ALL THE HELP, RESOURCES, AND PERSONAL SUPPORT YOU AND YOUR STUDENTS NEED!

www.wileyplus.com/resources

2-Minute Tutorials and all of the resources you & your students need to get started.

Student support from an experienced student user.

Collaborate with your colleagues, find a mentor, attend virtual and live events, and view resources.
www.WhereFacultyConnect.com

Pre-loaded, ready-to-use assignments and presentations. Created by subject matter experts.

Technical Support 24/7
FAQs, online chat, and phone support.
www.wileyplus.com/support

Your *WileyPLUS* Account Manager. Personal training and implementation support.

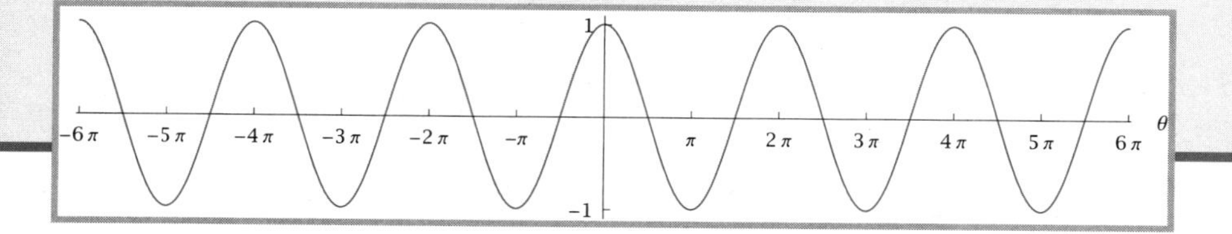

Algebra & Trigonometry

with
Student Solutions Manual

Sheldon Axler
San Francisco State University

JOHN WILEY & SONS, INC.

VICE PRESIDENT AND EXECUTIVE PUBLISHER	Laurie Rosatone
ACQUISITIONS EDITOR	Joanna Dingle
PROJECT EDITOR	Ellen Keohane
ASSISTANT EDITOR	Beth Pearson
DEVELOPMENT EDITOR	Anne Scanlan-Rohrer
SENIOR PRODUCTION EDITOR	Ken Santor
PHOTO EDITOR	Mary Ann Price
MARKETING MANAGER	Jonathan Cottrell
DESIGN DIRECTOR	Madelyn Lesure
MEDIA EDITOR	Melissa Edwards
MEDIA ASSISTANT	Lisa Sabatini

This book was typeset in pdfLaTeX by the author. Printing and binding by RR Donnelley, Jefferson City. Cover printed by Phoenix Color Corporation.

About the Cover The diagram on the cover shows a graph of the cosine function.

This book is printed on acid free paper. ∞

ISBN-13 978-0470-58579-5 (hardcover)
ISBN-13 978-0470-47081-7 (softcover)
ISBN-13 978-0470-47082-4 (binder ready)

Printed in the United States of America

10 9 8 7 6 5 4 3 2 1

About the Author

Sheldon Axler is Dean of the College of Science & Engineering at San Francisco State University, where he joined the faculty as Chair of the Mathematics Department in 1997.

Axler was valedictorian of his high school in Miami, Florida. He received his AB from Princeton University with highest honors, followed by a PhD in Mathematics from the University of California at Berkeley.

As a postdoctoral Moore Instructor at MIT, Axler received a university-wide teaching award. Axler was then an assistant professor, associate professor, and professor at Michigan State University, where he received the first J. Sutherland Frame Teaching Award and the Distinguished Faculty Award.

Axler received the Lester R. Ford Award for expository writing from the Mathematical Association of America in 1996. In addition to publishing numerous research papers, Axler is the author of *Linear Algebra Done Right* (which has been adopted as a textbook at over 240 universities and colleges), *College Algebra*, and *Precalculus: A Prelude to Calculus*; he is also co-author of *Harmonic Function Theory* (a graduate/research-level book).

Axler has served as Editor-in-Chief of the *Mathematical Intelligencer* and as Associate Editor of the *American Mathematical Monthly*. He has been a member of the Council of the American Mathematical Society and a member of the Board of Trustees of the Mathematical Sciences Research Institute. Axler currently serves on the editorial board of Springer's series Undergraduate Texts in Mathematics, Graduate Texts in Mathematics, and Universitext.

Contents

Preface to the Instructor

Goals

This book aims to provide students with the algebraic and trigonometric skills and understanding needed for other coursework and for participating as an educated citizen in a complex society.

Mathematics faculty frequently complain that many students do not read the textbook. When doing homework, a typical student may look only at the relevant section of the textbook or the student solutions manual for an example similar to the homework problem at hand. The student reads enough of that example to imitate the procedure and then does the homework problem. Little understanding may take place.

In contrast, this book is designed to be read by students. The writing style and layout are meant to induce students to read and understand the material. Explanations are more plentiful than typically found in algebra and trigonometry books. Examples of the concepts make the ideas concrete whenever possible.

Exercises and Problems

Each exercise has a unique correct answer, usually a number or a function; most problems have multiple correct answers, usually explanations or examples.

Students learn mathematics by actively working on a wide range of exercises and problems. Ideally, a student who reads and understands the material in a section of this book should be able to do the exercises and problems in that section without further help. However, some of the exercises require application of the ideas in a context that students may not have seen before; many students will need help with these exercises. This help is available from the complete worked-out solutions to all the odd-numbered exercises that appear at the end of each section.

Because the worked-out solutions were written solely by the author of the textbook, students can expect a consistent approach to the material. Furthermore, students will save money by not having to purchase a separate student solutions manual.

This book contains what is usually a separate book called the student solutions manual.

The exercises (but not the problems) occur in pairs, so that an odd-numbered exercise is followed by an even-numbered exercise whose solution uses the same ideas and techniques. A student stumped by an even-numbered exercise should be able to tackle it after reading the worked-out solution to the corresponding odd-numbered exercise. This arrangement allows the text to focus more centrally on explanations of the material and examples of the concepts.

Most students will read the student solutions manual when they are assigned homework, even though they are reluctant to read the main text. The integration of the student solutions manual within this book should encourage students to drift over and also read the main text. To reinforce this tendency, the worked-out solutions to the odd-numbered exercises at the end of each section are intentionally typeset with a slightly less appealing style (smaller type, two-column format, and not right justified) than the main text. The reader-friendly appearance of the main text might nudge students to spend some time there.

Exercises and problems in this book vary greatly in difficulty and purpose. Some exercises and problems are designed to hone algebraic manipulation skills; other exercises and problems are designed to push students to genuine understanding beyond rote algorithmic calculation.

Some exercises and problems intentionally reinforce material from earlier in the book and require multiple steps. For example, Exercise 30 in Section 5.3 asks students to find all numbers x such that

$$\log_5(x+4) + \log_5(x+2) = 2.$$

To solve this exercise, students will need to use the formula for a sum of logarithms as well as the quadratic formula; they will also need to eliminate one of the potential solutions produced by the quadratic formula because it would lead to the evaluation of the logarithm of a negative number. Although such multi-step exercises require more thought than most exercises in the book, they allow students to see crucial concepts more than once, sometimes in unexpected contexts.

The Calculator Issue

The issue of whether and how calculators should be used by students has generated immense controversy.

To aid instructors in presenting the kind of course they want, the symbol [calculator] appears with exercises and problems that require students to use a calculator.

Some sections of this book have many exercises and problems designed for calculators (for example Section 5.4 on exponential growth), but some sections deal with material not as amenable to calculator use. The text seeks to provide students with both understanding and skills. Thus the book does not aim for an artificially predetermined percentage of exercises and problems in each section requiring calculator use.

Some exercises and problems that require a calculator are intentionally designed to make students realize that by understanding the material, they can overcome the limitations of calculators. As one example among many, Exercise 83 in Section 5.3 asks students to find the number of digits in the decimal expansion of 7^{4000}. Brute force with a calculator will not work with this problem because the number involved has too many digits. However, a few moments' thought should show students that they can solve this problem by using logarithms (and their calculators!).

Regardless of what level of calculator use an instructor expects, students should not turn to a calculator to compute something like log 1, *because then* log *has become just a button on the calculator.*

The calculator icon can be interpreted for some exercises, depending on the instructor's preference, to mean that the solution should be a decimal approximation rather than the exact answer. For example, Exercise 3 in Section 6.3 asks how much would need to be deposited in a bank account paying 4% interest compounded continuously so that at the end of 10 years the account would contain $10,000. The exact answer to this exercise is $10000/e^{0.4}$ dollars, but it may be more satisfying to the student (after obtaining the exact answer) to use a calculator to see that approximately $6,703 needs to be deposited.

For exercises such as the one described in the paragraph above, instructors can decide whether to ask for exact answers or decimal approximations or both (the worked-out solutions for the odd-numbered exercises will usually contain both). If an instructor asks for only an exact answer, then a calculator may not be needed despite the presence of the calculator icon.

Symbolic processing programs such as *Mathematica* and *Maple* offer appealing alternatives to hand-held calculators because of their ability to solve equations and deal with symbols as well as numbers. Furthermore, the larger size, better resolution, and color on a computer screen make graphs produced by such software more informative than graphs on a typical hand-held graphing calculator.

Your students may not use a symbolic processing program because of the complexity or expense of such software. However, easy-to-use free web-based symbolic programs are becoming available. Occasionally this book shows how students can use Wolfram|Alpha, which has almost no learning curve, to go beyond what can be done easily by hand.

Even if you do not tell your students about such free tools, knowledge about such web-based homework aids is likely to spread rapidly among students.

Distinctive Approaches

Half-life and Exponential Growth

Almost all algebra and trigonometry books present radioactive decay as an example of exponential decay. Amazingly, the typical algebra and trigonometry textbook states that if a radioactive isotope has half-life h, then the amount left at time t will equal $e^{-(t\ln 2)/h}$ times the amount present at time 0.

A much clearer formulation would state, as this textbook does, that the amount left at time t will equal $2^{-t/h}$ times the amount present at time 0. The unnecessary use of e and $\ln 2$ in this context may suggest to students that e and natural logarithms have only contrived and artificial uses, which is not the message a textbook should send. Using $2^{-t/h}$ helps students understand the concept of half-life, with a formula connected to the meaning of the concept.

Similarly, many algebra and trigonometry textbooks consider, for example, a colony of bacteria doubling in size every 3 hours, with the textbook then producing the formula $e^{(t \ln 2)/3}$ for the growth factor after t hours. The simpler and more natural formula $2^{t/3}$ seems not to be mentioned in such books. This book presents the more natural approach to such issues of exponential growth and decay.

Algebraic Properties of Logarithms

The base for logarithms in Chapter 5 is arbitrary. Most of the examples and motivation use logarithms base 2 or logarithms base 10. Students will see how the algebraic properties of logarithms follow easily from the properties of exponents.

The crucial concepts of e and natural logarithms are saved for Chapter 6. Thus students can concentrate in Chapter 5 on understanding logarithms (arbitrary base) and their properties without at the same time worrying about grasping concepts related to e. Similarly, when natural logarithms arise naturally in Chapter 6, students should be able to concentrate on issues surrounding e without at the same time learning properties of logarithms.

The initial separation of logarithms and e should help students master both concepts.

Area

Section 2.4 in this book builds the intuitive notion of area starting with squares, and then quickly derives formulas for the area of rectangles, triangles, parallelograms, and trapezoids. A discussion of the effects of stretching either horizontally or vertically easily leads to the familiar formula for the area enclosed by a circle. Similar ideas are then used to find the formula for the area inside an ellipse (without calculus!).

Section 6.1 deals with the question of estimating the area under parts of the curve $y = \frac{1}{x}$ by using rectangles. This easy nontechnical introduction, with its emphasis on ideas without the clutter of the notation of Riemann sums, gives students a taste of an important idea from calculus.

e, The Exponential Function, and the Natural Logarithm

Most algebra and trigonometry textbooks either present no motivation for e or motivate e via continuously compounding interest or through the limit of an indeterminate expression of the form 1^∞; these concepts are difficult for students at this level to understand.

Chapter 6 presents a clean and well-motivated approach to e and the natural logarithm. We do this by looking at the area (intuitively defined) under the curve $y = \frac{1}{x}$, above the x-axis, and between the lines $x = 1$ and $x = c$.

The approach taken here to the exponential function and the natural logarithm shows that a good understanding of these subjects need not wait until a calculus course.

A similar approach to e and the natural logarithm is common in calculus courses. However, this approach is not usually adopted in algebra and trigonometry textbooks. Using basic properties of area, the simple presentation given here shows how these ideas can come through clearly without the technicalities of calculus or Riemann sums.

The approach taken here also has the advantage that it easily leads, as we will see in Chapter 6, to the approximation $\ln(1+h) \approx h$ for $|h|$ small. Furthermore, the same methods show that if r is any number, then

$$(1 + \tfrac{r}{x})^x \approx e^r$$

for large values of x. A final bonus of this approach is that the connection between continuously compounding interest and e becomes a nice corollary of natural considerations concerning area.

Inverse Functions

The unifying concept of inverse functions is introduced in Section 3.4. This crucial idea has its first major use in this book in the definition of $y^{1/m}$ as the number x such that $x^m = y$ (in other words, the function $y \mapsto y^{1/m}$ is the inverse of the function $x \mapsto x^m$; see Section 5.1). The second major use of inverse functions occurs in the definition of $\log_b y$ as the number x such that $b^x = y$ (in other words, the function $y \mapsto \log_b y$ is the inverse of the function $x \mapsto b^x$; see Section 5.2).

Thus students should be comfortable with using inverse functions by the time they reach the inverse trigonometric functions (arccosine, arcsine, and arctangent) in Section 10.1. For students who go on to calculus, this familiarity with inverse functions should help when dealing with inverse operations such as anti-differentiation.

With the approach taken in this book, the statement "$\log_b y = x$ means $b^x = y$" is consistent with the notation used for inverse functions.

This book emphasizes that $f^{-1}(y) = x$ means $f(x) = y$. Thus this book states that to find $f^{-1}(y)$, solve the equation $f(x) = y$ for x.

In contrast, many books at this level unfortunately instruct the reader wanting to find f^{-1} to start with the equation $y = f(x)$, then "interchange the variables x and y to obtain $x = f(y)$", then solve for y in terms of x. This "interchange" method ends up with notation expressing f^{-1} as a function of x.

However, the "interchange" method makes no sense when trying to find the value of an inverse function at a specific number instead of at a variable name. Consider, for example, the problem of finding $f^{-1}(11)$ if f is the function defined by $f(x) = 2x + 3$. The student mechanically following the "interchange" method as it is stated in many books would start with the equation $11 = 2x + 3$ and then interchange x and 11, getting the equation $x = 2 \cdot 11 + 3$. This is, of course, completely wrong.

In contrast, this book does this problem by solving the equation $11 = 2x+3$ for x, getting $x = 4$ and concluding that $f^{-1}(11) = 4$.

The "interchange" method will also be confusing to students when the variables names have meaning. For example, in an applied problem the variables might be t (for time) and d (for distance) rather than x and y, and we might have a function that gives distance in terms of time: $d = f(t)$. The inverse function should then give time in terms of distance: $t = f^{-1}(d)$. Interchanging the variable names here would be quite confusing.

Trigonometry

This book defines $\cos\theta$ and $\sin\theta$ as the first and second coordinates of the radius of the unit circle corresponding to θ (see Section 9.3). In contrast to this definition using only one symbol, many books at this level require students to juggle at least four symbols—θ (or t), x, y, and P—to parse the definitions of the trigonometric functions. These books define $\cos\theta = x$, and students become accustomed to thinking of $\cos\theta$ as the x-coordinate. When students encounter $\cos x$, as often happens within a dozen pages of the initial definition, they think that $\cos x$ is the x-coordinate of ... oops, that is a different use of x. No wonder so many students struggle with trigonometric functions.

This book defines sine and cosine in one section, then defines the tangent function (and the other three trigonometric functions that have less importance) in another section. This gentle approach contrasts with most books that define all six trigonometric functions on the same page. Students have difficulty assimilating so many definitions simultaneously.

This book emphasizes cos, sin, tan *and places little emphasis on* sec, csc, cot.

What to Cover

Different instructors will want to cover different sections of this book. Many instructors will want to cover Chapter 1 (The Real Numbers), even though it should be review, because it deals with familiar topics in a deeper fashion than students may have previously seen.

Some instructors will cover Section 4.3 (Rational Functions) only lightly because graphing rational functions, and in particular finding local minima and maxima, is better done with calculus. Many instructors will prefer to skip Chapter 8 (Sequences, Series, and Limits), leaving that material to a calculus course.

The inverse trigonometric identities (Section 10.2) are given more space in this book than in most books at this level. This material is included not so much for its intrinsic importance but as a way for students to obtain a deeper understanding of the trigonometric functions. Instructors can skip this material or cover it lightly.

Comments Welcome

I seek your help in making this a better book. Please send me your comments and your suggestions for improvements. Thanks!

Sheldon Axler
San Francisco State University

e-mail: algebra@axler.net
web site: algebraTrig.axler.net
Twitter: @AxlerAlgebra

WileyPLUS

WileyPLUS is an innovative online environment for effective teaching and learning.

A Research-based Design. *WileyPLUS* provides an online environment that integrates relevant resources, including the entire digital textbook, in an easy-to-navigate framework that helps students study more effectively.

- *WileyPLUS* adds structure by organizing textbook content into smaller, more manageable chunks.
- Related media, examples, and sample practice items reinforce learning objectives.
- Innovative features such as calendars, visual progress tracking, and self-evaluation tools improve time management and strengthen areas of weakness.

One-on-one Engagement. With *WileyPLUS* for *College Algebra* students receive 24/7 access to resources that promote positive learning outcomes. Students engage with related examples and sample practice items, including:

- videos;
- Guided Online (GO) tutorial exercises;
- concept questions.

Measurable Outcomes. Throughout each study session, students can assess their progress and gain immediate feedback. *WileyPLUS* provides precise reporting of strengths and weaknesses, as well as individualized quizzes, so that students are confident they are spending their time on the right things. With *WileyPLUS*, students always know the outcome of their efforts.

WileyPLUS provides reliable, customizable resources that reinforce course goals inside and outside the classroom. Pre-created materials and activities help instructors optimize their time.

Customizable Course Plan. *WileyPLUS* comes with a pre-created Course Plan designed specifically for this book. Simple drag-and-drop tools make it easy to assign the course plan as-is or modify it to reflect your course syllabus.

Pre-created Activity Types Include:

- questions;
- readings and resources;
- print tests;
- concept mastery;
- project.

Course Materials and Assessment Content:

- lecture slides;
- Instructor's Solutions Manual;
- question assignments: selected exercises coded algorithmically with hints, links to text, whiteboard/show-work feature, and instructor-controlled problem-solving help;
- proficiency exams;
- computerized testbank;
- printable testbank.

Gradebook. *WileyPLUS* provides instant access to reports on trends in class performance, student use of course materials, and progress towards learning objectives, helping inform decisions and drive classroom discussions.

Learn More: www.wileyplus.com

Acknowledgments

As usual in a textbook, as opposed to a research article, little attempt has been made to provide proper credit to the original creators of the ideas presented in this book. Where possible, I have tried to improve on standard approaches to this material. However, the absence of a reference does not imply originality on my part. I thank the many mathematicians who have created and refined our beautiful subject.

Most of the results in this book belong to the common heritage of mathematics, created over thousands of years by clever and curious people.

I chose Wiley as the publisher of this book because of the company's commitment to excellence. The people at Wiley have made outstanding contributions to this project, providing astute editorial advice, superb design expertise, high-level production skill, and insightful marketing savvy. I am truly grateful to the following Wiley folks, all of whom helped make this a better and more successful book than it would have been otherwise: Jonathan Cottrell, Joanna Dingle, Melissa Edwards, Jessica Jacobs, Ellen Keohane, Madelyn Lesure, Beth Pearson, Mary Ann Price, Laurie Rosatone, Lisa Sabatini, Ken Santor, Anne Scanlan-Rohrer, Jennifer Wreyford.

Celeste Hernandez, the accuracy checker, and Katrina Avery, the copy editor, excelled at catching mathematical and linguistic errors.

The instructors and students who used the earlier versions of this book provided wonderfully useful feedback. Numerous reviewers gave me terrific suggestions as the book progressed through various stages of development. I am grateful to all the class testers and reviewers whose names are listed on the following page, with special thanks to Michael Price.

Like most mathematicians, I owe thanks to Donald Knuth, who invented TeX, and to Leslie Lamport, who invented LaTeX, which I used to typeset this book. I am grateful to the authors of the many open-source LaTeX packages I used to improve the appearance of the book, especially to Hàn Thế Thành for pdfLaTeX, Robert Schlicht for microtype, and Frank Mittelbach for multicol.

Thanks also to Wolfram Research for producing *Mathematica*, which is the software I used to create the graphics in this book.

My awesome partner Carrie Heeter deserves considerable credit for her wise advice and continual encouragement throughout the long book-writing process.

Many thanks to all of you!

Sheldon

Class Testers and Reviewers

- Vladimir Akis, California State University, Los Angeles
- LaVerne Chambers Alan, Crichton College
- Aaron Altose, Cuyahoga Community College
- George Anastassiou, University of Memphis
- Karen Anglin, Blinn College Brenham
- Jan Archibald, Ventura College
- Vinod Arya, University of North Texas at Dallas
- Carlos Barron, Mountain View College
- Jamey Bass, City College of San Francisco
- Jaromir J. Becan, University of Texas at San Antonio
- Jeff Berg, Arapahoe Community College
- Matt Bertens, City College of San Francisco
- Nicoleta Bila, Fayetteville State University
- Arlene Blasius, SUNY College at Old Westbury
- Andrea Blum, Suffolk County Community College
- Valerie Bouagnon, DePaul University
- Brian Brock, San Jacinto College
- Connie Buller, Metropolitan Community College of Omaha
- Michael Butros, Victor Valley College
- Jennifer Cabaniss, Central Texas College
- Debananda Chakraborty, State University of New York at Buffalo
- Denise Chellsen, Cuesta College
- Sharon Christensen, Cameron University
- De Cook, Northwest Florida State College
- Kathy Cousins-Cooper, North Carolina A&T University
- Christopher Danielson, Minnesota State University Mankato
- Hilary Davies, University of Alaska Anchorage
- Michelle DeDeo, University of North Florida
- Luis Carlos Diaz, LaRoche College
- Deanna Dick, Alvin College
- David Dwiggins, University of Memphis
- Kathleen Ebert, Alfred State University
- Gay Ellis, Missouri State University
- Joan Evans, Texas Southern University
- Mike Everett, Santa Ana College
- Don Faust, Northern Michigan University
- Judy Fethe, Pellissippi State Technical Community College
- Anne Fine, East Central University
- Jack Flaherty, Brunswick Community College
- Patricia Foard, South Plains College
- Pari Ford, University of Nebraska Kearney
- Heng Fu, Thomas Nelson Community College
- Lee R. Gibson, University of Louisville
- Renu Gupta, Louisiana State University at Alexandria
- Joan Haig, University of Alaska Anchorage
- Daniel Harned, Lansing Community College
- Bud Hart, Oregon Institute of Technology
- Cheryl Hawker, Eastern Illinois University
- Max Hibbs, Blinn College Brenham
- Jada Hill, Richland College
- James Hilsenbeck, University of Texas at Brownsville
- Sarah Holliday, Southern Polytechnic State University

- Dwight Horan, Wentworth Institute of Technology
- Kerry Johnson, Missouri Southern State University
- Susan Jordan, Arkansas Tech University
- Basant Karna, Marshall University
- Alexander Kasiukov, Suffolk County Community College
- Brianna Kurtz, Daytona State College
- Grant Lathrom, Missouri Southern State University
- Kiseop Lee, University of Louisville
- Max Lee, Westchester Community College
- Scott Lewis, Utah Valley University
- Gary Lippman, California State University East Bay
- William Livingston, Missouri Southern State University
- Crepin M. Mahop, Howard University
- Syrous Marivani, Louisiana State University at Alexandria
- Mary Barone Martin, Middle Tennessee State University
- Eric Matsuoka, Leeward Community College
- Mike McCraith, Cuyahoga Community College
- Margaret Michener, University of Nebraska Kearney
- David Miller, West Virginia University
- Juan Carlos Molina, Austin Community College
- Hojin Moon, California State University Long Beach
- Bette Nelson, Alvin Community College
- Ron Palcic, Johnson County Community College
- Priti Patel, Tarrant County Community College Southeast
- Mary Beth Pattengale, Sierra College
- Vic Perera, Kent State University Trumbull
- Sandy Poinsett, College of Southern Maryland
- Michael Price, University of Oregon
- Adrienne Riley, California Polytechnic State University, San Luis Obispo
- Mike Rosenthal, Florida International University
- Daniel T. Russow, Arizona Western College
- Alan Saleski, Loyola University Chicago
- Rebecca Schantz, East Central College
- Jiajia Seffrood, Leeward Community College
- Mayada Shahroki, Lone Star College CyFair
- Sumanthi Shankar, Community College of Denver
- Robert Shea, Central Texas College
- Linda Snellings-Neal, Wright State University
- Shannon Solis, San Jacinto College, Central Campus
- Paul Sontag, University of Cincinnati
- Jacqueline Stone, University of Maryland
- Padma Tanniru, San Jose State University
- Marvelyn Tapp, University of Arkansas at Little Rock
- Mary Ann Teel, University of North Texas
- Jennie Thompson, Leeward Community College
- Jean Thornton, Western Kentucky University
- Michael van Opstall, University of Utah
- Sara Weiss, Richland College
- Nathanial Wiggins, San Jacinto College, North Campus

Preface to the Student

This book will help provide you with the algebraic and trigonometric skills and understanding needed for other coursework and for participating as an educated citizen in a complex society.

To learn this material well, you will need to spend serious time reading this book. You cannot expect to absorb mathematics the way you devour a novel. If you read through a section of this book in less than an hour, then you are going too fast. You should pause to ponder and internalize each definition, often by trying to invent some examples in addition to those given in the book. For each result stated in the book, you should seek examples to show why each hypothesis is necessary. When steps in a calculation are left out in the book, you need to supply the missing pieces, which will require some writing on your part. These activities can be difficult when attempted alone; try to work with a group of a few other students.

Complete worked-out solutions to the odd-numbered exercises are given at the end of each section.

You will need to spend several hours per section doing the exercises and problems. Make sure that you can do all the exercises and most of the problems, not just the ones assigned for homework. By the way, the difference between an exercise and a problem in this book is that each exercise has a unique correct answer that is a mathematical object such as a number or a function. In contrast, the solutions to problems often consist of explanations or examples; thus most problems have multiple correct answers.

Have fun, and best wishes in your studies!

Sheldon Axler
San Francisco State University

web site: `algebraTrig.axler.net`
Twitter: `@AxlerAlgebra`

CHAPTER

1

The Parthenon, built in Athens over 2400 years ago. The ancient Greeks developed and used remarkably sophisticated mathematics.

The Real Numbers

Success in this course will require a good understanding of the basic properties of the real number system. Thus this book begins with a review of the real numbers.

The first section of this chapter starts with the construction of the real line. This section contains as an optional highlight the ancient Greek proof that no rational number has a square equal to 2. This beautiful result appears here not because you will need it, but because it should be seen by everyone at least once.

Although this chapter will be mostly review, a thorough grounding in the real number system will serve you well throughout this course and then for the rest of your life. You will need good algebraic manipulation skills; thus the second section of this chapter reviews the fundamental algebra of the real numbers. You will also need to feel comfortable working with inequalities and absolute values, which are reviewed in the last section of this chapter.

Even if your instructor decides to skip this chapter, you may want to read through it. Make sure you can do all the exercises.

1.1 *The Real Line*

LEARNING OBJECTIVES

By the end of this section you should be able to

- explain the correspondence between the system of real numbers and the real line;
- show that some real numbers are not rational.

The **integers** are the numbers

$$\dots, -3, -2, -1, 0, 1, 2, 3, \dots;$$

here the dots indicate that the numbers continue without end in each direction. The sum, difference, and product of any two integers are also integers.

The quotient of two integers is not necessarily an integer. Thus we extend arithmetic to the **rational numbers**, which are numbers of the form

The use of a horizontal bar to separate the numerator and denominator of a fraction was introduced by Arabic mathematicians about 900 years ago.

$$\frac{m}{n},$$

where m and n are integers and $n \neq 0$.

Division is the inverse of multiplication, in the sense that we want the equation

$$\frac{m}{n} \cdot n = m$$

to hold. In the equation above, if we take $n = 0$ and (for example) $m = 1$, we get the nonsensical equation $\frac{1}{0} \cdot 0 = 1$. This equation is nonsensical because multiplying anything by 0 should give 0, not 1. To get around this problem, we leave expressions such as $\frac{1}{0}$ undefined. In other words, division by 0 is prohibited.

Division by 0 is undefined.

The rational numbers form a terrifically useful system. We can add, multiply, subtract, and divide rational numbers (with the exception of division by 0) and stay within the system of rational numbers. Rational numbers suffice for all actual physical measurements, such as length and weight, of any desired accuracy.

However, geometry and algebra force us to consider an even richer system of numbers—the real numbers. To see why we need to go beyond the rational numbers, we will investigate the real line.

Construction of the Real Line

Imagine a horizontal line, extending without end in both directions. Pick a point on this line and label it 0. Pick another point to the right of 0 and label it 1, as in the figure below.

0 1

Two key points on the real line.

Once the points 0 and 1 have been chosen on the line, everything else is determined by thinking of the distance between 0 and 1 as one unit of length. For example, 2 is one unit to the right of 1. Then 3 is one unit to the right of 2, and so on. The negative integers correspond to moving to the left of 0. Thus -1 is one unit to the left of 0. Then -2 is one unit to the left of -1, and so on.

The symbol for zero was invented in India more than 1100 years ago.

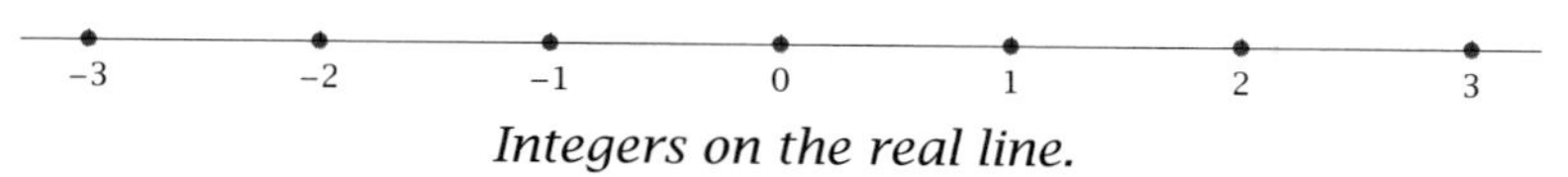

Integers on the real line.

If n is a positive integer, then $\frac{1}{n}$ is to the right of 0 by the length obtained by dividing the segment from 0 to 1 into n segments of equal length. Then $\frac{2}{n}$ is to the right of $\frac{1}{n}$ by the same length, and $\frac{3}{n}$ is to the right of $\frac{2}{n}$ by the same length again, and so on. The negative rational numbers are placed on the line similarly, but to the left of 0.

In this way, we associate with every rational number a point on the line. No figure can show the labels of all the rational numbers, because we can include only finitely many labels. The figure below shows the line with labels attached to a few of the points corresponding to rational numbers.

$-3 \quad -\frac{5}{2} \quad -2 \quad -\frac{115}{76} \quad -1 \quad -\frac{2}{3} \quad -\frac{1}{3} \quad 0 \quad \frac{1}{3} \quad \frac{2}{3} \quad 1 \quad \frac{12}{7} \quad 2 \quad \frac{257}{101} \quad 3$

Some rational numbers on the real line.

We will use the intuitive notion that the line has no gaps and that every conceivable distance can be represented by a point on the line. With these concepts in mind, we call the line shown above the **real line**. We think of each point on the real line as corresponding to a **real number**. The undefined intuitive notions (such as "no gaps") can be made precise using more advanced mathematics. In this book, we let our intuitive notions of the real line serve to define the system of real numbers.

Is Every Real Number Rational?

We know that every rational number corresponds to some point on the real line. Does every point on the real line correspond to some rational number? In other words, is every real number rational?

If more and more labels of rational numbers were placed on the figure above, the real line would look increasingly cluttered. Probably the first people to ponder these issues thought that the rational numbers fill up the entire real line. However, the ancient Greeks realized that this is not true. To see how they came to this conclusion, we make a brief detour into geometry.

Recall that for a right triangle, the sum of the squares of the lengths of the two sides that form the right angle equals the square of the length of the hypotenuse. The figure below illustrates this result, which is called the Pythagorean Theorem.

This theorem is named in honor of the Greek mathematician and philosopher Pythagoras who proved it over 2500 years ago. The Babylonians discovered this result a thousand years earlier than that.

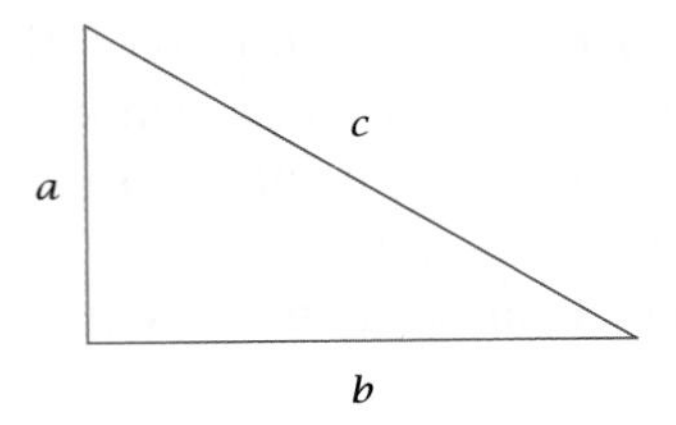

The Pythagorean Theorem for right triangles: $c^2 = a^2 + b^2$.

Now consider the special case where both sides that form the right angle have length 1, as in the figure below. In this case, the Pythagorean Theorem states that the length c of the hypotenuse has a square equal to 2.

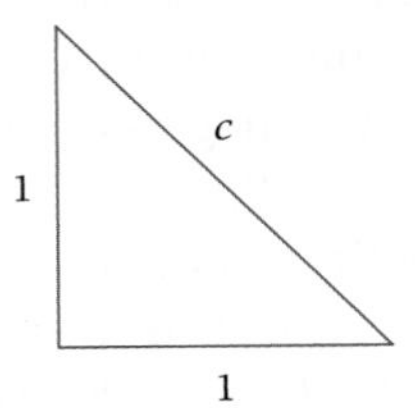

An isosceles right triangle. The Pythagorean Theorem implies that $c^2 = 2$.

Because we have constructed a line segment whose length c satisfies the equation $c^2 = 2$, a point to the right of 0 on the real line corresponds to c. In other words, there is a positive real number c whose square equals 2. This raises the question of whether there exists a rational number whose square equals 2.

We could try to find a rational number whose square equals 2 by experimentation. One striking example is

$$\left(\frac{99}{70}\right)^2 = \frac{9801}{4900};$$

here the numerator of the right side misses being twice the denominator by only 1. Although $\left(\frac{99}{70}\right)^2$ is close to 2, it is not exactly equal to 2.

Another example is $\frac{9369319}{6625109}$. The square of this rational number is approximately 1.9999999999992, which is very close to 2 but again is not exactly what we seek.

Because we have found rational numbers whose squares are very close to 2, you might suspect that with further cleverness we could find a rational number whose square equals 2. However, the ancient Greeks proved this is impossible. This course does not focus much on proofs. However, the Greek proof that there is no rational number whose square equals 2 is one of the great intellectual achievements of humanity. It should be experienced by every educated person. Thus this proof is presented below for your enrichment.

What follows is a proof by contradiction. We will start by assuming that there is a rational number whose square equals 2. Using that assumption, we will arrive at a contradiction. So our assumption must have been incorrect. Thus there is no rational number whose square equals 2.

Understanding the logical pattern of thinking that goes into this proof can be a valuable asset in dealing with complex issues.

No rational number has a square equal to 2.

Proof: Suppose there exist integers m and n such that

$$\left(\frac{m}{n}\right)^2 = 2.$$

By canceling any common factors, we can choose m and n to have no factors in common. In other words, $\frac{m}{n}$ is reduced to lowest terms.

The equation above is equivalent to the equation

$$m^2 = 2n^2.$$

This implies that m^2 is even; hence m is even. Thus $m = 2k$ for some integer k. Substituting $2k$ for m in the equation above gives

$$4k^2 = 2n^2,$$

or equivalently

$$2k^2 = n^2.$$

This implies that n^2 is even; hence n is even.

We have now shown that both m and n are even, contradicting our choice of m and n as having no factors in common. This contradiction means our original assumption that there is a rational number whose square equals 2 must be incorrect. Thus there do not exist integers m and n such that $\left(\frac{m}{n}\right)^2 = 2$.

"When you have excluded the impossible, whatever remains, however improbable, must be the truth."
—Sherlock Holmes

The notation $\sqrt{2}$ is used to denote the positive real number c such that $c^2 = 2$. As we saw earlier, the Pythagorean Theorem implies that there exists a real number $\sqrt{2}$ with the property that $\sqrt{2}^2 = 2$.

The result above implies that $\sqrt{2}$ is not a rational number. Thus not every real number is a rational number. In other words, not every point on the real line corresponds to a rational number.

Irrational numbers

A real number that is not rational is called an **irrational number**.

We have just shown that $\sqrt{2}$ is an irrational number. The real numbers π and e, which we will encounter in later chapters, are also irrational numbers.

Once we have found one irrational number, finding others is much easier, as shown in the example below.

EXAMPLE 1

The attitude of the ancient Greeks toward irrational numbers persists in our everyday use of "irrational" to mean "not based on reason".

Show that $3 + \sqrt{2}$ is an irrational number.

SOLUTION Suppose $3 + \sqrt{2}$ is a rational number. Because

$$\sqrt{2} = (3 + \sqrt{2}) - 3,$$

this implies that $\sqrt{2}$ is the difference of two rational numbers, which implies that $\sqrt{2}$ is a rational number, which is not true. Thus our assumption that $3 + \sqrt{2}$ is a rational number was incorrect. In other words, $3 + \sqrt{2}$ is an irrational number.

The next example provides another illustration of how to use one irrational number to generate another irrational number.

EXAMPLE 2

Show that $8\sqrt{2}$ is an irrational number.

SOLUTION Suppose $8\sqrt{2}$ is a rational number. Because

$$\sqrt{2} = \frac{8\sqrt{2}}{8},$$

this implies that $\sqrt{2}$ is the quotient of two rational numbers, which implies that $\sqrt{2}$ is a rational number, which is not true. Thus our assumption that $8\sqrt{2}$ is a rational number was incorrect. In other words, $8\sqrt{2}$ is an irrational number.

PROBLEMS

The problems in this section may be harder than typical problems found in the rest of this book.

1. Show that $\frac{6}{7} + \sqrt{2}$ is an irrational number.
2. Show that $5 - \sqrt{2}$ is an irrational number.
3. Show that $3\sqrt{2}$ is an irrational number.
4. Show that $\frac{3\sqrt{2}}{5}$ is an irrational number.
5. Show that $4 + 9\sqrt{2}$ is an irrational number.
6. Explain why the sum of a rational number and an irrational number is an irrational number.
7. Explain why the product of a nonzero rational number and an irrational number is an irrational number.
8. Suppose t is an irrational number. Explain why $\frac{1}{t}$ is also an irrational number.
9. Give an example of two irrational numbers whose sum is an irrational number.
10. Give an example of two irrational numbers whose sum is a rational number.
11. Give an example of three irrational numbers whose sum is a rational number.
12. Give an example of two irrational numbers whose product is an irrational number.
13. Give an example of two irrational numbers whose product is a rational number.

1.2 Algebra of the Real Numbers

LEARNING OBJECTIVES

By the end of this section you should be able to

- manipulate algebraic expressions using the commutative, associative, and distributive properties;
- recognize the order of algebraic operations and the role of parentheses;
- apply the crucial algebraic identities involving additive inverses and fractions;
- explain the importance of being careful about parentheses and the order of operations when using a calculator or computer.

The operations of addition, subtraction, multiplication, and division extend from the rational numbers to the real numbers. We can add, subtract, multiply, and divide any two real numbers and stay within the system of real numbers, again with the exception that division by 0 is prohibited.

Exercises woven throughout this book have been designed to sharpen your algebraic manipulation skills as we cover other topics.

In this section we review the basic algebraic properties of the real numbers. Because this material should indeed be review, no effort has been made to show how some of these properties follow from others. Instead, this section focuses on highlighting key properties that should become so familiar to you that you can use them comfortably and without effort.

Commutativity and Associativity

Commutativity is the formal name for the property stating that order does not matter in addition and multiplication:

Commutativity

$$a + b = b + a \quad \text{and} \quad ab = ba$$

Here (and throughout this section) a, b, and other variables denote either real numbers or expressions that take on values that are real numbers. For example, the commutativity of addition implies that $x^2 + \frac{x}{5} = \frac{x}{5} + x^2$.

Neither subtraction nor division is commutative because order does matter for those operations. For example, $5 - 3 \neq 3 - 5$, and $\frac{6}{2} \neq \frac{2}{6}$.

Associativity is the formal name for the property stating that grouping does not matter in addition and multiplication:

Associativity

$$(a + b) + c = a + (b + c) \quad \text{and} \quad (ab)c = a(bc)$$

Expressions inside parentheses should be calculated before further computation. For example, $(a + b) + c$ should be calculated by first adding a

and b, and then adding that sum to c. The associative property of addition asserts that this number will be the same as $a + (b + c)$, which should be calculated by first adding b and c, and then adding that sum to a.

Because of the associativity of addition, we can dispense with parentheses when adding three or more numbers, writing expressions such as

$$a + b + c + d$$

without worrying about how the terms are grouped. Similarly, because of the associative property of multiplication we do not need parentheses when multiplying together three or more numbers. Thus we can write expressions such as $abcd$ without specifying the order of multiplication or the grouping.

Neither subtraction nor division is associative because the grouping does matter for those operations. For example,

$$(9 - 6) - 2 = 3 - 2 = 1,$$

but

$$9 - (6 - 2) = 9 - 4 = 5,$$

which shows that subtraction is not associative.

The standard practice is to evaluate subtractions from left to right unless parentheses indicate otherwise. For example, $9 - 6 - 2$ should be interpreted to mean $(9 - 6) - 2$, which equals 1.

The Order of Algebraic Operations

Consider the expression

$$2 + 3 \cdot 7.$$

This expression contains no parentheses to guide us to which operation should be performed first. Should we first add 2 and 3, and then multiply the result by 7? If so, we would interpret the expression above as

$$(2 + 3) \cdot 7,$$

which equals 35.

Or to evaluate

$$2 + 3 \cdot 7$$

should we first multiply together 3 and 7, and then add 2 to that result. If so, we would interpret the expression above as

Note that $(2 + 3) \cdot 7$ does not equal $2 + (3 \cdot 7)$. Thus the order of these operations does matter.

$$2 + (3 \cdot 7),$$

which equals 23.

So does $2 + 3 \cdot 7$ equal $(2 + 3) \cdot 7$ or $2 + (3 \cdot 7)$? The answer to this question depends on custom rather than anything inherent in the mathematical situation. Every mathematically literate person would interpret $2 + 3 \cdot 7$ to mean $2 + (3 \cdot 7)$. In other words, people in the modern era have adopted the convention that multiplications should be performed before additions unless parentheses dictate otherwise. You need to become accustomed to this convention:

Multiplication and division before addition and subtraction

Unless parentheses indicate otherwise, products and quotients are calculated before sums and differences.

Thus, for example, $a + bc$ is interpreted to mean $a + (bc)$, although almost always we dispense with the parentheses and just write $a + bc$.

As another illustration of the principle above, consider the expression

$$4m + 3n + 11(p + q).$$

The correct interpretation of this expression is that 4 should be multiplied by m, 3 should be multiplied by n, 11 should be multiplied by $p + q$, and then the three numbers $4m$, $3n$, and $11(p + q)$ should be added together. In other words, the expression above equals

$$(4m) + (3n) + \big(11(p + q)\big).$$

The size of parentheses is sometimes used as an optional visual aid to indicate the order of operations. Smaller parentheses should be used for more inner parentheses. Thus expressions enclosed in smaller parentheses should usually be evaluated before expressions enclosed in larger parentheses.

The three newly added sets of parentheses in the expression above are unnecessary, although it is not incorrect to include them. However, the version of the same expression without the unnecessary parentheses is cleaner and easier to read.

When parentheses are enclosed within parentheses, expressions in the innermost parentheses are evaluated first.

Evaluate inner parentheses first

In an expression with parentheses inside parentheses, evaluate the innermost parentheses first and then work outward.

EXAMPLE 1

Evaluate the expression $2\big(6 + 3(1 + 4)\big)$.

SOLUTION Here the innermost parentheses surround $1 + 4$. Thus start by evaluating that expression, getting 5:

$$2\big(6 + 3\underbrace{(1 + 4)}_{5}\big) = 2(6 + 3 \cdot 5).$$

Now to evaluate the expression $6 + 3 \cdot 5$, first evaluate $3 \cdot 5$, getting 15, then add that to 6, getting 21. Multiplying by 2 completes our evaluation of this expression:

$$2\underbrace{(6 + 3\underbrace{(1+4)}_{5})}_{21} = 42.$$

The Distributive Property

The distributive property connects addition and multiplication, converting a product with a sum into a sum of two products.

Distributive property

$$a(b + c) = ab + ac$$

Because multiplication is commutative, the distributive property can also be written in the alternative form

$$(a + b)c = ac + bc.$$

The distributive property provides the justification for factoring expressions.

Sometimes you will need to use the distributive property to transform an expression of the form $a(b + c)$ into $ab + ac$, and sometimes you will need to use the distributive property in the opposite direction, transforming an expression of the form $ab + ac$ into $a(b + c)$. Because the distributive property is usually used to simplify an expression, the direction of the transformation depends on the context. The next example shows the use of the distributive property in both directions.

EXAMPLE 2 Simplify the expression $2(3m + x) + 5x$.

SOLUTION First use the distributive property to transform $2(3m + x)$ into $6m + 2x$:

$$2(3m + x) + 5x = 6m + 2x + 5x.$$

Now use the distributive property again, but in the other direction, to transform $2x + 5x$ to $(2 + 5)x$:

$$\begin{aligned} 6m + 2x + 5x &= 6m + (2 + 5)x \\ &= 6m + 7x. \end{aligned}$$

Putting all this together, we have used the distributive property (twice) to transform

$$2(3m + x) + 5x$$

into the simpler expression

$$6m + 7x.$$

One of the most common algebraic manipulations involves expanding a product of sums, as in the following example.

EXAMPLE 3

Expand $(a + b)(c + d)$.

SOLUTION Think of $(c + d)$ as a single number and then apply the distributive property to the expression above, getting

$$(a + b)(c + d) = a(c + d) + b(c + d).$$

Now apply the distributive property twice more, getting

$$(a + b)(c + d) = ac + ad + bc + bd.$$

After you use this formula several times, it will become so familiar that you can use it routinely without needing to pause. Note that every term in the first set of parentheses is multiplied by every term in the second set of parentheses.

If you are comfortable with the distributive property, there is no need to memorize the last formula from the example above, because you can always derive it again. Furthermore, by understanding how the identity above was obtained, you should have no trouble finding formulas for more complicated expressions such as $(a + b)(c + d + t)$.

An important special case of the identity above occurs when $c = a$ and $d = b$. In that case we have

$$(a + b)(a + b) = a^2 + ab + ba + b^2,$$

which, with a standard use of commutativity, becomes the identity

$$(a + b)^2 = a^2 + 2ab + b^2.$$

Additive Inverses and Subtraction

The **additive inverse**of a real number a is the number $-a$ such that

$$a + (-a) = \mathbf{0}.$$

The connection between subtraction and additive inverses is captured by the identity

$$a - b = a + (-b).$$

In fact, the equation above can be taken as the definition of subtraction.

You need to be comfortable using the following identities that involve additive inverses and subtraction:

Identities involving additive inverses and subtraction

$$-(-a) = a$$

$$-(a+b) = -a - b$$

$$(-a)(-b) = ab$$

$$(-a)b = a(-b) = -(ab)$$

$$(a-b)c = ac - bc$$

$$a(b-c) = ab - ac$$

EXAMPLE 4

Expand $(a+b)(a-b)$.

Be sure to distribute the minus signs correctly when using the distributive property, as shown here.

SOLUTION Start by thinking of $(a+b)$ as a single number and applying the distributive property. Then apply the distributive property twice more:

$$\begin{aligned}(a+b)(a-b) &= (a+b)a - (a+b)b \\ &= a^2 + ba - ab - b^2 \\ &= a^2 - b^2\end{aligned}$$

You need to become sufficiently comfortable with the following identities so that you can use them with ease.

Identities arising from the distributive property

$$(a+b)^2 = a^2 + 2ab + b^2$$

$$(a-b)^2 = a^2 - 2ab + b^2$$

$$(a+b)(a-b) = a^2 - b^2$$

EXAMPLE 5

Without using a calculator, evaluate 43×37.

SOLUTION

$$\begin{aligned}43 \times 37 &= (40+3)(40-3) \\ &= 40^2 - 3^2 \\ &= 1600 - 9 \\ &= 1591\end{aligned}$$

Multiplicative Inverses and the Algebra of Fractions

The **multiplicative inverse** of a real number $b \neq 0$ is the number $\frac{1}{b}$ such that

$$b \cdot \frac{1}{b} = 1.$$

The multiplicative inverse of b is sometimes called the **reciprocal** *of b.*

The connection between division and multiplicative inverses is captured by the identity

$$\frac{a}{b} = a \cdot \frac{1}{b}.$$

In fact, the equation above can be taken as the definition of division.

You need to be comfortable using various identities that involve multiplicative inverses and division. We start with the following identities:

Multiplication of fractions and cancellation

$$\frac{a}{b} \cdot \frac{c}{d} = \frac{ac}{bd} \qquad \frac{ac}{ad} = \frac{c}{d}$$

Assume that none of the denominators in this subsection equals 0.

The first identity above states that the product of two fractions can be computed by multiplying together the numerators and multiplying together the denominators. The second identity above, when used to transform $\frac{ac}{ad}$ into $\frac{c}{d}$, is the usual simplification of canceling a common factor from the numerator and denominator. When used in the other direction to transform $\frac{c}{d}$ into $\frac{ac}{ad}$, the second identity above becomes the familiar procedure of multiplying the numerator and denominator by the same factor.

Notice that the second identity above follows the first identity, as follows:

$$\frac{ac}{ad} = \frac{a}{a} \cdot \frac{c}{d} = 1 \cdot \frac{c}{d} = \frac{c}{d}.$$

EXAMPLE 6

Simplify the expression

$$\frac{3}{x^2 - 1} \cdot \frac{x+1}{x}.$$

SOLUTION Using the identities above, we have

$$\begin{aligned}\frac{3}{x^2 - 1} \cdot \frac{x+1}{x} &= \frac{3(x+1)}{(x^2-1)x}\\ &= \frac{3(x+1)}{(x+1)(x-1)x}\\ &= \frac{3}{(x-1)x}.\end{aligned}$$

Now we turn to the identity for adding two fractions:

The formula for adding fractions is more complicated than the formula for multiplying fractions.

Addition of fractions

$$\frac{a}{b} + \frac{c}{d} = \frac{ad + bc}{bd}$$

The derivation of the identity above is straightforward if we accept the formula for adding two fractions with the same denominator, which is

$$\frac{a}{b} + \frac{c}{b} = \frac{a + c}{b}.$$

To derive this formula, note that

$$\frac{a}{b} + \frac{c}{b} = a \cdot \frac{1}{b} + c \cdot \frac{1}{b} = (a + c) \cdot \frac{1}{b} = \frac{a + c}{b}.$$

For example, $\frac{2}{9} + \frac{5}{9} = \frac{7}{9}$, as you can visualize by thinking about a pie divided into 9 equal-size pieces, and then placing 2 pieces of the pie ($\frac{2}{9}$ of the pie) with 5 additional pieces ($\frac{5}{9}$ of the pie), getting 7 pieces of the pie ($\frac{7}{9}$ of the pie).

To obtain the formula for adding two fractions with different denominators, we use the multiplication identity to rewrite the fractions so that they have the same denominators:

$$\begin{aligned} \frac{a}{b} + \frac{c}{d} &= \frac{a}{b} \cdot \frac{d}{d} + \frac{b}{b} \cdot \frac{c}{d} \\ &= \frac{ad}{bd} + \frac{bc}{bd} \\ &= \frac{ad + bc}{bd}. \end{aligned}$$

Never, ever, make the mistake of thinking that $\frac{a}{b} + \frac{c}{d}$ equals $\frac{a+c}{b+d}$.

EXAMPLE 7

Write the sum

$$\frac{2}{w(w+1)} + \frac{3}{w^2}$$

as a single fraction.

SOLUTION Using the identity for adding fractions gives

$$\begin{aligned} \frac{2}{w(w+1)} + \frac{3}{w^2} &= \frac{w^2}{w^2} \cdot \frac{2}{w(w+1)} + \frac{3}{w^2} \cdot \frac{w(w+1)}{w(w+1)} \\ &= \frac{2w^2 + 3w(w+1)}{w^3(w+1)} \\ &= \frac{5w^2 + 3w}{w^3(w+1)} \\ &= \frac{w(5w+3)}{ww^2(w+1)} \\ &= \frac{5w+3}{w^2(w+1)}. \end{aligned}$$

REMARK Sometimes when adding two fractions, it is easier to use a common multiple of the two denominators that is simpler than the product of the two denominators. For example, the two denominators in this example are $w(w+1)$ and w^2. Their product is $w^3(w+1)$, which is the denominator used in the calculation above. However, $w^2(w+1)$ is also a common multiple of the two denominators. Here is the calculation using $w^2(w+1)$ as the denominator:

$$\begin{aligned}\frac{2}{w(w+1)}+\frac{3}{w^2} &= \frac{w}{w}\cdot\frac{2}{w(w+1)}+\frac{3}{w^2}\cdot\frac{w+1}{w+1}\\ &= \frac{2w}{w^2(w+1)}+\frac{3(w+1)}{w^2(w+1)}\\ &= \frac{2w+3(w+1)}{w^2(w+1)}\\ &= \frac{5w+3}{w^2(w+1)}.\end{aligned}$$

If you can easily find a common multiple that is simpler than the product of the two denominators, then using it will mean that less cancellation is necessary to simplify your final result.

The two methods produced the same answer. In general, either method will work fine.

Now we look at the identity for dividing by a fraction:

Here the size of the fraction bars are used to indicate that

$$\frac{a}{\frac{b}{c}}$$

should be interpreted to mean $a/(b/c)$.

Division by a fraction

$$\frac{a}{\frac{b}{c}} = a\cdot\frac{c}{b}$$

This identity gives the key to unraveling fractions that involve fractions, as shown in the following example.

EXAMPLE 8

Simplify the expression

$$\frac{\frac{y}{x}}{\frac{b}{c}}.$$

SOLUTION The size of the fraction bars indicates that the expression to be simplified is $(y/x)/(b/c)$. We use the identity above (thinking of $\frac{y}{x}$ as a), which shows that dividing by $\frac{b}{c}$ is the same as multiplying by $\frac{c}{b}$. Thus we have

$$\begin{aligned}\frac{\frac{y}{x}}{\frac{b}{c}} &= \frac{y}{x}\cdot\frac{c}{b}\\ &= \frac{yc}{xb}.\end{aligned}$$

When faced with complicated expressions involving fractions that are themselves fractions, remember that division by a fraction is the same as multiplication by the fraction flipped over.

Finally, we conclude this subsection by recording some identities involving fractions and additive inverses:

Fractions and additive inverses

$$\frac{-a}{b} = \frac{a}{-b} = -\frac{a}{b} \qquad \frac{-a}{-b} = \frac{a}{b} \qquad \frac{a}{b} - \frac{c}{d} = \frac{ad - bc}{bd}$$

Symbolic Calculators

Over the last several decades, inexpensive electronic calculators and computers drastically changed the ease of doing calculations. Many numeric computations that previously required considerable technical skill can now be done with a few pushes of a button or clicks of a mouse.

This development led to a change in the computational skills that are genuinely useful for most people to know. Thus some computational techniques are no longer routinely taught in schools. For example, very few people today know how to compute by hand an approximate square root. A mathematically literate person in 1960 could reasonably quickly compute by hand that $\sqrt{3} \approx 1.73205$, but today almost everyone would use a calculator or computer to obtain this approximation instantly and easily.

The symbol ≈ means "approximately equal".

Although calculators have made certain computational skills less important, correct use of a calculator requires some understanding, particularly of the order of operations.

EXAMPLE 9

Use a calculator to evaluate

$$8.7 + 2.1 \times 5.9.$$

SOLUTION On a calculator, you might enter

[8.7] [+] [2.1] [×] [5.9]

The result of entering these items on a calculator depends upon the calculator!

and then on most calculators you need to press the [*enter*] button or the [=] button.

Some calculators give 63.72 as the result of entering the items above, while other calculators give 21.09 as the result. Calculators that give the result 63.72 work by first calculating the sum $8.7 + 2.1$, getting 10.8, then multiplying by 5.9 to get 63.72. Thus such calculators interpret the input above to mean $(8.7 + 2.1) \times 5.9$.

Other calculators will interpret the input above to mean $8.7 + (2.1 \times 5.9)$, which equals 21.09 and which is what the expression $8.7 + 2.1 \times 5.9$ should mean.

This example shows the importance of paying careful attention to the order of operations, even when using a calculator.

If your calculator gives the result 63.72 when the items above are entered, then a correct answer for the desired calculation can be obtained by changing the order so that you enter

[2.1] [×] [5.9] [+] [8.7]

and then the [*enter*] button or the [=] button; this sequence of items should be interpreted by all calculators to mean $(2.1 \times 5.9) + 8.7$, giving the correct answer 21.09.

Make sure you know how your calculator interprets the kind of input shown in the example above, or you may get answers from your calculator that do not reflect the problem you have in mind.

Many calculators have parentheses buttons, which are often needed to control the order in which operations are performed. For example, to evaluate $(3.49 + 4.58)(5.67 + 6.76)$ on a calculator with parentheses buttons you can enter

[(] [3.49] [+] [4.58] [)] [×] [(] [5.67] [+] [6.76] [)]

and then the [*enter*] button or the [=] button, getting 100.3101.

Calculators and computers have evolved from being able to do arithmetic to having serious symbolic and graphing abilities. For now, we will focus on the ability of symbolic calculators to perform algebraic simplifications. Later chapters will illustrate the graphing abilities and additional algebraic capabilities of these modern machines.

Some calculators can do rational arithmetic, meaning that

$$\frac{1}{3} + \frac{2}{5}$$

gives the result $\frac{11}{15}$ *instead of* 0.7333333.

A symbolic calculator can handle symbols as well as numbers. Because of the many varieties of symbolic and graphing calculators available as either hand-held machines or as computer software, showing how all of them work would be impractical.

Thus the examples throughout this book that depend upon symbolic or graphing calculators will use Wolfram|Alpha, which is a free web-based symbolic and graphing calculator (and much more). You do not necessarily need to use Wolfram|Alpha to follow the examples; feel free to use your favorite calculator instead.

However, Wolfram|Alpha has some unusual capabilities. Thus even if you have your own symbolic calculator, you should take a look at Wolfram|Alpha so that you know the kind of computational power that can now be easily used. Here are some of the advantages of using Wolfram|Alpha at least occasionally:

- Wolfram|Alpha is free. Web access is needed to use Wolfram|Alpha, but many more people have web access than have symbolic calculators.
- Wolfram|Alpha is more powerful than standard symbolic calculators, allowing certain calculations that cannot be done with a hand-held calculator.
- Wolfram|Alpha is very easy to use. In particular, Wolfram|Alpha does not have the strict syntax requirements associated with many symbolic calculators.
- The *Show steps* feature of Wolfram|Alpha (discussed below) can be a terrific learning tool for students.
- The larger size, better resolution, and color on a computer screen make graphs produced by Wolfram|Alpha more informative than graphs on a typical hand-held graphing calculator.

The next example shows how to use Wolfram|Alpha as a web-based symbolic calculator.

EXAMPLE 10 Use Wolfram|Alpha to expand the expression $(x - 2y + 3z)^2$.

As usual on computers, a caret ^ is used to denote a power.

SOLUTION Point a web browser to `www.wolframalpha.com`. In the one-line entry box that appears near the top of the web page, enter

`expand (x - 2y + 3z)^2`

and then press the *enter* key on your keyboard or click the = box on the right side of the Wolfram|Alpha entry box, getting the result

$$x^2 - 4xy + 6xz + 4y^2 - 12yz + 9z^2.$$

The Show steps *option is not available for all Wolfram|Alpha results.*

If you want to see how this result was computed, click *Show steps* on the right side of the result box and you will see (in more detail that you probably want) how the distributive property was used multiple times to produce the result.

The next example is presented mainly to show that parentheses must sometimes be used with symbolic calculators even when parentheses do not appear in the usual mathematical notation.

EXAMPLE 11 Use Wolfram|Alpha to simplify the expression

$$\frac{\frac{1}{y-b} - \frac{1}{y}}{b}.$$

As shown here, Wolfram|Alpha does not mind if you insert extra spaces to make the input more easily readable by humans.

SOLUTION In a Wolfram|Alpha entry box, enter

`simplify ( 1/(y-b) - 1/y )/b`

to get the result

$$\frac{1}{y(y-b)}.$$

Make sure you understand why both sets of parentheses in the entry box above are needed. This example shows that even if you are using a machine for algebraic manipulation, you must have a good understanding of the order of operations.

Most of the exercises in this section can be done by Wolfram|Alpha or a symbolic calculator. However, you should want to acquire the basic algebraic manipulation skills and understanding needed to do these exercises. Very little skill or understanding will come from watching a machine do the exercises.

The best way to use Wolfram|Alpha or another symbolic calculator with the exercises is to check your answers and to experiment (and play!) by changing the input to see how the output varies.

EXERCISES

For Exercises 1–4, determine how many different values can arise by inserting one pair of parentheses into the given expression.

1. $19 - 12 - 8 - 2$
2. $3 - 7 - 9 - 5$
3. $6 + 3 \cdot 4 + 5 \cdot 2$
4. $5 \cdot 3 \cdot 2 + 6 \cdot 4$

For Exercises 5–22, expand the given expression.

5. $(x - y)(z + w - t)$
6. $(x + y - r)(z + w - t)$
7. $(2x + 3)^2$
8. $(3b + 5)^2$
9. $(2c - 7)^2$
10. $(4a - 5)^2$
11. $(x + y + z)^2$
12. $(x - 5y - 3z)^2$
13. $(x + 1)(x - 2)(x + 3)$
14. $(y - 2)(y - 3)(y + 5)$
15. $(a + 2)(a - 2)(a^2 + 4)$
16. $(b - 3)(b + 3)(b^2 + 9)$
17. $xy(x + y)(\frac{1}{x} - \frac{1}{y})$
18. $a^2z(z - a)(\frac{1}{z} + \frac{1}{a})$
19. $(t - 2)(t^2 + 2t + 4)$
20. $(m - 2)(m^4 + 2m^3 + 4m^2 + 8m + 16)$
21. $(n + 3)(n^2 - 3n + 9)$
22. $(y + 2)(y^4 - 2y^3 + 4y^2 - 8y + 16)$

For Exercises 23–50, simplify the given expression as much as possible.

23. $4(2m + 3n) + 7m$
24. $3(2m + 4(n + 5p)) + 6n$
25. $\frac{3}{4} + \frac{6}{7}$
26. $\frac{2}{5} + \frac{7}{8}$
27. $\frac{3}{4} \cdot \frac{14}{39}$
28. $\frac{2}{3} \cdot \frac{15}{22}$
29. $\frac{\frac{5}{7}}{\frac{2}{3}}$
30. $\frac{\frac{6}{5}}{\frac{7}{4}}$
31. $\frac{m+1}{2} + \frac{3}{n}$
32. $\frac{m}{3} + \frac{5}{n-2}$
33. $\frac{2}{3} \cdot \frac{4}{5} + \frac{3}{4} \cdot 2$
34. $\frac{3}{5} \cdot \frac{2}{7} + \frac{5}{4} \cdot 2$
35. $\frac{2}{5} \cdot \frac{m+3}{7} + \frac{1}{2}$
36. $\frac{3}{4} \cdot \frac{n-2}{5} + \frac{7}{3}$
37. $\frac{2}{x+3} + \frac{y-4}{5}$
38. $\frac{x-3}{4} - \frac{5}{y+2}$
39. $\frac{4t+1}{t^2} + \frac{3}{t}$
40. $\frac{5}{u^2} + \frac{1-2u}{u^3}$
41. $\frac{3}{v(v-2)} + \frac{v+1}{v^3}$
42. $\frac{w-1}{w^3} - \frac{2}{w(w-3)}$
43. $\frac{1}{x-y}\left(\frac{x}{y} - \frac{y}{x}\right)$
44. $\frac{1}{y}\left(\frac{1}{x-y} - \frac{1}{x+y}\right)$
45. $\frac{(x+a)^2 - x^2}{a}$
46. $\frac{\frac{1}{x+a} - \frac{1}{x}}{a}$
47. $\frac{\frac{x-2}{y}}{\frac{z}{x+2}}$
48. $\frac{\frac{x-4}{y+3}}{\frac{y-3}{x+4}}$
49. $\frac{\frac{a-t}{b-c}}{\frac{b+c}{a+t}}$
50. $\frac{\frac{r+m}{u-n}}{\frac{n+u}{m-r}}$

PROBLEMS

Some problems require considerably more thought than the exercises. Unlike exercises, problems usually have more than one correct answer.

51. Show that $(a + 1)^2 = a^2 + 1$ if and only if $a = 0$.
52. Explain why $(a + b)^2 = a^2 + b^2$ if and only if $a = 0$ or $b = 0$.
53. Show that $(a - 1)^2 = a^2 - 1$ if and only if $a = 1$.
54. Explain why $(a - b)^2 = a^2 - b^2$ if and only if $b = 0$ or $b = a$.
55. Explain how you could show that $51 \times 49 = 2499$ in your head by using the identity $(a + b)(a - b) = a^2 - b^2$.

56. Show that

$$a^3 + b^3 + c^3 - 3abc$$
$$= (a + b + c)(a^2 + b^2 + c^2 - ab - bc - ac).$$

57. Give an example to show that division does not satisfy the associative property.

58. The sales tax in San Francisco is 8.5%. Diners in San Francisco often compute a 17% tip on their before-tax restaurant bill by simply doubling the sales tax. For example, a \$64 dollar food and drink bill would come with a sales tax of \$5.44; doubling that amount would lead to a 17% tip of \$10.88 (which might be rounded up to \$11). Explain why this technique is an application of the associativity of multiplication.

59. A quick way to compute a 15% tip on a restaurant bill is first to compute 10% of the bill (by shifting the decimal point) and then add half of that amount for the total tip. For example, 15% of a \$43 restaurant bill is \$4.30 + \$2.15, which equals \$6.45. Explain why this technique is an application of the distributive property.

60. Suppose $b \neq 0$ and $d \neq 0$. Explain why

$$\frac{a}{b} = \frac{c}{d} \quad \text{if and only if} \quad ad = bc.$$

61. The first letters of the phrase "Please excuse my dear Aunt Sally" are used by some people to remember the order of operations: parentheses, exponents (which we will discuss in a later chapter), multiplication, division, addition, subtraction. Make up a catchy phrase that serves the same purpose but with exponents excluded.

62. (a) Verify that

$$\frac{16}{2} - \frac{25}{5} = \frac{16 - 25}{2 - 5}.$$

(b) From the example above you may be tempted to think that

$$\frac{a}{b} - \frac{c}{d} = \frac{a - c}{b - d}$$

provided none of the denominators equals 0. Give an example to show that this is not true.

63. Suppose $b \neq 0$ and $d \neq 0$. Explain why

$$\frac{a}{b} - \frac{c}{d} = \frac{ad - bc}{bd}.$$

WORKED-OUT SOLUTIONS *to Odd-numbered Exercises*

Do not read these worked-out solutions before first attempting to do the exercises yourself. Otherwise you may merely mimic the techniques shown here without understanding the ideas.

Best way to learn: Carefully read the section of the textbook, then do all the odd-numbered exercises (even if they have not been assigned) and check your answers here. If you get stuck on an exercise, reread the section of the textbook—then try the exercise again. If you are still stuck, then look at the worked-out solution here.

For Exercises 1–4, determine how many different values can arise by inserting one pair of parentheses into the given expression.

1. $19 - 12 - 8 - 2$

SOLUTION Here are the possibilities:

$19(-12 - 8 - 2) = -418$ $\quad$ $19 - (12 - 8) - 2 = 13$

$19(-12 - 8) - 2 = -382$ $\quad$ $19 - (12 - 8 - 2) = 17$

$19(-12) - 8 - 2 = -238$ $\quad$ $19 - 12 - 8(-2) = 23$

$(19 - 12) - 8 - 2 = -3$ $\quad$ $19 - 12(-8) - 2 = 113$

$19 - 12 - (8 - 2) = 1$ $\quad$ $19 - 12(-8 - 2) = 139$

Other possible ways to insert one pair of parentheses lead to values already included in the list above. For example,

$$(19 - 12 - 8) - 2 = -3.$$

Thus ten values are possible; they are −418, −382, −238, −3, 1, 13, 17, 23, 113, and 139.

3. $6 + 3 \cdot 4 + 5 \cdot 2$

SOLUTION Here are the possibilities:

$$(6 + 3 \cdot 4 + 5 \cdot 2) = 28$$
$$6 + (3 \cdot 4 + 5) \cdot 2 = 40$$
$$(6 + 3) \cdot 4 + 5 \cdot 2 = 46$$
$$6 + 3 \cdot (4 + 5 \cdot 2) = 48$$
$$6 + 3 \cdot (4 + 5) \cdot 2 = 60$$

Other possible ways to insert one pair of parentheses lead to values already included in the list above. For example,

$$(6 + 3 \cdot 4 + 5) \cdot 2 = 46.$$

Thus five values are possible; they are 28, 40, 46, 48, and 60.

For Exercises 5–22, expand the given expression.

5. $(x - y)(z + w - t)$

SOLUTION

$$\begin{aligned}(x - y)(z + w - t) \\ &= x(z + w - t) - y(z + w - t) \\ &= xz + xw - xt - yz - yw + yt\end{aligned}$$

7. $(2x + 3)^2$

SOLUTION

$$\begin{aligned}(2x + 3)^2 &= (2x)^2 + 2 \cdot (2x) \cdot 3 + 3^2 \\ &= 4x^2 + 12x + 9\end{aligned}$$

9. $(2c - 7)^2$

SOLUTION

$$\begin{aligned}(2c - 7)^2 &= (2c)^2 - 2 \cdot (2c) \cdot 7 + 7^2 \\ &= 4c^2 - 28c + 49\end{aligned}$$

11. $(x + y + z)^2$

SOLUTION

$$\begin{aligned}(x + y + z)^2 \\ &= (x + y + z)(x + y + z) \\ &= x(x + y + z) + y(x + y + z) + z(x + y + z) \\ &= x^2 + xy + xz + yx + y^2 + yz \\ &\quad + zx + zy + z^2 \\ &= x^2 + y^2 + z^2 + 2xy + 2xz + 2yz\end{aligned}$$

13. $(x + 1)(x - 2)(x + 3)$

SOLUTION

$$\begin{aligned}(x + 1)(x - 2)(x + 3) \\ &= ((x + 1)(x - 2))(x + 3) \\ &= (x^2 - 2x + x - 2)(x + 3) \\ &= (x^2 - x - 2)(x + 3) \\ &= x^3 + 3x^2 - x^2 - 3x - 2x - 6 \\ &= x^3 + 2x^2 - 5x - 6\end{aligned}$$

15. $(a + 2)(a - 2)(a^2 + 4)$

SOLUTION

$$\begin{aligned}(a + 2)(a - 2)(a^2 + 4) &= ((a + 2)(a - 2))(a^2 + 4) \\ &= (a^2 - 4)(a^2 + 4) \\ &= a^4 - 16\end{aligned}$$

17. $xy(x + y)(\frac{1}{x} - \frac{1}{y})$

SOLUTION

$$\begin{aligned}xy(x + y)\left(\frac{1}{x} - \frac{1}{y}\right) &= xy(x + y)\left(\frac{y}{xy} - \frac{x}{xy}\right) \\ &= xy(x + y)\left(\frac{y - x}{xy}\right) \\ &= (x + y)(y - x) \\ &= y^2 - x^2\end{aligned}$$

19. $(t - 2)(t^2 + 2t + 4)$

SOLUTION

$$(t-2)(t^2+2t+4) = t(t^2+2t+4) - 2(t^2+2t+4)$$
$$= t^3 + 2t^2 + 4t - 2t^2 - 4t - 8$$
$$= t^3 - 8$$

21. $(n+3)(n^2-3n+9)$

SOLUTION

$$(n+3)(n^2-3n+9)$$
$$= n(n^2-3n+9) + 3(n^2-3n+9)$$
$$= n^3 - 3n^2 + 9n + 3n^2 - 9n + 27$$
$$= n^3 + 27$$

For Exercises 23–50, simplify the given expression as much as possible.

23. $4(2m+3n)+7m$

SOLUTION

$$4(2m+3n)+7m = 8m + 12n + 7m$$
$$= 15m + 12n$$

25. $\frac{3}{4} + \frac{6}{7}$

SOLUTION $\frac{3}{4} + \frac{6}{7} = \frac{3}{4}\cdot\frac{7}{7} + \frac{6}{7}\cdot\frac{4}{4} = \frac{21}{28} + \frac{24}{28} = \frac{45}{28}$

27. $\frac{3}{4}\cdot\frac{14}{39}$

SOLUTION $\frac{3}{4}\cdot\frac{14}{39} = \frac{3\cdot 14}{4\cdot 39} = \frac{7}{2\cdot 13} = \frac{7}{26}$

29. $\frac{\frac{5}{7}}{\frac{2}{3}}$

SOLUTION $\frac{\frac{5}{7}}{\frac{2}{3}} = \frac{5}{7}\cdot\frac{3}{2} = \frac{5\cdot 3}{7\cdot 2} = \frac{15}{14}$

31. $\frac{m+1}{2} + \frac{3}{n}$

SOLUTION

$$\frac{m+1}{2} + \frac{3}{n} = \frac{m+1}{2}\cdot\frac{n}{n} + \frac{3}{n}\cdot\frac{2}{2}$$
$$= \frac{(m+1)n + 3\cdot 2}{2n}$$
$$= \frac{mn+n+6}{2n}$$

33. $\frac{2}{3}\cdot\frac{4}{5} + \frac{3}{4}\cdot 2$

SOLUTION

$$\frac{2}{3}\cdot\frac{4}{5} + \frac{3}{4}\cdot 2 = \frac{8}{15} + \frac{3}{2}$$
$$= \frac{8}{15}\cdot\frac{2}{2} + \frac{3}{2}\cdot\frac{15}{15}$$
$$= \frac{16+45}{30}$$
$$= \frac{61}{30}$$

35. $\frac{2}{5}\cdot\frac{m+3}{7} + \frac{1}{2}$

SOLUTION

$$\frac{2}{5}\cdot\frac{m+3}{7} + \frac{1}{2} = \frac{2m+6}{35} + \frac{1}{2}$$
$$= \frac{2m+6}{35}\cdot\frac{2}{2} + \frac{1}{2}\cdot\frac{35}{35}$$
$$= \frac{4m+12+35}{70}$$
$$= \frac{4m+47}{70}$$

37. $\frac{2}{x+3} + \frac{y-4}{5}$

SOLUTION

$$\frac{2}{x+3} + \frac{y-4}{5} = \frac{2}{x+3} \cdot \frac{5}{5} + \frac{y-4}{5} \cdot \frac{x+3}{x+3}$$

$$= \frac{2 \cdot 5 + (y-4)(x+3)}{5(x+3)}$$

$$= \frac{10 + yx + 3y - 4x - 12}{5(x+3)}$$

$$= \frac{xy - 4x + 3y - 2}{5(x+3)}$$

39. $\frac{4t+1}{t^2} + \frac{3}{t}$

SOLUTION

$$\frac{4t+1}{t^2} + \frac{3}{t} = \frac{4t+1}{t^2} + \frac{3}{t} \cdot \frac{t}{t}$$

$$= \frac{4t+1}{t^2} + \frac{3t}{t^2}$$

$$= \frac{7t+1}{t^2}$$

41. $\frac{3}{v(v-2)} + \frac{v+1}{v^3}$

SOLUTION

$$\frac{3}{v(v-2)} + \frac{v+1}{v^3} = \frac{v^2}{v^2} \cdot \frac{3}{v(v-2)} + \frac{v+1}{v^3} \cdot \frac{v-2}{v-2}$$

$$= \frac{3v^2}{v^3(v-2)} + \frac{v^2 - v - 2}{v^3(v-2)}$$

$$= \frac{4v^2 - v - 2}{v^3(v-2)}$$

43. $\frac{1}{x-y}\left(\frac{x}{y} - \frac{y}{x}\right)$

SOLUTION

$$\frac{1}{x-y}\left(\frac{x}{y} - \frac{y}{x}\right) = \frac{1}{x-y}\left(\frac{x}{y} \cdot \frac{x}{x} - \frac{y}{x} \cdot \frac{y}{y}\right)$$

$$= \frac{1}{x-y}\left(\frac{x^2 - y^2}{xy}\right)$$

$$= \frac{1}{x-y}\left(\frac{(x+y)(x-y)}{xy}\right)$$

$$= \frac{x+y}{xy}$$

45. $\frac{(x+a)^2 - x^2}{a}$

SOLUTION

$$\frac{(x+a)^2 - x^2}{a} = \frac{x^2 + 2xa + a^2 - x^2}{a}$$

$$= \frac{2xa + a^2}{a}$$

$$= 2x + a$$

47. $\frac{\frac{x-2}{y}}{\frac{z}{x+2}}$

SOLUTION

$$\frac{\frac{x-2}{y}}{\frac{z}{x+2}} = \frac{x-2}{y} \cdot \frac{x+2}{z}$$

$$= \frac{x^2 - 4}{yz}$$

49. $\frac{\frac{a-t}{b-c}}{\frac{b+c}{a+t}}$

SOLUTION

$$\frac{\frac{a-t}{b-c}}{\frac{b+c}{a+t}} = \frac{a-t}{b-c} \cdot \frac{a+t}{b+c}$$

$$= \frac{a^2 - t^2}{b^2 - c^2}$$

1.3 Inequalities, Intervals, and Absolute Value

LEARNING OBJECTIVES

By the end of this section you should be able to

- apply the algebraic properties involving positive and negative numbers;
- manipulate inequalities;
- use interval notation for the four types of intervals;
- use interval notation involving $-\infty$ and ∞;
- work with unions of intervals;
- manipulate and interpret expressions involving absolute value.

From now on, "number" means "real number" unless otherwise stated.

Positive and Negative Numbers

Positive and negative numbers

- A number is called **positive** if it is right of 0 on the real line.
- A number is called **negative** if it is left of 0 on the real line.

The words "positive" and "negative" have many uses in English in addition to their mathematical meaning. Some of these uses, such as in the phrase "photographic negative", are related to the mathematical meaning.

Every number is either right of 0, left of 0, or equal to 0. Thus every number is either positive, negative, or 0.

$-3 \quad -\frac{5}{2} \quad -2 \quad -\frac{115}{76} \quad -1 \quad -\frac{2}{3} \quad -\frac{1}{3} \quad 0 \quad \frac{1}{3} \quad \frac{2}{3} \quad 1 \quad \frac{12}{7} \quad 2 \quad \frac{257}{101} \quad 3$

negative numbers *positive numbers*

All of the following properties should already be familiar to you.

Algebraic properties of positive and negative numbers

- The sum of two positive numbers is positive.
- The sum of two negative numbers is negative.
- The additive inverseof a positive number is negative.
- The additive inverse of a negative number is positive.
- The product of two positive numbers is positive.
- The product of two negative numbers is positive.
- The product of a positive number and a negative number is negative.
- The multiplicative inverse of a positive number is positive.
- The multiplicative inverse of a negative number is negative.

Example:

$2 + 3 = 5$

$(-2) + (-3) = -5$

-2 *is negative*

$-(-2)$ *is positive*

$2 \cdot 3 = 6$

$(-2) \cdot (-3) = 6$

$2 \cdot (-3) = -6$

$\frac{1}{2}$ *is positive*

$\frac{1}{-2}$ *is negative*

Lesser and Greater

We say that a number a is **less than** a number b, written $a < b$, if a is left of b on the real line. Equivalently, $a < b$ if and only if $b - a$ is positive. In particular, b is positive if and only if $0 < b$.

a b

$a < b$.

We say that a is **less than or equal to** b, written $a \leq b$, if $a < b$ or $a = b$. Thus the statement $x < 4$ is true if x equals 3 but false if x equals 4, whereas the statement $x \leq 4$ is true if x equals 3 and also true if x equals 4.

We say that b is **greater than** a, written $b > a$, if b is right of a on the real line. Thus $b > a$ means the same as $a < b$. Similarly, we say that b is **greater than or equal to** a, written $b \geq a$, if $b > a$ or $b = a$. Thus $b \geq a$ means the same as $a \leq b$.

We now begin discussion of a series of simple but crucial properties of inequalities. The first property we will discuss is called transitivity.

Transitivity

If $a < b$ and $b < c$, then $a < c$.

For example, from the inequalities $\sqrt{15} < 4$ and $4 < \frac{21}{5}$ we can conclude that $\sqrt{15} < \frac{21}{5}$.

To see why transitivity holds, suppose $a < b$ and $b < c$. Then a is left of b on the real line and b is left of c. This implies that a is left of c, which means that $a < c$; see the figure below.

a b c

Transitivity: $a < b$ and $b < c$ implies that $a < c$.

Often multiple inequalities are written together as a single string of inequalities. Thus $a < b < c$ means the same thing as $a < b$ and $b < c$.

Our next result shows that we can add inequalities.

Addition of inequalities

If $a < b$ and $c < d$, then $a + c < b + d$.

For example, from the inequalities $\sqrt{8} < 3$ and $4 < \sqrt{17}$ we can conclude that $\sqrt{8} + 4 < 3 + \sqrt{17}$.

To see why this is true, note that if $a < b$ and $c < d$, then $b - a$ and $d - c$ are positive numbers. Because the sum of two positive numbers is positive, this implies that $(b - a) + (d - c)$ is positive. In other words, $(b + d) - (a + c)$ is positive. This means that $a + c < b + d$, as desired.

The next result states that we can multiply both sides of an inequality by a positive number and preserve the inequality. However, if both sides of an inequality are multiplied by a negative number, then the direction of the inequality must be reversed.

Multiplication of an inequality

Suppose $a < b$.

- If $c > 0$, then $ac < bc$.
- If $c < 0$, then $ac > bc$.

For example, from the inequality $\sqrt{7} < \sqrt{8}$ we can conclude that $3\sqrt{7} < 3\sqrt{8}$ and $(-3)\sqrt{7} > (-3)\sqrt{8}$.

To see why this is true, first suppose $c > 0$. We are assuming that $a < b$, which means that $b - a$ is positive. Because the product of two positive numbers is positive, this implies that $(b - a)c$ is positive. In other words, $bc - ac$ is positive, which means that $ac < bc$, as desired.

Now consider the case where $c < 0$. We are still assuming that $a < b$, which means that $b - a$ is positive. Because the product of a positive number and a negative number is negative, this implies that $(b - a)c$ is negative. In other words, $bc - ac$ is negative, which means that $ac > bc$, as desired.

An important special case of the result above is obtained by setting $c = -1$, which gives the following result:

Additive inverse and inequalities

If $a < b$, then $-a > -b$.

For example, from the inequality $2 < 3$ we can conclude that $-2 > -3$.

In other words, the direction of an inequality must be reversed when taking additive inverses of both sides.

The next result shows that the direction of an inequality must also be reversed when taking multiplicative inverses of both sides, unless one side is negative and the other side is positive.

Multiplicative inverse and inequalities

Suppose $a < b$.

- If a and b are both positive or both negative, then $\frac{1}{a} > \frac{1}{b}$.
- If $a < 0 < b$, then $\frac{1}{a} < \frac{1}{b}$.

For example, from the inequality $2 < 3$ we can conclude that $\frac{1}{2} > \frac{1}{3}$.

To see why this is true, first suppose a and b are both positive or both negative. In either case, ab is positive. Thus $\frac{1}{ab} > 0$. Thus we can multiply both sides of the inequality $a < b$ by $\frac{1}{ab}$, preserving the direction of the inequality. This gives

$$a \cdot \frac{1}{ab} < b \cdot \frac{1}{ab},$$

which is the same as $\frac{1}{b} < \frac{1}{a}$, or equivalently $\frac{1}{a} > \frac{1}{b}$, as desired.

The case where $a < 0 < b$ is even easier. In this case $\frac{1}{a}$ is negative and $\frac{1}{b}$ is positive. Thus $\frac{1}{a} < \frac{1}{b}$, as desired.

Intervals

We begin this subsection with an imprecise definition.

Set

A **set** is a collection of objects.

This definition is imprecise because the words "collection" and "objects" are vague.

The collection of positive numbers is an example of a set, as is the collection of odd negative integers. Most of the sets considered in this book are collections of real numbers, which at least removes some of the vagueness from the word "objects".

If a set contains only finitely many objects, then the objects in the set can be explicitly displayed between the symbols { }. For example, the set consisting of the numbers 4, $-\frac{17}{7}$, and $\sqrt{2}$ can be denoted by

$$\{4, -\tfrac{17}{7}, \sqrt{2}\}.$$

Sets can also be denoted by a property that characterizes objects of the set. For example, the set of real numbers greater than 2 can be denoted by

$$\{x : x > 2\}.$$

Here the notation $\{x : \dots\}$ should be read to mean "the set of real numbers x such that" and then whatever follows. There is no particular x here. The variable is simply a convenient device to describe a property, and the symbol used for the variable does not matter. Thus $\{x : x > 2\}$ and $\{y : y > 2\}$ and $\{t : t > 2\}$ all denote the same set, which can also be described (without mentioning any variables) as the set of real numbers greater than 2.

A special type of set occurs so often in mathematics that it gets its own name, which is given by the following definition.

Interval

An **interval** is a set of real numbers that contains all numbers between any two numbers in the set.

For example, the set of positive numbers is an interval because all numbers between any two positive numbers are positive. As a nonexample, the set of integers is not an interval because 0 and 1 are in this set, but $\frac{2}{3}$, which is between 0 and 1, is not in this set. As another nonexample, the set of rational numbers is not an interval, because 1 and 2 are in this set, but $\sqrt{2}$, which is between 1 and 2, is not in this set.

Intervals are so useful in mathematics that special notation has been designed for them. Suppose a and b are numbers with $a < b$. We define the following four intervals with endpoints a and b:

Intervals

- The **open interval** (a,b) with endpoints a and b is the set of numbers between a and b, not including either endpoint:

$$(a,b) = \{x : a < x < b\}.$$

The definition of $[a,b]$ also makes sense when $a = b$; the interval $[a,a]$ consists of the single number a.

- The **closed interval** $[a,b]$ with endpoints a and b is the set of numbers between a and b, including both endpoints:

$$[a,b] = \{x : a \leq x \leq b\}.$$

- The **half-open interval** $[a,b)$ with endpoints a and b is the set of numbers between a and b, including a but not including b:

$$[a,b) = \{x : a \leq x < b\}.$$

The term "half-closed" would make as much sense as "half-open".

- The **half-open interval** $(a,b]$ with endpoints a and b is the set of numbers between a and b, including b but not including a:

$$(a,b] = \{x : a < x \leq b\}.$$

With this notation, a parenthesis indicates that the corresponding endpoint is not included in the set, and a straight bracket indicates that the corresponding endpoint is included in the set. Thus the interval $(3,7]$ includes the numbers 4, $\sqrt{17}$, 5.49, and the endpoint 7 (along with many other numbers), but does not include the numbers 2 or 9 or the endpoint 3.

Sometimes we need to use intervals that extend arbitrarily far to the left or to the right on the real line. Suppose a is a real number. We define the following four intervals with endpoint a:

Intervals extending arbitrarily far

Example: $(0,\infty)$ denotes the set of positive numbers.

- The interval (a,∞) is the set of numbers greater than a:

$$(a,\infty) = \{x : x > a\}.$$

- The interval $[a,\infty)$ is the set of numbers greater than or equal to a:

$$[a,\infty) = \{x : x \geq a\}.$$

Example: $(-\infty,0)$ denotes the set of negative numbers.

- The interval $(-\infty,a)$ is the set of numbers less than a:

$$(-\infty,a) = \{x : x < a\}.$$

- The interval $(-\infty,a]$ is the set of numbers less than or equal to a:

$$(-\infty,a] = \{x : x \leq a\}.$$

Here the symbol ∞, called **infinity**, should be thought of simply as a notational convenience. Neither ∞ nor $-\infty$ is a real number; these symbols have no meaning in this context other than as notational shorthand. For example, the interval $(2, \infty)$ is defined to be the set of real numbers greater than 2 (note that ∞ is not mentioned in this definition). The notation $(2, \infty)$ is often used because writing $(2, \infty)$ is easier than writing $\{x : x > 2\}$.

As before, a parenthesis indicates that the corresponding endpoint is not included in the set, and a straight bracket indicates that the corresponding endpoint is included in the set. Thus the interval $(2, \infty)$ does not include the endpoint 2, but the interval $[2, \infty)$ does include the endpoint 2. Both of the intervals $(2, \infty)$ and $[2, \infty)$ include 2.5 and 98765 (along with many other numbers); neither of these intervals includes 1.5 or -857.

There do not exist intervals with a closed bracket adjacent to $-\infty$ or ∞. For example, $[-\infty, 2]$ and $[2, \infty]$ do not make sense because the closed brackets indicate that both endpoints should be included. The symbols $-\infty$ and ∞ can never be included in a set of real numbers because these symbols do not denote real numbers.

Some books use the notation $(-\infty, \infty)$ to denote the set of real numbers.

EXAMPLE 1

A party invitation states that guests can arrive any time after 4 pm on June 30. Let time be measured in hours from noon on June 30. If the invitation is taken literally, write an interval to represent acceptable times for guests to arrive.

SOLUTION Any time greater than 4 hours from noon on June 30 is acceptable. Thus the interval of acceptable times is $(4, \infty)$.

In later chapters we will occasionally find it useful to work with the union of two intervals. Here is the definition of union:

Union

The **union** of two sets A and B, denoted $A \cup B$, is the set of objects that are contained in at least one of the sets A and B.

Similarly, the union of three or more sets is the collection of objects that are contained in at least one of the sets.

Thus $A \cup B$ consists of the objects (usually numbers) that belong either to A or to B or to both A and B.

EXAMPLE 2

Write $(1, 5) \cup (3, 7]$ as an interval.

SOLUTION
The figure here shows that every number in the interval $(1, 7]$ is either in $(1, 5)$ or is in $(3, 7]$ or is in both $(1, 5)$ and $(3, 7]$. The figure shows that $(1, 5) \cup (3, 7] = (1, 7]$.

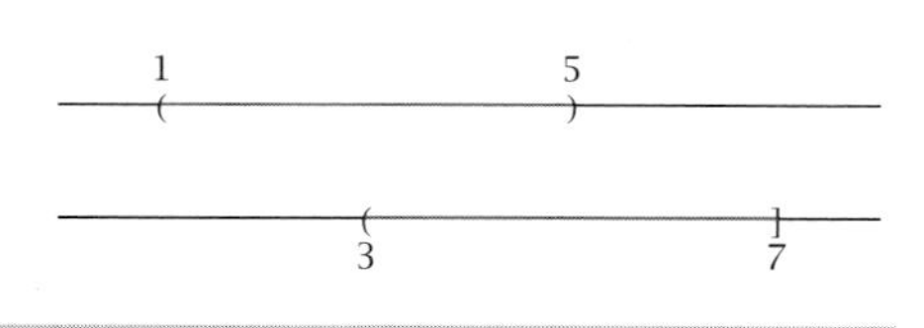

The next example goes in the other direction, starting with a set and then writing it as a union of intervals.

EXAMPLE 3 Write the set of nonzero real numbers as the union of two intervals.

SOLUTION The set of nonzero real numbers is the union of the set of negative numbers and the set of positive numbers. In other words, the set of nonzero real numbers equals $(-\infty, 0) \cup (0, \infty)$.

Absolute Value

The **absolute value**of a number is its distance from 0; here we are thinking of numbers as points on the real line. For example, the absolute value of $\frac{3}{2}$ equals $\frac{3}{2}$, as can be seen in the figure below. More interestingly, the absolute value of $-\frac{3}{2}$ equals $\frac{3}{2}$.

$-2 \quad -\frac{3}{2} \quad -1 \quad 0 \quad 1 \quad \frac{3}{2} \quad 2$

The absolute value of a number is its distance to 0.

The absolute value of a number b is denoted by $|b|$. Thus $|\frac{3}{2}| = \frac{3}{2}$ and $|-\frac{3}{2}| = \frac{3}{2}$. Here is the formal definition of absolute value:

Absolute value

The **absolute value** of a number b, denoted $|b|$, is defined by

$$|b| = \begin{cases} b & \text{if } b \geq 0 \\ -b & \text{if } b < 0. \end{cases}$$

This definition implies that $|b| \geq 0$ for every real number b.

For example, $-\frac{3}{2} < 0$, and thus by the formula above $|-\frac{3}{2}|$ equals $-(-\frac{3}{2})$, which equals $\frac{3}{2}$.

The concept of absolute value is fairly simple—just strip away the minus sign from any number that happens to have one. However, this rule can be applied only to numbers, not to expressions whose value is unknown. For example, if we encounter the expression $|-x|$, we cannot simplify this expression to x unless we know that $x \geq 0$. If x happens to be negative, then $|-x| = -x$; stripping away the negative sign would be incorrect in this case.

$|-x| = |x|$ regardless of the value of x, as you are asked to explain in Problem 72.

Inequalities involving absolute values can be written without using an absolute value, as shown in the following example.

EXAMPLE 4

(a) Write the inequality $|x| < 2$ without using an absolute value.

(b) Write the set $\{x : |x| < 2\}$ as an interval.

SOLUTION

(a) A number has absolute value less than 2 if only and only if its distance from 0 is less than 2, and this happens if and only if the number is between -2 and 2. Hence the inequality $|x| < 2$ could be written as

$$-2 < x < 2.$$

(b) The inequality above implies that the set $\{x : |x| < 2\}$ equals the open interval $(-2, 2)$.

In the next example, we end up with an interval not centered at 0.

EXAMPLE 5

$|x - 5|$ is the distance between x and 5. Thus $\{x : |x - 5| < 1\}$ is the set of points on the real line whose distance to 5 is less than 1.

(a) Write the inequality $|x - 5| < 1$ without using an absolute value.

(b) Write the set $\{x : |x - 5| < 1\}$ as an interval.

SOLUTION

(a) The absolute value of a number is less than 1 precisely when the number is between -1 and 1. Thus the inequality $|x - 5| < 1$ is equivalent to

$$-1 < x - 5 < 1.$$

Adding 5 to all three parts of the inequality above transforms it to the inequality

$$4 < x < 6.$$

(b) The inequality above implies that the set $\{x : |x - 5| < 1\}$ equals the open interval $(4, 6)$.

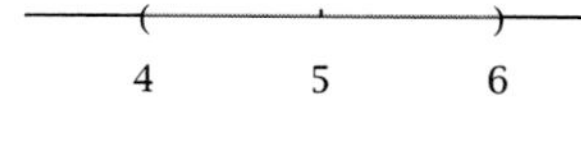

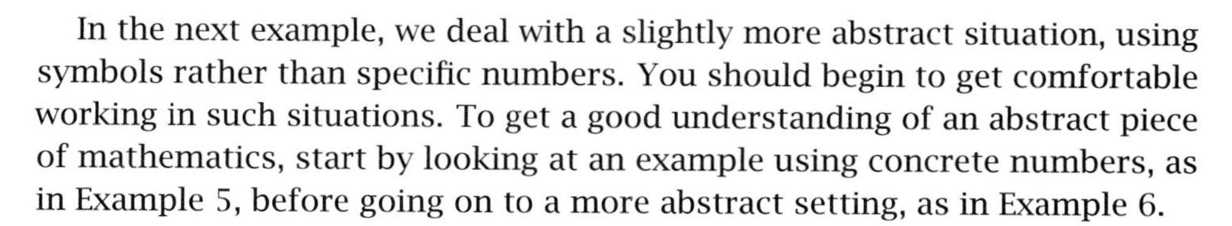

In the next example, we deal with a slightly more abstract situation, using symbols rather than specific numbers. You should begin to get comfortable working in such situations. To get a good understanding of an abstract piece of mathematics, start by looking at an example using concrete numbers, as in Example 5, before going on to a more abstract setting, as in Example 6.

EXAMPLE 6

$|x - b|$ is the distance between x and b. Thus $\{x : |x - b| < h\}$ is the set of points on the real line whose distance to b is less than h.

Suppose b is a real number and $h > 0$.

(a) Write the inequality $|x - b| < h$ without using an absolute value.

(b) Write the set $\{x : |x - b| < h\}$ as an interval.

SOLUTION

(a) The absolute value of a number is less than h precisely when the number is between $-h$ and h. Thus the inequality $|x - b| < h$ is equivalent to

$$-h < x - b < h.$$

Adding b to all three parts of the inequality above transforms it to the inequality

$$b - h < x < b + h.$$

(b) The inequality above implies that the set $\{x : |x - b| < h\}$ equals the open interval $(b - h, b + h)$.

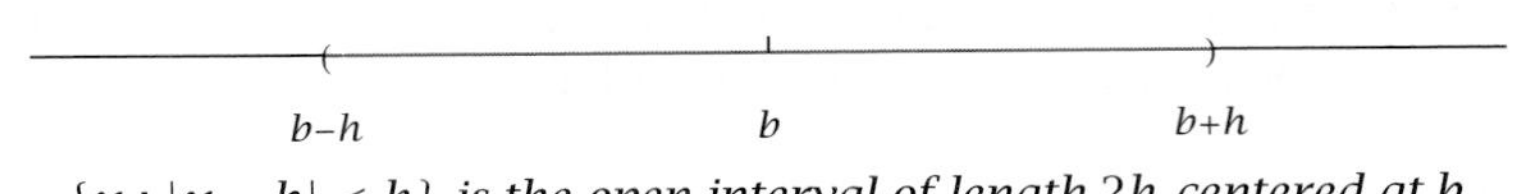

$\{x : |x - b| < h\}$ is the open interval of length $2h$ centered at b.

The set of numbers satisfying an inequality involving an absolute value may be the union of two intervals, as shown in the next example.

EXAMPLE 7

Ball bearings need to have extremely accurate sizes to work correctly. The ideal diameter of a particular ball bearing is 0.8 cm, but a ball bearing is declared acceptable if the error in the diameter size is less than 0.001 cm.

(a) Write an inequality using absolute values and the diameter d of a ball bearing (measured in cm, which is an abbreviation for centimeters) that gives the condition for a ball bearing to be unacceptable.

(b) Write the set of numbers satisfying the inequality produced in part (a) as the union of two intervals.

SOLUTION

(a) The error in the diameter size is $|d - 0.8|$. Thus a ball bearing with diameter d is unacceptable if

$$|d - 0.8| \geq 0.001.$$

(b) Because $0.8 - 0.001 = 0.799$ and $0.8 + .001 = 0.801$, the set of numbers d such that $|d - 0.8| \geq 0.001$ is $(-\infty, 0.799] \cup [0.801, \infty)$.

Equations involving absolute values must often be solved by considering multiple possibilities. Here is a simple example:

EXAMPLE 8

Find all numbers t such that

$$|3t - 4| = 10.$$

SOLUTION The equation $|3t - 4| = 10$ implies that $3t - 4 = 10$ or $3t - 4 = -10$. Solving these equations for t gives $t = \frac{14}{3}$ or $t = -2$. Substituting these values for t back into the original equation shows that both $\frac{14}{3}$ and -2 are indeed solutions.

As a more complicated example, consider the equation

$$|x - 3| + |x - 4| = 9.$$

The worked-out solution to Exercise 9 shows how to find the solutions without a symbolic calculator.

To find the solutions to this equation, think of the set of real numbers as the union of the three intervals $(-\infty, 3)$, $[3, 4]$, and $(4, \infty)$ and consider what the equation above becomes for x in each of those three intervals.

The next example shows how to use Wolfram|Alpha to find solutions to this equation.

EXAMPLE 9

Use Wolfram|Alpha to find all numbers x such that

$$|x-3| + |x-4| = 9.$$

SOLUTION In a Wolfram|Alpha entry box, enter

```
solve |x-3| + |x-4| = 9
```

to get the result that the solutions are $x = -1$ and $x = 8$ (ignore the Wolfram|Alpha output that deals with anything other than the real number solutions).

EXERCISES

1. Evaluate $|-4| + |4|$.
2. Evaluate $|5| + |-6|$.
3. Find all numbers with absolute value 9.
4. Find all numbers with absolute value 10.

In Exercises 5–18, find all numbers x satisfying the given equation.

5. $|2x - 6| = 11$
6. $|5x + 8| = 19$
7. $\left|\frac{x+1}{x-1}\right| = 2$
8. $\left|\frac{3x+2}{x-4}\right| = 5$
9. $|x-3| + |x-4| = 9$
10. $|x+1| + |x-2| = 7$
11. $|x-3| + |x-4| = 1$
12. $|x+1| + |x-2| = 3$
13. $|x-3| + |x-4| = \frac{1}{2}$
14. $|x+1| + |x-2| = 2$
15. $|x + 3| = x + 3$
16. $|x - 5| = 5 - x$
17. $|x| = x + 1$
18. $|x + 3| = x + 5$

In Exercises 19–28, write each union as a single interval.

19. $[2,7) \cup [5,20)$
20. $[-8,-3) \cup [-6,-1)$
21. $[-2,8] \cup (-1,4)$
22. $(-9,-2) \cup [-7,-5]$
23. $(3,\infty) \cup [2,8]$
24. $(-\infty,4) \cup (-2,6]$
25. $(-\infty,-3) \cup [-5,\infty)$
26. $(-\infty,-6] \cup (-8,12)$
27. $(-3,\infty) \cup [-5,\infty)$
28. $(-\infty,-10] \cup (-\infty,-8]$

29. A medicine is known to decompose and become ineffective if its temperature ever reaches 103 degrees Fahrenheit or more. Write an interval to represent the temperatures (in degrees Fahrenheit) at which the medicine is ineffective.

30. At normal atmospheric pressure, water boils at all temperatures of 100 degrees Celsius and higher. Write an interval to represent the temperatures (in degrees Celsius) at which water boils.

31. A shoelace manufacturer guarantees that its 33-inch shoelaces will be 33 inches long, with an error of at most 0.1 inch.

    (a) Write an inequality using absolute values and the length s of a shoelace that gives the condition that the shoelace does not meet the guarantee.

    (b) Write the set of numbers satisfying the inequality in part (a) as a union of two intervals.

32. A copying machine works with paper that is 8.5 inches wide, provided that the error in the paper width is less than 0.06 inch.

    (a) Write an inequality using absolute values and the length w of the paper that gives the condition that the paper's width fails the requirements of the copying machine.

    (b) Write the set of numbers satisfying the inequality in part (a) as a union of two intervals.

33. Give four examples of pairs of real numbers a and b such that

$$|a + b| = 2 \quad \text{and} \quad |a| + |b| = 8.$$

34. Give four examples of pairs of real numbers a and b such that

$$|a + b| = 3 \quad \text{and} \quad |a| + |b| = 11.$$

In Exercises 35–46, write each set as an interval or as a union of two intervals.

35. $\{x : |x - 4| < \frac{1}{10}\}$
36. $\{x : |x + 2| < \frac{1}{100}\}$
37. $\{x : |x + 4| < \frac{\varepsilon}{2}\}$; here $\varepsilon > 0$ [*Mathematicians often use the Greek letter* ε, *which is called* **epsilon**, *to denote a small positive number.*]
38. $\{x : |x - 2| < \frac{\varepsilon}{3}\}$; here $\varepsilon > 0$
39. $\{y : |y - a| < \varepsilon\}$; here $\varepsilon > 0$
40. $\{y : |y + b| < \varepsilon\}$; here $\varepsilon > 0$
41. $\{x : |3x - 2| < \frac{1}{4}\}$
42. $\{x : |4x - 3| < \frac{1}{5}\}$
43. $\{x : |x| > 2\}$
44. $\{x : |x| > 9\}$
45. $\{x : |x - 5| \geq 3\}$
46. $\{x : |x + 6| \geq 2\}$

The* intersection *of two sets of numbers consists of all numbers that are in both sets. If* A *and* B *are sets, then their intersection is denoted by* $A \cap B$*. In Exercises 47–56, write each intersection as a single interval.

47. $[2, 7) \cap [5, 20)$
48. $[-8, -3) \cap [-6, -1)$
49. $[-2, 8] \cap (-1, 4)$
50. $(-9, -2) \cap [-7, -5]$
51. $(3, \infty) \cap [2, 8]$
52. $(-\infty, 4) \cap (-2, 6]$
53. $(-\infty, -3) \cap [-5, \infty)$
54. $(-\infty, -6] \cap (-8, 12)$
55. $(-3, \infty) \cap [-5, \infty)$
56. $(-\infty, -10] \cap (-\infty, -8]$

PROBLEMS

57. Suppose a and b are numbers. Explain why either $a < b$, $a = b$, or $a > b$.
58. Show that if $a < b$ and $c \leq d$, then $a + c < b + d$.
59. Show that if b is a positive number and $a < b$, then
$$\frac{a}{b} < \frac{a+1}{b+1}.$$
60. In contrast to Problem 62 in Section 1.2, show that there do not exist positive numbers a, b, c, and d such that
$$\frac{a}{b} + \frac{c}{d} = \frac{a+c}{b+d}.$$
61. Explain why every open interval containing 0 contains an open interval centered at 0.
62. Give an example of an open interval and a closed interval whose union equals the interval $(2, 5)$.
63. (a) True or false:

    If $a < b$ and $c < d$, then $c - b < d - a$.

    (b) Explain your answer to part (a). This means that if the answer to part (a) is "true", then you should explain why $c - b < d - a$ whenever $a < b$ and $c < d$; if the answer to part (a) is "false", then you should give an example of numbers a, b, c, and d such that $a < b$ and $c < d$ but $c - b \geq d - a$.
64. (a) True or false:

    If $a < b$ and $c < d$, then $ac < bd$.

    (b) Explain your answer to part (a). This means that if the answer to part (a) is "true", then you should explain why $ac < bd$ whenever $a < b$ and $c < d$; if the answer to part (a) is "false", then you should give an example of numbers a, b, c, and d such that $a < b$ and $c < d$ but $ac \geq bd$.
65. (a) True or false:

    If $0 < a < b$ and $0 < c < d$, then $\dfrac{a}{d} < \dfrac{b}{c}$.

    (b) Explain your answer to part (a). This means that if the answer to part (a) is "true", then you should explain why $\frac{a}{d} < \frac{b}{c}$ whenever $0 < a < b$ and $0 < c < d$; if the answer to part (a) is "false", then you should give an example of numbers a, b, c, and d such that $0 < a < b$ and $0 < c < d$ but
$$\frac{a}{d} \geq \frac{b}{c}.$$
66. Give an example of an open interval and a closed interval whose intersection equals the interval $(2, 5)$.
67. Give an example of an open interval and a closed interval whose union equals the interval $[-3, 7]$.

68. Give an example of an open interval and a closed interval whose intersection equals the interval $[-3, 7]$.

69. Explain why the equation

$$|8x - 3| = -2$$

has no solutions.

70. Explain why

$$|a^2| = a^2$$

for every real number a.

71. Explain why

$$|ab| = |a||b|$$

for all real numbers a and b.

72. Explain why

$$|-a| = |a|$$

for all real numbers a.

73. Explain why

$$\left|\frac{a}{b}\right| = \frac{|a|}{|b|}$$

for all real numbers a and b (with $b \neq 0$).

74. Give an example of a set of real numbers such that the average of any two numbers in the set is in the set, but the set is not an interval.

75. (a) Show that if $a \geq 0$ and $b \geq 0$, then $|a+b| = |a| + |b|$.

(b) Show that if $a \geq 0$ and $b < 0$, then $|a+b| \leq |a| + |b|$.

(c) Show that if $a < 0$ and $b \geq 0$, then $|a+b| \leq |a| + |b|$.

(d) Show that if $a < 0$ and $b < 0$, then $|a+b| = |a| + |b|$.

(e) Explain why the previous four items imply that

$$|a + b| \leq |a| + |b|$$

for all real numbers a and b.

76. Show that if a and b are real numbers such that

$$|a + b| < |a| + |b|,$$

then $ab < 0$.

77. Show that

$$\big||a| - |b|\big| \leq |a - b|$$

for all real numbers a and b.

WORKED-OUT SOLUTIONS *to Odd-numbered Exercises*

1. Evaluate $|-4| + |4|$.

SOLUTION $|-4| + |4| = 4 + 4 = 8$

3. Find all numbers with absolute value 9.

SOLUTION The only numbers whose absolute value equals 9 are 9 and -9.

In Exercises 5–18, find all numbers x satisfying the given equation.

5. $|2x - 6| = 11$

SOLUTION The equation $|2x - 6| = 11$ implies that $2x - 6 = 11$ or $2x - 6 = -11$. Solving these equations for x gives $x = \frac{17}{2}$ or $x = -\frac{5}{2}$.

7. $\left|\frac{x+1}{x-1}\right| = 2$

SOLUTION The equation $\left|\frac{x+1}{x-1}\right| = 2$ implies that $\frac{x+1}{x-1} = 2$ or $\frac{x+1}{x-1} = -2$. Solving these equations for x gives $x = 3$ or $x = \frac{1}{3}$.

9. $|x-3| + |x-4| = 9$

SOLUTION First, consider numbers x such that $x > 4$. In this case, we have $x - 3 > 0$ and $x - 4 > 0$, which implies that $|x - 3| = x - 3$ and $|x - 4| = x - 4$. Thus the original equation becomes

$$x - 3 + x - 4 = 9,$$

which can be rewritten as $2x - 7 = 9$, which can easily be solved to yield $x = 8$. Substituting 8 for x in the original equation shows that $x = 8$ is indeed a solution (make sure you do this check).

Second, consider numbers x such that $x < 3$. In this case, we have $x - 3 < 0$ and $x - 4 < 0$, which implies that $|x - 3| = 3 - x$ and $|x - 4| = 4 - x$. Thus the original equation becomes

$$3 - x + 4 - x = 9,$$

which can be rewritten as $7 - 2x = 9$, which can easily be solved to yield $x = -1$. Substituting -1 for x in the original equation shows that

$x = -1$ is indeed a solution (make sure you do this check).

Third, we need to consider the only remaining possibility, which is that $3 \le x \le 4$. In this case, we have $x - 3 \ge 0$ and $x - 4 \le 0$, which implies that $|x - 3| = x - 3$ and $|x - 4| = 4 - x$. Thus the original equation becomes

$$x - 3 + 4 - x = 9,$$

which can be rewritten as $1 = 9$, which holds for no values of x.

Thus we can conclude that 8 and -1 are the only values of x that satisfy the original equation.

11. $|x-3| + |x-4| = 1$

SOLUTION If $x > 4$, then the distance from x to 3 is bigger than 1, and thus $|x - 3| > 1$ and thus $|x - 3| + |x - 4| > 1$. Hence there are no solutions to the equation above with $x > 4$.

If $x < 3$, then the distance from x to 4 is bigger than 1, and thus $|x - 4| > 1$ and thus $|x - 3| + |x - 4| > 1$. Hence there are no solutions to the equation above with $x < 3$.

The only remaining possibility is that $3 \le x \le 4$. In this case, we have $x - 3 \ge 0$ and $x - 4 \le 0$, which implies that $|x - 3| = x - 3$ and $|x - 4| = 4 - x$, which implies that

$$|x - 3| + |x - 4| = (x - 3) + (4 - x) = 1.$$

Thus the set of numbers x such that $|x - 3| + |x - 4| = 1$ is the interval $[3, 4]$.

13. $|x-3| + |x-4| = \frac{1}{2}$

SOLUTION As we saw in the solution to Exercise 11, if $x > 4$ or $x < 3$, then $|x-3| + |x-4| > 1$, and in particular $|x - 3| + |x - 4| \neq \frac{1}{2}$.

We also saw in the solution to Exercise 11 that if $3 \le x \le 4$, then $|x - 3| + |x - 4| = 1$, and in particular $|x - 3| + |x - 4| \neq \frac{1}{2}$.

Thus there are no numbers x such that $|x - 3| + |x - 4| = \frac{1}{2}$.

15. $|x + 3| = x + 3$

SOLUTION Note that $|x + 3| = x + 3$ if and only if $x + 3 \ge 0$, which is equivalent to the inequality $x \ge -3$. Thus the set of numbers x such that $|x + 3| = x + 3$ is the interval $[-3, \infty)$.

17. $|x| = x + 1$

SOLUTION If $x \ge 0$, then $|x| = x$ and the equation above becomes the equation $x = x + 1$, which has no solutions.

If $x < 0$, then $|x| = -x$ and the equation above becomes the equation $-x = x + 1$, which has the solution $x = -\frac{1}{2}$. Substituting $-\frac{1}{2}$ for x in the equation above shows that $x = -\frac{1}{2}$ is indeed a solution to the equation.

Thus the only number x satisfying $|x| = x + 1$ is $-\frac{1}{2}$.

In Exercises 19–28, write each union as a single interval.

19. $[2, 7) \cup [5, 20)$

SOLUTION The first interval is $\{x : 2 \le x < 7\}$, which includes the left endpoint 2 but does not include the right endpoint 7. The second interval is $\{x : 5 \le x < 20\}$, which includes the left endpoint 5 but does not include the right endpoint 20. The set of numbers in at least one of these sets equals $\{x : 2 \le x < 20\}$, as can be seen below:

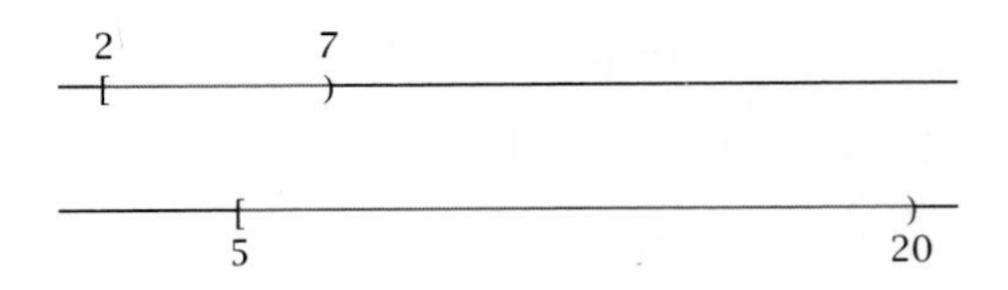

Thus $[2, 7) \cup [5, 20) = [2, 20)$.

21. $[-2, 8] \cup (-1, 4)$

SOLUTION The first interval is the set $\{x : -2 \le x \le 8\}$, which includes both endpoints. The second interval is $\{x : -1 < x < 4\}$, which does not include either endpoint. The set of numbers in at least one of these sets equals $\{x : -2 \le x \le 8\}$, as can be seen below:

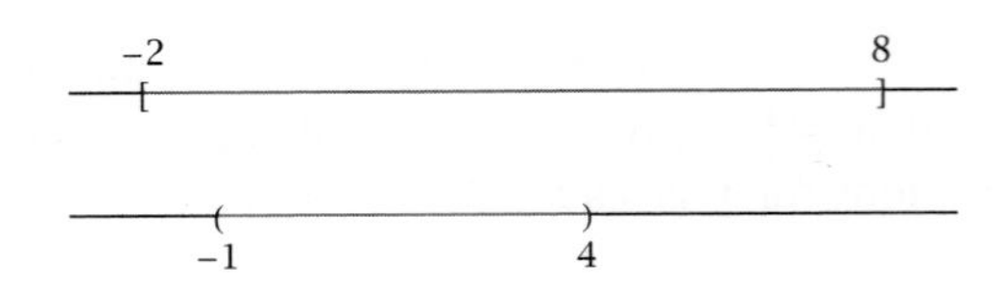

Thus $[-2,8] \cup (-1,4) = [-2,8]$.

23. $(3,\infty) \cup [2,8]$

SOLUTION The first interval is $\{x : 3 < x\}$, which does not include the left endpoint and which has no right endpoint. The second interval is $\{x : 2 \le x \le 8\}$, which includes both endpoints. The set of numbers in at least one of these sets equals $\{x : 2 \le x\}$, as can be seen below:

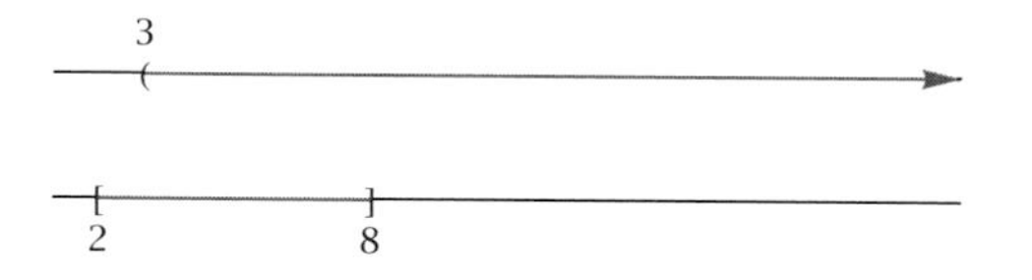

Thus $(3,\infty) \cup [2,8] = [2,\infty)$.

25. $(-\infty,-3) \cup [-5,\infty)$

SOLUTION The first interval is $\{x : x < -3\}$, which has no left endpoint and which does not include the right endpoint. The second interval is $\{x : -5 \le x\}$, which includes the left endpoint and which has no right endpoint. The set of numbers in at least one of these sets equals the entire real line, as can be seen below:

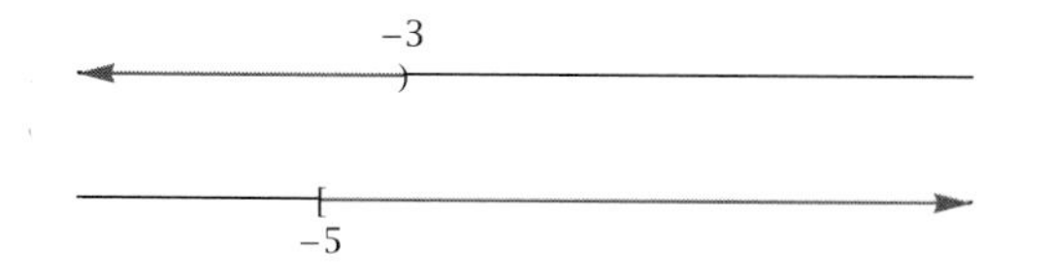

Thus $(-\infty,-3) \cup [-5,\infty) = (-\infty,\infty)$.

27. $(-3,\infty) \cup [-5,\infty)$

SOLUTION The first interval is $\{x : -3 < x\}$, which does not include the left endpoint and which has no right endpoint. The second interval is $\{x : -5 \le x\}$, which includes the left endpoint and which has no right endpoint. The set of numbers that are in at least one of these sets equals $\{x : -5 \le x\}$, as can be seen below:

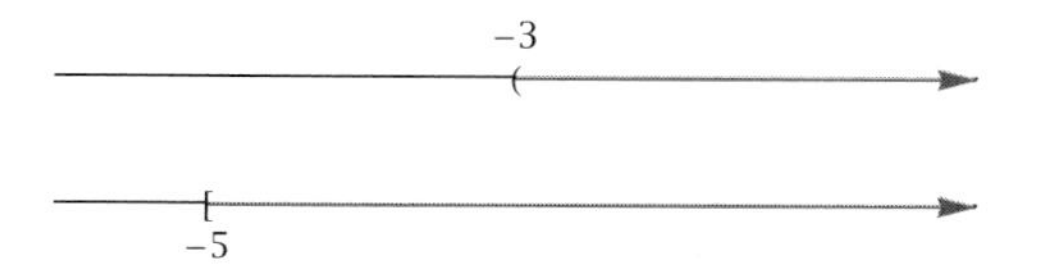

Thus $(-3,\infty) \cup [-5,\infty) = [-5,\infty)$.

29. A medicine is known to decompose and become ineffective if its temperature ever reaches 103 degrees Fahrenheit or more. Write an interval to represent the temperatures (in degrees Fahrenheit) at which the medicine is ineffective.

SOLUTION The medicine is ineffective at all temperatures of 103 degrees Fahrenheit or greater, which corresponds to the interval $[103,\infty)$.

31. A shoelace manufacturer guarantees that its 33-inch shoelaces will be 33 inches long, with an error of at most 0.1 inch.

(a) Write an inequality using absolute values and the length s of a shoelace that gives the condition that the shoelace does not meet the guarantee.

(b) Write the set of numbers satisfying the inequality in part (a) as a union of two intervals.

SOLUTION

(a) The error in the shoelace length is $|s - 33|$. Thus a shoelace with length s does not meet the guarantee if $|s - 33| > 0.1$.

(b) Because $33 - 0.1 = 32.9$ and $33 + 0.1 = 33.1$, the set of numbers s such that $|s - 33| > 0.1$ is $(-\infty, 32.9) \cup (33.1,\infty)$.

33. Give four examples of pairs of real numbers a and b such that

$$|a + b| = 2 \quad \text{and} \quad |a| + |b| = 8.$$

SOLUTION First consider the case where $a \ge 0$ and $b \ge 0$. In this case, we have $a + b \ge 0$. Thus the equations above become

$$a + b = 2 \quad \text{and} \quad a + b = 8.$$

There are no solutions to the simultaneous equations above, because $a + b$ cannot simultaneously equal both 2 and 8.

Next consider the case where $a < 0$ and $b < 0$. In this case, we have $a + b < 0$. Thus the equations above become

$$-a - b = 2 \quad \text{and} \quad -a - b = 8.$$

There are no solutions to the simultaneous equations above, because $-a - b$ cannot simultaneously equal both 2 and 8.

Now consider the case where $a \geq 0$, $b < 0$, and $a + b \geq 0$. In this case the equations above become

$$a + b = 2 \quad \text{and} \quad a - b = 8.$$

Solving these equations for a and b, we get $a = 5$ and $b = -3$.

Now consider the case where $a \geq 0$, $b < 0$, and $a + b < 0$. In this case the equations above become

$$-a - b = 2 \quad \text{and} \quad a - b = 8.$$

Solving these equations for a and b, we get $a = 3$ and $b = -5$.

Now consider the case where $a < 0$, $b \geq 0$, and $a + b \geq 0$. In this case the equations above become

$$a + b = 2 \quad \text{and} \quad -a + b = 8.$$

Solving these equations for a and b, we get $a = -3$ and $b = 5$.

Now consider the case where $a < 0$, $b \geq 0$, and $a + b < 0$. In this case the equations above become

$$-a - b = 2 \quad \text{and} \quad -a + b = 8.$$

Solving these equations for a and b, we get $a = -5$ and $b = 3$.

At this point, we have considered all possible cases. Thus the only solutions are $a = 5$, $b = -3$, or $a = 3$, $b = -5$, or $a = -3$, $b = 5$, or $a = -5$, $b = 3$.

In Exercises 35–46, write each set as an interval or as a union of two intervals.

35. $\{x : |x - 4| < \frac{1}{10}\}$

SOLUTION The inequality $|x - 4| < \frac{1}{10}$ is equivalent to the inequality

$$-\tfrac{1}{10} < x - 4 < \tfrac{1}{10}.$$

Add 4 to all parts of this inequality, getting

$$4 - \tfrac{1}{10} < x < 4 + \tfrac{1}{10},$$

which is the same as

$$\tfrac{39}{10} < x < \tfrac{41}{10}.$$

Thus $\{x : |x - 4| < \frac{1}{10}\} = (\frac{39}{10}, \frac{41}{10})$.

37. $\{x : |x + 4| < \frac{\varepsilon}{2}\}$; here $\varepsilon > 0$

SOLUTION The inequality $|x + 4| < \frac{\varepsilon}{2}$ is equivalent to the inequality

$$-\frac{\varepsilon}{2} < x + 4 < \frac{\varepsilon}{2}.$$

Add -4 to all parts of this inequality, getting

$$-4 - \frac{\varepsilon}{2} < x < -4 + \frac{\varepsilon}{2}.$$

Thus $\{x : |x + 4| < \frac{\varepsilon}{2}\} = (-4 - \frac{\varepsilon}{2}, -4 + \frac{\varepsilon}{2})$.

39. $\{y : |y - a| < \varepsilon\}$; here $\varepsilon > 0$

SOLUTION The inequality $|y - a| < \varepsilon$ is equivalent to the inequality

$$-\varepsilon < y - a < \varepsilon.$$

Add a to all parts of this inequality, getting

$$a - \varepsilon < y < a + \varepsilon.$$

Thus $\{y : |y - a| < \varepsilon\} = (a - \varepsilon, a + \varepsilon)$.

41. $\{x : |3x - 2| < \frac{1}{4}\}$

SOLUTION The inequality $|3x - 2| < \frac{1}{4}$ is equivalent to the inequality

$$-\tfrac{1}{4} < 3x - 2 < \tfrac{1}{4}.$$

Add 2 to all parts of this inequality, getting

$$\tfrac{7}{4} < 3x < \tfrac{9}{4}.$$

Now divide all parts of this inequality by 3, getting

$$\tfrac{7}{12} < x < \tfrac{3}{4}.$$

Thus $\{x : |3x - 2| < \frac{1}{4}\} = (\frac{7}{12}, \frac{3}{4})$.

43. $\{x : |x| > 2\}$

SOLUTION The inequality $|x| > 2$ means that $x > 2$ or $x < -2$. Thus $\{x : |x| > 2\} = (-\infty, -2) \cup (2, \infty)$.

45. $\{x : |x - 5| \geq 3\}$

SOLUTION The inequality $|x - 5| \geq 3$ means that $x - 5 \geq 3$ or $x - 5 \leq -3$. Adding 5 to both sides of these equalities shows that $x \geq 8$ or $x \leq 2$. Thus $\{x : |x - 5| \geq 3\} = (-\infty, 2] \cup [8, \infty)$.

The* intersection *of two sets of numbers consists of all numbers that are in both sets. If A and B are sets, then their intersection is denoted by $A \cap B$. In Exercises 47–56, write each intersection as a single interval.

47. $[2, 7) \cap [5, 20)$

    SOLUTION The first interval is the set $\{x : 2 \le x < 7\}$, which includes the left endpoint 2 but does not include the right endpoint 7. The second interval is the set $\{x : 5 \le x < 20\}$, which includes the left endpoint 5 but does not include the right endpoint 20. The set of numbers that are in both these sets equals $\{x : 5 \le x < 7\}$, as can be seen below:

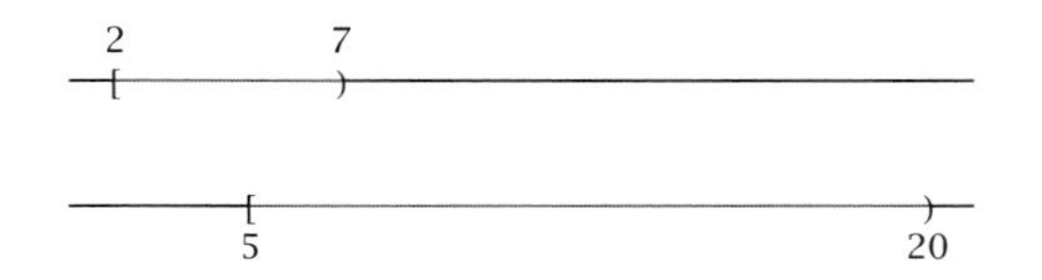

    Thus $[2, 7) \cap [5, 20) = [5, 7)$.

49. $[-2, 8] \cap (-1, 4)$

    SOLUTION The first interval is the set $\{x : -2 \le x \le 8\}$, which includes both endpoints. The second interval is the set $\{x : -1 < x < 4\}$, which includes neither endpoint. The set of numbers that are in both these sets equals $\{x : -1 < x < 4\}$, as can be seen below:

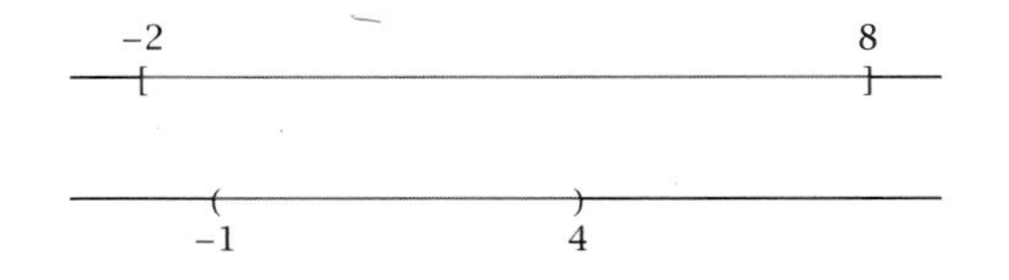

    Thus $[-2, 8] \cap (-1, 4) = (-1, 4)$.

51. $(3, \infty) \cap [2, 8]$

    SOLUTION The first interval is $\{x : 3 < x\}$, which does not include the left endpoint and which has no right endpoint. The second interval is $\{x : 2 \le x \le 8\}$, which includes both endpoints. The set of numbers in both sets equals $\{x : 3 < x \le 8\}$, as can be seen below:

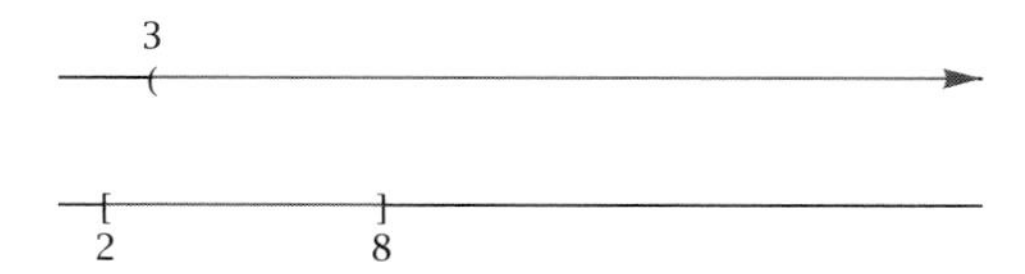

    Thus $(3, \infty) \cap [2, 8] = (3, 8]$.

53. $(-\infty, -3) \cap [-5, \infty)$

    SOLUTION The first interval is $\{x : x < -3\}$, which has no left endpoint and which does not include the right endpoint. The second interval is $\{x : -5 \le x\}$, which includes the left endpoint and which has no right endpoint. The set of numbers in both sets equals $\{x : -5 \le x < -3\}$, as can be seen below:

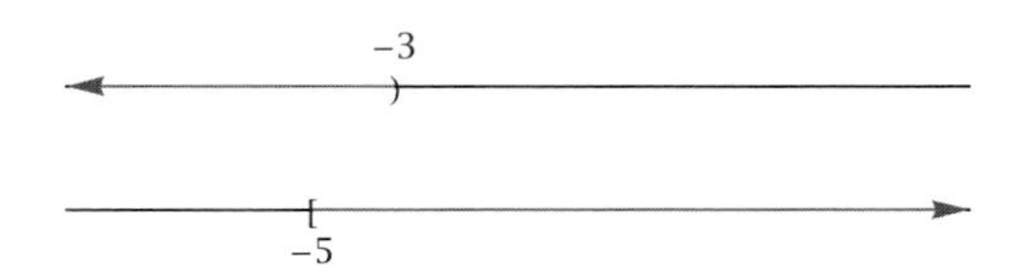

    Thus $(-\infty, -3) \cap [-5, \infty) = [-5, -3)$.

55. $(-3, \infty) \cap [-5, \infty)$

    SOLUTION The first interval is $\{x : -3 < x\}$, which does not include the left endpoint and which has no right endpoint. The second interval is $\{x : -5 \le x\}$, which includes the left endpoint and which has no right endpoint. The set of numbers in both sets equals $\{x : -3 < x\}$, as can be seen below:

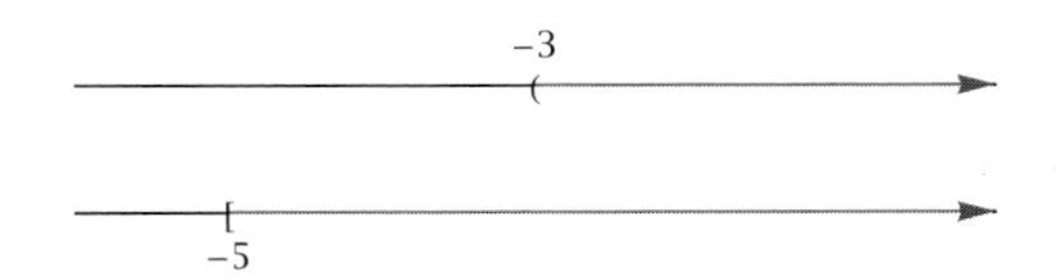

    Thus $(-3, \infty) \cap [-5, \infty) = (-3, \infty)$.

CHAPTER SUMMARY

To check that you have mastered the most important concepts and skills covered in this chapter, make sure that you can do each item in the following list:

- Explain the correspondence between the system of real numbers and the real line.
- Simplify algebraic expressions using the commutative, associative, and distributive properties.
- List the order of algebraic operations.
- Explain how parentheses are used to alter the order of algebraic operations.
- Use the algebraic identities involving additive inverses and multiplicative inverses.
- Manipulate inequalities.
- Use interval notation for open intervals, closed intervals, and half-open intervals.
- Use interval notation involving $-\infty$ and ∞, with the understanding that $-\infty$ and ∞ are not real numbers.
- Write inequalities involving an absolute value without using an absolute value.
- Compute the union of intervals.

To review a chapter, go through the list above to find items that you do not know how to do, then reread the material in the chapter about those items. Then try to answer the chapter review questions below without looking back at the chapter.

CHAPTER REVIEW QUESTIONS

1. Explain how the points on the real line correspond to the set of real numbers.
2. Show that $7 - 6\sqrt{2}$ is an irrational number.
3. What is the commutative property for addition?
4. What is the commutative property for multiplication?
5. What is the associative property for addition?
6. What is the associative property for multiplication?
7. Expand $(t + w)^2$.
8. Expand $(u - v)^2$.
9. Expand $(x - y)(x + y)$.
10. Expand $(a + b)(x - y - z)$.
11. Expand $(a + b - c)^2$.
12. Simplify the expression $\dfrac{\frac{1}{t-b} - \frac{1}{t}}{b}$.
13. Find all real numbers x such that $|3x - 4| = 5$.
14. Give an example of two numbers x and y such that $|x + y|$ does not equal $|x| + |y|$.
15. Suppose $0 < a < b$ and $0 < c < d$. Explain why $ac < bd$.
16. Write the set $\{t : |t - 3| < \frac{1}{4}\}$ as an interval.
17. Write the set $\{w : |5w + 2| < \frac{1}{3}\}$ as an interval.
18. Explain why the sets $\{x : |8x - 5| < 2\}$ and $\{t : |5 - 8t| < 2\}$ are the same set.
19. Write $[-5, 6) \cup [-1, 9)$ as an interval.
20. Write $(-\infty, 4] \cup (3, 8]$ as an interval.
21. Explain why $[7, \infty]$ is not an interval of real numbers.
22. Write the set $\{t : |2t + 7| \ge 5\}$ as a union of two intervals.
23. Suppose you put \$5.21 into a jar on June 22. Then you added one penny to the jar every day until the jar contained \$5.95. Is the set $\{5.21, 5.22, 5.23, \dots, 5.95\}$ of all amounts of money (measured in dollars) that were in the jar during the summer an interval?
24. Is the set of all real numbers x such that $x^2 > 3$ an interval? Explain your answer.

CHAPTER

2

René Descartes, who invented the coordinate system we use to graph equations, explaining his work to Queen Christina of Sweden (from an 18th-century painting by Dumesnil).

Combining Algebra and Geometry

Analytic geometry, which combines algebra and geometry, provides a tremendously powerful tool for visualizing equations with two variables.

We begin this chapter with a description of the coordinate plane, including a discussion of distance and length. Then we turn our attention to lines and their slopes, which are simple concepts that have immense importance.

We then investigate quadratic expressions. We will see how to complete the square and solve quadratic equations. Quadratic expressions will also lead us to the conic sections (ellipses, parabolas, and hyperbolas).

We conclude the chapter by learning methods for computing the area of triangles, trapezoids, circles, and ellipses.

2.1 The Coordinate Plane

LEARNING OBJECTIVES

By the end of this section you should be able to

- locate points in the coordinate plane;
- graph equations with two variables in the coordinate plane, possibly using technology;
- compute the distance between two points;
- compute the circumference of a circle.

Coordinates

Recall how the real line is constructed: We start with a horizontal line, pick a point on it that we label 0, pick a point to the right of 0 that we label 1, and then we label other points using the scale determined by 0 and 1 (see Section 1.1 to review the construction of the real line).

The coordinate plane is constructed in a similar fashion, but using a horizontal and a vertical line rather than just a horizontal line.

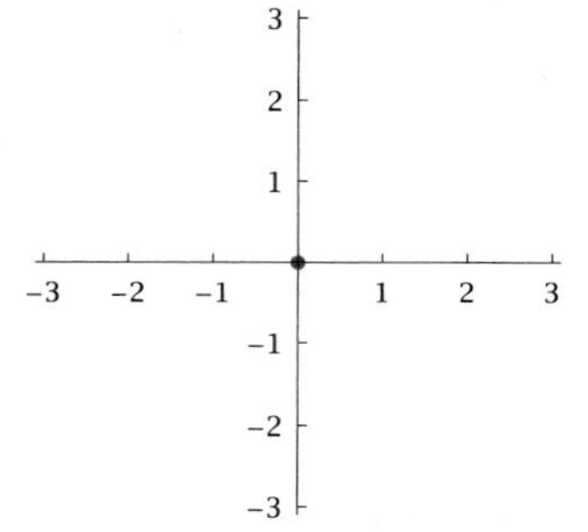

The coordinate plane, with a dot at the origin.

The coordinate plane

- The **coordinate plane** is constructed by starting with a horizontal line and a vertical line in a plane. These lines are called the **coordinate axes**.
- The intersection point of the coordinate axes is called the **origin**; it receives a label of 0 on both axes.
- On the horizontal axis, pick a point to the right of the origin and label it 1. Then label other points on the horizontal axis using the scale determined by the origin and 1.
- Similarly, on the vertical axis, pick a point above the origin and label it 1. Then label other points on the vertical axis using the scale determined by the origin and 1.

Sometimes it is important to use the same scale on both axes, as is done in the figure above. Other times it may be more convenient to use two different scales on the two axes.

A point in the plane is identified with its coordinates, which are written as an ordered pair of numbers surrounded by parentheses, as described below.

Coordinates

- The first coordinate indicates the horizontal distance from the origin, with positive numbers corresponding to points right of the origin and negative numbers corresponding to points left of the origin.
- The second coordinate indicates the vertical distance from the origin, with positive numbers corresponding to points above the origin and negative numbers corresponding to points below the origin.

The plane with this system of labeling is often called the **Cartesian plane** *in honor of the French mathematician René Descartes (1596–1650), who described this technique in his 1637 book* Discourse on Method.

EXAMPLE 1

Locate on a coordinate plane the following points:

(a) $(2, 1)$; (b) $(-1, 2.5)$; (c) $(-2.5, -2.5)$; (d) $(3, -2)$.

SOLUTION

(a) The point $(2, 1)$ can be located by starting at the origin, moving 2 units to the right along the horizontal axis, and then moving up 1 unit; see the figure below.

(b) The point $(-1, 2.5)$ can be located by starting at the origin, moving 1 unit to the left along the horizontal axis, and then moving up 2.5 units; see the figure below.

(c) The point $(-2.5, -2.5)$ can be located by starting at the origin, moving 2.5 units to the left along the horizontal axis, and then moving down 2.5 units; see the figure below.

(d) The point $(3, -2)$ can be located by starting at the origin, moving 3 units to the right along the horizontal axis, and then moving down 2 units; see the figure below.

The notation $(-1, 2.5)$ *could denote either the point with coordinates* $(-1, 2.5)$ *or the open interval* $(-1, 2.5)$. *You should be able to tell from the context which meaning is intended.*

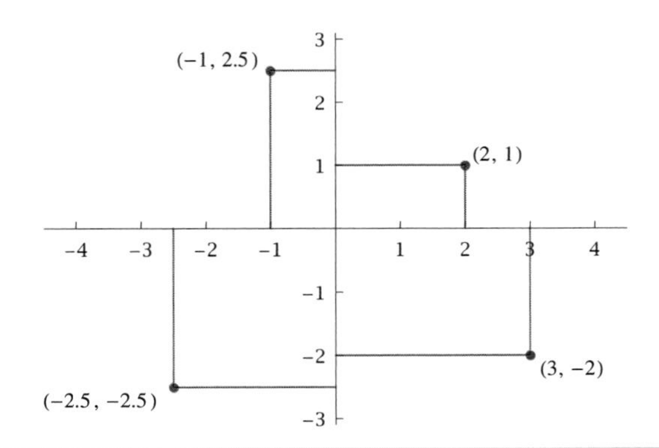

These coordinates are sometimes called **rectangular coordinates** *because each point's coordinates are determined by a rectangle, as shown in this figure.*

The horizontal axis is often called the x-**axis** and the vertical axis is often called the y-**axis**. In this case, the coordinate plane can be called the xy-plane. However, other variables can also be used, depending on the problem at hand.

Regardless of the names of the axes, remember that the first coordinate corresponds to horizontal distance from the origin and the second coordinate corresponds to vertical distance from the origin.

If the horizontal axis has been labeled the x-axis, then the first coordinate of a point is often called the x-**coordinate**. Similarly, if the vertical axis has been labeled the y-axis, then the second coordinate is often called the y-**coordinate**.

The potential confusion of this terminology becomes apparent when we want to consider a point whose coordinates are (y, x); here y is the x-coordinate and x is the y-coordinate. Furthermore, always calling the first coordinate the x-coordinate will lead to confusion when the horizontal axis is labeled with another variable such as t or θ.

Graphs of Equations

The coordinate plane allows us to visualize the set of points satisfied by equations with two variables.

Graph of an equation

The **graph** of an equation with two variables is the set of points in the corresponding coordinate plane satisfied by the equation.

EXAMPLE 2

The graph of the equation $4x^4 + y^2 = 2$ is shown in the margin.

(a) Where does this graph intersect the x-axis?

(b) Where does this graph intersect the y-axis?

(c) Is the point $(\frac{1}{2}, \frac{7}{5})$ on this graph?

(d) Find four points on this graph that are not on either of the coordinate axes.

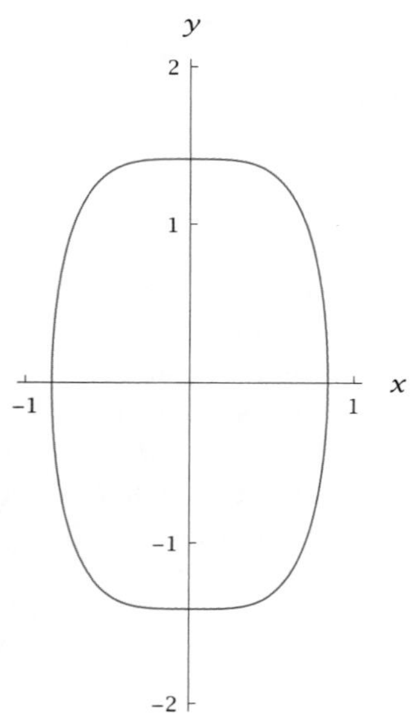

The graph of the equation $4x^4 + y^2 = 2$.

SOLUTION

(a) The x-axis is the set of points where $y = 0$. Thus to find the points where the graph intersects the x-axis, substitute $y = 0$ in the equation $4x^4 + y^2 = 2$ to obtain the equation $4x^4 = 2$, which implies that $x^4 = \frac{1}{2}$, which implies that $x = 1/2^{1/4} \approx 0.8$ or $x = -1/2^{1/4} \approx -0.8$ (we will discuss fractional exponents in detail in Section 5.1, but you are probably already familiar with this concept from a previous course).

Thus the points where the graph intersects the x-axis are approximately $(0.8, 0)$ and $(-0.8, 0)$. Make sure you can locate these points on the graph in the margin.

(b) The y-axis is the set of points where $x = 0$. Thus to find the points where the graph intersects the y-axis, substitute $x = 0$ in the equation $4x^4 + y^2 = 2$ to obtain the equation $y^2 = 2$, which implies that $y = \sqrt{2} \approx 1.4$ or $y = -\sqrt{2} \approx -1.4$.

Thus the points where the graph intersects the y-axis are approximately $(0, 1.4)$ and $(0, -1.4)$. Make sure you can locate these points on the graph in the margin.

(c) Asking whether $(\frac{1}{2}, \frac{7}{5})$ is on the graph of the equation $4x^4 + y^2 = 2$ is equivalent to asking whether $4(\frac{1}{2})^4 + (\frac{7}{5})^2$ equals 2. A little arithmetic shows that $4(\frac{1}{2})^4 + (\frac{7}{5})^2$ equals $\frac{221}{100}$, which equals 2.21, which is close but not equal to 2. Thus the point $(\frac{1}{2}, \frac{7}{5})$ is not on the graph of $4x^4 + y^2 = 2$.

The red dot in the figure in the margin shows the point $(\frac{1}{2}, \frac{7}{5})$. As you can see, it is not on the graph of $4x^4 + y^2 = 2$, although it is close. The vertical line segment has endpoints at $(\frac{1}{2}, 0)$ and $(\frac{1}{2}, \frac{7}{5})$, showing that the red dot has first coordinate equal to $\frac{1}{2}$.

(d) To find some points on the graph not on the coordinate axes, substitute $y = 1$ in the equation $4x^4 + y^2 = 2$ to obtain the equation $4x^4 = 1$, which implies that $x = 1/4^{1/4} \approx 0.7$ or $x = -1/4^{1/4} \approx -0.7$. Thus $(1/4^{1/4}, 1)$ and $(-1/4^{1/4}, 1)$, which are approximately $(0.7, 1)$ and $(-0.7, 1)$, are points on the graph.

Because $(-y)^2 = y^2$, the points $(1/4^{1/4}, -1)$ and $(-1/4^{1/4}, -1)$, which are approximately $(0.7, -1)$ and $(-0.7, -1)$, are also on the graph.

The blue dots in the figure in the margin show the four points on the graph that we have just found. The top horizontal line segment has endpoints at $(-1/4^{1/4}, 1)$ and $(1/4^{1/4}, 1)$, showing that these two points have second coordinate equal to 1. The bottom horizontal line segment has endpoints at $(-1/4^{1/4}, -1)$ and $(1/4^{1/4}, -1)$, showing that these two points have second coordinate equal to -1.

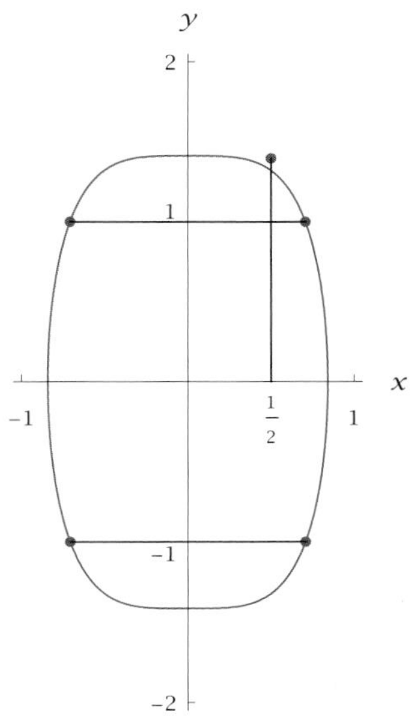

The graph of the equation $4x^4 + y^2 = 2$, *along with one point not on the graph and four points on the graph.*

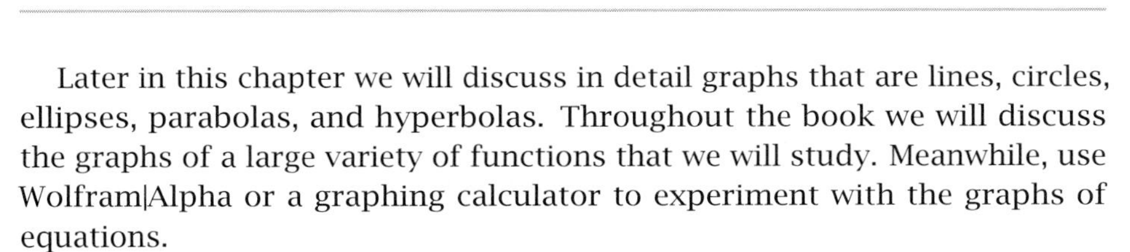

Later in this chapter we will discuss in detail graphs that are lines, circles, ellipses, parabolas, and hyperbolas. Throughout the book we will discuss the graphs of a large variety of functions that we will study. Meanwhile, use Wolfram|Alpha or a graphing calculator to experiment with the graphs of equations.

The next example uses Wolfram|Alpha in the solution, but you could use a graphing calculator or any other technology instead.

EXAMPLE 3

(a) Graph the equation $2t^3 + z = z^3 + t$.

(b) Graph the equation $2t^3 + z = z^3 + t$ for t in the interval $[-6, 6]$.

(c) Determine whether or not the point $(4.5, 5.5)$ is on the graph of $2t^3 + z = z^3 + t$.

SOLUTION

(a) Point a web browser to `www.wolframalpha.com`. In the one-line entry box that appears near the top of the web page, enter

```
graph 2t^3 + z = z^3 + t
```

and then press the *enter* key on your keyboard or click the = box on the right side of the Wolfram|Alpha entry box, getting a graph that should look like the one shown in the margin here.

Wolfram|Alpha selects the variable lowest in alphabetical order to use as the horizontal axis and the variable highest in alphabetical order to use as the vertical axis, as has been done here with t and z and as is usually what you want to do when the variables are x and y.

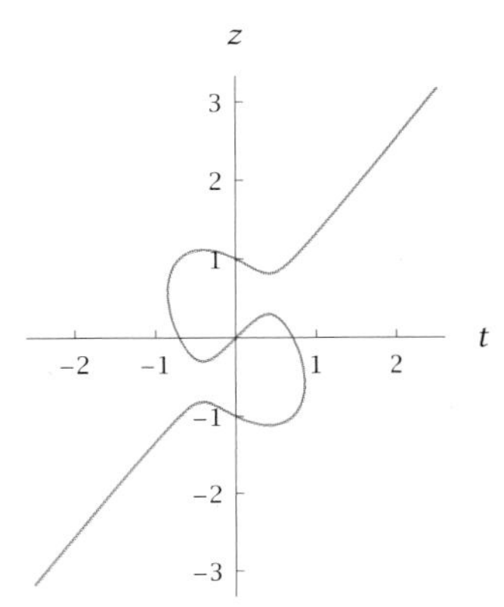

The graph of the equation $2t^3 + z = z^3 + t$.

(b) Unlike the previous graph we examined, the graph of $2t^3 + z = z^3 + t$ includes points arbitrarily far left, right, up, and down. Thus the figure shown in the margin above is not the entire graph of $2t^3 + z = z^3 + t$. No figure could show the entire graph of this equation.

Wolfram|Alpha usually chooses an interesting part of the graph to show, but you may want to zoom in to see more detail or zoom out to see a bigger picture. For example, to zoom out to see the graph for t in the interval $[-6, 6]$, type

```
graph 2t^3 + z = z^3 + t from t =-6 to 6
```

in a Wolfram|Alpha entry box to obtain the graph shown below.

Here the horizontal and vertical axes have different scales, which can distort the shape of the graph. However, the use of different scales is often necessary to prevent the graph from becoming too large.

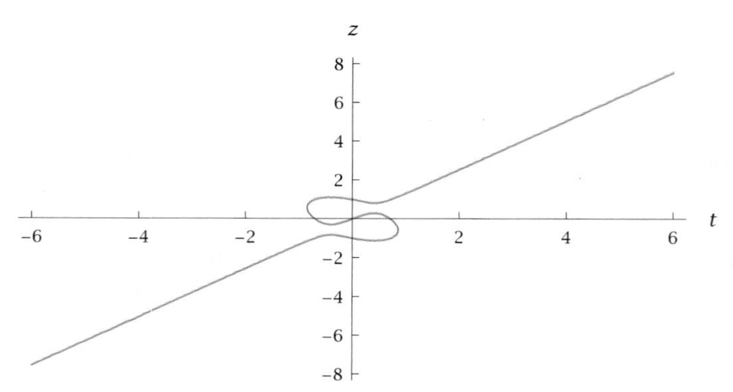

The graph of $2t^3 + z = z^3 + t$ *for* t *in the interval* $[-6, 6]$.

(c) From the graph above, it appears possible that the point $(4.5, 5.5)$ may be on the graph of $2t^3 + z = z^3 + t$. To check whether or not this is true, we substitute $t = 4.5$ and $z = 5.5$ into the equation and check whether we get a true statement. To do this check using Wolfram|Alpha, type

```
is 2 * 4.5^3 + 5.5 = 5.5^3 + 4.5
```

in a Wolfram|Alpha entry box to obtain the answer False. Thus the point $(4.5, 5.5)$ is not on the graph of $2t^3 + z = z^3 + t$.

Distance Between Two Points

We start gently with some concrete examples before getting to the formula for the distance between two points.

EXAMPLE 4 Find the distance between the point $(4, 3)$ and the origin.

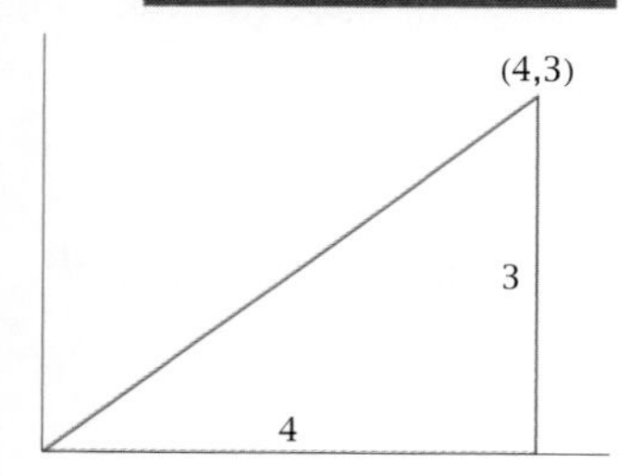

SOLUTION The distance between the point $(4, 3)$ and the origin is the length of the hypotenuse in the right triangle shown here. By the Pythagorean Theorem, this hypotenuse has length $\sqrt{4^2 + 3^2}$, which equals $\sqrt{25}$, which equals 5.

Here is another example, this time with neither of the points being the origin.

EXAMPLE 5

Find the distance between the points $(5, 6)$ and $(2, 1)$.

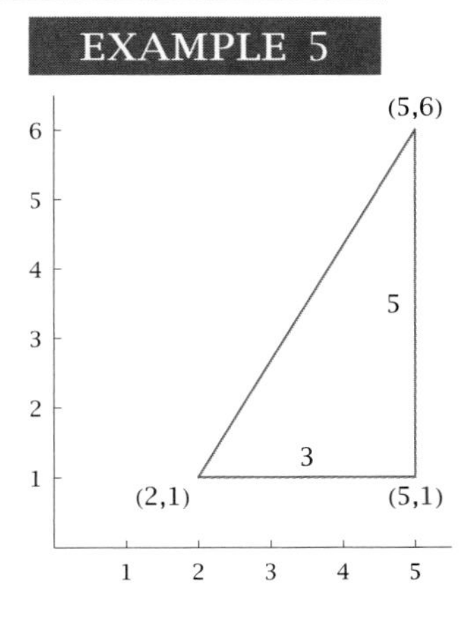

SOLUTION The distance between the points $(5, 6)$ and $(2, 1)$ is the length of the hypotenuse in the right triangle shown here. The horizontal side of this triangle has length $5 - 2$, which equals 3, and the vertical side of this triangle has length $6 - 1$, which equals 5. By the Pythagorean Theorem, the hypotenuse has length $\sqrt{3^2 + 5^2}$, which equals $\sqrt{34}$.

More generally, to find the formula for the distance between two points (x_1, y_1) and (x_2, y_2), consider the right triangle in the figure below:

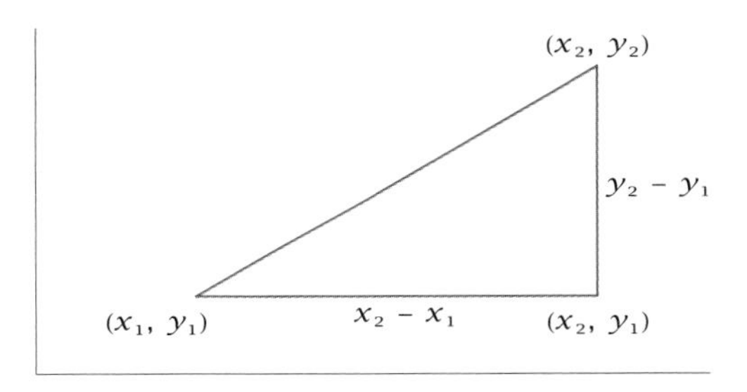

The length of the hypotenuse equals the distance between (x_1, y_1) and (x_2, y_2).

Starting with the points (x_1, y_1) and (x_2, y_2) in the figure above, make sure you understand why the third point in the triangle (the vertex at the right angle) has coordinates (x_2, y_1). Also, verify that the horizontal side of the triangle has length $x_2 - x_1$ and the vertical side of the triangle has length $y_2 - y_1$, as indicated in the figure above. The Pythagorean Theorem then gives the length of the hypotenuse, leading to the following formula:

Distance between two points

The distance between the points (x_1, y_1) and (x_2, y_2) is

$$\sqrt{(x_2 - x_1)^2 + (y_2 - y_1)^2}.$$

As a special case of this formula, the distance between a point (x, y) and the origin is $\sqrt{x^2 + y^2}$.

Using the formula above, we can now find the distance between two points without drawing a figure.

EXAMPLE 6

Find the distance between the points $(3, 1)$ and $(-4, -99)$.

SOLUTION The distance between these two points is $\sqrt{(3 - (-4))^2 + (1 - (-99))^2}$, which equals $\sqrt{7^2 + 100^2}$, which equals $\sqrt{49 + 10000}$, which equals $\sqrt{10049}$.

Length, Perimeter, and Circumference

The **length** of a line segment is the distance between the two endpoints. For example, the length of the line segment connecting the points $(-1,4)$ and $(2,6)$ equals

$$\sqrt{(2-(-1))^2+(6-4)^2},$$

which equals $\sqrt{13}$.

Defining the length of a path or curve in the coordinate plane is more complicated. A rigorous definition requires calculus, so we use the following intuitive definition:

Length

The **length** of a path or curve can be determined by placing a string on the path or curve and then measuring the length of the string when it is straightened into a line segment.

EXAMPLE 7

Find the length of the path shown here consisting of the line segment connecting $(-2,2)$ with $(5,3)$ followed by the line segment connecting $(5,3)$ with $(2,1)$.

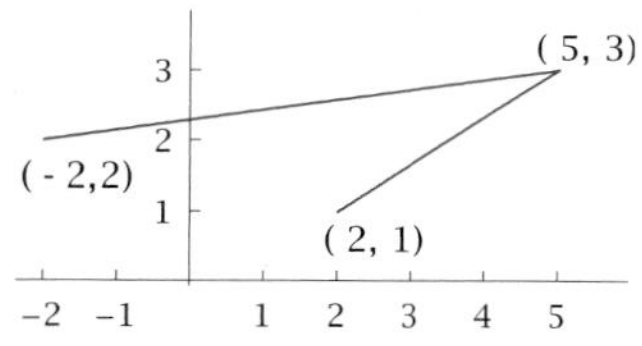

If a path consists of line segments, then the length of the path is the sum of the lengths of the line segments.

SOLUTION The first line segment has length

$$\sqrt{(5-(-2))^2+(3-2)^2},$$

which equals $\sqrt{50}$. The second line segment has length

$$\sqrt{(2-5)^2+(1-3)^2},$$

which equals $\sqrt{13}$. Thus this path has length $\sqrt{50}+\sqrt{13}$.

You are probably already familiar with two other words that are used to denote the lengths of certain paths that begin and end at the same point. One word probably would have been enough, but the two following words are commonly used:

Perimeter and circumference

- The **perimeter** of a polygon is the length of the path that surrounds the polygon.
- The **circumference** of a region is the length of the curve that surrounds the region.

EXAMPLE 8

(a) What is the perimeter of an equilateral triangle with sides of length s?

(b) What is the perimeter of a square with sides of length s?

(c) What is the perimeter of a rectangle with width w and height h?

SOLUTION

(a) An equilateral triangle with sides of length s has perimeter $3s$.

(b) A square with sides of length s has perimeter $4s$.

(c) A rectangle with width w and height h has perimeter $2w + 2h$.

The perimeter of an equilateral triangle is proportional to the length of one of its sides (the ratio is 3) and the perimeter of a square is proportional to the length of one of its sides (the ratio is 4). Thus it is reasonable to believe that the circumference of a circle is proportional to its diameter.

Physical experiments confirm this belief. For example, suppose you have a very accurate ruler that can measure lengths with 0.01-inch accuracy. If you place a string on top of a circle with diameter 1 inch, then straighten the string to a line segment, you will find that the string has length about 3.14 inches. Similarly, if you place a string on top of a circle with diameter 2 inches, then straighten the string to a line segment, you will find that the string has length about 6.28 inches. Thus the circumference of a circle with diameter two inches is twice the circumference of a circle with diameter 1 inch.

Just for fun, here are the first 504 digits of π:

3.141592653589793238462643383279502884197169399375105820974944592307816406286208998628034825342117067982148086513282306647093844609550582231725359408128481117450284102701938521105559644622948954930381964428810975665933446128475648233786783165271201909145648566923460348610454326648213393607260249141273724587006606315588174881520920962829254091715364367892590360011330530548820466521384146951941511609433057270365759591953092186117381932611793105118548074462379962749567351885752724891227938183011949129833

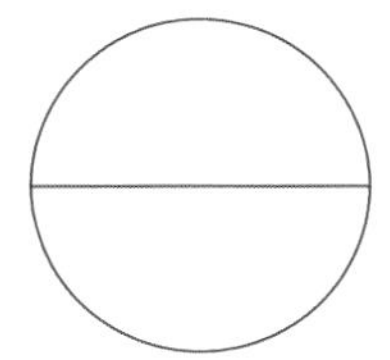

The circle on the left has been straightened into the line segment on the right. A measurement shows that this line segment is approximately 3.14 *times as long as the diameter of the circle, which is shown above in red.*

Similarly, you will find that for any circle you measure, the ratio of the circumference to the diameter is approximately 3.14. The exact value of this ratio is so important that it gets its own symbol:

> π
>
> The ratio of the circumference to the diameter of a circle is called π.

Does the decimal expansion of π contain one thousand consecutive 4's? No one knows, but mathematicians suspect that the answer is "yes".

It turns out that π is an irrational number (see Problem 53 in Section 9.4). For most practical purposes, 3.14 is a good approximation of π—the error is about 0.05%. If more accurate computations are needed, then 3.1416 is an even better approximation—the error is about 0.0002%.

The remarkable approximation $\pi \approx \frac{355}{113}$ was discovered over 1500 years ago by the Chinese mathematician Zu Chongzhi.

A fraction that approximates π well is $\frac{22}{7}$ (notice how page 22 is numbered in this book)—the error is about 0.04%. A fraction that approximates π even better is $\frac{355}{113}$—the error is extremely small at about 0.000008%.

Keep in mind that π is not equal to 3.14 or 3.1416 or $\frac{22}{7}$ or $\frac{355}{113}$. All of these are useful approximations, but π is an irrational number that cannot be represented exactly as a decimal number or as a fraction.

We have defined π to be the number such that a circle with diameter d has circumference πd. Because the diameter of a circle is equal to twice the radius, we have the following formula:

> ***Circumference of a circle***
>
> A circle with radius r has circumference $2\pi r$.

EXAMPLE 9

Suppose you want to design a 400-meter track consisting of two half-circles connected by parallel line segments. Suppose also that you want the total length of the curved part of the track to equal the total length of the straight part of the track. What dimensions should the track have?

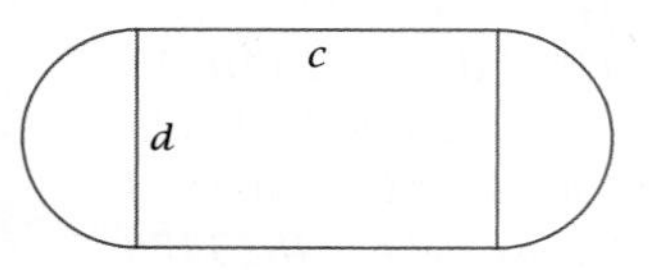

SOLUTION We want the total length of the straight part of the track to be 200 meters. Thus each of the two straight pieces must be 100 meters long. Hence we take $c = 100$ meters in the figure in the margin.

We also want the total length of the two half-circles to be 200 meters. Thus we want $\pi d = 200$. Hence we take $d = \frac{200}{\pi} \approx 63.66$ meters in the figure in the margin.

EXERCISES

For Exercises 1–8, give the coordinates of the specified point using the figure below:

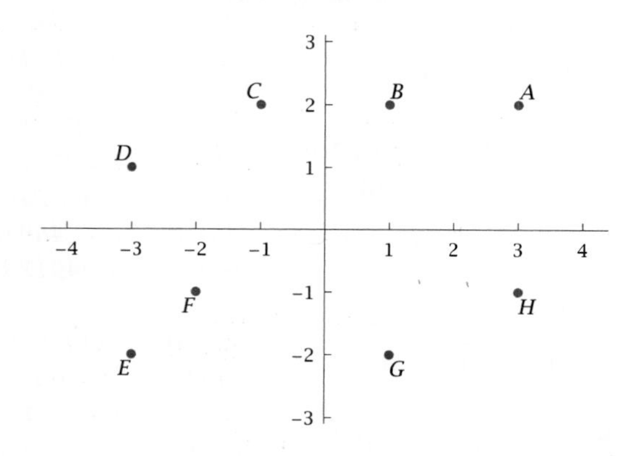

1. A
2. B
3. C
4. D
5. E
6. F
7. G
8. H

9. Sketch a coordinate plane showing the following four points, their coordinates, and the rectangles determined by each point (as in Example 1): $(1, 2)$, $(-2, 2)$, $(-3, -1)$, $(2, -3)$.

10. Sketch a coordinate plane showing the following four points, their coordinates, and the rectangles determined by each point (as in Example 1): $(2.5, 1)$, $(-1, 3)$, $(-1.5, -1.5)$, $(1, -3)$.

11. Find the distance between the points $(3, -2)$ and $(-1, 4)$.

12. Find the distance between the points $(-4, -7)$ and $(-8, -5)$.

13. Find two choices for t such that the distance between $(2, -1)$ and $(t, 3)$ equals 7.

14. Find two choices for t such that the distance between $(3, -2)$ and $(1, t)$ equals 5.

15. Find two choices for b such that $(4, b)$ has distance 5 from $(3, 6)$.

16. Find two choices for b such that $(b, -1)$ has distance 4 from $(3, 2)$.

17. Find two points on the horizontal axis whose distance from $(3, 2)$ equals 7.

18. Find two points on the horizontal axis whose distance from $(1, 4)$ equals 6.

19. A ship sails north for 2 miles and then west for 5 miles. How far is the ship from its starting point?

20. A ship sails east for 7 miles and then south for 3 miles. How far is the ship from its starting point?

Use the following information concerning Manhattan (New York City) for Exercises 21–22:

- ***Avenues in Manhattan run roughly north-south; streets run east-west.***
- ***For much of its length, Broadway runs diagonally across the grid formed by avenues and streets.***
- ***The distance between consecutive avenues in Manhattan is 922 feet.***
- ***The distance between consecutive streets in Manhattan is 260 feet.***

21. Suppose you walk from the corner of Central Park at 8th Avenue and 59th Street to 10th Avenue and 71st Street.

    (a) What is the length of your path if you walk along 8th Avenue to 71st Street, then along 71st Street to 10th Avenue?

    (b) What is the length of your path if you walk along Broadway, which goes in a straight line from 8th Avenue and 59th Street to 10th Avenue and 71st Street?

    (c) How much shorter is the direct path along Broadway than walking along 8th Avenue and 71st Street?

    (d) At the normal city walking speed of 250 feet per minute, how much time would you save by walking along Broadway as compared to walking along 8th Avenue and 71st Street?

22. Suppose you walk from 5th Avenue and 24th Street to Times Square at 7th Avenue and 42nd Street.

    (a) What is the length of your path if you walk along 5th Avenue to 42nd Street, then along 42nd Street to 7th Avenue?

    (b) What is the length of your path if you walk along Broadway, which goes in a straight line from 5th Avenue and 24th Street to 7th Avenue and 42nd Street?

    (c) How much shorter is the direct path along Broadway than walking along 5th Avenue and 42nd Street?

    (d) At the normal city walking speed of 250 feet per minute, how much time would you save by walking along Broadway as compared to walking along 5th Avenue and 42nd Street?

23. Find two points on the vertical axis whose distance from $(5, -1)$ equals 8.

24. Find two points on the vertical axis whose distance from $(2, -4)$ equals 5.

25. Find the perimeter of the triangle that has vertices at $(1, 2)$, $(5, -3)$, and $(-4, -1)$.

26. Find the perimeter of the triangle that has vertices at $(-3, 1)$, $(4, -2)$, and $(5, -1)$.

27. Find the radius of a circle that has circumference 12 inches.

28. Find the radius of a circle that has circumference 20 feet.

29. Find the radius of a circle that has circumference 8 more than its diameter.

30. Find the radius of a circle that has circumference 12 more than its diameter.

31. Find an equation whose graph in the xy-plane is the set of points whose distance from the origin is 3.

32. Find an equation whose graph in the xy-plane is the set of points whose distance from the origin is 5.

33. Find the length of the graph of the equation $x^2 + y^2 = 9$.

34. Find the length of the graph of the equation $x^2 + y^2 = 25$.

35. Find an equation whose graph in the rt-plane is the set of points whose distance from $(3, 2)$ is 2.

36. Find an equation whose graph in the bc-plane is the set of points whose distance from $(-2, 1)$ is 3.

37. Find two points that have distance 2 from the origin and distance 3 from $(3, 0)$.

38. Find two points that have distance 3 from the origin and distance 2 from $(2, 0)$.

39. Suppose you want to design a 400-meter track consisting of two half-circles connected by parallel line segments. Suppose also that you want the total length of the curved part of the track to equal half the total length of the straight part of the track. What dimensions should the track have?

40. Suppose you want to design a 200-meter indoor track consisting of two half-circles connected by parallel line segments. Suppose also that you want the total length of the curved part of the track to equal three-fourths the total length of the straight part of the track. What dimensions should the track have?

41. Suppose a rope is just long enough to cover the equator of the Earth. About how much longer would the rope need to be so that it could be suspended seven feet above the entire equator?

42. Suppose a satellite is in orbit one hundred miles above the equator of the Earth. About how much further does the satellite travel in one orbit than would a person traveling once around the equator on the surface of the Earth?

PROBLEMS

Some problems require considerably more thought than the exercises. Unlike exercises, problems often have more than one correct answer.

43. Find two points, one on the horizontal axis and one on the vertical axis, such that the distance between these two points equals 15.

44. Explain why there does not exist a point on the horizontal axis whose distance from $(5, 4)$ equals 3.

45. Use Wolfram|Alpha or a calculator to find the distance between the points $(-21, -15)$ and $(17, 28)$. [In Wolfram|Alpha, you can do this by typing `distance from (-21, -15) to (17, 28)` in an entry box. Note that in addition to the distance in both exact and approximate form, you get a figure showing the two points. Experiment with finding the distance between other pairs of points, and notice the placement of the points on the figures produced by Wolfram|Alpha.]

46. Find six distinct points whose distance from the origin equals 3.

47. Find six distinct points whose distance from $(3, 1)$ equals 4.

48. Graph the equation $x^4 + y^4 = 1$.

49. (a) Graph the equation $y^3 - 3y = x^2$ for x in the interval $[-3, 3]$.
    (b) What is unusual about this graph as compared to other graphs we have examined?

50. Show that a square whose diagonal has length d has perimeter $2\sqrt{2}d$.

51. The figure below illustrates an isosceles right triangle with legs of length 1, along with one-fourth of a circle centered at the right-angle vertex of the triangle. Using the result that the shortest path between two points is a line segment, explain why this figure shows that $2\sqrt{2} < \pi$.

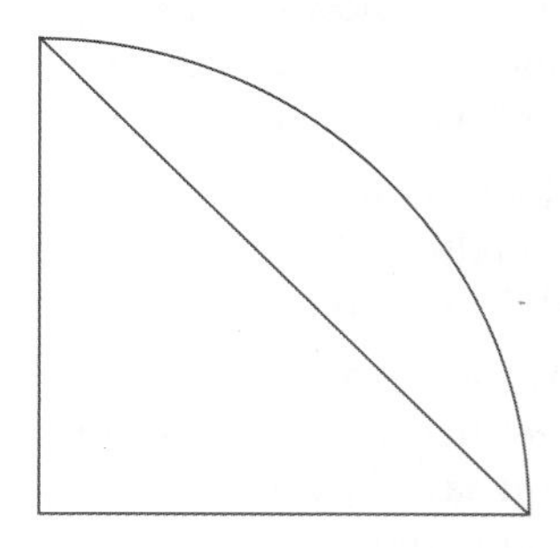

WORKED-OUT SOLUTIONS *to Odd-numbered Exercises*

Do not read these worked-out solutions before first attempting to do the exercises yourself. Otherwise you may merely mimic the techniques shown here without understanding the ideas.

Best way to learn: Carefully read the section of the textbook, then do all the odd-numbered exercises (even if they have not been assigned) and check your answers here. If you get stuck on an exercise, reread the section of the textbook—then try the exercise again. If you are still stuck, then look at the worked-out solution here.

For Exercises 1–8, give the coordinates of the specified point using the figure below:

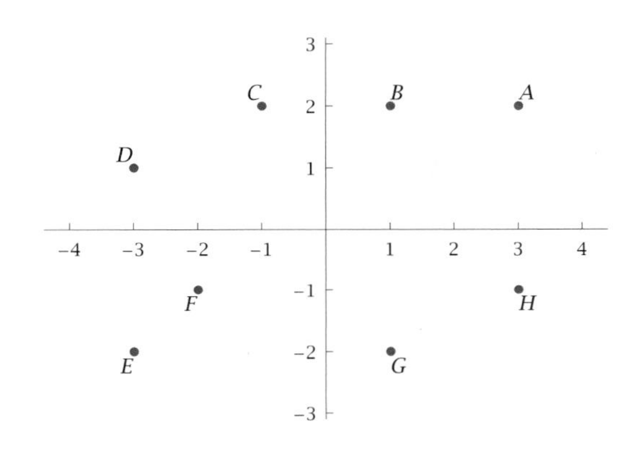

1. A

SOLUTION To get to the point A starting at the origin, we must move 3 units right and 2 units up. Thus A has coordinates $(3, 2)$.

Numbers obtained from a figure should be considered approximations. Thus the actual coordinates of A might be $(3.01, 1.98)$.

3. C

SOLUTION To get to the point C starting at the origin, we must move 1 unit left and 2 units up. Thus C has coordinates $(-1, 2)$.

5. E

SOLUTION To get to the point E starting at the origin, we must move 3 units left and 2 units down. Thus E has coordinates $(-3, -2)$.

7. G

SOLUTION To get to the point G starting at the origin, we must move 1 unit right and 2 units down. Thus G has coordinates $(1, -2)$.

9. Sketch a coordinate plane showing the following four points, their coordinates, and the rectangles determined by each point (as in Example 1): $(1, 2)$, $(-2, 2)$, $(-3, -1)$, $(2, -3)$.

SOLUTION

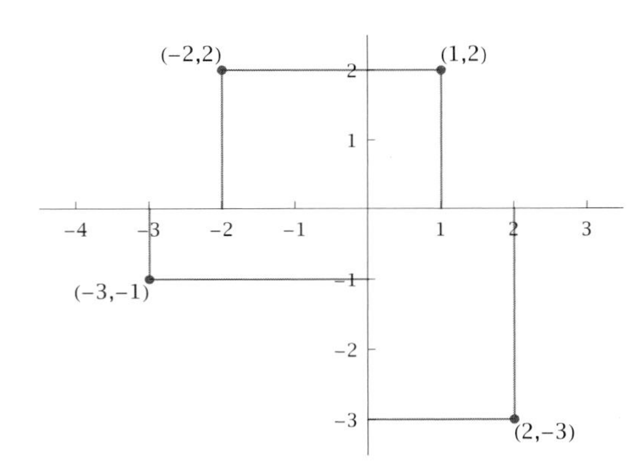

11. Find the distance between the points $(3, -2)$ and $(-1, 4)$.

SOLUTION The distance between the points $(3, -2)$ and $(-1, 4)$ equals

$$\sqrt{(-1-3)^2 + (4-(-2))^2},$$

which equals $\sqrt{(-4)^2 + 6^2}$, which equals $\sqrt{16 + 36}$, which equals $\sqrt{52}$, which can be simplified as follows:

$$\sqrt{52} = \sqrt{4 \cdot 13} = \sqrt{4} \cdot \sqrt{13} = 2\sqrt{13}.$$

Thus the distance between the points $(3, -2)$ and $(-1, 4)$ equals $2\sqrt{13}$.

13. Find two choices for t such that the distance between $(2, -1)$ and $(t, 3)$ equals 7.

SOLUTION The distance between $(2, -1)$ and $(t, 3)$ equals

$$\sqrt{(t-2)^2 + 16}.$$

We want this to equal 7, which means that we must have

$$(t-2)^2 + 16 = 49.$$

Subtracting 16 from both sides of the equation above gives

$$(t-2)^2 = 33,$$

which implies that $t - 2 = \pm\sqrt{33}$. Thus $t = 2 + \sqrt{33}$ or $t = 2 - \sqrt{33}$.

15. Find two choices for b such that $(4, b)$ has distance 5 from $(3, 6)$.

SOLUTION The distance between $(4, b)$ and $(3, 6)$ equals

$$\sqrt{1 + (6-b)^2}.$$

We want this to equal 5, which means that we must have

$$1 + (6-b)^2 = 25.$$

Subtracting 1 from both sides of the equation above gives

$$(6-b)^2 = 24,$$

which implies that $6 - b = \pm\sqrt{24}$. Thus $b = 6 - \sqrt{24}$ or $b = 6 + \sqrt{24}$.

17. Find two points on the horizontal axis whose distance from $(3, 2)$ equals 7.

SOLUTION A typical point on the horizontal axis has coordinates $(x, 0)$. The distance from this point to $(3, 2)$ is $\sqrt{(x-3)^2 + (0-2)^2}$. Thus we need to solve the equation

$$\sqrt{(x-3)^2 + 4} = 7.$$

Squaring both sides of the equation above, and then subtracting 4 from both sides gives

$$(x-3)^2 = 45.$$

Thus $x - 3 = \pm\sqrt{45} = \pm 3\sqrt{5}$. Thus $x = 3 \pm 3\sqrt{5}$. Hence the two points on the horizontal axis whose distance from $(3, 2)$ equals 7 are $(3 + 3\sqrt{5}, 0)$ and $(3 - 3\sqrt{5}, 0)$.

19. A ship sails north for 2 miles and then west for 5 miles. How far is the ship from its starting point?

SOLUTION The figure below shows the path of the ship. The length of the red line is the distance of the ship from its starting point. By the Pythagorean Theorem, this distance is $\sqrt{2^2 + 5^2}$ miles, which equals $\sqrt{29}$ miles.

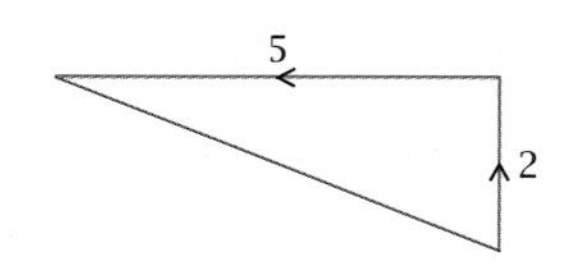

We have assumed that the surface of the Earth is part of a plane rather than part of a sphere. For distances of less than a few hundred miles, this is a good approximation.

Use the following information concerning Manhattan (New York City) for Exercises 21–22:

- ***Avenues in Manhattan run roughly north-south; streets run east-west.***
- ***For much of its length, Broadway runs diagonally across the grid formed by avenues and streets.***
- ***The distance between consecutive avenues in Manhattan is 922 feet.***
- ***The distance between consecutive streets in Manhattan is 260 feet.***

21. Suppose you walk from the corner of Central Park at 8th Avenue and 59th Street to 10th Avenue and 71st Street.

(a) What is the length of your path if you walk along 8th Avenue to 71st Street, then along 71st Street to 10th Avenue?

(b) What is the length of your path if you walk along Broadway, which goes in a straight line from 8th Avenue and 59th Street to 10th Avenue and 71st Street?

(c) How much shorter is the direct path along Broadway than walking along 8th Avenue and 71st Street?

(d) At the normal city walking speed of 250 feet per minute, how much time would you save by walking along Broadway as compared to walking along 8th Avenue and 71st Street?

SOLUTION

(a) Walking along 8th Avenue from 59th Street to 71st Street is 12 blocks (because $71 - 59 = 12$), each of which is 260 feet, for a total of 12×260 feet, which equals 3120 feet.

Walking along 71st Street from 8th Avenue to 10th Avenue is 2 blocks, each of which is 922 feet, for a total of 1844 feet.

Thus the total path consists of 3120 feet plus 1844 feet, which equals 4964 feet.

(b) The length of the path along Broadway is the length of the hypotenuse of a right triangle whose other sides have length 3120 feet and 1844 feet, as calculated in part (a). Thus the length of the path along Broadway is

$$\sqrt{3120^2 + 1844^2} = \sqrt{13134736} \approx 3624 \text{ feet.}$$

(c) Subtracting the results of part (b) from the result of part (a), we see that the path along Broadway is about 1340 feet shorter. Because a mile contains 5280 feet, the path along Broadway is about a quarter-mile shorter ($\frac{1340}{5280} \approx 0.25$).

(d) Part (c) shows that the path along Broadway is about 1340 feet shorter. Thus the amount of time saved is $\frac{1340}{250}$ minutes, which equals a bit more than 5 minutes.

23. Find two points on the vertical axis whose distance from $(5, -1)$ equals 8.

SOLUTION A typical point on the vertical axis has coordinates $(0, y)$. The distance from this point to $(5, -1)$ is $\sqrt{(0-5)^2 + (y-(-1))^2}$. Thus we need to solve the equation

$$\sqrt{25 + (y+1)^2} = 8.$$

Squaring both sides of the equation above, and then subtracting 25 from both sides gives

$$(y+1)^2 = 39.$$

Thus $y + 1 = \pm\sqrt{39}$. Thus $y = -1 \pm \sqrt{39}$. Hence the two points on the vertical axis whose distance from $(5, -1)$ equals 8 are $(0, -1 + \sqrt{39})$ and $(0, -1 - \sqrt{39})$.

25. Find the perimeter of the triangle that has vertices at $(1, 2)$, $(5, -3)$, and $(-4, -1)$.

SOLUTION The perimeter of the triangle equals the sum of the lengths of the three sides of the triangle. Thus we find the lengths of those three sides.

The side of the triangle connecting the vertices $(1, 2)$ and $(5, -3)$ has length

$$\sqrt{(5-1)^2 + (-3-2)^2} = \sqrt{41}.$$

The side of the triangle connecting the vertices $(5, -3)$ and $(-4, -1)$ has length

$$\sqrt{(-4-5)^2 + (-1-(-3))^2} = \sqrt{85}.$$

The side of the triangle connecting the vertices $(-4, -1)$ and $(1, 2)$ has length

$$\sqrt{(1-(-4))^2 + (2-(-1))^2} = \sqrt{34}.$$

Thus the perimeter of the triangle equals $\sqrt{41} + \sqrt{85} + \sqrt{34}$.

27. Find the radius of a circle that has circumference 12 inches.

SOLUTION Let r denote the radius of this circle in inches. Thus $2\pi r = 12$, which implies that $r = \frac{6}{\pi}$ inches.

29. Find the radius of a circle that has circumference 8 more than its diameter.

SOLUTION Let r denote the radius of this circle. Thus the circle has circumference $2\pi r$ and has diameter $2r$. Because the circumference is 8 more than diameter, we have $2\pi r = 2r + 8$. Thus $(2\pi - 2)r = 8$, which implies that $r = \frac{4}{\pi - 1}$.

31. Find an equation whose graph in the xy-plane is the set of points whose distance from the origin is 3.

SOLUTION The distance from a point (x, y) to the origin is $\sqrt{x^2 + y^2}$. Thus the equation we seek is

$$\sqrt{x^2 + y^2} = 3.$$

To write this equation without using square roots, square both sides to get the equivalent equation

$$x^2 + y^2 = 9.$$

33. Find the length of the graph of the equation $x^2 + y^2 = 9$.

SOLUTION From Exercise 31, we see that the graph of $x^2 + y^2 = 9$ is the set of points in the xy-plane whose distance from the origin is 3. In other words, the graph of $x^2 + y^2 = 9$ is the circle of radius 3 centered at the origin. The length (or circumference) of this graph is $2\pi \cdot 3$, which equals 6π.

35. Find an equation whose graph in the rt-plane is the set of points whose distance from $(3, 2)$ is 2.

SOLUTION The distance from a point (r, t) to $(3, 2)$ is $\sqrt{(r-3)^2 + (t-2)^2}$. Thus the equation we seek is

$$\sqrt{(r-3)^2 + (t-2)^2} = 2.$$

To write this equation without using square roots, square both sides to get the equivalent equation

$$(r-3)^2 + (t-2)^2 = 4.$$

37. Find two points that have distance 2 from the origin and distance 3 from $(3, 0)$.

SOLUTION Suppose (x, y) has distance 2 from the origin and distance 3 from $(3, 0)$. Thus

$$\sqrt{x^2 + y^2} = 2 \quad \text{and} \quad \sqrt{(x-3)^2 + y^2} = 3.$$

Squaring both sides of these equations gives

$$x^2 + y^2 = 4 \quad \text{and} \quad x^2 - 6x + 9 + y^2 = 9.$$

Subtracting the second equation from the first equation gives

$$6x = 4.$$

Thus $x = \frac{2}{3}$. Substituting this value of x into the equation $x^2 + y^2 = 4$ gives the equation $y^2 = \frac{32}{9}$. Thus $y = \pm\frac{\sqrt{32}}{3} = \pm\frac{4\sqrt{2}}{3}$. Hence the two points we seek are $(\frac{2}{3}, \frac{4\sqrt{2}}{3})$ and $(\frac{2}{3}, -\frac{4\sqrt{2}}{3})$.

39. Suppose you want to design a 400-meter track consisting of two half-circles connected by parallel line segments. Suppose also that you want the total length of the curved part of the track to equal half the total length of the straight part of the track. What dimensions should the track have?

SOLUTION Let t equal the total length of the straight part of the track in meters. We want the curved part of the track to have total length $\frac{t}{2}$. Thus we want

$$t + \frac{t}{2} = 400.$$

Solving this equation for t, we get $t = \frac{800}{3}$ meters. Thus each of the two straight pieces must be $\frac{400}{3}$ meters long. Hence we take $c = \frac{400}{3} \approx 133.33$ meters in the figure in the below.

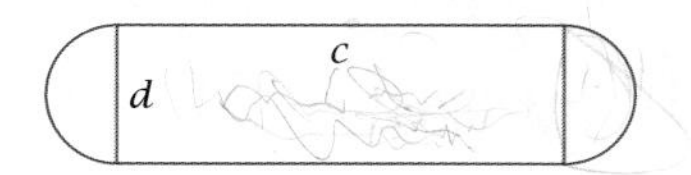

We also want the total length of the two half-circles to be $\frac{400}{3}$ meters. Thus we want $\pi d = \frac{400}{3}$. Hence we take $d = \frac{400}{3\pi} \approx 42.44$ meters in the figure above.

41. Suppose a rope is just long enough to cover the equator of the Earth. About how much longer would the rope need to be so that it could be suspended seven feet above the entire equator?

SOLUTION Assume that the equator of the Earth is a circle. This assumption is close enough to being correct to answer a question that requires only an approximation.

Assume that the radius of the Earth is r, measured in feet (note that we do not need to know the value of r for this exercise). For a rope to cover the equator, it needs to have length $2\pi r$ feet. For a rope to be suspended seven feet above the equator, it would need to have length $2\pi(r + 7)$ feet, which equals $(2\pi r + 14\pi)$ feet. In other words, to be suspended seven feet above the equator, the rope would need to be only 14π feet longer than a rope covering the equator. Because $14\pi \approx 14 \cdot \frac{22}{7} = 44$, the rope would need to be about 44 feet longer than a rope covering the equator.

2.2 *Lines*

LEARNING OBJECTIVES

By the end of this section you should be able to

- find the slope of a line;
- find the equation of a line given its slope and a point on it;
- find the equation of a line given two points on it;
- determine whether or not two lines are parallel;
- find the equation of a line perpendicular to a given line and containing a given point;
- find the midpoint of a line segment.

Slope

Consider a line in the coordinate plane, along with four points (x_1, y_1), (x_2, y_2), (x_3, y_3), and (x_4, y_4) on the line. Draw two right triangles with horizontal and vertical edges as in the figure below:

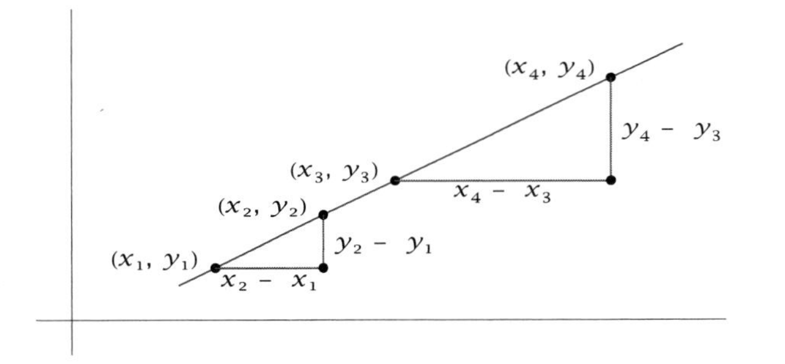

Similar triangles.

In this figure, each side of the larger triangle has twice the length of the corresponding side of the smaller triangle.

The two right triangles in the figure above are similar because their angles are equal. Thus the ratios of the corresponding sides of the two triangles above are equal. Specifically, taking the ratio of the vertical side and horizontal side for each triangle, we have

$$\frac{y_2 - y_1}{x_2 - x_1} = \frac{y_4 - y_3}{x_4 - x_3}.$$

The equation above states that for any pair of points (x_1, y_1) and (x_2, y_2) on the line, the ratio $\frac{y_2-y_1}{x_2-x_1}$ does not depend on the particular pair of points chosen on the line. If we choose another pair of points on the line, say (x_3, y_3) and (x_4, y_4) instead of (x_1, y_1) and (x_2, y_2), then the difference of second coordinates divided by the difference of first coordinates remains the same, as shown by the equation above.

Thus the ratio $\frac{y_2-y_1}{x_2-x_1}$ is a constant depending only on the line and not on the particular points (x_1, y_1) and (x_2, y_2) chosen on the line. This constant is called the **slope** of the line.

Slope

If (x_1, y_1) and (x_2, y_2) are any two points on a line, with $x_1 \neq x_2$, then the **slope** of the line is

$$\frac{y_2 - y_1}{x_2 - x_1}.$$

EXAMPLE 1

Find the slope of the line containing the points $(2, 1)$ and $(5, 3)$.

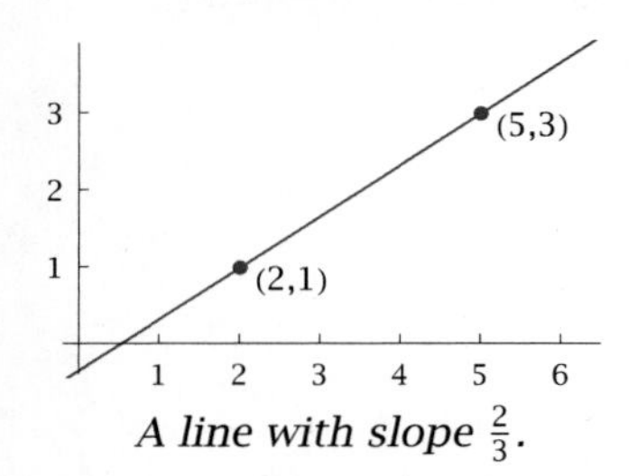

A line with slope $\frac{2}{3}$.

SOLUTION The line containing $(2, 1)$ and $(5, 3)$ is shown here. The slope of this line is $\frac{3-1}{5-2}$, which equals $\frac{2}{3}$.

A line with positive slope slants up from left to right; a line with negative slope slants down from left to right. Lines whose slopes have larger absolute value are steeper than lines whose slopes have smaller absolute value. This figure shows some lines and their slopes; the same scale has been used on both axes.

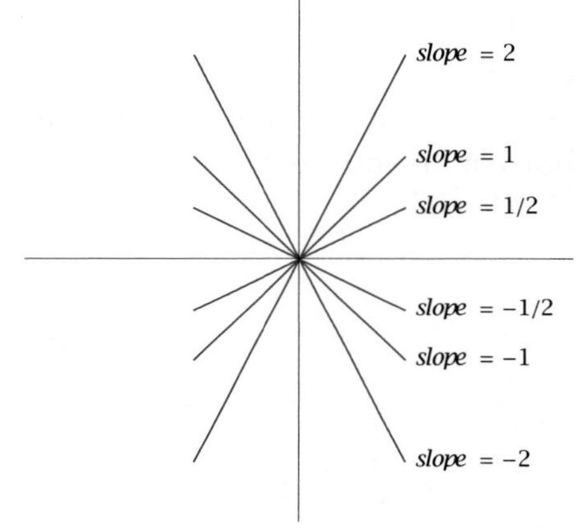

In the figure above, the horizontal axis has slope 0, as does every horizontal line. Vertical lines, including the vertical axis, do not have a slope, because a vertical line does not contain two points (x_1, y_1) and (x_2, y_2) with $x_1 \neq x_2$.

The Equation of a Line

Consider a line with slope m, and suppose (x_1, y_1) is a point on this line. Let (x, y) denote a typical point on the line, as shown here.

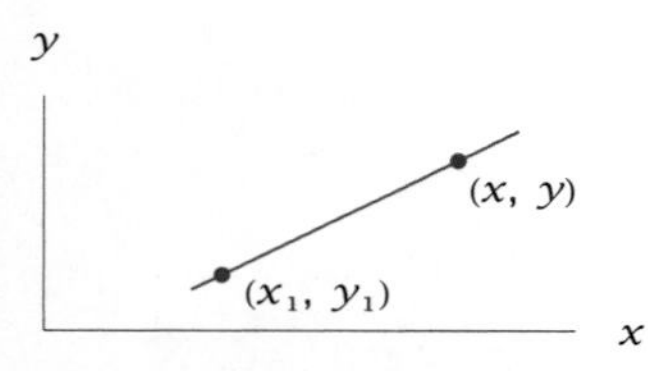

A line with slope $\frac{y - y_1}{x - x_1}$.

Because this line has slope m, we have

$$\frac{y - y_1}{x - x_1} = m.$$

Multiplying both sides of the equation above by $x - x_1$, we get the following formula:

The equation of a line, given its slope and one point on it

The line in the xy-plane that has slope m and contains the point (x_1, y_1) is given by the equation

$$y - y_1 = m(x - x_1).$$

The symbol m is often used to denote the slope of a line.

The equation above can be solved for y to get an equation for the line in the form $y = mx + b$, where m and b are constants.

EXAMPLE 2

Find the equation of the line in the xy-plane that has slope $\frac{1}{2}$ and contains $(4, 1)$.

SOLUTION In this case the equation displayed above becomes

$$y - 1 = \tfrac{1}{2}(x - 4).$$

Adding 1 to both sides and simplifying, we get

$$y = \tfrac{1}{2}x - 1.$$

As a check for possible errors, if we take $x = 4$ in the equation above, we get $y = 1$. Thus the point $(4, 1)$ is indeed on this line.

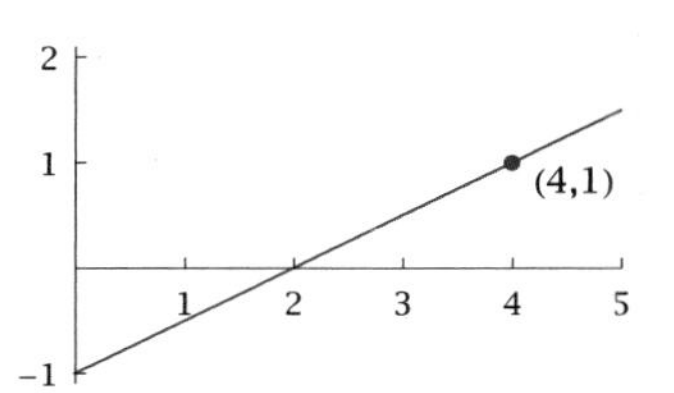

The line with slope $\frac{1}{2}$ that contains the point $(4, 1)$.

As a special case of finding the equation of a line when given its slope and one point on it, suppose we want to find the equation of the line in the xy-plane with slope m that intersects the y-axis at the point $(0, b)$. In this case, the formula above becomes

$$y - b = m(x - 0).$$

*The point where a line intersects the y-axis is often called the **y-intercept**.*

Solving this equation for y, we have the following result:

The equation of a line, given its slope and y-intercept

The line in the xy-plane with slope m that intersects the y-axis at $(0, b)$ is given by the equation

$$y = mx + b.$$

EXAMPLE 3

Find the equation of the line in the xy-plane that has slope $\frac{1}{2}$ and intersects the y-axis at $(0, -1)$.

SOLUTION The formula above shows that the desired equation is

$$y = \tfrac{1}{2}x - 1.$$

This line is shown in Example 2, where you can see that the line intersects the y-axis at $(0, -1)$.

If a line contains the origin, then $b = 0$ in the formula above. For example, the line in the xy-plane that has slope 2 and contains the origin is given by the equation $y = 2x$. The figure below Example 1 shows the line $y = 2x$ and several other lines containing the origin.

Suppose now that we want to find the equation of the line containing two specific points. We can reduce this problem to a problem we have already solved by computing the slope of the line and then using the formula in the box above.

Specifically, suppose we want to find the equation of the line containing the points (x_1, y_1) and (x_2, y_2), where $x_1 \neq x_2$. This line has slope $(y_2 - y_1)/(x_2 - x_1)$. Thus our formula for the equation of a line when given its slope and one point on it gives the following result:

The equation of a line, given two points on it

The line in the xy-plane that contains the points (x_1, y_1) and (x_2, y_2), where $x_1 \neq x_2$, is given by the equation

$$y - y_1 = \left(\frac{y_2 - y_1}{x_2 - x_1}\right)(x - x_1).$$

EXAMPLE 4 Find the equation of the line in the xy-plane that contains the points $(2, 4)$ and $(5, 1)$.

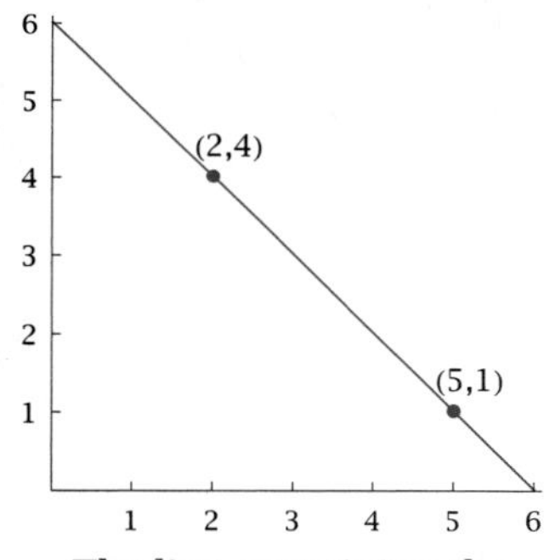

The line containing the points $(2, 4)$ and $(5, 1)$.

SOLUTION In this case the equation above becomes

$$y - 4 = \left(\frac{1 - 4}{5 - 2}\right)(x - 2).$$

Solving this equation for y, we get

$$y = -x + 6.$$

As a check, if we take $x = 2$ in the equation above, we get $y = 4$, and if we take $x = 5$ in the equation above, we get $y = 1$; thus the points $(2, 4)$ and $(5, 1)$ are indeed on this line.

Conversion between different units of measurement is usually done by an equation that represents one unit as a suitable multiple of another unit. For example, a pound is officially defined to be exactly 0.45359237 kilograms. Thus the equation that gives the weight k in kilograms for a object weighing p pounds is

The formula $p = 2.2k$ is often used, but it is an approximation rather than an exact formula.

$$k = 0.45359237p.$$

Conversion between temperature scales is unusual because the zero temperature on one scale does not correspond to the zero temperature on another scale. Most other quantities such weights, lengths, and currency have the same zero point regardless of the units used. For example, without knowing the conversion rate, you know that 0 centimeters is the same length as 0 inches.

The next example shows how to find a formula for converting from Celsius temperatures to Fahrenheit temperatures. The solution to the next example also shows that rather than memorizing the formula, for example, of the equation of a line given two points on it, sometimes it is simpler just to use the available information to find the constants m and b that characterize the line $y = mx + b$.

EXAMPLE 5

Find an equation that gives the temperature F on the Fahrenheit scale corresponding to temperature C on the Celsius scale.

SOLUTION We seek an equation of the form

$$F = mC + b$$

for some constants m and b.

To find m and b, we start by recalling that the freezing temperature of water equals 0 degrees Celsius and 32 degrees Fahrenheit. Plugging $C = 0$ and $F = 32$ into the equation above gives $32 = b$. Now that we know that $b = 32$, the equation above can be rewritten as

$$F = mC + 32.$$

Recall now that the boiling point of water equals 100 degrees Celsius and 212 degrees Fahrenheit. Plugging $C = 100$ and $F = 212$ into the equation above and then solving for m shows that $m = \frac{9}{5}$. Thus the formula we seek is

$$F = \tfrac{9}{5}C + 32.$$

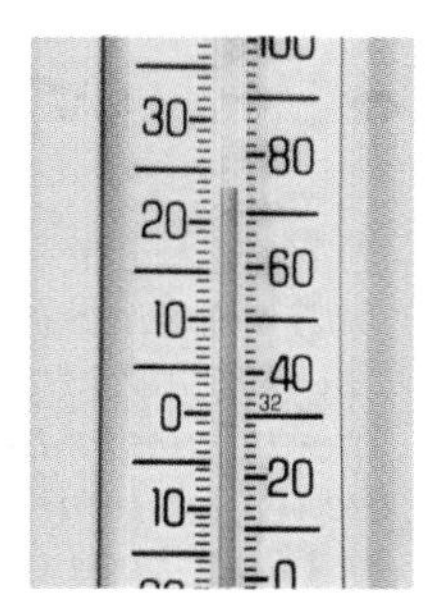

This thermometer shows Celsius degrees on the left, Fahrenheit degrees on the right.

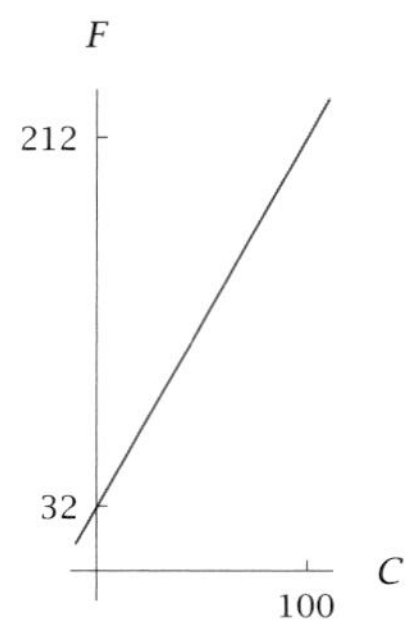

The graph of $F = \frac{9}{5}C + 32$ *on the interval* $[-10, 110]$.

Parallel Lines

Consider two parallel lines in the coordinate plane, as shown in the figure below:

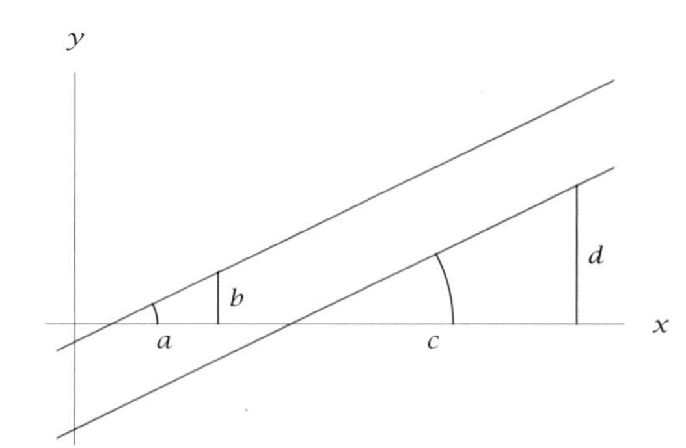

Parallel lines.

Because the two lines are parallel, the corresponding angles in the two triangles above are equal (as shown by the arcs in the figure above), and thus the two right triangles are similar. This implies that

$$\frac{b}{a} = \frac{d}{c}.$$

Because $\frac{b}{a}$ is the slope of the top line and $\frac{d}{c}$ is the slope of the bottom line, we conclude that these parallel lines have the same slope.

The logic used in the paragraph above is reversible. Specifically, suppose instead of starting with the assumption that the two lines in the figure above are parallel, we start with the assumption that the two lines have the same slope. Thus $\frac{b}{a} = \frac{d}{c}$, which implies that the two right triangles in the figure above are similar. Hence the two lines make equal angles with the horizontal axis, as shown by the arcs in the figure, which implies that the two lines are parallel.

The figure and reasoning given above do not work if both lines are horizontal or both lines are vertical. But horizontal lines all have slope 0, and the slope is not defined for vertical lines. Thus we can summarize our characterization of parallel lines as follows:

The phrase "if and only if", when connecting two statements, means that the two statements are either both true or both false. For example, $x + 1 > 6$ if and only if $x > 5$.

Parallel lines

Two nonvertical lines in the coordinate plane are parallel if and only if they have the same slope.

EXAMPLE 6

(a) Are the lines in the xy-plane given by the following equations parallel?

$$y = 4x - 5 \quad \text{and} \quad y = 4x + 18$$

(b) Are the lines in the xy-plane given by the following equations parallel?

$$y = 6x + 5 \quad \text{and} \quad y = 7x + 5$$

SOLUTION

(a) These lines are parallel because they have the same slope (which equals 4).

(b) These lines are not parallel because their slopes are not equal—the first line has slope 6 and the second line has slope 7.

Perpendicular Lines

Before beginning our treatment of perpendicular lines, we take a brief detour to make clear the geometry of a line with negative slope. A line with negative slope slants down from left to right. The figure below shows a line with negative slope; to avoid clutter the coordinate axes are not shown.

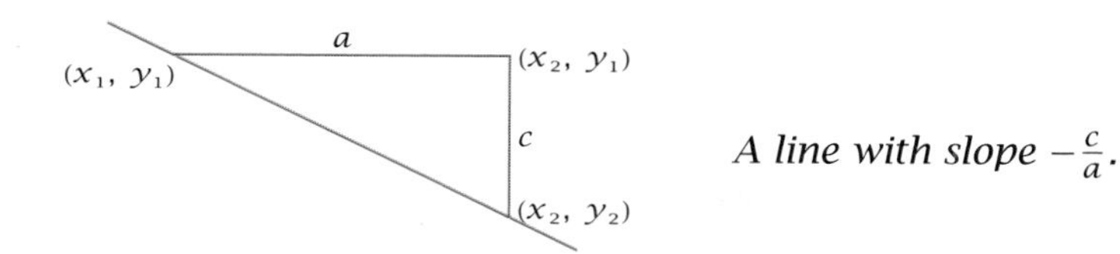

A line with slope $-\frac{c}{a}$.

In the figure above, a is the length of the horizontal line segment and c is the length of the vertical line segment. Of course a and c are positive numbers, because lengths are positive. In terms of the coordinates as shown in the figure above, we have $a = x_2 - x_1$ and $c = y_1 - y_2$. The slope of this line equals $(y_2 - y_1)/(x_2 - x_1)$, which equals $-c/a$.

The following result gives a useful characterization of perpendicular lines:

Perpendicular lines

Two nonvertical lines are perpendicular if and only if the product of their slopes equals -1.

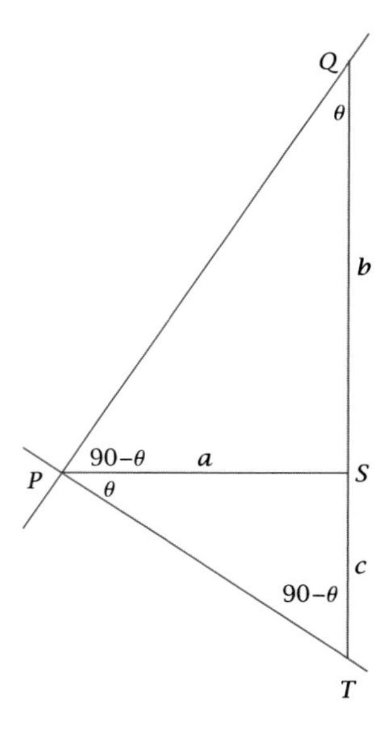

To explain why the result above holds, consider two perpendicular lines as shown in blue in the figure here. In addition to the two perpendicular lines in blue, the figure shows the horizontal line segment PS and the vertical line segment QT, which intersect at S.

We assume that the angle PQT is θ degrees. To check that the other three labeled angles in the figure are labeled correctly, first note that the two angles labeled $90 - \theta$ are each the third angle in a right triangle (the right triangles are PSQ and QPT), one of whose angles is θ. Consideration of the right angle QPT now shows that angle TPS is θ degrees, as labeled.

The line containing the points P and Q has slope b/a, as can be seen from the figure. Furthermore, our brief discussion of lines with negative slope shows that the line containing the points P and T has slope $-c/a$.

Consider the right triangles PSQ and TSP in the figure. These triangles have the same angles, and thus they are similar. Thus the ratios of corresponding sides are equal. Specifically, we have

$$\frac{b}{a} = \frac{a}{c}.$$

Multiplying both sides of this equation by $-c/a$, we get

$$\left(\frac{b}{a}\right) \cdot \left(-\frac{c}{a}\right) = -1.$$

As we have already seen, the first quantity on the left above is the slope of the line containing the points P and Q, and the second quantity is the slope of the line containing the points P and T. Thus we can conclude that the product of the slopes of these two perpendicular lines equals -1, as desired.

The logic used above is reversible. Specifically, suppose instead of starting with the assumption that the two lines in blue are perpendicular, we start

Numbers m_1 and m_2 such that $m_1 m_2 = -1$ are sometimes called **negative reciprocals** *of each other.*

with the assumption that the product of their slopes equals -1. This implies that $\frac{b}{a} = \frac{a}{c}$, which implies that the two right triangles PSQ and TSP are similar; thus these two triangles have the same angles. This implies that the angles are labeled correctly in the figure above (assuming that we start by declaring that angle PQS measures θ degrees). This then implies that angle QPT measures 90°. Thus the two lines in blue are perpendicular, as desired.

EXAMPLE 7

Show the lines in the xy-plane given by the following equations are perpendicular:

$$y = 4x - 5 \quad \text{and} \quad y = -\tfrac{1}{4}x + 18$$

SOLUTION The first line has slope 4; the second line has slope $-\frac{1}{4}$. The product of these slopes is $4 \cdot (-\frac{1}{4})$, which equals -1. Because the product of the two slopes equals -1, the two lines are perpendicular.

To show that two lines are perpendicular, we only need to know the slopes of the lines, not their full equations, as shown by the following example.

EXAMPLE 8

Show that the line containing the points $(1, -2)$ and $(3, 3)$ is perpendicular to the line containing the points $(9, -1)$ and $(4, 1)$.

SOLUTION The line containing $(1, -2)$ and $(3, 3)$ has slope $\frac{3-(-2)}{3-1}$, which equals $\frac{5}{2}$. Also, the line containing $(9, -1)$ and $(4, 1)$ has slope $\frac{1-(-1)}{4-9}$, which equals $-\frac{2}{5}$. Because the product $\frac{5}{2} \cdot (-\frac{2}{5})$ equals -1, the two lines are perpendicular.

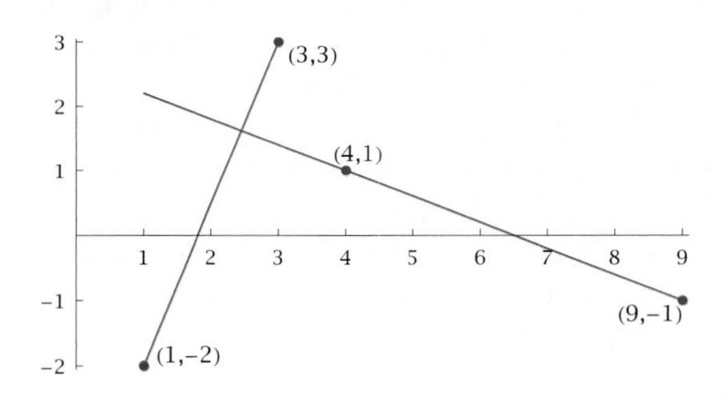

The line containing $(1, -2)$ and $(3, 3)$ is perpendicular to the line containing $(9, -1)$ and $(4, 1)$.

Midpoints

This subsection begins with an intuitive definition of the midpoint of a line segment:

> ***Midpoint***
>
> The **midpoint** of a line segment is the point on the line segment that lies halfway between the two endpoints.

As you might guess, the first coordinate of the midpoint of a line segment is the average of the first coordinates of the endpoints. Similarly, the second coordinate of the midpoint is the average of the second coordinates of the endpoints. Here is the formal statement of this formula:

Midpoint

The midpoint of the line segment connecting (x_1, y_1) and (x_2, y_2) equals

$$\left(\frac{x_1 + x_2}{2}, \frac{y_1 + y_2}{2}\right).$$

Problems 51–53 at the end of this section will lead you to an explanation of why this formula for the midpoint is correct.

The next example illustrates the use of the formula above.

EXAMPLE 9

(a) Find the midpoint of the line segment connecting $(1,3)$ and $(5,9)$.

(b) Verify that the distance between the midpoint found in (a) and the first endpoint $(1,3)$ equals the distance between the midpoint found in (a) and the second endpoint $(5,9)$.

(c) Verify that the midpoint found in (a) lies on the line connecting $(1,3)$ and $(5,9)$.

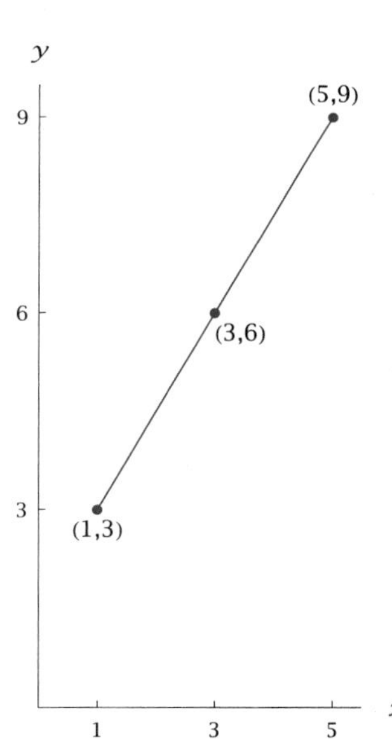

The point $(3,6)$ is the midpoint of the line segment connecting $(1,3)$ and $(5,9)$.

SOLUTION

(a) Using the formula above, we see that the midpoint of the line segment connecting $(1,3)$ and $(5,9)$ equals

$$\left(\frac{1+5}{2}, \frac{3+9}{2}\right),$$

which equals $(3,6)$.

(b) First compute the distance between the midpoint and the endpoint $(1,3)$:

$$\text{distance between } (3,6) \text{ and } (1,3) = \sqrt{(3-1)^2 + (6-3)^2} = \sqrt{2^2 + 3^2}$$
$$= \sqrt{13}.$$

Next compute the distance between the midpoint and the endpoint $(5,9)$:

$$\text{distance between } (3,6) \text{ and } (5,9) = \sqrt{(3-5)^2 + (6-9)^2} = \sqrt{(-2)^2 + (-3)^2}$$
$$= \sqrt{13}.$$

As expected, these two distances are equal—the distance between the midpoint and either endpoint is $\sqrt{13}$.

(c) First compute the slope of the line containing the midpoint and the endpoint $(1,3)$:

$$\text{slope of line containing } (3,6) \text{ and } (1,3) = \frac{6-3}{3-1} = \frac{3}{2}.$$

Next compute the slope of the line containing the midpoint and the endpoint $(5,9)$:

$$\text{slope of line containing } (3,6) \text{ and } (5,9) = \frac{6-9}{3-5} = \frac{-3}{-2} = \frac{3}{2}.$$

As expected, these two slopes are equal. In other words, the line segment from $(1,3)$ to $(3,6)$ points in the same direction as the line segment from $(3,6)$ to $(5,9)$. Thus these three points all lie on the same line.

EXERCISES

1. What are the coordinates of the unlabeled vertex of the smaller of the two right triangles in the figure at the beginning of this section?
2. What are the coordinates of the unlabeled vertex of the larger of the two right triangles in the figure at the beginning of this section?
3. Find the slope of the line that contains the points $(3,4)$ and $(7,13)$.
4. Find the slope of the line that contains the points $(2,11)$ and $(6,-5)$.
5. Find a number t such that the line containing the points $(1,t)$ and $(3,7)$ has slope 5.
6. Find a number c such that the line containing the points $(c,4)$ and $(-2,9)$ has slope -3.
7. Suppose the tuition per semester at Euphoria State University is \$525 plus \$200 for each class unit taken.
   (a) Find an equation that gives the tuition t in dollars for taking u class units.
   (b) Find the total tuition for accumulating 120 units over 8 semesters.
   (c) Find the total tuition for accumulating 120 units over 12 semesters.
8. Suppose the tuition per semester at Luxim University is \$900 plus \$850 for each class unit taken.
   (a) Find an equation that gives the tuition t in dollars for taking u class units.
   (b) Find the total tuition for accumulating 120 units over 8 semesters.
   (c) Find the total tuition for accumulating 120 units over 10 semesters.
9. Suppose your cell phone company offers two calling plans. The pay-per-call plan charges \$14 per month plus 3 cents for each minute. The unlimited-calling plan charges a flat rate of \$29 per month for unlimited calls.
   (a) Find an equation that gives the cost c in dollars for making m minutes of phone calls per month on the pay-per-call plan.
   (b) How many minutes per month must you use for the unlimited-calling plan to become cheaper?
10. Suppose your cell phone company offers two calling plans. The pay-per-call plan charges \$11 per month plus 4 cents for each minute. The unlimited-calling plan charges a flat rate of \$25 per month for unlimited calls.
    (a) Find an equation that gives the cost c in dollars for making m minutes of phone calls per month on the pay-per-call plan.
    (b) How many minutes per month must you use for the unlimited-calling plan to become cheaper?
11. Find the equation of the line in the xy-plane with slope 2 that contains the point $(7,3)$.
12. Find the equation of the line in the xy-plane with slope -4 that contains the point $(-5,-2)$.
13. Find the equation of the line that contains the points $(2,-1)$ and $(4,9)$.
14. Find the equation of the line that contains the points $(-3,2)$ and $(-5,7)$.
15. Find a number t such that the point $(3,t)$ is on the line containing the points $(7,6)$ and $(14,10)$.
16. Find a number t such that the point $(-2,t)$ is on the line containing the points $(5,-2)$ and $(10,-8)$.

17. Find a formula for the number of seconds s in d days.

18. Find a formula for the number of seconds s in w weeks.

19. Find a formula for the number of inches I in M miles.

20. Find a formula for the number of miles M in F feet.

21. Find a formula for the number of kilometers k in M miles.
[*The exact conversion between the English measurement system and the metric system is given by the equation* 1 *inch* = 2.54 *centimeters.*]

22. Find a formula for the number of miles M in m meters.

23. Find a formula for the number of inches I in c centimeters.

24. Find a formula for the number of meters m in F feet.

25. Find a number c such that the point $(c, 13)$ is on the line containing the points $(-4, -17)$ and $(6, 33)$.

26. Find a number c such that the point $(c, -19)$ is on the line containing the points $(2, 1)$ and $(4, 9)$.

27. Find a number t such that the point $(t, 2t)$ is on the line containing the points $(3, -7)$ and $(5, -15)$.

28. Find a number t such that the point $(t, \frac{t}{2})$ is on the line containing the points $(2, -4)$ and $(-3, -11)$.

29. Find the equation of the line in the xy-plane that contains the point $(3, 2)$ and that is parallel to the line $y = 4x - 1$.

30. Find the equation of the line in the xy-plane that contains the point $(-4, -5)$ and that is parallel to the line $y = -2x + 3$.

31. Find the equation of the line that contains the point $(2, 3)$ and that is parallel to the line containing the points $(7, 1)$ and $(5, 6)$.

32. Find the equation of the line that contains the point $(-4, 3)$ and that is parallel to the line containing the points $(3, -7)$ and $(6, -9)$.

33. Find a number t such that the line containing the points $(t, 2)$ and $(3, 5)$ is parallel to the line containing the points $(-1, 4)$ and $(-3, -2)$.

34. Find a number t such that the line containing the points $(-3, t)$ and $(2, -4)$ is parallel to the line containing the points $(5, 6)$ and $(-2, 4)$.

35. Find the intersection in the xy-plane of the lines $y = 5x + 3$ and $y = -2x + 1$.

36. Find the intersection in the xy-plane of the lines $y = -4x + 5$ and $y = 5x - 2$.

37. Find a number b such that the three lines in the xy-plane given by the equations $y = 2x + b$, $y = 3x - 5$, and $y = -4x + 6$ have a common intersection point.

38. Find a number m such that the three lines in the xy-plane given by the equations $y = mx + 3$, $y = 4x + 1$, and $y = 5x + 7$ have a common intersection point.

39. Find the equation of the line in the xy-plane that contains the point $(4, 1)$ and that is perpendicular to the line whose equation is $y = 3x + 5$.

40. Find the equation of the line in the xy-plane that contains the point $(-3, 2)$ and that is perpendicular to the line whose equation is $y = -5x + 1$.

41. Find a number t such that the line in the xy-plane containing the points $(t, 4)$ and $(2, -1)$ is perpendicular to the line $y = 6x - 7$.

42. Find a number t such that the line in the xy-plane containing the points $(-3, t)$ and $(4, 3)$ is perpendicular to the line $y = -5x + 999$.

43. Find a number t such that the line containing the points $(4, t)$ and $(-1, 6)$ is perpendicular to the line that contains the points $(3, 5)$ and $(1, -2)$.

44. Find a number t such that the line containing the points $(t, -2)$ and $(-3, 5)$ is perpendicular to the line that contains the points $(4, 7)$ and $(1, 11)$.

45. Find the midpoint of the line segment connecting $(-3, 4)$ and $(5, 7)$.

46. Find the midpoint of the line segment connecting $(6, -5)$ and $(-3, -8)$.

47. Find numbers x and y such that $(-2, 5)$ is the midpoint of the line segment connecting $(3, 1)$ and (x, y).

48. Find numbers x and y such that $(3, -4)$ is the midpoint of the line segment connecting $(-2, 5)$ and (x, y).

PROBLEMS

49. The Kelvin temperature scale is defined by $K = C + 273.15$, where K is the temperature on the Kelvin scale and C is the temperature on the Celsius scale. (Thus -273.15 degrees Celsius, which is the temperature at which all molecular movement ceases and thus is the lowest possible temperature, corresponds to 0 on the Kelvin scale.)

    (a) Find an equation that gives the temperature F on the Fahrenheit scale corresponding to temperature K on the Kelvin scale.

    (b) Explain why the graph of the equation from part (a) is parallel to the graph of the equation obtained in Example 5.

50. Find the equation of the line in the xy-plane that has slope m and intersects the x-axis at $(c, 0)$.

51. Suppose (x_1, y_1) and (x_2, y_2) are the endpoints of a line segment.

    (a) Show that the distance between the point $(\frac{x_1+x_2}{2}, \frac{y_1+y_2}{2})$ and the endpoint (x_1, y_1) equals half the length of the line segment.

    (b) Show that the distance between the point $(\frac{x_1+x_2}{2}, \frac{y_1+y_2}{2})$ and the endpoint (x_2, y_2) equals half the length of the line segment.

52. Suppose (x_1, y_1) and (x_2, y_2) are the endpoints of a line segment.

    (a) Show that the line containing the point $(\frac{x_1+x_2}{2}, \frac{y_1+y_2}{2})$ and the endpoint (x_1, y_1) has slope $\frac{y_2-y_1}{x_2-x_1}$.

    (b) Show that the line containing the point $(\frac{x_1+x_2}{2}, \frac{y_1+y_2}{2})$ and the endpoint (x_2, y_2) has slope $\frac{y_2-y_1}{x_2-x_1}$.

    (c) Explain why parts (a) and (b) of this problem imply that the point $(\frac{x_1+x_2}{2}, \frac{y_1+y_2}{2})$ lies on the line containing the endpoints (x_1, y_1) and (x_2, y_2).

53. Explain why the two previous problems imply that $(\frac{x_1+x_2}{2}, \frac{y_1+y_2}{2})$ is the midpoint of the line segment with endpoints (x_1, y_1) and (x_2, y_2).

54. Find six distinct points on the circle with center $(2, 3)$ and radius 5.

55. Find six distinct points on the circle with center $(4, 1)$ and circumference 3.

56. We used similar triangles to show that the product of the slopes of two perpendicular lines equals -1. The steps below outline an alternative proof that avoids the use of similar triangles but uses more algebra instead. Use the figure below, which is the same as the figure used earlier except that there is now no need to label the angles.

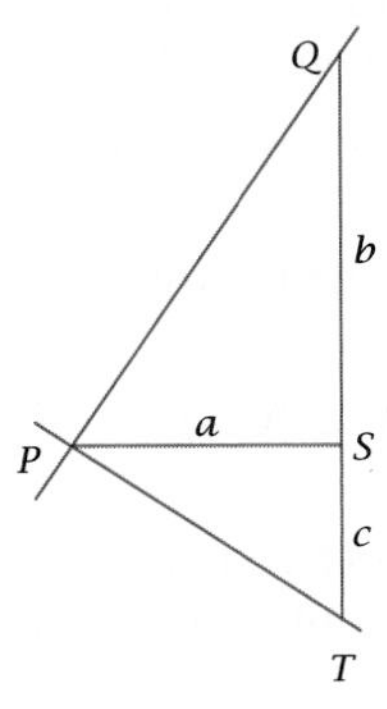

QP is perpendicular to PT.

    (a) Apply the Pythagorean Theorem to triangle PSQ to find the length of the line segment PQ in terms of a and b.

    (b) Apply the Pythagorean Theorem to triangle PST to find the length of the line segment PT in terms of a and c.

    (c) Apply the Pythagorean Theorem to triangle QPT to find the length of the line segment QT in terms of the lengths of the line segments of PQ and PT calculated in the first two parts of this problem.

    (d) As can be seen from the figure, the length of the line segment QT equals $b + c$. Thus set the formula for length of the line segment QT, as calculated in the previous part of this problem, equal to $b + c$, and solve the resulting equation for c in terms of a and b.

    (e) Use the result in the previous part of this problem to show that the slope of the line containing P and Q times the slope of the line containing P and T equals -1.

57. Suppose a and b are nonzero numbers. Where does the line in the xy-plane given by the equation

$$\frac{x}{a} + \frac{y}{b} = 1$$

intersect the coordinate axes?

58. Show that the points $(-84, -14)$, $(21, 1)$, and $(98, 12)$ lie on a line.

59. Show that the points $(-8, -65)$, $(1, 52)$, and $(3, 77)$ do not lie on a line.

60. Change just one of the six numbers in the problem above so that the resulting three points do lie on a line.

61. Show that for every number t, the point $(5 - 3t, 7 - 4t)$ is on the line containing the points $(2, 3)$ and $(5, 7)$.

WORKED-OUT SOLUTIONS *to Odd-numbered Exercises*

1. What are the coordinates of the unlabeled vertex of the smaller of the two right triangles in the figure at the beginning of this section?

SOLUTION Drawing vertical and horizontal lines from the point in question to the coordinate axes shows that the coordinates of the point are (x_2, y_1).

3. Find the slope of the line that contains the points $(3, 4)$ and $(7, 13)$.

SOLUTION The line containing the points $(3, 4)$ and $(7, 13)$ has slope

$$\frac{13 - 4}{7 - 3},$$

which equals $\frac{9}{4}$.

5. Find a number t such that the line containing the points $(1, t)$ and $(3, 7)$ has slope 5.

SOLUTION The slope of the line containing the points $(1, t)$ and $(3, 7)$ equals

$$\frac{7 - t}{3 - 1},$$

which equals $\frac{7-t}{2}$. We want this slope to equal 5. Thus we must find a number t such that

$$\frac{7 - t}{2} = 5.$$

Solving this equation for t, we get $t = -3$.

7. Suppose the tuition per semester at Euphoria State University is \$525 plus \$200 for each class unit taken.

   (a) Find an equation that gives the tuition t in dollars for taking u class units.

   (b) Find the total tuition for accumulating 120 units over 8 semesters.

   (c) Find the total tuition for accumulating 120 units over 12 semesters.

SOLUTION

(a) The tuition t in dollars for taking u class units is given by the equation

$$t = 200u + 525.$$

(b) Taking 120 units over 8 semesters will cost $200 \times 120 + 8 \times 525$ dollars, which equals \$28,200.

(c) Taking 120 units over 12 semesters will cost $200 \times 120 + 12 \times 525$ dollars, which equals \$30,300.

9. Suppose your cell phone company offers two calling plans. The pay-per-call plan charges \$14 per month plus 3 cents for each minute. The unlimited-calling plan charges a flat rate of \$29 per month for unlimited calls.

   (a) Find an equation that gives the cost c in dollars for making m minutes of phone calls per month on the pay-per-call plan.

   (b) How many minutes per month must you use for the unlimited-calling plan to become cheaper?

SOLUTION

(a) As usual, units must be consistent throughout any calculation. Thus we think of 3 cents as 0.03 dollars. Hence the cost c in dollars for making m minutes of phone calls per month is given by the equation

$$c = 0.03m + 14.$$

(b) Setting $c = 29$ in the equation above gives the equation $29 = 0.03m + 14$. Thus

$$m = \tfrac{29-14}{0.03} = \tfrac{15}{0.03} = \tfrac{1500}{3} = 500.$$

Thus the two plans cost an equal amount if 500 minutes per month are used. If more than 500 minutes per month are used, then the unlimited-calling plan is cheaper.

11. Find the equation of the line in the xy-plane with slope 2 that contains the point $(7, 3)$.

SOLUTION If (x, y) denotes a typical point on the line with slope 2 that contains the point $(7, 3)$, then

$$\frac{y-3}{x-7} = 2.$$

Multiplying both sides of this equation by $x - 7$ and then adding 3 to both sides gives the equation

$$y = 2x - 11.$$

CHECK The line whose equation is $y = 2x - 11$ has slope 2. We should also check that the point $(7, 3)$ is on this line. In other words, we need to verify the alleged equation

$$3 \stackrel{?}{=} 2 \cdot 7 - 11.$$

Simple arithmetic shows that this is indeed true.

13. Find the equation of the line that contains the points $(2, -1)$ and $(4, 9)$.

SOLUTION The line that contains the points $(2, -1)$ and $(4, 9)$ has slope

$$\frac{9-(-1)}{4-2},$$

which equals 5. Thus if (x, y) denotes a typical point on this line, then

$$\frac{y-9}{x-4} = 5.$$

Multiplying both sides of this equation by $x - 4$ and then adding 9 to both sides gives the equation

$$y = 5x - 11.$$

CHECK We need to check that both $(2, -1)$ and $(4, 9)$ are on the line whose equation is $y = 5x - 11$. In other words, we need to verify the alleged equations

$$-1 \stackrel{?}{=} 5 \cdot 2 - 11 \quad \text{and} \quad 9 \stackrel{?}{=} 5 \cdot 4 - 11.$$

Simple arithmetic shows that both alleged equations are indeed true.

15. Find a number t such that the point $(3, t)$ is on the line containing the points $(7, 6)$ and $(14, 10)$.

SOLUTION First we find the equation of the line containing the points $(7, 6)$ and $(14, 10)$. To do this, note that the line containing those two points has slope

$$\frac{10-6}{14-7},$$

which equals $\frac{4}{7}$. Thus if (x, y) denotes a typical point on this line, then

$$\frac{y-6}{x-7} = \frac{4}{7}.$$

Multiplying both sides of this equation by $x - 7$ and then adding 6 gives the equation

$$y = \tfrac{4}{7}x + 2.$$

Now we can find a number t such that the point $(3, t)$ is on the line given by the equation above. To do this, in the equation above replace x by 3 and y by t, getting

$$t = \tfrac{4}{7} \cdot 3 + 2.$$

Performing the arithmetic to compute the right side, we get $t = \frac{26}{7}$.

CHECK We should check that all three points $(7, 6)$, $(14, 10)$, and $(3, \frac{26}{7})$ are on the line $y = \frac{4}{7}x + 2$. In other words, we need to verify the alleged equations

$$6 \stackrel{?}{=} \tfrac{4}{7} \cdot 7 + 2, \quad 10 \stackrel{?}{=} \tfrac{4}{7} \cdot 14 + 2, \quad \tfrac{26}{7} \stackrel{?}{=} \tfrac{4}{7} \cdot 3 + 2.$$

Simple arithmetic shows that all three alleged equations are indeed true.

17. Find a formula for the number of seconds s in d days.

SOLUTION Each minute has 60 seconds, and each hour has 60 minutes. Thus each hour has

60×60 seconds, or 3600 seconds. Each day has 24 hours; thus each day has 24×3600 seconds, or 86400 seconds. Thus

$$s = 86400d.$$

19. Find a formula for the number of inches I in M miles.

SOLUTION Each foot has 12 inches, and each mile has 5280 feet. Thus each mile has 5280×12 inches, or 63360 inches. Thus

$$I = 63360M.$$

21. Find a formula for the number of kilometers k in M miles.

SOLUTION Multiplying both sides of the equation

$$1 \text{ inch} = 2.54 \text{ centimeters}$$

by 12 gives

$$\begin{aligned} 1 \text{ foot} &= 12 \times 2.54 \text{ centimeters} \\ &= 30.48 \text{ centimeters}. \end{aligned}$$

Multiplying both sides of the equation above by 5280 gives

$$\begin{aligned} 1 \text{ mile} &= 5280 \times 30.48 \text{ centimeters} \\ &= 160934.4 \text{ centimeters} \\ &= 1609.344 \text{ meters} \\ &= 1.609344 \text{ kilometers}. \end{aligned}$$

Multiplying both sides of the equation above by a number M shows that M miles $=$ $1.609344M$ kilometers. In other words,

$$k = 1.609344M.$$

[*The formula above is exact. However, the approximation* $k = 1.61M$ *is often used.*]

23. Find a formula for the number of inches I in c centimeters.

SOLUTION Dividing both sides of the equation

$$1 \text{ inch} = 2.54 \text{ centimeters}$$

by 2.54 gives

$$1 \text{ centimeter} = \frac{1}{2.54} \text{ inches}.$$

Multiplying both sides of the equation above by a number c shows that c centimeters $=$ $\frac{c}{2.54}$ inches. In other words,

$$I = \frac{c}{2.54}.$$

25. Find a number c such that the point $(c, 13)$ is on the line containing the points $(-4, -17)$ and $(6, 33)$.

SOLUTION First we find the equation of the line containing the points $(-4, -17)$ and $(6, 33)$. To do this, note that the line containing those two points has slope

$$\frac{33 - (-17)}{6 - (-4)},$$

which equals 5. Thus if (x, y) denotes a typical point on this line, then

$$\frac{y - 33}{x - 6} = 5.$$

Multiplying both sides of this equation by $x - 6$ and then adding 33 gives the equation

$$y = 5x + 3.$$

Now we can find a number c such that the point $(c, 13)$ is on the line given by the equation above. To do this, in the equation above replace x by c and y by 13, getting

$$13 = 5c + 3.$$

Solving this equation for c, we get $c = 2$.

CHECK We should check that the three points $(-4, -17)$, $(6, 33)$, and $(2, 13)$ are all on the line whose equation is $y = 5x + 3$. In other words, we need to verify the alleged equations

$$-17 \stackrel{?}{=} 5 \cdot (-4) + 3, \quad 33 \stackrel{?}{=} 5 \cdot 6 + 3, \quad 13 \stackrel{?}{=} 5 \cdot 2 + 2.$$

Simple arithmetic shows that all three alleged equations are indeed true.

27. Find a number t such that the point $(t, 2t)$ is on the line containing the points $(3, -7)$ and $(5, -15)$.

SOLUTION First we find the equation of the line containing the points $(3, -7)$ and $(5, -15)$. To do this, note that the line containing those two points has slope

$$\frac{-7-(-15)}{3-5},$$

which equals -4. Thus if (x, y) denotes a point on this line, then

$$\frac{y-(-7)}{x-3} = -4.$$

Multiplying both sides of this equation by $x - 3$ and then subtracting 7 gives the equation

$$y = -4x + 5.$$

Now we can find a number t such that the point $(t, 2t)$ is on the line given by the equation above. To do this, in the equation above replace x by t and y by $2t$, getting

$$2t = -4t + 5.$$

Solving this equation for t, we get $t = \frac{5}{6}$.

CHECK We should check that the three points $(3, -7)$, $(5, -15)$, and $(\frac{5}{6}, 2 \cdot \frac{5}{6})$ are all on the line whose equation is $y = -4x + 5$. In other words, we need to verify the alleged equations

$$-7 \stackrel{?}{=} -4 \cdot 3 + 5, \quad -15 \stackrel{?}{=} -4 \cdot 5 + 5, \quad \frac{5}{3} \stackrel{?}{=} -4 \cdot \frac{5}{6} + 5.$$

Simple arithmetic shows that all three alleged equations are indeed true.

29. Find the equation of the line in the xy-plane that contains the point $(3, 2)$ and that is parallel to the line $y = 4x - 1$.

SOLUTION The line in the xy-plane whose equation is $y = 4x - 1$ has slope 4. Thus each line parallel to it also has slope 4 and hence has the form

$$y = 4x + b$$

for some constant b.

Thus we need to find a constant b such that the point $(3, 2)$ is on the line given by the equation above. Replacing x by 3 and replacing y by 2 in the equation above, we have

$$2 = 4 \cdot 3 + b.$$

Solving this equation for b, we get $b = -10$. Thus the line that we seek is described by the equation

$$y = 4x - 10.$$

31. Find the equation of the line that contains the point $(2, 3)$ and that is parallel to the line containing the points $(7, 1)$ and $(5, 6)$.

SOLUTION The line containing the points $(7, 1)$ and $(5, 6)$ has slope

$$\frac{6-1}{5-7},$$

which equals $-\frac{5}{2}$. Thus each line parallel to it also has slope $-\frac{5}{2}$ and hence has the form

$$y = -\tfrac{5}{2}x + b$$

for some constant b.

Thus we need to find a constant b such that the point $(2, 3)$ is on the line given by the equation above. Replacing x by 2 and replacing y by 3 in the equation above, we have

$$3 = -\tfrac{5}{2} \cdot 2 + b.$$

Solving this equation for b, we get $b = 8$. Thus the line that we seek is described by the equation

$$y = -\tfrac{5}{2}x + 8.$$

33. Find a number t such that the line containing the points $(t, 2)$ and $(3, 5)$ is parallel to the line containing the points $(-1, 4)$ and $(-3, -2)$.

SOLUTION The line containing the points $(-1, 4)$ and $(-3, -2)$ has slope

$$\frac{4-(-2)}{-1-(-3)},$$

which equals 3. Thus each line parallel to it also has slope 3.

The line containing the points $(t, 2)$ and $(3, 5)$ has slope

$$\frac{5-2}{3-t},$$

which equals $\frac{3}{3-t}$. From the paragraph above, we want this slope to equal 3. In other words, we need to solve the equation

$$\frac{3}{3-t} = 3.$$

Dividing both sides of the equation above by 3 and then multiplying both sides by $3 - t$ gives the equation $1 = 3 - t$. Thus $t = 2$.

35. Find the intersection in the xy-plane of the lines $y = 5x + 3$ and $y = -2x + 1$.

SOLUTION Setting the two right sides of the equations above equal to each other, we get

$$5x + 3 = -2x + 1.$$

To solve this equation for x, add $2x$ to both sides and then subtract 3 from both sides, getting $7x = -2$. Thus $x = -\frac{2}{7}$.

To find the value of y at the intersection point, we can plug the value $x = -\frac{2}{7}$ into either of the equations of the two lines. Choosing the first equation, we have $y = -5 \cdot \frac{2}{7} + 3$, which implies that $y = \frac{11}{7}$. Thus the two lines intersect at the point $(-\frac{2}{7}, \frac{11}{7})$.

CHECK As a check, we can substitute the value $x = -\frac{2}{7}$ into the equation for the second line and see if that also gives the value $y = \frac{11}{7}$. In other words, we need to verify the alleged equation

$$\frac{11}{7} \stackrel{?}{=} -2(-\frac{2}{7}) + 1.$$

Simple arithmetic shows that this is true. Thus we indeed have the correct solution.

37. Find a number b such that the three lines in the xy-plane given by the equations $y = 2x + b$, $y = 3x - 5$, and $y = -4x + 6$ have a common intersection point.

SOLUTION The unknown b appears in the first equation; thus our first step will be to find the point of intersection of the last two lines. To do this, we set the right sides of the last two equations equal to each other, getting

$$3x - 5 = -4x + 6.$$

To solve this equation for x, add $4x$ to both sides and then add 5 to both sides, getting $7x = 11$. Thus $x = \frac{11}{7}$. Substituting this value of x into the equation $y = 3x - 5$, we get

$$y = 3 \cdot \frac{11}{7} - 5.$$

Thus $y = -\frac{2}{7}$.

At this stage, we have shown that the lines given by the equations $y = 3x - 5$ and $y = -4x + 6$ intersect at the point $(\frac{11}{7}, -\frac{2}{7})$. We want the line given by the equation $y = 2x + b$ also to contain this point. Thus we set $x = \frac{11}{7}$ and $y = -\frac{2}{7}$ in this equation, getting

$$-\frac{2}{7} = 2 \cdot \frac{11}{7} + b.$$

Solving this equation for b, we get $b = -\frac{24}{7}$.

CHECK As a check that the line given by the equation $y = -4x + 6$ contains the point $(\frac{11}{7}, -\frac{2}{7})$, we can substitute the value $x = \frac{11}{7}$ into the equation for that line and see if it gives the value $y = -\frac{2}{7}$. In other words, we need to verify the alleged equation

$$-\frac{2}{7} \stackrel{?}{=} -4 \cdot \frac{11}{7} + 6.$$

Simple arithmetic shows that this is true. Thus we indeed found the correct point of intersection.

We chose the line whose equation is given by $y = -4x + 6$ for this check because the other two lines had been used in direct calculations in our solution.

39. Find the equation of the line in the xy-plane that contains the point $(4, 1)$ and that is perpendicular to the line whose equation is $y = 3x + 5$.

SOLUTION The line in the xy-plane whose equation is $y = 3x + 5$ has slope 3. Thus every line perpendicular to it has slope $-\frac{1}{3}$. Hence the equation of the line that we seek has the form

$$y = -\tfrac{1}{3}x + b$$

for some constant b. We want the point $(4, 1)$ to be on this line. Substituting $x = 4$ and $y = 1$ into the equation above, we have

$$1 = -\tfrac{1}{3} \cdot 4 + b.$$

Solving this equation for b, we get $b = \frac{7}{3}$. Thus the equation of the line that we seek is

$$y = -\tfrac{1}{3}x + \tfrac{7}{3}.$$

41. Find a number t such that the line in the xy-plane containing the points $(t, 4)$ and $(2, -1)$ is perpendicular to the line $y = 6x - 7$.

SOLUTION The line in the xy-plane whose equation is $y = 6x - 7$ has slope 6. Thus every line perpendicular to it has slope $-\frac{1}{6}$. Thus we want the line containing the points $(t, 4)$ and $(2, -1)$ to have slope $-\frac{1}{6}$. In other words, we want

$$\frac{4 - (-1)}{t - 2} = -\frac{1}{6}.$$

Solving this equation for t, we get $t = -28$.

43. Find a number t such that the line containing the points $(4, t)$ and $(-1, 6)$ is perpendicular to the line that contains the points $(3, 5)$ and $(1, -2)$.

SOLUTION The line containing the points $(3, 5)$ and $(1, -2)$ has slope

$$\frac{5 - (-2)}{3 - 1},$$

which equals $\frac{7}{2}$. Thus every line perpendicular to it has slope $-\frac{2}{7}$. Thus we want the line containing the points $(4, t)$ and $(-1, 6)$ to have slope $-\frac{2}{7}$. In other words, we want

$$\frac{t - 6}{4 - (-1)} = -\frac{2}{7}.$$

Solving this equation for t, we get $t = \frac{32}{7}$.

45. Find the midpoint of the line segment connecting $(-3, 4)$ and $(5, 7)$.

SOLUTION The midpoint of the line segment connecting $(-3, 4)$ and $(5, 7)$ is

$$\left(\frac{-3 + 5}{2}, \frac{4 + 7}{2}\right),$$

which equals $(1, \frac{11}{2})$.

47. Find numbers x and y such that $(-2, 5)$ is the midpoint of the line segment connecting $(3, 1)$ and (x, y).

SOLUTION The midpoint of the line segment connecting $(3, 1)$ and (x, y) is

$$\left(\frac{3 + x}{2}, \frac{1 + y}{2}\right).$$

We want this to equal $(-2, 5)$. Thus we must solve the equations

$$\frac{3 + x}{2} = -2 \quad \text{and} \quad \frac{1 + y}{2} = 5.$$

Solving these equations gives $x = -7$ and $y = 9$.

2.3 *Quadratic Expressions and Conic Sections*

LEARNING OBJECTIVES

By the end of this section you should be able to

- rewrite a quadratic expression using the completing-the-square technique;
- find the point on a line closest to a given point;
- derive and use the quadratic formula;
- find the center and radius of a circle from its equation;
- find the vertex of a parabola;
- recognize and work with equations of ellipses and hyperbolas.

Completing the Square

Suppose you want to find the numbers x such that

$$x^2 - 5 = 0.$$

The solution is easy to obtain. Simply add 5 to both sides of the equation above, getting $x^2 = 5$, and then conclude that $x = \pm\sqrt{5}$.

Here ± indicates that we can choose either the plus sign or the minus sign.

Now consider the harder problem of finding the numbers x such that

$$x^2 + 6x - 4 = 0.$$

Nothing obvious works here. Adding either 4 or $4 - 6x$ to both sides of this equation does not produce a new equation that leads to the solution.

However, a technique called **completing the square** can be used to deal with equations such as the one above. The key to this technique is the identity

$$(x + t)^2 = x^2 + 2tx + t^2.$$

Be sure that you are thoroughly familiar with this crucial identity.

The next example illustrates the technique of completing the square.

EXAMPLE 1

Find the numbers x such that

$$x^2 + 6x - 4 = 0.$$

SOLUTION The idea here is that we want the $6x$ term in the equation above to match term $2tx$ term in the expansion above for $(x + t)^2$. In other words, we want $2t = 6$, and hence we choose $t = 3$.

When $(x + 3)^2$ is expanded, we get $x^2 + 6x + 9$. The x^2 and $6x$ terms match the corresponding terms in the expression $x^2 + 6x$, but the expansion of $(x + 3)^2$ has an extra constant term of 9. Thus we subtract 9, rewriting $x^2 + 6x$ in the equation above as $(x + 3)^2 - 9$, getting

$$(x + 3)^2 - 9 - 4 = 0.$$

Now add 13 to both sides of the equation above, getting $(x + 3)^2 = 13$, which implies that $x + 3 = \pm\sqrt{13}$. Finally, add -3 to both sides, concluding that $x = -3 \pm \sqrt{13}$.

The general formula for the substitution to make when completing the square is shown below. Do not memorize this formula. You only need to remember that the coefficient of the x term will need to be divided by 2, and then the appropriate constant will need to be subtracted to get a correct identity.

Completing the square

$$x^2 + bx = \left(x + \frac{b}{2}\right)^2 - \left(\frac{b}{2}\right)^2$$

For example, if $b = -10$, then the identity above becomes

$$x^2 - 10x = (x - 5)^2 - 25.$$

Note that the term that is subtracted is always positive because $(\frac{b}{2})^2$ is positive regardless of whether b is positive or negative.

The next example shows the usefulness of completing the square.

EXAMPLE 2

(a) What value of x makes $3x^2 - 5x + 4$ as small as possible?

(b) What is the smallest value of $3x^2 - 5x + 4$?

SOLUTION

(a) We factor out the coefficient 3 from the x^2 and x terms and then apply the completing-the-square identity, as follows:

The new expression has fractions making it look more cumbersome than the original expression. However, this new expression allows us to answer the questions above.

$$\begin{aligned} 3x^2 - 5x + 4 &= 3[x^2 - \tfrac{5}{3}x] + 4 \\ &= 3[(x - \tfrac{5}{6})^2 - \tfrac{25}{36}] + 4 \\ &= 3(x - \tfrac{5}{6})^2 - \tfrac{25}{12} + 4 \\ &= 3(x - \tfrac{5}{6})^2 + \tfrac{23}{12} \end{aligned}$$

The term $(x - \frac{5}{6})^2$ in the last expression is positive for all values of x except for $x = \frac{5}{6}$, which makes $(x - \frac{5}{6})^2$ equal to 0. Thus $3x^2 - 5x + 4$ is as small as possible when $x = \frac{5}{6}$.

(b) We could substitute the value $x = \frac{5}{6}$ into the expression $3x^2 - 5x + 4$ to find the smallest value of this expression. However, without doing any work we can see from the last equation above that this expression equals $\frac{23}{12}$ when $x = \frac{5}{6}$. Thus $\frac{23}{12}$ is the smallest value of $3x^2 - 5x + 4$.

The next example illustrates another application of the completing-the-square technique.

EXAMPLE 3

Find the point on the line $y = 2x - 1$ that is closest to the point $(2, 1)$.

SOLUTION A typical point on the line $y = 2x - 1$ has coordinates $(x, 2x - 1)$. The distance between this point and $(2, 1)$ equals

$$\sqrt{(x-2)^2 + (2x-1-1)^2},$$

which with a bit of algebra (do it!) can be rewritten as

$$\sqrt{5x^2 - 12x + 8}.$$

We want to make the quantity above as small as possible, which means that we need to make $5x^2 - 12x$ as small as possible. This can be done by completing the square:

$$\begin{aligned} 5x^2 - 12x &= 5[x^2 - \tfrac{12}{5}x] \\ &= 5[(x - \tfrac{6}{5})^2 - \tfrac{36}{25}]. \end{aligned}$$

The last quantity will be as small as possible when $x = \frac{6}{5}$. Plugging $x = \frac{6}{5}$ into the equation $y = 2x - 1$ gives $y = \frac{7}{5}$.

Thus $(\frac{6}{5}, \frac{7}{5})$ is the point on the line $y = 2x - 1$ that is closest to the point $(2, 1)$. The picture in the margin shows that $(\frac{6}{5}, \frac{7}{5})$ is indeed a plausible solution.

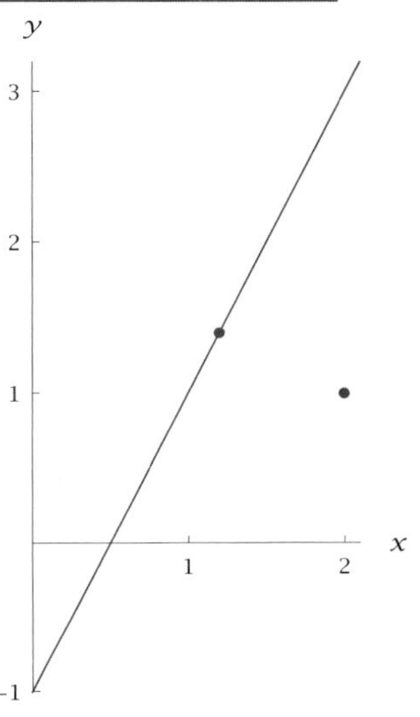

The line $y = 2x - 1$, the point $(2, 1)$, and the point on the line closest to $(2, 1)$.

The Quadratic Formula

Quadratic expression

A **quadratic expression** has the form

$$ax^2 + bx + c,$$

where $a \neq 0$.

Consider the quadratic expression $ax^2 + bx + c$, where $a \neq 0$. Factor out a from the first two terms and then complete the square, as follows:

We will follow the same pattern as in previous concrete examples and complete the square with an arbitrary quadratic expression. This will allow us to derive the quadratic formula.

$$\begin{aligned} ax^2 + bx + c &= a\Big[x^2 + \frac{b}{a}x\Big] + c \\ &= a\Big[\Big(x + \frac{b}{2a}\Big)^2 - \frac{b^2}{4a^2}\Big] + c \\ &= a\Big(x + \frac{b}{2a}\Big)^2 - \frac{b^2}{4a} + c \\ &= a\Big(x + \frac{b}{2a}\Big)^2 - \frac{b^2 - 4ac}{4a}. \end{aligned}$$

Suppose now that we want to find the numbers x, such that $ax^2+bx+c = 0$. Setting the last expression equal to 0, we get

$$a\left(x + \frac{b}{2a}\right)^2 = \frac{b^2 - 4ac}{4a},$$

and then dividing both sides by a gives

$$\left(x + \frac{b}{2a}\right)^2 = \frac{b^2 - 4ac}{4a^2}.$$

Regardless of the value of x, the left side of the last equation is a positive number or 0. Thus if the right side is negative, the equation does not hold for any real number x. In other words, if $b^2 - 4ac < 0$, then the equation $ax^2 + bx + c = 0$ has no real solutions.

If $b^2 - 4ac \geq 0$, then we can take the square root of both sides of the last equation, getting

$$x + \frac{b}{2a} = \pm\frac{\sqrt{b^2 - 4ac}}{2a},$$

and then adding $-\frac{b}{2a}$ to both sides gives

By completing the square, we have derived the quadratic formula!

$$x = \frac{-b \pm \sqrt{b^2 - 4ac}}{2a}.$$

Quadratic formula

Consider the equation

$$ax^2 + bx + c = 0,$$

where a, b, and c are real numbers with $a \neq 0$.

- If $b^2 - 4ac < 0$, then the equation above has no (real) solutions.
- If $b^2 - 4ac = 0$, then the equation above has one solution:

$$x = -\frac{b}{2a}.$$

- If $b^2 - 4ac > 0$, then the equation above has two solutions:

$$x = \frac{-b \pm \sqrt{b^2 - 4ac}}{2a}.$$

The quadratic formula often is useful in problems that do not initially seem to involve quadratic expressions. The following example illustrates how a simply stated problem can lead to a quadratic expression.

EXAMPLE 4

Find two numbers whose sum equals 7 and whose product equals 8.

SOLUTION Let's call the two numbers s and t. We want

$$s + t = 7 \quad \text{and} \quad st = 8.$$

Solving the first equation for s, we have $s = 7 - t$. Substituting this expression for s into the second equation gives $(7 - t)t = 8$, which is equivalent to the equation

$$t^2 - 7t + 8 = 0.$$

Using the quadratic formula to solve this equation for t gives

$$t = \frac{7 \pm \sqrt{7^2 - 4 \cdot 8}}{2} = \frac{7 \pm \sqrt{17}}{2}.$$

Let's choose the solution $t = \frac{7+\sqrt{17}}{2}$. Plugging this value of t into the equation $s = 7 - t$ then gives $s = \frac{7-\sqrt{17}}{2}$.

You should verify that if we had chosen $t = \frac{7-\sqrt{17}}{2}$, then we would have ended up with the same pair of numbers.

Thus two numbers whose sum equals 7 and whose product equals 8 are $\frac{7-\sqrt{17}}{2}$ and $\frac{7+\sqrt{17}}{2}$.

REMARK To check that this solution is correct, note that

$$\frac{7 - \sqrt{17}}{2} + \frac{7 + \sqrt{17}}{2} = \frac{14}{2} = 7$$

and

$$\frac{7 - \sqrt{17}}{2} \cdot \frac{7 + \sqrt{17}}{2} = \frac{7^2 - \sqrt{17}^2}{4} = \frac{49 - 17}{4} = \frac{32}{4} = 8.$$

Circles

The set of points that have distance 3 from the origin in a coordinate plane is the circle with radius 3 centered at the origin. The next example shows how to find an equation that describes this circle.

EXAMPLE 5

Find an equation that describes the circle with radius 3 centered at the origin in the xy-plane.

SOLUTION Recall that the distance from a point (x, y) to the origin is $\sqrt{x^2 + y^2}$. Hence a point (x, y) has distance 3 from the origin if and only if

$$\sqrt{x^2 + y^2} = 3.$$

Squaring both sides, we get

$$x^2 + y^2 = 9.$$

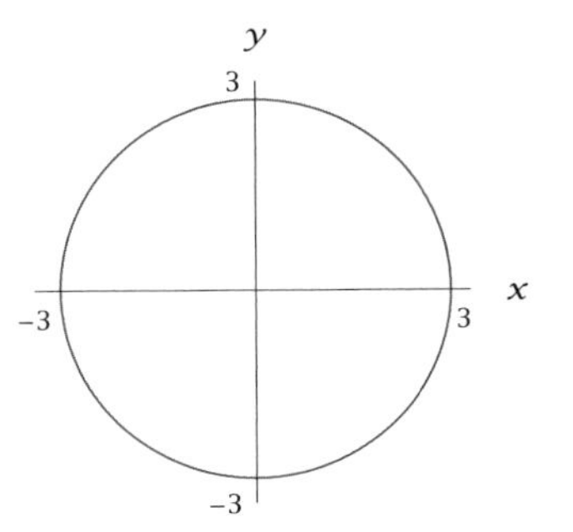

The circle of radius 3 centered at the origin.

More generally, suppose r is a positive number. Using the same reasoning as above, we see that

$$x^2 + y^2 = r^2$$

is the equation of the circle with radius r centered at the origin in the xy-plane.

We can also consider circles centered at points other than the origin.

EXAMPLE 6

Find the equation of the circle in the xy-plane centered at $(2, 1)$ with radius 5.

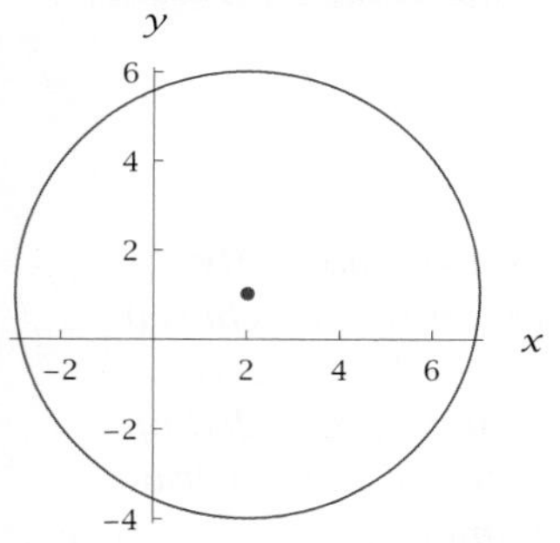

The circle centered at $(2, 1)$ with radius 5.

SOLUTION This circle is the set of points whose distance from $(2, 1)$ equals 5. In other words, the circle centered at $(2, 1)$ with radius 5 is the set of points (x, y) satisfying the equation

$$\sqrt{(x-2)^2 + (y-1)^2} = 5.$$

Squaring both sides, we can more conveniently describe this circle as the set of points (x, y) such that

$$(x-2)^2 + (y-1)^2 = 25.$$

Using the same reasoning as in the example above, we get the following more general result:

Equation of a circle

The circle with center (h, k) and radius r is the set of points (x, y) satisfying the equation

$$(x-h)^2 + (y-k)^2 = r^2.$$

For example, the equation $(x-3)^2 + (y+5)^2 = 7$ describes the circle in the xy-plane with radius $\sqrt{7}$ centered at $(3, -5)$.

Sometimes the equation of a circle may be in a form in which the radius and center are not obvious. You may then need to complete the square to find the radius and center. The following example illustrates this procedure:

EXAMPLE 7

Find the radius and center of the circle in the xy-plane described by

$$x^2 + 4x + y^2 - 6y = 12.$$

SOLUTION Completing the square, we have

$$\begin{aligned} 12 &= x^2 + 4x + y^2 - 6y \\ &= (x+2)^2 - 4 + (y-3)^2 - 9 \\ &= (x+2)^2 + (y-3)^2 - 13. \end{aligned}$$

Here the completing-the-square technique has been applied separately to the x and y variables.

Adding 13 to the first and last sides of the equation above shows that

$$(x+2)^2 + (y-3)^2 = 25.$$

Thus we have a circle with radius 5 centered at $(-2, 3)$.

Ellipses

Ellipse

Stretching a circle horizontally and/or vertically produces a curve called an **ellipse**.

EXAMPLE 8

Find an equation describing the ellipse in the xy-plane produced by stretching a circle of radius 1 centered at the origin horizontally by a factor of 5 and vertically by a factor of 3.

SOLUTION To find an equation describing this ellipse, consider a typical point (u, v) on the circle of radius 1 centered at the origin. Thus $u^2 + v^2 = 1$.

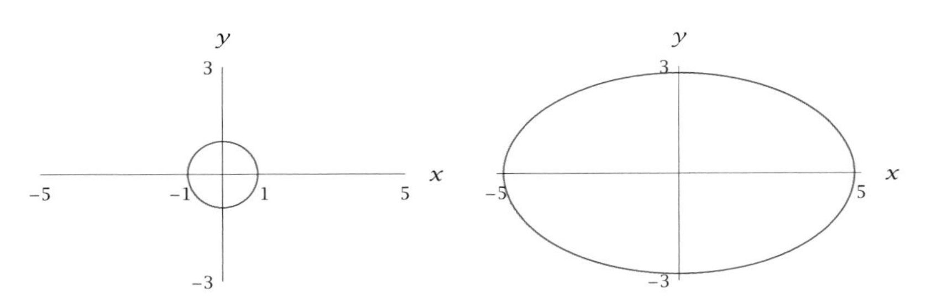

Stretching horizontally by a factor of 5 *and vertically by a factor of* 3 *transforms the circle on the left into the ellipse on the right.*

The German mathematician Johannes Kepler, who in 1609 published his discovery that the orbits of the planets are ellipses, not circles or combinations of circles as had been previously thought.

Stretching horizontally by a factor of 5 and stretching vertically by a factor of 3 transforms the point (u, v) to the point $(5u, 3v)$. Rewrite the equation $u^2 + v^2 = 1$ in terms of this new point, getting

$$\frac{(5u)^2}{25} + \frac{(3v)^2}{9} = 1.$$

Write the transformed point $(5u, 3v)$ as (x, y), thus setting $x = 5u$ and $y = 3v$, getting

$$\frac{x^2}{25} + \frac{y^2}{9} = 1,$$

which is the equation of the ellipse shown above on the right.

The points $(\pm 5, 0)$ and $(0, \pm 3)$ satisfy this equation and thus lie on the ellipse, as can also be seen from the figure above.

More generally, suppose a and b are positive numbers. Suppose the circle of radius 1 centered at the origin is stretched horizontally by a factor of a and stretched vertically by a factor of b. Using the same reasoning as above (just replace 5 by a and replace 3 by b), we see that the equation of the resulting ellipse in the xy-plane is

Viewing an ellipse as a stretched circle will lead us to the formula for the area inside an ellipse in Section 2.4.

$$\frac{x^2}{a^2} + \frac{y^2}{b^2} = 1.$$

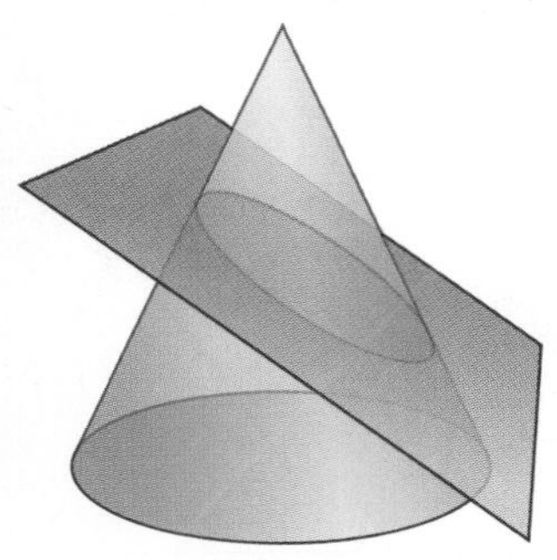

The ancient Greeks discovered that the intersection of a cone and an appropriately positioned plane is an ellipse.

The points $(\pm a, 0)$ and $(0, \pm b)$ satisfy this equation and lie on the ellipse.

Planets have orbits that are ellipses, but you may be surprised to learn that the sun is *not* located at the center of those orbits. Instead, the sun is located at what is called a **focus** of each elliptical orbit. The plural of focus is **foci**, which are defined as follows:

Foci of an ellipse

The **foci** of an ellipse are two points with the property that the sum of the distances from the foci to a point on the ellipse is a constant independent of the point on the ellipse.

As we will see in the next example, the foci for the ellipse $\frac{x^2}{25} + \frac{y^2}{9} = 1$ are the points $(-4, 0)$ and $(4, 0)$. This example shows how to verify that a pair of points are foci for an ellipse.

EXAMPLE 9

(a) Find a formula in terms of x for the distance from a typical point (x, y) on the ellipse $\frac{x^2}{25} + \frac{y^2}{9} = 1$ to the point $(4, 0)$.

(b) Find a formula in terms of x for the distance from a typical point (x, y) on the ellipse $\frac{x^2}{25} + \frac{y^2}{9} = 1$ to the point $(-4, 0)$.

(c) Show that $(4, 0)$ and $(-4, 0)$ are foci of the ellipse $\frac{x^2}{25} + \frac{y^2}{9} = 1$.

Isaac Newton showed that the equations of gravity imply that a planetary orbit around a star is an ellipse with the star at one of the foci. For example, if units are chosen so that the orbit of a planet is the ellipse $\frac{x^2}{25} + \frac{y^2}{9} = 1$, then the star must be located at either $(4, 0)$ or $(-4, 0)$.

SOLUTION

(a) The distance from (x, y) to the point $(4, 0)$ is $\sqrt{(x-4)^2 + y^2}$, which equals

$$\sqrt{x^2 - 8x + 16 + y^2}.$$

We want an answer solely in terms of x, assuming that (x, y) lies on the ellipse $\frac{x^2}{25} + \frac{y^2}{9} = 1$. Solving the ellipse equation for y^2, we have

$$y^2 = 9 - \tfrac{9}{25}x^2.$$

Substituting this expression for y^2 into the expression above shows that the distance from (x, y) to $(4, 0)$ equals

$$\sqrt{25 - 8x + \tfrac{16}{25}x^2},$$

which equals $\sqrt{(5 - \frac{4}{5}x)^2}$, which equals $5 - \frac{4}{5}x$.

(b) The distance from (x, y) to the point $(-4, 0)$ is $\sqrt{(x+4)^2 + y^2}$, which equals

$$\sqrt{x^2 + 8x + 16 + y^2}.$$

Now proceed as in the solution to part (a), with $-8x$ replaced by $8x$, concluding that the distance from (x, y) to $(-4, 0)$ is $5 + \frac{4}{5}x$.

(c) As we know from the previous two parts of this example, if (x, y) is a point on the ellipse $\frac{x^2}{25} + \frac{y^2}{9} = 1$, then

$$\text{distance from } (x, y) \text{ to } (4, 0) \text{ is } 5 - \tfrac{4}{5}x$$

and

$$\text{distance from } (x, y) \text{ to } (-4, 0) \text{ is } 5 + \tfrac{4}{5}x.$$

Adding these distances, we see that the sum equals 10, which is a constant independent of the point (x, y) on the ellipse. Thus $(4, 0)$ and $(-4, 0)$ are foci of this ellipse.

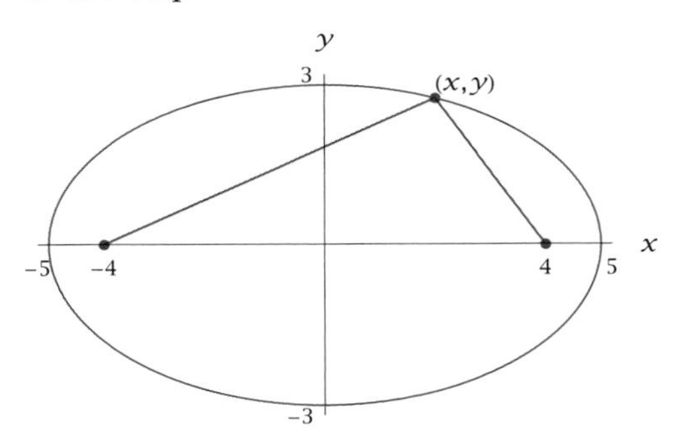

For every point (x, y) on the ellipse $\frac{x^2}{25} + \frac{y^2}{9} = 1$, the sum of the lengths of the two red line segments equals 10.

No pair of points other than $(4, 0)$ and $(-4, 0)$ has the property that the sum of the distances to points on this ellipse is constant.

The next result generalizes the example above. The verification of this result is outlined in Problems 72–77. To do this verification, simply use the ideas from the example above.

Formula for the foci of an ellipse

- If $a > b > 0$, then the foci of the ellipse $\frac{x^2}{a^2} + \frac{y^2}{b^2} = 1$ are the points

$$(\sqrt{a^2 - b^2}, 0) \quad \text{and} \quad (-\sqrt{a^2 - b^2}, 0).$$

- If $b > a > 0$, then the foci of the ellipse $\frac{x^2}{a^2} + \frac{y^2}{b^2} = 1$ are the points

$$(0, \sqrt{b^2 - a^2}) \quad \text{and} \quad (0, -\sqrt{b^2 - a^2}).$$

This result does not deal with the easy case $a = b$, when the ellipse is a circle centered at the origin. In this case, the foci can be thought of as two copies of the origin.

Parabolas

A kicked or thrown football follows a path shaped like the curve in the margin. You may be tempted to think that this curve is half of an ellipse. However, the equations of gravity show that objects thrown into the air follow a path that is part of a parabola, not part of an ellipse. Thus we now turn our attention to parabolas and the equations that define them.

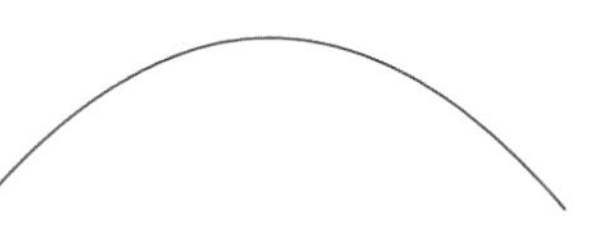

The path of a football.

Parabolas can be defined geometrically, but for our purposes it is simpler to look at a class of parabolas that are easy to define algebraically. For now, we will restrict our attention to parabolas in the xy-plane that have a vertical line of symmetry:

Parabolas

The graph of an equation of the form

$$y = ax^2 + bx + c,$$

where $a \neq 0$, is called a **parabola**.

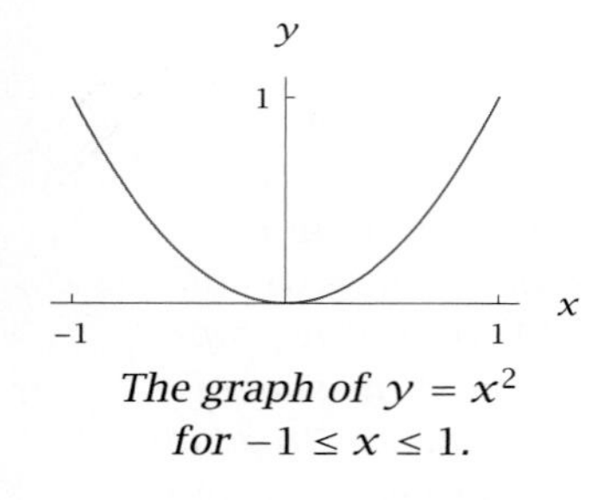

The graph of $y = x^2$ for $-1 \le x \le 1$.

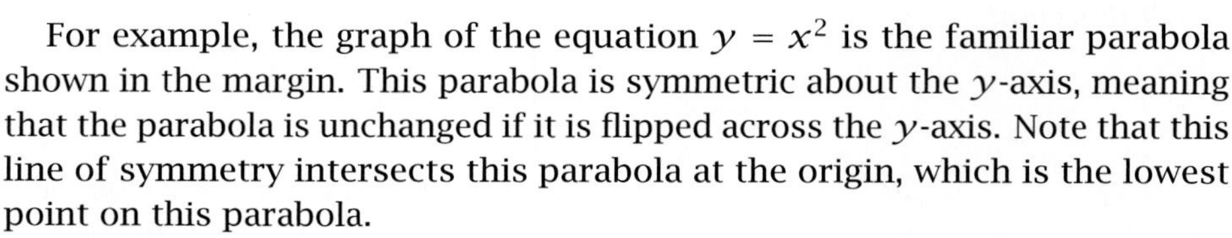

For example, the graph of the equation $y = x^2$ is the familiar parabola shown in the margin. This parabola is symmetric about the y-axis, meaning that the parabola is unchanged if it is flipped across the y-axis. Note that this line of symmetry intersects this parabola at the origin, which is the lowest point on this parabola.

So that you can become familiar with the shape of parabolas, the figure below shows two more typical parabolas.

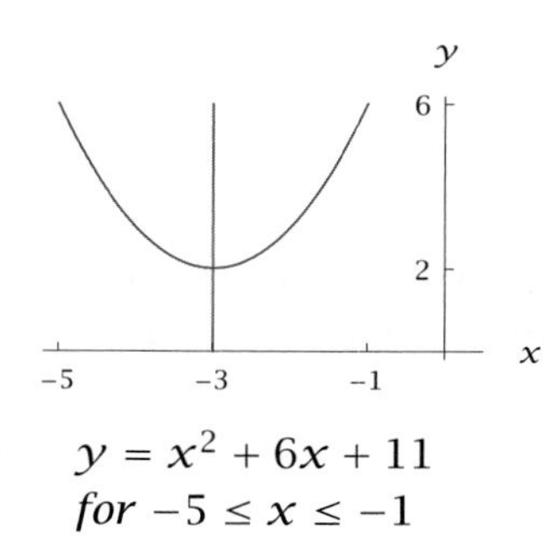

$y = x^2 + 6x + 11$
for $-5 \le x \le -1$

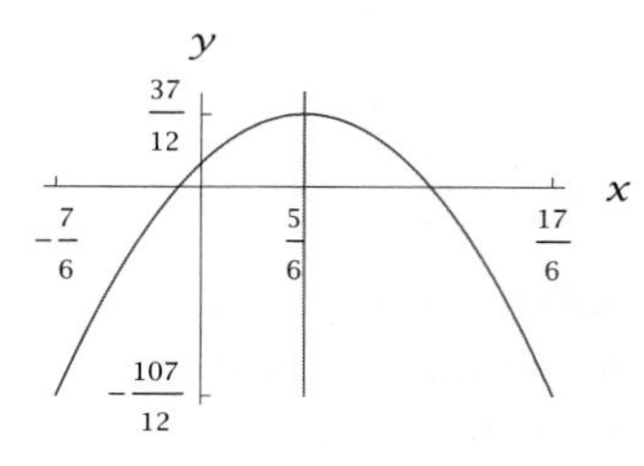

$y = -3x^2 + 5x + 1$
for $-\frac{7}{6} \le x \le \frac{17}{6}$

The parabola on the left is symmetric about the line $x = -3$ (shown in red), which intersects the parabola at its lowest point. The parabola on the right is symmetric about the line $x = \frac{5}{6}$ (shown in red), which intersects the parabola at its highest point.

Every parabola is symmetric about some line. The point where this line of symmetry intersects the parabola is sufficiently important to deserve its own name:

Vertex

The **vertex** of a parabola is the point where the line of symmetry of the parabola intersects the parabola.

For example, the figure above shows that the vertex of the parabola $y = x^2 + 6x + 11$ on the left is $(-3, 2)$, which is the lowest point on the graph. The figure above also shows that the vertex of the parabola $y = -3x^2 + 5x + 1$ is $(\frac{5}{6}, \frac{37}{12})$, which is the highest point on the graph.

The parabolas above exhibit the typical behavior described by the following result:

Parabola direction

Consider the parabola given by the equation

$$y = ax^2 + bx + c,$$

where $a \neq 0$.

- If $a > 0$, then the parabola opens upward and the vertex of the parabola is the lowest point on the graph.
- If $a < 0$, then the parabola opens downward and the vertex of the parabola is the highest point on the graph.

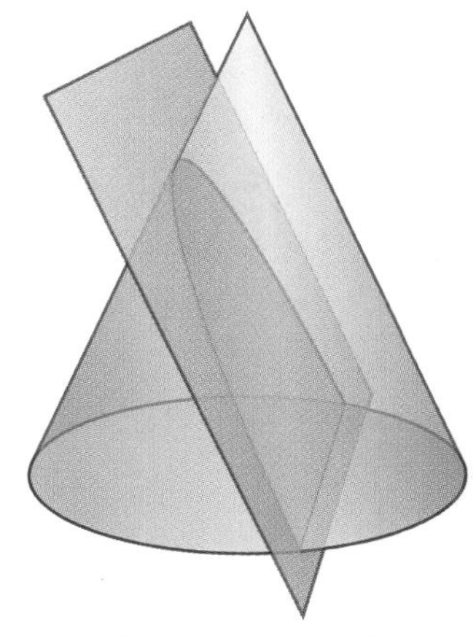

The ancient Greeks discovered that the intersection of a cone and an appropriately positioned plane is a parabola.

When we deal with function transformations in Section 3.2, we will see why the result above holds and we will learn how to use function transformations to obtain the graph of a parabola by appropriate transformations of the graph of $y = x^2$. Thus for now we give only one example showing how to find the vertex of a parabola.

EXAMPLE 10

(a) For what value of x does $x^2 + 6x + 11$ attain its minimum value?

(b) Find the vertex of the graph of $y = x^2 + 6x + 11$.

SOLUTION

(a) First complete the square to rewrite $x^2 + 6x + 11$, as follows:

$$\begin{aligned} x^2 + 6x + 11 &= (x + 3)^2 - 9 + 11 \\ &= (x + 3)^2 + 2 \end{aligned}$$

The last expression shows that $x^2 + 6x + 11$ takes on its minimum value when $x = -3$, because $(x + 3)^2$ is positive for all values of x except $x = -3$.

(b) The last expression above shows that $x^2 + 6x + 11$ equals 2 when $x = -3$. Thus the vertex of the graph of $y = x^2 + 6x + 11$ is the point $(-3, 2)$, as shown in the graph of this parabola above.

A comet whose orbit lies on a hyperbola will come near earth at most once. A comet whose orbit is an ellipse will return periodically.

Hyperbolas

Some comets and all planets travel in orbits that are ellipses. However, many comets have orbits that are not ellipses but instead lie on hyperbolas, which is another category of curves. Thus we now turn our attention to hyperbolas and the equations that define them.

Hyperbolas can be defined geometrically, but we will restrict attention to hyperbolas in the xy-plane that can be defined in the following simple fashion:

Hyperbolas

The graph of an equation of the form

$$\frac{y^2}{b^2} - \frac{x^2}{a^2} = c,$$

where a, b, and c are nonzero numbers, is called a **hyperbola**.

The figure below shows the graph of the hyperbola $\frac{y^2}{16} - \frac{x^2}{9} = 1$ in blue for $-6 \le x \le 6$. Note that this hyperbola consists of two branches rather than the single curve we get for ellipses and parabolas. Some key properties of this graph, which are typical of hyperbolas, are discussed in the next example.

EXAMPLE 11

Consider the hyperbola

$$\frac{y^2}{16} - \frac{x^2}{9} = 1.$$

(a) Explain why this hyperbola intersects the y-axis at $(0, 4)$ and $(0, -4)$.

(b) Explain why the hyperbola contains no points (x, y) with $|y| < 4$.

(c) Explain why points (x, y) on the hyperbola for large values of x are near the line $y = \frac{4}{3}x$ or the line $y = -\frac{4}{3}x$.

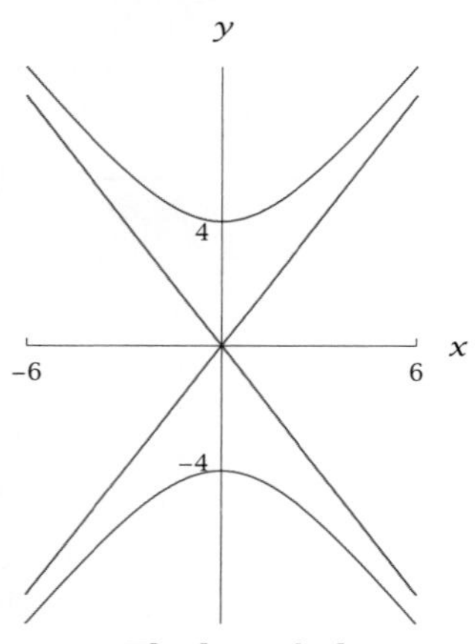

The hyperbola

$$\frac{y^2}{16} - \frac{x^2}{9} = 1$$

in blue for $-6 \le x \le 6$, along with the lines $y = \frac{4}{3}x$ and $y = -\frac{4}{3}x$ in red.

SOLUTION

(a) If $x = 0$, then the equation defining the hyperbola shows that $\frac{y^2}{16} = 1$. Thus $y = \pm 4$. Hence the hyperbola intersects the y-axis, which is defined by $x = 0$, at $(0, 4)$ and $(0, -4)$.

(b) If (x, y) is a point on the hyperbola, then

$$y^2 = 16(1 + \frac{x^2}{9}) \ge 16,$$

which implies that $|y| \ge 4$. Thus the hyperbola contains no points (x, y) with $|y| < 4$.

(c) The equation defining this parabola can be rewritten in the form

$$\frac{y^2}{x^2} = \frac{16}{9} + \frac{16}{x^2}$$

if $x \ne 0$. If x is large, then the equation above implies that

$$\frac{y^2}{x^2} \approx \frac{16}{9}.$$

Taking square roots then shows that $y \approx \pm\frac{4}{3}x$.

Each branch of a hyperbola may appear to be shaped like a parabola, but these curves are not parabolas. As one notable difference, a parabola cannot have the behavior described in part (c) of the example above.

Compare the following definition of the foci of a hyperbola with the definition earlier in this section of the foci of an ellipse.

Foci of a hyperbola

The **foci** of a hyperbola are two points with the property that the difference of the distances from the foci to a point on the hyperbola is a constant independent of the point on the hyperbola.

As we will see in the next example, the foci for the hyperbola $\frac{y^2}{16} - \frac{x^2}{9} = 1$ are the points $(0, -5)$ and $(0, 5)$. This example shows how to verify that a pair of points are foci for a hyperbola.

Problems 81–84 show why the graph of $y = \frac{1}{x}$ is also called a hyperbola.

(a) Find a formula in terms of y for the distance from a typical point (x, y) with $y > 0$ on the hyperbola $\frac{y^2}{16} - \frac{x^2}{9} = 1$ to the point $(0, -5)$.

(b) Find a formula in terms of y for the distance from a typical point (x, y) with $y > 0$ on the hyperbola $\frac{y^2}{16} - \frac{x^2}{9} = 1$ to the point $(0, 5)$.

(c) Show that $(0, -5)$ and $(0, 5)$ are foci of the hyperbola $\frac{y^2}{16} - \frac{x^2}{9} = 1$.

SOLUTION

(a) The distance from (x, y) to the point $(0, -5)$ is $\sqrt{x^2 + (y+5)^2}$, which equals

$$\sqrt{x^2 + y^2 + 10y + 25}.$$

We want an answer solely in terms of y, assuming that (x, y) lies on the hyperbola $\frac{y^2}{16} - \frac{x^2}{9} = 1$ and $y > 0$. Solving the hyperbola equation for x^2, we have

$$x^2 = \tfrac{9}{16}y^2 - 9.$$

Substituting this expression for x^2 into the expression above shows that the distance from (x, y) to $(0, -5)$ equals

$$\sqrt{\tfrac{25}{16}y^2 + 10y + 16},$$

which equals $\sqrt{(\frac{5}{4}y + 4)^2}$, which equals $\frac{5}{4}y + 4$.

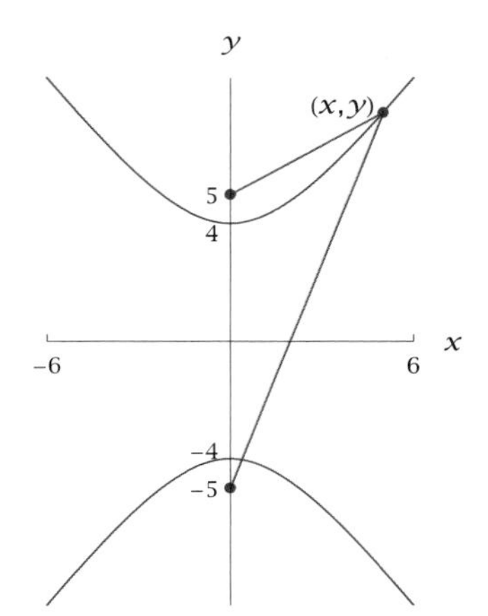

For every point (x, y) on the hyperbola $\frac{y^2}{16} - \frac{x^2}{9} = 1$, the difference of the lengths of the two red line segments equals 8.

(b) The distance from (x, y) to the point $(0, 5)$ is $\sqrt{x^2 + (y-5)^2}$, which equals

$$\sqrt{x^2 + y^2 - 10y + 25}.$$

Now proceed as in the solution to part (a), with $10y$ replaced by $-10y$, concluding that the distance from (x, y) to $(0, 5)$ is $\frac{5}{4}y - 4$.

(c) As we know from the previous two parts of this example, if (x, y) is a point on the hyperbola $\frac{y^2}{16} - \frac{x^2}{9} = 1$ and $y > 0$, then

$$\text{distance from } (x, y) \text{ to } (0, -5) \text{ is } \tfrac{5}{4}y + 4$$

and

$$\text{distance from } (x, y) \text{ to } (0, 5) \text{ is } \tfrac{5}{4}y - 4.$$

Subtracting these distances, we see that the difference equals 8, which is a constant independent of the point (x, y) on the hyperbola. Thus $(0, -5)$ and $(0, 5)$ are foci of this hyperbola.

No pair of points other than $(0, -5)$ and $(0, 5)$ has the property that the difference of the distances to points on this hyperbola is constant.

Isaac Newton showed that a comet's orbit around a star lies either on an ellipse or on a parabola (rare) or on a hyperbola with the star at one of the foci. For example, if units are chosen so that the orbit of a comet is the upper branch of the hyperbola $\frac{y^2}{16} - \frac{x^2}{9} = 1$, then the star must be located at $(0, 5)$.

The next result generalizes the example above. The verification of this result is outlined in Problems 78–80. To do this verification, simply use the ideas from the example above.

Formula for the foci of a hyperbola

If a and b are nonzero numbers, then the foci of the hyperbola $\frac{y^2}{b^2} - \frac{x^2}{a^2} = 1$ are the points

$$(0, -\sqrt{a^2 + b^2}) \quad \text{and} \quad (0, \sqrt{a^2 + b^2}).$$

Sometimes when a new comet is discovered, there are not enough observations to determine whether the comet is in an elliptical orbit or in a hyperbolic orbit. The distinction is important, because a comet in a hyperbolic orbit will disappear and never again be visible from earth.

EXERCISES

For Exercises 1–12, use the following information: If an object is thrown straight up into the air from height H ***feet at time 0 with initial velocity*** V ***feet per second, then at time*** t ***seconds the height of the object is***

$$-16.1t^2 + Vt + H$$

feet. This formula uses only gravitational force, ignoring air friction. It is valid only until the object hits the ground or some other object.

Some notebook computers have a sensor that detects sudden changes in motion and stops the notebook's hard drive, protecting it from damage.

1. Suppose a notebook computer is accidentally knocked off a shelf that is six feet high. How long before the computer hits the ground?

2. Suppose a notebook computer is accidentally knocked off a desk that is three feet high. How long before the computer hits the ground?

3. Suppose the motion detection/protection mechanism of a notebook computer takes 0.3 seconds to work after the computer starts to fall. What is the minimum height from which the notebook computer can fall and have the protection mechanism work?

4. Suppose the motion detection/protection mechanism of a notebook computer takes 0.4 seconds to work after the computer starts to fall. What is the minimum height from which the notebook computer can fall and have the protection mechanism work?

5. Suppose a ball is tossed straight up into the air from height 5 feet with initial velocity 20 feet per second.
   (a) How long before the ball hits the ground?
   (b) How long before the ball reaches its maximum height?
   (c) What is the ball's maximum height?

6. Suppose a ball is tossed straight up into the air from height 4 feet with initial velocity 40 feet per second.
   (a) How long before the ball hits the ground?
   (b) How long before the ball reaches its maximum height?
   (c) What is the ball's maximum height?

7. Suppose a ball is tossed straight up into the air from height 5 feet. What should be the initial velocity to have the ball stay in the air for 4 seconds?

8. Suppose a ball is tossed straight up into the air from height 4 feet. What should be the initial velocity to have the ball stay in the air for 3 seconds?

9. Suppose a ball is tossed straight up into the air from height 5 feet. What should be the initial velocity to have the ball reach its maximum height after 1 second?

10. Suppose a ball is tossed straight up into the air from height 4 feet. What should be the initial velocity to have the ball reach its maximum height after 2 seconds?

11. Suppose a ball is tossed straight up into the air from height 5 feet. What should be the initial velocity to have the ball reach a height of 50 feet?

12. Suppose a ball is tossed straight up into the air from height 4 feet. What should be the initial velocity to have the ball reach a height of 70 feet?

13. Find the equation of the circle in the xy-plane centered at $(3, -2)$ with radius 7.

14. Find the equation of the circle in the xy-plane centered at $(-4, 5)$ with radius 6.

15. Find two choices for b such that $(5, b)$ is on the circle with radius 4 centered at $(3, 6)$.

16. Find two choices for b such that $(b, 4)$ is on the circle with radius 3 centered at $(-1, 6)$.

17. Find all numbers x such that

$$\frac{x-1}{x+3} = \frac{2x-1}{x+2}.$$

18. Find all numbers x such that

$$\frac{3x+2}{x-2} = \frac{2x-1}{x-1}.$$

19. Find two numbers w such that the points $(3, 1)$, $(w, 4)$, and $(5, w)$ all lie on a straight line.

20. Find two numbers r such that the points $(-1, 4)$, $(r, 2r)$, and $(1, r)$ all lie on a straight line.

21. The graph of the equation

$$x^2 - 6x + y^2 + 8y = -25$$

contains exactly one point. Find the coordinates of that point.

22. The graph of the equation

$$x^2 + 5x + y^2 - 3y = -\tfrac{17}{2}$$

contains exactly one point. Find the coordinates of that point.

23. Find the point on the line $y = 3x + 1$ in the xy-plane that is closest to the point $(2, 4)$.

24. Find the point on the line $y = 2x - 3$ in the xy-plane that is closest to the point $(5, 1)$.

25. Find a number t such that the distance between $(2, 3)$ and $(t, 2t)$ is as small as possible.

26. Find a number t such that the distance between $(-2, 1)$ and $(3t, 2t)$ is as small as possible.

27. Find the length of the graph of the curve defined by

$$y = \sqrt{9 - x^2}$$

with $-3 \le x \le 3$.

28. Find the length of the graph of the curve defined by

$$y = \sqrt{25 - x^2}$$

with $0 \le x \le 5$.

29. Find the two points where the circle of radius 2 centered at the origin intersects the circle of radius 3 centered at $(3, 0)$.

30. Find the two points where the circle of radius 3 centered at the origin intersects the circle of radius 4 centered at $(5, 0)$.

31. Find the equation of the circle in the xy-plane centered at the origin with circumference 9.

32. Find the equation of the circle in the xy-plane centered at $(3, 7)$ with circumference 5.

33. Find the equation of the circle centered at the origin in the uv-plane that has twice the circumference of the circle whose equation equals

$$u^2 + v^2 = 10.$$

34. Find the equation of the circle centered at the origin in the tw-plane that has three times the circumference of the circle whose equation equals

$$t^2 + w^2 = 5.$$

For Exercises 35 and 36, find the following information about the circles in the xy-plane described by the given equation:

(a) ***center***
(b) ***radius***
(c) ***diameter***
(d) ***circumference***

35. $x^2 - 8x + y^2 + 2y = -14$

36. $x^2 + 5x + y^2 - 6y = 3$

37. Find the intersection of the line containing the points $(2, 3)$ and $(4, 7)$ and the circle with radius $\sqrt{15}$ centered at $(3, -3)$.

38. Find the intersection of the line containing the points $(3, 4)$ and $(1, 8)$ and the circle with radius $\sqrt{3}$ centered at $(2, 9)$.

For Exercises 39–44, find the vertex of the graph of the given equation.

39. $y = 7x^2 - 12$

40. $y = -9x^2 - 5$

41. $y = (x - 2)^2 - 3$

42. $y = (x + 3)^2 + 4$

43. $y = (2x - 5)^2 + 6$

44. $y = (7x + 3)^2 + 5$

In Exercises 45–48, for the given equation:

(a) ***Write the right side of the equation in the form*** $k(x + t)^2 + r$.

(b) ***Find the value of*** x ***where the right side of the equation attains its minimum value or its maximum value.***

(c) ***Find the minimum or maximum value of the right side of the equation.***

(d) ***Find the vertex of the graph of the equation.***

45. $y = x^2 + 7x + 12$

46. $y = 5x^2 + 2x + 1$

47. $y = -2x^2 + 5x - 2$

48. $y = -3x^2 + 5x - 1$

49. Find a constant c such that the graph of $y = x^2 + 6x + c$ has its vertex on the x-axis.

50. Find a constant c such that the graph of $y = x^2 + 5x + c$ in the xy-plane has its vertex on the line $y = x$.

51. Find two numbers whose sum equals 10 and whose product equals 7.

52. Find two numbers whose sum equals 6 and whose product equals 4.

53. Find two positive numbers whose difference equals 3 and whose product equals 20.

54. Find two positive numbers whose difference equals 4 and whose product equals 15.

55. Find the minimum value of $x^2 - 6x + 2$.

56. Find the minimum value of $3x^2 + 5x + 1$.

57. Find the maximum value of $7 - 2x - x^2$.

58. Find the maximum value of $9 + 5x - 4x^2$.

PROBLEMS

59. Show that
$$(a + b)^2 = a^2 + b^2$$
if and only if $a = 0$ or $b = 0$.

60. Explain why the graph of the equation
$$x^2 + 4x + y^2 - 10y = -30$$
contains no points.

61. Suppose
$$2x^2 + 3x + c > 0$$
for every real number x. Show that $c > \frac{9}{8}$.

62. Suppose
$$3x^2 + bx + 7 > 0$$
for every real number x. Show that $|b| < 2\sqrt{21}$.

63. Suppose
$$at^2 + 5t + 4 > 0$$
for every real number t. Show that $a > \frac{25}{16}$.

64. Suppose $a \neq 0$ and $b^2 \geq 4ac$. Verify by direct substitution that if
$$x = \frac{-b \pm \sqrt{b^2 - 4ac}}{2a},$$
then $ax^2 + bx + c = 0$.

65. Suppose $a \neq 0$ and $b^2 \geq 4ac$. Verify by direct calculation that
$$ax^2 + bx + c =$$
$$a\left(x - \frac{-b + \sqrt{b^2 - 4ac}}{2a}\right)\left(x - \frac{-b - \sqrt{b^2 - 4ac}}{2a}\right).$$

66. Find a number φ such that in the figure below, the yellow rectangle is similar to the large rectangle formed by the union of the blue square and the yellow rectangle.

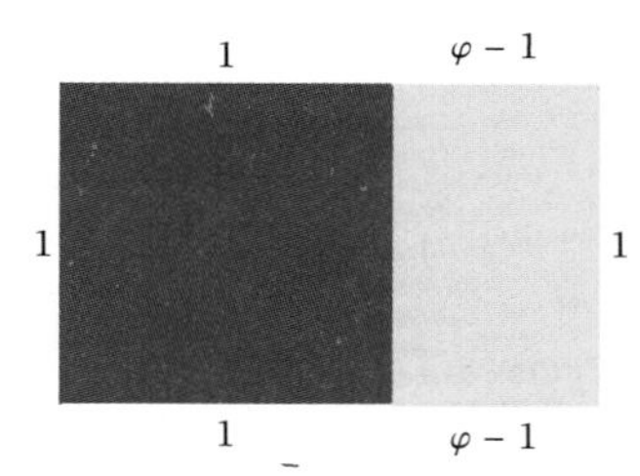

The number φ that solves this problem is called the **golden ratio** (the symbol φ is the Greek letter phi). Rectangles whose ratio between the length of the long side and the length of the short side equals φ are supposedly the most aesthetically pleasing rectangles. The large rectangle formed by the union of the blue square and the yellow rectangle has the golden ratio, as does the yellow rectangle. Many works of art feature rectangles with the golden ratio.

67. Suppose a, b, and c are numbers with $a \neq 0$. Show that the vertex of the graph of

$$y = ax^2 + bx + c$$

is the point $\left(-\frac{b}{2a}, \frac{4ac-b^2}{4a}\right)$.

68. Suppose a, b, and c are numbers with $a \neq 0$ and that only one number t satisfies the equation

$$at^2 + bt + c = 0.$$

Show that t is the first coordinate of the vertex of the graph of $y = ax^2 + bx + c$ and that the second coordinate of the vertex equals 0.

69. Suppose a, b, and c are numbers such that exactly two real numbers t satisfy the equation $at^2 + bt + c = 0$. Show that the average of these two real numbers is the first coordinate of the vertex of the graph of $y = ax^2 + bx + c$.

70. Suppose b and c are numbers such that the equation

$$x^2 + bx + c = 0$$

has no real solutions. Explain why the equation

$$x^2 + bx - c = 0$$

has two real solutions.

71. Show that there do not exist two real numbers whose sum is 7 and whose product is 13.

72. Suppose $a > b > 0$. Find a formula in terms of x for the distance from a typical point (x, y) on the ellipse $\frac{x^2}{a^2} + \frac{y^2}{b^2} = 1$ to the point $(\sqrt{a^2 - b^2}, 0)$.

73. Suppose $a > b > 0$. Find a formula in terms of x for the distance from a typical point (x, y) on the ellipse $\frac{x^2}{a^2} + \frac{y^2}{b^2} = 1$ to the point $(-\sqrt{a^2 - b^2}, 0)$.

74. Suppose $a > b > 0$. Use the results of the two previous problems to show that $(\sqrt{a^2 - b^2}, 0)$ and $(-\sqrt{a^2 - b^2}, 0)$ are foci of the ellipse

$$\frac{x^2}{a^2} + \frac{y^2}{b^2} = 1.$$

75. Suppose $b > a > 0$. Find a formula in terms of y for the distance from a typical point (x, y) on the ellipse $\frac{x^2}{a^2} + \frac{y^2}{b^2} = 1$ to the point $(0, \sqrt{b^2 - a^2})$.

76. Suppose $b > a > 0$. Find a formula in terms of y for the distance from a typical point (x, y) on the ellipse $\frac{x^2}{a^2} + \frac{y^2}{b^2} = 1$ to the point $(0, -\sqrt{b^2 - a^2})$.

77. Suppose $b > a > 0$. Use the results of the two previous problems to show that $(0, \sqrt{b^2 - a^2})$ and $(0, -\sqrt{b^2 - a^2})$ are foci of the ellipse

$$\frac{x^2}{a^2} + \frac{y^2}{b^2} = 1.$$

78. Suppose a and b are nonzero numbers. Find a formula in terms of y for the distance from a typical point (x, y) with $y > 0$ on the hyperbola $\frac{y^2}{b^2} - \frac{x^2}{a^2} = 1$ to the point $(0, -\sqrt{a^2 + b^2})$.

79. Suppose a and b are nonzero numbers. Find a formula in terms of y for the distance from a typical point (x, y) with $y > 0$ on the hyperbola $\frac{y^2}{b^2} - \frac{x^2}{a^2} = 1$ to the point $(0, \sqrt{a^2 + b^2})$.

80. Suppose a and b are nonzero numbers. Use the results of the two previous problems to show that $(0, -\sqrt{a^2 + b^2})$ and $(0, \sqrt{a^2 + b^2})$ are foci of the hyperbola

$$\frac{y^2}{b^2} - \frac{x^2}{a^2} = 1.$$

81. Suppose $x > 0$. Show that the distance from $(x, \frac{1}{x})$ to the point $(-\sqrt{2}, -\sqrt{2})$ is $x + \frac{1}{x} + \sqrt{2}$. [*See Example 7 in Section 4.1 for a graph of* $y = \frac{1}{x}$.]

82. Suppose $x > 0$. Show that the distance from $(x, \frac{1}{x})$ to the point $(\sqrt{2}, \sqrt{2})$ is $x + \frac{1}{x} - \sqrt{2}$.

83. Suppose $x > 0$. Show that the distance from $(x, \frac{1}{x})$ to $(-\sqrt{2}, -\sqrt{2})$ minus the distance from $(x, \frac{1}{x})$ to $(\sqrt{2}, \sqrt{2})$ equals $2\sqrt{2}$.

84. Explain why the result of the previous problem justifies calling the curve $y = \frac{1}{x}$ a hyperbola with foci at $(-\sqrt{2}, -\sqrt{2})$ and $(\sqrt{2}, \sqrt{2})$.

85. Explain why `graph x^2/25 + y^2/9 = 1` in Wolfram|Alpha produces what appears to be a circle rather than a typical ellipse-shape as shown in Examples 8 and 9.
[*Hint:* Notice the scales on the two axes.]

WORKED-OUT SOLUTIONS *to Odd-numbered Exercises*

For Exercises 1–12, use the following information: If an object is thrown straight up into the air from height H feet at time 0 with initial velocity V feet per second, then at time t seconds the height of the object is

$$-16.1t^2 + Vt + H$$

feet. This formula uses only gravitational force, ignoring air friction. It is valid only until the object hits the ground or some other object.

Some notebook computers have a sensor that detects sudden changes in motion and stops the notebook's hard drive, protecting it from damage.

1. Suppose a notebook computer is accidentally knocked off a shelf that is six feet high. How long before the computer hits the ground?

SOLUTION In this case, we have $V = 0$ and $H = 6$ in the formula above. We want to know the time t at which the height equals 0. In other words, we need to solve the equation

$$-16.1t^2 + 6 = 0.$$

The solutions to the equation are $t = \pm\sqrt{\frac{6}{16.1}}$. The negative value makes no sense here, and thus $t = \sqrt{\frac{6}{16.1}} \approx 0.61$ seconds.

3. Suppose the motion detection/protection mechanism of a notebook computer takes 0.3 seconds to work after the computer starts to fall. What is the minimum height from which the notebook computer can fall and have the protection mechanism work?

SOLUTION We want to find the initial height H such that the notebook computer will hit the ground (meaning have height 0) after 0.3 seconds. In other words, we want

$$-16.1 \times 0.3^2 + H = 0.$$

Thus $H = 16.1 \times 0.3^2 = 1.449$ feet.

5. Suppose a ball is tossed straight up into the air from height 5 feet with initial velocity 20 feet per second.

(a) How long before the ball hits the ground?

(b) How long before the ball reaches its maximum height?

(c) What is the ball's maximum height?

SOLUTION

(a) The ball hits the ground when

$$-16.1t^2 + 20t + 5 = 0.$$

The quadratic formula shows that $t \approx 1.46$ seconds (the other solution produced by the quadratic formula has been discarded because it is negative).

(b) Completing the square, we have

$$\begin{aligned}
-16.1t^2 &+ 20t + 5 \\
&= -16.1\left[t^2 - \frac{20}{16.1}t\right] + 5 \\
&= -16.1\left[\left(t - \frac{10}{16.1}\right)^2 - \left(\frac{10}{16.1}\right)^2\right] + 5 \\
&= -16.1\left(t - \frac{10}{16.1}\right)^2 + \frac{100}{16.1} + 5.
\end{aligned}$$

The expression above gives the height of the ball at time t. Thus the ball reaches its maximum height when $t = \frac{10}{16.1} \approx 0.62$ seconds.

(c) The solution to part (b) shows that the maximum height of the ball is $\frac{100}{16.1} + 5 \approx 11.2$ feet.

7. Suppose a ball is tossed straight up into the air from height 5 feet. What should be the initial velocity to have the ball stay in the air for 4 seconds?

SOLUTION Suppose the initial velocity of the ball is V. Then the height of the ball at time t is

$$-16.1t^2 + Vt + 5.$$

We want $t = 4$ when the ball hits the ground, meaning its height is 0. In other words, we want

$$-16.1 \times 4^2 + 4V + 5 = 0.$$

Solving this equation for V gives $V = 63.15$ feet per second.

9. Suppose a ball is tossed straight up into the air from height 5 feet. What should be the initial velocity to have the ball reach its maximum height after 1 second?

SOLUTION Suppose the initial velocity of the ball is V. Then the height of the ball at time t is

$$\begin{aligned} -16.1t^2 &+ Vt + 5 \\ &= -16.1\Big[t^2 - \frac{Vt}{16.1}\Big] + 5 \\ &= -16.1\Big[\Big(t - \frac{V}{32.2}\Big)^2 - \Big(\frac{V}{32.2}\Big)^2\Big] + 5 \\ &= -16.1\Big(t - \frac{V}{32.2}\Big)^2 + \frac{V^2}{64.4} + 5. \end{aligned}$$

The ball reaches its maximum height when $t - \frac{V}{32.2} = 0$. We want this to happen when $t = 1$. In other words, we want $1 - \frac{V}{32.2} = 0$, which implies that $V = 32.2$ feet per second.

11. Suppose a ball is tossed straight up into the air from height 5 feet. What should be the initial velocity to have the ball reach a height of 50 feet?

SOLUTION Suppose the initial velocity of the ball is V. As can be seen from the solution to Exercise 9, the maximum height of the ball is $\frac{V^2}{64.4} + 5$, which we want to equal 50. Solving the equation

$$\frac{V^2}{64.4} + 5 = 50$$

for V gives $V \approx 53.83$ feet per second.

13. Find the equation of the circle in the xy-plane centered at $(3, -2)$ with radius 7.

SOLUTION The equation of this circle is

$$(x-3)^2 + (y+2)^2 = 49.$$

15. Find two choices for b such that $(5, b)$ is on the circle with radius 4 centered at $(3, 6)$.

SOLUTION The equation of the circle with radius 4 centered at $(3, 6)$ is

$$(x-3)^2 + (y-6)^2 = 16.$$

The point $(5, b)$ is on this circle if and only if

$$(5-3)^2 + (b-6)^2 = 16,$$

which is equivalent to the equation $(b-6)^2 = 12$. Thus

$$b - 6 = \pm\sqrt{12} = \pm\sqrt{4 \cdot 3} = \pm\sqrt{4}\sqrt{3} = \pm 2\sqrt{3}.$$

Thus $b = 6 + 2\sqrt{3}$ or $b = 6 - 2\sqrt{3}$.

17. Find all numbers x such that

$$\frac{x-1}{x+3} = \frac{2x-1}{x+2}.$$

SOLUTION The equation above is equivalent to the equation

$$(x-1)(x+2) = (2x-1)(x+3).$$

Expanding and then collecting all terms on one side leads to the equation

$$x^2 + 4x - 1 = 0.$$

The quadratic formula now shows that $x = -2 \pm \sqrt{5}$.

19. Find two numbers w such that the points $(3, 1)$, $(w, 4)$, and $(5, w)$ all lie on a straight line.

SOLUTION The slope of the line containing $(3, 1)$ and $(w, 4)$ is $\frac{3}{w-3}$. The slope of the line containing $(3, 1)$ and $(5, w)$ is $\frac{w-1}{2}$. We need these two slopes to be equal, which means that

$$\frac{3}{w-3} = \frac{w-1}{2}.$$

Thus $2 \cdot 3 = (w-1)(w-3)$, which means that

$$w^2 - 4w - 3 = 0.$$

The quadratic formula now shows that $w = 2 \pm \sqrt{7}$.

21. The graph of the equation

$$x^2 - 6x + y^2 + 8y = -25$$

contains exactly one point. Find the coordinates of that point.

SOLUTION We complete the square to rewrite the equation above:

$$\begin{aligned} -25 &= x^2 - 6x + y^2 + 8y \\ &= (x-3)^2 - 9 + (y+4)^2 - 16 \\ &= (x-3)^2 + (y+4)^2 - 25. \end{aligned}$$

Thus the original equation can be rewritten as $(x-3)^2 + (y+4)^2 = 0$. The only solution to this equation is $x = 3$ and $y = -4$. Thus $(3, -4)$ is the only point on the graph of this equation.

23. Find the point on the line $y = 3x + 1$ in the xy-plane that is closest to the point $(2, 4)$.

SOLUTION A typical point on the line $y = 3x + 1$ in the xy-plane has coordinates $(x, 3x + 1)$. The distance between this point and $(2, 4)$ equals

$$\sqrt{(x-2)^2 + (3x + 1 - 4)^2},$$

which with a bit of algebra can be rewritten as

$$\sqrt{10x^2 - 22x + 13}.$$

We want to make the quantity above as small as possible, which means that we need to make $10x^2 - 22x$ as small as possible. This can be done by completing the square:

$$\begin{aligned} 10x^2 - 22x &= 10\left[x^2 - \frac{11}{5}x\right] \\ &= 10\left[\left(x - \frac{11}{10}\right)^2 - \frac{121}{100}\right]. \end{aligned}$$

The last quantity will be as small as possible when $x = \frac{11}{10}$. Plugging $x = \frac{11}{10}$ into the equation $y = 3x + 1$ gives $y = \frac{43}{10}$. Thus $(\frac{11}{10}, \frac{43}{10})$ is the point on the line $y = 3x + 1$ that is closest to the point $(2, 4)$.

25. Find a number t such that the distance between $(2, 3)$ and $(t, 2t)$ is as small as possible.

SOLUTION The distance between $(2, 3)$ and $(t, 2t)$ equals

$$\sqrt{(t-2)^2 + (2t-3)^2}.$$

We want to make this as small as possible, which happens when

$$(t-2)^2 + (2t-3)^2$$

is as small as possible. Note that

$$(t-2)^2 + (2t-3)^2 = 5t^2 - 16t + 13.$$

This will be as small as possible when $5t^2 - 16t$ is as small as possible. To find when that happens, we complete the square:

$$\begin{aligned} 5t^2 - 16t &= 5\left[t^2 - \tfrac{16}{5}t\right] \\ &= 5\left[\left(t - \tfrac{8}{5}\right)^2 - \tfrac{64}{25}\right]. \end{aligned}$$

This quantity is made smallest when $t = \frac{8}{5}$.

27. Find the length of the graph of the curve defined by

$$y = \sqrt{9 - x^2}$$

with $-3 \le x \le 3$.

SOLUTION Squaring both sides of the equation $y = \sqrt{9 - x^2}$ and then adding x^2 to both sides gives the equation $x^2 + y^2 = 9$, which is the equation of the circle of radius 3 centered at the origin. However, the equation $y = \sqrt{9 - x^2}$ implies that $y \ge 0$, and thus we have only the top half of the circle.

The entire circle of radius 3 has circumference 6π. Thus the graph of $y = \sqrt{9 - x^2}$, which is half of the circle, has length 3π.

29. Find the two points where the circle of radius 2 centered at the origin intersects the circle of radius 3 centered at $(3, 0)$.

SOLUTION The equations of these two circles are

$$x^2 + y^2 = 4 \quad \text{and} \quad (x-3)^2 + y^2 = 9.$$

Subtracting the first equation from the second equation, we get

$$(x-3)^2 - x^2 = 5,$$

which simplifies to the equation $-6x + 9 = 5$, whose solution is $x = \frac{2}{3}$. Plugging this value of x into either of the equations above and solving for y gives $y = \pm\frac{4\sqrt{2}}{3}$. Thus the two circles intersect at the points $(\frac{2}{3}, \frac{4\sqrt{2}}{3})$ and $(\frac{2}{3}, -\frac{4\sqrt{2}}{3})$.

31. Find the equation of the circle in the xy-plane centered at the origin with circumference 9.

SOLUTION Let r denote the radius of this circle. Then $2\pi r = 9$, which implies that $r = \frac{9}{2\pi}$. Thus the equation of the circle is

$$x^2 + y^2 = \frac{81}{4\pi^2}.$$

33. Find the equation of the circle centered at the origin in the uv-plane that has twice the circumference of the circle whose equation equals

$$u^2 + v^2 = 10.$$

SOLUTION The equation given above describes a circle centered at the origin whose radius equals $\sqrt{10}$. Because the circumference is proportional to the radius, if we want a circle with twice the circumference then we need to double the radius. Thus the circle we seek has radius $2\sqrt{10}$. Because $(2\sqrt{10})^2 = 2^2 \cdot \sqrt{10}^2 = 40$, the equation we seek is

$$u^2 + v^2 = 40.$$

For Exercises 35 and 36, find the following information about the circles in the xy-plane described by the given equation:

(a) *center*	**(c)** *diameter*
(b) *radius*	**(d)** *circumference*

35. $x^2 - 8x + y^2 + 2y = -14$

SOLUTION Completing the square, we can rewrite the left side of this equation as follows:

$$\begin{aligned} x^2 - 8x + y^2 + 2y &= (x - 4)^2 - 16 + (y + 1)^2 - 1 \\ &= (x - 4)^2 + (y + 1)^2 - 17. \end{aligned}$$

Substituting this expression into the left side of the original equation and then adding 17 to both sides shows that the original equation is equivalent to the equation

$$(x - 4)^2 + (y + 1)^2 = 3.$$

(a) The equation above shows that this circle has center $(4, -1)$.

(b) The equation above shows that this circle has radius $\sqrt{3}$.

(c) Because the diameter is twice the radius, this circle has diameter $2\sqrt{3}$.

(d) Because the circumference is 2π times the radius, this circle has circumference $2\pi\sqrt{3}$.

37. Find the intersection of the line containing the points $(2, 3)$ and $(4, 7)$ and the circle with radius $\sqrt{15}$ centered at $(3, -3)$.

SOLUTION First we find the equation of the line containing the points $(2, 3)$ and $(4, 7)$. This line will have slope $\frac{7-3}{4-2}$, which equals 2. Thus the equation of this line will have the form $y = 2x + b$. Because $(2, 3)$ is on this line, we can substitute $x = 2$ and $y = 3$ into the last equation and then solve for b, getting $b = -1$. Thus the equation of the line containing the points $(2, 3)$ and $(4, 7)$ is

$$y = 2x - 1.$$

The equation of the circle with radius $\sqrt{15}$ centered at $(3, -3)$ is

$$(x - 3)^2 + (y + 3)^2 = 15.$$

To find the intersection of the circle and the line, we replace y by $2x - 1$ in the equation above, getting

$$(x - 3)^2 + (2x + 2)^2 = 15.$$

Expanding the terms in the equation above and then collecting terms gives the equation

$$5x^2 + 2x - 2 = 0.$$

Using the quadratic formula, we then find that

$$x = \frac{-1 + \sqrt{11}}{5} \quad \text{or} \quad x = \frac{-1 - \sqrt{11}}{5}.$$

Substituting these values of x into the equation $y = 2x - 1$ shows that the line intersects the circle in the points

$$\left(\frac{-1 + \sqrt{11}}{5}, \frac{-7 + 2\sqrt{11}}{5}\right)$$

and

$$\left(\frac{-1 - \sqrt{11}}{5}, \frac{-7 - 2\sqrt{11}}{5}\right).$$

For Exercises 39–44, find the vertex of the graph of the given equation.

39. $y = 7x^2 - 12$

SOLUTION The value of $7x^2 - 12$ is minimized when $x = 0$. When $x = 0$, the value of $7x^2 - 12$ equals -12. Thus the vertex of the graph of $y = 7x^2 - 12$ is $(0, -12)$.

41. $y = (x-2)^2 - 3$

SOLUTION The value of $(x-2)^2 - 3$ is minimized when $x - 2 = 0$. When $x = 2$, the value of $(x-2)^2 - 3$ equals -3. Thus the vertex of the graph of $y = (x-2)^2 - 3$ is $(2, -3)$.

43. $y = (2x-5)^2 + 6$

SOLUTION The value of $(2x-5)^2 + 6$ is minimized when $2x - 5 = 0$. This happens when $x = \frac{5}{2}$, which makes the value of $(2x-5)^2 + 6$ equal to 6. Thus the vertex of the graph of $y = (2x-5)^2 + 6$ is $(\frac{5}{2}, 6)$.

In Exercises 45–48, for the given equation:

(a) ***Write the right side of the equation in the form $k(x+t)^2 + r$.***

(b) ***Find the value of x where the right side of the equation attains its minimum value or its maximum value.***

(c) ***Find the minimum or maximum value of the right side of the equation.***

(d) ***Find the vertex of the graph of the equation.***

45. $y = x^2 + 7x + 12$

SOLUTION

(a) By completing the square, we can write

$$x^2 + 7x + 12 = (x + \tfrac{7}{2})^2 - \tfrac{49}{4} + 12$$
$$= (x + \tfrac{7}{2})^2 - \tfrac{1}{4}.$$

(b) The expression above shows that the value of the right side is minimized when $x = -\frac{7}{2}$.

(c) The solution to part (a) shows that the minimum value of this expression is $-\frac{1}{4}$, which occurs when $x = -\frac{7}{2}$.

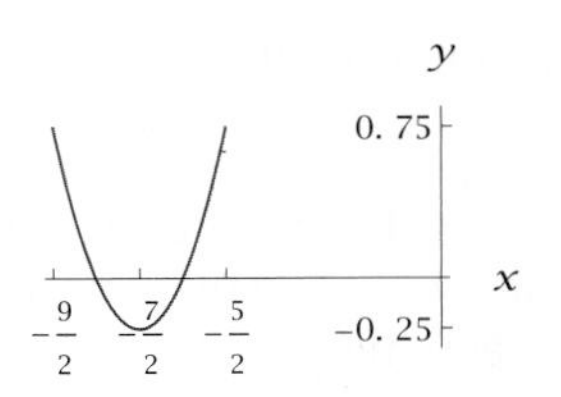

(d) The figure above shows that the vertex of the graph of $y = x^2 + 7x + 12$ is the point $(-\frac{7}{2}, -\frac{1}{4})$, as follows from the solutions to parts (b) and (c) and as is shown in the graph above.

47. $y = -2x^2 + 5x - 2$

SOLUTION

(a) By completing the square, we can write

$$-2x^2 + 5x - 2 = -2[x^2 - \tfrac{5}{2}x] - 2$$
$$= -2[(x - \tfrac{5}{4})^2 - \tfrac{25}{16}] - 2$$
$$= -2(x - \tfrac{5}{4})^2 + \tfrac{25}{8} - 2$$
$$= -2(x - \tfrac{5}{4})^2 + \tfrac{9}{8}.$$

(b) The expression above shows that the value of $-2x^2 + 5x - 2$ is maximized when $x = \frac{5}{4}$.

(c) The solution to part (a) shows that the maximum value of this expression is $\frac{9}{8}$, which occurs when $x = \frac{5}{4}$.

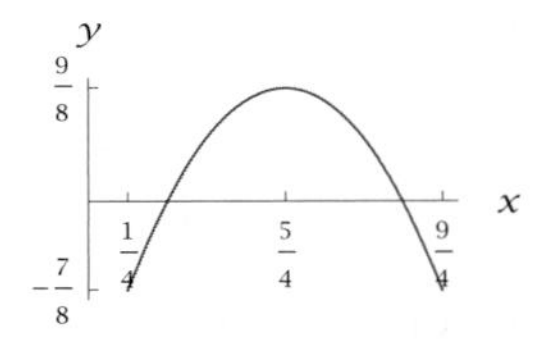

(d) The figure above shows that the vertex of the graph of $y = -2x^2 + 5x - 2$ is the point $(\frac{5}{4}, \frac{9}{8})$, as follows from the solutions to parts (b) and (c) and as is shown in the graph above.

49. Find a constant c such that the graph of $y = x^2 + 6x + c$ has its vertex on the x-axis.

SOLUTION First we find the vertex of the graph of $y = x^2 + 6x + c$. To do this, complete the square:

$$x^2 + 6x + c = (x+3)^2 - 9 + c.$$

Thus the value of $x^2 + 6x + c$ is minimized when $x = -3$. When $x = -3$, the value of $x^2 + 6x + c$

equals $-9+c$. Thus the vertex of $y = x^2+6x+c$ is $(-3, -9+c)$.

The x-axis consists of the points whose second coordinate equals 0. Thus the vertex of the graph of $y = x^2 + 6x + c$ will be on the x-axis when $-9 + c = 0$, or equivalently when $c = 9$.

51. Find two numbers whose sum equals 10 and whose product equals 7.

SOLUTION Let's call the two numbers s and t. We want

$$s + t = 10 \quad \text{and} \quad st = 7.$$

Solving the first equation for s, we have $s = 10 - t$. Substituting this expression for s into the second equation gives $(10 - t)t = 7$, which is equivalent to the equation

$$t^2 - 10t + 7 = 0.$$

Using the quadratic formula to solve this equation for t gives

$$t = \frac{10 \pm \sqrt{10^2 - 4 \cdot 7}}{2} = \frac{10 \pm \sqrt{72}}{2}$$

$$= \frac{10 \pm \sqrt{36 \cdot 2}}{2} = 5 \pm 3\sqrt{2}.$$

Let's choose the solution $t = 5 + 3\sqrt{2}$. Plugging this value of t into the equation $s = 10 - t$ then gives $s = 5 - 3\sqrt{2}$.

Thus two numbers whose sum equals 10 and whose product equals 7 are $5 - 3\sqrt{2}$ and $5 + 3\sqrt{2}$.

CHECK To check that this solution is correct, note that

$$(5 - 3\sqrt{2}) + (5 + 3\sqrt{2}) = 10$$

and

$$(5-3\sqrt{2})(5+3\sqrt{2}) = 5^2 - 3^2\sqrt{2}^2 = 25 - 9 \cdot 2 = 7.$$

53. Find two positive numbers whose difference equals 3 and whose product equals 20.

SOLUTION Let's call the two numbers s and t. We want

$$s - t = 3 \quad \text{and} \quad st = 20.$$

Solving the first equation for s, we have $s = t + 3$. Substituting this expression for s into the second equation gives $(t + 3)t = 20$, which is equivalent to the equation

$$t^2 + 3t - 20 = 0.$$

Using the quadratic formula to solve this equation for t gives

$$t = \frac{-3 \pm \sqrt{3^2 + 4 \cdot 20}}{2} = \frac{-3 \pm \sqrt{89}}{2}.$$

Choosing the minus sign in the plus-or-minus expression above would lead to a negative value for t. Because this exercise requires that t be positive, we choose $t = \frac{-3+\sqrt{89}}{2}$. Plugging this value of t into the equation $s = t + 3$ then gives $s = \frac{3+\sqrt{89}}{2}$.

Thus two numbers whose difference equals 3 and whose product equals 20 are $\frac{3+\sqrt{89}}{2}$ and $\frac{-3+\sqrt{89}}{2}$.

CHECK To check that this solution is correct, note that

$$\frac{3 + \sqrt{89}}{2} - \frac{-3 + \sqrt{89}}{2} = \frac{6}{2} = 3$$

and

$$\frac{3 + \sqrt{89}}{2} \cdot \frac{-3 + \sqrt{89}}{2} = \frac{\sqrt{89} + 3}{2} \cdot \frac{\sqrt{89} - 3}{2}$$

$$= \frac{\sqrt{89}^2 - 3^2}{4} = \frac{80}{4} = 20.$$

55. Find the minimum value of $x^2 - 6x + 2$.

SOLUTION By completing the square, we can write

$$x^2 - 6x + 2 = (x - 3)^2 - 9 + 2$$

$$= (x - 3)^2 - 7.$$

The expression above shows that the minimum value of $x^2 - 6x + 2$ is -7 (and that this minimum value occurs when $x = 3$).

57. Find the maximum value of $7 - 2x - x^2$.

SOLUTION By completing the square, we can write

$$7 - 2x - x^2 = -[x^2 + 2x] + 7$$

$$= -[(x + 1)^2 - 1] + 7$$

$$= -(x + 1)^2 + 8.$$

The expression above shows that the maximum value of $7 - 2x - x^2$ is 8 (and that this maximum value occurs when $x = -1$).

2.4 *Area*

LEARNING OBJECTIVES

By the end of this section you should be able to

- compute the areas of squares, rectangles, parallelograms, triangles, and trapezoids;
- explain how area changes when the coordinate axes are stretched;
- compute the area inside a circle;
- compute the area inside an ellipse.

You probably already have a good intuitive notion of area. In this section we will try to strengthen this intuition and build a good understanding of the formulas for the area of simple regions.

Squares, Rectangles, and Parallelograms

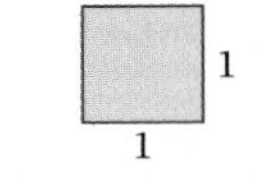

A 1*-by-*1 *square.*

The most primitive notion of area is that a 1-by-1 square has area 1. If we can decompose a region into 1-by-1 squares, then the area of that region is the number of 1-by-1 squares into which it can be decomposed, as shown in the figure below:

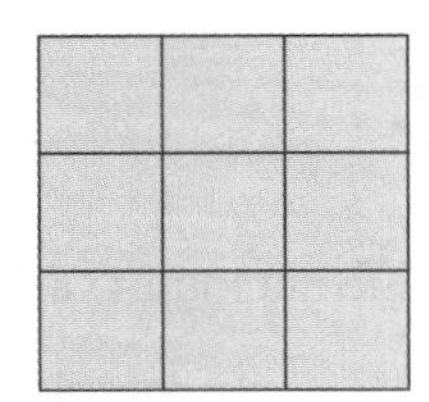

A 3*-by-*3 *square can be decomposed into nine* 1*-by-*1 *squares. Thus a* 3*-by-*3 *square has area* 9.

The expression m^2 is called "m squared" because a square whose sides have length m has area m^2.

If m is a positive integer, then an m-by-m square can be decomposed into m^2 squares of size 1-by-1. Thus it is no surprise that the area of an m-by-m square equals m^2.

The same formula holds for squares whose side length is not necessarily an integer, as shown below:

Four $\frac{1}{2}$-by-$\frac{1}{2}$ squares fill up a 1*-by-*1 *square. Thus each $\frac{1}{2}$-by-$\frac{1}{2}$ square has area $\frac{1}{4}$.*

More generally, we have the following formula:

> ***Area of a square***
>
> A square whose sides have length ℓ has area ℓ^2.

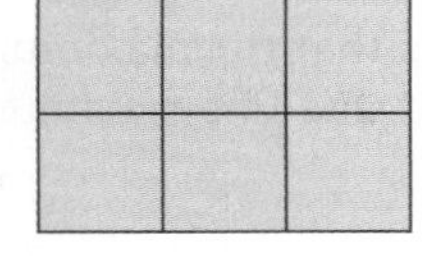

Consider a rectangle with base 3 and height 2, as shown here. This 3-by-2 rectangle can be decomposed into six 1-by-1 squares. Thus this rectangle has area 6.

Similarly, if b and h are positive integers, then a rectangle with base b and height h can be composed into bh squares of size 1-by-1, showing that the rectangle has area bh. More generally, the same formula is valid even if the base and height are not integers.

Use the same units for the base and the height. The unit of measurement for area is then the square of the unit used for these lengths. For example, a rectangle with base 3 feet and height 2 feet has area 6 square feet.

Area of a rectangle

A rectangle with base b and height h has area bh.

In the special case where the base equals the height, the formula for the area of a rectangle becomes the formula for the area of a square.

A **parallelogram** is a quadrilateral (a four-sided polygon) in which both pairs of opposite sides are parallel, as shown here.

To find the area of a parallelogram, select one of the sides and call its length the **base**. The opposite side of the parallelogram will have the same length. The **height** of the parallelogram is then defined to be the length of a line segment that connects these two sides and is perpendicular to both of them. Thus in the figure shown here, the parallelogram has base b and both vertical line segments have length equal to the height h.

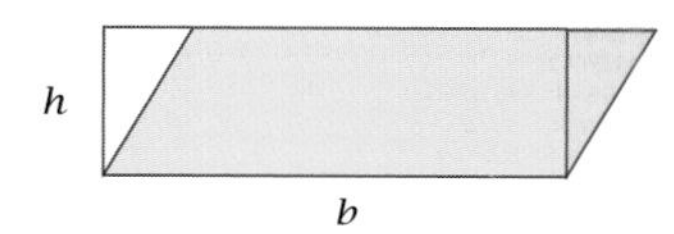

The yellow region is a parallelogram with base b and height h. The area of the parallelogram is the same as the area of the rectangle (outlined in red) with base b and height h.

The two small triangles in the figure above have the same size and thus the same area. The rectangle in the figure above could be obtained from the parallelogram by moving the triangle on the right to the position of the triangle on the left. This shows that the parallelogram and the rectangle above have the same area. Because the area of the rectangle equals bh, we thus have the following formula for the area of a parallelogram:

Area of a parallelogram

A parallelogram with base b and height h has area bh.

Triangles and Trapezoids

To find the area of a triangle, select one of the sides and call its length the **base**. The **height** of the triangle is then defined to be the length of the perpendicular line segment that connects the opposite vertex to the side determining the base, as shown in the figure in the margin.

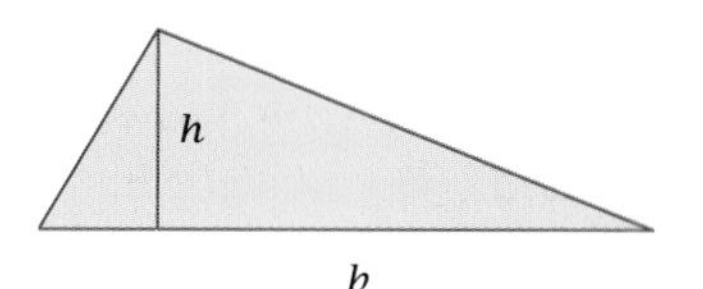

A triangle with base b and height h.

To derive the formula for the area of a triangle with base b and height h, draw two line segments, each parallel to and the same length as one of the sides of the triangle, to form a parallelogram as in the figure below:

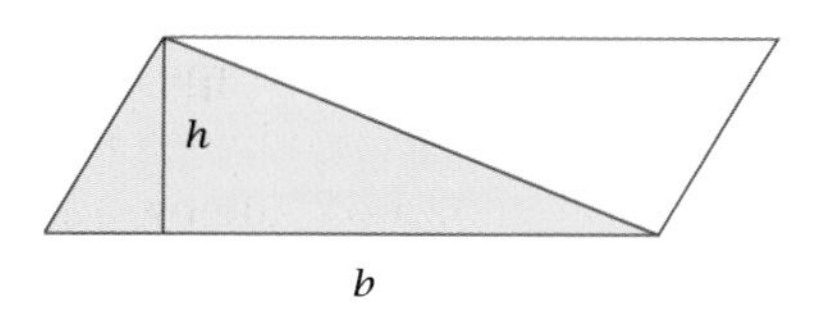

The triangle has been extended to a parallelogram by adjoining a second triangle with the same side lengths as the original triangle.

The parallelogram above has base b and height h and hence has area bh. The original triangle has area equal to half the area of the parallelogram. Thus we obtain the following formula:

> ***Area of a triangle***
>
> A triangle with base b and height h has area $\frac{1}{2}bh$.

EXAMPLE 1

Find the area of the triangle whose vertices are $(1, 0)$, $(9, 0)$, and $(7, 3)$.

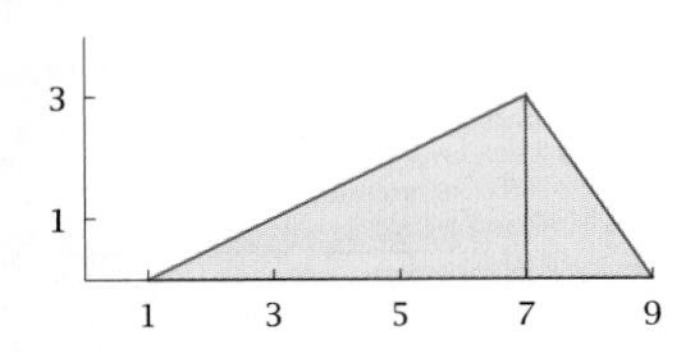

SOLUTION Choose the side connecting $(1, 0)$ and $(9, 0)$ as the base of this triangle. Thus this triangle has base $9 - 1$, which equals 8.

The height of this triangle is the length of the red line shown here; this height equals the second coordinate of the vertex $(7, 3)$. In other words, this triangle has height 3.

Thus this triangle has area $\frac{1}{2} \cdot 8 \cdot 3$, which equals 12.

---

a

b

A right triangle with area $\frac{1}{2}ab$.

Consider the special case where our triangle happens to be a right triangle,with the right angle between sides of length a and b. Choosing b to be the base of the triangle, we see that the height of this triangle equals a. Thus in this case the area of the triangle equals $\frac{1}{2}ab$.

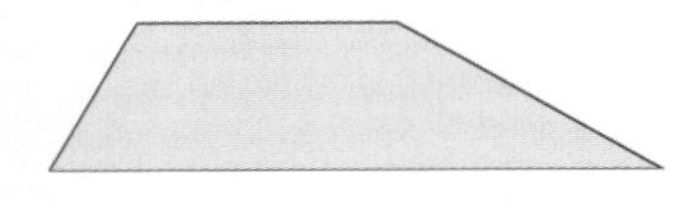

A **trapezoid** is a quadrilateral that has at least one pair of parallel sides, as for example shown here. The lengths of a pair of opposite parallel sides are called the **bases,** which are denoted below by b_1 and b_2. The **height** of the trapezoid, denoted h below, is then defined to be the length of a line segment that connects these two sides and that is perpendicular to both of them.

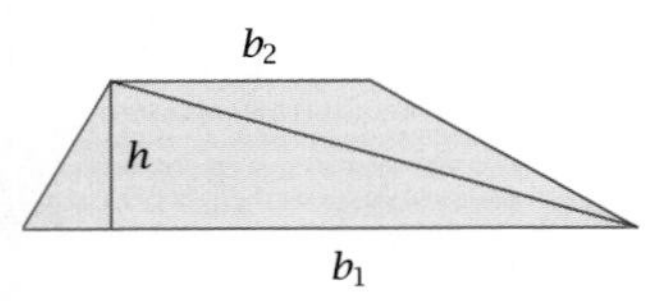

A trapezoid with bases b_1 and b_2 and height h.

The diagonal in the figure here divides the trapezoid into two triangles. The lower triangle has base b_1 and height h; thus the lower triangle has area $\frac{1}{2}b_1h$. The upper triangle has base b_2 and height h; thus the upper triangle has area $\frac{1}{2}b_2h$. The area of the trapezoid is the sum of the areas of these two triangles. Thus the area of the trapezoid equals $\frac{1}{2}b_1h + \frac{1}{2}b_2h$. Factoring out the $\frac{1}{2}$ and the h in this expression gives the following formula:

> ***Area of a trapezoid***
>
> A trapezoid with bases b_1, b_2 and height h has area $\frac{1}{2}(b_1 + b_2)h$.

Note that $\frac{1}{2}(b_1 + b_2)$ is just the average of the two bases of the trapezoid. In the special case where the trapezoid is a parallelogram, the two bases are equal and we are back to the familiar formula that the area of a parallelogram equals the base times the height.

EXAMPLE 2

Find the area of the region in the xy-plane under the line $y = 2x$, above the x-axis, and between the lines $x = 2$ and $x = 5$.

SOLUTION

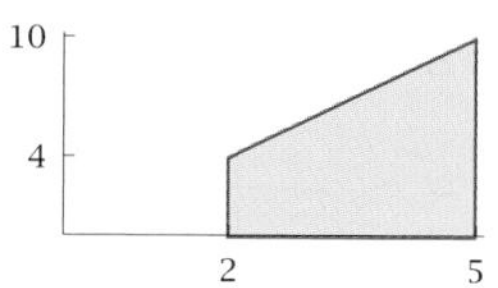

The line $x = 2$ intersects the line $y = 2x$ at the point $(2, 4)$. The line $x = 5$ intersects the line $y = 2x$ at the point $(5, 10)$.

Thus the region in question is the trapezoid shown above. The parallel sides of this trapezoid (the two vertical sides) have lengths 4 and 10, and thus this trapezoid has bases 4 and 10. As can be seen from the figure above, this trapezoid has height 3 (note that in this trapezoid, the height is the length of the horizontal side). Thus the area of this trapezoid is $\frac{1}{2} \cdot (4 + 10) \cdot 3$, which equals 21.

Stretching

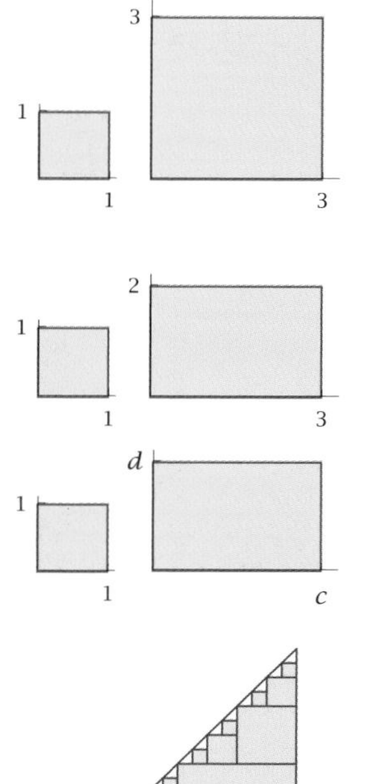

Suppose a square whose sides have length 1 has its sides tripled in length, resulting in a square whose sides have length 3, as shown here. You can think of this transformation as stretching both vertically and horizontally by a factor of 3. This transformation increases the area of the square by a factor of 9.

Consider now the transformation that stretches horizontally by a factor of 3 and stretches vertically by a factor of 2. This transformation changes a square whose sides have length 1 into a rectangle with base 3 and height 2, as shown here. Thus the area has been increased by a factor of 6.

More generally, suppose c, d are positive numbers, and consider the transformation that stretches horizontally by a factor of c and stretches vertically by a factor of d. This transformation changes a square whose sides have length 1 into a rectangle with base c and height d, as shown here. Thus the area has been increased by a factor of cd.

We need not restrict our attention to squares. The transformation that stretches horizontally by a factor of c and stretches vertically by a factor of d will change any region into a new region whose area has been changed by a factor of cd. This result follows from the result for squares, because any region can be approximated by a union of squares, as shown here for a triangle. Here is the formal statement of this result:

Area Stretch Theorem

Suppose R is a region in the coordinate plane and c, d are positive numbers. Let R' be the region obtained from R by stretching horizontally by a factor of c and stretching vertically by a factor of d. Then

the area of R' equals cd times the area of R.

Circles and Ellipses

We will now derive the formula for the area inside a circle of radius r. You are surely already familiar with this formula. The goal here is to explain why it is true.

Consider the region inside a circle of radius 1 centered at the origin. If we stretch both horizontally and vertically by a factor of r, this region becomes the region inside the circle of radius r centered at the origin, as shown in the figure below for $r = 2$.

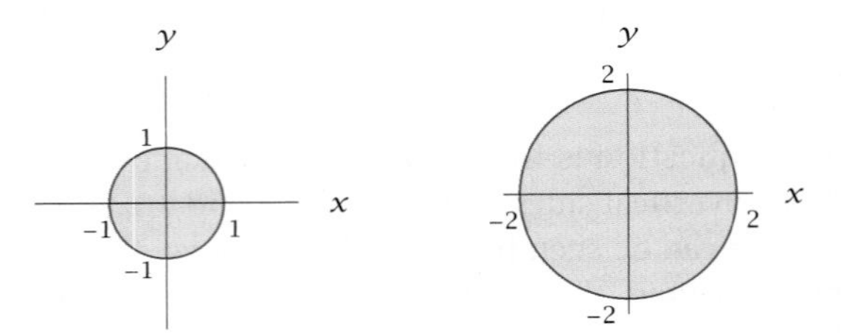

Stretching both horizontally and vertically by a factor of 2 *transforms a circle of radius* 1 *into a circle of radius* 2.

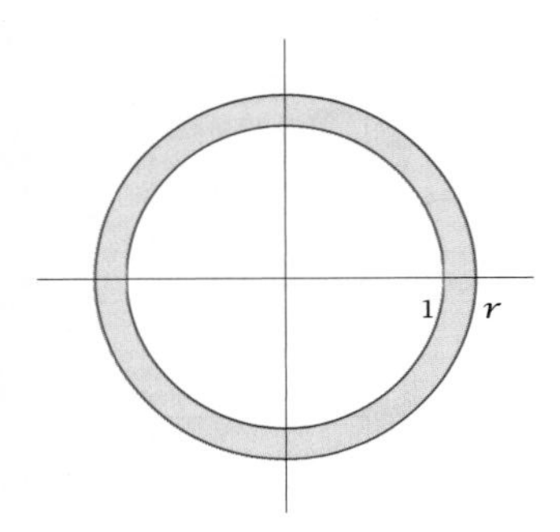

Let p denote the area inside a circle of radius 1. The Area Stretch Theorem implies that the area inside a circle of radius r equals $r^2 p$, which we write in the more familiar form pr^2. We need to find the value of p.

To find p, consider a circle of radius 1 surrounded by a slightly larger circle with radius r, as shown in the margin. Cut out the region between the two circles, then cut a slit in it and unwind it into the shape of a trapezoid (this requires a tiny bit of distortion) as shown below.

The upper base of the trapezoid is the circumference of the circle of radius 1*; the bottom base is the circumference of the circle of radius* r.

The trapezoid has height $r - 1$, which is the distance between the two original circles. The trapezoid has bases $2\pi r$ and 2π, corresponding to the circumferences of the two circles. Thus the trapezoid has area

$$\tfrac{1}{2}(2\pi r + 2\pi)(r - 1),$$

which equals $\pi(r + 1)(r - 1)$, which equals $\pi(r^2 - 1)$.

Our derivation of the formula for the area inside a circle shows the intimate connection between the area and the circumference of a circle.

The area inside the larger circle equals the area inside the circle of radius 1 plus the area of the region between the two circles. In other words, the area inside the larger circle equals $p + \pi(r^2 - 1)$. The area inside the larger circle also equals pr^2, because the larger circle has radius r. Thus we have

$$pr^2 = p + \pi(r^2 - 1).$$

Subtracting p from both sides, we get

$$p(r^2 - 1) = \pi(r^2 - 1).$$

Thus $p = \pi$. In other words, the area inside a circle of radius r equals πr^2.

We have derived the following formula:

Pie and π: The area of this pie with radius 4 inches is 16π square inches.

Area inside a circle

The area inside a circle of radius r is πr^2.

Thus to find the area inside a circle, we must first find the radius of the circle. Finding the radius sometimes requires a preliminary algebraic manipulation such as completing the square, as shown in the following example.

EXAMPLE 3

Consider the circle described by the equation

$$x^2 - 8x + y^2 + 6y = 4.$$

(a) Find the center of this circle.

(b) Find the radius of this circle.

(c) Find the circumference of this circle.

(d) Find the area inside this circle.

SOLUTION To obtain the desired information about the circle, we put its equation in a standard form. This can be done by completing the square:

$$\begin{aligned} 4 &= x^2 - 8x + y^2 + 6y \\ &= (x-4)^2 - 16 + (y+3)^2 - 9 \\ &= (x-4)^2 + (y+3)^2 - 25. \end{aligned}$$

Adding 25 to the first and last sides above shows that the circle is described by the equation

$$(x-4)^2 + (y+3)^2 = 29.$$

(a) The equation above shows that the center of the circle is $(4, -3)$.

(b) The equation above shows that the radius of the circle is $\sqrt{29}$.

(c) Because the circle has radius $\sqrt{29}$, its circumference is $2\sqrt{29}\pi$.

(d) Because the circle has radius $\sqrt{29}$, its area is 29π.

Do not make the mistake of thinking that this circle has radius 29.

In Section 2.3 we saw that the ellipse

$$\frac{x^2}{25} + \frac{y^2}{9} = 1$$

is obtained from the circle of radius 1 centered at the origin by stretching horizontally by a factor of 5 and stretching vertically by factor of 3.

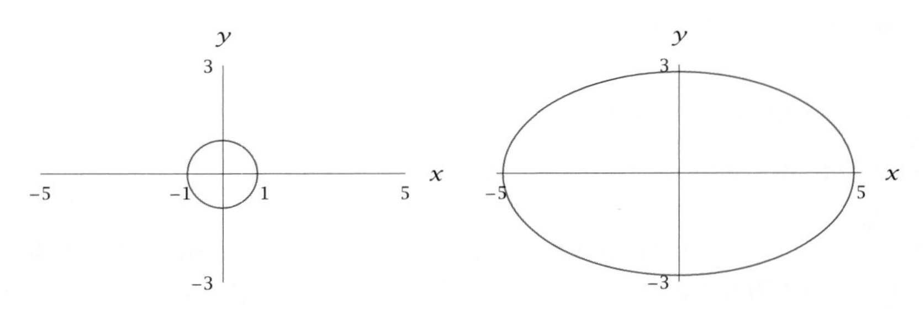

Stretching horizontally by a factor of 5 *and vertically by a factor of* 3 *transforms the circle on the left into the ellipse on the right.*

In addition to discovering that the orbits of planets are ellipses, Kepler also discovered that a line joining a planet to the sun sweeps out equal areas in equal times.

Because $5 \cdot 3 = 15$, the Area Stretch Theorem tells us that the area inside this ellipse equals 15 times the area inside the circle of radius 1. Because the area inside a circle of radius 1 is π, we conclude that the area inside this ellipse is 15π.

More generally, suppose a and b are positive numbers. Suppose the circle of radius 1 centered at the origin is stretched horizontally by a factor of a and stretched vertically by a factor of b. As we saw in Section 2.3, the equation of the resulting ellipse in the xy-plane is

$$\frac{x^2}{a^2} + \frac{y^2}{b^2} = 1.$$

The Area Stretch Theorem now gives us the following formula:

Area inside an ellipse

Suppose a and b are positive numbers. Then the area inside the ellipse

$$\frac{x^2}{a^2} + \frac{y^2}{b^2} = 1$$

is πab.

EXAMPLE 4 Find the area inside the ellipse

$$4x^2 + 5y^2 = 3.$$

SOLUTION To put the equation of this ellipse in the form given by the area formula, begin by dividing both sides by 3, and then force the equation into the desired form, as follows:

$$\begin{aligned} 1 &= \tfrac{4}{3}x^2 + \tfrac{5}{3}y^2 \\ &= \frac{x^2}{\frac{3}{4}} + \frac{y^2}{\frac{3}{5}} \\ &= \frac{x^2}{\left(\frac{\sqrt{3}}{2}\right)^2} + \frac{y^2}{\sqrt{\frac{3}{5}}^{\,2}}. \end{aligned}$$

Thus the area inside the ellipse is $\pi \cdot \frac{\sqrt{3}}{2} \cdot \sqrt{\frac{3}{5}}$, which equals $\frac{3\sqrt{5}}{10}\pi$.

Ellipses need not be centered at the origin. For example, the equation

$$\frac{(x-5)^2}{9} + \frac{(y-7)^2}{16} = 1$$

represents an ellipse centered at a point $(5, 7)$. This ellipse is obtained by shifting the ellipse whose equation is $\frac{x^2}{9} + \frac{y^2}{16} = 1$ right 5 units and up 7 units. The formula above tells us that the area inside the ellipse $\frac{x^2}{9} + \frac{y^2}{16} = 1$ is 12π, and thus the area inside the ellipse $\frac{(x-5)^2}{9} + \frac{(y-7)^2}{16} = 1$ is also 12π.

"The universe cannot be read until we have learned the language and become familiar with the characters in which it is written. It is written in mathematical language, and the letters are triangles, circles and other geometrical figures, without which means it is humanly impossible to comprehend a single word."
—Galileo

More generally, if a and b are positive numbers, then the equation

$$\frac{(x-h)^2}{a^2} + \frac{(y-k)^2}{b^2} = 1$$

represents an ellipse centered at a point (h, k). This ellipse is obtained by shifting the ellipse whose equation is $\frac{x^2}{a^2} + \frac{y^2}{b^2} = 1$. Thus the area inside the ellipse $\frac{(x-h)^2}{a^2} + \frac{(y-k)^2}{b^2} = 1$ is πab.

EXERCISES

1. Find the area of a triangle that has two sides of length 6 and one side of length 10.

2. Find the area of a triangle that has two sides of length 6 and one side of length 4.

3. (a) Find the distance from the point $(2, 3)$ to the line containing the points $(-2, -1)$ and $(5, 4)$.

   (b) Use the information from part (a) to find the area of the triangle whose vertices are $(2, 3)$, $(-2, -1)$, and $(5, 4)$.

4. (a) Find the distance from the point $(3, 4)$ to the line containing the points $(1, 5)$ and $(-2, 2)$.

   (b) Use the information from part (a) to find the area of the triangle whose vertices are $(3, 4)$, $(1, 5)$, and $(-2, 2)$.

5. Find the area of the triangle whose vertices are $(2, 0)$, $(9, 0)$, and $(4, 5)$.

6. Find the area of the triangle whose vertices are $(-3, 0)$, $(2, 0)$, and $(4, 3)$.

7. Suppose $(2, 3)$, $(1, 1)$, and $(7, 1)$ are three vertices of a parallelogram, two of whose sides are shown here.

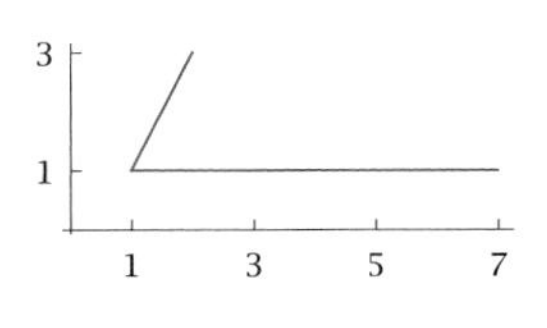

   (a) Find the fourth vertex of this parallelogram.

   (b) Find the area of this parallelogram.

8. Suppose $(3,4)$, $(2,1)$, and $(6,1)$ are three vertices of a parallelogram, two of whose sides are shown here.

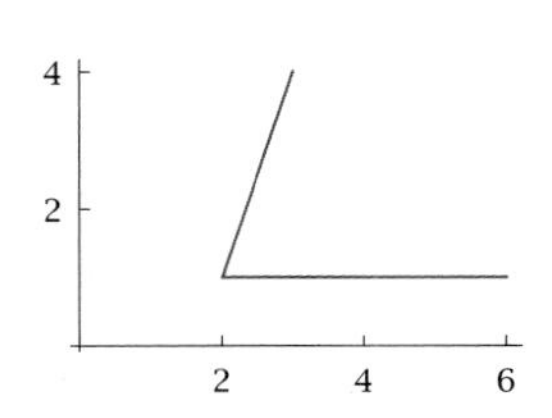

   (a) Find the fourth vertex of this parallelogram.

   (b) Find the area of this parallelogram.

9. Find the area of this trapezoid, whose vertices are $(1,1)$, $(7,1)$, $(5,3)$, and $(2,3)$.

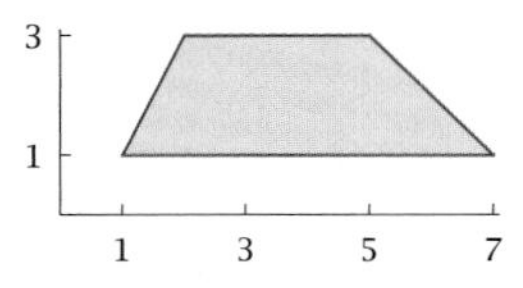

10. Find the area of this trapezoid, whose vertices are $(2,1)$, $(6,1)$, $(8,4)$, and $(1,4)$.

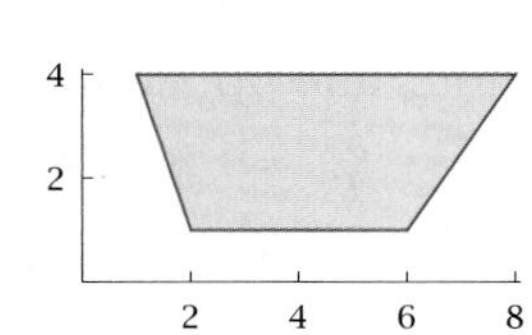

11. Find the area of the region in the xy-plane under the line $y = \frac{x}{2}$, above the x-axis, and between the lines $x = 2$ and $x = 6$.

12. Find the area of the region in the xy-plane under the line $y = 3x + 1$, above the x-axis, and between the lines $x = 1$ and $x = 5$.

13. Let $f(x) = |x|$. Find the area of the region in the xy-plane under the graph of f, above the x-axis, and between the lines $x = -2$ and $x = 5$.

14. Let $f(x) = |2x|$. Find the area of the region in the xy-plane under the graph of f, above the x-axis, and between the lines $x = -3$ and $x = 4$.

15. Find the area inside a circle with diameter 7.

16. Find the area inside a circle with diameter 9.

17. Find the area inside a circle with circumference 5 feet.

18. Find the area inside a circle with circumference 7 yards.

19. Find the area inside the circle whose equation is
$$x^2 - 6x + y^2 + 10y = 1.$$

20. Find the area inside the circle whose equation is
$$x^2 + 5x + y^2 - 3y = 1.$$

21. Find a number t such that the area inside the circle
$$3x^2 + 3y^2 = t$$
is 8.

22. Find a number t such that the area inside the circle
$$5x^2 + 5y^2 = t$$
is 2.

Use the following information for Exercises 23–28: A standard DVD disk has a 12-cm diameter (cm is the abbreviation for centimeters). The hole in the center of the disk has a 0.75-cm radius. About 50.2 megabytes of data can be stored on each square cm of usable surface of the DVD disk.

23. What is the area of a DVD disk, not counting the hole?

24. What is the area of a DVD disk, not counting the hole and the unusable circular ring with width 1.5 cm that surrounds the hole?

25. A movie company is manufacturing DVD disks containing one of its movies. Part of the surface of each DVD disk will be made usable by coating with laser-sensitive material a circular ring whose inner radius is 2.25 cm from the center of the disk. What is the minimum outer radius of this circular ring if the movie requires 3100 megabytes of data storage?

26. A movie company is manufacturing DVD disks containing one of its movies. Part of the surface of each DVD disk will be made usable by coating with laser-sensitive material a circular ring whose inner radius is 2.25 cm from the center of the disk. What is the minimum outer radius of this circular ring if the movie requires 4200 megabytes of data storage?

27. Suppose a movie company wants to store data for extra features in a circular ring on a DVD disk. If the circular ring has outer radius 5.9 cm and 200 megabytes of data storage is needed, what is the maximum inner radius of the circular ring for the extra features?

28. Suppose a movie company wants to store data for extra features in a circular ring on a DVD disk. If the circular ring has outer radius 5.9 cm and 350 megabytes of data storage is needed, what is the maximum inner radius of the circular ring for the extra features?

29. Find the area of the region in the xy-plane under the curve $y = \sqrt{4 - x^2}$ (with $-2 \le x \le 2$) and above the x-axis.

30. Find the area of the region in the xy-plane under the curve $y = \sqrt{9 - x^2}$ (with $-3 \le x \le 3$) and above the x-axis.

31. Using the answer from Exercise 29, find the area of the region in the xy-plane under the curve $y = 3\sqrt{4 - x^2}$ (with $-2 \le x \le 2$) and above the x-axis.

32. Using the answer from Exercise 30, find the area of the region in the xy-plane under the curve $y = 5\sqrt{9 - x^2}$ (with $-3 \le x \le 3$) and above the x-axis.

33. Using the answer from Exercise 29, find the area of the region in the xy-plane under the curve $y = \sqrt{4 - \frac{x^2}{9}}$ (with $-6 \le x \le 6$) and above the x-axis.

34. Using the answer from Exercise 30, find the area of the region in the xy-plane under the curve $y = \sqrt{9 - \frac{x^2}{16}}$ (with $-12 \le x \le 12$) and above the x-axis.

35. Find the area of the region in the xy-plane under the curve

$$y = 1 + \sqrt{4 - x^2},$$

above the x-axis, and between the lines $x = -2$ and $x = 2$.

36. Find the area of the region in the xy-plane under the curve

$$y = 2 + \sqrt{9 - x^2},$$

above the x-axis, and between the lines $x = -3$ and $x = 3$.

In Exercises 37–44, find the area inside the ellipse in the xy-plane determined by the given equation.

37. $\frac{x^2}{7} + \frac{y^2}{16} = 1$

38. $\frac{x^2}{9} + \frac{y^2}{5} = 1$

39. $2x^2 + 3y^2 = 1$

40. $10x^2 + 7y^2 = 1$

41. $3x^2 + 2y^2 = 7$

42. $5x^2 + 9y^2 = 3$

43. $3x^2 + 4x + 2y^2 + 3y = 2$

44. $4x^2 + 2x + 5y^2 + y = 2$

45. Find a positive number c such that the area inside the ellipse

$$2x^2 + cy^2 = 5$$

is 3.

46. Find a positive number c such that the area inside the ellipse

$$cx^2 + 7y^2 = 3$$

is 2.

47. Find numbers a and b such that $a > b$, $a + b = 15$, and the area inside the ellipse

$$\frac{x^2}{a^2} + \frac{y^2}{b^2} = 1$$

is 36π.

48. Find numbers a and b such that $a > b$, $a + b = 5$, and the area inside the ellipse

$$\frac{x^2}{a^2} + \frac{y^2}{b^2} = 1$$

is 3π.

49. Find a number t such that the area inside the ellipse

$$4x^2 + 9y^2 = t$$

is 5.

50. Find a number t such that the area inside the ellipse

$$2x^2 + 3y^2 = t$$

is 7.

PROBLEMS

51. Explain why a square yard contains 9 square feet.

52. Explain why a square foot contains 144 square inches.

53. Find a formula that gives the area of a square in terms of the length of the diagonal of the square.

54. Find a formula that gives the area of a square in terms of the perimeter.

55. Suppose a and b are positive numbers. Draw a figure of a square whose sides have length $a + b$. Partition this square into a square whose sides have length a, a square whose sides have length b, and two rectangles in a way that illustrates the identity

$$(a + b)^2 = a^2 + 2ab + b^2.$$

56. Find an example of a parallelogram whose area equals 10 and whose perimeter equals 16 (give the coordinates for all four vertices of your parallelogram).

57. Show that an equilateral triangle with sides of length r has area $\frac{\sqrt{3}}{4}r^2$.

58. Show that an equilateral triangle with area A has sides of length $\frac{2\sqrt{A}}{3^{1/4}}$.

59. Suppose $0 < a < b$. Show that the area of the region under the line $y = x$, above the x-axis, and between the lines $x = a$ and $x = b$ is $\frac{b^2 - a^2}{2}$.

60. Show that the area inside a circle with circumference c is $\frac{c^2}{4\pi}$.

61. Find a formula that gives the area inside a circle in terms of the diameter of the circle.

62. In ancient China and Babylonia, the area inside a circle was said to be one-half the radius times the circumference. Show that this formula agrees with our formula for the area inside a circle.

63. Suppose a, b, and c are positive numbers. Show that the area inside the ellipse

$$ax^2 + by^2 = c$$

is $\pi \frac{c}{\sqrt{ab}}$.

64. Consider the following figure, which is drawn accurately to scale:

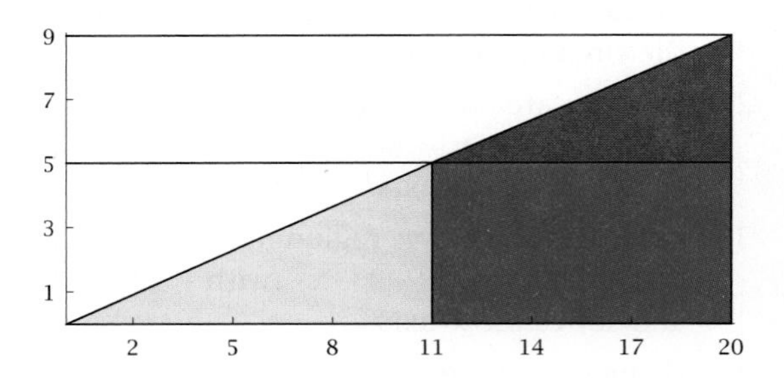

(a) Show that the right triangle whose vertices are $(0, 0)$, $(20, 0)$, and $(20, 9)$ has area 90.

(b) Show that the yellow right triangle has area 27.5.

(c) Show that the blue rectangle has area 45.

(d) Show that the red right triangle has area 18.

(e) Add the results of parts (b), (c), and (d), showing that the area of the colored region is 90.5.

(f) Seeing the figure above, most people expect that parts (a) and (e) will have the same result. Yet in part (a) we found area 90, and in part (e) we found area 90.5. Explain why these results differ.

65. The figure below illustrates a circle of radius 1 enclosed within a square. By comparing areas, explain why this figure shows that $\pi < 4$.

WORKED-OUT SOLUTIONS *to Odd-numbered Exercises*

1. Find the area of a triangle that has two sides of length 6 and one side of length 10.

SOLUTION By the Pythagorean Theorem (see figure below), the height of this triangle equals $\sqrt{6^2 - 5^2}$, which equals $\sqrt{11}$.

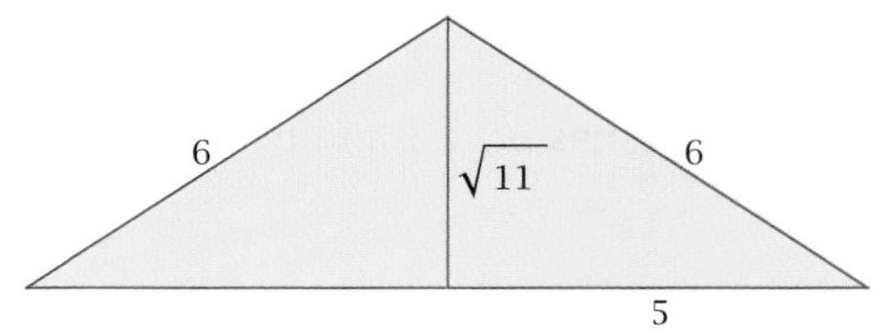

A triangle that has two sides of length 6 *and one side of length* 10.

Thus the area of this triangle equals $5\sqrt{11}$.

3. (a) Find the distance from the point $(2,3)$ to the line containing the points $(-2,-1)$ and $(5,4)$.

   (b) Use the information from part (a) to find the area of the triangle whose vertices are $(2,3)$, $(-2,-1)$, and $(5,4)$.

SOLUTION

(a) To find the distance from the point $(2,3)$ to the line containing the points $(-2,-1)$ and $(5,4)$, we first find the equation of the line containing the points $(-2,-1)$ and $(5,4)$. The slope of this line equals

$$\frac{4-(-1)}{5-(-2)},$$

which equals $\frac{5}{7}$. Thus the equation of the line containing the points $(-2,-1)$ and $(5,4)$ is

$$\frac{y-4}{x-5} = \frac{5}{7},$$

which can be rewritten as

$$y = \tfrac{5}{7}x + \tfrac{3}{7}.$$

To find the distance from the point $(2,3)$ to the line containing the points $(-2,-1)$ and $(5,4)$, we want to find the equation of the line containing the point $(2,3)$ that is perpendicular to the line containing the points $(-2,-1)$ and $(5,4)$. The equation of this line is

$$\frac{y-3}{x-2} = -\frac{7}{5},$$

which can be rewritten as

$$y = -\tfrac{7}{5}x + \tfrac{29}{5}.$$

To find where this line intersects the line containing the points $(-2,-1)$ and $(5,4)$, we need to solve the equation

$$\tfrac{5}{7}x + \tfrac{3}{7} = -\tfrac{7}{5}x + \tfrac{29}{5}.$$

Simple algebra shows that the solution to this equation is $x = \frac{94}{37}$. Plugging this value of x into the equation of either line shows that $y = \frac{83}{37}$. Thus the two lines intersect at the point $(\frac{94}{37}, \frac{83}{37})$.

Thus the distance from the point $(2,3)$ to the line containing the points $(-2,-1)$ and $(5,4)$ is the distance from the point $(2,3)$ to the point $(\frac{94}{37}, \frac{83}{37})$. This distance equals

$$\sqrt{(2-\tfrac{94}{37})^2 + (3-\tfrac{83}{37})^2},$$

which equals $\sqrt{\frac{32}{37}}$, which equals $4\sqrt{\frac{2}{37}}$.

(b) We will consider the line segment connecting the points $(-2,-1)$, and $(5,4)$ to be the base of this triangle. In part (a), we found that the height of this triangle equals $4\sqrt{\frac{2}{37}}$.

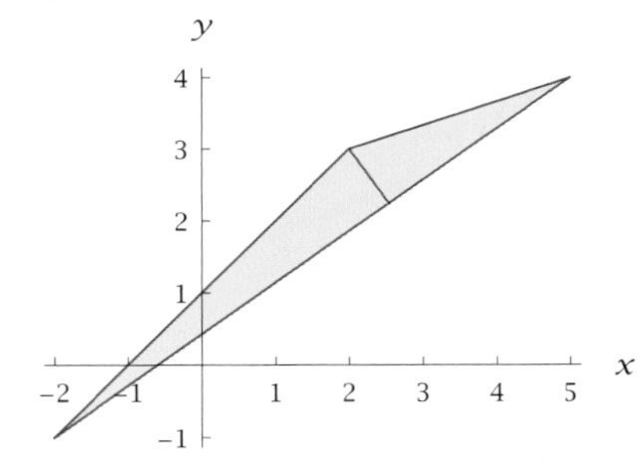

The triangle with vertices $(2,3)$, $(-2,-1)$, *and* $(5,4)$, *with a line segment showing its height.*

The base of the triangle is the distance between the points $(-2,-1)$ and $(5,4)$. This distance equals $\sqrt{74}$. Thus the area of the triangle (one-half the base times the height) equals

$$\tfrac{1}{2}\sqrt{74}\left(4\sqrt{\tfrac{2}{37}}\right),$$

which equals 4.

[There are easier ways to find the area of this triangle, but the technique used here gives you practice with several important concepts.]

5. Find the area of the triangle whose vertices are $(2, 0)$, $(9, 0)$, and $(4, 5)$.

SOLUTION Choose the side connecting $(2, 0)$ and $(9, 0)$ as the base of this triangle. Thus the triangle below has base $9 - 2$, which equals 7.

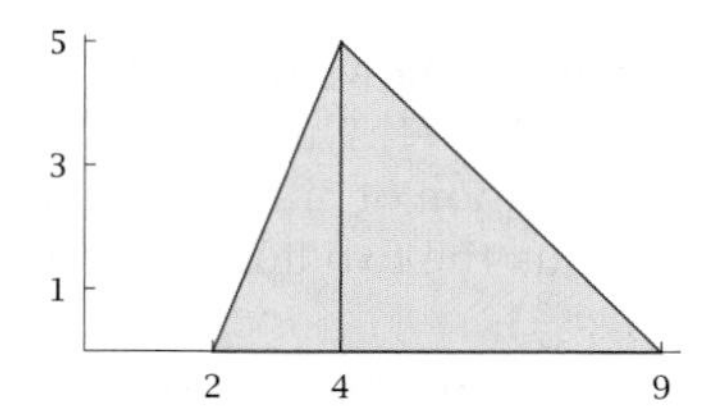

The height of this triangle is the length of the red line shown here; this height equals the second coordinate of the vertex $(4, 5)$. In other words, this triangle has height 5.

Thus this triangle has area $\frac{1}{2} \cdot 7 \cdot 5$, which equals $\frac{35}{2}$.

7. Suppose $(2, 3)$, $(1, 1)$, and $(7, 1)$ are three vertices of a parallelogram, two of whose sides are shown here.

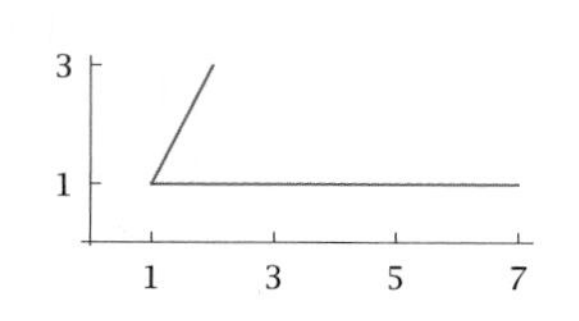

   (a) Find the fourth vertex of this parallelogram.

   (b) Find the area of this parallelogram.

SOLUTION

(a) Consider the horizontal side of the parallelogram connecting the points $(1, 1)$ and $(7, 1)$. This side has length 6. Thus the opposite side, which connects the point $(2, 3)$ and the fourth vertex, must also be horizontal and have length 6. Thus the second coordinate of the fourth vertex is the same as the second coordinate of $(2, 3)$, and the first coordinate of the fourth vertex is obtained by adding 6 to the first coordinate of $(2, 3)$. Hence the fourth vertex equals $(8, 3)$.

(b) The base of this parallelogram is the length of the side connecting the points $(1, 1)$ and $(7, 1)$, which equals 6. The height of this parallelogram is the length of a vertical line segment connecting the two horizontal sides. Because one of the horizontal sides lies on the line $y = 1$ and the other horizontal side lies on the line $y = 3$, a vertical line segment connecting these two sides will have length 2. Thus the parallelogram has height 2. Because this parallelogram has base 6 and height 2, it has area 12.

9. Find the area of this trapezoid, whose vertices are $(1, 1)$, $(7, 1)$, $(5, 3)$, and $(2, 3)$.

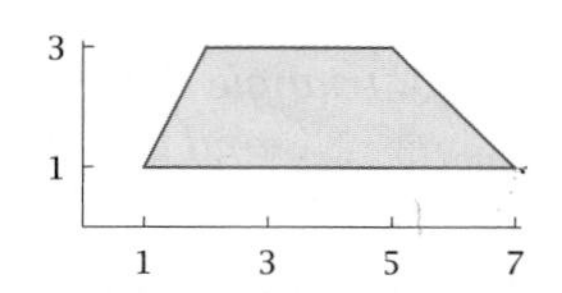

SOLUTION One base of this trapezoid is the length of the side connecting the points $(1, 1)$ and $(7, 1)$, which equals 6. The other base of this trapezoid is the length of the side connecting the points $(5, 3)$ and $(2, 3)$, which equals 3.

The height of this trapezoid is the length of a vertical line segment connecting the two horizontal sides. Because one of the horizontal sides lies on the line $y = 1$ and the other horizontal side lies on the line $y = 3$, a vertical line segment connecting these two sides will have length 2. Thus the trapezoid has height 2.

Because this trapezoid has bases 6 and 3 and has height 2, it has area $\frac{1}{2}(6 + 3) \cdot 2$, which equals 9.

11. Find the area of the region in the xy-plane under the line $y = \frac{x}{2}$, above the x-axis, and between the lines $x = 2$ and $x = 6$.

SOLUTION The line $x = 2$ intersects the line $y = \frac{x}{2}$ at the point $(2, 1)$. The line $x = 6$ intersects the line $y = \frac{x}{2}$ at the point $(6, 3)$.

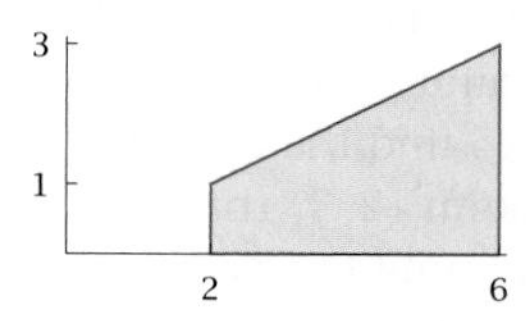

Thus the region in question is the trapezoid shown above. The parallel sides of this trapezoid (the two vertical sides) have lengths 1 and 3, and thus this trapezoid has bases 1 and 3. As can be seen from the figure above, this trapezoid has height 4. Thus the area of this trapezoid is $\frac{1}{2} \cdot (1 + 3) \cdot 4$, which equals 8.

13. Let $f(x) = |x|$. Find the area of the region in the xy-plane under the graph of f, above the x-axis, and between the lines $x = -2$ and $x = 5$.

SOLUTION

The region under consideration is the union of two triangles, as shown here.

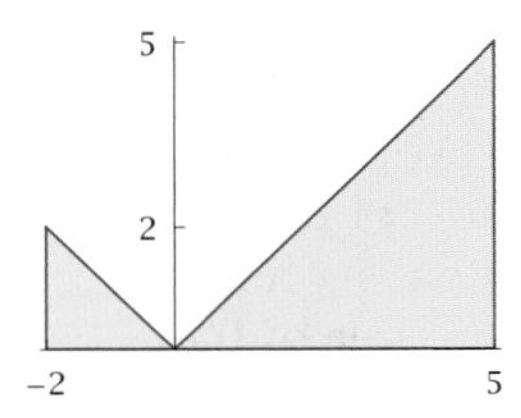

One of the triangles has base 2 and height 2 and thus has area 2. The other triangle has base 5 and height 5 and thus has area $\frac{25}{2}$. Thus the area of the region under consideration equals $2 + \frac{25}{2}$, which equals $\frac{29}{2}$.

15. Find the area inside a circle with diameter 7.

SOLUTION A circle with diameter 7 has radius $\frac{7}{2}$. Thus the area inside this circle is $\pi(\frac{7}{2})^2$, which equals $\frac{49\pi}{4}$.

17. Find the area inside a circle with circumference 5 feet.

SOLUTION Let r denote the radius of this circle in feet. Thus $2\pi r = 5$, which implies that $r = \frac{5}{2\pi}$. Thus the area inside this circle is $\pi(\frac{5}{2\pi})^2$ square feet, which equals $\frac{25}{4\pi}$ square feet.

19. Find the area inside the circle whose equation is

$$x^2 - 6x + y^2 + 10y = 1.$$

SOLUTION To find the radius of the circle given by the equation above, we complete the square, as follows:

$$\begin{aligned} 1 &= x^2 - 6x + y^2 + 10y \\ &= (x-3)^2 - 9 + (y+5)^2 - 25 \\ &= (x-3)^2 + (y+5)^2 - 34. \end{aligned}$$

Adding 34 to both sides of this equation gives

$$(x-3)^2 + (y+5)^2 = 35.$$

Thus we see that this circle is centered at $(3, -5)$ (which is irrelevant for this exercise) and that it has radius $\sqrt{35}$. Thus the area inside this circle equals $\pi\sqrt{35}^2$, which equals 35π.

21. Find a number t such that the area inside the circle

$$3x^2 + 3y^2 = t$$

is 8.

SOLUTION Rewriting the equation above as

$$x^2 + y^2 = \left(\sqrt{\tfrac{t}{3}}\right)^2,$$

we see that this circle has radius $\sqrt{\frac{t}{3}}$. Thus the area inside this circle is $\pi\left(\sqrt{\frac{t}{3}}\right)^2$, which equals $\frac{\pi t}{3}$. We want this area to equal 8, which means we need to solve the equation $\frac{\pi t}{3} = 8$. Thus $t = \frac{24}{\pi}$.

Use the following information for Exercises 23–28:

23. What is the area of a DVD disk, not counting the hole?

SOLUTION The DVD disk has radius 6 cm (because the diameter is 12 cm). The area inside a circle with radius 6 cm is 36π square cm.

The area of the hole is $0.75^2\pi$ square cm, which is 0.5625π square cm.

Subtracting 0.5625π from 36π, we see that the DVD disk has area 34.4375π square cm, which is approximately 111.33 square cm.

25. A movie company is manufacturing DVD disks containing one of its movies. Part of the surface of each DVD disk will be made usable by coating with laser-sensitive material a circular ring whose inner radius is 2.25 cm from the center of the disk. What is the minimum outer radius of this circular ring if the movie requires 3100 megabytes of data storage?

SOLUTION Because 50.2 megabytes of data can be stored in each square cm of usable surface, the usable surface must have area at least $\frac{3100}{50.2}$ square cm, or about 61.753 square cm.

If the outer radius of the circular ring of usable area is r, then the usable area will be $\pi r^2 - 2.25^2\pi$. Thus we solve the equation

$$\pi r^2 - 2.25^2\pi = 61.753$$

for r, getting $r \approx 4.97$ cm.

27. Suppose a movie company wants to store data for extra features in a circular ring on a DVD disk. If the circular ring has outer radius 5.9 cm and 200 megabytes of data storage is needed, what is the maximum inner radius of the circular ring for the extra features?

SOLUTION Because 50.2 megabytes of data can be stored in each square cm of usable surface, the usable surface must have area at least $\frac{200}{50.2}$ square cm, or about 3.984 square cm.

If the inner radius of the circular ring of usable area is r, then the usable area will be $5.9^2\pi - \pi r^2$. Thus we solve the equation

$$5.9^2\pi - \pi r^2 = 3.984$$

for r, getting $r \approx 5.79$ cm.

29. Find the area of the region in the xy-plane under the curve $y = \sqrt{4 - x^2}$ (with $-2 \le x \le 2$) and above the x-axis.

SOLUTION

Square both sides of the equation $y = \sqrt{4 - x^2}$ and then add x^2 to both sides.

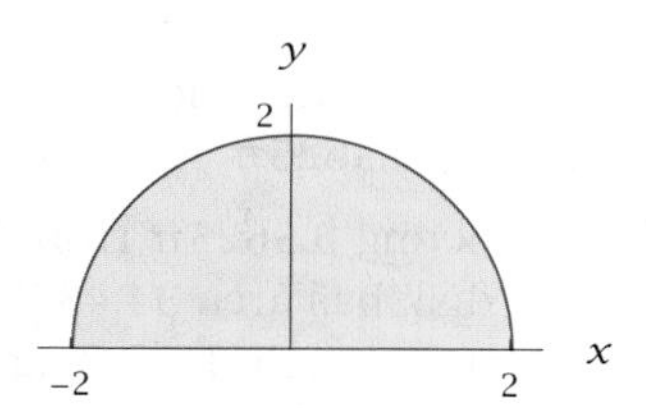

This gives the equation $x^2 + y^2 = 4$, which is the equation of a circle of radius 2 centered at the origin. However, the equation $y = \sqrt{4 - x^2}$ forces y to be nonnegative, and thus we have only the top half of the circle. Thus the region in question, which is shown above, has half the area inside a circle of radius 2. Hence the area of this region is $\frac{1}{2}\pi \cdot 2^2$, which equals 2π.

31. Using the answer from Exercise 29, find the area of the region in the xy-plane under the curve $y = 3\sqrt{4 - x^2}$ (with $-2 \le x \le 2$) and above the x-axis.

SOLUTION

The region in this exercise is obtained from the region in Exercise 29 by stretching vertically by a factor of 3. Thus by the Area Stretch Theorem, the area of this region is 3 times the area of the region in Exercise 29. Thus this region has area 6π.

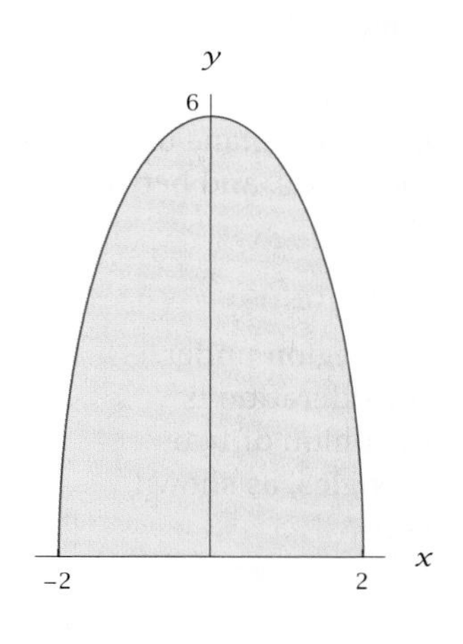

33. Using the answer from Exercise 29, find the area of the region in the xy-plane under the curve $y = \sqrt{4 - \frac{x^2}{9}}$ (with $-6 \le x \le 6$) and above the x-axis.

SOLUTION Define a function f with domain the interval $[-2, 2]$ by $f(x) = \sqrt{4 - x^2}$. Define a function h with domain the interval $[-6, 6]$ by $h(x) = f(\frac{x}{3})$. Thus

$$h(x) = f(\tfrac{x}{3}) = \sqrt{4 - (\tfrac{x}{3})^2} = \sqrt{4 - \tfrac{x^2}{9}}.$$

Hence the graph of h is obtained by horizontally stretching the graph of f by a factor of 3 (see Section 3.2). Thus the region in this exercise is obtained from the region in Exercise 29 by stretching horizontally by a factor of 3.

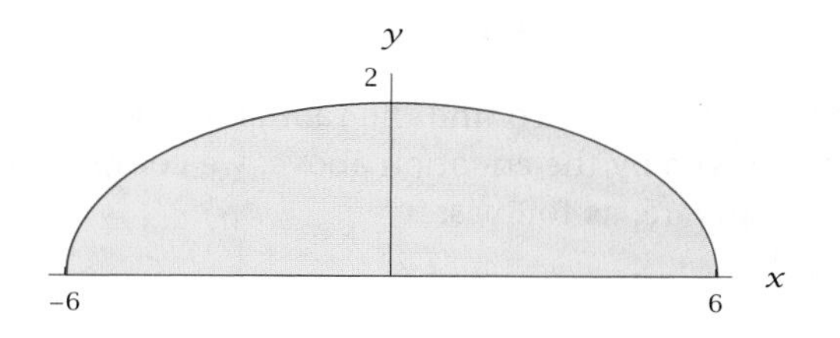

Thus by the Area Stretch Theorem, this region has area 6π.

35. Find the area of the region in the xy-plane under the curve

$$y = 1 + \sqrt{4 - x^2},$$

above the x-axis, and between the lines $x = -2$ and $x = 2$.

SOLUTION The curve $y = 1 + \sqrt{4 - x^2}$ is obtained by shifting the curve $y = \sqrt{4 - x^2}$ up 1 unit.

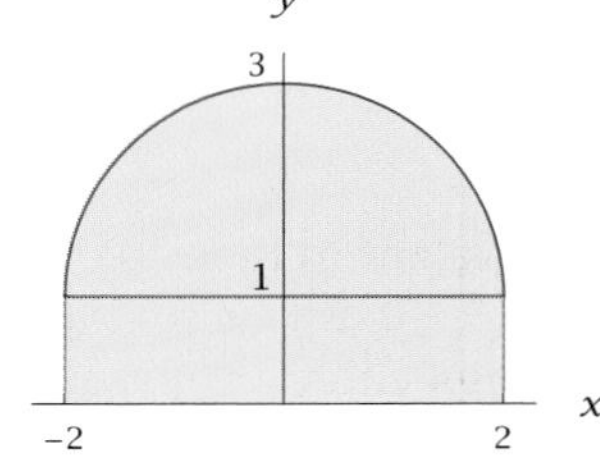

Thus we have the region above, which should be compared to the region shown in the solution to Exercise 29.

To find the area of this region, we break it into two parts. One part consists of the rectangle shown above that has base 4 and height 1 (and thus has area 4); the other part is obtained by shifting the region in Exercise 29 up 1 unit (and thus has area 2π, which is the area of the region in Exercise 29). Adding together the areas of these two parts, we conclude that the region shown above has area $4 + 2\pi$.

In Exercises 37–44, find the area inside the ellipse in the xy-plane determined by the given equation.

37. $\dfrac{x^2}{7} + \dfrac{y^2}{16} = 1$

SOLUTION Rewrite the equation of this ellipse as

$$\frac{x^2}{\sqrt{7}^2} + \frac{y^2}{4^2} = 1.$$

Thus the area inside this ellipse is $4\sqrt{7}\pi$.

39. $2x^2 + 3y^2 = 1$

SOLUTION Rewrite the equation of this ellipse in the form given by the area formula, as follows:

$$\begin{aligned} 1 &= 2x^2 + 3y^2 \\ &= \frac{x^2}{\frac{1}{2}} + \frac{y^2}{\frac{1}{3}} \\ &= \frac{x^2}{\sqrt{\frac{1}{2}}^2} + \frac{y^2}{\sqrt{\frac{1}{3}}^2}. \end{aligned}$$

Thus the area inside the ellipse is $\pi \cdot \sqrt{\frac{1}{2}} \cdot \sqrt{\frac{1}{3}}$, which equals $\frac{\pi}{\sqrt{6}}$. Multiplying numerator and denominator by $\sqrt{6}$, we see that we could also express this area as $\frac{\sqrt{6}\pi}{6}$.

41. $3x^2 + 2y^2 = 7$

SOLUTION To put the equation of the ellipse in the form given by the area formula, begin by dividing both sides by 7, and then force the equation into the desired form, as follows:

$$\begin{aligned} 1 &= \tfrac{3}{7}x^2 + \tfrac{2}{7}y^2 \\ &= \frac{x^2}{\frac{7}{3}} + \frac{y^2}{\frac{7}{2}} \\ &= \frac{x^2}{\sqrt{\frac{7}{3}}^2} + \frac{y^2}{\sqrt{\frac{7}{2}}^2}. \end{aligned}$$

Thus the area inside the ellipse is $\pi \cdot \sqrt{\frac{7}{3}} \cdot \sqrt{\frac{7}{2}}$, which equals $\frac{7\pi}{\sqrt{6}}$. Multiplying numerator and denominator by $\sqrt{6}$, we see that we could also express this area as $\frac{7\sqrt{6}\pi}{6}$.

43. $3x^2 + 4x + 2y^2 + 3y = 2$

SOLUTION To put the equation of this ellipse in a standard form, we complete the square, as follows:

$$\begin{aligned} 2 &= 3x^2 + 4x + 2y^2 + 3y \\ &= 3[x^2 + \tfrac{4}{3}x] + 2[y^2 + \tfrac{3}{2}y] \\ &= 3[(x + \tfrac{2}{3})^2 - \tfrac{4}{9}] + 2[(y + \tfrac{3}{4})^2 - \tfrac{9}{16}] \\ &= 3(x + \tfrac{2}{3})^2 - \tfrac{4}{3} + 2(y + \tfrac{3}{4})^2 - \tfrac{9}{8} \\ &= 3(x + \tfrac{2}{3})^2 + 2(y + \tfrac{3}{4})^2 - \tfrac{59}{24}. \end{aligned}$$

Adding $\frac{59}{24}$ to both sides of this equation gives

$$3(x + \tfrac{2}{3})^2 + 2(y + \tfrac{3}{4})^2 = \tfrac{107}{24}.$$

Now multiplying both sides of this equation by $\frac{24}{107}$ gives

$$\tfrac{72}{107}\left(x+\tfrac{2}{3}\right)^2+\tfrac{48}{107}\left(y+\tfrac{3}{4}\right)^2=1.$$

We rewrite this equation in the form

$$\frac{\left(x+\frac{2}{3}\right)^2}{\left(\frac{\sqrt{107}}{\sqrt{72}}\right)^2}+\frac{\left(y+\frac{3}{4}\right)^2}{\left(\frac{\sqrt{107}}{\sqrt{48}}\right)^2}=1.$$

Thus the area inside this ellipse is

$$\pi\frac{\sqrt{107}}{\sqrt{72}}\frac{\sqrt{107}}{\sqrt{48}}.$$

Because $\sqrt{72}\sqrt{48}=\sqrt{36\cdot 2}\sqrt{16\cdot 3}=6\sqrt{2}\cdot 4\sqrt{3}$, this equals

$$\pi\frac{107}{24\sqrt{6}}.$$

Multiplying numerator and denominator by $\sqrt{6}$ also allows us to express this area as

$$\pi\frac{107\sqrt{6}}{144}.$$

45. Find a positive number c such that the area inside the ellipse

$$2x^2+cy^2=5$$

is 3.

SOLUTION To put the equation of the ellipse in the form given by the area formula, begin by dividing both sides by 5, and then force the equation into the desired form, as follows:

$$\begin{aligned}1&=\tfrac{2}{5}x^2+\tfrac{c}{5}y^2\\&=\frac{x^2}{\frac{5}{2}}+\frac{y^2}{\frac{5}{c}}\\&=\frac{x^2}{\sqrt{\frac{5}{2}}^2}+\frac{y^2}{\sqrt{\frac{5}{c}}^2}.\end{aligned}$$

Thus the area inside the ellipse is $\pi\cdot\sqrt{\frac{5}{2}}\cdot\sqrt{\frac{5}{c}}$, which equals $\frac{5\pi}{\sqrt{2c}}$. We want this area to equal 3, so we must solve the equation $\frac{5\pi}{\sqrt{2c}}=3$. Squaring both sides and then solving for c gives $c=\frac{25\pi^2}{18}$.

47. Find numbers a and b such that $a>b$, $a+b=15$, and the area inside the ellipse

$$\frac{x^2}{a^2}+\frac{y^2}{b^2}=1$$

is 36π.

SOLUTION The area inside the ellipse is πab. Thus we need to solve the simultaneous equations

$$a+b=15 \quad\text{and}\quad ab=36.$$

The first equation can be rewritten as $b=15-a$, and this value for b can then be substituted into the second equation, giving the equation

$$a(15-a)=36.$$

This is equivalent to the equation $a^2-15a+36=0$, whose solutions (which can be found through either factoring or the quadratic formula) are $a=3$ and $a=12$. Choosing $a=3$ gives $b=15-a=12$, which violates the condition that $a>b$. Choosing $a=12$ gives $b=15-12=3$. Thus the only solution to this exercise is $a=12$, $b=3$.

49. Find a number t such that the area inside the ellipse

$$4x^2+9y^2=t$$

is 5.

SOLUTION Dividing the equation above by t, we have

$$\begin{aligned}1&=\tfrac{4}{t}x^2+\tfrac{9}{t}y^2\\&=\frac{x^2}{\frac{t}{4}}+\frac{y^2}{\frac{t}{9}}\\&=\frac{x^2}{\left(\frac{\sqrt{t}}{2}\right)^2}+\frac{y^2}{\left(\frac{\sqrt{t}}{3}\right)^2}.\end{aligned}$$

Thus the area inside this ellipse is

$$\pi\frac{\sqrt{t}}{2}\cdot\frac{\sqrt{t}}{3},$$

which equals $\frac{\pi t}{6}$. Hence we want $\frac{\pi t}{6}=5$, which means that $t=\frac{30}{\pi}$.

CHAPTER SUMMARY

To check that you have mastered the most important concepts and skills covered in this chapter, make sure that you can do each item in the following list:

- Locate points on the coordinate plane.
- Compute the distance between two points.
- Find the equation of a line given its slope and a point on it.
- Find the equation of a line given two points on it.
- Find the equation of a line parallel to a given line and containing a given point.
- Find the equation of a line perpendicular to a given line and containing a given point.
- Find the midpoint of a line segment.
- Use the completing-the-square technique with quadratic expressions.
- Solve quadratic equations.
- Find the equation of a circle, given its center and radius.
- Find the vertex of a parabola.
- Compute the area of triangles and trapezoids.
- Compute the area inside a circle or ellipse.
- Explain how area changes when stretching either horizontally or vertically or both.

To review a chapter, go through the list above to find items that you do not know how to do, then reread the material in the chapter about those items. Then try to answer the chapter review questions below without looking back at the chapter.

CHAPTER REVIEW QUESTIONS

1. Find the distance between the points $(5, -6)$ and $(-2, -4)$.

2. Find two points, one on the horizontal axis and one on the vertical axis, such that the distance between these two points equals 21.

3. Find the perimeter of the parallelogram whose vertices are $(2, 1)$, $(7, 1)$, $(10, 3)$, and $(5, 3)$.

4. Find the perimeter of the triangle whose vertices are $(1, 2)$, $(6, 2)$, and $(7, 5)$.

5. Find the perimeter of the trapezoid whose vertices are $(2, 3)$, $(8, 3)$, $(9, 5)$, and $(-1, 5)$.

6. Explain how to find the slope of a line if given the coordinates of two points on the line.

7. Given the slopes of two lines, how can you determine whether or not the lines are parallel?

8. Given the slopes of two lines, how can you determine whether or not the lines are perpendicular?

9. Find a number t such that the line containing the points $(3, -5)$ and $(-4, t)$ has slope -6.

10. Find the equation of the line in the xy-plane that has slope -4 and contains the point $(3, -7)$.

11. Find the equation of the line in the xy-plane that contains the points $(-6, 1)$ and $(-1, -8)$.

12. Find the equation of the line in the xy-plane that is perpendicular to the line $y = 6x - 7$ and that contains the point $(-2, 9)$.

13. Find a line segment that is not parallel to either of the coordinate axes and that has $(-3, 5)$ as its midpoint.

14. Find the vertex of the graph of the equation
$$y = 5x^2 + 2x + 3.$$

15. Give an example of numbers a, b, and c such that the graph of
$$y = ax^2 + bx + c$$
has its vertex at the point $(-4, 7)$.

16. Find a number c such that the equation

$$x^2 + cx + 3 = 0$$

has exactly one solution.

17. Find a number x such that

$$\frac{x+1}{x-2} = 3x.$$

18. Find the equation of the circle in the xy-plane centered at $(-4, 3)$ that has radius 6.

19. Find the center, radius, and circumference of the circle in the xy-plane described by

$$x^2 - 8x + y^2 + 10y = 2.$$

Also, find the area inside this circle.

20. Find the area of a triangle that has two sides of length 8 and one side of length 3.

21. Find the area of the parallelogram whose vertices are $(2, 1)$, $(7, 1)$, $(10, 3)$, and $(5, 3)$.

22. Find the area of the triangle whose vertices are $(1, 2)$, $(6, 2)$, and $(7, 5)$.

23. Find the area of the trapezoid whose vertices are $(2, 3)$, $(8, 3)$, $(9, 5)$, and $(-1, 5)$.

24. Find a number t such that the area inside the circle

$$x^2 + 6x + y^2 - 8y = t$$

is 11.

25. Find the area inside the ellipse

$$3x^2 + 2y^2 = 5.$$

26. Suppose a newly discovered planet is orbiting a far-away star and that units and a coordinate system have been chosen so that planet's orbit is described by the equation

$$\frac{x^2}{29} + \frac{y^2}{20} = 1.$$

What are the two possible locations of the star?

CHAPTER

3

Euclid explaining geometry (from The School of Athens, *painted by Raphael around 1510).*

Functions and Their Graphs

Functions lie at the heart of modern mathematics. We begin this chapter by introducing the notion of a function. Three key objects associated with each function are its graph, domain, and range. In the second section of this chapter, we will see how algebraic transformations of a function change these three key objects.

The third section of this chapter deals with the composition of functions. As we will see, the operation of composition allows us to express complicated functions in terms of simpler functions.

Inverse functions and their graphs become the center of attention in the last two sections of this chapter. Inverse functions will be key tools later in this book, for example in our treatment of roots and logarithms.

3.1 Functions

LEARNING OBJECTIVES

By the end of this section you should be able to

- evaluate functions defined by formulas;
- work with graphical as well as algebraic representations of functions;
- apply the vertical line test to determine if a curve is the graph of some function;
- determine the domain and range of a function, either algebraically or from a graph;
- work with functions defined by tables.

Definition and Examples

Functions and their domains

A **function** associates every number in some set of real numbers, called the **domain** of the function, with exactly one real number.

Although we do not need to do so in this book, functions can be defined more generally to deal with objects other than real numbers.

We usually denote functions by letters such as f, g, and h. If f is a function and x is a number in the domain of f, then the number that f associates with x is denoted by $f(x)$ and is called the value of f at x.

EXAMPLE 1

Suppose a function f is defined by the formula

$$f(x) = x^2$$

for every real number x. Evaluate each of the following:

(a) $f(3)$ (b) $f(-\frac{1}{2})$ (c) $f(1+t)$ (d) $f\left(\frac{x+5}{\pi}\right)$

The use of informal language when discussing functions is acceptable if the meaning is clear. For example, a textbook or your instructor might refer to "the function x^2" or "the function $f(x) = x^2$". Both these phrases are shorthand for the more formally correct "the function f defined by $f(x) = x^2$".

SOLUTION Here the domain of f is the set of real numbers, and f is the function that associates every real number with its square. To evaluate f at any number, we simply square that number, as shown by the solutions below:

(a) $f(3) = 3^2 = 9$

(b) $f(-\frac{1}{2}) = (-\frac{1}{2})^2 = \frac{1}{4}$

(c) $f(1+t) = (1+t)^2 = 1 + 2t + t^2$

(d) $f\left(\frac{x+5}{\pi}\right) = \left(\frac{x+5}{\pi}\right)^2 = \frac{x^2 + 10x + 25}{\pi^2}$

A function need not be defined by a single algebraic expression, as shown by the following example.

EXAMPLE 2

The U. S. 2010 federal income tax for a single person with taxable income x dollars (this is the net income after allowable deductions and exemptions) is $g(x)$ dollars, where g is the function defined by federal law as follows:

$$g(x) = \begin{cases} 0.1x & \text{if } 0 \le x \le 8375 \\ 0.15x - 418.75 & \text{if } 8375 < x \le 34000 \\ 0.25x - 3818.75 & \text{if } 34000 < x \le 82400 \\ 0.28x - 6290.75 & \text{if } 82400 < x \le 171850 \\ 0.33x - 14883.25 & \text{if } 171850 < x \le 373650 \\ 0.35x - 22356.25 & \text{if } 373650 < x. \end{cases}$$

(a) What is the 2010 federal income tax for single person whose taxable income that year was \$20,000?

(b) What is the 2010 federal income tax for single person whose taxable income that year was \$40,000?

SOLUTION

(a) Because 20000 is between 8375 and 34000, use the second line of the definition of g:

$$\begin{aligned} g(20000) &= 0.15 \times 20000 - 418.75 \\ &= 2581.25. \end{aligned}$$

Thus the 2010 federal income tax for a single person with \$20,000 taxable income that year is \$2,581.25.

(b) Because 40000 is between 34000 and 82400, use the third line of the definition of g:

$$\begin{aligned} g(40000) &= 0.25 \times 40000 - 3818.75 \\ &= 6181.25. \end{aligned}$$

Thus the 2010 federal income tax for a single person with \$40,000 taxable income that year is \$6,181.25.

The next example shows that using the flexibility offered by functions can be quicker than dealing with single algebraic expressions.

EXAMPLE 3

Give an example of a function h whose domain is the set of positive numbers and such that $h(1) = 10$, $h(3) = 2$, and $h(9) = 26$.

SOLUTION The hard way to find a function h with the required properties is to search for a function defined by a single algebraic expression. The easy way to come up with an example is to define h to have the desired values at 1, 3, and 9 and to be something simple at other values. For example, h could be defined as follows:

$$h(x) = \begin{cases} 10 & \text{if } x = 1 \\ 2 & \text{if } x = 3 \\ 26 & \text{if } x = 9 \\ 0 & \text{if } x \text{ is a positive number other than 1, 3, or 9.} \end{cases}$$

The function h defined by $h(x) = x^2 - 8x + 17$ for all positive numbers x provides another correct solution (as you can verify). However, finding this algebraic expression requires serious effort.

Instead of 0, we could have used any number to define $h(x)$ when x is a positive number other than 1, 3, or 9. Or we could have done something more complicated such as defining $h(x)$ to be x^4 when x is a positive number other than 1, 3, or 9.

You might sometimes find it useful to think of a function f as a machine:

Functions as machines

A function f can be visualized as a machine that takes an input x and produces an output $f(x)$.

This machine might work using a formula, or it might work in a more mysterious fashion, in which case it is sometimes called a "black box".

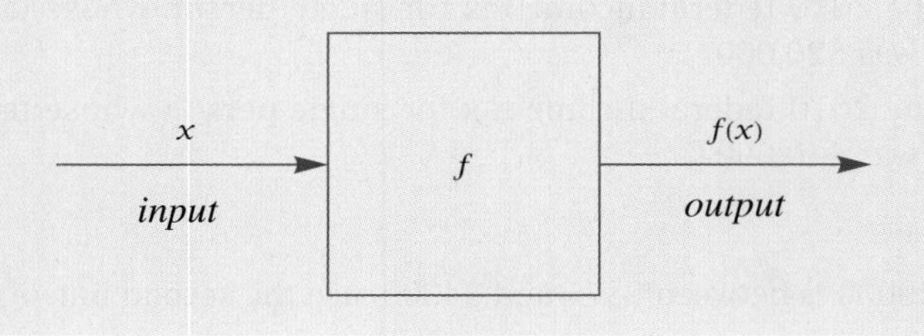

For example, if f is the function whose domain is the interval $[-4, 6]$, with f defined by the formula $f(x) = x^2$ for every x in the interval $[-4, 6]$, then giving input 3 to this machine produces output 9. The same input must always produce the same output; thus inputting 3 to this machine at a later time must again produce the output 9. Although each input has just one output, a given output may arise from more than one input. For example, the inputs -3 and 3 both produce the output 9 for this function.

When thinking of a function as the machine pictured above, the domain of the function is the set of numbers that the machine accepts as allowable inputs.

What if the number 8 is input to the machine described in the paragraph above? Because 8 is not in the domain of this function f, the machine does not produce an output for this input; the machine should produce an error message stating that 8 is not an allowable input.

Although the variable x is commonly used to denote the input for a function, other symbols can also be used:

Notation

The variable used for inputs when defining a function is irrelevant.

Distinguishing between f and $f(x)$ is usually worthwhile and helps lead to better understanding. Use f to denote a function and $f(x)$ to denote the value of a function f at a number x.

For example, consider the function f such that

$$f(x) = 3x$$

for every real number x. There is no particular x here. The symbol x is simply a placeholder to indicate that f associates any number with 3 times that number. We could have defined the same function f by using the formula $f(y) = 3y$ for every real number y. Or we could have used the formula $f(t) = 3t$ for every real number t. All these formulas show that $f(2) = 6$ and that $f(2w + 5) = 3(2w + 5)$.

The Graph of a Function

A function can be visualized by considering its graph.

The graph of a function

The **graph** of a function f is the set of points of the form $(x, f(x))$ as x varies over the domain of f.

Thus in the xy-plane, the graph of a function f is the set of points (x, y) satisfying the equation $y = f(x)$.

EXAMPLE 4

Suppose f is the function with domain $[0,3]$ defined by $f(x) = x^2 - 2x + 3$. Sketch the graph of f.

SOLUTION The graph of f is the set of points (x, y) in the xy-plane such that x is in the interval $[0,3]$ and

$$y = x^2 - 2x + 3.$$

To sketch the parabola defined by the equation above, we complete the square:

$$\begin{aligned} y &= x^2 - 2x + 3 \\ &= (x-1)^2 - 1 + 3 \\ &= (x-1)^2 + 2. \end{aligned}$$

Thus this parabola has vertex at the point $(1,2)$. We also see that if $x = 0$, then $y = 3$, producing the point $(0,3)$, and if $x = 3$, then $y = 6$, producing the point $(3,6)$. This information leads to the sketch shown in the margin.

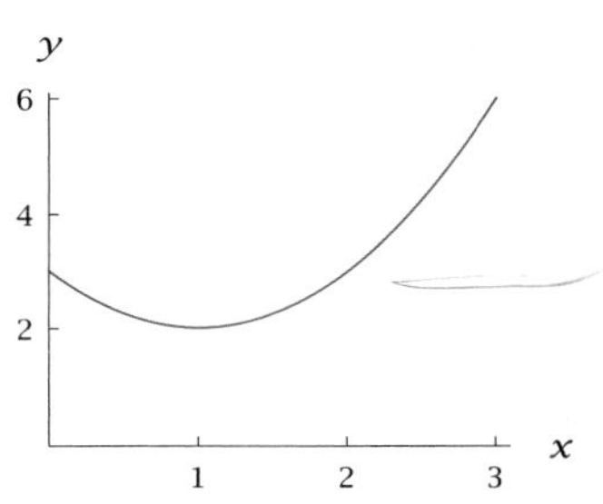

The graph of the function f with domain $[0,3]$ *defined by* $f(x) = x^2 - 2x + 3$.

Sketching the graph of a complicated function usually requires the aid of a computer or calculator. The next example uses Wolfram|Alpha in the solution, but you could use a graphing calculator or any other technology instead.

EXAMPLE 5

Let f be the function defined by

$$f(x) = \frac{4(5x - x^2 - 2)}{x^2 + 2}.$$

(a) Sketch the graph of f on the interval $[1,4]$.

(b) Sketch the graph of f on the interval $[-4,4]$.

SOLUTION

(a) Point a web browser to `www.wolframalpha.com`. In the one-line entry box that appears near the top of the web page, enter

```
graph 4(5x - x^2 - 2)/(x^2 + 2) from x=1 to 4
```

and then press the `enter` key on your keyboard or click the `=` box on the right side of the Wolfram|Alpha entry box, getting a graph that should allow you to sketch a figure like the one shown here.

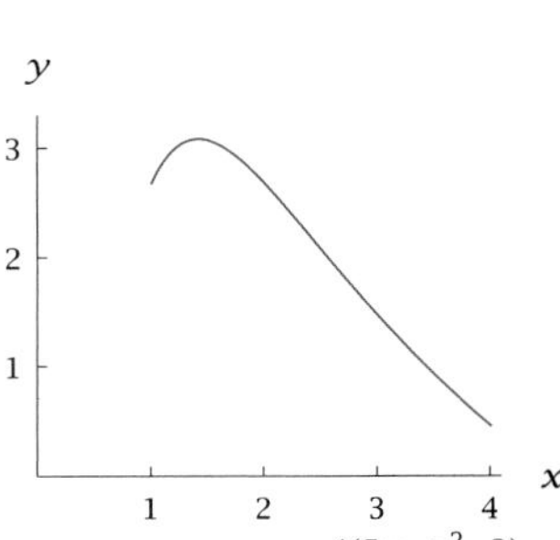

The graph of $\frac{4(5x-x^2-2)}{x^2+2}$ *on the interval* $[1,4]$.

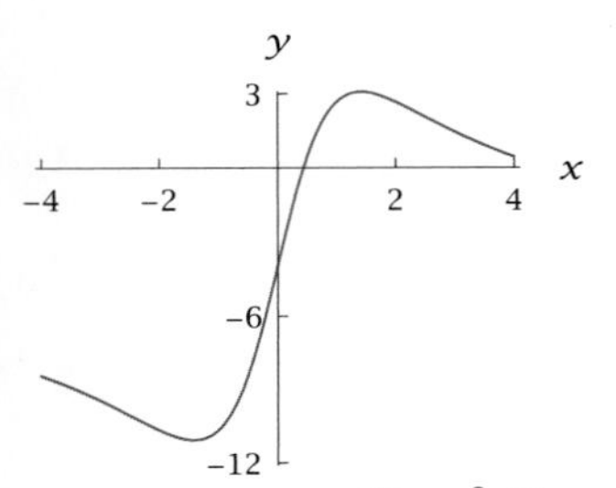

The graph of $\frac{4(5x-x^2-2)}{x^2+2}$ *on the interval* $[-4, 4]$.

(b) In the Wolfram|Alpha entry box from part (a), change 1 to -4, producing a graph like the one shown here.

The horizontal and vertical axes have different scales in this graph. The graph would become too large in the vertical direction if we used the same scale on both axis. Using different scales on the two axes makes the size of the graph more appropriate, but be aware that it changes the apparent shape of the curve. Specifically, the part of the graph on the interval $[1, 4]$ appears flatter in part (b) than the graph in part (a).

Sometimes the only information we have about a function is a sketch of its graph. The next example illustrates the procedure for finding approximate values of a function from a sketch of its graph.

EXAMPLE 6

The web site for the 4-mile HillyView race shows this graph of the function f, where $f(x)$ is the altitude in feet when the race path is x miles from the starting line. Estimate the altitude when the race path is three miles from the starting line.

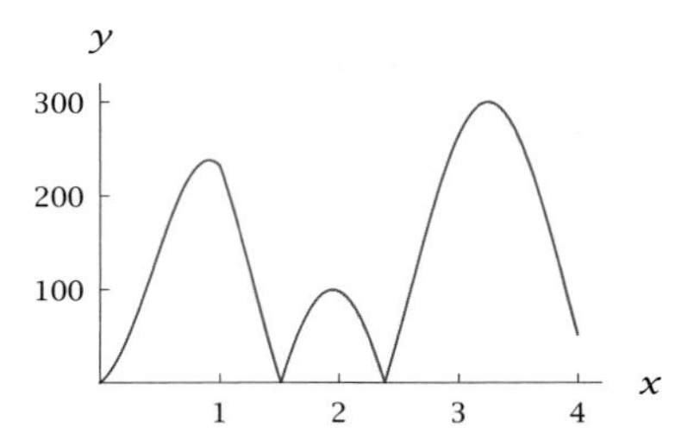

SOLUTION We need to estimate the value of $f(3)$. To do this, draw a vertical line segment from the point 3 on the x-axis until it intersects the graph. The length of that line segment will equal $f(3)$, as shown below on the left.

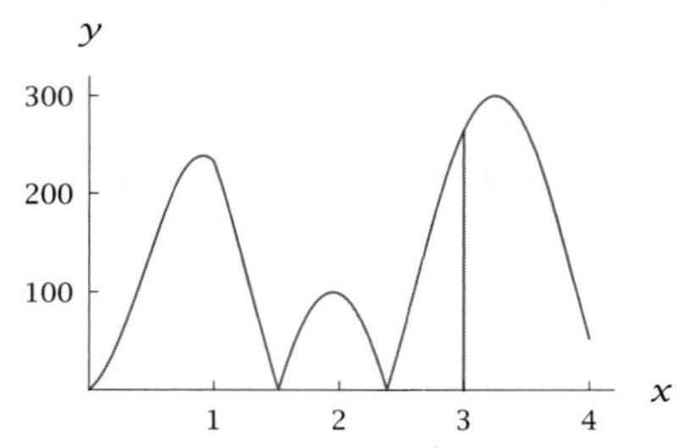

Vertical line segment has length $f(3)$.

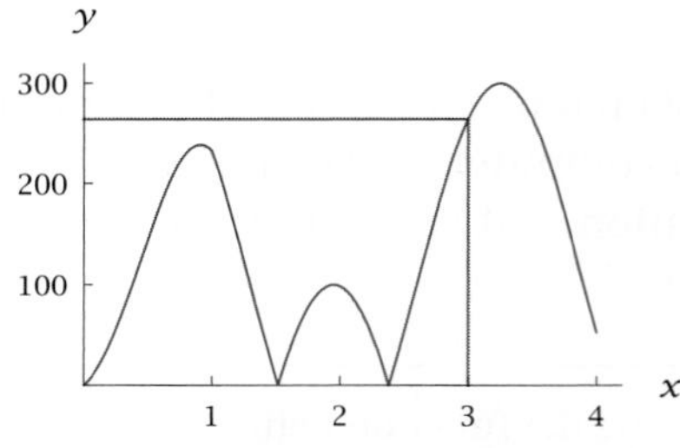

$f(3)$ *is approximately* 260.

Usually the easiest way to estimate the length of the vertical line segment shown above on the left is to draw the horizontal line shown above on the right. The point where this horizontal line intersects the y-axis then gives the length of the vertical line segment.

From the figure on the right, we see that $f(3)$ is a bit more than halfway between 200 and 300. Thus 260 is a good estimate of $f(3)$. In other words, the altitude is about 260 feet when the race path is three miles from the starting line.

The procedure used in the example above can be summarized as follows:

Finding values of a function from its graph

To evaluate $f(b)$ given only the graph of f in the xy-plane,

(a) find the point where the vertical line $x = b$ intersects the graph of f;

(b) draw a horizontal line from that point to the y-axis;

(c) the intersection of that horizontal line with the y-axis gives the value of $f(b)$.

Not every curve in the plane is the graph of some function, as illustrated by the following example.

EXAMPLE 7

Is this curve the graph of some function?

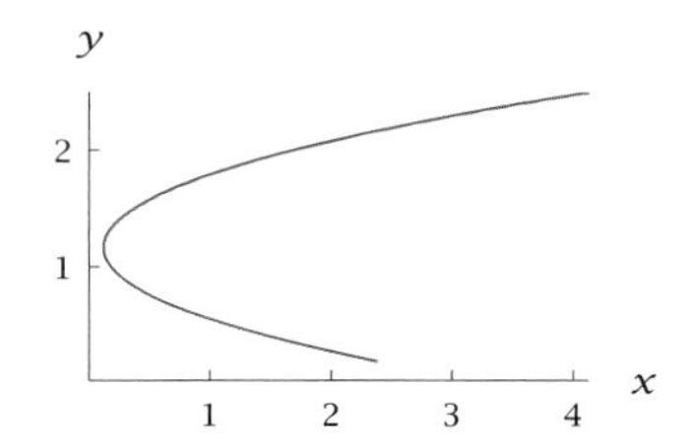

SOLUTION If this curve were the graph of some function f, then we could find the value of $f(1)$ by seeing where the line $x = 1$ intersects the curve. However, the figure below shows that the line $x = 1$ intersects the curve in two points. The definition of a function requires that $f(1)$ be a single number, not a pair of numbers. Thus this curve is not the graph of any function.

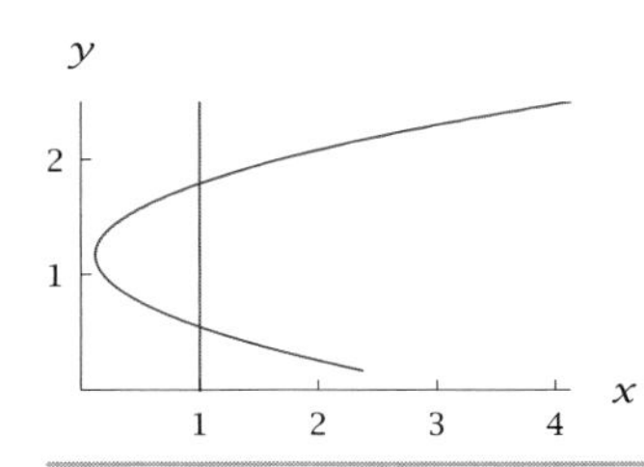

The line $x = 1$ intersects the curve in two points. Thus this curve is not the graph of a function.

More generally, any curve in the coordinate plane that intersects some vertical line in more than one point cannot be the graph of a function. Conversely, a curve in the plane that intersects each vertical line in at most one point is the graph of some function f, with the values of f determined as Example 6.

The condition for a curve in the coordinate plane to be the graph of some function can be summarized as follows:

Vertical line test

A curve in the coordinate plane is the graph of some function if and only if every vertical line intersects the curve in at most one point.

The vertical line test shows, for example, that no function has a graph that is a circle.

The Domain of a Function

Two functions are equal if and only if they have the same domain and the same value at every number in that domain.

Although the domain of a function is a formal part of characterizing the function, often we are loose about the domain of a function. Usually the domain is clear from the context or from a formula defining a function. Use the following informal rule when the domain is not specified:

Domain not specified

If a function is defined by a formula, with no domain specified, then the domain is assumed to be the set of all real numbers for which the formula makes sense and produces a real number.

The next three examples illustrate this informal rule that you can use when the domain of a function is not explicitly stated.

EXAMPLE 8 Find the domain of the function f defined by

$$f(x) = (3x - 1)^2.$$

SOLUTION No domain has been specified, but the formula above makes sense for all real numbers x. Thus unless the context indicates otherwise, we assume that the domain for this function is the set of real numbers.

The following example shows that avoiding division by 0 can determine the domain of a function.

EXAMPLE 9 Find the domain of the function h defined by

$$h(x) = \frac{x - 4}{x^2 - 2x - 2}.$$

SOLUTION No domain has been specified, but the formula above does not make sense when the denominator equals 0, which would lead to division by 0. The quadratic formula shows that

$$x^2 - 2x - 2 = 0$$

when $x = 1 \pm \sqrt{3}$. Thus unless the context indicates otherwise, we assume that the domain for this function is the set $\{x : x \neq 1 \pm \sqrt{3}\}$. Using interval notation, we can write the domain as

$$(-\infty, 1 - \sqrt{3}) \cup (1 - \sqrt{3}, 1 + \sqrt{3}) \cup (1 + \sqrt{3}, \infty).$$

The following example illustrates the requirement of the informal rule that the formula must produce a real number.

EXAMPLE 10

Find the domain of the function g defined by

$$g(x) = \sqrt{|x| - 5}.$$

SOLUTION No domain has been specified, but the formula above produces a real number only for numbers x with absolute value greater than or equal to 5. Thus unless the context indicates otherwise, we assume that the domain for this function is $(-\infty, -5] \cup [5, \infty)$.

The next example shows how the domain of a function can be determined from its graph.

EXAMPLE 11

Suppose all we know about a function f is the sketch of its graph shown here.

(a) Is 0.5 in the domain of f?

(b) Is 2.5 in the domain of f?

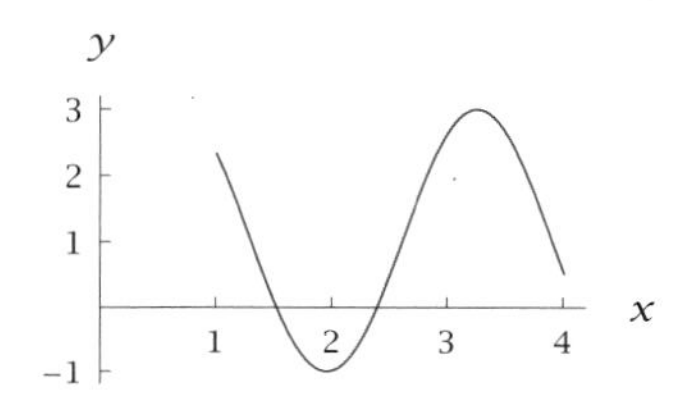

SOLUTION Recall that the graph of f consists of all points of the form $(b, f(b))$ as b varies over the domain of f. Thus the line $x = b$ in the xy-plane intersects the graph of f if and only if b is in the domain of f. The figure below, in addition to the graph of f, contains the lines $x = 0.5$ and $x = 2.5$:

Unlikely but possible: Perhaps a tiny hole in this graph, too small for us to see, implies that 2.5 is not in the domain of this function. Thus caution must be used when working with graphs. However, do not be afraid to draw reasonable conclusions that would be valid unless something weird is happening.

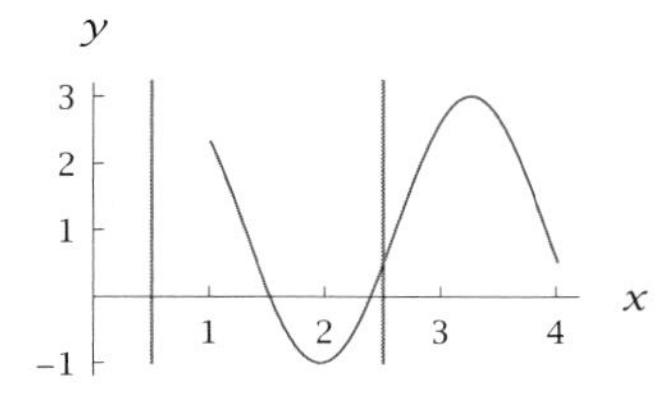

The vertical lines that intersect the graph correspond to numbers in the domain.

(a) The figure above shows that the line $x = 0.5$ does not intersect the graph of f. Thus 0.5 is not in the domain of f.

(b) The line $x = 2.5$ does intersect the graph of f. Thus 2.5 is in the domain of f.

The technique used above can be summarized as follows:

Determining the domain from the graph

A number b is in the domain of a function f if and only if the line $x = b$ in the xy-plane intersects the graph of f.

The Range of a Function

Another set associated with a function, along with the domain, is the range. The range of a function is the set of all values taken on by the function. Here is the precise definition:

> ### *Range*
>
> The **range** of a function f is the set of all numbers y such that $f(x) = y$ for at least one x in the domain of f.

Some books use the word **image** *instead of* **range**.

In other words, if we think of a function f as the machine below, then the range of f is the set of numbers that the machine produces as outputs (and the domain is the set of allowable inputs).

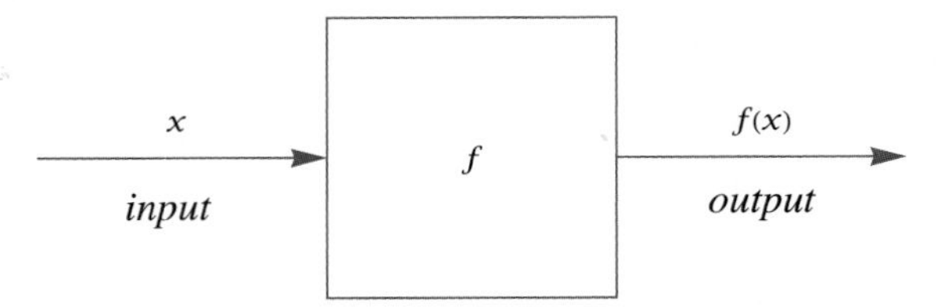

The set of inputs acceptable by this machine is the domain of f, and the set of outputs is the range of f.

EXAMPLE 12

Suppose the domain of f is the interval $[2, 5]$, with f defined on this interval by the equation $f(x) = 3x + 1$.

(a) Is 10 is the range of f?

(b) Is 19 in the range of f?

SOLUTION

(a) We need to determine whether the equation

$$3x + 1 = 10$$

has a solution in the interval $[2, 5]$, which is the domain of f. The only solution to the equation above is $x = 3$, which is in $[2, 5]$. Thus 10 is in the range of f.

(b) We need to determine whether the equation

$$3x + 1 = 19$$

has a solution in $[2, 5]$, which is the domain of f. The only solution to the equation above is $x = 6$, which is not in $[2, 5]$. Thus 19 is not in the range of f.

For a number y to be in the range of a function f, there is no requirement that the equation $f(x) = y$ have just one solution x in the domain of f. The requirement is that there be at least one solution. The next example shows that multiple solutions can easily arise.

EXAMPLE 13

Suppose the domain of g is the interval $[1, 20]$, with g defined on this interval by the equation $g(x) = |x - 5|$. Is 2 in the range of g?

SOLUTION We need to determine whether the equation

$$|x - 5| = 2$$

This equation implies that $x - 5 = 2$ or $x - 5 = -2$.

has at least one solution x in the interval $[1, 20]$. The equation above has two solutions, $x = 7$ and $x = 3$, both of which are in the domain of g. Thus 2 is in the range of g.

The range of a function can be determined by the horizontal lines that intersect the graph of the function, as shown by the next example.

EXAMPLE 14

Suppose f is the function with domain $[1, 4]$ whose graph is shown in the margin.

(a) Is 1.5 in the range of f?

(b) Is 4 in the range of f?

(c) Make a reasonable guess of the range of f.

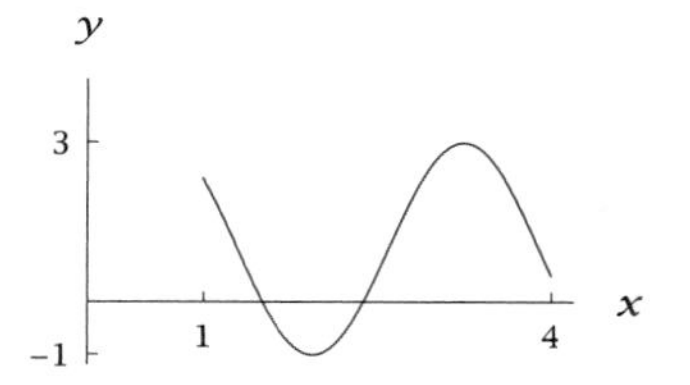

SOLUTION

(a) The figure below shows that the line $y = 1.5$ intersects the graph of f in three points. Thus 1.5 is in the range of f.

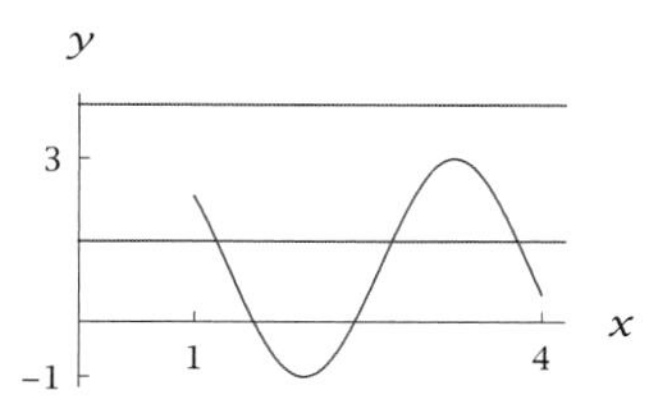

The horizontal lines that intersect the graph correspond to numbers in the range.

The lower red line shows that the equation $f(x) = 1.5$ has three solutions x in the domain of f. We need one or more such solutions for 1.5 to be in the range of f.

(b) The figure above shows that the line $y = 4$ does not intersect the graph of f. In other words, the equation $f(x) = 4$ has no solutions x in the domain of f. Thus 4 is not in the range of f.

(c) By drawing horizontal lines, we can see that the range of this function appears to be the interval $[-1, 3]$. The actual range of this function might be slightly different—we would not be able to notice the difference from the sketch of the graph if the range were actually equal to the interval $[-1.02, 3.001]$.

The technique used above can be summarized as follows:

Determining the range from the graph

A number c is in the range of a function f if and only if the horizontal line $y = c$ in the xy-plane intersects the graph of f.

Functions via Tables

If the domain of a function consists of only finitely many numbers, then all the values of a function can be listed in a table.

EXAMPLE 15

Describe the function f whose domain consists of the three numbers $\{2, 7, 13\}$ and whose values are given by the following table:

x	$f(x)$
2	3
7	$\sqrt{2}$
13	-4

SOLUTION For this function we have

$$f(2) = 3, \quad f(7) = \sqrt{2}, \quad \text{and} \quad f(13) = -4.$$

The equations above give a complete description of the function f.

Thinking about why the result below holds should be a good review of the concepts of domain and range.

Domain and range from a table

Suppose a function has only finitely many numbers in its domain and all the values of the function are displayed in a table. Then

- the domain of the function is the set of numbers that appear in the left column of the table;
- the range of the function is the set of numbers that appear in the right column of the table.

EXAMPLE 16

x	$f(x)$
1	6
2	6
3	-7
5	6

Suppose f is the function completely determined by the table in the margin.

(a) What is the domain of f?

(b) What is the range of f?

SOLUTION

(a) The left column of the table contains the numbers 1, 2, 3, and 5. Thus the domain of f is the set $\{1, 2, 3, 5\}$.

(b) The right column of the table contains only two distinct numbers, -7 and 6. Thus the range of f is the set $\{-7, 6\}$.

EXERCISES

For Exercises 1–12, assume that

$$f(x) = \frac{x+2}{x^2+1}$$

for every real number x. Evaluate and simplify each of the following expressions.

1. $f(0)$
2. $f(1)$
3. $f(-1)$
4. $f(-2)$
5. $f(2a)$
6. $f(\frac{b}{3})$
7. $f(2a+1)$
8. $f(3a-1)$
9. $f(x^2+1)$
10. $f(2x^2+3)$
11. $f(\frac{a}{b}-1)$
12. $f(\frac{2a}{b}+3)$

For Exercises 13–18, assume that

$$g(x) = \frac{x-1}{x+2}.$$

13. Find a number b such that $g(b) = 4$.
14. Find a number b such that $g(b) = 3$.
15. Evaluate and simplify the expression $\frac{g(x)-g(2)}{x-2}$.
16. Evaluate and simplify the expression $\frac{g(x)-g(3)}{x-3}$.
17. Evaluate and simplify the expression $\frac{g(a+t)-g(a)}{t}$.
18. Evaluate and simplify the expression $\frac{g(x+b)-g(x-b)}{2b}$.
19. Using the tax function given in Example 2, find the 2010 federal income tax on a single person whose taxable income that year was $45,000.
20. Using the tax function given in Example 2, find the 2010 federal income tax on a single person whose taxable income that year was $90,000.
21. Shown below is the graph of a function f.

    (a) What is the domain of f?

    (b) What is the range of f?

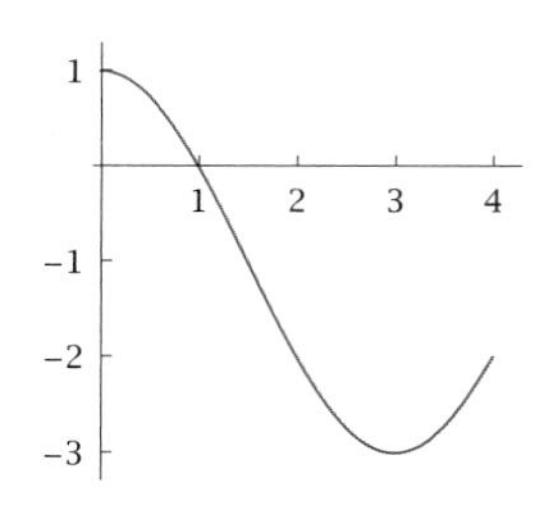

22. Shown below is the graph of a function f.

    (a) What is the domain of f?

    (b) What is the range of f?

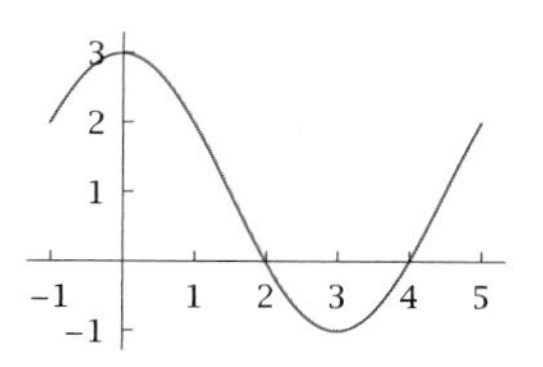

For Exercises 23–34, assume that f is the function with domain $[-4, 4]$ whose graph is shown below:

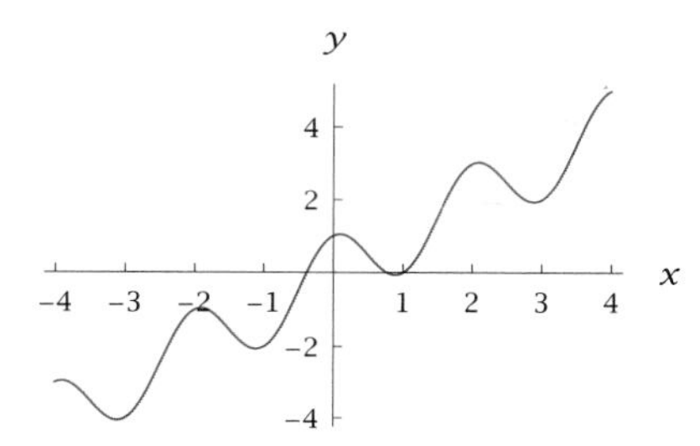

The graph of f.

23. Estimate the value of $f(-4)$.
24. Estimate the value of $f(-3)$.
25. Estimate the value of $f(-2)$.
26. Estimate the value of $f(-1)$.
27. Estimate the value of $f(2)$.
28. Estimate the value of $f(0)$.
29. Estimate the value of $f(4)$.
30. Estimate the value of $f(3)$.
31. Estimate a number b such that $f(b) = 4$.
32. Estimate a negative number b such that $f(b) = 0.5$.
33. How many values of x satisfy the equation $f(x) = \frac{1}{2}$?
34. How many values of x satisfy the equation $f(x) = -3.5$?

For Exercises 35–46, assume that g is the function with domain $[-4,4]$ whose graph is shown below:

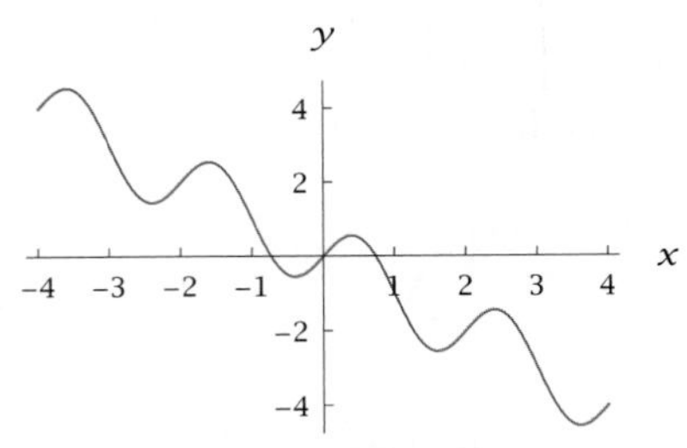

The graph of g.

35. Estimate the value of $g(-4)$.
36. Estimate the value of $g(-3)$.
37. Estimate the value of $g(-2)$.
38. Estimate the value of $g(-1)$.
39. Estimate the value of $g(2)$.
40. Estimate the value of $g(1)$.
41. Estimate the value of $g(2.5)$.
42. Estimate the value of $g(1.5)$.
43. Estimate a number b such that $g(b) = 3.5$.
44. Estimate a number b such that $g(b) = -3.5$.
45. How many values of x satisfy the equation $g(x) = -2$?
46. How many values of x satisfy the equation $g(x) = 0$?

For Exercises 47–54, assume that f is the function defined by

$$f(x) = \begin{cases} 2x+9 & \text{if } x < 0 \\ 3x-10 & \text{if } x \ge 0. \end{cases}$$

47. Evaluate $f(1)$.
48. Evaluate $f(2)$.
49. Evaluate $f(-3)$.
50. Evaluate $f(-4)$.
51. Evaluate $f(|x|+1)$.
52. Evaluate $f(|x-5|+2)$.
53. Find two different values of x such that $f(x) = 0$.
54. Find two different values of x such that $f(x) = 4$.

For Exercises 55–58, find a number b such that the function f equals the function g.

55. The function f has domain the set of positive numbers and is defined by $f(x) = 5x^2 - 7$; the function g has domain (b, ∞) and is defined by $g(x) = 5x^2 - 7$.
56. The function f has domain the set of numbers with absolute value less than 4 and is defined by $f(x) = \frac{3}{x+5}$; the function g has domain the interval $(-b, b)$ and is defined by $g(x) = \frac{3}{x+5}$.
57. Both f and g have domain $\{3, 5\}$, with f defined on this domain by the formula $f(x) = x^2 - 3$ and g defined on this domain by the formula $g(x) = \frac{18}{x} + b(x-3)$.
58. Both f and g have domain $\{-3, 4\}$, with f defined on this domain by the formula $f(x) = 3x + 5$ and g defined on this domain by the formula $g(x) = 15 + \frac{8}{x} + b(x-4)$.

For Exercises 59–64, a formula has been given defining a function f but no domain has been specified. Find the domain of each function f, assuming that the domain is the set of real numbers for which the formula makes sense and produces a real number.

59. $f(x) = \frac{2x+1}{3x-4}$
60. $f(x) = \frac{4x-9}{7x+5}$
61. $f(x) = \frac{\sqrt{x-5}}{x-7}$
62. $f(x) = \frac{\sqrt{2x+3}}{x-6}$
63. $f(x) = \sqrt{|x-6|-1}$
64. $f(x) = \sqrt{|x+5|-3}$

For Exercises 65–70, suppose h is defined by

$$h(t) = |t| + 1.$$

65. What is the range of h if the domain of h is the interval $(1, 4]$?
66. What is the range of h if the domain of h is the interval $[-8, -3)$?
67. What is the range of h if the domain of h is the interval $[-3, 5]$?
68. What is the range of h if the domain of h is the interval $[-8, 2]$?
69. What is the range of h if the domain of h is the set of positive numbers?
70. What is the range of h if the domain of h is the set of negative numbers?

71. Sketch the graph of the function f whose domain is the set of five numbers $\{-2, -1, 0, 1, 2\}$ and whose values are defined by the following table:

x	$f(x)$
-2	1
-1	3
0	-1
1	-2
2	3

72. Sketch the graph of the function f whose domain is the set of five numbers $\{-1, 0, 1, 2, 4\}$ and whose values are defined by the following table:

x	$f(x)$
-1	-2
0	2
1	0
2	1
4	-1

For Exercises 73–78, assume that g and h are the functions completely defined by the tables below:

x	$g(x)$
-3	-1
-1	1
1	2.5
3	-2

x	$h(x)$
-4	2
-2	-3
2	-1.5
3	1

73. What is the domain of g?
74. What is the domain of h?
75. What is the range of g?
76. What is the range of h?
77. Draw the graph of g.
78. Draw the graph of h.

For Exercises 79–86, assume that f and g are functions completely defined by the following tables:

x	$f(x)$
3	13
4	-5
6	$\frac{3}{5}$
7.3	-5

x	$g(x)$
3	3
8	$\sqrt{7}$
8.4	$\sqrt{7}$
12.1	$-\frac{2}{7}$

79. Evaluate $f(6)$.
80. Evaluate $g(8)$.
81. What is the domain of f?
82. What is the domain of g?
83. What is the range of f?
84. What is the range of g?
85. Find two different values of x such that $f(x) = -5$.
86. Find two different values of x such that $g(x) = \sqrt{7}$.
87. Find all functions (displayed as tables) whose domain is the set $\{2, 9\}$ and whose range is the set $\{4, 6\}$.
88. Find all functions (displayed as tables) whose domain is the set $\{5, 8\}$ and whose range is the set $\{1, 3\}$.
89. Find all functions (displayed as tables) whose domain is $\{1, 2, 4\}$ and whose range is $\{-2, 1, \sqrt{3}\}$.
90. Find all functions (displayed as tables) whose domain is $\{-1, 0, \pi\}$ and whose range is $\{-3, \sqrt{2}, 5\}$.
91. Find all functions (displayed as tables) whose domain is $\{3, 5, 9\}$ and whose range is $\{2, 4\}$.
92. Find all functions (displayed as tables) whose domain is $\{0, 2, 8\}$ and whose range is $\{6, 9\}$.

PROBLEMS

Some problems require considerably more thought than the exercises. Unlike exercises, problems often have more than one correct answer.

93. Sketch the graph of a function whose domain equals the interval $[1, 3]$ and whose range equals the interval $[-2, 4]$.
94. Sketch the graph of a function whose domain is the interval $[0, 4]$ and whose range is the set of two numbers $\{2, 3\}$.
95. Give an example of a function whose domain is $\{2, 5, 7\}$ and whose range is $\{-2, 3, 4\}$.
96. Give an example of a function whose domain is $\{3, 4, 7, 9\}$ and whose range is $\{-1, 0, 3\}$.
97. Find two different functions whose domain is $\{3, 8\}$ and whose range is $\{-4, 1\}$.

98. Explain why there does not exist a function whose domain is $\{-1, 0, 3\}$ and whose range is $\{3, 4, 7, 9\}$.

99. Give an example of a function f whose domain is the set of real numbers and such that the values of $f(-1)$, $f(0)$, and $f(2)$ are given by the following table:

x	$f(x)$
-1	$\sqrt{2}$
0	$\frac{17}{3}$
2	-5

100. Suppose the only information you know about a function f is that the domain of f is the set of real numbers and that $f(1) = 1$, $f(2) = 4$, $f(3) = 9$, and $f(4) = 16$. What can you say about the value of $f(5)$?
[*Hint:* The answer to this problem is not "25". The shortest correct answer is just one word.]

101. Give an example of two different functions f and g, both of which have the set of real numbers as their domain, such that $f(x) = g(x)$ for every rational number x.

102. Give an example of a function whose domain equals the set of real numbers and whose range equals the set $\{-1, 0, 1\}$.

103. Explain why no function has a graph that is an ellipse.

104. Give an example of a function whose domain equals the set of real numbers and whose range equals the set of integers.

105. Give an example of a function whose domain equals $[0, 1]$ and whose range equals $(0, 1)$.

106. Give an example of a function whose domain equals $(0, 1)$ and whose range equals $[0, 1]$.

107. Give an example of a function whose domain is the set of positive integers and whose range is the set of positive even integers.

108. Give an example of a function whose domain is the set of positive even integers and whose range is the set of positive odd integers.

109. Give an example of a function whose domain is the set of integers and whose range is the set of positive integers.

110. Give an example of a function whose domain is the set of positive integers and whose range is the set of integers.

For Problems 111–116, use the following information: If an object is thrown straight up into the air from height H feet at time 0 with initial velocity V feet per second, then at time t seconds the height of the object is

$$-16.1t^2 + Vt + H$$

feet. This formula uses only gravitational force, ignoring air friction. It is valid only until the object hits the ground or some other object.

111. Find a function b such that $b(V)$ is the length of time in seconds that a ball takes to reach its maximum height when thrown straight up with initial velocity V feet per second.

112. Find a function d such that $d(H)$ is the maximum height in feet the ball reaches when thrown straight up with initial velocity 50 feet per second from height H feet.

113. Find a function c such that $c(V)$ is the maximum height in feet the ball reaches when thrown straight up with initial velocity V feet per second from height 5 feet.

114. Find a function F such that $F(y)$ is the initial velocity in feet per second needed to make the maximum height of the ball be y feet when thrown from a height of 5 feet.

115. Find a function f such that $f(V)$ is the length of time in seconds that a ball stays in the air when thrown straight up with initial velocity V feet per second from height 5 feet.

116. Find a function g such that $g(H)$ is the length of time in seconds that a ball stays in the air when thrown straight up with initial velocity 50 feet per second from height H feet.

WORKED-OUT SOLUTIONS *to Odd-numbered Exercises*

Do not read these worked-out solutions before first attempting to do the exercises yourself. Otherwise you may merely mimic the techniques shown here without understanding the ideas.

Best way to learn: Carefully read the section of the textbook, then do all the odd-numbered exercises (even if they have not been assigned) and check your answers here. If you get stuck on an exercise, reread the section of the textbook—then try the exercise again. If you are still stuck, then look at the worked-out solution here.

For Exercises 1–12, assume that

$$f(x) = \frac{x+2}{x^2+1}$$

for every real number x. Evaluate and simplify each of the following expressions.

1. $f(0)$

SOLUTION $f(0) = \frac{0+2}{0^2+1} = \frac{2}{1} = 2$

3. $f(-1)$

SOLUTION $f(-1) = \dfrac{-1+2}{(-1)^2+1} = \dfrac{1}{1+1} = \dfrac{1}{2}$

5. $f(2a)$

SOLUTION $f(2a) = \dfrac{2a+2}{(2a)^2+1} = \dfrac{2a+2}{4a^2+1}$

7. $f(2a+1)$

SOLUTION

$$f(2a+1) = \frac{(2a+1)+2}{(2a+1)^2+1} = \frac{2a+3}{4a^2+4a+2}$$

9. $f(x^2+1)$

SOLUTION

$$f(x^2+1) = \frac{(x^2+1)+2}{(x^2+1)^2+1} = \frac{x^2+3}{x^4+2x^2+2}$$

11. $f(\frac{a}{b}-1)$

SOLUTION We have

$$f(\tfrac{a}{b}-1) = \frac{(\frac{a}{b}-1)+2}{(\frac{a}{b}-1)^2+1} = \frac{\frac{a}{b}+1}{\frac{a^2}{b^2}-2\frac{a}{b}+2}$$

$$= \frac{ab+b^2}{a^2-2ab+2b^2},$$

where the last expression was obtained by multiplying the numerator and denominator of the previous expression by b^2.

For Exercises 13–18, assume that

$$g(x) = \frac{x-1}{x+2}.$$

13. Find a number b such that $g(b) = 4$.

SOLUTION We want to find a number b such that

$$\frac{b-1}{b+2} = 4.$$

Multiply both sides of the equation above by $b+2$, getting

$$b - 1 = 4b + 8.$$

Now solve this equation for b, getting $b = -3$.

15. Evaluate and simplify the expression $\frac{g(x)-g(2)}{x-2}$.

SOLUTION We begin by evaluating the numerator:

$$\begin{aligned} g(x) - g(2) &= \frac{x-1}{x+2} - \frac{1}{4} \\ &= \frac{4(x-1)-(x+2)}{4(x+2)} \\ &= \frac{4x-4-x-2}{4(x+2)} \\ &= \frac{3x-6}{4(x+2)} \\ &= \frac{3(x-2)}{4(x+2)}. \end{aligned}$$

Thus

$$\begin{aligned} \frac{g(x)-g(2)}{x-2} &= \frac{3(x-2)}{4(x+2)} \cdot \frac{1}{x-2} \\ &= \frac{3}{4(x+2)}. \end{aligned}$$

17. Evaluate and simplify the expression $\frac{g(a+t)-g(a)}{t}$.

SOLUTION We begin by computing the numerator:

$$\begin{aligned} g(a+t)-g(a) \\ &= \frac{(a+t)-1}{(a+t)+2} - \frac{a-1}{a+2} \\ &= \frac{(a+t-1)(a+2)-(a-1)(a+t+2)}{(a+t+2)(a+2)} \\ &= \frac{3t}{(a+t+2)(a+2)}. \end{aligned}$$

Thus

$$\frac{g(a+t)-g(a)}{t} = \frac{3}{(a+t+2)(a+2)}.$$

19. Using the tax function given in Example 2, find the 2010 federal income tax on a single person whose taxable income that year was \$45,000.

SOLUTION Because 45000 is between 34000 and 82400, use the third line of the definition of g in Example 2:

$$\begin{aligned} g(45000) &= 0.25 \times 45000 - 3818.75 \\ &= 7431.25. \end{aligned}$$

Thus the 2010 federal income tax for a single person with \$40,000 taxable income that year is \$7,431.25.

21. Shown below is the graph of a function f.

(a) What is the domain of f?

(b) What is the range of f?

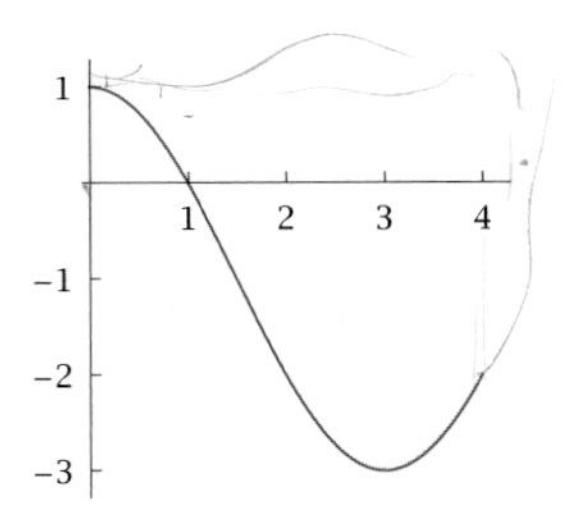

SOLUTION

(a) From the figure, it appears that the domain of f is $[0, 4]$.
The word "appears" is used here because a figure cannot provide precision. The actual domain of f might be $[0, 4.001]$ or $[0, 3.99]$ or $(0, 4)$.

(b) From the figure, it appears that the range of f is $[-3, 1]$.

For Exercises 23–34, assume that f is the function with domain $[-4, 4]$ whose graph is shown below:

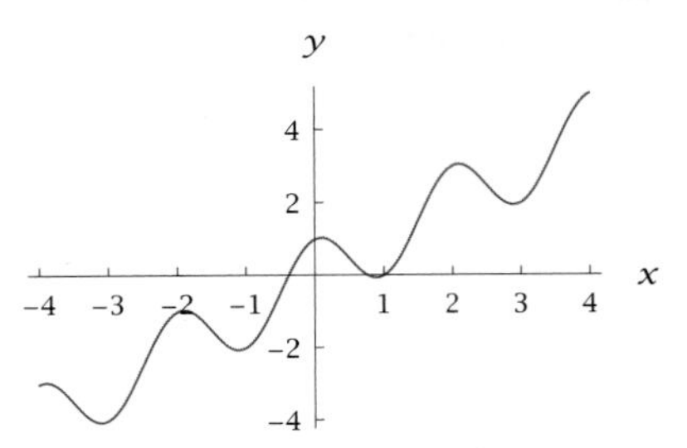

The graph of f.

23. Estimate the value of $f(-4)$.

SOLUTION To estimate the value of $f(-4)$, draw a vertical line from the point -4 on the x-axis to the graph, as shown below:

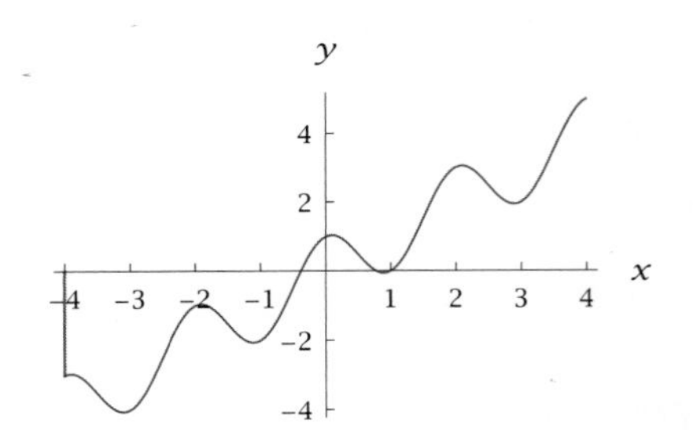

Then draw a horizontal line from where the vertical line intersects the graph to the y-axis:

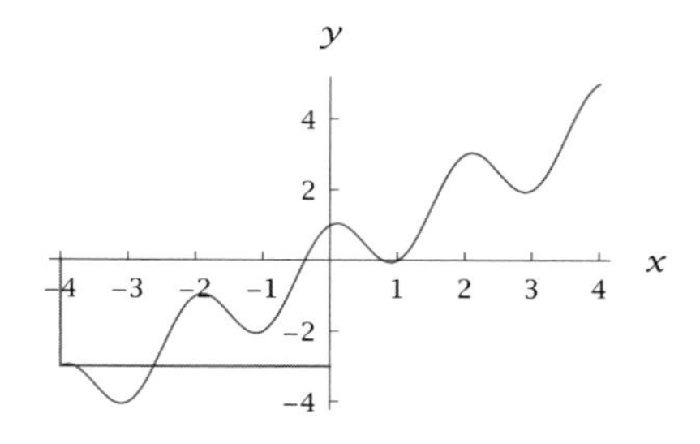

The intersection of the horizontal line with the y-axis gives the value of $f(-4)$. Thus we see that $f(-4) \approx -3$ (the symbol $\approx$ means "is approximately equal to", which is the best that can be done when using a graph).

25. Estimate the value of $f(-2)$.

SOLUTION To estimate the value of $f(-2)$, draw a vertical line from the point -2 on the x-axis to the graph, as shown below:

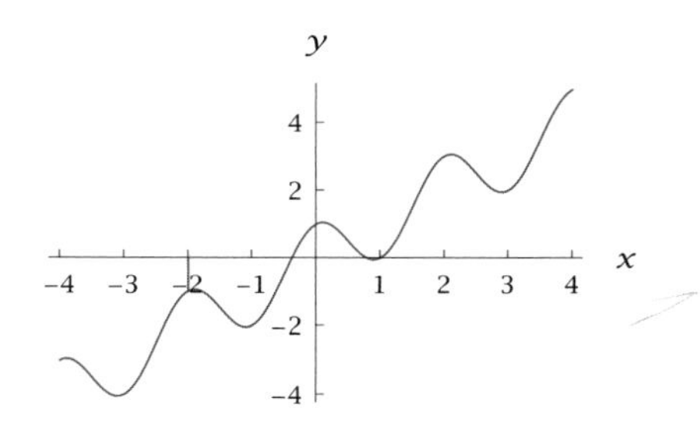

Then draw a horizontal line from where the vertical line intersects the graph to the y-axis:

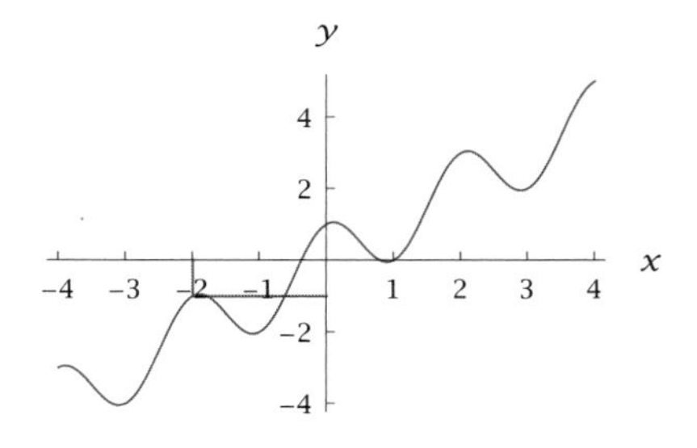

The intersection of the horizontal line with the y-axis gives the value of $f(-2)$. Thus we see that $f(-2) \approx -1$.

27. Estimate the value of $f(2)$.

SOLUTION To estimate the value of $f(2)$, draw a vertical line from the point 2 on the x-axis to the graph, as shown below:

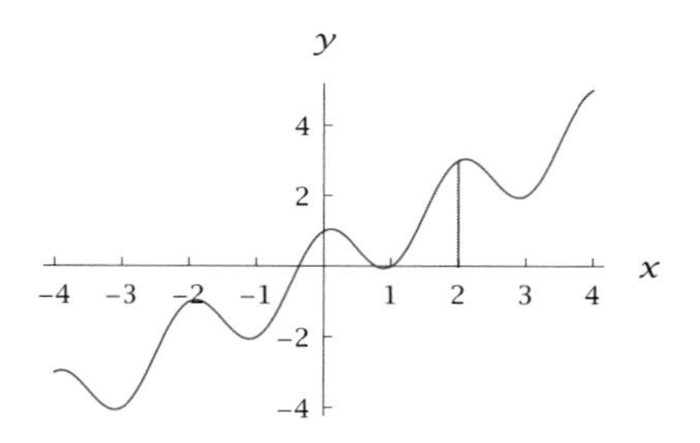

Then draw a horizontal line from where the vertical line intersects the graph to the y-axis:

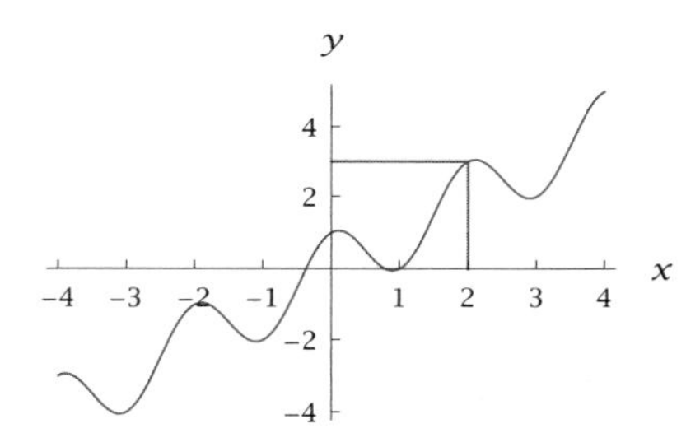

The intersection of the horizontal line with the y-axis gives the value of $f(2)$. Thus we see that $f(2) \approx 3$.

29. Estimate the value of $f(4)$.

SOLUTION To estimate the value of $f(4)$, draw a vertical line from the point 4 on the x-axis to the graph, as shown below:

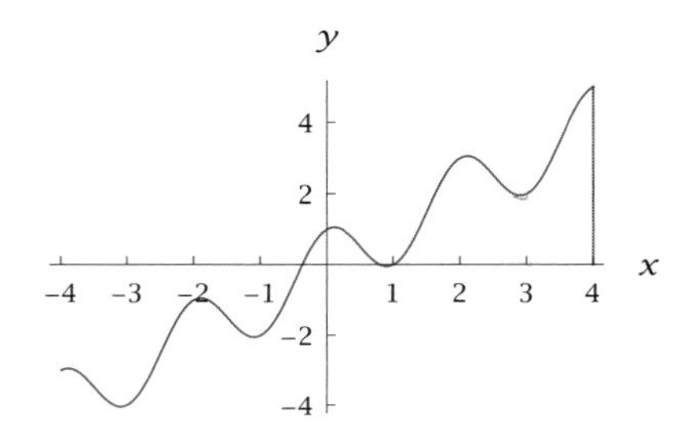

Then draw a horizontal line from where the vertical line intersects the graph to the y-axis:

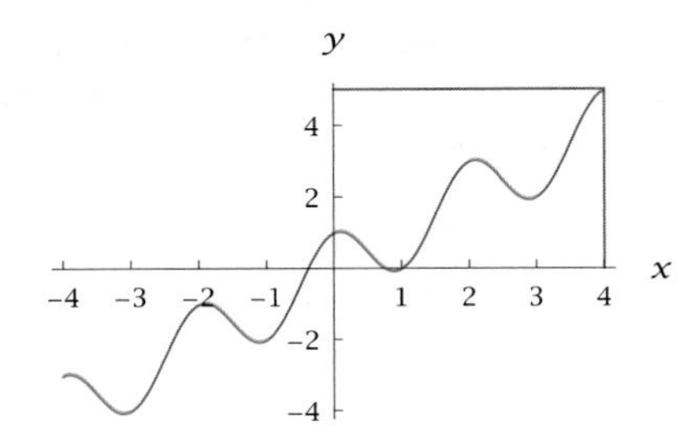

The intersection of the horizontal line with the y-axis gives the value of $f(4)$. Thus we see that $f(4) \approx 5$.

31. Estimate a number b such that $f(b) = 4$.

SOLUTION Draw the horizontal line $y = 4$, as shown below:

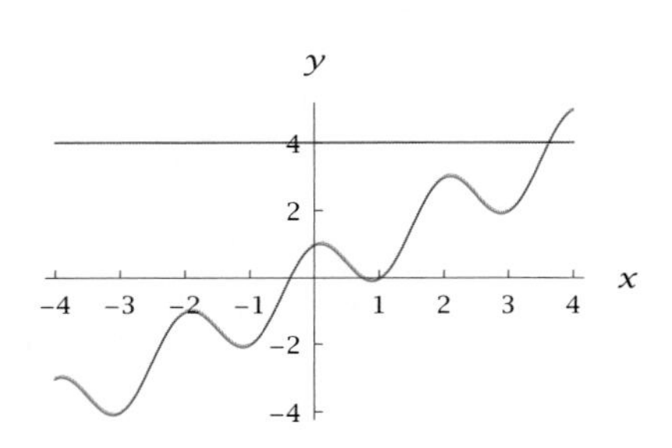

Then draw a vertical line from where this horizontal line intersects the graph to the x-axis:

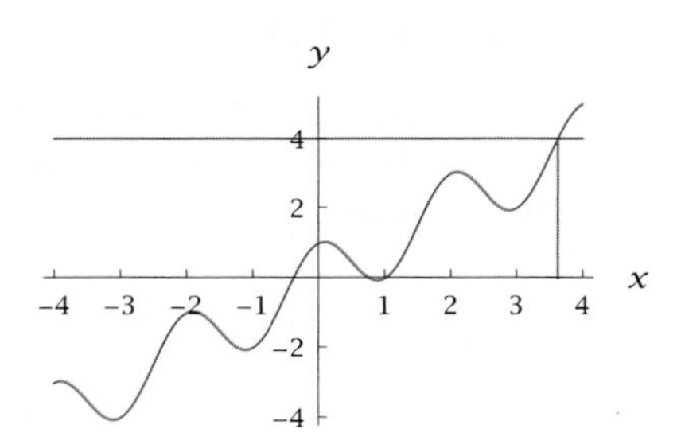

The intersection of the vertical line with the x-axis gives the value of b such that $f(b) = 4$. Thus we see that $b \approx 3.6$.

33. How many values of x satisfy the equation $f(x) = \frac{1}{2}$?

SOLUTION Draw the horizontal line $y = \frac{1}{2}$, as shown below. This horizontal line intersects the graph in three points. Thus there exist three values of x such that $f(x) = \frac{1}{2}$.

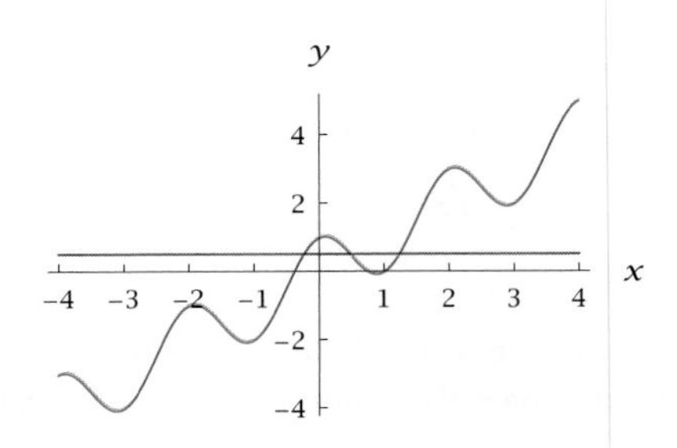

For Exercises 35–46, assume that g is the function with domain $[-4, 4]$ whose graph is shown below:

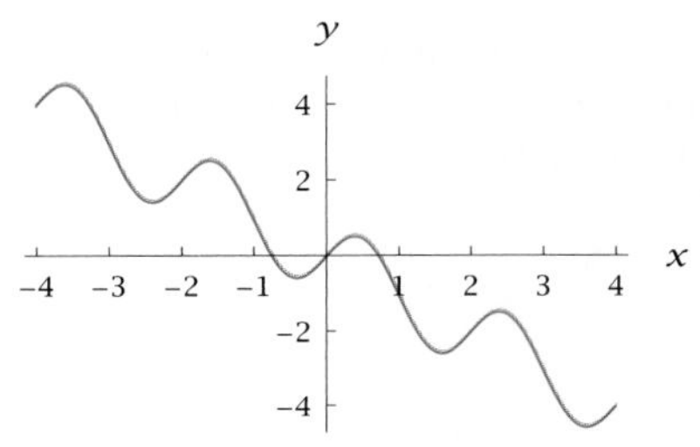

The graph of g.

35. Estimate the value of $g(-4)$.

SOLUTION To estimate the value of $g(-4)$, draw a vertical line from the point -4 on the x-axis to the graph, as shown below:

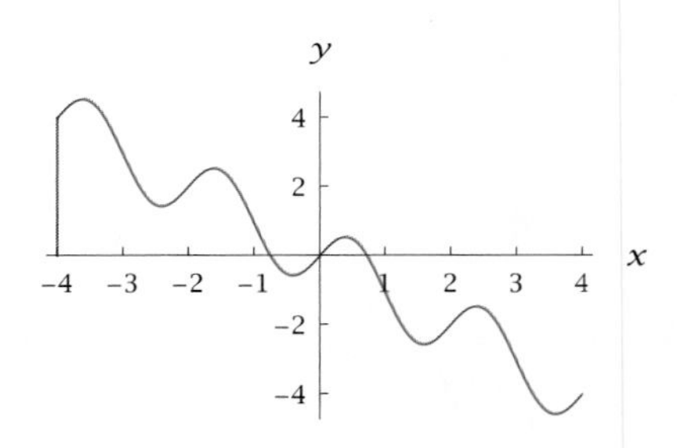

Then draw a horizontal line from where the vertical line intersects the graph to the y-axis:

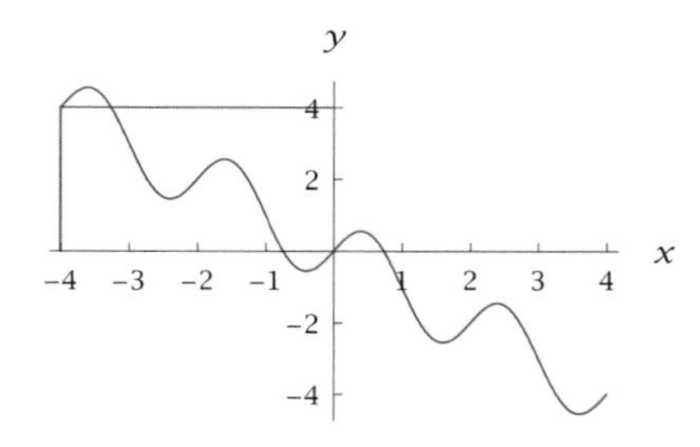

The intersection of the horizontal line with the y-axis gives the value of $g(-4)$. Thus we see that $g(-4) \approx 4$.

37. Estimate the value of $g(-2)$.

SOLUTION To estimate the value of $g(-2)$, draw a vertical line from the point -2 on the x-axis to the graph, as shown below:

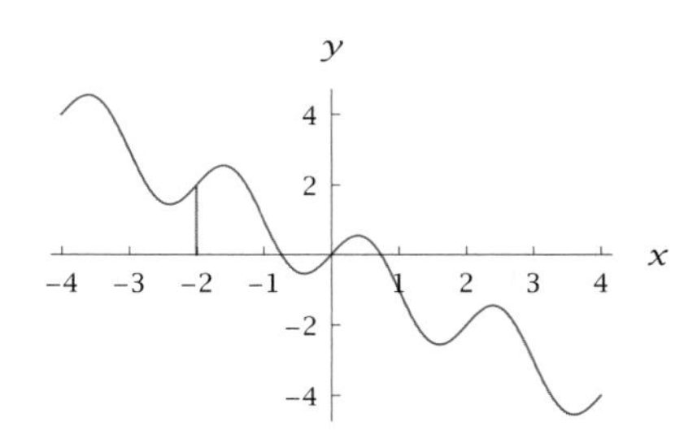

Then draw a horizontal line from where the vertical line intersects the graph to the y-axis:

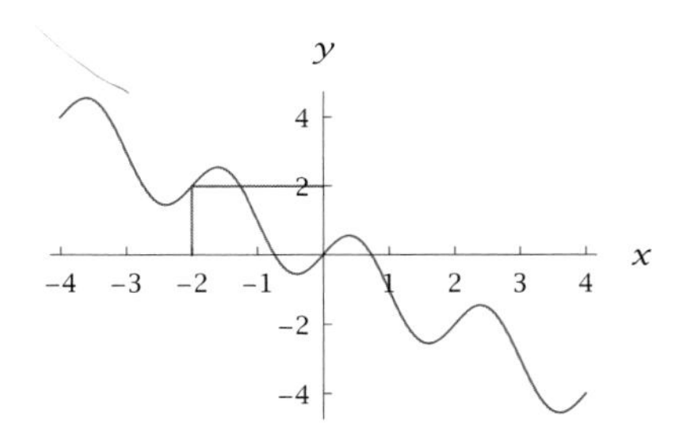

The intersection of the horizontal line with the y-axis gives the value of $g(-2)$. Thus we see that $g(-2) \approx 2$.

39. Estimate the value of $g(2)$.

SOLUTION To estimate the value of $g(2)$, draw a vertical line from the point 2 on the x-axis to the graph, as shown below:

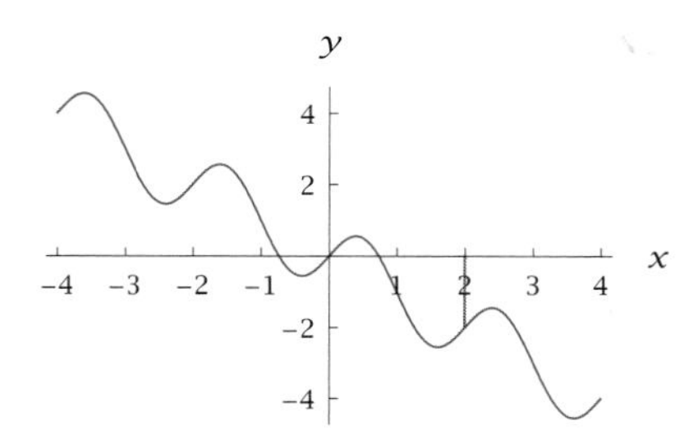

Then draw a horizontal line from where the vertical line intersects the graph to the y-axis:

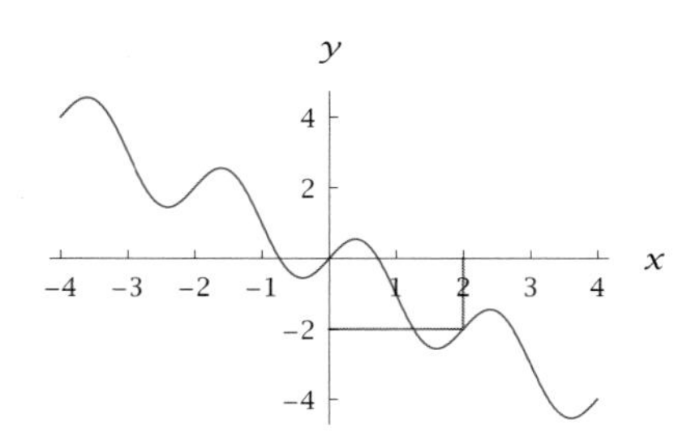

The intersection of the horizontal line with the y-axis gives the value of $g(2)$. Thus $g(2) \approx -2$.

41. Estimate the value of $g(2.5)$.

SOLUTION To estimate the value of $g(2.5)$, draw a vertical line from the point 2.5 on the x-axis to the graph, as shown below:

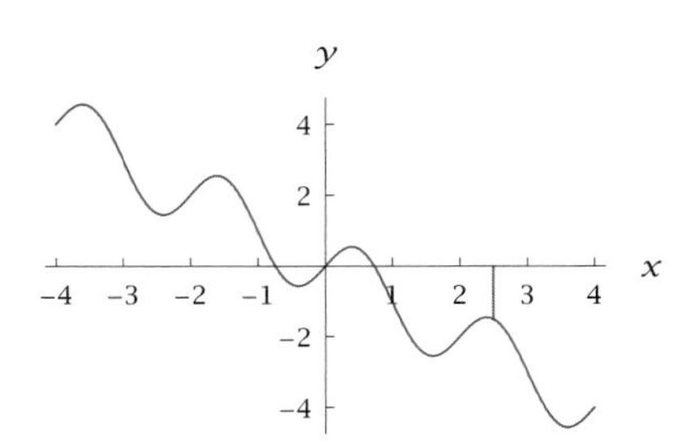

Then draw a horizontal line from where the vertical line intersects the graph to the y-axis:

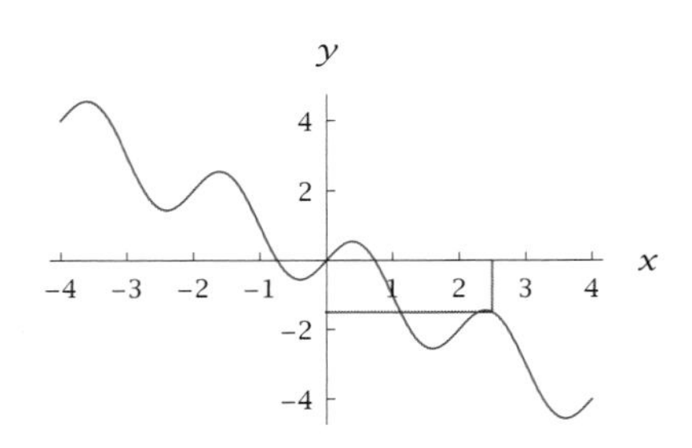

The intersection of the horizontal line with the y-axis gives the value of $g(2.5)$. Thus we see that $g(2.5) \approx -1.5$.

43. Estimate a number b such that $g(b) = 3.5$.

SOLUTION Draw the horizontal line $y = 3.5$, as shown below:

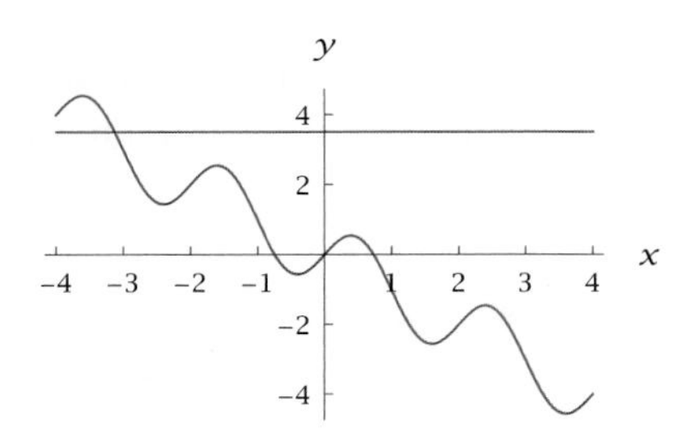

Then draw a vertical line from where this horizontal line intersects the graph to the x-axis:

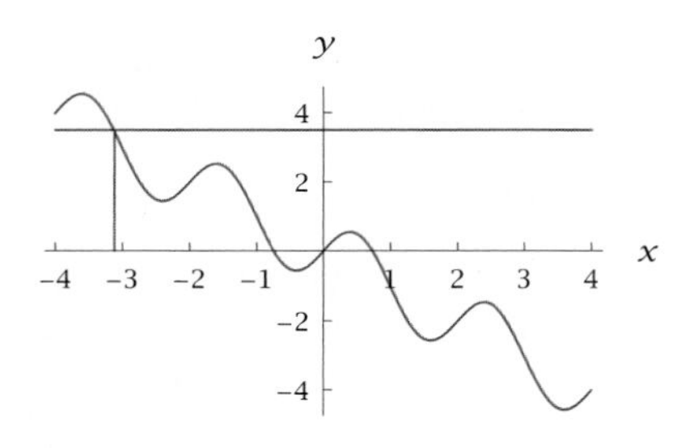

The intersection of the vertical line with the x-axis gives the value of b such that $g(b) = 3.5$. Thus we see that $b \approx -3.1$.

45. How many values of x satisfy the equation $g(x) = -2$?

SOLUTION Draw the horizontal line $y = -2$, as shown here. This horizontal line intersects the graph in three points. Thus there exist three values of x such that $g(x) = -2$.

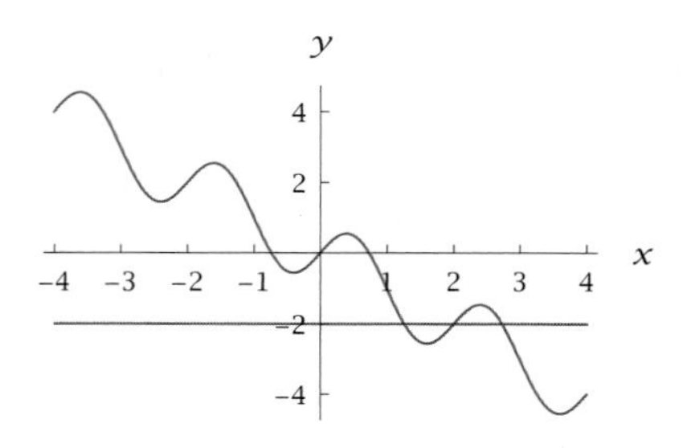

For Exercises 47–54, assume that f is the function defined by

$$f(x) = \begin{cases} 2x + 9 & \textit{if } x < 0 \\ 3x - 10 & \textit{if } x \geq 0. \end{cases}$$

47. Evaluate $f(1)$.

SOLUTION Because $1 \geq 0$, we have

$$f(1) = 3 \cdot 1 - 10 = -7.$$

49. Evaluate $f(-3)$.

SOLUTION Because $-3 < 0$, we have

$$f(-3) = 2(-3) + 9 = 3.$$

51. Evaluate $f(|x| + 1)$.

SOLUTION Because $|x| + 1 \geq 1 > 0$, we have

$$f(|x| + 1) = 3(|x| + 1) - 10 = 3|x| - 7.$$

53. Find two different values of x such that $f(x) = 0$.

SOLUTION If $x < 0$, then $f(x) = 2x + 9$. We want to find x such that $f(x) = 0$, which means that we need to solve the equation $2x + 9 = 0$ and hope that the solution satisfies $x < 0$. Subtracting 9 from both sides of $2x + 9 = 0$ and then dividing both sides by 2 gives $x = -\frac{9}{2}$. This value of x satisfies the inequality $x < 0$, and we do indeed have $f(-\frac{9}{2}) = 0$.

If $x \geq 0$, then $f(x) = 3x - 10$. We want to find x such that $f(x) = 0$, which means that we need to solve the equation $3x - 10 = 0$ and hope that the solution satisfies $x \geq 0$. Adding 10 to both sides of $3x - 10 = 0$ and then dividing both sides by 3 gives $x = \frac{10}{3}$. This value of x satisfies the inequality $x \geq 0$, and we do indeed have $f(\frac{10}{3}) = 0$.

For Exercises 55–58, find a number b such that the function f equals the function g.

55. The function f has domain the set of positive numbers and is defined by $f(x) = 5x^2 - 7$; the function g has domain (b, ∞) and is defined by $g(x) = 5x^2 - 7$.

SOLUTION For two functions to be equal, they must at least have the same domain. Because the domain of f is the set of positive numbers, which equals the interval $(0, \infty)$, we must have $b = 0$.

57. Both f and g have domain $\{3, 5\}$, with f defined on this domain by the formula $f(x) = x^2 - 3$ and g defined on this domain by the formula $g(x) = \frac{18}{x} + b(x - 3)$.

SOLUTION Note that

$$f(3) = 3^2 - 3 = 6 \quad \text{and} \quad f(5) = 5^2 - 3 = 22.$$

Also,

$$g(3) = \tfrac{18}{3} + b(3 - 3) = 6 \quad \text{and} \quad g(5) = \tfrac{18}{5} + 2b.$$

Thus regardless of the choice of b, we have $f(3) = g(3)$. To make the function f equal the function g, we must also have $f(5) = g(5)$, which means that we must have

$$22 = \tfrac{18}{5} + 2b.$$

Solving this equation for b, we get $b = \frac{46}{5}$.

For Exercises 59–64, a formula has been given defining a function f but no domain has been specified. Find the domain of each function f, assuming that the domain is the set of real numbers for which the formula makes sense and produces a real number.

59. $f(x) = \frac{2x+1}{3x-4}$

SOLUTION The formula above does not make sense when $3x - 4 = 0$, which would lead to division by 0. The equation $3x - 4 = 0$ is equivalent to $x = \frac{4}{3}$. Thus the domain of f is the set of real numbers not equal to $\frac{4}{3}$. In other words, the domain of f equals $\{x : x \neq \frac{4}{3}\}$, which could also be written as $(-\infty, \frac{4}{3}) \cup (\frac{4}{3}, \infty)$.

61. $f(x) = \frac{\sqrt{x-5}}{x-7}$

SOLUTION The formula above does not make sense when $x < 5$ because we cannot take the square root of a negative number. The formula above also does not make sense when $x = 7$, which would lead to division by 0. Thus the domain of f is the set of real numbers greater than or equal to 5 and not equal to 7. In other words, the domain of f equals $\{x : x \geq 5 \text{ and } x \neq 7\}$, which could also be written as $[5, 7) \cup (7, \infty)$.

63. $f(x) = \sqrt{|x - 6| - 1}$

SOLUTION Because we cannot take the square root of a negative number, we must have $|x - 6| - 1 \geq 0$. This inequality is equivalent to $|x - 6| \geq 1$, which means that $x - 6 \geq 1$ or $x - 6 \leq -1$. Adding 6 to both sides of these inequalities, we see that the formula above makes sense only when $x \geq 7$ or $x \leq 5$. In other words, the domain of f equals $\{x : x \leq 5 \text{ or } x \geq 7\}$, which could also be written as $(-\infty, 5] \cup [7, \infty)$.

For Exercises 65–70, suppose h is defined by

$$h(t) = |t| + 1.$$

65. What is the range of h if the domain of h is the interval $(1, 4]$?

SOLUTION For each number t in the interval $(1, 4]$, we have $h(t) = t + 1$. Thus the range of h is obtained by adding 1 to each number in the interval $(1, 4]$. This implies that the range of h is the interval $(2, 5]$.

67. What is the range of h if the domain of h is the interval $[-3, 5]$?

SOLUTION For each number t in the interval $[-3, 0)$, we have $h(t) = -t + 1$, and for each number t in the interval $[0, 5]$ we have $h(t) = t + 1$. Thus the range of h consists of the numbers obtained by multiplying each number in the interval $[-3, 0)$ by -1 and then adding 1 (this produces the interval $(1, 4]$), along with the numbers obtained by adding 1 to each number in the interval $[0, 5]$ (this produces the interval $[1, 6]$). This implies that the range of h is the interval $[1, 6]$.

69. What is the range of h if the domain of h is the set of positive numbers?

SOLUTION For each positive number t we have $h(t) = t + 1$. Thus the range of h is the set obtained by adding 1 to each positive number. Hence the range of h is the interval $(1, \infty)$.

71. Sketch the graph of the function f whose domain is the set of five numbers $\{-2, -1, 0, 1, 2\}$ and whose values are defined by the following table:

x	$f(x)$
-2	1
-1	3
0	-1
1	-2
2	3

SOLUTION

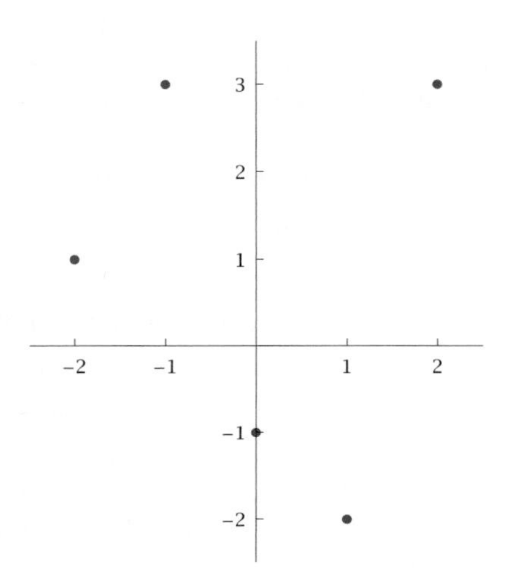

For Exercises 73–78, assume that g and h are the functions completely defined by the tables below:

x	$g(x)$
-3	-1
-1	1
1	2.5
3	-2

x	$h(x)$
-4	2
-2	-3
2	-1.5
3	1

73. What is the domain of g?

SOLUTION The domain of g is the set of numbers in the first column of the table defining g. Thus the domain of g is the set $\{-3, -1, 1, 3\}$.

75. What is the range of g?

SOLUTION The range of g is the set of numbers that appear in the second column of the table defining g. Thus the range of g is the set $\{-1, 1, 2.5, -2\}$.

77. Draw the graph of g.

SOLUTION The graph of g consists of the four points with coordinates $(-3, -1)$, $(-1, 1)$, $(1, 2.5)$, $(3, -2)$, as shown below:

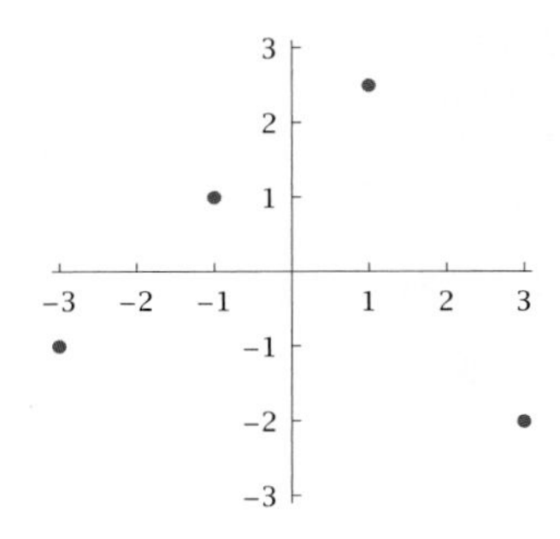

For Exercises 79–86, assume that f and g are functions completely defined by the following tables:

x	$f(x)$
3	13
4	-5
6	$\frac{3}{5}$
7.3	-5

x	$g(x)$
3	3
8	$\sqrt{7}$
8.4	$\sqrt{7}$
12.1	$-\frac{2}{7}$

79. Evaluate $f(6)$.

SOLUTION Looking at the table, we see that $f(6) = \frac{3}{5}$.

81. What is the domain of f?

SOLUTION The domain of f is the set of numbers in the first column of the table defining f. Thus the domain of f is the set $\{3, 4, 6, 7.3\}$.

83. What is the range of f?

SOLUTION The range of f is the set of numbers that appear in the second column of the table defining f. Numbers that appear more than once in the second column need to be listed only once when finding the range. Thus the range of f is the set $\{13, -5, \frac{3}{5}\}$.

85. Find two different values of x such that $f(x) = -5$.

SOLUTION Looking at the table, we see that $f(4) = -5$ and $f(7.3) = -5$.

87. Find all functions (displayed as tables) whose domain is the set $\{2, 9\}$ and whose range is the set $\{4, 6\}$.

SOLUTION Because we seek functions f whose domain is the set $\{2, 9\}$, the first column of the table for any such function must have 2 appear once and must have 9 appear once. In other words, the table must start like this:

x	$f(x)$
2	
9	

or this

x	$f(x)$
9	
2	

The order of the rows in a table that defines a function does not matter. For convenience, we choose the first possibility above.

Because the range must be the set $\{4, 6\}$, the second column must contain 4 and 6. There are only two slots in which to put these numbers in the first table above, and thus each one must appear exactly once in the second column. Thus there are only two functions whose domain is the set $\{2, 9\}$ and whose range is the set $\{4, 6\}$; these functions are given by the following two tables:

x	$f(x)$
2	4
9	6

x	$f(x)$
2	6
9	4

The first function above is the function f defined by $f(2) = 4$ and $f(9) = 6$; the second function above is the function f defined by $f(2) = 6$ and $f(9) = 4$.

89. Find all functions (displayed as tables) whose domain is $\{1, 2, 4\}$ and whose range is $\{-2, 1, \sqrt{3}\}$.

SOLUTION Because we seek functions f whose domain is $\{1, 2, 4\}$, the first column of the table for any such function must have 1 appear once, must have 2 appear once, and must have 4 appear once. The order of the rows in a table that defines a function does not matter. For convenience, we put the first column in numerical order 1, 2, 4.

Because the range must be $\{-2, 1, \sqrt{3}\}$, the second column must contain -2, 1, and $\sqrt{3}$. There are only three slots in which to put these three numbers, and thus each one must appear exactly once in the second column. There are six ways in which these three numbers can be ordered. Thus the six functions whose domain is $\{1, 2, 4\}$ and whose range is $\{-2, 1, \sqrt{3}\}$ are given by the following tables:

x	$f(x)$
1	-2
2	1
4	$\sqrt{3}$

x	$f(x)$
1	-2
2	$\sqrt{3}$
4	1

x	$f(x)$
1	1
2	-2
4	$\sqrt{3}$

x	$f(x)$
1	1
2	$\sqrt{3}$
4	-2

x	$f(x)$
1	$\sqrt{3}$
2	-2
4	1

x	$f(x)$
1	$\sqrt{3}$
2	1
4	-2

91. Find all functions (displayed as tables) whose domain is $\{3, 5, 9\}$ and whose range is $\{2, 4\}$.

SOLUTION Because we seek functions f whose domain is $\{3, 5, 9\}$, the first column of the table for any such function must have 3 appear once, must have 5 appear once, and must have 9 appear once. The order of the rows in a table that defines a function does not matter. For convenience, we put the first column in numerical order 3, 5, 9.

Because the range must be $\{2, 4\}$, the second column must contain 2 and 4. There are three slots in which to put these three numbers, and thus one of them must be repeated. There are six ways to do this. Thus the six functions whose domain is $\{3, 5, 9\}$ and whose range is $\{2, 4\}$ are given by the following tables:

x	$f(x)$
3	2
5	2
9	4

x	$f(x)$
3	2
5	4
9	2

x	$f(x)$
3	4
5	2
9	2

x	$f(x)$
3	4
5	4
9	2

x	$f(x)$
3	4
5	2
9	4

x	$f(x)$
3	2
5	4
9	4

3.2 Function Transformations and Graphs

LEARNING OBJECTIVES

By the end of this section you should be able to

- work with the vertical function transformations, which shift the graph up or down, stretch the graph vertically, or flip the graph across the horizontal axis;
- work with the horizontal function transformations, which shift the graph left or right, stretch the graph horizontally, or flip the graph across the vertical axis;
- combine multiple function transformations;
- determine the domain, range, and graph of a transformed function;
- recognize even functions and odd functions.

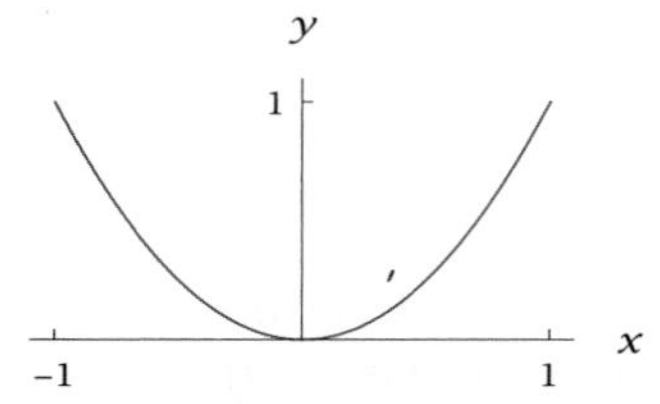

The graph of $f(x) = x^2$, with domain $[-1, 1]$. The range of f is $[0, 1]$.

In this section we investigate various transformations of functions and learn the effect of such transformations on the domain, range, and graph of a function. To illustrate these ideas, throughout this section we will use the function f defined by $f(x) = x^2$, with the domain of f being the interval $[-1, 1]$. Thus the graph of f is part of a familiar parabola.

Vertical Transformations: Shifting, Stretching, and Flipping

This subsection focuses on vertical function transformations, which change the vertical shape or location of the graph of a function. Because vertical function transformations affect the graph only vertically, the vertical function transformations do not change the domain of the function.

We begin with an example showing the procedure for shifting the graph of a function up.

EXAMPLE 1

Define a function g by

$$g(x) = f(x) + 1,$$

where f is the function defined by $f(x) = x^2$, with the domain of f the interval $[-1, 1]$.

(a) Find the domain of g. (b) Find the range of g. (c) Sketch the graph of g.

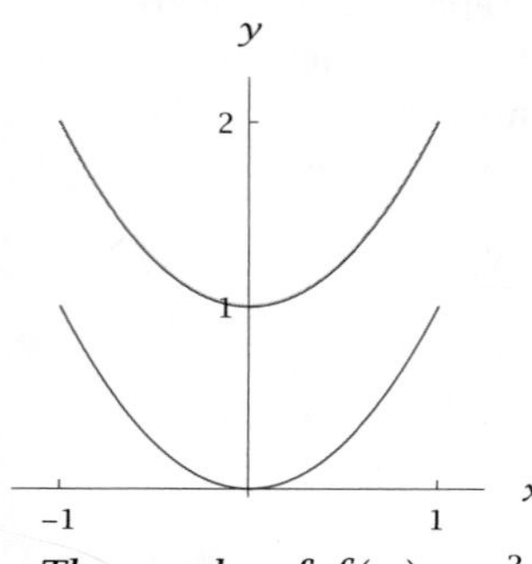

The graphs of $f(x) = x^2$ (blue) and $g(x) = x^2 + 1$ (red), each with domain $[-1, 1]$.

SOLUTION

(a) The formula defining g shows that $g(x)$ is defined precisely when $f(x)$ is defined. In other words, the domain of g equals the domain of f. Thus the domain of g is the interval $[-1, 1]$.

(b) Recall that the range of g is the set of values taken on by $g(x)$ as x varies over the domain of g. Because $g(x)$ equals $f(x) + 1$, we see that the range of g is obtained by adding 1 to each number in the range of f. Thus the range of g is the interval $[1, 2]$.

(c) A typical point on the graph of f has the form (x, x^2), where x is in the interval $[-1, 1]$. Because $g(x) = x^2 + 1$, a typical point on the graph of g has the form $(x, x^2 + 1)$, where x is in the interval $[-1, 1]$. In other words, each point on the graph of g is obtained by adding 1 to the second coordinate of a point on the graph of f. Thus the graph of g is obtained by shifting the graph of f up 1 unit, as shown above.

Shifting the graph of a function down follows a similar pattern, with a minus sign replacing the plus sign, as shown in the following example.

EXAMPLE 2

Define a function h by

$$h(x) = f(x) - 1,$$

where f is the function defined by $f(x) = x^2$, with the domain of f the interval $[-1, 1]$.

(a) Find the domain of h. (b) Find the range of h. (c) Sketch the graph of h.

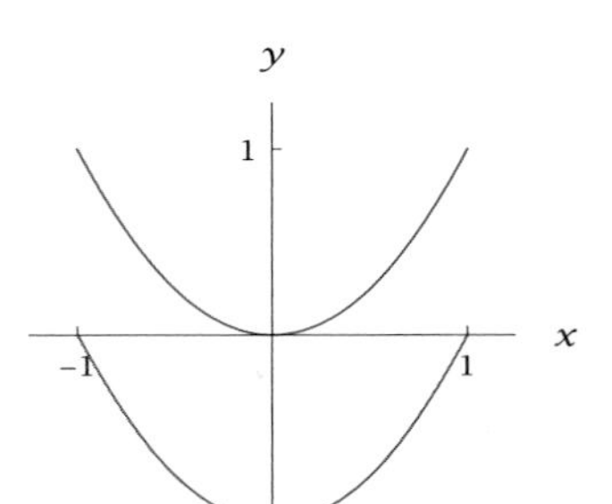

The graphs of $f(x) = x^2$ (blue) and $h(x) = x^2 - 1$ (red), each with domain $[-1, 1]$.

SOLUTION

(a) The formula above shows that $h(x)$ is defined precisely when $f(x)$ is defined. In other words, the domain of h equals the domain of f. Thus the domain of h is the interval $[-1, 1]$.

(b) Because $h(x)$ equals $f(x) - 1$, we see that the range of h is obtained by subtracting 1 from each number in the range of f. Thus the range of h is the interval $[-1, 0]$.

(c) Because $h(x) = x^2 - 1$, a typical point on the graph of h has the form $(x, x^2 - 1)$, where x is in the interval $[-1, 1]$. Thus the graph of h is obtained by shifting the graph of f down 1 unit, as shown here.

We could have used any positive number a instead of 1 in these examples when defining $g(x)$ as $f(x) + 1$ and defining $h(x)$ as $f(x) - 1$. Similarly, there is nothing special about the particular function f that we used. Thus the following results hold in general:

Shifting a graph up or down

Suppose f is a function and $a > 0$. Define functions g and h by

$$g(x) = f(x) + a \quad \text{and} \quad h(x) = f(x) - a.$$

Then

- the graph of g is obtained by shifting the graph of f up a units;
- the graph of h is obtained by shifting the graph of f down a units.

Instead of memorizing the conclusions in all the result boxes in this section, try to understand how these conclusions arise. Then you can figure out what you need depending on the problem at hand.

The procedure for stretching the graph of a function vertically is illustrated by the following example:

EXAMPLE 3

Define functions g and h by

$$g(x) = 2f(x) \quad \text{and} \quad h(x) = \tfrac{1}{2}f(x),$$

where f is the function defined by $f(x) = x^2$, with the domain of f the interval $[-1, 1]$.

(a) Find the domain of g and the domain of h.

(b) Find the range of g.

(c) Find the range of h.

(d) Sketch the graphs of g and h.

SOLUTION

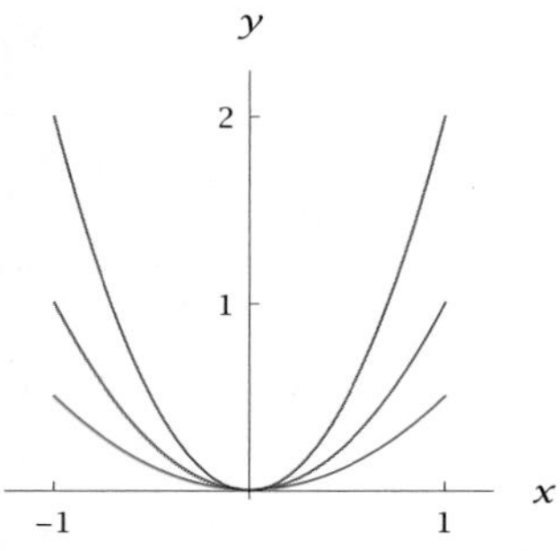

The graphs of $f(x) = x^2$ (blue) and $g(x) = 2x^2$ (red) and $h(x) = \frac{1}{2}x^2$ (green), each with domain $[-1, 1]$.

(a) The formulas defining g and h show that $g(x)$ and $h(x)$ are defined precisely when $f(x)$ is defined. In other words, the domain of g and the domain of h both equal the domain of f. Thus the domain of g and the domain of h both equal the interval $[-1, 1]$.

(b) Because $g(x)$ equals $2f(x)$, we see that the range of g is obtained by multiplying each number in the range of f by 2. Thus the range of g is the interval $[0, 2]$.

(c) Because $h(x)$ equals $\frac{1}{2}f(x)$, we see that the range of h is obtained by multiplying each number in the range of f by $\frac{1}{2}$. Thus the range of h is the interval $[0, \frac{1}{2}]$.

(d) For each x in the interval $[-1, 1]$, the point $(x, 2x^2)$ is on the graph of g and the point $(x, \frac{1}{2}x^2)$ is on the graph of h. Thus the graph of g is obtained by vertically stretching the graph of f by a factor of 2, and the graph of h is obtained by vertically stretching the graph of f by a factor of $\frac{1}{2}$, as shown here.

Perhaps the word "shrink" would be more appropriate here.

In the last part of the example above, we noted that the graph of h is obtained by vertically stretching the graph of f by a factor of $\frac{1}{2}$. This terminology may seem a bit strange because the word "stretch" often has the connotation of something getting larger. However, we will find it convenient to use the word "stretch" in the wider sense of multiplying by some positive number, which might be less than 1.

We could have used any positive number c instead of 2 or $\frac{1}{2}$ in the example above. Similarly, there is nothing special about the particular function f that we used. Thus the following result holds in general:

Stretching a graph vertically

Suppose f is a function and $c > 0$. Define a function g by

$$g(x) = cf(x).$$

Then the graph of g is obtained by vertically stretching the graph of f by a factor of c.

The procedure for flipping the graph of a function across the horizontal axis is illustrated by the following example. Flipping a graph across the horizontal axis changes only the vertical aspect of the graph. Thus flipping across the horizontal axis is indeed a vertical function transformation.

EXAMPLE 4

Define a function g by

$$g(x) = -f(x),$$

where f is the function defined by $f(x) = x^2$, with the domain of f the interval $[-1, 1]$.

(a) Find the domain of g. (b) Find the range of g. (c) Sketch the graph of g.

SOLUTION

(a) The formula defining g shows that $g(x)$ is defined precisely when $f(x)$ is defined. In other words, the domain of g equals the domain of f. Thus the domain of g is the interval $[-1, 1]$.

(b) Because $g(x)$ equals $-f(x)$, we see that the values taken on by g are the negatives of the values taken on by f. Thus the range of g is the interval $[-1, 0]$.

(c) Note that $g(x) = -x^2$ for each x in the interval $[-1, 1]$. For each point (x, x^2) on the graph of f, the point $(x, -x^2)$ is on the graph of g. Thus the graph of g is obtained by flipping the graph of f across the horizontal axis, as shown here.

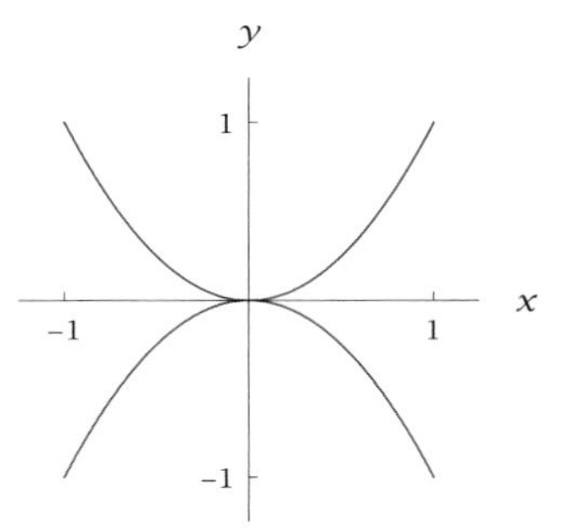

The graphs of $f(x) = x^2$ (blue) and $g(x) = -x^2$ (red), each with domain $[-1, 1]$.

The following result holds for every function f:

Flipping a graph across the horizontal axis

Suppose f is a function. Define a function g by

$$g(x) = -f(x).$$

Then the graph of g is obtained by flipping the graph of f across the horizontal axis.

Many books use the word reflecting *instead of* flipping. *However,* flipping *seems to be a simpler and a more accurate description of how the red graph above is obtained from the blue graph.*

Horizontal Transformations: Shifting, Stretching, Flipping

Now we focus on horizontal function transformations, which change the horizontal shape or location of the graph of a function. Because horizontal function transformations affect the graph only horizontally, the horizontal function transformations do not change the range of the function.

We begin with an example showing the procedure for shifting the graph of a function to the left.

Vertical transformations work pretty much as you would expect. As you will soon see, the actions of horizontal transformation are less intuitive.

EXAMPLE 5

Define a function g by

$$g(x) = f(x + 1),$$

where f is the function defined by $f(x) = x^2$, with the domain of f the interval $[-1, 1]$.

(a) Find the domain of g. (b) Find the range of g. (c) Sketch the graph of g.

SOLUTION

(a) The formula defining g shows that $g(x)$ is defined precisely when $f(x+1)$ is defined, which means that $x+1$ must be in the interval $[-1,1]$, which means that x must be in the interval $[-2,0]$. Thus the domain of g is the interval $[-2,0]$.

(b) Because $g(x)$ equals $f(x+1)$, we see that the values taken on by g are the same as the values taken on by f. Thus the range of g equals the range of f, which is the interval $[0,1]$.

(c) Note that $g(x) = (x+1)^2$ for each x in the interval $[-2,0]$. For each point (x, x^2) on the graph of f, the point $(x-1, x^2)$ is on the graph of g (because $g(x-1) = x^2$). Thus the graph of g is obtained by shifting the graph of f left 1 unit:

Example 1 with $g(x) = x^2 + 1$ differs from this example with $g(x) = (x+1)^2$. In the earlier example, the graph was shifted up; in this example, the graph is shifted left. The domains and ranges also behave differently in these two examples.

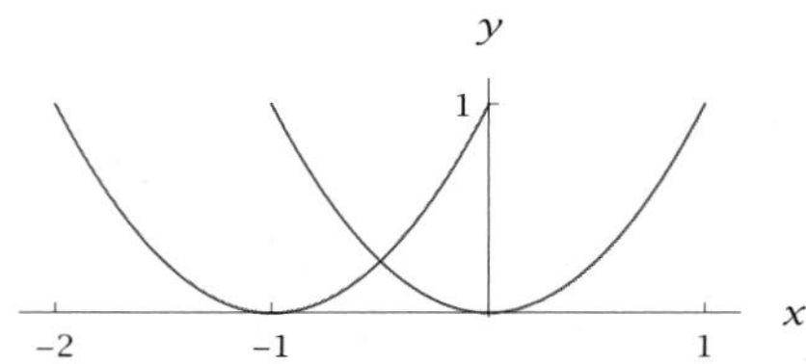

The graphs of $f(x) = x^2$ (blue, with domain $[-1,1]$) and $g(x) = (x+1)^2$ (red, with domain $[-2,0]$). The graph of g is obtained by shifting the graph of f left 1 unit.

Suppose we define a function h by

$$h(x) = f(x-1),$$

where f is again the function defined by $f(x) = x^2$, with the domain of f the interval $[-1,1]$. Then everything works as in the example above, except that the domain and graph of h are obtained by shifting the domain and graph of f right 1 unit (instead of left 1 unit as in the example above).

More generally, we could have used any positive number b instead of 1 in these examples when defining $g(x)$ as $f(x+1)$ and defining $h(x)$ as $f(x-1)$. Similarly, there is nothing special about the particular function f that we used. Thus the following results hold in general:

Shifting a graph left or right

Suppose f is a function and $b > 0$. Define functions g and h by

$$g(x) = f(x+b) \quad \text{and} \quad h(x) = f(x-b).$$

Then

- the graph of g is obtained by shifting the graph of f left b units;
- the graph of h is obtained by shifting the graph of f right b units.

The next example shows the procedure for horizontally stretching the graph of a function.

EXAMPLE 6

Define functions g and h by

$$g(x) = f(2x) \quad \text{and} \quad h(x) = f(\tfrac{1}{2}x),$$

where f is the function defined by $f(x) = x^2$, with the domain of f the interval $[-1, 1]$.

(a) Find the domain of g.

(b) Find the domain of h.

(c) Find the range of g and the range of h.

(d) Sketch the graphs of g and h.

SOLUTION

(a) The formula defining g shows that $g(x)$ is defined precisely when $f(2x)$ is defined, which means that $2x$ must be in the interval $[-1, 1]$, which means that x must be in the interval $[-\frac{1}{2}, \frac{1}{2}]$. Thus the domain of g is the interval $[-\frac{1}{2}, \frac{1}{2}]$.

(b) The formula defining h shows that $h(x)$ is defined precisely when $f(\frac{1}{2}x)$ is defined, which means that $\frac{1}{2}x$ must be in the interval $[-1, 1]$, which means that x must be in the interval $[-2, 2]$. Thus the domain of h is the interval $[-2, 2]$.

(c) The formulas defining g and h show that the values taken on by g and h are the same as the values taken on by f. Thus the range of g and the range of h both equal the range of f, which is the interval $[0, 1]$.

(d) For each point (x, x^2) on the graph of f, the point $(\frac{x}{2}, x^2)$ is on the graph of g (because $g(\frac{x}{2}) = x^2$) and the point $(2x, x^2)$ is on the graph of h (because $h(2x) = x^2$) . Thus the graph of g is obtained by horizontally stretching the graph of f by a factor of $\frac{1}{2}$, and the graph of h is obtained by horizontally stretching the graph of f by a factor of 2, as shown here.

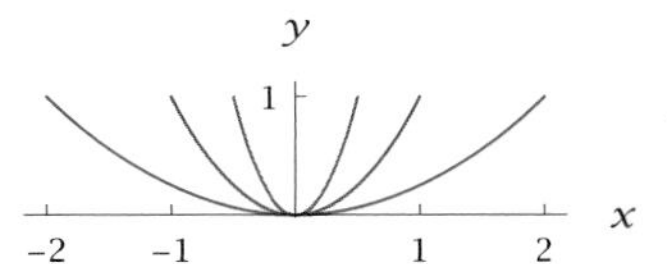

The graphs of $f(x) = x^2$ (blue, with domain $[-1, 1]$) and $g(x) = (2x)^2$ (red, with domain $[-\frac{1}{2}, \frac{1}{2}]$) and $h(x) = (\frac{1}{2}x)^2$ (green, with domain $[-2, 2]$).

We could have used any positive number c instead of 2 or $\frac{1}{2}$ when defining $g(x)$ as $f(2x)$ and defining $h(x)$ as $f(\frac{1}{2}x)$ in the example above. Similarly, there is nothing special about the particular function f that we used. Thus the following result holds in general:

Stretching a graph horizontally

Suppose f is a function and $c > 0$. Define a function g by

$$g(x) = f(cx).$$

Then the graph of g is obtained by horizontally stretching the graph of f by a factor of $\frac{1}{c}$.

The procedure for flipping the graph of a function across the vertical axis is illustrated by the following example. Flipping a graph across the vertical axis changes only the horizontal aspect of the graph. Thus flipping across the vertical axis is indeed a horizontal function transformation.

To show the ideas more clearly, in the next example we change the domain of f to the interval $[\frac{1}{2}, 1]$.

EXAMPLE 7

Define a function g by

$$g(x) = f(-x),$$

where f is the function defined by $f(x) = x^2$, with the domain of f the interval $[\frac{1}{2}, 1]$.

(a) Find the domain of g. (b) Find the range of g. (c) Sketch the graph of g.

SOLUTION

(a) The formula defining g shows that $g(x)$ is defined precisely when $f(-x)$ is defined, which means that $-x$ must be in the interval $[\frac{1}{2}, 1]$, which means that x must be in the interval $[-1, -\frac{1}{2}]$. Thus the domain of g is the interval $[-1, -\frac{1}{2}]$.

(b) Because $g(x)$ equals $f(-x)$, we see that the values taken on by g are the same as the values taken on by f. Thus the range of g equals the range of f, which is the interval $[\frac{1}{4}, 1]$.

(c) Note that $g(x) = (-x)^2 = x^2$ for each x in the interval $[-1, -\frac{1}{2}]$. For each point (x, x^2) on the graph of f, the point $(-x, x^2)$ is on the graph of g (because $g(-x) = x^2$). Thus the graph of g is obtained by flipping the graph of f across the vertical axis:

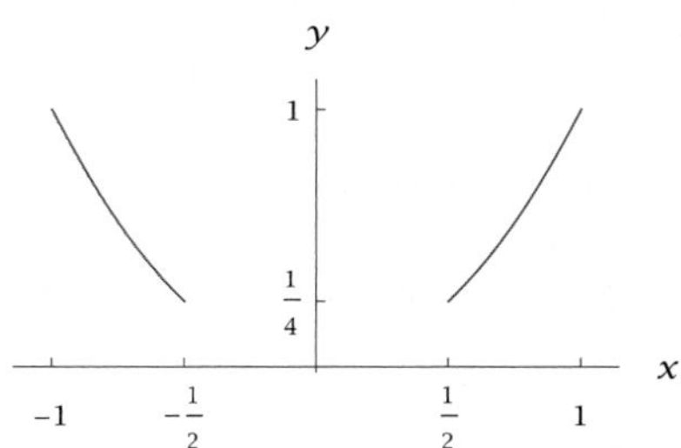

The graphs of $f(x) = x^2$ (blue, with domain $[\frac{1}{2}, 1]$) and $g(x) = (-x)^2 = x^2$ (red, with domain $[-1, -\frac{1}{2}]$). The graph of g is obtained by flipping the graph of f across the vertical axis.

The following result holds for every function f:

The domain of g is obtained by multiplying each number in the domain of f by -1.

Flipping a graph across the vertical axis

Suppose f is a function. Define a function g by

$$g(x) = f(-x).$$

Then the graph of g is obtained by flipping the graph of f across the vertical axis.

Combinations of Vertical Function Transformations

When dealing with combinations of vertical function transformations, the order in which those transformations are applied can be crucial. The following simple procedure provides a method for finding the graph when dealing with combinations of vertical function transformations.

Combinations of vertical function transformations

To obtain the graph of a function defined by combinations of vertical function transformations, apply the transformations in the same order as the corresponding operations when evaluating the function.

EXAMPLE 8

Define a function g by

$$g(x) = -2f(x) + 1,$$

where f is the function defined by $f(x) = x^2$, with the domain of f the interval $[-1, 1]$.

(a) List the order of the operations used to evaluate $g(x)$ after $f(x)$ has been evaluated.

(b) Find the domain of g.

(c) Find the range of g.

(d) Sketch the graph of g.

SOLUTION

(a) Because $g(x) = -2f(x) + 1$, operations should be done in the following order to evaluate $g(x)$:

1. Multiply $f(x)$ by 2.
2. Multiply the number obtained in the previous step by -1.
3. Add 1 to the number obtained in the previous step.

The order of steps 1 and 2 could be interchanged. However, the operation of adding 1 must be the last step.

Steps 1 and 2 above could be combined into a single step: multiply $f(x)$ by -2. However, we have separated this operation into steps 1 and 2 because the basic function transformation that stretches a graph vertically involves multiplication by a positive number.

(b) The formula defining g shows that $g(x)$ is defined precisely when $f(x)$ is defined. In other words, the domain of g equals the domain of f. Thus the domain of g is the interval $[-1, 1]$.

(c) The range of g is obtained by applying the operations from the answer to part (a), in the same order, to the range of f, which is the interval $[0, 1]$:

1. Multiplying each number in $[0, 1]$ by 2 gives the interval $[0, 2]$.
2. Multiplying each number in $[0, 2]$ by -1 gives the interval $[-2, 0]$.
3. Adding 1 to each number in $[-2, 0]$ gives the interval $[-1, 1]$, which is the range of g.

(d) Applying function transformations in the same order as in the answer to part (a), we see that the graph of g is obtained from the graph of f by vertically stretching the graph of f by a factor of 2, then flipping the resulting graph across the horizontal axis, then shifting the resulting graph up 1 unit, producing the graph shown here.

The operations listed in the solutions to part (a) of this example and the next example are the same, differing only in the order. However, the different order produces different graphs.

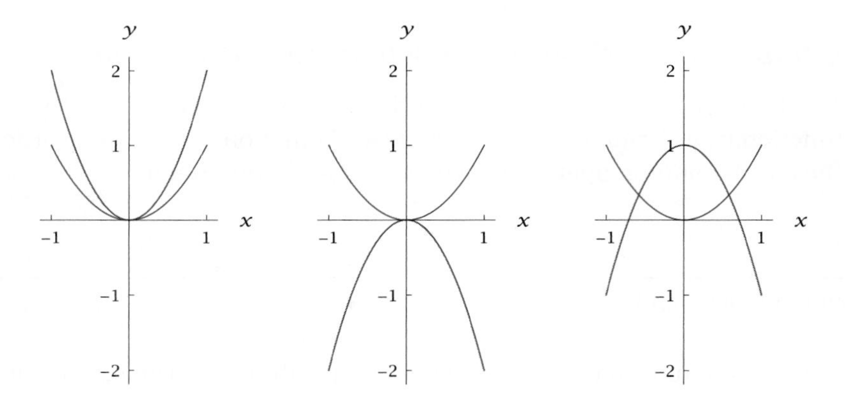

The graphs of $f(x) = x^2$ (blue) and $2f(x)$ (red, left) and $-2f(x)$ (red, center) and $-2f(x) + 1$ (red, right), each with domain $[-1, 1]$.

Comparing the example above to the next example shows the importance of applying the operations in the appropriate order.

EXAMPLE 9

Define a function h by

$$h(x) = -2(f(x) + 1),$$

where f is the function defined by $f(x) = x^2$, with the domain of f the interval $[-1, 1]$.

(a) List the order of the operations used to evaluate $h(x)$ after $f(x)$ has been evaluated.

(b) Find the domain of h.

(c) Find the range of h.

(d) Sketch the graph of h.

SOLUTION

(a) Because $h(x) = -2(f(x) + 1)$, operations should be done in the following order to evaluate $h(x)$:

1. Add 1 to $f(x)$.
2. Multiply the number obtained in the previous step by 2.
3. Multiply the number obtained in the previous step by -1.

(b) The formula defining h shows that $h(x)$ is defined precisely when $f(x)$ is defined. In other words, the domain of h equals the domain of f. Thus the domain of h is the interval $[-1, 1]$.

(c) The range of h is obtained by applying the operations from the answer to part (a), in the same order, to the range of f, which is the interval $[0, 1]$:

1. Adding 1 to each number in $[0, 1]$ gives the interval $[1, 2]$.
2. Multiplying each number in $[1, 2]$ by 2 gives the interval $[2, 4]$.
3. Multiplying each number in $[2, 4]$ by -1 gives the interval $[-4, -2]$, which is the range of h.

(d) Applying function transformations in the same order as in the answer to part (a), we see that the graph of g is obtained by shifting the graph of f up 1 unit, then vertically stretching the resulting graph by a factor of 2, then flipping the resulting graph across the horizontal axis, producing the graph shown here.

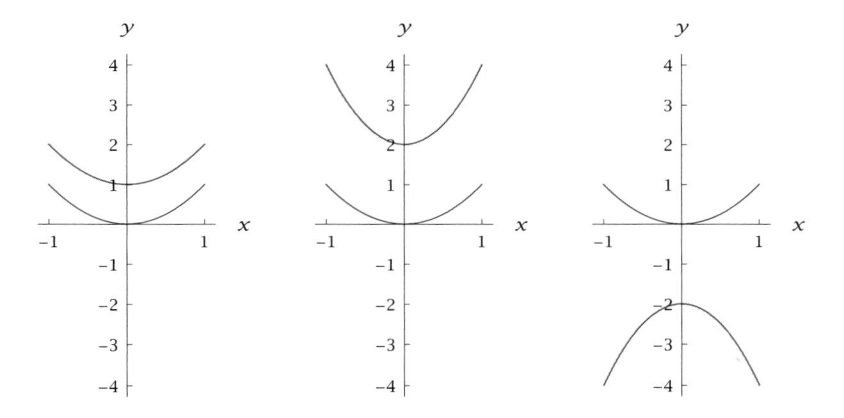

Notice how the graph of $-2(f(x) + 1)$ in this example differs from the graph of $-2f(x) + 1$ in the previous example.

The graphs of $f(x) = x^2$ (blue) and $f(x) + 1$ (red, left) and $2(f(x) + 1)$ (red, center) and $-2(f(x) + 1)$ (red, right), each with domain $[-1, 1]$.

When dealing with a combination of a vertical function transformation and a horizontal function transformation, the transformations can be applied in either order. For a combination of multiple vertical function transformations and a single horizontal function transformation, be sure to apply the vertical function transformations in the proper order; the horizontal function transformation can be applied either after or before the vertical function transformations.

Combinations of multiple horizontal function transformations, possibly combined with vertical function transformations, are more complicated. For those who are interested in these kinds of multiple function transformations, the worked-out solutions to some of the odd-numbered exercises in this section provide examples of the proper technique.

Even Functions

Suppose $f(x) = x^2$ for every real number x. Notice that

$$f(-x) = (-x)^2 = x^2 = f(x).$$

This property is sufficiently important that we give it a name.

Even functions

A function f is called **even** if

$$f(-x) = f(x)$$

for every x in the domain of f.

For the equation $f(-x) = f(x)$ to hold for every x in the domain of f, the expression $f(-x)$ must make sense. Thus $-x$ must be in the domain of f for every x in the domain of f. For example, a function whose domain is the interval $[-3, 5]$ cannot possibly be an even function, but a function whose domain is the interval $(-4, 4)$ may or may not be an even function.

As we have already observed, x^2 is an even function. Here is another simple example:

EXAMPLE 10

Show that the function f defined by $f(x) = |x|$ for every real number x is an even function.

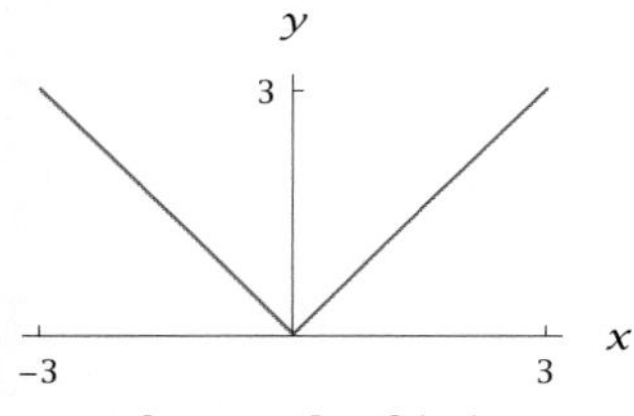

The graph of $|x|$ on the interval $[-3, 3]$.

SOLUTION This function is even because

$$f(-x) = |-x| = |x| = f(x)$$

for every real number x.

Suppose f is an even function. As we know, flipping the graph of f across the vertical axis gives the graph of the function h defined by $h(x) = f(-x)$. Because f is even, we actually have $h(x) = f(-x) = f(x)$, which implies that $h = f$. In other words, flipping the graph of f across the vertical axis gives us back the graph of f. Thus the graph of an even function is symmetric about the vertical axis. This symmetry can be seen, for example, in the graph shown above of $|x|$ on the interval $[-3, 3]$. Here is the statement of the result in general:

The graph of an even function

A function is even if and only if its graph is unchanged when flipped across the vertical axis.

Odd Functions

Consider now the function defined by $f(x) = x^3$ for every real number x. Notice that

$$f(-x) = (-x)^3 = -(x^3) = -f(x).$$

This property is sufficiently important that we also give it a name.

Odd functions

A function f is called **odd** if

$$f(-x) = -f(x)$$

for every x in the domain of f.

As was the case for even functions, for a function to be odd $-x$ must be in the domain of f for every x in the domain of f, because otherwise there is no possibility that the equation $f(-x) = -f(x)$ can hold for every x in the domain of f.

As we have already observed, x^3 is an odd function. Here is another simple example:

EXAMPLE 11

Show that the function f defined by $f(x) = \frac{1}{x}$ for every real number $x \neq 0$ is an odd function.

SOLUTION This function is odd because

$$f(-x) = \frac{1}{-x} = -\frac{1}{x} = -f(x)$$

for every real number $x \neq 0$.

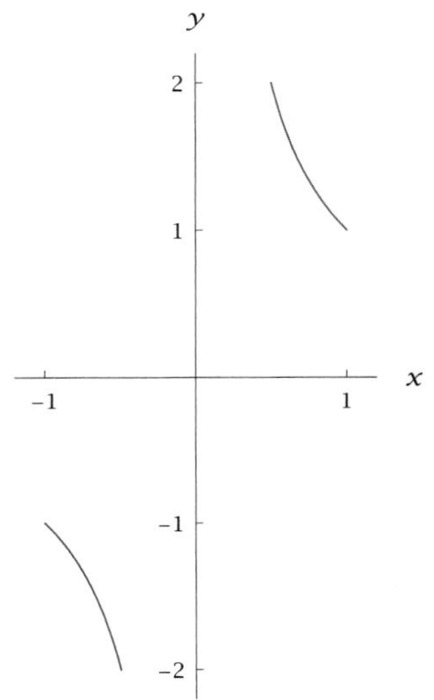

The graph of $\frac{1}{x}$ on $[-1, -\frac{1}{2}] \cup [\frac{1}{2}, 1]$.

Suppose f is an odd function. If x is a number in the domain of f, then $(x, f(x))$ is a point on the graph of f. Because $f(-x) = -f(x)$, the point $(-x, -f(x))$ also is on the graph of f. In other words, flipping a point $(x, f(x))$ on the graph of f across the origin gives a point $(-x, -f(x))$ that is also on the graph of f. Thus the graph of an odd function is symmetric about the origin. This symmetry can be seen, for example, in the graph shown here of $\frac{1}{x}$ on $[-1, -\frac{1}{2}] \cup [\frac{1}{2}, 1]$. Here is the statement of the result in general:

The graph of an odd function

A function is odd if and only if its graph is unchanged when flipped across the origin.

EXERCISES

For Exercises 1–14, assume that f is the function defined on the interval $[1,2]$ by the formula $f(x) = \frac{4}{x^2}$. Thus the domain of f is the interval $[1,2]$, the range of f is the interval $[1,4]$, and the graph of f is shown here.

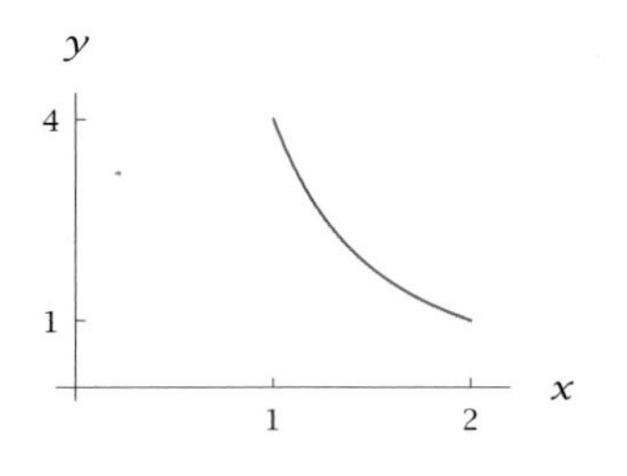

The graph of f.

For each function g described below:

- **(a)** ***Sketch the graph of g.***
- **(b)** ***Find the domain of g (the endpoints of this interval should be shown on the horizontal axis of your sketch of the graph of g).***
- **(c)** ***Give a formula for g.***
- **(d)** ***Find the range of g (the endpoints of this interval should be shown on the vertical axis of your sketch of the graph of g).***

1. The graph of g is obtained by shifting the graph of f up 1 unit.
2. The graph of g is obtained by shifting the graph of f up 3 units.
3. The graph of g is obtained by shifting the graph of f down 3 units.
4. The graph of g is obtained by shifting the graph of f down 2 units.
5. The graph of g is obtained by vertically stretching the graph of f by a factor of 2.
6. The graph of g is obtained by vertically stretching the graph of f by a factor of 3.
7. The graph of g is obtained by shifting the graph of f left 3 units.
8. The graph of g is obtained by shifting the graph of f left 4 units.
9. The graph of g is obtained by shifting the graph of f right 1 unit.
10. The graph of g is obtained by shifting the graph of f right 3 units.
11. The graph of g is obtained by horizontally stretching the graph of f by a factor of 2.
12. The graph of g is obtained by horizontally stretching the graph of f by a factor of $\frac{1}{2}$.
13. The graph of g is obtained by flipping the graph of f across the horizontal axis.
14. The graph of g is obtained by flipping the graph of f across the vertical axis.

For Exercises 15–50, assume that f is a function whose domain is the interval $[1,5]$, whose range is the interval $[1,3]$, and whose graph is the figure below.

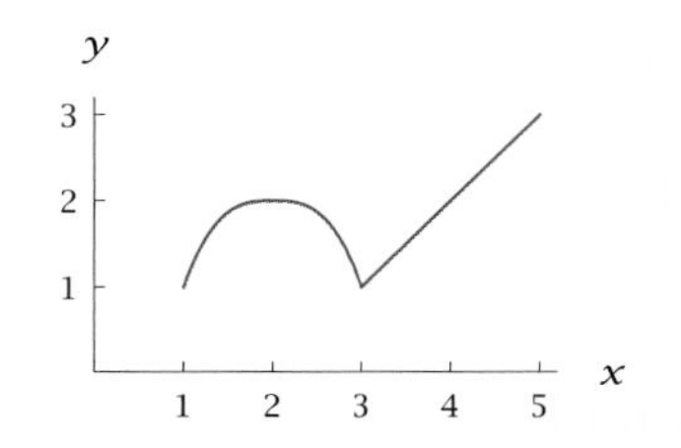

The graph of f.

For each given function g:

- **(a)** ***Find the domain of g.***
- **(b)** ***Find the range of g.***
- **(c)** ***Sketch the graph of g.***

15. $g(x) = f(x) + 1$
16. $g(x) = f(x) + 3$
17. $g(x) = f(x) - 3$
18. $g(x) = f(x) - 5$
19. $g(x) = 2f(x)$
20. $g(x) = \frac{1}{2}f(x)$
21. $g(x) = f(x+2)$
22. $g(x) = f(x+3)$
23. $g(x) = f(x-1)$
24. $g(x) = f(x-2)$
25. $g(x) = f(2x)$
26. $g(x) = f(3x)$
27. $g(x) = f(\frac{x}{2})$
28. $g(x) = f(\frac{5x}{8})$
29. $g(x) = 2f(x) + 1$
30. $g(x) = 3f(x) + 2$
31. $g(x) = \frac{1}{2}f(x) - 1$
32. $g(x) = \frac{2}{3}f(x) - 2$
33. $g(x) = 3 - f(x)$
34. $g(x) = 2 - f(x)$
35. $g(x) = -f(x-1)$
36. $g(x) = -f(x-3)$
37. $g(x) = f(x+1) + 2$
38. $g(x) = f(x+2) + 1$
39. $g(x) = f(2x) + 1$
40. $g(x) = f(3x) + 2$
41. $g(x) = f(2x+1)$
42. $g(x) = f(3x+2)$

43. $g(x) = 2f(\frac{x}{2} + 1)$

44. $g(x) = 3f(\frac{2x}{5} + 2)$

45. $g(x) = 2f(\frac{x}{2} + 1) - 3$

46. $g(x) = 3f(\frac{2x}{5} + 2) + 1$

47. $g(x) = 2f(\frac{x}{2} + 3)$

48. $g(x) = 3f(\frac{2x}{5} - 2)$

49. $g(x) = 6 - 2f(\frac{x}{2} + 3)$

50. $g(x) = 1 - 3f(\frac{2x}{5} - 2)$

51. Suppose g is an even function whose domain is $[-2, -1] \cup [1, 2]$ and whose graph on the interval $[1, 2]$ is the graph used in the instructions for Exercises 1–14. Sketch the graph of g on $[-2, -1] \cup [1, 2]$.

52. Suppose g is an even function whose domain is $[-5, -1] \cup [1, 5]$ and whose graph on the interval $[1, 5]$ is the graph used in the instructions for Exercises 15–50. Sketch the graph of g on $[-5, -1] \cup [1, 5]$.

53. Suppose h is an odd function whose domain is $[-2, -1] \cup [1, 2]$ and whose graph on the interval $[1, 2]$ is the graph used in the instructions for Exercises 1–14. Sketch the graph of h on $[-2, -1] \cup [1, 2]$.

54. Suppose h is an odd function whose domain is $[-5, -1] \cup [1, 5]$ and whose graph on the interval $[1, 5]$ is the graph used in the instructions for Exercises 15–50. Sketch the graph of h on $[-5, -1] \cup [1, 5]$.

For Exercises 55–58, suppose f is a function whose domain is the interval $[-5, 5]$ and

$$f(x) = \frac{x}{x + 3}$$

for every x in the interval $[0, 5]$.

55. Suppose f is an even function. Evaluate $f(-2)$.

56. Suppose f is an even function. Evaluate $f(-3)$.

57. Suppose f is an odd function. Evaluate $f(-2)$.

58. Suppose f is an odd function. Evaluate $f(-3)$.

PROBLEMS

59. Find the only function whose domain is the set of real numbers and that is both even and odd.

60. Show that if f is an odd function such that 0 is in the domain of f, then $f(0) = 0$.

61. Show that the sum of two even functions (with the same domain) is an even function.

62. Show that the product of two even functions (with the same domain) is an even function.

63. Show that the product of two odd functions (with the same domain) is an even function.

For Problems 64–66, suppose that to provide additional funds for higher education, the federal government adopts a new income tax plan that consists of the 2010 income tax plus an additional \$100 per taxpayer. Let g be the function such that $g(x)$ is the 2010 federal income tax for a single person with taxable income x dollars, and let h be the corresponding function for the new income tax plan.

64. Is h obtained from g by a vertical function transformation or by a horizontal function transformation?

65. Write a formula for $h(x)$ in terms of $g(x)$.

66. Using the explicit formula for $g(x)$ given in Example 2 in Section 3.1, give an explicit formula for $h(x)$.

For Problems 67–69, suppose that to pump more money into the economy during a recession, the federal government adopts a new income tax plan that makes income taxes 90% of the 2010 income tax. Let g be the function such that $g(x)$ is the 2010 federal income tax for a single person with taxable income x dollars, and let h be the corresponding function for the new income tax plan.

67. Is h obtained from g by a vertical function transformation or by a horizontal function transformation?

68. Write a formula for $h(x)$ in terms of $g(x)$.

69. Using the explicit formula for $g(x)$ given in Example 2 in Section 3.1, give an explicit formula for $h(x)$.

70. The result box following Example 2 could have been made more complete by including explicit information about the domain and range of the functions g and h. For example, the more complete result box might have looked like the one shown here:

Shifting a graph up or down

Suppose f is a function and $a > 0$. Define functions g and h by

$$g(x) = f(x) + a \quad \text{and} \quad h(x) = f(x) - a.$$

Then

- g and h have the same domain as f;
- the range of g is obtained by adding a to every number in the range of f;
- the range of h is obtained by subtracting a from every number in the range of f;
- the graph of g is obtained by shifting the graph of f up a units;
- the graph of h is obtained by shifting the graph of f down a units.

Construct similar complete result boxes, including explicit information about the domain and range of the functions g and h, for each of the other five result boxes in this section that deal with function transformations.

71. Show that the function f defined by $f(x) = mx + b$ is an odd function if and only if $b = 0$.

72. Show that the function f defined by $f(x) = mx + b$ is an even function if and only if $m = 0$.

73. Show that the function f defined by $f(x) = ax^2 + bx + c$ is an even function if and only if $b = 0$.

74. True or false: If f is an odd function whose domain is the set of real numbers and a function g is defined by

$$g(x) = \begin{cases} f(x) & \text{if } x \geq 0 \\ -f(x) & \text{if } x < 0, \end{cases}$$

then g is an even function. Explain your answer.

75. True or false: If f is an even function whose domain is the set of real numbers and a function g is defined by

$$g(x) = \begin{cases} f(x) & \text{if } x \geq 0 \\ -f(x) & \text{if } x < 0, \end{cases}$$

then g is an odd function. Explain your answer.

76. (a) True or false: Just as every integer is either even or odd, every function whose domain is the set of integers is either an even function or an odd function.

    (b) Explain your answer to part (a). This means that if the answer is "true", then you should explain why every function whose domain is the set of integers is either an even function or an odd function; if the answer is "false", then you should give an example of a function whose domain is the set of integers but that is neither even nor odd.

77. (a) True or false: The product of an even function and an odd function (with the same domain) is an odd function.

    (b) Explain your answer to part (a). This means that if the answer is "true", then explain why the product of every even function and every odd function (with the same domain) is an odd function; if the answer is "false", then give an example of an even function f and an odd function g (with the same domain) such that fg is not an odd function.

78. (a) True or false: The sum of an even function and an odd function (with the same domain) is an odd function.

    (b) Explain your answer to part (a). This means that if the answer is "true", then explain why the sum of every even function and every odd function (with the same domain) is an odd function; if the answer is "false", then give an example of an even function f and an odd function g (with the same domain) such that $f + g$ is not an odd function.

WORKED-OUT SOLUTIONS *to Odd-numbered Exercises*

For Exercises 1–14, assume that f is the function defined on the interval $[1,2]$ by the formula $f(x) = \frac{4}{x^2}$. Thus the domain of f is the interval $[1,2]$, the range of f is the interval $[1,4]$, and the graph of f is shown here.

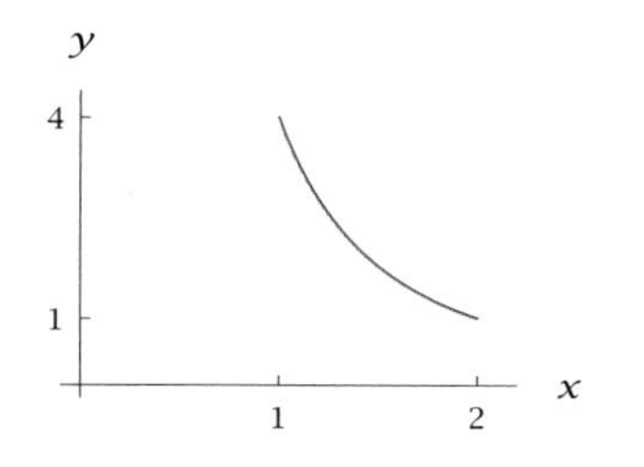

The graph of f.

For each function g described below:

(a) ***Sketch the graph of g.***

(b) ***Find the domain of g (the endpoints of this interval should be shown on the horizontal axis of your sketch of the graph of g).***

(c) ***Give a formula for g.***

(d) ***Find the range of g (the endpoints of this interval should be shown on the vertical axis of your sketch of the graph of g).***

1. The graph of g is obtained by shifting the graph of f up 1 unit.

SOLUTION

(a)

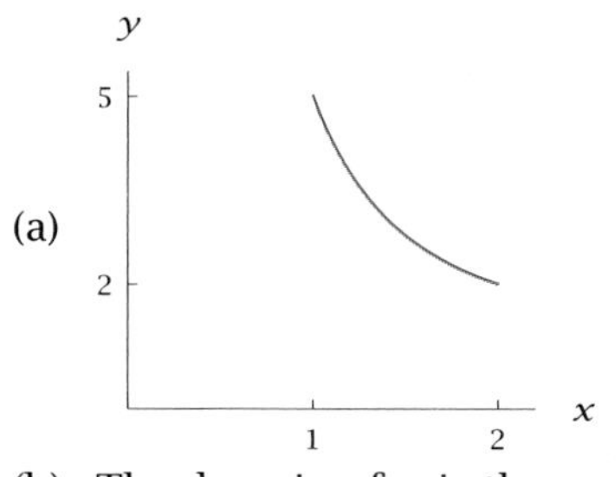

Shifting the graph of f up 1 unit gives this graph.

(b) The domain of g is the same as the domain of f. Thus the domain of g is the interval $[1,2]$.

(c) Because the graph of g is obtained by shifting the graph of f up 1 unit, we have $g(x) = f(x) + 1$. Thus

$$g(x) = \frac{4}{x^2} + 1$$

for each number x in the interval $[1,2]$.

(d) The range of g is obtained by adding 1 to each number in the range of f. Thus the range of g is the interval $[2,5]$.

3. The graph of g is obtained by shifting the graph of f down 3 units.

SOLUTION

(a)

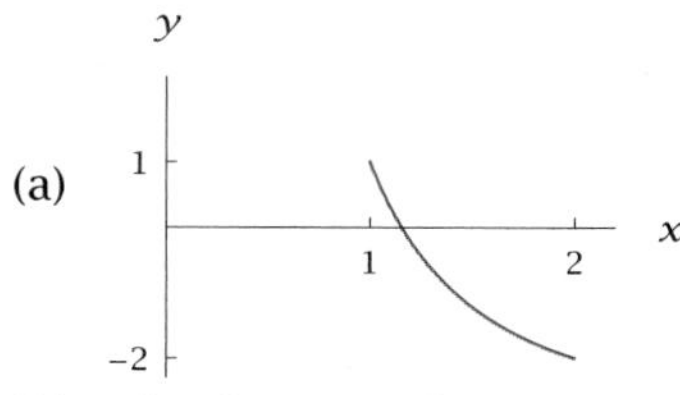

Shifting the graph of f down 3 units gives this graph.

(b) The domain of g is the same as the domain of f. Thus the domain of g is the interval $[1,2]$.

(c) Because the graph of g is obtained by shifting the graph of f down 3 units, we have $g(x) = f(x) - 3$. Thus

$$g(x) = \frac{4}{x^2} - 3$$

for each number x in the interval $[1,2]$.

(d) The range of g is obtained by subtracting 3 from each number in the range of f. Thus the range of g is the interval $[-2,1]$.

5. The graph of g is obtained by vertically stretching the graph of f by a factor of 2.

SOLUTION

(a)

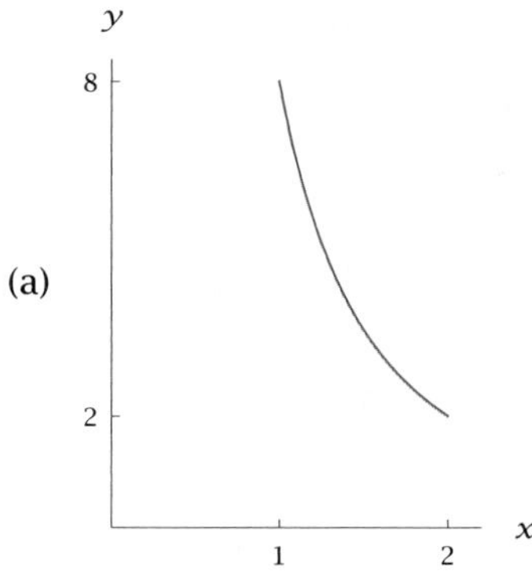

Vertically stretching the graph of f by a factor of 2 gives this graph.

(b) The domain of g is the same as the domain of f. Thus the domain of g is the interval $[1,2]$.

(c) Because the graph of g is obtained by vertically stretching the graph of f by a factor of 2, we have $g(x) = 2f(x)$. Thus

$$g(x) = \frac{8}{x^2}$$

for each number x in the interval $[1, 2]$.

(d) The range of g is obtained by multiplying every number in the range of f by 2. Thus the range of g is the interval $[2, 8]$.

7. The graph of g is obtained by shifting the graph of f left 3 units.

SOLUTION

(a)

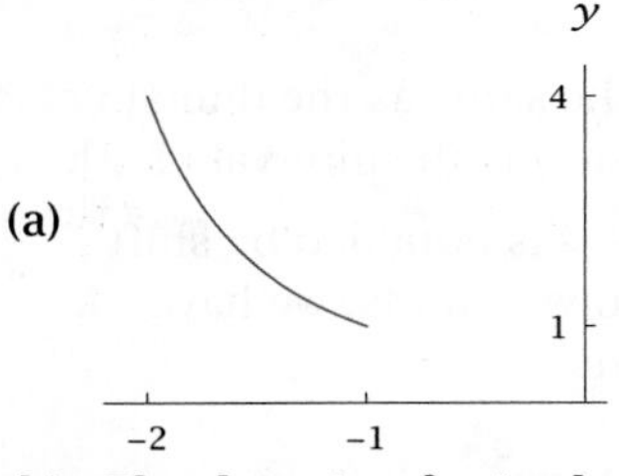

Shifting the graph of f left 3 units gives this graph.

(b) The domain of g is obtained by subtracting 3 from every number in domain of f. Thus the domain of g is the interval $[-2, -1]$.

(c) Because the graph of g is obtained by shifting the graph of f left 3 units, we have $g(x) = f(x+3)$. Thus

$$g(x) = \frac{4}{(x+3)^2}$$

for each number x in the interval $[-2, -1]$.

(d) The range of g is the same as the range of f. Thus the range of g is the interval $[1, 4]$.

9. The graph of g is obtained by shifting the graph of f right 1 unit.

SOLUTION

(a)

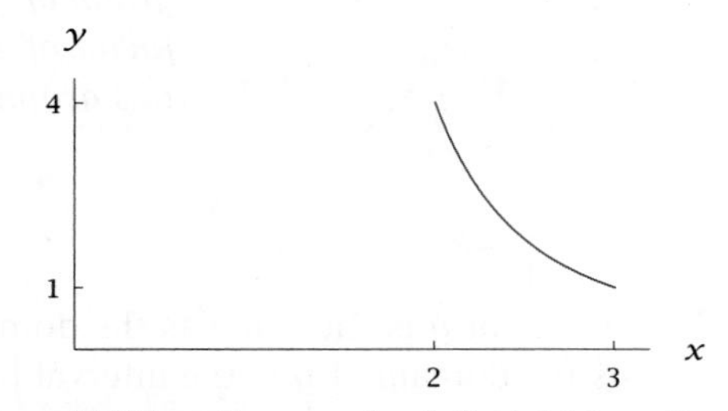

Shifting the graph of f right 1 unit gives this graph.

(b) The domain of g is obtained by adding 1 to every number in domain of f. Thus the domain of g is the interval $[2, 3]$.

(c) Because the graph of g is obtained by shifting the graph of f right 1 unit, we have $g(x) = f(x-1)$. Thus

$$g(x) = \frac{4}{(x-1)^2}$$

for each number x in the interval $[2, 3]$.

(d) The range of g is the same as the range of f. Thus the range of g is the interval $[1, 4]$.

11. The graph of g is obtained by horizontally stretching the graph of f by a factor of 2.

SOLUTION

(a)

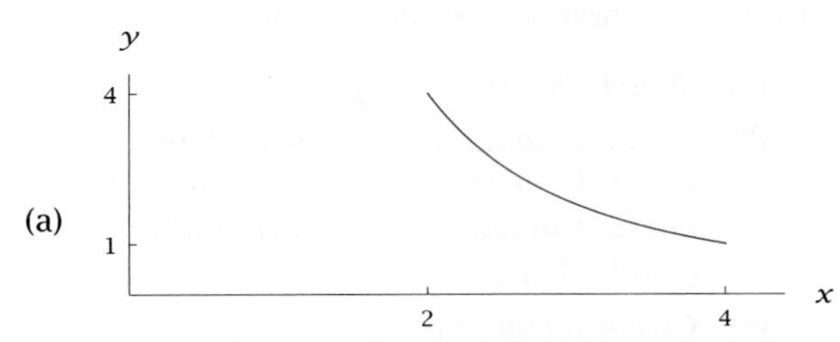

Horizontally stretching the graph of f by a factor of 2 gives this graph.

(b) The domain of g is obtained by multiplying every number in the domain of f by 2. Thus the domain of g is the interval $[2, 4]$.

(c) Because the graph of g is obtained by horizontally stretching the graph of f by a factor of 2, we have $g(x) = f(x/2)$. Thus

$$g(x) = \frac{4}{(x/2)^2} = \frac{16}{x^2}$$

for each number x in the interval $[2, 4]$.

(d) The range of g is the same as the range of f. Thus the range of g is the interval $[1, 4]$.

13. The graph of g is obtained by flipping the graph of f across the horizontal axis.

SOLUTION

(a)

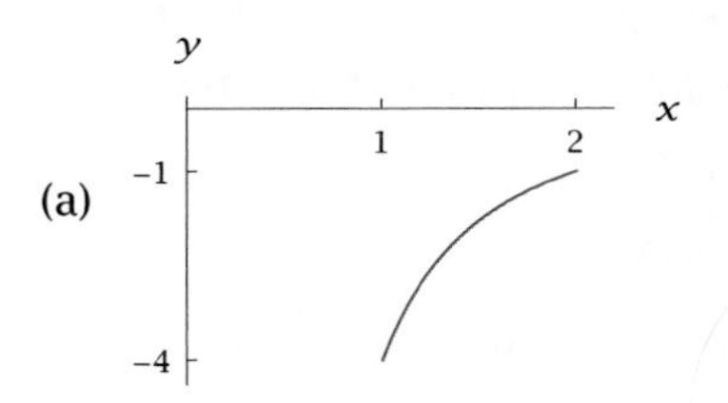

Flipping the graph of f across the horizontal axis gives this graph.

(b) The domain of g is the same as the domain of f. Thus the domain of g is the interval $[1, 2]$.

(c) Because the graph of g is obtained by flipping the graph of f across the horizontal axis, we have $g(x) = -f(x)$. Thus

$$g(x) = -\frac{4}{x^2}$$

for each number x in the interval $[1, 2]$.

(d) The range of g is obtained by multiplying every number in the range of f by -1. Thus the range of g is the interval $[-4, -1]$.

For Exercises 15–50, assume that f is a function whose domain is the interval $[1, 5]$, whose range is the interval $[1, 3]$, and whose graph is the figure below.

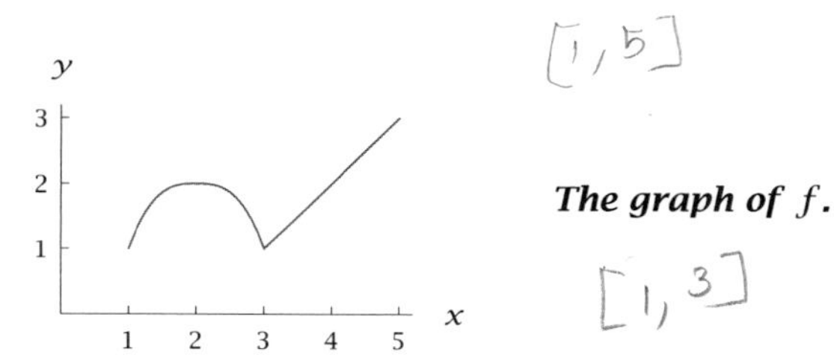

The graph of f.

[1, 5]

[1, 3]

For each given function g:

(a) ***Find the domain of g.***

(b) ***Find the range of g.***

(c) ***Sketch the graph of g.***

15. $g(x) = f(x) + 1$

SOLUTION

(a) Note that $g(x)$ is defined precisely when $f(x)$ is defined. In other words, the function g has the same domain as f. Thus the domain of g is the interval $[1, 5]$.

(b) The range of g is obtained by adding 1 to every number in the range of f. Thus the range of g is the interval $[2, 4]$.

(c) The graph of g, shown here, is obtained by shifting the graph of f up 1 unit.

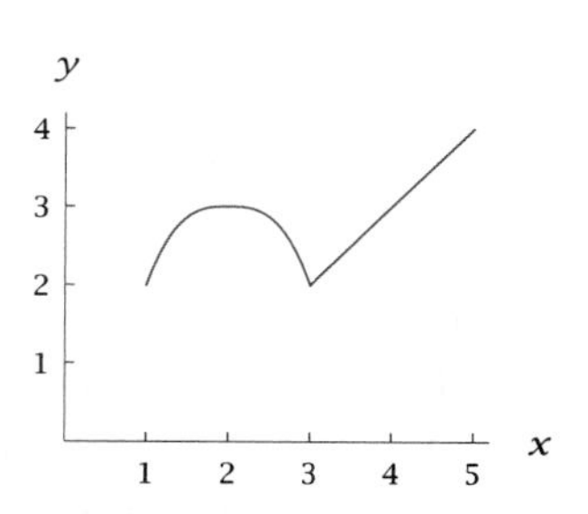

17. $g(x) = f(x) - 3$

SOLUTION

(a) Note that $g(x)$ is defined precisely when $f(x)$ is defined. In other words, the function g has the same domain as f. Thus the domain of g is the interval $[1, 5]$.

(b) The range of g is obtained by subtracting 3 from each number in the range of f. Thus the range of g is the interval $[-2, 0]$.

(c) The graph of g, shown here, is obtained by shifting the graph of f down 3 units.

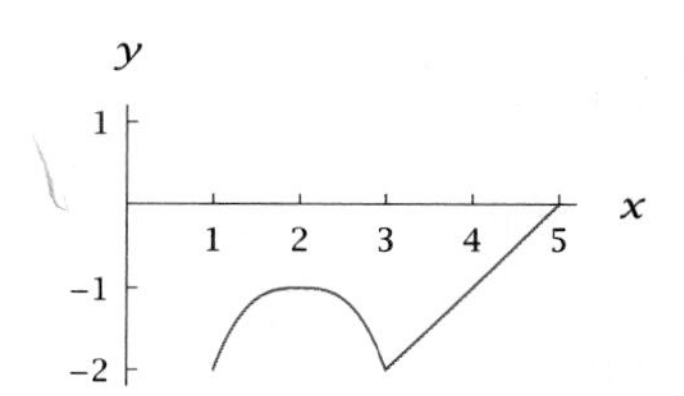

19. $g(x) = 2f(x)$

SOLUTION

(a) Note that $g(x)$ is defined precisely when $f(x)$ is defined. In other words, the function g has the same domain as f. Thus the domain of g is the interval $[1, 5]$.

(b) The range of g is obtained by multiplying each number in the range of f by 2. Thus the range of g is the interval $[2, 6]$.

(c) The graph of g, shown here, is obtained by vertically stretching the graph of f by a factor of 2.

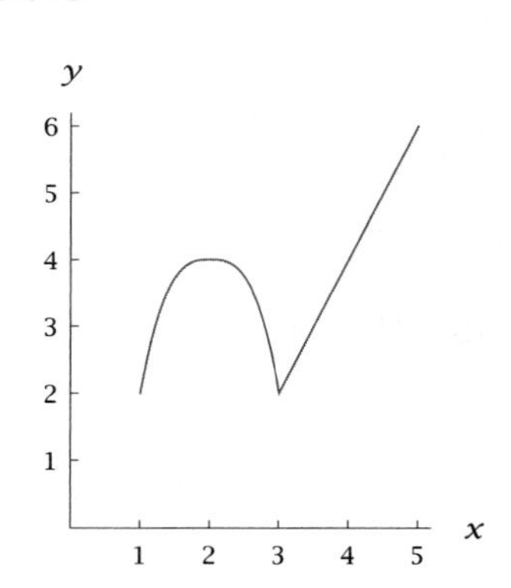

21. $g(x) = f(x + 2)$

SOLUTION

(a) Note that $g(x)$ is defined when $x + 2$ is in the interval $[1, 5]$, which means that x must be in the interval $[-1, 3]$. Thus the domain of g is the interval $[-1, 3]$.

(b) The range of g is the same as the range of f. Thus the range of g is the interval $[1, 3]$.

(c) The graph of g, shown here, is obtained by shifting the graph of f left 2 units.

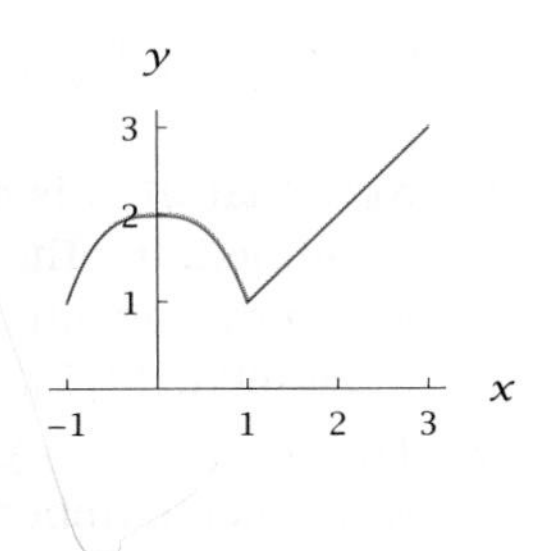

23. $g(x) = f(x - 1)$

SOLUTION

(a) Note that $g(x)$ is defined when $x - 1$ is in the interval $[1, 5]$, which means that x must be in the interval $[2, 6]$. Thus the domain of g is the interval $[2, 6]$.

(b) The range of g is the same as the range of f. Thus the range of g is the interval $[1, 3]$.

(c) The graph of g, shown here, is obtained by shifting the graph of f right 1 unit.

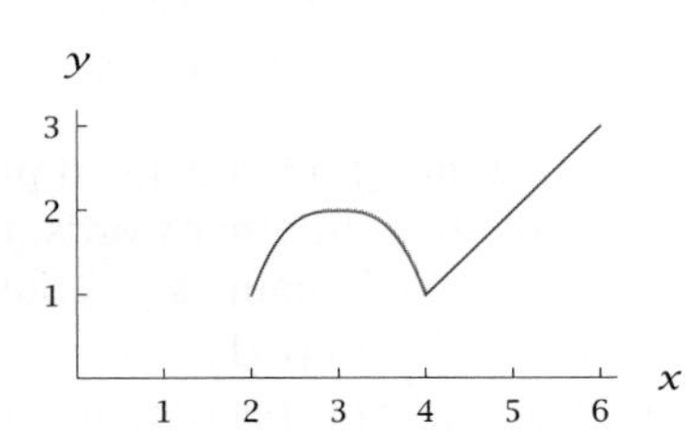

25. $g(x) = f(2x)$

SOLUTION

(a) Note that $g(x)$ is defined when $2x$ is in the interval $[1, 5]$, which means that x must be in the interval $[\frac{1}{2}, \frac{5}{2}]$. Thus the domain of g is the interval $[\frac{1}{2}, \frac{5}{2}]$.

(b) The range of g is the same as the range of f. Thus the range of g is the interval $[1, 3]$.

(c) The graph of g, shown here, is obtained by horizontally stretching the graph of f by a factor of $\frac{1}{2}$.

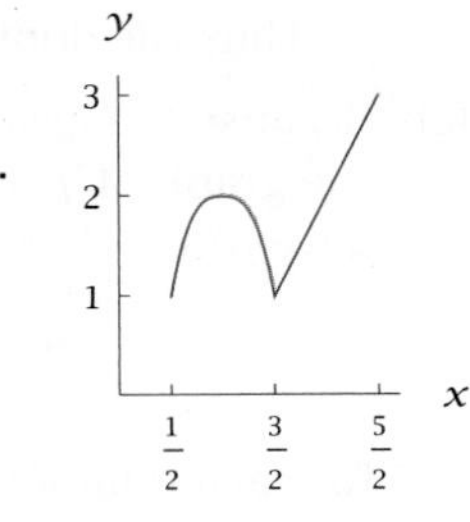

27. $g(x) = f(\frac{x}{2})$

SOLUTION

(a) Note that $g(x)$ is defined when $\frac{x}{2}$ is in the interval $[1, 5]$, which means that x must be in the interval $[2, 10]$. Thus the domain of g is the interval $[2, 10]$.

(b) The range of g is the same as the range of f. Thus the range of g is the interval $[1, 3]$.

(c) The graph of g, shown below, is obtained by horizontally stretching the graph of f by a factor of 2.

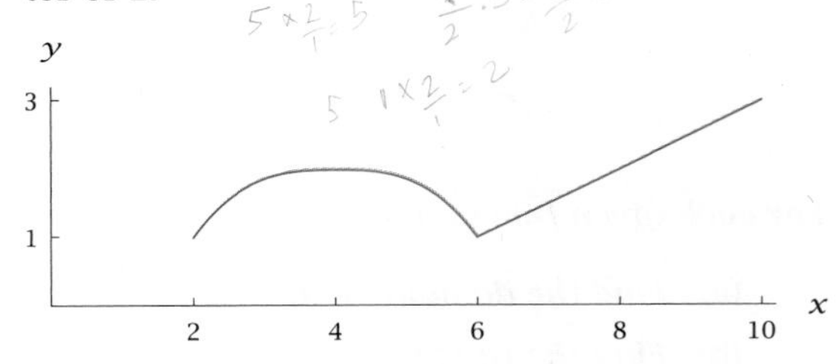

29. $g(x) = 2f(x) + 1$

SOLUTION

(a) Note that $g(x)$ is defined precisely when $f(x)$ is defined. In other words, the function g has the same domain as f. Thus the domain of g is the interval $[1, 5]$.

(b) The range of g is obtained by multiplying each number in the range of f by 2, which gives the interval $[2, 6]$, and then adding 1 to each number in this interval, which gives the interval $[3, 7]$. Thus the range of g is the interval $[3, 7]$.

(c) Note that $g(x)$ is evaluated by evaluating $f(x)$, then multiplying by 2, and then adding 1. Applying function transformations in the same order, we see that the graph of g, shown here, is obtained by vertically stretching the graph of f by a factor of 2, then shifting up 1 unit.

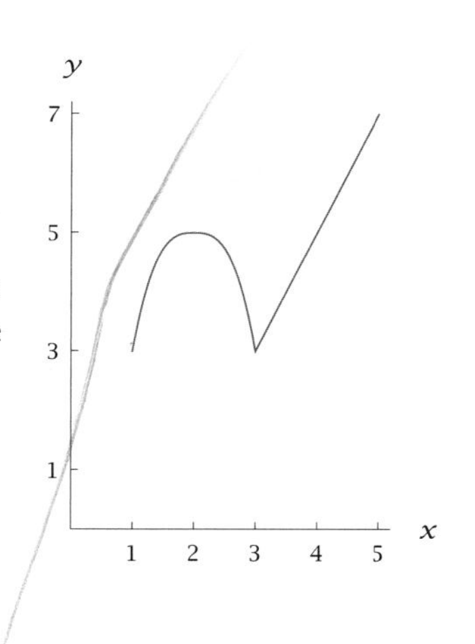

31. $g(x) = \frac{1}{2}f(x) - 1$

SOLUTION

(a) Note that $g(x)$ is defined precisely when $f(x)$ is defined. In other words, the function g has the same domain as f. Thus the domain of g is the interval $[1, 5]$.

(b) The range of g is obtained by multiplying each number in the range of f by $\frac{1}{2}$, which gives the interval $[\frac{1}{2}, \frac{3}{2}]$, and then subtracting 1 from each number in this interval, which gives the interval $[-\frac{1}{2}, \frac{1}{2}]$. Thus the range of g is the interval $[-\frac{1}{2}, \frac{1}{2}]$.

(c) Note that $g(x)$ is evaluated by evaluating $f(x)$, then multiplying by $\frac{1}{2}$, and then subtracting 1. Applying function transformations in the same order, we see that the graph of g, shown here, is obtained by vertically stretching the graph of f by a factor of $\frac{1}{2}$, then shifting down 1 unit.

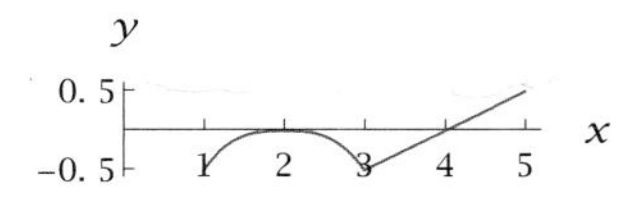

33. $g(x) = 3 - f(x)$

SOLUTION

(a) Note that $g(x)$ is defined precisely when $f(x)$ is defined. In other words, the function g has the same domain as f. Thus the domain of g is the interval $[1, 5]$.

(b) The range of g is obtained by multiplying each number in the range of f by -1, which gives the interval $[-3, -1]$, and then adding 3 to each number in this interval, which gives the interval $[0, 2]$. Thus the range of g is the interval $[0, 2]$.

(c) Note that $g(x)$ is evaluated by evaluating $f(x)$, then multiplying by -1, and then adding 3. Applying function transformations in the same order, we see that the graph of g, shown here, is obtained by flipping the graph of f across the horizontal axis, then shifting up 3 units.

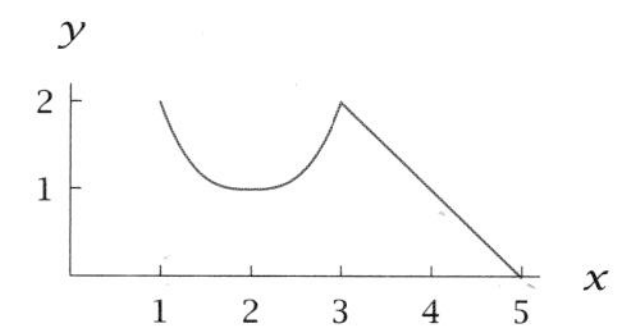

35. $g(x) = -f(x - 1)$

SOLUTION

(a) Note that $g(x)$ is defined when $x - 1$ is in the interval $[1, 5]$, which means that x must be in the interval $[2, 6]$. Thus the domain of g is the interval $[2, 6]$.

(b) The range of g is obtained by multiplying each number in the range of f by -1. Thus the range of g is the interval $[-3, -1]$.

(c) The graph of g, shown here, is obtained by shifting the graph of f right 1 unit, then flipping across the horizontal axis.

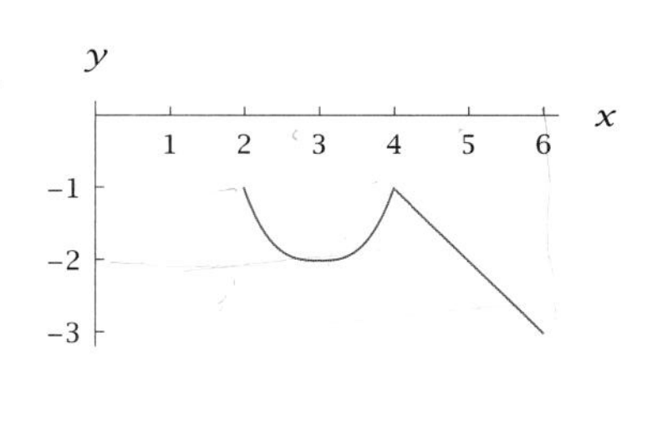

37. $g(x) = f(x + 1) + 2$

SOLUTION

(a) Note that $g(x)$ is defined when $x + 1$ is in the interval $[1, 5]$, which means that x must be in the interval $[0, 4]$. Thus the domain of g is the interval $[0, 4]$.

(b) The range of g is obtained by adding 2 to each number in the range of f. Thus the range of g is the interval $[3, 5]$.

(c) The graph of g, shown here, is obtained by shifting the graph of f left 1 unit, then shifting up 2 units.

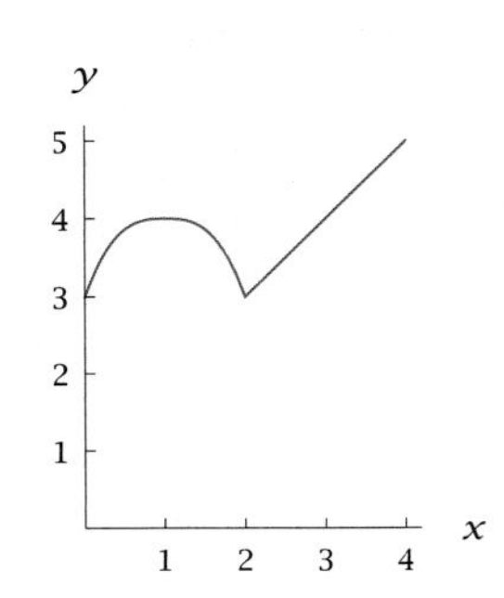

39. $g(x) = f(2x) + 1$

SOLUTION

(a) Note that $g(x)$ is defined when $2x$ is in the interval $[1, 5]$, which means that x must be in the interval $[\frac{1}{2}, \frac{5}{2}]$. Thus the domain of g is the interval $[\frac{1}{2}, \frac{5}{2}]$.

(b) The range of g is obtained by adding 1 to each number in the range of f. Thus the range of g is the interval $[2, 4]$.

(c) The graph of g, shown here, is obtained by horizontally stretching the graph of f by a factor of $\frac{1}{2}$, then shifting up 1 unit.

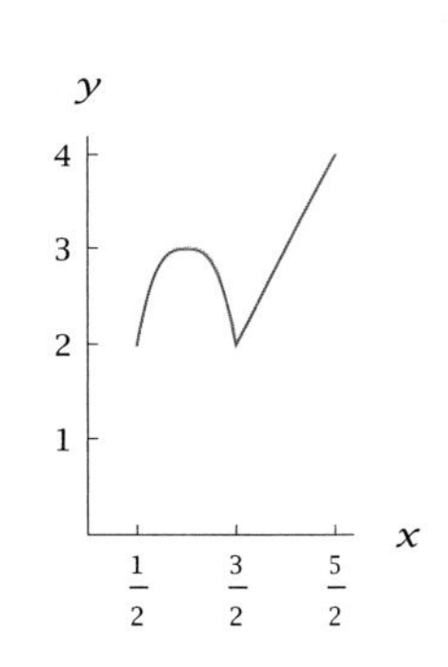

41. $g(x) = f(2x + 1)$

SOLUTION

(a) Note that $g(x)$ is defined when $2x + 1$ is in the interval $[1, 5]$, which means that

$$1 \le 2x + 1 \le 5.$$

Adding -1 to each part of the inequality above gives $0 \le 2x \le 4$, and then dividing each part by 2 produces $0 \le x \le 2$. Thus the domain of g is the interval $[0, 2]$.

(b) The range of g equals the range of f. Thus the range of g is the interval $[1, 3]$.

(c) Define a function h by $h(x) = f(2x)$. The graph of h is obtained by horizontally stretching the graph of f by a factor of $\frac{1}{2}$. Note that

$$g(x) = f(2x + 1) = f(2(x + \tfrac{1}{2})) = h(x + \tfrac{1}{2}).$$

Thus the graph of g is obtained by shifting the graph of h left $\frac{1}{2}$ unit.

Putting this together, we see that the graph of g, shown here, is obtained by horizontally stretching the graph of f by a factor of $\frac{1}{2}$, then shifting left $\frac{1}{2}$ unit.

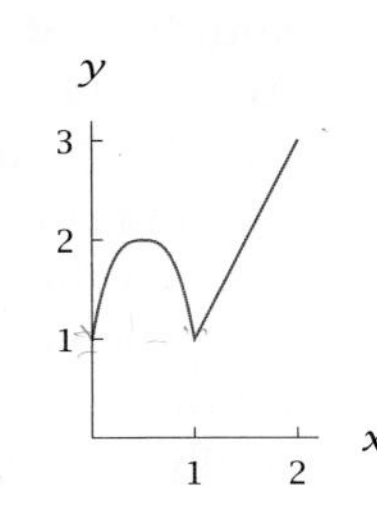

43. $g(x) = 2f(\frac{x}{2} + 1)$

SOLUTION

(a) Note that $g(x)$ is defined when $\frac{x}{2} + 1$ is in the interval $[1, 5]$, which means that

$$1 \le \tfrac{x}{2} + 1 \le 5.$$

Adding -1 to each part of the inequality above gives $0 \le \frac{x}{2} \le 4$, and then multiplying each part by 2 produces $0 \le x \le 8$. Thus the domain of g is the interval $[0, 8]$.

(b) The range of g is obtained by multiplying each number in the range of f by 2. Thus the range of g is the interval $[2, 6]$.

(c) Define a function h by

$$h(x) = f(\tfrac{x}{2}).$$

The graph of h is obtained by horizontally stretching the graph of f by a factor of 2. Note that

$$g(x) = 2f(\tfrac{x}{2} + 1) = 2f(\tfrac{x+2}{2}) = 2h(x + 2).$$

Thus the graph of g is obtained from the graph of h by shifting left 2 units and stretching vertically by a factor of 2.

Putting this together, we see that the graph of g, shown below, is obtained by horizontally stretching the graph of f by a factor of 2, shifting left 2 units, and then vertically stretching by a factor of 2.

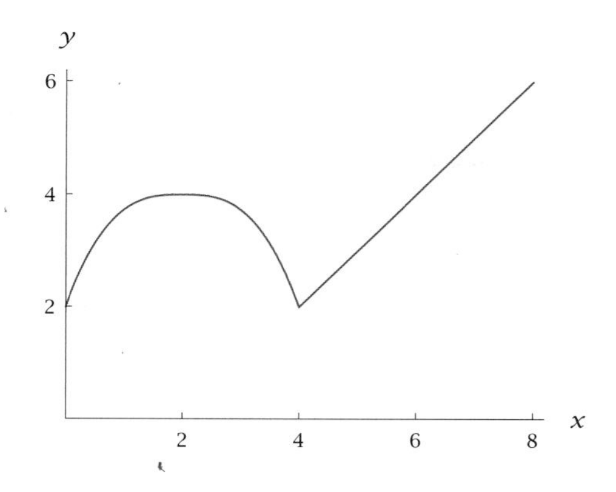

45. $g(x) = 2f(\frac{x}{2} + 1) - 3$

SOLUTION

(a) Note that $g(x)$ is defined when $\frac{x}{2} + 1$ is in the interval $[1, 5]$, which means that

$$1 \leq \frac{x}{2} + 1 \leq 5.$$

Adding -1 to each part of this inequality gives the inequality $0 \leq \frac{x}{2} \leq 4$, and then multiplying by 2 gives $0 \leq x \leq 8$. Thus the domain of g is the interval $[0, 8]$.

(b) The range of g is obtained by multiplying each number in the range of f by 2 and then subtracting 3. Thus the range of g is the interval $[-1, 3]$.

(c) The graph of g, shown below, is obtained by shifting the graph obtained in the solution to Exercise 43 down 3 units.

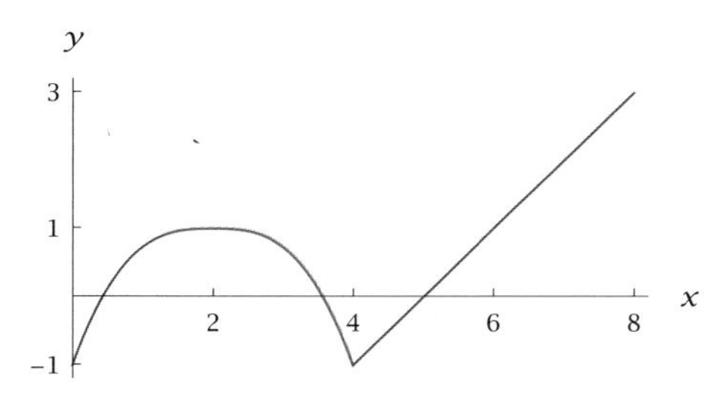

47. $g(x) = 2f(\frac{x}{2} + 3)$

SOLUTION

(a) Note that $g(x)$ is defined when $\frac{x}{2} + 3$ is in the interval $[1, 5]$, which means that

$$1 \leq \frac{x}{2} + 3 \leq 5.$$

Adding -3 to each part of this inequality gives the inequality $-2 \leq \frac{x}{2} \leq 2$, and then multiplying by 2 gives $-4 \leq x \leq 4$. Thus the domain of g is the interval $[-4, 4]$.

(b) Define a function h by

$$h(x) = f(\tfrac{x}{2}).$$

The graph of h is obtained by horizontally stretching the graph of f by a factor of 2. Note that

$$g(x) = 2f(\tfrac{x}{2} + 3) = 2f(\tfrac{x+6}{2}) = 2h(x + 6).$$

Thus the graph of g is obtained from the graph of h by shifting left 6 units and stretching vertically by a factor of 2.

Putting this together, we see that the graph of g, shown below, is obtained by horizontally stretching the graph of f by a factor of 2, shifting left 6 units, and then vertically stretching by a factor of 2.

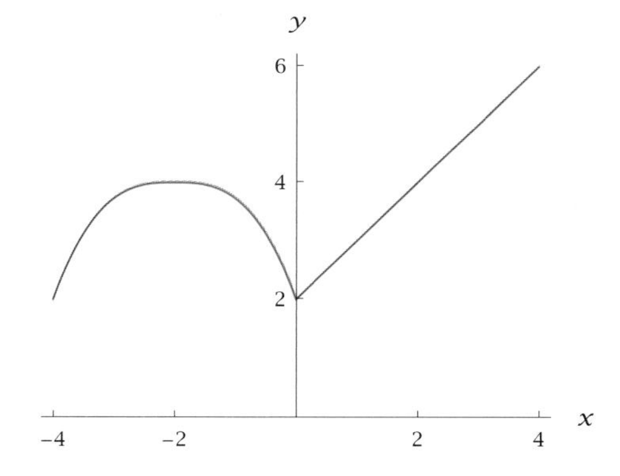

49. $g(x) = 6 - 2f(\frac{x}{2} + 3)$

SOLUTION

(a) Note that $g(x)$ is defined when $\frac{x}{2} + 3$ is in the interval $[1, 5]$, which means that

$$1 \leq \frac{x}{2} + 3 \leq 5.$$

Adding -3 to each part of this inequality gives the inequality $-2 \leq \frac{x}{2} \leq 2$, and then multiplying by 2 gives $-4 \leq x \leq 4$. Thus the domain of g is the interval $[-4, 4]$.

(b) The range of g is obtained by multiplying each number in the range of f by -2, giving the interval $[-6, -2]$, and then adding 6 to each number in this interval, which gives the interval $[0, 4]$.

(c) The graph of g, shown below, is obtained by flipping across the horizontal axis the graph obtained in the solution to Exercise 47, then shifting up by 6 units.

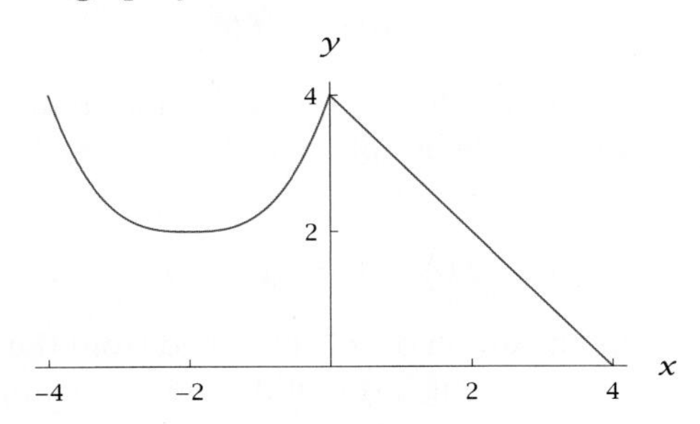

51. Suppose g is an even function whose domain is $[-2,-1] \cup [1,2]$ and whose graph on the interval $[1,2]$ is the graph used in the instructions for Exercises 1–14. Sketch the graph of g on $[-2,-1] \cup [1,2]$.

SOLUTION Because g is an even function, its graph is unchanged when flipped across the vertical axis. Thus we can find the graph of g on the interval $[-2,-1]$ by flipping the graph on the interval $[1,2]$ across the vertical axis, producing the following graph of g:

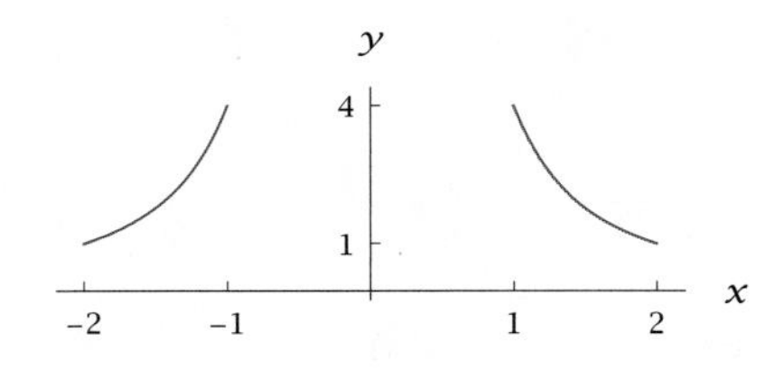

53. Suppose h is an odd function whose domain is $[-2,-1] \cup [1,2]$ and whose graph on the interval $[1,2]$ is the graph used in the instructions for Exercises 1–14. Sketch the graph of h on $[-2,-1] \cup [1,2]$.

SOLUTION Because h is an odd function, its graph is unchanged when flipped across the origin. Thus we can find the graph of h on the interval $[-2,-1]$ by flipping the graph on the interval $[1,2]$ across the origin, producing the following graph of h:

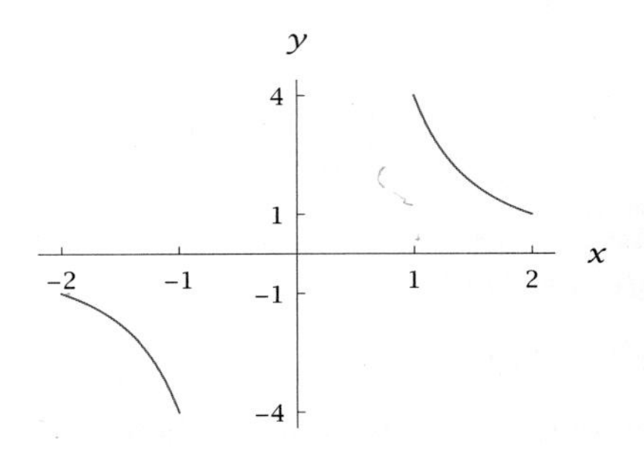

For Exercises 55–58, suppose f is a function whose domain is the interval $[-5,5]$ and

$$f(x) = \frac{x}{x+3}$$

for every x in the interval $[0,5]$.

55. Suppose f is an even function. Evaluate $f(-2)$.

SOLUTION Because 2 is in the interval $[0,5]$, we can use the formula above to evaluate $f(2)$. We have

$$f(2) = \tfrac{2}{2+3} = \tfrac{2}{5}.$$

Because f is an even function, we have

$$f(-2) = f(2) = \tfrac{2}{5}.$$

57. Suppose f is an odd function. Evaluate $f(-2)$.

SOLUTION Because f is an odd function, we have

$$f(-2) = -f(2) = -\tfrac{2}{5}.$$

3.3 *Composition of Functions*

LEARNING OBJECTIVES

By the end of this section you should be able to

- combine two functions using the usual algebraic operations;
- compute the composition of two functions;
- write a complicated function as the composition of simpler functions;
- express a function transformation as a composition of functions.

Combining Two Functions

Suppose f and g are functions. We define a new function, called the **sum** of f and g and denoted by $f + g$, by letting $f + g$ be the function whose value at a number x is given by the equation

$$(f + g)(x) = f(x) + g(x).$$

For $f(x) + g(x)$ to make sense, both $f(x)$ and $g(x)$ must make sense. Thus the domain of $f + g$ is the intersection of the domain of f and the domain of g.

Similarly, we can define the difference, product, and quotient of two functions in the expected manner. Here are the formal definitions:

Algebra of functions

Suppose f and g are functions. Then the **sum**, **difference**, **product**, and **quotient** of f and g are the functions denoted $f + g$, $f - g$, fg, and $\frac{f}{g}$ defined by

$$(f + g)(x) = f(x) + g(x)$$

$$(f - g)(x) = f(x) - g(x)$$

$$(fg)(x) = f(x)g(x)$$

$$\Big(\frac{f}{g}\Big)(x) = \frac{f(x)}{g(x)}.$$

Addition and multiplication of functions are commutative and associative operations; subtraction and division of functions have neither of these properties.

The domain of the first three functions above is the intersection of the domain of f and the domain of g. To avoid division by 0, the domain of $\frac{f}{g}$ is the set of numbers x such that x is in the domain of f and x is in the domain of g and $g(x) \neq 0$.

EXAMPLE 1

Suppose f and g are the functions defined by

$$f(x) = \sqrt{x - 3} \quad \text{and} \quad g(x) = \sqrt{8 - x}.$$

(a) What is the domain of $f + g$?

(b) Find a formula for $(f + g)(x)$.

(c) Evaluate $(f + g)(5)$.

SOLUTION

(a) Negative numbers do not have square roots in the real number system. Thus the domain of f is the interval $[3, \infty)$ and the domain of g is the interval $(-\infty, 8]$. The domain of $f + g$ is the intersection of these two domains, which is the interval $[3, 8]$.

(b) Using the definition of $f + g$, we have

$$\begin{aligned}(f + g)(x) &= f(x) + g(x) \\ &= \sqrt{x - 3} + \sqrt{8 - x}.\end{aligned}$$

(c) Using the definition of $f + g$, we have

$$\begin{aligned}(f + g)(5) &= f(5) + g(5) \\ &= \sqrt{5 - 3} + \sqrt{8 - 5} \\ &= \sqrt{2} + \sqrt{3}.\end{aligned}$$

Definition of Composition

Now we turn to a new way of combining two functions. The concept of the composition of functions allows us to write complicated functions in terms of simpler functions. This idea has applications throughout wide areas of mathematics.

EXAMPLE 2

The domain of h is the interval $[-3, \infty)$.

Consider the function h defined by

$$h(x) = \sqrt{x + 3}.$$

Thus, for example, $h(2) = \sqrt{5}$. The value of $h(x)$ is computed by carrying out two steps: first add 3 to x, and then take the square root of that sum. Write h in terms of two simpler functions that correspond to these two steps.

SOLUTION Define

$$f(x) = \sqrt{x} \quad \text{and} \quad g(x) = x + 3.$$

Then

$$h(x) = \sqrt{x + 3} = \sqrt{g(x)} = f(g(x)).$$

In the last term above, $f(g(x))$, we evaluate f at $g(x)$. This kind of construction occurs so often that it has been given a name and notation:

Composition

If f and g are functions, then the **composition** of f and g, denoted $f \circ g$, is the function defined by

$$(f \circ g)(x) = f(g(x)).$$

In evaluating $(f \circ g)(x)$, first we evaluate $g(x)$, then we evaluate $f(g(x))$.

The domain of $f \circ g$ is the set of numbers x such that $f(g(x))$ makes sense. For $f(g(x))$ to make sense, x must be in the domain of g [so that $g(x)$ will be defined] and $g(x)$ must be in the domain of f [so that $f(g(x))$ will be defined]. Thus the domain of $f \circ g$ is the set of numbers x in the domain of g such that $g(x)$ is in the domain of f.

EXAMPLE 3

Suppose

$$f(x) = \frac{1}{x-4} \quad \text{and} \quad g(x) = x^2.$$

(a) Evaluate $(f \circ g)(3)$.

(b) Find a formula for the composition $f \circ g$.

(c) What is the domain of $f \circ g$?

SOLUTION

(a) Using the definition of composition, we have

$$\begin{aligned}(f \circ g)(3) &= f(g(3))\\ &= f(9)\\ &= \frac{1}{5}.\end{aligned}$$

(b) Using the definition of composition, we have

$$\begin{aligned}(f \circ g)(x) &= f(g(x))\\ &= f(x^2)\\ &= \frac{1}{x^2-4}.\end{aligned}$$

(c) The domains of f and g were not specified, which means we are assuming that each domain is the set of numbers where the formulas defining these functions make sense. Thus the domain of f equals the set of real numbers except 4, and the domain of g equals the set of real numbers.

From part (b), we see that $f(g(x))$ makes sense provided $x^2 \neq 4$. Thus the domain of $f \circ g$ equals the set of all real numbers except -2 and 2.

In Section 3.1 we saw that a function g can be thought of as a machine that takes an input x and produces an output $g(x)$. The composition $f \circ g$ can then be thought of as the machine that feeds the output of the g machine into the f machine:

Here $g(x)$ is the output of the g machine and $g(x)$ is also the input of the f machine.

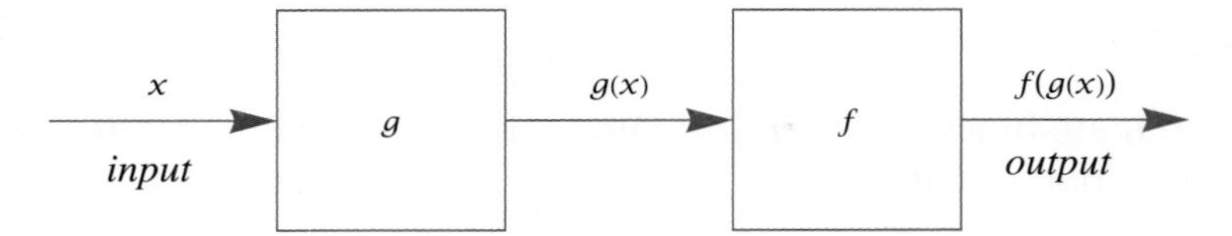

The composition $f \circ g$ as the combination of two machines.

EXAMPLE 4

(a) Suppose your cell phone company charges \$0.05 per minute plus \$0.47 for each call to China. Find a function p that gives the amount charged by your cell phone company for a call to China as a function of the number of minutes.

(b) Suppose the tax on cell phone bills is 6% plus \$0.01 for each call. Find a function t that gives your total cost, including tax, for a call to China as a function of the amount charged by your cell phone company.

(c) Explain why the composition $t \circ p$ gives your total cost, including tax, of making a cell phone call to China as a function of the number of minutes.

(d) Compute a formula for $t \circ p$.

(e) What is your total cost for a ten-minute call to China?

SOLUTION

*In this example, the functions are called p and t to help you remember that p is the function giving the amount charged by your **phone** company and t is the function giving the **total** cost, including tax.*

(a) For a call of m minutes to China, the amount $p(m)$ in dollars charged by your cell phone company will be

$$p(m) = 0.05m + 0.47,$$

consisting of \$0.05 times the number of minutes plus \$0.47. Most cell phone companies round up any extra seconds to the next highest minute. Thus for this application only positive integer values of m will be used.

(b) For an amount of y dollars charged by your phone company for a phone call, your total cost in dollars, including tax, will be

$$t(y) = 1.06y + 0.01.$$

(c) Your total cost for a call of m minutes to China is computed by first calculating the amount $p(m)$ charged by the phone company, and then calculating the total cost (including tax) of $t(p(m))$, which equals $(t \circ p)(m)$.

(d) The composition $t \circ p$ is given by the formula

$$\begin{aligned}(t \circ p)(m) &= t(p(m)) \\ &= t(0.05m + 0.47) \\ &= 1.06(0.05m + 0.47) + 0.01 \\ &= 0.053m + 0.5082.\end{aligned}$$

(e) Using the formula above from part (d), we have

$$(t \circ p)(10) = 0.053 \times 10 + 0.5082 = 1.0383.$$

Thus your total cost for a ten-minute call to China is \$1.04.

Order Matters in Composition

Composition is not commutative. In other words, it is not necessarily true that $f \circ g = g \circ f$, as can be shown by choosing almost any pair of functions.

EXAMPLE 5

Suppose

$$f(x) = 1 + x \quad \text{and} \quad g(x) = x^2.$$

(a) Evaluate $(f \circ g)(4)$.

(b) Evaluate $(g \circ f)(4)$.

(c) Find a formula for the composition $f \circ g$.

(d) Find a formula for the composition $g \circ f$.

SOLUTION

(a) Using the definition of composition, we have

$$\begin{aligned}(f \circ g)(4) &= f(g(4)) \\ &= f(16) \\ &= 17.\end{aligned}$$

(b) Using the definition of composition, we have

$$\begin{aligned}(g \circ f)(4) &= g(f(4)) \\ &= g(5) \\ &= 25.\end{aligned}$$

The solutions to (a) and (b) show that $(f \circ g)(4) \neq (g \circ f)(4)$ for these functions f and g.

(c) Using the definition of composition, we have

$$\begin{aligned}(f \circ g)(x) &= f(g(x)) \\ &= f(x^2) \\ &= 1 + x^2.\end{aligned}$$

(d) Using the definition of composition, we have

$$\begin{aligned}(g \circ f)(x) &= g(f(x)) \\ &= g(1 + x) \\ &= (1 + x)^2 \\ &= 1 + 2x + x^2.\end{aligned}$$

Never, ever make the mistake of thinking that $(1 + x)^2$ equals $1 + x^2$.

The example above is typical, meaning that for most functions f and g we have $f \circ g \neq g \circ f$. However, the identity function that we now define does commute (with respect to composition) with all other functions.

Identity function

The **identity function** is the function I defined by

$$I(x) = x$$

for every number x.

If f is any function and x is any number in the domain of f, then

$$(f \circ I)(x) = f(I(x)) = f(x) \quad \text{and} \quad (I \circ f)(x) = I(f(x)) = f(x).$$

Thus we have the following result, which explains why I is called the identity function.

The function I is the identity for composition

If f is any function, then $f \circ I = I \circ f = f$.

Decomposing Functions

Computing the composition of two functions is usually a straightforward application of the definition of composition. Less straightforward is the process of starting with a function and writing it as the composition of two simpler functions. The following example illustrates the process.

EXAMPLE 6

Suppose

$$T(x) = \left|\frac{x^2 - 3}{x^2 - 7}\right|.$$

Write T as the composition of two simpler functions. In other words, find two functions f and g, each of them simpler than T, such that $T = f \circ g$.

Typically a function can be decomposed into the composition of other functions in many different ways.

SOLUTION The problem here is that there is no rigorous definition of "simpler". Certainly it is easy to write T as the composition of two functions, because $T = T \circ I$, where I is the identity function, but that decomposition is unlikely to be useful.

Because evaluating an absolute value is the last operation done in computing $T(x)$, one reasonable possibility is to define

$$f(x) = |x| \quad \text{and} \quad g(x) = \frac{x^2 - 3}{x^2 - 7}.$$

Both potential solutions discussed here are correct. Choosing one or the other may depend on the context or on one's taste. Also, see Example 7, where T is decomposed into three simpler functions.

You should verify that with these definitions of f and g, we indeed have $T = f \circ g$. Furthermore, both f and g seem to be simpler functions than T.

Because x appears in the formula defining T only in the expression x^2, another reasonable possibility is to define

$$f(x) = \left|\frac{x - 3}{x - 7}\right| \quad \text{and} \quad g(x) = x^2.$$

Again you should verify that with these definitions of f and g, we have $T = f \circ g$; both f and g seem to be simpler functions than T.

Composing More than Two Functions

Although composition is not commutative, it is associative.

Composition is associative

If f, g, and h are functions, then

$$(f \circ g) \circ h = f \circ (g \circ h).$$

To prove the associativity of composition, note that

$$((f \circ g) \circ h)(x) = (f \circ g)(h(x)) = f(g(h(x)))$$

and

$$(f \circ (g \circ h))(x) = f((g \circ h)(x)) = f(g(h(x))).$$

The equations above show that the functions $(f \circ g) \circ h$ and $f \circ (g \circ h)$ have the same value at every number x in their domain. Thus $(f \circ g) \circ h = f \circ (g \circ h)$.

Because composition is associative, we can dispense with the parentheses and simply write $f \circ g \circ h$, which is the function whose value at a number x is $f(g(h(x)))$.

The domain of $f \circ g \circ h$ is the set of numbers x in the domain of h such that $h(x)$ is in the domain of g and $g(h(x))$ is in the domain of f.

EXAMPLE 7

Suppose

$$T(x) = \left|\frac{x^2 - 3}{x^2 - 7}\right|.$$

Write T as the composition of three simpler functions.

SOLUTION We want to choose reasonably simple functions f, g, and h such that $T = f \circ g \circ h$. Probably the best choice here is to take

$$f(x) = |x|, \quad g(x) = \frac{x - 3}{x - 7}, \quad h(x) = x^2.$$

With these choices, we have

$$\begin{aligned} f(g(h(x))) &= f(g(x^2)) \\ &= f\Big(\frac{x^2 - 3}{x^2 - 7}\Big) \\ &= \left|\frac{x^2 - 3}{x^2 - 7}\right|, \end{aligned}$$

as desired.

Here is how to come up with the choices made above for f, g, and h: Because x appears in the formula defining T only in the expression x^2, we start by taking $h(x) = x^2$. To make $(g \circ h)(x)$ equal $\frac{x^2-3}{x^2-7}$, we then take $g(x) = \frac{x-3}{x-7}$. Finally, because evaluating an absolute value is the last operation done in computing $T(x)$, we take $f(x) = |x|$.

Function Transformations as Compositions

All the function transformations discussed in Section 3.2 can be considered to be compositions with linear functions, which we now define.

Linear functions

The term linear function *is used because the graph of any such function is a line.*

A **linear function** is a function h of the form

$$h(x) = mx + b,$$

where m and b are constants.

Vertical function transformations can be expressed as compositions with a linear function on the left, as shown in the next example.

EXAMPLE 8

Suppose f is a function. Define a function g by

$$g(x) = -2f(x) + 1.$$

(a) Write g as the composition of a linear function with f.

(b) Describe how the graph of g is obtained from the graph of f.

SOLUTION

(a) Define a linear function h by

$$h(x) = -2x + 1.$$

For x in the domain of f, we have

$$\begin{aligned} g(x) &= -2f(x) + 1 \\ &= h(f(x)) \\ &= (h \circ f)(x). \end{aligned}$$

Thus $g = h \circ f$.

See Example 8 in Section 3.2 for a figure showing this function transformation.

(b) As discussed in the solution to Example 8 in Section 3.2, the graph of g is obtained by vertically stretching the graph of f by a factor of 2, then flipping the resulting graph across the horizontal axis, then shifting the resulting graph up 1 unit.

Horizontal function transformations can be expressed as compositions with a linear function on the right, as shown in the next example.

EXAMPLE 9

Suppose f is a function. Define a function g by

$$g(x) = f(2x).$$

(a) Write g as the composition of a linear function with f.

(b) Describe how the graph of g is obtained from the graph of f.

SOLUTION

(a) Define a linear function h by

$$h(x) = 2x.$$

For x in the domain of f, we have

$$\begin{aligned} g(x) &= f(2x) \\ &= f(h(x)) \\ &= (f \circ h)(x). \end{aligned}$$

Thus $g = f \circ h$.

(b) As discussed in the solution to Example 6 in Section 3.2, the graph of g is obtained by horizontally stretching the graph of f by a factor of $\frac{1}{2}$.

See Example 6 in Section 3.2 for a figure showing this function transformation.

Combinations of vertical function transformations and horizontal function transformations can be expressed as compositions with a linear function on the left and a linear function on the right, as shown in the next example.

EXAMPLE 10

Suppose f is a function. Define a function g by

$$g(x) = f(2x) + 1.$$

(a) Write g as the composition of a linear function, f, and another linear function.

(b) Describe how the graph of g is obtained from the graph of f.

SOLUTION

(a) Define linear functions h and p by

$$h(x) = x + 1 \quad \text{and} \quad p(x) = 2x.$$

For x in the domain of g, we have

$$\begin{aligned} g(x) &= f(2x) + 1 \\ &= h(f(2x)) \\ &= h(f(p(x))) \\ &= (h \circ f \circ p)(x). \end{aligned}$$

Thus $g = h \circ f \circ p$.

See the solution to Exercise 39 in Section 3.2 for a figure showing this function transformation.

(b) As discussed in the solution to Exercise 39 in Section 3.2, the graph of g is obtained by horizontally stretching the graph of f by a factor of $\frac{1}{2}$, then shifting up 1 unit.

EXERCISES

For Exercises 1–10, evaluate the indicated expression assuming that f, g, and h are the functions completely defined by the tables below:

x	$f(x)$
1	4
2	1
3	2
4	2

x	$g(x)$
1	2
2	4
3	1
4	3

x	$h(x)$
1	3
2	3
3	4
4	1

1. $(f \circ g)(1)$
2. $(f \circ g)(3)$
3. $(g \circ f)(1)$
4. $(g \circ f)(3)$
5. $(f \circ f)(2)$
6. $(f \circ f)(4)$
7. $(g \circ g)(4)$
8. $(g \circ g)(2)$
9. $(f \circ g \circ h)(2)$
10. $(h \circ g \circ f)(2)$

For Exercises 11–24, evaluate the indicated expression assuming that

$$f(x) = \sqrt{x}, \quad g(x) = \frac{x+1}{x+2}, \quad h(x) = |x - 1|.$$

11. $(f \circ g)(4)$
12. $(f \circ g)(5)$
13. $(g \circ f)(4)$
14. $(g \circ f)(5)$
15. $(f \circ h)(-3)$
16. $(f \circ h)(-15)$
17. $(f \circ g \circ h)(0)$
18. $(h \circ g \circ f)(0)$
19. $(f \circ g)(0.23)$
20. $(f \circ g)(3.85)$
21. $(g \circ f)(0.23)$
22. $(g \circ f)(3.85)$
23. $(h \circ f)(0.3)$
24. $(h \circ f)(0.7)$

In Exercises 25–30, for the given functions f and g find formulas for (a) $f \circ g$ and (b) $g \circ f$. Simplify your results as much as possible.

25. $f(x) = x^2 + 1, \quad g(x) = \dfrac{1}{x}$
26. $f(x) = (x+1)^2, \quad g(x) = \dfrac{3}{x}$
27. $f(x) = \dfrac{x-1}{x+1}, \quad g(x) = x^2 + 2$
28. $f(x) = \dfrac{x+2}{x-3}, \quad g(x) = \dfrac{1}{x+1}$
29. $f(x) = \dfrac{x-1}{x^2+1}, \quad g(x) = \dfrac{x+3}{x+4}$
30. $f(x) = \dfrac{x-2}{x+3}, \quad g(x) = \dfrac{1}{(x+2)^2}$
31. Find a number b such that $f \circ g = g \circ f$, where $f(x) = 2x + b$ and $g(x) = 3x + 4$.
32. Find a number c such that $f \circ g = g \circ f$, where $f(x) = 5x - 2$ and $g(x) = cx - 3$.
33. Suppose
$$h(x) = \left(\frac{x^2+1}{x-1} - 1\right)^3.$$
   (a) If $f(x) = x^3$, then find a function g such that $h = f \circ g$.
   (b) If $f(x) = (x-1)^3$, then find a function g such that $h = f \circ g$.
34. Suppose
$$h(x) = \sqrt{\frac{1}{x^2+1} + 2}.$$
   (a) If $f(x) = \sqrt{x}$, then find a function g such that $h = f \circ g$.
   (b) If $f(x) = \sqrt{x+2}$, then find a function g such that $h = f \circ g$.
35. Suppose
$$h(x) = 2 + \sqrt{\frac{1}{x^2+1}}.$$
   (a) If $g(x) = \frac{1}{x^2+1}$, then find a function f such that $h = f \circ g$.
   (b) If $g(x) = x^2$, then find a function f such that $h = f \circ g$.
36. Suppose
$$h(x) = \left(\frac{x^2+1}{x-1} - 1\right)^3.$$
   (a) If $g(x) = \frac{x^2+1}{x-1} - 1$, then find a function f such that $h = f \circ g$.
   (b) If $g(x) = \frac{x^2+1}{x-1}$, then find a function f such that $h = f \circ g$.

In Exercises 37–40, find functions f and g, each simpler than the given function h, such that $h = f \circ g$.

37. $h(x) = (x^2 - 1)^2$

38. $h(x) = \sqrt{x^2 - 1}$

39. $h(x) = \dfrac{3}{2 + x^2}$

40. $h(x) = \dfrac{2}{3 + \sqrt{1 + x}}$

In Exercises 41–42, find functions f, g, and h, each simpler than the given function T, such that $T = f \circ g \circ h$.

41. $T(x) = \dfrac{4}{5 + x^2}$

42. $T(x) = \sqrt{4 + x^2}$

43. Suppose the graph of f is a parabola with vertex at $(3, 2)$. Suppose $g(x) = 4x + 5$. What are the coordinates of the vertex of the graph of $f \circ g$?

44. Suppose the graph of f is a parabola with vertex at $(-5, 4)$. Suppose $g(x) = 3x - 1$. What are the coordinates of the vertex of the graph of $f \circ g$?

45. Suppose the graph of f is a parabola with vertex at $(3, 2)$. Suppose $g(x) = 4x + 5$. What are the coordinates of the vertex of the graph of $g \circ f$?

46. Suppose the graph of f is a parabola with vertex at $(-5, 4)$. Suppose $g(x) = 3x - 1$. What are the coordinates of the vertex of the graph of $g \circ f$?

47. Suppose the graph of f is a parabola with vertex at (t, s). Suppose $g(x) = ax + b$, where a and b are constants with $a \neq 0$. What are the coordinates of the vertex of the graph of $f \circ g$?

48. Suppose the graph of f is a parabola with vertex at (t, s). Suppose $g(x) = ax + b$, where a and b are constants with $a \neq 0$. What are the coordinates of the vertex of the graph of $g \circ f$?

PROBLEMS

For Problems 49–54, suppose you are exchanging currency in the London airport. The currency exchange service there only makes transactions in which one of the two currencies is British pounds, but you want to exchange dollars for Euros. Thus you first need to exchange dollars for British pounds, then exchange British pounds for Euros. At the time you want to make the exchange, the function f for exchanging dollars for British pounds is given by the formula

$$f(d) = 0.66d - 1$$

and the function g for exchanging British pounds for Euros is given by the formula

$$g(p) = 1.23p - 2.$$

The subtraction of 1 or 2 in the number of British pounds or Euros that you receive is the fee charged by the currency exchange service for each transaction.

49. Is the function describing the exchange of dollars for Euros $f \circ g$ or $g \circ f$?

50. Explain your answer to the previous problem in terms of which function is evaluated first when computing a value for a composition (the function on the left or the function on the right?).

51. Find a formula for the function given by your answer to Problem 49.

52. How many Euros would you receive for exchanging \$100 after going through this two-step exchange process?

53. How many Euros would you receive for exchanging \$200 after going through this two-step exchange process?

54. Which process ends up giving you more Euros: exchanging \$100 for Euros twice or exchanging \$200 for Euros once?

55. Show that the composition of two linear functions is a linear function.

56. Show that if f and g are linear functions, then the graphs of $f \circ g$ and $g \circ f$ have the same slope.

57. Suppose $f(x) = ax + b$ and $g(x) = cx + d$, where a, b, c, and d are constants. Show that $f \circ g = g \circ f$ if and only if $d(a - 1) = b(c - 1)$.

58. Show that the graphs of two linear functions f and g are perpendicular if and only if the graph of $f \circ g$ has slope -1.
59. Suppose f and g are functions. Show that the composition $f \circ g$ has the same domain as g if and only if the range of g is contained in the domain of f.
60. Show that if f is a constant function and g is any function, then $f \circ g$ and $g \circ f$ are both constant functions.
61. Give an example of three functions f, g, and h, none of which is a constant function, such that $f \circ g = f \circ h$ but g is not equal to h.
62. Give an example of three functions f, g, and h, none of which is a constant function, such that $f \circ h = g \circ h$ but f is not equal to g.

A quadratic function *is a function* g *of the form*

$$g(x) = ax^2 + bx + c,$$

where a, b, *and* c *are constants with* $a \neq 0$.

63. Show that if f is a nonconstant linear function and g is a quadratic function, then $f \circ g$ and $g \circ f$ are both quadratic functions.
64. Suppose g is an even function and f is any function. Show that $f \circ g$ is an even function.
65. Suppose f is an even function and g is an odd function. Show that $f \circ g$ is an even function.
66. Suppose f and g are both odd functions. Is the composition $f \circ g$ even, odd, or neither? Explain.
67. Show that if f, g, and h are functions, then

$$(f + g) \circ h = f \circ h + g \circ h.$$

68. Find functions f, g, and h such that

$$f \circ (g + h) \neq f \circ g + f \circ h.$$

WORKED-OUT SOLUTIONS *to Odd-numbered Exercises*

For Exercises 1–10, evaluate the indicated expression assuming that f, g, and h are the functions completely defined by the tables below:

x	$f(x)$
1	4
2	1
3	2
4	2

x	$g(x)$
1	2
2	4
3	1
4	3

x	$h(x)$
1	3
2	3
3	4
4	1

1. $(f \circ g)(1)$

SOLUTION $(f \circ g)(1) = f(g(1)) = f(2) = 1$

3. $(g \circ f)(1)$

SOLUTION $(g \circ f)(1) = g(f(1)) = g(4) = 3$

5. $(f \circ f)(2)$

SOLUTION $(f \circ f)(2) = f(f(2)) = f(1) = 4$

7. $(g \circ g)(4)$

SOLUTION $(g \circ g)(4) = g(g(4)) = g(3) = 1$

9. $(f \circ g \circ h)(2)$

SOLUTION

$$\begin{aligned}(f \circ g \circ h)(2) &= f(g(h(2))) \\ &= f(g(3)) = f(1) = 4\end{aligned}$$

For Exercises 11–24, evaluate the indicated expression assuming that

$$f(x) = \sqrt{x}, \quad g(x) = \frac{x+1}{x+2}, \quad h(x) = |x - 1|.$$

11. $(f \circ g)(4)$

SOLUTION

$$(f \circ g)(4) = f(g(4)) = f\left(\frac{4+1}{4+2}\right) = f\left(\frac{5}{6}\right) = \sqrt{\frac{5}{6}}$$

13. $(g \circ f)(4)$

SOLUTION

$$\begin{aligned}(g \circ f)(4) &= g(f(4)) \\ &= g(\sqrt{4}) = g(2) = \frac{2+1}{2+2} = \frac{3}{4}\end{aligned}$$

15. $(f \circ h)(-3)$

SOLUTION

$$\begin{aligned}(f \circ h)(-3) &= f(h(-3)) = f(|-3-1|)\\ &= f(|-4|) = f(4) = \sqrt{4} = 2\end{aligned}$$

17. $(f \circ g \circ h)(0)$

SOLUTION

$$\begin{aligned}(f \circ g \circ h)(0) &= f(g(h(0)))\\ &= f(g(1)) = f\left(\frac{2}{3}\right) = \sqrt{\frac{2}{3}}\end{aligned}$$

19. $(f \circ g)(0.23)$

SOLUTION

$$\begin{aligned}(f \circ g)(0.23) &= f(g(0.23)) = f\left(\frac{0.23+1}{0.23+2}\right)\\ &\approx f(0.55157) = \sqrt{0.55157} \approx 0.74268\end{aligned}$$

21. $(g \circ f)(0.23)$

SOLUTION

$$\begin{aligned}(g \circ f)(0.23) &= g(f(0.23)) = g(\sqrt{0.23})\\ &\approx g(0.47958) = \frac{0.47958+1}{0.47958+2}\\ &\approx 0.59671\end{aligned}$$

23. $(h \circ f)(0.3)$

SOLUTION

$$\begin{aligned}(h \circ f)(0.3) &= h(f(0.3)) = h(\sqrt{0.3})\\ &\approx h(0.547723) = |0.547723 - 1|\\ &= |-0.452277| = 0.452277\end{aligned}$$

In Exercises 25–30, for the given functions f and g find formulas for (a) $f \circ g$ and (b) $g \circ f$. Simplify your results as much as possible.

25. $f(x) = x^2 + 1, \quad g(x) = \dfrac{1}{x}$

SOLUTION

(a)
$$\begin{aligned}(f \circ g)(x) &= f(g(x))\\ &= f\left(\frac{1}{x}\right)\\ &= \left(\frac{1}{x}\right)^2 + 1\\ &= \frac{1}{x^2} + 1\end{aligned}$$

(b)
$$\begin{aligned}(g \circ f)(x) &= g(f(x))\\ &= g(x^2+1)\\ &= \frac{1}{x^2+1}\end{aligned}$$

27. $f(x) = \dfrac{x-1}{x+1}, \quad g(x) = x^2 + 2$

SOLUTION

(a)
$$\begin{aligned}(f \circ g)(x) &= f(g(x))\\ &= f(x^2+2)\\ &= \frac{(x^2+2)-1}{(x^2+2)+1}\\ &= \frac{x^2+1}{x^2+3}\end{aligned}$$

(b)
$$\begin{aligned}(g \circ f)(x) &= g(f(x))\\ &= g\left(\frac{x-1}{x+1}\right)\\ &= \left(\frac{x-1}{x+1}\right)^2 + 2\end{aligned}$$

29. $f(x) = \dfrac{x-1}{x^2+1}, \quad g(x) = \dfrac{x+3}{x+4}$

SOLUTION

(a) We have

$$\begin{aligned}(f \circ g)(x) &= f(g(x))\\ &= f\left(\frac{x+3}{x+4}\right)\\ &= \frac{\frac{x+3}{x+4} - 1}{\left(\frac{x+3}{x+4}\right)^2 + 1}\\ &= \frac{(x+3)(x+4) - (x+4)^2}{(x+3)^2 + (x+4)^2}\\ &= \frac{x^2+7x+12-x^2-8x-16}{x^2+6x+9+x^2+8x+16}\\ &= \frac{-x-4}{2x^2+14x+25}.\end{aligned}$$

In going from the third line above to the fourth line, both numerator and denominator were multiplied by $(x+4)^2$.

(b) We have

$$\begin{aligned}(g \circ f)(x) &= g(f(x)) \\ &= g\Big(\frac{x-1}{x^2+1}\Big) \\ &= \frac{\frac{x-1}{x^2+1}+3}{\frac{x-1}{x^2+1}+4} \\ &= \frac{x-1+3(x^2+1)}{x-1+4(x^2+1)} \\ &= \frac{3x^2+x+2}{4x^2+x+3}.\end{aligned}$$

In going from the third line above to the fourth line, both numerator and denominator were multiplied by x^2+1.

31. Find a number b such that $f \circ g = g \circ f$, where $f(x) = 2x + b$ and $g(x) = 3x + 4$.

SOLUTION We will compute $(f \circ g)(x)$ and $(g \circ f)(x)$, then set those two expressions equal to each other and solve for b. We begin with $(f \circ g)(x)$:

$$\begin{aligned}(f \circ g)(x) &= f(g(x)) = f(3x+4) \\ &= 2(3x+4) + b = 6x + 8 + b.\end{aligned}$$

Next we compute $(g \circ f)(x)$:

$$\begin{aligned}(g \circ f)(x) &= g(f(x)) = g(2x+b) \\ &= 3(2x+b) + 4 = 6x + 3b + 4.\end{aligned}$$

Looking at the expressions for $(f \circ g)(x)$ and $(g \circ f)(x)$, we see that they will equal each other if

$$8 + b = 3b + 4.$$

Solving this equation for b, we get $b = 2$.

33. Suppose

$$h(x) = \Big(\frac{x^2+1}{x-1} - 1\Big)^3.$$

(a) If $f(x) = x^3$, then find a function g such that $h = f \circ g$.

(b) If $f(x) = (x-1)^3$, then find a function g such that $h = f \circ g$.

SOLUTION

(a) We want the following equation to hold: $h(x) = f(g(x))$. Replacing h and f with the formulas for them, we have

$$\Big(\frac{x^2+1}{x-1} - 1\Big)^3 = (g(x))^3.$$

Looking at the equation above, we see that we want to choose

$$g(x) = \frac{x^2+1}{x-1} - 1.$$

(b) We want the following equation to hold: $h(x) = f(g(x))$. Replacing h and f with the formulas for them, we have

$$\Big(\frac{x^2+1}{x-1} - 1\Big)^3 = (g(x)-1)^3.$$

Looking at the equation above, we see that we want to choose

$$g(x) = \frac{x^2+1}{x-1}.$$

35. Suppose

$$h(x) = 2 + \sqrt{\frac{1}{x^2+1}}.$$

(a) If $g(x) = \frac{1}{x^2+1}$, then find a function f such that $h = f \circ g$.

(b) If $g(x) = x^2$, then find a function f such that $h = f \circ g$.

SOLUTION

(a) We want the following equation to hold: $h(x) = f(g(x))$. Replacing h and g with the formulas for them, we have

$$2 + \sqrt{\frac{1}{x^2+1}} = f\Big(\frac{1}{x^2+1}\Big).$$

Looking at the equation above, we see that we want to choose $f(x) = 2 + \sqrt{x}$.

(b) We want the following equation to hold: $h(x) = f(g(x))$. Replacing h and g with the formulas for them, we have

$$2 + \sqrt{\frac{1}{x^2+1}} = f(x^2).$$

Looking at the equation above, we see that we want to choose

$$f(x) = 2 + \sqrt{\frac{1}{x+1}}.$$

In Exercises 37–40, find functions f and g, each simpler than the given function h, such that $h = f \circ g$.

37. $h(x) = (x^2 - 1)^2$

    SOLUTION The last operation performed in the computation of $h(x)$ is squaring. Thus the most natural way to write h as a composition of two functions f and g is to choose $f(x) = x^2$, which then suggests that we choose $g(x) = x^2 - 1$.

39. $h(x) = \dfrac{3}{2 + x^2}$

    SOLUTION The last operation performed in the computation of $h(x)$ is dividing 3 by a certain expression. Thus the most natural way to write h as a composition of two functions f and g is to choose $f(x) = \frac{3}{x}$, which then requires that we choose $g(x) = 2 + x^2$.

In Exercises 41–42, find functions f, g, and h, each simpler than the given function T, such that $T = f \circ g \circ h$.

41. $T(x) = \dfrac{4}{5 + x^2}$

    SOLUTION A good solution is to take

$$f(x) = \frac{4}{x}, \quad g(x) = 5 + x, \quad h(x) = x^2.$$

43. Suppose the graph of f is a parabola with vertex at $(3, 2)$. Suppose $g(x) = 4x + 5$. What are the coordinates of the vertex of the graph of $f \circ g$?

    SOLUTION Note that

$$(f \circ g)(x) = f(g(x)) = f(4x + 5).$$

Because $f(x)$ attains its minimum or maximum value when $x = 3$, we see from the equation above that $(f \circ g)(x)$ attains its minimum or maximum value when $4x + 5 = 3$. Solving this equation for x, we see that $(f \circ g)(x)$ attains its minimum or maximum value when $x = -\frac{1}{2}$. The equation displayed above shows that this minimum or maximum value of $(f \circ g)(x)$ is the same as the minimum or maximum value of f, which equals 2. Thus the vertex of the graph of $f \circ g$ is $(-\frac{1}{2}, 2)$.

45. Suppose the graph of f is a parabola with vertex at $(3, 2)$. Suppose $g(x) = 4x + 5$. What are the coordinates of the vertex of the graph of $g \circ f$?

    SOLUTION Note that

$$(g \circ f)(x) = g(f(x)) = 4f(x) + 5.$$

Because $f(x)$ attains its minimum or maximum value (which equals 2) when $x = 3$, we see from the equation above that $(g \circ f)(x)$ also attains its minimum or maximum value when $x = 3$. We have

$$(g \circ f)(3) = g(f(3)) = 4f(3) + 5 = 4 \cdot 2 + 5 = 13.$$

Thus the vertex of the graph of $g \circ f$ is $(3, 13)$.

47. Suppose the graph of f is a parabola with vertex at (t, s). Suppose $g(x) = ax + b$, where a and b are constants with $a \neq 0$. What are the coordinates of the vertex of the graph of $f \circ g$?

    SOLUTION Note that

$$(f \circ g)(x) = f(ax + b).$$

Because $f(x)$ attains its minimum or maximum value when $x = t$, we see from the equation above that $(f \circ g)(x)$ attains its minimum or maximum value when $ax + b = t$. Thus $(f \circ g)(x)$ attains its minimum or maximum value when $x = \frac{t-b}{a}$. The equation displayed above shows that this minimum or maximum value of $(f \circ g)(x)$ is the same as the minimum or maximum value of f, which equals s. Thus the vertex of the graph of $f \circ g$ is $(\frac{t-b}{a}, s)$.

3.4 *Inverse Functions*

LEARNING OBJECTIVES

By the end of this section you should be able to

- determine which functions have inverses;
- find a formula for an inverse function (when possible);
- use the composition of a function and its inverse to check that an inverse function has been correctly found;
- find the domain and range of an inverse function.

The Inverse Problem

The concept of an inverse function will play a key role in this book in defining roots and logarithms. To motivate this concept, we begin with some simple examples.

Suppose f is the function defined by $f(x) = 3x$. Given a value of x, we can find the value of $f(x)$ by using the formula defining f. For example, taking $x = 5$, we see that $f(5)$ equals 15.

In the inverse problem, we are given the value of $f(x)$ and asked to find the value of x. The following example illustrates the idea of the inverse problem:

EXAMPLE 1

Suppose f is the function defined by

$$f(x) = 3x.$$

(a) Find x such that $f(x) = 6$.

(b) Find x such that $f(x) = 300$.

(c) For each number y, find a number x such that $f(x) = y$.

SOLUTION

(a) Solving the equation $3x = 6$ for x, we get $x = 2$.

(b) Solving the equation $3x = 300$ for x, we get $x = 100$.

(c) Solving the equation $3x = y$ for x, we get $x = \frac{y}{3}$.

Inverse functions will be defined more precisely after we work through some examples.

For each number y, part (c) of the example above asks for the number x such that $f(x) = y$. That number x is called $f^{-1}(y)$ (pronounced "f inverse of y"). The example above shows that if $f(x) = 3x$, then $f^{-1}(6) = 2$ and $f^{-1}(300) = 100$ and, more generally, $f^{-1}(y) = \frac{y}{3}$ for every number y.

To see how inverse functions can arise in real-world problems, suppose you know that a temperature of x degrees Celsius corresponds to $\frac{9}{5}x + 32$ degrees Fahrenheit (we derived this formula in Example 5 in Section 2.2). In other words, you know that the function f that converts the Celsius temperature scale to the Fahrenheit temperature scale is given by the formula

$$f(x) = \tfrac{9}{5}x + 32.$$

For example, because $f(20) = 68$, this formula shows that 20 degrees Celsius corresponds to 68 degrees Fahrenheit.

If you are given a temperature on the Fahrenheit scale and asked to convert it to Celsius, then you are facing the problem of finding the inverse of the function above, as shown in the following example.

EXAMPLE 2

The Fahrenheit temperature scale was invented in the 18th century by the German physicist and engineer Daniel Gabriel Fahrenheit.

(a) Convert 95 degrees Fahrenheit to the Celsius scale.

(b) For each temperature y on the Fahrenheit scale, what is the corresponding temperature on the Celsius scale?

SOLUTION Let

$$f(x) = \tfrac{9}{5}x + 32.$$

Thus x degrees Celsius corresponds to $f(x)$ degrees Fahrenheit.

(a) We need to find x such that $f(x) = 95$. Solving the equation $\frac{9}{5}x + 32 = 95$ for x, we get $x = 35$. Thus 35 degrees Celsius corresponds to 95 degrees Fahrenheit.

(b) For each number y, we need to find x such that $f(x) = y$. Solving the equation $\frac{9}{5}x + 32 = y$ for x, we get $x = \frac{5}{9}(y - 32)$. Thus $\frac{5}{9}(y - 32)$ degrees Celsius corresponds to y degrees Fahrenheit.

The Celsius temperature scale is named in honor of the 18th century Swedish astronomer Anders Celsius, who originally proposed a temperature scale with 0 as the boiling point of water and 100 as the freezing point. Later this was reversed, giving us the familiar scale in which higher numbers correspond to hotter temperatures.

In the example above we have $f(x) = \frac{9}{5}x + 32$. For each number y, part (b) of the example above asks for the number x such that $f(x) = y$. We call that number $f^{-1}(y)$. Part (a) of the example above shows that $f^{-1}(95) = 35$; part (b) shows more generally that

$$f^{-1}(y) = \tfrac{5}{9}(y - 32).$$

In this example, the function f converts from Celsius to Fahrenheit, and the function f^{-1} goes in the other direction, converting from Fahrenheit to Celsius.

One-to-one Functions

To see the difficulties that can arise with inverse problems, consider the function f, with domain the set of real numbers, defined by the formula

$$f(x) = x^2.$$

Suppose we are told that x is a number such that $f(x) = 16$, and we are asked to find the value of x. Of course $f(4) = 16$, but also $f(-4) = 16$. Thus with the information given we have no way to determine a unique value of x such that $f(x) = 16$. Hence in this case an inverse function does not exist.

The difficulty with the lack of a unique solution to an inverse problem can often be fixed by changing the domain. For example, consider the function g, with domain the set of positive numbers, defined by the formula

$$g(x) = x^2.$$

Note that g is defined by the same formula as f in the previous paragraph, but these two functions are not the same because they have different domains. Now if we are told that x is a number in the domain of g such that $g(x) = 16$ and we are asked to find x, we can assert that $x = 4$. More generally, given any positive number y, we can ask for the number x in the domain of g such that $g(x) = y$. This number x, which depends on y, is denoted $g^{-1}(y)$, and is given by the formula

$$g^{-1}(y) = \sqrt{y}.$$

We saw earlier that the function f defined by $f(x) = x^2$ (and with domain equal to the set of real numbers) does not have an inverse because, in particular, the equation $f(x) = 16$ has more than one solution. A function is called **one-to-one** if this situation does not arise.

As we will soon see, functions that are one-to-one are precisely the functions that have inverses.

One-to-one

A function f is called **one-to-one** if for each number y in the range of f there is exactly one number x in the domain of f such that $f(x) = y$.

For example, the function f, with domain the set of real numbers, defined by $f(x) = x^2$ is not one-to-one because there are two distinct numbers x in the domain of f such that $f(x) = 16$ (we could have used any positive number instead of 16 to show that f is not one-to-one). In contrast, the function g, with domain the set of positive numbers, defined by $g(x) = x^2$ is one-to-one.

The Definition of an Inverse Function

We are now ready to give the formal definition of an inverse function.

Definition of f^{-1}

Suppose f is a one-to-one function.

- If y is in the range of f, then $f^{-1}(y)$ is defined to be the number x such that $f(x) = y$.
- The function f^{-1} is called the **inverse function** of f.

Short version:

- $f^{-1}(y) = x$ means $f(x) = y$.

EXAMPLE 3

Suppose $f(x) = 2x + 3$.

(a) Evaluate $f^{-1}(11)$.

(b) Find a formula for $f^{-1}(y)$.

SOLUTION

(a) To evaluate $f^{-1}(11)$, we must find the number x such that $f(x) = 11$. In other words, we must solve the equation $2x + 3 = 11$. The solution to this equation is $x = 4$. Thus $f(4) = 11$, and hence $f^{-1}(11) = 4$.

(b) Fix a number y. To find a formula for $f^{-1}(y)$, we must find the number x such that $f(x) = y$. In other words, we must solve the equation

$$2x + 3 = y$$

for x. The solution to this equation is $x = \frac{y-3}{2}$. Thus $f(\frac{y-3}{2}) = y$, and hence $f^{-1}(y) = \frac{y-3}{2}$.

If f is a one-to-one function, then for each y in the range of f we have a uniquely defined number $f^{-1}(y)$. Thus f^{-1} is itself a function.

The inverse function is not defined for a function that is not one-to-one.

Think of f^{-1} as undoing whatever f does. This list gives some examples of a function f and its inverse f^{-1}.

f	f^{-1}
$f(x) = x + 2$	$f^{-1}(y) = y - 2$
$f(x) = 3x$	$f^{-1}(y) = \frac{y}{3}$
$f(x) = x^2$	$f^{-1}(y) = \sqrt{y}$
$f(x) = \sqrt{x}$	$f^{-1}(y) = y^2$

The first entry in the list above shows that if f is the function that adds 2 to a number, then f^{-1} is the function that subtracts 2 from a number.

The second entry in the list above shows that if f is the function that multiplies a number by 3, then f^{-1} is the function that divides a number by 3.

Similarly, the third entry in the list above shows that if f is the function that squares a number, then f^{-1} is the function that takes the square root of a number (here the domain of f is assumed to be the nonnegative numbers, so that we have a one-to-one function).

Finally, the fourth entry in the list above shows that if f is the function that takes the square root of a number, then f^{-1} is the function that squares a number (here the domain of f is assumed to be the nonnegative numbers, because the square root of a negative number is not defined as a real number).

In Section 3.1 we saw that a function f can be thought of as a machine that takes an input x and produces an output $f(x)$. Similarly, we can think of f^{-1} as a machine that takes an input $f(x)$ and produces an output x.

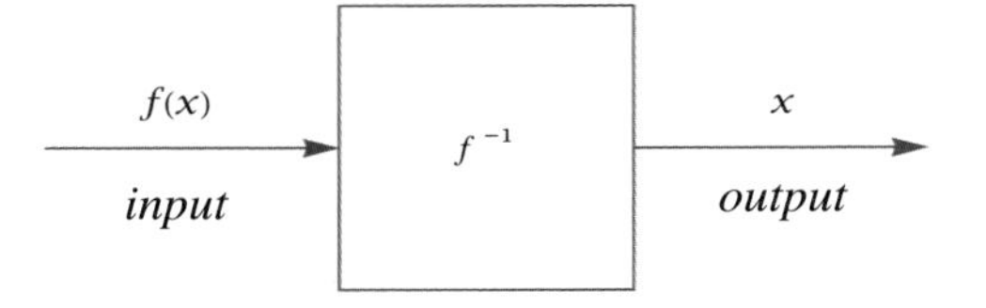

Thought of as a machine, f^{-1} reverses the action of f.

The procedure for finding a formula for an inverse function can be described as follows:

Finding a formula for an inverse function

Suppose f is a one-to-one function. To find a formula for $f^{-1}(y)$, solve the equation $f(x) = y$ for x in terms of y.

EXAMPLE 4

Suppose f is the function whose domain is the set of positive numbers, with

$$f(x) = \frac{x^2 + 2x}{2x + 6}$$

for every $x > 0$. Find a formula for f^{-1}.

This function f as defined here is one-to-one. However, this function would not be one-to-one if the domain were the set of real numbers.

SOLUTION To find a formula for $f^{-1}(y)$, we need to solve the equation

$$\frac{x^2 + 2x}{2x + 6} = y$$

for x in terms of y. This can be done be multiplying both sides of the equation above by $2x + 6$ and then using the quadratic formula to solve for x in terms of y (think of y as a constant and x as the unknown). Or you could use a computer (enter `solve (x^2 + 2x)/(2x + 6) = y for x` in a Wolfram|Alpha entry box) or a symbolic calculator. Using any of these methods, you should see that

$$x = \sqrt{y^2 + 4y + 1} + y - 1 \quad \text{or} \quad x = -\sqrt{y^2 + 4y + 1} + y - 1.$$

Because the domain of f is the set of positive numbers, the equation $f(x) = y$ along with the specific formula defining f show that both x and y must be positive. However, the second possibility given above for x is a negative number. Thus x must be given by the first possibility above. In other words,

$$f^{-1}(y) = \sqrt{y^2 + 4y + 1} + y - 1$$

for every $y > 0$.

The Domain and Range of an Inverse Function

The domain and range of a one-to-one function are nicely related to the domain and range of its inverse. To understand this relationship, consider a one-to-one function f. Note that $f^{-1}(y)$ is defined precisely when y is in the range of f. Thus the domain of f^{-1} equals the range of f.

Similarly, because f^{-1} reverses the action of f, a moment's thought shows that the range of f^{-1} equals the domain of f. We can summarize the relationship between the domains and ranges of functions and their inverses as follows:

Domain and range of an inverse function

If f is a one-to-one function, then

- the domain of f^{-1} equals the range of f;
- the range of f^{-1} equals the domain of f.

EXAMPLE 5

Suppose the domain of f is the interval $[0, 2]$, with f defined on this domain by the equation $f(x) = x^2$.

(a) What is the range of f?

(b) Find a formula for the inverse function f^{-1}.

(c) What is the domain of the inverse function f^{-1}?

(d) What is the range of the inverse function f^{-1}?

SOLUTION

(a) The range of f is the interval $[0, 4]$ because that interval is equal to the set of squares of numbers in the interval $[0, 2]$.

(b) Suppose y is in the range of f, which is the interval $[0, 4]$. To find a formula for $f^{-1}(y)$, we have to solve for x the equation $f(x) = y$. In other words, we have to solve the equation $x^2 = y$ for x. The solution x must be in the domain of f, which is $[0, 2]$, and in particular x must be nonnegative. Thus we have $x = \sqrt{y}$. In other words, $f^{-1}(y) = \sqrt{y}$.

(c) The domain of the inverse function f^{-1} is the interval $[0, 4]$, which is the range of f.

(d) The range of the inverse function f^{-1} is the interval $[0, 2]$, which is the domain of f.

This example illustrates how the inverse function interchanges the domain and range of the original function.

The Composition of a Function and Its Inverse

The following example will help motivate our next result.

EXAMPLE 6

Suppose f is the function whose domain is the set of real numbers, with f defined by $f(x) = 2x + 3$.

(a) Find a formula for $f \circ f^{-1}$.

(b) Find a formula for $f^{-1} \circ f$.

SOLUTION As we saw in Example 3, $f^{-1}(y) = \frac{y-3}{2}$. Thus we have the following:

(a) $(f \circ f^{-1})(y) = f(f^{-1}(y)) = f(\frac{y-3}{2}) = 2(\frac{y-3}{2}) + 3 = y$

(b) $(f^{-1} \circ f)(x) = f^{-1}(f(x)) = f^{-1}(2x + 3) = \frac{(2x+3)-3}{2} = x$

Similar equations hold for the composition of any one-to-one function and its inverse:

The composition of a function and its inverse

Suppose f is a one-to-one function. Then

- $f(f^{-1}(y)) = y$ for every y in the range of f;
- $f^{-1}(f(x)) = x$ for every x in the domain of f.

To see why these results hold, first suppose y is a number in the range of f. Let $x = f^{-1}(y)$. Then $f(x) = y$. Thus

$$f(f^{-1}(y)) = f(x) = y,$$

as claimed above.

To verify the second conclusion in the box above, suppose x is a number in the domain of f. Let $y = f(x)$ Then $f^{-1}(y) = x$. Thus

$$f^{-1}(f(x)) = f^{-1}(y) = x,$$

as claimed.

Recall that I is the identity function defined by $I(x) = x$ (where we have left the domain vague), or we could equally well define I by the equation $I(y) = y$. The results in the box above could be expressed by the equations

$$f \circ f^{-1} = I \quad \textbf{and} \quad f^{-1} \circ f = I.$$

Here the I in the first equation above has domain equal to the range of f (which equals the domain of f^{-1}), and the I in the second equation above has the same domain as f. The equations above explain why the terminology "inverse" is used for the inverse function: f^{-1} is the inverse of f under composition in the sense that the composition of f and f^{-1} in either order gives the identity function.

The figure below illustrates the equation $f^{-1} \circ f = I$, thinking of f and f^{-1} as machines.

Here we start with x as input and end with x as output. Thus this figure illustrates the equation $f^{-1} \circ f = I$.

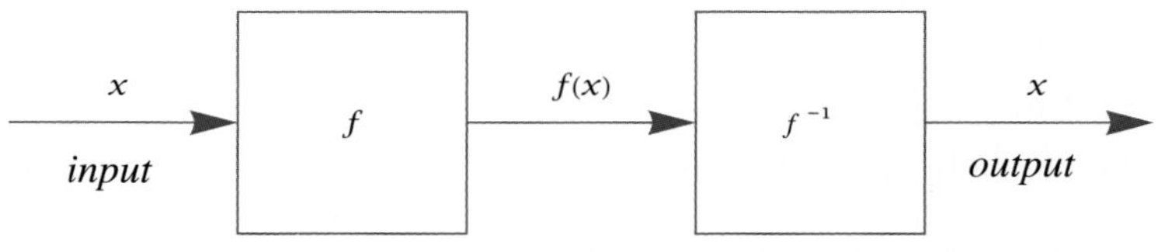

Here we start with x as the input. The first machine produces output $f(x)$, which then becomes the input for the second machine. When $f(x)$ is input into the second machine, the output is x because the second machine based on f^{-1} reverses the action of f.

Suppose you need to compute the inverse of a function f. As discussed earlier, to find a formula for f^{-1} you need to solve the equation $f(x) = y$ for x in terms of y. Once you have obtained a formula for f^{-1}, a good way to check that you have the correct formula is to verify one or both of the equations in the box above.

EXAMPLE 7

Suppose $f(x) = \frac{9}{5}x + 32$, which is the formula for converting the Celsius temperature scale to the Fahrenheit scale. We computed earlier that the inverse to this function is given by the formula $f^{-1}(y) = \frac{5}{9}(y - 32)$. Check that this is correct by verifying that $f(f^{-1}(y)) = y$ for every real number y.

SOLUTION To check that we have the right formula for f^{-1}, we compute as follows:

$$\begin{aligned} f(f^{-1}(y)) &= f\big(\tfrac{5}{9}(y - 32)\big) \\ &= \tfrac{9}{5}\big(\tfrac{5}{9}(y - 32)\big) + 32 \\ &= (y - 32) + 32 \\ &= y. \end{aligned}$$

To be doubly safe that we are not making an algebraic manipulation error, we could also verify that $f^{-1}(f(x)) = x$ for every real number x. However, one check is usually good enough.

Thus $f(f^{-1}(y)) = y$, which means that our formula for f^{-1} is correct. If our computation of $f(f^{-1}(y))$ had simplified to anything other than y, we would know that we had made a mistake in computing f^{-1}.

Comments About Notation

The notation $y = f(x)$ leads naturally to the notation $f^{-1}(y)$. Recall, however, that in defining a function the variable is simply a placeholder. Thus we could use other letters, including x, as the variable for the inverse function. For example, consider the function f, with domain equal to the set of positive numbers, defined by the equation

$$f(x) = x^2.$$

As we know, the inverse function is given by the formula

$$f^{-1}(y) = \sqrt{y}.$$

However, the inverse function could also be characterized by the formula

$$f^{-1}(x) = \sqrt{x}.$$

Other letters could also be used as the placeholder. For example, we might also characterize the inverse function by the formula

$$f^{-1}(t) = \sqrt{t}.$$

Do not confuse $f^{-1}(y)$ with $[f(y)]^{-1}$.

The notation f^{-1} for the inverse of a function (which means the inverse under composition) should not be confused with the multiplicative inverse $\frac{1}{f}$. In other words, $f^{-1} \neq \frac{1}{f}$. However, if the exponent -1 is placed anywhere other than immediately after a function symbol, then it should probably be interpreted as a multiplicative inverse.

EXAMPLE 8

Suppose $f(x) = x^2 - 1$, with the domain of f being the set of positive numbers.

(a) Evaluate $f^{-1}(8)$.

(b) Evaluate $[f(8)]^{-1}$.

(c) Evaluate $f(8^{-1})$.

SOLUTION

(a) To evaluate $f^{-1}(8)$, we must find a positive number x such that $f(x) = 8$. In other words, we must solve the equation $x^2 - 1 = 8$. The solution to this equation is $x = 3$. Thus $f(3) = 8$, and hence $f^{-1}(8) = 3$.

(b)
$$[f(8)]^{-1} = \frac{1}{f(8)} = \frac{1}{8^2 - 1} = \frac{1}{63}$$

(c)
$$f(8^{-1}) = f\left(\tfrac{1}{8}\right) = \left(\tfrac{1}{8}\right)^2 - 1 = \frac{1}{64} - 1 = -\frac{63}{64}$$

When dealing with real-world problems, you may want to choose the notation to reflect the context. The next example illustrates this idea, with the use of the variable d to denote distance and t to denote time.

EXAMPLE 9

Suppose you ran a marathon (26.2 miles) in exactly 4 hours. Let f be the function with domain $[0, 26.2]$ such that $f(d)$ is the number of minutes since the start of the race at which you reached distance d miles from the starting line.

(a) What is the range of f?

(b) What is the domain of the inverse function f^{-1}?

(c) What is the meaning of $f^{-1}(t)$ for a number t in the domain of f^{-1}?

SOLUTION

(a) Because 4 hours equals 240 minutes, the range of f is the interval $[0, 240]$.

(b) As usual, the domain of f^{-1} is the range of f. Thus the domain of f^{-1} is the interval $[0, 240]$.

(c) The function f^{-1} reverses the roles of the input and the output as compared to the function f. Thus $f^{-1}(t)$ is the distance in miles you had run from the starting line at time t minutes after the start of the race.

EXERCISES

For Exercises 1–8, check your answer by evaluating the appropriate function at your answer.

1. Suppose $f(x) = 4x + 6$. Evaluate $f^{-1}(5)$.
2. Suppose $f(x) = 7x - 5$. Evaluate $f^{-1}(-3)$.
3. Suppose $g(x) = \frac{x+2}{x+1}$. Evaluate $g^{-1}(3)$.
4. Suppose $g(x) = \frac{x-3}{x-4}$. Evaluate $g^{-1}(2)$.
5. Suppose $f(x) = 3x + 2$. Find a formula for f^{-1}.
6. Suppose $f(x) = 8x - 9$. Find a formula for f^{-1}.
7. Suppose $h(t) = \frac{1+t}{2-t}$. Find a formula for h^{-1}.
8. Suppose $h(t) = \frac{2-3t}{4+5t}$. Find a formula for h^{-1}.
9. Suppose $f(x) = 2 + \frac{x-5}{x+6}$.
   (a) Evaluate $f^{-1}(4)$.
   (b) Evaluate $[f(4)]^{-1}$.
   (c) Evaluate $f(4^{-1})$.
10. Suppose $h(x) = 3 - \frac{x+4}{x-7}$.
    (a) Evaluate $h^{-1}(9)$.
    (b) Evaluate $[h(9)]^{-1}$.
    (c) Evaluate $h(9^{-1})$.
11. Suppose $h(x) = x^2 + 3x + 4$, with the domain of h being the set of positive numbers. Evaluate $h^{-1}(7)$.
12. Suppose $h(x) = x^2 + 2x - 5$, with the domain of h being the set of positive numbers. Evaluate $h^{-1}(4)$.
13. Suppose $h(x) = 5x^2 + 7$, where the domain of h is the set of positive numbers. Find a formula for h^{-1}.
14. Suppose $h(x) = 3x^2 - 4$, where the domain of h is the set of positive numbers. Find a formula for h^{-1}.

For each of the functions f given in Exercises 15–24:

(a) Find the domain of f.

(b) Find the range of f.

(c) Find a formula for f^{-1}.

(d) Find the domain of f^{-1}.

(e) Find the range of f^{-1}.

You can check your solutions to part (c) by verifying that $f^{-1} \circ f = I$ and $f \circ f^{-1} = I$ (recall that I is the function defined by $I(x) = x$).

15. $f(x) = 3x + 5$
16. $f(x) = 2x - 7$
17. $f(x) = \frac{1}{3x+2}$
18. $f(x) = \frac{4}{5x-3}$
19. $f(x) = \frac{2x}{x+3}$
20. $f(x) = \frac{3x-2}{4x+5}$
21. $f(x) = \begin{cases} 3x & \text{if } x < 0 \\ 4x & \text{if } x \ge 0 \end{cases}$
22. $f(x) = \begin{cases} 2x & \text{if } x < 0 \\ x^2 & \text{if } x \ge 0 \end{cases}$
23. $f(x) = x^2 + 8$, where the domain of f equals $(0, \infty)$.
24. $f(x) = 2x^2 + 5$, where the domain of f equals $(0, \infty)$.
25. Suppose $f(x) = x^5 + 2x^3$. Which of the numbers listed below equals $f^{-1}(8.10693)$?

    1.1, 1.2, 1.3, 1.4

    [For this particular function, it is not possible to find a formula for $f^{-1}(y)$.]
26. Suppose $f(x) = 3x^5 + 4x^3$. Which of the numbers listed below equals $f^{-1}(0.28672)$?

    0.2, 0.3, 0.4, 0.5

    [For this particular function, it is not possible to find a formula for $f^{-1}(y)$.]

For Exercises 27–28, use the U. S. 2010 federal income tax function for a single person as defined in Example 2 of Section 3.1.

27. What is the taxable income of a single person who paid \$10,000 in federal taxes for 2010?

28. What is the taxable income of a single person who paid \$20,000 in federal taxes for 2010?

For Exercises 29–36, suppose f and g are functions whose domain is the interval $[1, \infty)$, with

$$f(x) = x^2 + 3x + 5 \quad \textit{and} \quad g(x) = x^2 + 4x + 7.$$

29. What is the range of f?

30. What is the range of g?

31. Find a formula for f^{-1}.

32. Find a formula for g^{-1}.

33. What is the domain of f^{-1}?

34. What is the domain of g^{-1}?

35. What is the range of f^{-1}?

36. What is the range of g^{-1}?

37. Suppose $g(x) = x^7 + x^3$. Evaluate

$$\left(g^{-1}(4)\right)^7 + \left(g^{-1}(4)\right)^3 + 1.$$

38. Suppose $g(x) = 8x^9 + 7x^3$. Evaluate

$$8\left(g^{-1}(5)\right)^9 + 7\left(g^{-1}(5)\right)^3 - 3.$$

PROBLEMS

39. The exact number of meters in y yards is $f(y)$, where f is the function defined by

$$f(y) = 0.9144y.$$

 (a) Find a formula for $f^{-1}(m)$.

 (b) What is the meaning of $f^{-1}(m)$?

40. The exact number of kilometers in M miles is $f(M)$, where f is the function defined by

$$f(M) = 1.609344M.$$

 (a) Find a formula for $f^{-1}(k)$.

 (b) What is the meaning of $f^{-1}(k)$?

41. A temperature F degrees Fahrenheit corresponds to $g(F)$ degrees on the Kelvin temperature scale, where

$$g(F) = \tfrac{5}{9}F + 255.37.$$

 (a) Find a formula for $g^{-1}(K)$.

 (b) What is the meaning of $g^{-1}(K)$?

 (c) Evaluate $g^{-1}(0)$? (This is absolute zero, the lowest possible temperature, because all molecular activity stops at 0 degrees Kelvin.)

42. Suppose g is the federal income tax function given by Example 2 of Section 3.1. What is the meaning of the function g^{-1}?

43. Suppose f is the function whose domain is the set of real numbers, with f defined on this domain by the formula

$$f(x) = |x + 6|.$$

Explain why f is not a one-to-one function.

44. Suppose g is the function whose domain is the interval $[-2, 2]$, with g defined on this domain by the formula

$$g(x) = (5x^2 + 3)^{7777}.$$

Explain why g is not a one-to-one function.

45. Show that if f is the linear function defined by $f(x) = mx + b$, where $m \neq 0$, then f is a one-to-one function.

46. Show that if f is the linear function defined by $f(x) = mx + b$, where $m \neq 0$, then the inverse function f^{-1} is defined by the formula $f^{-1}(y) = \frac{1}{m}y - \frac{b}{m}$.

47. Consider the function h whose domain is the interval $[-4, 4]$, with h defined on this domain by the formula

$$h(x) = (2 + x)^2.$$

Does h have an inverse? If so, find it, along with its domain and range. If not, explain why not.

48. Consider the function h whose domain is the interval $[-3, 3]$, with h defined on this domain by the formula

$$h(x) = (3 + x)^2.$$

Does h have an inverse? If so, find it, along with its domain and range. If not, explain why not.

49. Suppose f is a one-to-one function. Explain why the inverse of the inverse of f equals f. In other words, explain why

$$(f^{-1})^{-1} = f.$$

50. The function f defined by

$$f(x) = x^5 + x^3$$

is one-to-one (here the domain of f is the set of real numbers). Compute $f^{-1}(y)$ for four different values of y of your choice.
[For this particular function, it is not possible to find a formula for $f^{-1}(y)$.]

51. Suppose f is a function whose domain equals $\{2, 4, 7, 8, 9\}$ and whose range equals $\{-3, 0, 2, 6, 7\}$. Explain why f is a one-to-one function.

52. Suppose f is a function whose domain equals $\{2, 4, 7, 8, 9\}$ and whose range equals $\{-3, 0, 2, 6\}$. Explain why f is not a one-to-one function.

53. Show that the composition of two one-to-one functions is a one-to-one function.

54. Give an example to show that the sum of two one-to-one functions is not necessarily a one-to-one function.

55. Give an example to show that the product of two one-to-one functions is not necessarily a one-to-one function.

56. Give an example of a function f such that the domain of f and the range of f both equal the set of integers, but f is not a one-to-one function.

57. Give an example of a one-to-one function whose domain equals the set of integers and whose range equals the set of positive integers.

WORKED-OUT SOLUTIONS *to Odd-numbered Exercises*

For Exercises 1–8, check your answer by evaluating the appropriate function at your answer.

1. Suppose $f(x) = 4x + 6$. Evaluate $f^{-1}(5)$.

SOLUTION We need to find a number x such that $f(x) = 5$. In other words, we need to solve the equation

$$4x + 6 = 5.$$

This equation has solution $x = -\frac{1}{4}$. Thus $f^{-1}(5) = -\frac{1}{4}$.

CHECK To check that $f^{-1}(5) = -\frac{1}{4}$, we need to verify that $f(-\frac{1}{4}) = 5$. We have

$$f(-\tfrac{1}{4}) = 4(-\tfrac{1}{4}) + 6 = 5,$$

as desired.

3. Suppose $g(x) = \dfrac{x+2}{x+1}$. Evaluate $g^{-1}(3)$.

SOLUTION We need to find a number x such that $g(x) = 3$. In other words, we need to solve the equation

$$\frac{x+2}{x+1} = 3.$$

Multiplying both sides of this equation by $x + 1$ gives the equation

$$x + 2 = 3x + 3,$$

which has solution $x = -\frac{1}{2}$. Thus $g^{-1}(3) = -\frac{1}{2}$.

CHECK To check that $g^{-1}(3) = -\frac{1}{2}$, we need to verify that $g(-\frac{1}{2}) = 3$. We have

$$g(-\tfrac{1}{2}) = \frac{-\frac{1}{2} + 2}{-\frac{1}{2} + 1} = \frac{\frac{3}{2}}{\frac{1}{2}} = 3,$$

as desired.

5. Suppose $f(x) = 3x + 2$. Find a formula for f^{-1}.

SOLUTION For each number y, we need to find a number x such that $f(x) = y$. In other words, we need to solve the equation

$$3x + 2 = y$$

for x in terms of y. Subtracting 2 from both sides of the equation above and then dividing both sides by 3 gives

$$x = \frac{y-2}{3}.$$

Thus

$$f^{-1}(y) = \frac{y-2}{3}$$

for every number y.

CHECK To check that $f^{-1}(y) = \frac{y-2}{3}$, we need to verify that $f(\frac{y-2}{3}) = y$. We have

$$f(\tfrac{y-2}{3}) = 3(\frac{y-2}{3}) + 2 = y,$$

as desired.

7. Suppose $h(t) = \frac{1+t}{2-t}$. Find a formula for h^{-1}.

SOLUTION For each number y, we need to find a number t such that $h(t) = y$. In other words, we need to solve the equation

$$\frac{1+t}{2-t} = y$$

for t in terms of y. Multiplying both sides of this equation by $2-t$ and then collecting all the terms with t on one side gives

$$t + yt = 2y - 1.$$

Rewriting the left side as $(1+y)t$ and then dividing both sides by $1+y$ gives

$$t = \frac{2y-1}{1+y}.$$

Thus

$$h^{-1}(y) = \frac{2y-1}{1+y}$$

for every number $y \neq -1$.

CHECK To check that $h^{-1}(y) = \frac{2y-1}{1+y}$, we need to verify that $h(\frac{2y-1}{1+y}) = y$. We have

$$h(\tfrac{2y-1}{1+y}) = \frac{1 + \frac{2y-1}{1+y}}{2 - \frac{2y-1}{1+y}}.$$

Multiplying numerator and denominator of the expression on the right by $1+y$ gives

$$h(\tfrac{2y-1}{1+y}) = \frac{1+y+2y-1}{2+2y-2y+1} = \frac{3y}{3} = y,$$

as desired.

9. Suppose $f(x) = 2 + \frac{x-5}{x+6}$.

(a) Evaluate $f^{-1}(4)$.

(b) Evaluate $[f(4)]^{-1}$.

(c) Evaluate $f(4^{-1})$.

SOLUTION

(a) We need to find a number x such that $f(x) = 4$. In other words, we need to solve the equation

$$2 + \frac{x-5}{x+6} = 4.$$

Subtracting 2 from both sides and then multiplying both sides by $x+6$ gives the equation

$$x - 5 = 2x + 12,$$

which has solution $x = -17$. Thus $f^{-1}(4) = -17$.

(b) Note that

$$f(4) = 2 + \tfrac{4-5}{4+6} = \tfrac{20}{10} - \tfrac{1}{10} = \tfrac{19}{10}.$$

Thus $[f(4)]^{-1} = \frac{10}{19}$.

(c)

$$f(4^{-1}) = f(\tfrac{1}{4}) = 2 + \frac{\frac{1}{4} - 5}{\frac{1}{4} + 6} = \frac{31}{25}$$

11. Suppose $h(x) = x^2 + 3x + 4$, with the domain of h being the set of positive numbers. Evaluate $h^{-1}(7)$.

SOLUTION We need to find a positive number x such that $h(x) = 7$. In other words, we need to find a positive solution to the equation

$$x^2 + 3x + 4 = 7,$$

which is equivalent to the equation

$$x^2 + 3x - 3 = 0.$$

The quadratic formula shows that the equation above has solutions

$$x = \frac{-3+\sqrt{21}}{2} \quad \text{and} \quad x = \frac{-3-\sqrt{21}}{2}.$$

Because the domain of h is the set of positive numbers, the value of x that we seek must be positive. The second solution above is negative; thus it can be discarded, giving $h^{-1}(7) = \frac{-3+\sqrt{21}}{2}$.

CHECK To check that $h^{-1}(7) = \frac{-3+\sqrt{21}}{2}$, we must verify that $h(\frac{-3+\sqrt{21}}{2}) = 7$. We have

$$\begin{aligned} h(\tfrac{-3+\sqrt{21}}{2}) &= (\tfrac{-3+\sqrt{21}}{2})^2 + 3(\tfrac{-3+\sqrt{21}}{2}) + 4 \\ &= \tfrac{15-3\sqrt{21}}{2} + \tfrac{-9+3\sqrt{21}}{2} + 4 \\ &= 7, \end{aligned}$$

as desired.

13. Suppose $h(x) = 5x^2 + 7$, where the domain of h is the set of positive numbers. Find a formula for h^{-1}.

SOLUTION For each number y, we need to find a number x such that $h(x) = y$. In other words, we need to solve the equation

$$5x^2 + 7 = y$$

for x in terms of y. Subtracting 7 from both sides of the equation above and then dividing both sides by 5 and taking square roots gives

$$x = \sqrt{\frac{y-7}{5}},$$

where we chose the positive square root because x is required to be a positive number. Thus

$$h^{-1}(y) = \sqrt{\frac{y-7}{5}}$$

for every number $y > 7$ (the restriction that $y > 7$ is required to ensure that we get a positive number when evaluating the formula above).

For each of the functions f given in Exercises 15–24:

- **(a)** ***Find the domain of f.***
- **(b)** ***Find the range of f.***
- **(c)** ***Find a formula for f^{-1}.***
- **(d)** ***Find the domain of f^{-1}.***
- **(e)** ***Find the range of f^{-1}.***

You can check your solutions to part (c) by verifying that $f^{-1} \circ f = I$ and $f \circ f^{-1} = I$ (recall that I is the function defined by $I(x) = x$).

15. $f(x) = 3x + 5$

SOLUTION

(a) The expression $3x + 5$ makes sense for all real numbers x. Thus the domain of f is the set of real numbers.

(b) To find the range of f, we need to find the numbers y such that

$$y = 3x + 5$$

for some x in the domain of f. In other words, we need to find the values of y such that the equation above can be solved for a real number x. Solving the equation above for x, we get

$$x = \frac{y-5}{3}.$$

The expression above on the right makes sense for every real number y. Thus the range of f is the set of real numbers.

(c) The expression above shows that f^{-1} is given by the formula

$$f^{-1}(y) = \frac{y-5}{3}.$$

(d) The domain of f^{-1} equals the range of f. Thus the domain of f^{-1} is the set of real numbers.

(e) The range of f^{-1} equals the domain of f. Thus the range of f^{-1} is the set of real numbers.

17. $f(x) = \dfrac{1}{3x+2}$

SOLUTION

(a) The expression $\frac{1}{3x+2}$ makes sense except when $3x + 2 = 0$. Solving this equation for x gives $x = -\frac{2}{3}$. Thus the domain of f is the set $\{x : x \neq -\frac{2}{3}\}$.

(b) To find the range of f, we need to find the numbers y such that

$$y = \frac{1}{3x+2}$$

for some x in the domain of f. In other words, we need to find the values of y such that the equation above can be solved for a real number $x \neq -\frac{2}{3}$. To solve this equation for x, multiply both sides by $3x + 2$, getting

$$3xy + 2y = 1.$$

Now subtract $2y$ from both sides, then divide by $3y$, getting

$$x = \frac{1 - 2y}{3y}.$$

The expression above on the right makes sense for every real number $y \neq 0$ and produces a number $x \neq -\frac{2}{3}$ (because the equation $-\frac{2}{3} = \frac{1-2y}{3y}$ leads to nonsense, as you can verify if you try to solve it for y). Thus the range of f is the set $\{y : y \neq 0\}$.

(c) The expression above shows that f^{-1} is given by the formula

$$f^{-1}(y) = \frac{1 - 2y}{3y}.$$

(d) The domain of f^{-1} equals the range of f. Thus the domain of f^{-1} is the set $\{y : y \neq 0\}$.

(e) The range of f^{-1} equals the domain of f. Thus the range of f^{-1} is the set $\{x : x \neq -\frac{2}{3}\}$.

19. $f(x) = \dfrac{2x}{x + 3}$

SOLUTION

(a) The expression $\frac{2x}{x+3}$ makes sense except when $x = -3$. Thus the domain of f is the set $\{x : x \neq -3\}$.

(b) To find the range of f, we need to find the numbers y such that

$$y = \frac{2x}{x + 3}$$

for some x in the domain of f. In other words, we need to find the values of y such that the equation above can be solved for a real number $x \neq -3$. To solve this equation for x, multiply both sides by $x + 3$, getting

$$xy + 3y = 2x.$$

Now subtract xy from both sides, getting

$$3y = 2x - xy = x(2 - y).$$

Dividing by $2 - y$ gives

$$x = \frac{3y}{2 - y}.$$

The expression above on the right makes sense for every real number $y \neq 2$ and produces a number $x \neq -3$ (because the equation $-3 = \frac{3y}{2-y}$ leads to nonsense, as you can verify if you try to solve it for y). Thus the range of f is the set $\{y : y \neq 2\}$.

(c) The expression above shows that f^{-1} is given by the formula

$$f^{-1}(y) = \frac{3y}{2 - y}.$$

(d) The domain of f^{-1} equals the range of f. Thus the domain of f^{-1} is the set $\{y : y \neq 2\}$.

(e) The range of f^{-1} equals the domain of f. Thus the range of f^{-1} is the set $\{x : x \neq -3\}$.

21. $f(x) = \begin{cases} 3x & \text{if } x < 0 \\ 4x & \text{if } x \geq 0 \end{cases}$

SOLUTION

(a) The expression defining $f(x)$ makes sense for all real numbers x. Thus the domain of f is the set of real numbers.

(b) To find the range of f, we need to find the numbers y such that $y = f(x)$ for some real number x. From the definition of f, we see that if $y < 0$, then $y = f(\frac{y}{3})$, and if $y \geq 0$, then $y = f(\frac{y}{4})$. Thus every real number y is in the range of f. In other words, the range of f is the set of real numbers.

(c) From the paragraph above, we see that f^{-1} is given by the formula

$$f^{-1}(y) = \begin{cases} \frac{y}{3} & \text{if } y < 0 \\ \frac{y}{4} & \text{if } y \geq 0. \end{cases}$$

(d) The domain of f^{-1} equals the range of f. Thus the domain of f^{-1} is the set of real numbers.

(e) The range of f^{-1} equals the domain of f. Thus the range of f^{-1} is the set of real numbers.

23. $f(x) = x^2 + 8$, where the domain of f equals $(0, \infty)$.

SOLUTION

(a) As part of the definition of the function f, the domain has been specified to be the interval $(0, \infty)$, which is the set of positive numbers.

(b) To find the range of f, we need to find the numbers y such that

$$y = x^2 + 8$$

for some x in the domain of f. In other words, we need to find the values of y such that the equation above can be solved for a positive number x. To solve this equation for x, subtract 8 from both sides and then take square roots of both sides, getting

$$x = \sqrt{y - 8},$$

where we chose the positive square root of $y - 8$ because x is required to be a positive number.

The expression above on the right makes sense and produces a positive number x for every number $y > 8$. Thus the range of f is the interval $(8, \infty)$.

(c) The expression above shows that f^{-1} is given by the formula

$$f^{-1}(y) = \sqrt{y - 8}.$$

(d) The domain of f^{-1} equals the range of f. Thus the domain of f^{-1} is the interval $(8, \infty)$.

(e) The range of f^{-1} equals the domain of f. Thus the range of f^{-1} is the interval $(0, \infty)$, which is the set of positive numbers.

25. Suppose $f(x) = x^5 + 2x^3$. Which of the numbers listed below equals $f^{-1}(8.10693)$?

$$1.1, \quad 1.2, \quad 1.3, \quad 1.4$$

SOLUTION First we test whether or not $f^{-1}(8.10693)$ equals 1.1 by checking whether or not $f(1.1)$ equals 8.10693. Using a calculator, we find that

$$f(1.1) = 4.27251,$$

which means that $f^{-1}(8.10693) \neq 1.1$.

Next we test whether or not $f^{-1}(8.10693)$ equals 1.2 by checking whether or not $f(1.2)$ equals 8.10693. Using a calculator, we find that

$$f(1.2) = 5.94432,$$

which means that $f^{-1}(8.10693) \neq 1.2$.

Next we test whether or not $f^{-1}(8.10693)$ equals 1.3 by checking whether or not $f(1.3)$ equals 8.10693. Using a calculator, we find that

$$f(1.3) = 8.10693,$$

which means that $f^{-1}(8.10693) = 1.3$.

For Exercises 27–28, use the U. S. 2010 federal income tax function for a single person as defined in Example 2 of Section 3.1.

27. What is the taxable income of a single person who paid \$10,000 in federal taxes for 2010?

SOLUTION Let g be the income tax function as defined in Example 2 of Section 3.1. We need to evaluate $g^{-1}(10000)$. Letting $t = g^{-1}(10000)$, this means that we need to solve the equation $g(t) = 10000$ for t. Determining which formula to apply requires a bit of experimentation. Using the definition of g, we can calculate that $g(8375) = 837.5$, $g(34000) = 4681.25$, and $g(82400) = 16781.3$. Because 10000 is between 4681.25 and 16781.3, this means that t is between 34000 and 82400. Thus $g(t) = 0.25t - 3818.75$. Solving the equation

$$0.25t - 3818.75 = 10000$$

for t, we get $t = 55275$. Thus a single person whose federal tax bill was \$10,000 had a taxable income of \$55,275.

For Exercises 29–36, suppose f and g are functions whose domain is the interval $[1, \infty)$, with

$$f(x) = x^2 + 3x + 5 \quad \textbf{\textit{and}} \quad g(x) = x^2 + 4x + 7.$$

29. What is the range of f?

SOLUTION To find the range of f, we need to find the numbers y such that

$$y = x^2 + 3x + 5$$

for some x in the domain of f. In other words, we need to find the values of y such that the equation above can be solved for a number $x \geq 1$. To solve this equation for x, subtract y from both sides, getting the equation

$$x^2 + 3x + (5 - y) = 0.$$

Using the quadratic equation to solve this equation for x, we get

$$x = \frac{-3 \pm \sqrt{3^2 - 4(5 - y)}}{2} = \frac{-3 \pm \sqrt{4y - 11}}{2}.$$

Choosing the negative sign in the equation above would give a negative value for x, which is not possible because x is required to be in the domain of f, which is the interval $[1, \infty)$. Thus we must have

$$x = \frac{-3 + \sqrt{4y - 11}}{2}.$$

Because x is required to be in the domain of f, which is the interval $[1, \infty)$, we must have

$$\frac{-3 + \sqrt{4y - 11}}{2} \geq 1.$$

Multiplying both sides of this inequality by 2 and then adding 3 to both sides gives the inequality

$$\sqrt{4y - 11} \geq 5.$$

Thus $4y - 11 \geq 25$, which implies that $4y \geq 36$, which implies that $y \geq 9$. Thus the range of f is the interval $[9, \infty)$.

31. Find a formula for f^{-1}.

SOLUTION The expression derived in the solution to Exercise 29 shows that f^{-1} is given by the formula

$$f^{-1}(y) = \frac{-3 + \sqrt{4y - 11}}{2}.$$

33. What is the domain of f^{-1}?

SOLUTION The domain of f^{-1} equals the range of f. Thus the domain of f^{-1} is the interval $[9, \infty)$.

35. What is the range of f^{-1}?

SOLUTION The range of f^{-1} equals the domain of f. Thus the range of f^{-1} is the interval $[1, \infty)$.

37. Suppose $g(x) = x^7 + x^3$. Evaluate

$$(g^{-1}(4))^7 + (g^{-1}(4))^3 + 1.$$

SOLUTION We are asked to evaluate $g(g^{-1}(4)) + 1$. Because $g(g^{-1}(4)) = 4$, the quantity above equals 5.

3.5 A Graphical Approach to Inverse Functions

LEARNING OBJECTIVES

By the end of this section you should be able to

- sketch the graph of f^{-1} from the graph of f;
- apply the horizontal line test to determine whether a function has an inverse;
- recognize increasing functions and decreasing functions;
- compute an inverse function for a function defined by a table.

The Graph of an Inverse Function

We begin with an example that illustrates how the graph of an inverse function is related to the graph of the original function.

EXAMPLE 1

Suppose f is the function with domain $[0, 2]$ defined by $f(x) = x^2$. What is the relationship between the graph of f and the graph of f^{-1}?

SOLUTION

The graph of f is part of the familiar parabola defined by the curve $y = x^2$. The range of f is the interval $[0, 4]$. The inverse function f^{-1} has domain $[0, 4]$, with $f^{-1}(x) = \sqrt{x}$.

The graphs of f and f^{-1} are shown below. These graphs are symmetric about the line $y = x$, meaning that we could obtain either graph by flipping the other graph across this line.

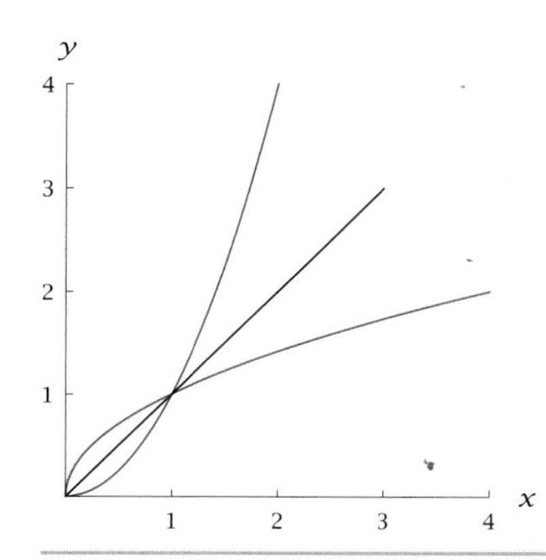

The graph of x^2 (blue) and the graph of its inverse $\sqrt{x}$ (red) are symmetric about the line $y = x$ (black).

The relationship noted above between the graph of x^2 and the graph of its inverse $\sqrt{x}$ holds in general for the graph of any one-to-one function and the graph of its inverse. Suppose, for example, that the point $(2, 1)$ is on the graph of some one-to-one function f. This means that $f(2) = 1$, which is equivalent to the equation $f^{-1}(1) = 2$, which means that $(1, 2)$ is on the graph of f^{-1}. The figure in the margin shows that the point $(1, 2)$ can be obtained by flipping the point $(2, 1)$ across the line $y = x$.

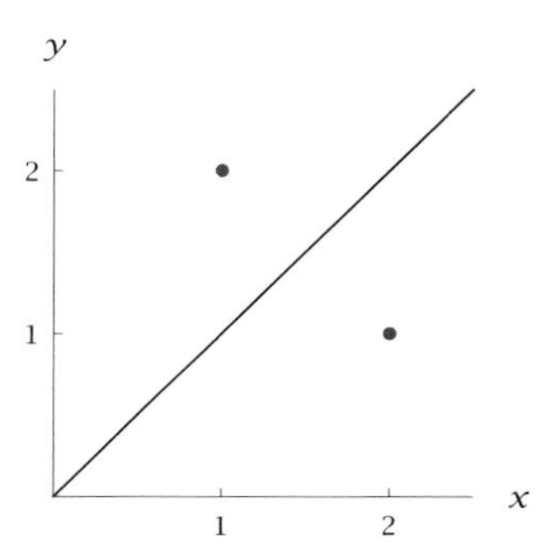

Flipping the point $(2, 1)$ (blue) across the line $y = x$ gives the point $(1, 2)$ (red).

More generally, a point (a, b) is on the graph of a one-to-one function f if and only if (b, a) is on the graph of its inverse function f^{-1}. In other words, the graph of f^{-1} can be obtained by interchanging the first and second coordinates of each point on the graph of f. Interchanging first and second coordinates amounts to flipping across the line through the origin with slope 1 (which is the line $y = x$ if we are working in the xy-plane).

The graph of a one-to-one function and its inverse

- A point (a, b) is on the graph of a one-to-one function if and only if (b, a) is on the graph of its inverse function.
- The graph of a one-to-one function and the graph of its inverse are symmetric about the line through the origin with slope 1.
- Each graph can be obtained from the other by flipping across the line through the origin with slope 1.

Sometimes an explicit formula cannot be found for f^{-1} because the equation $f(x) = y$ cannot be solved for x even though f is a one-to-one function. However, even in such cases we can obtain the graph of f^{-1}.

EXAMPLE 2

Suppose f is the function with domain $[0, 1]$ defined by $f(x) = \frac{1}{2}x^5 + \frac{3}{2}x^3$. Sketch the graph of f^{-1}.

The graph of $f(x) = \frac{1}{2}x^5 + \frac{3}{2}x^3$.

SOLUTION The graph of f is shown here in the margin; it was produced by a computer program that can graph a function if given a formula for the function.

Even though f is a one-to-one function, neither humans nor computers can solve the equation

$$\tfrac{1}{2}x^5 + \tfrac{3}{2}x^3 = y$$

for x in terms of y. Thus in this case there is no formula for f^{-1} that a computer can use to produce the graph of f^{-1}.

However, we can find the graph of f^{-1} by flipping the graph of f across the line $y = x$, as shown below:

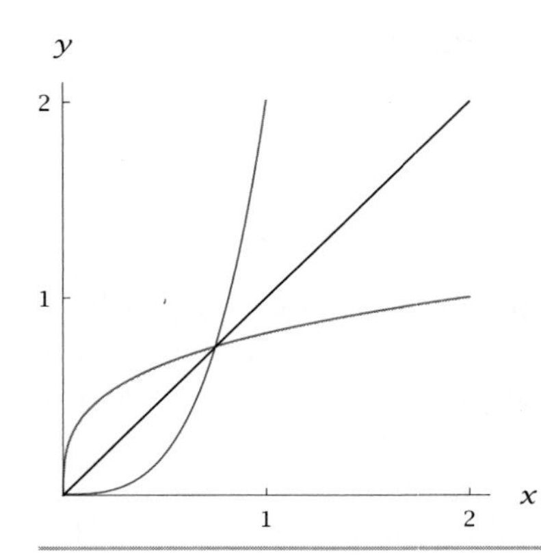

The graph of $f(x) = \frac{1}{2}x^5 + \frac{3}{2}x^3$ (blue) and the graph of its inverse (red), which is obtained by flipping across the line $y = x$.

Graphical Interpretation of One-to-One

The graph of a function can be used to determine whether or not the function is one-to-one (and thus whether or not the function has an inverse). The example below illustrates the idea.

EXAMPLE 3

Suppose f is the function with domain $[1, 4]$ whose graph is shown here in the margin. Is f a one-to-one function?

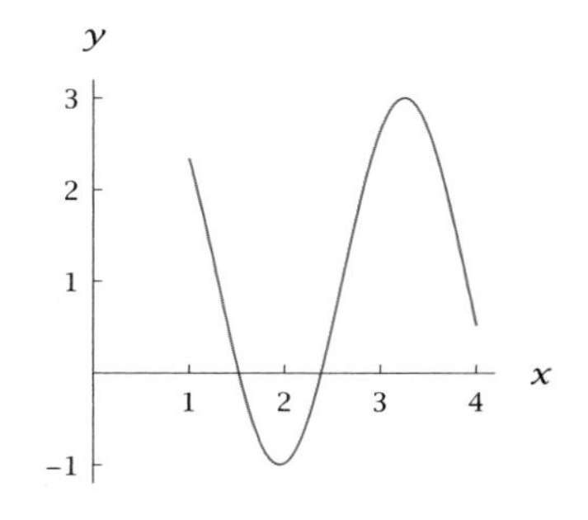

SOLUTION For f to be one-to-one, for each number y there must be at most one number x such that $f(x) = y$. Draw the line $y = 2$ on the same coordinate plane as the graph, as shown below.

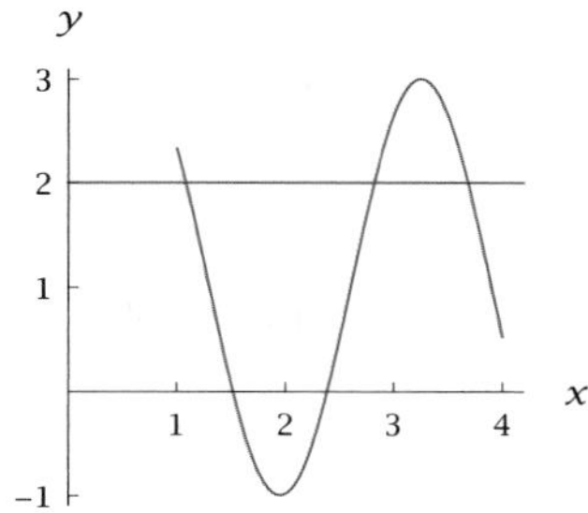

The line $y = 2$ intersects the graph of f in three points, as shown here. Thus there are three numbers x in the domain of f such that $f(x) = 2$. Hence f is not a one-to-one function.

The method used in the example above can be used with the graph of any function. Here is the formal statement of the resulting test:

Horizontal line test

A function is one-to-one if and only if every horizontal line intersects the graph of the function in at most one point.

The functions that have inverses are precisely the one-to-one functions. Thus the horizontal line test can be used to determine whether or not a function has an inverse.

When using the horizontal line test, be careful about its correct interpretation: If you find even one horizontal line that intersects the graph in more than one point, then the function is not one-to-one. However, finding one horizontal line that intersects the graph in at most one point does not imply anything concerning whether or not the function is one-to-one. For the function to be one-to-one, *every* horizontal line must intersect the graph in at most one point.

EXAMPLE 4

Suppose f is the function with domain $[-2, 2]$ whose graph is shown here in the margin. Is f a one-to-one function?

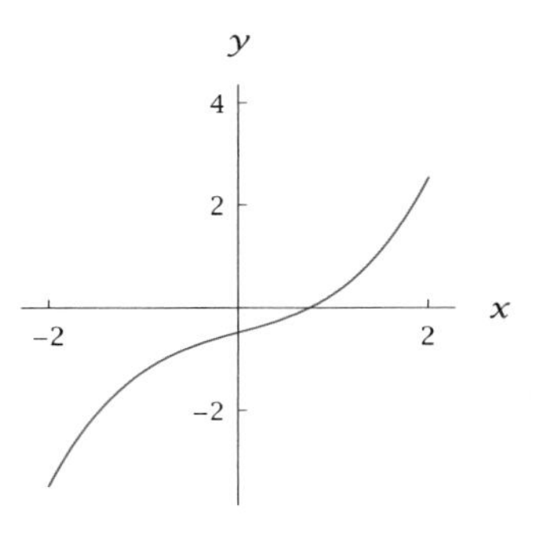

SOLUTION For f to be one-to-one, each horizontal line must intersect the graph of f in at most one point. The figure below shows the graph of f along with the horizontal lines $y = 1$ and $y = 3$.

The line $y = 1$ intersects the graph of f in one point, and the line $y = 3$ intersects the graph in zero points, as shown here. Furthermore, the figure shows that each horizontal line intersects the graph in at most one point. Hence f is a one-to-one function.

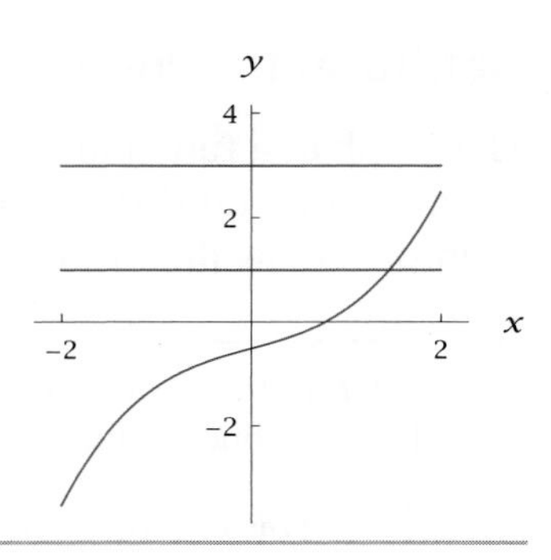

Increasing and Decreasing Functions

The domain of the function shown here is the interval $[1, 6]$. On the interval $[1, 3]$, the graph of this function gets higher from left to right; thus we say that this function is **increasing** on the interval $[1, 3]$. On the interval $[3, 6]$, the graph of this function gets lower from left to right; thus we say that this function is **decreasing** on the interval $[3, 6]$. Here are the formal definitions:

1 3 6

Increasing on an interval

A function f is called **increasing** on an interval if $f(a) < f(b)$ whenever $a < b$ and a, b are in the interval.

Decreasing on an interval

A function f is called **decreasing** on an interval if $f(a) > f(b)$ whenever $a < b$ and a, b are in the interval.

EXAMPLE 5

The function f whose graph is shown here has domain $[-1, 6]$.

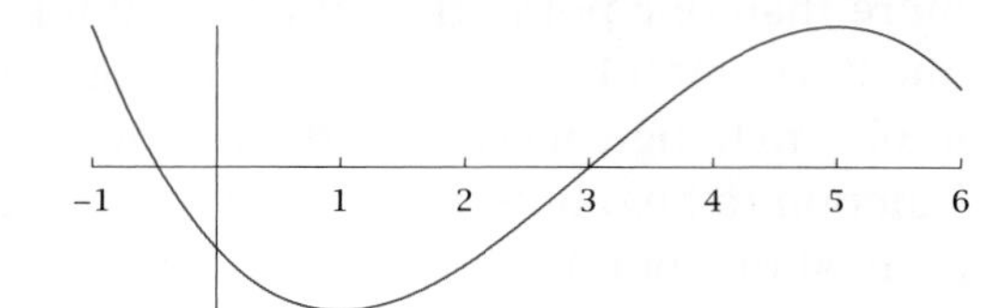

(a) Find the largest interval on which f is increasing.

(b) Find the largest interval on which f is decreasing.

(c) Find the largest interval containing 6 on which f is decreasing.

SOLUTION

(a) We see that $[1, 5]$ is the largest interval on which f is increasing.

(b) We see that $[-1, 1]$ is the largest interval on which f is decreasing.

(c) We see that $[5, 6]$ is the largest interval containing 6 on which f is decreasing.

A function is called **increasing** if its graph gets higher from left to right on its entire domain. Here is the formal definition:

Sometimes the terms "increasing" and "decreasing" are used without referring to an interval, as explained here.

Increasing functions

A function f is called **increasing** if $f(a) < f(b)$ whenever $a < b$ and a, b are in the domain of f.

Similarly, a function is called **decreasing** if its graph gets lower from left to right on its entire domain, as defined below:

Decreasing functions

A function f is called **decreasing** if $f(a) > f(b)$ whenever $a < b$ and a, b are in the domain of f.

EXAMPLE 6

Shown below are the graphs of three functions; each function is graphed on its entire domain.

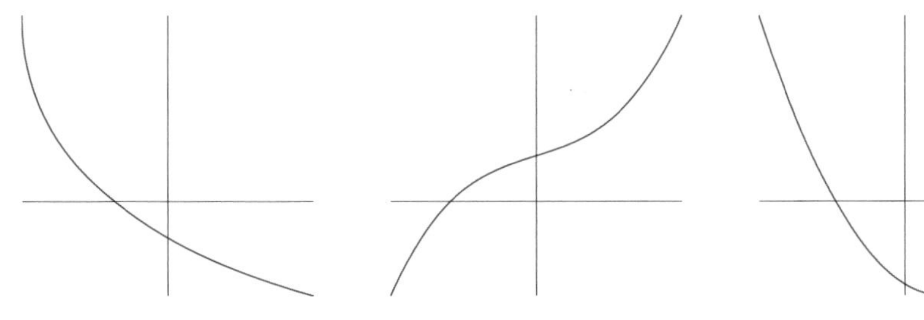

The graph of f. *The graph of g.* *The graph of h.*

(a) Is f increasing, decreasing, or neither?

(b) Is g increasing, decreasing, or neither?

(c) Is h increasing, decreasing, or neither?

SOLUTION

(a) The graph of f gets lower from left to right on its entire domain. Thus f is decreasing.

(b) The graph of g gets higher from left to right on its entire domain. Thus g is increasing.

(c) The graph of h gets lower from left to right on part of its domain and gets higher from left to right on another part of its domain. Thus h is neither increasing nor decreasing.

Every horizontal line intersects the graph of an increasing function in at most one point, and similarly for the graph of a decreasing function. Thus we have the following result:

This result implies that a function that is either increasing or decreasing has an inverse.

Increasing and decreasing functions are one-to-one

- Every increasing function is one-to-one.
- Every decreasing function is one-to-one.

The result above raises the question of whether every one-to-one function must be increasing or decreasing. The graph shown in the margin answers this question. Specifically, this function is one-to-one because each horizontal line intersects the graph in at most one point. However, this function is neither increasing nor decreasing.

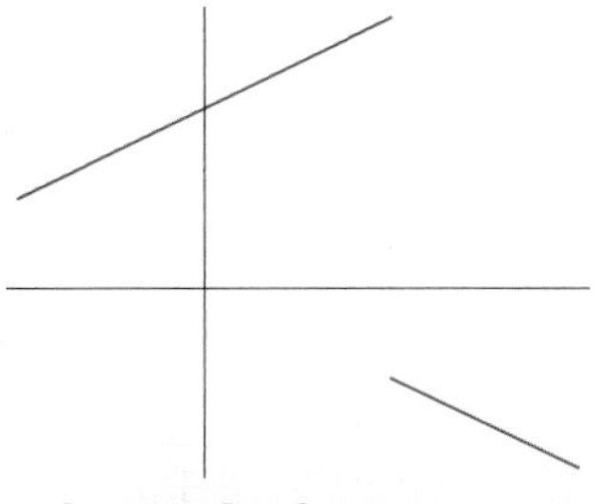

The graph of a one-to-one function that is neither increasing nor decreasing.

The graph in the example shown in the margin is not one connected piece—you cannot sketch it without lifting your pencil from the paper. A one-to-one function whose graph consists of just one connected piece must be either increasing or decreasing. However, a rigorous explanation of why this result holds requires tools from calculus.

Suppose f is an increasing function and a and b are numbers in the domain of f with $a < b$. Thus $f(a) < f(b)$. Recall that $f(a)$ and $f(b)$ are numbers in the domain of f^{-1}. We have

$$f^{-1}(f(a)) < f^{-1}(f(b))$$

because $f^{-1}(f(a)) = a$ and $f^{-1}(f(b)) = b$. The inequality above shows that f^{-1} is an increasing function.

In other words, we have just shown that the inverse of an increasing function is increasing. The figure of a function and its inverse in Example 2 illustrates this result graphically. A similar result holds for decreasing functions.

Inverses of increasing and decreasing functions

- The inverse of an increasing function is increasing.
- The inverse of a decreasing function is decreasing.

Inverse Functions via Tables

For functions whose domain consists of only finitely many numbers, tables provide good insight into the notion of an inverse function.

EXAMPLE 7

Suppose f is the function whose domain is the set of four numbers $\{\sqrt{2}, 8, 17, 18\}$, with the values of f given in the table shown here in the margin.

(a) What is the range of f?

(b) Explain why f is a one-to-one function.

(c) What is the table for the function f^{-1}?

x	$f(x)$
$\sqrt{2}$	3
8	−5
17	6
18	1

SOLUTION

(a) The range of f is the set of numbers appearing in the second column of the table defining f. Thus the range of f is the set $\{3, -5, 6, 1\}$.

(b) A function is one-to-one if and only if each number in its range corresponds to only one number in the domain. This means that a function defined by a table is one-to-one if and only if no number is repeated in the second column of the table defining the function. Because the second column of the table above contains no repetitions, we conclude that f is a one-to-one function.

(c) Suppose we want to evaluate $f^{-1}(3)$. This means that we need to find a number x such that $f(x) = 3$. Looking in the table above at the column labeled $f(x)$, we see that $f(\sqrt{2}) = 3$. Thus $f^{-1}(3) = \sqrt{2}$, which means that in the table for f^{-1} the positions of $\sqrt{2}$ and 3 should be interchanged from their positions in the table for f.

More generally, the table for f^{-1} is obtained by interchanging the columns in the table for f, producing the table shown here.

y	$f^{-1}(y)$
3	$\sqrt{2}$
−5	8
6	17
1	18

The ideas used in the example above apply to any function defined by a table, as summarized below.

Inverse functions via tables

Suppose f is a function defined by a table. Then:

- f is one-to-one if and only if the table defining f has no repetitions in the second column.
- If f is one-to-one, then the table for f^{-1} is obtained by interchanging the columns of the table defining f.

EXERCISES

For Exercises 1–12, use the following graphs:

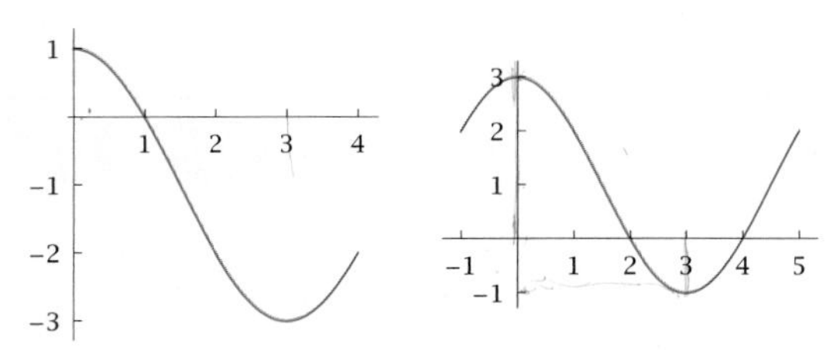

The graph of f. ***The graph of g.***

Here f has domain $[0,4]$ and g has domain $[-1,5]$.

1. What is the largest interval contained in the domain of f on which f is increasing?
2. What is the largest interval contained in the domain of g on which g is increasing?
3. Let F denote the function obtained from f by restricting the domain to the interval in Exercise 1. What is the domain of F^{-1}?
4. Let G denote the function obtained from g by restricting the domain to the interval in Exercise 2. What is the domain of G^{-1}?
5. With F as in Exercise 3, what is the range of F^{-1}?
6. With G as in Exercise 4, what is the range of G^{-1}?
7. What is the largest interval contained in the domain of f on which f is decreasing?
8. What is the largest interval contained in the domain of g on which g is decreasing?
9. Let H denote the function obtained from f by restricting the domain to the interval in Exercise 7. What is the domain of H^{-1}?
10. Let J denote the function obtained from g by restricting the domain to the interval in Exercise 8. What is the domain of J^{-1}?
11. With H as in Exercise 9, what is the range of H^{-1}?
12. With J as in Exercise 10, what is the range of J^{-1}?
13. Suppose
$$f(x) = x^2 - 6x + 11.$$
Find the smallest number b such that f is increasing on the interval $[b, \infty)$.
14. Suppose
$$f(x) = x^2 + 8x + 5.$$
Find the smallest number b such that f is increasing on the interval $[b, \infty)$.

For Exercises 15–38 suppose f and g are functions, each with domain of four numbers, with f and g defined by the tables below:

x	$f(x)$
1	4
2	5
3	2
4	3

x	$g(x)$
2	3
3	2
4	4
5	1

15. What is the domain of f?
16. What is the domain of g?
17. What is the range of f?
18. What is the range of g?
19. Sketch the graph of f.
20. Sketch the graph of g.
21. Give the table of values for f^{-1}.
22. Give the table of values for g^{-1}.
23. What is the domain of f^{-1}?
24. What is the domain of g^{-1}?
25. What is the range of f^{-1}?
26. What is the range of g^{-1}?
27. Sketch the graph of f^{-1}.
28. Sketch the graph of g^{-1}.
29. Give the table of values for $f^{-1} \circ f$.
30. Give the table of values for $g^{-1} \circ g$.
31. Give the table of values for $f \circ f^{-1}$.
32. Give the table of values for $g \circ g^{-1}$.
33. Give the table of values for $f \circ g$.
34. Give the table of values for $g \circ f$.
35. Give the table of values for $(f \circ g)^{-1}$.
36. Give the table of values for $(g \circ f)^{-1}$.
37. Give the table of values for $g^{-1} \circ f^{-1}$.
38. Give the table of values for $f^{-1} \circ g^{-1}$.

PROBLEMS

39. Suppose f is the function whose domain is the interval $[-2, 2]$, with f defined by the following formula:

$$f(x) = \begin{cases} -\frac{x}{3} & \text{if } -2 \le x < 0 \\ 2x & \text{if } 0 \le x \le 2. \end{cases}$$

    (a) Sketch the graph of f.

    (b) Explain why the graph of f shows that f is not a one-to-one function.

    (c) Give an explicit example of two distinct numbers a and b such that $f(a) = f(b)$.

40. Show that a linear function is increasing if and only if the slope of its graph is positive.

41. Show that a linear function is decreasing if and only if the slope of its graph is negative.

42. Draw the graph of a function that is increasing on the interval $[-2, 0]$ and decreasing on the interval $[0, 2]$.

43. Draw the graph of a function that is decreasing on the interval $[-2, 1]$ and increasing on the interval $[1, 5]$.

44. Give an example of an increasing function whose domain is the interval $[0, 1]$ but whose range does not equal the interval $[f(0), f(1)]$.

45. Show that the sum of two increasing functions is increasing.

46. Give an example of two increasing functions whose product is not increasing.
[*Hint:* There are no such examples where both functions are positive everywhere.]

47. Give an example of two decreasing functions whose product is increasing.

48. Show that the composition of two increasing functions is increasing.

49. Explain why it is important as a matter of social policy that the income tax function g in Example 2 of Section 3.1 be an increasing function.

50. The graph of the income tax function in Example 2 of Section 3.1 consists of six different line segments. The slopes of these line segments increase going from left to right on the graph. What is the social policy behind this increase in slopes, in terms of the amount of extra tax corresponding to an extra dollar of income at various income levels?

51. Suppose the income tax function in Example 2 of Section 3.1 is changed so that

$$g(x) = 0.15x - 450 \quad \text{if } 8375 < x \le 34000,$$

with the other parts of the definition of g left unchanged. Show that if this change is made, then the income tax function g would no longer be an increasing function.

52. Suppose the income tax function in Example 2 of Section 3.1 is changed so that

$$g(x) = 0.14x - 418.75 \quad \text{if } 8375 < x \le 34000,$$

with the other parts of the definition of g left unchanged. Show that if this change is made, then the income tax function g would no longer be an increasing function.

53. Explain why an even function whose domain contains a nonzero number cannot be a one-to-one function.

54. The solutions to Exercises 35 and 37 are the same, suggesting that

$$(f \circ g)^{-1} = g^{-1} \circ f^{-1}.$$

Explain why the equation above holds whenever f and g are one-to-one functions such that the range of g equals the domain of f.

WORKED-OUT SOLUTIONS *to Odd-numbered Exercises*

For Exercises 1–12, use the following graphs:

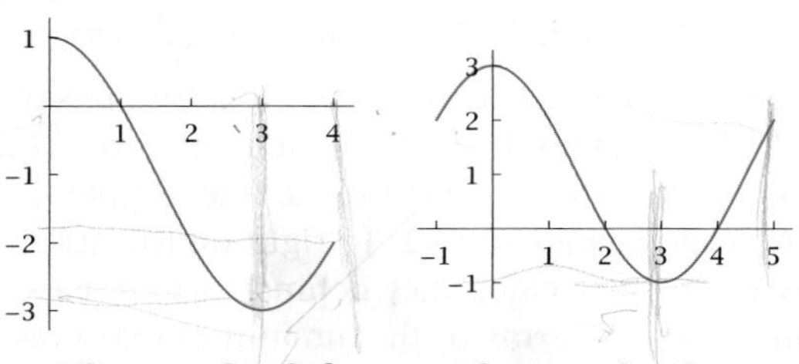

The graph of f. ***The graph of g.***

Here f has domain $[0, 4]$ and g has domain $[-1, 5]$.

1. What is the largest interval contained in the domain of f on which f is increasing?

   SOLUTION As can be seen from the graph, $[3, 4]$ is the largest interval on which f is increasing.

   As usual when obtaining information solely from graphs, this answer (as well as the answers to the other parts of this exercise) should be considered an approximation. An expanded graph at a finer scale might show that $[2.99, 4]$ or $[3.01, 4]$ would be a more accurate answer than $[3, 4]$.

3. Let F denote the function obtained from f by restricting the domain to the interval in Exercise 1. What is the domain of F^{-1}?

   SOLUTION The domain of F^{-1} equals the range of F. Because F is the function f with domain restricted to the interval $[3, 4]$, we see from the graph above that the range of F is the interval $[-3, -2]$. Thus the domain of F^{-1} is the interval $[-3, -2]$.

5. With F as in Exercise 3, what is the range of F^{-1}?

   SOLUTION The range of F^{-1} equals the domain of F. Thus the range of F^{-1} is the interval $[3, 4]$.

7. What is the largest interval contained in the domain of f on which f is decreasing?

   SOLUTION As can be seen from the graph, $[0, 3]$ is the largest interval on which f is decreasing.

9. Let H denote the function obtained from f by restricting the domain to the interval in Exercise 7. What is the domain of H^{-1}?

   SOLUTION The domain of H^{-1} equals the range of H. Because H is the function f with domain restricted to the interval $[0, 3]$, we see from the graph above that the range of H is the interval $[-3, 1]$. Thus the domain of H^{-1} is the interval $[-3, 1]$.

11. With H as in Exercise 9, what is the range of H^{-1}?

    SOLUTION The range of H^{-1} equals the domain of H. Thus the range of H^{-1} is the interval $[0, 3]$.

13. Suppose
    $$f(x) = x^2 - 6x + 11.$$
    Find the smallest number b such that f is increasing on the interval $[b, \infty)$.

    SOLUTION The graph of f is a parabola shaped like this:

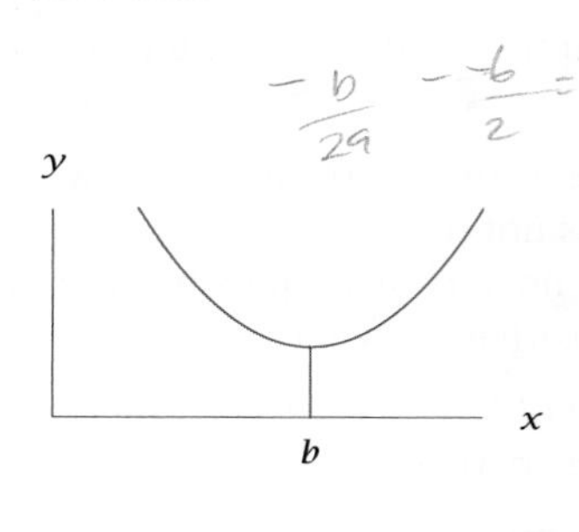

The largest interval on which f is increasing is $[b, \infty)$, where b is the first coordinate of the vertex of the graph of f.

As can be seen from the figure above, the smallest number b such that f is increasing on the interval $[b, \infty)$ is the first coordinate of the vertex of the graph of f. To find this number, we complete the square:

$$\begin{aligned} x^2 - 6x + 11 &= (x-3)^2 - 9 + 11 \\ &= (x-3)^2 + 2 \end{aligned}$$

The equation above shows that the first coordinate of the vertex of the parabola is 3. Thus we take $b = 3$.

For Exercises 15–38 suppose f and g are functions, each with domain of four numbers, with f and g defined by the tables below:

x	$f(x)$
1	4
2	5
3	2
4	3

x	$g(x)$
2	3
3	2
4	4
5	1

15. What is the domain of f?

SOLUTION The domain of f equals the set of numbers in the left column of the table defining f. Thus the domain of f equals $\{1, 2, 3, 4\}$.

17. What is the range of f?

SOLUTION The range of f equals the set of numbers in the right column of the table defining f. Thus the range of f equals $\{2, 3, 4, 5\}$.

19. Sketch the graph of f.

SOLUTION The graph of f consists of all points of the form $(x, f(x))$ as x varies over the domain of f. Thus the graph of f, shown below, consists of the four points $(1, 4)$, $(2, 5)$, $(3, 2)$, and $(4, 3)$.

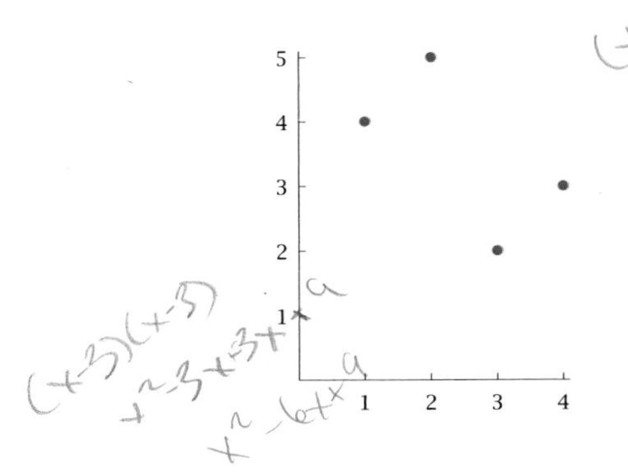

21. Give the table of values for f^{-1}.

SOLUTION The table for the inverse of a function is obtained by interchanging the two columns of the table for the function (after which one can, if desired, reorder the rows, as has been done below):

y	$f^{-1}(y)$
2	3
3	4
4	1
5	2

23. What is the domain of f^{-1}?

SOLUTION The domain of f^{-1} equals the range of f. Thus the domain of f^{-1} is the set $\{2, 3, 4, 5\}$.

25. What is the range of f^{-1}?

SOLUTION The range of f^{-1} equals the domain of f. Thus the range of f^{-1} is the set $\{1, 2, 3, 4\}$.

27. Sketch the graph of f^{-1}.

SOLUTION The graph of f^{-1} consists of all points of the form $(x, f^{-1}(x))$ as x varies over the domain of f^{-1}. Thus the graph of f^{-1}, shown below, consists of the four points $(4, 1)$, $(5, 2)$, $(2, 3)$, and $(3, 4)$.

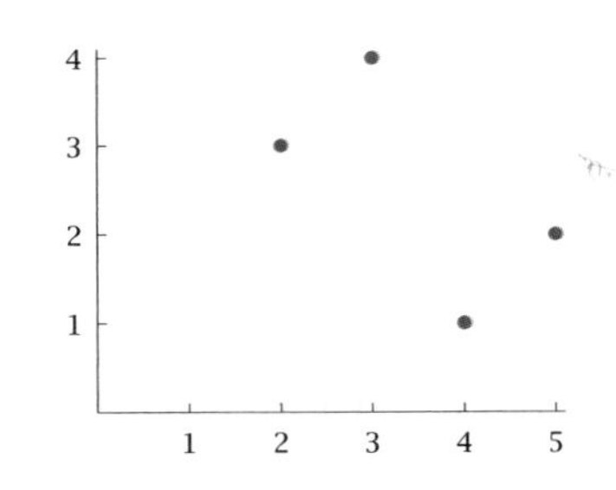

29. Give the table of values for $f^{-1} \circ f$.

SOLUTION We know that $f^{-1} \circ f$ is the identity function on the domain of f; thus no computations are necessary. However, because this function f has only four numbers in its domain, it may be instructive to compute $(f^{-1} \circ f)(x)$ for each value of x in the domain of f. Here is that computation:

$$(f^{-1} \circ f)(1) = f^{-1}(f(1)) = f^{-1}(4) = 1$$

$$(f^{-1} \circ f)(2) = f^{-1}(f(2)) = f^{-1}(5) = 2$$

$$(f^{-1} \circ f)(3) = f^{-1}(f(3)) = f^{-1}(2) = 3$$

$$(f^{-1} \circ f)(4) = f^{-1}(f(4)) = f^{-1}(3) = 4$$

Thus, as expected, the table of values for $f^{-1} \circ f$ is as shown below:

x	$(f^{-1} \circ f)(x)$
1	1
2	2
3	3
4	4

31. Give the table of values for $f \circ f^{-1}$.

SOLUTION We know that $f \circ f^{-1}$ is the identity function on the range of f (which equals the domain of f^{-1}); thus no computations are necessary. However, because this function f has only four numbers in its range, it may be instructive to compute $(f \circ f^{-1})(y)$ for each value of y in the range of f. Here is that computation:

$$(f \circ f^{-1})(2) = f(f^{-1}(2)) = f(3) = 2$$

$$(f \circ f^{-1})(3) = f(f^{-1}(3)) = f(4) = 3$$

$$(f \circ f^{-1})(4) = f(f^{-1}(4)) = f(1) = 4$$

$$(f \circ f^{-1})(5) = f(f^{-1}(5)) = f(2) = 5$$

Thus, as expected, the table of values for $f \circ f^{-1}$ is as shown here.

y	$(f \circ f^{-1})(y)$
2	2
3	3
4	4
5	5

33. Give the table of values for $f \circ g$.

SOLUTION We need to compute $(f \circ g)(x)$ for every x in the domain of g. Here is that computation:

$$(f \circ g)(2) = f(g(2)) = f(3) = 2$$

$$(f \circ g)(3) = f(g(3)) = f(2) = 5$$

$$(f \circ g)(4) = f(g(4)) = f(4) = 3$$

$$(f \circ g)(5) = f(g(5)) = f(1) = 4$$

Thus the table of values for $f \circ g$ is as shown here.

x	$(f \circ g)(x)$
2	2
3	5
4	3
5	4

35. Give the table of values for $(f \circ g)^{-1}$.

SOLUTION The table of values for $(f \circ g)^{-1}$ is obtained by interchanging the two columns of the table for $(f \circ g)$ (after which one can, if desired, reorder the rows, as has been done below).

Thus the table for $(f \circ g)^{-1}$ is as shown here.

y	$(f \circ g)^{-1}(y)$
2	2
3	4
4	5
5	3

37. Give the table of values for $g^{-1} \circ f^{-1}$.

SOLUTION We need to compute $(g^{-1} \circ f^{-1})(y)$ for every y in the domain of f^{-1}. Here is that computation:

$$(g^{-1} \circ f^{-1})(2) = g^{-1}(f^{-1}(2)) = g^{-1}(3) = 2$$

$$(g^{-1} \circ f^{-1})(3) = g^{-1}(f^{-1}(3)) = g^{-1}(4) = 4$$

$$(g^{-1} \circ f^{-1})(4) = g^{-1}(f^{-1}(4)) = g^{-1}(1) = 5$$

$$(g^{-1} \circ f^{-1})(5) = g^{-1}(f^{-1}(5)) = g^{-1}(2) = 3$$

Thus the table of values for $g^{-1} \circ f^{-1}$ is as shown here.

y	$(g^{-1} \circ f^{-1})(y)$
2	2
3	4
4	5
5	3

CHAPTER SUMMARY

To check that you have mastered the most important concepts and skills covered in this chapter, make sure that you can do each item in the following list:

- Explain the concept of a function, including its domain.
- Define the range of a function.
- Explain the relationship between a function and its graph.
- Determine the domain and range of a function from its graph.
- Use the vertical line test to determine if a set is the graph of some function.
- Determine whether a function transformation shifts the graph up, down, left, or right.
- Determine whether a function transformation stretches the graph vertically or horizontally.
- Determine whether a function transformation flips the graph vertically or horizontally.
- Determine the domain, range, and graph of a transformed function.
- Compute the composition of two functions.
- Write a complicated function as the composition of simpler functions.
- Explain the concept of an inverse function.
- Explain which functions have inverses.
- Find a formula for an inverse function (when possible).
- Sketch the graph of f^{-1} from the graph of f.
- Use the horizontal line test to determine whether a function has an inverse.
- Construct a table of values of f^{-1} from a table of values of f.
- Recognize from a graph whether a function is increasing or decreasing or neither on an interval.

To review a chapter, go through the list above to find items that you do not know how to do, then reread the material in the chapter about those items. Then try to answer the chapter review questions below without looking back at the chapter.

CHAPTER REVIEW QUESTIONS

1. Suppose f is a function. Explain what it means to say that $\frac{3}{2}$ is in the domain of f.
2. Suppose f is a function. Explain what it means to say that $\frac{3}{2}$ is in the range of f.
3. Write the domain of $\dfrac{x^5 + 2}{x^2 + 7x - 1}$ as a union of intervals.
4. Give an example of a function whose domain consists of five numbers and whose range consists of three numbers.
5. Give an example of a function whose domain is the set of real numbers and whose range is not an interval.
6. Explain how to find the domain of a function from its graph.
7. Explain how to find the range of a function from its graph.
8. Suppose f is defined by $f(x) = x^2$.
   (a) What is the range of f if the domain of f is the interval $[1, 3]$?
   (b) What is the range of f if the domain of f is the interval $[-2, 3]$?
   (c) What is the range of f if the domain of f is the set of positive numbers?
   (d) What is the range of f if the domain of f is the set of negative numbers?
   (e) What is the range of f if the domain of f is the set of real numbers?

9. Explain how to use the vertical line test to determine whether or not a set in the plane is the graph of some function.

10. Sketch a curve in the coordinate plane that is not the graph of any function.

For Questions 11–18, assume that f is the function defined on the interval $[1,3]$ by the formula

$$f(x) = \frac{1}{x^2 - 3x + 3}.$$

The domain of f is the interval $[1,3]$, the range of f is the interval $[\frac{1}{3}, \frac{4}{3}]$, and the graph of f is shown below.

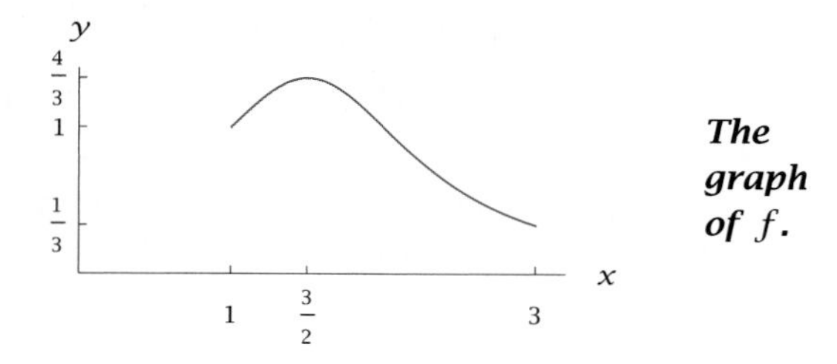

The graph of f.

For each function g described below:

- **(a)** ***Sketch the graph of g.***
- **(b)** ***Find the domain of g (the endpoints of this interval should be shown on the horizontal axis of your sketch of the graph of g).***
- **(c)** ***Give a formula for g.***
- **(d)** ***Find the range of g (the endpoints of this interval should be shown on the vertical axis of your sketch of the graph of g).***

11. The graph of g is obtained by shifting the graph of f up 2 units.

12. The graph of g is obtained by shifting the graph of f down 2 units.

13. The graph of g is obtained by shifting the graph of f left 2 units.

14. The graph of g is obtained by shifting the graph of f right 2 units.

15. The graph of g is obtained by vertically stretching the graph of f by a factor of 3.

16. The graph of g is obtained by horizontally stretching the graph of f by a factor of 2.

17. The graph of g is obtained by flipping the graph of f across the horizontal axis.

18. The graph of g is obtained by flipping the graph of f across the vertical axis.

19. Suppose f is a function with domain $[1,3]$ and range $[2,5]$. Define functions g and h by

$$g(x) = 3f(x) \quad \text{and} \quad h(x) = f(4x).$$

   (a) What is the domain of g?

   (b) What is the range of g?

   (c) What is the domain of h?

   (d) What is the range of h?

20. Suppose the domain of f is the set of four numbers $\{1,2,3,4\}$, with f defined by the table shown here. Draw the graph of f.

x	$f(x)$
1	2
2	3
3	−1
4	1

21. Show that the sum of two odd functions (with the same domain) is an odd function.

22. Define the composition of two functions.

23. Suppose f and g are functions. Explain why the domain of $f \circ g$ is contained in the domain of g.

24. Suppose f and g are functions. Explain why the range of $f \circ g$ is contained in the range of f.

25. Suppose f is a function and g is a linear function. Explain why the domain of $g \circ f$ is the same as the domain of f.

26. Suppose f is a function and g is a nonconstant linear function. Explain why the range of $f \circ g$ is the same as the range of f.

27. Suppose $f(x) = \frac{x^2+3}{5x^2-9}$. Find two functions g and h, each simpler than f, such that $f = g \circ h$.

28. Explain how to use the horizontal line test to determine whether or not a function is one-to-one.

For Questions 29–32, suppose

$$h(x) = |2x + 3| + x^2 \quad \textbf{\textit{and}} \quad f(x) = 3x - 5.$$

29. Evaluate $(h \circ f)(3)$.
30. Evaluate $(f \circ h)(-4)$.
31. Find a formula for $h \circ f$.
32. Find a formula for $f \circ h$.
33. Suppose $f(x) = \frac{2x+1}{3x-4}$. Evaluate $f^{-1}(-5)$.
34. Suppose $g(x) = 3 + \frac{x}{2x-3}$. Find a formula for g^{-1}.
35. Suppose f is a one-to-one function. Explain the relationship between the graph of f and the graph of f^{-1}.
36. Suppose f is a one-to-one function. Explain the relationship between the domain and range of f and the domain and range of f^{-1}.
37. Explain the different meanings of the notations $f^{-1}(x)$, $[f(x)]^{-1}$, and $f(x^{-1})$.
38. The function f defined by

$$f(x) = x^5 + 2x^3 + 2$$

is one-to-one (here the domain of f is the set of real numbers). Compute $f^{-1}(y)$ for four different values of y of your choice.

39. Draw the graph of a function that is decreasing on the interval $[1, 2]$ and increasing on the interval $[2, 5]$.
40. Make up a table that defines a one-to-one function whose domain consists of five numbers. Then sketch the graph of this function and its inverse.

CHAPTER

4

Statue of the Persian mathematician and poet Omar Khayyam, whose algebra book written in 1070 contained the first serious study of cubic polynomials.

Polynomial and Rational Functions

We begin this chapter by reviewing the properties of integer exponents, in preparation for dealing with polynomial functions. We will see why x^0 is defined to be 1 and why x^{-m} is defined to be $\frac{1}{x^m}$.

Then we deal with polynomials, one of the most important classes of functions. We will look at the connection between the zeros of a polynomial and its linear factors. We will also examine the behavior of the graphs of polynomials.

Next we turn to rational functions, which are ratios of polynomials. Unlike polynomials, rational functions can have asymptotes in their graphs.

This chapter concludes with a section on complex numbers, including a discussion of complex zeros of polynomials.

4.1 Integer Exponents

LEARNING OBJECTIVES

By the end of this section you should be able to

- explain why x^0 is defined to equal 1 (for $x \neq 0$);
- explain why x^{-m} is defined to equal $\frac{1}{x^m}$ (for m a positive integer and $x \neq 0$);
- manipulate and simplify expressions involving integer exponents.

Positive Integer Exponents

Multiplication by a positive integer is repeated addition, in the sense that if x is a real number and m is a positive integer, then mx equals the sum with x appearing m times:

$$mx = \underbrace{x + x + \cdots + x}_{x \text{ appears } m \text{ times}}.$$

Just as multiplication by a positive integer is defined as repeated addition, positive integer exponents denote repeated multiplication:

Positive integer exponents

If x is a real number and m is a positive integer, then x^m is defined to be the product with x appearing m times:

$$x^m = \underbrace{x \cdot x \cdot \;\cdots\; \cdot x}_{x \text{ appears } m \text{ times}}.$$

EXAMPLE 1 Evaluate $\left(\frac{1}{2}\right)^3$.

SOLUTION

$$\left(\tfrac{1}{2}\right)^3 = \tfrac{1}{2} \cdot \tfrac{1}{2} \cdot \tfrac{1}{2} = \tfrac{1}{8}$$

If m is a positive integer, then we can define a function f by

$$f(x) = x^m.$$

For $m = 1$, the graph of the function defined by $f(x) = x$ is a line through the origin with slope 1. For $m = 2$, the graph of the function defined by $f(x) = x^2$ is the familiar parabola with vertex at the origin, as shown here.

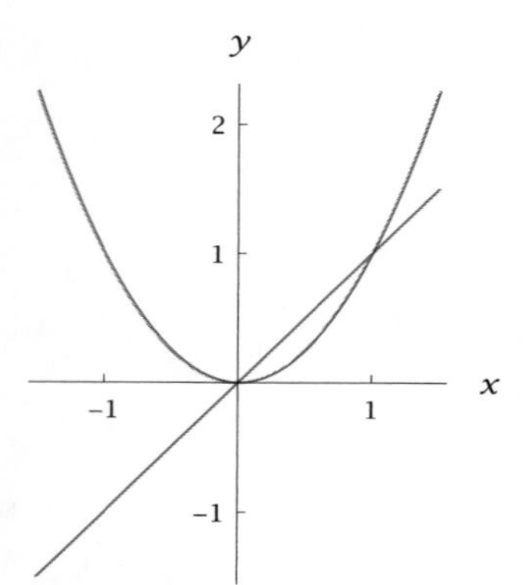

The graphs of x (blue) and x^2 (red) on $[-1.5, 1.5]$.

The graphs of x^3, x^4, x^5, and x^6 are shown below, separated into two groups according to their shape. Note that x^3 and x^5 are increasing functions, but x^4 and x^6 are decreasing on the interval $(-\infty, 0]$ and increasing on the interval $[0, \infty)$.

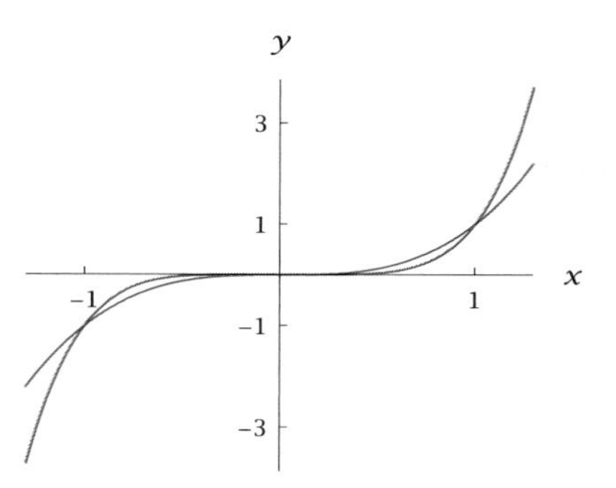

The graphs of x^3 (blue) and x^5 (red) on $[-1.3, 1.3]$.

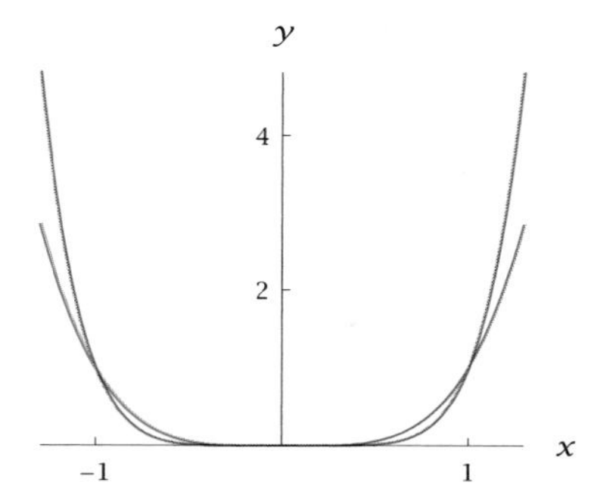

The graphs of x^4 (blue) and x^6 (red) on $[-1.3, 1.3]$.

Although the graphs of x^4 and x^6 have a parabola-type shape, these graphs are not true parabolas.

We have now seen graphs of x^m for $m = 1, 2, 3, 4, 5, 6$. For larger odd values of m, the graph of x^m has roughly the same shape as the graphs of x^3 and x^5; for larger even values of m, the graph of x^m has roughly the same shape as the graphs of x^2, x^4, and x^6.

A cube whose edges have length L has volume L^3. A ball with radius r has volume $\frac{4}{3}r^3$. Thus expressions involving an exponent of 3 often appear in questions connected with volume.

EXAMPLE 2

The NBA (National Basketball Association) states that a professional basketball should have a circumference of 29.5 inches. What is the volume of an NBA basketball?

SOLUTION First we find the radius r of an NBA basketball. Because the circumference is 29.5 inches, we have $2\pi r = 29.5$. Thus $r = \frac{29.5}{2\pi}$. Hence the volume of an NBA basketball is $\frac{4}{3}\pi\left(\frac{29.5}{2\pi}\right)^3$ cubic inches, which we can simplify as follows:

$$\begin{aligned}\frac{4}{3}\pi\left(\frac{29.5}{2\pi}\right)^3 &= \frac{4}{3}\pi \cdot \frac{29.5}{2\pi} \cdot \frac{29.5}{2\pi} \cdot \frac{29.5}{2\pi}\\ &= \frac{(29.5)^3}{6\pi^2}\\ &\approx 433.5.\end{aligned}$$

Thus the volume of an NBA basketball is approximately 433.5 cubic inches.

Properties of Exponents

The properties of positive integers exponents follow from the definition of x^m as repeated multiplication.

EXAMPLE 3

Suppose x is a real number and m and n are positive integers. Explain why

$$x^m x^n = x^{m+n}.$$

SOLUTION We have

$$x^m x^n = \underbrace{x \cdot x \cdot \cdots \cdot x}_{x \text{ appears } m \text{ times}} \cdot \underbrace{x \cdot x \cdot \cdots \cdot x}_{x \text{ appears } n \text{ times}}.$$

Thus $x^m x^n = x^{m+n}$ (because x appears a total of $m + n$ times in the product above).

The expression x^m is called the mth **power** *of x.*

Taking $n = m$ in the example above, we see that $x^m x^m = x^{m+m}$, which can be rewritten as $(x^m)^2 = x^{2m}$. The next example generalizes this equation, replacing 2 by any positive integer.

EXAMPLE 4

Suppose x is a real number and m and n are positive integers. Explain why

$$(x^m)^n = x^{mn}.$$

SOLUTION We have

$$(x^m)^n = \underbrace{x^m \cdot x^m \cdot \cdots \cdot x^m}_{x^m \text{ appears } n \text{ times}}.$$

Each x^m on the right side of the equation above equals the product with x appearing m times, and x^m appears n times. Thus x appears a total of mn times in the product above, which shows that $(x^m)^n = x^{mn}$.

The next example provides a formula for raising the product of two numbers to a power.

EXAMPLE 5

Suppose x and y are real numbers and m is a positive integer. Explain why

$$(xy)^m = x^m y^m.$$

SOLUTION We have

$$(xy)^m = \underbrace{(xy) \cdot (xy) \cdot \cdots \cdot (xy)}_{(xy) \text{ appears } m \text{ times}}.$$

Because of the associativity and commutativity of multiplication, the product above can be rearranged to show that

$$(xy)^m = \underbrace{x \cdot x \cdot \cdots \cdot x}_{x \text{ appears } m \text{ times}} \cdot \underbrace{y \cdot y \cdot \cdots \cdot y}_{y \text{ appears } m \text{ times}}.$$

Thus we see that $(xy)^m = x^m y^m$.

The three previous examples show that positive integer exponents obey the following rules. Soon we will extend these rules to integer exponents that are not necessarily positive.

Properties of positive integer exponents

Suppose x and y are numbers and m and n are positive integers. Then

$$x^m x^n = x^{m+n},$$

$$(x^m)^n = x^{mn},$$

$$x^m y^m = (xy)^m.$$

Defining x^0

We have defined x^m as the product with x appearing m times. This definition makes sense only when m is a positive integer. To define x^m for other values of m, we will choose definitions so that the properties listed above for positive integer exponents continue to hold.

We begin by considering how x^0 might be defined. Recall that if x is a real number and m and n are positive integers, then

$$x^m x^n = x^{m+n}.$$

We would like to choose the definition of x^0 so that the equation above holds even if $m = 0$. In other words, we would like to define x^0 so that

$$x^0 x^n = x^{0+n}.$$

Rewriting this equation as

$$x^0 x^n = x^n,$$

we see that if $x \neq 0$, then we have no choice but to define x^0 to equal 1.

The paragraph above shows how we should define x^0 for $x \neq 0$, but what happens when $x = 0$? Unfortunately, finding a definition for 0^0 that preserves other properties of exponents turns out to be impossible. Two conflicting tendencies point to different possible definitions for 0^0:

- The equation $x^0 = 1$, valid for all $x \neq 0$, suggests that we should define 0^0 to equal 1.
- The equation $0^m = 0$, valid for all positive integers m, suggests that we should define 0^0 to equal 0.

If we choose to define 0^0 to equal 1, as suggested by the first point above, then we violate the equation $0^m = 0$ suggested by the second point. If we choose to define 0^0 to equal 0, as suggested by the second point above, then we violate the equation $x^0 = 1$ suggested by the first point. Either way, we cannot maintain the consistency of our algebraic properties involving exponents.

To solve this dilemma, we leave 0^0 undefined rather than choose a definition that will violate some of our algebraic properties. Mathematics takes a similar position with respect to division by 0: The equations $x \cdot \frac{y}{x} = y$ and $0 \cdot \frac{y}{x} = 0$ cannot both be satisfied if $x = 0$ and $y = 1$, regardless of how we define $\frac{1}{0}$. Thus $\frac{1}{0}$ is left undefined.

In summary, here is our definition of x^0:

For example, $4^0 = 1$.

Definition of x^0

- If $x \neq 0$, then $x^0 = 1$.
- The expression 0^0 is undefined.

Negative Integer Exponents

As with the definition of a zero exponent, we will let consistency with previous algebraic properties force upon us the meaning of negative integer exponents.

At this stage, we have defined x^m whenever $x \neq 0$ and m is a positive integer or zero. We now turn our attention to defining the meaning of negative integers exponents.

Recall that if $x \neq 0$ and m and n are nonnegative integers, then

$$x^m x^n = x^{m+n}.$$

We would like to choose the meaning of negative integer exponents so that the equation above holds whenever m and n are integers (including the possibility that one or both of m and n might be negative). In the equation above, if we take $n = -m$, we get

$$x^m x^{-m} = x^{m+(-m)}.$$

Because $x^0 = 1$, this equation can be rewritten as

$$x^m x^{-m} = 1.$$

Thus we see that we have no choice but to define x^{-m} to equal the multiplicative inverse of x^m.

To avoid division by 0, we cannot allow x to equal 0 in this definition. Thus if m is a positive integer, then 0^{-m} is undefined.

Negative integer exponents

If $x \neq 0$ and m is a positive integer, then x^{-m} is defined to be the multiplicative inverse of x^m:

$$x^{-m} = \frac{1}{x^m}.$$

EXAMPLE 6

Evaluate 3^{-2}.

SOLUTION

$$3^{-2} = \frac{1}{3^2} = \frac{1}{9}$$

We can gain some insight into the behavior of the function x^m, for m a negative integer, by looking at its graph.

EXAMPLE 7

Compare the graphs of x^{-1} and x^{-2}.

SOLUTION The figures here show that the absolute value of $\frac{1}{x}$ and $\frac{1}{x^2}$ are both large for values of x near 0. Conversely, $\frac{1}{x}$ and $\frac{1}{x^2}$ are both near 0 for x with large absolute value.

Both $\frac{1}{x}$ and $\frac{1}{x^2}$ are decreasing on $(0, \infty)$.

However, $\frac{1}{x}$ is decreasing on $(-\infty, 0)$, but $\frac{1}{x^2}$ is increasing on $(-\infty, 0)$. Another difference between these graphs is that the graph of $\frac{1}{x^2}$ lies entirely above the x-axis.

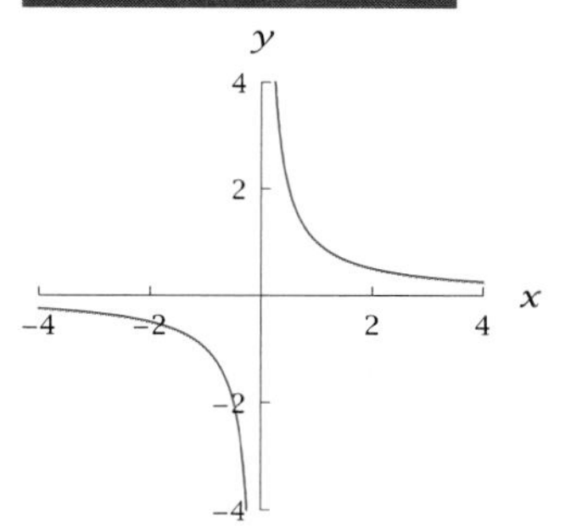

The graph of $\frac{1}{x}$ on $[-4, -\frac{1}{4}] \cup [\frac{1}{4}, 4]$.

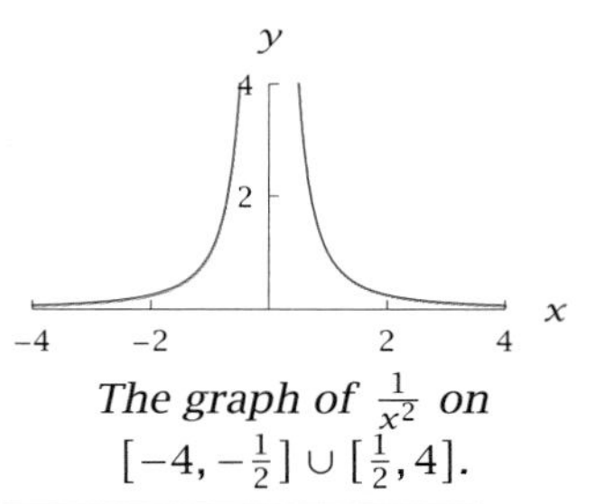

The graph of $\frac{1}{x^2}$ on $[-4, -\frac{1}{2}] \cup [\frac{1}{2}, 4]$.

In general, if m is a positive integer, then the graph of $\frac{1}{x^m}$ behaves like the graph of $\frac{1}{x}$ if m is odd and like the graph of $\frac{1}{x^2}$ if m is even. Larger values of m correspond to functions whose graphs get closer to the x-axis more rapidly for large values of x and closer to the vertical axis more rapidly for values of x near 0.

Manipulations with Exponents

For $x \neq 0$, we have now defined x^m for all integer values of m. Our previous identities involving positive integer exponents hold for all integers. The next example shows how to verify these identities when negative exponents are involved.

EXAMPLE 8

This identity holds for all integers.

Show that

$$x^m x^n = x^{m+n}$$

if m and n are negative integers and $x \neq 0$.

SOLUTION Suppose m and n are negative integers. Then there exist positive integers p and q such that $m = -p$ and $n = -q$. Now

$$\begin{aligned} x^m x^n &= x^{-p} x^{-q} \\ &= \frac{1}{x^p} \frac{1}{x^q} \\ &= \frac{1}{x^{p+q}} \\ &= x^{-(p+q)} \\ &= x^{(-p)+(-q)} \\ &= x^{m+n}, \end{aligned}$$

as desired.

The box below lists the key properties of integer exponents.

Algebraic properties of exponents

Suppose x and y are nonzero numbers and m and n are integers. Then

$$x^m x^n = x^{m+n},$$

$$x^m y^m = (xy)^m,$$

$$(x^m)^n = x^{mn},$$

$$x^0 = 1,$$

$$x^{-m} = \frac{1}{x^m},$$

$$\frac{x^m}{x^n} = x^{m-n},$$

$$\frac{x^m}{y^m} = \left(\frac{x}{y}\right)^m.$$

These properties can be verified using the properties of positive integer exponents and techniques similar to what was done in the example above.

EXAMPLE 9 Simplify the expression

$$\left(\frac{(x^{-3}y^4)^2}{x^{-9}y^3}\right)^2.$$

To manipulate fractions that involve exponents, keep in mind that an exponent changes sign when we move it from the numerator to the denominator or from the denominator to the numerator.

SOLUTION First we simplify the numerator inside the large parentheses. We have

$$(x^{-3}y^4)^2 = x^{-6}y^8.$$

The expression inside the large parentheses is thus equal to

$$\frac{x^{-6}y^8}{x^{-9}y^3}.$$

Now bring the terms in the denominator to the numerator, getting

$$\begin{aligned}\frac{x^{-6}y^8}{x^{-9}y^3} &= x^{-6}x^9y^8y^{-3}\\ &= x^{-6+9}y^{8-3}\\ &= x^3y^5.\end{aligned}$$

Thus the expression inside the large parentheses equals x^3y^5. Squaring that expression, we get

$$\left(\frac{(x^{-3}y^4)^2}{x^{-9}y^3}\right)^2 = x^6y^{10}.$$

EXERCISES

For Exercises 1–6, evaluate the given expression. Do not use a calculator.

1. $2^5 - 5^2$
2. $4^3 - 3^4$
3. $\dfrac{3^{-2}}{2^{-3}}$
4. $\dfrac{2^{-6}}{6^{-2}}$
5. $(\frac{2}{3})^{-4}$
6. $(\frac{5}{4})^{-3}$

The numbers in Exercises 7–14 are too large to be handled by a calculator. These exercises require an understanding of the concepts.

7. Write 9^{3000} as a power of 3.
8. Write 27^{4000} as a power of 3.
9. Write 5^{4000} as a power of 25.
10. Write 2^{3000} as a power of 8.
11. Write $2^5 \cdot 8^{1000}$ as a power of 2.
12. Write $5^3 \cdot 25^{2000}$ as a power of 5.
13. Write $2^{100} \cdot 4^{200} \cdot 8^{300}$ as a power of 2.
14. Write $3^{500} \cdot 9^{200} \cdot 27^{100}$ as a power of 3.

For Exercises 15–20, simplify the given expression by writing it as a power of a single variable.

15. $x^5(x^2)^3$
16. $y^4(y^3)^5$
17. $y^4(y^2(y^5)^2)^3$
18. $x(x^4(x^3)^2)^5$
19. $t^4(t^3(t^{-2})^5)^4$
20. $w^3(w^4(w^{-3})^6)^2$
21. Write $\dfrac{8^{1000}}{2^5}$ as a power of 2.
22. Write $\dfrac{25^{2000}}{5^3}$ as a power of 5.
23. Find integers m and n such that $2^m \cdot 5^n = 16000$.
24. Find integers m and n such that $2^m \cdot 5^n = 0.0032$.

For Exercises 25–32, simplify the given expression.

25. $\dfrac{(x^2)^3 y^8}{x^5(y^4)^3}$
26. $\dfrac{x^{11}(y^3)^2}{(x^3)^5(y^2)^4}$
27. $\dfrac{(x^{-2})^3 y^8}{x^{-5}(y^4)^{-3}}$
28. $\dfrac{x^{-11}(y^3)^{-2}}{(x^{-3})^5(y^2)^4}$
29. $\dfrac{(x^2y^4)^3}{(x^5y^2)^{-4}}$
30. $\dfrac{(x^2y^4)^{-3}}{(x^5y^{-2})^4}$
31. $\left(\dfrac{(x^2y^{-5})^{-4}}{(x^5y^{-2})^{-3}}\right)^2$
32. $\left(\dfrac{(x^{-3}y^5)^{-4}}{(x^{-5}y^{-2})^{-3}}\right)^{-2}$

33. A regulation ping-pong ball has a 4 cm diameter. Find the volume of a ping-pong ball.
34. A regulation tennis ball has a 2.5 inch diameter. Find the volume of a tennis ball.
35. A regulation baseball is defined to have a circumference of 9 inches. Find the volume of a baseball.
36. A regulation soccer ball is defined to have a circumference of 69 cm. Find the volume of a soccer ball.

For Exercises 37–44, find a formula for $f \circ g$ given the indicated functions f and g.

37. $f(x) = x^2$, $g(x) = x^3$
38. $f(x) = x^5$, $g(x) = x^4$
39. $f(x) = 4x^2$, $g(x) = 5x^3$
40. $f(x) = 3x^5$, $g(x) = 2x^4$
41. $f(x) = 4x^{-2}$, $g(x) = 5x^3$
42. $f(x) = 3x^{-5}$, $g(x) = 2x^4$
43. $f(x) = 4x^{-2}$, $g(x) = -5x^{-3}$
44. $f(x) = 3x^{-5}$, $g(x) = -2x^{-4}$

For Exercises 45–54, sketch the graph of the given function f on the interval $[-1.3, 1.3]$.

45. $f(x) = x^3 + 1$
46. $f(x) = x^4 + 2$
47. $f(x) = x^4 - 1.5$
48. $f(x) = x^3 - 0.5$
49. $f(x) = 2x^3$
50. $f(x) = 3x^4$
51. $f(x) = -2x^4$
52. $f(x) = -3x^3$
53. $f(x) = -2x^4 + 3$
54. $f(x) = -3x^3 + 4$

For Exercises 55–64, sketch the graph of the given function f on the domain $[-3, -\frac{1}{3}] \cup [\frac{1}{3}, 3]$.

55. $f(x) = \dfrac{1}{x} + 1$
56. $f(x) = \dfrac{1}{x^2} + 2$
57. $f(x) = \dfrac{1}{x^2} - 2$
58. $f(x) = \dfrac{1}{x} - 3$
59. $f(x) = \dfrac{2}{x}$
60. $f(x) = \dfrac{3}{x^2}$
61. $f(x) = -\dfrac{2}{x^2}$
62. $f(x) = -\dfrac{3}{x}$
63. $f(x) = -\dfrac{2}{x^2} + 3$
64. $f(x) = -\dfrac{3}{x} + 4$

PROBLEMS

Some problems require considerably more thought than the exercises. Unlike exercises, problems often have more than one correct answer.

65. Suppose m is a positive integer. Explain why 10^m, when written out in the usual decimal notation, is the digit 1 followed by m 0's.

66. (a) Verify that $2^4 = 4^2$.
    (b) Part (a) might lead someone to guess that we have a commutative operation here. However, for most choices of integers m and n, the inequality $m^n \neq n^m$ holds. For example, show that $2^3 \neq 3^2$ (which shows that we do not have commutativity).

67. (a) Verify that $(2^2)^2 = 2^{(2^2)}$.
    (b) Part (a) might lead someone to guess that we have an associative operation here. However, show that $(3^3)^3 \neq 3^{(3^3)}$ (which shows that we do not have associativity).

68. Suppose m is odd. Show that the function f defined by $f(x) = x^m$ is an odd function.

69. Suppose m is even. Show that the function f defined by $f(x) = x^m$ is an even function.

70. Suppose m and n are integers. Define functions f and g by $f(x) = x^m$ and $g(x) = x^n$. Explain why
$$(f \circ g)(x) = x^{mn}.$$

71. Suppose x is a real number and m, n, and p are positive integers. Explain why
$$x^{m+n+p} = x^m x^n x^p.$$

72. Suppose x is a real number and m, n, and p are positive integers. Explain why
$$((x^m)^n)^p = x^{mnp}.$$

73. Suppose x, y, and z are real numbers and m is a positive integer. Explain why
$$x^m y^m z^m = (xyz)^m.$$

74. Suppose x and y are real numbers, with $y \neq 0$, and m is a positive integer. Explain why
$$\frac{x^m}{y^m} = \left(\frac{x}{y}\right)^m.$$

75. Complete the verification begun in this section that
$$x^m x^n = x^{m+n}$$
for all $x \neq 0$ and all integers m and n.

[*We have already verified the identities in this problem and the next two problems when m and n are positive integers. The point of these problems is to verify these identities when one (or both) of m and n is negative or zero.*]

76. Show that if $x \neq 0$ and m and n are integers, then
$$(x^m)^n = x^{mn}.$$

77. Show that if x and y are nonzero real numbers and m is an integer, then
$$(xy)^m = x^m y^m.$$

78. Show that if $x \neq 0$, then
$$|x^n| = |x|^n$$
for all integers n.

Fermat's Last Theorem states that if n is an integer greater than 2, then there do not exist positive integers x, y, and z such that
$$x^n + y^n = z^n.$$

Fermat's Last Theorem was not proved until 1994, although mathematicians had been trying to find a proof for centuries.

79. Use Fermat's Last Theorem to show that if n is an integer greater than 2, then there do not exist positive rational numbers x and y such that
$$x^n + y^n = 1.$$
[*Hint:* Use proof by contradiction: Assume that there exist rational numbers $x = \frac{m}{p}$ and $y = \frac{q}{r}$ such that $x^n + y^n = 1$; then show that this assumption leads to a contradiction of Fermat's Last Theorem.]

80. Use Fermat's Last Theorem to show that if n is an integer greater than 2, then there do not exist positive rational numbers x, y, and z such that
$$x^n + y^n = z^n.$$
[*The equation $3^2 + 4^2 = 5^2$ shows the necessity of the hypothesis that $n > 2$.*]

WORKED-OUT SOLUTIONS *to Odd-numbered Exercises*

Do not read these worked-out solutions before first attempting to do the exercises yourself. Otherwise you may merely mimic the techniques shown here without understanding the ideas.

Best way to learn: Carefully read the section of the textbook, then do all the odd-numbered exercises (even if they have not been assigned) and check your answers here. If you get stuck on an exercise, reread the section of the textbook—then try the exercise again. If you are still stuck, then look at the worked-out solution here.

For Exercises 1–6, evaluate the given expression. Do not use a calculator.

1. $2^5 - 5^2$

SOLUTION $2^5 - 5^2 = 32 - 25 = 7$

3. $\dfrac{3^{-2}}{2^{-3}}$

SOLUTION $\dfrac{3^{-2}}{2^{-3}} = \dfrac{2^3}{3^2} = \dfrac{8}{9}$

5. $(\frac{2}{3})^{-4}$

SOLUTION $(\frac{2}{3})^{-4} = (\frac{3}{2})^4 = \frac{3^4}{2^4} = \frac{81}{16}$

The numbers in Exercises 7–14 are too large to be handled by a calculator. These exercises require an understanding of the concepts.

7. Write 9^{3000} as a power of 3.

SOLUTION $9^{3000} = (3^2)^{3000} = 3^{6000}$

9. Write 5^{4000} as a power of 25.

SOLUTION
$$\begin{aligned} 5^{4000} &= 5^{2\cdot 2000} \\ &= (5^2)^{2000} \\ &= 25^{2000} \end{aligned}$$

11. Write $2^5 \cdot 8^{1000}$ as a power of 2.

SOLUTION
$$\begin{aligned} 2^5 \cdot 8^{1000} &= 2^5 \cdot (2^3)^{1000} \\ &= 2^5 \cdot 2^{3000} \\ &= 2^{3005} \end{aligned}$$

13. Write $2^{100} \cdot 4^{200} \cdot 8^{300}$ as a power of 2.

SOLUTION
$$\begin{aligned} 2^{100} \cdot 4^{200} \cdot 8^{300} &= 2^{100} \cdot (2^2)^{200} \cdot (2^3)^{300} \\ &= 2^{100} \cdot 2^{400} \cdot 2^{900} \\ &= 2^{1400} \end{aligned}$$

For Exercises 15–20, simplify the given expression by writing it as a power of a single variable.

15. $x^5(x^2)^3$

SOLUTION $x^5(x^2)^3 = x^5x^6 = x^{11}$

17. $y^4(y^2(y^5)^2)^3$

SOLUTION
$$\begin{aligned} y^4(y^2(y^5)^2)^3 &= y^4(y^2y^{10})^3 \\ &= y^4(y^{12})^3 \\ &= y^4y^{36} \\ &= y^{40} \end{aligned}$$

19. $t^4(t^3(t^{-2})^5)^4$

SOLUTION
$$\begin{aligned} t^4(t^3(t^{-2})^5)^4 &= t^4(t^3t^{-10})^4 \\ &= t^4(t^{-7})^4 \\ &= t^4t^{-28} \\ &= t^{-24} \end{aligned}$$

21. Write $\dfrac{8^{1000}}{2^5}$ as a power of 2.

SOLUTION
$$\frac{8^{1000}}{2^5} = \frac{(2^3)^{1000}}{2^5} = \frac{2^{3000}}{2^5} = 2^{2995}$$

23. Find integers m and n such that $2^m \cdot 5^n = 16000$.

SOLUTION Note that

$$\begin{aligned} 16000 &= 16 \cdot 1000 \\ &= 2^4 \cdot 10^3 \\ &= 2^4 \cdot (2 \cdot 5)^3 \\ &= 2^4 \cdot 2^3 \cdot 5^3 \\ &= 2^7 \cdot 5^3. \end{aligned}$$

Thus if we want to find integers m and n such that $2^m \cdot 5^n = 16000$, we should choose $m = 7$ and $n = 3$.

For Exercises 25–32, simplify the given expression.

25. $\dfrac{(x^2)^3 y^8}{x^5 (y^4)^3}$

SOLUTION
$$\frac{(x^2)^3 y^8}{x^5 (y^4)^3} = \frac{x^6 y^8}{x^5 y^{12}} = \frac{x^{6-5}}{y^{12-8}} = \frac{x}{y^4}$$

27. $\dfrac{(x^{-2})^3 y^8}{x^{-5} (y^4)^{-3}}$

SOLUTION

$$\frac{(x^{-2})^3 y^8}{x^{-5} (y^4)^{-3}} = \frac{x^{-6} y^8}{x^{-5} y^{-12}} = \frac{y^{8+12}}{x^{6-5}} = \frac{y^{20}}{x}$$

29. $\dfrac{(x^2 y^4)^3}{(x^5 y^2)^{-4}}$

SOLUTION
$$\frac{(x^2 y^4)^3}{(x^5 y^2)^{-4}} = \frac{x^6 y^{12}}{x^{-20} y^{-8}} = x^{6+20} y^{12+8} = x^{26} y^{20}$$

31. $\left(\dfrac{(x^2 y^{-5})^{-4}}{(x^5 y^{-2})^{-3}}\right)^2$

SOLUTION
$$\left(\frac{(x^2 y^{-5})^{-4}}{(x^5 y^{-2})^{-3}}\right)^2 = \frac{(x^2 y^{-5})^{-8}}{(x^5 y^{-2})^{-6}} = \frac{x^{-16} y^{40}}{x^{-30} y^{12}} = x^{30-16} y^{40-12} = x^{14} y^{28}$$

33. A regulation ping-pong ball has a 4 cm diameter. Find the volume of a ping-pong ball.

SOLUTION Because the diameter of a ping-pong ball is 4 cm, the radius is 2 cm. Thus the volume of a ping-pong ball is $\frac{4}{3}\pi \cdot 2^3$ cubic centimeters, which we can simplify as follows:

$$\begin{aligned} \tfrac{4}{3}\pi \cdot 2^3 &= \tfrac{4}{3}\pi \cdot 8 \\ &= \tfrac{32\pi}{3} \\ &\approx 33.5. \end{aligned}$$

Thus the volume of a ping-pong ball is approximately 33.5 cubic centimeters.

35. A regulation baseball is defined to have a circumference of 9 inches. Find the volume of a baseball.

SOLUTION First we find the radius r of a baseball. Because the circumference is 9 inches, we have $2\pi r = 9$. Thus $r = \frac{9}{2\pi}$. Hence the volume of a baseball is $\frac{4}{3}\pi(\frac{9}{2\pi})^3$ cubic inches, which we can simplify as follows:

$$\begin{aligned} \tfrac{4}{3}\pi\left(\tfrac{9}{2\pi}\right)^3 &= \tfrac{4}{3}\pi \cdot \tfrac{9}{2\pi} \cdot \tfrac{9}{2\pi} \cdot \tfrac{9}{2\pi} \\ &= \tfrac{243}{2\pi^2} \\ &\approx 12.3. \end{aligned}$$

Thus the volume of a baseball is approximately 12.3 cubic inches.

For Exercises 37–44, find a formula for $f \circ g$ given the indicated functions f and g.

37. $f(x) = x^2$, $g(x) = x^3$

SOLUTION

$$(f \circ g)(x) = f(g(x)) = f(x^3) = (x^3)^2 = x^6$$

39. $f(x) = 4x^2$, $g(x) = 5x^3$

SOLUTION

$$(f \circ g)(x) = f(g(x)) = f(5x^3)$$
$$= 4(5x^3)^2 = 4 \cdot 5^2(x^3)^2 = 100x^6$$

41. $f(x) = 4x^{-2}$, $g(x) = 5x^3$

SOLUTION

$$(f \circ g)(x) = f(g(x)) = f(5x^3)$$
$$= 4(5x^3)^{-2} = 4 \cdot 5^{-2}(x^3)^{-2} = \tfrac{4}{25}x^{-6}$$

43. $f(x) = 4x^{-2}$, $g(x) = -5x^{-3}$

SOLUTION

$$(f \circ g)(x) = f(g(x)) = f(-5x^{-3})$$
$$= 4(-5x^{-3})^{-2} = 4(-5)^{-2}(x^{-3})^{-2} = \tfrac{4}{25}x^6$$

For Exercises 45–54, sketch the graph of the given function f on the interval $[-1.3, 1.3]$.

45. $f(x) = x^3 + 1$

SOLUTION Shift the graph of x^3 up 1 unit, getting this graph:

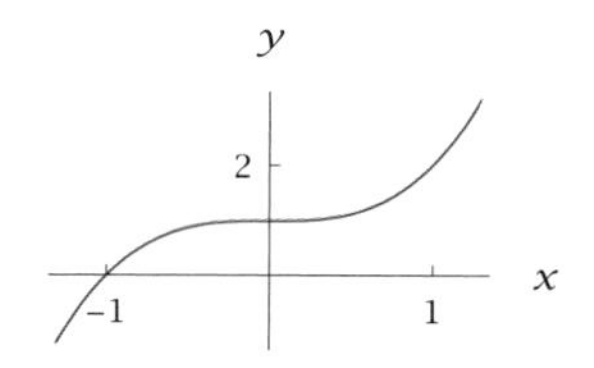

The graph of $x^3 + 1$.

47. $f(x) = x^4 - 1.5$

SOLUTION Shift the graph of x^4 down 1.5 units, getting this graph:

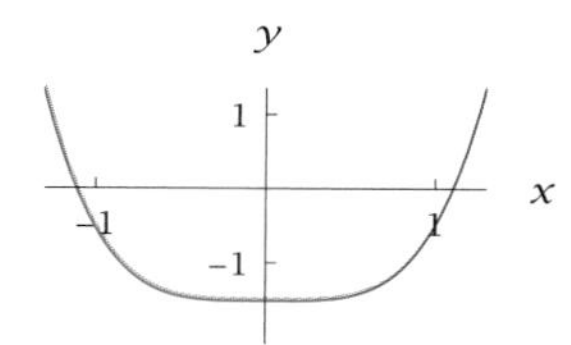

The graph of $x^4 - 1.5$.

49. $f(x) = 2x^3$

SOLUTION Vertically stretch the graph of x^3 by a factor of 2, getting this graph:

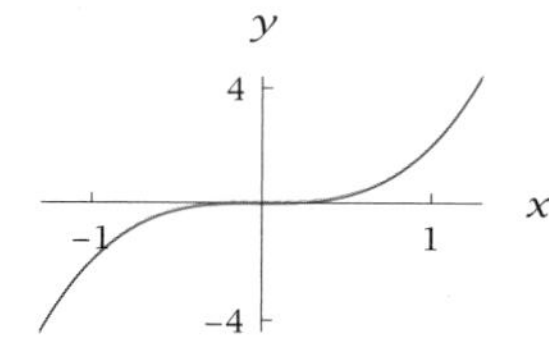

The graph of $2x^3$.

51. $f(x) = -2x^4$

SOLUTION Vertically stretch the graph of x^4 by a factor of 2 and then flip across the x-axis, getting this graph:

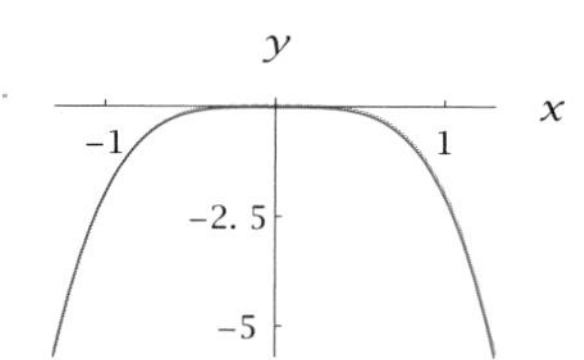

The graph of $-2x^4$.

53. $f(x) = -2x^4 + 3$

SOLUTION Vertically stretch the graph of x^4 by a factor of 2, then flip across the x-axis, and then shift up by 3 units, getting this graph:

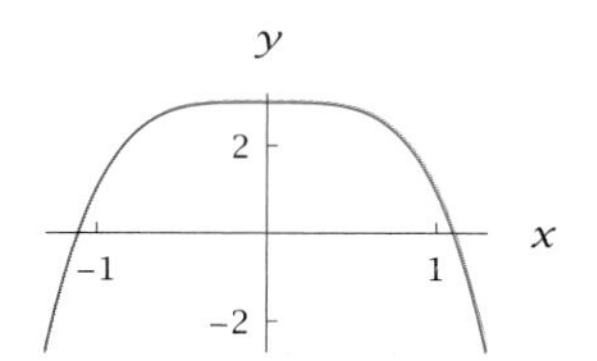

The graph of $-2x^4 + 3$.

For Exercises 55–64, sketch the graph of the given function f on the domain $[-3, -\frac{1}{3}] \cup [\frac{1}{3}, 3]$.

55. $f(x) = \dfrac{1}{x} + 1$

SOLUTION Shift the graph of $\frac{1}{x}$ up 1 unit, getting this graph:

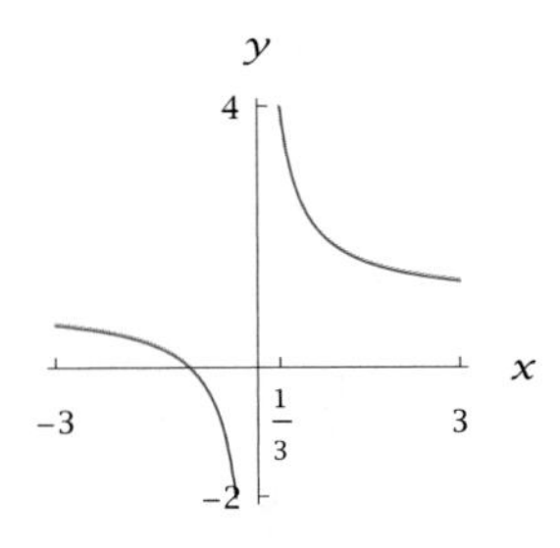

The graph of $\frac{1}{x} + 1$.

57. $f(x) = \dfrac{1}{x^2} - 2$

SOLUTION Shift the graph of $\frac{1}{x^2}$ down 2 units, getting this graph:

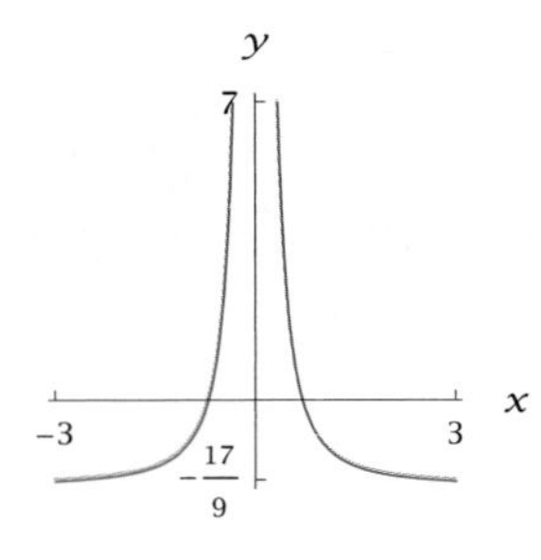

The graph of $\frac{1}{x^2} - 2$.

59. $f(x) = \dfrac{2}{x}$

SOLUTION Vertically stretch the graph of $\frac{1}{x}$ by a factor of 2, getting this graph:

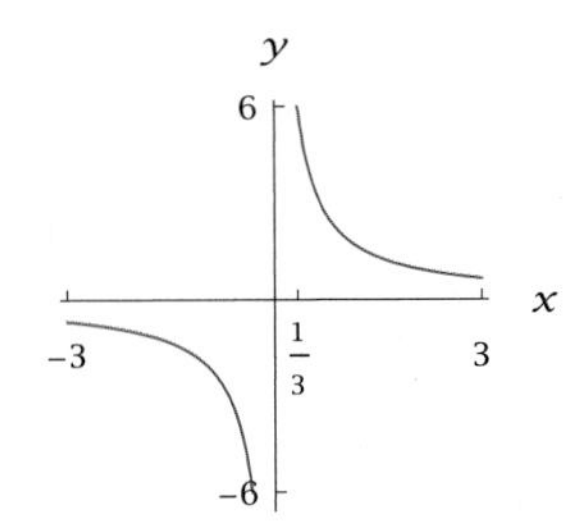

The graph of $\frac{2}{x}$.

61. $f(x) = -\dfrac{2}{x^2}$

SOLUTION Vertically stretch the graph of $\frac{1}{x^2}$ by a factor of 2 and then flip across the x-axis, getting this graph:

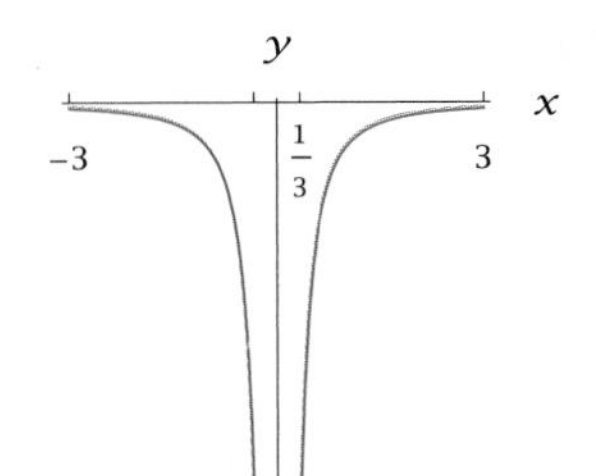

The graph of $-\frac{2}{x^2}$.

63. $f(x) = -\dfrac{2}{x^2} + 3$

SOLUTION Vertically stretch the graph of $\frac{1}{x^2}$ by a factor of 2, then flip across the x-axis, and then shift up by 3 units, getting this graph:

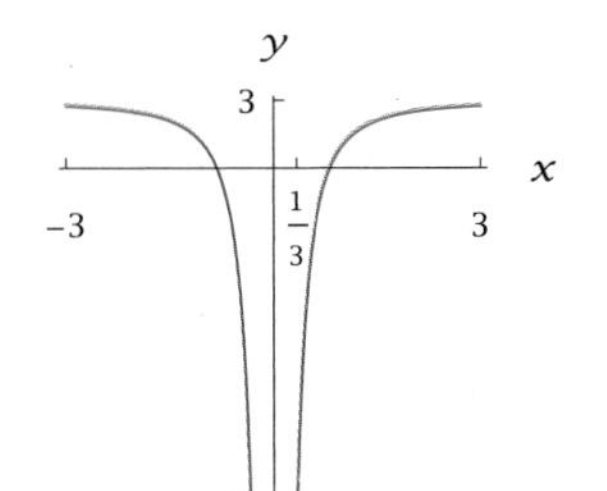

The graph of $-\frac{2}{x^2} + 3$.

4.2 *Polynomials*

LEARNING OBJECTIVES

By the end of this section you should be able to

- recognize the degree of a polynomial;
- manipulate polynomials algebraically;
- explain the connection between factorization and the zeros of a polynomial;
- determine the behavior of $p(x)$ when p is a polynomial and $|x|$ is large.

We have previously looked at linear functions and quadratic functions, which are among the simplest polynomials. In this section we will deal with more general polynomials. We begin with the definition of a polynomial.

Polynomials

A **polynomial** is a function p such that

$$p(x) = a_0 + a_1x + a_2x^2 + \cdots + a_nx^n,$$

where n is a nonnegative integer and $a_0, a_1, a_2, \dots, a_n$ are constants.

Because the expression defining a polynomial makes sense for every real number, you should assume that the domain of a polynomial is the set of real numbers unless another domain has been specified.

For example, the function p defined by

$$p(x) = 3 - 7x^5 + 2x^6$$

is a polynomial. Here, in terms of the definition above, we have $a_0 = 3$, $a_1 = a_2 = a_3 = a_4 = 0$, $a_5 = -7$, and $a_6 = 2$.

The Degree of a Polynomial

The highest exponent that appears in the expression defining a polynomial plays a key role in determining the behavior of the polynomial. Thus the following definition is useful.

Degree of a polynomial

Suppose p is a polynomial defined by

$$p(x) = a_0 + a_1x + a_2x^2 + \cdots + a_nx^n.$$

If $a_n \neq 0$, then we say that p has **degree** n. The degree of p is denoted by $\deg p$.

The numbers $a_0, a_1, a_2, \dots, a_n$ are called the ***coefficients*** *of the polynomial p.*

EXAMPLE 1

(a) Give an example of a polynomial of degree 0. Describe its graph.

(b) Give an example of a polynomial of degree 1. Describe its graph.

(c) Give an example of a polynomial of degree 2. Describe its graph.

(d) Give an example of a polynomial of degree 7.

SOLUTION

Polynomials of degree 0 are constant functions.

(a) The function p defined by

$$p(x) = 4$$

is a polynomial of degree 0. Its graph is a horizontal line.

Polynomials of degree 1 are linear functions.

(b) The function p defined by

$$p(x) = 2 + \tfrac{1}{2}x$$

is a polynomial of degree 1. Its graph is a nonhorizontal line.

Polynomials of degree 2 are quadratic functions.

(c) The function p defined by

$$p(x) = -3 + 5x^2$$

is a polynomial of degree 2. Its graph is a parabola.

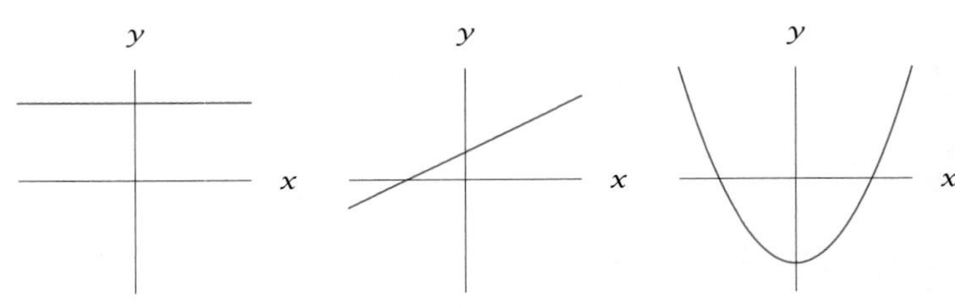

The graph of a polynomial of degree 0 (left), a polynomial of degree 1 (center), and a polynomial of degree 2 (right).

(d) The function p defined by

$$p(x) = 13 + 12x - x^3 - 9x^4 + 3x^7$$

is a polynomial of degree 7.

The Algebra of Polynomials

Two polynomials can be added or subtracted, producing another polynomial. Specifically, if p and q are polynomials then the polynomial $p + q$ is defined by

$$(p + q)(x) = p(x) + q(x)$$

and the polynomial $p - q$ is defined by

$$(p - q)(x) = p(x) - q(x).$$

Suppose p and q are polynomials defined by

EXAMPLE 2

$$p(x) = 2 - 7x^2 + 5x^3 \quad \text{and} \quad q(x) = 1 + 9x + x^2 + 5x^3.$$

(a) What is $\deg p$?

(b) What is $\deg q$?

(c) Find a formula for $p + q$.

(d) What is $\deg(p + q)$?

(e) Find a formula for $p - q$.

(f) What is $\deg(p - q)$?

Polynomial addition is commutative and associative. In other words, $p + q = q + p$ and $(p+q)+r = p+(q+r)$ for all polynomials p, q, and r.

SOLUTION

(a) The term with highest exponent that appears in the expression defining p is $5x^3$. Thus $\deg p = 3$.

(b) The term with highest exponent that appears in the expression defining q is $5x^3$. Thus $\deg q = 3$.

(c) Adding together the expressions defining p and q, we have

$$(p + q)(x) = 3 + 9x - 6x^2 + 10x^3.$$

(d) The term with highest exponent that appears in the expression above for $p + q$ is $10x^3$. Thus $\deg(p + q) = 3$.

(e) Subtracting the expression defining q from the expression defining p, we have

$$(p - q)(x) = 1 - 9x - 8x^2.$$

(f) The term with highest exponent that appears in the expression above for $p - q$ is $-8x^2$. Thus $\deg(p - q) = 2$.

More generally, we have the following result:

The constant polynomial p defined by $p(x) = 0$ for every number x has no nonzero coefficients. Thus the degree of this polynomial is undefined. Sometimes it is convenient to write $\deg 0 = -\infty$ to avoid trivial exceptions to various results.

Degree of the sum and difference of two polynomials

If p and q are nonzero polynomials, then

$$\deg(p + q) \leq \text{maximum}\{\deg p, \deg q\}$$

and

$$\deg(p - q) \leq \text{maximum}\{\deg p, \deg q\}.$$

This result holds because neither $p + q$ nor $p - q$ can contain an exponent larger than the largest exponent that appears in p or q.

Due to cancellation, the degree of $p + q$ or the degree of $p - q$ can be less than the maximum of the degree of p and the degree of q, as shown in part (f) of the example above.

Two polynomials can be multiplied together, producing another polynomial. Specifically, if p and q are polynomials, then the polynomial pq is defined by

$$(pq)(x) = p(x) \cdot q(x).$$

EXAMPLE 3

Suppose p and q are polynomials defined by

$$p(x) = 2 - 3x^2 \quad \text{and} \quad q(x) = 4x + 7x^5.$$

(a) What is $\deg p$?

(b) What is $\deg q$?

(c) Find a formula for pq.

(d) What is $\deg(pq)$?

Polynomial multiplication is commutative and associative. In other words, $pq = qp$ and $(pq)r = p(qr)$ for all polynomials p, q, and r.

SOLUTION

(a) The term with highest exponent that appears in the expression defining p is $-3x^2$. Thus $\deg p = 2$.

(b) The term with highest exponent that appears in the expression defining q is $7x^5$. Thus $\deg q = 5$.

(c)
$$\begin{aligned}(pq)(x) &= (2 - 3x^2)(4x + 7x^5)\\ &= 8x - 12x^3 + 14x^5 - 21x^7\end{aligned}$$

(d) The term with highest exponent that appears in the expression above for pq is $-21x^7$. Thus $\deg(pq) = 7$.

More generally, we have the following result:

Degree of the product of two polynomials

If p and q are nonzero polynomials, then

$$\deg(pq) = \deg p + \deg q.$$

This equality holds because when the highest-exponent term $x^{\deg p}$ in p is multiplied by the highest-exponent term $x^{\deg q}$ in q, we get $x^{\deg p + \deg q}$.

Zeros and Factorization of Polynomials

The zeros of a function are also sometimes called the **roots** *of the function.*

Zeros of a function

A number r is called a **zero** of a function p if $p(r) = 0$.

EXAMPLE 4

(a) Show that $\frac{3}{4}$ is a zero of the function p defined by $p(x) = 3 - 4x$.

(b) More generally, what are the zeros of the polynomial q of degree 1 defined by $q(x) = ax + b$, where a and b are constants with $a \neq 0$?

SOLUTION

(a) To show that $\frac{3}{4}$ is a zero of p, we must verify that $p(\frac{3}{4}) = 0$. This is easy to do:

$$p(\tfrac{3}{4}) = 3 - 4 \cdot \tfrac{3}{4} = 3 - 3 = 0.$$

(b) To find all zeros of q, we must find all numbers x such that $q(x) = 0$. In other words, we must solve the equation $ax + b = 0$. The only solution to this equation is $x = -\frac{b}{a}$. Thus $-\frac{b}{a}$ is the only zero of q.

The quadratic formula (see Section 2.3) can be used to find the zeros of a polynomial of degree 2, as summarized below.

Zeros of polynomials of degree 2

Suppose p is defined by $p(x) = ax^2 + bx + c$, where $a \neq 0$. Then

- p has two zeros, which equal $\frac{-b+\sqrt{b^2-4ac}}{2a}$ and $\frac{-b-\sqrt{b^2-4ac}}{2a}$, if $b^2 - 4ac > 0$.
- p has one zero equal to $-\frac{b}{2a}$ if $b^2 - 4ac = 0$;
- p has no (real) zeros if $b^2 - 4ac < 0$;

The cubic formula, which was discovered in the 16th century, is presented below for your amusement only. Do not memorize it.

Consider the cubic polynomial $p(x) = ax^3 + bx^2 + cx + d$, where $a \neq 0$. Set

$$u = \frac{bc}{6a^2} - \frac{b^3}{27a^3} - \frac{d}{2a}$$

and then set

$$v = u^2 + \left(\frac{c}{3a} - \frac{b^2}{9a^2}\right)^3.$$

Suppose $v \geq 0$. Then

$$-\frac{b}{3a} + \sqrt[3]{u + \sqrt{v}} + \sqrt[3]{u - \sqrt{v}}$$

is a zero of p.

Just as there is a quadratic formula to find the zeros of a polynomial of degree 2, there is a cubic formula to find the zeros of a polynomial of degree 3 and a quartic formula to find the zeros of a polynomial of degree 4. However, these complicated formulas are not of great practical value, and most mathematicians do not know these formulas.

Remarkably, mathematicians have proved that no formula exists for the zeros of polynomials of degree 5 or higher. But computers and calculators can use clever numerical methods to find good approximations to the zeros of any polynomial, even when exact zeros cannot be found. For example, no one will ever be able to give an exact formula for a zero of the polynomial p defined by

$$p(x) = x^5 - 5x^4 - 6x^3 + 17x^2 + 4x - 7.$$

However, a computer or symbolic calculator that can find approximate zeros of polynomials will produce the following five approximate zeros of p:

$$-1.87278 \quad -0.737226 \quad 0.624418 \quad 1.47289 \quad 5.51270$$

(for example, enter `solve x^5 - 5 x^4 - 6 x^3 + 17 x^2 + 4 x - 7 = 0` in a Wolfram|Alpha entry box to find the zeros of this polynomial p).

The zeros of a function have a graphical interpretation. Specifically, suppose p is a function and r is a zero of p. Then $p(r) = 0$ and thus $(r, 0)$ is on the graph of p; note that $(r, 0)$ is also on the horizontal axis. We conclude that each zero of p corresponds to a point where the graph of p intersects the horizontal axis.

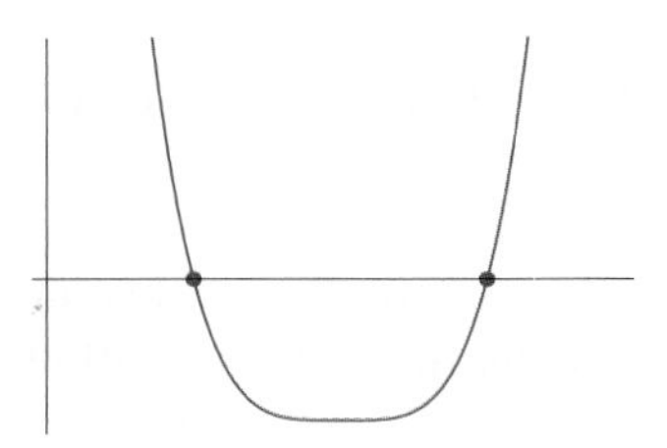

The zeros of a function correspond to the points where the graph of the function intersects the horizontal axis. Thus the function whose graph is shown here has two zeros.

EXAMPLE 5

Explain why the polynomial p defined by $p(x) = x^2 + 1$ has no (real) zeros.

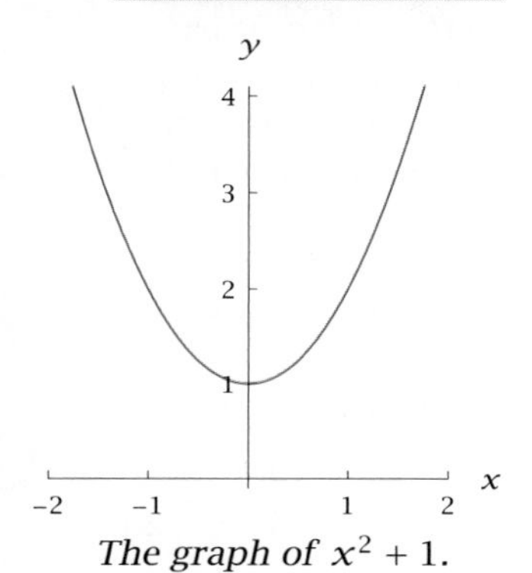

The graph of $x^2 + 1$.

SOLUTION In this case, the equation $p(x) = 0$ leads to the equation $x^2 = -1$, which has no real solutions because the square of a real number cannot be negative. Thus this polynomial p has no (real) zeros.

The graph in the margin also shows that p has no (real) zeros, because the graph does not intersect the horizontal axis.

Complex numbers were invented to provide solutions to the equation $x^2 = -1$. This book deals mostly with real numbers, but Section 4.4 will introduce complex numbers and complex zeros.

We now turn to the intimate connection between finding the zeros of a polynomial and finding factors of the polynomial of the form $x - r$.

Factor

Suppose p is a polynomial and r is a real number. Then $x - r$ is called a **factor** of $p(x)$ if there exists a polynomial G such that

$$p(x) = (x - r)G(x)$$

for every real number x.

EXAMPLE 6

Suppose p is the polynomial of degree 4 defined by

$$p(x) = (x - 2)(x - 5)(x^2 + 1).$$

(a) Explain why $x - 2$ is a factor of $p(x)$.

(b) Explain why $x - 5$ is a factor of $p(x)$.

(c) Show that 2 and 5 are zeros of p.

(d) Show that p has no (real) zeros except 2 and 5.

SOLUTION

(a) Because

$$p(x) = (x - 2)\big((x - 5)(x^2 + 1)\big),$$

we see that $x - 2$ is a factor of $p(x)$.

(b) Because

$$p(x) = (x - 5)\big((x - 2)(x^2 + 1)\big),$$

we see that $x - 5$ is a factor of $p(x)$.

(c) If 2 is substituted for x in the expression $p(x) = (x - 2)\big((x - 5)(x^2 + 1)\big)$, then without doing any computation we see that $p(2) = 0$. Thus 2 is a zero of p.

Similarly, setting x to equal 5 in the expression $p(x) = (x-5)\big((x-2)(x^2+1)\big)$, we easily see that $p(5) = 0$. Thus 5 is a zero of p.

(d) If x is a real number other than 2 or 5, then none of the numbers $x - 2$, $x - 5$, and $x^2 + 1$ equals 0 and thus $p(x) \neq 0$. Hence p has no (real) zeros except 2 and 5.

The example above provides a good illustration of the following result.

This result shows that the problem of finding the zeros of a polynomial is really the same as the problem of finding factors of the polynomial of the form $x - r$.

Zeros and factors of a polynomial

Suppose p is a polynomial and r is a real number. Then r is a zero of p if and only if $x - r$ is a factor of $p(x)$.

To see why the "if" part of the result above holds, suppose p is a polynomial and r is a real number such that $x - r$ is a factor of $p(x)$. Then there is a polynomial G such that $p(x) = (x - r)G(x)$. Thus $p(r) = (r - r)G(r) = 0$, and hence r is a zero of p.

The next section will provide an explanation of why the "only if" part of the result above holds.

The result above implies that to find the zeros of a polynomial, we need only factor the polynomial. However, there is no easy way to factor a typical polynomial other than by finding its zeros.

Without using a computer or calculator, you should be able to quickly factor expressions such as $x^2 + 2x + 1$, which equals $(x + 1)^2$, and $x^2 - 9$, which equals $(x + 3)(x - 3)$. However, for most quadratic and higher-degree polynomials (not counting the artificial examples often found in textbooks) it will be easier to use the quadratic formula or a computer or calculator to find zeros than to try to recognize factors of the form $x - r$.

A polynomial of degree 1 has exactly one zero (because the equation $ax + b = 0$ has exactly one solution $x = -\frac{b}{a}$). We know from the quadratic formula that a polynomial of degree 2 has at most two zeros. More generally, we have the following result:

Number of zeros of a polynomial

A nonzero polynomial has at most as many zeros as its degree.

Thus, for example, a polynomial of degree 15 has at most 15 zeros. This result holds because each (real) zero of a polynomial p corresponds to at least one term $x - r_j$ in a factorization of the form

$$p(x) = (x - r_1)(x - r_2)\dots(x - r_m)G(x),$$

where G is a polynomial with no (real) zeros. If the polynomial p had more zeros than its degree, then the right side of the equation above would have a higher degree than the left side, which would be a contradiction.

The Behavior of a Polynomial Near $\pm\infty$

Important: Always remember that neither ∞ nor $-\infty$ is a real number.

We now turn to an investigation of the behavior of a polynomial near ∞ and near $-\infty$. To say that x is near ∞ is just an informal way of saying that x is very large. Similarly, to say that x is near $-\infty$ is just an informal way of saying that x is negative and $|x|$ is very large. The phrase "very large" has no precise meaning; even its informal meaning can depend on the context. Our focus will be on determining whether a polynomial takes on positive or negative values near ∞ and near $-\infty$.

EXAMPLE 7

Let p be the polynomial defined by

$$p(x) = x^5 - 99999x^4 - 9999x^3 - 999x^2 - 99x - 9.$$

Is $p(x)$ positive or is $p(x)$ negative for x near ∞? In other words, if x is very large, is $p(x) > 0$ or is $p(x) < 0$?

SOLUTION If x is positive, then the x^5 term in $p(x)$ is also positive but the other terms in $p(x)$ are all negative. If $x > 1$, then x^5 is larger than x^4, but perhaps the $-99999x^4$ term along with the other terms will still make $p(x)$ negative. To get a feeling for the behavior of p, we can collect some evidence by evaluating $p(x)$ for some values of x, as in the table below:

The evidence in this table indicates that $p(x)$ is negative for positive values of x.

x	$p(x)$
1	−111104
10	−1009989899
100	−9999908999909
1000	−99008999999099009
10000	−899999999099900990009

From the table above, it appears that $p(x)$ is negative when x is positive, and more decisively negative for larger values of x, as shown in the graph below.

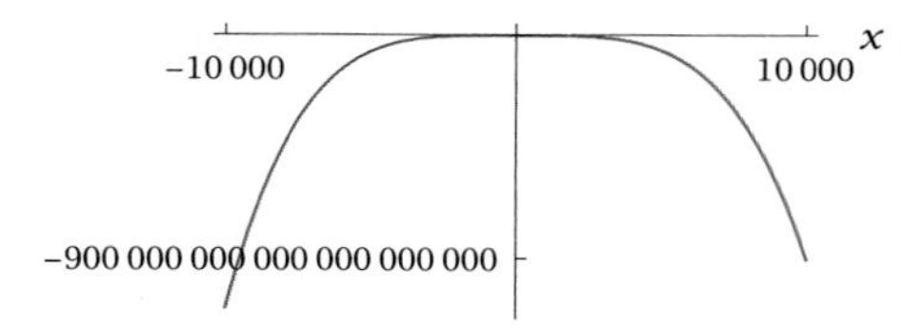

The graph of $x^5 - 99999x^4 - 9999x^3 - 999x^2 - 99x - 9$ on the interval $[-10000, 10000]$.

However, a bit of thought shows that this first impression is wrong. To see this, factor out x^5, the highest-degree term in the expression defining p, getting

$$p(x) = x^5\left(1 - \frac{99999}{x} - \frac{9999}{x^2} - \frac{999}{x^3} - \frac{99}{x^4} - \frac{9}{x^5}\right)$$

for all $x \neq 0$. If x is a very large number, say $x > 10^{10}$, then the five negative terms in the expression above are all very small. Thus if $x > 10^{10}$, then the expression in parentheses above is approximately 1. This means that $p(x)$ behaves like x^5 for very large values of x. In particular, this analysis implies that $p(x)$ is positive for very large values of x, unlike what we expected from the table and graph above.

Extending the table above to larger values of x, we find that $p(x)$ is positive for large values of x, as expected from the previous paragraph:

x	$p(x)$
100000	90000990009990099991
1000000	900000990000999000999900999991
10000000	9900000999000099990009999900999991
100000000	9990000099990000999990009999990099999991

Graphing the function over an interval 100 times larger than the interval used in the previous graph shows that $p(x)$ is indeed positive for large values of x.

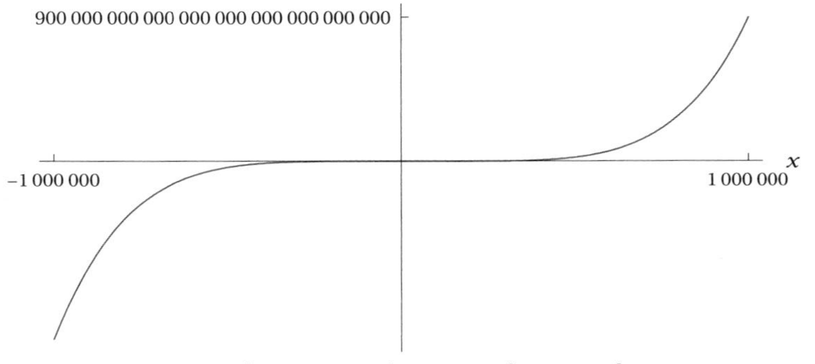

The graph of $x^5 - 99999x^4 - 9999x^3 - 999x^2 - 99x - 9$ *on the interval* $[-1000000, 1000000]$.

For very large values of x*, the polynomial* $p(x)$ *behaves like* x^5*. Thus* $p(x)$ *is positive for* x *near* ∞.

In general, the same procedure as used in the example above works with any polynomial:

Behavior of a polynomial near $\pm\infty$

- To determine the behavior of a polynomial near ∞ or near $-\infty$, factor out the term with highest degree.
- If cx^n is the term with highest degree of a polynomial p, then $p(x)$ behaves like cx^n when $|x|$ is very large.

EXAMPLE 8

Suppose

$$p(x) = 14 - 888x + 77777x^4 - 5x^6.$$

(a) Is $p(x)$ positive or negative for x near ∞?

(b) Is $p(x)$ positive or negative for x near $-\infty$?

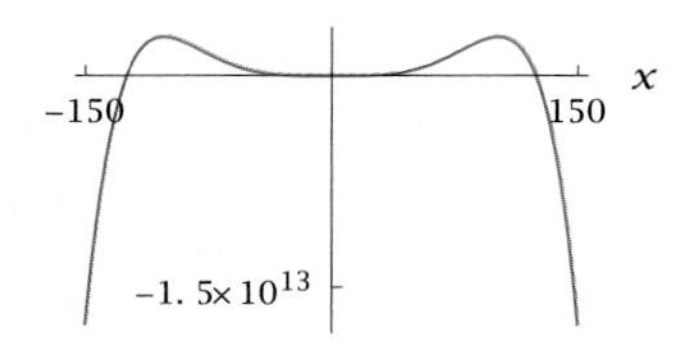

The graph of $14 - 888x + 77777x^4 - 5x^6$ *on* $[-150, 150]$.

SOLUTION The term with highest degree in $p(x)$ is $-5x^6$. Factoring out this term, we have

$$p(x) = -5x^6\left(1 - \frac{77777}{5x^2} + \frac{888}{5x^5} - \frac{14}{5x^6}\right).$$

If $|x|$ in very large, then the expression in parentheses above is approximately 1. Thus $p(x)$ behaves like $-5x^6$ when x is near ∞ or when x is near $-\infty$.

(a) If $x > 0$, then $-5x^6$ is negative. Thus $p(x)$ is negative when x is near ∞.

(b) If $x < 0$, then $-5x^6$ is negative. Thus $p(x)$ is negative when x is near $-\infty$.

The next example illustrates a method for discovering that an interval contains a zero.

EXAMPLE 9

Let p be the polynomial defined by

$$p(x) = x^5 + x^2 - 1.$$

Explain why p has a zero in the interval $(0, 1)$.

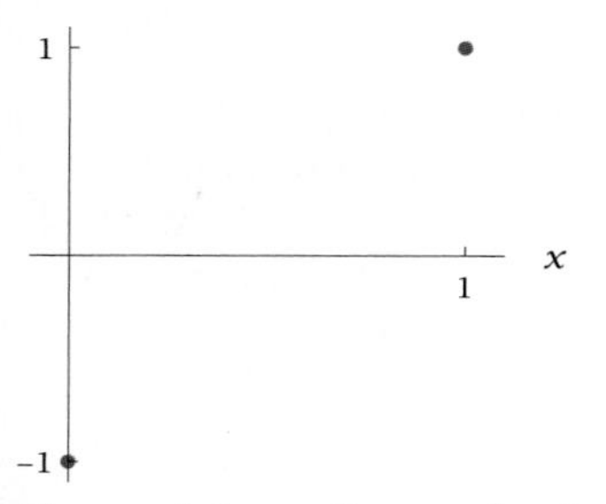

Two points on the graph of $x^5 + x^2 - 1$.

SOLUTION Because $p(0) = -1$ and $p(1) = 1$, the points $(0, -1)$ and $(1, 1)$ are on the graph of p (see figure in margin). The graph of p is a curve connecting these two points. As is intuitively clear (and can be rigorously proved using advanced mathematics), this curve must cross the horizontal axis at some point where the first coordinate is between 0 and 1. Thus p has a zero in the interval $(0, 1)$.

More generally, the reasoning used in the example above leads to the following result.

This result is a special case of what is called the Intermediate Value Theorem.

Zero in an interval

Suppose p is a polynomial and a and b are real numbers such that one of the numbers $p(a)$, $p(b)$ is negative and the other is positive. Then p has a zero in the interval (a, b).

The result above holds not just for polynomials but also for what are called **continuous** functions on an interval. A function whose graph consists of just one connected curve is continuous.

As an example of a function for which the result above fails, consider the function f defined by $f(x) = \frac{1}{x}$. Note that $f(-1)$ is negative and $f(1)$ is positive, but f does not have a zero on the interval $(-1, 1)$. This function f is not a polynomial; its graph (see Example 7 in Section 4.1) consists of two connected curves, and $f(x)$ is not defined for $x = 0$.

Suppose p is a polynomial with odd degree n. For simplicity, let's assume that x^n is the term of p with highest degree (if we care only about the zeros of p, we can always satisfy this condition by dividing p by the coefficient of the term with highest degree). We know that $p(x)$ behaves like x^n when $|x|$ is very large. Because n is an odd positive integer, this implies that $p(x)$ is negative for x near $-\infty$ and positive for x near ∞. Thus there is a negative number a such that $p(a)$ is negative and a positive number b such that $p(b)$ is positive. The result above now implies that p has a zero in the interval (a, b).

Our examination of the behavior near ∞ and near $-\infty$ of a polynomial with odd degree has led us to the following conclusion:

Zeros for polynomials with odd degree

Every polynomial with odd degree has at least one (real) zero.

If we work with complex numbers rather than real numbers, then every nonconstant polynomial has a zero. Section 4.4 includes a discussion of this result.

Graphs of Polynomials

Computers can draw graphs of polynomials better than humans. However, some human thought is usually needed to select on appropriate interval on which to graph a polynomial.

EXAMPLE 10

Let p be the polynomial defined by

$$p(x) = x^4 - 4x^3 - 2x^2 + 13x + 12.$$

Find an interval that does a good job of illustrating the key features of the graph of p.

SOLUTION If we ask a computer to graph this polynomial on the interval $[-2, 2]$, we obtain the following graph:

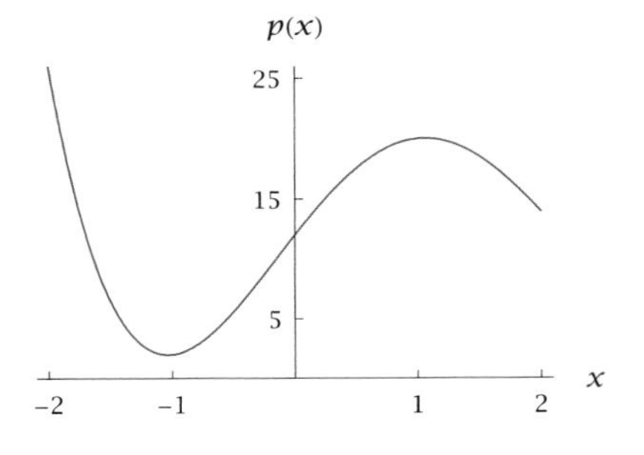

The graph of $x^4 - 4x^3 - 2x^2 + 13x + 12$ *on the interval* $[-2, 2]$.

Because $p(x)$ behaves like x^4 for very large values of x, we see that the graph above does not depict enough of the features of p. Often a bit of experimentation is needed to find an appropriate interval to illustrate the key features of the graph. For this polynomial p, the interval $[-2, 4]$ works well, as shown below:

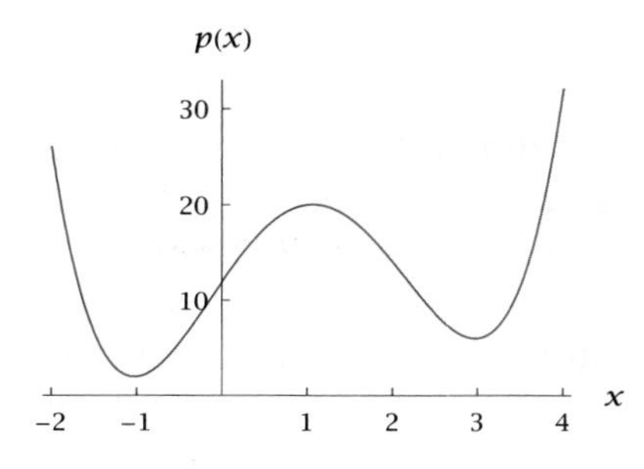

The graph of $x^4 - 4x^3 - 2x^2 + 13x + 12$ *on the interval* $[-2, 4]$.

The figure above shows the graph of p beginning to look like the graph of x^4 when $|x|$ is large. Thus the interval $[-2, 4]$ provides a more complete representation of the behavior of p than does $[-2, 2]$, which was the first interval we used.

We also now see that the graph of p above contains three points that might be thought of as either the top of a peak (at $x \approx 1$) or the bottom of a valley (at $x \approx -1$ and $x \approx 3$).

To search for additional behavior of p, we might try graphing p on a much larger interval, as follows:

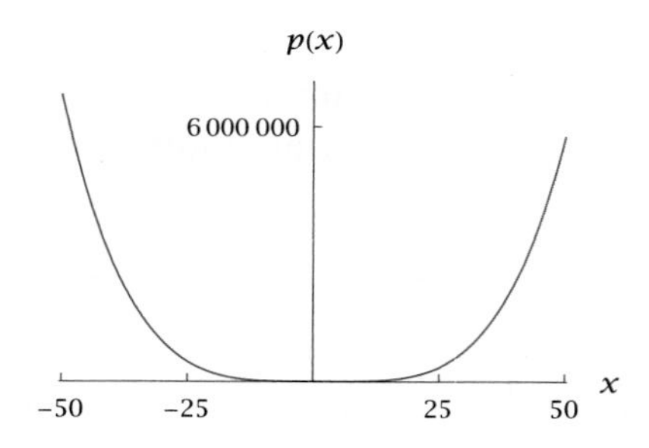

The graph of $x^4 - 4x^3 - 2x^2 + 13x + 12$ *on the interval* $[-50, 50]$.

The graph above shows no peaks or valleys, even though we know that it contains a total of at least three peaks and valleys. What happened here is that the scale needed to display the graph on the interval $[-50, 50]$ made the peaks and valleys so small that we cannot see them. Thus using this large interval hid some key features that were visible when we used the interval $[-2, 4]$.

Conclusion: A good choice for graphing this function is the interval $[-2, 4]$.

The following result is often useful for helping to determine whether any additional peaks or valleys in a graph remain to be discovered:

A good explanation of why this result holds requires more advanced mathematics than would be appropriate here.

Peaks and valleys for the graph of a polynomial

The graph of a polynomial p can have a total of at most $\deg p - 1$ peaks and valleys.

For example, the result above implies that the graph of the fourth-degree polynomial $x^4 - 4x^3 - 2x^2 + 13x + 12$ can have a total of at most three peaks and valleys. We discovered a total of three peaks and valleys when graphing this function on the interval $[-2, 4]$ in Example 10. Thus we need not worry that any remaining peaks or valleys are lurking undiscovered elsewhere.

EXERCISES

Suppose

$$p(x) = x^2 + 5x + 2,$$

$$q(x) = 2x^3 - 3x + 1,$$

$$s(x) = 4x^3 - 2.$$

In Exercises 1–18, write the indicated expression as a sum of terms, each of which is a constant times a power of x.

1. $(p + q)(x)$
2. $(p - q)(x)$
3. $(3p - 2q)(x)$
4. $(4p + 5q)(x)$
5. $(pq)(x)$
6. $(ps)(x)$
7. $(p(x))^2$
8. $(q(x))^2$
9. $(p(x))^2 s(x)$
10. $(q(x))^2 s(x)$
11. $(p \circ q)(x)$
12. $(q \circ p)(x)$
13. $(p \circ s)(x)$
14. $(s \circ p)(x)$
15. $(q \circ (p + s))(x)$
16. $((q + p) \circ s)(x)$
17. $\dfrac{q(2 + x) - q(2)}{x}$
18. $\dfrac{s(1 + x) - s(1)}{x}$
19. Find all real numbers x such that
$$x^6 - 8x^3 + 15 = 0.$$
20. Find all real numbers x such that
$$x^6 - 3x^3 - 10 = 0.$$
21. Find all real numbers x such that
$$x^4 - 2x^2 - 15 = 0.$$
22. Find all real numbers x such that
$$x^4 + 5x^2 - 14 = 0.$$
23. Factor $x^8 - y^8$ as nicely as possible.
24. Factor $x^{16} - y^8$ as nicely as possible.
25. Find a number b such that 3 is a zero of the polynomial p defined by
$$p(x) = 1 - 4x + bx^2 + 2x^3.$$
26. Find a number c such that -2 is a zero of the polynomial p defined by
$$p(x) = 5 - 3x + 4x^2 + cx^3.$$
27. Find a polynomial p of degree 3 such that -1, 2, and 3 are zeros of p and $p(0) = 1$.
28. Find a polynomial p of degree 3 such that -2, -1, and 4 are zeros of p and $p(1) = 2$.
29. Find all choices of b, c, and d such that 1 and 4 are the only zeros of the polynomial p defined by
$$p(x) = x^3 + bx^2 + cx + d.$$
30. Find all choices of b, c, and d such that -3 and 2 are the only zeros of the polynomial p defined by
$$p(x) = x^3 + bx^2 + cx + d.$$

PROBLEMS

31. Give an example of two polynomials of degree 4 whose sum has degree 3.
32. Give an example of two polynomials of degree 4 whose sum has degree 2.
33. Find a polynomial p of degree 2 with integer coefficients such that 2.1 and 4.1 are zeros of p.
34. Find a polynomial p of degree 3 with integer coefficients such that 2.1, 3.1, and 4.1 are zeros of p.
35. Show that if p and q are nonzero polynomials with $\deg p < \deg q$, then
$$\deg(p + q) = \deg q.$$
36. Give an example of polynomials p and q such that $\deg(pq) = 8$ and $\deg(p + q) = 5$.
37. Give an example of polynomials p and q such that $\deg(pq) = 8$ and $\deg(p + q) = 2$.

38. Suppose $q(x) = 2x^3 - 3x + 1$.

(a) Show that the point $(2, 11)$ is on the graph of q.

(b) Show that the slope of a line containing $(2, 11)$ and a point on the graph of q very close to $(2, 11)$ is approximately 21.

[*Hint:* Use the result of Exercise 17.]

39. Suppose $s(x) = 4x^3 - 2$.

(a) Show that the point $(1, 2)$ is on the graph of s.

(b) Give an estimate for the slope of a line containing $(1, 2)$ and a point on the graph of s very close to $(1, 2)$.

[*Hint:* Use the result of Exercise 18.]

40. Give an example of polynomials p and q of degree 3 such that $p(1) = q(1)$, $p(2) = q(2)$, and $p(3) = q(3)$, but $p(4) \neq q(4)$.

41. Suppose p and q are polynomials of degree 3 such that $p(1) = q(1)$, $p(2) = q(2)$, $p(3) = q(3)$, and $p(4) = q(4)$. Explain why $p = q$.

For Problems 42–43, let p be the polynomial defined by

$$p(x) = x^6 - 87x^4 - 92x + 2.$$

42. (a) Use a computer or calculator to sketch a graph of p on the interval $[-5, 5]$.

(b) Is $p(x)$ positive or negative for x near ∞?

(c) Is $p(x)$ positive or negative for x near $-\infty$?

(d) Explain why the graph from part (a) does not accurately show the behavior of $p(x)$ for large values of x.

43. (a) Evaluate $p(-2)$, $p(-1)$, $p(0)$, and $p(1)$.

(b) Explain why the results from part (a) imply that p has a zero in the interval $(-2, -1)$ and p has a zero in the interval $(0, 1)$.

(c) Show that p has at least four zeros in the interval $[-10, 10]$.
[*Hint:* We already know from part (*b*) that p has at least two zeros is the interval $[-10, 10]$. You can show the existence of other zeros by finding integers n such that one of the numbers $p(n)$, $p(n + 1)$ is positive and the other is negative.]

44. A new snack shop on campus finds that the number of students following it on *Twitter* at the end of each of its first five weeks in business is 23, 89, 223, 419, and 647. A clever employee discovers that the number of students following the new snack shop on *Twitter* after w weeks is $p(w)$, where p is defined by

$$p(w) = 7 + 3w + 5w^2 + 9w^3 - w^4.$$

Indeed, with p defined as above, we have $p(1) = 23$, $p(2) = 89$, $p(3) = 223$, $p(4) = 419$, and $p(5) = 647$. Explain why the polynomial p defined above cannot give accurate predictions for the number of followers on *Twitter* for weeks far into the future.

45. A textbook states that the rabbit population on a small island is observed to be

$$1000 + 120t - 0.4t^4,$$

where t is the time in months since observations of the island began. Explain why the formula above cannot correctly give the number of rabbits on the island for large values of t.

46. Verify that

$$(x + y)^3 = x^3 + 3x^2y + 3xy^2 + y^3.$$

47. Verify that

$$x^3 - y^3 = (x - y)(x^2 + xy + y^2).$$

48. Verify that

$$x^3 + y^3 = (x + y)(x^2 - xy + y^2).$$

49. Verify that

$$x^5 - y^5 = (x - y)(x^4 + x^3y + x^2y^2 + xy^3 + y^4).$$

50. Verify that

$$x^4 + 1 = (x^2 + \sqrt{2}x + 1)(x^2 - \sqrt{2}x + 1).$$

51. Write the polynomial $x^4 + 16$ as the product of two polynomials of degree 2.
[*Hint:* Use the result from the previous problem with x replaced by $\frac{x}{2}$.]

52. Show that

$$(a + b)^3 = a^3 + b^3$$

if and only if $a = 0$ or $b = 0$ or $a = -b$.

53. Suppose d is a real number. Show that

$$(d+1)^4 = d^4 + 1$$

if and only if $d = 0$.

54. Without doing any calculations or using a calculator, explain why

$$x^2 + 87559743x - 787727821$$

has no integer zeros.
[*Hint:* If x is an odd integer, is the expression above even or odd? If x is an even integer, is the expression above even or odd?]

55. Suppose M and N are odd integers. Explain why

$$x^2 + Mx + N$$

has no integer zeros.

56. Suppose M and N are odd integers. Explain why

$$x^2 + Mx + N$$

has no rational zeros.

57. Suppose $p(x) = 3x^7 - 5x^3 + 7x - 2$.

    (a) Show that if m is a zero of p, then

    $$\frac{2}{m} = 3m^6 - 5m^2 + 7.$$

    (b) Show that the only possible integer zeros of p are -2, -1, 1, and 2.

    (c) Show that no integer is a zero of p.

58. Suppose a, b, and c are integers and that

$$p(x) = ax^3 + bx^2 + cx + 9.$$

Explain why every zero of p that is an integer is contained in the set $\{-9, -3, -1, 1, 3, 9\}$.

59. Suppose $p(x) = 2x^5 + 5x^4 + 2x^3 - 1$. Show that -1 is the only integer zero of p.

60. Suppose $p(x) = a_0 + a_1x + \cdots + a_nx^n$, where $a_1, a_2, \dots, a_n$ are integers. Suppose m is a nonzero integer that is a zero of p. Show that a_0/m is an integer.
[*This result shows that to find integer zeros of a polynomial, we need only look at divisors of its constant term.*]

61. Suppose $p(x) = 2x^6 + 3x^5 + 5$.

    (a) Show that if $\frac{M}{N}$ is a zero of p, then

    $$2M^6 + 3M^5N + 5N^6 = 0.$$

    (b) Show that if M and N are integers with no common factors and $\frac{M}{N}$ is a zero of p, then $5/M$ and $2/N$ are integers.

    (c) Show that the only possible rational zeros of p are -5, -1, $-\frac{1}{2}$, and $-\frac{5}{2}$.

    (d) Show that no rational number is a zero of p.

62. Suppose $p(x) = 2x^4 + 9x^3 + 1$.

    (a) Show that if $\frac{M}{N}$ is a zero of p, then

    $$2M^4 + 9M^3N + N^4 = 0.$$

    (b) Show that if M and N are integers with no common factors and $\frac{M}{N}$ is a zero of p, then $M = -1$ or $M = 1$.

    (c) Show that if M and N are integers with no common factors and $\frac{M}{N}$ is a zero of p, then $N = -2$ or $N = 2$ or $N = -1$ or $N = 1$.

    (d) Show that $-\frac{1}{2}$ is the only rational zero of p.

63. Suppose $p(x) = a_0 + a_1x + \cdots + a_nx^n$, where $a_1, a_2, \dots, a_n$ are integers. Suppose M and N are nonzero integers with no common factors and $\frac{M}{N}$ is a zero of p. Show that a_0/M and a_n/N are integers.
[*This result shows that to find rational zeros of a polynomial, we need only look at fractions whose numerator is a divisor of the constant term and whose denominator is a divisor of the coefficient of highest degree. This result is called the* **Rational Roots Theorem**.]

64. Give an example of a polynomial of degree 5 that has exactly two zeros.

65. Give an example of a polynomial of degree 8 that has exactly three zeros.

66. Give an example of a polynomial p of degree 4 such that $p(7) = 0$ and $p(x) \geq 0$ for all real numbers x.

67. Give an example of a polynomial p of degree 6 such that $p(0) = 5$ and $p(x) \geq 5$ for all real numbers x.

68. Give an example of a polynomial p of degree 8 such that $p(2) = 3$ and $p(x) \geq 3$ for all real numbers x.

69. Explain why there does not exist a polynomial p of degree 7 such that $p(x) \geq -100$ for every real number x.

70. Explain why the composition of two polynomials is a polynomial.

71. Show that if p and q are nonzero polynomials, then
$$\deg(p \circ q) = (\deg p)(\deg q).$$

72. In the first figure in the solution to Example 7, the graph of the polynomial p clearly lies below the x-axis for x in the interval $[5000, 10000]$. Yet in the second figure in the same solution, the graph of p seems to be on or above the x-axis for all values of p in the interval $[0, 1000000]$. Explain.

73. Explain why the polynomial p defined by
$$p(x) = x^6 + 7x^5 - 2x - 3$$
has a zero in the interval $(0, 1)$.

WORKED-OUT SOLUTIONS *to Odd-numbered Exercises*

Suppose

$$p(x) = x^2 + 5x + 2,$$
$$q(x) = 2x^3 - 3x + 1,$$
$$s(x) = 4x^3 - 2.$$

In Exercises 1–18, write the indicated expression as a sum of terms, each of which is a constant times a power of x.

1. $(p + q)(x)$

SOLUTION

$$\begin{aligned}(p+q)(x) &= (x^2 + 5x + 2) + (2x^3 - 3x + 1)\\ &= 2x^3 + x^2 + 2x + 3\end{aligned}$$

3. $(3p - 2q)(x)$

SOLUTION

$$\begin{aligned}(3p - 2q)(x) &= 3(x^2 + 5x + 2) - 2(2x^3 - 3x + 1)\\ &= 3x^2 + 15x + 6 - 4x^3 + 6x - 2\\ &= -4x^3 + 3x^2 + 21x + 4\end{aligned}$$

5. $(pq)(x)$

SOLUTION

$$\begin{aligned}(pq)(x) &= (x^2 + 5x + 2)(2x^3 - 3x + 1)\\ &= x^2(2x^3 - 3x + 1)\\ &\quad + 5x(2x^3 - 3x + 1) + 2(2x^3 - 3x + 1)\\ &= 2x^5 - 3x^3 + x^2 + 10x^4 - 15x^2\\ &\quad + 5x + 4x^3 - 6x + 2\\ &= 2x^5 + 10x^4 + x^3 - 14x^2 - x + 2\end{aligned}$$

7. $(p(x))^2$

SOLUTION

$$\begin{aligned}(p(x))^2 &= (x^2 + 5x + 2)(x^2 + 5x + 2)\\ &= x^2(x^2 + 5x + 2) + 5x(x^2 + 5x + 2)\\ &\quad + 2(x^2 + 5x + 2)\\ &= x^4 + 5x^3 + 2x^2 + 5x^3 + 25x^2\\ &\quad + 10x + 2x^2 + 10x + 4\\ &= x^4 + 10x^3 + 29x^2 + 20x + 4\end{aligned}$$

9. $(p(x))^2 s(x)$

SOLUTION Using the expression that we computed for $(p(x))^2$ in the solution to Exercise 7, we have

$$\begin{aligned}(p(x))^2 s(x) &= (x^4 + 10x^3 + 29x^2 + 20x + 4)(4x^3 - 2)\\ &= 4x^3(x^4 + 10x^3 + 29x^2 + 20x + 4)\\ &\quad - 2(x^4 + 10x^3 + 29x^2 + 20x + 4)\\ &= 4x^7 + 40x^6 + 116x^5 + 80x^4 + 16x^3\\ &\quad - 2x^4 - 20x^3 - 58x^2 - 40x - 8\\ &= 4x^7 + 40x^6 + 116x^5 + 78x^4\\ &\quad - 4x^3 - 58x^2 - 40x - 8.\end{aligned}$$

11. $(p \circ q)(x)$

SOLUTION

$$\begin{aligned}(p \circ q)(x) &= p(q(x))\\ &= p(2x^3 - 3x + 1)\\ &= (2x^3 - 3x + 1)^2 + 5(2x^3 - 3x + 1) + 2\\ &= (4x^6 - 12x^4 + 4x^3 + 9x^2 - 6x + 1)\\ &\quad + (10x^3 - 15x + 5) + 2\\ &= 4x^6 - 12x^4 + 14x^3 + 9x^2 - 21x + 8\end{aligned}$$

13. $(p \circ s)(x)$

SOLUTION

$$\begin{aligned}(p \circ s)(x) &= p(s(x))\\ &= p(4x^3 - 2)\\ &= (4x^3 - 2)^2 + 5(4x^3 - 2) + 2\\ &= (16x^6 - 16x^3 + 4) + (20x^3 - 10) + 2\\ &= 16x^6 + 4x^3 - 4\end{aligned}$$

15. $(q \circ (p + s))(x)$

SOLUTION

$$\begin{aligned}(q \circ (p + s))(x) &= q((p + s)(x))\\ &= q(p(x) + s(x))\\ &= q(4x^3 + x^2 + 5x)\\ &= 2(4x^3 + x^2 + 5x)^3 - 3(4x^3 + x^2 + 5x) + 1\\ &= 2(4x^3 + x^2 + 5x)^2(4x^3 + x^2 + 5x)\\ &\quad - 12x^3 - 3x^2 - 15x + 1\\ &= 2(16x^6 + 8x^5 + 41x^4 + 10x^3 + 25x^2)\\ &\quad \times (4x^3 + x^2 + 5x) - 12x^3 - 3x^2 - 15x + 1\\ &= 128x^9 + 96x^8 + 504x^7 + 242x^6 + 630x^5\\ &\quad + 150x^4 + 238x^3 - 3x^2 - 15x + 1\end{aligned}$$

17. $\dfrac{q(2 + x) - q(2)}{x}$

SOLUTION

$$\begin{aligned}&\frac{q(2 + x) - q(2)}{x}\\ &= \frac{2(2 + x)^3 - 3(2 + x) + 1 - (2 \cdot 2^3 - 3 \cdot 2 + 1)}{x}\\ &= \frac{2x^3 + 12x^2 + 21x}{x}\\ &= 2x^2 + 12x + 21\end{aligned}$$

19. Find all real numbers x such that

$$x^6 - 8x^3 + 15 = 0.$$

SOLUTION This equation involves x^3 and x^6; thus we make the substitution $x^3 = y$. Squaring both sides of the equation $x^3 = y$ gives $x^6 = y^2$. With these substitutions, the equation above becomes

$$y^2 - 8y + 15 = 0.$$

This new equation can now be solved either by factoring the left side or by using the quadratic formula. Let's factor the left side, getting

$$(y - 3)(y - 5) = 0.$$

Thus $y = 3$ or $y = 5$ (the same result could have been obtained by using the quadratic formula).

Substituting x^3 for y now shows that $x^3 = 3$ or $x^3 = 5$. Thus $x = 3^{1/3}$ or $x = 5^{1/3}$.

21. Find all real numbers x such that

$$x^4 - 2x^2 - 15 = 0.$$

SOLUTION This equation involves x^2 and x^4; thus we make the substitution $x^2 = y$. Squaring both sides of the equation $x^2 = y$ gives $x^4 = y^2$. With these substitutions, the equation above becomes

$$y^2 - 2y - 15 = 0.$$

This new equation can now be solved either by factoring the left side or by using the quadratic formula. Let's use the quadratic formula, getting

$$y = \frac{2 \pm \sqrt{4 + 60}}{2} = \frac{2 \pm 8}{2}.$$

Thus $y = 5$ or $y = -3$ (the same result could have been obtained by factoring).

Substituting x^2 for y now shows that $x^2 = 5$ or $x^2 = -3$. The equation $x^2 = 5$ implies that $x = \sqrt{5}$ or $x = -\sqrt{5}$. The equation $x^2 = -3$ has no solutions in the real numbers. Thus the only solutions to our original equation $x^4 - 2x^2 - 15 = 0$ are $x = \sqrt{5}$ or $x = -\sqrt{5}$.

23. Factor $x^8 - y^8$ as nicely as possible.

SOLUTION

$$\begin{aligned} x^8 - y^8 &= (x^4 - y^4)(x^4 + y^4) \\ &= (x^2 - y^2)(x^2 + y^2)(x^4 + y^4) \\ &= (x - y)(x + y)(x^2 + y^2)(x^4 + y^4) \end{aligned}$$

25. Find a number b such that 3 is a zero of the polynomial p defined by

$$p(x) = 1 - 4x + bx^2 + 2x^3.$$

SOLUTION Note that

$$\begin{aligned} p(3) &= 1 - 4 \cdot 3 + b \cdot 3^2 + 2 \cdot 3^3 \\ &= 43 + 9b. \end{aligned}$$

We want $p(3)$ to equal 0. Thus we solve the equation $0 = 43 + 9b$, getting $b = -\frac{43}{9}$.

27. Find a polynomial p of degree 3 such that -1, 2, and 3 are zeros of p and $p(0) = 1$.

SOLUTION If p is a polynomial of degree 3 and -1, 2, and 3 are zeros of p, then

$$p(x) = c(x + 1)(x - 2)(x - 3)$$

for some constant c. We have $p(0) = c(0 + 1)(0 - 2)(0 - 3) = 6c$. Thus to make $p(0) = 1$ we must choose $c = \frac{1}{6}$. Thus

$$p(x) = \frac{(x + 1)(x - 2)(x - 3)}{6},$$

which by multiplying together the terms in the numerator can also be written in the form

$$p(x) = 1 + \frac{x}{6} - \frac{2x^2}{3} + \frac{x^3}{6}.$$

29. Find all choices of b, c, and d such that 1 and 4 are the only zeros of the polynomial p defined by

$$p(x) = x^3 + bx^2 + cx + d.$$

SOLUTION Because 1 and 4 are zeros of p, there is a polynomial q such that

$$p(x) = (x - 1)(x - 4)q(x).$$

Because p has degree 3, the polynomial q must have degree 1. Thus q has a zero, which must equal 1 or 4 because those are the only zeros of p. Furthermore, the coefficient of x in the polynomial q must equal 1 because the coefficient of x^3 in the polynomial p equals 1.

Thus $q(x) = x - 1$ or $q(x) = x - 4$. In other words, $p(x) = (x - 1)^2(x - 4)$ or $p(x) = (x - 1)(x - 4)^2$. Multiplying out these expressions, we see that $p(x) = x^3 - 6x^2 + 9x - 4$ or $p(x) = x^3 - 9x^2 + 24x - 16$.

Thus $b = -6$, $c = 9$, $d = -4$ or $b = -9$, $c = 24$, $c = -16$.

4.3 *Rational Functions*

LEARNING OBJECTIVES

By the end of this section you should be able to

- manipulate rational functions algebraically;
- decompose a rational function into a polynomial plus a rational function whose numerator has degree less than its denominator;
- determine the behavior of a rational function near $\pm\infty$;
- recognize the asymptotes of the graph of a rational function.

Ratios of Polynomials

Just as a rational number is the ratio of two integers, a **rational function** is the ratio of two polynomials:

Rational functions

A **rational function** is a function r such that

$$r(x) = \frac{p(x)}{q(x)},$$

where p and q are polynomials, with $q \neq 0$.

For example, the function r defined by

$$r(x) = \frac{2x^3 + 7x + 1}{x^4 + 3}$$

is a rational function.

Every polynomial is also a rational function because a polynomial can be written as the ratio of itself with the constant polynomial 1.

Unless some other domain has been specified, you should assume that the domain of a rational function is the set of real numbers where the expression defining the rational function makes sense. In the example from the paragraph above, the expression defining r makes sense for every real number; thus the domain of that rational function is the set of real numbers.

Because division by 0 is not defined, the domain of a rational function $\frac{p}{q}$ must exclude all zeros of q, as shown in the following example.

EXAMPLE 1

Find the domain of the rational function r defined by

$$r(x) = \frac{3x^5 + x^4 - 6x^3 - 2}{x^2 - 9}.$$

SOLUTION The denominator of the expression above is 0 if $x = 3$ or $x = -3$. Thus unless stated otherwise, we would assume that the domain of r is the set of numbers other than 3 and -3. In other words, the domain of r is $(-\infty, -3) \cup (-3, 3) \cup (3, \infty)$.

The Algebra of Rational Functions

Two rational functions can be added or subtracted, producing another rational function. Specifically, if r and s are rational functions then the rational function $r + s$ is defined by

Your algebraic manipulation skills can be sharpened by exercises involving the addition and subtraction of rational functions.

$$(r + s)(x) = r(x) + s(x)$$

and the rational function $r - s$ is defined by

$$(r - s)(x) = r(x) - s(x).$$

The procedure for adding or subtracting rational functions is the same as for adding or subtracting rational numbers—multiply numerator and denominator by the same factor to get common denominators.

EXAMPLE 2 Suppose

$$r(x) = \frac{2x}{x^2 + 1} \quad \text{and} \quad s(x) = \frac{3x + 2}{x^3 + 5}.$$

Write $r + s$ as the ratio of two polynomials.

SOLUTION

$$\begin{aligned}(r + s)(x) &= \frac{2x}{x^2 + 1} + \frac{3x + 2}{x^3 + 5} \\ &= \frac{(2x)(x^3 + 5)}{(x^2 + 1)(x^3 + 5)} + \frac{(3x + 2)(x^2 + 1)}{(x^2 + 1)(x^3 + 5)} \\ &= \frac{(2x)(x^3 + 5) + (3x + 2)(x^2 + 1)}{(x^2 + 1)(x^3 + 5)} \\ &= \frac{2x^4 + 3x^3 + 2x^2 + 13x + 2}{x^5 + x^3 + 5x^2 + 5}\end{aligned}$$

Two rational functions can be multiplied or divided, producing another rational function (except that division by the constant rational function 0 is not defined). Specifically, if r and s are rational functions then the rational function rs is defined by

$$(rs)(x) = r(x) \cdot s(x)$$

The quotient $\frac{r}{s}$ is not defined at the zeros of s.

and the rational function $\frac{r}{s}$ is defined by

$$\left(\frac{r}{s}\right)(x) = \frac{r(x)}{s(x)}.$$

The procedure for multiplying or dividing rational functions is the same as for multiplying or dividing rational numbers. In particular, dividing by a rational function $\frac{p}{q}$ is the same as multiplying by $\frac{q}{p}$.

EXAMPLE 3

Suppose

$$r(x) = \frac{2x}{x^2 + 1} \quad \text{and} \quad s(x) = \frac{3x + 2}{x^3 + 5}.$$

Write $\frac{r}{s}$ as the ratio of two polynomials.

SOLUTION

$$\begin{aligned}
\left(\frac{r}{s}\right)(x) &= \frac{\frac{2x}{x^2+1}}{\frac{3x+2}{x^3+5}} \\
&= \frac{2x}{x^2+1} \cdot \frac{x^3+5}{3x+2} \\
&= \frac{2x(x^3+5)}{(x^2+1)(3x+2)} \\
&= \frac{2x^4+10x}{3x^3+2x^2+3x+2}
\end{aligned}$$

Note that dividing by $\frac{3x+2}{x^3+5}$ is the same as multiplying by $\frac{x^3+5}{3x+2}$.

Division of Polynomials

Sometimes it is useful to express a rational number as an integer plus a rational number for which the numerator is less than the denominator. For example, $\frac{17}{3} = 5 + \frac{2}{3}$.

Similarly, sometimes it is useful to express a rational function as a polynomial plus a rational function for which the degree of the numerator is less than the degree of the denominator. For example,

$$\frac{x^5 - 7x^4 + 3x^2 + 6x + 4}{x^2} = (x^3 - 7x^2 + 3) + \frac{6x + 4}{x^2};$$

here the numerator of the rational function $\frac{6x+4}{x^2}$ has degree 1 and the denominator has degree 2.

To consider a more complicated example, suppose we want to express the rational function

$$\frac{x^5 + 6x^3 + 11x + 7}{x^2 + 4}$$

as a polynomial plus a rational function of the form $\frac{ax+b}{x^2+4}$, where a and b are constants. Entering `simplify (x^5 + 6x^3 + 11x + 7) / (x^2 + 4)` in a Wolfram|Alpha entry box produces the desired result. The next example shows how you could obtain the result by hand, giving you some insight into how a computer could get the result.

A procedure similar to long division of integers can be used with polynomials. However, mechanistic use of that procedure offers little insight into why it works. The procedure presented here, which is really just long division in slight disguise, has the advantage that its use leads to understanding its validity.

EXAMPLE 4 Write

$$\frac{x^5 + 6x^3 + 11x + 7}{x^2 + 4}$$

in the form $G(x) + \frac{ax+b}{x^2+4}$, where G is a polynomial and a, b are constants.

The idea throughout this procedure is to concentrate on the highest-degree term in the numerator.

SOLUTION The highest-degree term in the numerator is x^5; the denominator equals $x^2 + 4$. To get an x^5 term from $x^2 + 4$, we multiply $x^2 + 4$ by x^3. Thus we write

$$x^5 = x^3(x^2 + 4) - 4x^3.$$

The $-4x^3$ term above is the adjustment term that cancels the $4x^3$ term that arises when $x^3(x^2 + 4)$ is expanded to $x^5 + 4x^3$. Using the equation above, we write

$$\frac{x^5 + 6x^3 + 11x + 7}{x^2 + 4} = \frac{x^3(x^2 + 4) - 4x^3 + 6x^3 + 11x + 7}{x^2 + 4}$$

$$= x^3 + \frac{2x^3 + 11x + 7}{x^2 + 4}.$$

Again, we concentrate on the highest-degree term in the numerator.

The highest-degree term remaining in the numerator is now $2x^3$. We repeat the technique used above. Specifically, to get a $2x^3$ term from $x^2 + 4$, we multiply $x^2 + 4$ by $2x$. Thus we write

$$2x^3 = 2x(x^2 + 4) - 8x.$$

The $-8x$ term above is the adjustment term that cancels the $8x$ term that arises when $2x(x^2 + 4)$ is expanded to $2x^3 + 8x$. Using the equations above, we write

$$\frac{x^5 + 6x^3 + 11x + 7}{x^2 + 4} = x^3 + \frac{2x^3 + 11x + 7}{x^2 + 4}$$

$$= x^3 + \frac{(2x)(x^2 + 4) - 8x + 11x + 7}{x^2 + 4}$$

$$= x^3 + 2x + \frac{3x + 7}{x^2 + 4}.$$

Thus we have written $\frac{x^5+6x^3+11x+7}{x^2+4}$ in the desired form.

The procedure carried out in the example above can be applied to the ratio of any two polynomials:

Procedure for dividing polynomials

(a) Express the highest-degree term in the numerator as a single term times the denominator, plus whatever adjustment terms are necessary.

(b) Simplify the quotient using the numerator as rewritten in part (a).

(c) Repeat steps (a) and (b) on the remaining rational function until the degree of the numerator is less than the degree of the denominator or the numerator is 0.

The result of the procedure above is the decomposition of a rational function into a polynomial plus a rational function for which the degree of the numerator is less than the degree of the denominator (or the numerator is 0):

Division of polynomials

If p and q are polynomials, with $q \neq 0$, then there exist polynomials G and R such that

$$\frac{p}{q} = G + \frac{R}{q}$$

and $\deg R < \deg q$ or $R = 0$.

The symbol R is used because this term is analogous to the remainder term in division of integers.

Multiplying both sides of the equation in the box above by q gives a useful alternative way to state the conclusion:

Division of polynomials

If p and q are polynomials, with $q \neq 0$, then there exist polynomials G and R such that

$$p = qG + R$$

and $\deg R < \deg q$ or $R = 0$.

As a special case of the result above, fix a real number r and let q be the polynomial defined by $q(x) = x - r$. Because $\deg q = 1$, we will have $\deg R = 0$ or $R = 0$ in the result above; either way, R will be a constant polynomial. In other words, the result above implies that if p is a polynomial, then there exist a polynomial G and a constant c such that

$$p(x) = (x - r)G(x) + c$$

for every real number x. Taking $x = r$ in the equation above, we get $p(r) = c$, and thus the equation above can be rewritten as

$$p(x) = (x - r)G(x) + p(r).$$

Recall that r is called a zero of p if and only if $p(r) = 0$, which happens if and only if the equation above can be rewritten as $p(x) = (x - r)G(x)$. Thus we now see why the following result from the previous section holds:

Zeros and factors of a polynomial

Suppose p is a polynomial and r is a real number. Then r is a zero of p if and only if $x - r$ is a factor of $p(x)$.

The Behavior of a Rational Function Near $\pm\infty$

We now turn to an investigation of the behavior of a rational function near ∞ and near $-\infty$. Recall that to determine the behavior of a polynomial near ∞ or near $-\infty$, we factored out the term with highest degree. The procedure is the same for rational functions, except that the term of highest degree should be separately factored out of the numerator and the denominator. The next example illustrates this procedure.

EXAMPLE 5

Suppose

$$r(x) = \frac{9x^5 - 2x^3 + 1}{x^8 + x + 1}.$$

Discuss the behavior of $r(x)$ for x near ∞ and for x near $-\infty$.

Remember that neither ∞ nor $-\infty$ is a real number. To say that x is near ∞ is just an informal way of saying that x is very large. Similarly, to say that x is near $-\infty$ is an informal way of saying that x is negative and $|x|$ is very large.

SOLUTION The term of highest degree in the numerator is $9x^5$; the term of highest degree in the denominator is x^8. Factoring out these terms, and considering only values of x near ∞ or near $-\infty$, we have

$$\begin{aligned} r(x) &= \frac{9x^5(1 - \frac{2}{9x^2} + \frac{1}{9x^5})}{x^8(1 + \frac{1}{x^7} + \frac{1}{x^8})} \\ &= \frac{9}{x^3} \cdot \frac{(1 - \frac{2}{9x^2} + \frac{1}{9x^5})}{(1 + \frac{1}{x^7} + \frac{1}{x^8})} \\ &\approx \frac{9}{x^3}. \end{aligned}$$

For $|x|$ very large, $(1 - \frac{2}{9x^2} + \frac{1}{9x^5})$ and $(1 + \frac{1}{x^7} + \frac{1}{x^8})$ are both very close to 1, which explains how we got the approximation above.

The calculation above indicates that $r(x)$ should behave like $\frac{9}{x^3}$ for x near ∞ or near $-\infty$. In particular, if x is near ∞, then $\frac{9}{x^3}$ is positive but very close to 0; thus $r(x)$ has the same behavior. If x is near $-\infty$, then $\frac{9}{x^3}$ is negative but very close to 0; thus $r(x)$ has the same behavior. As the graph below shows, for this function we do not even need to take $|x|$ particularly large to see this behavior.

As can be seen here, the graph of a rational function can be unexpectedly beautiful and complex.

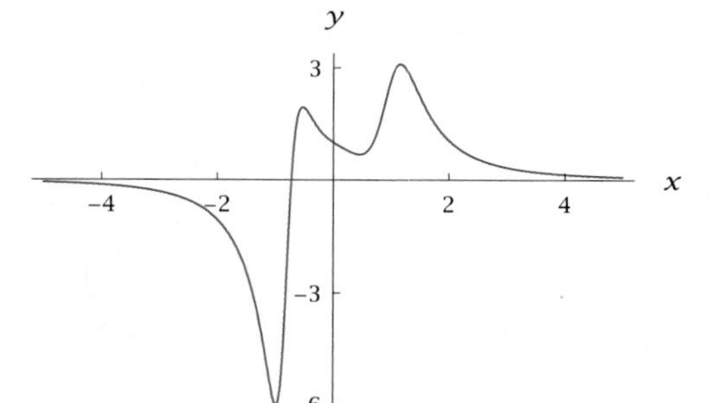

The graph of $\frac{9x^5-2x^3+1}{x^8+x+1}$ on the interval $[-5, 5]$. The values of this function are positive but close to 0 for x near 5, and negative but close to 0 for x near -5.

In general, the same procedure used in the example above works with any rational function:

Behavior of a Rational Function Near $\pm\infty$

To determine the behavior of a rational function near ∞ or near $-\infty$, separately factor out the term with highest degree in the numerator and the denominator.

EXAMPLE 6

The percentage survival rate of a seedling of a particular tropical tree can be modeled by the equation

$$s(d) = \frac{15d^2}{d^2 + 40},$$

where a seedling has an $s(d)$ percent chance of survival if it is d meters distance from its parent tree.

(a) At approximately what distance is a seedling's survival rate 10%?

(b) What survival rate does this model predict for seedlings growing very far from their parent tree?

SOLUTION

(a) We need to find d such that $\frac{15d^2}{d^2 + 40} = 10$. Multiplying both sides of this equation by $d^2 + 40$ and then subtracting $10d^2$ from both sides leads to the equation $5d^2 = 400$. Thus $d^2 = 80$, which means that $d = \sqrt{80} \approx 9$. Thus a seedling should be about 9 meters from its parent in order to have a 10% chance of survival.

(b) We need to estimate $s(d)$ for large values of d. If d is a large number, then

$$\begin{aligned} s(d) &= \frac{15d^2}{d^2 + 40} \\ &= \frac{15d^2}{d^2(1 + \frac{40}{d^2})} \\ &= \frac{15}{1 + \frac{40}{d^2}} \\ &\approx 15. \end{aligned}$$

Thus the model predicts that seedlings growing very far from their parent tree have approximately a 15% chance of survival.

The graph of $\frac{15d^2}{d^2+40}$ (blue) and the line $y = 15$ (red) on the interval $[0, 50]$. As can be seen here, if d is large then $s(d) \approx 15$.

The line $y = 15$ plays a special role in understanding the behavior of the graph above. Such lines are sufficiently important to have a name. Although the definition below is not precise (because "arbitrarily close" is vague), its meaning should be clear to you.

Asymptote

A line is called an **asymptote** of a graph if the graph becomes and stays arbitrarily close to the line in at least one direction along the line.

Thus the line $y = 15$ is an asymptote of the graph of $y = \frac{15d^2}{d^2+40}$, as can be seen from the graph in Example 6. Also, the x-axis (which is the line $y = 0$) is an asymptote of the graph of $y = \frac{9x^5-2x^3+1}{x^8+x+1}$, as we saw in Example 5.

EXAMPLE 7

Suppose

$$r(x) = \frac{3x^6 - 9x^4 + 5}{2x^6 + 4x + 3}.$$

Find an asymptote of the graph of r.

SOLUTION The graph below shows that the line $y = \frac{3}{2}$ appears to be an asymptote of the graph of r.

Here again we see that the graph of a rational function can be unexpectedly beautiful and complex.

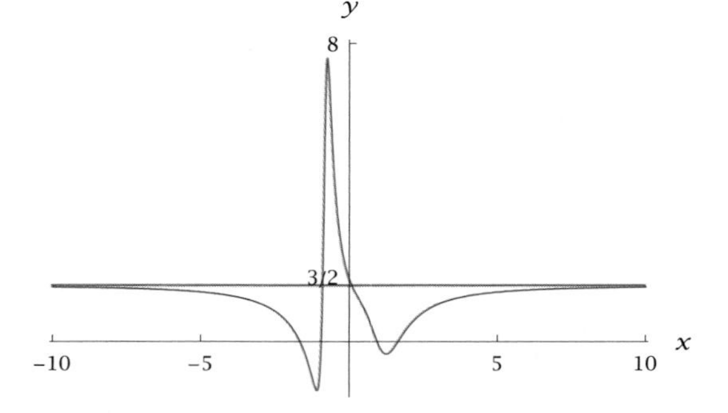

The graph of $\frac{3x^6-9x^4+5}{2x^6+4x+3}$ *(blue) and the line* $y = \frac{3}{2}$ *(red) on the interval* $[-10, 10]$.

To verify that we indeed have an asymptote here, we will investigate the behavior of $r(x)$ for x near $\pm\infty$. The term of highest degree in the numerator is $3x^6$; the term of highest degree in the denominator is $2x^6$. Factoring out these terms, and considering only values of x near ∞ or near $-\infty$, we have

$$\begin{aligned} r(x) &= \frac{3x^6(1 - \frac{3}{x^2} + \frac{5}{3x^6})}{2x^6(1 + \frac{2}{x^5} + \frac{3}{2x^6})} \\ &= \frac{3}{2} \cdot \frac{(1 - \frac{3}{x^2} + \frac{5}{3x^6})}{(1 + \frac{2}{x^5} + \frac{3}{2x^6})} \\ &\approx \frac{3}{2}. \end{aligned}$$

For $|x|$ very large, $(1 - \frac{3}{x^2} + \frac{5}{3x^6})$ and $(1 + \frac{2}{x^5} + \frac{3}{2x^6})$ are both very close to 1, which explains how we got the approximation above.

The calculation above indicates that $r(x)$ should equal approximately $\frac{3}{2}$ for x near ∞ or near $-\infty$. As the graph above shows, for this function we do not even need to take $|x|$ particularly large to see this behavior.

So far we have looked at the behavior of a rational function whose numerator has smaller degree than its denominator, and we have looked rational functions whose numerator and denominator have equal degree. The next example illustrates the behavior near $\pm\infty$ of a rational function whose numerator has larger degree than its denominator.

EXAMPLE 8

Suppose

$$r(x) = \frac{4x^{10} - 2x^3 + 3x + 15}{2x^6 + x^5 + 1}.$$

Discuss the behavior of $r(x)$ for x near ∞ and for x near $-\infty$.

SOLUTION The term of highest degree in the numerator is $4x^{10}$; the term of highest degree in the denominator is $2x^6$. Factoring out these terms, and considering only values of x near ∞ or near $-\infty$, we have

$$\begin{aligned} r(x) &= \frac{4x^{10}\left(1 - \frac{1}{2x^7} + \frac{3}{4x^9} + \frac{15}{4x^{10}}\right)}{2x^6\left(1 + \frac{1}{2x} + \frac{1}{2x^6}\right)} \\ &= 2x^4 \cdot \frac{\left(1 - \frac{1}{2x^7} + \frac{3}{4x^9} + \frac{15}{4x^{10}}\right)}{\left(1 + \frac{1}{2x} + \frac{1}{2x^6}\right)} \\ &\approx 2x^4. \end{aligned}$$

For $|x|$ very large, $\left(1 - \frac{1}{2x^7} + \frac{3}{4x^9} + \frac{15}{4x^{10}}\right)$ and $\left(1 + \frac{1}{2x} + \frac{1}{2x^6}\right)$ are both very close to 1, which explains how we got the approximation above.

The calculation above indicates that $r(x)$ should behave like $2x^4$ for x near ∞ or near $-\infty$. In particular, $r(x)$ should be positive and large for x near ∞ or near $-\infty$. As the following graph shows, for this function we do not even need to take $|x|$ particularly large to see this behavior.

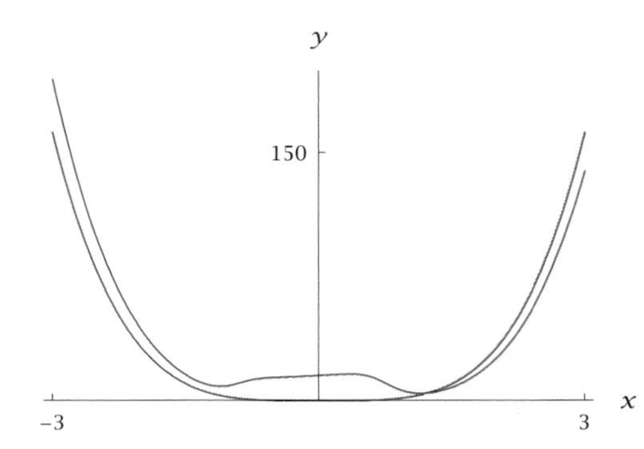

The graphs of $\frac{4x^{10}-2x^3+3x+15}{2x^6+x^5+1}$ *(blue) and* $2x^4$ *(red) on the interval* $[-3, 3]$.

The graph of r *looks like the graph of* $2x^4$ *for large values of* x.

Graphs of Rational Functions

Just as with polynomials, the task of graphing a rational function can be performed better by computers than by humans. We have already seen the graphs of several rational functions and discussed the behavior of rational functions near $\pm\infty$.

The graph of a rational function can look strikingly different from the graph of a polynomial in one important aspect that we have not yet discussed, as shown in the following example.

EXAMPLE 9

Suppose r is the rational function defined by

$$\frac{x^2+5}{x^3-2x^2-x+2}.$$

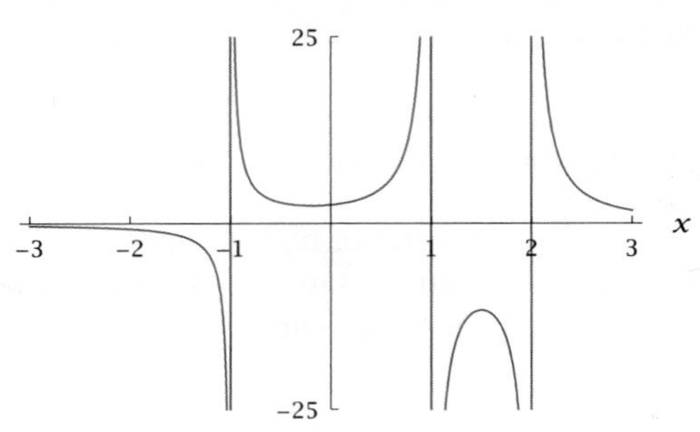

The graph of this function on the interval $[-3, 3]$ is shown here, truncated on the vertical axis to the interval $[-25, 25]$.

Discuss the asymptotes of the graph of this function.

SOLUTION Because the numerator of $r(x)$ has degree less than the denominator, $r(x)$ is close to 0 for x near ∞ and for x near $-\infty$. Thus the x-axis is an asymptote of the graph of r, as shown above.

The red lines are the vertical asymptotes of the graph of r. The x-axis is also an asymptote of this graph.

The strikingly different behavior of the graph above as compared to previous graphs that we have seen occurs near $x = -1$, $x = 1$, and $x = 2$; those three lines are shown above in red. To understand what is happening here, verify that the denominator of $r(x)$ is zero if $x = -1$, $x = 1$, or $x = 2$. Thus the numbers -1, 1, and 2 are not in the domain of r, because division by 0 is not defined.

For values of x very close to $x = -1$, $x = 1$, or $x = 2$, the denominator of $r(x)$ is very close to 0, but the numerator is always at least 5. Dividing a number larger than 5 by a number very close to 0 produces a number with very large absolute value, which explains the behavior of the graph of r near $x = -1$, $x = 1$, and $x = 2$. In other words, the lines $x = -1$, $x = 1$, and $x = 2$ are asymptotes of the graph of r.

We conclude this section by stating a result about the maximum number of peaks and valleys than can appear in the graph of a rational function. An explanation of why this result holds requires calculus.

Peaks and valleys for the graph of a rational function

The graph of a rational function $\frac{p}{q}$, where p and q are polynomials, can have a total of most $\deg p + \deg q - 1$ peaks and valleys.

EXERCISES

For Exercises 1–4, write the domain of the given function r as a union of intervals.

1. $r(x) = \dfrac{5x^3 - 12x^2 + 13}{x^2 - 7}$
2. $r(x) = \dfrac{x^5 + 3x^4 - 6}{2x^2 - 5}$
3. $r(x) = \dfrac{4x^7 + 8x^2 - 1}{x^2 - 2x - 6}$
4. $r(x) = \dfrac{6x^9 + x^5 + 8}{x^2 + 4x + 1}$

Suppose

$$r(x) = \frac{3x + 4}{x^2 + 1},$$

$$s(x) = \frac{x^2 + 2}{2x - 1},$$

$$t(x) = \frac{5}{4x^3 + 3}.$$

In Exercises 5–22, write the indicated expression as a ratio, with the numerator and denominator each written as a sum of terms of the form cx^m.

5. $(r + s)(x)$
6. $(r - s)(x)$
7. $(s - t)(x)$
8. $(s + t)(x)$
9. $(3r - 2s)(x)$
10. $(4r + 5s)(x)$
11. $(rs)(x)$
12. $(rt)(x)$
13. $(r(x))^2$
14. $(s(x))^2$
15. $(r(x))^2 t(x)$
16. $(s(x))^2 t(x)$
17. $(r \circ s)(x)$
18. $(s \circ r)(x)$
19. $(r \circ t)(x)$
20. $(t \circ r)(x)$
21. $\frac{s(1+x)-s(1)}{x}$
22. $\frac{t(x-1)-t(-1)}{x}$

For Exercises 23–28, suppose

$$r(x) = \frac{x + 1}{x^2 + 3} \quad \textit{and} \quad s(x) = \frac{x + 2}{x^2 + 5}.$$

23. What is the domain of r?
24. What is the domain of s?
25. Find two distinct numbers x such that $r(x) = \frac{1}{4}$.
26. Find two distinct numbers x such that $s(x) = \frac{1}{8}$.
27. What is the range of r?
28. What is the range of s?

In Exercises 29–34, write each expression in the form $G(x) + \frac{R(x)}{q(x)}$, where q is the denominator of the given expression and G and R are polynomials with $\deg R < \deg q$.

29. $\dfrac{2x + 1}{x - 3}$
30. $\dfrac{4x - 5}{x + 7}$
31. $\dfrac{x^2}{3x - 1}$
32. $\dfrac{x^2}{4x + 3}$
33. $\dfrac{x^6 + 3x^3 + 1}{x^2 + 2x + 5}$
34. $\dfrac{x^6 - 4x^2 + 5}{x^2 - 3x + 1}$

35. Find a constant c such that $r(10^{100}) \approx 6$, where
$$r(x) = \frac{cx^3 + 20x^2 - 15x + 17}{5x^3 + 4x^2 + 18x + 7}.$$

36. Find a constant c such that $r(2^{1000}) \approx 5$, where
$$r(x) = \frac{3x^4 - 2x^3 + 8x + 7}{cx^4 - 9x + 2}.$$

37. A bicycle company finds that its average cost per bicycle for producing n thousand bicycles is $a(n)$ dollars, where
$$a(n) = 700\frac{4n^2 + 3n + 50}{16n^2 + 3n + 35}.$$
What will be the approximate cost per bicycle when the company is producing many bicycles?

38. A bicycle company finds that its average cost per bicycle for producing n thousand bicycles is $a(n)$ dollars, where
$$a(n) = 800\frac{3n^2 + n + 40}{16n^2 + 2n + 45}.$$
What will be the approximate cost per bicycle when the company is producing many bicycles?

For Exercises 39–42, find the asymptotes of the graph of the given function r.

39. $r(x) = \dfrac{6x^4 + 4x^3 - 7}{2x^4 + 3x^2 + 5}$
40. $r(x) = \dfrac{6x^6 - 7x^3 + 3}{3x^6 + 5x^4 + x^2 + 1}$
41. $r(x) = \dfrac{3x + 1}{x^2 + x - 2}$
42. $r(x) = \dfrac{9x + 5}{x^2 - x - 6}$

43. Suppose you start driving a car on a chilly fall day. As you drive, the heater in the car makes the temperature inside the car $F(t)$ degrees Fahrenheit at time t minutes after you started driving, where

$$F(t) = 40 + \frac{30t^3}{t^3 + 100}.$$

(a) What was the temperature in the car when you started driving?

(b) What was the approximate temperature in the car ten minutes after you started driving?

(c) What will be the approximate temperature in the car after you have been driving for a long time?

44. Suppose you start driving a car on a hot summer day. As you drive, the air conditioner in the car makes the temperature inside the car $F(t)$ degrees Fahrenheit at time t minutes after you started driving, where

$$F(t) = 90 - \frac{18t^2}{t^2 + 65}.$$

(a) What was the temperature in the car when you started driving?

(b) What was the approximate temperature in the car 15 minutes after you started driving?

(c) What will be the approximate temperature in the car after you have been driving for a long time?

PROBLEMS

45. Suppose $s(x) = \dfrac{x^2 + 2}{2x - 1}$.

(a) Show that the point $(1, 3)$ is on the graph of s.

(b) Show that the slope of a line containing $(1, 3)$ and a point on the graph of s very close to $(1, 3)$ is approximately -4.

[*Hint:* Use the result of Exercise 21.]

46. Suppose $t(x) = \dfrac{5}{4x^3 + 3}$.

(a) Show that the point $(-1, -5)$ is on the graph of t.

(b) Give an estimate for the slope of a line containing $(-1, -5)$ and a point on the graph of t very close to $(-1, -5)$.

[*Hint:* Use the result of Exercise 22.]

47. Explain how the result in the previous section for the maximum number of peaks and valleys in the graph of a polynomial is a special case of the result in this section for the maximum number of peaks and valleys in the graph of a rational function.

48. Explain why the composition of a polynomial and a rational function (in either order) is a rational function.

49. Explain why the composition of two rational functions is a rational function.

50. Suppose p is a polynomial and r is a number. Explain why there is a polynomial G such that

$$\frac{p(x) - p(r)}{x - r} = G(x)$$

for every number $x \neq r$.

51. Suppose p is a nonzero polynomial with at least one (real) zero. Explain why

- there exist real numbers $r_1, r_2, \dots, r_m$ and a polynomial G such that G has no (real) zeros and

$$p(x) = (x - r_1)(x - r_2)\dots(x - r_m)G(x)$$

for every real number x;

- each of the numbers $r_1, r_2, \dots, r_m$ is a zero of p;
- p has no zeros other than $r_1, r_2, \dots, r_m$.

WORKED-OUT SOLUTIONS *to Odd-numbered Exercises*

For Exercises 1–4, write the domain of the given function r as a union of intervals.

1. $r(x) = \dfrac{5x^3 - 12x^2 + 13}{x^2 - 7}$

SOLUTION Because we have no other information about the domain of r, we assume that the domain of r is the set of numbers where the expression defining r makes sense, which means where the denominator is not 0. The denominator of the expression defining r is 0 if $x = -\sqrt{7}$ or $x = \sqrt{7}$. Thus the domain of r is the set of numbers other than $-\sqrt{7}$ and $\sqrt{7}$. In other words, the domain of r is $(-\infty, -\sqrt{7}) \cup (-\sqrt{7}, \sqrt{7}) \cup (\sqrt{7}, \infty)$.

3. $r(x) = \dfrac{4x^7 + 8x^2 - 1}{x^2 - 2x - 6}$

SOLUTION To find where the expression defining r does not make sense, apply the quadratic formula to the equation $x^2 - 2x - 6 = 0$, getting $x = 1 - \sqrt{7}$ or $x = 1 + \sqrt{7}$. Thus the domain of r is the set of numbers other than $1 - \sqrt{7}$ and $1 + \sqrt{7}$. In other words, the domain of r is $(-\infty, 1 - \sqrt{7}) \cup (1 - \sqrt{7}, 1 + \sqrt{7}) \cup (1 + \sqrt{7}, \infty)$.

Suppose

$$r(x) = \frac{3x + 4}{x^2 + 1},$$

$$s(x) = \frac{x^2 + 2}{2x - 1},$$

$$t(x) = \frac{5}{4x^3 + 3}.$$

In Exercises 5–22, write the indicated expression as a ratio, with the numerator and denominator each written as a sum of terms of the form cx^m.

5. $(r + s)(x)$

SOLUTION

$$(r + s)(x) = \frac{3x + 4}{x^2 + 1} + \frac{x^2 + 2}{2x - 1}$$

$$= \frac{(3x + 4)(2x - 1)}{(x^2 + 1)(2x - 1)} + \frac{(x^2 + 2)(x^2 + 1)}{(x^2 + 1)(2x - 1)}$$

$$= \frac{(3x + 4)(2x - 1) + (x^2 + 2)(x^2 + 1)}{(x^2 + 1)(2x - 1)}$$

$$= \frac{6x^2 - 3x + 8x - 4 + x^4 + x^2 + 2x^2 + 2}{2x^3 - x^2 + 2x - 1}$$

$$= \frac{x^4 + 9x^2 + 5x - 2}{2x^3 - x^2 + 2x - 1}$$

7. $(s - t)(x)$

SOLUTION

$$(s - t)(x) = \frac{x^2 + 2}{2x - 1} - \frac{5}{4x^3 + 3}$$

$$= \frac{(x^2 + 2)(4x^3 + 3)}{(2x - 1)(4x^3 + 3)} - \frac{5(2x - 1)}{(2x - 1)(4x^3 + 3)}$$

$$= \frac{(x^2 + 2)(4x^3 + 3) - 5(2x - 1)}{(2x - 1)(4x^3 + 3)}$$

$$= \frac{4x^5 + 8x^3 + 3x^2 - 10x + 11}{8x^4 - 4x^3 + 6x - 3}$$

9. $(3r - 2s)(x)$

SOLUTION

$$(3r - 2s)(x) = 3\Big(\frac{3x + 4}{x^2 + 1}\Big) - 2\Big(\frac{x^2 + 2}{2x - 1}\Big)$$

$$= \frac{9x + 12}{x^2 + 1} - \frac{2x^2 + 4}{2x - 1}$$

$$= \frac{(9x + 12)(2x - 1)}{(x^2 + 1)(2x - 1)} - \frac{(2x^2 + 4)(x^2 + 1)}{(x^2 + 1)(2x - 1)}$$

$$= \frac{(9x + 12)(2x - 1) - (2x^2 + 4)(x^2 + 1)}{(x^2 + 1)(2x - 1)}$$

$$= \frac{18x^2 - 9x + 24x - 12 - 2x^4 - 6x^2 - 4}{2x^3 - x^2 + 2x - 1}$$

$$= \frac{-2x^4 + 12x^2 + 15x - 16}{2x^3 - x^2 + 2x - 1}$$

11. $(rs)(x)$

SOLUTION

$$(rs)(x) = \frac{3x+4}{x^2+1} \cdot \frac{x^2+2}{2x-1}$$

$$= \frac{(3x+4)(x^2+2)}{(x^2+1)(2x-1)}$$

$$= \frac{3x^3+4x^2+6x+8}{2x^3-x^2+2x-1}$$

13. $(r(x))^2$

SOLUTION

$$(r(x))^2 = \Big(\frac{3x+4}{x^2+1}\Big)^2$$

$$= \frac{(3x+4)^2}{(x^2+1)^2}$$

$$= \frac{9x^2+24x+16}{x^4+2x^2+1}$$

15. $(r(x))^2 t(x)$

SOLUTION Using the expression that we computed for $(r(x))^2$ in the solution to Exercise 13, we have

$$(r(x))^2 t(x) = \frac{9x^2+24x+16}{x^4+2x^2+1} \cdot \frac{5}{4x^3+3}$$

$$= \frac{5(9x^2+24x+16)}{(x^4+2x^2+1)(4x^3+3)}$$

$$= \frac{45x^2+120x+80}{4x^7+8x^5+3x^4+4x^3+6x^2+3}.$$

17. $(r \circ s)(x)$

SOLUTION We have

$$(r \circ s)(x) = r(s(x))$$

$$= r\Big(\frac{x^2+2}{2x-1}\Big)$$

$$= \frac{3\big(\frac{x^2+2}{2x-1}\big)+4}{\big(\frac{x^2+2}{2x-1}\big)^2+1}$$

$$= \frac{3\frac{(x^2+2)}{(2x-1)}+4}{\frac{(x^2+2)^2}{(2x-1)^2}+1}.$$

Multiplying the numerator and denominator of the expression above by $(2x-1)^2$ gives

$$(r \circ s)(x) = \frac{3(x^2+2)(2x-1)+4(2x-1)^2}{(x^2+2)^2+(2x-1)^2}$$

$$= \frac{6x^3+13x^2-4x-2}{x^4+8x^2-4x+5}.$$

19. $(r \circ t)(x)$

SOLUTION We have

$$(r \circ t)(x) = r(t(x))$$

$$= r\Big(\frac{5}{4x^3+3}\Big)$$

$$= \frac{3\big(\frac{5}{4x^3+3}\big)+4}{\big(\frac{5}{4x^3+3}\big)^2+1}$$

$$= \frac{\frac{15}{4x^3+3}+4}{\frac{25}{(4x^3+3)^2}+1}.$$

Multiplying the numerator and denominator of the expression above by $(4x^3+3)^2$ gives

$$(r \circ t)(x) = \frac{15(4x^3+3)+4(4x^3+3)^2}{25+(4x^3+3)^2}$$

$$= \frac{64x^6+156x^3+81}{16x^6+24x^3+34}.$$

21. $\frac{s(1+x)-s(1)}{x}$

SOLUTION Note that $s(1) = 3$. Thus

$$\frac{s(1+x)-s(1)}{x} = \frac{\frac{(1+x)^2+2}{2(1+x)-1}-3}{x}$$

$$= \frac{\frac{x^2+2x+3}{2x+1}-3}{x}.$$

Multiplying the numerator and denominator of the expression above by $2x+1$ gives

$$\frac{s(1+x)-s(1)}{x} = \frac{x^2+2x+3-6x-3}{x(2x+1)}$$

$$= \frac{x^2-4x}{x(2x+1)}$$

$$= \frac{x-4}{2x+1}.$$

For Exercises 23–28, suppose

$$r(x) = \frac{x+1}{x^2+3} \quad \textit{and} \quad s(x) = \frac{x+2}{x^2+5}.$$

23. What is the domain of r?

SOLUTION The denominator of the expression defining r is a nonzero number for every real number x, and thus the expression defining r makes sense for every real number x. Because we have no other indication of the domain of r, we thus assume that the domain of r is the set of real numbers.

25. Find two distinct numbers x such that $r(x) = \frac{1}{4}$.

SOLUTION We need to solve the equation

$$\frac{x+1}{x^2+3} = \frac{1}{4}$$

for x. Multiplying both sides by x^2+3 and then multiplying both sides by 4 and collecting all the terms on one side, we have

$$x^2-4x-1=0.$$

Using the quadratic formula, we get the solutions $x = 2-\sqrt{5}$ and $x = 2+\sqrt{5}$.

27. What is the range of r?

SOLUTION To find the range of r, we must find all numbers y such that

$$\frac{x+1}{x^2+3} = y$$

for at least one number x. Thus we will solve the equation above for x and then determine for which numbers y we get an expression for x that makes sense. Multiplying both sides of the equation above by x^2+3 and then collecting terms gives

$$yx^2 - x + (3y-1) = 0.$$

If $y = 0$, then this equation has the solution $x = -1$. If $y \neq 0$, then use the quadratic formula to solve the equation above for x, getting

$$x = \frac{1+\sqrt{1+4y-12y^2}}{2y}$$

or

$$x = \frac{1-\sqrt{1+4y-12y^2}}{2y}.$$

These expressions for x make sense precisely when $1+4y-12y^2 \geq 0$. Completing the square, we can rewrite this inequality as

$$-12\big((y-\tfrac{1}{6})^2-\tfrac{1}{9}\big) \geq 0.$$

Thus we must have $(y-\frac{1}{6})^2 \leq \frac{1}{9}$, which is equivalent to $-\frac{1}{3} \leq y-\frac{1}{6} \leq \frac{1}{3}$. Adding $\frac{1}{6}$ to each side of these inequalities gives $-\frac{1}{6} \leq y \leq \frac{1}{2}$.

Thus the range of r is the interval $[-\frac{1}{6}, \frac{1}{2}]$.

In Exercises 29–34, write each expression in the form $G(x) + \frac{R(x)}{q(x)}$, where q is the denominator of the given expression and G and R are polynomials with $\deg R < \deg q$.

29. $\dfrac{2x+1}{x-3}$

SOLUTION $\dfrac{2x+1}{x-3} = \dfrac{2(x-3)+6+1}{x-3}$

$$= 2 + \frac{7}{x-3}$$

31. $\dfrac{x^2}{3x-1}$

SOLUTION $\dfrac{x^2}{3x-1} = \dfrac{\frac{x}{3}(3x-1)+\frac{x}{3}}{3x-1}$

$$= \frac{x}{3} + \frac{\frac{x}{3}}{3x-1}$$

$$= \frac{x}{3} + \frac{\frac{1}{9}(3x-1)+\frac{1}{9}}{3x-1}$$

$$= \frac{x}{3} + \frac{1}{9} + \frac{1}{9(3x-1)}$$

33. $\dfrac{x^6+3x^3+1}{x^2+2x+5}$

SOLUTION

$$\frac{x^6 + 3x^3 + 1}{x^2 + 2x + 5}$$

$$= \frac{x^4(x^2 + 2x + 5) - 2x^5 - 5x^4 + 3x^3 + 1}{x^2 + 2x + 5}$$

$$= x^4 + \frac{-2x^5 - 5x^4 + 3x^3 + 1}{x^2 + 2x + 5}$$

$$= x^4 + \frac{(-2x^3)(x^2 + 2x + 5)}{x^2 + 2x + 5}$$

$$+ \frac{4x^4 + 10x^3 - 5x^4 + 3x^3 + 1}{x^2 + 2x + 5}$$

$$= x^4 - 2x^3 + \frac{-x^4 + 13x^3 + 1}{x^2 + 2x + 5}$$

$$= x^4 - 2x^3 + \frac{(-x^2)(x^2 + 2x + 5)}{x^2 + 2x + 5}$$

$$+ \frac{2x^3 + 5x^2 + 13x^3 + 1}{x^2 + 2x + 5}$$

$$= x^4 - 2x^3 - x^2 + \frac{15x^3 + 5x^2 + 1}{x^2 + 2x + 5}$$

$$= x^4 - 2x^3 - x^2$$

$$+ \frac{15x(x^2 + 2x + 5) - 30x^2 - 75x + 5x^2 + 1}{x^2 + 2x + 5}$$

$$= x^4 - 2x^3 - x^2 + 15x + \frac{-25x^2 - 75x + 1}{x^2 + 2x + 5}$$

$$= x^4 - 2x^3 - x^2 + 15x$$

$$+ \frac{-25(x^2 + 2x + 5) + 50x + 125 - 75x + 1}{x^2 + 2x + 5}$$

$$= x^4 - 2x^3 - x^2 + 15x - 25 + \frac{-25x + 126}{x^2 + 2x + 5}$$

35. Find a constant c such that $r(10^{100}) \approx 6$, where

$$r(x) = \frac{cx^3 + 20x^2 - 15x + 17}{5x^3 + 4x^2 + 18x + 7}.$$

SOLUTION Because 10^{100} is a very large number, we need to estimate the value of $r(x)$ for very large values of x. The highest-degree term in the numerator of r is cx^3 (unless we choose $c = 0$); the highest-degree term in the denominator of r is $5x^3$. Factoring out these terms and considering only very large values of x, we have

$$\begin{aligned} r(x) &= \frac{cx^3(1 + \frac{20}{cx} - \frac{15}{cx^2} + \frac{17}{cx^3})}{5x^3(1 + \frac{4}{5x} + \frac{18}{5x^2} + \frac{7}{5x^3})} \\ &= \frac{c}{5} \cdot \frac{(1 + \frac{20}{cx} - \frac{15}{cx^2} + \frac{17}{cx^3})}{(1 + \frac{4}{5x} + \frac{18}{5x^2} + \frac{7}{5x^3})} \\ &\approx \frac{c}{5}. \end{aligned}$$

For x very large, $(1 + \frac{20}{cx} - \frac{15}{cx^2} + \frac{17}{cx^3})$ and $(1 + \frac{4}{5x} + \frac{18}{5x^2} + \frac{7}{5x^3})$ are both very close to 1, which explains how we got the approximation above.

The approximation above shows that $r(10^{100}) \approx \frac{c}{5}$. Hence we want to choose c so that $\frac{c}{5} = 6$. Thus we take $c = 30$.

37. A bicycle company finds that its average cost per bicycle for producing n thousand bicycles is $a(n)$ dollars, where

$$a(n) = 700\frac{4n^2 + 3n + 50}{16n^2 + 3n + 35}.$$

What will be the approximate cost per bicycle when the company is producing many bicycles?

SOLUTION For n large we have

$$\begin{aligned} a(n) &= 700\frac{4n^2 + 3n + 50}{16n^2 + 3n + 35} \\ &= 700\frac{4n^2(1 + \frac{3}{4n} + \frac{25}{2n^2})}{16n^2(1 + \frac{3}{16n} + \frac{35}{16n^2})} \\ &= 700\frac{4(1 + \frac{3}{4n} + \frac{25}{2n^2})}{16(1 + \frac{3}{16n} + \frac{35}{16n^2})} \\ &\approx 700 \cdot \frac{4}{16} \\ &= 175. \end{aligned}$$

Thus the average cost to produce each bicycle will be about \$175 when the company is producing many bicycles.

For Exercises 39–42, find the asymptotes of the graph of the given function r.

39. $r(x) = \dfrac{6x^4 + 4x^3 - 7}{2x^4 + 3x^2 + 5}$

SOLUTION The denominator of this rational function is never 0, so we only need to worry about the behavior of r near $\pm\infty$. For $|x|$ very large, we have

$$\begin{aligned} r(x) &= \frac{6x^4 + 4x^3 - 7}{2x^4 + 3x^2 + 5} \\ &= \frac{6x^4 (1 + \frac{2}{3x} - \frac{7}{6x^4})}{2x^4 (1 + \frac{3}{2x^2} + \frac{5}{2x^4})} \\ &\approx 3. \end{aligned}$$

Thus the line $y = 3$ is an asymptote of the graph of r, as shown below:

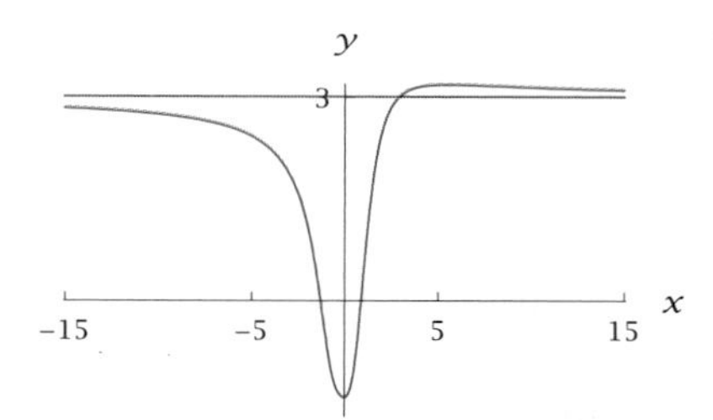

The graph of $\dfrac{6x^4 + 4x^3 - 7}{2x^4 + 3x^2 + 5}$ *on the interval* $[-15, 15]$.

41. $r(x) = \dfrac{3x + 1}{x^2 + x - 2}$

SOLUTION The denominator of this rational function is 0 when

$$x^2 + x - 2 = 0.$$

Solving this equation either by factoring or using the quadratic formula, we get $x = -2$ or $x = 1$. Because the degree of the numerator is less than the degree of the denominator, the value of this function is close to 0 when $|x|$ is large. Thus the asymptotes of the graph of r are the lines $x = -2$, $x = 1$, and $y = 0$, as shown below:

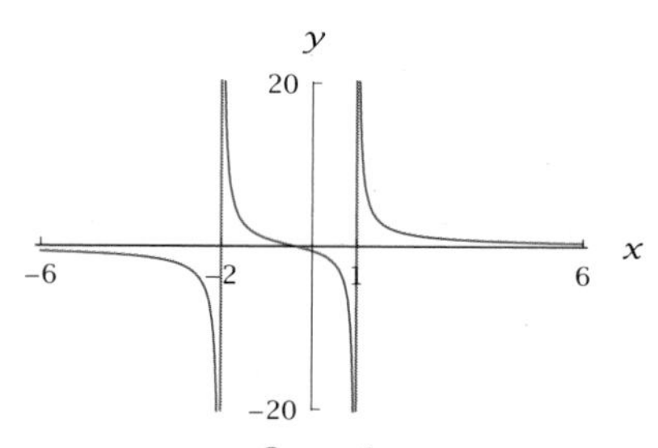

The graph of $\dfrac{3x + 1}{x^2 + x - 2}$ *on the interval* $[-6, 6]$*, truncated on the vertical axis to the interval* $[-20, 20]$.

43. Suppose you start driving a car on a chilly fall day. As you drive, the heater in the car makes the temperature inside the car $F(t)$ degrees Fahrenheit at time t minutes after you started driving, where

$$F(t) = 40 + \frac{30t^3}{t^3 + 100}.$$

(a) What was the temperature in the car when you started driving?

(b) What was the approximate temperature in the car ten minutes after you started driving?

(c) What will be the approximate temperature in the car after you have been driving for a long time?

SOLUTION

(a) Because $F(0) = 40$, the temperature in the car was 40 degrees Fahrenheit when you started driving.

(b) Because $F(10) \approx 67.3$, the temperature in the car was approximately 67.3 degrees Fahrenheit ten minutes after you started driving.

(c) Suppose t is a large number. Then

$$\begin{aligned} F(t) &= 40 + \frac{30t^3}{t^3 + 100} \\ &= 40 + \frac{30t^3}{t^3(1 + \frac{100}{t^3})} \\ &\approx 40 + 30 \\ &= 70. \end{aligned}$$

Thus the temperature in the car will be approximately 70 degrees Fahrenheit after you have been driving for a long time.

4.4 *Complex Numbers*

LEARNING OBJECTIVES

By the end of this section you should be able to

- add, multiply, and divide complex numbers;
- compute the complex conjugate of a complex number;
- solve quadratic equations using complex numbers;
- explain why nonreal roots of real polynomials come in pairs;
- explain the Fundamental Theorem of Algebra.

The Complex Number System

The real number system provides a powerful context for solving a broad array of problems. Calculus takes place mostly within the real number system. However, some important mathematical problems cannot be solved within the real number system. This section provides an introduction to the complex number system, which is a remarkably useful extension of the real number system.

Consider the equation

$$x^2 = -1.$$

The equation above has no solutions within the system of real numbers, because the square of a real number is either positive or zero.

The symbol i was first used to denote $\sqrt{-1}$ by the Swiss mathematician Leonard Euler in 1777.

Thus mathematicians invented a "number", called i, that provides a solution to the equation above. You can think of i as a symbol with the property that $i^2 = -1$. Numbers such as $2 + 3i$ are called **complex numbers**. We say that 2 is the **real part** of $2 + 3i$ and that 3 is the **imaginary part** of $2 + 3i$. More generally, we have the following definitions:

Complex numbers

- The symbol i has the property that

$$i^2 = -1.$$

- A **complex number** is a number of the form $a + bi$, where a and b are real numbers.
- If $z = a + bi$, where a and b are real numbers, then a is called the **real part** of z and b is called the **imaginary part** of z.

The complex number $4 + 0i$ is considered to be the same as the real number 4. More generally, if a is a real number, then the complex number $a + 0i$ is considered to be the same as the real number a. Thus every real number is also a complex number.

Arithmetic with Complex Numbers

The sum and difference of two complex numbers are defined as follows:

Stating the definition of complex addition in words rather than symbols, we could say that the real part of the sum is the sum of the real parts and the imaginary part of the sum is the sum of the imaginary parts.

Addition and subtraction of complex numbers

Suppose a, b, c, and d are real numbers. Then

- $(a + bi) + (c + di) = (a + c) + (b + d)i$;
- $(a + bi) - (c + di) = (a - c) + (b - d)i$.

EXAMPLE 1

(a) Evaluate $(2 + 3i) + (4 + 5i)$.

(b) Evaluate $(6 + 3i) - (2 + 8i)$.

SOLUTION

(a)

$$\begin{aligned}(2 + 3i) + (4 + 5i) &= (2 + 4) + (3 + 5)i \\ &= 6 + 8i\end{aligned}$$

(b)

$$\begin{aligned}(6 + 3i) - (2 + 8i) &= (6 - 2) + (3 - 8)i \\ &= 4 - 5i\end{aligned}$$

In the last solution above, note that we have written $4 + (-5)i$ in the equivalent form $4 - 5i$.

The product of two complex numbers is computed by using the property $i^2 = -1$ and by assuming that we can apply the usual properties of arithmetic (commutativity, associativity, and distributive property). The following example illustrates the idea.

EXAMPLE 2

Evaluate $(3i)(5i)$.

SOLUTION The commutative and associative properties of multiplication state that order and grouping do not matter. Thus we can rewrite $(3i)(5i)$ as $(3 \cdot 5)(i \cdot i)$ and then complete the calculation as follows:

$$\begin{aligned}(3i)(5i) &= (3 \cdot 5)(i \cdot i) \\ &= 15i^2 \\ &= 15(-1) \\ &= -15.\end{aligned}$$

After you become accustomed to working with complex numbers, you will do calculations such as the one above more quickly without the intermediate steps:

$$(3i)(5i) = 15i^2 = -15.$$

The next example shows how to compute a more complicated product of complex numbers. Again, the idea is to use the property $i^2 = -1$ and the usual rules of arithmetic, starting with the distributive property.

EXAMPLE 3

Evaluate $(2 + 3i)(4 + 5i)$.

SOLUTION

$$\begin{aligned}(2 + 3i)(4 + 5i) &= 2(4 + 5i) + (3i)(4 + 5i) \\ &= 2 \cdot 4 + 2 \cdot (5i) + (3i) \cdot 4 + (3i) \cdot (5i) \\ &= 8 + 10i + 12i - 15 \\ &= -7 + 22i\end{aligned}$$

More generally, we have the following formula for multiplication of complex numbers:

Do not memorize this formula. Instead, when you need to compute the product of complex numbers, just use the property $i^2 = -1$ and the usual rules of arithmetic.

Multiplication of complex numbers

Suppose a, b, c, and d are real numbers. Then

$$(a + bi)(c + di) = (ac - bd) + (ad + bc)i.$$

Do not make the mistake of thinking that the real part of the product of two complex numbers equals the product of the real parts (but see Problem 49). In this respect, products do not act like sums.

Complex Conjugates and Division of Complex Numbers

Division of a complex number by a real number behaves as you might expect. In keeping with the philosophy that arithmetic with complex numbers should obey the same algebraic rules as arithmetic with real numbers, division by (for example) 3 should be the same as multiplication by $\frac{1}{3}$, and we already know how to do multiplication involving complex numbers. The following simple example illustrates this idea.

EXAMPLE 4

Evaluate $\frac{5 + 6i}{3}$.

SOLUTION

$$\begin{aligned}\frac{5+6i}{3} &= \frac{1}{3}(5+6i)\\ &= \frac{1}{3}\cdot 5 + \frac{1}{3}(6i)\\ &= \frac{5}{3} + 2i\end{aligned}$$

Thus we see that to divide a complex number by a real number, simply divide the real and imaginary parts of the complex number by the real number to obtain the real and imaginary parts of the quotient.

Division by a nonreal complex number is more complicated. Consider, for example, how to divide a complex number by $2 + 3i$. This should be the same as multiplying by $\frac{1}{2+3i}$, but what is $\frac{1}{2+3i}$? Again using the principle that complex arithmetic should obey the same rules as real arithmetic, $\frac{1}{2+3i}$ is the number such that

$$(2+3i)\Big(\frac{1}{2+3i}\Big) = 1.$$

If you are just becoming acquainted with complex numbers, you might guess that $\frac{1}{2+3i}$ equals $\frac{1}{2} + \frac{1}{3}i$ or perhaps $\frac{1}{2} - \frac{1}{3}i$. However, neither of these guesses is correct, because neither $(2+3i)(\frac{1}{2} + \frac{1}{3}i)$ nor $(2+3i)(\frac{1}{2} - \frac{1}{3}i)$ equals 1 (as you should verify by actually doing the multiplications).

Complex numbers were first used by 16th century Italian mathematicians who were trying to solve cubic equations. Several more centuries passed before most mathematicians became comfortable with using complex numbers.

Thus we take a slight detour to discuss the complex conjugate, which will be useful in computing the quotient of two complex numbers.

Complex conjugate

Suppose a and b are real numbers. The complex conjugate of $a + bi$, denoted $\overline{a+bi}$, is defined by

$$\overline{a+bi} = a - bi.$$

For example,

$$\overline{2+3i} = 2 - 3i \quad \text{and} \quad \overline{2-3i} = 2 + 3i.$$

The next example hints at the usefulness of complex conjugates. Note the use of the key identity $(x+y)(x-y) = x^2 - y^2$.

EXAMPLE 5

Evaluate $(2+3i)(\overline{2+3i})$.

SOLUTION

$$\begin{aligned}(2+3i)(\overline{2+3i}) &= (2+3i)(2-3i)\\ &= 2^2 - (3i)^2\\ &= 4 - (-9)\\ &= 13\end{aligned}$$

EXAMPLE 6 Show that $\frac{1}{2+3i} = \frac{2}{13} - \frac{3}{13}i$.

Geometric interpretations of the complex number system and complex addition, subtraction, multiplication, division, and complex conjugation will be presented in Section 11.4.

SOLUTION The previous example shows that $(2+3i)(2-3i) = 13$. Dividing both sides of this equation by 13, we see that

$$(2+3i)\left(\frac{2}{13} - \frac{3}{13}i\right) = 1.$$

Thus $\frac{2}{13} - \frac{3}{13}i$ is the complex number that when multiplied by $2 + 3i$ gives 1. This means that

$$\frac{1}{2+3i} = \frac{2}{13} - \frac{3}{13}i.$$

The next example shows the general procedure for dividing by complex numbers. The idea is to multiply by 1, expressed as the ratio of the complex conjugate of the denominator with itself.

EXAMPLE 7 Evaluate $\frac{3+4i}{2-5i}$.

SOLUTION

$$\begin{aligned}\frac{3+4i}{2-5i} &= \frac{3+4i}{2-5i} \cdot \frac{2+5i}{2+5i}\\ &= \frac{(3+4i)(2+5i)}{(2-5i)(2+5i)}\\ &= \frac{(6-20)+(15+8)i}{2^2+5^2}\\ &= \frac{-14+23i}{29}\\ &= -\frac{14}{29} + \frac{23}{29}i\end{aligned}$$

More generally, we have the following formula for division of complex numbers:

Do not memorize this formula. Instead, when you need to compute the quotient of complex numbers, just multiply numerator and denominator by the complex conjugate of the denominator and then compute, as in the example above.

Division of complex numbers

Suppose a, b, c, and d are real numbers, with $c + di \neq 0$. Then

$$\frac{a+bi}{c+di} = \frac{ac+bd}{c^2+d^2} + \frac{bc-ad}{c^2+d^2}i.$$

Complex conjugation interacts well with algebraic operations. Specifically, the following properties hold:

Properties of complex conjugation

Suppose w and z are complex numbers. Then

- $\overline{\overline{z}} = z$;
- $\overline{w + z} = \overline{w} + \overline{z}$;
- $\overline{w - z} = \overline{w} - \overline{z}$;
- $\overline{w \cdot z} = \overline{w} \cdot \overline{z}$;
- $\overline{z^n} = (\overline{z})^n$ for every positive integer n;
- $\overline{\left(\frac{w}{z}\right)} = \frac{\overline{w}}{\overline{z}}$ if $z \neq 0$;
- $\frac{z + \overline{z}}{2}$ equals the real part of z;
- $\frac{z - \overline{z}}{2i}$ equals the imaginary part of z.

The expression $\overline{z}$ is pronounced "z-bar".

EXAMPLE 8

Verify the last two properties for $z = 5 + 3i$.

SOLUTION We have $\overline{z} = 5 - 3i$. Thus

$$\frac{z + \overline{z}}{2} = \frac{(5 + 3i) + (5 - 3i)}{2} = \frac{10}{2} = 5 = \text{the real part of } z$$

and

$$\frac{z - \overline{z}}{2i} = \frac{(5 + 3i) - (5 - 3i)}{2i} = \frac{6i}{2i} = 3 = \text{the imaginary part of } z.$$

To verify the properties above in general, write the complex numbers w and z in terms of their real and imaginary parts and then compute.

EXAMPLE 9

Show that if w and z are complex numbers, then $\overline{w + z} = \overline{w} + \overline{z}$.

SOLUTION Suppose $w = a + bi$ and $z = c + di$, where a, b, c, and d are real numbers. Then

$$\begin{aligned}
\overline{w + z} &= \overline{(a + bi) + (c + di)} \\
&= \overline{(a + c) + (b + d)i} \\
&= (a + c) - (b + d)i \\
&= (a - bi) + (c - di) \\
&= \overline{w} + \overline{z}.
\end{aligned}$$

Zeros and Factorization of Polynomials, Revisited

In Section 2.3 we saw that the equation $ax^2 + bx + c = 0$ has solutions

$$x = \frac{-b \pm \sqrt{b^2 - 4ac}}{2a}$$

provided $b^2 - 4ac \geq 0$. If we are willing to consider solutions that are complex numbers, then the formula above is valid (with the same derivation) without the restriction that $b^2 - 4ac \geq 0$.

The following example illustrates how the quadratic formula can be used to find complex zeros of quadratic functions.

EXAMPLE 10

Find the complex numbers z such that $z^2 - 2z + 5 = 0$.

Note that $\sqrt{-16}$ simplifies to $\pm 4i$, which is correct because $(\pm 4i)^2 = -16$.

SOLUTION Using the quadratic formula, we have

$$z = \frac{2 \pm \sqrt{4 - 4 \cdot 5}}{2} = \frac{2 \pm \sqrt{-16}}{2} = \frac{2 \pm 4i}{2} = 1 \pm 2i.$$

In the example above, the quadratic polynomial has two zeros, namely $-1 + 2i$ and $-1 - 2i$, that are complex conjugates of each other. This behavior is not a coincidence, even for higher-degree polynomials where the quadratic formula plays no role, as shown by the following example.

EXAMPLE 11

Let p be the polynomial defined by

$$p(z) = z^{12} - 6z^{11} + 13z^{10} + 2z^2 - 12z + 26.$$

Suppose you have been told (accurately) that $3 + 2i$ is a zero of p. Show that $3 - 2i$ is a zero of p.

SOLUTION We have been told that $p(3 + 2i) = 0$, which can be written as

$$0 = (3 + 2i)^{12} - 6(3 + 2i)^{11} + 13(3 + 2i)^{10} + 2(3 + 2i)^2 - 12(3 + 2i) + 26.$$

We need to verify that $p(3 - 2i) = 0$. This could be done by a long computation that would involve evaluating $(3 - 2i)^{12}$ and the other terms of $p(3 - 2i)$. However, we can get the desired result without calculation by taking the complex conjugate of both sides of the equation above and then using the properties of complex conjugation, getting

$$\begin{aligned} 0 &= \overline{(3 + 2i)^{12} - 6(3 + 2i)^{11} + 13(3 + 2i)^{10} + 2(3 + 2i)^2 - 12(3 + 2i) + 26} \\ &= \overline{(3 + 2i)^{12}} - \overline{6(3 + 2i)^{11}} + \overline{13(3 + 2i)^{10}} + \overline{2(3 + 2i)^2} - \overline{12(3 + 2i) + 26} \\ &= (3 - 2i)^{12} - 6(3 - 2i)^{11} + 13(3 - 2i)^{10} + 2(3 - 2i)^2 - 12(3 - 2i) + 26 \\ &= p(3 - 2i). \end{aligned}$$

The technique used in the example above can be used more generally to give the following result:

The complex conjugate of a zero is a zero

Suppose p is a polynomial with real coefficients. If z is a complex number that is a zero of p, then $\overline{z}$ is also a zero of p.

This result states that nonreal zeros of a polynomial with real coefficients come in pairs. In other words, if a and b are real numbers and $a + bi$ is a zero of such a polynomial, then so is $a - bi$.

We can think about five increasingly large numbers systems—the positive integers, the integers, the rational numbers, the real numbers, the complex numbers—with each successive number system viewed as an extension of the previous system to allow new kinds of equations to be solved:

- The equation $x + 2 = 0$ leads to the negative number $x = -2$. More generally, the equation $x + m = 0$, where m is a nonnegative integer, leads to the set of integers.
- The equation $5x = 3$ leads to the fraction $x = \frac{3}{5}$. More generally, the equation $nx = m$, where m and n are integers with $n \neq 0$, leads to the set of rational numbers.
- The equation $x^2 = 2$ leads to the irrational numbers $x = \pm\sqrt{2}$. More generally, the notion that the real line contains no holes leads to the set of real numbers.
- The equation $x^2 = -1$ leads to the complex numbers $x = \pm i$. More generally, the quadratic equation $x^2 + bx + c = 0$, where b and c are real numbers, leads to the set of complex numbers.

The progression above makes it reasonable to guess that we need to add new kinds of numbers to solve polynomial equations of higher degree. For example, there is no obvious solution within the complex number system to the equation $x^4 = -1$. Do we need to invent yet another new kind of number to solve this equation? And then yet another new kind of number to solve sixth-degree equations, and so on?

Somewhat surprisingly, rather than a continuing sequence of new kinds of numbers, we can stay within the complex numbers and still be assured that polynomial equations of every degree have solutions. We will soon state this result more precisely. However, first we turn to the example below, which shows that a solution to the equation $x^4 = -1$ does indeed exist within the complex number system.

EXAMPLE 12

Verify that

$$\left(\frac{\sqrt{2}}{2} + \frac{\sqrt{2}}{2}i\right)^4 = -1.$$

SOLUTION We have

$$\left(\frac{\sqrt{2}}{2}+\frac{\sqrt{2}}{2}i\right)^2=\left(\frac{\sqrt{2}}{2}\right)^2+2\cdot\frac{\sqrt{2}}{2}\cdot\frac{\sqrt{2}}{2}i-\left(\frac{\sqrt{2}}{2}\right)^2=i.$$

Thus

$$\left(\frac{\sqrt{2}}{2}+\frac{\sqrt{2}}{2}i\right)^4=\left(\left(\frac{\sqrt{2}}{2}+\frac{\sqrt{2}}{2}i\right)^2\right)^2=i^2=-1.$$

The German mathematician Carl Friedrich Gauss proved the Fundamental Theorem of Algebra in 1799, when he was 22 years old.

The next result is so important that it is called the **Fundamental Theorem of Algebra**. The proof of this result requires techniques from advanced mathematics and thus cannot be given here.

Fundamental Theorem of Algebra

Suppose p is a polynomial of degree $n \geq 1$. Then there exist complex numbers $r_1, r_2, \ldots, r_n$ and a constant c such that

$$p(z) = c(z - r_1)(z - r_2)\ldots(z - r_n)$$

for every complex number z.

The complex numbers $r_1, r_2, \ldots, r_n$ in the factorization above are not necessarily distinct. For example, if $p(z) = z^2 - 2z + 1$, then $c = 1$, $r_1 = 1$, and $r_2 = 1$.

The following remarks may help lead to a better understanding of the Fundamental Theorem of Algebra:

- The factorization above shows that $p(r_1) = p(r_2) = \cdots = p(r_n) = 0$. Thus each of the numbers $r_1, r_2, \ldots, r_n$ is a zero of p. Furthermore, p has no other zeros, as can be seen from the factorization above.

- The constant c is the coefficient of z^n in the expression $p(z)$. Thus if z^n has coefficient 1, then $c = 1$.

- In the statement of the Fundamental Theorem of Algebra, we have not specified whether the polynomial p has real coefficients or complex coefficients. The result is true either way. However, even if all the coefficients are real, then the numbers $r_1, r_2, \ldots, r_n$ cannot necessarily be assumed to be real numbers. For example, if $p(z) = z^2 + 1$, then the factorization promised by the Fundamental Theorem of Algebra is $p(z) = (z - i)(z + i)$.

Section 11.4 shows how to compute fractional powers using complex numbers.

- The Fundamental Theorem of Algebra is an existence theorem. It does not tell us how to find the zeros of p or how to factor p. Thus, for example, although we are assured that the equation $z^6 = -1$ has a solution in the complex number system (because the polynomial $z^6 + 1$ must have a complex zero), the Fundamental Theorem of Algebra does not tell us how to find a solution (but for this specific polynomial, see Problem 43).

EXERCISES

For Exercises 1–34, write each expression in the form $a + bi$, where a and b are real numbers.

1. $(4 + 2i) + (3 + 8i)$
2. $(5 + 7i) + (4 + 6i)$
3. $(5 + 3i) - (2 + 9i)$
4. $(9 + 2i) - (6 + 7i)$
5. $(6 + 2i) - (9 - 7i)$
6. $(1 + 3i) - (6 - 5i)$
7. $(2 + 3i)(4 + 5i)$
8. $(5 + 6i)(2 + 7i)$
9. $(2 + 3i)(4 - 5i)$
10. $(5 + 6i)(2 - 7i)$
11. $(4 - 3i)(2 - 6i)$
12. $(8 - 4i)(2 - 3i)$
13. $(3 + 4i)^2$
14. $(6 + 5i)^2$
15. $(5 - 2i)^2$
16. $(4 - 7i)^2$
17. $(4 + \sqrt{3}i)^2$
18. $(5 + \sqrt{6}i)^2$
19. $(\sqrt{5} - \sqrt{7}i)^2$
20. $(\sqrt{11} - \sqrt{3}i)^2$
21. $(2 + 3i)^3$
22. $(4 + 3i)^3$
23. $(1 + \sqrt{3}i)^3$
24. $\left(\frac{1}{2} - \frac{\sqrt{3}}{2}i\right)^3$
25. i^{8001}
26. i^{1003}
27. $\overline{8 + 3i}$
28. $\overline{-7 + \frac{2}{3}i}$
29. $\overline{-5 - 6i}$
30. $\overline{\frac{5}{3} - 9i}$
31. $\dfrac{1 + 2i}{3 + 4i}$
32. $\dfrac{5 + 6i}{2 + 3i}$
33. $\dfrac{4 + 3i}{5 - 2i}$
34. $\dfrac{3 - 4i}{6 - 5i}$
35. Find two complex numbers z that satisfy the equation $z^2 + 4z + 6 = 0$.
36. Find two complex numbers z that satisfy the equation $2z^2 + 4z + 5 = 0$.
37. Find a complex number whose square equals $5 + 12i$.
38. Find a complex number whose square equals $21 - 20i$.
39. Find two complex numbers whose sum equals 7 and whose product equals 13.
[*Compare to Problem 71 in Section 2.3.*]
40. Find two complex numbers whose sum equals 5 and whose product equals 11.

PROBLEMS

41. Write out a table showing the values of i^n with n ranging over the integers from 1 to 12. Describe the pattern that emerges.
42. Verify that
$$(\sqrt{3} + i)^6 = -64.$$
43. Explain why the previous problem implies that
$$\left(\frac{\sqrt{3}}{2} + \frac{1}{2}i\right)^6 = -1.$$
44. Show that addition of complex numbers is commutative, meaning that
$$w + z = z + w$$
for all complex numbers w and z.
[*Hint:* Show that
$$(a + bi) + (c + di) = (c + di) + (a + bi)$$
for all real numbers a, b, c, and d.]
45. Show that addition of complex numbers is associative, meaning that
$$u + (w + z) = (u + w) + z$$
for all complex numbers u, w, and z.
46. Show that multiplication of complex numbers is commutative, meaning that
$$wz = zw$$
for all complex numbers w and z.
47. Show that multiplication of complex numbers is associative, meaning that
$$u(wz) = (uw)z$$
for all complex numbers u, w, and z.
48. Show that addition and multiplication of complex numbers satisfy the distributive property, meaning that
$$u(w + z) = uw + uz$$
for all complex numbers u, w, and z.

49. Suppose w and z are complex numbers such that the real part of wz equals the real part of w times the real part of z. Explain why either w or z must be a real number.

50. Suppose z is a complex number. Show that z is a real number if and only if $z = \overline{z}$.

51. Suppose z is a complex number. Show that $\overline{z} = -z$ if and only if the real part of z equals 0.

52. Show that $\overline{\overline{z}} = z$ for every complex number z.

53. Show that $\overline{w - z} = \overline{w} - \overline{z}$ for all complex numbers w and z.

54. Show that $\overline{w \cdot z} = \overline{w} \cdot \overline{z}$ for all complex numbers w and z.

55. Show that $\overline{z^n} = (\overline{z})^n$ for every complex number z and every positive integer n.

56. Show that if $a + bi \neq 0$, then

$$\frac{1}{a + bi} = \frac{a - bi}{a^2 + b^2}.$$

57. Suppose w and z are complex numbers, with $z \neq 0$. Show that $\overline{\left(\frac{w}{z}\right)} = \frac{\overline{w}}{\overline{z}}$.

58. Suppose z is a complex number. Show that $\frac{z + \overline{z}}{2}$ equals the real part of z.

59. Suppose z is a complex number. Show that $\frac{z - \overline{z}}{2i}$ equals the imaginary part of z.

60. Show that if p is a polynomial with real coefficients, then

$$p(\overline{z}) = \overline{p(z)}$$

for every complex number z.

61. Explain why the result in the previous problem implies that if p is a polynomial with real coefficients and z is a complex number that is a zero of p, then $\overline{z}$ is also a zero of p.

62. Suppose f is a quadratic function with real coefficients and no real zeros. Show that the average of the two complex zeros of f is the first coordinate of the vertex of the graph of f.

63. Suppose

$$f(x) = ax^2 + bx + c,$$

where $a \neq 0$ and $b^2 < 4ac$. Verify by direct substitution into the formula above that

$$f\left(\frac{-b + \sqrt{4ac - b^2}\,i}{2a}\right) = 0$$

and

$$f\left(\frac{-b - \sqrt{4ac - b^2}\,i}{2a}\right) = 0.$$

64. Suppose $a \neq 0$ and $b^2 < 4ac$. Verify by direct calculation that

$$ax^2 + bx + c =$$

$$a\left(x - \frac{-b + \sqrt{4ac - b^2}\,i}{2a}\right)\left(x - \frac{-b - \sqrt{4ac - b^2}\,i}{2a}\right).$$

WORKED-OUT SOLUTIONS *to Odd-numbered Exercises*

For Exercises 1–34, write each expression in the form a + bi, where a and b are real numbers.

1. $(4 + 2i) + (3 + 8i)$

SOLUTION

$$\begin{aligned}(4 + 2i) + (3 + 8i) &= (4 + 3) + (2 + 8)i \\ &= 7 + 10i\end{aligned}$$

3. $(5 + 3i) - (2 + 9i)$

SOLUTION

$$\begin{aligned}(5 + 3i) - (2 + 9i) &= (5 - 2) + (3 - 9)i \\ &= 3 - 6i\end{aligned}$$

5. $(6 + 2i) - (9 - 7i)$

SOLUTION

$$\begin{aligned}(6 + 2i) - (9 - 7i) &= (6 - 9) + (2 + 7)i \\ &= -3 + 9i\end{aligned}$$

7. $(2 + 3i)(4 + 5i)$

SOLUTION

$$\begin{aligned}(2 + 3i)(4 + 5i) &= (2 \cdot 4 - 3 \cdot 5) + (2 \cdot 5 + 3 \cdot 4)i \\ &= -7 + 22i\end{aligned}$$

9. $(2+3i)(4-5i)$

SOLUTION

$$\begin{aligned}(2+3i)(4-5i) &= (2\cdot 4 + 3\cdot 5) + \big(2\cdot(-5) + 3\cdot 4\big)i \\ &= 23 + 2i\end{aligned}$$

11. $(4-3i)(2-6i)$

SOLUTION

$$\begin{aligned}&(4-3i)(2-6i) \\ &\quad = (4\cdot 2 - 3\cdot 6) + \big(4\cdot(-6) + (-3)\cdot 2\big)i \\ &\quad = -10 - 30i\end{aligned}$$

13. $(3+4i)^2$

SOLUTION

$$\begin{aligned}(3+4i)^2 &= 3^2 + 2\cdot 3\cdot 4i + (4i)^2 \\ &= 9 + 24i - 16 \\ &= -7 + 24i\end{aligned}$$

15. $(5-2i)^2$

SOLUTION

$$\begin{aligned}(5-2i)^2 &= 5^2 - 2\cdot 5\cdot 2i + (2i)^2 \\ &= 25 - 20i - 4 \\ &= 21 - 20i\end{aligned}$$

17. $(4+\sqrt{3}i)^2$

SOLUTION

$$\begin{aligned}(4+\sqrt{3}i)^2 &= 4^2 + 2\cdot 4\cdot\sqrt{3}i + (\sqrt{3}i)^2 \\ &= 16 + 8\sqrt{3}i - 3 \\ &= 13 + 8\sqrt{3}i\end{aligned}$$

19. $(\sqrt{5}-\sqrt{7}i)^2$

SOLUTION

$$\begin{aligned}(\sqrt{5}-\sqrt{7}i)^2 &= \sqrt{5}^2 - 2\cdot\sqrt{5}\cdot\sqrt{7}i + (\sqrt{7}i)^2 \\ &= 5 - 2\sqrt{35}i - 7 \\ &= -2 - 2\sqrt{35}i\end{aligned}$$

21. $(2+3i)^3$

SOLUTION First we compute $(2+3i)^2$:

$$\begin{aligned}(2+3i)^2 &= 2^2 + 2\cdot 2\cdot 3i + (3i)^2 \\ &= 4 + 12i - 9 \\ &= -5 + 12i.\end{aligned}$$

Now

$$\begin{aligned}(2+3i)^3 &= (2+3i)^2(2+3i) \\ &= (-5+12i)(2+3i) \\ &= (-10-36) + (-15+24)i \\ &= -46 + 9i.\end{aligned}$$

23. $(1+\sqrt{3}i)^3$

SOLUTION First we compute $(1+\sqrt{3}i)^2$:

$$\begin{aligned}(1+\sqrt{3}i)^2 &= 1^2 + 2\cdot 1\cdot\sqrt{3}i + (\sqrt{3}i)^2 \\ &= 1 + 2\sqrt{3}i - 3 \\ &= -2 + 2\sqrt{3}i.\end{aligned}$$

Now

$$\begin{aligned}(1+\sqrt{3}i)^3 &= (1+\sqrt{3}i)^2(1+\sqrt{3}i) \\ &= (-2+2\sqrt{3}i)(1+\sqrt{3}i) \\ &= (-2-2\sqrt{3}^2) + (-2\sqrt{3}+2\sqrt{3})i \\ &= -8.\end{aligned}$$

25. i^{8001}

SOLUTION $i^{8001} = i^{8000}i = (i^2)^{4000}i$
$= (-1)^{4000}i = i$

27. $\overline{8+3i}$

SOLUTION $\overline{8+3i} = 8 - 3i$

29. $\overline{-5-6i}$

SOLUTION $\overline{-5-6i} = -5 + 6i$

31. $\dfrac{1+2i}{3+4i}$

SOLUTION
$$\frac{1+2i}{3+4i} = \frac{1+2i}{3+4i}\cdot\frac{3-4i}{3-4i}$$
$$= \frac{(1+2i)(3-4i)}{(3+4i)(3-4i)}$$
$$= \frac{(3+8)+(-4+6)i}{3^2+4^2}$$
$$= \frac{11+2i}{25}$$
$$= \frac{11}{25} + \frac{2}{25}i$$

33. $\dfrac{4+3i}{5-2i}$

SOLUTION
$$\frac{4+3i}{5-2i} = \frac{4+3i}{5-2i}\cdot\frac{5+2i}{5+2i}$$
$$= \frac{(4+3i)(5+2i)}{(5-2i)(5+2i)}$$
$$= \frac{(20-6)+(8+15)i}{5^2+2^2}$$
$$= \frac{14+23i}{29} = \frac{14}{29} + \frac{23}{29}i$$

35. Find two complex numbers z that satisfy the equation $z^2 + 4z + 6 = 0$.

SOLUTION By the quadratic formula, we have
$$z = \frac{-4 \pm \sqrt{4^2 - 4\cdot 6}}{2}$$
$$= \frac{-4 \pm \sqrt{-8}}{2}$$
$$= \frac{-4 \pm \sqrt{8}i}{2}$$
$$= \frac{-4 \pm \sqrt{4\cdot 2}i}{2}$$
$$= \frac{-4 \pm 2\sqrt{2}i}{2}$$
$$= -2 \pm \sqrt{2}i.$$

37. Find a complex number whose square equals $5 + 12i$.

SOLUTION We seek real numbers a and b such that
$$5 + 12i = (a+bi)^2 = (a^2 - b^2) + 2abi.$$
The equation above implies that
$$a^2 - b^2 = 5 \quad\text{and}\quad 2ab = 12.$$
Solving the last equation for b, we have $b = \frac{6}{a}$. Substituting this value of b in the first equation gives
$$a^2 - \frac{36}{a^2} = 5.$$
Multiplying both sides of the equation above by a^2 and then moving all terms to one side produces the equation
$$0 = (a^2)^2 - 5a^2 - 36.$$
Think of a^2 as the unknown in the equation above. We can solve for a^2 either by factorization or by using the quadratic formula. For this particular equation, factorization is easy; we have
$$0 = (a^2)^2 - 5a^2 - 36 = (a^2 - 9)(a^2 + 4).$$
The equation above shows that we must choose $a^2 = 9$ or $a^2 = -4$. However, the equation $a^2 = -4$ is not satisfied for any real number a, and thus we must choose $a^2 = 9$, which implies that $a = 3$ or $a = -3$. Choosing $a = 3$, we can now solve for b in the original equation $2ab = 12$, getting $b = 2$.

Thus $3 + 2i$ is our candidate for a solution. Checking, we have
$$(3+2i)^2 = 3^2 + 2\cdot 3\cdot 2i - 2^2 = 5 + 12i,$$
as desired.

The other correct solution is $-3 - 2i$, which we would have obtained by choosing $a = -3$.

39. Find two complex numbers whose sum equals 7 and whose product equals 13.

SOLUTION Let's call the two numbers w and z. We want
$$w + z = 7 \quad\text{and}\quad wz = 13.$$

Solving the first equation for w, we have $w = 7 - z$. Substituting this expression for w into the second equation gives $(7 - z)z = 13$, which is equivalent to the equation

$$z^2 - 7z + 13 = 0.$$

Using the quadratic formula to solve this equation for z gives

$$z = \frac{7 \pm \sqrt{7^2 - 4 \cdot 13}}{2} = \frac{7 \pm \sqrt{-3}}{2} = \frac{7 \pm \sqrt{3}i}{2}.$$

Let's choose the solution $z = \frac{7+\sqrt{3}i}{2}$. Plugging this value of z into the equation $w = 7 - z$ then gives $w = \frac{7-\sqrt{3}i}{2}$.

Thus two complex numbers whose sum equals 7 and whose product equals 13 are $\frac{7-\sqrt{3}i}{2}$ and $\frac{7+\sqrt{3}i}{2}$.

CHECK To check that this solution is correct, note that

$$\frac{7 - \sqrt{3}i}{2} + \frac{7 + \sqrt{3}i}{2} = \frac{14}{2} = 7$$

and

$$\begin{aligned}\frac{7 - \sqrt{3}i}{2} \cdot \frac{7 + \sqrt{3}i}{2} &= \frac{7^2 + \sqrt{3}^2}{4} \\ &= \frac{49 + 3}{4} = \frac{52}{4} = 13.\end{aligned}$$

CHAPTER SUMMARY

To check that you have mastered the most important concepts and skills covered in this chapter, make sure that you can do each item in the following list:

- Manipulate and simplify expressions involving integer exponents.
- Explain how x^0 and x^{-m} are defined.
- Explain the connection between the linear factors of a polynomial and its zeros.
- Determine the behavior of a polynomial near $-\infty$ and near ∞.
- Compute the sum, difference, product, and quotient of two rational functions (and thus of two polynomials).
- Write a rational function as the sum of a polynomial and a rational function whose numerator has smaller degree than its denominator.
- Determine the behavior of a rational function near $-\infty$ and near ∞.
- Perform arithmetic involving addition, subtraction, multiplication, division, and complex conjugation of complex numbers.

To review a chapter, go through the list above to find items that you do not know how to do, then reread the material in the chapter about those items. Then try to answer the chapter review questions below without looking back at the chapter.

CHAPTER REVIEW QUESTIONS

1. Write $\dfrac{3^{800} \cdot 9^{30}}{27^7}$ as a power of 3.
2. Write $y^{-5}(y^6(y^3)^4)^2$ as a power of y.
3. Simplify the expression
$$\frac{(t^3 w^5)^{-3}}{(t^{-3} w^2)^4}.$$
4. Explain why 3^0 is defined to equal 1.
5. Explain why 3^{-44} is defined to equal $\frac{1}{3^{44}}$.
6. Sketch the graph of the function f defined by $f(x) = -5x^4 + 7$ on the interval $[-1, 1]$.
7. Sketch the graph of the function f defined by
$$f(x) = -\frac{4}{x} + 6$$
on $[-2, -\frac{1}{2}] \cup [\frac{1}{2}, 2]$.
8. Give an example of two polynomials of degree 9 whose sum has degree 4.
9. Find a polynomial whose zeros are -3, 2, and 5.
10. Find a polynomial p such that $p(-1) = 0$, $p(4) = 0$, and $p(2) = 3$.
11. Explain why $x^7 + 9999x^6 - 88x^5 + 77x^4 - 6x^3 + 55$ is negative for negative values of x with very large absolute value.
12. Write
$$\frac{3x^{80} + 2}{x^6 - 5} + \frac{x^7 + 10}{x^2 + 9}$$
as a ratio, with the numerator and denominator each written as a sum of terms of the form cx^m.
13. Write
$$\frac{3x^{80} + 2}{x^6 - 5} \Big/ \frac{x^7 + 10}{x^2 + 9}$$
as a ratio, with the numerator and denominator each written as a sum of terms of the form cx^m.
14. Suppose
$$r(x) = \frac{300x^{80} + 299}{x^{76} - 101} \quad \text{and} \quad s(x) = \frac{x^7 + 1}{x^2 + 9}.$$
Which is larger, $r(10^{100})$ or $s(10^{100})$?
15. Write the domain of $\dfrac{x^5 + 2}{x^2 + 7x - 1}$ as a union of intervals.

16. Write
$$\frac{4x^5 - 2x^4 + 3x^2 + 1}{2x^2 - 1}$$
in the form $G(x) + \frac{R(x)}{2x^2 - 1}$, where G is a polynomial and R is a linear function.

17. Find the asymptotes of the graph of the function f defined by
$$f(x) = \frac{3x^2 + 5x + 1}{x^2 + 7x + 10}.$$

18. Give an example of a rational function whose asymptotes in the xy-plane are the lines $x = 2$ and $x = 5$.

19. Give an example of a rational function whose asymptotes in the xy-plane are the lines $x = 2$, $x = 5$, and $y = 3$.

20. Write
$$\frac{-2 + 5i}{4 + 3i}$$
in the form $a + bi$, where a and b are real numbers.

21. Verify that
$$\left(-\frac{1}{2} + \frac{\sqrt{3}}{2}i\right)^3 = 1.$$

CHAPTER

5

Starry Night, painted by Vincent Van Gogh in 1889. The brightness of a star as seen from Earth is measured using a logarithmic scale.

Exponents and Logarithms

This chapter focuses on understanding exponents and logarithms, along with applications of these crucial concepts.

The algebraic properties of integer exponents will lead us to the definition of $x^{1/m}$ for m a positive integer. As we will see, the function that takes the mth root of a number is simply the inverse of the function that raises a number to the mth power. The algebraic properties of exponents will then provide a natural definition of rational number exponents. From there we will finally reach the notion of real exponents.

Logarithms will be defined as inverse functions of exponential functions. We will see that the important algebraic properties of logarithms follow directly from the algebraic properties of exponents.

This chapter contains numerous applications showing how exponents and logarithms are used to model radioactive decay, earthquake intensity, sound intensity, star brightness, population growth, and compound interest.

5.1 *Exponents and Exponential Functions*

LEARNING OBJECTIVES

By the end of this section you should be able to

- explain why $x^{1/m}$ is defined to equal the number whose m^{th} power equals x;
- explain why $x^{n/m}$ is defined to equal $(x^{1/m})^n$;
- manipulate and simplify expressions involving exponents;
- work with exponential functions.

So far we have defined the meaning of integer exponents. We will begin this section by making sense of the expression $x^{1/m}$, where m is a positive integer. From there we will progress to rational exponents, then to real exponents, and then to exponential functions.

Roots

Suppose x is a real number and m is a positive integer. How should we define $x^{1/m}$? To answer this question, we will let the algebraic properties of exponents force the definition upon us, as we did when we defined the meaning of negative integer exponents in Section 4.1.

Recall that if x is a real number and m and n are positive integers, then

$$(x^n)^m = x^{nm}.$$

We would like the equation above to hold even when m and n are not positive integers. In particular, if we take n equal to $1/m$, the equation above becomes

$$(x^{1/m})^m = x.$$

Thus we see that we should define $x^{1/m}$ to be a number that when raised to the mth power gives x.

EXAMPLE 1

How should $8^{1/3}$ be defined?

The expression x^3 is called the **cube** *of x.*

SOLUTION Taking $x = 8$ and $m = 3$ in the equation above, we get

$$(8^{1/3})^3 = 8.$$

Thus $8^{1/3}$ should be defined to be a number that when cubed gives 8. The only such number is 2; thus we should define $8^{1/3}$ to equal 2.

Similarly, $(-8)^{1/3}$ should be defined to equal -2, because -2 is the only number that when cubed gives -8. The next example shows that special care must be used when defining $x^{1/m}$ if m is an even integer.

EXAMPLE 2

How should $9^{1/2}$ be defined?

SOLUTION In the equation $(x^{1/m})^m = x$, take $x = 9$ and $m = 2$ to get

$$(9^{1/2})^2 = 9.$$

Thus $9^{1/2}$ should be defined to be a number that when squared gives 9. Both 3 and -3 have a square equal to 9; thus we have a choice. When this happens, we will always choose the positive possibility. Thus $9^{1/2}$ is defined to equal 3.

The next example shows the problem that arises when trying to define $x^{1/m}$ if x is negative and m is an even integer.

EXAMPLE 3

How should $(-9)^{1/2}$ be defined?

SOLUTION In the equation $(x^{1/m})^m = x$, take $x = -9$ and $m = 2$ to get

$$((-9)^{1/2})^2 = -9.$$

Thus $(-9)^{1/2}$ should be defined to be a number that when squared gives -9. But no such real number exists, because no real number has a square that is negative. Hence we leave $(-9)^{1/2}$ undefined when working only with real numbers. Similarly, we left $1/0$ and 0^0 undefined because no possible definition could preserve the necessary algebraic properties.

Complex numbers were invented so that meaning could be given to expressions such as $(-9)^{1/2}$, but we restrict our attention here to real numbers.

With the experience of the previous examples, we are now ready to give the formal definition of $x^{1/m}$.

Roots

If m is a positive integer and x is a real number, then $x^{1/m}$ is defined to be the real number satisfying the equation

$$(x^{1/m})^m = x,$$

subject to the following conditions:

- If $x < 0$ and m is an even integer, then $x^{1/m}$ is undefined.
- If $x > 0$ and m is an even integer, then $x^{1/m}$ is chosen to be the positive number satisfying the equation above.

The number $x^{1/m}$ is called the mth **root** of x. Thus the mth root of x is the number that when raised to the mth power gives x, with the understanding that if m is even and x is positive, then we choose the positive number with this property.

The number $x^{1/2}$ is called the **square root** of x, and $x^{1/3}$ is called the **cube root** of x. For example, the square root of $\frac{16}{9}$ equals $\frac{4}{3}$, and the cube root of 125 equals 5. The notation $\sqrt{x}$ denotes the square root of x, and the notation $\sqrt[3]{x}$ denotes the cube root of x. For example, $\sqrt{9} = 3$ and $\sqrt[3]{\frac{1}{8}} = \frac{1}{2}$. More generally, the notation $\sqrt[m]{x}$ denotes the mth root of x.

Notation for roots

$$\sqrt{x} = x^{1/2};$$

$$\sqrt[m]{x} = x^{1/m}.$$

The expression $\sqrt{2}$ cannot be simplified any further—there does not exist a rational number whose square equals 2 (see Section 1.1). Thus the expression $\sqrt{2}$ is usually left simply as $\sqrt{2}$, unless a numeric calculation is needed. The key property of $\sqrt{2}$ is that

$$(\sqrt{2})^2 = 2.$$

Make sure that you understand why the equation above holds as a consequence of our definitions.

Suppose x and y are positive numbers. Then

$$(\sqrt{x}\sqrt{y})^2 = (\sqrt{x})^2(\sqrt{y})^2 = xy.$$

Thus $\sqrt{x}\sqrt{y}$ is a positive number whose square equals xy. Our definition of square root now implies that $\sqrt{xy}$ equals $\sqrt{x}\sqrt{y}$, so we have the following result.

Never, ever, make the mistake of thinking that $\sqrt{x} + \sqrt{y}$ equals $\sqrt{x+y}$.

Square root of a product

If x and y are positive numbers, then

$$\sqrt{x}\sqrt{y} = \sqrt{xy}.$$

The result above is used twice in the next example, once in each direction.

EXAMPLE 4 Simplify $\sqrt{2}\sqrt{6}$.

SOLUTION

$$\begin{aligned}\sqrt{2}\sqrt{6} &= \sqrt{12}\\ &= \sqrt{4\cdot 3}\\ &= \sqrt{4}\sqrt{3}\\ &= 2\sqrt{3}\end{aligned}$$

The example below should help solidify your understanding of square roots.

EXAMPLE 5

Show that $\sqrt{7 + 4\sqrt{3}} = 2 + \sqrt{3}$.

SOLUTION No one knows a nice way to simplify an expression of the form $\sqrt{a + b\sqrt{c}}$. Thus we have no way to work with the left side of the equation above. However, to say that the square root of $7 + 4\sqrt{3}$ equals $2 + \sqrt{3}$ means that the square of $2 + \sqrt{3}$ equals $7 + 4\sqrt{3}$. Thus to verify the equation above, we will square the right side and see if we get $7 + 4\sqrt{3}$. Here is the calculation:

$$\begin{aligned}(2 + \sqrt{3})^2 &= 2^2 + 2 \cdot 2 \cdot \sqrt{3} + \sqrt{3}^2 \\ &= 4 + 4\sqrt{3} + 3 \\ &= 7 + 4\sqrt{3}.\end{aligned}$$

Thus $\sqrt{7 + 4\sqrt{3}} = 2 + \sqrt{3}$.

The key point to understand in the definition of $x^{1/m}$ is that the mth root function is simply the inverse function of the mth power function. Although we did not use this language when we defined m^{th} roots, we could have done so, because we defined $y^{1/m}$ as the number that makes the equation $(y^{1/m})^m = y$ hold, exactly as done in the definition of an inverse function (see Section 3.4). Here is a restatement of mth roots in terms of inverse functions:

Roots as inverse functions

Suppose m is a positive integer and f is the function defined by

$$f(x) = x^m,$$

with the domain of f being the set of real numbers if m is odd and the domain of f being $[0, \infty)$ if m is even. Then the inverse function f^{-1} is given by the formula

$$f^{-1}(y) = y^{1/m}.$$

If m is even, the domain of f is restricted to $[0, \infty)$ to obtain a one-to-one function.

Because the function $x^{1/m}$ is the inverse of the function x^m, we can obtain the graph of $x^{1/m}$ by flipping the graph of x^m across the line $y = x$, as is the case with any one-to-one function and its inverse. For the case when $m = 2$, we already did this, obtaining the graph of $\sqrt{x}$ by flipping the graph of x^2 across the line $y = x$; see Section 3.5. Here is the corresponding graph for x^3 and its inverse $\sqrt[3]{x}$:

The inverse of an increasing function is increasing. Thus the function $x^{1/m}$ is increasing for every positive integer m.

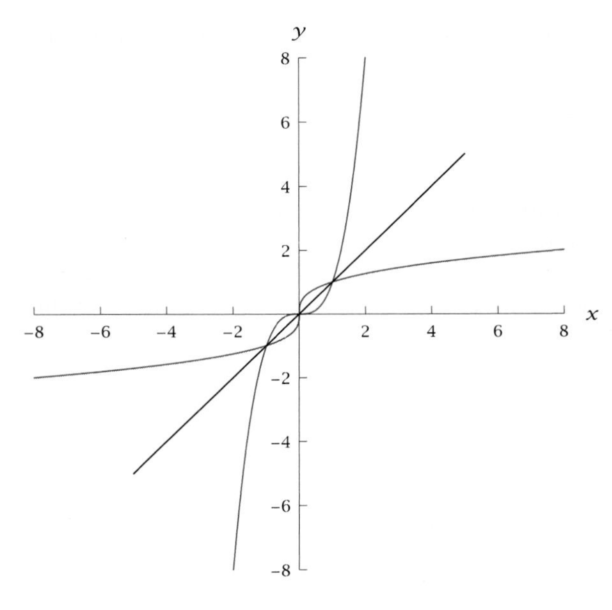

The graphs of x^3 *(blue) on the interval* $[-2, 2]$ *and its inverse function* $\sqrt[3]{x}$ *(red) on the interval* $[-8, 8]$.

Rational Exponents

Having defined the meaning of exponents of the form $1/m$, where m is a positive integer, we will now find it easy to define the meaning of rational exponents.

Recall from Section 4.1 that if n and p are positive integers, then

$$x^{np} = (x^p)^n$$

for every real number x. If we assume that the equation above should hold even when p is not a positive integer, we are led to the meaning of rational exponents. Specifically, suppose m is a positive integer and we take $p = 1/m$ in the equation above, getting

$$x^{n/m} = (x^{1/m})^n.$$

The left side of the equation above does not yet make sense, because we have not yet given meaning to a rational exponent. But the right side of the equation above does make sense, because we have defined $x^{1/m}$ and we have defined the nth power of every number. Thus we can use the right side of the equation above to define the left side, which we now do.

The phrase "whenever this makes sense" excludes the case where $x < 0$ *and* m *is even (because then* $x^{1/m}$ *is undefined) and the case where* $x = 0$ *and* $n \le 0$ *(because then* 0^n *is undefined).*

Rational exponents

If n/m is a fraction in reduced form, where m and n are integers and $m > 0$, then $x^{n/m}$ is defined by the equation

$$x^{n/m} = (x^{1/m})^n$$

whenever this makes sense.

EXAMPLE 6

(a) Evaluate $16^{3/2}$.

(b) Evaluate $27^{2/3}$.

(c) Evaluate $16^{-3/4}$.

(d) Evaluate $27^{-4/3}$.

SOLUTION

(a) $16^{3/2} = (16^{1/2})^3 = 4^3 = 64$

(b) $27^{2/3} = (27^{1/3})^2 = 3^2 = 9$

(c) $16^{-3/4} = (16^{1/4})^{-3} = 2^{-3} = \frac{1}{2^3} = \frac{1}{8}$

(d) $27^{-4/3} = (27^{1/3})^{-4} = 3^{-4} = \frac{1}{3^4} = \frac{1}{81}$

Real Exponents

At this stage we have defined the meaning of a rational exponent, but an expression such as $7^{\sqrt{2}}$ has not yet been defined. Nevertheless, the following example should make sense to you as the only reasonable (rational?) way to think of an irrational exponent.

EXAMPLE 7

Find an approximation for $7^{\sqrt{2}}$.

SOLUTION Because $\sqrt{2}$ is approximately 1.414, we expect that $7^{\sqrt{2}}$ should be approximately $7^{1.414}$ (which has been defined, because 1.414 is a rational number). A calculator shows that $7^{1.414} \approx 15.66638$.

If we use a better approximation of $\sqrt{2}$, then we should get a better approximation of $7^{\sqrt{2}}$. For example, 1.41421356 is a better rational approximation of $\sqrt{2}$ than 1.414. A calculator shows that

$$7^{1.41421356} \approx 15.67289,$$

which turns out to be correct for the first five digits after the decimal point in the decimal expansion of $7^{\sqrt{2}}$.

We could continue this process by taking rational approximations as close as we wish to $\sqrt{2}$, thus getting approximations as accurate as we wish to $7^{\sqrt{2}}$.

The example above gives the idea for defining the meaning of an irrational exponent:

Although rational numbers suffice for all physical quantities, we want to define functions f such as $f(x) = 2^x$ whose domain is the set of all real numbers. Thus we must define the meaning of irrational exponents.

Irrational exponents

Suppose $b > 0$ and x is an irrational number. Then b^x is the number that is approximated by numbers of the form b^r as r takes on rational values that approximate x.

The definition of b^x above does not have the level of rigor expected of a mathematical definition, but the idea should be clear from the example above. A rigorous approach to this question would take us beyond material appropriate for a college algebra course. Thus we will rely on our intuitive sense of the loose definition given above.

The box below summarizes the key algebraic properties of exponents. As we extended the meaning of exponents to larger classes of numbers, the definitions were chosen so that these algebraic properties were preserved.

> ***Algebraic properties of exponents***
>
> Suppose a and b are positive numbers and x and y are real numbers. Then
>
> $$b^x b^y = b^{x+y}, \qquad b^{-x} = \frac{1}{b^x},$$
>
> $$(b^x)^y = b^{xy}, \qquad \frac{a^x}{a^y} = a^{x-y},$$
>
> $$a^x b^x = (ab)^x,$$
>
> $$b^0 = 1, \qquad \frac{a^x}{b^x} = \left(\frac{a}{b}\right)^x.$$

Exponential Functions

Now that we have defined b^x for every positive number b and every real number x, we can define a function f by $f(x) = b^x$. These functions (one for each number b) are sufficiently important to have their own name:

> ***Exponential function***
>
> Suppose b is a positive number, with $b \neq 1$. Then the **exponential function** with base b is the function f defined by
>
> $$f(x) = b^x.$$

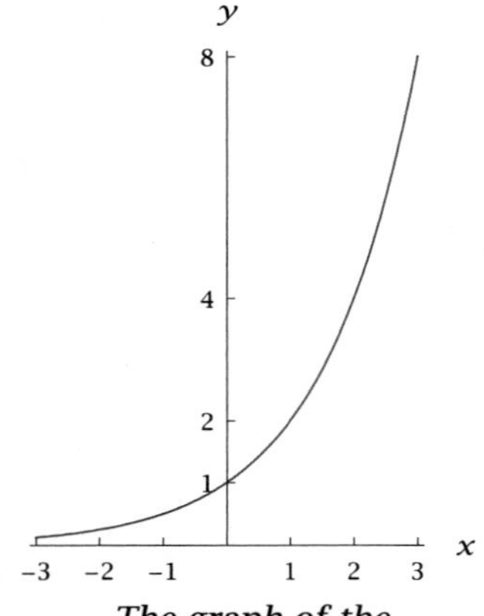

The graph of the exponential function 2^x on the interval $[-3, 3]$.

For example, taking $b = 2$, we have the exponential function f with base 2 defined by $f(x) = 2^x$. The domain of this function is the set of real numbers, and the range is the set of positive numbers. The graph of this function on the interval $[-3, 3]$ is shown in the margin.

Be careful to distinguish the graph of the function 2^x from the graph of the function x^2. These graphs have different shapes, as shown below.

The function g defined by $g(x) = x^2$ is not an exponential function. Remember that for an exponential function such as the function f defined by $f(x) = 2^x$, the variable appears in the exponent.

The graphs of 2^x (blue) and x^2 (red) on the interval $[-5.5, 5.5]$. The graph of 2^x gets very close to the x-axis for negative values of x with large absolute value. The graph of x^2 is a parabola with its vertex at the origin.

The potential base $b = 1$ is excluded from the definition of an exponential function because we do not want to make exceptions for this base. For example, it is convenient (and true) to say that the range of every exponential function is the set of positive numbers. But the function f defined by $f(x) = 1^x$ has the property that $f(x) = 1$ for every real number x; thus the range of this function is the set $\{1\}$ rather than the set of positive numbers. To exclude this kind of exception, we do not call this function f an exponential function.

EXERCISES

For Exercises 1–12, expand the expression.

1. $(2 + \sqrt{3})^2$
2. $(3 + \sqrt{2})^2$
3. $(2 - 3\sqrt{5})^2$
4. $(3 - 5\sqrt{2})^2$
5. $(2 + \sqrt{3})^4$
6. $(3 + \sqrt{2})^4$
7. $(3 + \sqrt{x})^2$
8. $(5 + \sqrt{x})^2$
9. $(3 - \sqrt{2x})^2$
10. $(5 - \sqrt{3x})^2$
11. $(1 + 2\sqrt{3x})^2$
12. $(3 + 2\sqrt{5x})^2$

For Exercises 13–20, evaluate the indicated quantities. Do not use a calculator because otherwise you will not gain the understanding that these exercises should help you attain.

13. $25^{3/2}$
14. $8^{5/3}$
15. $32^{3/5}$
16. $81^{3/4}$
17. $32^{-4/5}$
18. $8^{-5/3}$
19. $(-8)^{7/3}$
20. $(-27)^{4/3}$

For Exercises 21–32, find a formula for the inverse function f^{-1} of the indicated function f.

21. $f(x) = x^9$
22. $f(x) = x^{12}$
23. $f(x) = x^{1/7}$
24. $f(x) = x^{1/11}$
25. $f(x) = x^{-2/5}$
26. $f(x) = x^{-17/7}$
27. $f(x) = \frac{x^4}{81}$
28. $f(x) = 32x^5$
29. $f(x) = 6 + x^3$
30. $f(x) = x^6 - 5$
31. $f(x) = 4x^{3/7} - 1$
32. $f(x) = 7 + 8x^{5/9}$

For Exercises 33–38, find a formula for $(f \circ g)(x)$ assuming that f and g are the indicated functions.

33. $f(x) = x^{1/2}$ and $g(x) = x^{3/7}$
34. $f(x) = x^{5/3}$ and $g(x) = x^{4/9}$
35. $f(x) = 3 + x^{5/4}$ and $g(x) = x^{2/7}$
36. $f(x) = x^{2/3} - 7$ and $g(x) = x^{9/16}$
37. $f(x) = 5x^{\sqrt{2}}$ and $g(x) = x^{\sqrt{8}}$
38. $f(x) = 7x^{\sqrt{12}}$ and $g(x) = x^{\sqrt{3}}$

For Exercises 39–46, find all real numbers x that satisfy the indicated equation.

39. $x - 5\sqrt{x} + 6 = 0$
40. $x - 7\sqrt{x} + 12 = 0$
41. $x - \sqrt{x} = 6$
42. $x - \sqrt{x} = 12$
43. $x^{2/3} - 6x^{1/3} = -8$
44. $x^{2/3} + 3x^{1/3} = 10$
45. $x^4 - 3x^2 = 10$
46. $x^4 - 8x^2 = -15$
47. Suppose x is a number such that $3^x = 4$. Evaluate 3^{-2x}.
48. Suppose x is a number such that $2^x = \frac{1}{3}$. Evaluate 2^{-4x}.
49. Suppose x is a number such that $2^x = 5$. Evaluate 8^x.
50. Suppose x is a number such that $3^x = 5$. Evaluate $\left(\frac{1}{9}\right)^x$.

For Exercises 51–56, evaluate the indicated quantities assuming that f and g are the functions defined by

$$f(x) = 2^x \quad \textit{and} \quad g(x) = \frac{x+1}{x+2}.$$

51. $(f \circ g)(-1)$
52. $(g \circ f)(0)$
53. $(f \circ g)(0)$
54. $(g \circ f)(\frac{3}{2})$
55. $(f \circ f)(\frac{1}{2})$
56. $(f \circ f)(\frac{3}{5})$
57. What is the domain of the function $(3 + x)^{1/4}$?
58. What is the domain of the function $(1 + x^2)^{1/8}$?
59. Find an integer m such that
$$\left((3 + 2\sqrt{5})^2 - m\right)^2$$
is an integer.
60. Find an integer m such that
$$\left((5 - 2\sqrt{3})^2 - m\right)^2$$
is an integer.

PROBLEMS

Some problems require considerably more thought than the exercises. Unlike exercises, problems often have more than one correct answer.

61. Sketch the graph of the functions $\sqrt{x} + 1$ and $\sqrt{x+1}$ on the interval $[0, 4]$.

62. Explain why the spoken phrase "the square root of x plus one" could be interpreted in two different ways that would not give the same result.

63. Sketch the graph of the functions $2x^{1/3}$ and $(2x)^{1/3}$ on the interval $[0, 8]$.

64. Sketch the graphs of the functions $x^{1/4}$ and $x^{1/5}$ on the interval $[0, 81]$.

65. Show that $\sqrt{5} \cdot 5^{3/2} = 25$.

66. Show that $(3^{\sqrt{2}})^{\sqrt{2}} = 9$.

67. Show that $\sqrt{2}^3 \sqrt{8}^3 = 64$.

68. Show that $3^{3/2}12^{3/2} = 216$.

69. Show that $\sqrt{2+\sqrt{3}} = \sqrt{\frac{3}{2}} + \sqrt{\frac{1}{2}}$.

70. Show that $\sqrt{2-\sqrt{3}} = \sqrt{\frac{3}{2}} - \sqrt{\frac{1}{2}}$.

71. Show that $\sqrt{9-4\sqrt{5}} = \sqrt{5} - 2$.

72. Show that $(23 - 8\sqrt{7})^{1/2} = 4 - \sqrt{7}$.

73. Make up a problem similar in form to the problem above, without duplicating anything in this book.

74. Show that $(99 + 70\sqrt{2})^{1/3} = 3 + 2\sqrt{2}$.

75. Show that $(-37 + 30\sqrt{3})^{1/3} = -1 + 2\sqrt{3}$.

76. Show that if x and y are positive numbers with $x \neq y$, then
$$\frac{x-y}{\sqrt{x}-\sqrt{y}} = \sqrt{x} + \sqrt{y}.$$

77. Explain why
$$10^{100}(\sqrt{10^{200}+1} - 10^{100})$$
is approximately equal to $\frac{1}{2}$.

78. Explain why the equation $\sqrt{x^2} = x$ is not valid for all real numbers x and should be replaced by the equation $\sqrt{x^2} = |x|$.

79. Explain why the equation $\sqrt{x^8} = x^4$ is valid for all real numbers x, with no necessity for using absolute value.

80. Show that if x and y are positive numbers, then
$$\sqrt{x+y} < \sqrt{x} + \sqrt{y}.$$
[*In particular, if x and y are positive numbers, then $\sqrt{x+y} \neq \sqrt{x} + \sqrt{y}$.*]

81. Show that if $0 < x < y$, then
$$\sqrt{y} - \sqrt{x} < \sqrt{y-x}.$$

82. One of the graphs in this section suggests that
$$\sqrt{x} < \sqrt[3]{x} \quad \text{if} \quad 0 < x < 1$$
and
$$\sqrt{x} > \sqrt[3]{x} \quad \text{if} \quad x > 1.$$
Explain why each of these inequalities holds.

83. Suppose x is a positive number. Using only the definitions of roots and integer exponents, explain why
$$(x^{1/2})^3 = (x^{1/4})^6.$$

84. Suppose x is a positive number and n is a positive integer. Using only the definitions of roots and integer exponents, explain why
$$(x^{1/2})^n = (x^{1/4})^{2n}.$$

85. Suppose x is a positive number and n and p are positive integers. Using only the definitions of roots and integer exponents, explain why
$$(x^{1/2})^n = (x^{1/(2p)})^{np}.$$

86. Suppose x is a positive number and m, n, and p are positive integers. Using only the definitions of roots and integer exponents, explain why
$$(x^{1/m})^n = (x^{1/(mp)})^{np}.$$

87. Using the result from the problem above, explain why the definition of a positive number raised to a positive rational power gives the same result even if the positive rational number is not expressed in reduced form.

88. Using the result that $\sqrt{2}$ is irrational (proved in Section 1.1), show that $2^{5/2}$ is irrational.

89. Using the result that $\sqrt{2}$ is irrational, explain why $2^{1/6}$ is irrational.

90. Suppose you have a calculator that can only compute square roots. Explain how you could use this calculator to compute $7^{1/8}$.

91. Suppose you have a calculator that can only compute square roots and can multiply. Explain how you could use this calculator to compute $7^{3/4}$.

92. Give an example of three irrational numbers x, y, and z such that

$$xyz$$

is a rational number.

93. Give an example of three irrational numbers x, y, and z such that

$$(x^y)^z$$

is a rational number.

94. Is the function f defined by $f(x) = 2^x$ for every real number x an even function, an odd function, or neither?

95. What is wrong with the following string of equalities, which seems to show that $-1 = 1$?

$$-1 = i \cdot i = \sqrt{-1}\sqrt{-1} = \sqrt{(-1)(-1)} = \sqrt{1} = 1$$

96. Explain why the graph of $y = 8^x$ can be obtained by horizontally stretching the graph of $y = 2^x$ by a factor of $\frac{1}{3}$.

97. Explain why shifting the graph of $y = 2^x$ left 3 units produces the same graph as vertically stretching the graph of $y = 2^x$ by a factor of 8.

WORKED-OUT SOLUTIONS *to Odd-numbered Exercises*

Do not read these worked-out solutions before first attempting to do the exercises yourself. Otherwise you may merely mimic the techniques shown here without understanding the ideas.

Best way to learn: Carefully read the section of the textbook, then do all the odd-numbered exercises (even if they have not been assigned) and check your answers here. If you get stuck on an exercise, reread the section of the textbook—then try the exercise again. If you are still stuck, then look at the worked-out solution here.

For Exercises 1–12, expand the expression.

1. $(2 + \sqrt{3})^2$

SOLUTION

$$\begin{aligned}(2 + \sqrt{3})^2 &= 2^2 + 2 \cdot 2 \cdot \sqrt{3} + \sqrt{3}^2 \\ &= 4 + 4\sqrt{3} + 3 \\ &= 7 + 4\sqrt{3}\end{aligned}$$

3. $(2 - 3\sqrt{5})^2$

SOLUTION

$$\begin{aligned}(2 - 3\sqrt{5})^2 &= 2^2 - 2 \cdot 2 \cdot 3 \cdot \sqrt{5} + 3^2 \cdot \sqrt{5}^2 \\ &= 4 - 12\sqrt{5} + 9 \cdot 5 \\ &= 49 - 12\sqrt{5}\end{aligned}$$

5. $(2 + \sqrt{3})^4$

SOLUTION Note that

$$(2 + \sqrt{3})^4 = \left((2 + \sqrt{3})^2\right)^2.$$

Thus first we need to compute $(2 + \sqrt{3})^2$. We already did that in Exercise 1, getting

$$(2 + \sqrt{3})^2 = 7 + 4\sqrt{3}.$$

Thus

$$\begin{aligned}(2 + \sqrt{3})^4 &= \left((2 + \sqrt{3})^2\right)^2 \\ &= (7 + 4\sqrt{3})^2 \\ &= 7^2 + 2 \cdot 7 \cdot 4 \cdot \sqrt{3} + 4^2 \cdot \sqrt{3}^2 \\ &= 49 + 56\sqrt{3} + 16 \cdot 3 \\ &= 97 + 56\sqrt{3}.\end{aligned}$$

7. $(3 + \sqrt{x})^2$

SOLUTION

$$\begin{aligned}(3 + \sqrt{x})^2 &= 3^2 + 2 \cdot 3 \cdot \sqrt{x} + \sqrt{x}^2 \\ &= 9 + 6\sqrt{x} + x\end{aligned}$$

9. $(3 - \sqrt{2x})^2$

SOLUTION

$$\begin{aligned}(3 - \sqrt{2x})^2 &= 3^2 - 2 \cdot 3 \cdot \sqrt{2x} + \sqrt{2x}^2 \\ &= 9 - 6\sqrt{2x} + 2x\end{aligned}$$

11. $(1 + 2\sqrt{3x})^2$

SOLUTION

$$\begin{aligned}(1 + 2\sqrt{3x})^2 &= 1^2 + 2 \cdot 2 \cdot \sqrt{3x} + 2^2 \cdot \sqrt{3x}^2 \\ &= 1 + 4\sqrt{3x} + 4 \cdot 3x \\ &= 1 + 4\sqrt{3x} + 12x\end{aligned}$$

For Exercises 13–20, evaluate the indicated quantities. Do not use a calculator because otherwise you will not gain the understanding that these exercises should help you attain.

13. $25^{3/2}$

SOLUTION $25^{3/2} = (25^{1/2})^3 = 5^3 = 125$

15. $32^{3/5}$

SOLUTION $32^{3/5} = (32^{1/5})^3 = 2^3 = 8$

17. $32^{-4/5}$

SOLUTION $32^{-4/5} = (32^{1/5})^{-4} = 2^{-4} = \dfrac{1}{2^4} = \dfrac{1}{16}$

19. $(-8)^{7/3}$

SOLUTION

$$\begin{aligned}(-8)^{7/3} &= ((-8)^{1/3})^7 \\ &= (-2)^7 \\ &= -128\end{aligned}$$

For Exercises 21–32, find a formula for the inverse function f^{-1} of the indicated function f.

21. $f(x) = x^9$

SOLUTION By the definition of roots, the inverse of f is the function f^{-1} defined by

$$f^{-1}(y) = y^{1/9}.$$

23. $f(x) = x^{1/7}$

SOLUTION By the definition of roots, $f = g^{-1}$, where g is the function defined by $g(y) = y^7$. Thus $f^{-1} = (g^{-1})^{-1} = g$. In other words,

$$f^{-1}(y) = y^7.$$

25. $f(x) = x^{-2/5}$

SOLUTION To find a formula for f^{-1}, we solve the equation $x^{-2/5} = y$ for x. Raising both sides of this equation to the power $-\frac{5}{2}$, we get $x = y^{-5/2}$. Hence

$$f^{-1}(y) = y^{-5/2}.$$

27. $f(x) = \frac{x^4}{81}$

SOLUTION To find a formula for f^{-1}, we solve the equation $\frac{x^4}{81} = y$ for x. Multiplying both sides by 81 and then raising both sides of this equation to the power $\frac{1}{4}$, we get $x = (81y)^{1/4} = 81^{1/4}y^{1/4} = 3y^{1/4}$. Hence

$$f^{-1}(y) = 3y^{1/4}.$$

29. $f(x) = 6 + x^3$

SOLUTION To find a formula for f^{-1}, we solve the equation $6 + x^3 = y$ for x. Subtracting 6 from both sides and then raising both sides of this equation to the power $\frac{1}{3}$, we get $x = (y - 6)^{1/3}$. Hence

$$f^{-1}(y) = (y - 6)^{1/3}.$$

31. $f(x) = 4x^{3/7} - 1$

SOLUTION To find a formula for f^{-1}, we solve the equation $4x^{3/7} - 1 = y$ for x. Adding 1 to both sides, then dividing both sides by 4, and then raising both sides of this equation to the power $\frac{7}{3}$, we get $x = (\frac{y+1}{4})^{7/3}$. Hence

$$f^{-1}(y) = \left(\frac{y+1}{4}\right)^{7/3}.$$

For Exercises 33–38, find a formula for $(f \circ g)(x)$ assuming that f and g are the indicated functions.

33. $f(x) = x^{1/2}$ and $g(x) = x^{3/7}$

SOLUTION

$$\begin{aligned}(f \circ g)(x) = f(g(x)) = f(x^{3/7}) &= (x^{3/7})^{1/2} \\ &= x^{3/14}\end{aligned}$$

35. $f(x) = 3 + x^{5/4}$ and $g(x) = x^{2/7}$

SOLUTION

$$\begin{aligned}(f \circ g)(x) = f(g(x)) = f(x^{2/7}) &= 3 + (x^{2/7})^{5/4} \\ &= 3 + x^{5/14}\end{aligned}$$

37. $f(x) = 5x^{\sqrt{2}}$ and $g(x) = x^{\sqrt{8}}$

SOLUTION

$$\begin{aligned}(f \circ g)(x) &= f(g(x)) = f(x^{\sqrt{8}}) \\ &= 5\left(x^{\sqrt{8}}\right)^{\sqrt{2}} = 5x^{\sqrt{16}} = 5x^4\end{aligned}$$

For Exercises 39–46, find all real numbers x that satisfy the indicated equation.

39. $x - 5\sqrt{x} + 6 = 0$

SOLUTION This equation involves $\sqrt{x}$; thus we make the substitution $\sqrt{x} = y$. Squaring both sides of the equation $\sqrt{x} = y$ gives $x = y^2$. With these substitutions, the equation above becomes

$$y^2 - 5y + 6 = 0.$$

Factoring the left side gives

$$(y-2)(y-3) = 0.$$

Thus $y = 2$ or $y = 3$ (as could also have been discovered by using the quadratic formula).

Substituting $\sqrt{x}$ for y now shows that $\sqrt{x} = 2$ or $\sqrt{x} = 3$. Thus $x = 4$ or $x = 9$.

41. $x - \sqrt{x} = 6$

SOLUTION This equation involves $\sqrt{x}$; thus we make the substitution $\sqrt{x} = y$. Squaring both sides of the equation $\sqrt{x} = y$ gives $x = y^2$. Making these substitutions and subtracting 6 from both sides, we have

$$y^2 - y - 6 = 0.$$

The quadratic formula gives

$$y = \frac{1 \pm \sqrt{1+24}}{2} = \frac{1 \pm 5}{2}.$$

Thus $y = 3$ or $y = -2$ (the same result could have been obtained by factoring).

Substituting $\sqrt{x}$ for y now shows that $\sqrt{x} = 3$ or $\sqrt{x} = -2$. The first possibility corresponds to the solution $x = 9$. There are no real numbers x such that $\sqrt{x} = -2$. Thus $x = 9$ is the only solution to this equation.

43. $x^{2/3} - 6x^{1/3} = -8$

SOLUTION This equation involves $x^{1/3}$ and $x^{2/3}$; thus we make the substitution $x^{1/3} = y$. Squaring both sides of the equation $x^{1/3} = y$ gives $x^{2/3} = y^2$. Making these substitutions and adding 8 to both sides, we have

$$y^2 - 6y + 8 = 0.$$

Factoring the left side gives

$$(y-2)(y-4) = 0.$$

Thus $y = 2$ or $y = 4$ (as could also have been discovered by using the quadratic formula).

Substituting $x^{1/3}$ for y now shows that $x^{1/3} = 2$ or $x^{1/3} = 4$. Thus $x = 2^3$ or $x = 4^3$. In other words, $x = 8$ or $x = 64$.

45. $x^4 - 3x^2 = 10$

SOLUTION This equation involves x^2 and x^4; thus we make the substitution $x^2 = y$. Squaring both sides of the equation $x^2 = y$ gives $x^4 = y^2$. Making these substitutions and subtracting 10 from both sides, we have

$$y^2 - 3y - 10 = 0.$$

Factoring the left side gives

$$(y-5)(y+2) = 0.$$

Thus $y = 5$ or $y = -2$ (as could also have been discovered by using the quadratic formula).

Substituting x^2 for y now shows that $x^2 = 5$ or $x^2 = -2$. The first of these equations implies that $x = \sqrt{5}$ or $x = -\sqrt{5}$; the second equation is not satisfied by any real value of x. In other words, the original equation implies that $x = \sqrt{5}$ or $x = -\sqrt{5}$.

47. Suppose x is a number such that $3^x = 4$. Evaluate 3^{-2x}.

SOLUTION

$$\begin{aligned} 3^{-2x} &= (3^x)^{-2} \\ &= 4^{-2} \\ &= \frac{1}{4^2} \\ &= \frac{1}{16} \end{aligned}$$

49. Suppose x is a number such that $2^x = 5$. Evaluate 8^x.

SOLUTION

$$\begin{aligned} 8^x &= (2^3)^x \\ &= 2^{3x} \\ &= (2^x)^3 \\ &= 5^3 \\ &= 125 \end{aligned}$$

For Exercises 51–56, evaluate the indicated quantities assuming that f and g are the functions defined by

$$f(x) = 2^x \quad \textit{and} \quad g(x) = \frac{x+1}{x+2}.$$

51. $(f \circ g)(-1)$

SOLUTION

$$(f \circ g)(-1) = f(g(-1)) = f(0) = 2^0 = 1$$

53. $(f \circ g)(0)$

SOLUTION

$$(f \circ g)(0) = f(g(0)) = f(\tfrac{1}{2}) = 2^{1/2} \approx 1.414$$

55. $(f \circ f)(\frac{1}{2})$

SOLUTION

$$\begin{aligned} (f \circ f)(\tfrac{1}{2}) = f(f(\tfrac{1}{2})) &= f(2^{1/2}) \\ &\approx f(1.41421) \\ &= 2^{1.41421} \\ &\approx 2.66514 \end{aligned}$$

57. What is the domain of the function $(3 + x)^{1/4}$?

SOLUTION For $(3 + x)^{1/4}$ to be defined as a real number, we must have $3 + x \geq 0$, which is equivalent to the inequality $x \geq -3$. Thus the domain of the function $(3 + x)^{1/4}$ is the interval $[-3, \infty)$.

59. Find an integer m such that

$$\left((3 + 2\sqrt{5})^2 - m\right)^2$$

is an integer.

SOLUTION First we evaluate $(3 + 2\sqrt{5})^2$:

$$\begin{aligned} (3 + 2\sqrt{5})^2 &= 3^2 + 2 \cdot 3 \cdot 2 \cdot \sqrt{5} + 2^2 \cdot \sqrt{5}^2 \\ &= 9 + 12\sqrt{5} + 4 \cdot 5 \\ &= 29 + 12\sqrt{5}. \end{aligned}$$

Thus

$$\left((3 + 2\sqrt{5})^2 - m\right)^2 = (29 + 12\sqrt{5} - m)^2.$$

If we choose $m = 29$, then we have

$$\begin{aligned} \left((3 + 2\sqrt{5})^2 - m\right)^2 &= (12\sqrt{5})^2 \\ &= 12^2 \cdot \sqrt{5}^2 \\ &= 12^2 \cdot 5, \end{aligned}$$

which is an integer. Any choice other than $m = 29$ will leave a term involving $\sqrt{5}$ when $(29 + 12\sqrt{5} - m)^2$ is expanded. Thus $m = 29$ is the only solution to this exercise.

5.2 *Logarithms as Inverses of Exponential Functions*

LEARNING OBJECTIVES

By the end of this section you should be able to

- evaluate logarithms in simple cases;
- use logarithms to find inverses of functions involving b^x;
- compute the number of digits in a positive integer from its common logarithm;
- apply the formula for the logarithm of a power;
- model radioactive decay using half-life.

Logarithms Base 2

Consider the exponential function f defined by $f(x) = 2^x$. The domain of f is the set of real numbers, and the range of f is the set of positive numbers. The table in the margin gives the value of 2^x for some choices of x.

x	2^x
-3	$\frac{1}{8}$
-2	$\frac{1}{4}$
-1	$\frac{1}{2}$
0	1
1	2
2	4
3	8

Each time x increases by 1, the value of 2^x doubles; this happens because $2^{x+1} = 2 \cdot 2^x$.

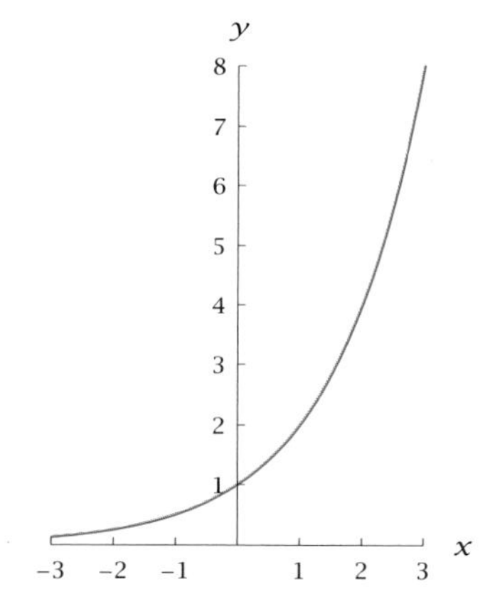

The graph of $y = 2^x$ on the interval $[-3, 3]$.

The graph of $y = 2^x$, as shown above, differs from the graph of $y = x^2$, which is a parabola. Recall that the square root function is the inverse of the function x^2 (with the domain of x^2 restricted to $[0, \infty)$ in order to obtain a one-to-one function). In this section we will define a new function, called the logarithm base 2, that is the inverse of the exponential function 2^x.

Logarithm base 2

For each positive number y the **logarithm** base 2 of y, denoted $\log_2 y$, is defined to be the number x such that $2^x = y$.

For example, $\log_2 8$ equals 3 because $2^3 = 8$. Similarly, $\log_2 \frac{1}{32} = -5$ because $2^{-5} = \frac{1}{32}$. The table in the margin gives the value of $\log_2 y$ for some choices of y. To help your understanding of logarithms, you should verify that each value of $\log_2 y$ in the table is correct.

y	$\log_2 y$
$\frac{1}{8}$	-3
$\frac{1}{4}$	-2
$\frac{1}{2}$	-1
1	0
2	1
4	2
8	3

The definition of $\log_2 y$ as the number such that

$$2^{\log_2 y} = y$$

means that if f is the function defined by $f(x) = 2^x$, then the inverse function of f is given by the formula $f^{-1}(y) = \log_2 y$. Thus the table shown above giving values of $\log_2 y$ is obtained by interchanging the two columns of the earlier table giving the values of 2^x, as always happens with a function and its inverse.

"Algebra as far as the quadratic equation and the use of logarithms are often of value."
—Thomas Jefferson

Because a function and its inverse interchange domains and ranges, the domain of the function f^{-1} defined by $f^{-1}(y) = \log_2 y$ is the set of positive numbers, and the range of this function is the set of real numbers. Expressions such as $\log_2 0$ and $\log_2(-1)$ make no sense because there does not exist a number x such that $2^x = 0$, nor does there exist a number x such that $2^x = -1$.

The figure below shows part of the graph of $\log_2 x$. Because the function $\log_2 x$ is the inverse of the function 2^x, flipping the graph of $y = 2^x$ across the line $y = x$ gives the graph of $y = \log_2 x$.

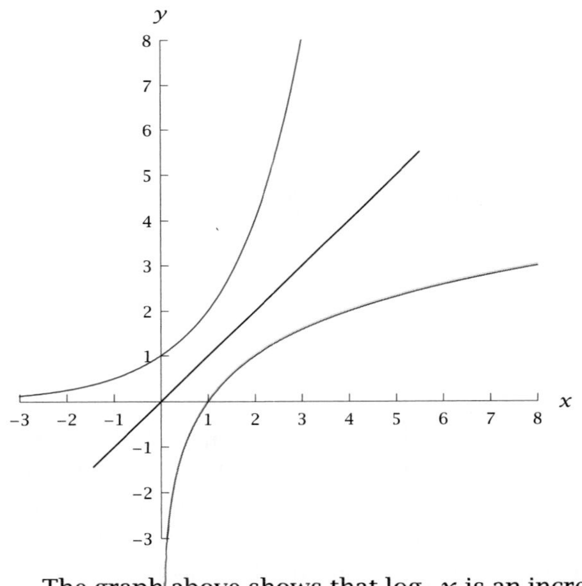

The graph of $\log_2 x$ *(red) on the interval* $[\frac{1}{8}, 8]$ *is obtained by flipping the graph of* 2^x *(blue) on the interval* $[-3, 3]$ *across the line* $y = x$.

The graph above shows that $\log_2 x$ is an increasing function. This behavior is expected, because 2^x is an increasing function and the inverse of an increasing function is increasing.

If x is a real number, then by definition of the logarithm the equation $\log_2 2^x = t$ means that $2^t = 2^x$, which implies that $t = x$. In other words,

$$\log_2 2^x = x.$$

If $f(x) = 2^x$, then $f^{-1}(y) = \log_2 y$ and the equation displayed above could be rewritten as $(f^{-1} \circ f)(x) = x$, which is an equation that always holds for a function and its inverse.

Logarithms with Any Base

We now take up the topic of defining logarithms with bases other than 2. No new ideas are needed for this more general situation—we will simply replace 2 by a positive number $b \neq 1$. Here is the formal definition:

Logarithms

Suppose b and y are positive numbers, with $b \neq 1$.

- The **logarithm** base b of y, denoted $\log_b y$, is defined to be the number x such that $b^x = y$.

Short version:

- $\log_b y = x$ means $b^x = y$.

The base $b = 1$ is excluded because $1^x = 1$ for every real number x.

EXAMPLE 1

(a) Evaluate $\log_{10} 1000$.

(b) Evaluate $\log_7 49$.

(c) Evaluate $\log_3 \frac{1}{81}$.

SOLUTION

(a) Because $10^3 = 1000$, we have $\log_{10} 1000 = 3$.

(b) Because $7^2 = 49$, we have $\log_7 49 = 2$.

(c) Because $3^{-4} = \frac{1}{81}$, we have $\log_3 \frac{1}{81} = -4$.

Two important identities follow immediately from the definition:

The logarithm of 1 and the logarithm of the base

If b is a positive number with $b \neq 1$, then

- $\log_b 1 = 0$;
- $\log_b b = 1$.

Logs have many uses, and the word "log" has more than one meaning.

The first identity holds because $b^0 = 1$; the second holds because $b^1 = b$.

The definition of $\log_b y$ as the number such that $b^{\log_b y} = y$ has the following consequence:

Logarithms as inverse functions

Suppose b is a positive number with $b \neq 1$ and f is the exponential function defined by $f(x) = b^x$. Then the inverse function of f is given by the formula

$$f^{-1}(y) = \log_b y.$$

If $y \leq 0$, then $\log_b y$ is not defined.

Because a function and its inverse interchange domains and ranges, the domain of the function f^{-1} defined by $f^{-1}(y) = \log_b y$ is the set of positive numbers, and the range of this function is the set of real numbers.

EXAMPLE 2 Suppose f is the function defined by $f(x) = 3 \cdot 5^{x-7}$. Find a formula for f^{-1}.

SOLUTION To find a formula for $f^{-1}(y)$, we solve the equation $3 \cdot 5^{x-7} = y$ for x. Dividing by 3, we have $5^{x-7} = \frac{y}{3}$. Thus $x - 7 = \log_5 \frac{y}{3}$, which implies that $x = 7 + \log_5 \frac{y}{3}$. Hence

$$f^{-1}(y) = 7 + \log_5 \tfrac{y}{3}.$$

Most applications of logarithms involve bases bigger than 1.

Because the function $\log_b x$ is the inverse of the function b^x, flipping the graph of $y = b^x$ across the line $y = x$ gives the graph of $y = \log_b x$. As we will see, if $b > 1$ then $\log_b x$ is an increasing function (because b^x is an increasing function) and the shape of the graph of $\log_b x$ is similar to the shape of the graph of $\log_2 x$ obtained earlier.

The definition of logarithm implies the two equations displayed below. Be sure that you are comfortable with these equations and understand why they hold. Note that if f is defined by $f(x) = b^x$, then $f^{-1}(y) = \log_b y$. The equations below could then be written in the form $(f \circ f^{-1})(y) = y$ and $(f^{-1} \circ f)(x) = x$, which are equations that always hold for a function and its inverse.

Inverse properties of logarithms

If b and y are positive numbers, with $b \neq 1$, and x is a real number, then

$$b^{\log_b y} = y \quad \text{and} \quad \log_b b^x = x.$$

In applications of logarithms, the most commonly used values for the base are 10, 2, and the number e (which we will discuss in Chapter 6). The use of a logarithm with base 10 is so frequent that it gets a special name:

John Napier, the Scottish mathematician who invented logarithms around 1614.

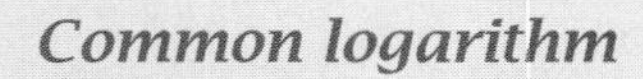

Common logarithm

- The logarithm base 10 is called the **common logarithm**.
- To simplify notation, sometimes logarithms base 10 are written without the base. If no base is displayed, then the base is assumed to be 10. In other words,

$$\log y = \log_{10} y.$$

Thus, for example, $\log 10000 = 4$ (because $10^4 = 10000$) and $\log \frac{1}{100} = -2$ (because $10^{-2} = \frac{1}{100}$). If your calculator has a button labeled "log", then it will compute the logarithm base 10, which is often just called the logarithm.

Common Logarithms and the Number of Digits

Note that 10^1 is a two-digit number, 10^2 is a three-digit number, 10^3 is a four-digit number, and so on. In general, 10^{n-1} is an n-digit number. Because 10^n, which consists of 1 followed by n zeros, is the smallest positive integer with $n + 1$ digits, we see that every integer in the interval $[10^{n-1}, 10^n)$ has n digits. Because $\log 10^{n-1} = n - 1$ and $\log 10^n = n$, this implies that an n-digit positive integer has a logarithm in the interval $[n - 1, n)$.

Digits and logarithms

The logarithm of an n-digit positive integer is in the interval $[n - 1, n)$.

Alternative statement: If M is a positive integer with n digits, then $n - 1 \le \log M < n$.

The conclusion above is often useful in making estimates. For example, without using a calculator we can see that the number 123456789, which has nine digits, has a logarithm between 8 and 9 (the actual value is about 8.09).

The next example shows how to use the conclusion above to determine the number of digits in a number from its logarithm.

EXAMPLE 3

Suppose M is a positive integer such that $\log M \approx 73.1$. How many digits does M have?

SOLUTION Because 73.1 is in the interval $[73, 74)$, we can conclude that M is a 74-digit number.

Always round up the logarithm of a number to determine the number of digits. Here $\log M \approx 73.1$ is rounded up to show that M has 74 digits.

Logarithm of a Power

We will use the formula $(b^r)^t = b^{tr}$ to derive a formula for the logarithm of a power. First we look at an example.

To motivate the formula for the logarithm of a power, we note that

$$\log (10^3)^4 = \log 10^{12} = 12 \quad \text{and} \quad \log 10^3 = 3.$$

An expression such as $\log 10^{12}$ should be interpreted to mean $\log(10^{12})$, not $(\log 10)^{12}$.

Putting these equations together, we see that

$$\log (10^3)^4 = 4 \log 10^3.$$

More generally, logarithms convert powers to products, as we will now show.

Suppose b and y are positive numbers, with $b \neq 1$, and t is a real number. Then

$$\begin{aligned} \log_b y^t &= \log_b (b^{\log_b y})^t \\ &= \log_b b^{t \log_b y} \\ &= t \log_b y. \end{aligned}$$

In other words, we have the following formula for the logarithm of a power:

Logarithm of a power

If b and y are positive numbers, with $b \neq 1$, and t is a real number, then

$$\log_b y^t = t \log_b y.$$

The next example shows a nice application of the formula above.

EXAMPLE 4

How many digits does 3^{5000} have?

Your calculator cannot evaluate 3^{5000}. Thus the formula for the logarithm of a power is needed even though a calculator is being used.

SOLUTION We can answer this question by evaluating the common logarithm of 3^{5000}. Using the formula for the logarithm of a power and a calculator, we see that

$$\log 3^{5000} = 5000 \log 3 \approx 2385.61.$$

Thus 3^{5000} has 2386 digits.

In the era before calculators and computers existed, books of common logarithm tables were frequently used to compute powers of numbers. As an example of how this worked, consider how these books of logarithms would have been used to evaluate $1.7^{3.7}$. The key to performing this calculation is the formula

$$\log 1.7^{3.7} = 3.7 \log 1.7.$$

With the advent of calculators and computers, books of logarithms have essentially disappeared. However, your calculator is using logarithms and the formula $\log_b y^t = t \log_b y$ when you ask it to evaluate an expression such as $1.7^{3.7}$.

Let's assume that we have a book that gives the logarithms of the numbers from 1 to 10 in increments of 0.001, meaning that the book gives the logarithms of 1.001, 1.002, 1.003, and so on.

The idea is first to compute the right side of the equation above. To do that, we would look in the book of logarithms, getting $\log 1.7 \approx 0.230449$. Multiplying the last number by 3.7, we would conclude that the right side of the equation above is approximately 0.852661. Thus, according to the equation above, we have

$$\log 1.7^{3.7} \approx 0.852661.$$

Hence we can evaluate $1.7^{3.7}$ by finding a number whose logarithm equals 0.852661. To do this, we would look through our book of logarithms and find that the closest match is provided by the entry showing that $\log 7.123 \approx 0.852663$. Thus $1.7^{3.7} \approx 7.123$.

Although nowadays logarithms rarely are used directly by humans for computations such as evaluating $1.7^{3.7}$, logarithms are used by your calculator for such computations. Logarithms also have important uses in calculus and several other branches of mathematics. Furthermore, logarithms have several practical uses.

Radioactive Decay and Half-Life

Scientists have observed that starting with a large sample of radon atoms, after 92 hours one-half of the radon atoms will decay into polonium. After another 92 hours, one-half of the remaining radon atoms will also decay into polonium. In other words, after 184 hours, only one-fourth of the original radon atoms will be left. After another 92 hours, one-half of those remaining one-fourth of the original atoms will decay into polonium, leaving only one-eighth of the original radon atoms after 276 hours.

Pioneering work on radioactive decay was done by Marie Curie, the only person ever to win Nobel Prizes in both physics (1903) and chemistry (1911).

After t hours, the number of radon atoms will be reduced by half $t/92$ times. Thus after t hours, the number of radon atoms left will equal the original number of radon atoms divided by $2^{t/92}$. Here t need not be an integer multiple of 92. For example, our formula predicts that after five hours the original number of radon atoms will be divided by $2^{5/92}$. A calculator shows that $\frac{1}{2^{5/92}} \approx 0.963$. Indeed, observations verify that after five hours a sample of radon will contain 96.3% of the original number of radon atoms.

Because half of the atoms in any sample of radon will decay to polonium in 92 hours, we say that radon has a **half-life** of 92 hours. Some radon atoms exist for much less than 92 hours, and some radon atoms exist for much longer than 92 hours.

The half-life of any radioactive isotope is the length of time it takes for half the atoms in a large sample of the isotope to decay. The table below gives the approximate half-life for several radioactive isotopes (each isotope number shown in the table gives the total number of protons and neutrons in the variety of atom under consideration).

isotope	*half-life*
neon-18	2 seconds
nitrogen-13	10 minutes
radon-222	92 hours
polonium-210	138 days
cesium-137	30 years
carbon-14	5730 years
plutonium-239	24,110 years
uranium-238	4.5 billion years

Half-life of some radioactive isotopes.

Some of the isotopes in this table are human creations that do not exist in nature. For example, the nitrogen on Earth is almost entirely nitrogen-14 (7 protons and 7 neutrons), which is not radioactive and does not decay. The nitrogen-13 listed here has 7 protons and 6 neutrons; it can be created in a laboratory, but it is radioactive and half of it will decay within 10 minutes.

If a radioactive isotope has a half-life of h time units (here the time units might be seconds, minutes, hours, days, years, or whatever unit is appropriate), then after t time units the number of atoms of this isotope is reduced by half t/h times. Thus after t time units, the remaining number of atoms of the radioactive isotope will equal the original number of atoms divided by $2^{t/h}$. Because $\frac{1}{2^{t/h}} = 2^{-t/h}$, we have the following result:

Radioactive decay

If a radioactive isotope has half-life h, then the function modeling the number of atoms in a sample of this isotope is

$$a(t) = a_0 \cdot 2^{-t/h},$$

where a_0 is the number of atoms of the isotope in the sample at time 0.

The radioactive decay of carbon-14 leads to a clever way of determining the age of fossils, wood, and other remnants of plants and animals. Carbon-12, by far the most common form of carbon on Earth, is not radioactive and does not decay. Radioactive carbon-14 is produced regularly as cosmic rays hit the upper atmosphere. Radioactive carbon-14 then filters down to the lower atmosphere, where it is absorbed by all living organisms as part of the food or photosynthesis cycle. Carbon-14 accounts for about 10^{-10} percent of the carbon atoms in a living organism.

The 1960 Nobel Prize in Chemistry was awarded to Willard Libby for his invention of this carbon-14 dating method.

When an organism dies, it stops absorbing new carbon because it is no longer eating or engaging in photosynthesis. Thus no new carbon-14 is absorbed. The radioactive carbon-14 in the organism then decays, with half of it gone after 5730 years, as shown in the table above. By measuring the amount of carbon-14 as a percentage of the total amount of carbon in the remains of an organism, we can then determine how long ago it died.

EXAMPLE 5

Suppose a cat skeleton found in an old well has a ratio of carbon-14 to carbon-12 that is 61% of the corresponding ratio for living organisms. Approximately how long ago did the cat die?

The author's first cat.

SOLUTION If t denotes the number of years since the cat died, then

$$0.61 = 2^{-t/5730}.$$

To solve this equation for t, take the logarithm of both sides, getting

$$\log 0.61 = -\frac{t}{5730}\log 2.$$

Solve this equation for t, getting

$$t = -5730\,\frac{\log 0.61}{\log 2} \approx 4086.$$

Because there will be some errors in measuring the percentage of carbon-14 in the cat skeleton, we should not produce such a precise-looking estimate. Thus we might estimate that the skeleton is about 4100 years old. Or if we want to indicate even less precision, we might say that the cat died about four thousand years ago.

EXERCISES

For Exercises 1–16, evaluate the indicated expression. Do not use a calculator for these exercises.

1. $\log_2 64$
2. $\log_2 1024$
3. $\log_2 \frac{1}{128}$
4. $\log_2 \frac{1}{256}$
5. $\log_4 2$
6. $\log_8 2$
7. $\log_4 8$
8. $\log_8 128$
9. $\log 10000$
10. $\log \frac{1}{1000}$
11. $\log \sqrt{1000}$
12. $\log \frac{1}{\sqrt{10000}}$
13. $\log_2 8^{3.1}$
14. $\log_8 2^{6.3}$
15. $\log_{16} 32$
16. $\log_{27} 81$
17. Find a number y such that $\log_2 y = 7$.
18. Find a number t such that $\log_2 t = 8$.
19. Find a number y such that $\log_2 y = -5$.
20. Find a number t such that $\log_2 t = -9$.

For Exercises 21–28, find a number b such that the indicated equality holds.

21. $\log_b 64 = 1$
22. $\log_b 64 = 2$
23. $\log_b 64 = 3$
24. $\log_b 64 = 6$
25. $\log_b 64 = 12$
26. $\log_b 64 = 18$
27. $\log_b 64 = \frac{3}{2}$
28. $\log_b 64 = \frac{6}{5}$

For Exercises 33–44, find all numbers x such that the indicated equation holds.

29. $\log |x| = 2$
30. $\log |x| = 3$
31. $|\log x| = 2$
32. $|\log x| = 3$
33. $\log_3(5x + 1) = 2$
34. $\log_4(3x + 1) = -2$
35. $13 = 10^{2x}$
36. $59 = 10^{3x}$
37. $\dfrac{10^x + 1}{10^x + 2} = 0.8$
38. $\dfrac{10^x + 3.8}{10^x + 3} = 1.1$
39. $10^{2x} + 10^x = 12$
40. $10^{2x} - 3 \cdot 10^x = 18$
41. $3^x = 8$
42. $7^x = 5$
43. $6^{\sqrt{x}} = 2$
44. $5^{\sqrt{x}} = 9$
45. For $x = 5$ and $y = 2$, evaluate each of the following:
    (a) $\log x^y$ (b) $(\log x)^y$
    [*This exercise and the next one emphasize that* $\log x^y$ *does not equal* $(\log x)^y$.]
46. For $x = 2$ and $y = 3$, evaluate each of the following:
    (a) $\log x^y$ (b) $(\log x)^y$
47. Suppose y is such that $\log_2 y = 17.67$. Evaluate $\log_2 y^{100}$.
48. Suppose x is such that $\log_6 x = 23.41$. Evaluate $\log_6 x^{10}$.

For Exercises 49–66, find a formula for the inverse function f^{-1} of the indicated function f.

49. $f(x) = 3^x$
50. $f(x) = 4.7^x$
51. $f(x) = 2^{x-5}$
52. $f(x) = 9^{x+6}$
53. $f(x) = 6^x + 7$
54. $f(x) = 5^x - 3$
55. $f(x) = 4 \cdot 5^x$
56. $f(x) = 8 \cdot 7^x$
57. $f(x) = 2 \cdot 9^x + 1$
58. $f(x) = 3 \cdot 4^x - 5$
59. $f(x) = \log_8 x$
60. $f(x) = \log_3 x$
61. $f(x) = \log_4(3x + 1)$
62. $f(x) = \log_7(2x - 9)$
63. $f(x) = 5 + 3\log_6(2x + 1)$
64. $f(x) = 8 + 9\log_2(4x - 7)$
65. $f(x) = \log_x 13$
66. $f(x) = \log_{5x} 6$

For Exercises 67–74, find a formula for $(f \circ g)(x)$ assuming that f and g are the indicated functions.

67. $f(x) = \log_6 x$ and $g(x) = 6^{3x}$
68. $f(x) = \log_5 x$ and $g(x) = 5^{3+2x}$
69. $f(x) = 6^{3x}$ and $g(x) = \log_6 x$
70. $f(x) = 5^{3+2x}$ and $g(x) = \log_5 x$
71. $f(x) = \log_x 4$ and $g(x) = 10^x$
72. $f(x) = \log_{2x} 7$ and $g(x) = 10^x$
73. $f(x) = \log_x 4$ and $g(x) = 100^x$
74. $f(x) = \log_{2x} 7$ and $g(x) = 100^x$
75. Find a number n such that $\log_3(\log_5 n) = 1$.
76. Find a number n such that $\log_3(\log_2 n) = 2$.
77. Find a number m such that $\log_7(\log_8 m) = 2$.
78. Find a number m such that $\log_5(\log_6 m) = 3$.

79. Suppose N is a positive integer such that $\log N \approx 35.4$. How many digits does N have?

80. Suppose k is a positive integer such that $\log k \approx 83.2$. How many digits does k have?

81. Suppose m is a positive integer such that $\log m \approx 13.2$. How many digits does m^3 have?

82. Suppose M is a positive integer such that $\log M \approx 50.3$. How many digits does M^4 have?

83. How many digits does 7^{4000} have?

84. How many digits does 8^{4444} have?

85. Find an integer k such that 18^k has 357 digits.

86. Find an integer n such that 22^n has 222 digits.

87. Find an integer m such that m^{1234} has 1991 digits.

88. Find an integer N such that N^{4321} has 6041 digits.

89. Find the smallest integer n such that $7^n > 10^{100}$.

90. Find the smallest integer k such that $9^k > 10^{1000}$.

91. Find the smallest integer M such that $5^{1/M} < 1.01$.

92. Find the smallest integer m such that $8^{1/m} < 1.001$.

93. Suppose $\log_8(\log_7 m) = 5$. How many digits does m have?

94. Suppose $\log_5(\log_9 m) = 6$. How many digits does m have?

95. At the end of 2004, the largest known prime number was $2^{24036583} - 1$. How many digits does this prime number have?

[*A* **prime number** *is an integer greater than* 1 *that has no divisors other than itself and* 1.]

96. At the end of 2005, the largest known prime number was $2^{30402457} - 1$. How many digits does this prime number have?

97. About how many hours will it take for a sample of radon-222 to have only one-eighth as much radon-222 as the original sample?

98. About how many minutes will it take for a sample of nitrogen-13 to have only one sixty-fourth as much nitrogen-13 as the original sample?

99. About how many years will it take for a sample of cesium-137 to have only two-thirds as much cesium-137 as the original sample?

100. About how many years will it take for a sample of plutonium-239 to have only 1% as much plutonium-239 as the original sample?

101. Suppose a radioactive isotope is such that one-fifth of the atoms in a sample decay after three years. Find the half-life of this isotope.

102. Suppose a radioactive isotope is such that five-sixths of the atoms in a sample decay after four days. Find the half-life of this isotope.

103. Suppose the ratio of carbon-14 to carbon-12 in a mummified cat is 64% of the corresponding ratio for living organisms. About how long ago did the cat die?

104. Suppose the ratio of carbon-14 to carbon-12 in a fossilized wooden tool is 20% of the corresponding ratio for living organisms. About how old is the wooden tool?

PROBLEMS

105. Explain why $\log_3 100$ is between 4 and 5.

106. Explain why $\log_{40} 3$ is between $\frac{1}{4}$ and $\frac{1}{3}$.

107. Show that $\log_2 3$ is an irrational number. [*Hint:* Use proof by contradiction: Assume that $\log_2 3$ is equal to a rational number $\frac{m}{n}$; write out what this means, and think about even and odd numbers.]

108. Show that $\log 2$ is irrational.

109. Explain why logarithms with base 0 are not defined.

110. Explain why logarithms with a negative base are not defined.

111. Explain why $\log_5 \sqrt{5} = \frac{1}{2}$.

112. Explain why there does not exist an integer m such that 67^m has 9236 digits.

113. Do a web search to find the largest currently known prime number. Then calculate the number of digits in this number.
[*The discovery of a new largest known prime number usually gets some newspaper coverage, including a statement of the number of digits. Thus you can probably find on the web the number of digits in the largest currently known prime number; you are asked here to do the calculation to verify that the reported number of digits is correct.*]

114. Explain why the graph of $y = \log(x^4)$ (with $x > 0$) can be obtained by vertically stretching the graph of $y = \log x$ by a factor of 4.

WORKED-OUT SOLUTIONS *to Odd-numbered Exercises*

For Exercises 1–16, evaluate the indicated expression. Do not use a calculator for these exercises.

1. $\log_2 64$

SOLUTION If we let $x = \log_2 64$, then x is the number such that

$$64 = 2^x.$$

Because $64 = 2^6$, we see that $x = 6$. Thus $\log_2 64 = 6$.

3. $\log_2 \frac{1}{128}$

SOLUTION If we let $x = \log_2 \frac{1}{128}$, then x is the number such that

$$\tfrac{1}{128} = 2^x.$$

Because $\frac{1}{128} = \frac{1}{2^7} = 2^{-7}$, we see that $x = -7$. Thus $\log_2 \frac{1}{128} = -7$.

5. $\log_4 2$

SOLUTION Because $2 = 4^{1/2}$, we have $\log_4 2 = \frac{1}{2}$.

7. $\log_4 8$

SOLUTION Because $8 = 2 \cdot 4 = 4^{1/2} \cdot 4 = 4^{3/2}$, we have $\log_4 8 = \frac{3}{2}$.

9. $\log 10000$

SOLUTION

$$\begin{aligned} \log 10000 &= \log 10^4 \\ &= 4 \end{aligned}$$

11. $\log \sqrt{1000}$

SOLUTION

$$\begin{aligned} \log \sqrt{1000} &= \log 1000^{1/2} \\ &= \log (10^3)^{1/2} \\ &= \log 10^{3/2} \\ &= \tfrac{3}{2} \end{aligned}$$

13. $\log_2 8^{3.1}$

SOLUTION

$$\begin{aligned} \log_2 8^{3.1} &= \log_2 (2^3)^{3.1} \\ &= \log_2 2^{9.3} \\ &= 9.3 \end{aligned}$$

15. $\log_{16} 32$

SOLUTION

$$\begin{aligned} \log_{16} 32 &= \log_{16} 2^5 \\ &= \log_{16} (2^4)^{5/4} \\ &= \log_{16} 16^{5/4} \\ &= \tfrac{5}{4} \end{aligned}$$

17. Find a number y such that $\log_2 y = 7$.

SOLUTION The equation $\log_2 y = 7$ implies that

$$y = 2^7 = 128.$$

19. Find a number y such that $\log_2 y = -5$.

SOLUTION The equation $\log_2 y = -5$ implies that

$$y = 2^{-5} = \tfrac{1}{32}.$$

For Exercises 21–28, find a number b such that the indicated equality holds.

21. $\log_b 64 = 1$

SOLUTION The equation $\log_b 64 = 1$ implies that

$$b^1 = 64.$$

Thus $b = 64$.

23. $\log_b 64 = 3$

SOLUTION The equation $\log_b 64 = 3$ implies that

$$b^3 = 64.$$

Because $4^3 = 64$, this implies that $b = 4$.

25. $\log_b 64 = 12$

SOLUTION The equation $\log_b 64 = 12$ implies that

$$b^{12} = 64.$$

Thus

$$\begin{aligned} b &= 64^{1/12} \\ &= (2^6)^{1/12} \\ &= 2^{6/12} \\ &= 2^{1/2} \\ &= \sqrt{2}. \end{aligned}$$

27. $\log_b 64 = \frac{3}{2}$

SOLUTION The equation $\log_b 64 = \frac{3}{2}$ implies that

$$b^{3/2} = 64.$$

Raising both sides of this equation to the $2/3$ power, we get

$$\begin{aligned} b &= 64^{2/3} \\ &= (2^6)^{2/3} \\ &= 2^4 \\ &= 16. \end{aligned}$$

For Exercises 33–44, find all numbers x such that the indicated equation holds.

29. $\log |x| = 2$

SOLUTION The equation $\log |x| = 2$ is equivalent to the equation

$$|x| = 10^2 = 100.$$

Thus the two values of x satisfying this equation are $x = 100$ and $x = -100$.

31. $|\log x| = 2$

SOLUTION The equation $|\log x| = 2$ means that $\log x = 2$ or $\log x = -2$, which means that $x = 10^2 = 100$ or $x = 10^{-2} = \frac{1}{100}$.

33. $\log_3(5x + 1) = 2$

SOLUTION The equation $\log_3(5x + 1) = 2$ implies that $5x + 1 = 3^2 = 9$. Thus $5x = 8$, which implies that $x = \frac{8}{5}$.

35. $13 = 10^{2x}$

SOLUTION The equation $13 = 10^{2x}$ implies that $2x = \log 13$. Thus $x = \frac{\log 13}{2}$, which is approximately equal to 0.557.

37. $\dfrac{10^x + 1}{10^x + 2} = 0.8$

SOLUTION Multiplying both sides of the equation above by $10^x + 2$, we get

$$10^x + 1 = 0.8 \cdot 10^x + 1.6.$$

Solving this equation for 10^x gives $10^x = 3$, which means that $x = \log 3 \approx 0.477121$.

39. $10^{2x} + 10^x = 12$

SOLUTION Note that $10^{2x} = (10^x)^2$. This suggests that we let $y = 10^x$. Then the equation above can be rewritten as

$$y^2 + y - 12 = 0.$$

The solutions to this equation (which can be found either by using the quadratic formula or by factoring) are $y = -4$ and $y = 3$. Thus $10^x = -4$ or $10^x = 3$. However, there is no real number x such that $10^x = -4$ (because 10^x is positive for every real number x), and thus we must have $10^x = 3$. Thus $x = \log 3 \approx 0.477121$.

41. $3^x = 8$

SOLUTION The equation $3^x = 8$ implies that

$$x = \log_3 8 = \frac{\log 8}{\log 3} \approx 1.89279.$$

43. $6^{\sqrt{x}} = 2$

SOLUTION The equation $6^{\sqrt{x}} = 2$ implies that

$$\sqrt{x} = \log_6 2 = \frac{\log 2}{\log 6}.$$

Thus

$$x = \left(\frac{\log 2}{\log 6}\right)^2 \approx 0.149655.$$

45. For $x = 5$ and $y = 2$, evaluate each of the following:

(a) $\log x^y$ (b) $(\log x)^y$

SOLUTION

(a) $\log 5^2 = \log 25 \approx 1.39794$

(b) $(\log 5)^2 \approx (0.69897)^2 \approx 0.48856$

47. Suppose y is such that $\log_2 y = 17.67$. Evaluate $\log_2 y^{100}$.

SOLUTION
$$\begin{aligned} \log_2 y^{100} &= 100 \log_2 y \\ &= 100 \cdot 17.67 \\ &= 1767 \end{aligned}$$

For Exercises 49–66, find a formula for the inverse function f^{-1} of the indicated function f.

49. $f(x) = 3^x$

SOLUTION By definition of the logarithm, the inverse of f is the function f^{-1} defined by

$$f^{-1}(y) = \log_3 y.$$

51. $f(x) = 2^{x-5}$

SOLUTION To find a formula for $f^{-1}(y)$, we solve the equation $2^{x-5} = y$ for x. This equation means that $x - 5 = \log_2 y$. Thus $x = 5 + \log_2 y$. Hence

$$f^{-1}(y) = 5 + \log_2 y.$$

53. $f(x) = 6^x + 7$

SOLUTION To find a formula for $f^{-1}(y)$, we solve the equation $6^x + 7 = y$ for x. Subtract 7 from both sides, getting $6^x = y - 7$. This equation means that $x = \log_6(y - 7)$. Hence

$$f^{-1}(y) = \log_6(y - 7).$$

55. $f(x) = 4 \cdot 5^x$

SOLUTION To find a formula for $f^{-1}(y)$, we solve the equation $4 \cdot 5^x = y$ for x. Divide both sides by 4, getting $5^x = \frac{y}{4}$. This equation means that $x = \log_5 \frac{y}{4}$. Hence

$$f^{-1}(y) = \log_5 \tfrac{y}{4}.$$

57. $f(x) = 2 \cdot 9^x + 1$

SOLUTION To find a formula for $f^{-1}(y)$, we solve the equation $2 \cdot 9^x + 1 = y$ for x. Subtract 1 from both sides, then divide both sides by 2, getting $9^x = \frac{y-1}{2}$. This equation means that $x = \log_9 \frac{y-1}{2}$. Hence

$$f^{-1}(y) = \log_9 \tfrac{y-1}{2}.$$

59. $f(x) = \log_8 x$

SOLUTION By the definition of the logarithm, the inverse of f is the function f^{-1} defined by

$$f^{-1}(y) = 8^y.$$

61. $f(x) = \log_4(3x + 1)$

SOLUTION To find a formula for $f^{-1}(y)$, we solve the equation

$$\log_4(3x + 1) = y$$

for x. This equation means that $3x + 1 = 4^y$. Solving for x, we get $x = \frac{4^y - 1}{3}$. Hence

$$f^{-1}(y) = \frac{4^y - 1}{3}.$$

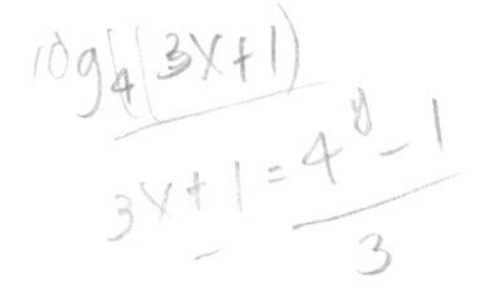

63. $f(x) = 5 + 3\log_6(2x + 1)$

SOLUTION To find a formula for $f^{-1}(y)$, we solve the equation

$$5 + 3\log_6(2x + 1) = y$$

for x. Subtracting 5 from both sides and then dividing by 3 gives

$$\log_6(2x + 1) = \frac{y - 5}{3}.$$

This equation means that $2x + 1 = 6^{(y-5)/3}$. Solving for x, we get $x = \frac{6^{(y-5)/3} - 1}{2}$. Hence

$$f^{-1}(y) = \frac{6^{(y-5)/3} - 1}{2}.$$

65. $f(x) = \log_x 13$

SOLUTION To find a formula for $f^{-1}(y)$, we solve the equation $\log_x 13 = y$ for x. This equation means that $x^y = 13$. Raising both sides to the power $\frac{1}{y}$, we get $x = 13^{1/y}$. Hence

$$f^{-1}(y) = 13^{1/y}.$$

For Exercises 67–74, find a formula for $(f \circ g)(x)$ assuming that f and g are the indicated functions.

67. $f(x) = \log_6 x$ and $g(x) = 6^{3x}$

SOLUTION

$$(f \circ g)(x) = f(g(x)) = f(6^{3x}) = \log_6 6^{3x} = 3x$$

69. $f(x) = 6^{3x}$ and $g(x) = \log_6 x$

SOLUTION

$$\begin{aligned}(f \circ g)(x) &= f(g(x)) = f(\log_6 x)\\ &= 6^{3\log_6 x} = (6^{\log_6 x})^3 = x^3\end{aligned}$$

71. $f(x) = \log_x 4$ and $g(x) = 10^x$

SOLUTION

$$\begin{aligned}(f \circ g)(x) = f(g(x)) &= f(10^x)\\ &= \log_{10^x} 4\\ &= \frac{\log 4}{\log 10^x}\\ &= \frac{\log 4}{x}\end{aligned}$$

73. $f(x) = \log_x 4$ and $g(x) = 100^x$

SOLUTION

$$\begin{aligned}(f \circ g)(x) = f(g(x)) &= f(100^x)\\ &= \log_{100^x} 4\\ &= \frac{\log 4}{\log 100^x}\\ &= \frac{\log 4}{\log 10^{2x}}\\ &= \frac{\log 4}{2x}\end{aligned}$$

75. Find a number n such that $\log_3(\log_5 n) = 1$.

SOLUTION The equation $\log_3(\log_5 n) = 1$ implies that $\log_5 n = 3$, which implies that $n = 5^3 = 125$.

77. Find a number m such that $\log_7(\log_8 m) = 2$.

SOLUTION The equation $\log_7(\log_8 m) = 2$ implies that

$$\log_8 m = 7^2 = 49.$$

The equation above now implies that

$$m = 8^{49}.$$

79. Suppose N is a positive integer such that $\log N \approx 35.4$. How many digits does N have?

SOLUTION Because 35.4 is in the interval $[35, 36)$, we can conclude that N is a 36-digit number.

81. Suppose m is a positive integer such that $\log m \approx 13.2$. How many digits does m^3 have?

SOLUTION Note that

$$\log(m^3) = 3\log m \approx 3 \times 13.2 = 39.6.$$

Because 39.6 is in the interval $[39, 40)$, we can conclude that m^3 is a 40-digit number.

83. How many digits does 7^{4000} have?

SOLUTION Using the formula for the logarithm of a power and a calculator, we have

$$\log 7^{4000} = 4000\log 7 \approx 3380.39.$$

Thus 7^{4000} has 3381 digits.

85. Find an integer k such that 18^k has 357 digits.

SOLUTION We want to find an integer k such that

$$356 \le \log 18^k < 357.$$

Using the formula for the logarithm of a power, we can rewrite the inequalities above as

$$356 \le k \log 18 < 357.$$

Dividing by log 18 gives

$$\frac{356}{\log 18} \le k < \frac{357}{\log 18}.$$

Using a calculator, we see that $\frac{356}{\log 18} \approx 283.6$ and $\frac{357}{\log 18} \approx 284.4$. Thus the only possible choice is to take $k = 284$.

Again using a calculator, we see that

$$\log 18^{284} = 284 \log 18 \approx 356.5.$$

Thus 18^{284} indeed has 357 digits.

87. Find an integer m such that m^{1234} has 1991 digits.

SOLUTION We want to find an integer m such that

$$1990 \le \log m^{1234} < 1991.$$

Using the formula for the logarithm of a power, we can rewrite the inequalities above as

$$1990 \le 1234 \log m < 1991.$$

Dividing by 1234 gives

$$\frac{1990}{1234} \le \log m < \frac{1991}{1234}.$$

Thus

$$10^{1990/1234} \le m < 10^{1991/1234}.$$

Using a calculator, we see that $10^{1990/1234} \approx 40.99$ and $10^{1991/1234} \approx 41.06$. Thus the only possible choice is to take $m = 41$.

Again using a calculator, we see that

$$\log 41^{1234} = 1234 \log 41 \approx 1990.18.$$

Thus 41^{1234} indeed has 1991 digits.

89. Find the smallest integer n such that $7^n > 10^{100}$.

SOLUTION Suppose $7^n > 10^{100}$. Taking the common logarithm of both sides, we have

$$\log 7^n > \log 10^{100},$$

which can be rewritten as

$$n \log 7 > 100.$$

This implies that

$$n > \frac{100}{\log 7} \approx 118.33.$$

The smallest integer that is bigger than 118.33 is 119. Thus we take $n = 119$.

91. Find the smallest integer M such that $5^{1/M} < 1.01$.

SOLUTION Suppose $5^{1/M} < 1.01$. Taking the common logarithm of both sides, we have

$$\log 5^{1/M} < \log 1.01,$$

which can be rewritten as

$$\frac{\log 5}{M} < \log 1.01.$$

This implies that

$$M > \frac{\log 5}{\log 1.01} \approx 161.7.$$

The smallest integer that is bigger than 161.7 is 162. Thus we take $M = 162$.

93. Suppose $\log_8(\log_7 m) = 5$. How many digits does m have?

SOLUTION The equation $\log_8(\log_7 m) = 5$ implies that

$$\log_7 m = 8^5 = 32768.$$

The equation above now implies that

$$m = 7^{32768}.$$

To compute the number of digits that m has, note that

$$\log m = \log 7^{32768} = 32768 \log 7 \approx 27692.2.$$

Thus m has 27693 digits.

95. At the end of 2004, the largest known prime number was $2^{24036583} - 1$. How many digits does this prime number have?

SOLUTION To calculate the number of digits in $2^{24036583} - 1$, we need to evaluate $\log(2^{24036583} - 1)$. However, $2^{24036583} - 1$ is too large to evaluate directly on a calculator, and no formula exists for the logarithm of the difference of two numbers.

The trick here is to note that $2^{24036583}$ and $2^{24036583} - 1$ have the same number of digits, as we will now see. Although it is possible for a number and the number minus 1 to have a different number of digits (for example, 100 and 99 do not have the same number of digits), this happens only if the larger of the two numbers consists of 1 followed by a bunch of 0's and the smaller of the two numbers consists of all 9's. Here are three different ways to see that this situation does not apply to $2^{24036583}$ and $2^{24036583} - 1$ (pick whichever explanation seems easiest to you): (a) $2^{24036583}$ cannot end in a 0 because all positive integer powers of 2 end in either 2, 4, 6, or 8; (b) $2^{24036583}$ cannot end in a 0 because then it would be divisible by 5, but $2^{24036583}$ is divisible only by integer powers of 2; (c) $2^{24036583} - 1$ cannot consist of all 9's because then it would be divisible by 9, which is not possible for a prime number.

Now that we know that $2^{24036583}$ and $2^{24036583} - 1$ have the same number of digits, we can calculate the number of digits by taking the logarithm of $2^{24036583}$ and using the formula for the logarithm of a power. We have

$$\log 2^{24036583} = 24036583 \log 2 \approx 7235732.5.$$

Thus $2^{24036583}$ has 7235733 digits; hence $2^{24036583} - 1$ also has 7235733 digits.

97. About how many hours will it take for a sample of radon-222 to have only one-eighth as much radon-222 as the original sample?

SOLUTION The half-life of radon-222 is about 92 hours, as shown in the chart in this section. To reduce the number of radon-222 atoms to one-eighth the original number, we need 3 half-lives (because $2^3 = 8$). Thus it will take 276 hours (because $92 \times 3 = 276$) to have only one-eighth as much radon-222 as the original sample.

99. About how many years will it take for a sample of cesium-137 to have only two-thirds as much cesium-137 as the original sample?

SOLUTION The half-life of cesium-137 is about 30 years, as shown in the chart in this section. Thus if we start with a atoms of cesium-137 at time 0, then after t years there will be

$$a \cdot 2^{-t/30}$$

atoms left. We want this to equal $\frac{2}{3}a$. Thus we must solve the equation

$$a \cdot 2^{-t/30} = \tfrac{2}{3}a.$$

To solve this equation for t, divide both sides by a and then take the logarithm of both sides, getting

$$-\tfrac{t}{30} \log 2 = \log \tfrac{2}{3}.$$

Now multiply both sides by -1, replace $-\log \frac{2}{3}$ by $\log \frac{3}{2}$, and then solve for t, getting

$$t = 30 \frac{\log \frac{3}{2}}{\log 2} \approx 17.5.$$

Thus two-thirds of the original sample will be left after approximately 17.5 years.

101. Suppose a radioactive isotope is such that one-fifth of the atoms in a sample decay after three years. Find the half-life of this isotope.

SOLUTION Let h denote the half-life of this isotope, measured in years. If we start with a sample of a atoms of this isotope, then after 3 years there will be

$$a \cdot 2^{-3/h}$$

atoms left. We want this to equal $\frac{4}{5}a$. Thus we must solve the equation

$$a \cdot 2^{-3/h} = \tfrac{4}{5}a.$$

To solve this equation for h, divide both sides by a and then take the logarithm of both sides, getting

$$-\tfrac{3}{h} \log 2 = \log \tfrac{4}{5}.$$

Now multiply both sides by -1, replace $-\log \frac{4}{5}$ by $\log \frac{5}{4}$, and then solve for h, getting

$$h = 3\frac{\log 2}{\log \frac{5}{4}} \approx 9.3.$$

Thus the half-life of this isotope is approximately 9.3 years.

103. Suppose the ratio of carbon-14 to carbon-12 in a mummified cat is 64% of the corresponding ratio for living organisms. About how long ago did the cat die?

SOLUTION The half-life of carbon-14 is 5730 years. If we start with a sample of a atoms of carbon-14, then after t years there will be

$$a \cdot 2^{-t/5730}$$

atoms left. We want to find t such that this equals $0.64a$. Thus we must solve the equation

$$a \cdot 2^{-t/5730} = 0.64a.$$

To solve this equation for t, divide both sides by a and then take the logarithm of both sides, getting

$$-\frac{t}{5730}\log 2 = \log 0.64.$$

Now solve for t, getting

$$t = -5730\frac{\log 0.64}{\log 2} \approx 3689.$$

Thus the cat died about 3689 years ago. Carbon-14 cannot be measured with extreme accuracy. Thus it is better to estimate that the cat died about 3700 years ago (because a number such as 3689 conveys more accuracy than will be present in such measurements).

The author's second cat.

5.3 *Applications of Logarithms*

LEARNING OBJECTIVES

By the end of this section you should be able to

- apply the formula for the logarithm of a product;
- apply the formula for the logarithm of a quotient;
- model earthquake intensity with the logarithmic Richter magnitude scale;
- model sound intensity with the logarithmic decibel scale;
- model star brightness with the logarithmic apparent magnitude scale;
- apply the change of base formula for logarithms.

Logarithm of a Product

To motivate the formula for the logarithm of a product, we note that

$$\log(10^2 10^3) = \log 10^5 = 5$$

and that

$$\log 10^2 = 2 \quad \text{and} \quad \log 10^3 = 3.$$

Putting these equations together, we see that

$$\log(10^2 10^3) = \log 10^2 + \log 10^3.$$

More generally, logarithms convert products to sums, as we will now show.

Never, ever, make the mistake of thinking that $\log_b(x + y)$ equals $\log_b x + \log_b y$. There is no nice formula for $\log_b(x + y)$.

Suppose b, x, and y are positive numbers, with $b \neq 1$. Then

$$\begin{aligned} \log_b(xy) &= \log_b(b^{\log_b x} b^{\log_b y}) \\ &= \log_b b^{\log_b x + \log_b y} \\ &= \log_b x + \log_b y. \end{aligned}$$

In other words, we have the following nice formula for the logarithm of a product:

Logarithm of a product

If b, x, and y are positive numbers, with $b \neq 1$, then

$$\log_b(xy) = \log_b x + \log_b y.$$

Never, ever, make the mistake of thinking that $\log_b(xy)$ equals the product $(\log_b x)(\log_b y)$.

EXAMPLE 1

Use the information that $\log 3 \approx 0.477$ to evaluate $\log 30000$.

SOLUTION

$$\begin{aligned}\log 30000 &= \log(10^4 \cdot 3)\\ &= \log(10^4) + \log 3\\ &= 4 + \log 3\\ &\approx 4.477.\end{aligned}$$

Logarithm of a Quotient

As we have seen, the formula $\log_b(xy) = \log_b x + \log_b y$ arises naturally from the formula $b^s b^t = b^{s+t}$. Similarly, we will use the formula $b^s/b^t = b^{s-t}$ to derive a formula for the logarithm of a quotient. First we look at an example.

To motivate the formula for the logarithm of a quotient, we note that

$$\log \frac{10^8}{10^3} = \log 10^5 = 5$$

and that

$$\log 10^8 = 8 \quad \text{and} \quad \log 10^3 = 3.$$

Putting these equations together, we see that

$$\log \frac{10^8}{10^3} = \log 10^8 - \log 10^3.$$

More generally, logarithms convert quotients to differences, as we will now show.

Suppose b, x, and y are positive numbers, with $b \neq 1$. Then

$$\begin{aligned}\log_b \frac{x}{y} &= \log_b \frac{b^{\log_b x}}{b^{\log_b y}}\\ &= \log_b b^{\log_b x - \log_b y}\\ &= \log_b x - \log_b y.\end{aligned}$$

Never, ever, make the mistake of thinking that $\log_b(x-y)$ equals $\log_b x - \log_b y$. There is no nice formula for $\log_b(x-y)$.

In other words, we have the following formula for the logarithm of a quotient:

Logarithm of a quotient

If b, x, and y are positive numbers, with $b \neq 1$, then

$$\log_b \frac{x}{y} = \log_b x - \log_b y.$$

Never, ever, make the mistake of thinking that $\log_b \frac{x}{y}$ equals $\frac{\log_b x}{\log_b y}$.

EXAMPLE 2 Use the information that $\log 7 \approx 0.845$to evaluate $\log \frac{1000}{7}$.

SOLUTION

$$\log \tfrac{1000}{7} = \log 1000 - \log 7$$
$$= 3 - \log 7$$
$$\approx 2.155$$

As a special case of the formula for the logarithm of a quotient, take $x = 1$ in the formula above for the logarithm of a quotient, getting

$$\log_b \frac{1}{y} = \log_b 1 - \log_b y.$$

Recalling that $\log_b 1 = 0$, we get the following result:

Logarithm of a multiplicative inverse

If b and y are positive numbers, with $b \neq 1$, then

$$\log_b \frac{1}{y} = -\log_b y.$$

Earthquakes and the Richter Scale

The intensity of an earthquake is measured by the size of the seismic waves generated by the earthquake. These numbers vary across such a huge scale that earthquakes are usually reported using the Richter magnitude scale, which is a logarithmic scale using common logarithms (base 10).

Richter magnitude scale

An earthquake with seismic waves of size S has **Richter magnitude**

$$\log \frac{S}{S_0},$$

where S_0 is the size of the seismic waves corresponding to what has been declared to be an earthquake with Richter magnitude 0.

The size of the seismic wave is roughly proportional to the amount of ground shaking.

A few points will help clarify this definition:

- The value of S_0 was set in 1935 by the American seismologist Charles Richter as approximately the size of the smallest seismic waves that could be measured at that time.

- The units used to measure S and S_0 do not matter (provided the same units are used for S and S_0) because any change in the scale of these units disappears in the ratio $\frac{S}{S_0}$.
- An increase of earthquake intensity by a factor of 10 corresponds to an increase of 1 in Richter magnitude, as can be seen from the equation

$$\log \frac{10S}{S_0} = \log 10 + \log \frac{S}{S_0} = 1 + \log \frac{S}{S_0}.$$

EXAMPLE 3

The world's most intense recorded earthquake struck Chile in 1960; it had Richter magnitude 9.5. The most intense recorded earthquake in the United States struck Alaska in 1964; it had Richter magnitude 9.2. Approximately how many times more intense was the 1960 earthquake in Chile than the 1964 earthquake in Alaska?

SOLUTION Let S_C denote the size of the seismic waves from the 1960 earthquake in Chile and let S_A denote the size of the seismic waves from the 1964 earthquake in Alaska. Thus

$$9.5 = \log \frac{S_C}{S_0} \quad \text{and} \quad 9.2 = \log \frac{S_A}{S_0}.$$

Subtracting the second equation from the first equation, we get

$$0.3 = \log \frac{S_C}{S_0} - \log \frac{S_A}{S_0} = \log\Big(\frac{S_C}{S_0} \Big/ \frac{S_A}{S_0}\Big) = \log \frac{S_C}{S_A}.$$

Thus

$$\frac{S_C}{S_A} = 10^{0.3} \approx 2.$$

In other words, the 1960 earthquake in Chile was approximately twice as intense as the 1964 earthquake in Alaska.

As this example shows, even small differences in the Richter magnitude can correspond to large differences in intensity.

Sound Intensity and Decibels

The intensity of a sound is the amount of energy carried by the sound through each unit of area. The human ear can perceive sound over an enormous range of intensities.

The ratio of the intensity of the sound level that causes pain to the intensity of the quietest sound that we can hear is over one trillion. Working with such large numbers can be inconvenient. Thus sound is usually measured in decibels, which is a logarithmic scale using common logarithms.

Decibel scale for sound

A sound with intensity E has

$$10 \log \frac{E}{E_0}$$

decibels, where E_0 is the intensity of an extremely quiet sound at the threshold of human hearing.

The factor of 10 that appears in the definition of the decibel scale can be a minor nuisance. The "deci" part of the name "decibel" comes from this factor of 10.

A few points will help clarify this definition:

- The value of E_0 is 10^{-12} watts per square meter.
- The intensity of sound is usually measured in watts per square meter, but the units used to measure E and E_0 do not matter (provided the same units are used for E and E_0) because any change in the scale of these units disappears in the ratio $\frac{E}{E_0}$.
- Multiplying sound intensity by a factor of 10 corresponds to adding 10 to the decibel measurement, as can be seen from the equation

$$10\log\frac{10E}{E_0} = 10\log 10 + 10\log\frac{E}{E_0} = 10 + 10\log\frac{E}{E_0}.$$

EXAMPLE 4

Because of worries about potential hearing damage, France passed a law limiting iPods and other MP3 players to a maximum possible volume of 100 decibels. Assuming that normal conversation has a sound level of 65 decibels, how many more times intense than normal conversation is the sound of an iPod operating at the French maximum of 100 decibels?

SOLUTION Let E_F denote the sound intensity of 100 decibels allowed in France and let E_C denote the sound intensity of normal conversation. Thus

$$100 = 10\log\frac{E_F}{E_0} \quad \text{and} \quad 65 = 10\log\frac{E_C}{E_0}.$$

Subtracting the second equation from the first equation, we get

$$35 = 10\log\frac{E_F}{E_0} - 10\log\frac{E_C}{E_0}.$$

Thus

$$3.5 = \log\frac{E_F}{E_0} - \log\frac{E_C}{E_0} = \log\left(\frac{E_F}{E_0}\Big/\frac{E_C}{E_0}\right) = \log\frac{E_F}{E_C}.$$

Thus

$$\frac{E_F}{E_C} = 10^{3.5} \approx 3162.$$

In other words, an iPod operating at the maximum legal French volume of 100 decibels produces sound about three thousand times more intense than normal conversation.

The increase in sound intensity by a factor of more than 3000 in the last example is not as drastic as it seems because of how we perceive loudness:

Loudness

The human ear perceives each increase in sound by 10 decibels to be a doubling of loudness (even though the sound intensity has actually increased by a factor of 10).

EXAMPLE 5

By what factor has the loudness increased in going from normal speech at 65 decibels to an iPod at 100 decibels?

SOLUTION Here we have an increase of 35 decibels, so we have had an increase of 10 decibels 3.5 times. Thus the perceived loudness has doubled 3.5 times, which means that it has increased by a factor of $2^{3.5}$. Because $2^{3.5} \approx 11$, this means that an iPod operating at 100 decibels seems about 11 times louder than normal conversation.

Star Brightness and Apparent Magnitude

The ancient Greeks divided the visible stars into six groups based on their brightness. The brightest stars were called first-magnitude stars. The next brightest group of stars were called second-magnitude stars, and so on, until the sixth-magnitude stars consisted of the barely visible stars.

About two thousand years later, astronomers made the ancient Greek star magnitude scale more precise. The typical first-magnitude stars were about 100 times brighter than the typical sixth-magnitude stars. Because there are five steps in going from the first magnitude to the sixth magnitude, this means that with each magnitude the brightness should decrease by a factor of $100^{1/5}$.

Because $100^{1/5} \approx$ 2.512, each magnitude is approximately 2.512 times dimmer than the previous magnitude.

Originally the scale was defined so that Polaris (the North Star) had magnitude 2. If we let b_2 denote the brightness of Polaris, this would mean that a third-magnitude star has brightness $b_2/100^{1/5}$, a fourth-magnitude star has brightness $b_2/(100^{1/5})^2$, a fifth-magnitude star has brightness $b_2/(100^{1/5})^3$, and so on. Thus the brightness b of a star with magnitude m should be given by the equation

$$b = \frac{b_2}{(100^{1/5})^{(m-2)}} = b_2 100^{(2-m)/5} = b_2 100^{2/5} 100^{-m/5} = b_0 100^{-m/5},$$

where $b_0 = b_2 100^{2/5}$. If we divide both sides of the equation above by b_0 and then take logarithms we get

$$\log \frac{b}{b_0} = \log 100^{-m/5} = -\frac{m}{5} \log 100 = -\frac{2m}{5}.$$

Solving this equation for m leads to the following definition:

Apparent magnitude

An object with brightness b has **apparent magnitude**

$$\frac{5}{2} \log \frac{b_0}{b},$$

where b_0 is the brightness of an object with magnitude 0.

A few points will help clarify this definition:

- The term "apparent magnitude" is more accurate than "magnitude" because we are measuring how bright a star appears from Earth. A glowing luminous star might appear dim from Earth because it is very far away.
- Although this apparent magnitude scale was originally set up for stars, it can be applied to other objects such as the full moon.
- Although the value of b_0 was originally set so that Polaris (the North Star) would have apparent magnitude 2, the definition has changed slightly. With the current definition of b_0, Polaris has magnitude close to 2 but not exactly equal to 2.
- The units used to measure brightness do not matter (provided the same units are used for b and b_0) because any change in the scale of these units disappears in the ratio $\frac{b_0}{b}$.

EXAMPLE 6

Because of the lack of atmospheric interference, the Hubble telescope can see dimmer stars than Earth-based telescopes of the same size.

With good binoculars you can see stars with apparent magnitude 9. The Hubble telescope, which is in orbit around the Earth, can detect stars with apparent magnitude 30. How much better is the Hubble telescope than binoculars, measured in terms of the ratio of the brightness of stars that they can detect?

SOLUTION Let b_9 denote the brightness of a star with apparent magnitude 9 and let b_{30} denote the brightness of a star with apparent magnitude 30. Thus

$$9 = \frac{5}{2}\log\frac{b_0}{b_9} \quad \text{and} \quad 30 = \frac{5}{2}\log\frac{b_0}{b_{30}}.$$

Subtracting the first equation from the second equation, we get

$$21 = \frac{5}{2}\log\frac{b_0}{b_{30}} - \frac{5}{2}\log\frac{b_0}{b_9}.$$

Thus

$$\frac{42}{5} = \log\frac{b_0}{b_{30}} - \log\frac{b_0}{b_9} = \log\Big(\frac{b_0}{b_{30}}\Big/\frac{b_0}{b_9}\Big) = \log\frac{b_9}{b_{30}}.$$

Thus

$$\frac{b_9}{b_{30}} = 10^{42/5} = 10^{8.4} \approx 250,000,000.$$

Thus the Hubble telescope can detect stars 250 million times dimmer than stars we can see with binoculars.

Change of Base

If we want to use a calculator to evaluate something like $\log_2 73.9$, then we need a formula for converting logarithms from one base to another. Thus we now consider the relationship between logarithms with different bases.

To motivate the formula we will discover, first we look at an example. Note that $\log_2 64 = 6$ because $2^6 = 64$, and $\log_8 64 = 2$ because $8^2 = 64$. Thus

$$\log_2 64 = 3 \log_8 64.$$

Your calculator can probably evaluate logarithms for only two bases. One of these is probably the logarithm base 10 (the common logarithm, probably labeled "log" on your calculator), and the other is probably the logarithm base e (this is the natural logarithm that we will discuss in the next chapter; it is probably labeled "ln" on your calculator).

The relationship above holds if 64 is replaced by any positive number y. To see this, note that

$$\begin{aligned} y &= 8^{\log_8 y} \\ &= (2^3)^{\log_8 y} \\ &= 2^{3 \log_8 y}, \end{aligned}$$

which implies that

$$\log_2 y = 3 \log_8 y.$$

In other words, the logarithm base 2 equals 3 times the logarithm base 8. Note that $3 = \log_2 8$.

The relationship derived above holds more generally. To see this, suppose a, b, and y are positive numbers, with $a \neq 1$ and $b \neq 1$. Then

$$\begin{aligned} y &= b^{\log_b y} \\ &= (a^{\log_a b})^{\log_b y} \\ &= a^{(\log_a b)(\log_b y)}. \end{aligned}$$

The equation above implies that $\log_a y = (\log_a b)(\log_b y)$. Solving this equation for $\log_b y$ (so that both base a logarithms will be on the same side), we have the following formula for converting logarithms from one base to another:

Change of base for logarithms

If a, b, and y are positive numbers, with $a \neq 1$ and $b \neq 1$, then

$$\log_b y = \frac{\log_a y}{\log_a b}.$$

A special case of this formula, suitable for use with calculators, is to take $a = 10$, thus using common logarithms and getting the following formula:

Change of base with common logarithms

If b and y are positive numbers, with $b \neq 1$, then

$$\log_b y = \frac{\log y}{\log b}.$$

EXAMPLE 7

Evaluate $\log_2 73.9$.

SOLUTION Use a calculator with $b = 2$ and $y = 73.9$ in the formula above, getting

$$\log_2 73.9 = \frac{\log 73.9}{\log 2} \approx 6.2075.$$

The change of base formula for logarithms implies that the graph of the logarithm using any base can be obtained by vertically stretching the graph of the logarithm using any other base (assuming both bases are bigger than 1), as shown in the following example.

EXAMPLE 8

Sketch the graphs of $\log_2 x$ and $\log x$ on the interval $[\frac{1}{8}, 8]$. What is the relationship between these two graphs?

SOLUTION The change of base formula implies that $\log x = (\log 2)(\log_2 x)$. Because $\log 2 \approx 0.3$, this means that the graph of $\log x$ is obtained from the graph of $\log_2 x$ (sketched earlier) by stretching vertically by a factor of approximately 0.3.

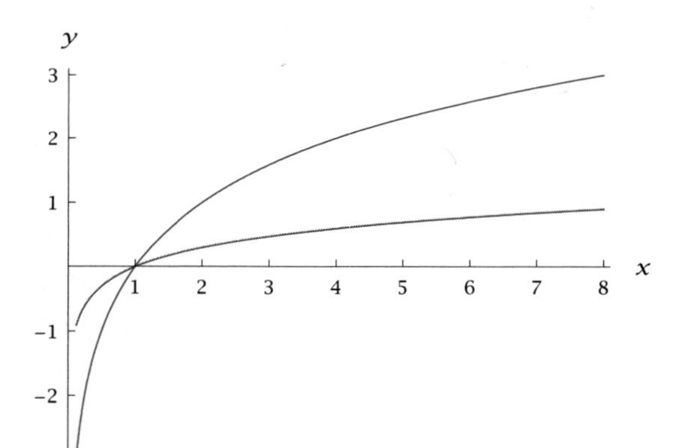

The graphs of $\log_2 x$ *(blue) and* $\log x$ *(red) on the interval* $[\frac{1}{8}, 8]$.

This change of base formula explains why the shape of thé graph of the logarithm function does not depend the base, as long as we consider only bases bigger than 1.

More generally, for fixed positive numbers a and b, neither which is 1, the change of base formula

$$\log_a x = (\log_a b) \log_b x$$

implies that the graph of $y = \log_a x$ can be obtained from the graph of $y = \log_b x$ by stretching vertically by a factor of $\log_a b$.

EXERCISES

1. For $x = 7$ and $y = 13$, evaluate each of the following:

   (a) $\log(x + y)$ (b) $\log x + \log y$

   [*This exercise and the next one emphasize that* $\log(x + y)$ *does not equal* $\log x + \log y$.]

2. For $x = 0.4$ and $y = 3.5$, evaluate each of the following:

   (a) $\log(x + y)$ (b) $\log x + \log y$

3. For $x = 3$ and $y = 8$, evaluate each of the following:

   (a) $\log(xy)$ (b) $(\log x)(\log y)$

   [*This exercise and the next one emphasize that* $\log(xy)$ *does not equal* $(\log x)(\log y)$.]

4. For $x = 1.1$ and $y = 5$, evaluate each of the following:

   (a) $\log(xy)$ (b) $(\log x)(\log y)$

5. For $x = 12$ and $y = 2$, evaluate each of the following:

   (a) $\log \frac{x}{y}$ (b) $\dfrac{\log x}{\log y}$

   [*This exercise and the next one emphasize that* $\log \frac{x}{y}$ *does not equal* $\frac{\log x}{\log y}$.]

6. For $x = 18$ and $y = 0.3$, evaluate each of the following:

   (a) $\log \frac{x}{y}$ (b) $\dfrac{\log x}{\log y}$

7. How many digits does $6^{700} \cdot 23^{1000}$ have?

8. How many digits does $5^{999} \cdot 17^{2222}$ have?

9. Suppose m and n are positive integers such that $\log m \approx 32.1$ and $\log n \approx 7.3$. How many digits does mn have?

10. Suppose m and n are positive integers such that $\log m \approx 41.3$ and $\log n \approx 12.8$. How many digits does mn have?

11. Suppose $\log a = 118.7$ and $\log b = 119.7$. Evaluate $\frac{b}{a}$.

12. Suppose $\log a = 203.4$ and $\log b = 205.4$. Evaluate $\frac{b}{a}$.

For Exercises 13–26, evaluate the given quantities assuming that

$$\log_3 x = 5.3 \quad \textit{and} \quad \log_3 y = 2.1,$$

$$\log_4 u = 3.2 \quad \textit{and} \quad \log_4 v = 1.3.$$

13. $\log_3(9xy)$
14. $\log_4(2uv)$
15. $\log_3 \frac{x}{3y}$
16. $\log_4 \frac{u}{8v}$
17. $\log_3 \sqrt{x}$
18. $\log_4 \sqrt{u}$
19. $\log_3 \dfrac{1}{\sqrt{y}}$
20. $\log_4 \dfrac{1}{\sqrt{v}}$
21. $\log_3(x^2y^3)$
22. $\log_4(u^3v^4)$
23. $\log_3 \dfrac{x^3}{y^2}$
24. $\log_4 \dfrac{u^2}{v^3}$
25. $\log_9 x^{10}$
26. $\log_2 u^{100}$

For Exercises 27–34, find all numbers x that satisfy the given equation.

27. $\log_7(x + 5) - \log_7(x - 1) = 2$
28. $\log_4(x + 4) - \log_4(x - 2) = 3$
29. $\log_3(x + 5) + \log_3(x - 1) = 2$
30. $\log_5(x + 4) + \log_5(x + 2) = 2$
31. $\dfrac{\log_6(15x)}{\log_6(5x)} = 2$
32. $\dfrac{\log_9(13x)}{\log_9(4x)} = 2$
33. $(\log(3x)) \log x = 4$
34. $(\log(6x)) \log x = 5$
35. How many more times intense is an earthquake with Richter magnitude 7 than an earthquake with Richter magnitude 5?
36. How many more times intense is an earthquake with Richter magnitude 6 than an earthquake with Richter magnitude 3?
37. The 1994 Northridge earthquake in Southern California, which killed several dozen people, had Richter magnitude 6.7. What would be the Richter magnitude of an earthquake that was 100 times more intense than the Northridge earthquake?
38. The 1995 earthquake in Kobe (Japan), which killed over 6000 people, had Richter magnitude 7.2. What would be the Richter magnitude of an earthquake that was 1000 times less intense than the Kobe earthquake?

39. The most intense recorded earthquake in the state of New York occurred in 1944; it had Richter magnitude 5.8. The most intense recorded earthquake in Minnesota occurred in 1975; it had Richter magnitude 5.0. Approximately how many times more intense was the 1944 earthquake in New York than the 1975 earthquake in Minnesota?

40. The most intense recorded earthquake in Wyoming occurred in 1959; it had Richter magnitude 6.5. The most intense recorded earthquake in Illinois occurred in 1968; it had Richter magnitude 5.3. Approximately how many times more intense was the 1959 earthquake in Wyoming than the 1968 earthquake in Illinois?

41. The most intense recorded earthquake in Texas occurred in 1931; it had Richter magnitude 5.8. If an earthquake were to strike Texas next year that was three times more intense than the current record in Texas, what would its Richter magnitude be?

42. The most intense recorded earthquake in Ohio occurred in 1937; it had Richter magnitude 5.4. If an earthquake were to strike Ohio next year that was 1.6 times more intense than the current record in Ohio, what would its Richter magnitude be?

43. Suppose you whisper at 20 decibels and normally speak at 60 decibels.

    (a) What is the ratio of the sound intensity of your normal speech to the sound intensity of your whisper?

    (b) How many times louder does your normal speech seem as compared to your whisper?

44. Suppose your vacuum cleaner produces a sound of 80 decibels and you normally speak at 60 decibels.

    (a) What is the ratio of the sound intensity of your vacuum cleaner to the sound intensity of your normal speech?

    (b) How many times louder does your vacuum cleaner seem as compared to your normal speech?

45. Suppose an airplane taking off makes a noise of 117 decibels and you normally speak at 63 decibels.

    (a) What is the ratio of the sound intensity of the airplane to the sound intensity of your normal speech?

    (b) How many times louder does the airplane seem than your normal speech?

46. Suppose your cell phone rings at a noise level of 74 decibels and you normally speak at 61 decibels.

    (a) What is the ratio of the sound intensity of your cell phone ring to the sound intensity of your normal speech?

    (b) How many times louder does your cell phone ring seem than your normal speech?

47. Suppose a television is playing softly at a sound level of 50 decibels. What decibel level would make the television sound eight times as loud?

48. Suppose a radio is playing loudly at a sound level of 80 decibels. What decibel level would make the radio sound one-fourth as loud?

49. Suppose a motorcycle produces a sound level of 90 decibels. What decibel level would make the motorcycle sound one-third as loud?

50. Suppose a rock band is playing loudly at a sound level of 100 decibels. What decibel level would make the band sound three-fifths as loud?

51. How many times brighter is a star with apparent magnitude 2 than a star with apparent magnitude 17?

52. How many times brighter is a star with apparent magnitude 3 than a star with apparent magnitude 23?

53. Sirius, the brightest star that can be seen from Earth (not counting the sun), has an apparent magnitude of -1.4. Vega, which was the North Star about 12,000 years ago (slight changes in Earth's orbit lead to changing North Stars every several thousand years), has an apparent magnitude of 0.03. How many times brighter than Vega is Sirius?

54. The full moon has an apparent magnitude of approximately -12.6. How many times brighter than Sirius is the full moon?

55. Neptune has an apparent magnitude of about 7.8. What is the apparent magnitude of a star that is 20 times brighter than Neptune?

56. What is the apparent magnitude of a star that is eight times brighter than Neptune?

For Exercises 57–64, evaluate the indicated quantities. Your calculator is unlikely to be able to evaluate logarithms using any of the bases in these exercises, so you will need to use an appropriate change of base formula.

57. $\log_2 13$

58. $\log_4 27$

59. $\log_{13} 9.72$

60. $\log_{17} 12.31$

61. $\log_9 0.23$

62. $\log_7 0.58$

63. $\log_{4.38} 7.1$

64. $\log_{5.06} 99.2$

PROBLEMS

65. Explain why
$$\log 500 = 3 - \log 2.$$

66. Explain why
$$\log \sqrt{0.07} = \frac{\log 7}{2} - 1.$$

67. Explain why
$$1 + \log x = \log(10x)$$
for every positive number x.

68. Explain why
$$2 - \log x = \log \tfrac{100}{x}$$
for every positive number x.

69. Explain why
$$(1 + \log x)^2 = \log(10x^2) + (\log x)^2$$
for every positive number x.

70. Explain why
$$\frac{1 + \log x}{2} = \log \sqrt{10x}$$
for every positive number x.

71. Explain why the graph of $y = \log(1000x)$ can be obtained by shifting the graph of $y = \log x$ up 3 units.

72. Explain why the graph of $y = \log(100x^3)$ can be obtained by vertically stretching the graph of $y = \log x$ by a factor of 3 and then shifting up 2 units.

73. Pretend that you are living in the time before calculators and computers existed, and that you have a book showing the logarithms of 1.001, 1.002, 1.003, and so on, up to the logarithm of 9.999. Explain how you would find the logarithm of 457.2, which is beyond the range of your book.

74. Explain why books of logarithm tables, which were frequently used before the era of calculators and computers, gave logarithms only for numbers between 1 and 10.

75. Derive the formula for the logarithm of a quotient by applying the formula for the logarithm of a product to $\log_b(y \cdot \frac{x}{y})$.
[*Sometimes seeing an alternative derivation can help increase your understanding.*]

76. Derive the formula $\log_b \frac{1}{y} = -\log_b y$ directly from the formula $1/b^t = b^{-t}$.

77. Show that an earthquake with Richter magnitude R has seismic waves of size $S_0 10^R$, where S_0 is the size of the seismic waves of an earthquake with Richter magnitude 0.

78. Do a web search to find the most intense earthquake in the United States in the last calendar year and the most intense earthquake in Japan in the last calendar year. Approximately how many times more intense was the larger of these two earthquakes than the smaller of the two?

79. Show that a sound with d decibels has intensity $E_0 10^{d/10}$, where E_0 is the intensity of a sound with 0 decibels.

80. Find at least three different web sites giving the apparent magnitude of Polaris (the North Star) accurate to at least two digits after the decimal point. If you find different values on different web sites (as the author did), then try to explain what could account for the discrepancy (and take this as a good lesson in the caution necessary when using the web as a source of scientific information).

81. Write a description of the logarithmic scale used for the pH scale, which measures acidity (this will probably require use of the library or the web).

82. Without doing any calculations, explain why the solutions to the equations in Exercises 31 and 32 are unchanged if we change the base for all the logarithms in those exercises to any positive number $b \neq 1$.

83. Explain why expressing a large positive integer in binary notation (base 2) should take approximately 3.3 times as many digits as expressing the same positive integer in standard decimal notation (base 10).
[*For example, this problem predicts that* 5×10^{12}, *which requires* 13 *digits to express in decimal notation, should require approximately* 13×3.3 *digits (which equals* 42.9 *digits) to express in binary notation. Expressing* 5×10^{12} *in binary notation actually requires* 43 *digits.*]

84. Suppose a and b are positive numbers, with $a \neq 1$ and $b \neq 1$. Show that

$$\log_a b = \frac{1}{\log_b a}.$$

85. Suppose b and y are positive numbers, with $b \neq 1$ and $b \neq \frac{1}{2}$. Show that

$$\log_{2b} y = \frac{\log_b y}{1 + \log_b 2}.$$

WORKED-OUT SOLUTIONS *to Odd-numbered Exercises*

1. For $x = 7$ and $y = 13$, evaluate each of the following:
(a) $\log(x + y)$ (b) $\log x + \log y$

SOLUTION

(a) $\log(7 + 13) = \log 20 \approx 1.30103$

(b)
$$\begin{aligned}\log 7 + \log 13 &\approx 0.845098 + 1.113943\\ &= 1.959041\end{aligned}$$

3. For $x = 3$ and $y = 8$, evaluate each of the following:
(a) $\log(xy)$ (b) $(\log x)(\log y)$

SOLUTION

(a) $\log(3 \cdot 8) = \log 24 \approx 1.38021$

(b)
$$\begin{aligned}(\log 3)(\log 8) &\approx (0.477121)(0.903090)\\ &\approx 0.430883\end{aligned}$$

5. For $x = 12$ and $y = 2$, evaluate each of the following:
(a) $\log \frac{x}{y}$ (b) $\dfrac{\log x}{\log y}$

SOLUTION

(a) $\log \frac{12}{2} = \log 6 \approx 0.778151$

(b) $\dfrac{\log 12}{\log 2} \approx \dfrac{1.079181}{0.301030} \approx 3.58496$

7. How many digits does $6^{700} \cdot 23^{1000}$ have?

SOLUTION Using the formulas for the logarithm of a product and the logarithm of a power, we have

$$\begin{aligned}\log(6^{700} \cdot 23^{1000}) &= \log 6^{700} + \log 23^{1000}\\ &= 700 \log 6 + 1000 \log 23\\ &\approx 1906.43.\end{aligned}$$

Thus $6^{700} \cdot 23^{1000}$ has 1907 digits.

9. Suppose m and n are positive integers such that $\log m \approx 32.1$ and $\log n \approx 7.3$. How many digits does mn have?

SOLUTION Note that

$$\begin{aligned}\log(mn) &= \log m + \log n\\ &\approx 32.1 + 7.3\\ &= 39.4.\end{aligned}$$

Thus mn has 40 digits.

11. Suppose $\log a = 118.7$ and $\log b = 119.7$. Evaluate $\frac{b}{a}$.

SOLUTION Note that

$$\begin{aligned}\log \tfrac{b}{a} &= \log b - \log a\\ &= 119.7 - 118.7\\ &= 1.\end{aligned}$$

Thus $\frac{b}{a} = 10$.

For Exercises 13–26, evaluate the given quantities assuming that

$$\log_3 x = 5.3 \quad \textit{and} \quad \log_3 y = 2.1,$$

$$\log_4 u = 3.2 \quad \textit{and} \quad \log_4 v = 1.3.$$

13. $\log_3(9xy)$

SOLUTION

$$\begin{aligned}\log_3(9xy) &= \log_3 9 + \log_3 x + \log_3 y\\ &= 2 + 5.3 + 2.1\\ &= 9.4\end{aligned}$$

15. $\log_3 \frac{x}{3y}$

SOLUTION

$$\begin{aligned}\log_3 \tfrac{x}{3y} &= \log_3 x - \log_3(3y)\\ &= \log_3 x - \log_3 3 - \log_3 y\\ &= 5.3 - 1 - 2.1\\ &= 2.2\end{aligned}$$

17. $\log_3 \sqrt{x}$

SOLUTION

$$\begin{aligned}\log_3 \sqrt{x} &= \log_3 x^{1/2}\\ &= \tfrac{1}{2}\log_3 x\\ &= \tfrac{1}{2} \times 5.3\\ &= 2.65\end{aligned}$$

19. $\log_3 \frac{1}{\sqrt{y}}$

SOLUTION

$$\begin{aligned}\log_3 \frac{1}{\sqrt{y}} &= \log_3 y^{-1/2}\\ &= -\tfrac{1}{2}\log_3 y\\ &= -\tfrac{1}{2} \times 2.1\\ &= -1.05\end{aligned}$$

21. $\log_3(x^2y^3)$

SOLUTION

$$\begin{aligned}\log_3(x^2y^3) &= \log_3 x^2 + \log_3 y^3\\ &= 2\log_3 x + 3\log_3 y\\ &= 2 \cdot 5.3 + 3 \cdot 2.1\\ &= 16.9\end{aligned}$$

23. $\log_3 \frac{x^3}{y^2}$

SOLUTION

$$\begin{aligned}\log_3 \frac{x^3}{y^2} &= \log_3 x^3 - \log_3 y^2\\ &= 3\log_3 x - 2\log_3 y\\ &= 3 \cdot 5.3 - 2 \cdot 2.1\\ &= 11.7\end{aligned}$$

25. $\log_9 x^{10}$

SOLUTION Because $\log_3 x = 5.3$, we see that $3^{5.3} = x$. This equation can be rewritten as $(9^{1/2})^{5.3} = x$, which can then be rewritten as $9^{2.65} = x$. In other words, $\log_9 x = 2.65$. Thus

$$\log_9 x^{10} = 10\log_9 x = 26.5.$$

For Exercises 27–34, find all numbers x that satisfy the given equation.

27. $\log_7(x+5) - \log_7(x-1) = 2$

SOLUTION Rewrite the equation as follows:

$$\begin{aligned}2 &= \log_7(x+5) - \log_7(x-1)\\ &= \log_7 \frac{x+5}{x-1}.\end{aligned}$$

Thus

$$\frac{x+5}{x-1} = 7^2 = 49.$$

We can solve the equation above for x, getting $x = \frac{9}{8}$.

29. $\log_3(x+5) + \log_3(x-1) = 2$

SOLUTION Rewrite the equation as follows:

$$\begin{aligned} 2 &= \log_3(x+5) + \log_3(x-1) \\ &= \log_3((x+5)(x-1)) \\ &= \log_3(x^2+4x-5). \end{aligned}$$

Thus

$$x^2 + 4x - 5 = 3^2 = 9,$$

which implies that

$$x^2 + 4x - 14 = 0.$$

We can solve the equation above using the quadratic formula, getting $x = 3\sqrt{2} - 2$ or $x = -3\sqrt{2} - 2$. However, both $x + 5$ and $x - 1$ are negative if $x = -3\sqrt{2} - 2$; because the logarithm of a negative number is undefined, we must discard this root of the equation above. We conclude that the only value of x satisfying the equation $\log_3(x+5) + \log_3(x-1) = 2$ is $x = 3\sqrt{2} - 2$.

31. $\dfrac{\log_6(15x)}{\log_6(5x)} = 2$

SOLUTION Rewrite the equation as follows:

$$\begin{aligned} 2 &= \frac{\log_6(15x)}{\log_6(5x)} \\ &= \frac{\log_6 15 + \log_6 x}{\log_6 5 + \log_6 x}. \end{aligned}$$

Solving this equation for $\log_6 x$ (the first step in doing this is to multiply both sides by the denominator $\log_6 5 + \log_6 x$), we get

$$\begin{aligned} \log_6 x &= \log_6 15 - 2\log_6 5 \\ &= \log_6 15 - \log_6 25 \\ &= \log_6 \tfrac{15}{25} \\ &= \log_6 \tfrac{3}{5}. \end{aligned}$$

Thus $x = \frac{3}{5}$.

33. $(\log(3x))\log x = 4$

SOLUTION Rewrite the equation as follows:

$$\begin{aligned} 4 &= (\log(3x))\log x \\ &= (\log x + \log 3)\log x \\ &= (\log x)^2 + (\log 3)(\log x). \end{aligned}$$

Letting $y = \log x$, we can rewrite the equation above as

$$y^2 + (\log 3)y - 4 = 0.$$

Use the quadratic formula to solve the equation above for y, getting

$$y \approx -2.25274 \quad \text{or} \quad y \approx 1.77562.$$

Thus

$$\log x \approx -2.25274 \quad \text{or} \quad \log x \approx 1.77562,$$

which means that

$$x \approx 10^{-2.25274} \approx 0.00558807$$

or

$$x \approx 10^{1.77562} \approx 59.6509.$$

35. How many more times intense is an earthquake with Richter magnitude 7 than an earthquake with Richter magnitude 5?

SOLUTION Here is an informal but accurate solution: Each increase of 1 in the Richter magnitude corresponds to an increase in the size of the seismic wave by a factor of 10. Thus an increase of 2 in the Richter magnitude corresponds to an increase in the size of the seismic wave by a factor of 10^2. Hence an earthquake with Richter magnitude 7 is 100 times more intense than an earthquake with Richter magnitude 5.

Here is a more formal explanation using logarithms: Let S_7 denote the size of the seismic waves from an earthquake with Richter magnitude 7 and let S_5 denote the size of the seismic waves from an earthquake with Richter magnitude 5. Thus

$$7 = \log \frac{S_7}{S_0} \quad \text{and} \quad 5 = \log \frac{S_5}{S_0}.$$

Subtracting the second equation from the first equation, we get

$$2 = \log \frac{S_7}{S_0} - \log \frac{S_5}{S_0} = \log\left(\frac{S_7}{S_0} \Big/ \frac{S_5}{S_0}\right) = \log \frac{S_7}{S_5}.$$

Thus

$$\frac{S_7}{S_5} = 10^2 = 100.$$

Hence an earthquake with Richter magnitude 7 is 100 times more intense than an earthquake with Richter magnitude 5.

37. The 1994 Northridge earthquake in Southern California, which killed several dozen people, had Richter magnitude 6.7. What would be the Richter magnitude of an earthquake that was 100 times more intense than the Northridge earthquake?

SOLUTION Each increase of 1 in the Richter magnitude corresponds to an increase in the intensity of the earthquake by a factor of 10. Hence an increase in intensity by a factor of 100 (which equals 10^2) corresponds to an increase of 2 is the Richter magnitude. Thus an earthquake that was 100 times more intense than the Northridge earthquake would have Richter magnitude $6.7 + 2$, which equals 8.7.

39. The most intense recorded earthquake in the state of New York occurred in 1944; it had Richter magnitude 5.8. The most intense recorded earthquake in Minnesota occurred in 1975; it had Richter magnitude 5.0. Approximately how many times more intense was the 1944 earthquake in New York than the 1975 earthquake in Minnesota?

SOLUTION Let S_N denote the size of the seismic waves from the 1944 earthquake in New York and let S_M denote the size of the seismic waves from the 1975 earthquake in Minnesota. Thus

$$5.8 = \log \frac{S_N}{S_0} \quad \text{and} \quad 5.0 = \log \frac{S_M}{S_0}.$$

Subtracting the second equation from the first equation, we get

$$0.8 = \log \frac{S_N}{S_0} - \log \frac{S_M}{S_0} = \log\left(\frac{S_N}{S_0} \Big/ \frac{S_M}{S_0}\right) = \log \frac{S_N}{S_M}.$$

Thus

$$\frac{S_N}{S_M} = 10^{0.8} \approx 6.3.$$

In other words, the 1944 earthquake in New York was approximately 6.3 times more intense than the 1975 earthquake in Minnesota.

41. The most intense recorded earthquake in Texas occurred in 1931; it had Richter magnitude 5.8. If an earthquake were to strike Texas next year that was three times more intense than the current record in Texas, what would its Richter magnitude be?

SOLUTION Let S_T denote the size of the seismic waves from the 1931 earthquake in Texas. Thus

$$5.8 = \log \frac{S_T}{S_0}.$$

An earthquake three times more intense would have Richter magnitude

$$\log \frac{3S_T}{S_0} = \log 3 + \log \frac{S_T}{S_0} \approx 0.477 + 5.8 = 6.277.$$

Because of the difficulty of obtaining accurate measurements, Richter magnitudes are usually reported with only one digit after the decimal place. Rounding off, we would thus say that an earthquake in Texas that was three times more intense than the current record would have Richter magnitude 6.3.

43. Suppose you whisper at 20 decibels and normally speak at 60 decibels.

(a) What is the ratio of the sound intensity of your normal speech to the sound intensity of your whisper?

(b) How many times louder does your normal speech seem as compared to your whisper?

SOLUTION

(a) Each increase of 10 decibels corresponds to multiplying the sound intensity by a factor of 10. Going from a 20-decibel whisper to 60-decibel normal speech means that the sound intensity has been increased by a factor of 10 four times. Because $10^4 = 10{,}000$, this means that the ratio of the sound intensity of your normal speech to the sound intensity of your whisper is 10,000.

(b) Each increase of 10 decibels results in a doubling of loudness. Here we have an increase of

40 decibels, so we have had an increase of 10 decibels four times. Thus the perceived loudness has increased by a factor of 2^4. Because $2^4 = 16$, this means that your normal conversation seems 16 times louder than your whisper.

45. Suppose an airplane taking off makes a noise of 117 decibels and you normally speak at 63 decibels.

(a) What is the ratio of the sound intensity of the airplane to the sound intensity of your normal speech?

(b) How many times louder does the airplane seem than your normal speech?

SOLUTION

(a) Let E_A denote the sound intensity of the airplane taking off and let E_S denote the sound intensity of your normal speech. Thus

$$117 = 10\log\frac{E_A}{E_0} \quad \text{and} \quad 63 = 10\log\frac{E_S}{E_0}.$$

Subtracting the second equation from the first equation, we get

$$54 = 10\log\frac{E_A}{E_0} - 10\log\frac{E_S}{E_0}.$$

Thus

$$5.4 = \log\frac{E_A}{E_0} - \log\frac{E_S}{E_0} = \log\Bigl(\frac{E_A}{E_0}\Big/\frac{E_S}{E_0}\Bigr) = \log\frac{E_A}{E_S}.$$

Thus

$$\frac{E_A}{E_S} = 10^{5.4} \approx 251{,}189.$$

In other words, the airplane taking off produces sound about 250 thousand times more intense than your normal speech.

(b) Each increase of 10 decibels results in a doubling of loudness. Here we have an increase of 54 decibels, so we have had an increase of 10 decibels 5.4 times. Thus the perceived loudness has increased by a factor of $2^{5.4}$. Because $2^{5.4} \approx 42$, this means that the airplane seems about 42 times louder than your normal speech.

47. Suppose a television is playing softly at a sound level of 50 decibels. What decibel level would make the television sound eight times as loud?

SOLUTION Each increase of ten decibels makes the television sound twice as loud. Because $8 = 2^3$, the sound level must double three times to make the television sound eight times as loud. Thus 30 decibels must be added to the sound level, raising it to 80 decibels.

49. Suppose a motorcycle produces a sound level of 90 decibels. What decibel level would make the motorcycle sound one-third as loud?

SOLUTION Each decrease of ten decibels makes the motorcycle sound half as loud. The sound level must be cut in half x times, where $\frac{1}{3} = (\frac{1}{2})^x$, to make the motorcycle sound one-third as loud. This equation can be rewritten as $2^x = 3$. Taking common logarithms of both sides gives $x\log 2 = \log 3$, which implies that

$$x = \frac{\log 3}{\log 2} \approx 1.585.$$

Thus the sound level must be decreased by ten decibels 1.585 times, meaning that the sound level must be reduced by 15.85 decibels. Because $90 - 15.85 = 74.15$, a sound level of 74.15 decibels would make the motorcycle sound one-third as loud.

51. How many times brighter is a star with apparent magnitude 2 than a star with apparent magnitude 17?

SOLUTION Every five magnitudes correspond to a change in brightness by a factor of 100. Thus a change in 15 magnitudes corresponds to a change in brightness by a factor of 100^3 (because $15 = 5 \times 3$). Because $100^3 = (10^2)^3 = 10^6$, a star with apparent magnitude 2 is one million times brighter than a star with apparent magnitude 17.

53. Sirius, the brightest star that can be seen from Earth (not counting the sun), has an apparent magnitude of -1.4. Vega, which was the North Star about 12,000 years ago (slight changes in Earth's orbit lead to changing North Stars every several thousand years), has an apparent magnitude of 0.03. How many times brighter than Vega is Sirius?

SOLUTION Let b_V denote the brightness of Vega and let b_S denote the brightness of Sirius.

Thus

$$0.03 = \frac{5}{2}\log\frac{b_0}{b_V} \quad\text{and}\quad -1.4 = \frac{5}{2}\log\frac{b_0}{b_S}.$$

Subtracting the second equation from the first equation, we get

$$1.43 = \frac{5}{2}\log\frac{b_0}{b_V} - \frac{5}{2}\log\frac{b_0}{b_S}.$$

Multiplying both sides by $\frac{2}{5}$, we get

$$0.572 = \log\frac{b_0}{b_V} - \log\frac{b_0}{b_S} = \log\Big(\frac{b_0}{b_V}\Big/\frac{b_0}{b_S}\Big)$$

$$= \log\frac{b_S}{b_V}.$$

Thus

$$\frac{b_S}{b_V} = 10^{0.572} \approx 3.7.$$

Thus Sirius is approximately 3.7 times brighter than Vega.

55. Neptune has an apparent magnitude of about 7.8. What is the apparent magnitude of a star that is 20 times brighter than Neptune?

SOLUTION Each decrease of apparent magnitude by 1 corresponds to brightness increase by a factor of $100^{1/5}$. If we decrease the magnitude by x, then the brightness increases by a factor of $(100^{1/5})^x$. For this exercise, we want $20 = (100^{1/5})^x$. To solve this equation for x, take logarithms of both sides, getting

$$\log 20 = x\log 100^{1/5} = \frac{2x}{5}.$$

Thus

$$x = \frac{5}{2}\log 20 \approx 3.25.$$

Because $7.8 - 3.25 = 4.55$, we conclude that a star 20 times brighter than Neptune has apparent magnitude approximately 4.55.

For Exercises 57–64, evaluate the indicated quantities. Your calculator is unlikely to be able to evaluate logarithms using any of the bases in these exercises, so you will need to use an appropriate change of base formula.

57. $\log_2 13$

SOLUTION $\log_2 13 = \dfrac{\log 13}{\log 2} \approx 3.70044$

59. $\log_{13} 9.72$

SOLUTION $\log_{13} 9.72 = \dfrac{\log 9.72}{\log 13} \approx 0.88664$

61. $\log_9 0.23$

SOLUTION $\log_9 0.23 = \dfrac{\log 0.23}{\log 9} \approx -0.668878$

63. $\log_{4.38} 7.1$

SOLUTION $\log_{4.38} 7.1 = \dfrac{\log 7.1}{\log 4.38} \approx 1.32703$

5.4 *Exponential Growth*

LEARNING OBJECTIVES

By the end of this section you should be able to

- describe the behavior of functions with exponential growth;
- model population growth;
- compute compound interest.

We begin this section with a story.

A Doubling Fable

A mathematician in ancient India invented the game of chess. Filled with gratitude for the remarkable entertainment of this game, the King offered the mathematician anything he wanted. The King expected the mathematician to ask for rare jewels or a majestic palace.

But the mathematician asked only that he be given one grain of rice for the first square on a chessboard, plus two grains of rice for the next square, plus four grains for the next square, and so on, doubling the amount for each square, until the 64th square on an 8-by-8 chessboard had been reached. The King was pleasantly surprised that the mathematician had asked for such a modest reward.

A bag of rice was opened, and first 1 grain was set aside, then 2, then 4, then 8, and so on. As the eighth square (the end of the first row of the chessboard) was reached, 128 grains of rice were counted out, and the King was secretly delighted to be paying such a small reward and also wondering at the foolishness of the mathematician.

As the 16th square was reached, 32,768 grains of rice were counted out, but this was still a small part of a bag of rice. But the 21st square required a full bag of rice, and the 24th square required eight bags of rice. This was more than the King had expected, but it was a trivial amount because the royal granary contained about 200,000 bags of rice to feed the kingdom during the coming winter.

As the 31st square was reached, over a thousand bags of rice were required and were delivered from the royal granary. Now the King was worried. By the 37th square, the royal granary was two-thirds empty. The 38th square would have required more bags of rice than were left, but the King stopped the process and ordered that the mathematician's head be chopped off as a warning about the greed induced by exponential growth.

To understand why the mathematician's seemingly modest request turned out to be so extravagant, note that the nth square of the chessboard required 2^{n-1} grains of rice. These numbers start slowly but grow rapidly, as shown in the table below:

n	2^n
10	1024
20	1048576
30	1073741824
40	1099511627776
50	1125899906842624
60	1152921504606846976

Powers of 2.

The 64th square of the chessboard would have required 2^{63} grains of rice. To get a rough estimate of the magnitude of this number, note that $2^{10} = 1024 \approx 10^3$. Thus

$$2^{63} = 2^3 \cdot 2^{60} = 8 \cdot (2^{10})^6 \approx 8 \cdot (10^3)^6 = 8 \cdot 10^{18} \approx 10^{19}.$$

When estimating large powers of 2, the approximation
$2^{10} \approx 1000$
often simplifies the calculation.

If each large bag contains a million (which equals 10^6) grains of rice, then the approximately 10^{19} grains of rice needed for the 64th square would have required approximately $10^{19}/10^6$ bags of rice, or approximately 10^{13} bags of rice. If we assume that ancient India had a population of about ten million (which equals 10^7), then each resident would have had to produce about $10^{13}/10^7$, which equals one million, bags of rice to satisfy the mathematician's request for the 64th square of the chessboard. Because it would have been impossible for each resident in India to produce a million bags of rice, the mathematician should not have been surprised at losing his head.

As x gets large, 2^x increases much faster than x^2. For example, 2^{63} equals 9223372036854775808 *but 63^2 equals only* 3969.

Functions with Exponential Growth

The function f defined by $f(x) = 2^x$ is an example of what is called a function with exponential growth. Other examples of functions of exponential growth are the functions g and h defined by $g(x) = 3 \cdot 5^x$ and $h(x) = 5 \cdot 7^{3x}$. More generally, we have the following definition:

Exponential growth

A function f is said to have **exponential growth** if f is of the form

$$f(x) = cb^{kx},$$

where c and k are positive constants and $b > 1$.

Functions with exponential growth increase rapidly. In fact, every function with exponential growth increases more rapidly than every polynomial, in the sense that if f is a function with exponential growth and p is any polynomial, then $f(x) > p(x)$ for all sufficiently large x. For example, $2^x > x^{1000}$ for all $x > 13747$ (Problem 35 shows that 13747 could not be replaced by 13746).

Functions with exponential growth increase so rapidly that graphing them in the usual manner can display too little information, as shown in the following example.

EXAMPLE 1 Discuss the graph of the function 9^x on the interval $[0, 8]$.

SOLUTION

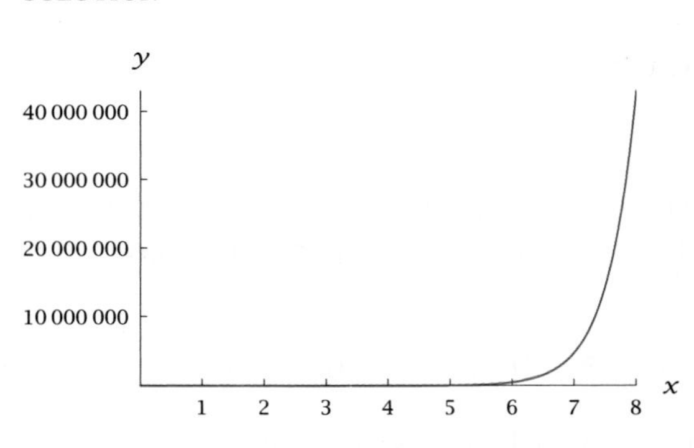

The graph of $y = 9^x$ on the interval $[0, 8]$.

The graph of the function 9^x on the interval $[0, 8]$ is shown above. In this graph, we cannot use the same scale on the x- and y-axes because 9^8 is larger than forty million. Due to the scale, the shape of the graph in the interval $[0, 5]$ gives little insight into the behavior of the function there. For example, this graph does not adequately distinguish between the values 9^2 (which equals 81) and 9^5 (which equals 59049).

Here we are taking the logarithm base 10, but the conclusion about the linearity of the logarithm of f would hold regardless of the base used for the logarithm.

Because the graphs of functions with exponential growth often do not provide sufficient visual information, data that is expected to have exponential growth is often graphed by taking the logarithm of the data. The advantage of this procedure is that if f is a function with exponential growth, then the logarithm of f is a linear function. For example, if $f(x) = 2^x$, then $\log f(x) = (\log 2)x$; thus the graph of $\log f$ is the line whose equation is $y = (\log 2)x$ (which is the line through the origin with slope $\log 2$).

More generally, if $f(x) = cb^{kx}$, then

$$\begin{aligned} \log f(x) &= \log c + \log b^{kx} \\ &= k(\log b)x + \log c. \end{aligned}$$

Here k, $\log b$, and $\log c$ are all constants; thus the function $\log f$ is indeed linear. If $k > 0$ and $b > 1$, as is required in the definition of exponential growth, then $k \log b > 0$, which implies that the line $y = \log f(x)$ has positive slope.

Logarithm of a function of exponential growth

A function f has exponential growth if and only if the graph of

$$y = \log f(x)$$

is a line with positive slope.

EXAMPLE 2

Moore's Law is named in honor of Gordon Moore, co-founder of Intel, who predicted in 1965 that computing power would follow a pattern of exponential growth.

Moore's Law is the phrase used to describe the observation that computing power has exponential growth, doubling roughly every 18 months. One standard measure of computing power is the number of transistors used per integrated circuit; the logarithm of this quantity (for common computer chips manufactured by Intel) is shown in the graph below for certain years between 1972 and 2010, with line segments connecting the data points:

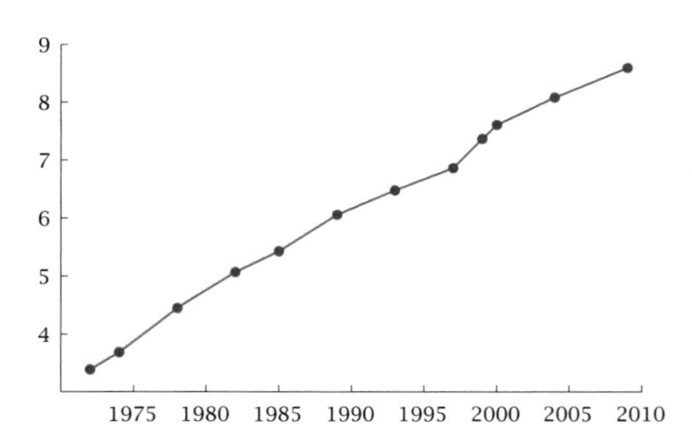

The logarithm of the number of transistors per integrated circuit. Moore's Law predicts exponential growth of computing power, which would make this graph a line.

Does this graph indicate that computing power has had exponential growth?

SOLUTION

The graph above of the logarithm of the number of transistors is roughly a line, as would be expected for a function with roughly exponential growth. Thus the graph above does indeed indicate that computing power has had exponential growth.

Real data, as in the graph above, rarely fits theoretical mathematical models perfectly. The graph above is not exactly a line, and thus we do not have exactly exponential growth. However, the graph above is close enough to a line so that a model of exponential growth can help explain what has happened to computing power over several decades.

Consider the function f with exponential growth defined by

$$f(x) = 5 \cdot 3^{2x}.$$

Because $3^{2x} = (3^2)^x = 9^x$, we can rewrite f in the form

$$f(x) = 5 \cdot 9^x.$$

More generally, suppose f is a function with exponential growth defined by

$$f(x) = cB^{kx}.$$

Because $B^{kx} = (B^k)^x$, if we let $b = B^k$ then we can rewrite f in the form

$$f(x) = cb^x.$$

In other words, by changing b we can, if we wish, always take $k = 1$ in the definition given earlier of a function with exponential growth.

Exponential growth, simpler form

Every function f with exponential growth can be written in the form

$$f(x) = cb^x,$$

where c is a positive constant and $b > 1$.

Consider now the function f with exponential growth defined by

$$f(x) = 3 \cdot 5^{7x}.$$

Because $5^{7x} = (2^{\log_2 5})^{7x} = 2^{7(\log_2 5)x}$, we can rewrite f in the form

$$f(x) = 3 \cdot 2^{kx},$$

where $k = 7(\log_2 5)$.

There is nothing special about the numbers 5 and 7 that appear in the paragraph above. The same procedure could be applied to any function f with exponential growth defined by $f(x) = cb^{kx}$. Thus we have the following result, which shows that by changing k we can, if we wish, always take $b = 2$ in the definition given earlier of a function with exponential growth.

Exponential growth, base 2

Every function f with exponential growth can be written in the form

$$f(x) = c2^{kx},$$

where c and k are positive constants.

For some applications, the most natural choice of the base is the number e that we will investigate in Chapter 6.

In the result above, there is nothing special about the number 2. The same result holds if 2 is replaced by 3 or 4 or any number bigger than 1. In other words, we can choose the base for a function with exponential growth to be whatever we wish (and then k needs to be suitably adjusted). You will often want to choose a value for the base that is related to the topic under consideration. We will soon consider population doubling models, where 2 is the most natural choice for the base.

EXAMPLE 3

Suppose f is a function with exponential growth such that $f(2) = 3$ and $f(5) = 7$.

(a) Find a formula for $f(x)$.

(b) Evaluate $f(17)$.

SOLUTION

(a) We will use the simpler form derived above. In other words, we can assume that

$$f(x) = cb^x.$$

We need to find c and b. We have

$$3 = f(2) = cb^2 \quad \text{and} \quad 7 = f(5) = cb^5.$$

Dividing the second equation by the first equation shows that $b^3 = \frac{7}{3}$. Thus $b = \left(\frac{7}{3}\right)^{1/3}$. Substituting this value for b into the first equation above gives

$$3 = c\left(\tfrac{7}{3}\right)^{2/3},$$

which implies that $c = 3\left(\frac{3}{7}\right)^{2/3}$. Thus

$$f(x) = 3\left(\tfrac{3}{7}\right)^{2/3}\left(\tfrac{7}{3}\right)^{x/3}.$$

(b) Using the formula above, we have

$$f(17) = 3\left(\tfrac{3}{7}\right)^{2/3}\left(\tfrac{7}{3}\right)^{17/3} \approx 207.494.$$

Population Growth

Populations of various organisms, ranging from bacteria to humans, often exhibit exponential growth. To illustrate this behavior, we will begin by considering bacteria. Bacteria are single-celled creatures that reproduce by absorbing some nutrients, growing, and then dividing in half—one bacterium cell becomes two bacteria cells.

EXAMPLE 4

Suppose a colony of bacteria in a petri dish has 700 cells at 1 pm. These bacteria reproduce at a rate that leads to doubling every three hours. How many bacteria cells will be in the petri dish at 9 pm on the same day?

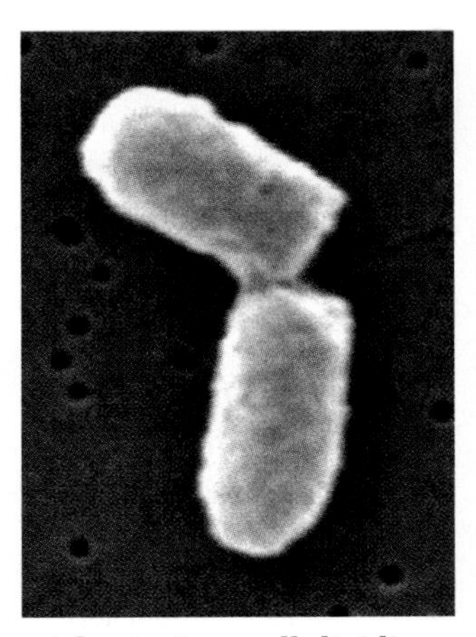

A bacterium cell dividing, as photographed by an electron microscope.

SOLUTION Because the number of bacteria cells doubles every three hours, at 4 pm there will be 1400 cells, at 7 pm there will be 2800 cells, and so on. In other words, in three hours the number of cells increases by a factor of two, in six hours the number of cells increases by a factor of four, in nine hours the number of cells increases by a factor of eight, and so on.

More generally, in t hours there are $t/3$ doubling periods. Hence in t hours the number of cells increases by a factor of $2^{t/3}$ and we should have $700 \cdot 2^{t/3}$ bacteria cells.

Thus at 9 pm, which is eight hours after 1 pm, our colony of bacteria should have $700 \cdot 2^{8/3}$ cells. However, this result should be thought of as an estimate rather than as an exact count. Actually, $700 \cdot 2^{8/3}$ is an irrational number (approximately equal to 4444.7), which makes no sense when counting bacteria cells. Thus we might predict that at 9 pm there would be about 4445 cells. Even better, because the real world rarely strictly adheres to formulas, we might expect between 4400 and 4500 cells at 9 pm.

Because functions with exponential growth increase so rapidly, they can be used to model real data for only limited time periods.

Although a function with exponential growth will often provide the best model for population growth for a certain time period, real population data cannot exhibit exponential growth for excessively long time periods. For example, the formula $700 \cdot 2^{t/3}$ derived above for our colony of bacteria predicts that after 10 days, which equals 240 hours, we would have about 10^{27} cells, which is far more than could fit in even a gigantic petri dish. The bacteria would have run out of space and nutrients long before reaching this population level.

Now we extend our example with bacteria to a more general situation. Suppose a population doubles every d time units (here the time units might be hours, days, years, or whatever unit is appropriate). Suppose also that at some specific time t_0 we know that the population is p_0. At time t there have been $t - t_0$ time units since time t_0. Thus at time t there have been $(t - t_0)/d$ doubling periods, and hence the population increases by a factor of $2^{(t-t_0)/d}$. This factor must be multiplied by the population at the starting time t_0. In other words, at time t we could expect a population of $p_0 \cdot 2^{(t-t_0)/d}$.

Exponential growth and doubling

If a population doubles every d time units, then the function p modeling this population growth is given by the formula

$$p(t) = p_0 \cdot 2^{(t-t_0)/d},$$

where p_0 is the population at time t_0.

The function p has exponential growth because we could rewrite p in the form

$$p(t) = (2^{-t_0/d} p_0) 2^{(1/d)t},$$

which corresponds to our definition of a function with exponential growth by taking $c = 2^{-t_0/d} p_0$, $b = 2$, and $k = 1/d$.

Human population data often follow patterns of exponential growth for decades or centuries. The graph below shows the logarithm of the world population for each year from 1950 to 2000:

This graph of the logarithm of world population looks remarkably like a straight line for these decades, showing that world population had exponential growth.

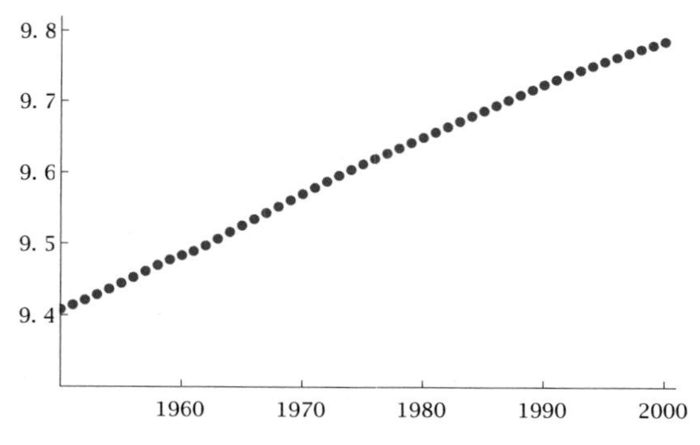

The logarithm of the world population each year from 1950 to 2000, as estimated by the U.S. Census Bureau.

EXAMPLE 5

The world population in mid-year 1950 was about 2.56 billion. During the period 1950–2000, world population increased at a rate that doubled the population approximately every 40 years.

(a) Find a formula that estimates the mid-year world population for 1950-2000.

(b) Using the formula from part (a), estimate the world population in mid-year 1955.

World population is now increasing at a slower rate, doubling about every 69 years.

SOLUTION

(a) Using the formula and data above, we see that the mid-year world population in the year y, expressed in billions, was approximately

$$2.56 \cdot 2^{(y-1950)/40}.$$

Here we are using y rather than t as the time variable.

(b) Taking $y = 1955$ in the formula above gives the estimate that the mid-year world population in 1955 was $2.56 \cdot 2^{(1955-1950)/40}$ billion, which is approximately 2.79 billion. The actual value was about 2.78 billion; thus the formula has good accuracy in this case.

Compound Interest

This Dilbert comic illustrates the power of compound interest. See the solution to Exercise 33 to learn whether this plan would work.

The computation of compound interest involves functions with exponential growth. We begin with a simple example.

EXAMPLE 6

Suppose you deposit \$8000 in a bank account that pays 5% annual interest. Assume that the bank pays interest once per year, at the end of each year, and that each year you place the interest in a cookie jar for safekeeping.

(a) How much will you have (original amount plus interest) at the end of one year?

(b) How much will you have (original amount plus interest) at the end of two years?

(c) How much will you have (original amount plus interest) at the end of t years?

SOLUTION

(a) Because 5% of \$8000 is \$400, at the end of the first year you will receive \$400 interest. Thus the total amount you will have at the end of one year is \$8400.

(b) You receive \$400 in interest at the end of the second year, bringing the amount in the cookie jar to \$800 and the total amount to \$8800.

(c) Because you receive \$400 interest each year, at the end of t years the cookie jar will contain $400t$ dollars. Thus the total amount you will have at the end of t years is $8000 + 400t$ dollars.

The symbol P comes from **principal**, *which is a fancy word for the initial amount.*

The situation in the example above, where interest is paid only on the original amount, is called **simple interest**. To generalize the example above, we can replace the \$8000 used in the example above with any initial amount P. Furthermore, we can replace the 5% annual interest with any annual interest rate r, expressed as a number rather than as a percent (thus 5% interest would correspond to $r = 0.05$). Each year the interest received will be rP. Thus after t years the total interest received will be rPt. Hence the total amount after t years will be $P + rPt$. Factoring out P from this expression, we have the following result:

Simple interest

If interest is paid once per year at annual interest rate r, with no interest paid on the interest, then after t years an initial amount P grows to

$$P(1 + rt).$$

The expression $P(1 + rt)$ that appears above is a linear function of t (assuming that the principal P and the interest rate r are constant). Thus when money grows with simple interest, linear functions arise naturally. We now turn to the more realistic situation of **compound interest**, meaning that interest is paid on the interest.

EXAMPLE 7

Suppose you deposit \$8000 in a bank account that pays 5% annual interest. Assume that the bank pays interest once per year, at the end of each year, and that each year the interest is deposited in the bank account.

(a) How much will you have at the end of one year?

(b) How much will you have at the end of two years?

(c) How much will you have at the end of three years?

(d) How much will you have at the end of t years?

SOLUTION

(a) Because 5% of \$8000 is \$400, at the end of the first year you will receive \$400 interest. Thus at the end of the first year the bank account will contain \$8400.

(b) At the end of the second year you will receive as interest 5% of \$8400, which equals \$420, which when added to the bank account gives a total of \$8820.

(c) At the end of the third year you will receive as interest 5% of \$8820, which equals \$441, which when added to the bank account gives a total of \$9261.

(d) Note that each year the amount in the bank account increases by a factor of 1.05. At the end of the first year you will have

$$8000 \times 1.05$$

dollars (which equals \$8400). At the end of two years, you will have the amount above multiplied by 1.05, which equals

$$8000 \times 1.05^2$$

dollars (which equals \$8820). At the end of three years, you will have the amount above again multiplied by 1.05, which equals

$$8000 \times 1.05^3$$

dollars (which equals \$9261). After t years, the original \$8000 will have grown to

$$8000 \times 1.05^t$$

dollars.

The table below summarizes the data for the two methods of computing interest that we have considered in the last two examples.

After the first year, compound interest produces a higher total than simple interest. This happens because with compound interest, interest is paid on the interest.

	simple interest		compound interest	
year	*interest*	*total*	*interest*	*total*
initial amount		\$8000		\$8000
1	\$400	\$8400	\$400	\$8400
2	\$400	\$8800	\$420	\$8820
3	\$400	\$9200	\$441	\$9261

Simple and compound interest, once per year, on \$8000 at 5%.

The compound interest computation done in part (d) of the last example can be extended to more general situations. To generalize the example above, we can replace the \$8000 used in the example above with any initial amount P. Furthermore, we can replace the 5% annual interest with any annual interest rate r, expressed as a number rather than as a percent. Each year the amount in the bank account increases by a factor of $1 + r$. Thus at the end of the first year the initial amount P will grow to $P(1 + r)$. At the end of two years, this will have grown to $P(1 + r)^2$. At the end of three years, this will have grown to $P(1 + r)^3$. More generally, we have the following result:

The variable t is used as a reminder that it denotes a time period.

Compound interest, once per year

If interest is compounded once per year at annual interest rate r, then after t years an initial amount P grows to

$$P(1+r)^t.$$

The expression $P(1+r)^t$ that appears above has exponential growth as a function of t (assuming that the principal P and the interest rate r are constant). Thus we see that when money grows with compound interest, functions with exponential growth arise naturally. Because functions with exponential growth increase rapidly, compound interest can lead to large amounts of money after long time periods.

EXAMPLE 8

Little historical evidence exists concerning the alleged sale of Manhattan. Most of the stories about this event should be considered legends.

Today Manhattan contains well-known New York City landmarks such as Times Square, the Empire State Building, Wall Street, and United Nations headquarters.

In 1626 Dutch settlers supposedly purchased from Native Americans the island of Manhattan for \$24. To determine whether or not this was a bargain, suppose \$24 earned 7% per year (a reasonable rate for a real estate investment), compounded once per year since 1626. How much would this investment be worth by 2011?

SOLUTION
Because $2011 - 1626 = 385$, the formula above shows that an initial amount of \$24 earning 7% per year compounded once per year would be worth

$$24(1.07)^{385}$$

dollars in 2011. A calculator shows that this is about five trillion dollars, which is more than the current assessed value of all land in Manhattan.

Interest is often compounded more than once per year. To see how this works, we now modify an earlier example. In our new example, interest will be paid and compounded twice per year rather than once per year. This means that instead of 5% interest being paid at the end of each year, the interest comes as two payments of 2.5% each year, with the 2.5% interest payments made at the end of every six months.

EXAMPLE 9

Suppose you deposit \$8000 in a bank account that pays 5% annual interest, compounded twice per year. How much will you have at the end of one year?

SOLUTION Because 2.5% of \$8000 equals \$200, at the end of the first six months \$200 will be deposited into the bank account; the bank account will then have a total of \$8200.

At the end of the second six months (in other words, at the end of the first year), 2.5% interest will be paid on the $8200 that was in the bank account for the previous six months. Because 2.5% of $8200 equals $205, the bank account will have $8405 at the end of the first year.

In the example above, the $8405 in the bank account at the end of the first year should be compared with the $8400 that would be in the bank account if interest had been paid only at the end of the year. The extra $5 arises because of the interest during the second six months on the interest earned at the end of the first six months.

As usual with compounding, the interest on the interest adds to the total.

Instead of compounding interest twice per year, as in the previous example, in the next example we will assume that interest is compounded four times per year. At 5% annual interest, this would mean that 1.25% interest will be paid at the end of every three months.

EXAMPLE 10

Suppose you deposit $8000 in a bank account that pays 5% annual interest, compounded four times per year. How much will you have at the end of one year?

SOLUTION Because 1.25% of $8000 equals $100, at the end of the first three months $100 will be deposited into the bank account; the bank account will then have a total of $8100.

At the end of the second three months (in other words, at the end of six months), 1.25% interest will be paid on the $8100 that was in the bank account for the previous three months. Because 1.25% of $8100 equals $101.25, the bank account will have $8201.25 at the end of the first six months.

A similar calculation shows that the bank account will have $8303.77 at the end of the first nine months and $8407.56 at the end of the first year.

Compare the results of the last two examples. Note that the bank account will contain $8405 at the end of the first year if compounding twice per year but $8407.56 if compounding four times per year.

If interest is compounded 12 times per year (at the end of each month), then we can expect a higher total than when interest is compounded 4 times per year. The table below shows the growth of $8000 at 5% interest for three years, with compounding either 1, 2, 4, or 12 times per year.

More frequent compounding leads to higher total amounts because more frequent interest payments give more time for interest to earn on the interest. In Chapter 6 we will discuss what happens when interest is compounded a huge number of times per year.

	times compounded per year			
year	1	2	4	12
initial amount	$8000	$8000	$8000	$8000
1	$8400	$8405	$8408	$8409
2	$8820	$8831	$8836	$8840
3	$9261	$9278	$9286	$9292

The growth of $8000 at 5% interest, rounded to the nearest dollar.

To find a formula for how money grows when compounded more than once per year, consider a bank account with annual interest rate r, compounded twice per year. Thus every six months, the amount in the bank account increases by a factor of $1 + \frac{r}{2}$. After t years, this will happen $2t$ times. Thus an initial amount P will grow to $P(1 + \frac{r}{2})^{2t}$ in t years.

More generally, suppose now that an annual interest rate r is compounded n times per year. Then n times per year, the amount in the bank account increases by a factor of $1 + \frac{r}{n}$. After t years, this will happen nt times, leading to the following result:

If the interest rate r and the compounding times per year n are fixed, then the function f defined by $f(t) = P(1 + \frac{r}{n})^{nt}$ is a function with exponential growth.

Compound interest, n times per year

If interest is compounded n times per year at annual interest rate r, then after t years an initial amount P grows to

$$P(1 + \tfrac{r}{n})^{nt}.$$

EXAMPLE 11

Suppose a bank account starting out with \$8000 receives 5% annual interest, compounded twelve times per year. How much will be in the bank account after three years?

SOLUTION Take $r = 0.05$, $n = 12$, $t = 3$, and $P = 8000$ in the formula above, which shows that after three years the amount in the bank account will be

$$8000(1 + \tfrac{0.05}{12})^{12 \cdot 3}$$

dollars. A calculator shows that this amount is approximately \$9292 (which is the last entry in the table above).

Advertisements from financial institutions often list the "APY" that you will earn on your money rather than the interest rate. The abbreviation "APY" denotes "annual percentage yield", which means the actual interest rate that you would receive at the end of one year after compounding.

For example, if a bank is paying 5% annual interest, compounded once per month (as is fairly common), then the bank can legally advertise that it pays an APY of 5.116%. Here the APY equals 5.116% because

$$1.05116 = (1 + \tfrac{0.05}{12})^{12}.$$

In other words, at 5% annual interest compounded twelve times per year, \$1000 will grow to \$1051.16. For a period of one year, this corresponds to simple annual interest of 5.116%.

EXERCISES

1. Without using a calculator or computer, give a rough estimate of 2^{83}.

2. Without using a calculator or computer, give a rough estimate of 2^{103}.

3. Without using a calculator or computer, determine which of the two numbers 2^{125} and $32 \cdot 10^{36}$ is larger.

4. Without using a calculator or computer, determine which of the two numbers 2^{400} and 17^{100} is larger.
[*Hint:* Note that $2^4 = 16$.]

For Exercises 5–8, suppose you deposit into a savings account one cent on January 1, two cents on January 2, four cents on January 3, and so on, doubling the amount of your deposit each day (assume that you use an electronic bank that is open every day of the year).

5. How much will you deposit on January 7?

6. How much will you deposit on January 11?

7. What is the first day that your deposit will exceed \$10,000?

8. What is the first day that your deposit will exceed \$100,000?

For Exercises 9–12, suppose you deposit into your savings account one cent on January 1, three cents on January 2, nine cents on January 3, and so on, tripling the amount of your deposit each day.

9. How much will you deposit on January 7?

10. How much will you deposit on January 11?

11. What is the first day that your deposit will exceed \$10,000?

12. What is the first day that your deposit will exceed \$100,000?

13. Suppose $f(x) = 7 \cdot 2^{3x}$. Find a constant b such that the graph of $\log_b f$ has slope 1.

14. Suppose $f(x) = 4 \cdot 2^{5x}$. Find a constant b such that the graph of $\log_b f$ has slope 1.

15. A colony of bacteria is growing exponentially, doubling in size every 100 minutes. How many minutes will it take for the colony of bacteria to triple in size?

16. A colony of bacteria is growing exponentially, doubling in size every 140 minutes. How many minutes will it take for the colony of bacteria to become five times its current size?

17. At current growth rates, the Earth's population is doubling about every 69 years. If this growth rate were to continue, about how many years will it take for the Earth's population to increase 50% from the present level?

18. At current growth rates, the Earth's population is doubling about every 69 years. If this growth rate were to continue, about how many years will it take for the Earth's population to become one-fourth larger than the current level?

19. Suppose a colony of bacteria starts with 200 cells and triples in size every four hours.

    (a) Find a function that models the population growth of this colony of bacteria.

    (b) Approximately how many cells will be in the colony after six hours?

20. Suppose a colony of bacteria starts with 100 cells and triples in size every two hours.

    (a) Find a function that models the population growth of this colony of bacteria.

    (b) Approximately how many cells will be in the colony after one hour?

21. Suppose \$700 is deposited in a bank account paying 6% interest per year, compounded 52 times per year. How much will be in the bank account at the end of 10 years?

22. Suppose \$8000 is deposited in a bank account paying 7% interest per year, compounded 12 times per year. How much will be in the bank account at the end of 100 years?

23. Suppose a bank account paying 4% interest per year, compounded 12 times per year, contains \$10,555 at the end of 10 years. What was the initial amount deposited in the bank account?

24. Suppose a bank account paying 6% interest per year, compounded four times per year, contains \$27,707 at the end of 20 years. What was the initial amount deposited in the bank account?

25. Suppose a savings account pays 6% interest per year, compounded once per year. If the savings account starts with \$500, how long would it take for the savings account to exceed \$2000?

26. Suppose a savings account pays 5% interest per year, compounded four times per year. If the savings account starts with \$600, how many years would it take for the savings account to exceed \$1400?

27. Suppose a bank wants to advertise that \$1000 deposited in its savings account will grow to \$1040 in one year. This bank compounds interest 12 times per year. What annual interest rate must the bank pay?

28. Suppose a bank wants to advertise that \$1000 deposited in its savings account will grow to \$1050 in one year. This bank compounds interest 365 times per year. What annual interest rate must the bank pay?

29. An advertisement for real estate published in the 28 July 2004 *New York Times* states:

> Did you know that the percent increase of the value of a home in Manhattan between the years 1950 and 2000 was 721%? Buy a home in Manhattan and invest in your future.

Suppose that instead of buying a home in Manhattan in 1950, someone had invested money in a bank account that compounds interest four times per year. What annual interest rate would the bank have to pay to equal the growth claimed in the ad?

30. Suppose that instead of buying a home in Manhattan in 1950, someone had invested money in a bank account that compounds interest once per month. What annual interest rate would the bank have to pay to equal the growth claimed in the ad from the previous exercise?

31. Suppose f is a function with exponential growth such that

$$f(1) = 3 \quad \text{and} \quad f(3) = 5.$$

Evaluate $f(8)$.

32. Suppose f is a function with exponential growth such that

$$f(2) = 3 \quad \text{and} \quad f(5) = 8.$$

Evaluate $f(10)$.

Exercises 33–34 will help you determine whether or not the Dilbert comic earlier in this section gives a reasonable method for turning a hundred dollars into a million dollars.

33. At 5% interest compounded once per year, how many years will it take to turn a hundred dollars into a million dollars?

34. At 5% interest compounded monthly, how long will it take to turn a hundred dollars into a million dollars?

PROBLEMS

35. Explain how you would use a calculator to verify that

$$2^{13746} < 13746^{1000}$$

but

$$2^{13747} > 13747^{1000},$$

and then actually use a calculator to verify both these inequalities.
[*The numbers involved in these inequalities have over four thousand digits. Thus some cleverness in using your calculator is required.*]

36. Show that

$$2^{10n} = (1.024)^n 10^{3n}.$$

[*This equality leads to the approximation* $2^{10n} \approx 10^{3n}$.]

37. Show that if f is a function with exponential growth, then so is the square root of f. More precisely, show that if f is a function with exponential growth, then so is the function g defined by $g(x) = \sqrt{f(x)}$.

38. Suppose f is a function with exponential growth and $f(0) = 1$. Explain why f can be represented by a formula of the form $f(x) = b^x$ for some $b > 1$.

39. Explain why every function f with exponential growth can be represented by a formula of the form $f(x) = c \cdot 3^{kx}$ for appropriate choices of c and k.

40. Find at least three newspaper articles that use the word "exponentially" (the easiest way to do this is to use the web site of a newspaper that allows searches of its articles). For each use of the word "exponentially" that you find in a newspaper article, discuss whether the word is used in its correct mathematical sense.
[*In a recent year the word "exponentially" appeared 87 times in the* New York Times.]

41. Suppose a bank pays annual interest rate r, compounded n times per year. Explain why the bank can advertise that its APY equals

$$\left(1 + \tfrac{r}{n}\right)^n - 1.$$

42. Find an advertisement in a newspaper or web site that gives the interest rate (before compounding), the frequency of compounding, and the APY. Determine whether or not the APY has been computed correctly.

43. Suppose f is a function with exponential growth. Show that there is a constant $b > 1$ such that

$$f(x+1) = bf(x)$$

for every x.

WORKED-OUT SOLUTIONS *to Odd-numbered Exercises*

1. Without using a calculator or computer, give a rough estimate of 2^{83}.

SOLUTION

$$2^{83} = 2^3 \cdot 2^{80} = 8 \cdot 2^{10 \cdot 8} = 8 \cdot (2^{10})^8$$
$$\approx 8 \cdot (10^3)^8 = 8 \cdot 10^{24} \approx 10^{25}$$

3. Without using a calculator or computer, determine which of the two numbers 2^{125} and $32 \cdot 10^{36}$ is larger.

SOLUTION Note that

$$\begin{aligned} 2^{125} &= 2^5 \cdot 2^{120} \\ &= 32 \cdot (2^{10})^{12} \\ &> 32 \cdot (10^3)^{12} \\ &= 32 \cdot 10^{36}. \end{aligned}$$

Thus 2^{125} is larger than $32 \cdot 10^{36}$.

For Exercises 5–8, suppose you deposit into a savings account one cent on January 1, two cents on January 2, four cents on January 3, and so on, doubling the amount of your deposit each day (assume that you use an electronic bank that is open every day of the year).

5. How much will you deposit on January 7?

SOLUTION On the nth day, 2^{n-1} cents are deposited. Thus on January 7, the amount deposited is 2^6 cents. In other words, \$0.64 will be deposited on January 7.

7. What is the first day that your deposit will exceed \$10,000?

SOLUTION On the nth day, 2^{n-1} cents are deposited. Because \$10,000 equals 10^6 cents, we need to find the smallest integer n such that

$$2^{n-1} > 10^6.$$

We can do a quick estimate by noting that

$$10^6 = (10^3)^2 < (2^{10})^2 = 2^{20}.$$

Thus taking $n - 1 = 20$, which is equivalent to taking $n = 21$, should be close to the correct answer.

To be more precise, note that the inequality $2^{n-1} > 10^6$ is equivalent to the inequality

$$\log 2^{n-1} > \log 10^6,$$

which can be rewritten as

$$(n-1)\log 2 > 6.$$

Dividing both sides by $\log 2$ and then adding 1 to both sides shows that this is equivalent to

$$n > 1 + \frac{6}{\log 2}.$$

A calculator shows that $1 + \frac{6}{\log 2} \approx 20.9$. Because 21 is the smallest integer bigger than 20.9, January 21 is the first day that the deposit will exceed \$10,000.

For Exercises 9–12, suppose you deposit into your savings account one cent on January 1, three cents on January 2, nine cents on January 3, and so on, tripling the amount of your deposit each day.

9. How much will you deposit on January 7?

SOLUTION On the nth day, 3^{n-1} cents are deposited. Thus on January 7, the amount deposited is 3^6 cents. Because $3^6 = 729$, we conclude that \$7.29 will be deposited on January 7.

11. What is the first day that your deposit will exceed \$10,000?

SOLUTION On the nth day, 3^{n-1} cents are deposited. Because \$10,000 equals 10^6 cents, we need to find the smallest integer n such that

$$3^{n-1} > 10^6.$$

This is equivalent to the inequality

$$\log 3^{n-1} > \log 10^6,$$

which can be rewritten as

$$(n-1)\log 3 > 6.$$

Dividing both sides by $\log 3$ and then adding 1 to both sides shows that this is equivalent to

$$n > 1 + \frac{6}{\log 3}.$$

A calculator shows that $1 + \frac{6}{\log 3} \approx 13.6$. Because 14 is the smallest integer bigger than 13.6, January 14 is the first day that the deposit will exceed \$10,000.

13. Suppose $f(x) = 7 \cdot 2^{3x}$. Find a constant b such that the graph of $\log_b f$ has slope 1.

SOLUTION Note that

$$\begin{aligned}\log_b f(x) &= \log_b 7 + \log_b 2^{3x}\\ &= \log_b 7 + 3(\log_b 2)x.\end{aligned}$$

Thus the slope of the graph of $\log_b f$ equals $3\log_b 2$, which equals 1 when $\log_b 2 = \frac{1}{3}$. Thus $b^{1/3} = 2$, which means that $b = 2^3 = 8$.

15. A colony of bacteria is growing exponentially, doubling in size every 100 minutes. How many minutes will it take for the colony of bacteria to triple in size?

SOLUTION Let $p(t)$ denote the number of cells in the colony of bacteria at time t, where t is measured in minutes. Then

$$p(t) = p_0 2^{t/100},$$

where p_0 is the number of cells at time 0. We need to find t such that $p(t) = 3p_0$. In other words, we need to find t such that

$$p_0 2^{t/100} = 3p_0.$$

Dividing both sides of the equation above by p_0 and then taking the logarithm of both sides gives

$$\frac{t}{100}\log 2 = \log 3.$$

Thus $t = 100\frac{\log 3}{\log 2}$, which is approximately 158.496. Thus the colony of bacteria will triple in size approximately every 158 minutes.

17. At current growth rates, the Earth's population is doubling about every 69 years. If this growth rate were to continue, about how many years will it take for the Earth's population to increase 50% from the present level?

SOLUTION Let $p(t)$ denote the Earth's population at time t, where t is measured in years starting from the present. Then

$$p(t) = p_0 2^{t/69},$$

where p_0 is the present population of the Earth. We need to find t such that $p(t) = 1.5p_0$. In other words, we need to find t such that

$$p_0 2^{t/69} = 1.5p_0.$$

Dividing both sides of the equation above by p_0 and then taking the logarithm of both sides gives

$$\frac{t}{69}\log 2 = \log 1.5.$$

Thus $t = 69\frac{\log 1.5}{\log 2}$, which is approximately 40.4. Thus the Earth's population, at current growth rates, would increase by 50% in approximately 40.4 years.

19. Suppose a colony of bacteria starts with 200 cells and triples in size every four hours.

    (a) Find a function that models the population growth of this colony of bacteria.

    (b) Approximately how many cells will be in the colony after six hours?

SOLUTION

(a) Let $p(t)$ denote the number of cells in the colony of bacteria at time t, where t is measured in hours. We know that $p(0) = 200$. In t hours, there are $t/4$ tripling periods; thus the number of cells increases by a factor of $3^{t/4}$. Hence

$$p(t) = 200 \cdot 3^{t/4}.$$

(b) After six hours, we could expect that there would be $p(6)$ cells of bacteria. Using the equation above, we have

$$p(6) = 200 \cdot 3^{6/4} = 200 \cdot 3^{3/2} \approx 1039.$$

21. Suppose \$700 is deposited in a bank account paying 6% interest per year, compounded 52 times per year. How much will be in the bank account at the end of 10 years?

SOLUTION With interest compounded 52 times per year at 6% per year, after 10 years \$700 will grow to

$$\$700(1 + \tfrac{0.06}{52})^{52 \cdot 10} \approx \$1275.$$

23. Suppose a bank account paying 4% interest per year, compounded 12 times per year, contains \$10,555 at the end of 10 years. What was the initial amount deposited in the bank account?

SOLUTION Let P denote the initial amount deposited in the bank account. With interest compounded 12 times per year at 4% per year, after 10 years P dollars will grow to

$$P(1 + \tfrac{0.04}{12})^{12 \cdot 10}$$

dollars, which we are told equals \$10,555. Thus we need to solve the equation

$$P(1 + \tfrac{0.04}{12})^{120} = \$10{,}555.$$

The solution to this equation is

$$P = \$10{,}555/(1 + \tfrac{0.04}{12})^{120} \approx \$7080.$$

25. Suppose a savings account pays 6% interest per year, compounded once per year. If the savings account starts with \$500, how long would it take for the savings account to exceed \$2000?

SOLUTION With 6% interest compounded once per year, a savings account starting with \$500 would have

$$500(1.06)^t$$

dollars after t years. We want this amount to exceed \$2000, which means that

$$500(1.06)^t > 2000.$$

Dividing both sides by 500 and then taking the logarithm of both sides gives

$$t \log 1.06 > \log 4.$$

Thus

$$t > \frac{\log 4}{\log 1.06} \approx 23.8.$$

Because interest is compounded only once per year, t needs to be an integer. The smallest integer larger than 23.8 is 24. Thus it will take 24 years for the amount in the savings account to exceed \$2000.

27. Suppose a bank wants to advertise that \$1000 deposited in its savings account will grow to \$1040 in one year. This bank compounds interest 12 times per year. What annual interest rate must the bank pay?

SOLUTION Let r denote the annual interest rate to be paid by the bank. At that interest rate, compounded 12 times per year, in one year \$1000 will grow to

$$1000(1 + \tfrac{r}{12})^{12}$$

dollars. We want this to equal \$1040, which means that we need to solve the equation

$$1000(1 + \tfrac{r}{12})^{12} = 1040.$$

To solve this equation, divide both sides by 1000 and then raise both sides to the power 1/12, getting

$$1 + \tfrac{r}{12} = 1.04^{1/12}.$$

Now subtract 1 from both sides and then multiply both sides by 12, getting

$$r = 12(1.04^{1/12} - 1) \approx 0.0393.$$

Thus the annual interest should be approximately 3.93%.

29. An advertisement for real estate published in the 28 July 2004 *New York Times* states:

> Did you know that the percent increase of the value of a home in Manhattan between the years 1950 and 2000 was 721%? Buy a home in Manhattan and invest in your future.

Suppose that instead of buying a home in Manhattan in 1950, someone had invested money in a bank account that compounds interest four times per year. What annual interest rate would the bank have to pay to equal the growth claimed in the ad?

SOLUTION An increase of 721% means that the final value is 821% of the initial value. Let r denote the interest rate the bank would have to pay for the 50 years from 1950 to 2000 to grow to 821% of the initial value. At that interest rate, compounded four times per year, in 50 years an initial amount of P dollars grows to

$$P(1 + \tfrac{r}{4})^{4\times 50}$$

dollars. We want this to equal 8.21 times the initial amount, which means that we need to solve the equation

$$P(1 + \tfrac{r}{4})^{200} = 8.21P.$$

To solve this equation, divide both sides by P and then raise both sides to the power $1/200$, getting

$$1 + \tfrac{r}{4} = 8.21^{1/200}.$$

Now subtract 1 from both sides and then multiply both sides by 4, getting

$$r = 4(8.21^{1/200} - 1) \approx 0.0423.$$

Thus the annual interest would need to be approximately 4.23% to equal the growth claimed in the ad.

[*Note that 4.23% is not a particularly high return for a long-term investment, contrary to the ad's implication.*]

31. Suppose f is a function with exponential growth such that

$$f(1) = 3 \quad \text{and} \quad f(3) = 5.$$

Evaluate $f(8)$.

SOLUTION We can assume that

$$f(x) = cb^x.$$

We need to find c and b. We have

$$3 = f(1) = cb \quad \text{and} \quad 5 = f(3) = cb^3.$$

Dividing the second equation by the first equation shows that $b^2 = \frac{5}{3}$. Thus $b = (\frac{5}{3})^{1/2}$. Substituting this value for b into the first equation above gives

$$3 = c(\tfrac{5}{3})^{1/2},$$

which implies that $c = 3(\frac{3}{5})^{1/2}$. Thus

$$f(x) = 3(\tfrac{3}{5})^{1/2}(\tfrac{5}{3})^{x/2}.$$

Using the formula above, we have

$$f(8) = 3(\tfrac{3}{5})^{1/2}(\tfrac{5}{3})^4 = \tfrac{625}{27}(\tfrac{3}{5})^{1/2} \approx 17.93.$$

Exercises 33–34 will help you determine whether or not the Dilbert comic earlier in this section gives a reasonable method for turning a hundred dollars into a million dollars.

33. At 5% interest compounded once per year, how many years will it take to turn a hundred dollars into a million dollars?

SOLUTION We want to find t so that

$$10^6 = 100 \times 1.05^t.$$

Thus $1.05^t = 10^4$. Take logarithms of both sides, getting $t \log 1.05 = 4$. Thus $t = \frac{4}{\log 1.05} \approx 188.8$. Because interest is compounded only once per year, we round up to the next year, concluding that it will take 189 years to turn a hundred dollars into a million dollars at 5% annual interest compounded once per year. Hence the comic is correct in stating that there will be at least a million dollars after 190 years. In fact, the comic could have used 189 years instead of 190 years.

CHAPTER SUMMARY

To check that you have mastered the most important concepts and skills covered in this chapter, make sure that you can do each item in the following list:

- Manipulate and simplify expressions involving exponents.
- Define logarithms.
- Use the change of base formula for logarithms.
- Use the formulas for the logarithm of a product, quotient, and power.
- Use common logarithms to determine how many digits a number has.
- Model population growth.
- Compute compound interest.
- Model radioactive decay using half-life.
- Use logarithmic scales for measuring earthquakes, sound, and stars.

To review a chapter, go through the list above to find items that you do not know how to do, then reread the material in the chapter about those items. Then try to answer the chapter review questions below without looking back at the chapter.

CHAPTER REVIEW QUESTIONS

1. Explain why $\sqrt{5}^2 = 5$.

2. Give an example of a number t such that $\sqrt{t^2} \neq t$.

3. Show that $(29 + 12\sqrt{5})^{1/2} = 3 + 2\sqrt{5}$.

4. Evaluate $32^{7/5}$.

5. Expand $(4 - 3\sqrt{5x})^2$.

6. What is the domain of the function f defined by $f(x) = x^{3/5}$?

7. What is the domain of the function f defined by $f(x) = (x - 5)^{3/4}$?

8. Find the inverse of the function f defined by
$$f(x) = 3 + 2x^{4/5}.$$

9. Find a formula for $(f \circ g)(x)$, where
$$f(x) = 3x^{\sqrt{32}} \quad \text{and} \quad g(x) = x^{\sqrt{2}}.$$

10. Explain how logarithms are defined.

11. What is the domain of the function f defined by $f(x) = \log_2(5x + 1)$?

12. What is the range of the function f defined by $f(x) = \log_7 x$?

13. Explain why
$$3^{\log_3 7} = 7.$$

14. Explain why
$$\log_5 5^{444} = 444.$$

15. Without using a calculator or computer, estimate the number of digits in 2^{1000}.

16. Find all numbers x such that
$$\log x + \log(x + 2) = 1.$$

17. Evaluate $\log_5 \sqrt{125}$.

18. Find a number b such that $\log_b 9 = -2$.

19. How many digits does 4^{7000} have?

20. At the time this book was written, the largest known prime number not of the form $2^n - 1$ was $19249 \cdot 2^{13018586} + 1$. How many digits does this prime number have?

21. Find the smallest integer m such that
$$8^m > 10^{500}.$$

22. Find the largest integer k such that
$$15^k < 11^{900}.$$

23. Explain why

$$\log 200 = 2 + \log 2.$$

24. Explain why

$$\log \sqrt{300} = 1 + \frac{\log 3}{2}.$$

25. Which of the expressions

$$\log x + \log y \quad \text{and} \quad (\log x)(\log y)$$

can be rewritten using only one log?

26. Which of the expressions

$$\log x - \log y \quad \text{and} \quad \frac{\log x}{\log y}$$

can be rewritten using only one log?

27. Find a formula for the inverse of the function f defined by

$$f(x) = 4 + 5\log_3(7x + 2).$$

28. Find a formula for $(f \circ g)(x)$, where

$$f(x) = 7^{4x} \quad \text{and} \quad g(x) = \log_7 x.$$

29. Find a formula for $(f \circ g)(x)$, where

$$f(x) = \log_2 x \quad \text{and} \quad g(x) = 2^{5x-9}.$$

30. Evaluate $\log_{3.2} 456$.

31. Suppose $\log_6 t = 4.3$. Evaluate $\log_6 t^{200}$.

32. Suppose $\log_7 w = 3.1$ and $\log_7 z = 2.2$. Evaluate

$$\log_7 \frac{49w^2}{z^3}.$$

33. Suppose \$7000 is deposited in a bank account paying 4% interest per year, compounded 12 times per year. How much will be in the bank account at the end of 50 years?

34. Suppose \$5000 is deposited in a bank account that compounds interest four times per year. The bank account contains \$9900 after 13 years. What is the annual interest rate for this bank account?

35. A colony that initially contains 100 bacteria cells is growing exponentially, doubling in size every 75 minutes. Approximately how many bacteria cells will the colony have after 6 hours?

36. A colony of bacteria is growing exponentially, doubling in size every 50 minutes. How many minutes will it take for the colony to become six times its current size?

37. A colony of bacteria is growing exponentially, increasing in size from 200 to 500 cells in 100 minutes. How many minutes does it take the colony to double in size?

38. Explain why a population cannot have exponential growth indefinitely.

39. About how many years will it take for a sample of cesium-137, which has a half-life of 30 years, to have only 3% as much cesium-137 as the original sample?

40. How many more times intense is an earthquake with Richter magnitude 6.8 than an earthquake with Richter magnitude 6.1?

41. Explain why adding ten decibels to a sound multiplies the intensity of the sound by a factor of 10.

42. Most stars have an apparent magnitude that is a positive number. However, four stars (not counting the sun) have an apparent magnitude that is a negative number. Explain how a star can have a negative magnitude.

CHAPTER

6

The St. Louis Gateway Arch, the tallest national monument in the United States. The shape of this arch comes directly from the exponential function involving e that we will learn about in this chapter.

e and the Natural Logarithm

We begin this chapter by learning how to use rectangles to estimate the area of a region bounded by a curve. These ideas will then lead us to the magical number e as well as to the natural logarithm and the exponential function.

Our approach to e and the natural logarithm via area will easily lead us to several important approximations that show the special properties of e. These approximations demonstrate why the natural logarithm deserves its name.

This chapter concludes by relooking at exponential growth through the lens of our new knowledge of e. We will see how e is used to model continuously compounded interest and continuous growth rates. Finally, e will be used to find a simple formula showing how long it takes to double your money.

6.1 Defining e and ln

LEARNING OBJECTIVES

By the end of this section you should be able to

- approximate the area under a curve using rectangles;
- explain the definition of e;
- explain the definition of the natural logarithm and its connection with area;
- work comfortably with the exponential and natural logarithm functions.

Estimating Area Using Rectangles

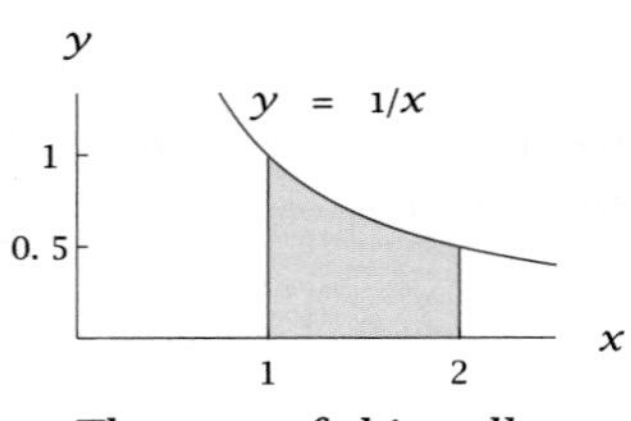

The area of this yellow region is denoted by area$(\frac{1}{x}, 1, 2)$.

The basic idea for calculating the area of a region bounded by a curve is to approximate the region by rectangles. We will illustrate this idea using the curve $y = \frac{1}{x}$, which will lead us to e (one of the most useful numbers in mathematics) and to the natural logarithm.

We begin by considering the yellow region shown here, whose area is denoted by area$(\frac{1}{x}, 1, 2)$. In other words, area$(\frac{1}{x}, 1, 2)$ equals the area of the region in the xy-plane under the curve $y = \frac{1}{x}$, above the x-axis, and between the lines $x = 1$ and $x = 2$.

The next example shows how to obtain a rough estimate of the area in the crudest possible fashion by using only one rectangle.

EXAMPLE 1

Show that

$$\text{area}(\tfrac{1}{x}, 1, 2) < 1$$

by enclosing the yellow region above in a single rectangle.

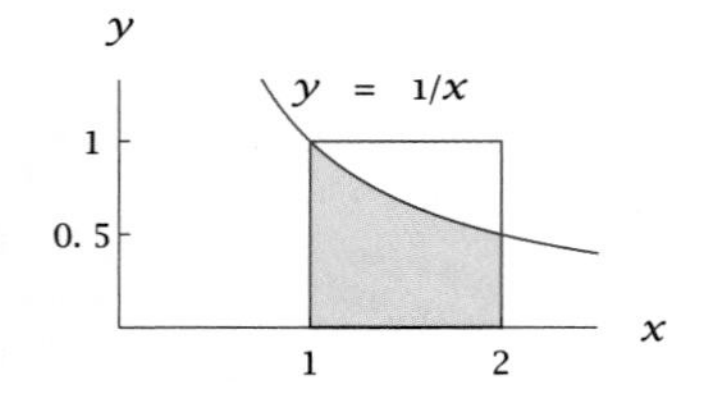

SOLUTION The smallest rectangle (with sides parallel to the coordinate axes) that contains the yellow region is the 1-by-1 square shown here.

Because the yellow region lies inside the 1-by-1 square, the figure here allows us to conclude that the area of the yellow region is less than 1. In other words,

$$\text{area}(\tfrac{1}{x}, 1, 2) < 1.$$

Now consider the yellow region shown below:

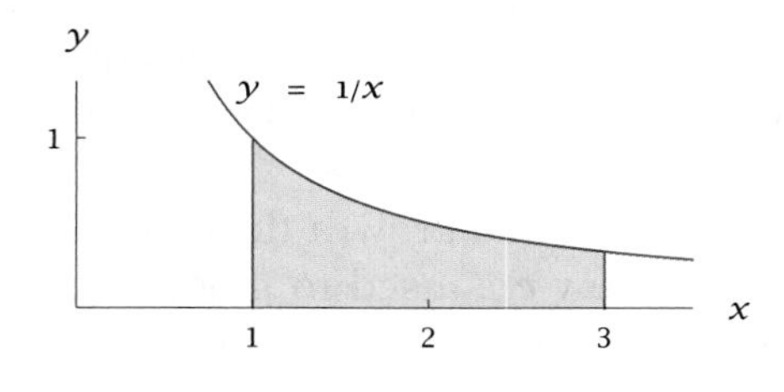

The area of this yellow region is denoted by area$(\frac{1}{x}, 1, 3)$.

The area of the yellow region above is denoted by area$(\frac{1}{x}, 1, 3)$. In other words, area$(\frac{1}{x}, 1, 3)$ equals the area of the region in the xy-plane under the curve $y = \frac{1}{x}$, above the x-axis, and between the lines $x = 1$ and $x = 3$.

The next example illustrates the procedure for approximating the area of a region by placing rectangles inside the region.

EXAMPLE 2

Show that area$(\frac{1}{x}, 1, 3) > 1$ by placing eight rectangles, each with the same size base, inside the yellow region above.

SOLUTION Place eight rectangles under the curve, as shown in the figure below:

Unlike the previous crude example, this time we use eight rectangles to get a more accurate estimate.

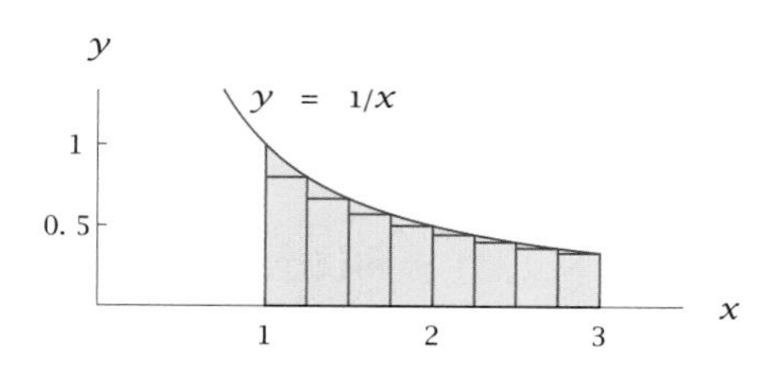

We have divided the interval $[1, 3]$ into eight intervals of equal size. The interval $[1, 3]$ has length 2. Thus the base of each rectangle has length $\frac{2}{8}$, which equals $\frac{1}{4}$.

The base of the first rectangle is the interval $[1, \frac{5}{4}]$. The figure above shows that the height of this first rectangle is $1/\frac{5}{4}$, which equals $\frac{4}{5}$. Because the first rectangle has base $\frac{1}{4}$ and height $\frac{4}{5}$, the area of the first rectangle equals $\frac{1}{4} \cdot \frac{4}{5}$, which equals $\frac{1}{5}$.

The base of the second rectangle is the interval $[\frac{5}{4}, \frac{3}{2}]$. The height of the second rectangle is $1/\frac{3}{2}$, which equals $\frac{2}{3}$. Thus the area of the second rectangle equals $\frac{1}{4} \cdot \frac{2}{3}$, which equals $\frac{1}{6}$.

The area of the third rectangle is computed in the same fashion. Specifically, the third rectangle has base $\frac{1}{4}$ and height $1/\frac{7}{4}$, which equals $\frac{4}{7}$. Thus the area of the third rectangle equals $\frac{1}{4} \cdot \frac{4}{7}$, which equals $\frac{1}{7}$.

The first three rectangles have area $\frac{1}{5}$, $\frac{1}{6}$, and $\frac{1}{7}$, as we have now computed. From this data, you might guess that the eight rectangles have area $\frac{1}{5}$, $\frac{1}{6}$, $\frac{1}{7}$, $\frac{1}{8}$, $\frac{1}{9}$, $\frac{1}{10}$, $\frac{1}{11}$, and $\frac{1}{12}$. This guess is correct, as you should verify using the same procedure as used above.

Thus the sum of the areas of all eight rectangles is

$$\frac{1}{5} + \frac{1}{6} + \frac{1}{7} + \frac{1}{8} + \frac{1}{9} + \frac{1}{10} + \frac{1}{11} + \frac{1}{12},$$

which equals $\frac{28271}{27720}$. Because these eight rectangles lie inside the yellow region, the area of the region is larger than the sum of the areas of the rectangles. Hence

$$\text{area}(\tfrac{1}{x}, 1, 3) > \tfrac{28271}{27720}.$$

The fraction on the right has a larger numerator than denominator; thus this fraction is larger than 1. Hence without further computation the inequality above shows that

$$\text{area}(\tfrac{1}{x}, 1, 3) > 1.$$

The inequality here goes in the opposite direction from the inequality in the previous example. Now we are placing the rectangles under the curve rather than above it.

In the example above, $\frac{28271}{27720}$ gives us an estimate for area$(\frac{1}{x}, 1, 3)$. If we want a more accurate estimate, we could use more and thinner rectangles under the curve.

The table below shows the sum of the areas of the rectangles under the curve for several different choices of the number of rectangles. Here we are assuming that the rectangles all have bases of the same size, as in the example above. The sums have been rounded off to five digits.

The sum of the areas of these rectangles was calculated with the aid of a computer.

number of rectangles	*sum of area of rectangles*
10	1.0349
100	1.0920
1000	1.0979
10000	1.0985
100000	1.0986

Estimates of area$(\frac{1}{x}, 1, 3)$.

The actual value of area$(\frac{1}{x}, 1, 3)$ is an irrational number whose first five digits are 1.0986, which agrees with the last entry in the table above.

In summary, we can get an accurate estimate of the area of the yellow region by dividing the interval $[1, 3]$ into many small intervals, then computing the sum of the areas of the corresponding rectangles that lie under the curve.

Defining e

The area under portions of the curve $y = \frac{1}{x}$ has some remarkable properties. To discuss these properties, we introduce the following notation, which we have already used for $c = 2$ and $c = 3$:

area$(\frac{1}{x}, 1, c)$

For $c > 1$, let area$(\frac{1}{x}, 1, c)$ denote the area of the yellow region below:

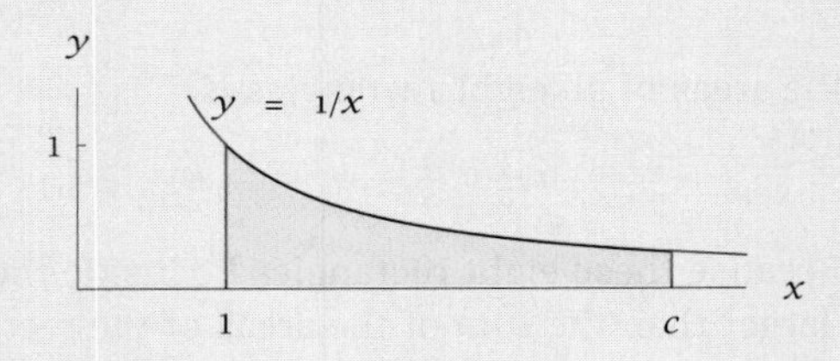

In other words, area$(\frac{1}{x}, 1, c)$ is the area of the region under the curve $y = \frac{1}{x}$, above the x-axis, and between the lines $x = 1$ and $x = c$.

To get a feeling for how area$(\frac{1}{x}, 1, c)$ depends on c, consider the following table:

c	$\text{area}(\frac{1}{x}, 1, c)$
2	0.693147
3	1.098612
4	1.386294
5	1.609438
6	1.791759
7	1.945910
8	2.079442
9	2.197225

Here values of $\text{area}(\frac{1}{x}, 1, c)$ *are rounded off to six digits after the decimal point.*

The table above agrees with the inequalities that we derived earlier in this section: $\text{area}(\frac{1}{x}, 1, 2) < 1$ and $\text{area}(\frac{1}{x}, 1, 3) > 1$.

Before reading the next paragraph, pause for a moment to see if you can discover a relationship between any entries in the table above.

If you look for a relationship between entries in the table above, most likely the first thing you will notice is that $\text{area}(\frac{1}{x}, 1, 4) = 2\,\text{area}(\frac{1}{x}, 1, 2)$. To see if any other such relationships lurk in the table, we now add a third column showing the ratio of $\text{area}(\frac{1}{x}, 1, c)$ to $\text{area}(\frac{1}{x}, 1, 2)$ and a fourth column showing the ratio of $\text{area}(\frac{1}{x}, 1, c)$ to $\text{area}(\frac{1}{x}, 1, 3)$ (both new columns rounded off to five digits after the decimal point):

c	$\text{area}(\frac{1}{x}, 1, c)$	$\dfrac{\text{area}(\frac{1}{x}, 1, c)}{\text{area}(\frac{1}{x}, 1, 2)}$	$\dfrac{\text{area}(\frac{1}{x}, 1, c)}{\text{area}(\frac{1}{x}, 1, 3)}$
2	0.693147	**1.00000**	0.63093
3	1.098612	1.58496	**1.00000**
4	1.386294	**2.00000**	1.26186
5	1.609438	2.32193	1.46497
6	1.791759	2.58496	1.63093
7	1.945910	2.80735	1.77124
8	2.079442	**3.00000**	1.89279
9	2.197225	3.16993	**2.00000**

The integer entries in the last two columns stand out. We already noted that $\text{area}(\frac{1}{x}, 1, 4) = 2\,\text{area}(\frac{1}{x}, 1, 2)$; the table above now shows the nice relationships $\text{area}(\frac{1}{x}, 1, 8) = 3\,\text{area}(\frac{1}{x}, 1, 2)$ and $\text{area}(\frac{1}{x}, 1, 9) = 2\,\text{area}(\frac{1}{x}, 1, 3)$. Because $4 = 2^2$ and $8 = 2^3$ and $9 = 3^2$, we write these equations more suggestively as

$$\text{area}(\tfrac{1}{x}, 1, 2^2) = 2\,\text{area}(\tfrac{1}{x}, 1, 2);$$

$$\text{area}(\tfrac{1}{x}, 1, 2^3) = 3\,\text{area}(\tfrac{1}{x}, 1, 2);$$

$$\text{area}(\tfrac{1}{x}, 1, 3^2) = 2\,\text{area}(\tfrac{1}{x}, 1, 3).$$

The equations above suggest the following remarkable formula:

An area formula

$$\text{area}(\tfrac{1}{x}, 1, c^t) = t\,\text{area}(\tfrac{1}{x}, 1, c)$$

for every $c > 1$ and every $t > 0$.

We already know that the formula above holds in three special cases. The formula above will be derived more generally in the next section. For now, assume that the evidence from the table above is sufficiently compelling to accept this formula.

The right side of the equation above would be simplified if c is such that $\text{area}(\frac{1}{x}, 1, c) = 1$. Thus we make the following definition:

Definition of e

e is the number such that

$$\text{area}(\tfrac{1}{x}, 1, e) = 1.$$

Earlier in this chapter we showed that $\text{area}(\frac{1}{x}, 1, 2)$ is less than 1 and that $\text{area}(\frac{1}{x}, 1, 3)$ is greater than 1. Thus for some number between 2 and 3, the area of the region we are considering must equal 1. That number is called e.

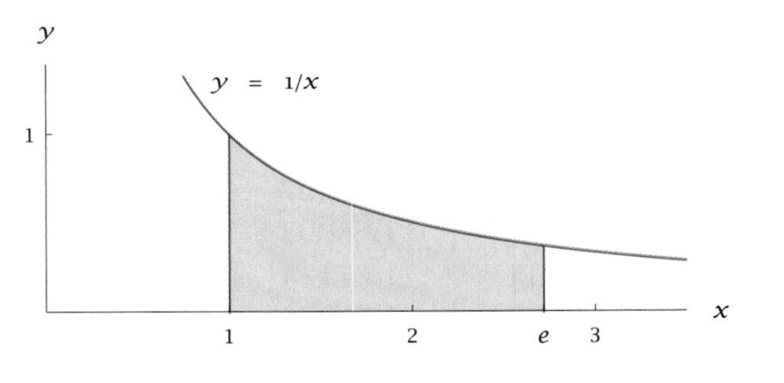

We define e to be the number such that the yellow region has area 1.

To attract mathematically skilled employees, Google once put up billboards around the country asking for the first 10-digit prime number found in consecutive digits of e. The solution, found with the aid of a computer, is 7427466391. These ten digits start in the 99th digit after the decimal point in the decimal representation of e.

The number e is given a special name because it is so useful in many parts of mathematics. We will see some applications of e later in this chapter.

It turns out that e is an irrational number. Here is a 40-digit approximation of e:

$$e \approx 2.718281828459045235360287471352662497757$$

For many practical purposes, 2.718 is a good approximation of e—the error is about 0.01%.

The fraction $\frac{19}{7}$ approximates e fairly well—the error is about 0.1%. The fraction $\frac{2721}{1001}$ approximates e even better—the error is about 0.000004%.

Keep in mind that e is not equal to 2.718 or $\frac{19}{7}$ or $\frac{2721}{1001}$. All of these are useful approximations, but e is an irrational number that cannot be represented exactly as a decimal number or as a fraction.

Defining the Natural Logarithm

The formula

$$\text{area}(\tfrac{1}{x}, 1, c^t) = t\,\text{area}(\tfrac{1}{x}, 1, c)$$

was introduced above. This formula should remind you of the behavior of logarithms with respect to powers. We will now see that the area under the curve $y = \frac{1}{x}$ is indeed intimately connected with a logarithm.

In the formula above, set c equal to e and use the equation $\text{area}(\frac{1}{x}, 1, e) = 1$ to see that

$$\text{area}(\tfrac{1}{x}, 1, e^t) = t.$$

for every positive number t.

Now consider a number $c > 1$. We can write c as a power of e in the usual fashion: $c = e^{\log_e c}$. Thus

$$\begin{aligned}\text{area}(\tfrac{1}{x}, 1, c) &= \text{area}(\tfrac{1}{x}, 1, e^{\log_e c}) \\ &= \log_e c,\end{aligned}$$

where the last equality comes from setting $t = \log_e c$ in the equation from the previous paragraph.

The logarithm with base e, which appeared above, is so useful that it has a special name and special notation.

Natural logarithm

For $c > 0$ the **natural logarithm** of c, denoted $\ln c$, is defined by

$$\ln c = \log_e c.$$

With this new notation, the equality $\text{area}(\frac{1}{x}, 1, c) = \log_e c$ derived above can be rewritten as follows:

As an indication of the usefulness of e and the natural logarithm, take a look at your calculator. It probably has buttons for e^x and $\ln x$.

Natural logarithms as areas

For $c > 1$, the natural logarithm of c is the area of the region below:

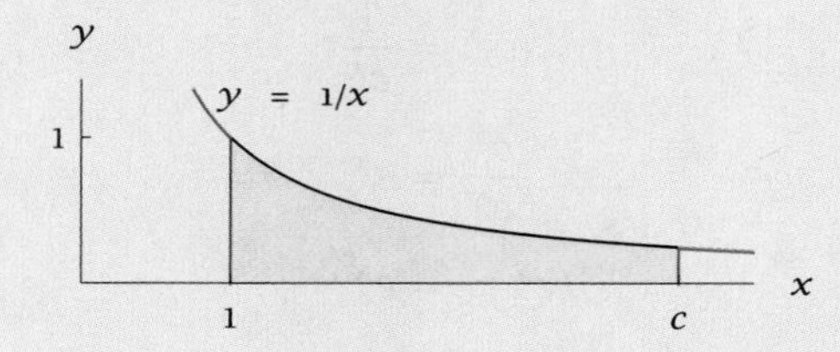

In other words,

$$\ln c = \text{area}(\tfrac{1}{x}, 1, c).$$

Properties of the Exponential Function and ln

The function whose value at a number x equals e^x is so important that it also has a special name.

The exponential function

The **exponential function** is the function f defined by

$$f(x) = e^x$$

for every real number x.

In Chapter 5 we defined the exponential function with base b to be the function whose value at x is b^x. Thus the exponential function defined above is just the exponential function with base e. In other words, if no base is mentioned, then assume that the base is e.

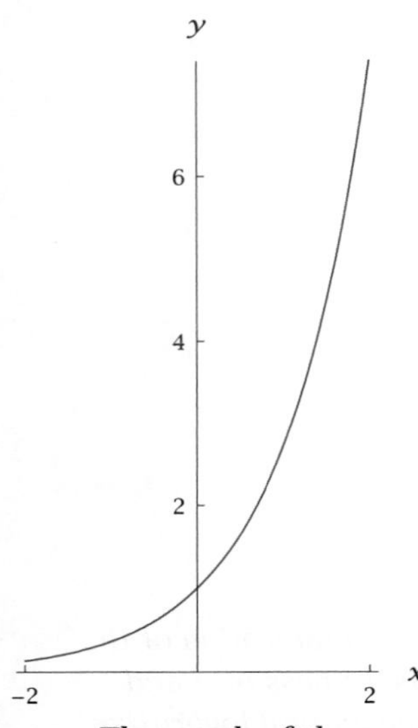

The graph of the exponential function e^x on $[-2, 2]$. The same scale is used on both axes to show the rapid growth of e^x as x increases.

The graph of the exponential function e^x looks similar to the graphs of the functions exponential functions 2^x or 3^x or any other exponential function with base $b > 1$. Specifically, e^x grows rapidly as x gets large, and e^x is close to 0 for negative values of x with large absolute value.

The domain of the exponential function is the set of real numbers, and the range of the exponential function is the set of positive numbers. Furthermore, the exponential function is an increasing function, as is every function of the form b^x for $b > 1$.

Powers of e have the same algebraic properties as powers of any number. Thus the identities listed below should already be familiar to you. They are included here as a review of key algebraic properties in the specific case of powers of e.

Properties of powers of e

$$e^0 = 1$$

$$e^1 = e$$

$$e^x e^y = e^{x+y}$$

$$e^{-x} = \frac{1}{e^x}$$

$$\frac{e^x}{e^y} = e^{x-y}$$

$$(e^x)^y = e^{xy}$$

The natural logarithm of a positive number x, denoted $\ln x$, equals $\log_e x$. Thus the graph of the natural logarithm looks similar to the graphs of the functions $\log_2 x$ or $\log x$ or $\log_b x$ for any number $b > 1$. Specifically, $\ln x$

grows slowly as x gets large. Furthermore, if x is a small positive number, then $\ln x$ is a negative number with large absolute value, as shown in the following figure:

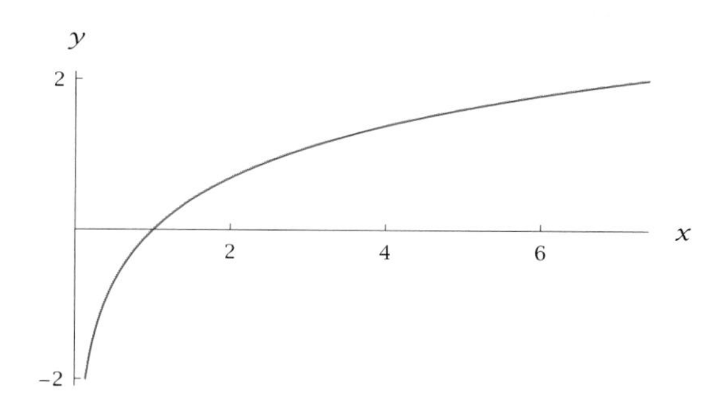

The graph of $\ln x$ on the interval $[e^{-2}, e^2]$. The same scale is used on both axes to show the slow growth of $\ln x$ and the rapid descent near 0 toward negative numbers with large absolute value.

The domain of $\ln x$ is the set of positive numbers, and the range of $\ln x$ is the set of real numbers. Furthermore, $\ln x$ is an increasing function because it is the inverse of the increasing function e^x.

Because the natural logarithm is the logarithm with base e, it has all the properties we saw earlier for logarithms with any base. For review, we summarize the key properties here. In the box below, we assume that x and y are positive numbers.

Recall that in this book, as in most college algebra books, $\log x$ means $\log_{10} x$. However, the natural logarithm is so important that many mathematicians use $\log x$ to denote the natural logarithm rather than the logarithm with base 10.

Properties of the natural logarithm

$$\ln 1 = 0$$

$$\ln e = 1$$

$$\ln(xy) = \ln x + \ln y$$

$$\ln \tfrac{1}{x} = -\ln x$$

$$\ln \tfrac{x}{y} = \ln x - \ln y$$

$$\ln x^t = t \ln x$$

The exponential function e^x and the natural logarithm $\ln x$ (which equals $\log_e x$) are the inverse functions for each other, just as the functions 2^x and $\log_2 x$ are the inverse functions for each other (in this statement, we could replace 2 by any positive number $b \neq 1$). Thus the exponential function and the natural logarithm exhibit the same behavior as any two functions that are the inverse functions for each other. For review, we summarize here the key properties connecting the exponential function and the natural logarithm:

Connections between the exponential function and the natural logarithm

- $\ln y = x$ means $e^x = y$.
- $\ln e^x = x$ for every real number x.
- $e^{\ln y} = y$ for every positive number y.

As usual for a function and its inverse, the graphs of the exponential function and the natural logarithm are symmetric to each other about the line $y = x$.

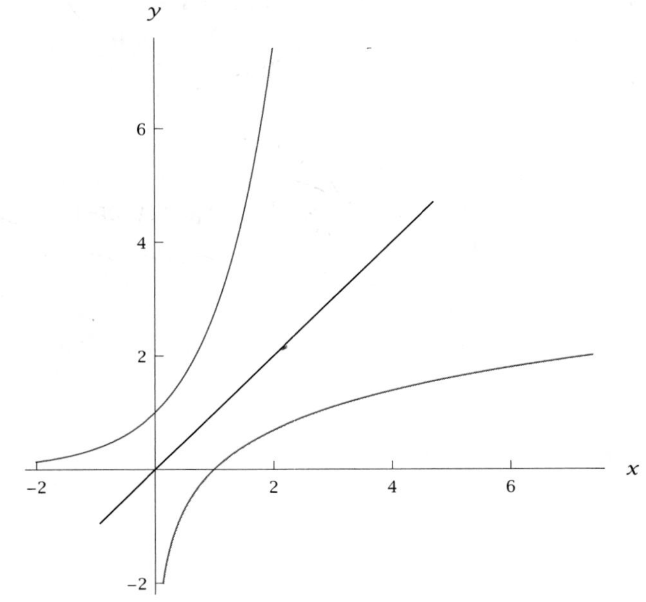

The figure here shows the graphs of e^x (blue) on $[-2, 2]$ and $\ln x$ (red) on $[e^{-2}, e^2]$. Each graph is obtained by flipping the other across the line $y = x$.

EXERCISES

1. For $x = 7$ and $y = 13$, evaluate each of the following:
   (a) $\ln(x + y)$ (b) $\ln x + \ln y$
   [*This exercise and the next one emphasize that* $\ln(x + y)$ *does not equal* $\ln x + \ln y$.]
2. For $x = 0.4$ and $y = 3.5$, evaluate each of the following:
   (a) $\ln(x + y)$ (b) $\ln x + \ln y$
3. For $x = 3$ and $y = 8$, evaluate each of the following:
   (a) $\ln(xy)$ (b) $(\ln x)(\ln y)$
   [*This exercise and the next one emphasize that* $\ln(xy)$ *does not equal* $(\ln x)(\ln y)$.]
4. For $x = 1.1$ and $y = 5$, evaluate each of the following:
   (a) $\ln(xy)$ (b) $(\ln x)(\ln y)$
5. For $x = 12$ and $y = 2$, evaluate each of the following:
   (a) $\ln \frac{x}{y}$ (b) $\frac{\ln x}{\ln y}$
   [*This exercise and the next one emphasize that* $\ln \frac{x}{y}$ *does not equal* $\frac{\ln x}{\ln y}$.]
6. For $x = 18$ and $y = 0.3$, evaluate each of the following:
   (a) $\ln \frac{x}{y}$ (b) $\frac{\ln x}{\ln y}$
7. Find a number y such that $\ln y = 4$.
8. Find a number c such that $\ln c = 5$.
9. Find a number x such that $\ln x = -2$.
10. Find a number x such that $\ln x = -3$.
11. Find a number t such that $\ln(2t + 1) = -4$.

12. Find a number w such that $\ln(3w - 2) = 5$.

13. Find all numbers y such that $\ln(y^2 + 1) = 3$.

14. Find all numbers r such that $\ln(2r^2 - 3) = -1$.

15. Find a number x such that $e^{3x-1} = 2$.

16. Find a number y such that $e^{4y-3} = 5$.

For Exercises 17–28, find all numbers x that satisfy the given equation.

17. $\ln(x + 5) - \ln(x - 1) = 2$

18. $\ln(x + 4) - \ln(x - 2) = 3$

19. $\ln(x + 5) + \ln(x - 1) = 2$

20. $\ln(x + 4) + \ln(x + 2) = 2$

21. $\dfrac{\ln(12x)}{\ln(5x)} = 2$

22. $\dfrac{\ln(11x)}{\ln(4x)} = 2$

23. $e^{2x} + e^x = 6$

24. $e^{2x} - 4e^x = 12$

25. $e^x + e^{-x} = 6$

26. $e^x + e^{-x} = 8$

27. $(\ln(3x)) \ln x = 4$

28. $(\ln(6x)) \ln x = 5$

29. Find the number c such that $\text{area}(\frac{1}{x}, 1, c) = 2$.

30. Find the number c such that $\text{area}(\frac{1}{x}, 1, c) = 3$.

31. Find the number t that makes e^{t^2+6t} as small as possible.
[*Here* e^{t^2+6t} *means* $e^{(t^2+6t)}$.]

32. Find the number t that makes e^{t^2+8t+3} as small as possible.

33. Find a number y such that
$$\frac{1 + \ln y}{2 + \ln y} = 0.9.$$

34. Find a number w such that
$$\frac{4 - \ln w}{3 - 5\ln w} = 3.6.$$

For Exercises 35–38, find a formula for $(f \circ g)(x)$ assuming that f and g are the indicated functions.

35. $f(x) = \ln x$ and $g(x) = e^{5x}$

36. $f(x) = \ln x$ and $g(x) = e^{4-7x}$

37. $f(x) = e^{2x}$ and $g(x) = \ln x$

38. $f(x) = e^{8-5x}$ and $g(x) = \ln x$

For each of the functions f given in Exercises 39–48:

(a) ***Find the domain of f.***

(b) ***Find the range of f.***

(c) ***Find a formula for f^{-1}.***

(d) ***Find the domain of f^{-1}.***

(e) ***Find the range of f^{-1}.***

You can check your solutions to part (c) by verifying that $f^{-1} \circ f = I$ and $f \circ f^{-1} = I$. (Recall that I is the function defined by $I(x) = x$.)

39. $f(x) = 2 + \ln x$

40. $f(x) = 3 - \ln x$

41. $f(x) = 4 - 5\ln x$

42. $f(x) = -6 + 7\ln x$

43. $f(x) = 3e^{2x}$

44. $f(x) = 5e^{9x}$

45. $f(x) = 4 + \ln(x - 2)$

46. $f(x) = 3 + \ln(x + 5)$

47. $f(x) = 5 + 6e^{7x}$

48. $f(x) = 4 - 2e^{8x}$

49. What is the area of the region under the curve $y = \frac{1}{x}$, above the x-axis, and between the lines $x = 1$ and $x = e^2$?

50. What is the area of the region under the curve $y = \frac{1}{x}$, above the x-axis, and between the lines $x = 1$ and $x = e^5$?

PROBLEMS

Some problems require considerably more thought than the exercises. Unlike exercises, problems often have more than one correct answer.

51. Verify that the last five rectangles in the figure in Example 2 have area $\frac{1}{8}$, $\frac{1}{9}$, $\frac{1}{10}$, $\frac{1}{11}$, and $\frac{1}{12}$.

52. Consider this figure:

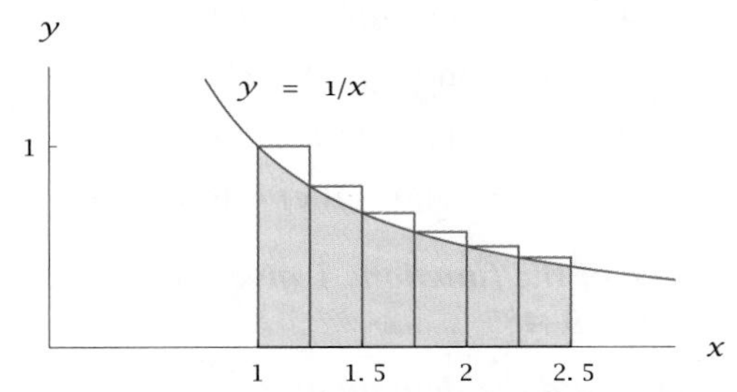

The region under the curve $y = \frac{1}{x}$, above the x-axis, and between the lines $x = 1$ and $x = 2.5$.

(a) Calculate the sum of the areas of all six rectangles shown in the figure above.

(b) Explain why the calculation you did in part (a) shows that

$$\text{area}(\tfrac{1}{x}, 1, 2.5) < 1.$$

(c) Explain why the inequality above shows that $e > 2.5$.

The following notation is used in Problems 53–56:* area$(x^2, 1, 2)$ *is the area of the region under the curve $y = x^2$, above the x-axis, and between the lines $x = 1$ and $x = 2$, as shown below.

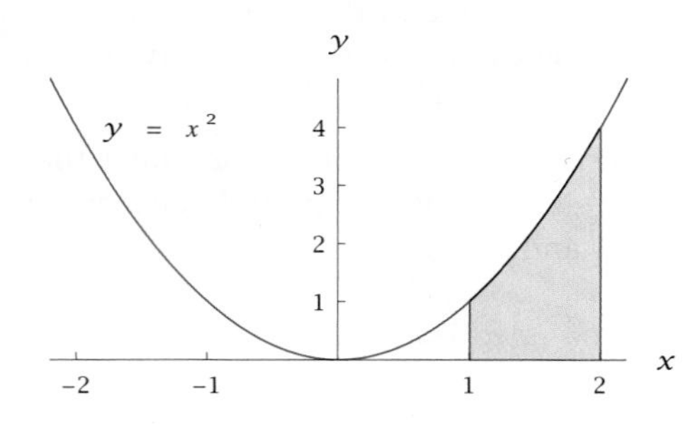

53. Using one rectangle, show that

$$1 < \text{area}(x^2, 1, 2).$$

54. Using one rectangle, show that

$$\text{area}(x^2, 1, 2) < 4.$$

55. Using four rectangles, show that

$$1.96 < \text{area}(x^2, 1, 2).$$

56. Using four rectangles, show that

$$\text{area}(x^2, 1, 2) < 2.72.$$

[*The two problems above show that* area$(x^2, 1, 2)$ *is in the interval* $[1.96, 2.72]$. *If we use the midpoint of that interval as an estimate, we get* area$(x^2, 1, 2) \approx \frac{1.96+2.72}{2} = 2.34$. *This is a very good estimate—the exact value of* area$(x^2, 1, 2)$ *is* $\frac{7}{3}$, *which is approximately* 2.33.]

57. Explain why

$$\ln x \approx 2.302585 \log x$$

for every positive number x.

58. Explain why the solution to part (b) of Exercise 5 in this section is the same as the solution to part (b) of Exercise 5 in Section 5.3.

59. Suppose c is a number such that area$(\frac{1}{x}, 1, c) > 1000$. Explain why $c > 2^{1000}$.

The functions* cosh *and* sinh *are defined by

$$\cosh x = \frac{e^x + e^{-x}}{2} \quad \text{and} \quad \sinh x = \frac{e^x - e^{-x}}{2}$$

for every real number x. For reasons that do not concern us here, these functions are called the* hyperbolic cosine *and* hyperbolic sine*; they are useful in engineering.

60. Show that cosh is an even function.

61. Show that sinh is an odd function.

62. Show that

$$(\cosh x)^2 - (\sinh x)^2 = 1$$

for every real number x.

63. Show that $\cosh x \geq 1$ for every real number x.

64. Show that

$$\cosh(x + y) = \cosh x \cosh y + \sinh x \sinh y$$

for all real numbers x and y.

65. Show that

$$\sinh(x + y) = \sinh x \cosh y + \cosh x \sinh y$$

for all real numbers x and y.

66. Show that

$$(\cosh x + \sinh x)^t = \cosh(tx) + \sinh(tx)$$

for all real numbers x and t.

67. Show that if x is very large, then

$$\cosh x \approx \sinh x \approx \frac{e^x}{2}.$$

68. Show that the range of sinh is the set of real numbers.

69. Show that sinh is a one-to-one function and that its inverse is given by the formula

$$(\sinh)^{-1}(y) = \ln(y + \sqrt{y^2 + 1})$$

for every real number y.

70. Show that the range of cosh is the interval $[1, \infty)$.

71. Suppose f is the function defined by

$$f(x) = \cosh x$$

for every $x \geq 0$. In other words, f is defined by the same formula as cosh, but the domain of f is the interval $[0, \infty)$ and the domain of cosh is the set of real numbers. Show that f is a one-to-one function and that its inverse is given by the formula

$$f^{-1}(y) = \ln(y + \sqrt{y^2 - 1})$$

for every $y \geq 1$.

72. Write a description of how the shape of the St. Louis Gateway Arch, whose picture appears on the opening page of this chapter, is related to the graph of $\cosh x$.
[*You should be able to find the necessary information using an appropriate web search.*]

WORKED-OUT SOLUTIONS *to Odd-numbered Exercises*

Do not read these worked-out solutions before first attempting to do the exercises yourself. Otherwise you may merely mimic the techniques shown here without understanding the ideas.

Best way to learn: Carefully read the section of the textbook, then do all the odd-numbered exercises (even if they have not been assigned) and check your answers here. If you get stuck on an exercise, reread the section of the textbook—then try the exercise again. If you are still stuck, then look at the worked-out solution here.

1. For $x = 7$ and $y = 13$, evaluate each of the following:
   (a) $\ln(x + y)$ (b) $\ln x + \ln y$

SOLUTION

(a) $\ln(7 + 13) = \ln 20 \approx 2.99573$

(b)
$$\ln 7 + \ln 13 \approx 1.94591 + 2.56495 = 4.51086$$

3. For $x = 3$ and $y = 8$, evaluate each of the following:
   (a) $\ln(xy)$ (b) $(\ln x)(\ln y)$

SOLUTION

(a) $\ln(3 \cdot 8) = \ln 24 \approx 3.17805$

(b)
$$(\ln 3)(\ln 8) \approx (1.09861)(2.07944) \approx 2.2845$$

5. For $x = 12$ and $y = 2$, evaluate each of the following:
   (a) $\ln \frac{x}{y}$ (b) $\frac{\ln x}{\ln y}$

SOLUTION

(a) $\ln \frac{12}{2} = \ln 6 \approx 1.79176$

(b) $\dfrac{\ln 12}{\ln 2} \approx \dfrac{2.48491}{0.693147} \approx 3.58496$

7. Find a number y such that $\ln y = 4$.

SOLUTION Recall that $\ln y$ is simply shorthand for $\log_e y$. Thus the equation $\ln y = 4$ can be rewritten as $\log_e y = 4$. The definition of a logarithm now implies that $y = e^4$.

9. Find a number x such that $\ln x = -2$.

SOLUTION Recall that $\ln x$ is simply shorthand for $\log_e x$. Thus the equation $\ln x = -2$ can be rewritten as $\log_e x = -2$. The definition of a logarithm now implies that $x = e^{-2}$.

11. Find a number t such that $\ln(2t + 1) = -4$.

SOLUTION The equation $\ln(2t + 1) = -4$ implies that

$$e^{-4} = 2t + 1.$$

Solving this equation for t, we get

$$t = \frac{e^{-4} - 1}{2}.$$

13. Find all numbers y such that $\ln(y^2 + 1) = 3$.

SOLUTION The equation $\ln(y^2 + 1) = 3$ implies that

$$e^3 = y^2 + 1.$$

Thus $y^2 = e^3 - 1$, which means that $y = \sqrt{e^3 - 1}$ or $y = -\sqrt{e^3 - 1}$.

15. Find a number x such that $e^{3x-1} = 2$.

SOLUTION The equation $e^{3x-1} = 2$ implies that

$$3x - 1 = \ln 2.$$

Solving this equation for x, we get

$$x = \frac{1 + \ln 2}{3}.$$

For Exercises 17–28, find all numbers x that satisfy the given equation.

17. $\ln(x + 5) - \ln(x - 1) = 2$

SOLUTION Our equation can be rewritten as follows:

$$\begin{aligned} 2 &= \ln(x + 5) - \ln(x - 1) \\ &= \ln \frac{x + 5}{x - 1}. \end{aligned}$$

Thus

$$\frac{x + 5}{x - 1} = e^2.$$

We can solve the equation above for x, getting $x = \dfrac{e^2 + 5}{e^2 - 1}$.

19. $\ln(x + 5) + \ln(x - 1) = 2$

SOLUTION Our equation can be rewritten as follows:

$$\begin{aligned} 2 &= \ln(x + 5) + \ln(x - 1) \\ &= \ln((x + 5)(x - 1)) \\ &= \ln(x^2 + 4x - 5). \end{aligned}$$

Thus

$$x^2 + 4x - 5 = e^2,$$

which implies that

$$x^2 + 4x - (e^2 + 5) = 0.$$

We can solve the equation above using the quadratic formula, getting $x = -2 + \sqrt{9 + e^2}$ or $x = -2 - \sqrt{9 + e^2}$. However, both $x + 5$ and $x - 1$ are negative if $x = -2 - \sqrt{9 + e^2}$; because the logarithm of a negative number is undefined, we must discard this root of the equation above. We conclude that the only value of x satisfying the equation $\ln(x + 5) + \ln(x - 1) = 2$ is $x = -2 + \sqrt{9 + e^2}$.

21. $\dfrac{\ln(12x)}{\ln(5x)} = 2$

SOLUTION Our equation can be rewritten as follows:

$$\begin{aligned} 2 &= \frac{\ln(12x)}{\ln(5x)} \\ &= \frac{\ln 12 + \ln x}{\ln 5 + \ln x}. \end{aligned}$$

Solving this equation for $\ln x$ (the first step in doing this is to multiply both sides by the denominator $\ln 5 + \ln x$), we get

$$\begin{aligned} \ln x &= \ln 12 - 2\ln 5 \\ &= \ln 12 - \ln 25 \\ &= \ln \tfrac{12}{25}. \end{aligned}$$

Thus $x = \frac{12}{25}$.

23. $e^{2x} + e^x = 6$

SOLUTION Note that $e^{2x} = (e^x)^2$. This suggests that we let $t = e^x$. Then the equation above can be rewritten as

$$t^2 + t - 6 = 0.$$

The solutions to this equation (which can be found either by using the quadratic formula or by factoring) are $t = -3$ and $t = 2$. Thus $e^x = -3$ or $e^x = 2$. However, there is no real number x such that $e^x = -3$ (because e^x is positive for every real number x), and thus we must have $e^x = 2$. Thus $x = \ln 2 \approx 0.693147$.

25. $e^x + e^{-x} = 6$

SOLUTION Let $t = e^x$. Then the equation above can be rewritten as

$$t + \frac{1}{t} - 6 = 0.$$

Multiply both sides by t, giving the equation

$$t^2 - 6t + 1 = 0.$$

The solutions to this equation (which can be found by using the quadratic formula) are $t = 3 - 2\sqrt{2}$ and $t = 3 + 2\sqrt{2}$. Thus $e^x = 3 - 2\sqrt{2}$ or $e^x = 3 + 2\sqrt{2}$. Thus the solutions to the original equation are $x = \ln(3 - 2\sqrt{2})$ and $x = \ln(3 + 2\sqrt{2})$.

27. $(\ln(3x))\ln x = 4$

SOLUTION Our equation can be rewritten as follows:

$$\begin{aligned} 4 &= (\ln(3x))\ln x \\ &= (\ln x + \ln 3)\ln x \\ &= (\ln x)^2 + (\ln 3)(\ln x). \end{aligned}$$

Letting $y = \ln x$, we can rewrite the equation above as

$$y^2 + (\ln 3)y - 4 = 0.$$

Use the quadratic formula to solve the equation above for y, getting

$$y \approx -2.62337 \quad \text{or} \quad y \approx 1.52476.$$

Thus

$$\ln x \approx -2.62337 \quad \text{or} \quad \ln x \approx 1.52476,$$

which means that

$$x \approx e^{-2.62337} \approx 0.072558$$

or

$$x \approx e^{1.52476} \approx 4.59403.$$

29. Find the number c such that $\text{area}(\frac{1}{x}, 1, c) = 2$.

SOLUTION Because $2 = \text{area}(\frac{1}{x}, 1, c) = \ln c$, we see that $c = e^2$.

31. Find the number t that makes e^{t^2+6t} as small as possible.

SOLUTION Because e^x is an increasing function of x, the number e^{t^2+6t} will be as small as possible when $t^2 + 6t$ is as small as possible. To find when $t^2 + 6t$ is as small as possible, we complete the square:

$$t^2 + 6t = (t + 3)^2 - 9.$$

The equation above shows that $t^2 + 6t$ is as small as possible when $t = -3$.

33. Find a number y such that

$$\frac{1 + \ln y}{2 + \ln y} = 0.9.$$

SOLUTION Multiplying both sides of the equation above by $2 + \ln y$ and then solving for $\ln y$ gives $\ln y = 8$. Thus $y = e^8 \approx 2980.96$.

For Exercises 35–38, find a formula for $(f \circ g)(x)$ assuming that f and g are the indicated functions.

35. $f(x) = \ln x$ and $g(x) = e^{5x}$

SOLUTION

$$(f \circ g)(x) = f(g(x)) = f(e^{5x}) = \ln e^{5x} = 5x$$

37. $f(x) = e^{2x}$ and $g(x) = \ln x$

SOLUTION

$$\begin{aligned} (f \circ g)(x) &= f(g(x)) = f(\ln x) \\ &= e^{2\ln x} = (e^{\ln x})^2 = x^2 \end{aligned}$$

For each of the functions f given in Exercises 39–48:

(a) ***Find the domain of f.***

(b) ***Find the range of f.***

(c) ***Find a formula for f^{-1}.***

(d) ***Find the domain of f^{-1}.***

(e) ***Find the range of*** f^{-1}.

You can check your solutions to part (c) by verifying that $f^{-1} \circ f = I$ ***and*** $f \circ f^{-1} = I$. ***(Recall that*** I ***is the function defined by*** $I(x) = x$.***)***

39. $f(x) = 2 + \ln x$

SOLUTION

(a) The expression $2 + \ln x$ makes sense for all positive numbers x. Thus the domain of f is the set of positive numbers.

(b) To find the range of f, we need to find the numbers y such that

$$y = 2 + \ln x$$

for some x in the domain of f. In other words, we need to find the values of y such that the equation above can be solved for a positive number x. To solve this equation for x, subtract 2 from both sides, getting $y - 2 = \ln x$, which implies that

$$x = e^{y-2}.$$

The expression above on the right makes sense for every real number y and produces a positive number x (because e raised to any power is positive). Thus the range of f is the set of real numbers.

(c) The expression above shows that f^{-1} is given by the expression

$$f^{-1}(y) = e^{y-2}.$$

(d) The domain of f^{-1} equals the range of f. Thus the domain of f^{-1} is the set of real numbers.

(e) The range of f^{-1} equals the domain of f. Thus the range of f^{-1} is the set of positive numbers.

41. $f(x) = 4 - 5\ln x$

SOLUTION

(a) The expression $4 - 5\ln x$ makes sense for all positive numbers x. Thus the domain of f is the set of positive numbers.

(b) To find the range of f, we need to find the numbers y such that

$$y = 4 - 5\ln x$$

for some x in the domain of f. In other words, we need to find the values of y such that the equation above can be solved for a positive number x. To solve this equation for x, subtract 4 from both sides, then divide both sides by -5, getting $\frac{4-y}{5} = \ln x$, which implies that

$$x = e^{(4-y)/5}.$$

The expression above on the right makes sense for every real number y and produces a positive number x (because e raised to any power is positive). Thus the range of f is the set of real numbers.

(c) The expression above shows that f^{-1} is given by the expression

$$f^{-1}(y) = e^{(4-y)/5}.$$

(d) The domain of f^{-1} equals the range of f. Thus the domain of f^{-1} is the set of real numbers.

(e) The range of f^{-1} equals the domain of f. Thus the range of f^{-1} is the set of positive numbers.

43. $f(x) = 3e^{2x}$

SOLUTION

(a) The expression $3e^{2x}$ makes sense for all real numbers x. Thus the domain of f is the set of real numbers.

(b) To find the range of f, we need to find the numbers y such that

$$y = 3e^{2x}$$

for some x in the domain of f. In other words, we need to find the values of y such that the equation above can be solved for a real number x. To solve this equation for x, divide both sides by 3, getting $\frac{y}{3} = e^{2x}$, which implies that $2x = \ln \frac{y}{3}$. Thus

$$x = \frac{\ln \frac{y}{3}}{2}.$$

The expression above on the right makes sense for every positive number y and produces a real number x. Thus the range of f is the set of positive numbers.

(c) The expression above shows that f^{-1} is given by the expression

$$f^{-1}(y) = \frac{\ln \frac{y}{3}}{2}.$$

(d) The domain of f^{-1} equals the range of f. Thus the domain of f^{-1} is the set of positive numbers.

(e) The range of f^{-1} equals the domain of f. Thus the range of f^{-1} is the set of real numbers.

45. $f(x) = 4 + \ln(x - 2)$

SOLUTION

(a) The expression $4 + \ln(x - 2)$ makes sense when $x > 2$. Thus the domain of f is the interval $(2, \infty)$.

(b) To find the range of f, we need to find the numbers y such that

$$y = 4 + \ln(x - 2)$$

for some x in the domain of f. In other words, we need to find the values of y such that the equation above can be solved for a number $x > 2$. To solve this equation for x, subtract 4 from both sides, getting $y - 4 = \ln(x - 2)$, which implies that $x - 2 = e^{y-4}$. Thus

$$x = 2 + e^{y-4}.$$

The expression above on the right makes sense for every real number y and produces a number $x > 2$ (because e raised to any power is positive). Thus the range of f is the set of real numbers.

(c) The expression above shows that f^{-1} is given by the expression

$$f^{-1}(y) = 2 + e^{y-4}.$$

(d) The domain of f^{-1} equals the range of f. Thus the domain of f^{-1} is the set of real numbers.

(e) The range of f^{-1} equals the domain of f. Thus the range of f^{-1} is the interval $(2, \infty)$.

47. $f(x) = 5 + 6e^{7x}$

SOLUTION

(a) The expression $5 + 6e^{7x}$ makes sense for all real numbers x. Thus the domain of f is the set of real numbers.

(b) To find the range of f, we need to find the numbers y such that

$$y = 5 + 6e^{7x}$$

for some x in the domain of f. In other words, we need to find the values of y such that the equation above can be solved for a real number x. To solve this equation for x, subtract 5 from both sides, then divide both sides by 6, getting $\frac{y-5}{6} = e^{7x}$, which implies that $7x = \ln \frac{y-5}{6}$. Thus

$$x = \frac{\ln \frac{y-5}{6}}{7}.$$

The expression above on the right makes sense for every $y > 5$ and produces a real number x. Thus the range of f is the interval $(5, \infty)$.

(c) The expression above shows that f^{-1} is given by the expression

$$f^{-1}(y) = \frac{\ln \frac{y-5}{6}}{7}.$$

(d) The domain of f^{-1} equals the range of f. Thus the domain of f^{-1} is the interval $(5, \infty)$.

(e) The range of f^{-1} equals the domain of f. Thus the range of f^{-1} is the set of real numbers.

49. What is the area of the region under the curve $y = \frac{1}{x}$, above the x-axis, and between the lines $x = 1$ and $x = e^2$?

SOLUTION The area of this region is $\ln e^2$, which equals 2.

6.2 *Approximations and area with e and* ln

LEARNING OBJECTIVES

By the end of this section you should be able to

- approximate $\ln(1+t)$ for small values of $|t|$;
- approximate e^t for small values of $|t|$;
- approximate $(1+\frac{r}{x})^x$ when x is much larger than $|r|$;
- explain the area formula that led to e and the natural logarithm.

Approximation of the Natural Logarithm

The next example leads to an important result.

EXAMPLE 1

Discuss the behavior of $\ln(1+t)$ for $|t|$ a small number.

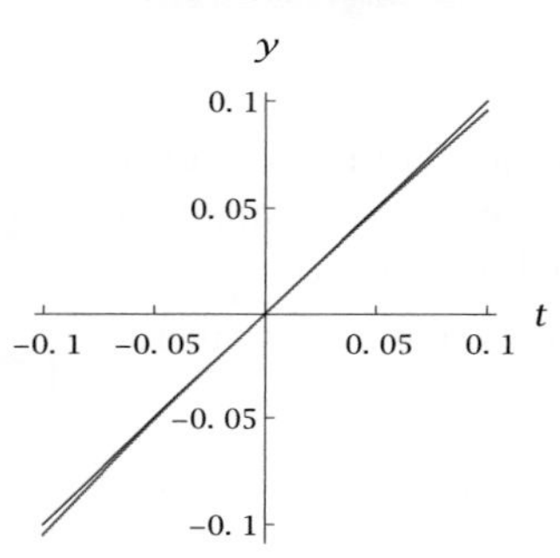

The graphs of $y = \ln(1+t)$ (blue) and $y = t$ (red) on $[-0.1, 0.1]$.

SOLUTION

The table shows the value of $\ln(1+t)$, rounded off to six significant digits, for some small values of $|t|$. This table leads us to guess that $\ln(1+t) \approx t$ if $|t|$ is a small number, with the approximation becoming more accurate as $|t|$ becomes smaller.

t	$\ln(1+t)$
0.05	0.0487902
0.005	0.00498754
0.0005	0.000499875
0.00005	0.0000499988
−0.05	−0.0512933
−0.005	−0.00501254
−0.0005	−0.000500125
−0.00005	−0.0000500013

The graph in the margin confirms that $\ln(1+t) \approx t$ if $|t|$ is small. At this scale, we cannot see the difference between $\ln(1+t)$ and t for t in the interval $[-0.05, 0.05]$.

To explain the behavior in the example above, suppose $t > 0$. Recall from the previous section that $\ln(1+t) = \text{area}(\frac{1}{x}, 1, 1+t)$. In other words, $\ln(1+t)$ equals the area of the region shown below on the left. If t is a small positive number, then the area of this region is approximately equal to the area of the rectangle shown below on the right. This rectangle has base t and height 1; thus the rectangle has area t. We conclude that $\ln(1+t) \approx t$.

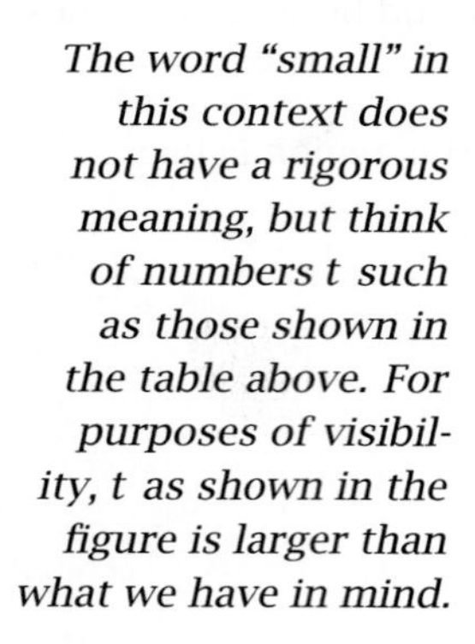

The word "small" in this context does not have a rigorous meaning, but think of numbers t such as those shown in the table above. For purposes of visibility, t as shown in the figure is larger than what we have in mind.

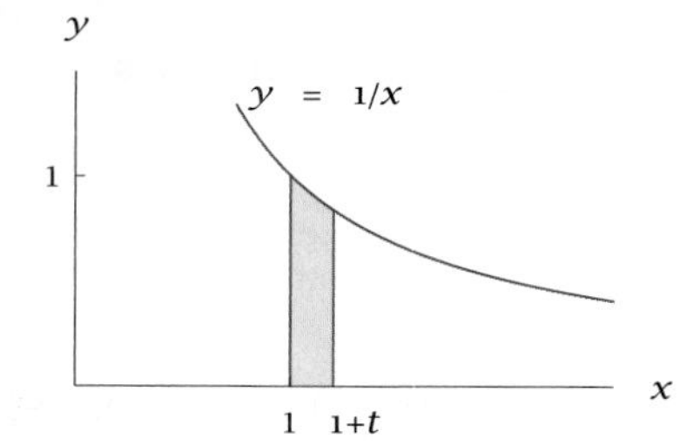

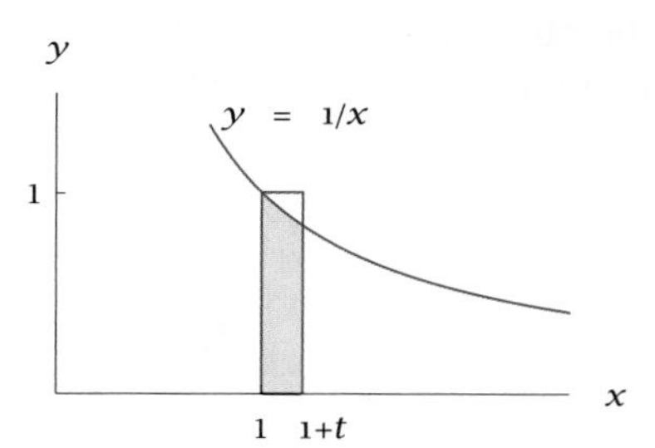

The area of the region on the left equals $\ln(1+t)$.
The rectangle on the right has area t.
Thus $\ln(1+t) \approx t$.

The result below demonstrates again why the natural logarithm deserves the name *natural.* No base for logarithms other than e produces such a nice approximation formula.

To remember whether $\ln(1+t)$ or $\ln t$ is approximately t for $|t|$ small, take $t = 0$ and recall that $\ln 1 = 0$; this should point you toward the correct approximation $\ln(1+t) \approx t$.

Approximation of the natural logarithm

If $|t|$ is small, then $\ln(1+t) \approx t$.

Consider now the figure below, where we assume that t is positive but not necessarily small. In this figure, $\ln(1+t)$ equals the area of the yellow region under the curve.

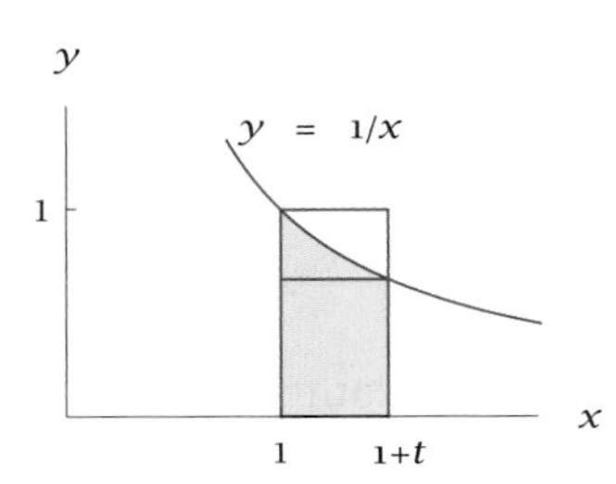

The area of the yellow region under the curve is greater than the area of the lower rectangle and is less than the area of the large rectangle.

The yellow region above contains the lower rectangle; thus the lower rectangle has a smaller area. The lower rectangle has base t and height $\frac{1}{1+t}$ and hence has area $\frac{t}{1+t}$. Thus

$$\frac{t}{1+t} < \ln(1+t).$$

The large rectangle in the figure above has base t and height 1 and thus has area t. The yellow region above is contained in the large rectangle; thus the large rectangle has a bigger area. In other words,

$$\ln(1+t) < t.$$

Putting together the inequalities from the previous two paragraphs, we have the result below.

Inequalities with the natural logarithm

If $t > 0$, then $\dfrac{t}{1+t} < \ln(1+t) < t$.

This result is valid for all positive numbers t, regardless of whether t is small or large.

If t is small, then $\frac{t}{1+t}$ and t are close to each other, showing that either one is a good estimate for $\ln(1+t)$. For small t, the estimate $\ln(1+t) \approx t$ is usually easier to use than the estimate $\ln(1+t) \approx \frac{t}{1+t}$. However, if we need an estimate that is either slightly too large or slightly too small, then the result above shows which one to use.

Approximations with the Exponential Function

Now we turn to approximations of e^x.

EXAMPLE 2 Discuss the behavior of e^x for $|x|$ a small number.

SOLUTION

The table shows the value of e^x, rounded off appropriatetly, for some small values of $|x|$. This table leads us to guess that $e^x \approx 1 + x$ if $|x|$ is a small number, with the approximation becoming more accurate as $|x|$ becomes smaller.

The graph in the margin confirms that $e^x \approx 1 + x$ if $|x|$ is small. At this scale, we cannot see the difference between e^x and $1 + x$ for x in the interval $[-0.05, 0.05]$.

x	e^x
0.05	1.051
0.005	1.00501
0.0005	1.0005001
0.00005	1.000050001
−0.05	0.951
−0.005	0.99501
−0.0005	0.9995001
−0.00005	0.999950001

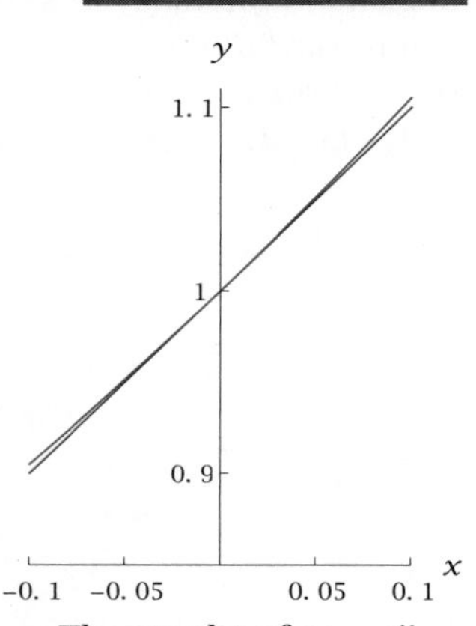

The graphs of $y = e^x$ (blue) and $y = 1 + x$ (red) on $[-0.1, 0.1]$. To save space, the horizontal axis has been drawn at $y = 0.85$ instead of at the usual location $y = 0$.

To explain the behavior in the example above, suppose $|x|$ is small. Then, as we already know, $x \approx \ln(1 + x)$. Thus

$$e^x \approx e^{\ln(1+x)}$$
$$= 1 + x.$$

Hence we have the following result:

> ***Approximation of the exponential function***
>
> If $|x|$ is small, then $e^x \approx 1 + x$.

Another useful approximation gives good estimates for e^r even when r is not small. As an example, consider the following table of values of $(1 + \frac{1}{x})^x$ for large values of x:

x	$(1 + \frac{1}{x})^x$
100	2.70481
1000	2.71692
10000	2.71815
100000	2.71827
1000000	2.71828

Values of $(1 + \frac{1}{x})^x$, rounded off to six digits.

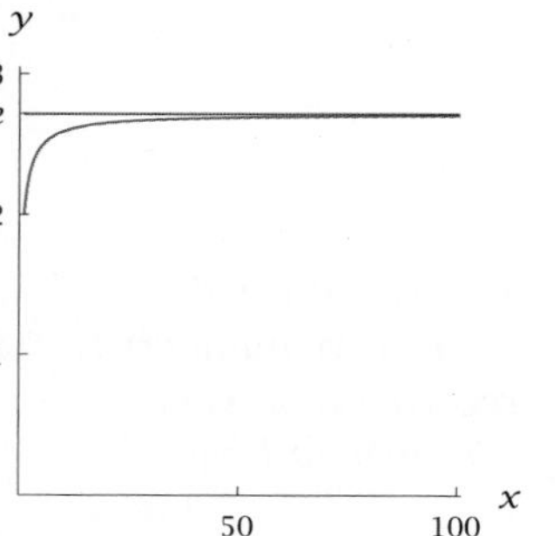

The graphs of $y = (1 + \frac{1}{x})^x$ (blue) and $y = e$ (red) on $[1, 100]$.

You may recognize the last entry in the table above as the value of e, rounded off to six digits. In other words, it appears that $(1 + \frac{1}{x})^x \approx e$ for large values of x. We will now see that an even more general approximation is valid.

Let r be any number, and suppose x is a number much larger than $|r|$. Thus $\left|\frac{r}{x}\right|$ is small. Then, as we already know, $e^{r/x} \approx 1 + \frac{r}{x}$. Thus

$$\begin{aligned} e^r &= (e^{r/x})^x \\ &\approx (1 + \tfrac{r}{x})^x. \end{aligned}$$

Hence we have the following result:

Approximation of the exponential function

If x is much larger than $|r|$, then

$$(1 + \tfrac{r}{x})^x \approx e^r.$$

Notice how e appears naturally in formulas that seem to have nothing to do with e. For example, $(1 + \frac{1}{1000000})^{1000000}$ is approximately e. If we had not discovered e through other means, we probably would have discovered it by investigating $(1 + \frac{1}{x})^x$ for large values of x.

For example, taking $r = 1$, this approximation shows that

$$(1 + \tfrac{1}{x})^x \approx e$$

for large values of x, confirming the results indicated by the table above.

An Area Formula

The area formula

$$\text{area}(\tfrac{1}{x}, 1, c^t) = t\,\text{area}(\tfrac{1}{x}, 1, c)$$

played a crucial role in the previous section, leading to the definitions of e and the natural logarithm. We found evidence that led us to this formula. Now we explain why this formula is true.

We start by introducing some slightly more general notation than was used in the previous section.

area$(\frac{1}{x}, b, c)$

For positive numbers b and c with $b < c$, let area$(\frac{1}{x}, b, c)$ denote the area of the yellow region below:

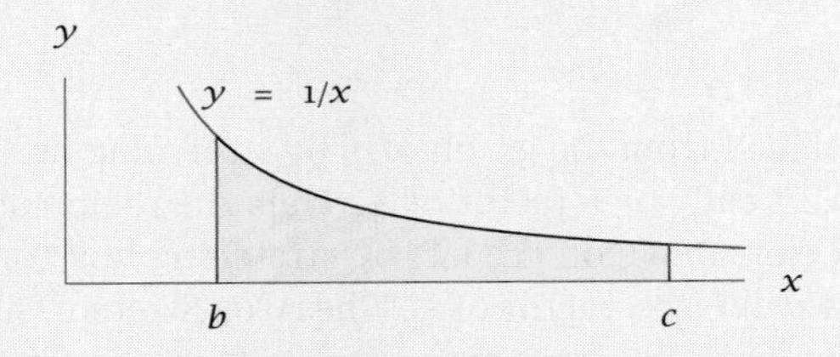

In other words, area$(\frac{1}{x}, b, c)$ is the area of the region under the curve $y = \frac{1}{x}$, above the x-axis, and between the lines $x = b$ and $x = c$.

The solution to the next example contains the key idea that will help us derive the area formula. In this example and the other results in the remainder of this section, we cannot use the equation $\text{area}(\frac{1}{x}, 1, c) = \ln c$. Using that equation would be circular reasoning because we are now trying to show that $\text{area}(\frac{1}{x}, 1, c^t) = t\,\text{area}(\frac{1}{x}, 1, c)$, which was used to show that $\text{area}(\frac{1}{x}, 1, c) = \ln c$.

EXAMPLE 3 Explain why $\text{area}(\frac{1}{x}, 1, 2) = \text{area}(\frac{1}{x}, 2, 4) = \text{area}(\frac{1}{x}, 4, 8)$.

SOLUTION We need to explain why the three regions below have the same area.

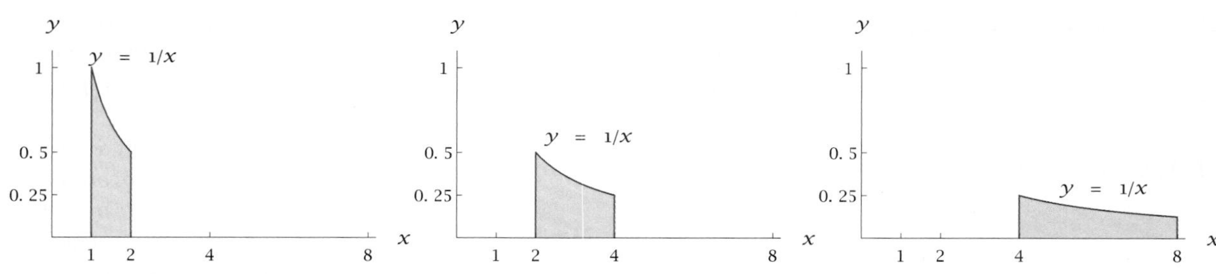

The region in the center is obtained from the region on the left by stretching horizontally by a factor of 2 and stretching vertically by a factor of $\frac{1}{2}$. Thus the Area Stretch Theorem implies that these two regions have the same area.

Similarly, the region on the right is obtained from the region on the left by stretching horizontally by a factor of 4 and stretching vertically by a factor of $\frac{1}{4}$. Thus the Area Stretch Theorem implies that these two regions have the same area.

Define a function f with domain $[1, 2]$ by

$$f(x) = \frac{1}{x}$$

and define a function g by

$$g(x) = \tfrac{1}{2} f(\tfrac{x}{2}) = \tfrac{1}{2} \frac{1}{\frac{x}{2}} = \frac{1}{x}.$$

Our results on function transformations (see Section 3.2) show that the graph of g is obtained from the graph of f by stretching horizontally by a factor of 2 and stretching vertically by a factor of $\frac{1}{2}$. In other words, the region above in the center is obtained from the region above on the left by stretching horizontally by a factor of 2 and stretching vertically by a factor of $\frac{1}{2}$. The Area Stretch Theorem (see Section 2.4) now implies that the area of the region in the center is $2 \cdot \frac{1}{2}$ times the area of the region on the left. Because $2 \cdot \frac{1}{2} = 1$, this implies that the two regions have the same area.

To show that the region above on the right has the same area as the region above on the left, follow the same procedure, but now define a function h by

$$h(x) = \tfrac{1}{4} f(\tfrac{x}{4}) = \tfrac{1}{4} \frac{1}{\frac{x}{4}} = \frac{1}{x}.$$

The graph of h is obtained from the graph of f by stretching horizontally by a factor of 4 and stretching vertically by a factor of $\frac{1}{4}$. Thus the region above on the right is obtained from the region above on the left by stretching horizontally by a factor of 4 and stretching vertically by a factor of $\frac{1}{4}$. The Area Stretch Theorem now implies that these two regions have the same area.

By inspecting a table of numbers in the previous section, we noticed that $\text{area}(\frac{1}{x}, 1, 2^3) = 3\,\text{area}(\frac{1}{x}, 1, 2)$. The next result explains why this is true.

EXAMPLE 4

Explain why $\text{area}(\frac{1}{x}, 1, 2^3) = 3\,\text{area}(\frac{1}{x}, 1, 2)$.

SOLUTION The idea here is to partition the region under the curve $y = \frac{1}{x}$, above the x-axis, and between the lines $x = 1$ and $x = 8$ into three regions, as shown below:

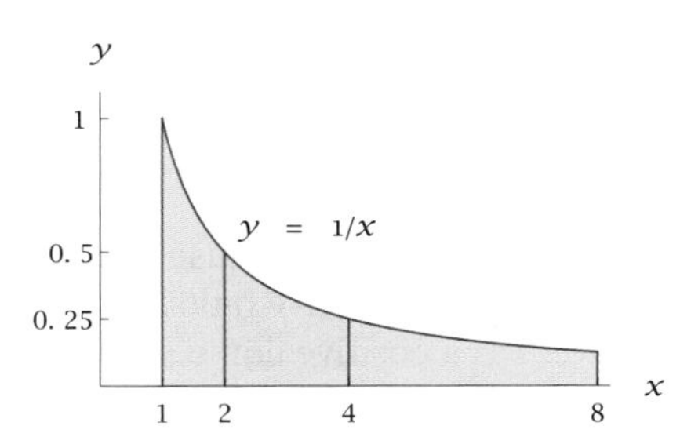

The previous example shows that each of these three regions has the same area. Thus $\text{area}(\frac{1}{x}, 1, 2^3) = 3\,\text{area}(\frac{1}{x}, 1, 2)$.

In the example above, there is nothing special about the number 2. We can replace 2 by any number $c > 1$, and using the same reasoning as in the two previous examples we can conclude that

$$\text{area}(\tfrac{1}{x}, 1, c^3) = 3\,\text{area}(\tfrac{1}{x}, 1, c).$$

If you learn calculus, you will encounter many of the ideas we have been discussing concerning area. The part of calculus called integral calculus *focuses on area.*

Furthermore, there is nothing special here about the number 3 in the equation above. Replacing 3 by any positive integer t, we can use the same reasoning to show that

$$\text{area}(\tfrac{1}{x}, 1, c^t) = t\,\text{area}(\tfrac{1}{x}, 1, c)$$

whenever $c > 1$ and t is a positive integer.

At this point, we have derived the desired area formula with the restriction that t must be a positive integer. If you have understood everything up to this point, this is an excellent achievement and a reasonable stopping place. If you want to understand the full area formula, then work through the following example, which removes the restriction that t be an integer.

EXAMPLE 5

Explain why

$$\text{area}(\tfrac{1}{x}, 1, c^t) = t\,\text{area}(\tfrac{1}{x}, 1, c)$$

for every $c > 1$ and every $t > 0$.

SOLUTION First we will verify the desired equation when t is a positive rational number. So suppose $t = \frac{m}{n}$, where m and n are positive integers. Using the restricted area formula that we have already derived, but replacing c by $c^{n/m}$ and replacing t by m, we have

$$\text{area}(\tfrac{1}{x}, 1, (c^{n/m})^m) = m\,\text{area}(\tfrac{1}{x}, 1, c^{n/m}).$$

Because $(c^{n/m})^m = c^n$, we can rewrite the equation above as

$$\text{area}(\tfrac{1}{x}, 1, c^n) = m\,\text{area}(\tfrac{1}{x}, 1, c^{n/m}).$$

By the restricted area formula that we have already derived, the left side of the equation above equals $n\,\text{area}(\frac{1}{x}, 1, c)$. Thus $n\,\text{area}(\frac{1}{x}, 1, c) = m\,\text{area}(\frac{1}{x}, 1, c^{n/m})$, which implies that

$$\text{area}(\tfrac{1}{x}, 1, c^{n/m}) = \frac{n}{m}\,\text{area}(\tfrac{1}{x}, 1, c).$$

In other words, we have now shown that

$$\text{area}(\tfrac{1}{x}, 1, c^t) = t\,\text{area}(\tfrac{1}{x}, 1, c)$$

whenever t is a positive rational number. Because every positive number can be approximated as closely as we like by a positive rational number, this implies that the equation above holds whenever t is a positive number. This completes the derivation of our area formula.

EXERCISES

For Exercises 1–14, estimate the indicated value without using a calculator.

1. $\ln 1.003$
2. $\ln 1.0007$
3. $\ln 0.993$
4. $\ln 0.9996$
5. $\ln 3.0012 - \ln 3$
6. $\ln 4.001 - \ln 4$
7. $e^{0.0013}$
8. $e^{0.00092}$
9. $e^{-0.0083}$
10. $e^{-0.00046}$
11. $\dfrac{e^9}{e^{8.997}}$
12. $\dfrac{e^5}{e^{4.984}}$
13. $\left(\dfrac{e^{7.001}}{e^7}\right)^2$
14. $\left(\dfrac{e^{8.0002}}{e^8}\right)^3$
15. Estimate the value of
$$\left(1 + \frac{3}{10^{100}}\right)^{(10^{100})}.$$
[*Your calculator will be unable to evaluate directly the expressions in this exercise and the next five exercises. Thus you will need to do more than button pushing for these exercises.*]
16. Estimate the value of
$$\left(1 + \frac{5}{10^{90}}\right)^{(10^{90})}.$$
17. Estimate the value of
$$\left(1 - \frac{4}{9^{80}}\right)^{(9^{80})}.$$
18. Estimate the value of
$$\left(1 - \frac{2}{8^{99}}\right)^{(8^{99})}.$$
19. Estimate the value of
$$(1 + 10^{-1000})^{2 \cdot 10^{1000}}.$$
20. Estimate the value of
$$(1 + 10^{-100})^{3 \cdot 10^{100}}.$$
21. Estimate the slope of the line containing the points
$$(5, \ln 5) \quad \text{and} \quad (5 + 10^{-100}, \ln(5 + 10^{-100})).$$
22. Estimate the slope of the line containing the points
$$(4, \ln 4) \quad \text{and} \quad (4 + 10^{-1000}, \ln(4 + 10^{-1000})).$$
23. Suppose t is a small positive number. Estimate the slope of the line containing the points $(4, e^4)$ and $(4 + t, e^{4+t})$.
24. Suppose r is a small positive number. Estimate the slope of the line containing the points $(7, e^7)$ and $(7 + r, e^{7+r})$.
25. Suppose r is a small positive number. Estimate the slope of the line containing the points $(e^2, 6)$ and $(e^{2+r}, 6 + r)$.

26. Suppose b is a small positive number. Estimate the slope of the line containing the points $(e^3, 5 + b)$ and $(e^{3+b}, 5)$.

27. Find a number r such that

$$\left(1 + \frac{r}{10^{90}}\right)^{(10^{90})} \approx 5.$$

28. Find a number r such that

$$\left(1 + \frac{r}{10^{75}}\right)^{(10^{75})} \approx 4.$$

29. Find the number c such that

$$\text{area}(\tfrac{1}{x}, 2, c) = 3.$$

30. Find the number c such that

$$\text{area}(\tfrac{1}{x}, 5, c) = 4.$$

PROBLEMS

31. Show that

$$\frac{1}{10^{20} + 1} < \ln(1 + 10^{-20}) < \frac{1}{10^{20}}.$$

32. (a) Using a calculator, verify that

$$\log(1 + t) \approx 0.434294t$$

for some small numbers t (for example, try $t = 0.001$ and then smaller values of t).

(b) Explain why the approximation above follows from the approximation $\ln(1 + t) \approx t$.

33. (a) Using a calculator or computer, verify that

$$2^t - 1 \approx 0.693147t$$

for some small numbers t (for example, try $t = 0.001$ and then smaller values of t).

(b) Explain why $2^t = e^{t \ln 2}$ for every number t.

(c) Explain why the approximation in part (a) follows from the approximation $e^t \approx 1 + t$.

34. Suppose x is a positive number.

(a) Explain why $x^t = e^{t \ln x}$ for every number t.

(b) Explain why

$$\frac{x^t - 1}{t} \approx \ln x$$

if t is close to 0.

[Part (b) of this problem gives another illustration of why the natural logarithm deserves the title "natural".]

35. (a) Using a calculator or computer, verify that

$$(1 + \tfrac{\ln 10}{x})^x \approx 10$$

for large values of x (for example, try $x = 1000$ and then larger values of x).

(b) Explain why the approximation above follows from the approximation $(1 + \frac{r}{x})^x \approx e^r$.

36. Using a calculator, discover a formula for a good approximation for

$$\ln(2 + t) - \ln 2$$

for small values of t (for example, try $t = 0.04$, $t = 0.02$, $t = 0.01$, and then smaller values of t). Then explain why your formula is indeed a good approximation.

37. Show that for every positive number c, we have

$$\ln(c + t) - \ln c \approx \tfrac{t}{c}$$

for small values of t.

38. Show that for every number c, we have

$$e^{c+t} - e^c \approx te^c$$

for small values of t.

39. Show that if $t > 0$, then $1 + t < e^t$.
[This problem and the next problem combine to show that

$$1 + t < e^t < (1 + t)^{1+t}$$

if $t > 0$.]

40. Show that if $t > 0$, then $e^t < (1 + t)^{1+t}$.

41. Show that if $x > 0$, then $(1 + \frac{1}{x})^x < e$.
[This problem and the next problem combine to show that

$$(1 + \tfrac{1}{x})^x < e < (1 + \tfrac{1}{x})^{x+1}$$

if $x > 0$.]

42. Show that if $x > 0$, then $e < (1 + \frac{1}{x})^{x+1}$.

43. (a) Show that

$$1.01^{100} < e < 1.01^{101}.$$

(b) Explain why

$$\frac{1.01^{100} + 1.01^{101}}{2}$$

is a reasonable estimate of e.

44. Show that

$$\text{area}(\tfrac{1}{x}, \tfrac{1}{b}, 1) = \text{area}(\tfrac{1}{x}, 1, b)$$

for every number $b > 1$.

45. Show that if $0 < a < 1$, then

$$\text{area}(\tfrac{1}{x}, a, 1) = -\ln a.$$

46. Show that

$$\text{area}(\tfrac{1}{x}, a, b) = \text{area}(\tfrac{1}{x}, 1, \tfrac{b}{a})$$

whenever $0 < a < b$.

47. Show that

$$\text{area}(\tfrac{1}{x}, a, b) = \ln \frac{b}{a}$$

whenever $0 < a < b$.

48. Show that $\sinh x \approx x$ if x is close to 0. [*The definition of* sinh *was given before Exercise 60 in Section 6.1.*]

WORKED-OUT SOLUTIONS *to Odd-numbered Exercises*

For Exercises 1–14, estimate the indicated value without using a calculator.

1. $\ln 1.003$

SOLUTION

$$\ln 1.003 = \ln(1 + 0.003) \approx 0.003$$

3. $\ln 0.993$

SOLUTION

$$\ln 0.993 = \ln(1 + (-0.007)) \approx -0.007$$

5. $\ln 3.0012 - \ln 3$

SOLUTION

$$\begin{aligned} \ln 3.0012 - \ln 3 = \ln \frac{3.0012}{3} &= \ln 1.0004 \\ &= \ln(1 + 0.0004) \\ &\approx 0.0004 \end{aligned}$$

7. $e^{0.0013}$

SOLUTION

$$e^{0.0013} \approx 1 + 0.0013 = 1.0013$$

9. $e^{-0.0083}$

SOLUTION

$$e^{-0.0083} \approx 1 + (-0.0083) = 0.9917$$

11. $\dfrac{e^9}{e^{8.997}}$

SOLUTION

$$\frac{e^9}{e^{8.997}} = e^{9-8.997} = e^{0.003} \approx 1 + 0.003 = 1.003$$

13. $\left(\dfrac{e^{7.001}}{e^7}\right)^2$

SOLUTION

$$\begin{aligned} \left(\frac{e^{7.001}}{e^7}\right)^2 = (e^{7.001-7})^2 &= (e^{0.001})^2 \\ &= e^{0.002} \\ &\approx 1 + 0.002 = 1.002 \end{aligned}$$

15. Estimate the value of

$$\left(1 + \frac{3}{10^{100}}\right)^{(10^{100})}.$$

SOLUTION $\left(1 + \dfrac{3}{10^{100}}\right)^{(10^{100})} \approx e^3 \approx 20.09$

17. Estimate the value of

$$\left(1 - \frac{4}{9^{80}}\right)^{(9^{80})}.$$

SOLUTION $\left(1 - \dfrac{4}{9^{80}}\right)^{(9^{80})} \approx e^{-4} \approx 0.01832$

19. Estimate the value of

$$(1 + 10^{-1000})^{2\cdot 10^{1000}}.$$

SOLUTION

$$\begin{aligned}(1 + 10^{-1000})^{2\cdot 10^{1000}} &= \Big((1 + 10^{-1000})^{10^{1000}}\Big)^2 \\ &= \Big((1 + \frac{1}{10^{1000}})^{10^{1000}}\Big)^2 \\ &\approx e^2 \\ &\approx 7.389\end{aligned}$$

21. Estimate the slope of the line containing the points

$$(5, \ln 5) \quad \text{and} \quad (5 + 10^{-100}, \ln(5 + 10^{-100})).$$

SOLUTION The slope of the line containing the points

$$(5, \ln 5) \quad \text{and} \quad (5 + 10^{-100}, \ln(5 + 10^{-100}))$$

is obtained in the usual way by taking the ratio of the difference of the second coordinates with the difference of the first coordinates:

$$\begin{aligned}\frac{\ln(5 + 10^{-100}) - \ln 5}{5 + 10^{-100} - 5} &= \frac{\ln(1 + \frac{1}{5}\cdot 10^{-100})}{10^{-100}} \\ &\approx \frac{\frac{1}{5}\cdot 10^{-100}}{10^{-100}} \\ &= \tfrac{1}{5}.\end{aligned}$$

Thus the slope of the line in question is approximately $\frac{1}{5}$.

23. Suppose t is a small positive number. Estimate the slope of the line containing the points $(4, e^4)$ and $(4 + t, e^{4+t})$.

SOLUTION The slope of the line containing $(4, e^4)$ and $(4 + t, e^{4+t})$ is obtained in the usual way by taking the ratio of the difference of the second coordinates to the difference of the first coordinates:

$$\begin{aligned}\frac{e^{4+t} - e^4}{4 + t - 4} &= \frac{e^4(e^t - 1)}{t} \\ &\approx \frac{e^4(1 + t - 1)}{t} \\ &= e^4 \\ &\approx 54.598\end{aligned}$$

Thus the slope of the line in question is approximately 54.598.

25. Suppose r is a small positive number. Estimate the slope of the line containing the points $(e^2, 6)$ and $(e^{2+r}, 6 + r)$.

SOLUTION The slope of the line containing $(e^2, 6)$ and $(e^{2+r}, 6 + r)$ is obtained in the usual way by taking the ratio of the difference of the second coordinates to the difference of the first coordinates:

$$\begin{aligned}\frac{6 + r - 6}{e^{2+r} - e^2} &= \frac{r}{e^2(e^r - 1)} \\ &\approx \frac{r}{e^2(1 + r - 1)} \\ &= \frac{1}{e^2} \\ &\approx 0.135\end{aligned}$$

Thus the slope of the line in question is approximately 0.135.

27. Find a number r such that

$$\Big(1 + \frac{r}{10^{90}}\Big)^{(10^{90})} \approx 5.$$

SOLUTION If r is not a huge number, then

$$\Big(1 + \frac{r}{10^{90}}\Big)^{(10^{90})} \approx e^r.$$

Thus we need to find a number r such that $e^r \approx 5$. This implies that $r \approx \ln 5 \approx 1.60944$.

29. Find the number c such that

$$\text{area}(\tfrac{1}{x}, 2, c) = 3.$$

SOLUTION We have

$$\begin{aligned}3 &= \text{area}(\tfrac{1}{x}, 2, c) \\ &= \text{area}(\tfrac{1}{x}, 1, c) - \text{area}(\tfrac{1}{x}, 1, 2) \\ &= \ln c - \ln 2 \\ &= \ln \tfrac{c}{2}.\end{aligned}$$

Thus $\frac{c}{2} = e^3$, which implies that $c = 2e^3 \approx 40.171$.

6.3 Exponential Growth Revisited

LEARNING OBJECTIVES

By the end of this section you should be able to

- explain the connection between continuous compounding and e;
- make computations using continuous compounding;
- make computations using continuous growth rates;
- estimate doubling time under continuous compounding.

Continuously Compounded Interest

Recall that if interest is compounded n times per year at annual interest rate r, then after t years an initial amount P grows to

$$P(1 + \tfrac{r}{n})^{nt};$$

see Section 5.4 to review the derivation of this formula.

More frequent compounding leads to a larger amount, because interest is earned on the interest more frequently. We could imagine compounding interest once per month ($n = 12$), or once per day ($n = 365$), or once per hour ($n = 365 \times 24 = 8760$), or once per minute ($n = 365 \times 24 \times 60 = 525600$), or once per second ($n = 365 \times 24 \times 60 \times 60 = 31536000$), or even more frequently.

To see what happens when interest is compounded very frequently, we need to consider what happens to the formula above when n is very large. Recall from the previous section that if n is much larger than r, then $(1 + \frac{r}{n})^n \approx e^r$. Thus

$$\begin{aligned} P(1 + \tfrac{r}{n})^{nt} &= P\big((1 + \tfrac{r}{n})^n\big)^t \\ &\approx P(e^r)^t \\ &= Pe^{rt}. \end{aligned}$$

This bank has been paying continuously compounded interest for many years.

In other words, if interest is compounded many times per year at annual interest rate r, then after t years an initial amount P grows to approximately Pe^{rt}. We can think of Pe^{rt} as the amount that we would have if interest were compounded continuously. This formula is actually shorter and cleaner than the formula involving compounding n times per year.

Many banks and other financial institutions use continuous compounding rather than compounding a specific number of times per year. Thus they use the formula derived above involving e, which we now restate as follows:

Continuous compounding

If interest is compounded continuously at annual interest rate r, then after t years an initial amount P grows to

$$Pe^{rt}.$$

This formula for continuous compounding gives another example of how e arises naturally.

Continuous compounding always produces a larger amount than compounding any specific number of times per year. However, for moderate initial amounts, moderate interest rates, and moderate time periods, the difference is not large, as shown in the following example.

EXAMPLE 1

Suppose \$10,000 is placed in a bank account that pays 5% annual interest.

(a) If interest is compounded continuously, how much will be in the bank account after 10 years?

(b) If interest is compounded four times per year, how much will be in the bank account after 10 years?

Continuous compounding indeed yields more in this example, as expected, but the difference is only about \$51 after 10 years.

SOLUTION

(a) The continuous compounding formula shows that \$10,000 compounded continuously for 10 years at 5% annual interest grows to become

$$\$10,000e^{0.05\times 10} \approx \$16,487.$$

(b) The compound interest formula shows that \$10,000 compounded four times per year for 10 years at 5% annual interest grows to become

$$\$10,000(1+\tfrac{0.05}{4})^{4\times 10} \approx \$16,436.$$

See Exercise 25 for an example of the dramatic difference continuous compounding can make over a very long time period.

Continuous Growth Rates

The model presented above of continuous compounding of interest can be applied to any situation with continuous growth at a fixed percentage. The units of time do not necessarily need to be years, but as usual the same time units must be used in all aspects of the model. Similarly, the quantity being measured need not be dollars; for example, this model works well for population growth over time intervals that are not too large.

Because continuous growth at a fixed percentage behaves the same as continuous compounding with money, the formulas are the same. Instead of referring to an annual interest rate that is compounded continuously, we use the term **continuous growth rate**. In other words, the continuous growth rate operates like an interest rate that is continuously compounded.

The continuous growth rate gives a good way to measure how fast something is growing. Again, the magic number e plays a special role. Our result above about continuous compounding can be restated to apply to more general situations, as follows:

Continuous growth rates

If a quantity has a continuous growth rate of r per unit time, then after t time units an initial amount P grows to

$$Pe^{rt}.$$

EXAMPLE 2

A continuous growth rate of 10% per hour does not imply that the colony increases by 10% after one hour. In one hour the colony increases in size by a factor of $e^{0.1}$, which is approximately 1.105, which is an increase of 10.5%.

Suppose a colony of bacteria has a continuous growth rate of 10% per hour.

(a) By what percent will the colony grow after five hours?

(b) How long will it take for the colony to grow to 250% of its initial size?

SOLUTION

(a) A continuous growth rate of 10% per hour means that we should set $r = 0.1$. If the colony starts at size P at time 0, then at time t (measured in hours) its size will be $Pe^{0.1t}$.

Thus after five hours the size of the colony will be $Pe^{0.5}$, which is an increase by a factor of $e^{0.5}$ over the initial size P. Because $e^{0.5} \approx 1.65$, this means that the colony will grow by about 65% after five hours.

(b) We want to find t such that

$$Pe^{0.1t} = 2.5P.$$

Dividing both sides by P, we see that $0.1t = \ln 2.5$. Thus $t = \frac{\ln 2.5}{0.1} \approx 9.16$. Because $0.16 \approx \frac{1}{6}$ and because one-sixth of an hour is 10 minutes, we conclude that it will take about 9 hours, 10 minutes for the colony to grow to 250% of its initial size.

Doubling Your Money

The following example shows how to compute how long it takes to double your money with continuous compounding.

EXAMPLE 3

How many years does it take for money to double at 5% annual interest compounded continuously?

SOLUTION After t years an initial amount P compounded continuously at 5% annual interest grows to $Pe^{0.05t}$. We want this to equal twice the initial amount. Thus we must solve the equation

$$Pe^{0.05t} = 2P,$$

which is equivalent to the equation $e^{0.05t} = 2$, which implies that $0.05t = \ln 2$. Thus

$$t = \frac{\ln 2}{0.05} \approx \frac{0.693}{0.05} = \frac{69.3}{5} \approx 13.9.$$

Hence the initial amount of money will double in about 13.9 years.

Suppose we want to know how long it takes money to double at 4% annual interest compounded continuously instead of 5%. Repeating the calculation above, but with 0.04 replacing 0.05, we see that money doubles in about $\frac{69.3}{4}$ years at 4% annual interest compounded continuously. More generally, money doubles in about $\frac{69.3}{R}$ years at R percent interest compounded continuously. Here R is expressed as a percent, rather than as a number. In other words, 5% interest corresponds to $R = 5$.

For quick estimates, usually it is best to round up the 69.3 appearing in the expression $\frac{69.3}{R}$ to 70. Using 70 instead of 69 is easier because 70 is evenly divisible by more numbers than 69 (some people even use 72 instead of 70, but using 70 gives more accurate results than using 72). Thus we have the following useful approximation formula:

Doubling time

At R percent annual interest compounded continuously, money doubles in approximately

$$\frac{70}{R}$$

years.

This approximation formula illustrates again the usefulness of the natural logarithm. The number 70 *appearing in this formula is really an approximation for* 69.3, *which is an approximation for* $100 \ln 2$.

For example, this formula shows that at 5% annual interest compounded continuously, money doubles in about $\frac{70}{5}$ years, which equals 14 years. This is close enough to the more precise estimate of 13.9 years that we obtained above. Furthermore, the computation using the $\frac{70}{R}$ estimate is easy enough to do without a calculator.

Instead of focusing on how long it takes money to double at a specified interest rate, we could ask what interest rate is required to make money double in a specified time period. Here is an example:

EXAMPLE 4

What annual interest rate is needed so that money will double in seven years when compounded continuously?

SOLUTION After seven years an initial amount P compounded continuously at R% annual interest grows to $Pe^{7R/100}$. We want this to equal twice the initial amount. Thus we must solve the equation

$$Pe^{7R/100} = 2P,$$

which is equivalent to the equation $e^{7R/100} = 2$, which implies that $\frac{7R}{100} = \ln 2$. Thus

$$R = \frac{100 \ln 2}{7} \approx \frac{69.3}{7} \approx 9.9.$$

Hence about 9.9% annual interest will make money double in seven years.

Suppose we want to know what annual interest rate is needed to double money in 11 years when compounded continuously. Repeating the calculation above, but with 11 replacing 7, we see that about $\frac{69.3}{11}$% annual interest would be needed. More generally, we see that to double money in t years, $\frac{69.3}{t}$ percent interest is needed.

For quick estimates, usually it is best to round up the 69.3 appearing in the expression $\frac{69.3}{t}$ to 70. Thus we have the following useful approximation formula:

Doubling rate

The annual interest rate needed for money to double in t years with continuous compounding is approximately

$$\frac{70}{t}$$

percent.

For example, this formula shows that for money to double in seven years when compounded continuously requires about $\frac{70}{7}$% annual interest, which equals 10%. This is close enough to the more precise estimate of 9.9% that we obtained above.

EXERCISES

1. How much would an initial amount of \$2000, compounded continuously at 6% annual interest, become after 25 years?
2. How much would an initial amount of \$3000, compounded continuously at 7% annual interest, become after 15 years?
3. How much would you need to deposit in a bank account paying 4% annual interest compounded continuously so that at the end of 10 years you would have \$10,000?
4. How much would you need to deposit in a bank account paying 5% annual interest compounded continuously so that at the end of 15 years you would have \$20,000?
5. Suppose a bank account that compounds interest continuously grows from \$100 to \$110 in two years. What annual interest rate is the bank paying?
6. Suppose a bank account that compounds interest continuously grows from \$200 to \$224 in three years. What annual interest rate is the bank paying?
7. Suppose a colony of bacteria has a continuous growth rate of 15% per hour. By what percent will the colony have grown after eight hours?
8. Suppose a colony of bacteria has a continuous growth rate of 20% per hour. By what percent will the colony have grown after seven hours?
9. Suppose a country's population increases by a total of 3% over a two-year period. What is the continuous growth rate for this country?
10. Suppose a country's population increases by a total of 6% over a three-year period. What is the continuous growth rate for this country?

11. Suppose the amount of the world's computer hard disk storage increases by a total of 200% over a four-year period. What is the continuous growth rate for the amount of the world's hard disk storage?

12. Suppose the number of cell phones in the world increases by a total of 150% over a five-year period. What is the continuous growth rate for the number of cell phones in the world?

13. Suppose a colony of bacteria has a continuous growth rate of 30% per hour. If the colony contains 8000 cells now, how many did it contain five hours ago?

14. Suppose a colony of bacteria has a continuous growth rate of 40% per hour. If the colony contains 7500 cells now, how many did it contain three hours ago?

15. Suppose a colony of bacteria has a continuous growth rate of 35% per hour. How long does it take the colony to triple in size?

16. Suppose a colony of bacteria has a continuous growth rate of 70% per hour. How long does it take the colony to quadruple in size?

17. About how many years does it take for money to double when compounded continuously at 2% per year?

18. About how many years does it take for money to double when compounded continuously at 10% per year?

19. About how many years does it take for \$200 to become \$800 when compounded continuously at 2% per year?

20. About how many years does it take for \$300 to become \$2,400 when compounded continuously at 5% per year?

21. How long does it take for money to triple when compounded continuously at 5% per year?

22. How long does it take for money to increase by a factor of five when compounded continuously at 7% per year?

23. Find a formula for estimating how long money takes to triple at R percent annual interest rate compounded continuously.

24. Find a formula for estimating how long money takes to increase by a factor of ten at R percent annual interest compounded continuously.

25. Suppose that one bank account pays 5% annual interest compounded once per year, and a second bank account pays 5% annual interest compounded continuously. If both bank accounts start with the same initial amount, how long will it take for the second bank account to contain twice the amount of the first bank account?

26. Suppose that one bank account pays 3% annual interest compounded once per year, and a second bank account pays 4% annual interest compounded continuously. If both bank accounts start with the same initial amount, how long will it take for the second bank account to contain 50% more than the first bank account?

27. Suppose a colony of 100 bacteria cells has a continuous growth rate of 30% per hour. Suppose a second colony of 200 bacteria cells has a continuous growth rate of 20% per hour. How long does it take for the two colonies to have the same number of bacteria cells?

28. Suppose a colony of 50 bacteria cells has a continuous growth rate of 35% per hour. Suppose a second colony of 300 bacteria cells has a continuous growth rate of 15% per hour. How long does it take for the two colonies to have the same number of bacteria cells?

29. Suppose a colony of bacteria has doubled in five hours. What is the approximate continuous growth rate of this colony of bacteria?

30. Suppose a colony of bacteria has doubled in two hours. What is the approximate continuous growth rate of this colony of bacteria?

31. Suppose a colony of bacteria has tripled in five hours. What is the continuous growth rate of this colony of bacteria?

32. Suppose a colony of bacteria has tripled in two hours. What is the continuous growth rate of this colony of bacteria?

PROBLEMS

33. Using compound interest, explain why

$$(1 + \tfrac{0.05}{n})^n < e^{0.05}$$

for every positive integer n.

34. Suppose that in Exercise 9 we had simply divided the 3% increase over two years by 2, getting 1.5% per year. Explain why this number is close to the more accurate answer of approximately 1.48% per year.

35. Suppose that in Exercise 11 we had simply divided the 200% increase over four years by 4, getting 50% per year. Explain why we should not be surprised that this number is not close to the more accurate answer of approximately 27.5% per year.

36. In Section 5.4 we saw that if a population doubles every d time units, then the function p modeling this population growth is given by the formula

$$p(t) = p_0 \cdot 2^{t/d},$$

where p_0 is the population at time 0. Some books do not use the formula above but instead use the formula

$$p(t) = p_0 e^{(t \ln 2)/d}.$$

Show that the two formulas above are really the same.

[*Which of the two formulas in this problem do you think is cleaner and easier to understand?*]

37. In Section 5.2 we saw that if a radioactive isotope has half-life h, then the function modeling the number of atoms in a sample of this isotope is

$$a(t) = a_0 \cdot 2^{-t/h},$$

where a_0 is the number of atoms of the isotope in the sample at time 0. Many books do not use the formula above but instead use the formula

$$a(t) = a_0 e^{-(t \ln 2)/h}.$$

Show that the two formulas above are really the same.

[*Which of the two formulas in this problem do you think is cleaner and easier to understand?*]

38. Explain why every function f with exponential growth (see Section 5.4 for the definition) can be written in the form

$$f(x) = ce^{kx},$$

where c and k are positive constants.

WORKED-OUT SOLUTIONS *to Odd-numbered Exercises*

1. How much would an initial amount of \$2000, compounded continuously at 6% annual interest, become after 25 years?

SOLUTION After 25 years, \$2000 compounded continuously at 6% annual interest would grow to $2000e^{0.06 \times 25}$ dollars, which equals $2000e^{1.5}$ dollars, which is approximately \$8963.

3. How much would you need to deposit in a bank account paying 4% annual interest compounded continuously so that at the end of 10 years you would have \$10,000?

SOLUTION We need to find P such that

$$10000 = Pe^{0.04 \times 10} = Pe^{0.4}.$$

Thus

$$P = \frac{10000}{e^{0.4}} \approx 6703.$$

In other words, the initial amount in the bank account should be $\frac{10000}{e^{0.4}}$ dollars, which is approximately \$6703.

5. Suppose a bank account that compounds interest continuously grows from \$100 to \$110 in two years. What annual interest rate is the bank paying?

SOLUTION Let r denote the annual interest rate paid by the bank. Then

$$110 = 100e^{2r}.$$

Dividing both sides of this equation by 100 gives $1.1 = e^{2r}$, which implies that $2r = \ln 1.1$, which is equivalent to

$$r = \frac{\ln 1.1}{2} \approx 0.0477.$$

Thus the annual interest is approximately 4.77%.

7. Suppose a colony of bacteria has a continuous growth rate of 15% per hour. By what percent will the colony have grown after eight hours?

SOLUTION A continuous growth rate of 15% per hour means that $r = 0.15$. If the colony starts at size P at time 0, then at time t (measured in hours) its size will be $Pe^{0.15t}$.

Because $0.15 \times 8 = 1.2$, after eight hours the size of the colony will be $Pe^{1.2}$, which is an increase by a factor of $e^{1.2}$ over the initial size P. Because $e^{1.2} \approx 3.32$, this means that the colony will be about 332% of its original size after eight hours. Thus the colony will have grown by about 232% after eight hours.

9. Suppose a country's population increases by a total of 3% over a two-year period. What is the continuous growth rate for this country?

SOLUTION A 3% increase means that we have 1.03 times as much as the initial amount. Thus $1.03P = Pe^{2r}$, where P is the country's population at the beginning of the measurement period and r is the country's continuous growth rate. Thus $e^{2r} = 1.03$, which means that $2r = \ln 1.03$. Thus $r = \frac{\ln 1.03}{2} \approx 0.0148$. Thus the country's continuous growth rate is approximately 1.48% per year.

11. Suppose the amount of the world's computer hard disk storage increases by a total of 200% over a four-year period. What is the continuous growth rate for the amount of the world's hard disk storage?

SOLUTION A 200% increase means that we have three times as much as the initial amount. Thus $3P = Pe^{4r}$, where P is amount of the world's hard disk storage at the beginning of the measurement period and r is the continuous growth rate. Thus $e^{4r} = 3$, which means that $4r = \ln 3$. Thus $r = \frac{\ln 3}{4} \approx 0.275$. Thus the continuous growth rate is approximately 27.5%.

13. Suppose a colony of bacteria has a continuous growth rate of 30% per hour. If the colony contains 8000 cells now, how many did it contain five hours ago?

SOLUTION Let P denote the number of cells at the initial time five hours ago. Thus we have $8000 = Pe^{0.3\times 5}$, or $8000 = Pe^{1.5}$. Thus

$$P = 8000/e^{1.5} \approx 1785.$$

15. Suppose a colony of bacteria has a continuous growth rate of 35% per hour. How long does it take the colony to triple in size?

SOLUTION Let P denote the initial size of the colony, and let t denote the time that it takes the colony to triple in size. Then $3P = Pe^{0.35t}$, which means that $e^{0.35t} = 3$. Thus $0.35t = \ln 3$, which implies that $t = \frac{\ln 3}{0.35} \approx 3.14$. Thus the colony triples in size in approximately 3.14 hours.

17. About how many years does it take for money to double when compounded continuously at 2% per year?

SOLUTION At 2% per year compounded continuously, money will double in approximately $\frac{70}{2}$ years, which equals 35 years.

19. About how many years does it take for \$200 to become \$800 when compounded continuously at 2% per year?

SOLUTION At 2% per year, money doubles in approximately 35 years. For \$200 to become \$800, it must double twice. Thus this will take about 70 years.

21. How long does it take for money to triple when compounded continuously at 5% per year?

SOLUTION To triple an initial amount P in t years at 5% annual interest compounded continuously, the following equation must hold:

$$Pe^{0.05t} = 3P.$$

Dividing both sides by P and then taking the natural logarithm of both sides gives $0.05t = \ln 3$. Thus $t = \frac{\ln 3}{0.05}$. Thus it would take $\frac{\ln 3}{0.05}$ years, which is about 22 years.

23. Find a formula for estimating how long money takes to triple at R percent annual interest rate compounded continuously.

SOLUTION To triple an initial amount P in t years at R percent annual interest compounded continuously, the following equation must hold:

$$Pe^{Rt/100} = 3P.$$

Dividing both sides by P and then taking the natural logarithm of both sides gives $Rt/100 = \ln 3$. Thus $t = \frac{100 \ln 3}{R}$. Because $\ln 3 \approx 1.10$, this shows that money triples in about $\frac{110}{R}$ years.

25. Suppose that one bank account pays 5% annual interest compounded once per year, and a second bank account pays 5% annual interest compounded continuously. If both bank accounts start with the same initial amount, how long will it take for the second bank account to contain twice the amount of the first bank account?

SOLUTION Suppose both bank accounts start with P dollars. After t years, the first bank account will contain $P(1.05)^t$ dollars and the second bank account will contain $Pe^{0.05t}$ dollars. Thus we need to solve the equation

$$\frac{Pe^{0.05t}}{P(1.05)^t} = 2.$$

The initial amount P drops out of this equation (as expected), and we can rewrite this equation as follows:

$$2 = \frac{e^{0.05t}}{1.05^t} = \frac{(e^{0.05})^t}{1.05^t} = \left(\frac{e^{0.05}}{1.05}\right)^t.$$

Taking the natural logarithm of the first and last terms above gives

$$\begin{aligned} \ln 2 = t \ln \frac{e^{0.05}}{1.05} &= t(\ln e^{0.05} - \ln 1.05) \\ &= t(0.05 - \ln 1.05), \end{aligned}$$

which we can then solve for t, getting

$$t = \frac{\ln 2}{0.05 - \ln 1.05}.$$

Using a calculator to evaluate the expression above, we see that t is approximately 573 years.

27. Suppose a colony of 100 bacteria cells has a continuous growth rate of 30% per hour. Suppose a second colony of 200 bacteria cells has a continuous growth rate of 20% per hour. How long does it take for the two colonies to have the same number of bacteria cells?

SOLUTION After t hours, the first colony contains $100e^{0.3t}$ bacteria cells and the second colony contains $200e^{0.2t}$ bacteria cells. Thus we need to solve the equation

$$100e^{0.3t} = 200e^{0.2t}.$$

Dividing both sides by 100 and then dividing both sides by $e^{0.2t}$ gives the equation

$$e^{0.1t} = 2.$$

Thus $0.1t = \ln 2$, which implies that

$$t = \frac{\ln 2}{0.1} \approx 6.93.$$

Thus the two colonies have the same number of bacteria cells in a bit less than 7 hours.

29. Suppose a colony of bacteria has doubled in five hours. What is the approximate continuous growth rate of this colony of bacteria?

SOLUTION The approximate formula for doubling the number of bacteria is the same as for doubling money. Thus if a colony of bacteria doubles in five hours, then it has a continuous growth rate of approximately (70/5)% per hour. In other words, this colony of bacteria has a continuous growth rate of approximately 14% per hour.

31. Suppose a colony of bacteria has tripled in five hours. What is the continuous growth rate of this colony of bacteria?

SOLUTION Let r denote the continuous growth rate of this colony of bacteria. If the colony initially contains P bacteria cells, then after five hours it will contain Pe^{5r} bacteria cells. Thus we need to solve the equation

$$Pe^{5r} = 3P.$$

Dividing both sides by P gives the equation $e^{5r} = 3$, which implies that $5r = \ln 3$. Thus

$$r = \frac{\ln 3}{5} \approx 0.2197.$$

Thus the continuous growth rate of this colony of bacteria is approximately 22% per hour.

CHAPTER SUMMARY

To check that you have mastered the most important concepts and skills covered in this chapter, make sure that you can do each item in the following list:

- Approximate the area under a curve using rectangles.
- Explain the definition of e.
- Explain the definition of the natural logarithm.
- Give at least one explanation of why the natural logarithm deserves to be called "natural".
- Approximate e^x and $\ln(1+x)$ for $|x|$ small.
- Compute continuously compounded interest.
- Estimate how long it takes to double money at a given interest rate.

To review a chapter, go through the list above to find items that you do not know how to do, then reread the material in the chapter about those items. Then try to answer the chapter review questions below without looking back at the chapter.

CHAPTER REVIEW QUESTIONS

1. What is the definition of e?

2. What are the domain and range of the function f defined by $f(x) = e^x$?

3. What is the definition of the natural logarithm?

4. What are the domain and range of the function g defined by $g(y) = \ln y$?

5. Find a number t such that $\ln(4t+3) = 5$.

6. Find a number t that makes $(e^{t+8})^t$ as small as possible.

7. Find a number w such that $e^{2w-7} = 6$.

8. Find a formula for the inverse of the function g defined by $g(x) = 8 - 3e^{5x}$.

9. Find a formula for the inverse of the function h defined by $h(x) = 1 - 5\ln(x+4)$.

10. Find the area of the region under the curve $y = \frac{1}{x}$, above the x-axis, and between the lines $x = 1$ and $x = e^2$.

11. Find a number c such that the area of the region under the curve $y = \frac{1}{x}$, above the x-axis, and between the lines $x = 1$ and $x = c$ is 45.

12. What is the area of the region under the curve $y = \frac{1}{x}$, above the x-axis, and between the lines $x = 3$ and $x = 5$?

13. Draw an appropriate figure and use it to explain why
$$\ln(1.0001) \approx 0.0001.$$

14. Estimate the slope of the line containing the points $(2, \ln(6 + 10^{-500}))$ and $(6, \ln 6)$.

15. Estimate the value of $\dfrac{e^{1000.002}}{e^{1000}}$.

16. Estimate the slope of the line containing the points $(6, e^{0.0002})$ and $(2, 1)$.

17. Estimate the value of
$$\left(1 - \frac{6}{788}\right)^{(788)}.$$

18. How much would an initial amount of \$12,000, compounded continuously at 6% annual interest, become after 20 years?

19. How much would you need to deposit in a bank account paying 6% annual interest compounded continuously so that at the end of 25 years you would have \$100,000?

20. Suppose a bank account that compounds interest continuously grows from \$2000 to \$2878.15 in seven years. What annual interest rate is the bank paying?

21. Approximately how many years does it take for money to double when compounded continuously at 5% per year?

22. Suppose a colony of bacteria has doubled in 10 hours. What is the approximate continuous growth rate of this colony of bacteria?

23. At 5% interest compounded continuously but paid monthly (as many banks do), how long will it take to turn a hundred dollars into a million dollars? (See the Dilbert comic in Section 5.4.)

CHAPTER

7

The standard method of solving systems of linear equations is called Gaussian elimination. This technique is named in honor of Carl Friedrich Gauss, pictured here on an old German 10-mark bill. Gauss used this method in a book published in 1809. However, the method was also used in a Chinese book published over 1600 years earlier.

Systems of Equations

This chapter provides only a taste of some topics in systems of equations. Proper treatment of these subjects requires a full book devoted just to them. The first course in linear algebra focuses on systems of linear equations and matrices.

In the first section of this chapter we look at equations and systems of equations. We will use graphical methods to find approximate solutions to a system of equations with two variables. We will also learn how to solve some systems of equations by substitution.

In the second section of this chapter we specialize to systems of linear equations. Unlike the situation with more general systems of equations, good methods exist for finding the solutions to systems of linear equations. We will learn about Gaussian elimination, the most efficient and most commonly used of these methods.

The third section of this chapter recasts Gaussian elimination in the context of matrices. This helps us understand how computers deal with large systems of linear equations.

We delve more deeply into matrix algebra in the last section of this chapter. Key topics here include matrix multiplication and the inverse of a matrix.

7.1 Equations and Systems of Equations

LEARNING OBJECTIVES

By the end of this section you should be able to

- find solutions of equations, when possible;
- graphically solve a system of equations with two variables;
- solve a system of equations by substitution.

Solving an Equation

In this chapter we will discuss solutions to systems of equations. Before we get to systems of equations, we get a firm foundation by briefly discussing what it means to solve a single equation.

Equations come in several types. The equation

$$2^3 - \log 100 = 6$$

contains no variables. This equation expresses an arithmetic relation between some numbers.

This identity expresses the commutative property of addition.

The equation

$$x + y = y + x$$

contains two variables. This equation is not meant to be solved, because it holds for all numbers x and y. An equation that holds regardless of the numerical values of the variables is called an **identity**.

The equation

$$\frac{1}{x+3} = 2$$

holds if and only if $x = -\frac{5}{2}$. This type of equation, which holds for some but not all values of the variables, is generally what we have in mind when discussing solving equations.

Solution to an Equation

- A **solution** to an equation is an assignment of values to the variables that satisfies the equation.
- To **solve** an equation means to find all solutions to the equation.

An equation might have no solutions, one solution, any finite number of solutions, or infinitely many solutions, as shown in the following example.

EXAMPLE 1

Find all solutions for each of the following equations.

(a) $3^x = -1$

(b) $10^x = 2$

(c) $x^2 + 3x - 5 = 0$

(d) $x^2 = 1.5^x$

(e) $x = \sqrt{x^2}$

SOLUTION

(a) The equation $3^x = -1$ has no solutions because $3^x > 0$ for every real number x.

In this book, assume unless stated otherwise that we seek solutions only from among the real numbers.

(b) This equation has one solution, which is $x = \log 2$. In fact, $\log 2$ is defined as the number x such that $10^x = 2$.

(c) The quadratic formula shows that this equation has two solutions, which are

$$x = \frac{-3 - \sqrt{29}}{2} \quad \text{and} \quad x = \frac{-3 + \sqrt{29}}{2}.$$

(d) No formula exists for finding exact solutions to this equation. To deal with such equations, use a computer or symbolic calculator to find approximate solutions. For example, if using Wolfram|Alpha type `solve x^2 = 1.5^x` into an entry box to learn that this equation has three solutions:

$$x \approx 1.3021, \quad x \approx -0.8429, \quad \text{and} \quad x \approx 12.4312.$$

(e) A real number x satisfies the equation $x = \sqrt{x^2}$ if and only if $x \geq 0$. Thus this equation has infinitely many solutions—every nonnegative number is a solution.

Sadly, the solution to part (d) above illustrates the most common situation faced when needing to solve an equation. Specifically, except for a small number of special types of equation, most equations cannot be solved exactly. Something like the quadratic formula, which gives a beautiful and clean way to find exact solutions to quadratic equations, simply does not exist for most equations.

Only in textbooks do most equations have exact solutions that can be found without using technology. In the real world, you should expect that most of the time you will need to use a computer or symbolic calculator to find approximate solutions.

When you find what you think are the solutions to an equation, you should always check whether or not you have actual solutions. You may have made an error in doing a calculation. Even if you have been careful, you may get incorrect results because some algebraic manipulations are not reversible.

Check your solutions

Always check each of your solutions to an equation by substituting into the original equation.

The next example shows how you might get incorrect solutions even when no errors have been made. The checking procedure in the next example exposes the incorrect solution.

EXAMPLE 2

Find all solutions to the equation

$$\log(x + 1) + \log(x + 10) = 1.$$

SOLUTION Using the formula for the logarithm of a product, rewrite the equation above as

$$\log[(x+1)(x+10)] = 1,$$

which implies that

$$(x+1)(x+10) = 10.$$

Expand the left side of the equation above, then subtract 10 from both sides, getting

$$x(x+11) = 0.$$

The solutions to the equation above are $x = 0$ and $x = -11$.

CHECK To check whether $x = 0$ is an actual solution, substitute $x = 0$ into the original equation to see whether

$$\log(0+1) + \log(0+10) \stackrel{?}{=} 1.$$

Because $\log 1 = 0$ and $\log 10 = 1$, the equality above does indeed hold. Thus $x = 0$ is indeed a solution to the original equation.

The incorrect solution $x = -11$ that arose in this example shows the importance of always checking your solutions.

To check whether $x = -11$ is an actual solution, substitute $x = -11$ into the original equation to see whether

$$\log(-11+1) + \log(-11+10) \stackrel{?}{=} 1.$$

The two terms on the left side are $\log(-10)$ and $\log(-1)$, neither of which is defined. Thus we do not have an equality above. Hence $x = -11$ is not a solution to the original equation.

An equation can have more than one variable, as shown in the next example. Like an equation with just one variable, an equation with more than one variable can have no solutions, one solution, any finite number of solutions, or infinitely many solutions.

EXAMPLE 3

Find all solutions for each of the following equations.

(a) $x^2 + y^2 = -1$

(b) $(x-5)^2 + (y-3)^2 = 0$

(c) $x^2y^2 + (x-1)^2(y-1)^2 = 0$

(d) $x^2 + y^2 = 1$

SOLUTION

(a) This equation has no solutions because $x^2 + y^2 \geq 0$ for all real numbers x and y.

(b) This equation has one solution, which is $x = 5, y = 3$. Any other choice for x or y makes the left side of the equation positive.

(c) The equation $x^2y^2 + (x-1)^2(y-1)^2 = 0$ has two solutions, which are

$$x = 0, y = 1 \quad \text{and} \quad x = 1, y = 0.$$

To see this, note that for the equation to hold, both x^2y^2 and $(x-1)^2(y-1)^2$ must equal 0. The condition $x^2y^2 = 0$ implies that $x = 0$ or $y = 0$. If $x = 0$, then the equation $(x-1)^2(y-1)^2 = 0$ implies that $y = 1$. If $y = 0$, then the equation $(x-1)^2(y-1)^2 = 0$ implies that $x = 1$.

(d) The equation $x^2 + y^2 = 1$ has infinitely many solutions. Every point on the circle of radius 1 centered at the origin in the xy-plane corresponds to a solution to this equation.

The graph of $x^2 + y^2 = 1$. The graph of an equation with two variables is defined to correspond to the solutions to the equation.

Solving a System of Equations Graphically

Systems of equations

- A **system of equations** is a collection of equations, usually with two or more variables.
- A **solution** to a system of equations is an assignment of values to the variables that satisfies all the equations in the system.

For example,

$$x^5y^3 + xyz + z^7 = 10$$
$$x^2yz^3 + xy^5 + yz^7 = 11$$

is a system of two equations with three variables x, y, and z.

As with a single equation, except in some special cases no technique exists to find exact solutions to a system of equations.

For systems of equations with two variables, the following procedure can sometimes be used to estimate the solutions and give some insight into the system of equations.

Graphically solving a system of equations with two variables

(a) Label the two coordinate axes in a coordinate plane with the two variables.

(b) Graph each equation.

(c) The solutions to the system of equations correspond to the points where all the graphs from the previous step intersect.

The following example illustrates the graphical technique for estimating solutions to a system of equations.

EXAMPLE 4

Graphically estimate the solutions to the system of equations

$$x^2 - y^2 = 1$$
$$2x + y = 4.$$

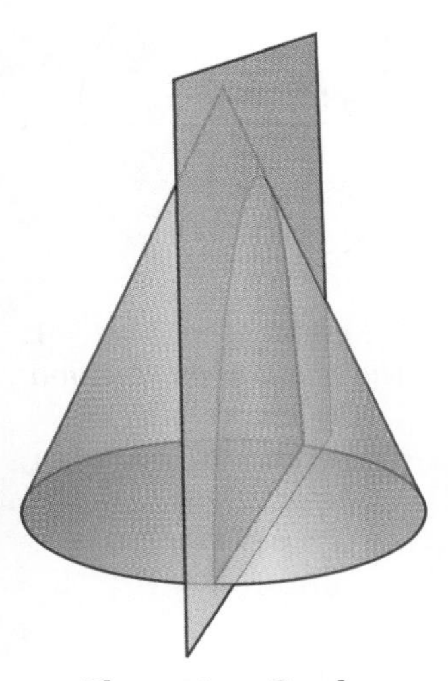

The ancient Greeks discovered that the intersection of a cone and an appropriately positioned plane is a hyperbola.

SOLUTION

A computer produced this graph of the points satisfying the two equations. Note that the set of points satisfying the equation $x^2 - y^2 = 1$ (in blue) is a hyperbola consisting of two curves.

The solution to this system of equations corresponds to the intersection of the two graphs. The figure here shows the two graphs intersect at two points.

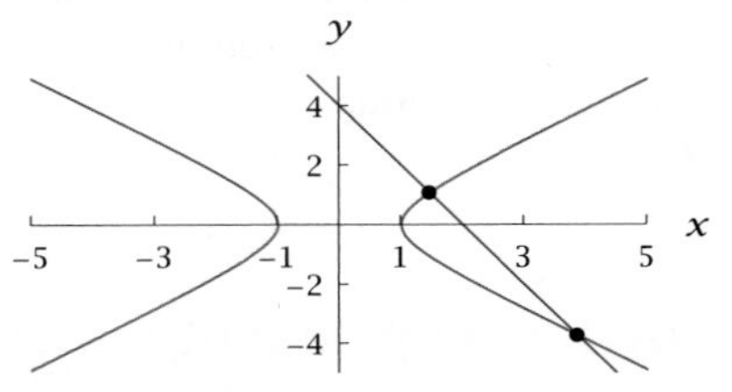

The points satisfying the equation $x^2 - y^2 = 1$ (blue) and $2x + y = 4$ (red).

One of the points of intersection has coordinates that appear to be approximately $(1.5, 1)$, and thus we estimate that one solution to this system of equations is $x \approx 1.5$, $y \approx 1$. The other point of intersection of the two graphs has coordinates that appear to be approximately $(4, -3.75)$, and thus we estimate that another solution to this system of equations is $x \approx 4$, $y \approx -3.75$.

A better estimate can be found by using a computer or graphing calculator to zoom in on a region containing one of the solutions. Here we have zoomed in on a region containing the first solution mentioned above. This figure shows that $x \approx 1.45$, $y \approx 1.1$ is a better estimate than our original approximation. Still better estimates could be found by zooming in further.

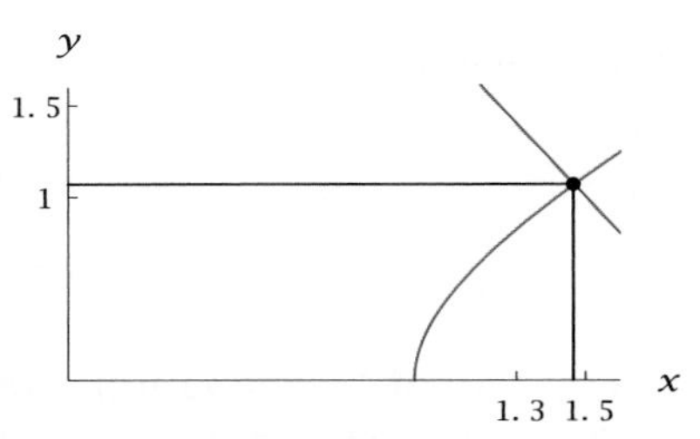

Zooming in. The vertical and horizontal lines from the intersection point to the coordinate axes help estimate the coordinates of the intersection point.

Solving a System of Equations by Substitution

One method for finding solutions to a system of equations is **substitution**.

Solving a system of equations by substitution

(a) Use an equation in the system of equations to solve for one of the variables in terms of the other variables.

(b) Substitute the expression found in the previous step into the other equations, resulting in a new system of equations with one less variable and one less equation.

(c) Repeat the first two steps until you can solve the remaining system.

(d) Substitute the values you have found into the previously obtained equations to get the complete solutions.

To get started with substitution, you must be able to solve for one of the variables in terms of the other variables. Substitution works well for some systems of equations, but for other systems of equations it may not be possible to solve for one variable in terms of the other variables.

EXAMPLE 5

Use substitution to find exact solutions to the system of equations

$$x^2 - y^2 = 1$$

$$2x + y = 4.$$

SOLUTION Solving the second equation for y gives

$$y = 4 - 2x.$$

Substituting this expression for y into the first equation gives

$$x^2 - (4 - 2x)^2 = 1,$$

which can be rewritten as

$$3x^2 - 16x + 17 = 0.$$

Using the quadratic formula to solve this equation gives

$$x = \frac{8 - \sqrt{13}}{3} \quad \text{or} \quad x = \frac{8 + \sqrt{13}}{3}.$$

Substituting these two values for x into the equation $y = 4 - 2x$ gives the following two exact solutions for the original system of equations:

$$x = \frac{8 - \sqrt{13}}{3},\ y = \frac{-4 + 2\sqrt{13}}{3} \quad \text{and} \quad x = \frac{8 + \sqrt{13}}{3},\ y = \frac{-4 - 2\sqrt{13}}{3}.$$

These exact solutions show that the estimates found using the graphical method in Example 4 are reasonable approximations.

Using a calculator to evaluate the exact solutions above shows that we have

$$x \approx 1.46482,\ y \approx 1.07037 \quad \text{and} \quad x \approx 3.86852,\ y \approx -3.73703.$$

EXERCISES

In Exercises 1–4, use graphs to find estimates for all solutions to the given system of equations.

1. $y = x^2 + 1$

   $y = e^x - 2$

2. $y = (x + 1)^2$

   $y = e^{x-2}$

3. $(x - 2)^2 + y^2 = 4$

   $y = \ln x$

4. $(x - 3)^2 + y^2 = 4$

   $y = 2 \ln x$

In Exercises 5–6, find all solutions to the given equation.

5. $x^6 + 8x^4 = -3$

6. $\frac{3}{x^2} + 7x^8 = -9$

In Exercises 7–20, find all solutions to the given system of equations.

7. $x^2 - 2y^2 = 3$
   $x + 2y = 1$

8. $x^2 + 3y^2 = 5$
   $x - 3y = 2$

9. $\frac{1}{x} - \frac{1}{y} = 2$
   $4x + y = 3$

10. $\frac{2}{x} - \frac{3}{y} = 1$
    $2x + y = -1$

11. $x^2 + y^3 = 11$
    $5x^2 - 2y^3 = 6$

12. $x^3 - y^2 = 3$
    $3x^3 - 2y^2 = 18$

13. $2^x + y^2 = 11$
    $3 \cdot 2^x - 2y^2 = 3$

14. $3^x - y^2 = 3$
    $2 \cdot 3^x + 3y^2 = 41$

15. $x + \log y = 7$
    $3x + 2\log y = 19$

16. $x + 2\log y = 6$
    $2x + 5\log y = 11$

17. $x^2 + \ln y = 7$
    $2x^2 - 3\ln y = 4$

18. $x^3 + 2\ln y = 3$
    $2x^3 - 3\ln y = 13$

19. $x^2 + y^3 = 9$
    $x^2y^3 = 20$

20. $x^2 + 2y^3 = 13$
    $x^2y^3 = 15$

21. An investment account starts with \$1000 and earns 5% annual interest, compounded once per year. An expense account starts at the same time with \$800 and has \$100 added to it each year.

    (a) Use a graph to determine the first year that the expense account is larger than the investment account.

    (b) Use a graph to determine the first year after your answer in part (a) at which the investment account again becomes larger than the expense account.

22. An investment account starts with \$1500 and earns 6% annual interest, compounded once per year. An expense account starts at the same time with \$1000 and has \$250 added to it each year.

    (a) Use a graph to determine the first year that the expense account is larger than the investment account.

    (b) Use a graph to determine the first year after your answer in part (a) at which the investment account again becomes larger than the expense account.

23. The outer boundary of a football stadium is described by the equation

$$75x^2 + 100y^2 = 3$$

where the units are kilometers measured from the center of the stadium. A bird flies over the stadium on the path described by the line

$$10x + 20y = 1.$$

    (a) Find the points at which the bird flies over the boundary of the stadium.

    (b) What is the distance the bird flies over the stadium?

24. The outer boundary of a football stadium is described by the equation

$$20x^2 + 30y^2 = 1$$

where the units are kilometers as measured from the center of the stadium. A bird flies over the stadium on the path described by the line

$$10x + 40y = 1.$$

(a) Find the points at which the bird flies over the boundary of the stadium.

(b) What is the distance the bird flies over the stadium?

PROBLEMS

Some problems require considerably more thought than the exercises.
Unlike exercises, problems often have more than one correct answer.

25. Give an example of an equation with one variable whose solutions are 2, 3, 5, 10, and 12.

26. Show that the equation

$$x^2y^2 + (x-1)^2(y-1)^2(y-2)^2 = 0$$

has exactly three solutions.

27. Find an equation with two variables that has exactly five solutions.

28. Give an example of a system of three equations with three variables that has exactly three solutions.

WORKED-OUT SOLUTIONS *to Odd-numbered Exercises*

Do not read these worked-out solutions before first attempting to do the exercises yourself. Otherwise you may merely mimic the techniques shown here without understanding the ideas.

Best way to learn: Carefully read the section of the textbook, then do all the odd-numbered exercises (even if they have not been assigned) and check your answers here. If you get stuck on an exercise, reread the section of the textbook—then try the exercise again. If you are still stuck, then look at the worked-out solution here.

In Exercises 1–4, use graphs to find estimates for all solutions to the given system of equations.

1.
$$y = x^2 + 1$$
$$y = e^x - 2$$

SOLUTION The graph of $y = x^2 + 1$ is obtained by shifting the familiar graph of $y = x^2$ up 1 unit. The graph of $y = e^x - 2$ is obtained by shifting the familiar graph of $y = e^x$ down 2 units.

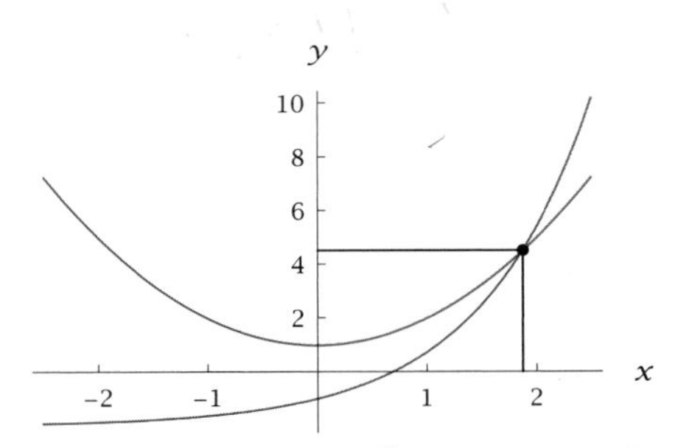

The graph of $y = x^2 + 1$ (blue) and the graph of $y = e^x - 2$ (red).

As can be seen from the figure above, these two graphs intersect in a point whose coordinates are approximately $(1.9, 4.5)$. Thus the solution to this system of equations is $x \approx 1.9, y \approx 4.5$.

3. $$(x-2)^2 + y^2 = 4$$
$$y = \ln x$$

SOLUTION The graph of $(x-2)^2 + y^2 = 4$ is the circle with radius 2 centered at $(2,0)$. The graph of $y = \ln x$ is the familiar graph of the natural logarithm.

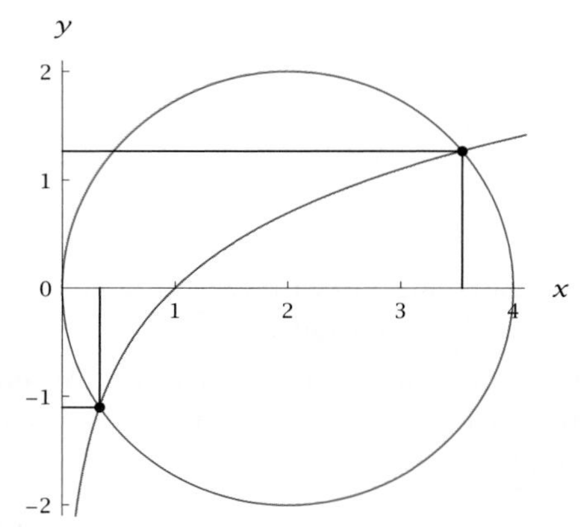

The graph of $(x-2)^2 + y^2 = 4$ (blue) and the graph of $y = \ln x$ (red).

As can be seen from the figure above, these two graphs intersect in two points whose coordinates are approximately $(3.5, 1.3)$ and $(0.3, -1.1)$. Thus the solutions to this system of equations are $x \approx 3.5, y \approx 1.3$ and $x \approx 0.3, y \approx -1.1$.

In Exercises 5–6, find all solutions to the given equation.

5. $x^6 + 8x^4 = -3$

SOLUTION If x is a real number, then x^6 and $8x^4$ are both nonnegative numbers. Thus the left side of the equation above is a nonnegative number, and hence does not equal -3. Thus this equation has no solutions in the real numbers.

In Exercises 7–20, find all solutions to the given system of equations.

7. $$x^2 - 2y^2 = 3$$
$$x + 2y = 1$$

SOLUTION Solve the second equation for x, getting $x = 1 - 2y$. Substitute this value for x into the first equation, getting $(1-2y)^2 - 2y^2 = 3$, which can be rewritten as

$$2y^2 - 4y - 2 = 0.$$

Divide both sides of this equation by 2, and then use the quadratic formula to get

$$y = 1 - \sqrt{2} \quad \text{or} \quad y = 1 + \sqrt{2}.$$

Now substitute these values for y into the equation $x = 1 - 2y$, getting the solutions

$$x = -1 + 2\sqrt{2}, \quad y = 1 - \sqrt{2}$$

and

$$x = -1 - 2\sqrt{2}, \quad y = 1 + \sqrt{2}.$$

9. $$\frac{1}{x} - \frac{1}{y} = 2$$
$$4x + y = 3$$

SOLUTION Solve the second equation for y, getting $y = 3 - 4x$. Substitute this value for y into the first equation, getting

$$\frac{1}{x} - \frac{1}{3-4x} = 2.$$

Multiply both sides of this equation by $x(3-4x)$, getting

$$(3-4x) - x = 2x(3-4x),$$

which can be rewritten as

$$8x^2 - 11x + 3 = 0.$$

Then use the quadratic formula to get

$$x = \frac{3}{8} \quad \text{or} \quad x = 1.$$

Now substitute these values for x into the equation $y = 3 - 4x$, getting the solutions

$$x = \frac{3}{8}, \quad y = \frac{3}{2}$$

and

$$x = 1, \quad y = -1.$$

11. $$x^2 + y^3 = 11$$
$$5x^2 - 2y^3 = 6$$

SOLUTION Note that x^2 is the only use of x in both equations. Thus we solve the first equation for x^2, getting

$$x^2 = 11 - y^3.$$

Substituting this value of x^2 into the second equation gives

$$5(11 - y^3) - 2y^3 = 6,$$

which is equivalent to the equation

$$7y^3 = 49.$$

Thus $y^3 = 7$, which means that $y = 7^{1/3}$. Substituting this value of y into the equation above for x^2 shows that $x^2 = 4$. Thus $x = \pm 2$. Hence the solutions to this system of equations are $x = 2, y = 7^{1/3}$ and $x = -2, y = 7^{1/3}$.

13.
$$2^x + y^2 = 11$$
$$3 \cdot 2^x - 2y^2 = 3$$

SOLUTION Note that 2^x is the only use of x in both equations. Thus we solve the first equation for 2^x, getting

$$2^x = 11 - y^2.$$

Substituting this value of 2^x into the second equation gives

$$3(11 - y^2) - 2y^2 = 3,$$

which is equivalent to the equation

$$5y^2 = 30.$$

Thus $y^2 = 6$, which means that $y = \pm\sqrt{6}$. Substituting these values of y into the equation above for 2^x shows that $2^x = 5$. Thus $x = \log_2 5$. Hence the solutions to this system of equations are $x = \log_2 5, y = \sqrt{6}$ and $x = \log_2 5, y = -\sqrt{6}$.

15.
$$x + \log y = 7$$
$$3x + 2\log y = 19$$

SOLUTION Solve the first equation for x, getting

$$x = 7 - \log y.$$

Substituting this value of x into the second equation gives

$$3(7 - \log y) + 2\log y = 19,$$

which is equivalent to the equation

$$\log y = 2.$$

Thus $y = 100$. Substituting this value of y into the equation above for x shows that $x = 5$. Hence the solution of this system of equations is $x = 5, y = 100$.

17.
$$x^2 + \ln y = 7$$
$$2x^2 - 3\ln y = 4$$

SOLUTION Note that x^2 is the only use of x in both equations. Thus we solve the first equation for x^2, getting

$$x^2 = 7 - \ln y.$$

Substituting this value of x^2 into the second equation gives

$$2(7 - \ln y) - 3\ln y = 4,$$

which is equivalent to the equation

$$5\ln y = 10.$$

Thus $\ln y = 2$, which means that $y = e^2$. Substituting this value of y into the equation above for x^2 shows that $x^2 = 5$. Thus $x = \pm\sqrt{5}$. Hence the solutions to this system of equations are $x = \sqrt{5}, y = e^2$ and $x = -\sqrt{5}, y = e^2$.

19.
$$x^2 + y^3 = 9$$
$$x^2 y^3 = 20$$

SOLUTION Note that x^2 is the only use of x in both equations. Thus we solve the first equation for x^2, getting

$$x^2 = 9 - y^3.$$

Substituting this value of x^2 into the second equation gives

$$(9 - y^3)y^3 = 20,$$

which is equivalent to the equation

$$y^6 - 9y^3 + 20 = 0,$$

which we rewrite as

$$(y^3)^2 - 9y^3 + 20 = 0.$$

Use the quadratic formula to solve for y^3 (or factor), getting $y^3 = 4$ or $y^3 = 5$. Thus $y = 4^{1/3}$ or $y = 5^{1/3}$. Substituting these values for y into the equation above for x^2 shows that $x^2 = 5$ or $x^2 = 4$. Thus $x = \pm\sqrt{5}$ (corresponding to $y = 4^{1/3}$) or $x = \pm 2$ (corresponding to $y = 5^{1/3}$). Hence the solutions to this system of equations are $x = \sqrt{5}, y = 4^{1/3}$ and $x = -\sqrt{5}, y = 4^{1/3}$ and $x = 2, y = 5^{1/3}$ and $x = -2, y = 5^{1/3}$.

21. An investment account starts with \$1000 and earns 5% annual interest, compounded once per year. An expense account starts at the same time with \$800 and has \$100 added to it each year.

(a) Use a graph to determine the first year that the expense account is larger than the investment account.

(b) Use a graph to determine the first year after your answer in part (a) at which the investment account again becomes larger than the expense account.

SOLUTION The amount in the investment account after t years will be $1000 \cdot 1.05^t$. The amount in the expense account after t years will be $800 + 100t$. The graph below plots these expressions for t in the interval $[0, 30]$.

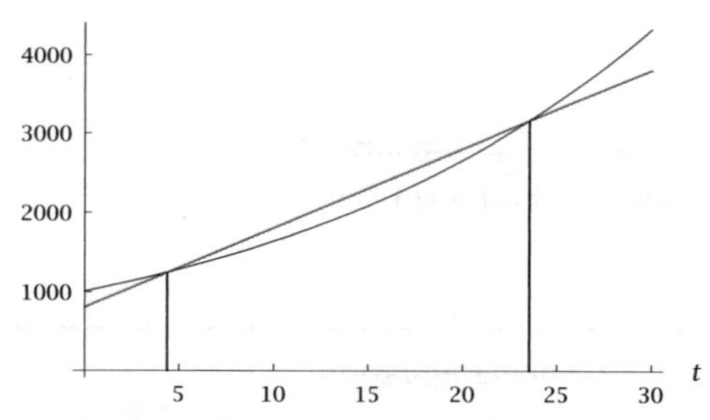

The graphs of $1000 \cdot 1.05^t$ *(blue) and* $800 + 100t$ *(red) on the interval* $[0, 30]$.

(a) The figure above shows that the two graphs first intersect with t between 4 and 5. Because money is added to both accounts only at the end of each year, we round up to the next year. Thus the expense account first becomes larger than the investment account after 5 years.

(b) The figure above shows that the two graphs next intersect with t between 23 and 24. Because money is added to both accounts only at the end of each year, we round up to the next year. Thus the investment account again becomes larger than the expense account after 24 years.

23. The outer boundary of a football stadium is described by the equation

$$75x^2 + 100y^2 = 3$$

where the units are kilometers measured from the center of the stadium. A bird flies over the stadium on the path described by the line

$$10x + 20y = 1.$$

(a) Find the points at which the bird flies over the boundary of the stadium.

(b) What is the distance the bird flies over the stadium?

SOLUTION

(a) The points at which the bird flies over the stadium correspond to the solutions to the system of equations given by the two equations above. To find solutions to this system of equations, solve the second equation for x, getting $x = \frac{1-20y}{10}$. Substitute this expression for x into the first equation, getting

$$\tfrac{3}{4}(1 - 40y + 400y^2) + 100y^2 = 3,$$

which can be rewritten as

$$1600y^2 - 120y - 9 = 0.$$

The quadratic formula now gives the solutions

$$y \approx 0.121353 \quad \text{and} \quad y \approx -0.0463525.$$

Substituting these values for y into the equation $x = \frac{1-20y}{10}$ shows that the bird flies over the boundary of the stadium at approximately the points

$$(-0.1427, 0.1214) \quad \text{and} \quad (0.1927, -0.04635).$$

(b) The distance the bird flies over the stadium is the distance between the two points above. This distance is

$$\sqrt{(-0.1427 - 0.1927)^2 + (0.1214 + 0.04635)^2},$$

which is approximately 0.375 kilometers.

7.2 *Solving Systems of Linear Equations*

LEARNING OBJECTIVES

By the end of this section you should be able to

- work with systems of linear equations;
- use Gaussian elimination to find the solutions to a system of linear equations.

Linear Equations: How Many Solutions?

Examples 4 and 5 in the previous section showed that the system of equations

$$x^2 - y^2 = 1$$
$$2x + y = 4$$

has exactly two solutions.

We will now turn our attention to systems of what are called linear equations. As we will see, a system of linear equations cannot have exactly two solutions, unlike the example in the paragraph above.

We begin extremely gently by looking at a linear equation with one variable.

Linear equation with one variable

A **linear equation with one variable** is an equation of the form

$$ax = b,$$

where a and b are constants.

For example, $3x = 15$ is a linear equation with one variable.

Although we have used x as the variable above, the variable might be called something other than x.

Remarkably, even the simple case of a linear equation with one variable gives us insight into the behavior we will encounter when we examine systems of linear equations with more variables.

EXAMPLE 1

Suppose a and b are constants. How many solutions are there to the linear equation $ax = b$?

SOLUTION A quick but incorrect response to this question would be that we must have $x = \frac{b}{a}$ and thus there is exactly one solution to the equation $ax = b$. However, more care is needed to deal with the case where $a = 0$.

For example, suppose $a = 0$ and $b = 1$. Our equation then becomes $0x = 1$. This equation is satisfied by no value of x. Thus in this case our equation has no solutions. More generally, if $a = 0$ and $b \neq 0$, then our equation has no solutions.

The other case to consider is where $a = 0$ and $b = 0$. In this case our equation becomes $0x = 0$. This equation is satisfied by every value of x. Thus in this case our equation has infinitely many solutions.

Larger systems of linear equations will also have only these three possibilities for the number of solutions.

In summary, the number of solutions depends on a and b:

- If $a \neq 0$, then the equation $ax = b$ has exactly one solution ($x = \frac{b}{a}$).
- If $a = 0$ and $b \neq 0$, then the equation $ax = b$ has no solutions.
- If $a = 0$ and $b = 0$, then the equation $ax = b$ has infinitely many solutions.

Continuing with our gentle approach, we now look at linear equations with two variables.

Linear equation with two variables

A **linear equation with two variables** is an equation of the form

$$ax + by = c,$$

where a, b, and c are constants.

For example, $5x - 3y = 7$ is a linear equation with two variables.

The terminology **linear equation** arose because unless a and b are both 0, the set of points (x, y) satisfying the equation $ax + by = c$ forms a line in the xy-plane.

Although we have used x and y as the variables above, the variables might be called something other than x and y.

The next example illustrates the behavior of a system of two linear equations with two variables.

EXAMPLE 2

Suppose a, b, and c are constants. How many solutions are there to the following system of linear equations?

$$2x + 3y = 6$$

$$ax + by = c$$

SOLUTION The graphical method gives the best insight into this question. The set of points that satisfy the equation $2x + 3y = 6$ forms the blue line shown below in the margin.

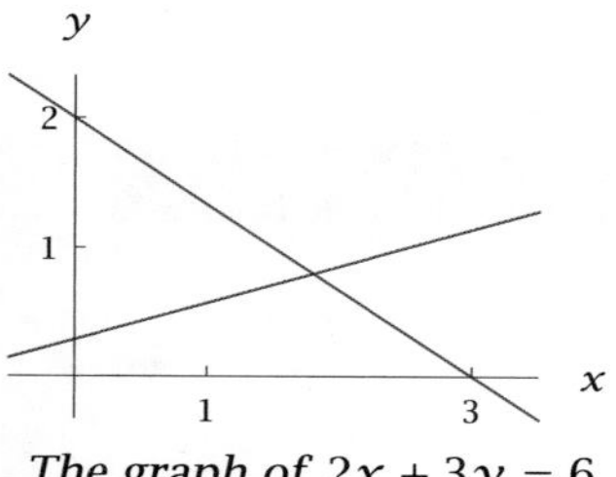

The graph of $2x + 3y = 6$ (blue). Most lines intersect the blue line at exactly one point.

A quick but incorrect response to this question would be that the set of points that satisfy the equation $ax + by = c$ forms a line, that two lines intersect at just one point, and thus there is exactly one solution to our system of equations, as shown in the figure in the margin.

Although the reasoning in the paragraph above is correct most of the time, more care is needed to deal with special cases. One special case occurs, for example, if $a = 0$, $b = 0$, and $c = 1$. In this case, the second equation in our system of equations becomes $0x + 0y = 1$. This equation is satisfied by no values of x and y. Thus no numbers x and y can satisfy both equations in our system of equations, and hence in this case our system of equations has no solutions.

Another special case occurs, for example, if $a = 4$, $b = 6$, and $c = 7$. In this case the set of points satisfying the second equation (which is $4x + 6y = 7$) is a line parallel to the line corresponding to the first equation. Because these lines are parallel, they do not intersect, as shown in the figure in the margin. Thus in this case, the system of equations has no solutions.

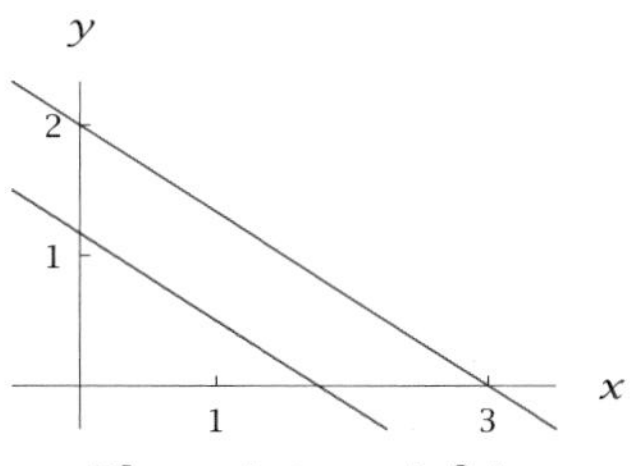

The points satisfying $2x + 3y = 6$ (blue) and the points satisfying $4x + 6y = 7$ (red). These parallel lines do not intersect.

Yet another special case occurs, for example, if $a = 6$, $b = 9$, and $c = 18$. In this case, our system of equations is

$$2x + 3y = 6$$
$$6x + 9y = 18.$$

Dividing both sides of the second equation by 3 produces the equation $2x + 3y = 6$, which is the same as the first equation. Thus the line determined by the second equation is the same as the line determined by the first equation. Hence in this case our system of equations has infinitely many solutions.

Finally, we have one more special case to consider. If $a = 0$, $b = 0$, and $c = 0$, then the second equation in our system of equations is

$$0x + 0y = 0.$$

This equation is satisfied for all values of x and y and thus in this case the second equation places no restrictions on x and y. The set of solutions to our system of equations in this case corresponds to the line determined by the first equation. Thus in this case our system of equations has infinitely many solutions.

In summary, the number of solutions depends upon a, b, and c:

Thus for this system of equations, there are either no solutions, one solution, or infinitely many solutions.

- If the graph of $ax + by = c$ is a line not parallel to or identical to the graph of $2x + 3y = 6$, then the system of equations has exactly one solution (see Problems 17 and 18 for that solution).
- If $a = b = 0$ and $c \neq 0$, then the system of equations has no solutions.
- If the graph of $ax + by = c$ is a line parallel to (but not identical to) the graph of $2x + 3y = 6$, then the system of equations has no solutions.
- If the graph of $ax + by = c$ is the same as the graph of $2x + 3y = 6$, then the system of equations has infinitely many solutions.
- If $a = b = c = 0$, then the system of equations has infinitely many solutions.

There is nothing special about the numbers 2, 3, and 6 that appear in the first equation of the example above. The same reasoning as used in the example above shows that for any system of two linear equations with two variables, there is either exactly one solution, no solutions, or infinitely many solutions.

In fact, as we will see in Section 7.3, the same conclusion (exactly one solution, no solutions, or infinitely many solutions) holds for every system of linear equations with any number of variables. Before we can understand why this result holds, we need to see how to solve a system of linear equations.

Systems of Linear Equations

Having defined linear equations with one variable and linear equations with two variables, our next step is to define linear equations with three variables.

For example, $4x - 5y + 3z = 7$ is a linear equation with three variables.

Linear equation with three variables

A **linear equation with three variables** is an equation of the form

$$ax + by + cz = d,$$

where a, b, c, and d are constants.

Although we have used x, y, and z as the variables above, the variables might be called something else.

Even though the set of points (x, y, z) satisfying an equation such as $4x - 5y + 3z = 7$ forms a plane rather than a line in three-dimensional space, this equation is still called a linear equation because its form is similar to the form of a linear equation with two variables.

The notion of a **linear equation** in any number of variables generalizes the ideas we have been discussing.

Linear equation

A **linear equation** has one side consisting of a sum of terms, each of which is a constant times a variable, and the other side is a constant.

EXAMPLE 3

A large fruit basket contains apples, bananas, oranges, plums, and strawberries. Each apple weighs 180 grams, each banana weighs 120 grams, each orange weighs 140 grams, each plum weighs 65 grams, and each strawberry weighs 25 grams. The total weight of the fruit basket is 2465 grams, including the basket itself, which weighs 500 grams. Write an equation connecting the number of apples, bananas, oranges, plums, and strawberries in the fruit basket to the total weight of the fruit in the fruit basket.

SOLUTION Most real-world problems do not come with variables already selected. Thus our first step is to assign variable names, as follows:

a = number of apples in the fruit basket;
b = number of bananas in the fruit basket;
r = number of oranges in the fruit basket;
p = number of plums in the fruit basket;
s = number of strawberries in the fruit basket.

Try to choose variable names that remind you of the meaning of the variables. Here, for example, we have chosen the first letter of each fruit name, except that the first letter of *oranges* looks too much like a zero and thus the second letter (and dominant sound) was used instead.

Each apple weighs 180 grams. Thus the a apples in the fruit basket weigh a total of $180a$ grams.

Each banana weighs 120 grams. Thus the b bananas in the fruit basket weigh a total of $120b$ grams.

Each orange weighs 140 grams. Thus the r oranges in the fruit basket weigh a total of $140r$ grams.

Each plum weighs 65 grams. Thus the p plums in the fruit basket weigh a total of $65p$ grams.

Each strawberry weighs 25 grams. Thus the s strawberries in the fruit basket weigh a total of $25s$ grams.

The total weight of the fruit basket is 2465 grams, including the basket itself, which weighs 500 grams. Thus the fruit in the fruit basket weights 1965 grams. Hence we have the following equation:

$$180a + 120b + 140r + 65p + 25s = 1965.$$

The equation above is a linear equation with five variables (a, b, r, p, and s).

You might want to avoid choosing e and i as variable names because these symbols have specific meanings in mathematics.

In the rest of this section and throughout the next section we will be dealing with systems of linear equations.

System of linear equations

A **system of linear equations** is a system of equations, each of which is a linear equation.

How can we find the solutions to a system of linear equations such as the one above? One method that works for any system of linear equations is substitution, which was discussed in Section 7.1.

EXAMPLE 4

Explain how substitution could be used to solve the following system of linear equations:

$$x + 2y + 3z = 4$$
$$3x - 3y + 4z = 1$$
$$2x + y - z = 7.$$

Here we have a system of three linear equations with three variables (x, y, and z).

SOLUTION To solve this system of equations, we could solve the first equation for x in terms of y and z (getting $x = 4 - 2y - 3z$) and then substitute this value for x into the last two equations. The last two equations would then become a system of two linear equations with two variables (y and z). We could solve one of these equations for y in terms of z and substitute that value for y into the other equation. This procedure would give us a single linear equation with one variable z. That equation could easily be solved for z. The value of z would then give us the value of y, and the values of y and z would then give us the value of x (from $x = 4 - 2y - 3z$).

Systems of linear equations arise in numerous contexts, usually with many more variables than the two or three variables used in examples in textbooks. Mathematical models of a sector of the economy or of an industrial production process typically involve dozens, hundreds, or thousands of variables.

Google computers solve a very large system of linear equations to rank the results of web searches.

The substitution technique discussed in the example above could be used to solve any system of linear equations. However, substitution is not as efficient for solving typical large systems of linear equations as another technique called Gaussian elimination that we will soon learn. Thus substitution is not used much in practice.

Another method for solving systems of linear equations is called Cramer's rule. You may have learned Cramer's rule in high school. However, Cramer's rule is used only in textbooks, not in real-world applications. Thus Cramer's rule will not be discussed here, except to note the following points of comparison with Gaussian elimination:

As we will see, you can understand why Gaussian elimination works. In contrast, Cramer's rule is usually presented without motivation or explanation of why it works.

- Gaussian elimination finds the solutions for all systems of linear equations. In contrast, Cramer's rule can be used only on systems of linear equations for which the number of equations equals the number of variables. Even for such systems of equations, Cramer's Rule works only when there is exactly one solution.
- Gaussian elimination is much faster and much more efficient than Cramer's rule for large systems of linear equations. No one would use Cramer's rule even on a relatively small system of ten linear equations with ten variables.
- Gaussian elimination is the standard tool used by computers to solve systems of linear equations.

Systems of linear equations are usually solved by computers (using Gaussian elimination) rather than by hand calculation. For example, to solve the system of linear equations in Example 4 you could type

```
solve x + 2y + 3z = 4, 3x - 3y + 4z = 1, 2x + y - z = 7
```

into a Wolfram|Alpha entry box (or use other software).

Nevertheless, as an educated citizen you should have some understanding of the process used by computers to solve systems of linear equations. Furthermore, you should be aware that you can use technology to solve even very large systems of linear equations quickly. Thus we now turn our attention to Gaussian elimination.

Gaussian Elimination

A procedure called **Gaussian elimination** provides a very fast method for finding solutions to systems of linear equations. The idea of Gaussian elimination is that multiples of the first equation are added to the other equations to eliminate the first variable from those equations. Then multiples of the new second equation are added to the equations below it to eliminate the second variable from those equations. And so on.

We will start learning Gaussian elimination gently by initially dealing only with systems of linear equations that have exactly one solution. Go through the next example and several exercises to make sure that you understand how to use Gaussian elimination in that context. Then in the next section we will extend Gaussian elimination to work with all systems of linear equations.

Gaussian elimination is particularly well suited for implementation on a computer.

The basic method shown next breaks down in certain special cases that we will discuss in the next section, but it works perfectly with almost all systems of linear equations in which the number of equations equals the number of variables.

Gaussian elimination for a system of linear equations, basic method

(a) Add multiples of the first equation to the other equations to eliminate the first variable from the other equations.

(b) Add multiples of the new second equation to the equations below the second equation to eliminate the second variable from the equations below the second equation.

(c) Continue this process, at each stage starting one equation lower and eliminating the next variable from the equations below.

(d) When the process above can no longer continue, solve for the last variable. Then solve for the second-to-last variable, then the third-to-last variable, and so on.

EXAMPLE 5

Find all solutions to the following system of linear equations:

$$\begin{aligned} x + 2y + 3z &= 4 \\ 3x - 3y + 4z &= 1 \\ 2x + y - z &= 7. \end{aligned}$$

SOLUTION The first step in Gaussian elimination is to use the first equation to eliminate the first variable (which here is x) from the other equations. To carry out this procedure, note that the coefficient of x in the second equation is 3. Thus we add -3 times the first equation to the second equation, getting a new second equation: $-9y - 5z = -11$. Similarly, adding -2 times the first equation to the third equation gives a new second equation: $-3y - 7z = -1$. At this stage, our system of linear equations has been changed to

$$\begin{aligned} x + 2y + 3z &= 4 \\ -9y - 5z &= -11 \\ -3y - 7z &= -1. \end{aligned}$$

If the coefficient of a variable in an equation is b and the coefficient of the same variable in another equation is c, then add $-\frac{c}{b}$ times the first equation to the second equation to eliminate the variable.

The second step in Gaussian elimination is to use the second equation above to eliminate the second variable (which here is y) from the equations below the second equation. To carry out this procedure, note that the coefficient of y in the second equation above is -9 and the coefficient of y in the third equation is -3. Thus we add $-\frac{1}{3}$ times the second equation to the third equation, getting a new third equation: $-\frac{16}{3}z = \frac{8}{3}$. At this stage, our system of linear equations has been changed to

$$\begin{aligned} x + 2y + 3z &= 4 \\ -9y - 5z &= -11 \\ -\tfrac{16}{3}z &= \tfrac{8}{3}. \end{aligned}$$

This process of eliminating variables below an equation cannot continue further. Thus we now solve the last equation in the new system for z and then work our way back up the most recent system of equations (this is called **back substitution**). Specifically, solving the last equation for z, we have $z = -\frac{1}{2}$. Substituting $z = -\frac{1}{2}$ into the second equation in the new system above gives the new equation $-9y + \frac{5}{2} = -11$, which we solve for y, getting $y = \frac{3}{2}$. Finally, substituting $y = \frac{3}{2}$ and $z = -\frac{1}{2}$ into the first equation in the new system above gives the new equation $x + 3 - \frac{3}{2} = 4$, which we solve for x, getting $x = \frac{5}{2}$.

Thus the only solution to our original system of equations is $x = \frac{5}{2}$, $y = \frac{3}{2}$, $z = -\frac{1}{2}$.

CHECK To check whether $x = \frac{5}{2}$, $y = \frac{3}{2}$, $z = -\frac{1}{2}$ is an actual solution, substitute these values into the original equations to see whether

$$\begin{aligned} \tfrac{5}{2} + 2 \cdot \tfrac{3}{2} - 3 \cdot \tfrac{1}{2} &\stackrel{?}{=} 4 \\ 3 \cdot \tfrac{5}{2} - 3 \cdot \tfrac{3}{2} - 4 \cdot \tfrac{1}{2} &\stackrel{?}{=} 1 \\ 2 \cdot \tfrac{5}{2} + \tfrac{3}{2} + \tfrac{1}{2} &\stackrel{?}{=} 7. \end{aligned}$$

Simple arithmetic shows that all three equations above hold. Thus $x = \frac{5}{2}$, $y = \frac{3}{2}$, $z = -\frac{1}{2}$ is indeed a solution of the system of equations.

EXERCISES

In Exercises 1–12, use Gaussian elimination to find all solutions to the given system of equations.

1. $$\begin{aligned} x - 4y &= 3 \\ 3x + 2y &= 7 \end{aligned}$$

2. $$\begin{aligned} -x + 2y &= 4 \\ 2x - 7y &= -3 \end{aligned}$$

3. $$\begin{aligned} 2x + 3y &= 4 \\ 5x - 6y &= 1 \end{aligned}$$

4. $$\begin{aligned} -4x + 5y &= 7 \\ 7x - 8y &= 9 \end{aligned}$$

5. $$\begin{aligned} 3r + 5t &= 1 \\ r - 4t &= 6 \end{aligned}$$

6. $$\begin{aligned} 2r + 5w &= 1 \\ r - 4w &= 7 \end{aligned}$$

7.
$$x + 3y - 2z = 1$$
$$2x - 4y + 3z = -5$$
$$-3x + 5y - 4z = 0$$

8.
$$x - 2y - 3z = 4$$
$$-3x + 2y + 3z = -1$$
$$2x + 2y - 3z = -2$$

9.
$$2x - 3y - 5z = -4$$
$$5x + y + 3z = 0$$
$$3x - 2y + 4z = 8$$

10.
$$3x - 3y - z = 5$$
$$2x + y + 3z = -6$$
$$3x - 2y + 4z = 7$$

11.
$$2a + 3b - 4c = 5$$
$$a - 2b + 4c = 6$$
$$3a + 3b - 2c = 7$$

12.
$$3a + 3b - 4d = 5$$
$$a - 2b + 4d = 6$$
$$4a + 3b - 2d = 7$$

13. An ad for a snack consisting of peanuts and raisins states that one serving of the regular snack contains 15 peanuts and 15 raisins and has 84 calories. The lite version of the snack consists of 10 peanuts and 20 raisins per serving and, according to the ad, has 72 calories.

(a) How many calories are in each raisin?

(b) How many calories are in each peanut?

14. An ad for a snack consisting of chocolate-covered peanuts and golden raisins states that one serving of the regular snack contains 15 chocolate-covered peanuts and 20 golden raisins and has 151 calories. The lite version of the snack consists of 10 chocolate-covered peanuts and 25 golden raisins per serving and, according to the ad, has 124 calories.

(a) How many calories are in each golden raisin?

(b) How many calories are in each chocolate-covered peanut?

15. At an educational district's office, three types of employee wages are incorporated into the budget: specialists, managers, and directors. Employees of each type have the same salary across districts. One district employs five specialists, two managers, and a director with total annual budgeted salaries of \$550,000. Another district employs six specialists, three managers, and two directors with an annual salary budget of \$777,000. A third district has eight specialists, no managers, and two directors with an annual salary budget of \$676,000.

(a) What is the annual salary for each specialist?

(b) What is the annual salary for each manager?

(c) What is the annual salary for each director?

16. At a university, three types of employee wages are incorporated into each department's budget: technical staff, clerical staff, and budget officers. Employees of each type have the same salary across departments. The Physics Department employs three technical staff, two clerical staff, and a budget officer with total annual budgeted salaries of \$325,000. The Biology Department employs four technical staff, three clerical staff, and two budget officers with an annual salary budget of \$485,000. The Mathematics Department has no technical staff, four clerical staff, and two budget officers with an annual salary budget of \$260,000.

(a) What is the annual salary for each technical staff member?

(b) What is the annual salary for each clerical staff member?

(c) What is the annual salary for each budget officer?

PROBLEMS

The next two problems give an explicit formula for the solution to the system of equations in Example 2 when there is exactly one solution.

17. Show that the graph of $2x + 3y = 6$ and the graph of $ax + by = c$ intersect in exactly one point if and only if $3a \neq 2b$.

18. Show that if $3a \neq 2b$, then the solution to the system of equations

$$2x + 3y = 6$$

$$ax + by = c$$

is $x = \frac{6b-3c}{2b-3a}$, $y = \frac{2c-6a}{2b-3a}$.

WORKED-OUT SOLUTIONS *to Odd-numbered Exercises*

In Exercises 1–12, use Gaussian elimination to find all solutions to the given system of equations.

1.
$$x - 4y = 3$$
$$3x + 2y = 7$$

SOLUTION Add -3 times the first equation to the second equation, getting the system of linear equations

$$x - 4y = 3$$

$$14y = -2.$$

Solve the second equation for y, getting $y = -\frac{1}{7}$. Substitute this value for y in the first equation, getting the equation $x + \frac{4}{7} = 3$. Thus $x = \frac{17}{7}$. Hence the only solution to this system of linear equations is $x = \frac{17}{7}$, $y = -\frac{1}{7}$.

3.
$$2x + 3y = 4$$
$$5x - 6y = 1$$

SOLUTION Add $-\frac{5}{2}$ times the first equation to the second equation, getting the system of linear equations

$$2x + 3y = 4$$

$$-\tfrac{27}{2}y = -9.$$

Solve the second equation for y, getting $y = \frac{2}{3}$. Substitute this value for y in the first equation, getting the equation $2x + 2 = 4$. Thus $x = 1$. Hence the only solution to this system of linear equations is $x = 1$, $y = \frac{2}{3}$.

5.
$$3r + 5t = 1$$
$$r - 4t = 6$$

SOLUTION To simplify the arithmetic, we switch the order of the equations, getting

$$r - 4t = 6$$

$$3r + 5t = 1.$$

Add -3 times the first equation to the second equation, getting

$$r - 4t = 6$$

$$17t = -17.$$

The second equation shows that $t = -1$. Substitute this value for t into the first equation and solve for r, getting $r = 2$. Hence the only solution to this system of equations is $r = 2, t = -1$.

7.
$$x + 3y - 2z = 1$$
$$2x - 4y + 3z = -5$$
$$-3x + 5y - 4z = 0$$

SOLUTION Add -2 times the first equation to the second equation and add 3 times the first equation to the third equation, getting the system of linear equations

$$x + 3y - 2z = 1$$

$$-10y + 7z = -7$$

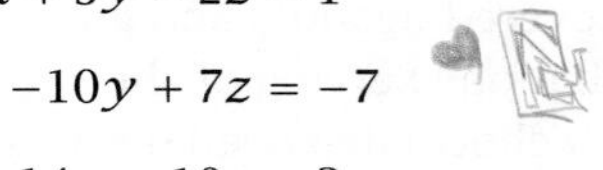

$$14y - 10z = 3.$$

Now add $\frac{14}{10}$ (which equals $\frac{7}{5}$) times the second equation to the third equation, getting the system of linear equations

$$x + 3y - 2z = 1$$
$$-10y + 7z = -7$$
$$-\frac{1}{5}z = -\frac{34}{5}.$$

Solve the third equation for z, getting $z = 34$. Substitute this value for z in the second equation, getting the equation $-10y + 7 \cdot 34 = -7$. Thus $y = \frac{49}{2}$. Substitute these values for x and y into the first equation, getting the equation $x + 3 \cdot \frac{49}{2} - 2 \cdot 34 = 1$. Thus $x = -\frac{9}{2}$. Hence the only solution to this system of linear equations is $x = -\frac{9}{2}$, $y = \frac{49}{2}$, $z = 34$.

9.
$$2x - 3y - 5z = -4$$
$$5x + y + 3z = 0$$
$$3x - 2y + 4z = 8$$

SOLUTION Add $-\frac{5}{2}$ times the first equation to the second equation and add $-\frac{3}{2}$ times the first equation to the third equation, getting the system of linear equations

$$2x - 3y - 5z = -4$$
$$\tfrac{17}{2}y + \tfrac{31}{2}z = 10$$
$$\tfrac{5}{2}y + \tfrac{23}{2}z = 14.$$

Now add $-\frac{2}{17} \cdot \frac{5}{2}$ (which equals $-\frac{5}{17}$) times the second equation to the third equation, getting the system of linear equations

$$2x - 3y - 5z = -4$$
$$\tfrac{17}{2}y + \tfrac{31}{2}z = 10$$
$$\tfrac{118}{17}z = \tfrac{188}{17}.$$

Solve the third equation for z, getting $z = \frac{188}{118} = \frac{94}{59}$. Substitute this value for z in the second equation and then solve for y, getting $y = -\frac{102}{59}$. Substitute these values for x and y into the first equation and then solve for x, getting $x = -\frac{36}{59}$. Hence the only solution to this system of linear equations is $x = -\frac{36}{59}$, $y = -\frac{102}{59}$, $z = \frac{94}{59}$.

11.
$$2a + 3b - 4c = 5$$
$$a - 2b + 4c = 6$$
$$3a + 3b - 2c = 7$$

SOLUTION To simplify the arithmetic, we switch the order of the first and second equations, getting

$$a - 2b + 4c = 6$$
$$2a + 3b - 4c = 5$$
$$3a + 3b - 2c = 7.$$

Add -2 times the first equation to the second equation, and add -3 times the first equation to the third equation, getting

$$a - 2b + 4c = 6$$
$$7b - 12c = -7$$
$$9b - 14c = -11.$$

Now add $-\frac{9}{7}$ times the second equation to the third equation, getting

$$a - 2b + 4c = 6$$
$$7b - 12c = -7$$
$$\tfrac{10}{7}c = -2.$$

Solving the last equation for c gives $c = -\frac{7}{5}$. Plugging this value for c into the second equation and then solving for b gives $b = -\frac{17}{5}$. Plugging these values for b and c into the first equation and then solving for a gives $a = \frac{24}{5}$. Hence the only solution to this equation is $a = \frac{24}{5}, b = -\frac{17}{5}, c = -\frac{7}{5}$.

13. An ad for a snack consisting of peanuts and raisins states that one serving of the regular snack contains 15 peanuts and 15 raisins and has 84 calories. The lite version of the snack consists of 10 peanuts and 20 raisins per serving and, according to the ad, has 72 calories.

(a) How many calories are in each raisin?

(b) How many calories are in each peanut?

SOLUTION Let p denote the number of calories in each peanut and let r denote the number of calories in each raisin. Then

$$15p + 15r = 84$$
$$10p + 20r = 72.$$

To solve this system of equations, add $-\frac{2}{3}$ times the first equation to the second equation, getting the new system of equations

$$15p + 15r = 84$$
$$10r = 16.$$

The last equation implies that $r = 1.6$. Plugging this value for r into the first equation and then solving for p shows that $p = 4$. Thus we have:

(a) Each raisin has 1.6 calories.

(b) Each peanut has 4 calories.

15. At an educational district's office, three types of employee wages are incorporated into the budget: specialists, managers, and directors. Employees of each type have the same salary across districts. One district employs five specialists, two managers, and a director with total annual budgeted salaries of \$550,000. Another district employs six specialists, three managers, and two directors with an annual salary budget of \$777,000. A third district has eight specialists, no managers, and two directors with an annual salary budget of \$676,000.

    (a) What is the annual salary for each specialist?

    (b) What is the annual salary for each manager?

    (c) What is the annual salary for each director?

    SOLUTION Use thousands of dollars as the units. Let s denote the annual salary of a specialist, let m denote the annual salary of a manager, and let d denote the annual salary of a director. Thus we have the following:

$$5s + 2m + d = 550$$
$$6s + 3m + 2d = 777$$
$$8s + 0m + 2d = 676.$$

Because the coefficient of d in the first equation is 1, we will reverse the order of the variables to make the arithmetic easier when doing Gaussian elimination. Thus we rewrite the system of equations above as

$$d + 2m + 5s = 550$$
$$2d + 3m + 6s = 777$$
$$2d + 0m + 8s = 676.$$

Add -2 times the first equation to the second equation, and add -2 times the first equation to the third equation, getting

$$d + 2m + 5s = 550$$
$$-m - 4s = -323$$
$$-4m - 2s = -424.$$

Now add -4 times the second equation to third equation, getting

$$d + 2m + 5s = 550$$
$$-m - 4s = -323$$
$$14s = 868.$$

Solving the last equation for s shows that $s = 62$. Plugging this value for s into the second equation above and then solving for m gives $m = 75$. Plugging these values for s and m into the first equation and then solving for d gives $d = 90$. Thus we have:

(a) Each specialist has \$62,000 annual salary.

(b) Each manager has \$75,000 annual salary.

(c) Each director has \$90,000 annual salary.

7.3 *Solving Systems of Linear Equations Using Matrices*

LEARNING OBJECTIVES

By the end of this section you should be able to

- represent a system of linear equations by a matrix;
- interpret a matrix as a system of linear equations;
- solve a system of linear equations using elementary row operations on a matrix.

Representing Systems of Linear Equations by Matrices

In the last example in the previous section, we used Gaussian elimination to solve a system of three linear equations with three variables. Real-world applications often involve many more linear equations and many more variables. Such systems of linear equations are solved by computers using Gaussian elimination.

Although computers can calculate quickly, efficiency becomes important when dealing with large systems of equations. Storing an equation such as

$$3x + 8y + 2z = 1$$

in a computer as "$3x + 8y + 2z = 1$" is highly inefficient, especially if we have dozens or hundreds or thousands of equations in that form. For example, if we know that we are dealing with linear equations with three variables x, y, and z, all that we really need to construct the equation above are the numbers 3, 8, 2 and 1, in that order. There is no need to list the variables repeatedly, nor do we need the plus signs connecting the terms above, nor do we need the equals sign.

Thus computers store systems of linear equations as matrices.

Matrices

- A **matrix** is a rectangular array of numbers. Usually a matrix is enclosed in straight brackets.
- A horizontal line of numbers within a matrix is called a **row**; a vertical line of numbers within a matrix is called a **column**.

Thus

$$\begin{bmatrix} 2 & 3 & 8 \\ -1 & 7 & 4 \end{bmatrix}$$

is a matrix with two rows and three columns. The examples in the margin of this page should help you learn to identify various parts of a matrix.

$$\begin{bmatrix} 2 & 3 & 8 \\ -1 & 7 & 4 \end{bmatrix}$$

Row 1 is red.

$$\begin{bmatrix} 2 & 3 & 8 \\ -1 & 7 & 4 \end{bmatrix}$$

Column 3 is red.

$$\begin{bmatrix} 2 & 3 & 8 \\ -1 & 7 & 4 \end{bmatrix}$$

Entry in row 1, column 3 is red.

Matrices have many uses within mathematics and in other fields. In this introductory look at matrices, we focus on their use in solving systems of linear equations.

The main idea in this subject is to represent a system of linear equations as a matrix and then manipulate the matrix. To represent a system of linear equations as a matrix, make the coefficients and the constant term from each equation into one row of the matrix, as shown in the following example.

EXAMPLE 1

Represent the system of linear equations

$$2x + 3y = 7$$

$$5x - 6y = 4$$

as a matrix.

The red bars are not part of the matrix. The red bars are simply a visual reminder about the placement of the equals sign when interpreting a row of the matrix as an equation.

SOLUTION The equation $2x + 3y = 7$ is represented as the row $\begin{bmatrix} 2 & 3 & | & 7 \end{bmatrix}$ and the equation $5x - 6y = 4$ is represented as the row $\begin{bmatrix} 5 & -6 & | & 4 \end{bmatrix}$. Thus the system of two linear equations above is represented by the matrix

$$\begin{bmatrix} 2 & 3 & | & 7 \\ 5 & -6 & | & 4 \end{bmatrix}.$$

Note that in the example above, the coefficients of x in the system of linear equations form column 1

$$\begin{bmatrix} 2 \\ 5 \end{bmatrix}$$

of the matrix. The coefficients of y form column 2

$$\begin{bmatrix} 3 \\ -6 \end{bmatrix}$$

of the matrix. The constant terms form the last column

$$\begin{bmatrix} 7 \\ 4 \end{bmatrix}$$

of the matrix.

The word "matrix" was first used to mean a rectangular array of numbers by the British mathematician James Sylvester, shown above, around 1850.

When representing a system of linear equations as a matrix, it is important to decide which symbol represents the first variable, which symbol represents the second variable, and so on. Furthermore, once that decision has been made, it is important to maintain consistency in the order of variables. For example, once we decide that x will denote the first variable and y will denote the second variable, then an equation such as $-6y + 5x = 4$ should be rewritten as $5x - 6y = 4$ so that it can be represented as the row $\begin{bmatrix} 5 & -6 & | & 4 \end{bmatrix}$.

In the next example we go in the other direction, interpreting a matrix as a system of linear equations.

EXAMPLE 2

Interpret the matrix

$$\left[\begin{array}{cc|c} -8 & 1 & -3 \\ 0 & 2 & 9 \end{array}\right]$$

as a system of linear equations.

SOLUTION To interpret a matrix as a system of linear equations, we need to have a symbol for the first variable, a symbol for the second variable, and so on. Sometimes the choice of symbols is dictated by the context. When the context does not suggest a choice of symbols, we are free to choose whatever symbols we want. In this case, we will choose x to denote the first variable and y to denote the second variable.

Thus the first row $\left[\begin{array}{cc|c} -8 & 1 & -3 \end{array}\right]$ is interpreted as the equation $-8x + y = -3$ and the second row $\left[\begin{array}{cc|c} 0 & 2 & 9 \end{array}\right]$ is interpreted as the equation $0x + 2y = 9$, which we rewrite as $2y = 9$. Thus the matrix above is interpreted as the following system of linear equations:

$$-8x + y = -3$$

$$2y = 9.$$

Gaussian Elimination with Matrices

The basic idea of using matrices to solve systems of linear equations is to represent the system of linear equations as a matrix and then perform the operations of Gaussian elimination on the matrix, at least until the stage of back substitution is reached.

Using matrices to solve systems of linear equations saves considerable time (for humans and computers) by not carrying along the names of the variables in each step.

The next example illustrates this idea. Furthermore, the next example shows how to deal with one of the special cases where the basic method of Gaussian elimination needs modification to work.

EXAMPLE 3

Use matrix operations to find all solutions to this system of linear equations:

$$2y + 3z = 5$$

$$x + y + z = 2$$

$$2x - y - 2z = -2.$$

SOLUTION First we represent this system of linear equations as the matrix

$$\left[\begin{array}{ccc|c} 0 & 2 & 3 & 5 \\ 1 & 1 & 1 & 2 \\ 2 & -1 & -2 & -2 \end{array}\right],$$

Some books use the term ***augmented matrix*** *to refer to a matrix whose last column consists of the constant terms in a system of equations.*

where we have made the natural choice of letting the first variable be x, the second variable be y, and the third variable be z.

Now we are ready to perform Gaussian elimination on the matrix. Normally the first step in Gaussian elimination is to add multiples of the first equation to the other equations to eliminate the first variable from the other equations. In terms of matrices, this translates to adding multiples of row 1 to the other rows to make the entries in column 1 equal to 0, except for the entry in row 1, column 1. However, the entry in the row 1, column 1 of the matrix above equals 0. Thus adding multiples of row 1 to the other rows cannot produce additional entries of 0 in column 1.

Interchanging two rows is one of what are called the elementary row operations.

The solution to this problem is easy. We will simply interchange the first two rows of the matrix above. This operation corresponds to rewriting the order of the equations in our system of linear equations so that the equation $x + y + z = 2$ comes first, with the equation $2y + 3z = 5$ second. The operation of changing the order of the equations (or equivalently interchanging two rows of the matrix) does not change the solutions to the system of linear equations. After interchanging the first two rows, we now have the matrix

$$\left[\begin{array}{ccc|c} 1 & 1 & 1 & 2 \\ 0 & 2 & 3 & 5 \\ 2 & -1 & -2 & -2 \end{array}\right].$$

Adding a multiple of one row to another row is another of the elementary row operations.

Now we can proceed with Gaussian elimination. The entry in row 2, column 1 is already 0, so we do not need to do anything to row 2. To make the entry in row 3, column 1 equal to 0 (equivalent to eliminating the first variable x), we add -2 times row 1 to row 3 (equivalent to adding -2 times one equation to another equation), getting the matrix

$$\left[\begin{array}{ccc|c} 1 & 1 & 1 & 2 \\ 0 & 2 & 3 & 5 \\ 0 & -3 & -4 & -6 \end{array}\right].$$

Now we need to make the entry in row 3, column 2 equal to 0 (equivalent to eliminating y from the third equation). We do this by adding $\frac{3}{2}$ times row 2 to row 3, getting the matrix

$$\left[\begin{array}{ccc|c} 1 & 1 & 1 & 2 \\ 0 & 2 & 3 & 5 \\ 0 & 0 & \frac{1}{2} & \frac{3}{2} \end{array}\right].$$

Multiplying a row by a nonzero constant is another of the elementary row operations.

The last row of the matrix above corresponds to the equation $\frac{1}{2}z = \frac{3}{2}$. Multiplying both sides of this equation by 2 corresponds to multiplying the last row of the matrix above by 2, giving the matrix

$$\left[\begin{array}{ccc|c} 1 & 1 & 1 & 2 \\ 0 & 2 & 3 & 5 \\ 0 & 0 & 1 & 3 \end{array}\right].$$

The last row of the matrix above corresponds to the equation $z = 3$. Having solved for the last variable z, we are ready to enter the back-substitution phase. Row 2 of the matrix above corresponds to the equation $2y + 3z = 5$. Substituting $z = 3$ into this equation gives $2y + 9 = 5$, which we easily solve for y, getting $y = -2$.

Finally, row 1 of the matrix above corresponds to the equation $x + y + z = 2$. Substituting $y = -2$ and $z = 3$ into this equation gives $x + 1 = 2$, which implies that $x = 1$.

In conclusion, we have shown that the only solution to our original system of equations is $x = 1$, $y = -2$, $z = 3$.

Careful study of the example above will lead to a good understanding of how matrix operations are used to solve systems of linear equations. Only three matrix operations are needed, as in the example above. These three operations are called **elementary row operations**:

Elementary row operations

Each of the following operations on a matrix is called an **elementary row operation**:

- adding a multiple of one row to another row;
- multiplying a row by a nonzero constant;
- interchanging two rows.

The constant 0 is excluded in the second elementary row operation because multiplying both sides of an equation by 0 results in a loss of information. For example, multiplying both sides of the equation $2x = 6$ by 0 produces the useless equation $0x = 0$.

These elementary row operations become easy to understand if you keep in mind that each of them corresponds to an operation on a system of linear equations. Thus the first elementary row operation corresponds to adding a multiple of one equation to another equation.

The second elementary row operation corresponds to multiplying an equation by a nonzero constant. The third elementary row operation corresponds to interchanging the order of two equations in a system of linear equations.

Each elementary row operation does not change the set of solutions to the corresponding system of linear equations. Thus performing a series of elementary row operations, as in the example above, does not change the set of solutions.

Although a full course (called linear algebra) is needed to deal carefully with these ideas, here is the main idea of using matrices to solve systems of linear equations:

Solving a system of linear equations with elementary row operations

(a) Represent the system of linear equations as a matrix.

(b) Perform elementary row operations on the matrix corresponding to the steps of Gaussian elimination until back substitution is easy or until the set of solutions becomes clear.

Systems of Linear Equations with No Solutions

So far we have used Gaussian elimination (or its equivalent formulation using elementary row operations) only on systems of linear equations that turned out to have exactly one solution. The next example shows how to deal with one of the special cases that we have not yet encountered.

EXAMPLE 4 Use matrix operations to find all solutions to this system of linear equations:

$$x + y + z = 3$$

$$x + 2y + 3z = 8$$

$$x + 4y + 7z = 10.$$

SOLUTION First we represent this system of linear equations as the matrix

$$\left[\begin{array}{ccc|c} 1 & 1 & 1 & 3 \\ 1 & 2 & 3 & 8 \\ 1 & 4 & 7 & 10 \end{array}\right],$$

where we have made the natural choice of letting the first variable be x, the second variable be y, and the third variable be z.

To start Gaussian elimination, we add -1 times row 1 to row 2 and also add -1 times row 1 to row 3, getting the matrix

$$\left[\begin{array}{ccc|c} 1 & 1 & 1 & 3 \\ 0 & 1 & 2 & 5 \\ 0 & 3 & 6 & 7 \end{array}\right].$$

Now we make the entry in row 3, column 2 equal to 0 by adding -3 times row 2 to row 3, getting the matrix

$$\left[\begin{array}{ccc|c} 1 & 1 & 1 & 3 \\ 0 & 1 & 2 & 5 \\ 0 & 0 & 0 & -8 \end{array}\right].$$

Look at the original system of linear equations. It is not at all obvious that it has no solutions. However, using Gaussian elimination on the matrix easily shows us that there are no solutions.

The last row of the matrix above corresponds to the equation

$$0x + 0y + 0z = -8.$$

Because the left side of the equation above equals 0 regardless of the values of x, y, and z, we see that there are no values of x, y, and z that satisfy this equation. Thus the original system of linear equations has no solutions.

We can summarize the experience of the example above as follows:

No solutions

If any stage of Gaussian elimination produces a row consisting of all 0's except for a nonzero entry in the last position, then the corresponding system of linear equations has no solutions.

Systems of Linear Equations with Infinitely Many Solutions

The next example illustrates yet another special case that we have not yet encountered when using Gaussian elimination.

EXAMPLE 5

Use matrix operations to find all solutions to this system of linear equations:

$$x + y + z = 3$$
$$x + 2y + 3z = 8$$
$$x + 4y + 7z = 18.$$

This system of linear equations is the same as in the previous example except now the constant term in the last equation is 18 rather than 10. However, now the solutions are quite different from what we found in the previous example.

SOLUTION First we represent this system of linear equations as the matrix

$$\left[\begin{array}{ccc|c} 1 & 1 & 1 & 3 \\ 1 & 2 & 3 & 8 \\ 1 & 4 & 7 & 18 \end{array}\right],$$

where we have made the natural choice of letting the first variable be x, the second variable be y, and the third variable be z.

To start Gaussian elimination, we add -1 times row 1 to row 2 and also add -1 times row 1 to row 3, getting the matrix

$$\left[\begin{array}{ccc|c} 1 & 1 & 1 & 3 \\ 0 & 1 & 2 & 5 \\ 0 & 3 & 6 & 15 \end{array}\right].$$

Now we make the entry in row 3, column 2 equal to 0 by adding -3 times row 2 to row 3, getting the matrix

$$\left[\begin{array}{ccc|c} 1 & 1 & 1 & 3 \\ 0 & 1 & 2 & 5 \\ 0 & 0 & 0 & 0 \end{array}\right].$$

The last row of the matrix above corresponds to the equation

$$0x + 0y + 0z = 0.$$

This equation is satisfied for all values of x, y, and z. In other words, this equation provides no information, and we can just ignore it.

Row 2 of the matrix above corresponds to the equation $y + 2z = 5$. Because the variable z cannot be eliminated, we simply solve this equation for y, getting

$$y = 5 - 2z.$$

Row 1 of the matrix above corresponds to the equation $x + y + z = 3$. Substituting $y = 5 - 2z$ into this equation gives $x + (5 - 2z) + z = 3$, which implies that

$$x = -2 + z.$$

Thus the solutions to our original system of linear equations are given by

$$x = -2 + z, \quad y = 5 - 2z.$$

Here z is any number, and then x and y are determined by the equations above. For example, taking $z = 0$, we have the solution $x = -2$, $y = 5$, $z = 0$. As another example, taking $z = 1$, we have the solution $x = -1$, $y = 3$, $z = 1$. Our original system of linear equations has one solution for each choice of z, showing that this system of linear equations has infinitely many solutions.

The example above shows that for some systems of linear equations, a complete description of the solutions consists of equations that express some of the variables in terms of the other variables.

Infinitely many solutions

In some systems of linear equations, Gaussian elimination leads to solving for some of the variables in terms of other variables. Such systems of linear equations have infinitely many solutions.

How Many Solutions, Revisited

At the beginning of Section 7.2, we saw that a linear equation with one variable has either no solutions, exactly one solution, or infinitely many solutions. We also saw that a system of two linear equations with two variables also has either no solutions, exactly one solution, or infinitely many solutions.

A bit of thought about how Gaussian elimination can end up shows that the same conclusion holds for every system of linear equations, regardless of the number of equations or the number of variables. If Gaussian elimination is applied to a system of linear equations and we do not end up with exactly one solution, then we must have either no solutions (corresponding to a row of the matrix consisting of all 0's except for a nonzero entry in the last position) or we get solutions with some variables solved only in terms of other variables (which gives infinitely many solutions).

The number of solutions to a system of linear equations

Each system of linear equations has either no solutions, one solution, or infinitely many solutions.

We conclude this section with a comment about notation for systems of equations with many variables. For a system of equations with three variables, it's fine to call the variables x, y, and z. However, once the number of variables exceeds a few dozen you will run out of letters. The common solution to this dilemma is use a single subscripted letter. Thus in a system of equations with 100 variables, the first variable could be denoted x_1, the second variable could be denoted x_2, and so on, up to the hundredth variable x_{100}.

The use of subscripted variables works particularly well on computers.

For example, the system of equations from Example 5 could be rewritten as follows if we call the variables x_1, x_2, and x_3 instead of x, y, and z:

$$x_1 + x_2 + x_3 = 3$$

$$x_1 + 2x_2 + 3x_3 = 8$$

$$x_1 + 4x_2 + 7x_3 = 18.$$

EXERCISES

In Exercises 1–4, represent the given system of linear equations as a matrix. Use alphabetical order for the variables.

1. $$\begin{aligned} 5x - 3y &= 2 \\ 4x + 7y &= -1 \end{aligned}$$

2. $$\begin{aligned} 8x + 6y &= -9 \\ 10x - \tfrac{3}{2}y &= 2 \end{aligned}$$

3. $$\begin{aligned} 8x + 6y - 5z &= -9 \\ 10x - \tfrac{3}{2}y + 4z &= 2 \\ x + 7y - 3z &= 11 \end{aligned}$$

4. $$\begin{aligned} 5x - 3y + \sqrt{2}z &= 2 \\ 4x + 7y - \sqrt{3}z &= -1 \\ -x + \tfrac{1}{3}y + 17z &= 6 \end{aligned}$$

In Exercises 5–8, interpret the given matrix as a system of linear equations. Use x for the first variable, y for the second variable, and (if needed) z for the third variable.

5. $$\left[\begin{array}{cc|c} 5 & -3 & 2 \\ -1 & \frac{1}{3} & 6 \end{array}\right]$$

6. $$\left[\begin{array}{cc|c} -7 & 4 & 23 \\ \frac{7}{3} & 31 & -5 \end{array}\right]$$

7. $$\left[\begin{array}{ccc|c} -7 & 4 & 23 & 6 \\ \frac{7}{3} & 31 & -5 & -11 \end{array}\right]$$

8. $$\left[\begin{array}{ccc|c} \sqrt{7} & 8 & 12 & -55 \\ 2 & -2\sqrt{3} & 15 & 1 \end{array}\right]$$

In Exercises 9–16, use Gaussian elimination to find all solutions to the given system of equations. For these exercises, work with matrices at least until the back-substitution stage is reached.

9. $$\begin{aligned} x - 2y + 3z &= -1 \\ 3x + 2y - 5z &= 3 \\ 2x - 5y + 2z &= 0 \end{aligned}$$

10. $$\begin{aligned} x + 3y + 2z &= 1 \\ 2x - 3y + 5z &= -2 \\ 3x + 4y - 7z &= 3 \end{aligned}$$

11. $$\begin{aligned} 3y + 2z &= 1 \\ x - 3y + 5z &= -2 \\ 3x + 4y - 7z &= 3 \end{aligned}$$

12. $$\begin{aligned} 2y + 3z &= 4 \\ -x + 4y + 3z &= -1 \\ 2x + 5y - 3z &= 0 \end{aligned}$$

13. $$\begin{aligned} x + 2y + 4z &= -3 \\ -2x + y + 3z &= 1 \\ -3x + 4y + 10z &= 4 \end{aligned}$$

14. $$\begin{aligned} -x - 3y + 5z &= 6 \\ 4x + 5y + 6z &= 7 \\ 2x - y + 16z &= 8 \end{aligned}$$

15. $$\begin{aligned} x + 2y + 4z &= -3 \\ -2x + y + 3z &= 1 \\ -3x + 4y + 10z &= -1 \end{aligned}$$

16. $$\begin{aligned} -x - 3y + 5z &= 6 \\ 4x + 5y + 6z &= 7 \\ 2x - y + 16z &= 19 \end{aligned}$$

17. Find a number b such that the system of linear equations
$$\begin{aligned} 2x + 3y &= 4 \\ 3x + by &= 7 \end{aligned}$$
has no solutions.

18. Find a number b such that the system of linear equations
$$\begin{aligned} 3x - 2y &= 1 \\ 4x + by &= 5 \end{aligned}$$
has no solutions.

19. Find a number b such that the system of linear equations

$$2x + 3y = 5$$
$$4x + 6y = b$$

has infinitely many solutions.

20. Find a number b such that the system of linear equations

$$3x - 2y = b$$
$$9x - 6y = 5$$

has infinitely many solutions.

PROBLEMS

21. Give an example of a system of three linear equations with two variables that has no solutions.

22. Give an example of a system of three linear equations with two variables that has exactly one solution.

23. Give an example of a system of three linear equations with two variables that has infinitely many solutions.

24. Give an example of a system of two linear equations with three variables that has no solutions.

25. Give an example of a system of two linear equations with three variables that has infinitely many solutions.

WORKED-OUT SOLUTIONS *to Odd-numbered Exercises*

In Exercises 1–4, represent the given system of linear equations as a matrix. Use alphabetical order for the variables.

1. $$5x - 3y = 2$$
$$4x + 7y = -1$$

SOLUTION The linear equation $5x - 3y = 2$ is represented as the row $\begin{bmatrix} 5 & -3 & | & 2 \end{bmatrix}$ and the linear equation $4x + 7y = -1$ is represented as the row $\begin{bmatrix} 4 & 7 & | & -1 \end{bmatrix}$. Thus the system of two linear equations above is represented by the matrix

$$\begin{bmatrix} 5 & -3 & | & 2 \\ 4 & 7 & | & -1 \end{bmatrix}.$$

3. $$8x + 6y - 5z = -9$$
$$10x - \tfrac{3}{2}y + 4z = 2$$
$$x + 7y - 3z = 11$$

SOLUTION The equation $8x + 6y - 5z = -9$ is represented as the row $\begin{bmatrix} 8 & 6 & -5 & | & -9 \end{bmatrix}$, the equation $10x - \frac{3}{2}y + 4z = 2$ is represented as the row $\begin{bmatrix} 10 & -\frac{3}{2} & 4 & | & 2 \end{bmatrix}$, and the equation $x + 7y - 3z = 11$ is represented as the row $\begin{bmatrix} 1 & 7 & -3 & | & 11 \end{bmatrix}$. Thus the system of three linear equations above is represented by the matrix

$$\begin{bmatrix} 8 & 6 & -5 & | & -9 \\ 10 & -\frac{3}{2} & 4 & | & 2 \\ 1 & 7 & -3 & | & 11 \end{bmatrix}.$$

In Exercises 5–8, interpret the given matrix as a system of linear equations. Use x for the first variable, y for the second variable, and (if needed) z for the third variable.

5. $$\begin{bmatrix} 5 & -3 & | & 2 \\ -1 & \frac{1}{3} & | & 6 \end{bmatrix}$$

SOLUTION Interpreting each row as the corresponding equation, we get the following system of linear equations:

$$5x - 3y = 2$$
$$-x + \tfrac{1}{3}y = 6.$$

7. $$\begin{bmatrix} -7 & 4 & 23 & | & 6 \\ \frac{7}{3} & 31 & -5 & | & -11 \end{bmatrix}$$

SOLUTION Interpreting each row as the corresponding equation, we get the following system of linear equations:

$$-7x + 4y + 23z = 6$$
$$\tfrac{7}{3}x + 31y - 5z = -11.$$

In Exercises 9–16, use Gaussian elimination to find all solutions to the given system of equations. For these exercises, work with matrices at least until the back-substitution stage is reached.

9. $$x - 2y + 3z = -1$$
$$3x + 2y - 5z = 3$$
$$2x - 5y + 2z = 0$$

SOLUTION First we represent this system of linear equations as the matrix

$$\left[\begin{array}{ccc|c} 1 & -2 & 3 & -1 \\ 3 & 2 & -5 & 3 \\ 2 & -5 & 2 & 0 \end{array}\right].$$

Add -3 times row 1 to row 2 and add -2 times row 1 to row 3, getting the matrix

$$\left[\begin{array}{ccc|c} 1 & -2 & 3 & -1 \\ 0 & 8 & -14 & 6 \\ 0 & -1 & -4 & 2 \end{array}\right].$$

Because all the entries in row 2 of the matrix above are divisible by 2, we can simplify a bit by multiplying row 2 of the matrix above by $\frac{1}{2}$, getting the matrix

$$\left[\begin{array}{ccc|c} 1 & -2 & 3 & -1 \\ 0 & 4 & -7 & 3 \\ 0 & -1 & -4 & 2 \end{array}\right].$$

To make the arithmetic in the next step a bit simpler, we now interchange rows 2 and 3, getting the matrix

$$\left[\begin{array}{ccc|c} 1 & -2 & 3 & -1 \\ 0 & -1 & -4 & 2 \\ 0 & 4 & -7 & 3 \end{array}\right].$$

Now add 4 times row 2 to row 3, getting the matrix

$$\left[\begin{array}{ccc|c} 1 & -2 & 3 & -1 \\ 0 & -1 & -4 & 2 \\ 0 & 0 & -23 & 11 \end{array}\right].$$

The last row of the matrix above corresponds to the equation $-23z = 11$, and thus $z = -\frac{11}{23}$.

Row 2 of the matrix above corresponds to the linear equation $-y - 4z = 2$. Substituting $z = -\frac{11}{23}$ into this equation gives $-y + \frac{44}{23} = 2$, which we can easily solve for y, getting $y = -\frac{2}{23}$.

Finally, row 1 of the matrix above corresponds to the equation $x - 2y + 3z = -1$. Substituting $y = -\frac{2}{23}$ and $z = -\frac{11}{23}$ into this equation and then solving for x gives $x = \frac{6}{23}$.

Thus the only solution to our original system of equations is $x = \frac{6}{23}$, $y = -\frac{2}{23}$, $z = -\frac{11}{23}$.

11. $$3y + 2z = 1$$
$$x - 3y + 5z = -2$$
$$3x + 4y - 7z = 3$$

SOLUTION First we represent this system of linear equations as the matrix

$$\left[\begin{array}{ccc|c} 0 & 3 & 2 & 1 \\ 1 & -3 & 5 & -2 \\ 3 & 4 & -7 & 3 \end{array}\right].$$

So that we can begin Gaussian elimination, we interchange the first two rows, getting the matrix

$$\left[\begin{array}{ccc|c} 1 & -3 & 5 & -2 \\ 0 & 3 & 2 & 1 \\ 3 & 4 & -7 & 3 \end{array}\right].$$

Add -3 times row 1 to row 3, getting the matrix

$$\left[\begin{array}{ccc|c} 1 & -3 & 5 & -2 \\ 0 & 3 & 2 & 1 \\ 0 & 13 & -22 & 9 \end{array}\right].$$

Now add $-\frac{13}{3}$ times row 2 to row 3, getting the matrix

$$\left[\begin{array}{ccc|c} 1 & -3 & 5 & -2 \\ 0 & 3 & 2 & 1 \\ 0 & 0 & -\frac{92}{3} & \frac{14}{3} \end{array}\right].$$

The last row of the matrix above corresponds to the equation $-\frac{92}{3}z = \frac{14}{3}$, and thus $z = -\frac{7}{46}$.

Row 2 of the matrix above corresponds to the linear equation $3y + 2z = 1$. Substituting $z = -\frac{7}{46}$ into this equation gives $3y - \frac{7}{23} = 1$, which we can easily solve for y, getting $y = \frac{10}{23}$.

Finally, row 1 of the matrix above corresponds to the equation $x - 3y + 5z = -2$. Substituting

$y = \frac{10}{23}$ and $z = -\frac{7}{46}$ into this equation and then solving for x gives $x = \frac{3}{46}$.

Thus the only solution to our original system of equations is $x = \frac{3}{46}$, $y = \frac{10}{23}$, $z = -\frac{7}{46}$.

13.
$$x + 2y + 4z = -3$$
$$-2x + y + 3z = 1$$
$$-3x + 4y + 10z = 4$$

SOLUTION First we represent this system of linear equations as the matrix

$$\left[\begin{array}{ccc|c} 1 & 2 & 4 & -3 \\ -2 & 1 & 3 & 1 \\ -3 & 4 & 10 & 4 \end{array}\right].$$

Add 2 times row 1 to row 2 and add 3 times row 1 to row 3, getting the matrix

$$\left[\begin{array}{ccc|c} 1 & 2 & 4 & -3 \\ 0 & 5 & 11 & -5 \\ 0 & 10 & 22 & -5 \end{array}\right].$$

Now add -2 times row 2 to row 3, getting the matrix

$$\left[\begin{array}{ccc|c} 1 & 2 & 4 & -3 \\ 0 & 5 & 11 & -5 \\ 0 & 0 & 0 & 5 \end{array}\right].$$

The last row of the matrix above corresponds to the equation

$$0x + 0y + 0z = 5,$$

which is not satisfied by any values of x, y, and z. Thus the original system of linear equations has no solutions.

15.
$$x + 2y + 4z = -3$$
$$-2x + y + 3z = 1$$
$$-3x + 4y + 10z = -1$$

SOLUTION First we represent this system of linear equations as the matrix

$$\left[\begin{array}{ccc|c} 1 & 2 & 4 & -3 \\ -2 & 1 & 3 & 1 \\ -3 & 4 & 10 & -1 \end{array}\right].$$

Add 2 times row 1 to row 2 and add 3 times row 1 to row 3, getting the matrix

$$\left[\begin{array}{ccc|c} 1 & 2 & 4 & -3 \\ 0 & 5 & 11 & -5 \\ 0 & 10 & 22 & -10 \end{array}\right].$$

Now add -2 times row 2 to row 3, getting the matrix

$$\left[\begin{array}{ccc|c} 1 & 2 & 4 & -3 \\ 0 & 5 & 11 & -5 \\ 0 & 0 & 0 & 0 \end{array}\right].$$

The last row of the matrix above corresponds to the equation

$$0x + 0y + 0z = 0,$$

which is satisfied for all values of x, y, and z. Thus this equation provides no information, and we can just ignore it.

Row 2 of the matrix above corresponds to the linear equation $5y + 11z = -5$. Solving this equation for y, we have

$$y = -1 - \tfrac{11}{5}z.$$

Row 1 of the matrix above corresponds to the linear equation $x + 2y + 4z = -3$. Substituting $y = -1 - \frac{11}{5}z$ into this equation and then solving for x gives

$$x = -1 + \tfrac{2}{5}z.$$

Thus the solutions to our original equation are given by

$$x = -1 + \tfrac{2}{5}z, \quad y = -1 - \tfrac{11}{5}z,$$

where z is any number (in particular, this system of linear equations has infinitely many solutions).

17. Find a number b such that the system of linear equations

$$2x + 3y = 4$$
$$3x + by = 7$$

has no solutions.

SOLUTION Represent this system of equations as the matrix

$$\left[\begin{array}{cc|c} 2 & 3 & 4 \\ 3 & b & 7 \end{array}\right].$$

Add $-\frac{3}{2}$ times row 1 to row 2, getting the matrix

$$\left[\begin{array}{cc|c} 2 & 3 & 4 \\ 0 & b - \frac{9}{2} & 1 \end{array}\right].$$

The last row corresponds to the equation $(b - \frac{9}{2})y = 1$. If we choose $b = \frac{9}{2}$, then this becomes the equation $0y = 1$, which has no solutions.

19. Find a number b such that the system of linear equations

$$2x + 3y = 5$$

$$4x + 6y = b$$

has infinitely many solutions.

SOLUTION Represent this system of equations as the matrix

$$\left[\begin{array}{cc|c} 2 & 3 & 5 \\ 4 & 6 & b \end{array}\right].$$

Add -2 times row 1 to row 2, getting the matrix

$$\left[\begin{array}{cc|c} 2 & 3 & 5 \\ 0 & 0 & b - 10 \end{array}\right].$$

The last row corresponds to the equation $0x + 0y = b - 10$. If we choose $b = 10$, then this becomes the equation $0x + 0y = 0$, which is satisfied by all values of x and y, which means that we can ignore it. Thus if $b = 10$, then our system of equations is equivalent to the equation $2x + 3y = 5$, which has infinitely many solutions (obtained by choosing any number y and then setting $x = \frac{5-3y}{2}$).

7.4 *Matrix Algebra*

LEARNING OBJECTIVES

By the end of this section you should be able to

- find the size of a matrix;
- determine when two matrices are equal;
- add and subtract matrices of the same size;
- compute the product of two matrices of appropriate sizes;
- find the inverse of an invertible matrix.

Matrix Size

We have been using matrices to represent systems of linear equations efficiently. Matrices have many other uses in mathematics and in other fields. Matrices arise naturally whenever we need to think about collections of numbers arranged in rows and columns. In this section we will deal with the basic properties of matrix algebra.

Recall that a matrix is a rectangular array of numbers, usually enclosed in straight brackets. The horizontal lines of numbers in a matrix are called rows; the vertical lines of numbers are called columns.

Size of a matrix

The **size of a matrix** is stated as the number of rows by the number of columns.

EXAMPLE 1

(a) What is the size of the matrix $\begin{bmatrix} 12 & \frac{7}{8} & 5.9 \\ 4 & 6 & -55 \end{bmatrix}$?

(b) Give an example of a 3-by-2 matrix.

SOLUTION

When dealing with rows and columns, remember that the row number always comes first.

(a) This matrix has 2 rows and 3 columns. Thus the size of this matrix is 2-by-3.

(b) A 3-by-2 matrix has 3 rows and 2 columns. Thus

$$\begin{bmatrix} 1 & 1 \\ 9 & 8 \\ 4 & 5 \end{bmatrix}$$

is a 3-by-2 matrix. Of course, there are many other choices for the entries of the matrix.

Now we turn to the issue of when two matrices are considered to be equal to each other.

Equality of matrices

Two matrices are **equal** if and only if

- they have have the same size, and
- in each corresponding position they have the same entries.

EXAMPLE 2

(a) Are the matrices

$$\begin{bmatrix} 7 & 5 \\ 2 & 9 \end{bmatrix} \quad \text{and} \quad \begin{bmatrix} 7 & 5 & 5 \\ 2 & 9 & 9 \end{bmatrix}$$

equal?

(b) Are the matrices

$$\begin{bmatrix} \sqrt{7} & 3 \\ -2 & 8 \end{bmatrix} \quad \text{and} \quad \begin{bmatrix} \sqrt{7} & 3 \\ -2 & 1 \end{bmatrix}$$

equal?

(c) Are the matrices

$$\begin{bmatrix} 1 & 1 & 1 \\ 2 & 2 & 2 \end{bmatrix} \quad \text{and} \quad \begin{bmatrix} 2 & 2 & 2 \\ 1 & 1 & 1 \end{bmatrix}$$

equal?

(d) Find numbers a, b, c, and d such that the matrices

$$\begin{bmatrix} 5 & 7 \\ 6 & d \end{bmatrix} \quad \text{and} \quad \begin{bmatrix} a & b^3 \\ c-4 & b+c \end{bmatrix}$$

are equal.

SOLUTION

(a) The first matrix has size 2-by-2, and the second matrix has size 2-by-3. Because these matrices do not have same size, they are not equal.

(b) These two matrices have the same size (2-by-2). They have the same entries in three of the four positions. However, the entry in row 2, column 2 of the first matrix is 8, while the entry in row 2, column 2 of the second matrix is 1. Because these matrices have different entries in row 2, column 2, they are not equal.

(c) These two matrices have the same size (2-by-3). Each of them contains three 1's and three 2's. However, the 1's and 2's appear in different positions in the two matrices. Thus these two matrices are not equal.

Order does not matter when listing elements of a set. However, positions of the entries do matter for matrices.

(d) For these two matrices to be equal, we must have

$$5 = a \qquad 7 = b^3$$
$$6 = c - 4 \qquad d = b + c$$

The first equation tells us that $a = 5$. The equation $7 = b^3$ tells us that $b = 7^{1/3}$. Solving the equation $6 = c - 4$ gives $c = 10$. Now that we know the values of b and c, the last equation becomes $d = 7^{1/3} + 10$.

Adding and Subtracting Matrices

We now turn to algebraic operations with matrices, beginning with matrix addition.

The sum of two matrices is defined only when the two matrices have the same size.

Matrix addition

The **sum of two matrices** of the same size is obtained by adding corresponding entries of the two matrices.

EXAMPLE 3

A small sandwich shop open for lunch on weekends tracks its sales by the hour and also by the kind of sandwich sold. The manager of the sandwich shop records this data in a matrix, with the rows corresponding to the hours and the columns corresponding to the kind of sandwich. The data for last Saturday was as follows:

	roast beef	tuna	turkey	veggie combo
11 am to noon	12	7	8	3
noon to 1 pm	31	15	13	9
1 pm to 2 pm	23	11	7	5

Let's use A to denote the matrix above (without the labels) for last Saturday, and let B be the matrix below with last Sunday's data:

Note the use here of single letters to denote entire matrices.

$$A = \begin{bmatrix} 12 & 7 & 8 & 3 \\ 31 & 15 & 13 & 9 \\ 23 & 11 & 7 & 5 \end{bmatrix} \quad \text{and} \quad B = \begin{bmatrix} 14 & 11 & 15 & 9 \\ 42 & 21 & 10 & 8 \\ 30 & 18 & 5 & 7 \end{bmatrix}.$$

For example, the entry in row 2, column 3 of B means that on Sunday in the time from noon to 1 pm (corresponding to row 2) the number of turkey sandwiches sold (corresponding to column 3) was 10.

Find the matrix that shows the sales by hour and by kind of sandwich for the entire past weekend.

SOLUTION The sales for the entire past weekend are obtained by adding together the sales for Saturday and the sales for Sunday. In other words, we seek the matrix $A + B$ obtained by adding corresponding entries of A and B. Simple arithmetic shows that

$$A + B = \begin{bmatrix} 26 & 18 & 23 & 12 \\ 73 & 36 & 23 & 17 \\ 53 & 29 & 12 & 12 \end{bmatrix}.$$

Thus, for example, the entry in row 3, column 2 of $A+B$ means that over the weekend in the time slot from 1 pm to 2 pm (corresponding to row 3) the number of tuna sandwiches sold (corresponding to column 2) was 29. The entry 29 comes from adding row 3, column 2 of A (showing that on Saturday from 1 pm to 2 pm there were 11 tuna sandwiches sold) to row 3, column 2 of B (showing that on Sunday from 1 pm to 2 pm there were 18 tuna sandwiches sold).

Matrix addition inherits the same commutative and associative properties enjoyed by regular addition of numbers. Thus we do not need to worry about order or grouping when dealing with matrix addition.

Commutativity and associativity of matrix addition

If A, B, and C are matrices of the same size, then

- $A + B = B + A$;
- $(A + B) + C = A + (B + C)$.

The definition of matrix subtraction should not be a surprise.

Matrix subtraction

The **difference of two matrices** of the same size is obtained by subtracting corresponding entries of the two matrices.

The difference of two matrices is defined only when the two matrices have the same size.

EXAMPLE 4

Using the same setting and data as in the previous example, find the matrix that shows the increases in sales from Saturday to Sunday for the past weekend by hour and by kind of sandwich.

SOLUTION The increases in sales from Saturday to Sunday of the past weekend are obtained by subtracting the Saturday sales from the Sunday sales. In other words, we seek the matrix $B - A$ obtained by subtracting entries of A from the corresponding entries of A. Simple arithmetic shows that

$$B - A = \begin{bmatrix} 2 & 4 & 7 & 6 \\ 11 & 6 & -3 & -1 \\ 7 & 7 & -2 & 2 \end{bmatrix}.$$

Thus, for example, the entry in row 2, column 1 of $B - A$ means that in the time slot from noon to 1 pm (corresponding to row 2) the number of roast beef sandwiches sold (corresponding to column 1) increased by 11 from Saturday to Sunday.

The negative entries in $B - A$ indicate a decrease from Saturday to Sunday. For example, the entry in row 2, column 4 of $B - A$ means that in the time slot from noon to 1 pm (corresponding to row 2) the number of veggie combo sandwiches sold (corresponding to column 4) decreased by 1 from Saturday to Sunday.

As with regular subtraction of numbers, matrix subtraction is neither commutative nor associative.

Multiplying a Matrix by a Number

We begin this topic by defining what it means to multiply a matrix by a number. The word **scalar** is a fancy word for *number*, and thus this operation is often called **scalar multiplication**.

Scalar multiplication

The product of a number and a matrix is obtained by multiplying each entry in the matrix by the number.

EXAMPLE 5

Using the same setting and data as in Example 3, find the matrix that shows the average sales by hour and by kind of sandwich for the entire past weekend.

SOLUTION The average sales for the entire weekend are obtained by averaging the sales for Saturday and Sunday. In other words, we seek the matrix $\frac{1}{2}(A+B)$ obtained by multiplying each entry in the matrix $A+B$ by $\frac{1}{2}$. Simple arithmetic using the matrix $A+B$ from Example 3 shows that

$$\tfrac{1}{2}(A+B) = \begin{bmatrix} 13 & 9 & 11.5 & 6 \\ 36.5 & 18 & 11.5 & 8.5 \\ 26.5 & 14.5 & 6 & 6 \end{bmatrix}.$$

Thus, for example, the entry 9 in row 1, column 2 of $\frac{1}{2}(A+B)$ means that over the weekend in the time slot from 11 am to noon (corresponding to row 1) the average number of tuna sandwiches sold (corresponding to column 2) was 9. The entry 9 comes from multiplying row 1, column 2 of $A+B$ by $\frac{1}{2}$.

The operation of multiplying a matrix by a number has the expected distributive properties.

Here there are two distributive properties—the first involves one number and two matrices; the second involves two numbers and one matrix.

Distributive properties

If A and B are matrices of the same size and s and t are numbers, then

- $s(A+B) = sA + sB$;
- $(s+t)A = sA + tA$.

Multiplying Matrices

Now we turn to the topic of multiplying together two matrices. You may expect that matrix multiplication is defined similarly to matrix addition. Specifically, a reasonable guess is that the product of two matrices (of the same size) is obtained by multiplying corresponding entries of the two matrices.

However, experience has shown that a different definition is much more useful. We begin by defining the product of a row of one matrix and the column of another matrix. In the following definition, the **length** of a row or of a column is the number of entries in that row or column. Thus in Example 6 the rows of A have length 2 and the columns of B also have length 2.

Row times column

The product of a row and a column of the same length is sum of the products of corresponding entries.

The product of a row of one length and a column of a different length is not defined.

EXAMPLE 6

Suppose

$$A = \begin{bmatrix} 5 & 4 \\ 3 & 7 \\ 8 & 6 \end{bmatrix} \quad \text{and} \quad B = \begin{bmatrix} 1 & 3 & 9 & 2 \\ 5 & 4 & 7 & 3 \end{bmatrix}.$$

(a) What is the product of row 1 of A and column 4 of B?

(b) What is the product of row 2 of A and column 1 of B?

SOLUTION

(a) Row 1 of A is $\begin{bmatrix} 5 & 4 \end{bmatrix}$. Column 4 of B is $\begin{bmatrix} 2 \\ 3 \end{bmatrix}$. Using the definition above, we have

$$\begin{aligned} \begin{bmatrix} 5 & 4 \end{bmatrix} \begin{bmatrix} 2 \\ 3 \end{bmatrix} &= 5 \cdot 2 + 4 \cdot 3 \\ &= 22. \end{aligned}$$

(b) Row 2 of A is $\begin{bmatrix} 3 & 7 \end{bmatrix}$. Column 1 of B is $\begin{bmatrix} 1 \\ 5 \end{bmatrix}$. Using the definition above, we have

$$\begin{aligned} \begin{bmatrix} 3 & 7 \end{bmatrix} \begin{bmatrix} 1 \\ 5 \end{bmatrix} &= 3 \cdot 1 + 7 \cdot 5 \\ &= 38. \end{aligned}$$

The next example illustrates why the definition above of the product of a row and a column can be useful.

EXAMPLE 7

Suppose the prices charged by the sandwich shop from Example 3 are as shown at the right. Using the data from Example 3, find the sandwich shop's total sales in dollars on Saturday from 11 am to noon.

roast beef	\$7
tuna	\$5
turkey	\$6
veggie combo	\$4

SOLUTION The number of sandwiches of each type sold between 11 am and noon on Saturday is shown in row 1 of the matrix A from Example 3, which is $\begin{bmatrix} 12 & 7 & 8 & 3 \end{bmatrix}$, where the sandwich types are listed in the same order as in the price table above (roast beef first, etc.).

The prices from the price table above can be written as the 4-by-1 price matrix at the right. Roast beef sales brought in $12 \cdot 7$ dollars (12 roast beef sandwiches at \$7 each), the tuna sales brought in $7 \cdot 5$ dollars (7 tuna sandwiches at \$5 each), and so on.

$$\begin{bmatrix} 7 \\ 5 \\ 6 \\ 4 \end{bmatrix}$$

Thus the shop's total sales from all four types of sandwiches between 11 am and noon on Saturday is the product of row 1 of the matrix A from Example 3 and column 1 of the price matrix (which has only one column). We have

$$\begin{bmatrix} 12 & 7 & 8 & 3 \end{bmatrix} \begin{bmatrix} 7 \\ 5 \\ 6 \\ 4 \end{bmatrix} = 12 \cdot 7 + 7 \cdot 5 + 8 \cdot 6 + 3 \cdot 4$$

$$= 179.$$

Thus the sandwich shop's total sales on Saturday from 11 am to noon was \$179.

Having defined the product of a row and a column, we can now define the product of two matrices.

The product of two matrices is defined only when the number of columns in the first matrix equals the number of rows in the second matrix.

Matrix multiplication

Suppose A is an m-by-n matrix and B is an n-by-p matrix. Then

- AB is an m-by-p matrix;
- the entry in row j, column k of AB equals row j of A times column k of B.

EXAMPLE 8

Suppose

$$A = \begin{bmatrix} 5 & 4 \\ 3 & 7 \\ 8 & 6 \end{bmatrix} \quad \text{and} \quad B = \begin{bmatrix} 1 & 3 & 9 & 2 \\ 5 & 4 & 7 & 3 \end{bmatrix}.$$

Find the matrix AB.

SOLUTION Here A is a 3-by-2 matrix and B is a 2-by-4 matrix. Thus AB is a 3-by-4 matrix.

The entry in row 1, column 1 of AB equals row 1 of A times column 1 of B. In other words, the entry in row 1, column 1 of AB equals $\begin{bmatrix} 5 & 4 \end{bmatrix} \begin{bmatrix} 1 \\ 5 \end{bmatrix}$, which equals $5 \cdot 1 + 4 \cdot 5$, which equals 25.

In Example 6 we computed that row 1 of A times column 4 of B is 22. Thus the entry in row 1, column 4 of AB is 22. We also computed in Example 6 that row 2 of A times column 1 of B is 38. Thus the entry in row 2, column 1 of AB is 38.

Computing the other entries of AB similarly, we get

$$AB = \begin{bmatrix} 25 & 31 & 73 & 22 \\ 38 & 37 & 76 & 27 \\ 38 & 48 & 114 & 34 \end{bmatrix}.$$

Warning: Matrix multiplication is not commutative. In the example above, we computed the product AB. However, the product BA is not defined for the matrices above because the number of columns of B (which is 4) is not equal to the number of rows of A (which is 3).

Even when the products AB and BA are both defined, they are not necessarily equal, as shown in the next example.

EXAMPLE 9

Suppose

$$A = \begin{bmatrix} 0 & 0 \\ 0 & 1 \end{bmatrix} \quad \text{and} \quad B = \begin{bmatrix} 0 & 1 \\ 0 & 0 \end{bmatrix}.$$

(a) What is the matrix AB?

(b) What is the matrix BA?

SOLUTION

(a) Using the definition of matrix multiplication, we have

$$\begin{aligned} AB &= \begin{bmatrix} 0 & 0 \\ 0 & 1 \end{bmatrix} \begin{bmatrix} 0 & 1 \\ 0 & 0 \end{bmatrix} \\ &= \begin{bmatrix} 0 & 0 \\ 0 & 0 \end{bmatrix}. \end{aligned}$$

(b) Using the definition of matrix multiplication, we have

$$\begin{aligned} BA &= \begin{bmatrix} 0 & 1 \\ 0 & 0 \end{bmatrix} \begin{bmatrix} 0 & 0 \\ 0 & 1 \end{bmatrix} \\ &= \begin{bmatrix} 0 & 1 \\ 0 & 0 \end{bmatrix}. \end{aligned}$$

Note that $AB \neq BA$ for these matrices A and B.

Although matrix multiplication is not commutative, it is associative.

Associativity of matrix multiplication

If A is an m-by-n matrix, B is an n-by-p matrix, and C is a p-by-q matrix, then

$$(AB)C = A(BC).$$

Because of associativity, we can delete the parentheses and simply write ABC.

Matrix multiplication also satisfies the distributive property.

Distributive property of matrix multiplication

- If A and B are m-by-n matrices and C is an n-by-p matrix, then

$$(A + B)C = AC + BC.$$

- If A is an m-by-n matrix and B and C are n-by-p matrices, then

$$A(B + C) = AB + AC.$$

We need two versions of the distributive property because matrix multiplication is not commutative.

The next example shows why the definition of matrix multiplication often leads to useful applications.

EXAMPLE 10

Let A be the matrix below showing the number of sandwiches sold (by time slot and type of sandwich) last Saturday by the sandwich shop in Example 3:

	roast beef	tuna	turkey	veggie combo
11 am to noon	12	7	8	3
noon to 1 pm	31	15	13	9
1 pm to 2 pm	23	11	7	5

$$A = \begin{bmatrix} 12 & 7 & 8 & 3 \\ 31 & 15 & 13 & 9 \\ 23 & 11 & 7 & 5 \end{bmatrix}.$$

The time-slot labels and sandwich-type labels above are not part of the matrix A.

Let P be the 4-by-1 matrix to the right showing the prices of each type of sandwich, listed in the same order as used for the columns of A (roast beef first, tuna second, etc.).

$$P = \begin{bmatrix} 7 \\ 5 \\ 6 \\ 4 \end{bmatrix}$$

Let C be the 1-by-3 matrix $C = \begin{bmatrix} 1 & 1 & 1 \end{bmatrix}$.

(a) What is the matrix AP?

(b) What is the meaning of the entries in the matrix AP?

(c) What is the matrix CA?

(d) What is the meaning of the entries in the matrix CA?

SOLUTION

Because matrix multiplication is used so often, techniques have been developed that allow computers to multiply even huge matrices very rapidly.

(a) Using the definition of matrix multiplication, we have

$$AP = \begin{bmatrix} 12 & 7 & 8 & 3 \\ 31 & 15 & 13 & 9 \\ 23 & 11 & 7 & 5 \end{bmatrix} \begin{bmatrix} 7 \\ 5 \\ 6 \\ 4 \end{bmatrix} = \begin{bmatrix} 179 \\ 406 \\ 278 \end{bmatrix}.$$

(b) The entry in row 1, column 1 of AP is obtained by multiplying row 1 of A (sales from 11 am to noon) by column 1 of P (the price of each type of sandwich). Thus the entry in row 1, column 1 of AP means that the total sales on Saturday from 11 am to noon was \$179. See Example 7 for more about this computation.

Similarly, the entry in row 2, column 1 of AP means that the total sales on Saturday from noon to 1 pm was \$406. Finally, the entry in row 3, column 1 of AP means that the total sales on Saturday from 1 pm to 2 pm was \$278.

(c) Using the definition of matrix multiplication, we have

$$CA = \begin{bmatrix} 1 & 1 & 1 \end{bmatrix} \begin{bmatrix} 12 & 7 & 8 & 3 \\ 31 & 15 & 13 & 9 \\ 23 & 11 & 7 & 5 \end{bmatrix} = \begin{bmatrix} 66 & 33 & 28 & 17 \end{bmatrix}.$$

(d) Because each entry of C is 1, multiplying row 1 of C (which has only one row) by column 1 of A just adds up the entries in column 1 of A. Thus the entry in row 1, column 1 of CA is the total number of roast beef sandwiches that were sold on Saturday. In other words, 66 roast beef sandwiches were sold on Saturday. Similarly, 33 tuna sandwiches were sold on Saturday, corresponding to row 1, column 2 of CA. The matrix CA also shows that 28 turkey sandwiches and 17 veggie combos were sold on Saturday.

The Inverse of a Matrix

Our focus now turns to the inverse of a matrix, which is another aspect of matrix algebra. As we will see, the inverse of a matrix is taken with respect to matrix multiplication. We begin with some definitions.

The diagonal of a square matrix

- A **square matrix** is a matrix that has the same number of rows as columns. Equivalently, a **square matrix** has size m-by-m for some positive integer m.
- The **diagonal** of a square matrix consists of the entries where the row number equals the column number.

$$\begin{bmatrix} 4 & -3 & 2 \\ 1 & 6 & -7 \\ 0 & 5 & -8 \end{bmatrix}$$

The diagonal entries of the matrix above are 4, 6, *and* −8.

For example, the diagonal entries of the 3-by-3 square matrix in the margin are printed in red.

Now we define some special square matrices.

Identity matrix

- For each positive integer m, the **identity matrix** I_m is the m-by-m matrix consisting of 1's on the diagonal and 0's elsewhere.
- If the context shows the size or there is no danger of confusion, then I_m can be written as I.

The first equation in the example below shows the identity matrix I_3, and the second equation shows the identity matrix I_2. Both of these identity matrices can also be denoted simply by I.

EXAMPLE 11

These two products show why the term identity matrix *is used: multiplying a matrix on either the right or the left by a suitably sized identity matrix produces the original matrix.*

Show that

$$\begin{bmatrix} 2 & 7 & 4 \\ 8 & 5 & 9 \end{bmatrix}\begin{bmatrix} 1 & 0 & 0 \\ 0 & 1 & 0 \\ 0 & 0 & 1 \end{bmatrix} = \begin{bmatrix} 2 & 7 & 4 \\ 8 & 5 & 9 \end{bmatrix}$$

and

$$\begin{bmatrix} 1 & 0 \\ 0 & 1 \end{bmatrix}\begin{bmatrix} 2 & 7 & 4 \\ 8 & 5 & 9 \end{bmatrix} = \begin{bmatrix} 2 & 7 & 4 \\ 8 & 5 & 9 \end{bmatrix}.$$

SOLUTION We begin with the first equation above. The entry in row 1, column 1 of the product is computed by multiplying the first row $\begin{bmatrix} 2 & 7 & 4 \end{bmatrix}$ by the first column of the identity matrix, which has 1 in the first entry and 0 in the other entries. Thus the entry in row 1, column 1 of the product in the first equation is $2 \cdot 1 + 7 \cdot 0 + 4 \cdot 0$, which equals 2, as shown in the right on the first equation.

Similarly, the entry in row 1, column 2 of the product in the first equation is $2 \cdot 0 + 7 \cdot 1 + 4 \cdot 0$, which equals 7, as shown in the right on the first equation.

The other four entries in the first product are computed similarly, as are all six entries in the second product. If you work through all these simple computations, you will have a good understanding of why multiplication by the identity matrix (on either side) works as it does.

The next result generalizes the previous example.

Identity matrix is a multiplicative identity

Suppose A is an m-by-n matrix. Then

$$AI_n = A \quad \text{and} \quad I_m A = A.$$

The conclusion above is often written in the less precise form

$$AI = A \quad \text{and} \quad IA = A.$$

Having defined the identity matrix, we can now define what it means for a matrix to be invertible.

The possibility of being invertible arises only for square matrices. If A is a matrix that is not square, then there cannot exist a matrix B such that $AB = I$ and $BA = I$.

Invertible matrix

A square matrix A is called **invertible** if there is a matrix B such that

$$AB = I \quad \text{and} \quad BA = I.$$

If A is an invertible matrix, we are tempted to call the matrix B that satisfies the equations above the inverse of A and denote it by A^{-1}. But suppose more than one matrix B satisfies the equations above, just as more than one number satisfies the equation $x^2 = 9$. If that happened, then the notation A^{-1} would be ambiguous. Fortunately all is well, as shown by the next result.

Inverse matrix

Suppose A is an invertible matrix. Then

- there is only one matrix B such that $AB = I$ and $BA = I$;
- this matrix B is called the **inverse** of A and is denoted A^{-1};
- the matrix A^{-1} is characterized by the equations $AA^{-1} = A^{-1}A = I$.

To see why there cannot be more than one choice for B in the equations above, suppose A, B, and C are matrices such that $AB = I$ and $CA = I$. Multiply both sides of the first equation by C on the left, getting $C(AB) = CI$. The associative property of matrix multiplication means we can rewrite this equation as $(CA)B = C$. However, $CA = I$, so the last equation becomes $C = B$. Thus we have shown that there cannot be more than one choice for the matrix B that satisfies the equations defining an invertible matrix.

EXAMPLE 12

Verify that $\begin{bmatrix} -4 & 2 \\ -3 & 1 \end{bmatrix}^{-1} = \begin{bmatrix} \frac{1}{2} & -1 \\ \frac{3}{2} & -2 \end{bmatrix}$.

We will see later how to calculate the inverse of an invertible matrix.

SOLUTION According to the definition of an inverse matrix, we need to show that

$$\begin{bmatrix} -4 & 2 \\ -3 & 1 \end{bmatrix}\begin{bmatrix} \frac{1}{2} & -1 \\ \frac{3}{2} & -2 \end{bmatrix} = \begin{bmatrix} 1 & 0 \\ 0 & 1 \end{bmatrix} \quad \text{and} \quad \begin{bmatrix} \frac{1}{2} & -1 \\ \frac{3}{2} & -2 \end{bmatrix}\begin{bmatrix} -4 & 2 \\ -3 & 1 \end{bmatrix} = \begin{bmatrix} 1 & 0 \\ 0 & 1 \end{bmatrix}.$$

The two equations above can be verified by applying the definition of matrix multiplication. For example, the entry in row 1, column 1 of the first product is

$$(-4)\tfrac{1}{2} + 2 \cdot \tfrac{3}{2},$$

which equals $-2 + 3$, which equals 1, as desired. The entry in row 1, column 2 of the first product is

$$(-4)(-1) + 2(-2),$$

which equals $4 + (-4)$, which equals 0, as desired. The verification of the other entries (two more for the first product and then four more for the second product) is left to the reader (do it!).

According to the definition, to verify that a matrix B is the inverse of a matrix A we must verify that $AB = I$ and $BA = I$. Matrix multiplication is not commutative, so in general AB is not equal to BA. However, the next result shows that in the special case where AB equals I, then BA also equals I.

Special case of commutativity

If A and B are square matrices such that $AB = I$, then $BA = I$.

This result can cut in half the work needed to verify the inverse of a matrix.

A proof of the result above would take us more deeply into linear algebra than is appropriate for a course at this level. Thus please accept this result as something that you will encounter again if you pursue more advanced mathematics.

EXAMPLE 13

Verify that $\begin{bmatrix} 1 & 2 & 3 \\ 5 & -1 & 6 \\ 7 & -2 & 8 \end{bmatrix}^{-1} = \begin{bmatrix} -4 & 22 & -15 \\ -2 & 13 & -9 \\ 3 & -16 & 11 \end{bmatrix}$.

SOLUTION By the special case of commutativity discussed above, we need only show that

$$\begin{bmatrix} 1 & 2 & 3 \\ 5 & -1 & 6 \\ 7 & -2 & 8 \end{bmatrix}\begin{bmatrix} -4 & 22 & -15 \\ -2 & 13 & -9 \\ 3 & -16 & 11 \end{bmatrix} = \begin{bmatrix} 1 & 0 & 0 \\ 0 & 1 & 0 \\ 0 & 0 & 1 \end{bmatrix}.$$

The equation above can be verified by applying the definition of matrix multiplication. For example, the entry in row 1, column 1 of the product is

$$1(-4) + 2(-2) + 3 \cdot 3,$$

which equals $-4 - 4 + 9$, which equals 1, as desired. The entry in row 1, column 2 of the product is

$$1 \cdot 22 + 2 \cdot 13 + 3(-16),$$

which equals $22 + 26 - 48$, which equals 0, as desired. The verification of the other seven entries is left to the reader (do it!).

In Section 7.3 we used matrices to solve systems of linear equations. The idea was to make the coefficients and the constant term from each equation into one row of a matrix. Then we performed elementary row operations on the matrix to solve the system of linear equations via Gaussian elimination.

Now we turn to another way to express a system of linear equations using matrices. We will make the coefficients of each equation into one row of a matrix, but now we will not include the constant term. Instead, the constant terms will form their own matrix with one column. The variables will also be put into a matrix with one column.

The next example shows how these ideas are used to rewrite a system of linear equations into a matrix equation. Instead of calling the variables x, y, and z, the variables here have been called x_1, x_2, and x_3 so that you will become comfortable with the subscript notation that is used when dealing with a large number of variables.

EXAMPLE 14 Consider the following system of equations:

$$\begin{aligned} x_1 + 2x_2 + 3x_3 &= 9 \\ 5x_1 - x_2 + 6x_3 &= -4 \\ 7x_1 - 2x_2 + 8x_3 &= 6. \end{aligned}$$

(a) Rewrite the system of equations above as a matrix equation.

(b) Use an inverse matrix to solve the matrix equation and find the values of x_1, x_2, and x_3.

SOLUTION

(a) Form a matrix A whose rows consist of the coefficients on the left side of the equations above. Form another matrix X with one column consisting of the variables. Finally, form another matrix B with one column consisting of the constant terms from the right side of the equations above. In each of these matrices, keep everything in the proper order as listed above. In other words, we have

$$A = \begin{bmatrix} 1 & 2 & 3 \\ 5 & -1 & 6 \\ 7 & -2 & 8 \end{bmatrix}, \quad X = \begin{bmatrix} x_1 \\ x_2 \\ x_3 \end{bmatrix}, \quad \text{and} \quad B = \begin{bmatrix} 9 \\ -4 \\ 6 \end{bmatrix}.$$

This matrix equation provides another reason why the definition of matrix multiplication, which at first seem strange, turns out to be so useful.

The system of linear equations above can now be expressed as the single matrix equation

$$AX = B.$$

To verify that the matrix equation above is the same as the original system of equations, use the definition of matrix multiplication. For example, the entry in row 1, column 1 of AX equals row 1 of A times column 1 of X (which has only one column). This product equals $x_1 + 2x_2 + 3x_3$. Setting that expression equal to the entry in row 1, column 1 of B gives the original first equation.

Similarly, the other two original equations arise by computing the entries in rows 2 and 3, column 1 of AX and setting them equal to the entries in rows 2 and 3, column 1 of B.

(b) To solve the matrix equation $AX = B$ for X, we multiply both sides of the equation by A^{-1} on the left, getting $A^{-1}AX = A^{-1}B$. Of course $A^{-1}A$ is the identity I, and IX equals X. Thus $X = A^{-1}B$. We know the entries in the matrix A^{-1} from Example 13. Thus

$$\begin{aligned} X &= A^{-1}B \\ &= \begin{bmatrix} -4 & 22 & -15 \\ -2 & 13 & -9 \\ 3 & -16 & 11 \end{bmatrix} \begin{bmatrix} 9 \\ -4 \\ 6 \end{bmatrix} \\ &= \begin{bmatrix} -214 \\ -124 \\ 157 \end{bmatrix}. \end{aligned}$$

The matrix equation above implies that $x_1 = -214$, $x_2 = -124$, and $x_3 = 157$.

The example above shows that if we know A^{-1}, then solving the matrix equation $AX = B$ can be done simply by matrix multiplication: $X = A^{-1}B$. However, this is not an efficient method for solving systems of linear equations because Gaussian elimination (as was presented in Sections 7.2 and 7.3) is faster than computing the inverse of a matrix.

Suppose you need to solve multiple systems of linear equations of the form $AX = B$ with the same matrix A (corresponding to the same coefficients) but many different choices of B (corresponding to many different choices of constant terms). In this case, computing A^{-1} (which only needs to be done once) is likely to be faster than redoing Gaussian elimination for many different choices of B. Thus the following procedure for finding the inverse of a square matrix may sometimes be useful.

Procedure for finding the inverse

Suppose A is a square matrix. Find the inverse of A as follows:

- Place an identity matrix of the same size to the right of A, creating a new matrix that has twice as many columns as A.
- Use elementary row operations on this larger matrix to transform the part originally occupied by A into the identity matrix (if this is not possible, then A is not invertible).
- The part of the larger matrix originally occupied by the identity matrix is A^{-1}.

EXAMPLE 15 Find the inverse of the matrix $\begin{bmatrix} 7 & 5 \\ 3 & 2 \end{bmatrix}$.

SOLUTION Create a larger matrix by placing the 2-by-2 identity matrix to the right of the original matrix, getting the matrix

The red bars are not part of the matrix. The red bars are simply a visual reminder about the location of the original matrix.

$$\left[\begin{array}{cc|cc} 7 & 5 & 1 & 0 \\ 3 & 2 & 0 & 1 \end{array}\right].$$

Now use elementary row operations on the matrix above. As the first step, multiply the first row by $\frac{1}{7}$ (to make the entry in row 1, column 1 equal to 1), getting

$$\left[\begin{array}{cc|cc} 1 & \frac{5}{7} & \frac{1}{7} & 0 \\ 3 & 2 & 0 & 1 \end{array}\right].$$

Next, add -3 times the first row to the second row, getting the matrix

$$\left[\begin{array}{cc|cc} 1 & \frac{5}{7} & \frac{1}{7} & 0 \\ 0 & -\frac{1}{7} & -\frac{3}{7} & 1 \end{array}\right].$$

The worked-out solution to Exercise 31 shows this process for finding the inverse of a 3-by-3 matrix.

Now multiply the second row by -7, getting

$$\left[\begin{array}{cc|cc} 1 & \frac{5}{7} & \frac{1}{7} & 0 \\ 0 & 1 & 3 & -7 \end{array}\right].$$

Next, add $-\frac{5}{7}$ times the second row to the first row (to make the entry in row 1, column 2 equal to 0), getting

$$\left[\begin{array}{cc|cc} 1 & 0 & -2 & 5 \\ 0 & 1 & 3 & -7 \end{array}\right].$$

The part of the matrix originally occupied by $\left[\begin{smallmatrix} 7 & 5 \\ 3 & 2 \end{smallmatrix}\right]$ has now been transformed into the 2-by-2 identity matrix. The part of the matrix originally occupied by the 2-by-2 identity matrix has been transformed into $\left[\begin{smallmatrix} -2 & 5 \\ 3 & -7 \end{smallmatrix}\right]$. Thus

$$\begin{bmatrix} 7 & 5 \\ 3 & 2 \end{bmatrix}^{-1} = \begin{bmatrix} -2 & 5 \\ 3 & -7 \end{bmatrix}.$$

CHECK To check that we have found the inverse correctly, use the definition of matrix multiplication to verify that

$$\begin{bmatrix} 7 & 5 \\ 3 & 2 \end{bmatrix}\begin{bmatrix} -2 & 5 \\ 3 & -7 \end{bmatrix} = \begin{bmatrix} 1 & 0 \\ 0 & 1 \end{bmatrix}.$$

A careful examination of the next example will show you why the procedure used above led to the inverse of the matrix. Once you understand why this procedure worked for this particular matrix, you should understand why it works for any square matrix that has an inverse.

EXAMPLE 16

Explain why the procedure in the previous example produced the inverse of the matrix $\begin{bmatrix} 7 & 5 \\ 3 & 2 \end{bmatrix}$.

SOLUTION Temporarily ignore the last column in the matrix $\left[\begin{array}{cc|cc} 7 & 5 & 1 & 0 \\ 3 & 2 & 0 & 1 \end{array}\right]$ that was used in the solution of the previous example. Looking only at the first three columns, we have

$$\left[\begin{array}{cc|c} 7 & 5 & 1 \\ 3 & 2 & 0 \end{array}\right].$$

We can think of the matrix above as representing the system of equations

$$7x + 5y = 1$$

$$3x + 2y = 0,$$

which could also be written as the matrix equation

$$\left[\begin{array}{cc|c} 7 & 5 & 1 \\ 3 & 2 & 0 \end{array}\right] \begin{bmatrix} x \\ y \end{bmatrix} = \begin{bmatrix} 1 \\ 0 \end{bmatrix}.$$

Our use of elementary row operations to change $\left[\begin{smallmatrix} 7 & 5 \\ 3 & 2 \end{smallmatrix}\right]$ to the 2-by-2 identity matrix solves this system of equations, producing the solution $x = -2, y = 3$ (corresponding to the third column in the last 2-by-4 matrix in the previous example). Thus

$$\begin{bmatrix} 7 & 5 \\ 3 & 2 \end{bmatrix} \begin{bmatrix} -2 \\ 3 \end{bmatrix} = \begin{bmatrix} 1 \\ 0 \end{bmatrix}.$$

Similarly, if we temporarily ignore the third column in the matrix $\left[\begin{array}{cc|cc} 7 & 5 & 1 & 0 \\ 3 & 2 & 0 & 1 \end{array}\right]$ that was used in the solution of the previous example, we have

$$\left[\begin{array}{cc|c} 7 & 5 & 0 \\ 3 & 2 & 1 \end{array}\right].$$

Think of the matrix above as representing a system of equations. Our use of elementary row operations to change $\left[\begin{smallmatrix} 7 & 5 \\ 3 & 2 \end{smallmatrix}\right]$ to the 2-by-2 identity matrix produces the solution $x = 5, y = -7$ (corresponding to the fourth column in the last 2-by-4 matrix in the previous example). Thus

$$\begin{bmatrix} 7 & 5 \\ 3 & 2 \end{bmatrix} \begin{bmatrix} 5 \\ -7 \end{bmatrix} = \begin{bmatrix} 0 \\ 1 \end{bmatrix}.$$

If you think for a moment about the definition of matrix multiplication, you will see that the equations

$$\begin{bmatrix} 7 & 5 \\ 3 & 2 \end{bmatrix} \begin{bmatrix} -2 \\ 3 \end{bmatrix} = \begin{bmatrix} 1 \\ 0 \end{bmatrix} \quad \text{and} \quad \begin{bmatrix} 7 & 5 \\ 3 & 2 \end{bmatrix} \begin{bmatrix} 5 \\ -7 \end{bmatrix} = \begin{bmatrix} 0 \\ 1 \end{bmatrix}$$

imply that

$$\begin{bmatrix} 7 & 5 \\ 3 & 2 \end{bmatrix} \begin{bmatrix} -2 & 5 \\ 3 & -7 \end{bmatrix} = \begin{bmatrix} 1 & 0 \\ 0 & 1 \end{bmatrix},$$

as desired.

EXERCISES

For Exercises 1–40, assume that

$$A = \begin{bmatrix} 2 & 5 \\ 3 & 4 \end{bmatrix}, \quad B = \begin{bmatrix} 3 & -2 \\ 5 & 4 \end{bmatrix}, \quad C = \begin{bmatrix} 8 & 6 \\ 1 & -7 \end{bmatrix},$$

$$D = \begin{bmatrix} 1 & 2 & 3 \\ 5 & -1 & 6 \\ 7 & -2 & 8 \end{bmatrix}, \quad E = \begin{bmatrix} 1 & -1 & 5 \\ 4 & -2 & 6 \\ 5 & -2 & 5 \end{bmatrix},$$

$$F = \begin{bmatrix} 3 & -2 & 6 \\ 4 & 1 & -8 \\ 5 & 0 & -7 \end{bmatrix}.$$

1. Evaluate $A + B$.
2. Evaluate $A + C$.
3. Evaluate $D + E$.
4. Evaluate $D + F$.
5. Evaluate $A - C$.
6. Evaluate $B - A$.
7. Evaluate $D - F$.
8. Evaluate $E - D$.
9. Evaluate $3A$.
10. Evaluate $4B$.
11. Evaluate $-2D$.
12. Evaluate $-5E$.
13. Find a matrix X such that $A + 2X = C$.
14. Find a matrix X such that $B + 3X = C$.
15. Find a matrix X such that $D - 2X = F$.
16. Find a matrix X such that $E - 3X = F$.
17. Evaluate AB.
18. Evaluate BC.
19. Evaluate BA.
20. Evaluate CB.
21. Evaluate DE.
22. Evaluate EF.
23. Evaluate ED.
24. Evaluate FE.
25. Evaluate ABC.
26. Evaluate CBA.
27. Evaluate DEF.
28. Evaluate FED.
29. Find A^{-1}.
30. Find B^{-1}.
31. Find D^{-1}.
32. Find E^{-1}.
33. Find a matrix X such that $AX = B$.
34. Find a matrix X such that $BX = C$.
35. Find a matrix X such that $XA = B$.
36. Find a matrix X such that $XB = C$.
37. Find a matrix X such that $DX = E$.
38. Find a matrix X such that $EX = F$.
39. Find a matrix X such that $XD = E$.
40. Find a matrix X such that $XE = F$.

PROBLEMS

For Problems 41–44, assume that two separate experimenters work for an ad company testing several different commercials on groups of consumers. The subjects are split into several groups, each of which is given a commercial that is considered either funny or dramatic, and also judged to be either conservative, liberal, or moderate in political tone. Mean ratings (possible values of 1 through 10) for consumers on each of the six possible commercials are given in the matrices below, matrix A for the first experimenter, matrix B for the second. Row 1 contains scores for the funny commercials, row 2 the scores for dramatic. Columns contain scores for, in order, the conservative, liberal, and moderate commercials.

$$A = \begin{bmatrix} 5 & 4.5 & 7 \\ 6 & 4.5 & 8 \end{bmatrix} \qquad B = \begin{bmatrix} 4.5 & 6 & 6 \\ 5 & 4 & 7 \end{bmatrix}$$

41. Interpret the meaning of the entry in row 2, column 1 of matrix B.
42. Compute $\frac{1}{2}(A + B)$.
43. Interpret the meaning of $\frac{1}{2}(A + B)$.
44. Which commercial type was most popular overall?

For Problems 45–49, represent a typical point (x, y) in the coordinate plane by the 1-by-2 matrix $\begin{bmatrix} x & y \end{bmatrix}$. Also, let

$$B = \begin{bmatrix} -1 & 0 \\ 0 & 1 \end{bmatrix} \quad \textit{and} \quad C = \begin{bmatrix} 1 & 0 \\ 0 & -1 \end{bmatrix}.$$

45. Consider the point $(4, 2)$, represented by the matrix $P = \begin{bmatrix} 4 & 2 \end{bmatrix}$.
    (a) Compute the product PB.
    (b) How does the point in the coordinate plane represented by PB relate to the original point $P = \begin{bmatrix} 4 & 2 \end{bmatrix}$?

46. Consider the point (b, c), represented by the matrix $Q = \begin{bmatrix} b & c \end{bmatrix}$.
    (a) Compute the product QB.
    (b) How does the point in the coordinate plane represented by QB relate to the original point $Q = \begin{bmatrix} b & c \end{bmatrix}$?

47. Consider the point $(-3, -5)$, represented by the matrix $P = \begin{bmatrix} -3 & -5 \end{bmatrix}$.
    (a) Compute the product PC.
    (b) How does the point in the coordinate plane represented by PC relate to the original point $P = \begin{bmatrix} -3 & -5 \end{bmatrix}$?

48. Consider the point (b, c), represented by the matrix $Q = \begin{bmatrix} b & c \end{bmatrix}$.
    (a) Compute the product QC.
    (b) How does the point in the coordinate plane represented by QC relate to the original point $Q = \begin{bmatrix} b & c \end{bmatrix}$?

49. Consider the point (b, c), represented by the matrix $Q = \begin{bmatrix} b & c \end{bmatrix}$.
    (a) Compute the product QBC.
    (b) How does the point in the coordinate plane represented by QBC relate to the original point $Q = \begin{bmatrix} b & c \end{bmatrix}$?

50. Give an example of 2-by-2 matrices A and B such that no entry in either matrix is 0 but

$$AB = \begin{bmatrix} 0 & 0 \\ 0 & 0 \end{bmatrix}.$$

51. Suppose A is a 3-by-3 matrix and B and C are 3-by-n matrices such that

$$AB = AC.$$

Show that if A is invertible, then $B = C$.

52. Suppose A is a 3-by-3 matrix and B and C are m-by-3 matrices such that

$$BA = CA.$$

Show that if A is invertible, then $B = C$.

For Problems 53–55, let

$$A = \begin{bmatrix} 2 & 0 & 0 \\ 0 & 7 & 0 \\ 0 & 0 & 5 \end{bmatrix}.$$

53. Suppose B is a 3-by-n matrix. Explain why AB is the matrix obtained from B by multiplying the row 1 of B by 2, multiplying row 2 of B by 7, and multiplying row 3 of B by 5.

54. Suppose B is an m-by-3 matrix. Explain why BA is the matrix obtained from B by multiplying the column 1 of B by 2, multiplying column 2 of B by 7, and multiplying column 3 of B by 5.

55. (a) Find A^{-1}.
    (b) Generalize the result in part (a) to a result about the inverse of any 3-by-3 matrix all of whose entries off the diagonal equal 0.

For Problems 56–58, let

$$A = \begin{bmatrix} 0 & 1 & 0 \\ 1 & 0 & 0 \\ 0 & 0 & 1 \end{bmatrix}.$$

56. Suppose B is a 3-by-n matrix. Explain why AB is the matrix obtained from B by interchanging rows 1 and 2.

57. Suppose B is an m-by-3 matrix. Explain why BA is the matrix obtained from B by interchanging columns 1 and 2.

58. Find A^{-1}.

For Problems 59–61, let

$$A = \begin{bmatrix} 1 & 0 & 0 \\ 4 & 1 & 0 \\ 0 & 0 & 1 \end{bmatrix}.$$

59. Suppose B is a 3-by-n matrix. Explain why AB is the matrix obtained from B by adding 4 times row 1 to row 2.

60. Suppose B is an m-by-3 matrix. Explain why BA is the matrix obtained from B by adding 4 times column 2 to column 1.

61. Find A^{-1}.

WORKED-OUT SOLUTIONS *to Odd-numbered Exercises*

For Exercises 1–40, assume that

$$A = \begin{bmatrix} 2 & 5 \\ 3 & 4 \end{bmatrix}, \quad B = \begin{bmatrix} 3 & -2 \\ 5 & 4 \end{bmatrix}, \quad C = \begin{bmatrix} 8 & 6 \\ 1 & -7 \end{bmatrix},$$

$$D = \begin{bmatrix} 1 & 2 & 3 \\ 5 & -1 & 6 \\ 7 & -2 & 8 \end{bmatrix}, \quad E = \begin{bmatrix} 1 & -1 & 5 \\ 4 & -2 & 6 \\ 5 & -2 & 5 \end{bmatrix},$$

$$F = \begin{bmatrix} 3 & -2 & 6 \\ 4 & 1 & -8 \\ 5 & 0 & -7 \end{bmatrix}.$$

1. Evaluate $A + B$.

SOLUTION

$$A + B = \begin{bmatrix} 2 & 5 \\ 3 & 4 \end{bmatrix} + \begin{bmatrix} 3 & -2 \\ 5 & 4 \end{bmatrix} = \begin{bmatrix} 5 & 3 \\ 8 & 8 \end{bmatrix}$$

3. Evaluate $D + E$.

SOLUTION

$$D + E = \begin{bmatrix} 1 & 2 & 3 \\ 5 & -1 & 6 \\ 7 & -2 & 8 \end{bmatrix} + \begin{bmatrix} 1 & -1 & 5 \\ 4 & -2 & 6 \\ 5 & -2 & 5 \end{bmatrix}$$

$$= \begin{bmatrix} 2 & 1 & 8 \\ 9 & -3 & 12 \\ 12 & -4 & 13 \end{bmatrix}$$

5. Evaluate $A - C$.

SOLUTION

$$A - C = \begin{bmatrix} 2 & 5 \\ 3 & 4 \end{bmatrix} - \begin{bmatrix} 8 & 6 \\ 1 & -7 \end{bmatrix} = \begin{bmatrix} -6 & -1 \\ 2 & 11 \end{bmatrix}$$

7. Evaluate $D - F$.

SOLUTION

$$D - F = \begin{bmatrix} 1 & 2 & 3 \\ 5 & -1 & 6 \\ 7 & -2 & 8 \end{bmatrix} - \begin{bmatrix} 3 & -2 & 6 \\ 4 & 1 & -8 \\ 5 & 0 & -7 \end{bmatrix}$$

$$= \begin{bmatrix} -2 & 4 & -3 \\ 1 & -2 & 14 \\ 2 & -2 & 15 \end{bmatrix}$$

9. Evaluate $3A$.

SOLUTION $3A = 3\begin{bmatrix} 2 & 5 \\ 3 & 4 \end{bmatrix} = \begin{bmatrix} 6 & 15 \\ 9 & 12 \end{bmatrix}$

11. Evaluate $-2D$.

SOLUTION

$$-2D = -2\begin{bmatrix} 1 & 2 & 3 \\ 5 & -1 & 6 \\ 7 & -2 & 8 \end{bmatrix} = \begin{bmatrix} -2 & -4 & -6 \\ -10 & 2 & -12 \\ -14 & 4 & -16 \end{bmatrix}$$

13. Find a matrix X such that $A + 2X = C$.

SOLUTION Subtracting A from both sides of the equation above gives $2X = C - A$. Multiplying both sides by $\frac{1}{2}$ then gives

$$X = \tfrac{1}{2}(C - A) = \tfrac{1}{2}\left(\begin{bmatrix} 8 & 6 \\ 1 & -7 \end{bmatrix} - \begin{bmatrix} 2 & 5 \\ 3 & 4 \end{bmatrix}\right)$$

$$= \tfrac{1}{2}\begin{bmatrix} 6 & 1 \\ -2 & -11 \end{bmatrix}$$

$$= \begin{bmatrix} 3 & \frac{1}{2} \\ -1 & -\frac{11}{2} \end{bmatrix}.$$

15. Find a matrix X such that $D - 2X = F$.

SOLUTION Subtracting D from both sides of the equation above gives $-2X = F - D$. Multiplying both sides by $-\frac{1}{2}$ then gives

$$X = \tfrac{1}{2}(D - F) = \tfrac{1}{2}\left(\begin{bmatrix} 1 & 2 & 3 \\ 5 & -1 & 6 \\ 7 & -2 & 8 \end{bmatrix} - \begin{bmatrix} 3 & -2 & 6 \\ 4 & 1 & -8 \\ 5 & 0 & -7 \end{bmatrix}\right)$$

$$= \tfrac{1}{2}\begin{bmatrix} -2 & 4 & -3 \\ 1 & -2 & 14 \\ 2 & -2 & 15 \end{bmatrix}$$

$$= \begin{bmatrix} -1 & 2 & -\frac{3}{2} \\ \frac{1}{2} & -1 & 7 \\ 1 & -1 & \frac{15}{2} \end{bmatrix}.$$

17. Evaluate AB.

SOLUTION

$$AB = \begin{bmatrix} 2 & 5 \\ 3 & 4 \end{bmatrix}\begin{bmatrix} 3 & -2 \\ 5 & 4 \end{bmatrix}$$

$$= \begin{bmatrix} 2 \cdot 3 + 5 \cdot 5 & 2 \cdot (-2) + 5 \cdot 4 \\ 3 \cdot 3 + 4 \cdot 5 & 3 \cdot (-2) + 4 \cdot 4 \end{bmatrix}$$

$$= \begin{bmatrix} 31 & 16 \\ 29 & 10 \end{bmatrix}$$

19. Evaluate BA.

SOLUTION $BA = \begin{bmatrix} 3 & -2 \\ 5 & 4 \end{bmatrix} \begin{bmatrix} 2 & 5 \\ 3 & 4 \end{bmatrix} = \begin{bmatrix} 0 & 7 \\ 22 & 41 \end{bmatrix}$

21. Evaluate DE.

SOLUTION

$$DE = \begin{bmatrix} 1 & 2 & 3 \\ 5 & -1 & 6 \\ 7 & -2 & 8 \end{bmatrix} \begin{bmatrix} 1 & -1 & 5 \\ 4 & -2 & 6 \\ 5 & -2 & 5 \end{bmatrix}$$

$$= \begin{bmatrix} 24 & -11 & 32 \\ 31 & -15 & 49 \\ 39 & -19 & 63 \end{bmatrix}$$

23. Evaluate ED.

SOLUTION

$$ED = \begin{bmatrix} 1 & -1 & 5 \\ 4 & -2 & 6 \\ 5 & -2 & 5 \end{bmatrix} \begin{bmatrix} 1 & 2 & 3 \\ 5 & -1 & 6 \\ 7 & -2 & 8 \end{bmatrix}$$

$$= \begin{bmatrix} 31 & -7 & 37 \\ 36 & -2 & 48 \\ 30 & 2 & 43 \end{bmatrix}$$

25. Evaluate ABC.

SOLUTION We evaluated AB in Exercise 17, so we will use that result, getting

$$ABC = (AB)C = \begin{bmatrix} 31 & 16 \\ 29 & 10 \end{bmatrix} \begin{bmatrix} 8 & 6 \\ 1 & -7 \end{bmatrix}$$

$$= \begin{bmatrix} 264 & 74 \\ 242 & 104 \end{bmatrix}.$$

27. Evaluate DEF.

SOLUTION We evaluated DE in Exercise 21, so we will use that result, getting

$$DEF = (DE)F = \begin{bmatrix} 24 & -11 & 32 \\ 31 & -15 & 49 \\ 39 & -19 & 63 \end{bmatrix} \begin{bmatrix} 3 & -2 & 6 \\ 4 & 1 & -8 \\ 5 & 0 & -7 \end{bmatrix}$$

$$= \begin{bmatrix} 188 & -59 & 8 \\ 278 & -77 & -37 \\ 356 & -97 & -55 \end{bmatrix}.$$

29. Find A^{-1}.

SOLUTION Create a larger matrix by placing the 2-by-2 identity matrix to the right of A, getting the matrix

$$\left[\begin{array}{cc|cc} 2 & 5 & 1 & 0 \\ 3 & 4 & 0 & 1 \end{array}\right].$$

Multiply the first row by $\frac{1}{2}$ (to make the entry in row 1, column 1 equal to 1), getting

$$\left[\begin{array}{cc|cc} 1 & \frac{5}{2} & \frac{1}{2} & 0 \\ 3 & 4 & 0 & 1 \end{array}\right].$$

Next, add -3 times the first row to the second row, getting the matrix

$$\left[\begin{array}{cc|cc} 1 & \frac{5}{2} & \frac{1}{2} & 0 \\ 0 & -\frac{7}{2} & -\frac{3}{2} & 1 \end{array}\right].$$

Now multiply the second row by $-\frac{2}{7}$, getting

$$\left[\begin{array}{cc|cc} 1 & \frac{5}{2} & \frac{1}{2} & 0 \\ 0 & 1 & \frac{3}{7} & -\frac{2}{7} \end{array}\right].$$

Next, add $-\frac{5}{2}$ times the second row to the first row (to make the entry in row 1, column 2 equal to 0), getting

$$\left[\begin{array}{cc|cc} 1 & 0 & -\frac{4}{7} & \frac{5}{7} \\ 0 & 1 & \frac{3}{7} & -\frac{2}{7} \end{array}\right].$$

The part of the matrix originally occupied by A has now been transformed into the 2-by-2 identity matrix. The part of the matrix originally occupied by the 2-by-2 identity matrix has been transformed into A^{-1}. Thus

$$A^{-1} = \begin{bmatrix} -\frac{4}{7} & \frac{5}{7} \\ \frac{3}{7} & -\frac{2}{7} \end{bmatrix}.$$

31. Find D^{-1}.

SOLUTION Create a larger matrix by placing the 3-by-3 identity matrix to the right of D, getting the matrix

$$\left[\begin{array}{ccc|ccc} 1 & 2 & 3 & 1 & 0 & 0 \\ 5 & -1 & 6 & 0 & 1 & 0 \\ 7 & -2 & 8 & 0 & 0 & 1 \end{array}\right].$$

Add -5 times the first row to the second row and add -7 times the first row to the third row, getting the matrix

$$\left[\begin{array}{ccc|ccc} 1 & 2 & 3 & 1 & 0 & 0 \\ 0 & -11 & -9 & -5 & 1 & 0 \\ 0 & -16 & -13 & -7 & 0 & 1 \end{array}\right].$$

Now multiply the second row by $-\frac{1}{11}$, getting

$$\left[\begin{array}{ccc|ccc} 1 & 2 & 3 & 1 & 0 & 0 \\ 0 & 1 & \frac{9}{11} & \frac{5}{11} & -\frac{1}{11} & 0 \\ 0 & -16 & -13 & -7 & 0 & 1 \end{array}\right].$$

Next, add -2 times the second row to the first row (to make the entry in row 1, column 2 equal to 0) and 16 times the second row to the third row, getting

$$\left[\begin{array}{ccc|ccc} 1 & 0 & \frac{15}{11} & \frac{1}{11} & \frac{2}{11} & 0 \\ 0 & 1 & \frac{9}{11} & \frac{5}{11} & -\frac{1}{11} & 0 \\ 0 & 0 & \frac{1}{11} & \frac{3}{11} & -\frac{16}{11} & 1 \end{array}\right].$$

Next, add -15 times the third row to the first row (to make the entry in row 1, column 3 equal to 0) and -9 times the third row to the second row, getting

$$\left[\begin{array}{ccc|ccc} 1 & 0 & 0 & -4 & 22 & -15 \\ 0 & 1 & 0 & -2 & 13 & -9 \\ 0 & 0 & \frac{1}{11} & \frac{3}{11} & -\frac{16}{11} & 1 \end{array}\right].$$

Finally, multiply the third row by 11, getting

$$\left[\begin{array}{ccc|ccc} 1 & 0 & 0 & -4 & 22 & -15 \\ 0 & 1 & 0 & -2 & 13 & -9 \\ 0 & 0 & 1 & 3 & -16 & 11 \end{array}\right].$$

The part of the matrix originally occupied by D has now been transformed into the 3-by-3 identity matrix. The part of the matrix originally occupied by the 3-by-3 identity matrix has been transformed into D^{-1}. Thus

$$D^{-1} = \begin{bmatrix} -4 & 22 & -15 \\ -2 & 13 & -9 \\ 3 & -16 & 11 \end{bmatrix}.$$

33. Find a matrix X such that $AX = B$.

SOLUTION Multiplying both sides of the equation above by A^{-1} on the left and using the result for A^{-1} from Exercise 29, we have

$$X = A^{-1}B = \begin{bmatrix} -\frac{4}{7} & \frac{5}{7} \\ \frac{3}{7} & -\frac{2}{7} \end{bmatrix} \begin{bmatrix} 3 & -2 \\ 5 & 4 \end{bmatrix}$$

$$= \begin{bmatrix} \frac{13}{7} & 4 \\ -\frac{1}{7} & -2 \end{bmatrix}.$$

35. Find a matrix X such that $XA = B$.

SOLUTION Multiplying both sides of the equation above by A^{-1} on the right and using the result for A^{-1} from Exercise 29, we have

$$X = BA^{-1} = \begin{bmatrix} 3 & -2 \\ 5 & 4 \end{bmatrix} \begin{bmatrix} -\frac{4}{7} & \frac{5}{7} \\ \frac{3}{7} & -\frac{2}{7} \end{bmatrix}$$

$$= \begin{bmatrix} -\frac{18}{7} & \frac{19}{7} \\ -\frac{8}{7} & \frac{17}{7} \end{bmatrix}.$$

37. Find a matrix X such that $DX = E$.

SOLUTION Multiplying both sides of the equation above by D^{-1} on the left and using the result for D^{-1} from Exercise 31, we have

$$X = D^{-1}E = \begin{bmatrix} -4 & 22 & -15 \\ -2 & 13 & -9 \\ 3 & -16 & 11 \end{bmatrix} \begin{bmatrix} 1 & -1 & 5 \\ 4 & -2 & 6 \\ 5 & -2 & 5 \end{bmatrix}$$

$$= \begin{bmatrix} 9 & -10 & 37 \\ 5 & -6 & 23 \\ -6 & 7 & -26 \end{bmatrix}.$$

39. Find a matrix X such that $XD = E$.

SOLUTION Multiplying both sides of the equation above by D^{-1} on the right and using the result for D^{-1} from Exercise 31, we have

$$X = ED^{-1} = \begin{bmatrix} 1 & -1 & 5 \\ 4 & -2 & 6 \\ 5 & -2 & 5 \end{bmatrix} \begin{bmatrix} -4 & 22 & -15 \\ -2 & 13 & -9 \\ 3 & -16 & 11 \end{bmatrix}$$

$$= \begin{bmatrix} 13 & -71 & 49 \\ 6 & -34 & 24 \\ -1 & 4 & -2 \end{bmatrix}.$$

CHAPTER SUMMARY

To check that you have mastered the most important concepts and skills covered in this chapter, make sure that you can do each item in the following list:

- Graphically solve a system of equations with two variables.
- Solve a system of equations by substitution.
- Solve a system of linear equations using Gaussian elimination and back substitution.
- Use elementary row operations to solve a system of linear equations.
- Add and subtract matrices of the same size.
- Compute the product of two matrices of appropriate sizes.
- Find the inverse of an invertible matrix.

To review a chapter, go through the list above to find items that you do not know how to do, then reread the material in the chapter about those items. Then try to answer the chapter review questions below without looking back at the chapter.

CHAPTER REVIEW QUESTIONS

1. Find all points where the line $2x + 3y = 1$ intersects the circle in the xy-plane centered at $(2, 1)$ with radius 3.

2. Find all solutions to the following system of equations:

$$\begin{aligned} 4x - 5y &= 3 \\ -7x + 6y &= 2. \end{aligned}$$

3. Find a number b such that the following system of equations has no solutions:

$$\begin{aligned} 4x - 5y &= 3 \\ -7x + by &= 2. \end{aligned}$$

4. Find numbers b and c such that the following system of equations has infinitely many solutions:

$$\begin{aligned} 4x - 5y &= 3 \\ -7x + by &= c. \end{aligned}$$

5. Find all solutions to the following system of equations:

$$\begin{aligned} x + 2y + 3z &= 4 \\ 4x + 5y + 6z &= 7 \\ 7x + 8y - 9z &= 1. \end{aligned}$$

6. Give an example of two different 3-by-3 matrices A and B such that $AB = BA$.

7. Give an example of two 3-by-3 matrices A and B such that $AB \neq BA$.

8. Evaluate

$$\begin{bmatrix} 3 & 2 & 1 \\ 1 & 1 & 1 \\ -2 & -3 & 5 \end{bmatrix} \begin{bmatrix} 4 & 0 & 1 \\ 2 & 9 & -3 \\ 5 & -6 & 4 \end{bmatrix}.$$

9. Find the inverse of the matrix $\begin{bmatrix} 1 & 3 & 5 \\ 2 & 4 & 6 \\ 0 & -1 & -3 \end{bmatrix}$.

10. Find a matrix X such that

$$\begin{bmatrix} 1 & 3 & 5 \\ 2 & 4 & 6 \\ 0 & -1 & -3 \end{bmatrix} X = \begin{bmatrix} 2 \\ 0 \\ -4 \end{bmatrix}.$$

11. Find a matrix Y such that

$$\begin{bmatrix} 1 & 3 & 5 \\ 2 & 4 & 6 \\ 0 & -1 & -3 \end{bmatrix} Y = \begin{bmatrix} 2 & 0 \\ 0 & 1 \\ -4 & 0 \end{bmatrix}.$$

12. Suppose A is an invertible matrix. Explain why

$$(A^{-1})^{-1} = A.$$

CHAPTER

8

Isaac Newton, as painted by Godfrey Kneller in 1689, two years after the publication of Newton's monumental book Principia. *This book uses series and limits to help explain the mathematics of planetary motion.*

Sequences, Series, and Limits

This chapter begins by considering sequences, which are lists of numbers. We particularly focus on the following special sequences:

- arithmetic sequences—consecutive terms have a constant difference;
- geometric sequences—consecutive terms have a constant ratio;
- recursively defined sequences—each term is defined by previous terms.

Then we consider series, which are sums of numbers. Here you will learn about summation notation, which is used in many parts of mathematics and statistics. We will derive formulas to evaluate arithmetic series and geometric series. We will also discuss the Binomial Theorem.

Finally, the chapter and the book conclude with an introduction to limits, one of the central ideas of calculus.

8.1 *Sequences*

LEARNING OBJECTIVES

By the end of this section you should be able to

- use sequence notation;
- compute the terms of an arithmetic sequence;
- compute the terms of a geometric sequence;
- compute the terms of a recursively defined sequence.

Introduction to Sequences

Sequences

A **sequence** is an ordered list of numbers.

For example, $7, \sqrt{3}, \frac{5}{2}$ is a sequence. The first term of this sequence is 7, the second term of this sequence is $\sqrt{3}$, and the third term of this sequence is $\frac{5}{2}$.

Sequences differ from sets in that order matters and repetitions are allowed in a sequence. For example, the sets $\{2, 3, 5\}$ and $\{5, 3, 2\}$ are the same, but the sequences $2, 3, 5$ and $5, 3, 2$ are not the same. As another example, the sets $\{8, 8, 4, 5\}$ and $\{8, 4, 5\}$ are the same, but the sequences $8, 8, 4, 5$ and $8, 4, 5$ are not the same.

A sequence might end, as is the case with all the sequences mentioned in the paragraphs above, or a sequence might continue indefinitely. A sequence that ends is called a **finite sequence**; a sequence that does not end is called an **infinite sequence**.

Recall that there is no real number named "infinity". The term "infinite sequence" can be regarded as abbreviation for the phrase "sequence that does not end".

An example of an infinite sequence is the sequence whose n^{th} term is $3n$. The first term of this sequence is 3, the second term of this sequence is 6, the third term of this sequence is 9, and so on. Because this sequence does not end, the entire sequence cannot be written down. Thus we write this sequence as

$$3, \ 6, \ 9, \ \ldots,$$

where the three dots indicate that the sequence continues without end.

When using the three-dot notation to designate a sequence, information should be given about how each term of the sequence is determined. Sometimes this is done by giving an explicit formula for the nth term of the sequence, as in the next example.

EXAMPLE 1

Each of the equations below gives a formula for the nth term of a sequence $a_1, a_2, \ldots$. Write each sequence below using the three-dot notation, giving the first four terms of the sequence. Furthermore, describe each sequence in words.

(a) $a_n = n$
(b) $a_n = 2n$
(c) $a_n = 2n - 1$
(d) $a_n = 3$
(e) $a_n = (-1)^n$
(f) $a_n = 2^{n-1}$

SOLUTION

(a) The sequence $a_1, a_2, \ldots$ defined by $a_n = n$ is $1, 2, 3, 4, \ldots$; this is the sequence of positive integers.

(b) The sequence $a_1, a_2, \ldots$ defined by $a_n = 2n$ is $2, 4, 6, 8 \ldots$; this is the sequence of even positive integers.

(c) The sequence $a_1, a_2, \ldots$ defined by $a_n = 2n - 1$ is $1, 3, 5, 7 \ldots$; this is the sequence of odd positive integers.

(d) The sequence $a_1, a_2, \ldots$ defined by $a_n = 3$ is $3, 3, 3, 3 \ldots$; this is the sequence of all 3's.

(e) The sequence $a_1, a_2, \ldots$ defined by $a_n = (-1)^n$ is $-1, 1, -1, 1 \ldots$; this is the sequence of alternating -1's and 1's, beginning with -1.

(f) The sequence $a_1, a_2, \ldots$ defined by $a_n = 2^{n-1}$ is $1, 2, 4, 8 \ldots$; this is the sequence of powers of 2, starting with 2 to the zeroth power.

Caution must be used when determining a sequence simply from the pattern of some of the terms, as illustrated by the following example.

EXAMPLE 2

What is the fifth term of the sequence $1, 4, 9, 16, \ldots$?

SOLUTION This is a trick question. You may reasonably suspect that the nth term of this sequence is n^2, which would imply that the fifth term equals 25.

However, the sequence whose n^{th} term equals

$$\frac{n^4 - 10n^3 + 39n^2 - 50n + 24}{4}$$

Problem 60 explains how this expression was obtained.

has as its first four terms $1, 4, 9, 16$, as you can verify. The fifth term of this sequence is 31, not 25.

Because we do not know whether the formula for the nth term of the sequence $1, 4, 9, 16, \ldots$ is given by n^2 or by the formula above or by some other formula, we cannot determine whether the fifth term of this sequence equals 25 or 31 or some other number.

One way out of the dilemma posed by the example above is to assume that the sequence is defined by the simplest possible means. Unless other information is given, you may need to make this assumption. This then raises another problem, because "simplest" is an imprecise notion and can be a matter of taste. However, in most cases almost everyone will agree on which among the many possibilities is the simplest. In Example 2 above, all reasonable people would agree that the expression n^2 is simpler than the expression $\frac{n^4-10n^3+39n^2-50n+24}{4}$.

Arithmetic Sequences

The sequence $1, 3, 5, 7 \ldots$ of odd positive integers has the property that the difference between any two consecutive terms is 2. Thus the difference between two consecutive terms is constant throughout the sequence. Sequences with this property are important enough to deserve their own name:

Arithmetic sequences

An **arithmetic sequence** is a sequence such that the difference between two consecutive terms is constant throughout the sequence.

When we consider the difference between consecutive terms in a sequence $a_1, a_2, \ldots$, we will subtract each term from its successor. In other words, we consider the difference $a_{n+1} - a_n$.

EXAMPLE 3

For each of the following sequences, determine whether or not the sequence is an arithmetic sequence. If the sequence is an arithmetic sequence, determine the difference between consecutive terms in the sequence.

An arithmetic sequence can be either an infinite sequence or a finite sequence. All the sequences in this example are infinite sequences except the last one.

(a) The sequence $1, 2, 3, 4, \ldots$ of positive integers.

(b) The sequence $-1, -2, -3, -4, \ldots$ of negative integers.

(c) The sequence $6, 8, 10, 12, \ldots$ of even positive integers starting with 6.

(d) The sequence $-1, 1, -1, 1, \ldots$ of alternating -1's and 1's.

(e) The sequence $1, 2, 4, 8, \ldots$ of powers of 2.

(f) The sequence $10, 15, 20, 25$.

SOLUTION

(a) The sequence $1, 2, 3, 4, \ldots$ of positive integers is an arithmetic sequence. The difference between any two consecutive terms is 1.

(b) The sequence $-1, -2, -3, -4, \ldots$ of negative integers is an arithmetic sequence. The difference between any two consecutive terms is -1.

(c) The sequence $6, 8, 10, 12, \ldots$ of even positive integers starting with 6 is an arithmetic sequence. The difference between any two consecutive terms is 2.

(d) The difference between consecutive terms of the sequence $-1, 1, -1, 1, \ldots$ oscillates between 2 and -2. Because the difference between consecutive terms of the sequence $-1, 1, -1, 1, \ldots$ is not constant, this sequence is not an arithmetic sequence.

(e) In the sequence $1, 2, 4, 8, \ldots$, the first two terms differ by 1, but the second and third terms differ by 2. Because the difference between consecutive terms of the sequence $1, 2, 4, 8, \ldots$ is not constant, this sequence is not an arithmetic sequence.

(f) In the finite sequence $10, 15, 20, 25$, the difference between any two consecutive terms is 5. Thus this sequence is an arithmetic sequence.

Consider an arithmetic sequence with first term b and difference d between consecutive terms. Each term of this sequence after the first term is obtained by adding d to the previous term. Thus this sequence is

$$b,\ b+d,\ b+2d,\ b+3d,\ \ldots.$$

The nth term of this sequence is obtained by adding d a total of $n-1$ times to the first term b. Thus we have the following result:

When used in the phrase "arithmetic sequence", the word "arithmetic" is pronounced differently from the word used to describe the subject you started learning in elementary school. You should be able to hear the difference when your instructor pronounces "arithmetic sequence".

Formula for an arithmetic sequence

The nth term of an arithmetic sequence with first term b and with difference d between consecutive terms is $b+(n-1)d$.

EXAMPLE 4

Suppose at the beginning of the year your iPod contains 53 songs and that you purchase four new songs each week to place on your iPod. Consider the sequence whose n^{th} term is the number of songs on your iPod at the beginning of the nth week of the year.

(a) What are the first four terms of this sequence?

(b) What is the 30th term of this sequence? In other words, how many songs will be on your iPod at the beginning of the 30th week?

SOLUTION

(a) The first four terms of this sequence are $53, 57, 61, 65$.

(b) To find the 30th term of this sequence, use the formula in the box above with $b = 53$, $n = 30$, and $d = 4$. Thus at the beginning of the 30th week the number of songs on the iPod will be $53 + (30 - 1) \cdot 4$, which equals 169.

Geometric Sequences

The sequence $1, 3, 9, 27\ldots$ of powers of 3 has the property that the ratio of any two consecutive terms is 3. Thus the ratio of two consecutive terms is constant throughout the sequence. Sequences with this property are important enough to deserve their own name:

Geometric sequences

A **geometric sequence** is a sequence such that the ratio of two consecutive terms is constant throughout the sequence.

When we consider the ratio of consecutive terms in a sequence $a_1, a_2, \ldots$, we will divide each term into its successor. In other words, we consider the ratio a_{n+1}/a_n.

EXAMPLE 5

A geometric sequence can be either an infinite sequence or a finite sequence. All the sequences in this example are infinite sequences except the last one.

For each of the following sequences, determine whether or not the sequence is a geometric sequence. If the sequence is a geometric sequence, determine the ratio of consecutive terms in the sequence.

(a) The sequence $16, 32, 64, 128, \ldots$ of the powers of 2 starting with 2^4.

(b) The sequence $3, 6, 12, 24, \ldots$ of 3 times the powers of 2 starting with $3 \cdot 2^0$.

(c) The sequence $-1, 1, -1, 1, \ldots$ of alternating -1's and 1's.

(d) The sequence $1, 4, 9, 16, \ldots$ of the squares of the positive integers.

(e) The sequence $2, 4, 6, 8, \ldots$ of even positive integers.

(f) The sequence $2, \frac{2}{3}, \frac{2}{9}, \frac{2}{27}$.

SOLUTION

(a) The sequence $16, 32, 64, 128, \ldots$ of the powers of 2 starting with 2^4 is a geometric sequence. The ratio of any two consecutive terms is 2.

(b) The sequence $3, 6, 12, 24, \ldots$ of 3 times the powers of 2 starting with $3 \cdot 2^0$ is a geometric sequence. The ratio of any two consecutive terms is 2.

(c) The sequence $-1, 1, -1, 1, \ldots$ of alternating -1's and 1's is a geometric sequence. The ratio of any two consecutive terms is -1.

(d) In the sequence $1, 4, 9, 16, \ldots$, the second and first terms have a ratio of 4, but the third and second terms have a ratio of $\frac{9}{4}$. Because the ratio of consecutive terms of the sequence $1, 4, 9, 16, \ldots$ is not constant, this sequence is not a geometric sequence.

(e) In the sequence $2, 4, 6, 8, \ldots$, the second and first terms have a ratio of 2, but the third and second terms have a ratio of $\frac{3}{2}$. Because the ratio of consecutive terms of the sequence $2, 4, 6, 8, \ldots$ is not constant, this sequence is not a geometric sequence.

(f) In the finite sequence $2, \frac{2}{3}, \frac{2}{9}, \frac{2}{27}$, the ratio of any two consecutive terms is $\frac{1}{3}$. Thus this sequence is a geometric sequence.

Consider a geometric sequence with first term b and ratio r of consecutive terms. Each term of this sequence after the first term is obtained by multiplying the previous term by r. Thus this sequence is

$$b,\ br,\ br^2,\ br^3,\ \ldots.$$

The nth term of this sequence is obtained by multiplying the first term b by r a total of $n - 1$ times. Thus we have the following result:

Formula for a geometric sequence

The nth term of a geometric sequence with first term b and with ratio r of consecutive terms is br^{n-1}.

EXAMPLE 6

Suppose at the beginning of the year \$1000 is deposited in a bank account that pays 5% interest per year, compounded once per year at the end of the year. Consider the sequence whose n^{th} term is the amount in the bank account at the beginning of the nth year.

(a) What are the first four terms of this sequence?

(b) What is the 20th term of this sequence? In other words, how much will be in the bank account at the beginning of the 20th year?

SOLUTION

(a) Each term of this sequence is obtained by multiplying the previous term by 1.05. Thus we have a geometric sequence whose first four terms are

$$\$1000, \quad \$1000 \cdot 1.05, \quad \$1000 \cdot (1.05)^2, \quad \$1000 \cdot (1.05)^3.$$

These four terms can be rewritten as \$1000, \$1050, \$1102.50, \$1157.63.

As this example shows, compound interest leads to geometric sequences.

(b) To find the 20th term of this sequence, use the formula in the box above with $b = \$1000$, $r = 1.05$, and $n = 20$. Thus at the beginning of the 20th year the amount of money in the bank account will be $\$1000 \cdot (1.05)^{19}$, which equals \$2526.95.

The next example shows how to deal with a geometric sequence when we have information about terms that are not consecutive.

EXAMPLE 7

Find the tenth term of a geometric sequence whose second term is 7 and whose fifth term is 35.

SOLUTION Let r denote the ratio of consecutive terms of this geometric sequence. To get from the second term of this sequence to the fifth term, we must multiply by r three times. Thus

$$7r^3 = 35.$$

Solving the equation above for r, we have $r = 5^{1/3}$.

To get from the fifth term of this sequence to the tenth term, we must multiply by r five times. Thus the tenth term of this sequence is $35r^5$. Now

$$\begin{aligned} 35r^5 &= 35(5^{1/3})^5 \\ &= 35 \cdot 5^{5/3} \\ &= 35 \cdot 5 \cdot 5^{2/3} \\ &= 105 \cdot 5^{2/3}. \end{aligned}$$

Thus the tenth term of this sequence is $105 \cdot 5^{2/3}$, which is approximately 307.022.

Recursively Defined Sequences

Sometimes the nth term of a sequence is defined by a formula involving n. For example, we might have the sequence $a_1, a_2, \ldots$ whose n^{th} term is defined by $a_n = 4 + 3n$. This is the arithmetic sequence

$$7,\ 10,\ 13,\ 16,\ 19,\ 22,\ \ldots$$

whose first term is 7, with a difference of 3 between consecutive terms.

Suppose we want to compute the seventh term of the sequence above, which has six terms displayed. To compute the seventh term we could use the formula $a_n = 4 + 3n$ to evaluate $a_7 = 4 + 3 \cdot 7$, or we could use the simpler method of adding 3 to the sixth term. Using this second viewpoint, we think of the sequence above as being defined by starting with 7 and then getting each later term by adding 3 to the previous term. In other words, we could think of this sequence as being defined by the equations

$$a_1 = 7 \quad \text{and} \quad a_{n+1} = a_n + 3 \text{ for } n \geq 1.$$

This viewpoint is sufficiently useful to deserve a name:

Sometimes this terminology is shortened to recursive sequence. *Actually, it is not the sequence that is recursive but its definition.*

Recursively defined sequences

A **recursively defined sequence** is a sequence in which each term from some point on is defined by using previous terms.

In the definition above, the phrase "from some point on" means that some terms at the beginning of the sequence will be defined explicitly rather than by using previous terms. In a recursively defined sequence, at least the first term must be defined explicitly because it has no previous terms.

EXAMPLE 8

Write the geometric sequence $6, 12, 24, 48 \ldots$ whose n^{th} term is defined by

$$a_n = 3 \cdot 2^n$$

as a recursively defined sequence.

SOLUTION Each term of this sequence is obtained by multiplying the previous term by 2. Thus the recursive definition of this sequence is given by the equations

$$a_1 = 6 \quad \text{and} \quad a_{n+1} = 2a_n \text{ for } n \geq 1.$$

If n is a positive integer, then $n!$ (pronounced "n factorial") is defined to be the product of the integers from 1 to n. Thus $1! = 1$, $2! = 2$, $3! = 6$, $4! = 24$, and so on.

EXAMPLE 9

Write the factorial sequence $1!, 2!, 3!, 4! \ldots$, whose n^{th} term is defined by $a_n = n!$, as a recursively defined sequence.

SOLUTION Note that $(n+1)!$ is the product of the integers from 1 to $n+1$. Thus $(n+1)!$ equals $n!$ times $n+1$. Hence the recursive definition of this sequence is given by the equations

$$a_1 = 1 \quad \text{and} \quad a_{n+1} = (n+1)a_n \text{ for } n \geq 1.$$

Perhaps the most famous recursively defined sequence is the Fibonacci sequence, which was defined by the Italian mathematician Leonardo Fibonacci over eight hundred years ago. Each term of the Fibonacci sequence is the sum of the two previous terms (except the first two terms, which are defined to equal 1). Thus the Fibonacci sequence has the recursive definition

$$a_1 = 1, \quad a_2 = 1, \quad \text{and} \quad a_{n+2} = a_n + a_{n+1} \text{ for } n \geq 1.$$

You may want to do a web search to learn about some of the ways in which the Fibonacci sequence arises in nature.

EXAMPLE 10

Find the first ten terms of the Fibonacci sequence.

SOLUTION The first two terms of the Fibonacci sequence are 1, 1.

The third term of the Fibonacci sequence is the sum of the first two terms; thus the third term is 2. The fourth term of the Fibonacci sequence is the sum of the second and third terms; thus the fourth term is 3. Continuing in this fashion, we get the first ten terms of the Fibonacci sequence:

$$1,\ 1,\ 2,\ 3,\ 5,\ 8,\ 13,\ 21,\ 34,\ 55$$

Leonardo Fibonacci, whose book written in 1202 introduced Europe to the Indian-Arabic decimal number system that we use today.

Recursive formulas provide a method for estimating square roots with remarkable accuracy. To estimate $\sqrt{c}$, the idea is to define a sequence recursively by letting a_1 be any crude estimate for $\sqrt{c}$; then use the recursive formula $a_{n+1} = \frac{1}{2}(\frac{c}{a_n} + a_n)$. The number a_n will be a good estimate for $\sqrt{c}$ even for small values of n; for larger values of n the estimate becomes extraordinarily accurate.

The example below illustrates this procedure to estimate $\sqrt{5}$. Note that we start with $a_1 = 2$, which means that we are using the crude estimate $\sqrt{5} \approx 2$.

EXAMPLE 11

Define a sequence recursively using the equations

$$a_1 = 2 \quad \text{and} \quad a_{n+1} = \frac{1}{2}\Big(\frac{5}{a_n} + a_n\Big) \text{ for } n \geq 1.$$

(a) Compute a_4. For how many digits after the decimal point does a_4 agree with $\sqrt{5}$?

(b) Compute a_7. For how many digits after the decimal point does a_7 agree with $\sqrt{5}$?

This recursive formula for computing square roots is a special case of a technique called Newton's method.

SOLUTION

(a) Using the recursive formula above and doing some simple arithmetic, we get $a_2 = \frac{9}{4}$, then $a_3 = \frac{161}{72}$, then $a_4 = \frac{51841}{23184}$.

Using a calculator, we see that

$$a_4 = \frac{51841}{23184} \approx 2.2360679779 \quad \text{and} \quad \sqrt{5} \approx 2.2360679774.$$

Thus a_4, which is computed with only a small amount of calculation, agrees with $\sqrt{5}$ for nine digits after the decimal point.

(b) A typical calculator cannot handle enough digits to compute a_7 exactly. However, a computer algebra system such as *Mathematica* or *Maple* can be used to compute that

$$a_5 = \frac{5374978561}{2403763488}, \quad a_6 = \frac{57780789062419261441}{25840354427429161536},$$

and

$$a_7 = \frac{6677239169351578707225356193679818792961}{2986152136938872067784669198846010266752}.$$

The number of accurate digits produced by this method roughly doubles with each recursion.

Even though computing a_7 requires only three more calculations after computing a_4, the value for a_7 calculated above agrees with $\sqrt{5}$ for 79 digits after the decimal point. This remarkable level of accuracy is typical of this recursive method for computing square roots.

EXERCISES

For Exercises 1–8, a formula is given for the nth term of a sequence $a_1, a_2, \ldots$.

(a) ***Write the sequence using the three-dot notation, giving the first four terms.***

(b) ***Give the 100th term of the sequence.***

1. $a_n = -n$
2. $a_n = \frac{1}{n}$
3. $a_n = 2 + 5n$
4. $a_n = 4n - 3$
5. $a_n = \sqrt{\frac{n}{n+1}}$
6. $a_n = \sqrt{\frac{2n-1}{3n-2}}$
7. $a_n = 3 + 2^n$
8. $a_n = 1 - \frac{1}{3^n}$

For Exercises 9–14, consider an arithmetic sequence with first term b and difference d between consecutive terms.

(a) ***Write the sequence using the three-dot notation, giving the first four terms of the sequence.***

(b) ***Give the 100th term of the sequence.***

9. $b = 2$, $d = 5$
10. $b = 7$, $d = 3$
11. $b = 4$, $d = -6$
12. $b = 8$, $d = -5$
13. $b = 0$, $d = \frac{1}{3}$
14. $b = -1$, $d = \frac{3}{2}$

For Exercises 15–18, suppose that at the beginning of the first day of a new year you have 3324 e-mail messages saved on your computer. At the end of each day you save only your 12 most important new e-mail messages along with the previously saved messages. Consider the sequence whose n^{th} term is the number of e-mail messages you have saved on your computer at the beginning of the nth day of the year.

15. What are the first, second, and third terms of this sequence?
16. What are the fourth, fifth, and sixth terms of this sequence?
17. What is the 100th term of this sequence? In other words, how many e-mail messages will you have saved on your computer at the beginning of the 100th day of the year?

18. What is the 250th term of this sequence? In other words, how many e-mail messages will you have saved on your computer at the beginning of the 250th day of the year?

For Exercises 19–24, consider a geometric sequence with first term b and ratio r of consecutive terms.

(a) ***Write the sequence using the three-dot notation, giving the first four terms.***

(b) ***Give the 100th term of the sequence.***

19. $b = 1, r = 5$
20. $b = 1, r = 4$
21. $b = 3, r = -2$
22. $b = 4, r = -5$
23. $b = 2, r = \frac{1}{3}$
24. $b = 5, r = \frac{2}{3}$
25. Find the fifth term of an arithmetic sequence whose second term is 8 and whose third term is 14.
26. Find the eighth term of an arithmetic sequence whose fourth term is 7 and whose fifth term is 4.
27. Find the first term of an arithmetic sequence whose second term is 19 and whose fourth term is 25.
28. Find the first term of an arithmetic sequence whose second term is 7 and whose fifth term is 11.
29. Find the 100th term of an arithmetic sequence whose tenth term is 5 and whose eleventh term is 8.
30. Find the 200th term of an arithmetic sequence whose fifth term is 23 and whose sixth term is 25.
31. Find the fifth term of a geometric sequence whose second term is 8 and whose third term is 14.
32. Find the eighth term of a geometric sequence whose fourth term is 7 and whose fifth term is 4.
33. Find the first term of a geometric sequence whose second term is 8 and whose fifth term is 27.
34. Find the first term of a geometric sequence whose second term is 64 and whose fifth term is 1.
35. Find the ninth term of a geometric sequence whose fourth term is 4 and whose seventh term is 5.
36. Find the tenth term of a geometric sequence whose second term is 3 and whose seventh term is 11.
37. Find the 100th term of a geometric sequence whose tenth term is 5 and whose eleventh term is 8.
38. Find the 400th term of a geometric sequence whose fifth term is 25 and whose sixth term is 27.

For Exercises 39–42, suppose that your annual salary at the beginning of your first year at a new company is \$38,000. Assume that your salary increases by 7% per year at the end of each year of employment. Consider the sequence whose n^{th} term is your salary at the beginning of your n^{th} year at this company.

39. What are the first, second, and third terms of this sequence?
40. What are the fourth, fifth, and sixth terms of this sequence?
41. What is the 10th term of this sequence? In other words, what will your salary be at the beginning of your 10th year at this company?
42. What is the 15th term of this sequence? In other words, what will your salary be at the beginning of your 15th year at this company?

For Exercises 43–46, give the first four terms of the specified recursively defined sequence.

43. $a_1 = 3$ and $a_{n+1} = 2a_n + 1$ for $n \geq 1$.
44. $a_1 = 2$ and $a_{n+1} = 3a_n - 5$ for $n \geq 1$.
45. $a_1 = 2$, $a_2 = 3$, and $a_{n+2} = a_n a_{n+1}$ for $n \geq 1$.
46. $a_1 = 4$, $a_2 = 7$, and $a_{n+2} = a_{n+1} - a_n$ for $n \geq 1$.

For Exercises 47–48, let $a_1, a_2, \ldots$ be the sequence defined by setting a_1 equal to the value shown below and for $n \geq 1$ letting

$$a_{n+1} = \begin{cases} \dfrac{a_n}{2} & \textit{if } a_n \textit{ is even;} \\ 3a_n + 1 & \textit{if } a_n \textit{ is odd.} \end{cases}$$

47. Suppose $a_1 = 3$. Find the smallest value of n such that $a_n = 1$.

[No one knows whether a_1 can be chosen to be a positive integer such that this recursively defined sequence does not contain any term equal to 1. You can become famous by finding such a choice for a_1. If you want to find out more about this problem, do a web search for "Collatz Problem".]

48. Suppose $a_1 = 7$. Find the smallest value of n such that $a_n = 1$.

For Exercises 49–54, consider the sequence whose n^{th} term a_n is given by the indicated formula.

(a) *Write the sequence using the three-dot notation, giving the first four terms of the sequence.*

(b) *Give a recursive definition of the specified sequence.*

49. $a_n = 5n - 3$
50. $a_n = 1 - 6n$
51. $a_n = 3(-2)^n$
52. $a_n = 5 \cdot 3^{-n}$
53. $a_n = 2^n n!$
54. $a_n = \frac{3^n}{n!}$
55. Define a sequence recursively by

$$a_1 = 3 \quad \text{and} \quad a_{n+1} = \frac{1}{2}\Big(\frac{7}{a_n} + a_n\Big) \text{ for } n \geq 1.$$

Find the smallest value of n such that a_n agrees with $\sqrt{7}$ for at least six digits after the decimal point.

56. Define a sequence recursively by

$$a_1 = 6 \quad \text{and} \quad a_{n+1} = \frac{1}{2}\Big(\frac{17}{a_n} + a_n\Big) \text{ for } n \geq 1.$$

Find the smallest value of n such that a_n agrees with $\sqrt{17}$ for at least four digits after the decimal point.

PROBLEMS

Some problems require considerably more thought than the exercises. Unlike exercises, problems often have more than one correct answer.

57. Explain why an infinite sequence is sometimes defined to be a function whose domain is the set of positive integers.

58. Find a sequence

$$3, -7, 18, 93, \ldots$$

whose 100th term equals 29.
[*Hint:* A correct solution to this problem can be obtained with no calculation.]

59. Find all infinite sequences that are both arithmetic and geometric sequences.

60. For Example 2, the author wanted to find a polynomial p such that

$$p(1) = 1,\ p(2) = 4,\ p(3) = 9,\ p(4) = 16,$$

and $p(5) = 31$. Carry out the following steps to see how that polynomial was obtained.

(a) Note that the polynomial

$$(x-2)(x-3)(x-4)(x-5)$$

is 0 for $x = 2, 3, 4, 5$ but is not zero for $x = 1$. By dividing the polynomial above by a suitable number, find a polynomial p_1 such that $p_1(1) = 1$ and

$$p_1(2) = p_1(3) = p_1(4) = p_1(5) = 0.$$

(b) Similarly, find a polynomial p_2 of degree 4 such that $p_2(2) = 1$ and

$$p_2(1) = p_2(3) = p_2(4) = p_2(5) = 0.$$

(c) Similarly, find polynomials p_j, for $j = 3, 4, 5$, such that each p_j satisfies $p_j(j) = 1$ and $p_j(k) = 0$ for values of k in $\{1, 2, 3, 4, 5\}$ other than j.

(d) Explain why the polynomial p defined by

$$p = p_1 + 4p_2 + 9p_3 + 16p_4 + 31p_5$$

satisfies

$$p(1) = 1,\ p(2) = 4,\ p(3) = 9,\ p(4) = 16,$$

and $p(5) = 31$.

61. Explain why the polynomial p defined by

$$p(x) = \frac{x^4 - 10x^3 + 39x^2 - 50x + 24}{4}$$

is the only polynomial of degree 4 such that $p(1) = 1$, $p(2) = 4$, $p(3) = 9$, $p(4) = 16$, and $p(5) = 31$.

WORKED-OUT SOLUTIONS *to Odd-numbered Exercises*

Do not read these worked-out solutions before first attempting to do the exercises yourself. Otherwise you may merely mimic the techniques shown here without understanding the ideas.

Best way to learn: Carefully read the section of the textbook, then do all the odd-numbered exercises (even if they have not been assigned) and check your answers here. If you get stuck on an exercise, reread the section of the textbook—then try the exercise again. If you are still stuck, then look at the worked-out solution here.

For Exercises 1–8, a formula is given for the nth term of a sequence $a_1, a_2, \dots$.

(a) ***Write the sequence using the three-dot notation, giving the first four terms.***

(b) ***Give the 100th term of the sequence.***

1. $a_n = -n$

SOLUTION

(a) The sequence $a_1, a_2, \dots$ defined by $a_n = -n$ is $-1, -2, -3, -4, \dots$.

(b) The 100th term of this sequence is -100.

3. $a_n = 2 + 5n$

SOLUTION

(a) The sequence $a_1, a_2, \dots$ defined by $a_n = 2 + 5n$ is $7, 12, 17, 22, \dots$.

(b) The 100th term of this sequence is $2 + 5 \cdot 100$, which equals 502.

5. $a_n = \sqrt{\frac{n}{n+1}}$

SOLUTION

(a) The sequence $a_1, a_2, \dots$ defined by $a_n = \sqrt{\frac{n}{n+1}}$ is $\sqrt{\frac{1}{2}}, \sqrt{\frac{2}{3}}, \sqrt{\frac{3}{4}}, \sqrt{\frac{4}{5}}, \dots$. Note that $\sqrt{\frac{3}{4}}$ has not been simplified to $\frac{\sqrt{3}}{2}$; similarly, $\sqrt{\frac{4}{5}}$ has not been simplified to $\frac{2}{\sqrt{5}}$. Making those simplifications would make it harder to discern the pattern in the sequence.

(b) The 100th term of this sequence is $\sqrt{\frac{100}{101}}$.

7. $a_n = 3 + 2^n$

SOLUTION

(a) The sequence $a_1, a_2, \dots$ defined by $a_n = 3 + 2^n$ is $5, 7, 11, 19, \dots$.

(b) The 100th term of this sequence is $3 + 2^{100}$.

For Exercises 9–14, consider an arithmetic sequence with first term b and difference d between consecutive terms.

(a) ***Write the sequence using the three-dot notation, giving the first four terms of the sequence.***

(b) ***Give the 100th term of the sequence.***

9. $b = 2$, $d = 5$

SOLUTION

(a) The arithmetic sequence with first term 2 and difference 5 between consecutive terms is $2, 7, 12, 17, \dots$.

(b) The 100th term of this sequence is $2 + 99 \cdot 5$, which equals 497.

11. $b = 4$, $d = -6$

SOLUTION

(a) The arithmetic sequence with first term 4 and difference -6 between consecutive terms is $4, -2, -8, -14, \dots$.

(b) The 100th term of this sequence is $4 + 99 \cdot (-6)$, which equals -590.

13. $b = 0$, $d = \frac{1}{3}$

SOLUTION

(a) The arithmetic sequence with first term 0 and difference $\frac{1}{3}$ between consecutive terms is $0, \frac{1}{3}, \frac{2}{3}, 1, \dots$.

(b) The 100th term of this sequence is $0 + 99 \cdot \frac{1}{3}$, which equals 33.

For Exercises 15–18, suppose that at the beginning of the first day of a new year you have 3324 e-mail messages saved on your computer. At the end of each day you save only your 12 most important new e-mail messages along with the previously saved messages. Consider the sequence whose n^{th} term is the number of e-mail messages you have saved on your computer at the beginning of the nth day of the year.

15. What are the first, second, and third terms of this sequence?

SOLUTION We have an arithmetic sequence whose first three terms are 3324, 3336, 3348; each term is the previous term plus 12.

17. What is the 100th term of this sequence? In other words, how many e-mail messages will you have saved on your computer at the beginning of the 100th day of the year?

SOLUTION The 100th term of this sequence is

$$3324 + (100 - 1) \cdot 12,$$

which equals 4512.

For Exercises 19–24, consider a geometric sequence with first term b and ratio r of consecutive terms.

(a) ***Write the sequence using the three-dot notation, giving the first four terms.***

(b) ***Give the 100th term of the sequence.***

19. $b = 1$, $r = 5$

SOLUTION

(a) The geometric sequence with first term 1 and ratio 5 of consecutive terms is $1, 5, 25, 125, \ldots$.

(b) The 100th term of this sequence is 5^{99}.

21. $b = 3$, $r = -2$

SOLUTION

(a) The geometric sequence with first term 3 and ratio -2 of consecutive terms is $3, -6, 12, -24, \ldots$.

(b) The 100th term of this sequence is $3 \cdot (-2)^{99}$, which equals $-3 \cdot 2^{99}$.

23. $b = 2$, $r = \frac{1}{3}$

SOLUTION

(a) The geometric sequence with first term 2 and ratio $\frac{1}{3}$ of consecutive terms is $2, \frac{2}{3}, \frac{2}{9}, \frac{2}{27}, \ldots$.

(b) The 100th term of this sequence is $2 \cdot \left(\frac{1}{3}\right)^{99}$, which equals $2/3^{99}$.

25. Find the fifth term of an arithmetic sequence whose second term is 8 and whose third term is 14.

SOLUTION Because the second term of this arithmetic sequence is 8 and the third term is 14, we see that the difference between consecutive terms is 6. Thus the fourth term is $14 + 6$, which equals 20, and the fifth term is $20 + 6$, which equals 26.

27. Find the first term of an arithmetic sequence whose second term is 19 and whose fourth term is 25.

SOLUTION Because the second term of this arithmetic sequence is 19 and the fourth term is 25, and because the fourth term is two terms away from the second term, we see that twice the difference between consecutive terms is 6. Thus the difference between consecutive terms is 3. Thus 19, which is the second term, is 3 more than the first term. This implies that the first term equals 16.

29. Find the 100th term of an arithmetic sequence whose tenth term is 5 and whose eleventh term is 8.

SOLUTION Because the tenth term of this arithmetic sequence is 5 and the eleventh term is 8, we see that the difference between consecutive terms is 3. To get from the eleventh term to the 100th term, we need to add 3 to the eleventh term $100 - 11$ times, which equals 89 times. Thus the 100th term is $8 + 89 \cdot 3$, which equals 275.

31. Find the fifth term of a geometric sequence whose second term is 8 and whose third term is 14.

SOLUTION The second term of this geometric sequence is 8, and the third term is 14. Hence the ratio of consecutive terms is $\frac{14}{8}$, which equals $\frac{7}{4}$. Thus the fourth term equals the third term times $\frac{7}{4}$. In other words, the fourth term is $14 \cdot \frac{7}{4}$, which equals $\frac{49}{2}$. Similarly, the fifth term is $\frac{49}{2} \cdot \frac{7}{4}$, which equals $\frac{343}{8}$.

33. Find the first term of a geometric sequence whose second term is 8 and whose fifth term is 27.

SOLUTION Let r denote the ratio of consecutive terms of this geometric sequence. Because the second term of this sequence is 8 and the fifth term is 27, and because the fifth term is three terms away from the second term, we have $8r^3 = 27$. Solving for r, we get $r = \frac{3}{2}$. Thus the ratio of consecutive terms is $\frac{3}{2}$. Thus 8, which is the second term, is $\frac{3}{2}$ times the first term. This implies that the first term equals $8 \cdot \frac{2}{3}$, which equals $\frac{16}{3}$.

35. Find the ninth term of a geometric sequence whose fourth term is 4 and whose seventh term is 5.

SOLUTION Let r denote the ratio of consecutive terms of this geometric sequence. To get from the fourth term of this sequence to the seventh term, we must multiply by r three times. Thus

$$4r^3 = 5.$$

Solving the equation above for r, we have $r = \left(\frac{5}{4}\right)^{1/3}$.

To get from the seventh term of this sequence to the ninth term, we must multiply by r twice. Thus the ninth term of this sequence is $5r^2$. Now

$$5r^2 = 5\left(\left(\tfrac{5}{4}\right)^{1/3}\right)^2 = 5\left(\tfrac{5}{4}\right)^{2/3} \approx 5.80199.$$

Thus the ninth term of this sequence is approximately 5.80199.

37. Find the 100th term of a geometric sequence whose tenth term is 5 and whose eleventh term is 8.

SOLUTION Because the tenth term of this geometric sequence is 5 and the eleventh term is 8, we see that the ratio of consecutive terms is $\frac{8}{5}$. To get from the eleventh term to the 100th term, we need to multiply the eleventh term by $\frac{8}{5}$ a total of $100 - 11$ times, which equals 89 times. Thus the 100th term is $8 \cdot \left(\frac{8}{5}\right)^{89}$, which equals $8 \cdot 1.6^{89}$, which is approximately 1.2×10^{19}.

For Exercises 39–42, suppose that your annual salary at the beginning of your first year at a new company is $38,000. Assume that your salary increases by 7% per year at the end of each year of employment. Consider the sequence whose n^{th} term is your salary at the beginning of your n^{th} year at this company.

39. What are the first, second, and third terms of this sequence?

SOLUTION We have here a geometric sequence, with first term 38000 and with ratio 1.07 of consecutive terms. Each term is 1.07 times the previous term. Thus the first three terms are

$$38000, 38000 \times 1.07, 38000 \times 1.07^2,$$

which equals

$$38000, 40660, 43506.2.$$

41. What is the 10th term of this sequence? In other words, what will your salary be at the beginning of your 10th year at this company?

SOLUTION The 10th term of this sequence is

$$38000 \times 1.07^9,$$

which equals 69861.45. In other words, your salary at the beginning of your 10th year will be almost $70,000.

For Exercises 43–46, give the first four terms of the specified recursively defined sequence.

43. $a_1 = 3$ and $a_{n+1} = 2a_n + 1$ for $n \geq 1$.

SOLUTION Each term after the first term is obtained by doubling the previous term and then adding 1. Thus the first four terms of this sequence are $3, 7, 15, 31$.

45. $a_1 = 2$, $a_2 = 3$, and $a_{n+2} = a_n a_{n+1}$ for $n \geq 1$.

SOLUTION Each term after the first two terms is the product of the two previous terms.

Thus the first four terms of this sequence are $2, 3, 6, 18$.

For Exercises 47–48, let $a_1, a_2, \ldots$ be the sequence defined by setting a_1 equal to the value shown below and for $n \geq 1$ letting

$$a_{n+1} = \begin{cases} \dfrac{a_n}{2} & \text{if } a_n \text{ is even;} \\ 3a_n + 1 & \text{if } a_n \text{ is odd.} \end{cases}$$

47. Suppose $a_1 = 3$. Find the smallest value of n such that $a_n = 1$.

SOLUTION Using the recursive formula above, starting with $a_1 = 3$ we compute terms of the sequence until one of them equals 1. The first eight terms of the sequence are

$$3,\ 10,\ 5,\ 16,\ 8,\ 4,\ 2,\ 1.$$

The eighth term of this sequence equals 1, with no earlier term equal to 1. Thus $n = 8$ is the smallest value of n such that $a_n = 1$.

For Exercises 49–54, consider the sequence whose n^{th} term a_n is given by the indicated formula.

(a) Write the sequence using the three-dot notation, giving the first four terms of the sequence.

(b) Give a recursive definition of the specified sequence.

49. $a_n = 5n - 3$

SOLUTION

(a) The sequence $a_1, a_2, \ldots$ defined by $a_n = 5n - 3$ is $2, 7, 12, 17, \ldots$.

(b) We have

$$\begin{aligned} a_{n+1} &= 5(n+1) - 3 = 5n + 5 - 3 = (5n - 3) + 5 \\ &= a_n + 5. \end{aligned}$$

Thus this sequence is defined by the equations

$$a_1 = 2 \quad \text{and} \quad a_{n+1} = a_n + 5 \text{ for } n \geq 1.$$

51. $a_n = 3(-2)^n$

SOLUTION

(a) The sequence $a_1, a_2, \ldots$ defined by $a_n = 3(-2)^n$ is $-6, 12, -24, 48, \ldots$.

(b) We have

$$a_{n+1} = 3(-2)^{n+1} = 3(-2)^n(-2) = -2a_n.$$

Thus this sequence is defined by the equations

$$a_1 = -6 \quad \text{and} \quad a_{n+1} = -2a_n \text{ for } n \geq 1.$$

53. $a_n = 2^n n!$

SOLUTION

(a) The sequence $a_1, a_2, \ldots$ defined by $a_n = 2^n n!$ is $2, 8, 48, 384, \ldots$.

(b) We have

$$\begin{aligned} a_{n+1} &= 2^{n+1}(n+1)! = 2 \cdot 2^n n!(n+1) \\ &= 2(n+1)2^n n! = 2(n+1)a_n. \end{aligned}$$

Thus this sequence is defined by the equations

$$a_1 = 2 \quad \text{and} \quad a_{n+1} = 2(n+1)a_n \text{ for } n \geq 1.$$

55. Define a sequence recursively by

$$a_1 = 3 \quad \text{and} \quad a_{n+1} = \frac{1}{2}\Big(\frac{7}{a_n} + a_n\Big) \text{ for } n \geq 1.$$

Find the smallest value of n such that a_n agrees with $\sqrt{7}$ for at least six digits after the decimal point.

SOLUTION A calculator shows that $\sqrt{7} \approx 2.6457513$. Using a calculator and the recursive formula above, we compute terms of the sequence until one of them agrees with $\sqrt{7}$ for at least six digits after the decimal point. The first four terms of the sequence are

$$3,\ 2.6666667,\ 2.6458333,\ 2.6457513.$$

The fourth term of this sequence agrees with $\sqrt{7}$ for at least six digits after the decimal point; no earlier term has this property. Thus $n = 4$ is the smallest value of n such that a_n agrees with $\sqrt{7}$ for at least six digits after the decimal point.

8.2 *Series*

LEARNING OBJECTIVES

By the end of this section you should be able to

- compute the sum of a finite arithmetic sequence;
- compute the sum of a finite geometric sequence;
- use summation notation;
- use the Binomial Theorem.

Sums of Sequences

A **series** is the sum of the terms of a sequence. For example, corresponding to the finite sequence $1, 4, 9, 16$ is the series $1 + 4 + 9 + 16$, which equals 30. In this section we will deal only with the series that arise from finite sequences; in the next section we will investigate the intricacies of infinite series.

We can refer to the terms of a series using the same terminology as for a sequence. For example, the series $1 + 4 + 9 + 16$ has first term 1, second term 4, and last term 16.

The three-dot notation for infinite sequences was introduced in the previous section. Now we want to extend that notation so that it can be used to indicate terms in a finite sequence or series that are not explicitly displayed. For example, consider the geometric sequence with 50 terms, where the mth term of this sequence is 2^m. We could denote this sequence by

$$2,\ 4,\ 8,\ \ldots,\ 2^{48},\ 2^{49},\ 2^{50}.$$

When three dots are used in a sequence, they are placed vertically at the same level as a comma. When three dots are used in a series, they are vertically centered with the plus sign.

Here the three dots denote the 44 terms of this sequence that are not explicitly displayed. Similarly, in the corresponding series

$$2 + 4 + 8 + \cdots + 2^{48} + 2^{49} + 2^{50},$$

the three dots denote the 44 terms that are not displayed.

Arithmetic Series

An **arithmetic series** is the sum obtained by adding up the terms of an arithmetic sequence. The next example provides our model for evaluating an arithmetic series.

EXAMPLE 1

Find the sum of all the odd numbers between 100 and 200.

SOLUTION We want to find the sum of the finite arithmetic sequence

$$101,\ 103,\ 105,\ \ldots,\ 195,\ 197,\ 199.$$

We could just add up the numbers above by brute force, but that will become tiresome when we need to deal with sequences that have 50,000 terms instead of 50 terms.

Thus we employ a trick. Let s denote the sum of all the odd numbers between 100 and 200. Our trick is to write out the sum defining s twice, but in reverse order the second time:

$$s = 101 + 103 + 105 + \cdots + 195 + 197 + 199$$

$$s = 199 + 197 + 195 + \cdots + 105 + 103 + 101.$$

Now add the two equations above, getting

$$2s = 300 + 300 + 300 + \cdots + 300 + 300 + 300.$$

The right side of the equation above consists of 50 terms, each equal to 300. Thus the equation above can be rewritten as $2s = 50 \cdot 300$. Solving for s, we have

$$s = 50 \cdot \frac{300}{2} = 50 \cdot 150 = 7500.$$

As you read this explanation of how to evaluate any arithmetic series, refer frequently to the concrete example above to help visualize the procedure.

The trick used in the example above works with any arithmetic series. Specifically, consider an arithmetic series with n terms and difference d between consecutive terms. Write the series twice, in reverse order the second time. With the series in the original order, each term is obtained by adding d to the previous term. With the series written in the reverse order, each term is obtained by subtracting d from the previous term. Thus when the two series are added, the addition of d and the subtraction of d cancel out; in the sum of the two series, all the terms are the same, equal to the sum of the first and last terms.

Thus twice the value of the series is equal to the number of terms times the sum of the first term and the last term. Dividing by 2, we obtain the following simple formula for evaluating an arithmetic series:

Arithmetic series

The sum of a finite arithmetic sequence equals the number of terms times the average of the first and last terms.

EXAMPLE 2

Evaluate the arithmetic series

$$3 + 8 + 13 + 18 + \cdots + 1003 + 1008.$$

SOLUTION The arithmetic sequence $3, 8, 13, 18, \ldots, 1003, 1008$ has first term 3 and a difference of 5 between consecutive terms. We need to determine the number n of terms in this sequence. Using the formula for the terms in an arithmetic sequence, we have

$$3 + (n - 1)5 = 1008.$$

Solving this equation for n gives $n = 202$. Thus this series has 202 terms.

The average of the first and last terms of this series is $\frac{3+1008}{2}$, which equals $\frac{1011}{2}$. The result in the box above now tells us that the arithmetic series

$$3 + 8 + 13 + 18 + \cdots + 1003 + 1008$$

equals $202 \cdot \frac{1011}{2}$, which equals $101 \cdot 1011$, which equals 102111.

To obtain the symbolic form of the formula in the box above, consider an arithmetic series with n terms, initial term b, and difference d between consecutive terms. The last term of this series is $b + (n - 1)d$. Thus the average of the first and last terms is

$$\frac{b + (b + (n - 1)d)}{2},$$

which equals $b + \frac{(n-1)d}{2}$. Hence we have the following symbolic version of the formula for evaluating an arithmetic series:

The version in the box above using just words is easier to remember than the symbolic version in the box below. However, the symbolic version is often useful.

Arithmetic series

The sum of an arithmetic sequence with n terms, initial term b, and difference d between consecutive terms is

$$n\left(b + \frac{(n-1)d}{2}\right).$$

In other words,

$$b + (b + d) + (b + 2d) + \cdots + (b + (n - 1)d) = n\left(b + \frac{(n-1)d}{2}\right).$$

EXAMPLE 3

For your summer exercise, you rode your bicycle every day, starting with 5 miles on the first day of summer. You increased the distance you rode by one-half mile each day. How many miles total did you ride over the 90 days of summer?

SOLUTION The number of miles you rode each day increased by a constant amount; thus we have an arithmetic sequence. To sum that arithmetic sequence using the formula above, we take the number of terms $n = 90$, the initial term $b = 5$, and the difference between consecutive terms $d = 0.5$. Using the formula above, the sum of this arithmetic sequence is

$$90\left(5 + \frac{89 \times 0.5}{2}\right),$$

which equals 2452.5. Thus your summer bicycle riding totaled 2452.5 miles.

Geometric Series

A **geometric series** is the sum obtained by adding up the terms of a geometric sequence. The next example provides our model for evaluating a geometric series.

EXAMPLE 4 Evaluate the geometric series

$$1 + 3 + 9 + \cdots + 3^{47} + 3^{48} + 3^{49}.$$

SOLUTION We could evaluate the series above by brute force, but that would become too difficult when we need to deal with geometric series that have 50,000 terms instead of 50 terms.

Thus we again employ a trick. Let s equal the sum above. Multiply s by 3, writing the resulting sum with terms aligned under the same terms of s, as follows:

$$s = 1 + 3 + 9 + \cdots + 3^{47} + 3^{48} + 3^{49}$$

$$3s = \quad 3 + 9 + \cdots + 3^{47} + 3^{48} + 3^{49} + 3^{50}.$$

Now subtract the first equation from the second equation, getting $2s = 3^{50} - 1$. Thus $s = (3^{50} - 1)/2$.

The trick used in the example above works with any geometric series. Specifically, consider a geometric series with n terms, starting with first term b, and with ratio r of consecutive terms. Let s equal the value of this geometric series. Multiply s by r, writing the resulting sum with terms aligned under the same terms of s, as follows:

$$s = b + br + br^2 + \cdots + br^{n-2} + br^{n-1}$$

$$rs = \quad br + br^2 + \cdots + br^{n-2} + br^{n-1} + br^n.$$

Now subtract the second equation from the first equation, getting $s - rs = b - br^n$, which can be rewritten as $(1 - r)s = b(1 - r^n)$. Dividing both sides by $1 - r$ gives the following formula:

The case $r = 1$ is excluded to avoid division by 0.

Geometric series

The sum of a geometric sequence with initial term b, ratio $r \neq 1$ of consecutive terms, and n terms is

$$b \cdot \frac{1 - r^n}{1 - r}.$$

In other words, if $r \neq 1$ then

$$b + br + br^2 + \cdots + br^{n-1} = b \cdot \frac{1 - r^n}{1 - r}.$$

EXAMPLE 5

Suppose tuition during your first year in college is \$12,000. You expect tuition to increase 6% per year, and you expect to take five years total to graduate. What is the total amount of tuition you should expect to pay in college?

SOLUTION The tuition each year after the first year is 1.06 times the previous year's tuition; thus we have a geometric sequence. To sum that sequence using the formula above, we take the initial term $b = 12,000$, the ratio $r = 1.06$ of consecutive terms, and the number of terms $n = 5$. Using the formula above, the sum of this geometric sequence is

$$12000 \cdot \frac{1 - 1.06^5}{1 - 1.06},$$

which equals 67645.1. Thus you should expect to pay a total of about \$67,645 in tuition during five years in college.

To express the formula $b + br + br^2 + \cdots + br^{n-1} = b \cdot \frac{1-r^n}{1-r}$ in words, first rewrite the right side the equation as

$$(b - br^n)/(1 - r).$$

The expression br^n would be the next term if we added one more term to the geometric sequence. Thus we have the following restatement of the formula for a geometric series:

Geometric series

The sum of a finite geometric sequence equals the first term minus what would be the term following the last term, divided by 1 minus the ratio of consecutive terms.

This box allows you to think about the formula for a geometric series in words instead of symbols.

EXAMPLE 6

Evaluate the geometric series

$$\frac{5}{3} + \frac{5}{9} + \frac{5}{27} + \cdots + \frac{5}{3^{20}}.$$

SOLUTION The first term of this geometric series is $\frac{5}{3}$. The ratio of consecutive terms is $\frac{1}{3}$. If we added one more term to this geometric series, the next term would be $\frac{5}{3^{21}}$. Using the box above, we see that

$$\begin{aligned}\frac{5}{3} + \frac{5}{9} + \frac{5}{27} + \cdots + \frac{5}{3^{20}} &= \frac{\frac{5}{3} - \frac{5}{3^{21}}}{1 - \frac{1}{3}}\\ &= \frac{\frac{5}{3} - \frac{5}{3^{21}}}{\frac{2}{3}}\\ &= \frac{5}{2} - \frac{5}{2 \cdot 3^{20}},\end{aligned}$$

where the last expression is obtained by multiplying the numerator and denominator of the previous expression by 3.

Summation Notation

The three-dot notation that we have been using has the advantage of presenting an easily understandable representation of a series. Another notation, called summation notation, is also often used for series. Summation notation has the advantage of explicitly displaying the formula used to compute the terms of the sequence. For some manipulations, summation notation works better than three-dot notation.

The following equation uses summation notation on the left side and three-dot notation on the right side:

The symbol Σ used in summation notation is an uppercase Greek sigma.

$$\sum_{k=1}^{99} k^2 = 1 + 4 + 9 + \cdots + 98^2 + 99^2.$$

In spoken language, the left side of the equation above becomes "the sum as k goes from 1 to 99 of k^2". This means that the first term of the series is obtained by starting with $k = 1$ and computing k^2 (which equals 1). The second term of the series is obtained by taking $k = 2$ and computing k^2 (which equals 4), and so on, until $k = 99$, giving the last term of the series (which is 99^2).

There is no specific k in the series above. We could have used j or m or any other letter, as long as we consistently use the same letter throughout the notation. Thus

$$\sum_{j=1}^{99} j^2 \quad \text{and} \quad \sum_{k=1}^{99} k^2 \quad \text{and} \quad \sum_{m=1}^{99} m^2$$

all denote the same series $1 + 4 + 9 + \cdots + 98^2 + 99^2$.

EXAMPLE 7

Write the geometric series

$$3 + 9 + 27 + \cdots + 3^{80}$$

using summation notation.

SOLUTION The kth term of this series is 3^k. Thus

$$3 + 9 + 27 + \cdots + 3^{80} = \sum_{k=1}^{80} 3^k.$$

You should also become comfortable translating in the other direction, from summation notation to either an explicit sum or the three-dot notation.

EXAMPLE 8

Write the series

$$\sum_{k=0}^{3} (k^2 - 1)2^k$$

as an explicit sum.

SOLUTION In this case the summation starts with $k = 0$. When $k = 0$, the expression $(k^2 - 1)2^k$ equals -1, so the first term of this series is -1.

When $k = 1$, the expression $(k^2 - 1)2^k$ equals 0, so the second term is 0. When $k = 2$, the expression $(k^2 - 1)2^k$ equals 12, so the third term is 12. When $k = 3$, the expression $(k^2 - 1)2^k$ equals 64, so the fourth term is 64. Thus

$$\sum_{k=0}^{3} (k^2 - 1)2^k = -1 + 0 + 12 + 64 = 75.$$

Usually the starting and ending values for a summation are written below and above the Σ. Sometimes to save vertical space this information appears alongside the Σ. For example, the sum above might be written as $\sum_{k=0}^{3}(k^2 - 1)2^k$.

Sometimes there is more than one convenient way to write a series using summation notation, as illustrated in the following example.

EXAMPLE 9

Suppose $r \neq 0$. Write the geometric series

$$1 + r + r^2 + \cdots + r^{n-1}$$

using summation notation.

SOLUTION This series has n terms. The kth term of this series is r^{k-1}. Thus

$$1 + r + r^2 + \cdots + r^{n-1} = \sum_{k=1}^{n} r^{k-1}.$$

We could also think of this series as the sum of powers of r, starting with r^0 (recall that $r^0 = 1$ provided $r \neq 0$) and ending with r^{n-1}. From this perspective, we could write

$$1 + r + r^2 + \cdots + r^{n-1} = \sum_{k=0}^{n-1} r^k.$$

Note that on the right side of the last equation, k starts at 0 and ends at $n - 1$.

Thus we have written this geometric series in two different ways using summation notation. Both are correct; the choice of which to use may depend on the context.

When using any kind of technology to evaluate a sum, you will probably need to convert the three-dot notation to the summation notation expected by the software, as shown in the next example.

EXAMPLE 10

(a) Evaluate $1 + 4 + 9 + \cdots + 98^2 + 99^2$.

(b) Find a formula for the sum of the first n squares $1 + 4 + 9 + \cdots + n^2$.

SOLUTION

(a) Enter `sum k^2 for k = 1 to 99` in a Wolfram|Alpha entry box to get

$$\sum_{k=1}^{99} k^2 = 328350.$$

(b) Enter `sum k^2 for k = 1 to n` in a Wolfram|Alpha entry box to get

$$\sum_{k=1}^{n} k^2 = \frac{n(n+1)(2n+1)}{6}$$

The solution here uses Wolfram|Alpha, but feel free to use your preferred technology instead.

The Binomial Theorem

We now turn to the question of expanding $(x + y)^n$ into a sum of terms.

EXAMPLE 11

Expand $(x + y)^n$ for $n = 0, 1, 2, 3, 4, 5$.

SOLUTION Each equation below (except the first) is obtained by multiplying both sides of the previous equation by $x + y$ and then expanding the right side.

Here the terms in each expansion have been written in decreasing order of powers of x.

$$(x + y)^0 = 1$$

$$(x + y)^1 = x + y$$

$$(x + y)^2 = x^2 + 2xy + y^2$$

$$(x + y)^3 = x^3 + 3x^2y + 3xy^2 + y^4$$

$$(x + y)^4 = x^4 + 4x^3y + 6x^2y^2 + 4xy^3 + y^4$$

$$(x + y)^5 = x^5 + 5x^4y + 10x^3y^2 + 10x^2y^3 + 5xy^4 + y^5$$

Some patterns are evident in the expansions above:

- Each expansion of $(x + y)^n$ above begins with x^n and ends with y^n.
- The second term in each expansion of $(x + y)^n$ is $nx^{n-1}y$, and the second-to-last term is nxy^{n-1}.
- Each term in the expansion of $(x + y)^n$ is a constant times $x^j y^k$, where j and k are nonnegative integers and $j + k = n$.

EXAMPLE 12

Center the expansions above, then find a pattern for computing the coefficients.

SOLUTION Centering the expansions above gives the following display.

$$
\begin{array}{lccccccccccc}
(x+y)^0 = & & & & & & 1 & & & & & \\
(x+y)^1 = & & & & & x & + & y & & & & \\
(x+y)^2 = & & & & x^2 & + & 2xy & + & y^2 & & & \\
(x+y)^3 = & & & x^3 & + & 3x^2y & + & 3xy^2 & + & y^4 & & \\
(x+y)^4 = & & x^4 & + & \boxed{4x^3y} & + & \boxed{6x^2y^2} & + & 4xy^3 & + & y^4 & \\
(x+y)^5 = & x^5 + 5x^4y & & + & & \boxed{10x^3y^2} & + & 10x^2y^3 & + & 5xy^4 + y^5 & &
\end{array}
$$

Consider, for example, the term $10x^3y^2$ (red) in the last row above. The coefficient diagonally to the left in the row above $10x^3y^2$ is 4 (blue) and the coefficient diagonally to the right above $10x^3y^2$ is 6 (blue), and furthermore $10 = 4 + 6$.

The same pattern holds throughout the expansions above. As you should verify, each coefficient that is not 1 is the sum of the coefficient diagonally to the left and the coefficient diagonally to the right in the row above.

In the previous example we discovered that the coefficient 10 in the term $10x^3y^2$ is the sum of the coefficient diagonally to the left and the coefficient diagonally to the right in the row above. The next example explains why this happens.

EXAMPLE 13

Explain why the coefficient of x^3y^2 in the expansion of $(x+y)^5$ is the sum of the coefficients of x^3y and x^2y^2 in the expansion of $(x+y)^4$.

SOLUTION Consider the following equations:

$$\begin{aligned}(x+y)^5 &= (x+y)^4(x+y)\\ &= (x^4 + \boxed{4x^3y} + \boxed{6x^2y^2} + 4xy^3 + y^4)(x+y).\end{aligned}$$

When the last line above is expanded using the distributive property, the only way that x^3y^2 terms can arise is when $4x^3y$ is multiplied by y (giving $4x^3y^2$) and when $6x^2y^2$ is multiplied by x (giving $6x^3y^2$). Adding together these two x^3y^2 terms gives $10x^3y^2$.

The same idea as used in the example explains why every coefficient in the expansion of $(x+y)^n$ can be computed by adding together two coefficients from the expansion of $(x+y)^{n-1}$.

The pattern of coefficients in the expansion of $(x+y)^n$ becomes more visible if we write just the coefficients from Example 12, leaving out the plus signs and the powers of x and y. We then have the following numbers, arranged in a triangular shape.

$$\begin{array}{ccccccccccc} &&&&&1&&&&& \\ &&&&1&&1&&&& \\ &&&1&&2&&1&&& \\ &&1&&3&&3&&1&& \\ &1&&4&&6&&4&&1& \\ 1&&5&&10&&10&&5&&1 \end{array}$$

Statue in France of Blaise Pascal. Pascal's triangle was used centuries before Pascal in India, Persia, China, and Italy.

This triangular array of numbers, with any number of rows, is called **Pascal's triangle**, in honor of Blaise Pascal, the 17th-century French mathematician who discovered properties of this triangle of numbers.

Pascal's triangle

- **Pascal's triangle** is a triangular array with n numbers in the nth row.
- The first and last entries in each row are 1.
- Each entry that is not a first or last entry in a row is the sum of the two entries diagonally above to the left and right.

The reasoning that was used in Example 13 now gives the following method for expanding $(x + y)^n$. Because row 1 of Pascal's triangle corresponds to $(x + y)^0$, row $n + 1$ of Pascal's triangle gives the coefficients of the expansion of $(x + y)^n$.

The coefficients correspond to terms in decreasing order of powers of x.

Coefficients of $(x + y)^n$

Suppose n is a nonnegative integer. Then row $n + 1$ of Pascal's triangle gives the coefficients in the expansion of $(x + y)^n$.

EXAMPLE 14 Use Pascal's triangle to find the expansion of $(x + y)^7$.

To expand $(x + y)^7$, we find the eighth row of Pascal's triangle.

SOLUTION We already know the sixth row of Pascal's triangle, corresponding to the expansion of $(x + y)^5$. It is listed as the first line below, and then below it we do the usual addition to get the seventh row, and then below it we do the usual addition to get the eighth row.

$$\begin{array}{ccccccccccccccc} & & 1 & & 5 & & 10 & & 10 & & 5 & & 1 & & \\ & 1 & & 6 & & 15 & & 20 & & 15 & & 6 & & 1 & \\ 1 & & 7 & & 21 & & 35 & & 35 & & 21 & & 7 & & 1 \end{array}$$

The eighth row of Pascal's triangle now allows us to expand $(x + y)^7$ easily:

$$(x + y)^7 = x^7 + 7x^6y + 21x^5y^2 + 35x^4y^3 + 35x^3y^4 + 21x^2y^5 + 7xy^6 + y^7.$$

Here is another example showing how Pascal's triangle can help speed up a calculation.

EXAMPLE 15 Use Pascal's triangle to simplify the expression $(2 + \sqrt{3})^5$.

SOLUTION The sixth row of Pascal's triangle is 1 5 10 10 5 1. Thus

$$\begin{aligned}(2 + \sqrt{3})^5 &= 2^5 + 5 \cdot 2^4\sqrt{3} + 10 \cdot 2^3\sqrt{3}^2 + 10 \cdot 2^2\sqrt{3}^3 + 5 \cdot 2\sqrt{3}^4 + \sqrt{3}^5 \\ &= 32 + 80\sqrt{3} + 80 \cdot 3 + 40 \cdot \sqrt{3}^2\sqrt{3} + 10 \cdot \sqrt{3}^2\sqrt{3}^2 + \sqrt{3}^2\sqrt{3}^2\sqrt{3} \\ &= 32 + 80\sqrt{3} + 80 \cdot 3 + 40 \cdot 3\sqrt{3} + 10 \cdot 3 \cdot 3 + 3 \cdot 3\sqrt{3} \\ &= 32 + 80\sqrt{3} + 240 + 120\sqrt{3} + 90 + 9\sqrt{3} \\ &= 362 + 209\sqrt{3}.\end{aligned}$$

The next example shows how the expansions we have been discussing can lead to useful approximations.

EXAMPLE 16

Explain why $1.0001^7 \approx 1.0007$.

See Problem 62 for a generalization of this result.

SOLUTION Using the expansion of $(x + y)^7$ from the previous example, with $x = 1$ and $y = 10^{-4}$, we have

$$\begin{aligned}1.0001^7 &= (1 + 10^{-4})^7 \\ &= 1 + 7 \cdot 10^{-4} + 21 \cdot 10^{-8} + 35 \cdot 10^{-12} \\ &\qquad + 35 \cdot 10^{-16} + 21 \cdot 10^{-20} + 7 \cdot 10^{-24} + 10^{-28} \\ &\approx 1 + 7 \cdot 10^{-4} \\ &= 1.0007,\end{aligned}$$

where the approximation in the second-to-last line is valid because the neglected terms are much smaller than 10^{-4}.

The exact value is

$$1.0001^7 = 1.0007002100350035002100070001,$$

which shows that our approximation is quite good.

Suppose we want to find the coefficient of $x^{97}y^3$ in the expansion of $(x + y)^{100}$. We could calculate this coefficient using Pascal's triangle, but only after much work to get to row 101. Thus we now discuss binomial coefficients, which will give us a direct way to compute the expansion of $(x + y)^n$ without first having to compute the expansion for lower powers.

Recall that $n!$ is defined to be the product of the integers from 1 to n. We will also find it convenient to define $0!$ to be 1. Thus we have the following.

$$0! = 1 \qquad 1! = 1 \qquad 2! = 2 \qquad 3! = 2 \cdot 3 = 6 \qquad 4! = 2 \cdot 3 \cdot 4 = 24$$

Now we can define binomial coefficients.

Binomial coefficient

Suppose n and k are nonnegative integers with $k \leq n$. The **binomial coefficient** $\binom{n}{k}$ is defined by

$$\binom{n}{k} = \frac{n!}{k!\,(n-k)!}.$$

EXAMPLE 17

(a) Evaluate $\binom{10}{3}$.

(b) Evaluate $\binom{n}{0}$.

(c) Evaluate $\binom{n}{1}$.

(d) Evaluate $\binom{n}{2}$.

(e) Evaluate $\binom{n}{3}$.

SOLUTION

(a) We do not actually need to compute 10! because the six terms from 7! in the denominator cancel the first six terms in the numerator, as shown below:

$$\binom{10}{3} = \frac{10!}{3!\,7!}$$

$$= \frac{2 \cdot 3 \cdot 4 \cdot 5 \cdot 6 \cdot 7 \cdot 8 \cdot 9 \cdot 10}{(2 \cdot 3)(2 \cdot 3 \cdot 4 \cdot 5 \cdot 6 \cdot 7)}$$

$$= \frac{8 \cdot 9 \cdot 10}{2 \cdot 3}$$

$$= 4 \cdot 3 \cdot 10$$

$$= 120.$$

The large cancellation of the numbers in red in this example often happens when working with binomial coefficients.

Despite their appearance as fractions, each binomial coefficient is an integer.

(b) $\binom{n}{0} = \frac{n!}{0!\,n!} = \frac{n!}{n!} = 1$

(c) $\binom{n}{1} = \frac{n!}{1!\,(n-1)!} = \frac{n!}{(n-1)!} = n$

(d) $\binom{n}{2} = \frac{n!}{2!\,(n-2)!} = \frac{(n-1)n}{2}$

(e) $\binom{n}{3} = \frac{n!}{3!\,(n-3)!} = \frac{(n-2)(n-1)n}{2 \cdot 3}$

The next result allows us to compute coefficients in the expansion of $(x+y)^n$ without using Pascal's triangle.

The word binomial *means an expression with two terms, just as* polynomial *refers to an expression with multiple terms. Thus* $x + y$ *is a binomial.*

Binomial Theorem

Suppose n is a positive integer. Then

$$(x+y)^n = \sum_{k=0}^{n} \binom{n}{k} x^{n-k} y^k.$$

Before explaining why the Binomial Theorem holds, let's look at two examples of its use.

EXAMPLE 18

Find the coefficient of $x^{97}y^3$ in the expansion of $(x+y)^{100}$.

SOLUTION According to the Binomial Theorem, the coefficient of $x^{97}y^3$ in the expansion of $(x+y)^{100}$ is $\binom{100}{3}$, which we compute as follows:

$$\binom{100}{3} = \frac{100!}{3!\,97!} = \frac{98 \cdot 99 \cdot 100}{2 \cdot 3} = 161700.$$

The next example confirms our previous observation that the first term of the expansion of $(x+y)^n$ is x^n and the second term is $nx^{n-1}y$.

EXAMPLE 19

Suppose $n \geq 4$. Explicitly display the first four terms in the expansion of $(x+y)^n$.

SOLUTION Using the Binomial Theorem and the results from Example 17, we have

$$\begin{aligned}(x+y)^n &= \tbinom{n}{0}x^n + \tbinom{n}{1}x^{n-1}y + \tbinom{n}{2}x^{n-2}y^2 + \tbinom{n}{3}x^{n-3}y^3 + \sum_{k=4}^{n}\tbinom{n}{k}x^{n-k}y^k \\ &= x^n + nx^{n-1}y + \tfrac{(n-1)n}{2}x^{n-2}y^2 + \tfrac{(n-2)(n-1)n}{2\cdot 3}x^{n-3}y^3 \\ &\quad + \sum_{k=4}^{n}\tbinom{n}{k}x^{n-k}y^k.\end{aligned}$$

The summation starts with $k = 4$ because the terms for $k = 0, 1, 2, 3$ are explicitly displayed.

The result in the next example provides the key to understanding why the Binomial Theorem holds.

EXAMPLE 20

Suppose n and k are positive integers with $k < n$. Show that

$$\binom{n-1}{k-1} + \binom{n-1}{k} = \binom{n}{k}.$$

SOLUTION

$$\begin{aligned}\binom{n-1}{k-1} + \binom{n-1}{k} &= \frac{(n-1)!}{(n-k)!\,(k-1)!} + \frac{(n-1)!}{(n-k-1)!\,k!} \\ &= \frac{(n-1)!}{(n-k-1)!\,(k-1)!}\Big(\frac{1}{n-k} + \frac{1}{k}\Big) \\ &= \frac{(n-1)!}{(n-k-1)!\,(k-1)!} \cdot \frac{n}{(n-k)k} \\ &= \frac{(n-1)!\,n}{(n-k-1)!\,(n-k)\,(k-1)!\,k} \\ &= \frac{n!}{(n-k)!\,k!} \\ &= \binom{n}{k}\end{aligned}$$

We know that row $n + 1$ of Pascal's triangle gives the coefficients in the expansion of $(x + y)^n$. Thus the next result implies that the Binomial Theorem holds.

Pascal's triangle and binomial coefficients

For k and n integers with $0 \le k \le n$, entry $k + 1$ in row $n + 1$ of Pascal's triangle is $\binom{n}{k}$.

A version of Pascal's triangle published in China in 1303. The fourth entry in row 8 is 34 rather than the correct value of $\binom{7}{3}$, which is 35. All other entries are correct. The error is probably a typo.

To see why the result above holds, consider the triangular array with entry $k + 1$ in row $n + 1$ equal to $\binom{n}{k}$. Thus the first three rows of this triangular array are

$$\begin{matrix} & & \binom{0}{0} & & \\ & \binom{1}{0} & & \binom{1}{1} & \\ \binom{2}{0} & & \binom{2}{1} & & \binom{2}{2}, \end{matrix} \qquad \text{which equals} \qquad \begin{matrix} & & 1 & & \\ & 1 & & 1 & \\ 1 & & 2 & & 1, \end{matrix}$$

which are the same as the first three rows of Pascal's triangle.

The entries in this triangular array diagonally above to the left and right of entry $k + 1$ in row $n + 1$ are entries k and $k + 1$ in row n, which are $\binom{n-1}{k-1}$ and $\binom{n-1}{k}$. The result in Example 20 shows that these two entries add up to $\binom{n}{k}$. In other words, this new triangular array is constructed just as Pascal's triangle is constructed, with each entry equal to the sum of the two entries diagonally above to the left and right.

Because this new triangular array and Pascal's triangle have the same first few rows and the same method for determining further rows, these two triangular arrays are equal.

EXERCISES

In Exercises 1–10, evaluate the arithmetic series.

1. $1 + 2 + 3 + \cdots + 98 + 99 + 100$
2. $1001 + 1002 + 1003 + \cdots + 2998 + 2999 + 3000$
3. $302 + 305 + 308 + \cdots + 6002 + 6005 + 6008$
4. $25 + 31 + 37 + \cdots + 601 + 607 + 613$
5. $200 + 195 + 190 + \cdots + 75 + 70 + 65$
6. $300 + 293 + 286 + \cdots + 55 + 48 + 41$
7. $\sum_{m=1}^{80} (4 + 5m)$
8. $\sum_{m=1}^{75} (2 + 3m)$
9. $\sum_{k=5}^{65} (4k - 1)$
10. $\sum_{k=10}^{900} (3k - 2)$
11. Find the sum of all the four-digit positive integers.
12. Find the sum of all the four-digit odd positive integers.
13. Find the sum of all the four-digit positive integers whose last digit equals 3.
14. Find the sum of all the four-digit positive integers that are evenly divisible by 5.
15. Evaluate $1 + 8 + 27 + \cdots + 998^3 + 999^3$.
16. Evaluate $1 + 16 + 81 + \cdots + 998^4 + 999^4$.

For Exercises 17–20, suppose you started an exercise program by riding your bicycle 10 miles on the first day and then you increased the distance you rode by one-quarter mile each day.

17. How many total miles did you ride after 50 days?

18. How many total miles did you ride after 70 days?

19. What is the first day on which the total number of miles you rode exceeded 2000?

20. What is the first day on which the total number of miles you rode exceeded 3000?

In Exercises 21–30, evaluate the geometric series.

21. $1 + 3 + 9 + \cdots + 3^{200}$

22. $1 + 2 + 4 + \cdots + 2^{100}$

23. $\frac{1}{4} + \frac{1}{16} + \frac{1}{64} + \cdots + \frac{1}{4^{50}}$

24. $\frac{1}{3} + \frac{1}{9} + \frac{1}{27} + \cdots + \frac{1}{3^{33}}$

25. $1 - \frac{1}{2} + \frac{1}{4} - \frac{1}{8} + \cdots + \frac{1}{2^{80}} - \frac{1}{2^{81}}$

26. $1 - \frac{1}{3} + \frac{1}{9} - \frac{1}{27} + \cdots + \frac{1}{3^{60}} - \frac{1}{3^{61}} + \frac{1}{3^{62}}$

27. $\sum_{k=1}^{40} \frac{3}{2^k}$

28. $\sum_{k=1}^{90} \frac{5}{7^k}$

29. $\sum_{m=3}^{77} (-5)^m$

30. $\sum_{m=5}^{91} (-2)^m$

In Exercises 31–34, write the series explicitly and evaluate the sum.

31. $\sum_{m=1}^{4} (m^2 + 5)$

32. $\sum_{m=1}^{5} (m^2 - 2m + 7)$

33. $\sum_{k=0}^{3} \log(k^2 + 2)$

34. $\sum_{k=0}^{4} \ln(2^k + 1)$

In Exercises 35–38, write the series using summation notation (starting with $k = 1$). Each series in Exercises 35–38 is either an arithmetic series or a geometric series.

35. $2 + 4 + 6 + \cdots + 100$

36. $1 + 3 + 5 + \cdots + 201$

37. $\frac{5}{9} + \frac{5}{27} + \frac{5}{81} + \cdots + \frac{5}{3^{40}}$

38. $\frac{7}{16} + \frac{7}{32} + \frac{7}{64} + \cdots + \frac{7}{2^{25}}$

39. Restate the symbolic version of the formula for evaluating an arithmetic series using summation notation.

40. Restate the symbolic version of the formula for evaluating a geometric series using summation notation.

For Exercises 41–44, consider the fable from the beginning of Section 5.4. In this fable, one grain of rice is placed on the first square of a chessboard, then two grains on the second square, then four grains on the third square, and so on, doubling the number of grains placed on each square.

41. Find the total number of grains of rice on the first 18 squares of the chessboard.

42. Find the total number of grains of rice on the first 30 squares of the chessboard.

43. Find the smallest number n such that the total number of grains of rice on the first n squares of the chessboard is more than 30,000,000.

44. Find the smallest number n such that the total number of grains of rice on the first n squares of the chessboard is more than 4,000,000,000.

For Exercises 45–46, use Pascal's triangle to simplify the indicated expression.

45. $(2 - \sqrt{3})^5$

46. $(3 - \sqrt{2})^6$

For Exercises 47–50, assume that n is a positive integer.

47. Evaluate $\binom{n}{n}$.

48. Evaluate $\binom{n}{n-1}$.

49. Evaluate $\binom{n}{n-2}$.

50. Evaluate $\binom{n}{n-3}$.

51. Evaluate $\binom{70}{3}$.

52. Evaluate $\binom{60}{4}$.

PROBLEMS

53. Explain why the polynomial factorization

$$1 - x^n = (1 - x)(1 + x + x^2 + \cdots + x^{n-1})$$

holds for every integer $n \geq 2$.

54. Show that

$$\tfrac{1}{2} + \tfrac{1}{3} + \cdots + \tfrac{1}{n} < \ln n$$

for every integer $n \geq 2$.
[*Hint:* Draw the graph of the curve $y = \frac{1}{x}$ in the xy-plane. Think of $\ln n$ as the area under part of this curve. Draw appropriate rectangles under the curve.]

55. Show that

$$\ln n < 1 + \tfrac{1}{2} + \cdots + \tfrac{1}{n-1}$$

for every integer $n \geq 2$.
[*Hint:* Draw the graph of the curve $y = \frac{1}{x}$ in the xy-plane. Think of $\ln n$ as the area under part of this curve. Draw appropriate rectangles above the curve.]

56. Show that the sum of a finite arithmetic sequence is 0 if and only if the last term equals the negative of the first term.

57. Use technology to find a formula for the sum of the first n cubes $1 + 8 + 27 + \cdots + n^3$.

58. Use technology to find a formula for the sum of the first n fourth powers $1 + 16 + 81 + \cdots + n^4$.

59. For $n = 0, 1, 2, 3, 4, 5$, show that the sum of the entries in row $n + 1$ of Pascal's triangle equals 2^n.

60. Show that

$$\sum_{k=0}^{n} \frac{n!}{k!(n-k)!} = 2^n$$

for every positive integer n.
[*Hint:* Expand $(1 + 1)^n$ using the Binomial Theorem.]

61. Suppose n is a positive integer. Show that the sum of the entries in row $n + 1$ of Pascal's triangle equals 2^n.
[*Hint:* Use the result in the previous problem.]

62. Suppose n is a positive integer. Show that if $|t|$ is small, then

$$(1 + t)^n \approx 1 + nt.$$

63. Suppose n is a positive integer and x is any number. Show that if $|t|$ is small and nonzero, then

$$\frac{(x + t)^n - x^n}{t} \approx nx^{n-1}.$$

WORKED-OUT SOLUTIONS *to Odd-numbered Exercises*

In Exercises 1–10, evaluate the arithmetic series.

1. $1 + 2 + 3 + \cdots + 98 + 99 + 100$

SOLUTION This series contains 100 terms.

The average of the first and last terms in this series is $\frac{1+100}{2}$, which equals $\frac{101}{2}$.

Thus $1 + 2 + \cdots + 99 + 100$ equals $100 \cdot \frac{101}{2}$, which equals $50 \cdot 101$, which equals 5050.

3. $302 + 305 + 308 + \cdots + 6002 + 6005 + 6008$

SOLUTION The difference between consecutive terms in this series is 3. We need to determine the number n of terms in this series. Using the formula for the terms of an arithmetic sequence, we have

$$302 + (n - 1)3 = 6008.$$

Subtracting 302 from both sides of this equation gives the equation $(n-1)3 = 5706$; dividing both sides by 3 then gives $n - 1 = 1902$. Thus $n = 1903$.

The average of the first and last terms in this series is $\frac{302+6008}{2}$, which equals 3155.

Thus $302 + 305 + \cdots + 6005 + 6008$ equals $1903 \cdot 3155$, which equals 6003965.

5. $200 + 195 + 190 + \cdots + 75 + 70 + 65$

SOLUTION The difference between consecutive terms in this series is -5. We need to determine the number n of terms in this series. Using the formula for the terms of an arithmetic sequence, we have

$$200 + (n - 1)(-5) = 65.$$

Subtracting 200 from both sides of this equation gives the equation $(n - 1)(-5) = -135$; dividing both sides by -5 then gives $n - 1 = 27$. Thus $n = 28$.

The average of the first and last terms in this series is $\frac{200+65}{2}$, which equals $\frac{265}{2}$.

Thus $200 + 195 + 190 + \cdots + 75 + 70 + 65$ equals $28 \cdot \frac{265}{2}$, which equals 3710.

7. $\sum_{m=1}^{80} (4 + 5m)$

SOLUTION Because $4 + 5 \cdot 1 = 9$ and $4 + 5 \cdot 80 = 404$, we have

$$\sum_{m=1}^{80} (4 + 5m) = 9 + 14 + 19 \cdots + 404.$$

Thus the first term of this arithmetic sequence is 9, the last term is 404, and we have 80 terms. Hence

$$\sum_{m=1}^{80} (4 + 5m) = 80 \cdot \frac{9 + 404}{2} = 16520.$$

9. $\sum_{k=5}^{65} (4k - 1)$

SOLUTION Because $4 \cdot 5 - 1 = 19$ and $4 \cdot 65 - 1 = 259$, we have

$$\sum_{k=5}^{65} (4k - 1) = 19 + 23 + 27 + \cdots + 259.$$

Thus the first term of this arithmetic sequence is 19, the last term is 259, and we have $65 - 5 + 1$ terms, or 61 terms. Hence

$$\sum_{k=5}^{65} (4k - 1) = 61 \cdot \frac{19 + 259}{2} = 8479.$$

11. Find the sum of all the four-digit positive integers.

SOLUTION We need to evaluate the arithmetic series

$$1000 + 1001 + 1002 + \cdots + 9999.$$

The number of terms in this arithmetic series is $9999 - 1000 + 1$, which equals 9000.

The average of the first and last terms is $\frac{1000+9999}{2}$, which equals $\frac{10999}{2}$.

Thus the sum of all the four-digit positive integers equals $9000 \cdot \frac{10999}{2}$, which equals 49495500.

13. Find the sum of all the four-digit positive integers whose last digit equals 3.

SOLUTION We need to evaluate the arithmetic series

$$1003 + 1013 + 1023 + \cdots + 9983 + 9993.$$

Consecutive terms in this series differ by 10. We need to determine the number n of terms in this series. Using the formula for the terms of an arithmetic sequence, we have

$$1003 + (n - 1)10 = 9993.$$

Subtracting 1003 from both sides of this equation gives the equation $(n - 1)10 = 8990$; dividing both sides by 10 then gives $n - 1 = 899$. Thus $n = 900$.

The average of the first and last terms is $\frac{1003+9993}{2}$, which equals 5498.

Thus the sum of all the four-digit positive integers whose last digit equals 3 is $900 \cdot 5498$, which equals 4948200.

15. Evaluate $1 + 8 + 27 + \cdots + 998^3 + 999^3$.

SOLUTION Enter `sum k^3 for k = 1 to 999` in a Wolfram|Alpha entry box to get

$$\sum_{k=1}^{999} k^3 = 249500250000.$$

Instead of Wolfram|Alpha, you could use your preferred technology to obtain the same result.

For Exercises 17–20, suppose you started an exercise program by riding your bicycle 10 miles on the first day and then you increased the distance you rode by one-quarter mile each day.

17. How many total miles did you ride after 50 days?

SOLUTION Let the number of miles you rode on the nth day be the nth term of a sequence. Then this is an arithmetic sequence, with first term 10 and difference 0.25 between consecutive terms. The sum of the first 50 terms of this arithmetic sequence is

$$50\left(10 + \frac{49 \times 0.25}{2}\right),$$

which equals 806.25. Thus you rode a total of 806.25 miles after 50 days.

19. What is the first day on which the total number of miles you rode exceeded 2000?

SOLUTION After n days, the total amount you rode is

$$n\left(10 + \frac{(n-1)}{8}\right).$$

To see when the total reaches 2000 miles, we solve the equation

$$n\left(10 + \frac{(n-1)}{8}\right) = 2000$$

for n (use the quadratic formula), getting a negative solution (which we discard because it makes no sense) and $n \approx 93.0151$. Thus the 94th day was the first day that the total number of miles you rode exceeded 2000.

In Exercises 21–30, evaluate the geometric series.

21. $1 + 3 + 9 + \cdots + 3^{200}$

SOLUTION The first term of this series is 1. If we added one more term to this series, the next term would be 3^{201}. The ratio of consecutive terms in this geometric series is 3. Thus

$$1 + 3 + 9 + \cdots + 3^{200} = \frac{1 - 3^{201}}{1 - 3} = \frac{3^{201} - 1}{2}.$$

23. $\frac{1}{4} + \frac{1}{16} + \frac{1}{64} + \cdots + \frac{1}{4^{50}}$

SOLUTION The first term of this series is $\frac{1}{4}$. If we added one more term to this series, the next term would be $1/4^{51}$. The ratio of consecutive terms in this geometric series is $\frac{1}{4}$. Thus

$$\frac{1}{4} + \frac{1}{16} + \frac{1}{64} + \cdots + \frac{1}{4^{50}} = \frac{\frac{1}{4} - \frac{1}{4^{51}}}{1 - \frac{1}{4}} = \frac{1 - \frac{1}{4^{50}}}{3},$$

where the last expression was obtained by multiplying the numerator and denominator of the previous expression by 4.

25. $1 - \frac{1}{2} + \frac{1}{4} - \frac{1}{8} + \cdots + \frac{1}{2^{80}} - \frac{1}{2^{81}}$

SOLUTION The first term of this series is 1. If we added one more term to this series, the next term would be $1/2^{82}$. The ratio of consecutive terms in this geometric series is $-\frac{1}{2}$. Thus

$$\begin{aligned} 1 - \frac{1}{2} + \frac{1}{4} - \frac{1}{8} + \cdots + \frac{1}{2^{80}} - \frac{1}{2^{81}} &= \frac{1 - \frac{1}{2^{82}}}{1 - (-\frac{1}{2})} \\ &= \frac{1 - \frac{1}{2^{82}}}{\frac{3}{2}} = \frac{2 - \frac{1}{2^{81}}}{3}, \end{aligned}$$

where the last expression was obtained by multiplying the numerator and denominator of the previous expression by 2.

27. $\sum_{k=1}^{40} \frac{3}{2^k}$

SOLUTION The first term of the series is $\frac{3}{2}$. If we added one more term to this geometric series, the next term would be $\frac{3}{2^{41}}$. The ratio of consecutive terms in this geometric series is $\frac{1}{2}$. Putting all this together, we have

$$\sum_{k=1}^{40} \frac{3}{2^k} = \frac{\frac{3}{2} - \frac{3}{2^{41}}}{1 - \frac{1}{2}} = 3 - \frac{3}{2^{40}}.$$

29. $\sum_{m=3}^{77} (-5)^m$

SOLUTION The first term of the series is $(-5)^3$, which equals -125. If we added one more term to this geometric series, the next term would be $(-5)^{78}$, which equals 5^{78}. The ratio of consecutive terms in this geometric series is -5. Putting all this together, we have

$$\sum_{m=3}^{77} (-5)^m = \frac{-125 - 5^{78}}{1 - (-5)} = -\frac{125 + 5^{78}}{6}.$$

In Exercises 31–34, write the series explicitly and evaluate the sum.

31. $\sum_{m=1}^{4} (m^2 + 5)$

SOLUTION When $m = 1$, the expression $m^2 + 5$ equals 6. When $m = 2$, the expression $m^2 + 5$

equals 9. When $m = 3$, the expression $m^2 + 5$ equals 14. When $m = 4$, the expression $m^2 + 5$ equals 21. Thus

$$\sum_{m=1}^{4} (m^2 + 5) = 6 + 9 + 14 + 21 = 50.$$

33. $\sum_{k=0}^{3} \log(k^2 + 2)$

SOLUTION When $k = 0$, the expression $\log(k^2 + 2)$ equals $\log 2$. When $k = 1$, the expression $\log(k^2 + 2)$ equals $\log 3$. When $k = 2$, the expression $\log(k^2 + 2)$ equals $\log 6$. When $k = 3$, the expression $\log(k^2 + 2)$ equals $\log 11$. Thus

$$\sum_{k=0}^{3} \log(k^2 + 2) = \log 2 + \log 3 + \log 6 + \log 11$$

$$= \log(2 \cdot 3 \cdot 6 \cdot 11) = \log 396.$$

In Exercises 35–38, write the series using summation notation (starting with $k = 1$). Each series in Exercises 35–38 is either an arithmetic series or a geometric series.

35. $2 + 4 + 6 + \cdots + 100$

SOLUTION The kth term of this sequence is $2k$. The last term corresponds to $k = 50$. Thus

$$2 + 4 + 6 + \cdots + 100 = \sum_{k=1}^{50} 2k.$$

37. $\frac{5}{9} + \frac{5}{27} + \frac{5}{81} + \cdots + \frac{5}{3^{40}}$

SOLUTION The kth term of this sequence is $\frac{5}{3^{k+1}}$. The last term corresponds to $k = 39$ (because when $k = 39$, the expression $\frac{5}{3^{k+1}}$ equals $\frac{5}{3^{40}}$). Thus

$$\frac{5}{9} + \frac{5}{27} + \frac{5}{81} + \cdots + \frac{5}{3^{40}} = \sum_{k=1}^{39} \frac{5}{3^{k+1}}.$$

39. Restate the symbolic version of the formula for evaluating an arithmetic series using summation notation.

SOLUTION Consider an arithmetic series with n terms, with an initial term b, and with difference d between consecutive terms. The kth term of this series is $b + (k - 1)d$. Thus the formula for evaluating an arithmetic series using summation notation is

$$\sum_{k=1}^{n} (b + (k - 1)d) = n(b + \tfrac{(n-1)d}{2}).$$

This could also be written in the form

$$\sum_{k=0}^{n-1} (b + kd) = n(b + \tfrac{(n-1)d}{2}).$$

For Exercises 41–44, consider the fable from the beginning of Section 5.4. In this fable, one grain of rice is placed on the first square of a chessboard, then two grains on the second square, then four grains on the third square, and so on, doubling the number of grains placed on each square.

41. Find the total number of grains of rice on the first 18 squares of the chessboard.

SOLUTION The total number of grains of rice on the first 18 squares of the chessboard is

$$1 + 2 + 4 + 8 + \cdots + 2^{17}.$$

This is a geometric series; the ratio of consecutive terms is 2. The term that would follow the last term is 2^{18}. Thus the sum of this series is

$$\frac{1 - 2^{18}}{1 - 2},$$

which equals $2^{18} - 1$.

43. Find the smallest number n such that the total number of grains of rice on the first n squares of the chessboard is more than 30,000,000.

SOLUTION The formula for a geometric series shows that the total number of grains of rice on the first n squares of the chessboard is $2^n - 1$. Thus we want to find the smallest integer n such that

$$2^n > 30000001.$$

Taking logs of both sides, we see that we need

$$n > \frac{\log 30000001}{\log 2} \approx 24.8.$$

The smallest integer satisfying this inequality is $n = 25$.

For Exercises 45–46, use Pascal's triangle to simplify the indicated expression.

45. $(2-\sqrt{3})^5$

SOLUTION The sixth row of Pascal's triangle is 1 5 10 10 5 1. Thus

$$\begin{aligned}(2-\sqrt{3})^5 &= 2^5 - 5\cdot 2^4\sqrt{3} + 10\cdot 2^3\sqrt{3}^2 - 10\cdot 2^2\sqrt{3}^3 \\ &\quad + 5\cdot 2\sqrt{3}^4 - \sqrt{3}^5 \\ &= 32 - 80\sqrt{3} + 80\cdot 3 - 40\cdot\sqrt{3}^2\sqrt{3} \\ &\quad + 10\cdot\sqrt{3}^2\sqrt{3}^2 - \sqrt{3}^2\sqrt{3}^2\sqrt{3} \\ &= 32 - 80\sqrt{3} + 80\cdot 3 - 40\cdot 3\sqrt{3} \\ &\quad + 10\cdot 3\cdot 3 - 3\cdot 3\sqrt{3} \\ &= 32 - 80\sqrt{3} + 240 - 120\sqrt{3} + 90 - 9\sqrt{3} \\ &= 362 - 209\sqrt{3}.\end{aligned}$$

For Exercises 47–50, assume that n is a positive integer.

47. Evaluate $\binom{n}{n}$.

SOLUTION

$$\binom{n}{n} = \frac{n!}{n!\,0!} = 1$$

49. Evaluate $\binom{n}{n-2}$.

SOLUTION

$$\binom{n}{n-2} = \frac{n!}{(n-2)!\,2!} = \frac{(n-1)n}{2}$$

51. Evaluate $\binom{70}{3}$.

SOLUTION

$$\binom{70}{3} = \frac{70!}{3!\,67!} = \frac{68\cdot 69\cdot 70}{2\cdot 3} = 54740$$

8.3 *Limits*

LEARNING OBJECTIVES

By the end of this section you should be able to

- recognize the limit of a sequence;
- use partial sums to evaluate an infinite series;
- compute the sum of an infinite geometric series;
- convert repeating decimals to fractions.

Introduction to Limits

Consider the sequence

$$1, \tfrac{1}{2}, \tfrac{1}{3}, \tfrac{1}{4}, \ldots;$$

here the nth term of the sequence is $\frac{1}{n}$. For all large values of n, the nth term of this sequence is close to 0. For example, all the terms after the one-millionth term of this sequence are within one-millionth of 0. We say that this sequence has limit 0. More generally, the following informal definition explains what it means for a sequence to have limit equal to some number L.

Limit of a sequence (less precise version)

A sequence has **limit** L if from some point on, all the terms of the sequence are very close to L.

This definition fails to be precise because the phrase "very close" is too vague. A more precise definition of limit will be given soon, but first we examine some examples to get a feel for what is meant by taking the limit of a sequence.

EXAMPLE 1

What is the limit of the sequence whose n^{th} term equals $\sqrt{n^2+n}-n$?

SOLUTION

The limit of a sequence depends on the behavior of the nth term for large values of n. The table to the right shows the values of the nth term of this sequence for some large values of n, calculated by a computer and rounded off to seven digits after the decimal point.

n	$\sqrt{n^2+n}-n$
1	0.4142136
10	0.4880885
100	0.4987562
1000	0.4998751
10000	0.4999875
100000	0.4999988
1000000	0.4999999

This table leads us to suspect that this sequence has limit $\frac{1}{2}$. This suspicion is correct, as can be shown by rewriting the nth term of this sequence as follows:

$$\sqrt{n^2+n}-n = \frac{1}{\sqrt{1+\frac{1}{n}}+1}.$$

See Problem 25 for a hint on how to derive this identity.

If n is very large, then $1 + \frac{1}{n}$ is very close to 1, and thus the right side of the equation above is very close to $\frac{1}{2}$. Hence the limit of the sequence in question is indeed equal to $\frac{1}{2}$.

Not every sequence has a limit, as shown by the following example:

EXAMPLE 2

Explain why the sequence whose n^{th} term equals $(-1)^{n-1}$ does not have a limit.

SOLUTION The sequence in question is the sequence of alternating 1's and -1's:

$$1, -1, 1, -1, \ldots.$$

A number that is very close to -1 must be negative, and a number that is very close to 1 must be positive; thus no number can be very close to both -1 and 1. Hence this sequence does not have a limit.

The next example shows why we need to be careful about the meaning of "very close".

EXAMPLE 3

What is the limit of the sequence all of whose terms equal 10^{-100}?

SOLUTION The sequence in question is the constant sequence

$$10^{-100}, 10^{-100}, 10^{-100}, \ldots.$$

The limit of this sequence is 10^{-100}. Note, however, that all the terms of this sequence are within one-billionth of 0. Thus if "very close" were defined to mean "within one-billionth", then the imprecise definition above might lead us to conclude incorrectly that this sequence has limit 0.

The example above shows that in our initial definition of limit, we cannot replace "very close to L" by "within one-billionth of L". For similar reasons, no single positive number, no matter how small, could be used to define "very close". This dilemma is solved by considering all positive numbers, including those that are very small (whatever that means). The following more precise definition of limit captures the notion that a sequence gets as close as we like to its limit if we go far enough out in the sequence:

As mentioned in Chapter 1, the Greek letter ε (epsilon) is often used when we are thinking about small positive numbers.

Limit of a sequence (more precise version)

A sequence has **limit** L if for every $\varepsilon > 0$, from some point on all terms of the sequence are within ε of L.

This definition means that for each possible choice of a positive number ε, there is some term of the sequence such that all following terms are within ε of L. How far out in the sequence we need to go (to have all the terms beyond there be within ε of L) can depend on ε.

For example, consider the sequence

$$-1,\ \tfrac{1}{2},\ -\tfrac{1}{3},\ \tfrac{1}{4},\ \dots;$$

here the nth term of the sequence equals $\frac{(-1)^n}{n}$. This sequence has limit 0. If we consider the choice $\varepsilon = 10^{-6}$, then all terms after the millionth term of this sequence are within ε of the limit 0. If we consider the choice $\varepsilon = 10^{-9}$, then all terms after the billionth term of this sequence are within ε of the limit 0. No matter how small we choose ε, we can go far enough out in the sequence (depending on ε) so that all the terms beyond there are within ε of 0.

Because the limit of a sequence depends only on what happens "from some point on", changing the first five terms or even the first five million terms does not affect the limit of a sequence. For example, consider the sequence

$$10,\ 100,\ 1000,\ 10000,\ 100000,\ \tfrac{1}{6},\ \tfrac{1}{7},\ \tfrac{1}{8},\ \tfrac{1}{9},\ \tfrac{1}{10},\ \dots;$$

here the nth term of the sequence equals 10^n if $n \le 5$ and equals $\frac{1}{n}$ if $n > 5$. Make sure you understand why the limit of this sequence equals 0.

The notation commonly used to denote the limit of a sequence is introduced below:

Limit notation

The notation

$$\lim_{n\to\infty} a_n = L$$

means that the sequence $a_1, a_2, \dots$ has limit L. We say that the limit of a_n as n goes to infinity equals L.

Once again, remember that ∞ is not a real number; it appears here to help convey the notion that only large values of n matter.

For example, we could write

$$\lim_{n\to\infty} \frac{1}{n} = 0;$$

we would say that the limit of $\frac{1}{n}$ as n goes to infinity equals 0. As another example from earlier in this section, we could write

$$\lim_{n\to\infty} \left(\sqrt{n^2+n} - n\right) = \frac{1}{2};$$

we would say that the limit of $\sqrt{n^2+n} - n$ as n goes to infinity equals $\frac{1}{2}$.

EXAMPLE 4

Evaluate $\lim_{n\to\infty} (1 + \frac{1}{n})^n$.

SOLUTION This is the sequence whose first five terms are

$$2,\ \left(\tfrac{3}{2}\right)^2,\ \left(\tfrac{4}{3}\right)^3,\ \left(\tfrac{5}{4}\right)^4,\ \left(\tfrac{6}{5}\right)^5.$$

A computer can tell us that the one-millionth term of this sequence is approximately 2.71828, which you should recognize as being approximately e. Indeed, in Section 6.2 we saw that $(1 + \frac{1}{n})^n \approx e$ for large values of n. The precise meaning of that approximation is that $\lim\limits_{n\to\infty} (1 + \frac{1}{n})^n = e$.

Consider the geometric sequence

$$\tfrac{1}{2},\ \tfrac{1}{4},\ \tfrac{1}{8},\ \tfrac{1}{16},\ \ldots.$$

Here the nth term equals $(\frac{1}{2})^n$, which is very small for large values of n. Thus this sequence has limit 0, which we can write as $\lim\limits_{n\to\infty} (\frac{1}{2})^n = 0$.

Similarly, multiplying any number with absolute value less than 1 by itself many times produces a number close to 0, as shown in the following example.

EXAMPLE 5

In the decimal expansion of 0.99^{100000}, how many zeros follow the decimal point before the first nonzero digit?

Even though 0.99 *is just slightly less than* 1, *raising it to a large power produces a very small number.*

SOLUTION Calculators cannot evaluate 0.99^{100000}, so take its common logarithm:

$$\log 0.99^{100000} = 100000 \log 0.99 \approx 100000 \cdot (-0.004365) = -436.5.$$

This means that 0.99^{100000} is between 10^{-437} and 10^{-436}. Thus 436 zeros follow the decimal point in the decimal expansion of 0.99^{100000} before the first nonzero digit.

The example above should help convince you that if r is any number with $|r| < 1$, then $\lim\limits_{n\to\infty} r^n = 0$.

Similarly, if $|r| > 1$, then r^n is very large for large values of n. Thus if $|r| > 1$, then the geometric sequence $r, r^2, r^3, \ldots$ does not have a limit.

If $r = -1$, then the geometric sequence $r, r^2, r^3, r^4 \ldots$ is the alternating sequence $-1, 1, -1, 1, \ldots$; this sequence does not have a limit. If $r = 1$, then the geometric sequence $r, r^2, r^3, r^4 \ldots$ is the constant sequence $1, 1, 1, 1, \ldots$; this sequence has limit 1.

Putting together the results above, we have the following summary concerning the limit of a geometric sequence:

Limit of a geometric sequence

Suppose r is a real number. Then the geometric sequence

$$r, r^2, r^3, \ldots$$

- has limit 0 if $|r| < 1$;
- has limit 1 if $r = 1$;
- does not have a limit if $r \leq -1$ or $r > 1$.

Infinite Series

Addition is initially defined as an operation that takes two numbers a and b and produces their sum $a + b$. We can find the sum of a finite sequence $a_1, a_2, \ldots, a_n$ by adding the first two terms a_1 and a_2, getting $a_1 + a_2$, then adding the third term, getting $a_1 + a_2 + a_3$, then adding the fourth term, getting $a_1 + a_2 + a_3 + a_4$, and so on. After n terms we will have found the sum for this finite sequence; this sum can be denoted

Because of the associative property, we do not need to worry about putting parentheses in these sums.

$$a_1 + a_2 + \cdots + a_n \quad \text{or} \quad \sum_{k=1}^{n} a_k.$$

Now consider an infinite sequence $a_1, a_2, \ldots$. What does it mean to find the sum of this infinite sequence? In other words, we want to attach a meaning to the infinite sum

$$a_1 + a_2 + a_3 + \cdots \quad \text{or} \quad \sum_{k=1}^{\infty} a_k.$$

Such sums are called **infinite series**.

The problem with trying to evaluate an infinite series by adding one term at a time is that the process will never terminate. Nevertheless, let's see what happens when we add one term at a time in a familiar geometric sequence.

EXAMPLE 6

What value should be assigned to the infinite sum $\sum_{k=1}^{\infty} \frac{1}{2^k}$?

SOLUTION We need to evaluate the infinite sum

$$\frac{1}{2} + \frac{1}{4} + \frac{1}{8} + \frac{1}{16} + \cdots.$$

The sum of the first two terms equals $\frac{3}{4}$. The sum of the first three terms equals $\frac{7}{8}$. The sum of the first four terms equals $\frac{15}{16}$. More generally, the sum of the first n terms equals $1 - \frac{1}{2^n}$, as can be shown by using the formula from the previous section for the sum of a finite geometric series.

Although the process of adding terms of this series never ends, we see that after adding a large number of terms the sum is close to 1. In other words, the limit of the sum of the first n terms is 1. Thus we declare that the infinite sum equals 1. Expressing all this in summation notation, we have

$$\sum_{k=1}^{\infty} \frac{1}{2^k} = \lim_{n\to\infty} \sum_{k=1}^{n} \frac{1}{2^k} = \lim_{n\to\infty} \left(1 - \frac{1}{2^n}\right) = 1.$$

The example above provides motivation for the formal definition of an infinite sum. To evaluate an infinite series, the idea is to add up the first n terms and then take the limit as n goes to infinity:

The numbers $\sum_{k=1}^{n} a_k$ are called the **partial sums** *of the infinite series. Thus the infinite sum is the limit of the sequence of partial sums.*

Infinite series

The infinite sum $\sum_{k=1}^{\infty} a_k$ is defined by

$$\sum_{k=1}^{\infty} a_k = \lim_{n\to\infty} \sum_{k=1}^{n} a_k$$

if this limit exists.

EXAMPLE 7

Evaluate the geometric series $\sum_{k=1}^{\infty} \frac{1}{10^k}$.

SOLUTION According to the definition above, we need to evaluate the partial sums $\sum_{k=1}^{n} \frac{1}{10^k}$ and then take the limit as n goes to infinity. Using the formula from the previous section for the sum of a finite geometric series, we have

$$\sum_{k=1}^{n} \frac{1}{10^k} = \frac{\frac{1}{10} - \frac{1}{10^{n+1}}}{1 - \frac{1}{10}} = \frac{1 - \frac{1}{10^n}}{9},$$

where the last expression is obtained by multiplying the numerator and denominator of the middle expression by 10. Thus

$$\sum_{k=1}^{\infty} \frac{1}{10^k} = \lim_{n\to\infty} \sum_{k=1}^{n} \frac{1}{10^k} = \lim_{n\to\infty} \frac{1 - \frac{1}{10^n}}{9} = \frac{1}{9}.$$

Some infinite sequences cannot be summed, because the limit of the sequence of partial sums does not exist. When this happens, the infinite sum is left undefined.

EXAMPLE 8

Explain why the infinite series $\sum_{k=1}^{\infty} (-1)^k$ is undefined.

SOLUTION We are trying to make sense of the infinite sum

$$-1 + 1 - 1 + 1 - 1 + \cdots.$$

Following the usual procedure for infinite sums, first we evaluate the partial sums $\sum_{k=1}^{n} (-1)^k$, getting

$$\sum_{k=1}^{n} (-1)^k = \begin{cases} -1 & \text{if } n \text{ is odd} \\ 0 & \text{if } n \text{ is even.} \end{cases}$$

Thus the sequence of partial sums is the alternating sequence of -1's and 0's. This sequence of partial sums does not have a limit. Thus the infinite sum is undefined.

We turn now to the problem of finding a formula for evaluating an infinite geometric series. Fix a number $r \neq 1$, and consider the geometric series

$$1 + r + r^2 + r^3 + \cdots;$$

here the ratio of consecutive terms is r. The sum of the first n terms is $1 + r + r^2 + \cdots + r^{n-1}$. The term following the last term would be r^n; thus by our formula for evaluating a geometric series we have

$$1 + r + r^2 + \cdots + r^{n-1} = \frac{1 - r^n}{1 - r}.$$

By definition, the infinite sum $1 + r + r^2 + r^3 + \cdots$ equals the limit (if it exists) of the partial sums above as n goes to infinity. We already know that the limit of r^n as n goes to infinity is 0 if $|r| < 1$ (and does not exist if $|r| > 1$). Thus we get the following beautiful formula:

Evaluating an infinite geometric series

If $|r| < 1$, then

$$1 + r + r^2 + r^3 + \cdots = \frac{1}{1 - r}.$$

If $|r| \geq 1$, then this infinite sum is not defined.

Any infinite geometric series can be reduced to the form above by factoring out the first term. The following example illustrates the procedure.

EXAMPLE 9

Evaluate the geometric series $\frac{7}{3} + \frac{7}{9} + \frac{7}{27} + \cdots$.

SOLUTION We factor out the first term $\frac{7}{3}$ and then apply the formula above, getting

$$\frac{7}{3} + \frac{7}{9} + \frac{7}{27} + \cdots = \frac{7}{3}\left(1 + \frac{1}{3} + \frac{1}{9} + \cdots\right) = \frac{7}{3} \cdot \frac{1}{1 - \frac{1}{3}} = \frac{7}{2}.$$

Decimals as Infinite Series

A **digit** is one of the numbers $0, 1, 2, 3, 4, 5, 6, 7, 8, 9$. Each real number t between 0 and 1 can be expressed as a decimal in the form

$$t = 0.d_1 d_2 d_3 \ldots,$$

where $d_1, d_2, d_3, \ldots$ is a sequence of digits. The interpretation of this representation is that

$$t = \frac{d_1}{10} + \frac{d_2}{100} + \frac{d_3}{1000} + \cdots,$$

which we can write in summation notation as

$$t = \sum_{k=1}^{\infty} \frac{d_k}{10^k}.$$

In other words, real numbers are represented by infinite series.

If from some point on each d_k equals 0, then we have what is called a **terminating decimal**; in this case we usually do not write the ending string of 0's.

EXAMPLE 10

Express 0.217 as a fraction.

SOLUTION In this case, the infinite series above becomes a finite series:

$$0.217 = \frac{2}{10} + \frac{1}{100} + \frac{7}{1000} = \frac{200}{1000} + \frac{10}{1000} + \frac{7}{1000} = \frac{217}{1000}$$

If the decimal representation of a number has a pattern that repeats from some point on, then we have what is called a **repeating decimal**.

EXAMPLE 11

Express

$$0.11111\ldots$$

as a fraction; here the digit 1 keeps repeating forever.

SOLUTION Using the interpretation of the decimal representation, we have

$$0.11111\ldots = \sum_{k=1}^{\infty} \frac{1}{10^k}.$$

The sum above is an infinite geometric series. As we saw in Example 7, this infinite geometric series equals $\frac{1}{9}$. Thus

$$0.11111\ldots = \frac{1}{9}.$$

Every irrational number has a nonrepeating decimal expansion.

Any repeating decimal can be converted to a fraction by evaluating an appropriate infinite geometric series. However, the technique used in the following example is usually easier.

EXAMPLE 12

Express

$$0.52473473473\ldots$$

as a fraction; here the digits 473 keep repeating forever.

SOLUTION Let

$$t = 0.52473473473\ldots.$$

The trick is to note that

$$1000t = 524.73473473473\ldots.$$

Subtracting the first equation above from the last equation, we get

$$999t = 524.21.$$

Thus

$$t = \frac{524.21}{999} = \frac{52421}{99900}.$$

Special Infinite Series

Advanced mathematics produces many beautiful special infinite series. We cannot derive the values for these infinite series here, but they are so pretty that you should at least see a few of them.

EXAMPLE 13

Evaluate $\sum_{k=1}^{\infty} \frac{1}{k!}$.

SOLUTION A computer calculation can give a partial sum that leads to a correct guess. Specifically,

$$\sum_{k=1}^{1000} \frac{1}{k!} \approx 1.718281828459.$$

You may recognize the digits after the decimal point as the digits after the decimal point of e. It is indeed true that this infinite sum equals $e - 1$. Adding 1 to both sides of the equation $\sum_{k=1}^{\infty} \frac{1}{k!} = e - 1$ gives the beautiful infinite series

$$1 + \frac{1}{1!} + \frac{1}{2!} + \frac{1}{3!} + \cdots = e.$$

This equation again shows how e magically appears throughout mathematics.

More generally, as you will learn if you take calculus, the following equation is true for every number x:

$$1 + \frac{x}{1!} + \frac{x^2}{2!} + \frac{x^3}{3!} + \cdots = e^x.$$

The next example shows again that the natural logarithm deserves the word "natural".

EXAMPLE 14

Evaluate $\sum_{k=1}^{\infty} \frac{(-1)^{k+1}}{k}$.

SOLUTION Once again a computer calculation can give a partial sum that leads to a correct guess. Specifically,

$$\sum_{k=1}^{100000} \frac{(-1)^{k+1}}{k} \approx 0.693142.$$

You may recognize the first five digits after the decimal point as the first five digits in the decimal expansion of ln 2. The infinite sum indeed equals ln 2. In other words, we have the following delightful equation:

$$1 - \frac{1}{2} + \frac{1}{3} - \frac{1}{4} + \frac{1}{5} - \frac{1}{6} + \cdots = \ln 2.$$

The next example presents another famous infinite series.

EXAMPLE 15 Evaluate $\sum_{k=1}^{\infty} \frac{1}{k^2}$.

SOLUTION A computer calculation can give a partial sum. Specifically,

$$\sum_{k=1}^{1000000} \frac{1}{k^2} \approx 1.64493.$$

Leonard Euler, the most important mathematician of the 18th century.

The value of this infinite series is hard to recognize even from this good approximation. In fact, the exact evaluation of this infinite sum was an unsolved problem for many years, but the Swiss mathematician Leonard Euler showed in 1735 that this infinite series equals $\frac{\pi^2}{6}$. In other words, we have the beautiful equation

$$1 + \frac{1}{4} + \frac{1}{9} + \frac{1}{16} + \cdots = \frac{\pi^2}{6}.$$

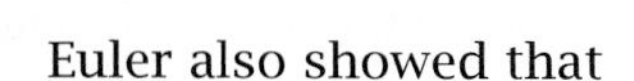

Euler also showed that

$$\sum_{k=1}^{\infty} \frac{1}{k^4} = \frac{\pi^4}{90} \quad \text{and} \quad \sum_{k=1}^{\infty} \frac{1}{k^6} = \frac{\pi^6}{945}.$$

The next example is presented to show that there are still unsolved problems in mathematics that are easy to state.

EXAMPLE 16 Evaluate $\sum_{k=1}^{\infty} \frac{1}{k^3}$.

SOLUTION A computer calculation can give a partial sum. Specifically,

$$\sum_{k=1}^{1000000} \frac{1}{k^3} \approx 1.2020569.$$

No one knows an exact expression for the infinite series $\sum_{k=1}^{\infty} \frac{1}{k^3}$. You will become famous if you find one!

EXERCISES

1. Evaluate $\lim_{n\to\infty} \frac{3n+5}{2n-7}$.
2. Evaluate $\lim_{n\to\infty} \frac{4n-2}{7n+6}$.
3. Evaluate $\lim_{n\to\infty} \frac{2n^2+5n+1}{5n^2-6n+3}$.
4. Evaluate $\lim_{n\to\infty} \frac{7n^2-4n+3}{3n^2+5n+9}$.
5. Evaluate $\lim_{n\to\infty} (1+\frac{3}{n})^n$.
6. Evaluate $\lim_{n\to\infty} (1-\frac{1}{n})^n$.
7. Evaluate $\lim_{n\to\infty} n(e^{1/n}-1)$.
8. Evaluate $\lim_{n\to\infty} n\ln(1+\frac{1}{n})$.
9. Evaluate $\lim_{n\to\infty} n(\ln(3+\frac{1}{n})-\ln 3)$.
10. Evaluate $\lim_{n\to\infty} n(\ln(7+\frac{1}{n})-\ln 7)$.
11. Find the smallest integer n such that $0.8^n < 10^{-100}$.
12. Find the smallest integer n such that $0.9^n < 10^{-200}$.
13. In the decimal expansion of 0.87^{1000}, how many zeros follow the decimal point before the first nonzero digit?
14. In the decimal expansion of 0.9^{9999}, how many zeros follow the decimal point before the first nonzero digit?
15. Evaluate $\sum_{k=1}^{\infty} \frac{3}{7^k}$.
16. Evaluate $\sum_{k=1}^{\infty} \frac{8}{5^k}$.
17. Evaluate $\sum_{m=2}^{\infty} \frac{5}{6^m}$.
18. Evaluate $\sum_{m=3}^{\infty} \frac{8}{3^m}$.
19. Express
$$0.23232323\ldots$$
as a fraction; here the digits 23 keep repeating forever.
20. Express
$$0.859859859\ldots$$
as a fraction; here the digits 859 keep repeating forever.
21. Express
$$8.237545454\ldots$$
as a fraction; here the digits 54 keep repeating forever.
22. Express
$$5.1372647264\ldots$$
as a fraction; here the digits 7264 keep repeating forever.

PROBLEMS

23. Give an example of a sequence that has limit 3 and whose first five terms are $2, 4, 6, 8, 10$.
24. Suppose you are given a sequence with limit L and that you change the sequence by adding 50 to the first 1000 terms, leaving the other terms unchanged. Explain why the new sequence also has limit L.
25. Show that
$$\sqrt{n^2+n}-n = \frac{1}{\sqrt{1+\frac{1}{n}}+1}.$$
[*Hint:* Multiply by $\sqrt{n^2+n}-n$ by $(\sqrt{n^2+n}+n)/(\sqrt{n^2+n}+n)$. Then factor n out of the numerator and denominator of the resulting expression.]
[*This identity was used in Example 1.*]
26. Which arithmetic sequences have a limit?
27. Suppose x is a positive number.
    (a) Explain why $x^{1/n} = e^{(\ln x)/n}$ for every nonzero number n.
    (b) Explain why
    $$n(x^{1/n}-1) \approx \ln x$$
    if n is very large.
    (c) Explain why
    $$\ln x = \lim_{n\to\infty} n(x^{1/n}-1).$$

[*A few books use the last equation above as the definition of the natural logarithm.*]

28. Find the only arithmetic sequence $a_1, a_2, a_3, \ldots$ such that the infinite sum $\sum_{k=1}^{\infty} a_k$ exists.

29. Show that if $|r| < 1$, then

$$\sum_{m=1}^{\infty} r^m = \frac{r}{1-r}.$$

30. Explain why 0.2 and the repeating decimal 0.199999... both represent the real number $\frac{1}{5}$.

31. Learn about Zeno's paradox (from a book, a friend, or a web search) and then relate the explanation of this ancient Greek problem to the infinite series

$$\frac{1}{2} + \frac{1}{4} + \frac{1}{8} + \frac{1}{16} + \cdots = 1.$$

32. Explain how the formula

$$e^x = 1 + \frac{x}{1!} + \frac{x^2}{2!} + \frac{x^3}{3!} + \cdots$$

leads to the approximation $e^x \approx 1 + x$ if $|x|$ is small (which we derived by another method in Section 6.2).

WORKED-OUT SOLUTIONS *to Odd-numbered Exercises*

1. Evaluate $\lim_{n\to\infty} \frac{3n+5}{2n-7}$.

SOLUTION Dividing numerator and denominator of this fraction by n, we see that

$$\frac{3n+5}{2n-7} = \frac{3+\frac{5}{n}}{2-\frac{7}{n}}.$$

If n is very large, then the numerator of the fraction on the right is close to 3 and the denominator is close to 2. Thus $\lim_{n\to\infty} \frac{3n+5}{2n-7} = \frac{3}{2}$.

3. Evaluate $\lim_{n\to\infty} \frac{2n^2+5n+1}{5n^2-6n+3}$.

SOLUTION Dividing numerator and denominator of this fraction by n^2, we see that

$$\frac{2n^2+5n+1}{5n^2-6n+3} = \frac{2+\frac{5}{n}+\frac{1}{n^2}}{5-\frac{6}{n}+\frac{3}{n^2}}.$$

If n is very large, then the numerator of the fraction on the right is close to 2 and the denominator is close to 5. Thus

$$\lim_{n\to\infty} \frac{2n^2+5n+1}{5n^2-6n+3} = \frac{2}{5}.$$

5. Evaluate $\lim_{n\to\infty} (1+\frac{3}{n})^n$.

SOLUTION The properties of the exponential function imply that if n is very large, then $(1+\frac{3}{n})^n \approx e^3$; see Section 6.2. Thus $\lim_{n\to\infty} (1+\frac{3}{n})^n = e^3$.

7. Evaluate $\lim_{n\to\infty} n(e^{1/n}-1)$.

SOLUTION Suppose n is very large. Then $\frac{1}{n}$ is very close to 0, which means that $e^{1/n} \approx 1 + \frac{1}{n}$. Thus $e^{1/n} - 1 \approx \frac{1}{n}$, which implies that

$$n(e^{1/n}-1) \approx 1.$$

Thus

$$\lim_{n\to\infty} n(e^{1/n}-1) = 1.$$

9. Evaluate $\lim_{n\to\infty} n(\ln(3+\frac{1}{n}) - \ln 3)$.

SOLUTION Note that

$$\ln(3+\tfrac{1}{n}) - \ln 3 = \ln(1+\tfrac{1}{3n}).$$

Suppose n is very large. Then $\frac{1}{3n}$ is very close to 0, which implies $\ln(1+\frac{1}{3n}) \approx \frac{1}{3n}$. Thus $n(\ln(3+\frac{1}{n}) - \ln 3) = n\ln(1+\frac{1}{3n}) \approx \frac{1}{3}$. Thus

$$\lim_{n\to\infty} n(\ln(3+\tfrac{1}{n}) - \ln 3) = \tfrac{1}{3}.$$

11. Find the smallest integer n such that $0.8^n < 10^{-100}$.

SOLUTION The inequality $0.8^n < 10^{-100}$ is equivalent to the inequality

$$\log 0.8^n < \log 10^{-100},$$

which can be rewritten as $n \log 0.8 < -100$. Because 0.8 is less than 1, we know that $\log 0.8$ is negative. Thus dividing by $\log 0.8$ reverses the direction of the inequality, changing the previous inequality into the inequality

$$n > \frac{-100}{\log 0.8} \approx 1031.9.$$

The smallest integer that is greater than 1031.9 is 1032. Thus 1032 is the smallest integer n such that $0.8^n < 10^{-100}$.

13. In the decimal expansion of 0.87^{1000}, how many zeros follow the decimal point before the first nonzero digit?

SOLUTION Taking a common logarithm, we have

$$\log 0.87^{1000} = 1000 \log 0.87 \approx -60.5.$$

This means that 0.87^{1000} is between 10^{-61} and 10^{-60}. Thus 60 zeros follow the decimal point in the decimal expansion of 0.87^{1000} before the first nonzero digit.

15. Evaluate $\sum_{k=1}^{\infty} \frac{3}{7^k}$.

SOLUTION

$$\begin{aligned}\sum_{k=1}^{\infty} \frac{3}{7^k} &= \frac{3}{7} + \frac{3}{7^2} + \frac{3}{7^3} + \cdots \\ &= \frac{3}{7}\left(1 + \frac{1}{7} + \frac{1}{7^2} + \cdots\right) \\ &= \frac{3}{7} \cdot \frac{1}{1 - \frac{1}{7}} \\ &= \frac{1}{2}\end{aligned}$$

17. Evaluate $\sum_{m=2}^{\infty} \frac{5}{6^m}$.

SOLUTION

$$\begin{aligned}\sum_{m=2}^{\infty} \frac{5}{6^m} &= \frac{5}{6^2} + \frac{5}{6^3} + \frac{5}{6^4} + \cdots \\ &= \frac{5}{36}\left(1 + \frac{1}{6} + \frac{1}{6^2} + \cdots\right) \\ &= \frac{5}{36} \cdot \frac{1}{1 - \frac{1}{6}} \\ &= \frac{1}{6}\end{aligned}$$

19. Express

$$0.23232323\ldots$$

as a fraction; here the digits 23 keep repeating forever.

SOLUTION Let

$$t = 0.23232323\ldots.$$

Note that

$$100t = 23.23232323\ldots.$$

Subtracting the first equation above from the last equation, we get

$$99t = 23.$$

Thus

$$t = \frac{23}{99}.$$

21. Express

$$8.237545454\ldots$$

as a fraction; here the digits 54 keep repeating forever.

SOLUTION Let

$$t = 8.237545454\ldots.$$

Note that

$$100t = 823.754545454\ldots.$$

Subtracting the first equation above from the last equation, we get

$$99t = 815.517.$$

Thus

$$t = \frac{815.517}{99} = \frac{815517}{99000} = \frac{90613}{11000}.$$

CHAPTER SUMMARY

To check that you have mastered the most important concepts and skills covered in this chapter, make sure that you can do each item in the following list:

- Compute the terms of an arithmetic sequence given any term and the difference between consecutive terms.
- Compute the terms of an arithmetic sequence given any two terms.
- Compute the terms of a geometric sequence given any term and the ratio of consecutive terms.
- Compute the terms of a geometric sequence given any two terms.
- Compute the terms of a recursively defined sequence given the equations defining the sequence.
- Compute the sum of a finite arithmetic sequence.
- Compute the sum of a finite geometric sequence.
- Work with summation notation.
- Use the Binomial Theorem.
- Explain the intuitive notion of limit.
- Compute the sum of an infinite geometric sequence.
- Convert a repeating decimal to a fraction.

To review a chapter, go through the list above to find items that you do not know how to do, then reread the material in the chapter about those items. Then try to answer the chapter review questions below without looking back at the chapter.

CHAPTER REVIEW QUESTIONS

1. Explain why a sequence whose first four terms are $41, 58, 75, 94$ is not an arithmetic sequence.

2. Give two different examples of arithmetic sequences whose fifth term equals 17.

3. Explain why a sequence whose first four terms are $24, 36, 54, 78$ is not a geometric sequence.

4. Give two different examples of geometric sequences whose fourth term equals 29.

5. Find a number t such that the finite sequence $1, 5, t$ is an arithmetic sequence.

6. Find a number t such that the finite sequence $1, 5, t$ is a geometric sequence.

7. Find the fifth term of the sequence defined recursively by the equations

$$a_1 = 2 \quad \text{and} \quad a_{n+1} = \frac{1}{a_n + 1} \text{ for } n \geq 1.$$

8. Give a recursive definition of the sequence whose n^{th} term equals $4^{-n}n!$.

9. Find the sum of all the three-digit even positive integers.

10. Evaluate $\sum_{j=1}^{22} (-5)^j$.

11. What is the coefficient of $x^{48}y^2$ in the expansion of $(x + y)^{50}$?

12. Evaluate $\lim_{n\to\infty} \frac{4n^2 + 1}{3n^2 - 5n}$.

13. Evaluate $\sum_{k=1}^{\infty} \frac{6}{12^k}$.

14. Evaluate $\sum_{m=3}^{\infty} \frac{5}{4^m}$.

15. Express

$$0.417898989\ldots$$

as a fraction; here the digits 89 keep repeating forever.

CHAPTER

9

The Transamerica Pyramid in San Francisco. Architects used trigonometry to design the unusual triangular faces of this building.

Trigonometric Functions

This chapter introduces the trigonometric functions. These remarkably useful functions appear in many parts of mathematics.

Trigonometric functions live most comfortably in the context of the unit circle. Thus this chapter begins with a careful examination of the unit circle, including a discussion of negative angles and angles greater than 360°.

Many formulas become simpler if angles are measured in radians rather than degrees. Hence we will become familiar with radians before defining the basic trigonometric functions—the cosine, sine, and tangent.

After defining the trigonometric functions in the context of the unit circle, we will see how these functions allow us to compute the measurements of right triangles. We will also dip into the vast sea of trigonometric identities.

9.1 *The Unit Circle*

LEARNING OBJECTIVES

By the end of this section you should be able to

- find points on the unit circle;
- find the radius of the unit circle corresponding to any angle, including negative angles and angles greater than 360°;
- compute the length of a circular arc;
- find the coordinates of the endpoint of the radius of the unit circle corresponding to any multiple of 30° or 45°.

The Equation of the Unit Circle

Trigonometry takes place most conveniently in the context of the unit circle. Thus we begin this chapter by acquainting ourselves with this crucial object.

The unit circle

The **unit circle** is the circle of radius 1 centered at the origin.

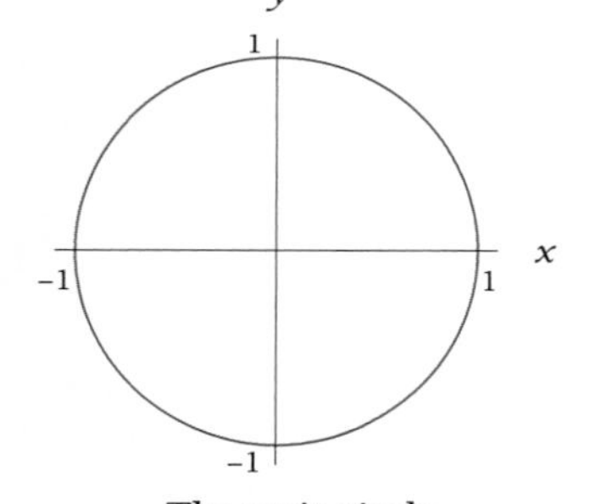

The unit circle.

The unit circle intersects the horizontal axis at the points $(1, 0)$ and $(-1, 0)$, and it intersects the vertical axis at the points $(0, 1)$ and $(0, -1)$, as shown in the figure here.

The unit circle in the xy-plane is described by the equation below. You should become thoroughly familiar with this equation.

Equation of the unit circle

The unit circle in the xy-plane is the set of points (x, y) such that

$$x^2 + y^2 = 1.$$

EXAMPLE 1 Find the points on the unit circle whose first coordinate equals $\frac{2}{3}$.

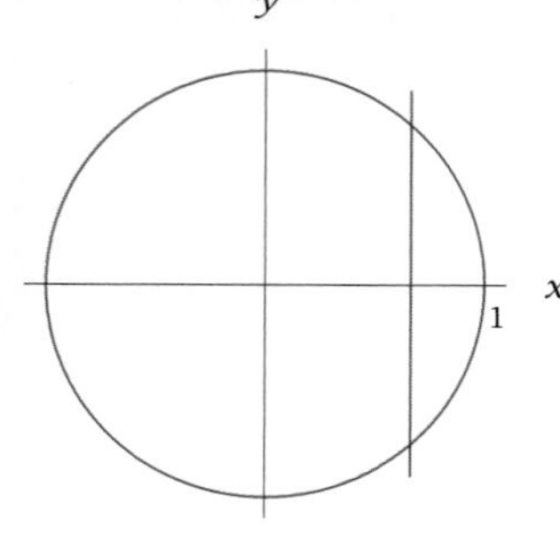

The unit circle and the line $x = \frac{2}{3}$.

SOLUTION We need to find the intersection of the unit circle and the line in the xy-plane whose equation is $x = \frac{2}{3}$, as shown here. To find this intersection, set x equal to $\frac{2}{3}$ in the equation $x^2 + y^2 = 1$ and then solve for y. In other words, we need to solve the equation

$$\left(\tfrac{2}{3}\right)^2 + y^2 = 1.$$

This simplifies to the equation $y^2 = \frac{5}{9}$, which implies that $y = \frac{\sqrt{5}}{3}$ or $y = -\frac{\sqrt{5}}{3}$. Thus the points on the unit circle whose first coordinate equals $\frac{2}{3}$ are $\left(\frac{2}{3}, \frac{\sqrt{5}}{3}\right)$ and $\left(\frac{2}{3}, -\frac{\sqrt{5}}{3}\right)$.

The next example shows how to find the coordinates of the points where the unit circle intersects the line through the origin with slope 1 (which in the xy-plane is described by the equation $y = x$).

EXAMPLE 2

Find the points on the unit circle whose two coordinates are equal.

SOLUTION We need to find the intersection of the unit circle and the line in the xy-plane whose equation is $y = x$. To find this intersection, set y equal to x in the equation $x^2 + y^2 = 1$ and then solve for x. In other words, we need to solve the equation

$$x^2 + x^2 = 1.$$

This simplifies to the equation $2x^2 = 1$, which implies that $x = \pm\frac{1}{\sqrt{2}} = \pm\frac{\sqrt{2}}{2}$. Thus the points on the unit circle whose coordinates are equal are $(\frac{\sqrt{2}}{2}, \frac{\sqrt{2}}{2})$ and $(-\frac{\sqrt{2}}{2}, -\frac{\sqrt{2}}{2})$.

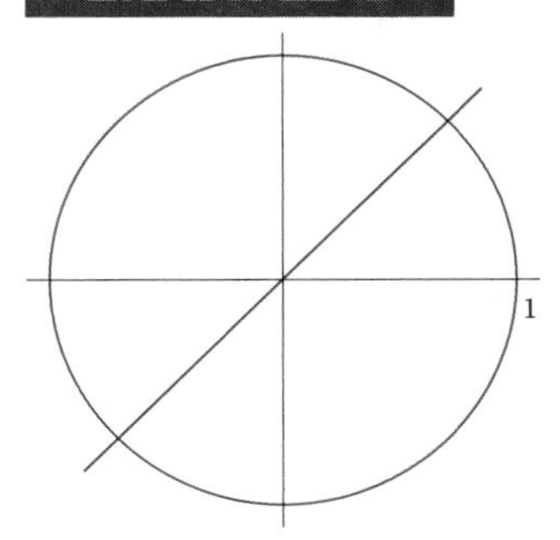

The unit circle and the line through the origin with slope 1.

Angles in the Unit Circle

The **positive horizontal axis**, which plays a special role in trigonometry, is the set of points on the horizontal axis that lie to the right of the origin. When we want to call attention to the positive horizontal axis, sometimes we draw it thicker than normal, as shown here.

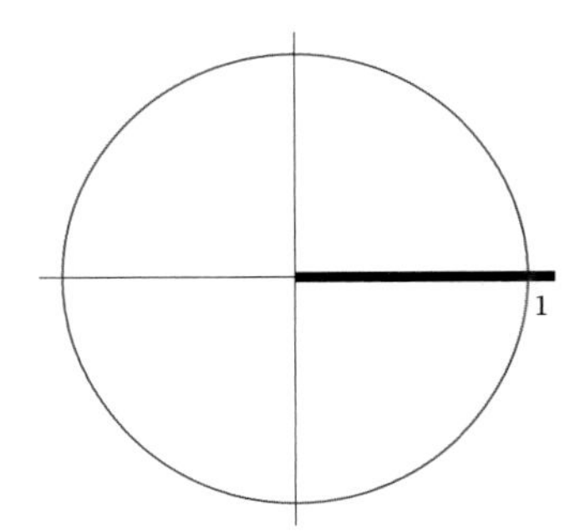

The unit circle with a thickened positive horizontal axis.

We will also occasionally refer to the negative horizontal axis, the positive vertical axis, and the negative vertical axis. These terms are sufficiently descriptive so that definitions are almost unneeded, but here are the formal definitions:

Positive and negative horizontal and vertical axes

- The **positive horizontal axis** is the set of points in the coordinate plane of the form $(x, 0)$, where $x > 0$.
- The **negative horizontal axis** is the set of points in the coordinate plane of the form $(x, 0)$, where $x < 0$.
- The **positive vertical axis** is the set of points in the coordinate plane of the form $(0, y)$, where $y > 0$.
- The **negative vertical axis** is the set of points in the coordinate plane of the form $(0, y)$, where $y < 0$.

We will be dealing with the angle between a radius of the unit circle and the positive horizontal axis, measured counterclockwise from the positive horizontal axis. **Counterclockwise** refers to the opposite direction from the motion of a clock's hands. For example, the figure here shows the radius of the unit circle whose endpoint is $(\frac{\sqrt{2}}{2}, \frac{\sqrt{2}}{2})$. This radius has a 45° angle with the positive horizontal axis.

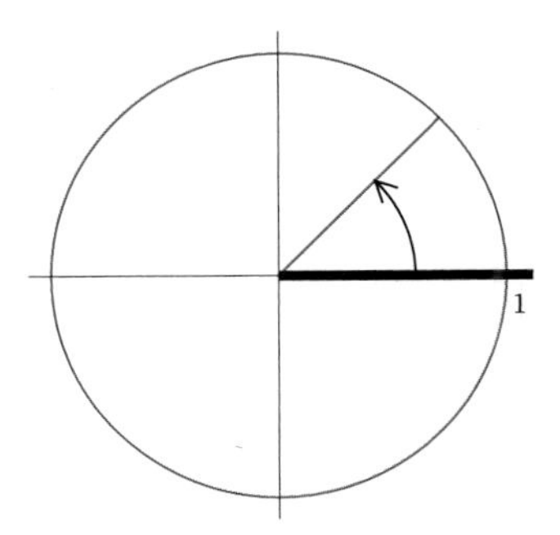

The radius that has a 45° angle with the positive horizontal axis. The arrow indicates the counter-clockwise direction.

The radius corresponding to an angle

For $\theta > 0$, the **radius of the unit circle corresponding to θ degrees** is the radius that has angle θ degrees with the positive horizontal axis, as measured counterclockwise from the positive horizontal axis.

EXAMPLE 3

Sketch the radius of the unit circle corresponding to each of the following angles: $90°$, $180°$, and $360°$.

SOLUTION The radius ending at $(0, 1)$ on the positive vertical axis has a $90°$ angle with the positive horizontal axis.

Similarly, the radius ending at $(-1, 0)$ on the negative horizontal axis has a $180°$ angle with the positive horizontal axis.

Going all the way around the circle corresponds to a $360°$ angle, getting us back to where we started with the radius ending at $(1, 0)$ on the positive horizontal axis.

The figure below shows each of these key angles and its corresponding radius:

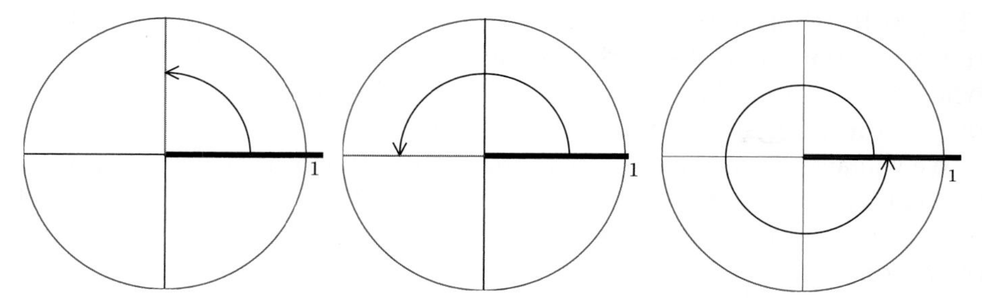

The radius corresponding to $90°$ *(left),* $180°$ *(center), and* $360°$ *(right).*

EXAMPLE 4

Sketch the radius of the unit circle corresponding to each of the following angles: $20°$, $100°$, and $200°$.

The radius corresponding to $100°$ lies slightly to the left of the positive vertical axis because $100°$ is slightly bigger than $90°$. The radius corresponding to $200°$ lies somewhat below the negative horizontal axis because $200°$ is somewhat bigger than $180°$.

SOLUTION The figure below shows each angle and its corresponding radius:

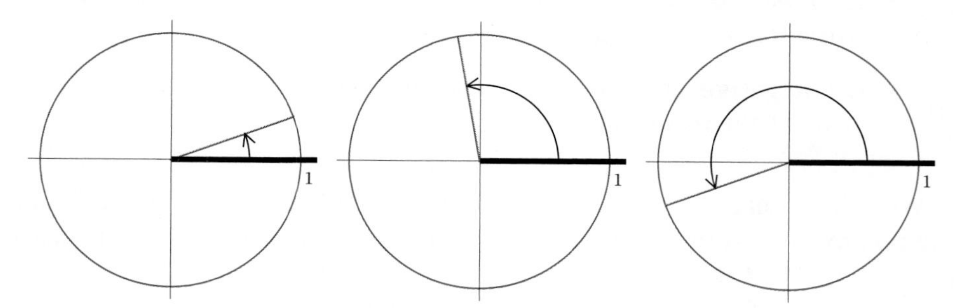

The radius corresponding to $20°$ *(left),* $100°$ *(center), and* $200°$ *(right).*

Negative Angles

Sometimes it is useful to consider the radius of the unit circle corresponding to a negative angle. By a negative angle we simply mean an angle that is measured clockwise from the positive horizontal axis.

> ***The radius corresponding to a negative angle***
>
> For $\theta < 0$, the **radius of the unit circle corresponding to θ degrees** is the radius that has angle $|\theta|$ degrees with the positive horizontal axis, as measured clockwise from the positive horizontal axis.

Clockwise *refers to the direction in which a clock's hands move, as shown by the arrows in Example 5.*

EXAMPLE 5

Sketch the radius of the unit circle corresponding to each of the following angles: $-30°$, $-60°$, and $-90°$.

SOLUTION

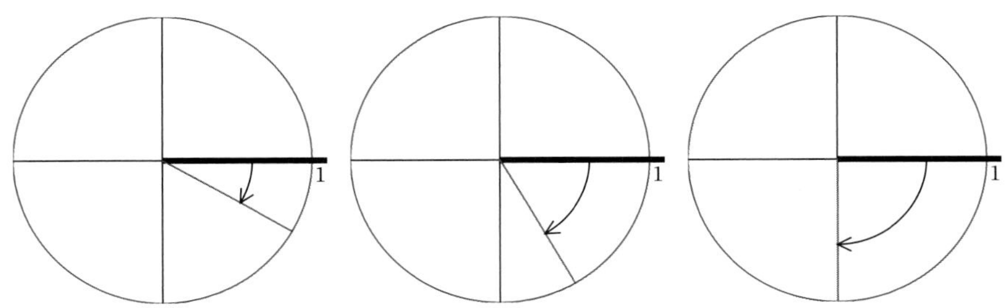

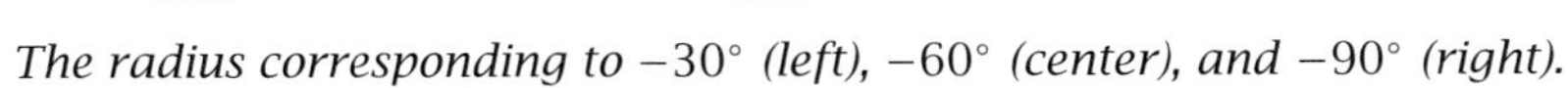

The radius corresponding to $-30°$ (left), $-60°$ (center), and $-90°$ (right).

The next example points out that a radius can correspond to more than one angle.

EXAMPLE 6

Does the radius below on the left correspond to a positive angle or to a negative angle?

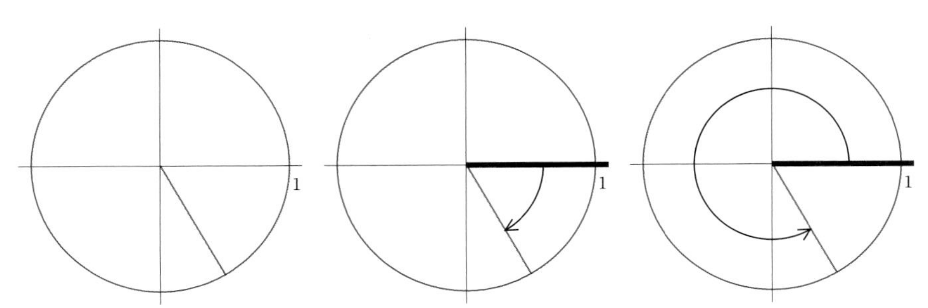

Does this radius correspond to $-60°$ (as in the center) or to $300°$ (as on the right)?

SOLUTION Measuring clockwise from the positive horizontal axis, the radius on the left corresponds to $-60°$. Or measuring counterclockwise from the positive horizontal axis, the same radius corresponds to $300°$. Which is correct?

Depending on the context, either of these interpretations could be correct. Each angle (positive or negative) corresponds to a unique radius of the unit circle, but a given radius corresponds to more than one angle. Thus given only the radius above on the left with no other information, it is not possible to determine whether it corresponds to a positive angle or to a negative angle.

In summary, the radius on the unit circle corresponding to an angle is determined as follows:

Positive and negative angles

- Angle measurements for a radius on the unit circle are made from the positive horizontal axis.
- Positive angles correspond to moving counterclockwise from the positive horizontal axis.
- Negative angles correspond to moving clockwise from the positive horizontal axis.

Angles Greater Than 360°

Sometimes we must consider angles with absolute value greater than $360°$. Such angles arise by starting at the positive horizontal axis and looping around the circle one or more times. The next example shows this procedure.

EXAMPLE 7

Consider the radius of the unit circle corresponding to $40°$. Discuss other angles that correspond to this radius.

SOLUTION Starting from the positive horizontal axis and moving counterclockwise, we could end up at the same radius by going completely around the circle ($360°$) and then continuing for another $40°$, for a total of $400°$ as shown in the center below. Or we could go completely around the circle twice ($720°$) and then continue counterclockwise for another $40°$ for a total of $760°$, as shown below on the right.

We could continue to add multiples of $360°$, showing that the same radius corresponds to $40 + 360n$ degrees for every positive integer n.

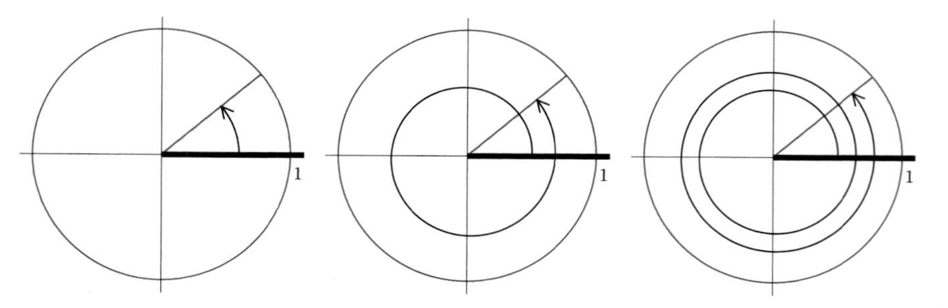

The same radius corresponds to $40°$ (left), $400°$ (center), $760°$ (right), and so on.

We can get another set of angles for the same radius by measuring clockwise from the positive horizontal axis. The figure below in the center shows that the radius corresponding to 40° also corresponds to −320°. Or we could go completely around the circle in the clockwise direction (−360°) and then continue clockwise to the radius (another −320°) for a total of −680°, as shown below on the right.

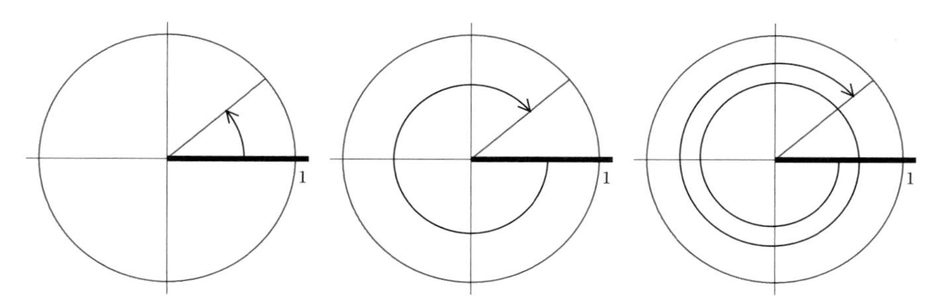

The same radius corresponds to 40° *(left),* −320° *(center),* −680° *(right), and so on.*

We could continue to subtract multiples of 360°*, showing that the same radius corresponds to* $40 + 360n$ *degrees for every negative integer* n.

We have defined the radius of the unit circle corresponding to any positive or negative angle. For completeness, we should also state explicitly that the radius corresponding to 0° is the radius along the positive horizontal axis. More generally, if n is any integer, then the radius corresponding to $360n$ degrees is the radius along the positive horizontal axis.

For each real number θ, there is one radius of the unit circle corresponding to θ degrees. However, for each radius of the unit circle, there are infinitely many angles corresponding to that radius. Here is the precise result:

Multiple choices for the angle corresponding to a radius

A radius of the unit circle corresponding to θ degrees also corresponds to $\theta + 360n$ degrees for every integer n.

Length of a Circular Arc

In Section 2.1 we saw that the length of a curve can be determined by placing a string on the curve and then measuring the length of the string when it is straightened into a line segment. We will now find a formula for the length of a circular arc, starting with a simple example.

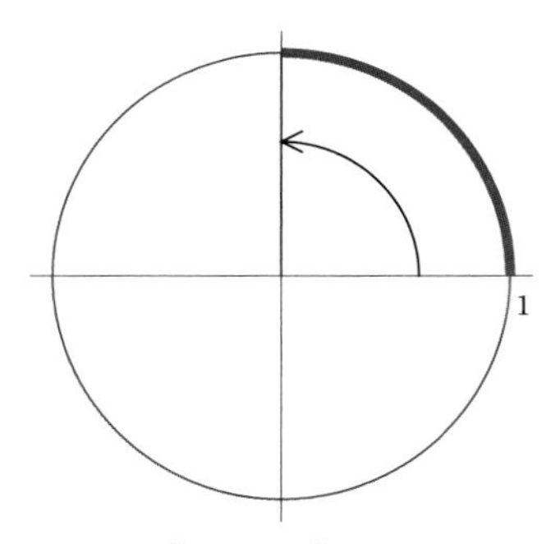

The circular arc corresponding to 90°.

EXAMPLE 8

What is the length of a circular arc on the unit circle corresponding to 90°?

SOLUTION The circular arc on the unit circle corresponding to 90° is shown above as the thickened part of the unit circle.

To find the length of this circular arc, recall that the length (circumference) of the entire unit circle equals 2π (from the familiar expression $2\pi r$ with $r = 1$). The circular arc here is one-fourth of the unit circle. Thus its length equals $\frac{2\pi}{4}$, which equals $\frac{\pi}{2}$.

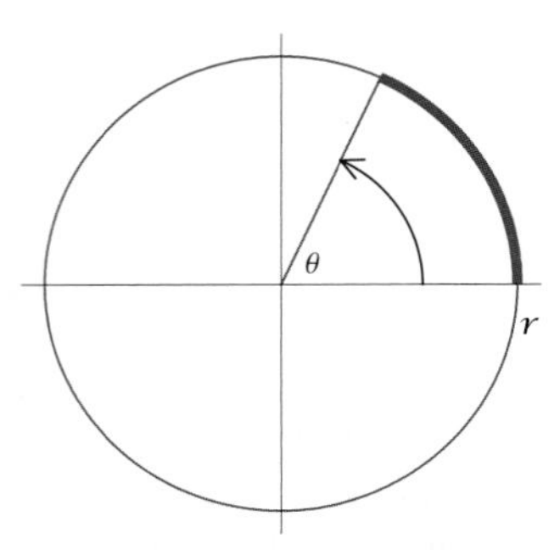

On a circle of radius r, the circular arc corresponding to θ degrees has length $\frac{\theta\pi r}{180}$.

More generally, suppose $0 < \theta \leq 360$ and consider a circular arc corresponding to θ degrees on a circle of radius r, as shown in the thickened part of the circle here. The length of this circular arc equals the fraction of the entire circle taken up by this circular arc times the circumference of the entire circle. In other words, the length of this circular arc equals $\frac{\theta}{360} \cdot 2\pi r$, which equals $\frac{\theta\pi r}{180}$.

In the following summary of the result that we derived above, we assume that $0 < \theta \leq 360$:

Length of a circular arc

An arc corresponding to θ degrees on a circle of radius r has length $\frac{\theta\pi r}{180}$.

In the special case where $\theta = 360$, the formula above states that a circle of radius r has length $2\pi r$, as expected.

Special Points on the Unit Circle

We know that the radius of the unit circle that corresponds to 45° has its endpoint at $(\frac{\sqrt{2}}{2}, \frac{\sqrt{2}}{2})$ [see Example 2]. The coordinates of the endpoint can also be explicitly found for the radius corresponding to 30° and for the radius corresponding to 60°. To do this, we first need to examine the dimensions of a right triangle with those angles.

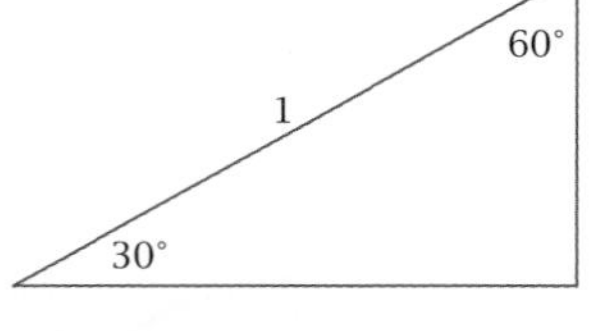

Consider a right triangle, one of whose angles is 30°. Because the angles of a triangle add up to 180°, the other angle of the triangle is 60°. Suppose this triangle has a hypotenuse of length 1, as shown in the figure here. Our goal is to find the lengths of the other two sides of the triangle.

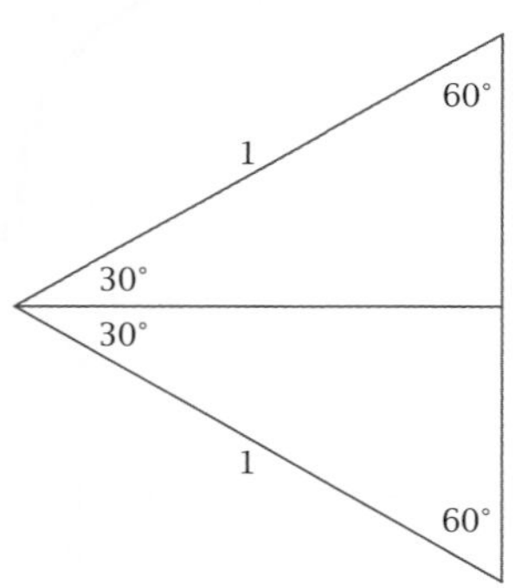

Because the large triangle is an equilateral triangle, the unlabeled vertical side of the large triangle has length 1. Thus in the top triangle, the vertical side has length $\frac{1}{2}$.

Flip the triangle across the base adjacent to the 30° angle, creating the figure shown here. Notice that all three angles in the large triangle are 60°. Thus the large triangle is an equilateral triangle. We already knew that two sides of this large triangle have length 1, as labeled here; now we know that the third side also has length 1.

Looking at the two smaller triangles here, note that each side opposite the 30° angle has half the length of the vertical side of the large triangle. Thus the vertical side in the top triangle has length $\frac{1}{2}$. The Pythagorean Theorem then implies that the horizontal side has length $\frac{\sqrt{3}}{2}$ (you should verify this). Thus the dimensions of this triangle are as in the figure shown below:

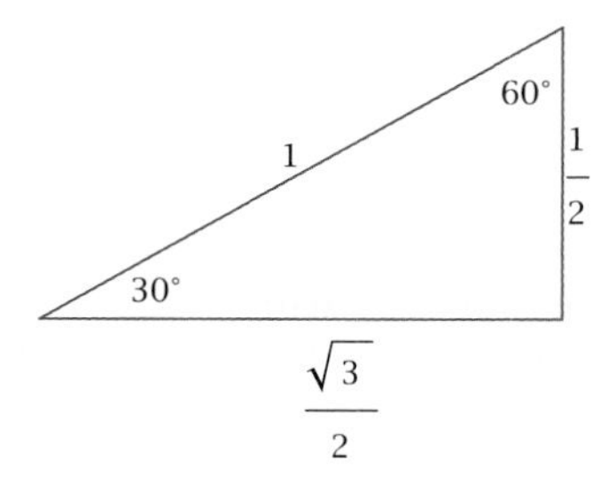

In a triangle with angles of 30°, 60°, and 90°, the side opposite the 30° angle has half the length of the hypotenuse.

In summary, we have shown the following:

Dimensions of a triangle with angles of **30°, 60°, *and*** **90°**

In a triangle with angles of 30°, 60°, and 90° and hypotenuse of length 1,

- the side opposite the 30° angle has length $\frac{1}{2}$;
- the side opposite the 60° angle has length $\frac{\sqrt{3}}{2}$.

EXAMPLE 9

(a) Find the coordinates of the endpoint of the radius of the unit circle corresponding to 30°.

(b) Find the coordinates of the endpoint of the radius of the unit circle corresponding to 60°.

Because of its origin in Latin, the plural of radius *is* radii.

SOLUTION

(a) The radius corresponding to 30° is shown below on the left. If we drop a perpendicular line segment from the endpoint of the radius to the horizontal axis, as shown below, we get a 30°-60°-90° triangle. The hypotenuse of that triangle is a radius of the unit circle and hence has length 1. Thus the side opposite the 60° angle has length $\frac{\sqrt{3}}{2}$ and the side opposite the 30° angle has length $\frac{1}{2}$. Hence the radius corresponding to 30° has its endpoint at $(\frac{\sqrt{3}}{2}, \frac{1}{2})$.

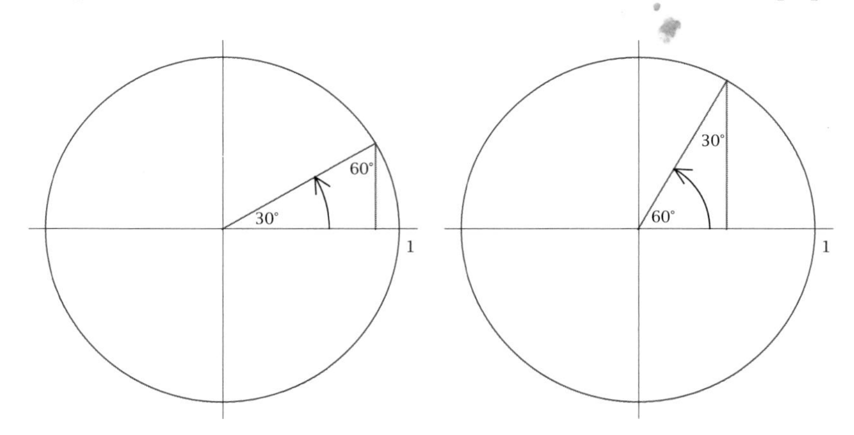

This radius has endpoint $(\frac{\sqrt{3}}{2}, \frac{1}{2})$. *This radius has endpoint* $(\frac{1}{2}, \frac{\sqrt{3}}{2})$.

(b) The radius corresponding to 60° is shown above on the right. The same reasoning as used in part (a) shows that this radius has its endpoint at $(\frac{1}{2}, \frac{\sqrt{3}}{2})$.

The table here shows the endpoint of the radius of the unit circle corresponding to some special angles. Exercises 35–40 ask you to extend this table to some other special angles. As we will see soon, the trigonometric functions were invented to extend this table to all angles.

angle	*endpoint of radius*
0°	$(1, 0)$
30°	$(\frac{\sqrt{3}}{2}, \frac{1}{2})$
45°	$(\frac{\sqrt{2}}{2}, \frac{\sqrt{2}}{2})$
60°	$(\frac{1}{2}, \frac{\sqrt{3}}{2})$
90°	$(0, 1)$
180°	$(-1, 0)$

EXERCISES

1. Find all numbers t such that $(\frac{1}{3}, t)$ is a point on the unit circle.
2. Find all numbers t such that $(\frac{3}{5}, t)$ is a point on the unit circle.
3. Find all numbers t such that $(t, -\frac{2}{5})$ is a point on the unit circle.
4. Find all numbers t such that $(t, -\frac{3}{7})$ is a point on the unit circle.
5. Find the points where the line through the origin with slope 3 intersects the unit circle.
6. Find the points where the line through the origin with slope 4 intersects the unit circle.

For Exercises 7–14, sketch the unit circle and the radius corresponding to the given angle. Include an arrow to show the direction in which the angle is measured from the positive horizontal axis.

7. 20°
8. 80°
9. 160°
10. 330°
11. 460°
12. −10°
13. −75°
14. −170°
15. What is the angle between the hour hand and the minute hand on a clock at 4 o'clock?
16. What is the angle between the hour hand and the minute hand on a clock at 5 o'clock?
17. What is the angle between the hour hand and the minute hand on a clock at 4:30?
18. What is the angle between the hour hand and the minute hand on a clock at 7:15?
19. What is the angle between the hour hand and the minute hand on a clock at 1:23?
20. What is the angle between the hour hand and the minute hand on a clock at 11:17?

For Exercises 21–24, give the answers to the nearest second.

21. At what time between 1 o'clock and 2 o'clock are the hour hand and the minute hand of a clock pointing in the same direction?
22. At what time between 4 o'clock and 5 o'clock are the hour hand and the minute hand of a clock pointing in the same direction?
23. Find two times between 1 o'clock and 2 o'clock when the hour hand and the minute hand of a clock are perpendicular.
24. Find two times between 4 o'clock and 5 o'clock when the hour hand and the minute hand of a clock are perpendicular.
25. Suppose an ant walks counterclockwise on the unit circle from the point $(1, 0)$ to the endpoint of the radius corresponding to 70°. How far has the ant walked?
26. Suppose an ant walks counterclockwise on the unit circle from the point $(1, 0)$ to the endpoint of the radius corresponding to 130°. How far has the ant walked?
27. What angle corresponds to a circular arc on the unit circle with length $\frac{\pi}{5}$?
28. What angle corresponds to a circular arc on the unit circle with length $\frac{\pi}{6}$?
29. What angle corresponds to a circular arc on the unit circle with length $\frac{5}{2}$?
30. What angle corresponds to a circular arc on the unit circle with length 1?

For Exercises 31–32, assume that the surface of the earth is a sphere with diameter 7926 miles.

31. Approximately how far does a ship travel when sailing along the equator in the Atlantic Ocean from longitude 20° west to longitude 30° west?
32. Approximately how far does a ship travel when sailing along the equator in the Pacific Ocean from longitude 170° west to longitude 120° west?
33. Find the lengths of both circular arcs on the unit circle connecting the points $(1, 0)$ and $(\frac{\sqrt{2}}{2}, \frac{\sqrt{2}}{2})$.
34. Find the lengths of both circular arcs on the unit circle connecting the points $(1, 0)$ and $(-\frac{\sqrt{2}}{2}, \frac{\sqrt{2}}{2})$.

For Exercises 35–40, find the endpoint of the radius of the unit circle corresponding to the given angle.

35. 120°
36. 240°
37. −30°
38. −150°
39. 390°
40. 510°

For Exercises 41–46, find the angle corresponding to the radius of the unit circle ending at the given point. Among the infinitely many possible correct solutions, choose the one with the smallest absolute value.

41. $\left(-\frac{1}{2}, \frac{\sqrt{3}}{2}\right)$
42. $\left(-\frac{\sqrt{3}}{2}, \frac{1}{2}\right)$
43. $\left(\frac{\sqrt{2}}{2}, -\frac{\sqrt{2}}{2}\right)$
44. $\left(\frac{1}{2}, -\frac{\sqrt{3}}{2}\right)$
45. $\left(-\frac{\sqrt{2}}{2}, -\frac{\sqrt{2}}{2}\right)$
46. $\left(-\frac{1}{2}, -\frac{\sqrt{3}}{2}\right)$
47. Find the lengths of both circular arcs on the unit circle connecting the point $\left(\frac{1}{2}, \frac{\sqrt{3}}{2}\right)$ and the endpoint of the radius corresponding to 130°.
48. Find the lengths of both circular arcs on the unit circle connecting the point $\left(\frac{\sqrt{3}}{2}, -\frac{1}{2}\right)$ and the endpoint of the radius corresponding to 50°.
49. Find the lengths of both circular arcs on the unit circle connecting the point $\left(-\frac{\sqrt{2}}{2}, -\frac{\sqrt{2}}{2}\right)$ and the endpoint of the radius corresponding to 125°.
50. Find the lengths of both circular arcs on the unit circle connecting the point $\left(-\frac{\sqrt{3}}{2}, -\frac{1}{2}\right)$ and the endpoint of the radius corresponding to 20°.
51. What is the slope of the radius of the unit circle corresponding to 30°?
52. What is the slope of the radius of the unit circle corresponding to 60°?

PROBLEMS

Some problems require considerably more thought than the exercises. Unlike exercises, problems often have more than one correct answer.

53. Suppose m is a real number. Find the points where the line through the origin with slope m intersects the unit circle.

For Problems 54–56, suppose a spider moves along the edge of a circular web at a distance of 3 cm from the center.

54. If the spider begins on the far right side of the web and creeps counterclockwise until it reaches the top of the web, approximately how far does it travel?
55. If the spider crawls along the edge of the web a distance of 2 cm, approximately what is the angle formed by the line segment from the center of the web to the spider's starting point and the line segment from the center of the web to the spider's finishing point?
56. Place the origin of the coordinate plane at the center of the web. What are the coordinates of the spider when it reaches the point directly southwest of the center?

Use the following information for Problems 57–62: A grad is a unit of measurement for angles that is sometimes used in surveying, especially in some European countries. A complete revolution once around a circle is 400 grads.

[These problems may help you work comfortably with angles in units other than degrees. In the next section we will introduce radians, the most important units used for angles.]

57. How many grads in a right angle?
58. The angles in a triangle add up to how many grads?
59. How many grads in each angle of an equilateral triangle?
60. Convert 37° to grads.
61. Convert 37 grads to degrees.
62. Discuss advantages and disadvantages of using grads as compared to degrees.
63. Verify that each of the following points is on the unit circle:

    (a) $\left(\frac{3}{5}, \frac{4}{5}\right)$ (b) $\left(\frac{5}{13}, \frac{12}{13}\right)$ (c) $\left(\frac{8}{17}, \frac{15}{17}\right)$
64. Show that if m and n are integers, not both zero, then

$$\left(\frac{m^2 - n^2}{m^2 + n^2}, \frac{2mn}{m^2 + n^2}\right)$$

is a point on the unit circle.
[*The result above shows that the unit circle contains infinitely many points for which both coordinates are rational.*]

WORKED-OUT SOLUTIONS *to Odd-numbered Exercises*

Do not read these worked-out solutions before first attempting to do the exercises yourself. Otherwise you may merely mimic the techniques shown here without understanding the ideas.

Best way to learn: Carefully read the section of the textbook, then do all the odd-numbered exercises (even if they have not been assigned) and check your answers here. If you get stuck on an exercise, reread the section of the textbook—then try the exercise again. If you are still stuck, then look at the worked-out solution here.

1. Find all numbers t such that $(\frac{1}{3}, t)$ is a point on the unit circle.

SOLUTION For $(\frac{1}{3}, t)$ to be a point on the unit circle means that the sum of the squares of the coordinates equals 1. In other words,

$$\left(\tfrac{1}{3}\right)^2 + t^2 = 1.$$

This simplifies to the equation $t^2 = \frac{8}{9}$, which implies that $t = \frac{\sqrt{8}}{3}$ or $t = -\frac{\sqrt{8}}{3}$. Because $\sqrt{8} = \sqrt{4 \cdot 2} = \sqrt{4} \cdot \sqrt{2} = 2\sqrt{2}$, we can rewrite this as $t = \frac{2\sqrt{2}}{3}$ or $t = -\frac{2\sqrt{2}}{3}$.

3. Find all numbers t such that $(t, -\frac{2}{5})$ is a point on the unit circle.

SOLUTION For $(t, -\frac{2}{5})$ to be a point on the unit circle means that the sum of the squares of the coordinates equals 1. In other words,

$$t^2 + \left(-\tfrac{2}{5}\right)^2 = 1.$$

This simplifies to the equation $t^2 = \frac{21}{25}$, which implies that $t = \frac{\sqrt{21}}{5}$ or $t = -\frac{\sqrt{21}}{5}$.

5. Find the points where the line through the origin with slope 3 intersects the unit circle.

SOLUTION The line through the origin with slope 3 is characterized by the equation $y = 3x$. Substituting this value for y into the equation for the unit circle ($x^2 + y^2 = 1$) gives

$$x^2 + (3x)^2 = 1,$$

which simplifies to the equation $10x^2 = 1$. Thus $x = \frac{\sqrt{10}}{10}$ or $x = -\frac{\sqrt{10}}{10}$. Using each of these values of x along with the equation $y = 3x$ gives the points $(\frac{\sqrt{10}}{10}, \frac{3\sqrt{10}}{10})$ and $(-\frac{\sqrt{10}}{10}, -\frac{3\sqrt{10}}{10})$ as the points of intersection of the line $y = 3x$ and the unit circle.

For Exercises 7–14, sketch the unit circle and the radius corresponding to the given angle. Include an arrow to show the direction in which the angle is measured from the positive horizontal axis.

7. 20°

SOLUTION

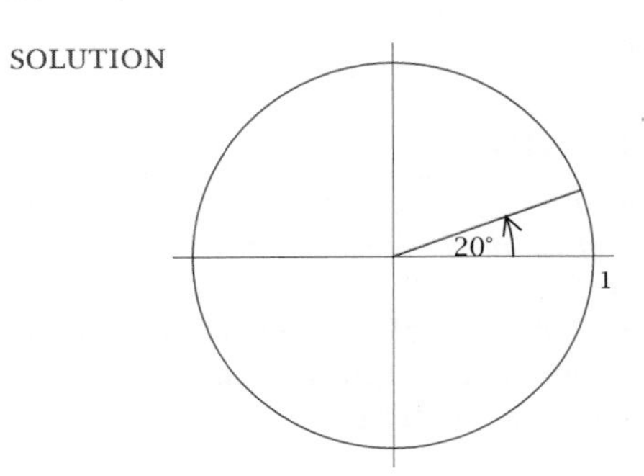

9. 160°

SOLUTION

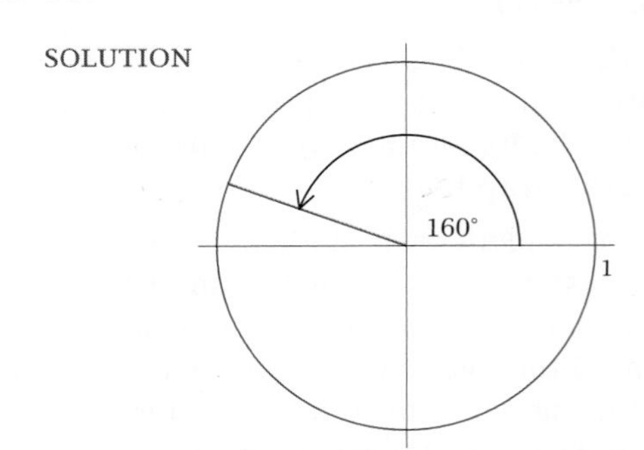

11. 460°

SOLUTION

460
1

13. -75°

SOLUTION

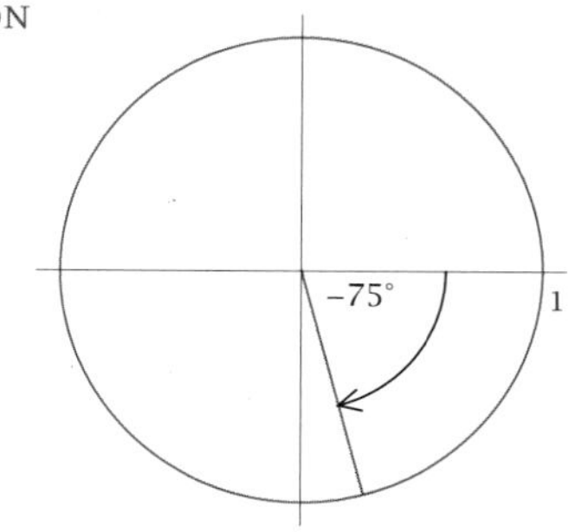

15. What is the angle between the hour hand and the minute hand on a clock at 4 o'clock?

SOLUTION At 4 o'clock, the minute hand is at 12 and the hour hand is at 4. The angle between those two clock hands takes up one-third of a circle. Because going around the entire circle is a 360° angle, the one-third of the circle between 12 and 4 forms an angle of 120° $(120 = \frac{1}{3} \cdot 360)$.

17. What is the angle between the hour hand and the minute hand on a clock at 4:30?

SOLUTION At 4:30, the hour and minute hand are separated by 1.5 numbers on the clock (because the minute hand is at 6 and the hour hand is halfway between 4 and 5). Because going around the entire circle is a 360° angle and the clock face has 12 numbers, the angle between two consecutive numbers on the clock is 30° $(30 = \frac{1}{12} \cdot 360)$. Thus the angle between the hour hand and the minute hand at 4:30 is 45° $(45 = 1.5 \times 30)$.

19. What is the angle between the hour hand and the minute hand on a clock at 1:23?

SOLUTION At 1:23, the hour hand is at position $1 + \frac{23}{60}$, which equals $\frac{83}{60}$.

In five minutes the minute hand moves one number on the face of the clock (for example, from five minutes after the hour to ten minutes after the hour, the minute hand moves from 1 to 2). Thus in one minute the minute hand moves $\frac{1}{5}$ number on the face of the clock. Thus at 23 minutes after the hour, the minute hand is at number $\frac{23}{5}$ on the face of the clock (because $\frac{23}{5} = 4 + \frac{3}{5}$, the minute hand is $\frac{3}{5}$ of the way between 4 and 5).

Thus at 1:23, the hour hand and minute hand differ by $\frac{23}{5} - \frac{83}{60}$, which equals $\frac{193}{60}$. Each number on the clock represents an angle of 30°, and thus the angle between the hour hand and minute hand at 1:23 is $\frac{193}{60} \cdot 30^\circ$, which equals $\frac{193}{2}^\circ$, which equals 96.5°.

For Exercises 21–24, give the answers to the nearest second.

21. At what time between 1 o'clock and 2 o'clock are the hour hand and the minute hand of a clock pointing in the same direction?

SOLUTION At m minutes after 1 o'clock, the minute hand will be at number $\frac{m}{5}$ and the hour hand will be at number $1 + \frac{m}{60}$ on the clock (see the solution to Exercise 19 for an explanation in the case when $m = 23$). We want the hour hand and the minute hand to be in the same location, which means that

$$\frac{m}{5} = 1 + \frac{m}{60}.$$

Solving the equation above for m gives $m = \frac{60}{11}$. Because $\frac{60}{11} = 5 + \frac{5}{11}$, the hour and minute hands point in the same direction at 1:05 plus $\frac{5}{11}$ of a minute. Now $\frac{5}{11}$ of a minute equals $60 \cdot \frac{5}{11}$ seconds, which is approximately 27.3 seconds. Thus the time between 1 o'clock and 2 o'clock at which the hour hand and minute hand point in the same direction is approximately 1:05:27.

23. Find two times between 1 o'clock and 2 o'clock when the hour hand and the minute hand of a clock are perpendicular.

SOLUTION At m minutes after 1 o'clock, the minute hand will be at number $\frac{m}{5}$ and the hour hand will be at number $1 + \frac{m}{60}$ on the clock (see the solution to Exercise 19 for an explanation in the case when $m = 23$).

The minute hand will make a 90° angle with the hour hand (measured clockwise from the hour hand) if

$$\frac{m}{5} = \left(1 + \frac{m}{60}\right) + 3,$$

where the addition of 3 numbers on the clock represents one-fourth of a rotation around the clock (because $\frac{3}{12} = \frac{1}{4}$). Solving the equation above for m gives $m = \frac{240}{11}$. Because $\frac{240}{11} = 21 + \frac{9}{11}$, the hour hand and the minute hand are perpendicular at 1:21 plus $\frac{9}{11}$ of a minute. Now $\frac{9}{11}$ of a minute equals $60 \cdot \frac{9}{11}$ seconds, which is approximately 49.1 seconds. Thus one time between 1 o'clock and 2 o'clock at which the hour hand and minute hand are perpendicular is approximately 1:21:49.

The minute hand will make a 270° angle with the hour hand (measured clockwise from the hour hand) if

$$\frac{m}{5} = \left(1 + \frac{m}{60}\right) + 9,$$

where the addition of 9 numbers on the clock represents three-fourths of a rotation around the clock (because $\frac{9}{12} = \frac{3}{4}$). Solving the equation above for m gives $m = \frac{600}{11}$. Because $\frac{600}{11} = 54 + \frac{6}{11}$, the hour hand and the minute hand are perpendicular at 1:54 plus $\frac{6}{11}$ of a minute. Now $\frac{6}{11}$ of a minute equals $60 \cdot \frac{6}{11}$ seconds, which is approximately 32.7 seconds. Thus the second time between 1 o'clock and 2 o'clock at which the hour hand and minute hand are perpendicular is approximately 1:54:33.

25. Suppose an ant walks counterclockwise on the unit circle from the point $(1, 0)$ to the endpoint of the radius corresponding to 70°. How far has the ant walked?

SOLUTION We need to find the length of the circular arc on the unit circle corresponding to 70°. This length equals $\frac{70\pi}{180}$, which equals $\frac{7\pi}{18}$.

27. What angle corresponds to a circular arc on the unit circle with length $\frac{\pi}{5}$?

SOLUTION Suppose θ degrees corresponds to an arc on the unit circle with length $\frac{\pi}{5}$. Thus $\frac{\theta\pi}{180} = \frac{\pi}{5}$. Solving this equation for θ, we get $\theta = 36$. Thus the angle in question is 36°.

29. What angle corresponds to a circular arc on the unit circle with length $\frac{5}{2}$?

SOLUTION Suppose θ degrees corresponds to an arc on the unit circle with length $\frac{5}{2}$. Thus $\frac{\theta\pi}{180} = \frac{5}{2}$. Solving this equation for θ, we get $\theta = \frac{450}{\pi}$. Thus the angle in question is $\frac{450}{\pi}$°, which is approximately equal to 143.2°.

For Exercises 31–32, assume that the surface of the earth is a sphere with diameter 7926 miles.

31. Approximately how far does a ship travel when sailing along the equator in the Atlantic Ocean from longitude 20° west to longitude 30° west?

SOLUTION Because the diameter of the earth is approximately 7926 miles, the radius of the earth is approximately 3963 miles. The arc along which the ship has sailed corresponds to 10°. Thus the length of this arc is approximately $\frac{10\pi \cdot 3963}{180}$ miles, which is approximately 692 miles.

33. Find the lengths of both circular arcs on the unit circle connecting the points $(1, 0)$ and $(\frac{\sqrt{2}}{2}, \frac{\sqrt{2}}{2})$.

SOLUTION The radius of the unit circle ending at the point $(\frac{\sqrt{2}}{2}, \frac{\sqrt{2}}{2})$ corresponds to 45°. One of the circular arcs connecting $(1, 0)$ and $(\frac{\sqrt{2}}{2}, \frac{\sqrt{2}}{2})$ is shown below as the thickened circular arc; the other circular arc connecting $(1, 0)$ and $(\frac{\sqrt{2}}{2}, \frac{\sqrt{2}}{2})$ is the unthickened part of the unit circle below.

The length of the thickened arc below is $\frac{45\pi}{180}$, which equals $\frac{\pi}{4}$. The entire unit circle has length 2π. Thus the length of the other circular arc below is $2\pi - \frac{\pi}{4}$, which equals $\frac{7\pi}{4}$.

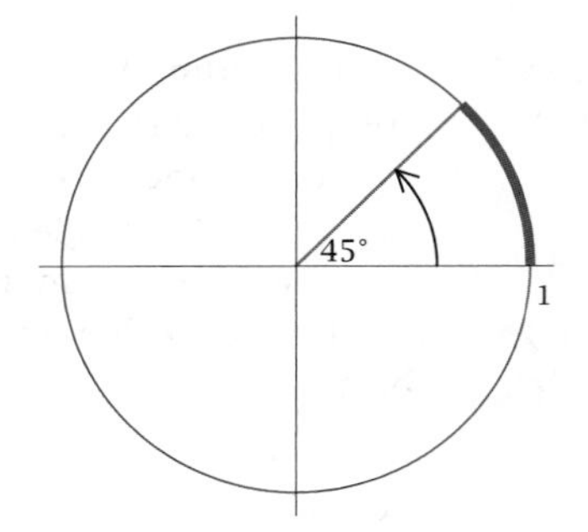

The thickened circular arc has length $\frac{\pi}{4}$. The other circular arc has length $\frac{7\pi}{4}$.

For Exercises 35–40, find the endpoint of the radius of the unit circle corresponding to the given angle.

35. 120°

SOLUTION The radius corresponding to 120° is shown below. The angle from this radius to the negative horizontal axis equals 180° − 120°, which equals 60° as shown in the figure below. Drop a perpendicular line segment from the endpoint of the radius to the horizontal axis, forming a right triangle as shown below. We already know that one angle of this right triangle is 60°; thus the other angle must be 30°, as labeled below:

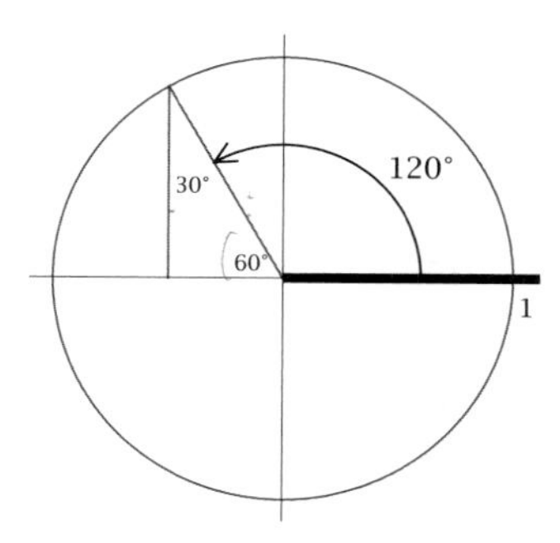

The side of the right triangle opposite the 30° angle has length $\frac{1}{2}$; the side of the right triangle opposite the 60° angle has length $\frac{\sqrt{3}}{2}$. Looking at the figure above, we see that the first coordinate of the endpoint of the radius is the negative of the length of the side opposite the 30° angle, and the second coordinate of the endpoint of the radius is the length of the side opposite the 60° angle. Thus the endpoint of the radius is $\left(-\frac{1}{2}, \frac{\sqrt{3}}{2}\right)$.

37. −30°

SOLUTION The radius corresponding to −30° is shown below. Draw a perpendicular line segment from the endpoint of the radius to the horizontal axis, forming a right triangle as shown below. We already know that one angle of this right triangle is 30°; thus the other angle must be 60°, as labeled below.

The side of the right triangle opposite the 30° angle has length $\frac{1}{2}$; the side of the right triangle opposite the 60° angle has length $\frac{\sqrt{3}}{2}$. Looking at the figure below, we see that the first coordinate of the endpoint of the radius is the length of the side opposite the 60° angle, and the second coordinate of the endpoint of the radius is the negative of the length of the side opposite the 30° angle. Thus the endpoint of the radius is $\left(\frac{\sqrt{3}}{2}, -\frac{1}{2}\right)$.

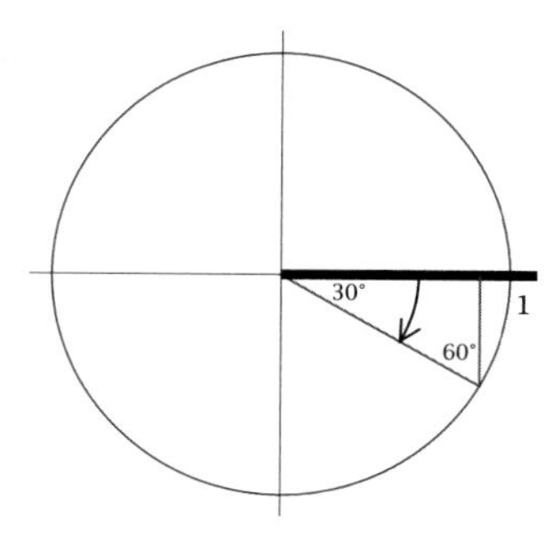

39. 390°

SOLUTION The radius corresponding to 390° is obtained by starting at the horizontal axis, making one complete counterclockwise rotation, and then continuing for another 30°. The resulting radius is shown below. Drop a perpendicular line segment from the endpoint of the radius to the horizontal axis, forming a right triangle as shown below. We already know that one angle of this right triangle is 30°; thus the other angle must be 60°, as labeled below.

The side of the right triangle opposite the 30° angle has length $\frac{1}{2}$; the side opposite the 60° angle has length $\frac{\sqrt{3}}{2}$. Looking at the figure below, we see that the first coordinate of the endpoint of the radius is the length of the side opposite the 60° angle, and the second coordinate of the endpoint of the radius is the length of the side opposite the 30° angle. Thus the endpoint of the radius is $\left(\frac{\sqrt{3}}{2}, \frac{1}{2}\right)$.

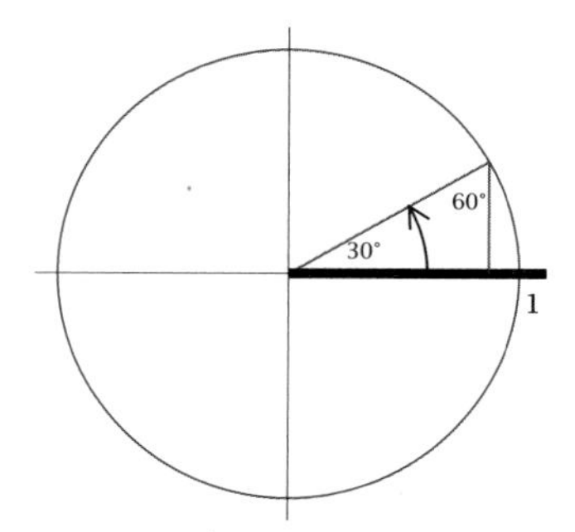

For Exercises 41–46, find the angle corresponding to the radius of the unit circle ending at the given point. Among the infinitely many possible correct solutions, choose the one with the smallest absolute value.

41. $\left(-\frac{1}{2}, \frac{\sqrt{3}}{2}\right)$

SOLUTION Draw the radius whose endpoint is $\left(-\frac{1}{2}, \frac{\sqrt{3}}{2}\right)$. Drop a perpendicular line segment from the endpoint of the radius to the horizontal axis, forming a right triangle. The hypotenuse of this right triangle is a radius of the unit circle and thus has length 1. The horizontal side has length $\frac{1}{2}$ and the vertical side of this triangle has length $\frac{\sqrt{3}}{2}$ because the endpoint of the radius is $\left(-\frac{1}{2}, \frac{\sqrt{3}}{2}\right)$.

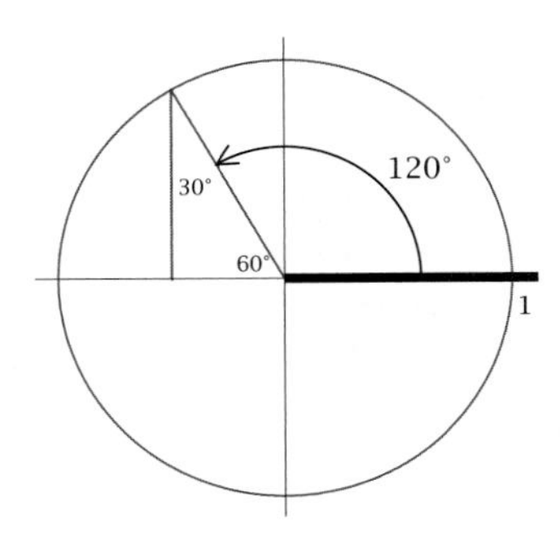

Thus we have a 30°-60°-90° triangle, with the 30° angle opposite the horizontal side of length $\frac{1}{2}$, as labeled above. Because $180 - 60 = 120$, the radius corresponds to 120°, as shown above.

In addition to corresponding to 120°, this radius also corresponds to 480°, 840°, and so on. This radius also corresponds to −240°, −600°, and so on. But of all the possible choices for this angle, the one with the smallest absolute value is 120°.

43. $\left(\frac{\sqrt{2}}{2}, -\frac{\sqrt{2}}{2}\right)$

SOLUTION Draw the radius whose endpoint is $\left(\frac{\sqrt{2}}{2}, -\frac{\sqrt{2}}{2}\right)$. Draw a perpendicular line segment from the endpoint of the radius to the horizontal axis, forming a right triangle. The hypotenuse of this right triangle is a radius of the unit circle and thus has length 1. The horizontal side has length $\frac{\sqrt{2}}{2}$ and the vertical side of this triangle also has length $\frac{\sqrt{2}}{2}$ because the endpoint of the radius is $\left(\frac{\sqrt{2}}{2}, -\frac{\sqrt{2}}{2}\right)$.

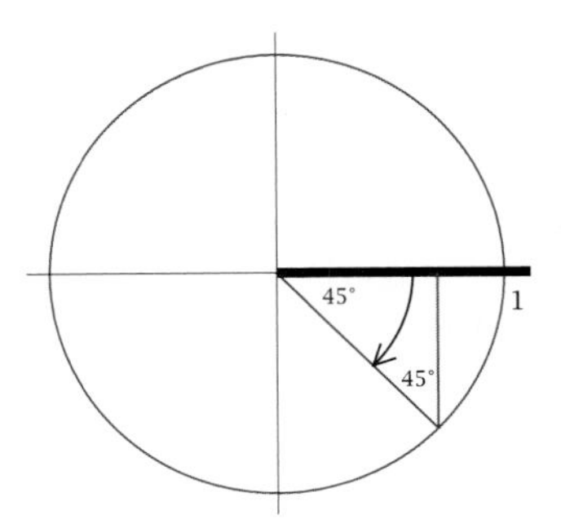

Thus we have here an isosceles right triangle, with two angles of 45° as labeled above. As can be seen from the figure above, the radius thus corresponds to −45°.

In addition to corresponding to −45°, this radius also corresponds to 315°, 675°, and so on. This radius also corresponds to −405°, −765°, and so on. But of all the possible choices for this angle, the one with the smallest absolute value is −45°.

45. $\left(-\frac{\sqrt{2}}{2}, -\frac{\sqrt{2}}{2}\right)$

SOLUTION Draw the radius whose endpoint is $\left(-\frac{\sqrt{2}}{2}, -\frac{\sqrt{2}}{2}\right)$. Draw a perpendicular line segment from the endpoint of the radius to the horizontal axis, forming a right triangle. The hypotenuse of this right triangle is a radius of the unit circle and thus has length 1. The horizontal side has length $\frac{\sqrt{2}}{2}$ and the vertical side of this triangle also has length $\frac{\sqrt{2}}{2}$ because the endpoint of the radius is $\left(-\frac{\sqrt{2}}{2}, -\frac{\sqrt{2}}{2}\right)$.

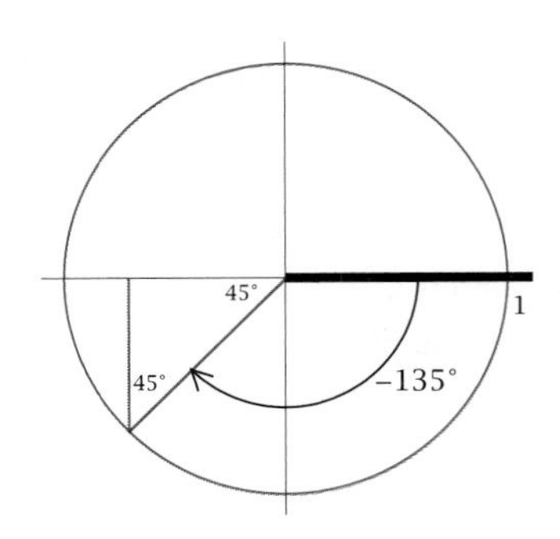

Thus we have here an isosceles right triangle, with two angles of 45° as labeled above. Because the radius makes a 45° angle with the negative horizontal axis, it corresponds to −135°, as shown here (because 135° = 180° − 45°).

In addition to corresponding to −135°, this radius also corresponds to 225°, 585°, and so on. This radius also corresponds to −495°, −855°, and so on. But of all the possible choices for this angle, the one with the smallest absolute value is −135°.

47. Find the lengths of both circular arcs on the unit circle connecting the point $(\frac{1}{2}, \frac{\sqrt{3}}{2})$ and the endpoint of the radius corresponding to 130°.

SOLUTION The radius of the unit circle ending at the point $(\frac{1}{2}, \frac{\sqrt{3}}{2})$ corresponds to 60°. One of the circular arcs connecting $(\frac{1}{2}, \frac{\sqrt{3}}{2})$ and the endpoint of the radius corresponding to 130° is shown below as the thickened circular arc; the other circular arc connecting these two points is the unthickened part of the unit circle below.

The thickened arc below corresponds to 70° (because 70° = 130° − 60°). Thus the length of the thickened arc below is $\frac{70\pi}{180}$, which equals $\frac{7\pi}{18}$. The entire unit circle has length 2π. Thus the length of the other circular arc below is $2\pi - \frac{7\pi}{18}$, which equals $\frac{29\pi}{18}$.

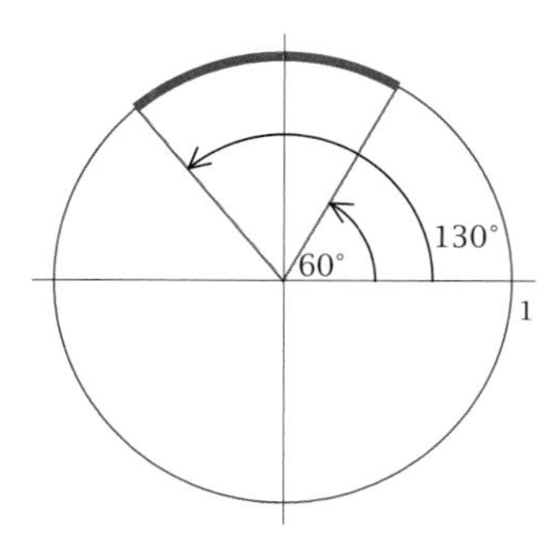

The thickened circular arc has length $\frac{7\pi}{18}$. The other circular arc has length $\frac{29\pi}{18}$.

49. Find the lengths of both circular arcs on the unit circle connecting the point $(-\frac{\sqrt{2}}{2}, -\frac{\sqrt{2}}{2})$ and the endpoint of the radius corresponding to 125°.

SOLUTION The radius of the unit circle ending at $(-\frac{\sqrt{2}}{2}, -\frac{\sqrt{2}}{2})$ corresponds to 225° (because 225° = 180° + 45°). One of the circular arcs connecting $(-\frac{\sqrt{2}}{2}, -\frac{\sqrt{2}}{2})$ and the endpoint of the radius corresponding to 125° is shown below as the thickened circular arc; the other circular arc connecting these two points is the unthickened part of the unit circle below.

The thickened arc below corresponds to 100° (because 100° = 225° − 125°). Thus the length of the thickened arc below is $\frac{100\pi}{180}$, which equals $\frac{5\pi}{9}$. The entire unit circle has length 2π. Thus the length of the other circular arc below is $2\pi - \frac{5\pi}{9}$, which equals $\frac{13\pi}{9}$.

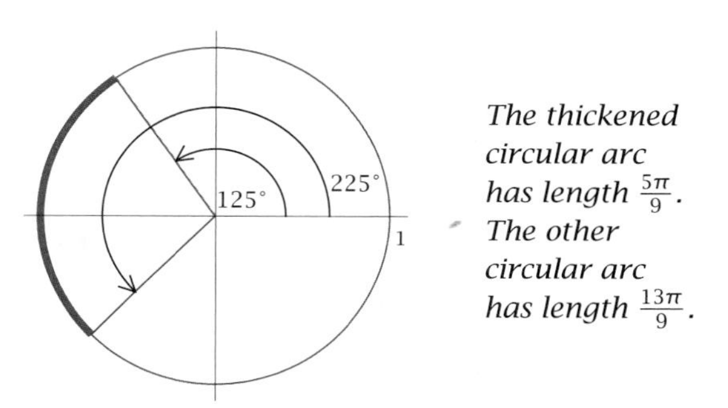

The thickened circular arc has length $\frac{5\pi}{9}$. The other circular arc has length $\frac{13\pi}{9}$.

51. What is the slope of the radius of the unit circle corresponding to 30°?

SOLUTION The radius of the unit circle corresponding to 30° has its initial point at $(0, 0)$ and its endpoint at $(\frac{\sqrt{3}}{2}, \frac{1}{2})$. Thus the slope of this radius is $(\frac{1}{2} - 0)/(\frac{\sqrt{3}}{2} - 0)$, which equals $\frac{1}{\sqrt{3}}$, which equals $\frac{\sqrt{3}}{3}$.

9.2 *Radians*

LEARNING OBJECTIVES

By the end of this section you should be able to

- convert radians to degrees and convert degrees to radians;
- find the length of a circular arc that is described by radians;
- find the area of a circular slice;
- find the coordinates of the endpoint of the radius of the unit circle corresponding to any multiple of $\frac{\pi}{6}$ radians or $\frac{\pi}{4}$ radians.

A Natural Unit of Measurement for Angles

We have been measuring angles in degrees, with 360° corresponding to a rotation through the entire circle. Hence 180° corresponds to a rotation through one-half the circle (thus generating a line), and 90° corresponds to a rotation through one-fourth the circle (thus generating a right angle).

The use of 360° to denote a complete rotation around the circle probably arose from trying to make one day's rotation of the Earth around the sun (or the sun around the Earth) correspond to 1°, as would be the case if the year had 360 days instead of 365 days.

There is nothing natural about the choice of 360 as the number of degrees in a complete circle. Mathematicians have introduced another unit of measurement for angles, called **radians**. Radians are often used rather than degrees because working with radians can lead to much nicer formulas than working with degrees.

The unit circle has circumference 2π. In other words, an ant walking around the unit circle once would walk a total distance of 2π. Because going around the circle once corresponds to traveling a distance of 2π, the following definition is a natural choice for a unit of measurement for angles. As we will see, this definition makes the length of an arc on the unit circle equal to the corresponding angle as measured in radians.

Radians

Radians are a unit of measurement for angles such that 2π radians correspond to a rotation through an entire circle.

Radians and degrees are two different units for measuring angles, just as feet and meters are two different units for measuring lengths.

EXAMPLE 1

Convert each of the following angles to degrees. Then sketch the radius of the unit circle corresponding to each angle.

(a) 2π radians; (b) π radians; (c) $\frac{\pi}{2}$ radians.

SOLUTION

(a) To translate between radians and degrees, note that a rotation through an entire circle equals 2π radians and also equals 360°. Thus

$$2\pi \text{ radians} = 360°.$$

(b) Rotation through half a circle equals π radians (because rotation through an entire circle equals 2π radians). Rotation through half a circle also equals 180°. Thus

$$\pi \textbf{ radians} = \mathbf{180^\circ}.$$

(c) Because rotation through an entire circle equals 2π radians, a right angle (which amounts to one-fourth of a circle) equals $\frac{2\pi}{4}$ radians, which equals $\frac{\pi}{2}$ radians. A right angle also equals 90°. Thus

$$\tfrac{\pi}{2} \textbf{ radians} = \mathbf{90^\circ}.$$

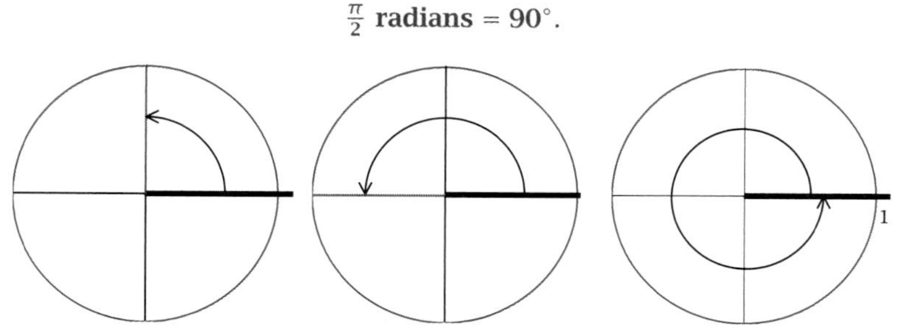

Angles of $\frac{\pi}{2}$ radians (left), π radians (center), and 2π radians (right) with the positive horizontal axis.

Try to think of the geometry of key angles directly in terms of radians instead of translating to degrees:

- *One complete rotation around a circle is 2π radians.*
- *The angles of a triangle add up to π radians.*
- *A right angle is $\frac{\pi}{2}$ radians.*

EXAMPLE 2

Convert each of the following angles to radians. Then sketch the radius of the unit circle corresponding to each angle.

(a) 45°; (b) 60°; (c) 30°.

SOLUTION

(a) If both sides of the equation $90^\circ = \frac{\pi}{2}$ radians are divided by 2, we get

$$\mathbf{45^\circ} = \tfrac{\pi}{4} \textbf{ radians}.$$

(b) If both sides of the equation $180^\circ = \pi$ radians are divided by 3, we get

$$\mathbf{60^\circ} = \tfrac{\pi}{3} \textbf{ radians}.$$

(c) If both sides of the equation $180^\circ = \pi$ radians are divided by 6, we get

$$\mathbf{30^\circ} = \tfrac{\pi}{6} \textbf{ radians}.$$

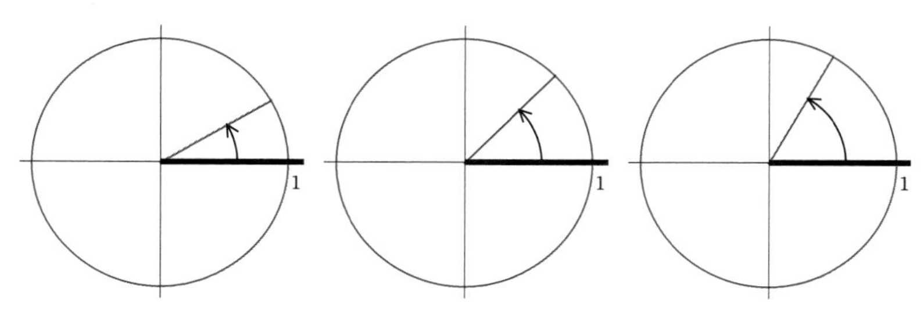

Angles of $\frac{\pi}{6}$ radians (left), $\frac{\pi}{4}$ radians (center), and $\frac{\pi}{3}$ radians (right) with the positive horizontal axis.

Additional examples of thinking of the geometry of key angles directly in terms of radians instead of translating to degrees:

- *Each angle of an equilateral triangle is $\frac{\pi}{3}$ radians.*
- *The line $y = x$ in the xy-plane makes a $\frac{\pi}{4}$ radian angle with the positive x-axis.*
- *In a right triangle with a hypotenuse of length 1 and another side with length $\frac{1}{2}$, the angle opposite the side with length $\frac{1}{2}$ is $\frac{\pi}{6}$ radians.*

The table below summarizes the translations between degrees and radians for some key angles. As you work more with radians, you will need to refer to this table less frequently because these translations will become part of your automatic vocabulary:

degrees	*radians*
30°	$\frac{\pi}{6}$ radians
45°	$\frac{\pi}{4}$ radians
60°	$\frac{\pi}{3}$ radians
90°	$\frac{\pi}{2}$ radians
180°	π radians
360°	2π radians

Translation between degrees and radians for commonly used angles.

To find a formula for converting any number of radians to degrees, start with the equation 2π radians = 360° and divide both sides by 2π, getting

Because $\frac{180}{\pi} \approx 57.3$, *one radian equals approximately* 57.3°.

$$1 \text{ radian} = \left(\frac{180}{\pi}\right)^\circ.$$

Multiply both sides of the equation above by any number θ to get the formula for converting radians to degrees:

Converting radians to degrees

$$\theta \text{ radians} = \left(\frac{180\theta}{\pi}\right)^\circ$$

To convert in the other direction (from degrees to radians), start with the equation 360° = 2π radians and divide both sides by 360, getting

$$1^\circ = \frac{\pi}{180} \text{ radians.}$$

Multiply both sides of the equation above by any number θ to get the formula for converting degrees to radians:

Converting degrees to radians

$$\theta^\circ = \frac{\theta\pi}{180} \text{ radians}$$

You should not need to memorize the two boxed formulas above for converting between radians and degrees. You need to remember only the defining equation 2π radians = 360°, from which you can derive the other formulas as needed. The following two examples illustrate this procedure (without using the two boxed formulas above).

EXAMPLE 3

Convert $\frac{7\pi}{90}$ radians to degrees.

SOLUTION Start with the equation

$$2\pi \text{ radians} = 360^\circ.$$

Divide both sides by 2 to obtain

$$\pi \text{ radians} = 180^\circ.$$

Now multiply both sides by $\frac{7}{90}$, obtaining

$$\tfrac{7\pi}{90} \text{ radians} = \tfrac{7}{90} \cdot 180^\circ = 14^\circ.$$

The next example illustrates the procedure for converting from degrees to radians.

EXAMPLE 4

Convert 10° to radians.

SOLUTION Start with the equation

$$360^\circ = 2\pi \text{ radians}.$$

Divide both sides by 360 to obtain

$$1^\circ = \frac{\pi}{180} \text{ radians}.$$

Now multiply both sides by 10, obtaining

$$10^\circ = \frac{10\pi}{180} \text{ radians} = \frac{\pi}{18} \text{ radians}.$$

Because $\frac{\pi}{18} \approx 0.1745$, this example shows that 10° is approximately 0.1745 radians.

The Radius Corresponding to an Angle

The radius corresponding to θ degrees was defined in Section 9.1. The definition is the same if using radians, except that the angle with the positive horizontal axis should be measured in radians rather than degrees. Thus here is the definition for a positive number of radians:

The radius corresponding to an angle

For $\theta > 0$, the **radius of the unit circle corresponding to θ radians** is the radius that has angle θ radians with the positive horizontal axis, as measured counterclockwise from the positive horizontal axis.

For example, the radius corresponding to $\frac{\pi}{2}$ radians, the radius corresponding to π radians, and the radius corresponding to 2π radians are shown in the solution to Example 1. Additionally, the solution to Example 2 shows the radius corresponding to $\frac{\pi}{6}$ radians, the radius corresponding to $\frac{\pi}{4}$ radians, and the radius corresponding to $\frac{\pi}{3}$ radians.

In Section 9.1 we introduced negative angles, which are measured clockwise from the positive horizontal axis. We can now think of such angles as being measured in radians instead of degrees. Thus we have the following definition for the radius corresponding to a negative number of radians:

The radius corresponding to a negative angle

For $\theta < 0$, the **radius of the unit circle corresponding to θ radians** is the radius that has angle $|\theta|$ radians with the positive horizontal axis, as measured clockwise from the positive horizontal axis.

EXAMPLE 5

Sketch the radius of the unit circle corresponding to each of the following angles: $-\frac{\pi}{4}$ radians, $-\frac{\pi}{2}$ radians, $-\pi$ radians.

SOLUTION

As we see here, negative angles are measured clockwise from the positive horizontal axis.

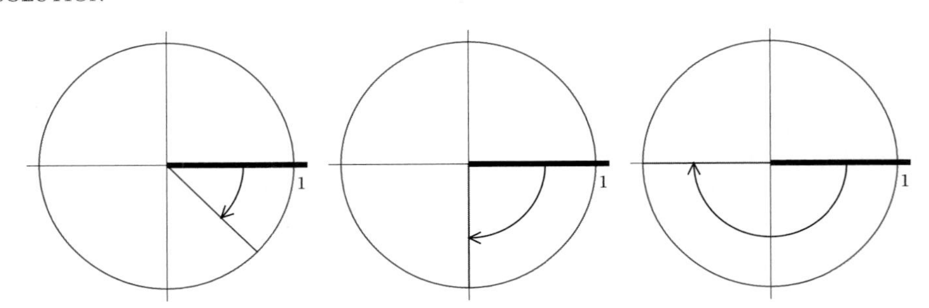

The radius corresponding to $-\frac{\pi}{4}$ radians (left), $-\frac{\pi}{2}$ radians (center), and $-\pi$ radians (right).

We have defined the radius of the unit circle corresponding to any positive or negative angle measured in radians. For completeness, we also state explicitly that the radius corresponding to 0 radians is the radius along the positive horizontal axis.

In the previous section we saw that we could obtain angles larger than 360° by starting at the positive horizontal axis and moving counterclockwise around the circle for more than a complete rotation. The same principle applies when working with radians, except that a complete counterclockwise rotation around the circle is measured as 2π radians rather than 360°. The next example illustrates this idea.

EXAMPLE 6

Sketch the radius of the unit circle corresponding to each of the following angles: π radians, 3π radians, 5π radians.

SOLUTION The radius corresponding to π radians is shown below on the left.

As shown in the center figure below, we end up at the same radius by moving counterclockwise completely around the circle (2π radians) and then continuing for another π radians, for a total of 3π radians.

The figure below on the right shows that we also end up at the same radius by moving counterclockwise around the circle twice (4π radians) and then continuing for another π radians for a total of 5π radians.

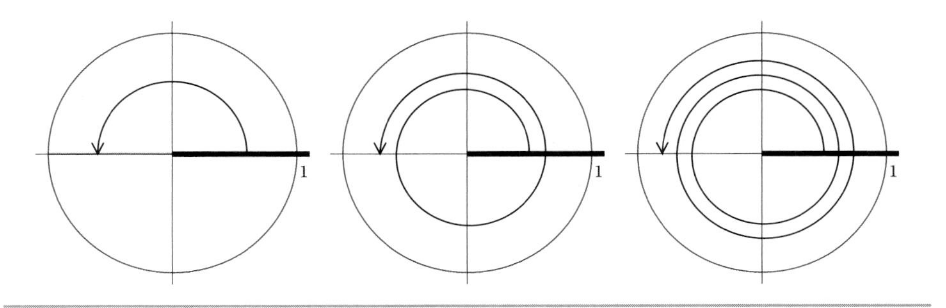

The same radius corresponds to π radians (left), 3π radians (center), 5π radians, and so on.

In the figure above, we could continue to add multiples of 2π, showing that the same radius corresponds to $\pi + 2\pi n$ radians for every positive integer n.

Adding negative multiples of 2π also leads to the same radius, as shown in the next example.

EXAMPLE 7

Sketch the radius of the unit circle corresponding to each of the following angles: π radians, $-\pi$ radians, -3π radians.

SOLUTION The radius corresponding to π radians is shown below on the left.

The figure below in the center shows that the radius corresponding to π radians also corresponds to $-\pi$ radians.

Or we could go completely around the circle in the clockwise direction (-2π radians) and then continue clockwise to the same radius (another $-\pi$ radians) for a total of -3π radians, as shown below on the right.

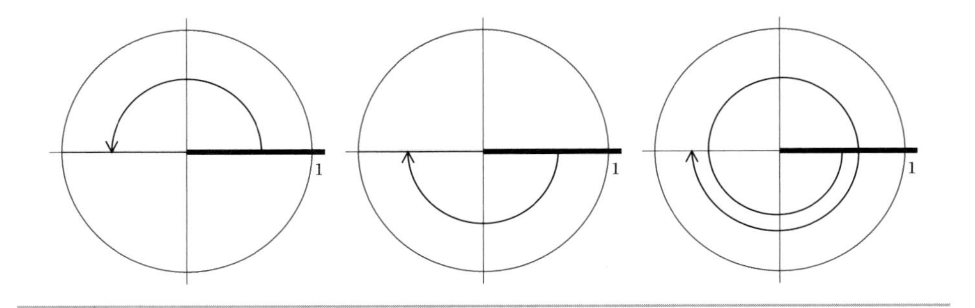

The same radius corresponds to π radians (left), $-\pi$ radians (center), -3π radians, and so on.

In the figure above, we could continue to subtract multiples of 2π radians, showing that the same radius corresponds to $\pi + 2\pi n$ radians for every integer n, positive or negative.

If we had started with any angle θ radians instead of π radians in the two previous examples, we would obtain the following result:

Multiple choices for the angle corresponding to a radius

A radius of the unit circle corresponding to θ radians also corresponds to $\theta + 2\pi n$ radians for every integer n.

Length of a Circular Arc

In the previous section, we found a formula for the length of a circular arc corresponding to an angle measured in degrees. We will now derive the formula that should be used when measuring angles in radians.

We begin by considering a circular arc on the unit circle corresponding to one radian (which is a bit more than $57°$), as shown here. The entire circle corresponds to 2π radians; thus the fraction of the circle contained in this circular arc is $\frac{1}{2\pi}$. Thus the length of this circular arc equals $\frac{1}{2\pi}$ times the circumference of the entire unit circle. In other words, the length of this circular arc equals $\frac{1}{2\pi} \cdot 2\pi$, which equals 1.

The circular arc on the unit circle corresponding to 1 *radian has length* 1.

Similarly, suppose $0 < \theta \leq 2\pi$. The fraction of the circle contained in a circular arc corresponding to θ radians is $\frac{\theta}{2\pi}$. Thus the length of a circular arc on the unit circle corresponding to θ radians is $\frac{\theta}{2\pi} \cdot 2\pi$. Hence we have the following result:

Length of a circular arc

If $0 < \theta \leq 2\pi$, then a circular arc on the unit circle corresponding to θ radians has length θ.

The formula above using radians is much cleaner than the corresponding formula using degrees (see Section 9.1). The formula above should not be a surprise, because we defined radians so that 2π radians equals the whole circle, which has length 2π for the unit circle. In fact, the definition of radians was chosen precisely to make this formula come out so nicely.

EXAMPLE 8

Suppose that the distance from the center of a wall clock to the endpoint of the minute hand is 1 foot.

(a) What time is it when the endpoint of the minute hand has traveled a distance of $\frac{\pi}{2}$ feet since 10 am?

(b) What time is it be when the endpoint of the minute hand has traveled a distance of 2π feet since 10 am?

(c) What time is it when the endpoint of the minute hand has traveled a distance of 3π feet since 10 am?

SOLUTION

(a) With feet as the units of measurement, the endpoint of the minute hand travels along the unit circle. Thus when the minute hand has traveled $\frac{\pi}{2}$ feet it makes an angle of $\frac{\pi}{2}$ radians with its initial position. Because $\frac{\pi}{2}$ radians is a right angle, when the minute hand has traveled a distance of $\frac{\pi}{2}$ feet since 10 am the time is 10:15 am.

(b) Traveling on the unit circle (where the units are feet) for a distance of 2π feet is exactly one full rotation around the circle. Thus when the minute hand has traveled a distance of 2π feet since 10 am, the time is 11 am.

(c) Traveling on the unit circle (where the units are feet) for a distance of 3π feet is 1.5 full rotations around the circle. Thus when the minute hand has traveled a distance of 3π feet since 10 am, the time is 11:30 am.

Area of a Slice

The following example will help us find a formula for the area of a slice inside a circle.

EXAMPLE 9

If a 14-inch pizza is cut into eight slices of equal size, what is the area of one slice? (Pizza sizes are measured in terms of the diameter of the pizza.)

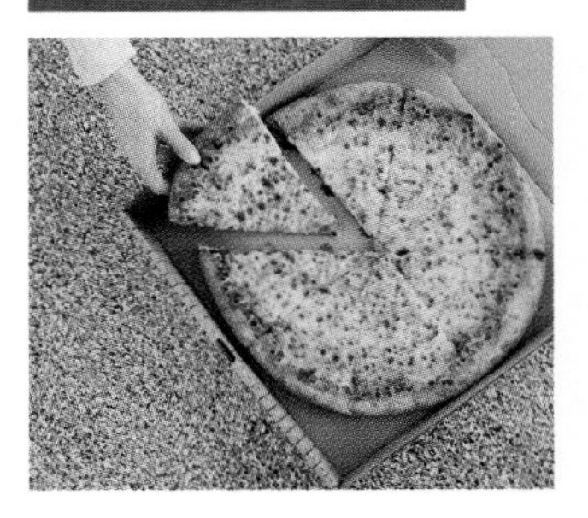

SOLUTION The diameter of the pizza is 14 inches; thus the radius of the pizza is 7 inches. Hence the entire pizza has area 49π square inches. One slice is one-eighth of the entire pizza. Thus a slice of this pizza has area $\frac{49\pi}{8}$ square inches, which is approximately 19.2 square inches.

To find the general formula for the area of a slice inside a circle, consider a circle of radius r. The area inside this circle is πr^2. The area of a slice with angle θ radians equals the fraction of the entire circle taken up by the slice times πr^2. The whole circle corresponds to 2π radians, and thus the fraction taken up by a slice with angle θ is $\frac{\theta}{2\pi}$. Putting all this together, we see that the area of the slice with angle θ radians is $(\frac{\theta}{2\pi})(\pi r^2)$, which equals $\frac{1}{2}\theta r^2$. Thus we have the following formula:

Area of a slice

A slice with angle θ radians inside a circle of radius r has area $\frac{1}{2}\theta r^2$.

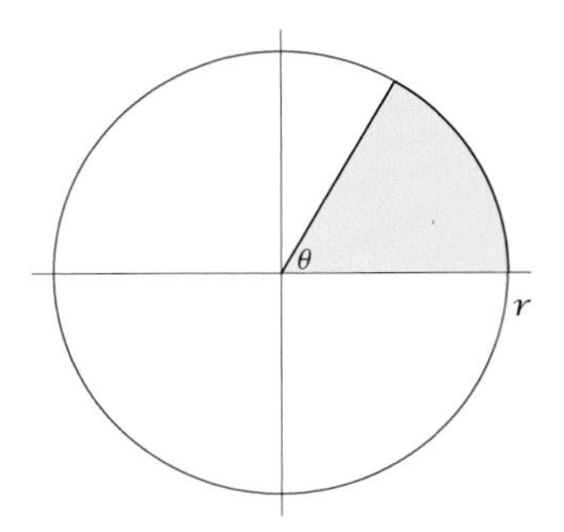

The yellow region has area $\frac{1}{2}\theta r^2$. This clean formula would not be as nice if the angle θ were measured in degrees instead of radians.

To test that the formula above is correct, we can let θ equal 2π radians, which means that the slice is the entire circle. The formula above tells us that the area should equal $\frac{1}{2}(2\pi)r^2$, which equals πr^2, which is indeed the area inside a circle of radius r.

Special Points on the Unit Circle

The table below shows the endpoint of the radius of the unit circle corresponding to some special angles. This is the same as the table in Section 9.1, except now we use radians rather than degrees.

angle	*endpoint of radius*
0 radians	$(1,0)$
$\frac{\pi}{6}$ radians	$(\frac{\sqrt{3}}{2}, \frac{1}{2})$
$\frac{\pi}{4}$ radians	$(\frac{\sqrt{2}}{2}, \frac{\sqrt{2}}{2})$
$\frac{\pi}{3}$ radians	$(\frac{1}{2}, \frac{\sqrt{3}}{2})$
$\frac{\pi}{2}$ radians	$(0,1)$
π radians	$(-1,0)$

Coordinates of the endpoint of the radius of the unit circle corresponding to some special angles.

The example below shows how to find the endpoints of the radius of the unit circle associated with additional special angles.

EXAMPLE 10

Find the coordinates of the endpoint of the radius of the unit circle corresponding to $\frac{14\pi}{3}$ radians.

SOLUTION Recall that integer multiples of 2π radians do not matter when locating the radius corresponding to an angle. Thus we write

$$\frac{14\pi}{3} = \frac{12\pi + 2\pi}{3} = 4\pi + \frac{2\pi}{3},$$

and we will use $\frac{2\pi}{3}$ radians instead of $\frac{14\pi}{3}$ radians in this problem.

At this point you may be more comfortable switching to degrees. Note that $\frac{2\pi}{3}$ radians equals $120°$. Thus now we need to solve the problem of finding the coordinates of the endpoint of the radius of the unit circle corresponding to $120°$.

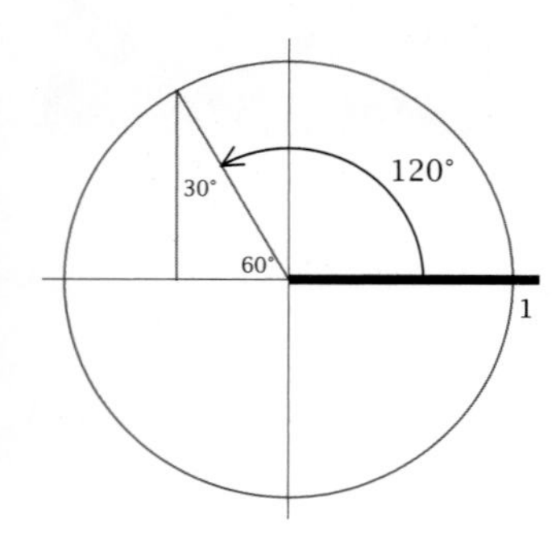

The radius corresponding to $120°$ is shown in the figure here. The angle from this radius to the negative horizontal axis equals $180° - 120°$, which equals $60°$ as shown in the figure. Drop a perpendicular line segment from the endpoint of the radius to the horizontal axis, forming a right triangle. We already know that one angle of this right triangle is $60°$; thus the other angle must be $30°$, as labeled in the figure.

The side of the right triangle opposite the $30°$ angle has length $\frac{1}{2}$; the side of the right triangle opposite the $60°$ angle has length $\frac{\sqrt{3}}{2}$. Looking at the figure, we see that the first coordinate of the endpoint of the radius is the negative of the length of the side opposite the $30°$ angle, and the second coordinate of the endpoint of the radius is the length of the side opposite the $60°$ angle. Thus the endpoint of the radius is $(-\frac{1}{2}, \frac{\sqrt{3}}{2})$.

EXERCISES

In Exercises 1–8, convert each angle to radians.

1. $15°$
2. $40°$
3. $-45°$
4. $-60°$
5. $270°$
6. $240°$
7. $1080°$
8. $1440°$

In Exercises 9–16, convert each angle to degrees.

9. 4π radians
10. 6π radians
11. $\frac{\pi}{9}$ radians
12. $\frac{\pi}{10}$ radians
13. 3 radians
14. 5 radians
15. $-\frac{2\pi}{3}$ radians
16. $-\frac{3\pi}{4}$ radians

For Exercises 17–24, sketch the unit circle and the radius corresponding to the given angle. Include an arrow to show the direction in which the angle is measured from the positive horizontal axis.

17. $\frac{5\pi}{18}$ radians
18. $\frac{1}{2}$ radian
19. 2 radians
20. 5 radians
21. $\frac{11\pi}{5}$ radians
22. $-\frac{\pi}{12}$ radians
23. -1 radian
24. $-\frac{8\pi}{9}$ radians

25. Suppose an ant walks counterclockwise on the unit circle from the point $(0, 1)$ to the endpoint of the radius corresponding to $\frac{5\pi}{4}$ radians. How far has the ant walked?

26. Suppose an ant walks counterclockwise on the unit circle from the point $(-1, 0)$ to the endpoint of the radius corresponding to 6 radians. How far has the ant walked?

27. Find the lengths of both circular arcs of the unit circle connecting the point $(1, 0)$ and the endpoint of the radius corresponding to 3 radians.

28. Find the lengths of both circular arcs of the unit circle connecting the point $(1, 0)$ and the endpoint of the radius corresponding to 4 radians.

29. Find the lengths of both circular arcs of the unit circle connecting the point $(\frac{\sqrt{2}}{2}, -\frac{\sqrt{2}}{2})$ and the point whose radius corresponds to 1 radian.

30. Find the lengths of both circular arcs of the unit circle connecting the point $(-\frac{\sqrt{2}}{2}, \frac{\sqrt{2}}{2})$ and the point whose radius corresponds to 2 radians.

31. For a 16-inch pizza, find the area of a slice with angle $\frac{3}{4}$ radians.

32. For a 14-inch pizza, find the area of a slice with angle $\frac{4}{5}$ radians.

33. Suppose a slice of a 12-inch pizza has an area of 20 square inches. What is the angle of this slice?

34. Suppose a slice of a 10-inch pizza has an area of 15 square inches. What is the angle of this slice?

35. Suppose a slice of pizza with an angle of $\frac{5}{6}$ radians has an area of 21 square inches. What is the diameter of this pizza?

36. Suppose a slice of pizza with an angle of 1.1 radians has an area of 25 square inches. What is the diameter of this pizza?

For each of the angles in Exercises 37–42, find the endpoint of the radius of the unit circle that corresponds to the given angle.

37. $\frac{5\pi}{6}$ radians
38. $\frac{7\pi}{6}$ radians
39. $-\frac{\pi}{4}$ radians
40. $-\frac{3\pi}{4}$ radians
41. $\frac{5\pi}{2}$ radians
42. $\frac{11\pi}{2}$ radians

For each of the angles in Exercises 43–46, find the slope of the radius of the unit circle that corresponds to the given angle.

43. $\frac{5\pi}{6}$ radians
44. $\frac{7\pi}{6}$ radians
45. $-\frac{\pi}{4}$ radians
46. $-\frac{3\pi}{4}$ radians

PROBLEMS

47. Find the formula for the length of a circular arc corresponding to θ radians on a circle of radius r.

48. Most dictionaries define acute angles and obtuse angles in terms of degrees. Restate these definitions in terms of radians.

49. Suppose the region bounded by the thickened radii and circular arc shown below is removed. Find a formula (in terms of θ) for the perimeter of the remaining region inside the unit circle.

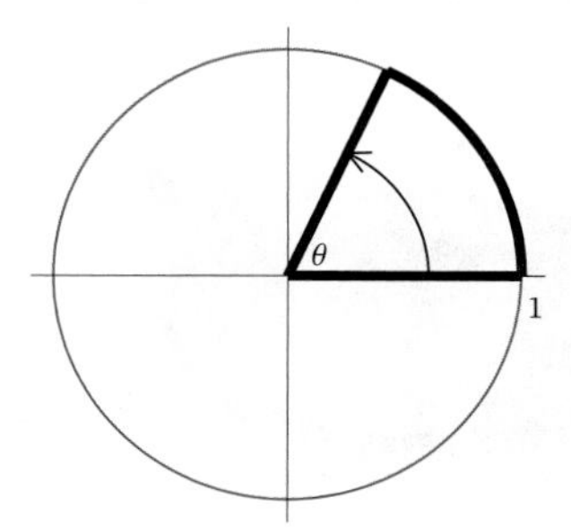

Assume that $0 < \theta < 2\pi$.

50. Find a formula for converting from radians to grads. [See the note before Problem 57 in Section 9.1 for the definition of grads.]

51. Find a formula for converting from grads to radians.

WORKED-OUT SOLUTIONS *to Odd-numbered Exercises*

In Exercises 1–8, convert each angle to radians.

1. 15°

SOLUTION Start with the equation

$$360° = 2\pi \text{ radians}.$$

Divide both sides by 360 to obtain

$$1° = \frac{\pi}{180} \text{ radians}.$$

Now multiply both sides by 15, obtaining

$$15° = \frac{15\pi}{180} \text{ radians} = \frac{\pi}{12} \text{ radians}.$$

3. −45°

SOLUTION Start with the equation

$$360° = 2\pi \text{ radians}.$$

Divide both sides by 360 to obtain

$$1° = \frac{\pi}{180} \text{ radians}.$$

Now multiply both sides by −45, obtaining

$$-45° = -\frac{45\pi}{180} \text{ radians} = -\frac{\pi}{4} \text{ radians}.$$

5. 270°

SOLUTION Start with the equation

$$360° = 2\pi \text{ radians}.$$

Divide both sides by 360 to obtain

$$1° = \frac{\pi}{180} \text{ radians}.$$

Now multiply both sides by 270, obtaining

$$270° = \frac{270\pi}{180} \text{ radians} = \frac{3\pi}{2} \text{ radians}.$$

7. 1080°

SOLUTION Start with the equation

$$360° = 2\pi \text{ radians}.$$

Divide both sides by 360 to obtain

$$1° = \frac{\pi}{180} \text{ radians}.$$

Now multiply both sides by 1080, obtaining

$$1080^\circ = \frac{1080\pi}{180} \text{ radians} = 6\pi \text{ radians.}$$

In Exercises 9–16, convert each angle to degrees.

9. 4π radians

SOLUTION Start with the equation

$$2\pi \text{ radians} = 360^\circ.$$

Multiply both sides by 2, obtaining

$$4\pi \text{ radians} = 2 \cdot 360^\circ = 720^\circ.$$

11. $\frac{\pi}{9}$ radians

SOLUTION Start with the equation

$$2\pi \text{ radians} = 360^\circ.$$

Divide both sides by 2 to obtain

$$\pi \text{ radians} = 180^\circ.$$

Now divide both sides by 9, obtaining

$$\frac{\pi}{9} \text{ radians} = \frac{180}{9}^\circ = 20^\circ.$$

13. 3 radians

SOLUTION Start with the equation

$$2\pi \text{ radians} = 360^\circ.$$

Divide both sides by 2π to obtain

$$1 \text{ radian} = \frac{180}{\pi}^\circ.$$

Now multiply both sides by 3, obtaining

$$3 \text{ radians} = 3 \cdot \frac{180}{\pi}^\circ = \frac{540}{\pi}^\circ.$$

15. $-\frac{2\pi}{3}$ radians

SOLUTION Start with the equation

$$2\pi \text{ radians} = 360^\circ.$$

Divide both sides by 2 to obtain

$$\pi \text{ radians} = 180^\circ.$$

Now multiply both sides by $-\frac{2}{3}$, obtaining

$$-\frac{2\pi}{3} \text{ radians} = -\frac{2}{3} \cdot 180^\circ = -120^\circ.$$

For Exercises 17–24, sketch the unit circle and the radius corresponding to the given angle. Include an arrow to show the direction in which the angle is measured from the positive horizontal axis.

17. $\frac{5\pi}{18}$ radians

SOLUTION

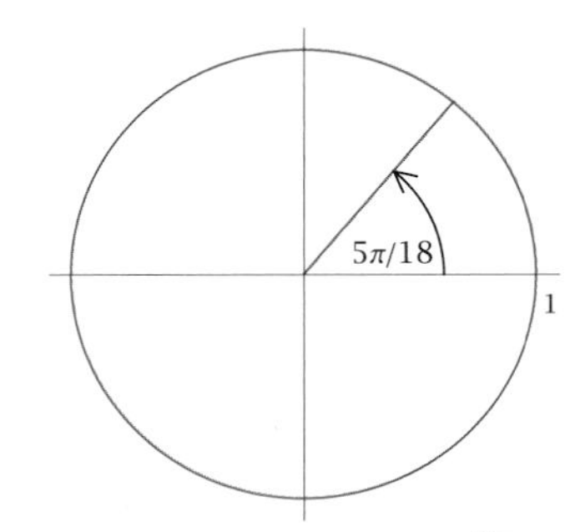

The radius corresponding to $\frac{5\pi}{18}$ radians, which equals 50°.

19. 2 radians

SOLUTION

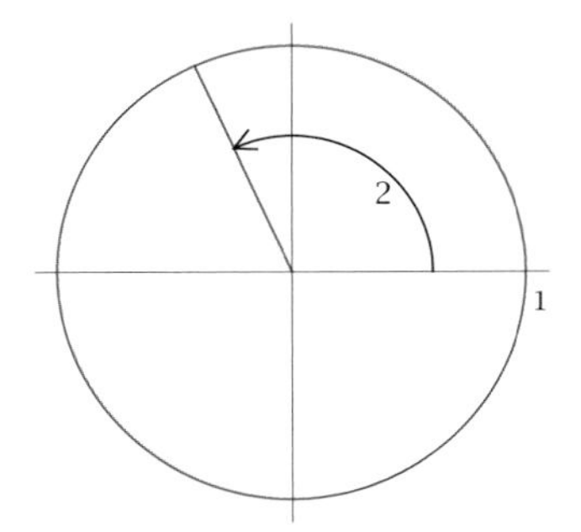

The radius corresponding to 2 radians, which approximately equals 114.6°.

21. $\frac{11\pi}{5}$ radians

SOLUTION

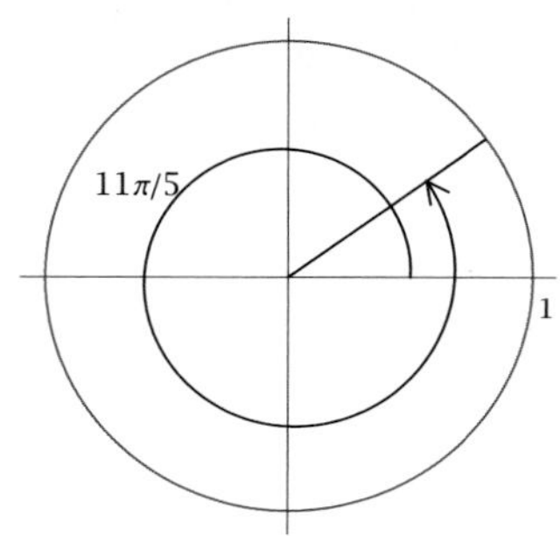

The radius corresponding to $\frac{11\pi}{5}$ radians, which equals $396°$.

23. -1 radian

SOLUTION

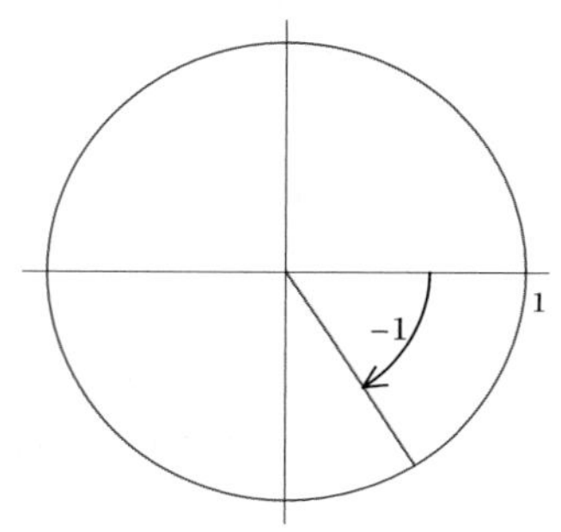

The radius corresponding to -1 radian, which approximately equals $-57.3°$.

25. Suppose an ant walks counterclockwise on the unit circle from the point $(0, 1)$ to the endpoint of the radius corresponding to $\frac{5\pi}{4}$ radians. How far has the ant walked?

SOLUTION The radius whose endpoint equals $(0, 1)$ corresponds to $\frac{\pi}{2}$ radians, which is the smaller angle shown below.

Because $\frac{5\pi}{4} = \pi + \frac{\pi}{4}$, the radius corresponding to $\frac{5\pi}{4}$ radians lies $\frac{\pi}{4}$ radians beyond the negative horizontal axis (halfway between the negative horizontal axis and the negative vertical axis). Thus the ant ends its walk at the endpoint of the radius corresponding to the larger angle shown below:

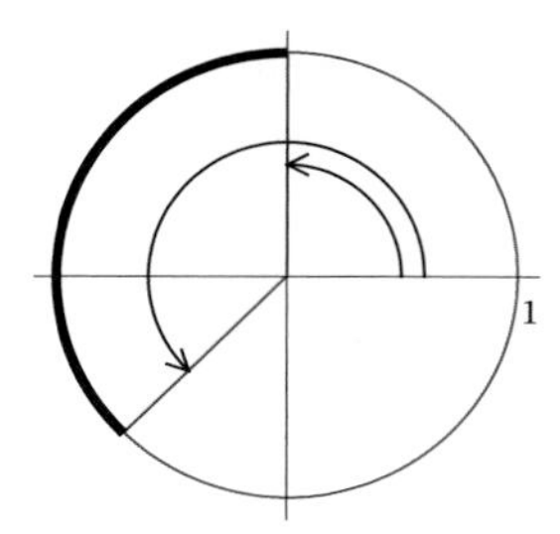

The ant walks along the thickened circular arc shown above. This circular arc corresponds to $\frac{5\pi}{4} - \frac{\pi}{2}$ radians, which equals $\frac{3\pi}{4}$ radians. Thus the distance walked by the ant is $\frac{3\pi}{4}$.

27. Find the lengths of both circular arcs of the unit circle connecting the point $(1, 0)$ and the endpoint of the radius corresponding to 3 radians.

SOLUTION Because 3 is a bit less than π, the radius corresponding to 3 radians lies a bit above the negative horizontal axis, as shown below. The thickened circular arc corresponds to 3 radians and thus has length 3. The entire unit circle has length 2π. Thus the length of the other circular arc is $2\pi - 3$, which is approximately 3.28.

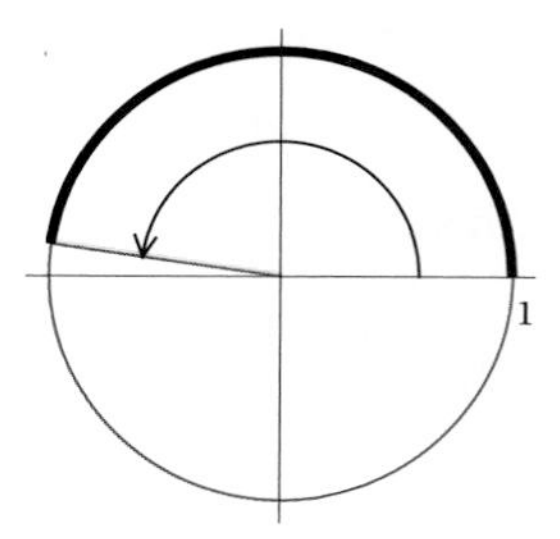

29. Find the lengths of both circular arcs of the unit circle connecting the point $(\frac{\sqrt{2}}{2}, -\frac{\sqrt{2}}{2})$ and the point whose radius corresponds to 1 radian.

SOLUTION The radius of the unit circle whose endpoint equals $(\frac{\sqrt{2}}{2}, -\frac{\sqrt{2}}{2})$ corresponds to $-\frac{\pi}{4}$ radians, as shown with the clockwise arrow below. The radius corresponding to 1 radian is shown with a counterclockwise arrow.

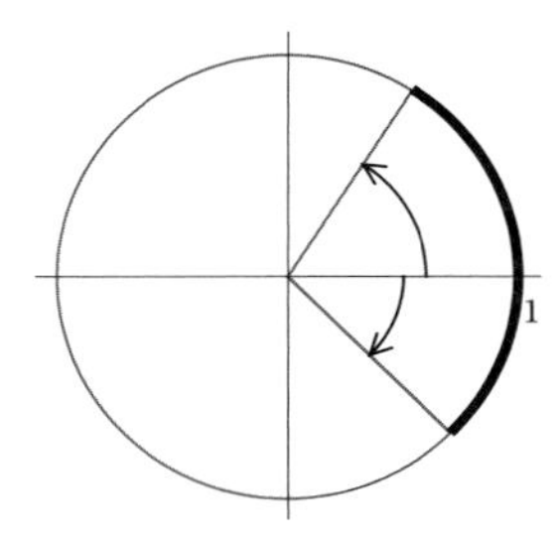

Thus the thickened circular arc above corresponds to $1 + \frac{\pi}{4}$ radians and thus has length $1 + \frac{\pi}{4}$, which is approximately 1.79. The entire unit circle has length 2π. Thus the length of the other circular arc below is $2\pi - (1 + \frac{\pi}{4})$, which equals $\frac{7\pi}{4} - 1$, which is approximately 4.50.

31. For a 16-inch pizza, find the area of a slice with angle $\frac{3}{4}$ radians.

SOLUTION Pizzas are measured by their diameters; thus this pizza has a radius of 8 inches. Thus the area of the slice is $\frac{1}{2} \cdot \frac{3}{4} \cdot 8^2$, which equals 24 square inches.

33. Suppose a slice of a 12-inch pizza has an area of 20 square inches. What is the angle of this slice?

SOLUTION This pizza has a radius of 6 inches. Let θ denote the angle of this slice, measured in radians. Then

$$20 = \tfrac{1}{2}\theta \cdot 6^2.$$

Solving this equation for θ, we get $\theta = \frac{10}{9}$ radians.

35. Suppose a slice of pizza with an angle of $\frac{5}{6}$ radians has an area of 21 square inches. What is the diameter of this pizza?

SOLUTION Let r denote the radius of this pizza. Thus

$$21 = \tfrac{1}{2} \cdot \tfrac{5}{6} r^2.$$

Solving this equation for r, we get $r = \sqrt{\frac{252}{5}} \approx 7.1$. Thus the diameter of the pizza is approximately 14.2 inches.

For each of the angles in Exercises 37–42, find the endpoint of the radius of the unit circle that corresponds to the given angle.

37. $\frac{5\pi}{6}$ radians

SOLUTION For this exercise it may be easier to convert to degrees. Thus we translate $\frac{5\pi}{6}$ radians to 150°.

The radius corresponding to 150° is shown below. The angle from this radius to the negative horizontal axis equals 180° − 150°, which equals 30° as shown in the figure below. Drop a perpendicular line segment from the endpoint of the radius to the horizontal axis, forming a right triangle as shown below. We already know that one angle of this right triangle is 30°; thus the other angle must be 60°, as labeled below.

The side of the right triangle opposite the 30° angle has length $\frac{1}{2}$; the side of the right triangle opposite the 60° angle has length $\frac{\sqrt{3}}{2}$. Looking at the figure below, we see that the first coordinate of the endpoint of the radius is the negative of the length of the side opposite the 60° angle, and the second coordinate of the endpoint of the radius is the length of the side opposite the 30° angle. Thus the endpoint of the radius is $(-\frac{\sqrt{3}}{2}, \frac{1}{2})$.

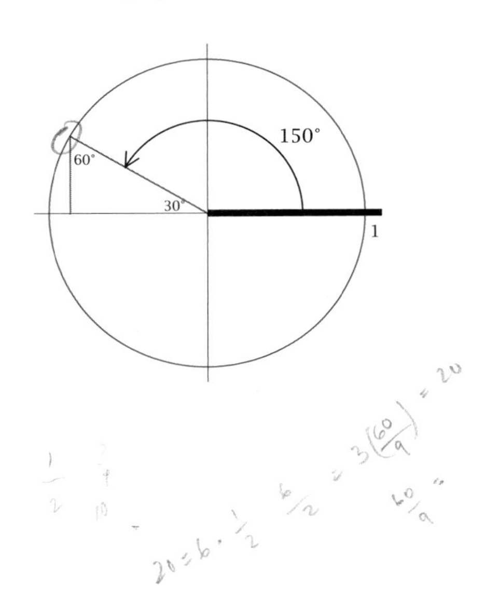

39. $-\frac{\pi}{4}$ radians

SOLUTION For this exercise it may be easier to convert to degrees. Thus we translate $-\frac{\pi}{4}$ radians to $-45°$.

The radius corresponding to $-45°$ is shown below. Draw a perpendicular line segment from the endpoint of the radius to the horizontal axis, forming a right triangle as shown below. We already know that one angle of this right triangle is $45°$; thus the other angle must also be $45°$, as labeled below.

The hypotenuse of this right triangle is a radius of the unit circle and thus has length 1. The other two sides each have length $\frac{\sqrt{2}}{2}$. Looking at the figure below, we see that the first coordinate of the endpoint of the radius is $\frac{\sqrt{2}}{2}$ and the second coordinate of the endpoint of the radius is $-\frac{\sqrt{2}}{2}$. Thus the endpoint of the radius is $(\frac{\sqrt{2}}{2}, -\frac{\sqrt{2}}{2})$.

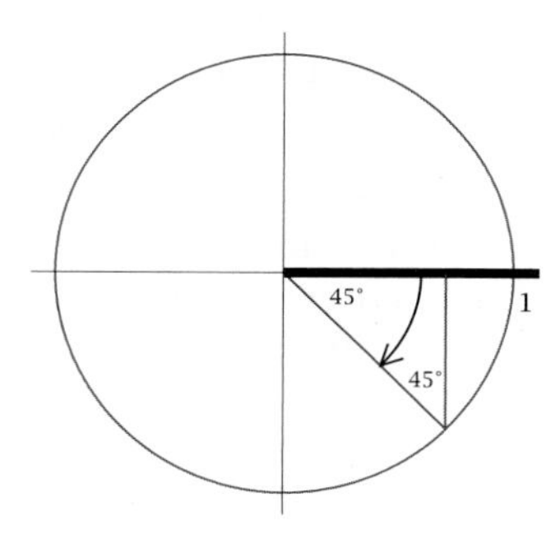

41. $\frac{5\pi}{2}$ radians

SOLUTION Note that $\frac{5\pi}{2} = 2\pi + \frac{\pi}{2}$. Thus the radius corresponding to $\frac{5\pi}{2}$ radians is obtained by starting at the horizontal axis, making one complete counterclockwise rotation (which is 2π radians), and then continuing for another $\frac{\pi}{2}$ radians. The resulting radius is shown below. Its endpoint is $(0, 1)$.

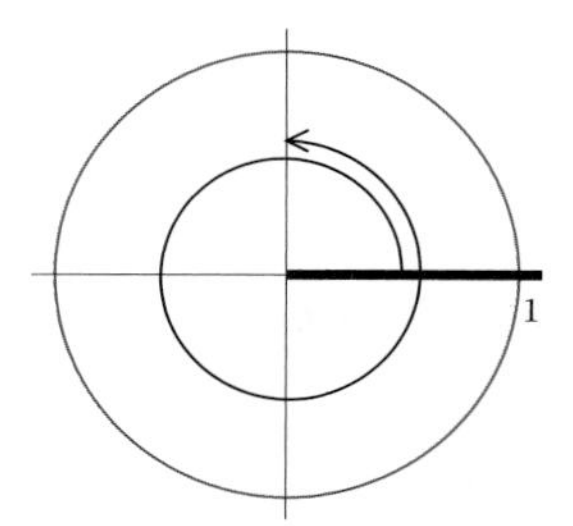

For each of the angles in Exercises 43–46, find the slope of the radius of the unit circle that corresponds to the given angle.

43. $\frac{5\pi}{6}$ radians

SOLUTION As we saw in the solution to Exercise 37, the endpoint of the radius corresponding to $\frac{5\pi}{6}$ radians is $(-\frac{\sqrt{3}}{2}, \frac{1}{2})$. Thus the slope of this radius is

$$\frac{\frac{1}{2}}{-\frac{\sqrt{3}}{2}},$$

which equals $-\frac{1}{\sqrt{3}}$, which equals $-\frac{\sqrt{3}}{3}$.

45. $-\frac{\pi}{4}$ radians

SOLUTION As we saw in the solution to Exercise 39, the endpoint of the radius corresponding to $-\frac{\pi}{4}$ radians is $(\frac{\sqrt{2}}{2}, -\frac{\sqrt{2}}{2})$. Thus the slope of this radius is

$$\frac{-\frac{\sqrt{2}}{2}}{\frac{\sqrt{2}}{2}},$$

which equals -1.

9.3 *Cosine and Sine*

LEARNING OBJECTIVES

By the end of this section you should be able to

- evaluate the cosine and sine of any multiple of 30° or 45° ($\frac{\pi}{6}$ radians or $\frac{\pi}{4}$ radians);
- determine whether the cosine (or sine) of an angle is positive or negative from the location of the corresponding radius;
- sketch the radius corresponding to θ if given either $\cos\theta$ or $\sin\theta$ and the sign of the other quantity;
- find $\cos\theta$ and $\sin\theta$ given one of these quantities and the quadrant of the corresponding radius.

Definition of Cosine and Sine

The table below shows the endpoint of the radius of the unit circle corresponding to some special angles. This table comes from tables in Sections 9.1 and 9.2.

θ (*radians*)	θ (*degrees*)	*endpoint of radius corresponding to* θ
0	$0°$	$(1, 0)$
$\frac{\pi}{6}$	$30°$	$(\frac{\sqrt{3}}{2}, \frac{1}{2})$
$\frac{\pi}{4}$	$45°$	$(\frac{\sqrt{2}}{2}, \frac{\sqrt{2}}{2})$
$\frac{\pi}{3}$	$60°$	$(\frac{1}{2}, \frac{\sqrt{3}}{2})$
$\frac{\pi}{2}$	$90°$	$(0, 1)$
π	$180°$	$(-1, 0)$
2π	$360°$	$(1, 0)$

Coordinates of the endpoint of the radius of the unit circle corresponding to some special angles.

We might consider extending the table above to other angles. For example, suppose we want to know the endpoint of the radius corresponding to $\frac{\pi}{18}$ radians (which equals 10°). Unfortunately, the coordinates of the endpoint of that radius do not have a nice form—neither coordinate is a rational number or even the square root of a rational number. The cosine and sine functions, which we are about to introduce, were invented to help us extend the table above to all angles.

The endpoint of the radius corresponding to $\frac{\pi}{18}$ radians is approximately $(0.9848, 0.1736)$.

Before introducing the cosine and sine functions, we explain a common assumption about notation in trigonometry:

Angles without units

If no units are given for an angle, then assume that the units are radians.

The figure below shows a radius of the unit circle corresponding to θ (here θ might be measured in either radians or degrees):

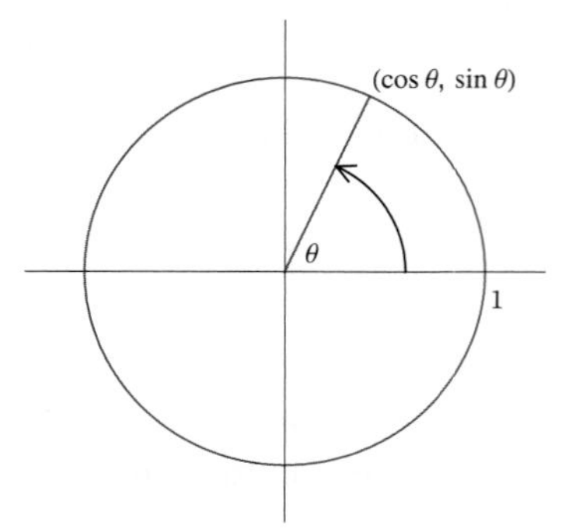

This figure defines the cosine and sine.

The endpoint of this radius is used to define the cosine and sine, as follows:

Cosine

The **cosine** of θ, denoted $\cos\theta$, is the first coordinate of the endpoint of the radius of the unit circle corresponding to θ.

Sine

The **sine** of θ, denoted $\sin\theta$, is the second coordinate of the endpoint of the radius of the unit circle corresponding to θ.

The two definitions above can be combined into a single statement, as follows:

Cosine and sine

The endpoint of the radius of the unit circle corresponding to θ has coordinates $(\cos\theta, \sin\theta)$.

EXAMPLE 1 Evaluate $\cos\frac{\pi}{2}$ and $\sin\frac{\pi}{2}$.

Here units are not specified for the angle $\frac{\pi}{2}$. Thus we assume that we are dealing with $\frac{\pi}{2}$ radians.

SOLUTION The radius corresponding to $\frac{\pi}{2}$ radians has endpoint $(0, 1)$. Thus

$$\cos\tfrac{\pi}{2} = 0 \quad \text{and} \quad \sin\tfrac{\pi}{2} = 1.$$

Equivalently, using degrees instead of radians we could write

$$\cos 90^\circ = 0 \quad \text{and} \quad \sin 90^\circ = 1.$$

The table below gives the cosine and sine of some special angles. This table is obtained from the table on the first page of this section by breaking the last column of that earlier table into two columns, with the first coordinate labeled as the cosine and the second coordinate labeled as the sine. Look at both tables and make sure that you understand what is going on here.

Most calculators can be set to work in either radians or degrees. Whenever you use a calculator to compute values of cosine or sine, be sure that your calculator is set to work in the appropriate units.

θ *(radians)*	θ *(degrees)*	$\cos\theta$	$\sin\theta$
0	$0°$	1	0
$\frac{\pi}{6}$	$30°$	$\frac{\sqrt{3}}{2}$	$\frac{1}{2}$
$\frac{\pi}{4}$	$45°$	$\frac{\sqrt{2}}{2}$	$\frac{\sqrt{2}}{2}$
$\frac{\pi}{3}$	$60°$	$\frac{1}{2}$	$\frac{\sqrt{3}}{2}$
$\frac{\pi}{2}$	$90°$	0	1
π	$180°$	-1	0
2π	$360°$	1	0

Cosine and sine of some special angles.

The table above giving the cosine and sine of special angles can be extended to other special angles, as shown in the next example.

EXAMPLE 2

Evaluate $\cos(-\frac{\pi}{2})$ and $\sin(-\frac{\pi}{2})$.

SOLUTION The radius corresponding to $-\frac{\pi}{2}$ radians has endpoint $(0, -1)$ as shown here. Thus

$$\cos(-\tfrac{\pi}{2}) = 0 \quad \text{and} \quad \sin(-\tfrac{\pi}{2}) = 1.$$

Using degrees instead of radians, we can also write

$$\cos(-90°) = 0 \quad \text{and} \quad \sin(-90°) = -1.$$

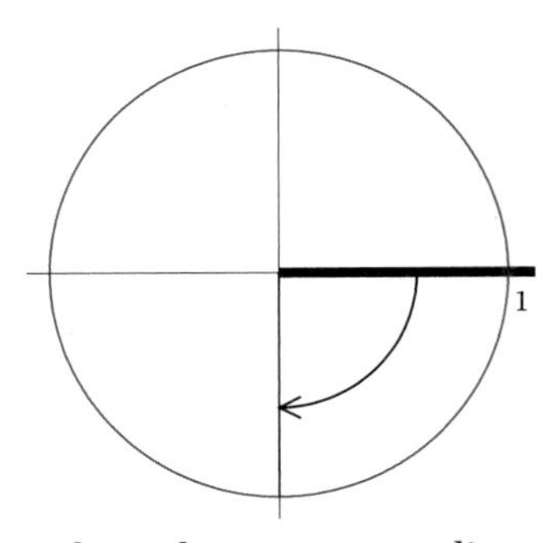

The radius corresponding to $-\frac{\pi}{2}$ radians has endpoint $(0, -1)$.

In addition to adding a row for $-\frac{\pi}{2}$ radians (which equals $-90°$), we could add many more entries to the table for the cosine and sine of special angles. Possibilities would include $\frac{2\pi}{3}$ radians (which equals $120°$), $\frac{5\pi}{6}$ radians (which equals $150°$), the negatives of all the angles already in the table, and so on. This would quickly become far too much information to memorize. Instead of memorizing the table above, concentrate on understanding the definitions of cosine and sine. Then you will be able to figure out the cosine and sine of any of the special angles, as needed.

To help remember that $\cos\theta$ is the first coordinate of the endpoint of the radius corresponding to θ and $\sin\theta$ is the second coordinate, keep cosine *and* sine *in alphabetical order.*

Similarly, do not become dependent on a calculator for evaluating the cosine and sine of special angles. If you need numeric values for $\cos 2$ or $\sin 17°$, then you will need to use a calculator. But if you get in the habit of using a calculator for evaluating expressions such as $\cos 0$ or $\sin(-180°)$, then cosine and sine will become simply buttons on your calculator and you will not be able to use these functions meaningfully.

The Signs of Cosine and Sine

The coordinate axes divide the coordinate plane into four regions, often called **quadrants**. The quadrant in which a radius lies determines whether the cosine and sine of the corresponding angle are positive or negative. The figure below shows the sign of the cosine and the sign of the sine in each of the four quadrants. Thus, for example, an angle corresponding to a radius lying in the region marked "$\cos\theta < 0, \sin\theta > 0$" will have a cosine that is negative and a sine that is positive.

There is no need to memorize this figure, because you can always reconstruct it if you understand the definitions of cosine and sine.

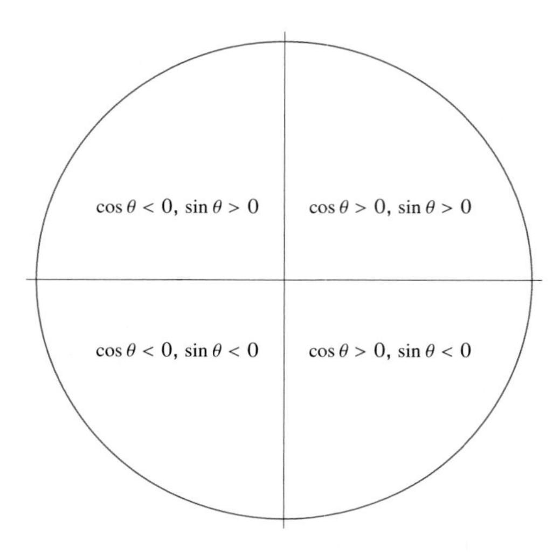

The quadrant in which a radius lies determines whether the cosine and sine of the corresponding angle are positive or negative.

Recall that the cosine of an angle is the first coordinate of the endpoint of the corresponding radius and the sine is the second coordinate. Thus the cosine is positive in the two quadrants where the first coordinate is positive and the cosine is negative in the two quadrants where the first coordinate is negative. Similarly, the sine is positive in the two quadrants where the second coordinate is positive and the sine is negative in the two quadrants where the second coordinate is negative.

The example below should help you understand how the quadrant determines the sign of the cosine and sine.

EXAMPLE 3

(a) Evaluate $\cos\frac{\pi}{4}$ and $\sin\frac{\pi}{4}$.

(b) Evaluate $\cos\frac{3\pi}{4}$ and $\sin\frac{3\pi}{4}$.

(c) Evaluate $\cos(-\frac{\pi}{4})$ and $\sin(-\frac{\pi}{4})$.

(d) Evaluate $\cos(-\frac{3\pi}{4})$ and $\sin(-\frac{3\pi}{4})$.

SOLUTION The four angles $\frac{\pi}{4}$, $\frac{3\pi}{4}$, $-\frac{\pi}{4}$, and $-\frac{3\pi}{4}$ radians (or, equivalently, $45°$, $135°$, $-45°$, and $-135°$) are shown below. Each coordinate of the radius corresponding to each of these angles is either $\frac{\sqrt{2}}{2}$ or $-\frac{\sqrt{2}}{2}$; the only issue to worry about in computing the cosine and sine of these angles is the sign.

The radius corresponding to $\frac{\pi}{4}$ radians, $\frac{3\pi}{4}$ radians, $-\frac{\pi}{4}$ radians, and $-\frac{3\pi}{4}$ radians (or, equivalently, 45°, 135°, −45°, and −135°).

(a) Both coordinates of the endpoint of the radius corresponding to $\frac{\pi}{4}$ radians are positive. Thus $\cos\frac{\pi}{4} = \sin\frac{\pi}{4} = \frac{\sqrt{2}}{2}$.

(b) The first coordinate of the endpoint of the radius corresponding to $\frac{3\pi}{4}$ radians is negative and the second coordinate of the endpoint is positive. Thus $\cos\frac{3\pi}{4} = -\frac{\sqrt{2}}{2}$ and $\sin\frac{3\pi}{4} = \frac{\sqrt{2}}{2}$.

(c) The first coordinate of the endpoint of the radius corresponding to $-\frac{\pi}{4}$ radians is positive and the second coordinate of the endpoint is negative. Thus we have $\cos(-\frac{\pi}{4}) = \frac{\sqrt{2}}{2}$ and $\sin(-\frac{\pi}{4}) = -\frac{\sqrt{2}}{2}$.

(d) Both coordinates of the endpoint of the radius corresponding to $-\frac{3\pi}{4}$ radians are negative. Thus we have $\cos(-\frac{3\pi}{4}) = \sin(-\frac{3\pi}{4}) = -\frac{\sqrt{2}}{2}$.

Quadrants can labeled by the descriptive terms upper/lower and right/left. For example, the radius corresponding to $\frac{3\pi}{4}$ radians (which equals 135°) lies in the upper-left quadrant.

The next example shows how to use information about the signs of the cosine and sine to locate the corresponding radius.

EXAMPLE 4

Sketch the radius of the unit circle corresponding to an angle θ such that $\cos\theta = 0.4$ and $\sin\theta < 0$.

SOLUTION Because $\cos\theta$ is positive and $\sin\theta$ is negative, the radius corresponding to θ lies in the lower-right quadrant. To find the endpoint of this radius, which has first coordinate 0.4, start with the point 0.4 on the horizontal axis and then move vertically down to reach a point on the unit circle. Then draw the radius from the origin to that point, as shown below.

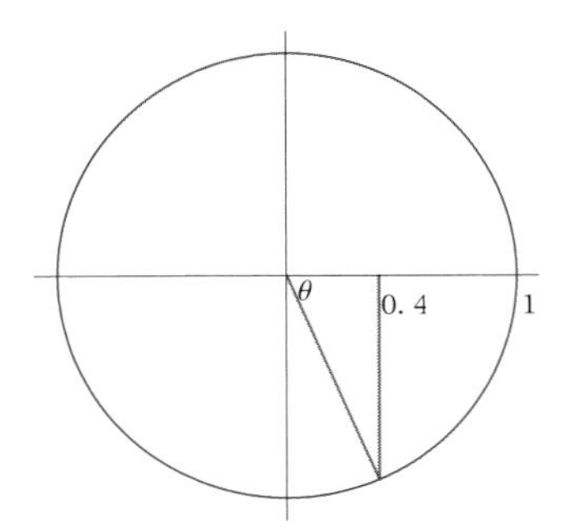

The radius corresponding to an angle θ such that $\cos\theta = 0.4$ and $\sin\theta < 0$.

The Key Equation Connecting Cosine and Sine

The figure defining cosine and sine is sufficiently important that we should look at it again:

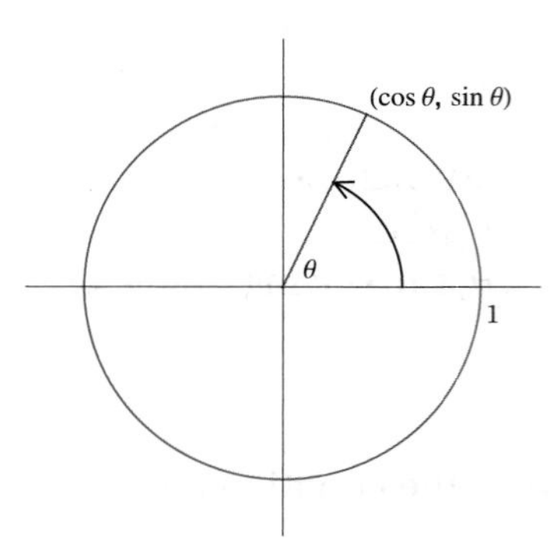

The point $(\cos\theta, \sin\theta)$ *is on the unit circle.*

By definition of cosine and sine, the point $(\cos\theta, \sin\theta)$ is on the unit circle, which is the set of points in the coordinate plane such that the sum of the squares of the coordinates equals 1. In the xy-plane, the unit circle is described by the equation

$$x^2 + y^2 = 1.$$

Thus the following crucial equation holds:

Relationship between cosine and sine

$$(\cos\theta)^2 + (\sin\theta)^2 = 1$$

for every angle θ.

Given either $\cos\theta$ or $\sin\theta$, the equation above can be used to solve for the other quantity, provided that we have enough additional information to determine the sign. The following example illustrates this procedure.

EXAMPLE 5

Suppose θ is an angle such that $\sin\theta = 0.6$, and suppose also that $\frac{\pi}{2} < \theta < \pi$. Evaluate $\cos\theta$.

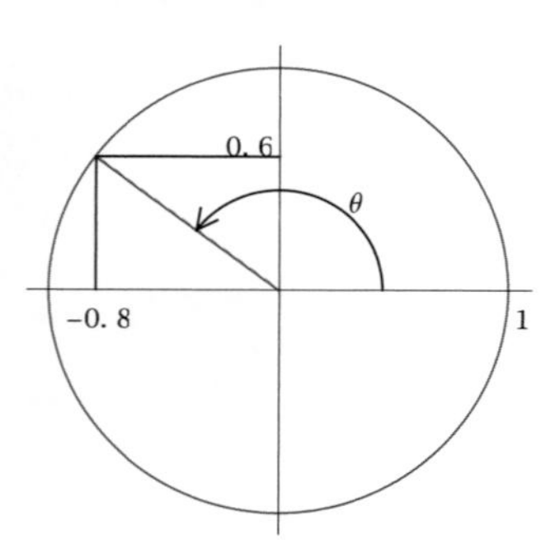

SOLUTION The equation above implies that

$$(\cos\theta)^2 + (0.6)^2 = 1.$$

Because $(0.6)^2 = 0.36$, this implies that

$$(\cos\theta)^2 = 0.64.$$

Thus $\cos\theta = 0.8$ or $\cos\theta = -0.8$. The additional information that $\frac{\pi}{2} < \theta < \pi$ implies that $\cos\theta$ is negative. Thus $\cos\theta = -0.8$.

The Graphs of Cosine and Sine

Before graphing the cosine and sine functions, we should think carefully about the domain and range of these functions. Recall that for each real number θ, there is a radius of the unit circle corresponding to θ.

Recall also that the coordinates of the endpoints of the radius corresponding to the angle θ are labeled $(\cos\theta, \sin\theta)$, thus defining the cosine and sine functions. These functions are defined for every real number θ. Thus the domain of both cosine and sine is the set of real numbers.

As we have already noted, a consequence of $(\cos\theta, \sin\theta)$ lying on the unit circle is the equation

$$(\cos\theta)^2 + (\sin\theta)^2 = 1.$$

Because $(\cos\theta)^2$ and $(\sin\theta)^2$ are both nonnegative, the equation above implies that

$$(\cos\theta)^2 \le 1 \quad \text{and} \quad (\sin\theta)^2 \le 1.$$

These inequalities can be used as a crude test of the plausibility of a result. For example, suppose you do a calculation involving θ and determine that $\cos\theta = 2$. Because the cosine of every angle is between -1 and 1, this is impossible. Thus you must have made a mistake in your calculation.

Thus $\cos\theta$ and $\sin\theta$ must both be between -1 and 1:

Cosine and sine are between -1 and 1

$$-1 \le \cos\theta \le 1 \quad \text{and} \quad -1 \le \sin\theta \le 1$$

for every angle θ.

These inequalities could also be written in the following form:

$$|\cos\theta| \le 1 \quad \text{and} \quad |\sin\theta| \le 1.$$

The first coordinates of the points of the unit circle are precisely the values of the cosine function. A figure of the unit circle shows that every point of the unit circle has a first coordinate in the interval $[-1, 1]$ (which is another way to derive the inequality $-1 \le \cos\theta \le 1$). Conversely, every number in the interval $[-1, 1]$ is the first coordinate of some point on the unit circle. Thus we can conclude that the range of the cosine function is the interval $[-1, 1]$. A similar conclusion holds for the sine function (use second coordinates instead of first coordinates).

We can summarize our results concerning the domain and range of the cosine and sine as follows:

Domain and range of cosine and sine

- The domain of both cosine and sine is the set of real numbers.
- The range of both cosine and sine is the interval $[-1, 1]$.

Because the domain of the cosine and the sine is the set of real numbers, we cannot show the graph of these functions on their entire domain. To understand what the graphs of these functions look like, we start by looking at the graph of cosine on the interval $[-6\pi, 6\pi]$:

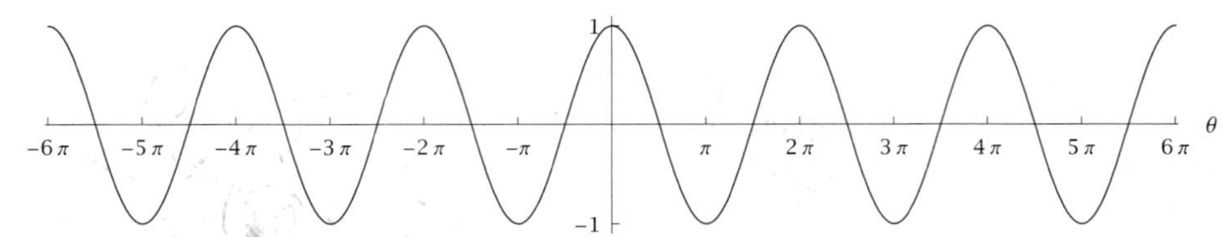

The graph of cosine on the interval $[-6\pi, 6\pi]$.

The graphs in this book were generated by the computer algebra software Mathematica.

Let's begin examining the graph above by noting that the point $(0, 1)$ is on the graph, as expected from the equation $\cos 0 = 1$. Note that the horizontal axis has been called the θ-axis.

Moving to the right along the θ-axis from the origin, we see that the graph crosses the θ-axis at the point $(\frac{\pi}{2}, 0)$, as expected from the equation $\cos \frac{\pi}{2} = 0$. Continuing further to the right, we see that the graph hits its lowest value when $\theta = \pi$, as expected from the equation $\cos \pi = -1$. The graph then crosses the θ-axis again at the point $(\frac{3\pi}{2}, 0)$, as expected from the equation $\cos \frac{3\pi}{2} = 0$. Then the graph hits its highest value again when $\theta = 2\pi$, as expected from the equation $\cos 2\pi = 1$.

The most striking feature of the graph above is its periodic nature—the graph repeats itself. To understand why the graph of cosine exhibits this periodic behavior, consider a radius of the unit circle starting along the positive horizontal axis and moving counterclockwise. As the radius moves, the first coordinate of its endpoint gives the value of the cosine of the corresponding angle. After the radius moves through an angle of 2π radians, it returns to its original position. Then it begins the cycle again, returning to its original position after moving through a total angle of 4π, and so on. Thus we see the periodic behavior of the graph of cosine.

In Section 9.6 we will examine the properties of cosine and its graph more deeply. For now, let's turn to the graph of sine. Here is the graph of sine on the interval $[-6\pi, 6\pi]$:

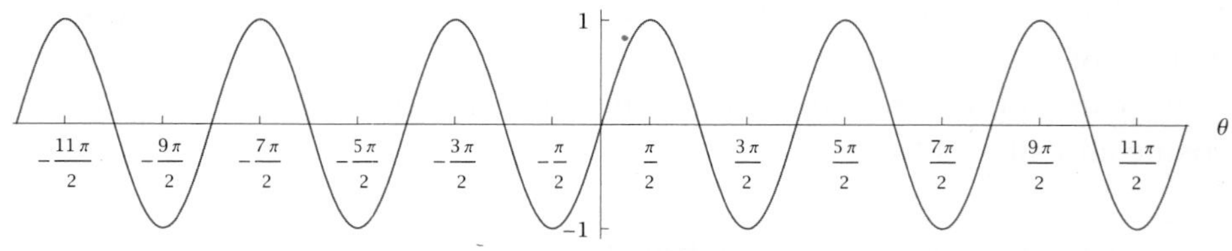

The graph of sine on the interval $[-6\pi, 6\pi]$.

The word sine *comes from the Latin word* sinus, *which means curve.*

This graph goes through the origin, as expected because $\sin 0 = 0$. Moving to the right along the θ-axis from the origin, we see that the graph hits its highest value when $\theta = \frac{\pi}{2}$, as expected because $\sin \frac{\pi}{2} = 1$. Continuing further to the right, we see that the graph crosses the θ-axis at the point $(\pi, 0)$, as expected because $\sin\pi = 0$. The graph then hits its lowest value when $\theta = \frac{3\pi}{2}$, as expected because $\sin\frac{3\pi}{2} = -1$. Then the graph crosses the θ-axis again at $(2\pi, 0)$, as expected because $\sin 2\pi = 0$.

Surely you have noticed that the graph of sine looks much like the graph of cosine. It appears that shifting one graph somewhat to the left or right produces the other graph. We will see that this is indeed the case when we delve more deeply into properties of cosine and sine in Section 9.6.

EXERCISES

Give exact values for the quantities in Exercises 1–10. Do not use a calculator for any of these exercises—otherwise you will likely get decimal approximations for some solutions rather than exact answers. More importantly, good understanding will come from working these exercises by hand.

1. (a) $\cos 3\pi$ (b) $\sin 3\pi$
2. (a) $\cos(-\frac{3\pi}{2})$ (b) $\sin(-\frac{3\pi}{2})$
3. (a) $\cos\frac{11\pi}{4}$ (b) $\sin\frac{11\pi}{4}$
4. (a) $\cos\frac{15\pi}{4}$ (b) $\sin\frac{15\pi}{4}$
5. (a) $\cos\frac{2\pi}{3}$ (b) $\sin\frac{2\pi}{3}$
6. (a) $\cos\frac{4\pi}{3}$ (b) $\sin\frac{4\pi}{3}$
7. (a) $\cos 210°$ (b) $\sin 210°$
8. (a) $\cos 300°$ (b) $\sin 300°$
9. (a) $\cos 360045°$ (b) $\sin 360045°$
10. (a) $\cos(-360030°)$ (b) $\sin(-360030°)$
11. Find the smallest number θ larger than 4π such that $\cos\theta = 0$.
12. Find the smallest number θ larger than 6π such that $\sin\theta = \frac{\sqrt{2}}{2}$.
13. Find the four smallest positive numbers θ such that $\cos\theta = 0$.
14. Find the four smallest positive numbers θ such that $\sin\theta = 0$.
15. Find the four smallest positive numbers θ such that $\sin\theta = 1$.
16. Find the four smallest positive numbers θ such that $\cos\theta = 1$.
17. Find the four smallest positive numbers θ such that $\cos\theta = -1$.
18. Find the four smallest positive numbers θ such that $\sin\theta = -1$.
19. Find the four smallest positive numbers θ such that $\sin\theta = \frac{1}{2}$.
20. Find the four smallest positive numbers θ such that $\cos\theta = \frac{1}{2}$.
21. Suppose $0 < \theta < \frac{\pi}{2}$ and $\cos\theta = \frac{2}{5}$. Evaluate $\sin\theta$.
22. Suppose $0 < \theta < \frac{\pi}{2}$ and $\sin\theta = \frac{3}{7}$. Evaluate $\cos\theta$.
23. Suppose $\frac{\pi}{2} < \theta < \pi$ and $\sin\theta = \frac{2}{9}$. Evaluate $\cos\theta$.
24. Suppose $\frac{\pi}{2} < \theta < \pi$ and $\sin\theta = \frac{3}{8}$. Evaluate $\cos\theta$.
25. Suppose $-\frac{\pi}{2} < \theta < 0$ and $\cos\theta = 0.1$. Evaluate $\sin\theta$.
26. Suppose $-\frac{\pi}{2} < \theta < 0$ and $\cos\theta = 0.3$. Evaluate $\sin\theta$.
27. Find the smallest number x such that
$$\sin(e^x) = 0.$$
28. Find the smallest number x such that
$$\cos(e^x + 1) = 0.$$
29. Find the smallest positive number x such that
$$\sin(x^2 + x + 4) = 0.$$
30. Find the smallest positive number x such that
$$\cos(x^2 + 2x + 6) = 0.$$
31. Let θ be the acute angle between the positive horizontal axis and the line with slope 3 through the origin. Evaluate $\cos\theta$ and $\sin\theta$.
32. Let θ be the acute angle between the positive horizontal axis and the line with slope 4 through the origin. Evaluate $\cos\theta$ and $\sin\theta$.

PROBLEMS

33. (a) Sketch a radius of the unit circle corresponding to an angle θ such that $\cos\theta = \frac{6}{7}$.

    (b) Sketch another radius, different from the one in part (a), also illustrating $\cos\theta = \frac{6}{7}$.

34. (a) Sketch a radius of the unit circle corresponding to an angle θ such that $\sin\theta = -0.8$.

    (b) Sketch another radius, different from the one in part (a), also illustrating $\sin\theta = -0.8$.

35. Find angles u and v such that $\cos u = \cos v$ but $\sin u \neq \sin v$.

36. Find angles u and v such that $\sin u = \sin v$ but $\cos u \neq \cos v$.

37. Show that $\ln(\cos\theta)$ is the average of $\ln(1-\sin\theta)$ and $\ln(1+\sin\theta)$ for every θ in the interval $(-\frac{\pi}{2}, \frac{\pi}{2})$.

38. Suppose you have borrowed two calculators from friends, but you do not know whether they are set to work in radians or degrees. Thus you ask each calculator to evaluate $\cos 3.14$. One calculator replies with an answer of -0.999999; the other calculator replies with an answer of 0.998499. Without further use of a calculator, how would you decide which calculator is using radians and which calculator is using degrees? Explain your answer.

39. Suppose you have borrowed two calculators from friends, but you do not know whether they are set to work in radians or degrees. Thus you ask each calculator to evaluate $\sin 1$. One calculator replies with an answer of 0.017452; the other calculator replies with an answer of 0.841471. Without further use of a calculator, how would you decide which calculator is using radians and which calculator is using degrees? Explain your answer.

40. Suppose m is a real number. Let θ be the acute angle between the positive horizontal axis and the line with slope m through the origin. Evaluate $\cos\theta$ and $\sin\theta$.

41. Explain why there does not exist a real number x such that $2^{\sin x} = \frac{3}{7}$.

42. Explain why $\pi^{\cos x} < 4$ for every real number x.

43. Explain why $\frac{1}{3} < e^{\sin x}$ for every number real number x.

44. Explain why the equation

$$(\sin x)^2 - 4\sin x + 4 = 0$$

has no solutions.

45. Explain why the equation

$$(\cos x)^{99} + 4\cos x - 6 = 0$$

has no solutions.

46. Explain why there does not exist a number θ such that $\log\cos\theta = 0.1$.

WORKED-OUT SOLUTIONS *to Odd-numbered Exercises*

Give exact values for the quantities in Exercises 1-10. Do not use a calculator for any of these exercises—otherwise you will likely get decimal approximations for some solutions rather than exact answers. More importantly, good understanding will come from working these exercises by hand.

1. (a) $\cos 3\pi$ (b) $\sin 3\pi$

SOLUTION Because $3\pi = 2\pi + \pi$, an angle of 3π radians (as measured counterclockwise from the positive horizontal axis) consists of a complete revolution around the circle (2π radians) followed by another π radians (180°), as shown below. The endpoint of the corresponding radius is $(-1, 0)$. Thus $\cos 3\pi = -1$ and $\sin 3\pi = 0$.

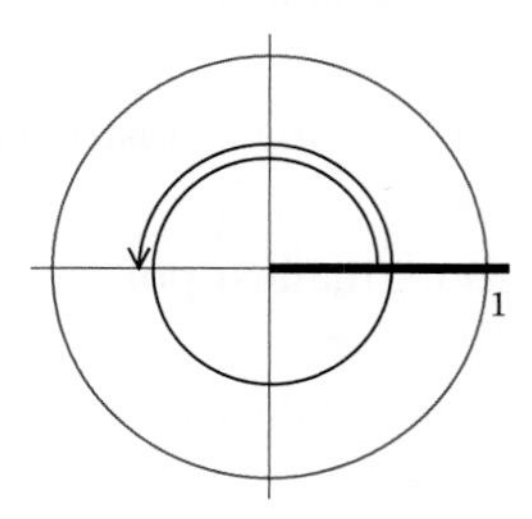

3. (a) $\cos \frac{11\pi}{4}$ (b) $\sin \frac{11\pi}{4}$

SOLUTION Because $\frac{11\pi}{4} = 2\pi + \frac{\pi}{2} + \frac{\pi}{4}$, an angle of $\frac{11\pi}{4}$ radians (as measured counterclockwise from the positive horizontal axis) consists of a complete revolution around the circle (2π radians) followed by another $\frac{\pi}{2}$ radians (90°), followed by another $\frac{\pi}{4}$ radians (45°), as shown below. Hence the endpoint of the corresponding radius is $(-\frac{\sqrt{2}}{2}, \frac{\sqrt{2}}{2})$. Thus $\cos \frac{11\pi}{4} = -\frac{\sqrt{2}}{2}$ and $\sin \frac{11\pi}{4} = \frac{\sqrt{2}}{2}$.

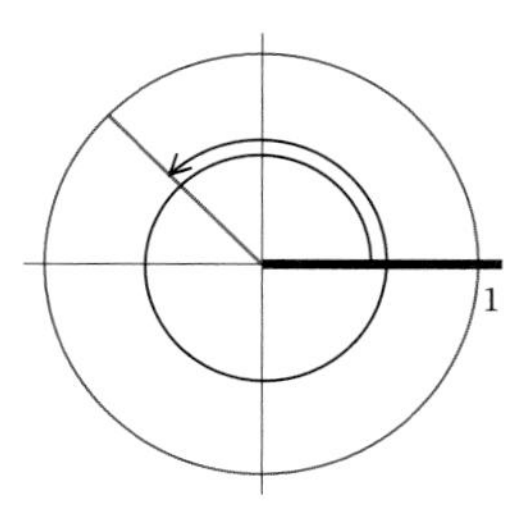

5. (a) $\cos \frac{2\pi}{3}$ (b) $\sin \frac{2\pi}{3}$

SOLUTION Because $\frac{2\pi}{3} = \frac{\pi}{2} + \frac{\pi}{6}$, an angle of $\frac{2\pi}{3}$ radians (as measured counterclockwise from the positive horizontal axis) consists of $\frac{\pi}{2}$ radians (90° radians) followed by another $\frac{\pi}{6}$ radians (30°), as shown below. The endpoint of the corresponding radius is $(-\frac{1}{2}, \frac{\sqrt{3}}{2})$. Thus $\cos \frac{2\pi}{3} = -\frac{1}{2}$ and $\sin \frac{2\pi}{3} = \frac{\sqrt{3}}{2}$.

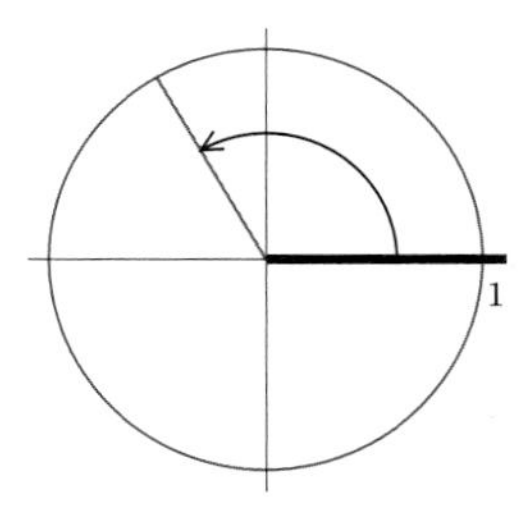

7. (a) $\cos 210°$ (b) $\sin 210°$

SOLUTION Because $210 = 180 + 30$, an angle of 210° (as measured counterclockwise from the positive horizontal axis) consists of 180° followed by another 30°, as shown below. The endpoint of the corresponding radius is $(-\frac{\sqrt{3}}{2}, -\frac{1}{2})$. Thus $\cos 210° = -\frac{\sqrt{3}}{2}$ and $\sin 210° = -\frac{1}{2}$.

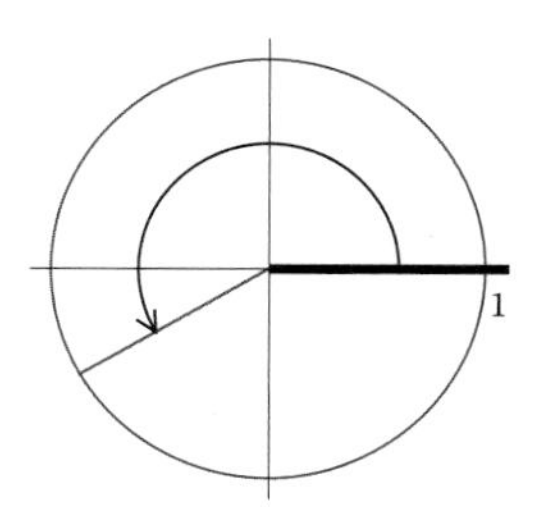

9. (a) $\cos 360045°$ (b) $\sin 360045°$

SOLUTION Because $360045 = 360 \times 1000 + 45$, an angle of 360045° (as measured counterclockwise from the positive horizontal axis) consists of 1000 complete revolutions around the circle followed by another 45°. The endpoint of the corresponding radius is $(\frac{\sqrt{2}}{2}, \frac{\sqrt{2}}{2})$. Thus

$$\cos 360045° = \frac{\sqrt{2}}{2} \quad \text{and} \quad \sin 360045° = \frac{\sqrt{2}}{2}.$$

11. Find the smallest number θ larger than 4π such that $\cos\theta = 0$.

SOLUTION Note that

$$0 = \cos \frac{\pi}{2} = \cos \frac{3\pi}{2} = \cos \frac{5\pi}{2} = \dots$$

and that the only numbers whose cosine equals 0 are of the form $\frac{(2n+1)\pi}{2}$, where n is an integer. The smallest number of this form larger than 4π is $\frac{9\pi}{2}$. Thus $\frac{9\pi}{2}$ is the smallest number larger than 4π whose cosine equals 0.

13. Find the four smallest positive numbers θ such that $\cos\theta = 0$.

SOLUTION Think of a radius of the unit circle whose endpoint is $(1, 0)$. If this radius moves counterclockwise, forming an angle of θ radians with the positive horizontal axis, the first coordinate of its endpoint first becomes 0 when θ equals $\frac{\pi}{2}$ (which equals 90°), then again when θ equals $\frac{3\pi}{2}$ (which equals 270°), then again when θ equals $\frac{5\pi}{2}$ (which equals 360° + 90°, or 450°), then again when θ equals $\frac{7\pi}{2}$ (which equals 360° + 270°, or 630°), and so on. Thus the four smallest positive numbers θ such that $\cos\theta = 0$ are $\frac{\pi}{2}$, $\frac{3\pi}{2}$, $\frac{5\pi}{2}$, and $\frac{7\pi}{2}$.

15. Find the four smallest positive numbers θ such that $\sin\theta = 1$.

SOLUTION Think of a radius of the unit circle whose endpoint is $(1, 0)$. If this radius moves counterclockwise, forming an angle of θ radians with the positive horizontal axis, then the second coordinate of its endpoint first becomes 1 when θ equals $\frac{\pi}{2}$ (which equals $90°$), then again when θ equals $\frac{5\pi}{2}$ (which equals $360° + 90°$, or $450°$), then again when θ equals $\frac{9\pi}{2}$ (which equals $2 \times 360° + 90°$, or $810°$), then again when θ equals $\frac{13\pi}{2}$ (which equals $3 \times 360° + 90°$, or $1170°$), and so on. Thus the four smallest positive numbers θ such that $\sin\theta = 1$ are $\frac{\pi}{2}$, $\frac{5\pi}{2}$, $\frac{9\pi}{2}$, and $\frac{13\pi}{2}$.

17. Find the four smallest positive numbers θ such that $\cos\theta = -1$.

SOLUTION Think of a radius of the unit circle whose endpoint is $(1, 0)$. If this radius moves counterclockwise, forming an angle of θ radians with the positive horizontal axis, the first coordinate of its endpoint first becomes -1 when θ equals π (which equals $180°$), then again when θ equals 3π (which equals $360° + 180°$, or $540°$), then again when θ equals 5π (which equals $2 \times 360° + 180°$, or $900°$), then again when θ equals 7π (which equals $3 \times 360° + 180°$, or $1260°$), and so on. Thus the four smallest positive numbers θ such that $\cos\theta = -1$ are π, 3π, 5π, and 7π.

19. Find the four smallest positive numbers θ such that $\sin\theta = \frac{1}{2}$.

SOLUTION Think of a radius of the unit circle whose endpoint is $(1, 0)$. If this radius moves counterclockwise, forming an angle of θ radians with the positive horizontal axis, the second coordinate of its endpoint first becomes $\frac{1}{2}$ when θ equals $\frac{\pi}{6}$ (which equals $30°$), then again when θ equals $\frac{5\pi}{6}$ (which equals $150°$), then again when θ equals $\frac{13\pi}{6}$ (which equals $360° + 30°$, or $390°$), then again when θ equals $\frac{17\pi}{6}$ (which equals $360° + 150°$, or $510°$), and so on. Thus the four smallest positive numbers θ such that $\sin\theta = \frac{1}{2}$ are $\frac{\pi}{6}$, $\frac{5\pi}{6}$, $\frac{13\pi}{6}$, and $\frac{17\pi}{6}$.

21. Suppose $0 < \theta < \frac{\pi}{2}$ and $\cos\theta = \frac{2}{5}$. Evaluate $\sin\theta$.

SOLUTION We know that

$$(\cos\theta)^2 + (\sin\theta)^2 = 1.$$

Thus

$$\begin{aligned}(\sin\theta)^2 &= 1 - (\cos\theta)^2\\ &= 1 - \left(\tfrac{2}{5}\right)^2\\ &= \frac{21}{25}.\end{aligned}$$

Because $0 < \theta < \frac{\pi}{2}$, we know that $\sin\theta > 0$. Thus taking square roots of both sides of the equation above gives

$$\sin\theta = \frac{\sqrt{21}}{5}.$$

23. Suppose $\frac{\pi}{2} < \theta < \pi$ and $\sin\theta = \frac{2}{9}$. Evaluate $\cos\theta$.

SOLUTION We know that

$$(\cos\theta)^2 + (\sin\theta)^2 = 1.$$

Thus

$$\begin{aligned}(\cos\theta)^2 &= 1 - (\sin\theta)^2\\ &= 1 - \left(\tfrac{2}{9}\right)^2\\ &= \frac{77}{81}.\end{aligned}$$

Because $\frac{\pi}{2} < \theta < \pi$, we know that $\cos\theta < 0$. Thus taking square roots of both sides of the equation above gives

$$\cos\theta = -\frac{\sqrt{77}}{9}.$$

25. Suppose $-\frac{\pi}{2} < \theta < 0$ and $\cos\theta = 0.1$. Evaluate $\sin\theta$.

SOLUTION We know that

$$(\cos\theta)^2 + (\sin\theta)^2 = 1.$$

Thus

$$\begin{aligned}(\sin\theta)^2 &= 1 - (\cos\theta)^2\\ &= 1 - (0.1)^2\\ &= 0.99.\end{aligned}$$

Because $-\frac{\pi}{2} < \theta < 0$, we know that $\sin\theta < 0$. Thus taking square roots of both sides of the equation above gives

$$\sin\theta = -\sqrt{0.99} \approx -0.995.$$

27. Find the smallest number x such that

$$\sin(e^x) = 0.$$

SOLUTION Note that e^x is an increasing function. Because e^x is positive for every real number x, and because π is the smallest positive number whose sine equals 0, we want to choose x so that $e^x = \pi$. Thus $x = \ln\pi$.

29. Find the smallest positive number x such that

$$\sin(x^2 + x + 4) = 0.$$

SOLUTION Note that $x^2 + x + 4$ is an increasing function on the interval $[0, \infty)$. If x is positive, then $x^2 + x + 4 > 4$. Because 4 is larger than π but less than 2π, the smallest number bigger than 4 whose sine equals 0 is 2π. Thus we want to choose x so that $x^2 + x + 4 = 2\pi$. In other words, we need to solve the equation

$$x^2 + x + (4 - 2\pi) = 0.$$

Using the quadratic formula, we see that the solutions to this equation are

$$x = \frac{-1 \pm \sqrt{8\pi - 15}}{2}.$$

A calculator shows that choosing the plus sign in the equation above gives $x \approx 1.0916$ and choosing the minus sign gives $x \approx -2.0916$. We seek only positive values of x, and thus we choose the plus sign in the equation above, getting $x \approx 1.0916$.

31. Let θ be the acute angle between the positive horizontal axis and the line with slope 3 through the origin. Evaluate $\cos\theta$ and $\sin\theta$.

SOLUTION From the solution to Exercise 5 in Section 9.1, we see that the endpoint of the relevant radius on the unit circle has coordinates $\left(\frac{\sqrt{10}}{10}, \frac{3\sqrt{10}}{10}\right)$. Thus

$$\cos\theta = \frac{\sqrt{10}}{10} \quad \text{and} \quad \sin\theta = \frac{3\sqrt{10}}{10}.$$

9.4 *More Trigonometric Functions*

LEARNING OBJECTIVES

By the end of this section you should be able to

- evaluate the tangent of any multiple of 30° or 45° ($\frac{\pi}{6}$ radians or $\frac{\pi}{4}$ radians);
- find the equation of the line making a given angle with the positive horizontal axis and containing a given point;
- sketch a radius of the unit circle corresponding to a given value of the tangent function;
- compute $\cos\theta$, $\sin\theta$, and $\tan\theta$ if given just one of these quantities and the location of the corresponding radius;
- evaluate $\sec\theta$, $\csc\theta$, and $\cot\theta$ as 1 divided by the value of one of the other trigonometric functions.

Section 9.3 introduced the cosine and the sine, the two most important trigonometric functions. This section introduces the tangent, another key trigonometric function, along with three more trigonometric functions.

Definition of Tangent

Recall that $\cos\theta$ and $\sin\theta$ are defined to be the first and second coordinates of the endpoint of the radius of the unit circle corresponding to θ. The ratio of these two numbers, with the cosine in the denominator, turns out to be sufficiently useful to deserve its own name.

Tangent

The **tangent** of an angle θ, denoted $\tan\theta$, is defined by

$$\tan\theta = \frac{\sin\theta}{\cos\theta}$$

provided that $\cos\theta \neq 0$.

Recall that the slope of the line segment connecting (x_1, y_1) and (x_2, y_2) is $\frac{y_2-y_1}{x_2-x_1}$.

The radius of the unit circle corresponding to θ has its initial point at $(0,0)$ and its endpoint at $(\cos\theta, \sin\theta)$. Thus the slope of this line segment equals $\frac{\sin\theta-0}{\cos\theta-0}$, which equals $\frac{\sin\theta}{\cos\theta}$, which equals $\tan\theta$. In other words, we have the following interpretation of the tangent of an angle:

Tangent as slope

$\tan\theta$ is the slope of the radius of the unit circle corresponding to θ.

The following figure illustrates how the cosine, sine, and tangent of an angle are defined:

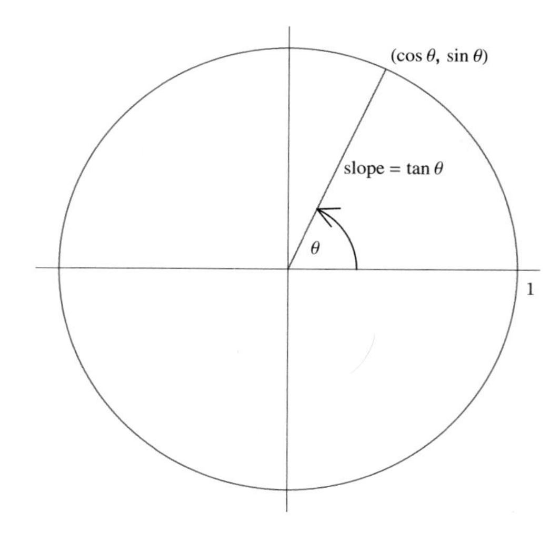

The radius corresponding to θ has slope $\tan\theta$.

Most of what you need to know about trigonometry can be derived from careful consideration of this figure.

EXAMPLE 1

Evaluate $\tan\frac{\pi}{4}$.

SOLUTION The radius corresponding to $\frac{\pi}{4}$ radians (which equals 45°) has its endpoint at $(\frac{\sqrt{2}}{2}, \frac{\sqrt{2}}{2})$. For this point, the second coordinate divided by the first coordinate equals 1. Thus

$$\tan\frac{\pi}{4} = \tan 45^\circ = 1.$$

The equation above is no surprise, because the line through the origin that makes a 45° angle with the positive horizontal axis has slope 1.

The table below gives the tangent of some special angles. This table is obtained from the table of cosines and sines of special angles in Section 9.3 simply by dividing the sine of each angle by its cosine.

θ (*radians*)	θ (*degrees*)	$\tan\theta$
0	0°	0
$\frac{\pi}{6}$	30°	$\frac{\sqrt{3}}{3}$
$\frac{\pi}{4}$	45°	1
$\frac{\pi}{3}$	60°	$\sqrt{3}$
$\frac{\pi}{2}$	90°	undefined
π	180°	0

Tangent of some special angles.

If you have trouble remembering whether $\tan\theta$ *equals* $\frac{\sin\theta}{\cos\theta}$ *or* $\frac{\cos\theta}{\sin\theta}$, *note that the wrong choice would lead to* $\tan 0$ *being undefined, which is not desirable.*

The table above shows that the tangent of $\frac{\pi}{2}$ radians (or equivalently the tangent of 90°) is not defined. The reason for this is that division by 0 is not defined.

Similarly, $\tan\theta$ is not defined for each angle θ such that $\cos\theta = 0$. In other words, $\tan\theta$ is not defined for $\theta = \pm\frac{\pi}{2}, \pm\frac{3\pi}{2}, \pm\frac{5\pi}{2}, \ldots$.

The tangent function allows us to find the equation of a line making a given angle with the positive horizontal axis and containing a given point.

EXAMPLE 2

Find the equation of the line in the xy-plane that contains the point $(3, 1)$ and makes an angle of $28°$ with the positive x-axis.

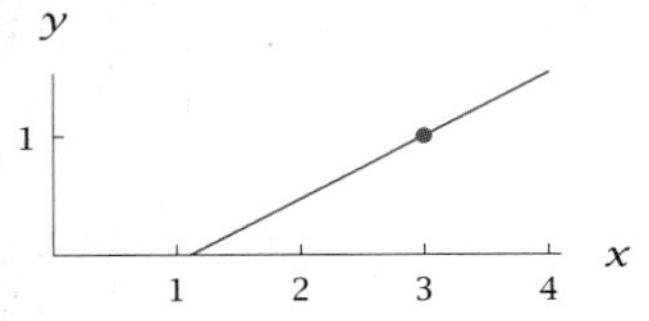

This line has a $28°$ angle with the positive x-axis and contains $(3, 1)$.

SOLUTION A line that makes an angle of $28°$ with the positive x-axis has slope $\tan 28°$. Thus the equation of the line is $y - 1 = (\tan 28°)(x - 3)$. Because $\tan 28° \approx 0.531709$, we could rewrite this as $y \approx 0.531709x - 0.59513$.

The Sign of Tangent

The quadrant in which a radius lies determines whether the tangent of the corresponding angle is positive or negative. The figure below shows the sign of the tangent in each of the four quadrants:

There is no need to memorize this figure, because you can always reconstruct it if you understand the definition of tangent.

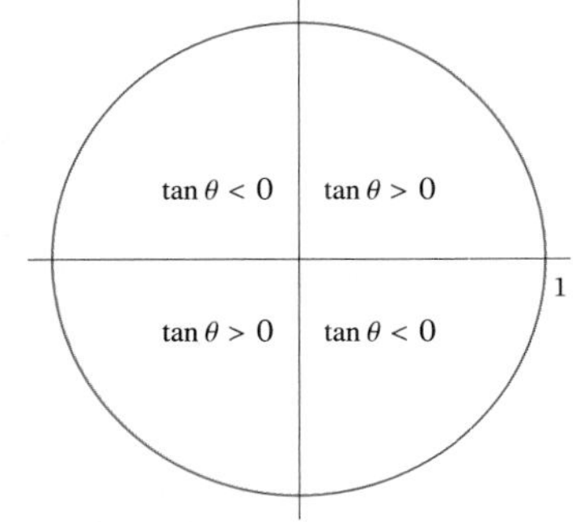

The quadrant in which a radius lies determines whether the tangent of the corresponding angle is positive or negative.

The tangent of an angle is the ratio of the second and first coordinates of the endpoint of the corresponding radius. Thus the tangent is positive in the quadrant where both coordinates are positive and also in the quadrant where both coordinates are negative. The tangent is negative in quadrants where one coordinate is positive and the other coordinate is negative.

EXAMPLE 3

Sketch the radius of the unit circle corresponding to an angle θ such that $\tan \theta = \frac{1}{2}$.

SOLUTION Because $\tan \theta$ is the slope of the radius of the unit circle corresponding to θ, we seek a radius with slope $\frac{1}{2}$. Because $\frac{1}{2} > 0$, any radius with slope $\frac{1}{2}$ must lie in either the upper-right quadrant or the lower-left quadrant.

One such radius is shown in the figure below on the left, and another such radius is shown below in the figure on the right.

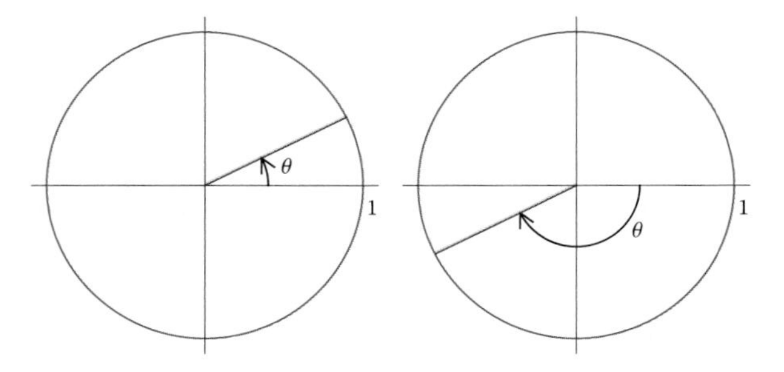

Two radii corresponding to two different values of θ such that $\tan \theta = \frac{1}{2}$. Each of these radii has slope $\frac{1}{2}$.

Connections Among Cosine, Sine, and Tangent

Given any one of $\cos\theta$ or $\sin\theta$ or $\tan\theta$, the equations

$$(\cos\theta)^2 + (\sin\theta)^2 = 1 \quad \text{and} \quad \tan\theta = \frac{\sin\theta}{\cos\theta}$$

can be used to solve for the other two quantities, provided that we have enough additional information to determine the sign. Suppose, for example, that we know $\cos\theta$ (and the quadrant in which the angle θ lies). Knowing $\cos\theta$, we can use the first equation above to calculate $\sin\theta$ (as we did in the previous section), and then we can use the second equation above to calculate $\tan\theta$.

The example below shows how to calculate $\cos\theta$ and $\sin\theta$ from $\tan\theta$ and the information about the quadrant of the angle.

EXAMPLE 4

Suppose $\pi < \theta < \frac{3\pi}{2}$ and $\tan\theta = 4$. Evaluate $\cos\theta$ and $\sin\theta$.

SOLUTION In solving such problems, a sketch can help us understand what is going on. In this case, we know that the angle θ is between π radians (which equals 180°) and $\frac{3\pi}{2}$ radians (which equals 270°). Furthermore, the corresponding radius has a fairly steep slope of 4. Thus the sketch here gives a good depiction of the situation.

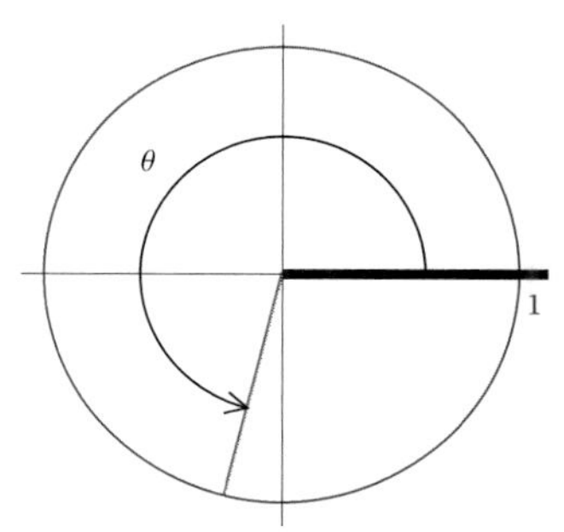

The angle between π and $\frac{3\pi}{2}$ whose tangent equals 4.

To solve this problem, rewrite the information that $\tan\theta = 4$ in the form

$$\frac{\sin\theta}{\cos\theta} = 4.$$

Multiplying both sides of this equation by $\cos\theta$, we get

$$\sin\theta = 4\cos\theta.$$

In the equation $(\cos\theta)^2 + (\sin\theta)^2 = 1$, substitute the expression above for $\sin\theta$, getting

$$(\cos\theta)^2 + (4\cos\theta)^2 = 1,$$

which is equivalent to the equation $17(\cos\theta)^2 = 1$. Thus $\cos\theta = \frac{1}{\sqrt{17}}$ or $\cos\theta = -\frac{1}{\sqrt{17}}$. A glance at the figure above shows that $\cos\theta$ is negative. Thus we must have $\cos\theta = -\frac{1}{\sqrt{17}}$.

The equation $\sin\theta = 4\cos\theta$ now implies that $\sin\theta = -\frac{4}{\sqrt{17}}$.

If we want to remove the square roots from the denominators, then our solution could be written in the form $\cos\theta = -\frac{\sqrt{17}}{17}$, $\sin\theta = -\frac{4\sqrt{17}}{17}$.

The Graph of Tangent

Before graphing the tangent function, we should think carefully about its domain and range. We have already noted that the tangent is defined for all real numbers except odd multiples of $\frac{\pi}{2}$.

The tangent of an angle is the slope of the corresponding radius of the unit circle. Because every real number is the slope of some radius of the unit circle, we see that every number is the tangent of some angle. In other words, the range of the tangent function equals the set of real numbers.

We can summarize our conclusions concerning the domain and range of the tangent as follows:

Domain and range of tangent

- The domain of the tangent function is the set of real numbers that are not odd multiples of $\frac{\pi}{2}$.
- The range of the tangent function is the set of real numbers.

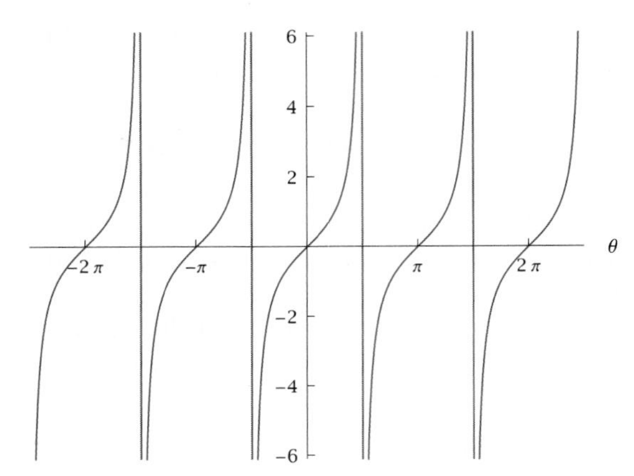

The graph of tangent on the interval $(-\frac{5}{2}\pi, \frac{5}{2}\pi)$.

This graph has been vertically truncated to show only values of the tangent that have absolute value less than 6.

Let's begin examining the graph above by noting that the graph goes through the origin, as expected from the equation $\tan 0 = 0$. Moving to the right along the θ-axis (the horizontal axis) from the origin, we see that the point $(\frac{\pi}{4}, 1)$ is on the graph, as expected from the equation $\tan \frac{\pi}{4} = 1$.

Continuing further to the right along the θ-axis toward the point where $\theta = \frac{\pi}{2}$, we see that as θ gets close to $\frac{\pi}{2}$, the values of $\tan \theta$ rapidly become very large. In fact, the values of $\tan \theta$ become too large to be shown on the figure above while maintaining a reasonable scale on the vertical axis.

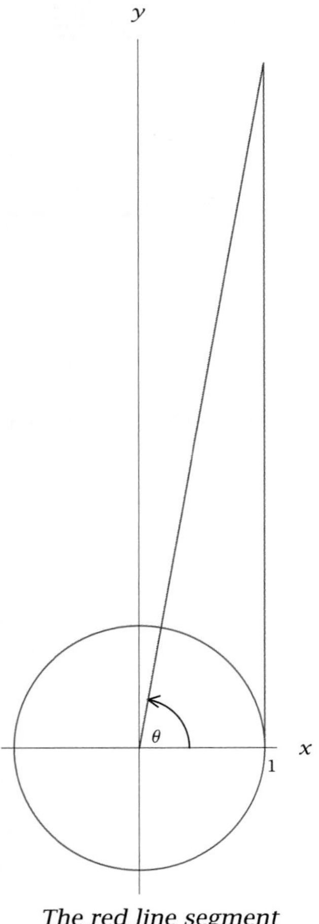

The red line segment has slope $\tan \theta$. *The blue line segment has length* $\tan \theta$.

To understand why $\tan \theta$ is large when θ is slightly less than a right angle, consider the figure in the margin, which shows an angle a bit less than $\frac{\pi}{2}$ radians. We know that the red line segment has slope $\tan \theta$. Thus the red line segment lies on the line $y = (\tan \theta)x$. Hence the point on the red line segment with $x = 1$ has $y = \tan \theta$. In other words, the blue line segment has length $\tan \theta$.

The blue line segment becomes very long when the red line segment comes close to making a right angle with the positive horizontal axis. Thus $\tan \theta$ becomes large when θ is just slightly less than a right angle.

This behavior can also be seen numerically as well as graphically. For example,

$$\sin(\tfrac{\pi}{2} - 0.01) \approx 0.99995 \quad \text{and} \quad \cos(\tfrac{\pi}{2} - 0.01) \approx 0.0099998.$$

Thus $\tan(\frac{\pi}{2} - 0.01)$, which is the ratio of the two numbers above, is approximately 100.

What's happening here is that if θ is a number just slightly less than $\frac{\pi}{2}$ (for example, θ might be $\frac{\pi}{2} - 0.01$ as in the example above), then $\sin\theta$ is just slightly less than 1 and $\cos\theta$ is just slightly more than 0. Thus the ratio $\frac{\sin\theta}{\cos\theta}$, which equals $\tan\theta$, will be large.

In addition to the behavior of the graph near the lines where θ is an odd multiple of $\frac{\pi}{2}$, another striking feature of the graph above is its periodic nature. We will discuss this property of the graph of the tangent in Section 9.6 when we examine the properties of the tangent more deeply.

Three More Trigonometric Functions

The three main trigonometric functions are cosine, sine, and tangent. Three more trigonometric functions are sometimes used. These functions are simply the multiplicative inverses of the functions we have already defined. Here are the formal definitions:

Secant

The **secant** of an angle θ, denoted $\sec\theta$, is defined by

$$\sec\theta = \frac{1}{\cos\theta}.$$

The secant, cosecant, and cotangent functions do not exist in France, in the sense that students there do not learn about these functions.

Cosecant

The **cosecant** of an angle θ, denoted $\csc\theta$, is defined by

$$\csc\theta = \frac{1}{\sin\theta}.$$

Cotangent

The **cotangent** of an angle θ, denoted $\cot\theta$, is defined by

$$\cot\theta = \frac{\cos\theta}{\sin\theta}.$$

In all three of these definitions, the function is not defined for values of θ that would result in a division by 0.

Because the cotangent is defined to be the cosine divided by the sine and the tangent is defined to be the sine divided by the cosine, we have the following consequence of the definitions:

Tangent and cotangent are multiplicative inverses.

If θ is an angle such that both $\tan\theta$ and $\cot\theta$ are defined, then

$$\cot\theta = \frac{1}{\tan\theta}.$$

The scientific calculator on an iPhone (obtained by rotating the standard calculator sideways) has buttons for cos, sin, *and* tan *but omits buttons for* sec, csc, *and* cot.

Many books place too much emphasis on the secant, cosecant, and cotangent. You will rarely need to know anything about these functions beyond their definitions. Whenever you do encounter one of these functions, simply replace it by its definition in terms of cosine, sine, and tangent and then use your knowledge of those more familiar functions. By concentrating on cosine, sine, and tangent rather than all six trigonometric functions, you will attain a better understanding with less clutter in your mind.

EXAMPLE 5

(a) Evaluate $\sec 60°$. (b) Evaluate $\csc\frac{\pi}{4}$. (c) Evaluate $\cot\frac{\pi}{3}$.

So that you will be comfortable with these functions in case you encounter them elsewhere, some of the exercises in this section require you to use the secant, cosecant, or cotangent. However, after this section we will rarely use these functions in this book.

SOLUTION

(a)

$$\begin{aligned}\sec 60° &= \frac{1}{\cos 60°}\\ &= \frac{1}{\frac{1}{2}}\\ &= 2\end{aligned}$$

Note that $|\sec\theta| \geq 1$ for all θ such that $\sec\theta$ is defined.

(b)

$$\begin{aligned}\csc\tfrac{\pi}{4} &= \frac{1}{\sin\frac{\pi}{4}}\\ &= \frac{1}{\frac{1}{\sqrt{2}}}\\ &= \sqrt{2}\\ &\approx 1.414\end{aligned}$$

Note that $|\csc\theta| \geq 1$ for all θ such that $\csc\theta$ is defined.

(c)

$$\begin{aligned}\cot\tfrac{\pi}{3} &= \frac{\cos\frac{\pi}{3}}{\sin\frac{\pi}{3}}\\ &= \frac{\frac{1}{2}}{\frac{\sqrt{3}}{2}}\\ &= \frac{1}{\sqrt{3}}\\ &\approx 0.577\end{aligned}$$

EXERCISES

1. Find the equation of the line in the xy-plane that goes through the origin and makes an angle of 0.7 radians with the positive x-axis.
2. Find the equation of the line in the xy-plane that goes through the origin and makes an angle of 1.2 radians with the positive x-axis.
3. Find the equation of the line in the xy-plane that contains the point $(3, 2)$ and makes an angle of $41°$ with the positive x-axis.
4. Find the equation of the line in the xy-plane that contains the point $(2, 5)$ and makes an angle of $73°$ with the positive x-axis.
5. Find a number t such that the line through the origin that contains the point $(4, t)$ makes a $22°$ angle with the positive horizontal axis.
6. Find a number w such that the line through the origin that contains the point $(7, w)$ makes a $17°$ angle with the positive horizontal axis.
7. Find the four smallest positive numbers θ such that $\tan\theta = 1$.
8. Find the four smallest positive numbers θ such that $\tan\theta = -1$.
9. Suppose $0 < \theta < \frac{\pi}{2}$ and $\cos\theta = \frac{1}{5}$. Evaluate:
   (a) $\sin\theta$ (b) $\tan\theta$
10. Suppose $0 < \theta < \frac{\pi}{2}$ and $\sin\theta = \frac{1}{4}$. Evaluate:
   (a) $\cos\theta$ (b) $\tan\theta$
11. Suppose $\frac{\pi}{2} < \theta < \pi$ and $\sin\theta = \frac{2}{3}$. Evaluate:
   (a) $\cos\theta$ (b) $\tan\theta$
12. Suppose $\frac{\pi}{2} < \theta < \pi$ and $\sin\theta = \frac{3}{4}$. Evaluate:
   (a) $\cos\theta$ (b) $\tan\theta$
13. Suppose $-\frac{\pi}{2} < \theta < 0$ and $\cos\theta = \frac{4}{5}$. Evaluate:
   (a) $\sin\theta$ (b) $\tan\theta$
14. Suppose $-\frac{\pi}{2} < \theta < 0$ and $\cos\theta = \frac{1}{5}$. Evaluate:
   (a) $\sin\theta$ (b) $\tan\theta$
15. Suppose $0 < \theta < \frac{\pi}{2}$ and $\tan\theta = \frac{1}{4}$. Evaluate:
   (a) $\cos\theta$ (b) $\sin\theta$
16. Suppose $0 < \theta < \frac{\pi}{2}$ and $\tan\theta = \frac{2}{3}$. Evaluate:
   (a) $\cos\theta$ (b) $\sin\theta$
17. Suppose $-\frac{\pi}{2} < \theta < 0$ and $\tan\theta = -3$. Evaluate:
   (a) $\cos\theta$ (b) $\sin\theta$
18. Suppose $-\frac{\pi}{2} < \theta < 0$ and $\tan\theta = -2$. Evaluate:
   (a) $\cos\theta$ (b) $\sin\theta$

Given that

$$\cos 15° = \frac{\sqrt{2+\sqrt{3}}}{2} \quad \textit{and} \quad \sin 22.5° = \frac{\sqrt{2-\sqrt{2}}}{2},$$

in Exercises 19–28 find exact expressions for the indicated quantities.
[These values for $\cos 15°$ and $\sin 22.5°$ will be derived in Examples 4 and 5 in Section 10.5.]

19. $\sin 15°$
20. $\cos 22.5°$
21. $\tan 15°$
22. $\tan 22.5°$
23. $\cot 15°$
24. $\cot 22.5°$
25. $\csc 15°$
26. $\csc 22.5°$
27. $\sec 15°$
28. $\sec 22.5°$

Suppose u and v are in the interval $(0, \frac{\pi}{2})$, with

$$\tan u = 2 \quad \textit{and} \quad \tan v = 3.$$

In Exercises 29–38, find exact expressions for the indicated quantities.

29. $\cot u$
30. $\cot v$
31. $\cos u$
32. $\cos v$
33. $\sin u$
34. $\sin v$
35. $\csc u$
36. $\csc v$
37. $\sec u$
38. $\sec v$
39. Find the smallest number x such that $\tan e^x = 0$.
40. Find the smallest number x such that $\tan e^x$ is undefined.

PROBLEMS

41. (a) Sketch a radius of the unit circle corresponding to an angle θ such that $\tan\theta = \frac{1}{7}$.
    (b) Sketch another radius, different from the one in part (a), also illustrating $\tan\theta = \frac{1}{7}$.
42. (a) Sketch a radius of the unit circle corresponding to an angle θ such that $\tan\theta = 7$.
    (b) Sketch another radius, different from the one in part (a), also illustrating $\tan\theta = 7$.
43. Suppose a radius of the unit circle corresponds to an angle whose tangent equals 5, and another radius of the unit circle corresponds to an angle whose tangent equals $-\frac{1}{5}$. Explain why these two radii are perpendicular to each other.
44. Explain why

$$\tan(\theta + \tfrac{\pi}{2}) = -\frac{1}{\tan\theta}$$

for every number θ that is not an integer multiple of $\frac{\pi}{2}$.

45. Explain why the previous problem excluded integer multiples of $\frac{\pi}{2}$ from the allowable values for θ.
46. Suppose you have borrowed two calculators from friends, but you do not know whether they are set to work in radians or degrees. Thus you ask each calculator to evaluate $\tan 89.9$. One calculator replies with an answer of -2.62; the other calculator replies with an answer of 572.96. Without further use of a calculator, how would you decide which calculator is using radians and which calculator is using degrees? Explain your answer.
47. Suppose you have borrowed two calculators from friends, but you do not know whether they are set to work in radians or degrees. Thus you ask each calculator to evaluate $\tan 1$. One calculator replies with an answer of 0.017455; the other calculator replies with an answer of 1.557408. Without further use of a calculator, how would you decide which calculator is using radians and which calculator is using degrees? Explain your answer.
48. Find a number θ such that the tangent of θ degrees is larger than 50000.
49. Find a positive number θ such that the tangent of θ degrees is less than -90000.
50. Explain why

$$|\sin\theta| \le |\tan\theta|$$

for all θ such that $\tan\theta$ is defined.

51. Suppose θ is not an odd multiple of $\frac{\pi}{2}$. Explain why the point $(\tan\theta, 1)$ is on the line containing the point $(\sin\theta, \cos\theta)$ and the origin.
52. Explain why $\log(\cot\theta) = -\log(\tan\theta)$ for every θ in the interval $(0, \frac{\pi}{2})$.
53. In 1768 the Swiss mathematician Johann Lambert proved that if θ is a rational number in the interval $(0, \frac{\pi}{2})$, then $\tan\theta$ is irrational. Use the equation $\tan\frac{\pi}{4} = 1$ to explain why this result implies that π is irrational.

    [*Lambert's result provided the first proof that π is irrational.*]

WORKED-OUT SOLUTIONS *to Odd-numbered Exercises*

1. Find the equation of the line in the xy-plane that goes through the origin and makes an angle of 0.7 radians with the positive x-axis.

SOLUTION A line that makes an angle of 0.7 radians with the positive x-axis has slope $\tan 0.7$. Thus the equation of the line is $y = (\tan 0.7)x$. Because $\tan 0.7 \approx 0.842288$, we could rewrite this as $y \approx 0.842288x$.

3. Find the equation of the line in the xy-plane that contains the point $(3, 2)$ and makes an angle of $41°$ with the positive x-axis.

SOLUTION A line that makes an angle of $41°$ with the positive x-axis has slope $\tan 41°$. Thus the equation of the line is $y - 2 = (\tan 41°)(x - 3)$. Because $\tan 41° \approx 0.869287$, we could rewrite this as $y \approx 0.869287x - 0.60786$.

5. Find a number t such that the line through the origin that contains the point $(4, t)$ makes a $22°$ angle with the positive horizontal axis.

SOLUTION The line through the origin that contains the point $(4, t)$ has slope $\frac{t}{4}$. Thus we want $\tan 22° = \frac{t}{4}$. Hence

$$t = 4 \tan 22° \approx 1.6161.$$

7. Find the four smallest positive numbers θ such that $\tan \theta = 1$.

SOLUTION Think of a radius of the unit circle whose endpoint is $(1, 0)$. If this radius moves counterclockwise, forming an angle of θ radians with the positive horizontal axis, then the first and second coordinates of its endpoint first become equal (which is equivalent to having $\tan \theta = 1$) when θ equals $\frac{\pi}{4}$ (which equals $45°$), then again when θ equals $\frac{5\pi}{4}$ (which equals $225°$), then again when θ equals $\frac{9\pi}{4}$ (which equals $360° + 45°$, or $405°$), then again when θ equals $\frac{13\pi}{4}$ (which equals $360° + 225°$, or $585°$), and so on.

Thus the four smallest positive numbers θ such that $\tan \theta = 1$ are $\frac{\pi}{4}$, $\frac{5\pi}{4}$, $\frac{9\pi}{4}$, and $\frac{13\pi}{4}$.

9. Suppose $0 < \theta < \frac{\pi}{2}$ and $\cos \theta = \frac{1}{5}$. Evaluate:
(a) $\sin \theta$ (b) $\tan \theta$

SOLUTION The figure below gives a sketch of the angle involved in this exercise:

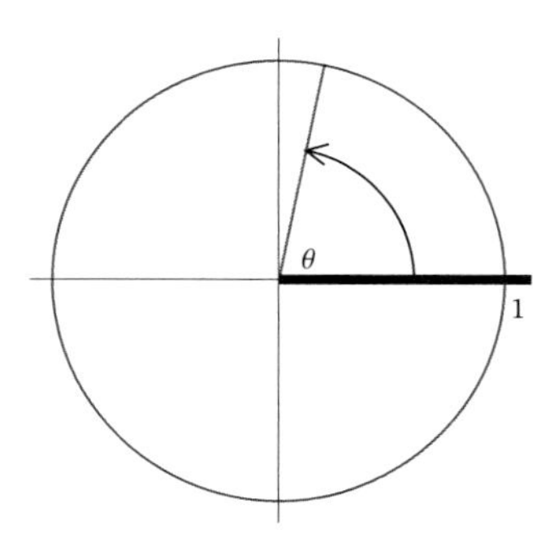

The angle between 0 and $\frac{\pi}{2}$ whose cosine equals $\frac{1}{5}$.

(a) We know that

$$(\cos \theta)^2 + (\sin \theta)^2 = 1.$$

Thus $\left(\frac{1}{5}\right)^2 + (\sin \theta)^2 = 1$. Solving this equation for $(\sin \theta)^2$ gives

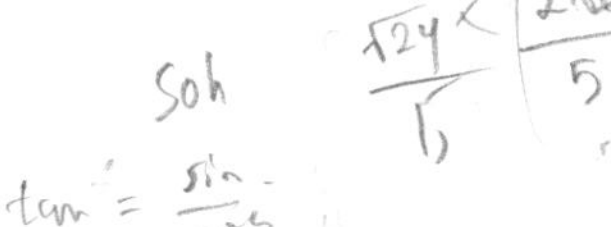

$$(\sin \theta)^2 = \frac{24}{25}.$$

The sketch above shows that $\sin \theta > 0$. Thus taking square roots of both sides of the equation above gives

$$\sin \theta = \frac{\sqrt{24}}{5} = \frac{\sqrt{4 \cdot 6}}{5} = \frac{2\sqrt{6}}{5}.$$

(b)

$$\tan \theta = \frac{\sin \theta}{\cos \theta} = \frac{\frac{2\sqrt{6}}{5}}{\frac{1}{5}} = 2\sqrt{6}.$$

11. Suppose $\frac{\pi}{2} < \theta < \pi$ and $\sin \theta = \frac{2}{3}$. Evaluate:
(a) $\cos \theta$ (b) $\tan \theta$

SOLUTION The figure below gives a sketch of the angle involved in this exercise:

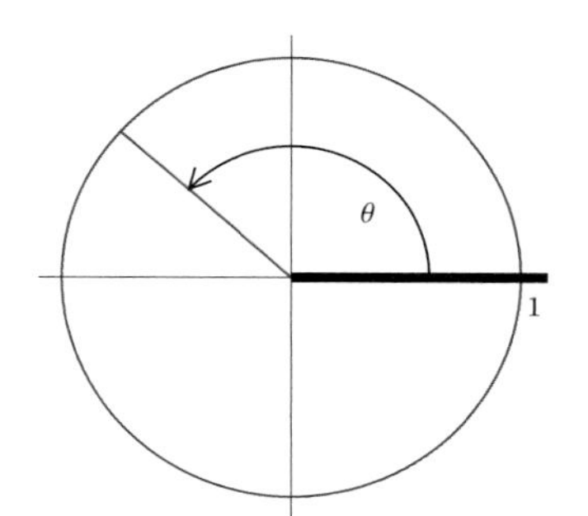

The angle between $\frac{\pi}{2}$ and π whose sine equals $\frac{2}{3}$.

(a) We know that

$$(\cos \theta)^2 + (\sin \theta)^2 = 1.$$

Thus $(\cos \theta)^2 + \left(\frac{2}{3}\right)^2 = 1$. Solving this equation for $(\cos \theta)^2$ gives

$$(\cos \theta)^2 = \frac{5}{9}.$$

The sketch above shows that $\cos \theta < 0$. Thus taking square roots of both sides of the equation above gives

$$\cos \theta = -\frac{\sqrt{5}}{3}.$$

(b)

$$\tan \theta = \frac{\sin \theta}{\cos \theta} = -\frac{\frac{2}{3}}{\frac{\sqrt{5}}{3}} = -\frac{2}{\sqrt{5}} = -\frac{2\sqrt{5}}{5}.$$

13. Suppose $-\frac{\pi}{2} < \theta < 0$ and $\cos \theta = \frac{4}{5}$. Evaluate:
(a) $\sin \theta$ (b) $\tan \theta$

SOLUTION The figure below gives a sketch of the angle involved in this exercise:

The angle between $-\frac{\pi}{2}$ and 0 whose cosine equals $\frac{4}{5}$.

(a) We know that

$$(\cos\theta)^2 + (\sin\theta)^2 = 1.$$

Thus $\left(\frac{4}{5}\right)^2 + (\sin\theta)^2 = 1$. Solving this equation for $(\sin\theta)^2$ gives

$$(\sin\theta)^2 = \frac{9}{25}.$$

The sketch above shows that $\sin\theta < 0$. Thus taking square roots of both sides of the equation above gives

$$\sin\theta = -\frac{3}{5}.$$

(b)

$$\tan\theta = \frac{\sin\theta}{\cos\theta} = -\frac{\frac{3}{5}}{\frac{4}{5}} = -\frac{3}{4}.$$

15. Suppose $0 < \theta < \frac{\pi}{2}$ and $\tan\theta = \frac{1}{4}$. Evaluate:

(a) $\cos\theta$ (b) $\sin\theta$

SOLUTION The figure below gives a sketch of the angle involved in this exercise:

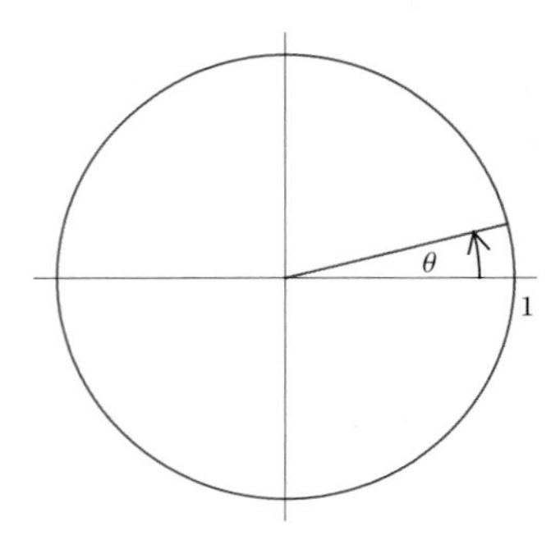

The angle between 0 and $\frac{\pi}{2}$ whose tangent equals $\frac{1}{4}$.

(a) Rewrite the equation $\tan\theta = \frac{1}{4}$ in the form $\frac{\sin\theta}{\cos\theta} = \frac{1}{4}$. Multiplying both sides of this equation by $\cos\theta$, we get

$$\sin\theta = \tfrac{1}{4}\cos\theta.$$

Substitute this expression for $\sin\theta$ into the equation $(\cos\theta)^2 + (\sin\theta)^2 = 1$, getting

$$(\cos\theta)^2 + \tfrac{1}{16}(\cos\theta)^2 = 1,$$

which is equivalent to

$$(\cos\theta)^2 = \frac{16}{17}.$$

The sketch above shows that $\cos\theta > 0$. Thus taking square roots of both sides of the equation above gives

$$\cos\theta = \frac{4}{\sqrt{17}} = \frac{4\sqrt{17}}{17}.$$

(b) We have already noted that $\sin\theta = \frac{1}{4}\cos\theta$. Thus

$$\sin\theta = \frac{\sqrt{17}}{17}.$$

17. Suppose $-\frac{\pi}{2} < \theta < 0$ and $\tan\theta = -3$. Evaluate:

(a) $\cos\theta$ (b) $\sin\theta$

SOLUTION The figure below gives a sketch of the angle involved in this exercise:

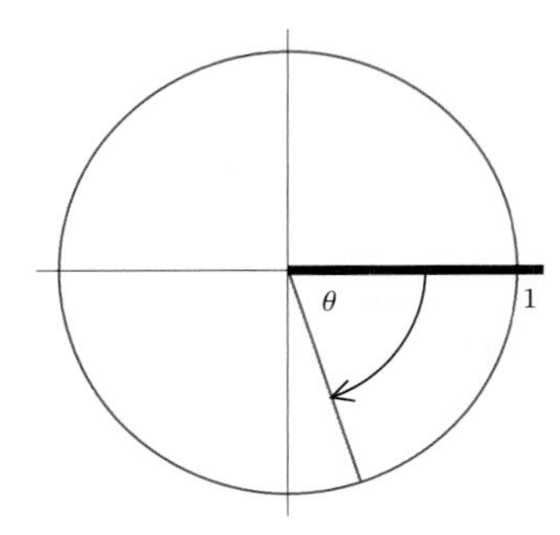

The angle between $-\frac{\pi}{2}$ and 0 whose tangent equals -3.

(a) Rewrite the equation $\tan\theta = -3$ in the form $\frac{\sin\theta}{\cos\theta} = -3$. Multiplying both sides of this equation by $\cos\theta$, we get

$$\sin\theta = -3\cos\theta.$$

Substitute this expression for $\sin\theta$ into the equation $(\cos\theta)^2 + (\sin\theta)^2 = 1$, getting

$$(\cos\theta)^2 + 9(\cos\theta)^2 = 1,$$

which is equivalent to

$$(\cos\theta)^2 = \frac{1}{10}.$$

The sketch above shows that $\cos\theta > 0$. Thus taking square roots of both sides of the equation above gives

$$\cos\theta = \frac{1}{\sqrt{10}} = \frac{\sqrt{10}}{10}.$$

(b) We have already noted that $\sin\theta = -3\cos\theta$. Thus

$$\sin\theta = -\frac{3\sqrt{10}}{10}.$$

Given that

$$\cos 15^\circ = \frac{\sqrt{2+\sqrt{3}}}{2} \quad \textit{and} \quad \sin 22.5^\circ = \frac{\sqrt{2-\sqrt{2}}}{2},$$

in Exercises 19–28 find exact expressions for the indicated quantities.

19. $\sin 15^\circ$

SOLUTION We know that

$$(\cos 15^\circ)^2 + (\sin 15^\circ)^2 = 1.$$

Thus

$$\begin{aligned}(\sin 15^\circ)^2 &= 1 - (\cos 15^\circ)^2 \\ &= 1 - \Big(\frac{\sqrt{2+\sqrt{3}}}{2}\Big)^2 \\ &= 1 - \frac{2+\sqrt{3}}{4} \\ &= \frac{2-\sqrt{3}}{4}.\end{aligned}$$

Because $\sin 15^\circ > 0$, taking square roots of both sides of the equation above gives

$$\sin 15^\circ = \frac{\sqrt{2-\sqrt{3}}}{2}.$$

21. $\tan 15^\circ$

SOLUTION

$$\begin{aligned}\tan 15^\circ &= \frac{\sin 15^\circ}{\cos 15^\circ} \\ &= \frac{\sqrt{2-\sqrt{3}}}{\sqrt{2+\sqrt{3}}} \\ &= \frac{\sqrt{2-\sqrt{3}}}{\sqrt{2+\sqrt{3}}} \cdot \frac{\sqrt{2-\sqrt{3}}}{\sqrt{2-\sqrt{3}}} \\ &= \frac{2-\sqrt{3}}{\sqrt{4-3}} \\ &= 2-\sqrt{3}\end{aligned}$$

23. $\cot 15^\circ$

SOLUTION

$$\begin{aligned}\cot 15^\circ &= \frac{1}{\tan 15^\circ} \\ &= \frac{1}{2-\sqrt{3}} \\ &= \frac{1}{2-\sqrt{3}} \cdot \frac{2+\sqrt{3}}{2+\sqrt{3}} \\ &= \frac{2+\sqrt{3}}{4-3} \\ &= 2+\sqrt{3}\end{aligned}$$

25. $\csc 15^\circ$

SOLUTION

$$\begin{aligned}\csc 15^\circ &= \frac{1}{\sin 15^\circ} \\ &= \frac{2}{\sqrt{2-\sqrt{3}}} \\ &= \frac{2}{\sqrt{2-\sqrt{3}}} \cdot \frac{\sqrt{2+\sqrt{3}}}{\sqrt{2+\sqrt{3}}} \\ &= \frac{2\sqrt{2+\sqrt{3}}}{\sqrt{4-3}} \\ &= 2\sqrt{2+\sqrt{3}}\end{aligned}$$

27. $\sec 15^\circ$

SOLUTION

$$\begin{aligned}\sec 15^\circ &= \frac{1}{\cos 15^\circ} \\ &= \frac{2}{\sqrt{2+\sqrt{3}}} \\ &= \frac{2}{\sqrt{2+\sqrt{3}}} \cdot \frac{\sqrt{2-\sqrt{3}}}{\sqrt{2-\sqrt{3}}} \\ &= \frac{2\sqrt{2-\sqrt{3}}}{\sqrt{4-3}} \\ &= 2\sqrt{2-\sqrt{3}}\end{aligned}$$

Suppose u and v are in the interval $(0, \frac{\pi}{2})$, with

$$\tan u = 2 \quad \textit{and} \quad \tan v = 3.$$

In Exercises 29–38, find exact expressions for the indicated quantities.

29. $\cot u$

SOLUTION
$$\cot u = \frac{1}{\tan u} = \frac{1}{2}$$

31. $\cos u$

SOLUTION We know that

$$2 = \tan u = \frac{\sin u}{\cos u}.$$

To find $\cos u$, make the substitution $\sin u = \sqrt{1 - (\cos u)^2}$ in the equation above (this substitution is valid because we know that $0 < u < \frac{\pi}{2}$ and thus $\sin u > 0$), getting

$$2 = \frac{\sqrt{1 - (\cos u)^2}}{\cos u}.$$

Now square both sides of the equation above, then multiply both sides by $(\cos u)^2$ and rearrange to get the equation

$$5(\cos u)^2 = 1.$$

Because $0 < u < \frac{\pi}{2}$, we see that $\cos u > 0$. Thus taking square roots of both sides of the equation above gives $\cos u = \frac{1}{\sqrt{5}}$, which can be rewritten as $\cos u = \frac{\sqrt{5}}{5}$.

33. $\sin u$

SOLUTION

$$\sin u = \sqrt{1 - (\cos u)^2} = \sqrt{1 - \frac{1}{5}} = \sqrt{\frac{4}{5}} = \frac{2}{\sqrt{5}} = \frac{2\sqrt{5}}{5}$$

35. $\csc u$

SOLUTION
$$\csc u = \frac{1}{\sin u} = \frac{\sqrt{5}}{2}$$

37. $\sec u$

SOLUTION
$$\sec u = \frac{1}{\cos u} = \sqrt{5}$$

39. Find the smallest number x such that $\tan e^x = 0$.

SOLUTION Note that e^x is an increasing function. Because e^x is positive for every real number x, and because π is the smallest positive number whose tangent equals 0, we want to choose x so that $e^x = \pi$. Thus $x = \ln \pi \approx 1.14473$.

9.5 *Trigonometry in Right Triangles*

LEARNING OBJECTIVES

By the end of this section you should be able to

- compute the cosine, sine, and tangent of any angle of a right triangle if given the lengths of any two sides of the triangle;
- compute the lengths of all three sides of a right triangle if given any angle (in addition to the right angle) and the length of any side.

Trigonometry originated in the study of triangles. In this section we study trigonometry in the context of right triangles. In the next chapter we will deal with general triangles.

The English word trigonometry *first appeared in 1614 in an English translation of a book written in Latin by the German mathematician Bartholomeo Pitiscus. A prominent crater on the moon is named for him.*

Trigonometric Functions via Right Triangles

Consider the radius of the unit circle corresponding to θ radians, where $0 < \theta < \frac{\pi}{2}$ (in degrees, this angle is between $0°$ and $90°$), as shown here.

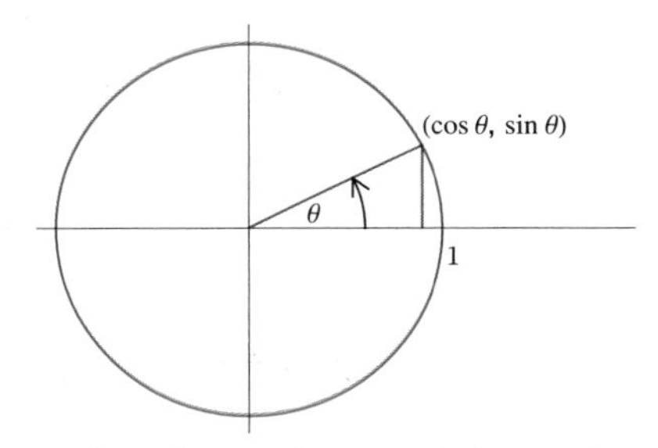

In the figure above, a vertical line segment has been dropped from the endpoint of the radius to the horizontal axis, producing a right triangle. The hypotenuse of this right triangle is a radius of the unit circle and hence has length 1. Because the endpoint of this radius has coordinates $(\cos\theta, \sin\theta)$, the horizontal side of the triangle has length $\cos\theta$ and the vertical side of the triangle has length $\sin\theta$. To get a clearer picture of what is going on, this triangle is displayed here, without the unit circle or the coordinate axes cluttering the figure (and for additional clarity, the scale has been enlarged).

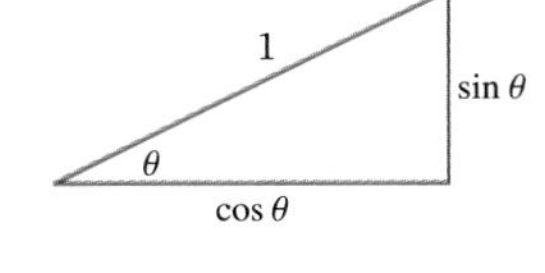

If we apply the Pythagorean Theorem to the triangle above, we get

$$(\cos\theta)^2 + (\sin\theta)^2 = 1,$$

which is a familiar equation.

Using the same angle θ as above, consider now a right triangle where one of the angles is θ but where the hypotenuse does not necessarily have length 1. Let c denote the length of the hypotenuse of this right triangle. Let a denote the length of the other side of the triangle adjacent to the angle θ, and let b denote the length of the side opposite the angle θ, as shown here.

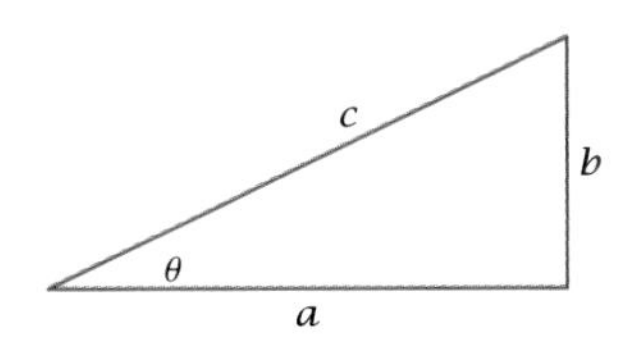

The two triangles shown above have the same angles. Thus these two triangles are similar. This similarity implies that the ratio of the lengths of any two sides of one of the triangles equals the ratio of the lengths of the corresponding sides of the other triangle.

For example, in our first triangle consider the horizontal side and the hypotenuse. These two sides have lengths $\cos\theta$ and 1. Thus the ratio of their lengths is $\frac{\cos\theta}{1}$, which equals $\cos\theta$. In our second triangle, the corresponding sides have lengths a and c. Their ratio (in the same order as used for the first triangle) is $\frac{a}{c}$. Setting these ratios from the two similar triangles equal to each other, we have

$$\cos\theta = \frac{a}{c}.$$

Similarly, in our first triangle above consider the vertical side and the hypotenuse. These two sides have lengths $\sin\theta$ and 1. Thus the ratio of their lengths is $\frac{\sin\theta}{1}$, which equals $\sin\theta$. In our second triangle, the corresponding sides have lengths b and c. Their ratio (in the same order as used for the first triangle) is $\frac{b}{c}$. Setting these ratios from the two similar triangles equal to each other, we have

$$\sin\theta = \frac{b}{c}.$$

Finally, in our first triangle consider the vertical side and the horizontal side. These two sides have lengths $\sin\theta$ and $\cos\theta$. Thus the ratio of their lengths is $\frac{\sin\theta}{\cos\theta}$, which equals $\tan\theta$. In our second triangle, the corresponding sides have lengths b and a. Their ratio (in the same order as used for the first triangle) is $\frac{b}{a}$. Setting these ratios from the two similar triangles equal to each other, we have

$$\tan\theta = \frac{b}{a}.$$

The last three equations displayed above form the basis of what is called right-triangle trigonometry. The box below restates these three equations using words rather than symbols.

Here the word "hypotenuse" is shorthand for "the length of the hypotenuse". Similarly, "adjacent side" is shorthand for "the length of the nonhypotenuse side adjacent to the angle θ". Finally, "opposite side" is shorthand for "the length of the side opposite the angle θ".

Right-triangle characterization of cosine, sine, and tangent

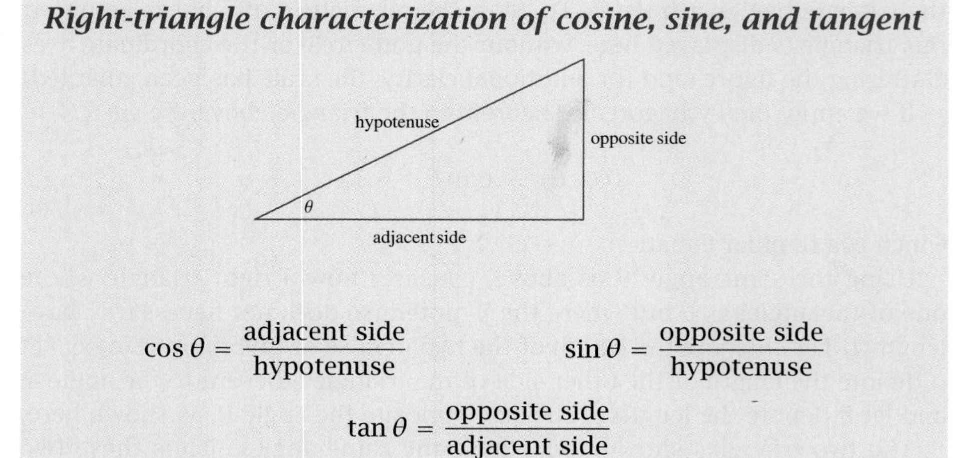

$$\cos\theta = \frac{\text{adjacent side}}{\text{hypotenuse}} \qquad \sin\theta = \frac{\text{opposite side}}{\text{hypotenuse}}$$

$$\tan\theta = \frac{\text{opposite side}}{\text{adjacent side}}$$

The figure and the equations in the box above capture the fundamentals of right-triangle trigonometry. Be sure that you thoroughly internalize the contents of the box above and that you can comfortably use these characterizations of the trigonometric functions.

Caution: The characterizations of $\cos\theta$, $\sin\theta$, *and* $\tan\theta$ *in the box above are valid only in right triangles, not in all triangles.*

Some books use the equations in the box above as the definitions of cosine, sine, and tangent. That approach makes sense only when θ is between 0 radians and $\frac{\pi}{2}$ radians (or between 0° and 90°), because there do not exist right triangles with angles bigger than $\frac{\pi}{2}$ radians (or 90°), just as there do not exist right triangles with negative angles. The characterizations of cosine, sine, and tangent given in the box above are highly useful, but keep in mind that the box above is valid only when θ is a positive angle less than $\frac{\pi}{2}$ radians (or 90°).

Two Sides of a Right Triangle

Given the lengths of any two sides of a right triangle, the Pythagorean Theorem allows us to find the length of the third side. Once we know the lengths of all three sides of a right triangle, we can find the cosine, sine, and tangent of any angle of the triangle. The example below illustrates this procedure.

EXAMPLE 1

Find the length of the hypotenuse and evaluate $\cos\theta$, $\sin\theta$, and $\tan\theta$ in this triangle.

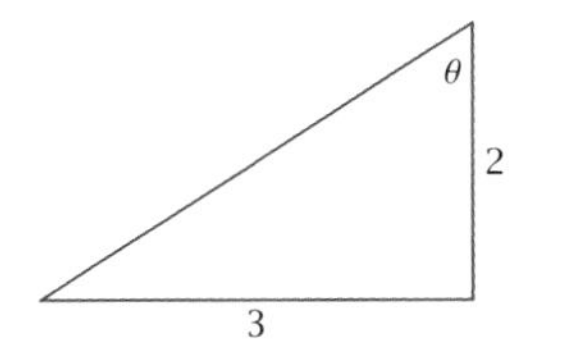

SOLUTION Let c denote the length of the hypotenuse of the triangle above. By the Pythagorean Theorem, we have

$$c^2 = 3^2 + 2^2.$$

Thus $c = \sqrt{13}$.

In this example the side opposite the angle θ *is the horizontal side of the triangle rather than the vertical side. This illustrates the usefulness of thinking in terms of opposite and adjacent sides rather than specific letters such as* a, b, *and* c.

Now $\cos\theta$ equals the length of the side adjacent to θ divided by the length of the hypotenuse. Thus

$$\cos\theta = \frac{\text{adjacent side}}{\text{hypotenuse}} = \frac{2}{\sqrt{13}} = \frac{2\sqrt{13}}{13}.$$

Similarly, $\sin\theta$ equals the length of the side opposite θ divided by the length of the hypotenuse. Thus

$$\sin\theta = \frac{\text{opposite side}}{\text{hypotenuse}} = \frac{3}{\sqrt{13}} = \frac{3\sqrt{13}}{13}.$$

Finally, $\tan\theta$ equals the length of the side opposite θ divided by the length of the side adjacent to θ. Thus

$$\tan\theta = \frac{\text{opposite side}}{\text{adjacent side}} = \frac{3}{2}.$$

In Section 10.1 we will see how to find the angle θ from the knowledge of either its cosine, its sine, or its tangent.

One Side and One Angle of a Right Triangle

Given the length of any side of a right triangle and any angle (in addition to the right angle), we can find the lengths of the other two sides of the triangle.

EXAMPLE 2

Find the lengths of the other two sides of this triangle.

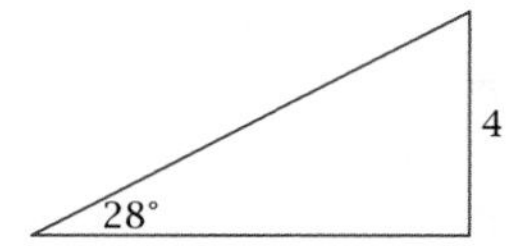

Real-world problems often do not come with labels attached. Thus sometimes the first step toward a solution is the assignment of appropriate labels.

SOLUTION The other two sides of the triangle are not labeled. Thus in the figure above, let a denote the length of the side adjacent to the 28° angle and let c denote the length of the hypotenuse. You may want to write these labels on the figure above.

Because we know the length of the side opposite the 28° angle, we will start with the sine. We have

$$\sin 28^\circ = \frac{\text{opposite side}}{\text{hypotenuse}} = \frac{4}{c}.$$

Solving for c, we get

$$c = \frac{4}{\sin 28^\circ} \approx 8.52,$$

where the approximation was obtained with the aid of a calculator.

Now we can find the length of the side adjacent to the 28° angle by using our characterization of the tangent. We have

$$\tan 28^\circ = \frac{\text{opposite side}}{\text{adjacent side}} = \frac{4}{a}.$$

Solving for a, we get

$$a = \frac{4}{\tan 28^\circ} \approx 7.52,$$

where the approximation was obtained with the aid of a calculator.

Trigonometry has a huge number of practical applications. The next example shows how trigonometry can be used to find the height of a building.

EXAMPLE 3

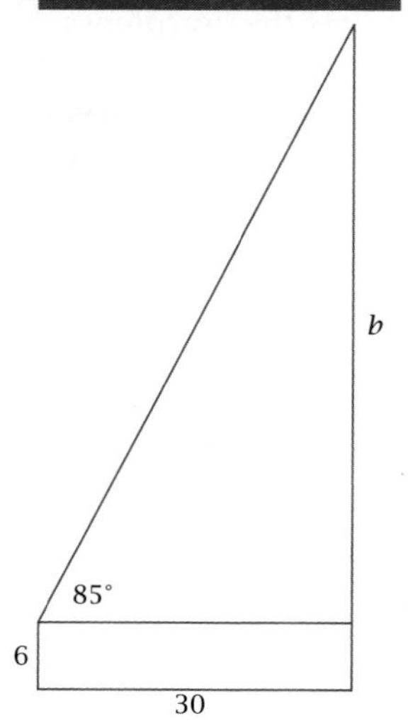

Standing 30 feet from the base of a tall building, you aim a laser pointer at the closest part of the top of the building. You measure that the laser pointer is 5° tilted from pointing straight up. The laser pointer is held 6 feet above the ground. How tall is the building?

SOLUTION In the sketch here, the rightmost vertical line represents the building and the hypotenuse represents the path of the laser beam. Because the laser pointer is 5° tilted from pointing straight up, the angle formed by the laser beam and a line parallel to the ground is 85°, as indicated in the sketch (which is not drawn to scale).

The side of the right triangle opposite the 85° angle has been labeled b. Thus the height of the building is $b + 6$.

We have $\tan 85^\circ = \frac{\text{opposite side}}{\text{adjacent side}} = \frac{b}{30}$. Solving this equation for b, we get

$$b = 30 \tan 85^\circ \approx 343.$$

Adding 6, we see that the height of the building is approximately 349 feet.

The next example illustrates another practical application of the ideas in this section.

EXAMPLE 4

A surveyor wishes to measure the distance between points A and B, but a canyon between A and B prevents a direct measurement. Thus the surveyor moves 500 meters perpendicular to the line AB to the point C and measures angle BCA as 78° (such angles can be measured with a tool called a *transit level*). What is the distance between the points A and B?

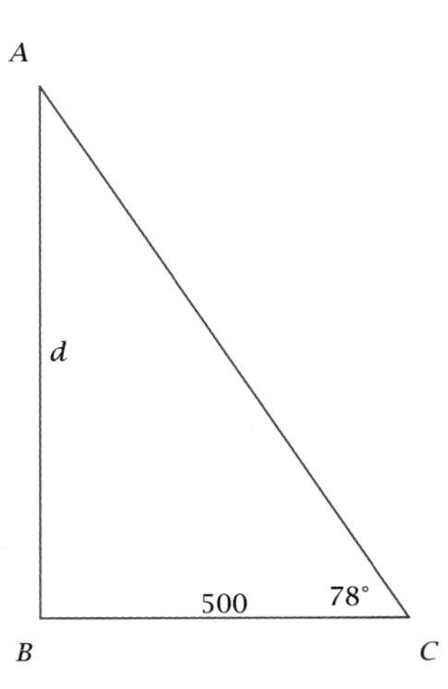

SOLUTION Let d denote the distance from A to B. From the figure (which is not drawn to scale) we have

$$\tan 78^\circ = \frac{\text{opposite side}}{\text{adjacent side}} = \frac{d}{500}.$$

Solving this equation for d, we get

$$d = 500 \tan 78^\circ \approx 2352.$$

Thus the distance between A and B is approximately 2352 meters.

EXERCISES

Use the right triangle below for Exercises 1–76. This triangle is not drawn to scale corresponding to the data in the exercises.

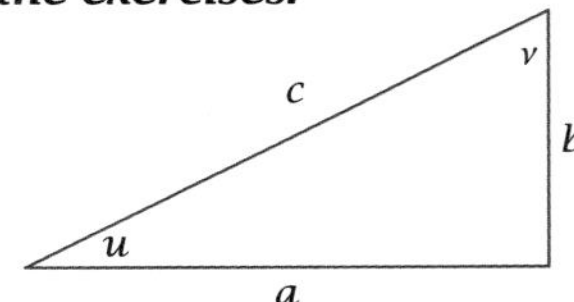

1. Suppose $a = 2$ and $b = 7$. Evaluate c.
2. Suppose $a = 3$ and $b = 5$. Evaluate c.
3. Suppose $a = 2$ and $b = 7$. Evaluate $\cos u$.
4. Suppose $a = 3$ and $b = 5$. Evaluate $\cos u$.
5. Suppose $a = 2$ and $b = 7$. Evaluate $\sin u$.
6. Suppose $a = 3$ and $b = 5$. Evaluate $\sin u$.
7. Suppose $a = 2$ and $b = 7$. Evaluate $\tan u$.
8. Suppose $a = 3$ and $b = 5$. Evaluate $\tan u$.
9. Suppose $a = 2$ and $b = 7$. Evaluate $\cos v$.
10. Suppose $a = 3$ and $b = 5$. Evaluate $\cos v$.
11. Suppose $a = 2$ and $b = 7$. Evaluate $\sin v$.
12. Suppose $a = 3$ and $b = 5$. Evaluate $\sin v$.
13. Suppose $a = 2$ and $b = 7$. Evaluate $\tan v$.
14. Suppose $a = 3$ and $b = 5$. Evaluate $\tan v$.
15. Suppose $b = 2$ and $c = 7$. Evaluate a.
16. Suppose $b = 4$ and $c = 6$. Evaluate a.
17. Suppose $b = 2$ and $c = 7$. Evaluate $\cos u$.
18. Suppose $b = 4$ and $c = 6$. Evaluate $\cos u$.
19. Suppose $b = 2$ and $c = 7$. Evaluate $\sin u$.
20. Suppose $b = 4$ and $c = 6$. Evaluate $\sin u$.
21. Suppose $b = 2$ and $c = 7$. Evaluate $\tan u$.
22. Suppose $b = 4$ and $c = 6$. Evaluate $\tan u$.
23. Suppose $b = 2$ and $c = 7$. Evaluate $\cos v$.
24. Suppose $b = 4$ and $c = 6$. Evaluate $\cos v$.
25. Suppose $b = 2$ and $c = 7$. Evaluate $\sin v$.
26. Suppose $b = 4$ and $c = 6$. Evaluate $\sin v$.
27. Suppose $b = 2$ and $c = 7$. Evaluate $\tan v$.
28. Suppose $b = 4$ and $c = 6$. Evaluate $\tan v$.
29. Suppose $a = 5$ and $u = 17°$. Evaluate b.
30. Suppose $b = 3$ and $v = 38°$. Evaluate a.
31. Suppose $a = 5$ and $u = 17°$. Evaluate c.
32. Suppose $b = 3$ and $v = 38°$. Evaluate c.
33. Suppose $u = 17°$. Evaluate $\cos v$.
34. Suppose $v = 38°$. Evaluate $\cos u$.
35. Suppose $u = 17°$. Evaluate $\sin v$.

36. Suppose $v = 38°$. Evaluate $\sin u$.

37. Suppose $u = 17°$. Evaluate $\tan v$.

38. Suppose $v = 38°$. Evaluate $\tan u$.

39. Suppose $c = 8$ and $u = 1$ radian. Evaluate a.

40. Suppose $c = 3$ and $v = 0.2$ radians. Evaluate a.

41. Suppose $c = 8$ and $u = 1$ radian. Evaluate b.

42. Suppose $c = 3$ and $v = 0.2$ radians. Evaluate b.

43. Suppose $u = 1$ radian. Evaluate $\cos v$.

44. Suppose $v = 0.2$ radians. Evaluate $\cos u$.

45. Suppose $u = 1$ radian. Evaluate $\sin v$.

46. Suppose $v = 0.2$ radians. Evaluate $\sin u$.

47. Suppose $u = 1$ radian. Evaluate $\tan v$.

48. Suppose $v = 0.2$ radians. Evaluate $\tan u$.

49. Suppose $c = 4$ and $\cos u = \frac{1}{5}$. Evaluate a.

50. Suppose $c = 5$ and $\cos u = \frac{2}{3}$. Evaluate a.

51. Suppose $c = 4$ and $\cos u = \frac{1}{5}$. Evaluate b.

52. Suppose $c = 5$ and $\cos u = \frac{2}{3}$. Evaluate b.

53. Suppose $\cos u = \frac{1}{5}$. Evaluate $\sin u$.

54. Suppose $\cos u = \frac{2}{3}$. Evaluate $\sin u$.

55. Suppose $\cos u = \frac{1}{5}$. Evaluate $\tan u$.

56. Suppose $\cos u = \frac{2}{3}$. Evaluate $\tan u$.

57. Suppose $\cos u = \frac{1}{5}$. Evaluate $\cos v$.

58. Suppose $\cos u = \frac{2}{3}$. Evaluate $\cos v$.

59. Suppose $\cos u = \frac{1}{5}$. Evaluate $\sin v$.

60. Suppose $\cos u = \frac{2}{3}$. Evaluate $\sin v$.

61. Suppose $\cos u = \frac{1}{5}$. Evaluate $\tan v$.

62. Suppose $\cos u = \frac{2}{3}$. Evaluate $\tan v$.

63. Suppose $b = 4$ and $\sin v = \frac{1}{3}$. Evaluate a.

64. Suppose $b = 2$ and $\sin v = \frac{3}{7}$. Evaluate a.

65. Suppose $b = 4$ and $\sin v = \frac{1}{3}$. Evaluate c.

66. Suppose $b = 2$ and $\sin v = \frac{3}{7}$. Evaluate c.

67. Suppose $\sin v = \frac{1}{3}$. Evaluate $\cos u$.

68. Suppose $\sin v = \frac{3}{7}$. Evaluate $\cos u$.

69. Suppose $\sin v = \frac{1}{3}$. Evaluate $\sin u$.

70. Suppose $\sin v = \frac{3}{7}$. Evaluate $\sin u$.

71. Suppose $\sin v = \frac{1}{3}$. Evaluate $\tan u$.

72. Suppose $\sin v = \frac{3}{7}$. Evaluate $\tan u$.

73. Suppose $\sin v = \frac{1}{3}$. Evaluate $\cos v$.

74. Suppose $\sin v = \frac{3}{7}$. Evaluate $\cos v$.

75. Suppose $\sin v = \frac{1}{3}$. Evaluate $\tan v$.

76. Suppose $\sin v = \frac{3}{7}$. Evaluate $\tan v$.

77. Find the perimeter of a right triangle that has hypotenuse of length 6 and a 40° angle.

78. Find the perimeter of a right triangle that has hypotenuse of length 8 and a 35° angle.

79. Suppose a 25-foot ladder is leaning against a wall, making a 63° angle with the ground (as measured from a perpendicular line from the base of the ladder to the wall). How high up the wall is the end of the ladder?

80. Suppose a 19-foot ladder is leaning against a wall, making a 71° angle with the ground (as measured from a perpendicular line from the base of the ladder to the wall). How high up the wall is the end of the ladder?

81. Suppose you need to find the height of a tall building. Standing 20 meters from the base of the building, you aim a laser pointer at the closest part of the top of the building. You measure that the laser pointer is 4° tilted from pointing straight up. The laser pointer is held 2 meters above the ground. How tall is the building?

82. Suppose you need to find the height of a tall building. Standing 15 meters from the base of the building, you aim a laser pointer at the closest part of the top of the building. You measure that the laser pointer is 7° tilted from pointing straight up. The laser pointer is held 2 meters above the ground. How tall is the building?

83. A surveyor wishes to measure the distance between points A and B, but buildings between A and B prevent a direct measurement. Thus the surveyor moves 50 meters perpendicular to the line AB to the point C and measures that angle BCA is 87°. What is the distance between the points A and B?

84. A surveyor wishes to measure the distance between points A and B, but a river between A and B prevent a direct measurement. Thus the surveyor moves 200 feet perpendicular to the line AB to the point C and measures that angle BCA is 81°. What is the distance between the points A and B?

For Exercises 85–90, assume that the surface of the earth is a sphere with radius 3963 miles. The* latitude *of a point P on the earth's surface is the angle between the line segment from the center of the earth to P and the line segment from the center of the earth to the point on the equator closest to P.

85. Dallas has latitude 32.8° north. Find the radius of the circle formed by the points with the same latitude as Dallas.

86. Cleveland has latitude 41.5° north. Find the radius of the circle formed by the points with the same latitude as Cleveland.

87. Suppose you travel east on the surface of the earth from Dallas (latitude 32.8° north, longitude 96.8° west), always staying at the same latitude as Dallas. You stop when reaching latitude 32.8° north, longitude 84.4° west (directly south of Atlanta). How far have your traveled?

88. Suppose you travel east on the surface of the earth from Cleveland (latitude 41.5° north, longitude 81.7° west), always staying at the same latitude as Cleveland. You stop when reaching latitude 41.5° north, longitude 75.1° west (directly north of Philadelphia). How far have your traveled?

89. How fast is Dallas moving due to the daily rotation of the earth about its axis?

90. How fast is Cleveland moving due to the daily rotation of the earth about its axis?

PROBLEMS

91. In doing several of the exercises in this section, you should have noticed a relationship between $\cos u$ and $\sin v$, along with a relationship between $\sin u$ and $\cos v$. What are these relationships? Explain why they hold.

92. In doing several of the exercises in this section, you should have noticed a relationship between $\tan u$ and $\tan v$. What is this relationship? Explain why it holds.

93. Find the lengths of all three sides of a right triangle that has perimeter 29 and has a 42° angle.

94. Find the latitude of your location, then find how fast you are moving due to the daily rotation of the earth about its axis.

WORKED-OUT SOLUTIONS *to Odd-numbered Exercises*

Use the right triangle below for Exercises 1–76. This triangle is not drawn to scale corresponding to the data in the exercises.

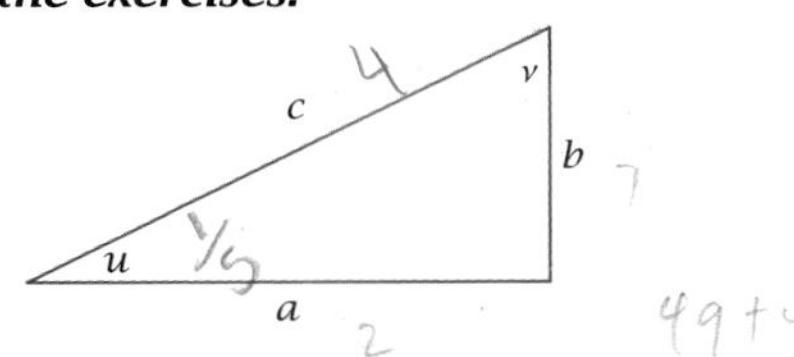

1. Suppose $a = 2$ and $b = 7$. Evaluate c.

SOLUTION The Pythagorean Theorem implies that $c^2 = 2^2 + 7^2$. Thus

$$c = \sqrt{2^2 + 7^2} = \sqrt{53}.$$

3. Suppose $a = 2$ and $b = 7$. Evaluate $\cos u$.

SOLUTION

$$\cos u = \frac{\text{adjacent side}}{\text{hypotenuse}} = \frac{a}{c} = \frac{2}{\sqrt{53}} = \frac{2\sqrt{53}}{53}$$

5. Suppose $a = 2$ and $b = 7$. Evaluate $\sin u$.

SOLUTION

$$\sin u = \frac{\text{opposite side}}{\text{hypotenuse}} = \frac{b}{c} = \frac{7}{\sqrt{53}} = \frac{7\sqrt{53}}{53}$$

7. Suppose $a = 2$ and $b = 7$. Evaluate $\tan u$.

SOLUTION

$$\tan u = \frac{\text{opposite side}}{\text{adjacent side}} = \frac{b}{a} = \frac{7}{2}$$

9. Suppose $a = 2$ and $b = 7$. Evaluate $\cos v$.

SOLUTION

$$\cos v = \frac{\text{adjacent side}}{\text{hypotenuse}} = \frac{b}{c} = \frac{7}{\sqrt{53}} = \frac{7\sqrt{53}}{53}$$

11. Suppose $a = 2$ and $b = 7$. Evaluate $\sin v$.

SOLUTION

$$\sin v = \frac{\text{opposite side}}{\text{hypotenuse}} = \frac{a}{c} = \frac{2}{\sqrt{53}} = \frac{2\sqrt{53}}{53}$$

13. Suppose $a = 2$ and $b = 7$. Evaluate $\tan v$.

SOLUTION

$$\tan v = \frac{\text{opposite side}}{\text{adjacent side}} = \frac{a}{b} = \frac{2}{7}$$

15. Suppose $b = 2$ and $c = 7$. Evaluate a.

SOLUTION The Pythagorean Theorem implies that $a^2 + 2^2 = 7^2$. Thus

$$a = \sqrt{7^2 - 2^2} = \sqrt{45} = \sqrt{9 \cdot 5} = \sqrt{9} \cdot \sqrt{5} = 3\sqrt{5}.$$

17. Suppose $b = 2$ and $c = 7$. Evaluate $\cos u$.

SOLUTION

$$\cos u = \frac{\text{adjacent side}}{\text{hypotenuse}} = \frac{a}{c} = \frac{3\sqrt{5}}{7}$$

19. Suppose $b = 2$ and $c = 7$. Evaluate $\sin u$.

SOLUTION

$$\sin u = \frac{\text{opposite side}}{\text{hypotenuse}} = \frac{b}{c} = \frac{2}{7}$$

21. Suppose $b = 2$ and $c = 7$. Evaluate $\tan u$.

SOLUTION

$$\tan u = \frac{\text{opposite side}}{\text{adjacent side}} = \frac{b}{a} = \frac{2}{3\sqrt{5}} = \frac{2\sqrt{5}}{15}$$

23. Suppose $b = 2$ and $c = 7$. Evaluate $\cos v$.

SOLUTION

$$\cos v = \frac{\text{adjacent side}}{\text{hypotenuse}} = \frac{b}{c} = \frac{2}{7}$$

25. Suppose $b = 2$ and $c = 7$. Evaluate $\sin v$.

SOLUTION

$$\sin v = \frac{\text{opposite side}}{\text{hypotenuse}} = \frac{a}{c} = \frac{3\sqrt{5}}{7}$$

27. Suppose $b = 2$ and $c = 7$. Evaluate $\tan v$.

SOLUTION

$$\tan v = \frac{\text{opposite side}}{\text{adjacent side}} = \frac{a}{b} = \frac{3\sqrt{5}}{2}$$

29. Suppose $a = 5$ and $u = 17°$. Evaluate b.

SOLUTION We have

$$\tan 17° = \frac{\text{opposite side}}{\text{adjacent side}} = \frac{b}{5}.$$

Solving for b, we get $b = 5 \tan 17° \approx 1.53$.

31. Suppose $a = 5$ and $u = 17°$. Evaluate c.

SOLUTION We have

$$\cos 17° = \frac{\text{adjacent side}}{\text{hypotenuse}} = \frac{5}{c}.$$

Solving for c, we get

$$c = \frac{5}{\cos 17°} \approx 5.23.$$

33. Suppose $u = 17°$. Evaluate $\cos v$.

SOLUTION Because $v = 90° - u$, we have $v = 73°$. Thus $\cos v = \cos 73° \approx 0.292$.

35. Suppose $u = 17°$. Evaluate $\sin v$.

SOLUTION $\sin v = \sin 73° \approx 0.956$

37. Suppose $u = 17°$. Evaluate $\tan v$.

SOLUTION $\tan v = \tan 73° \approx 3.27$

39. Suppose $c = 8$ and $u = 1$ radian. Evaluate a.

SOLUTION We have

$$\cos 1 = \frac{\text{adjacent side}}{\text{hypotenuse}} = \frac{a}{8}.$$

Solving for a, we get

$$a = 8 \cos 1 \approx 4.32.$$

When using a calculator to do the approximation above, be sure that your calculator is set to operate in radian mode.

41. Suppose $c = 8$ and $u = 1$ radian. Evaluate b.

SOLUTION We have

$$\sin 1 = \frac{\text{opposite side}}{\text{hypotenuse}} = \frac{b}{8}.$$

Solving for b, we get

$$b = 8\sin 1 \approx 6.73.$$

43. Suppose $u = 1$ radian. Evaluate $\cos v$.

SOLUTION Because $v = \frac{\pi}{2} - u$, we have $v = \frac{\pi}{2} - 1$. Thus

$$\cos v = \cos(\tfrac{\pi}{2} - 1) \approx 0.841.$$

45. Suppose $u = 1$ radian. Evaluate $\sin v$.

SOLUTION

$$\sin v = \sin(\tfrac{\pi}{2} - 1) \approx 0.540$$

47. Suppose $u = 1$ radian. Evaluate $\tan v$.

SOLUTION

$$\tan v = \tan(\tfrac{\pi}{2} - 1) \approx 0.642$$

49. Suppose $c = 4$ and $\cos u = \frac{1}{5}$. Evaluate a.

SOLUTION We have

$$\frac{1}{5} = \cos u = \frac{\text{adjacent side}}{\text{hypotenuse}} = \frac{a}{4}.$$

Solving this equation for a, we get

$$a = \frac{4}{5}.$$

51. Suppose $c = 4$ and $\cos u = \frac{1}{5}$. Evaluate b.

SOLUTION The Pythagorean Theorem implies that $(\frac{4}{5})^2 + b^2 = 4^2$. Thus

$$b = \sqrt{16 - \frac{16}{25}} = 4\sqrt{1 - \frac{1}{25}} = 4\sqrt{\frac{24}{25}} = \frac{8\sqrt{6}}{5}.$$

53. Suppose $\cos u = \frac{1}{5}$. Evaluate $\sin u$.

SOLUTION

$$\sin u = \sqrt{1 - (\cos u)^2} = \sqrt{1 - \frac{1}{25}}$$

$$= \sqrt{\frac{24}{25}} = \frac{2\sqrt{6}}{5}$$

55. Suppose $\cos u = \frac{1}{5}$. Evaluate $\tan u$.

SOLUTION

$$\tan u = \frac{\sin u}{\cos u} = \frac{\frac{2\sqrt{6}}{5}}{\frac{1}{5}} = 2\sqrt{6}$$

57. Suppose $\cos u = \frac{1}{5}$. Evaluate $\cos v$.

SOLUTION

$$\cos v = \frac{b}{c} = \sin u = \frac{2\sqrt{6}}{5}$$

59. Suppose $\cos u = \frac{1}{5}$. Evaluate $\sin v$.

SOLUTION

$$\sin v = \frac{a}{c} = \cos u = \frac{1}{5}$$

61. Suppose $\cos u = \frac{1}{5}$. Evaluate $\tan v$.

SOLUTION

$$\tan v = \frac{\sin v}{\cos v} = \frac{\frac{1}{5}}{\frac{2\sqrt{6}}{5}} = \frac{1}{2\sqrt{6}} = \frac{\sqrt{6}}{12}$$

63. Suppose $b = 4$ and $\sin v = \frac{1}{3}$. Evaluate a.

SOLUTION We have

$$\frac{1}{3} = \sin v = \frac{\text{opposite side}}{\text{hypotenuse}} = \frac{a}{c}.$$

Thus

$$c = 3a.$$

By the Pythagorean Theorem, we also have

$$c^2 = a^2 + 16.$$

Substituting $3a$ for c in this equation gives

$$9a^2 = a^2 + 16.$$

Solving the equation above for a shows that $a = \sqrt{2}$.

65. Suppose $b = 4$ and $\sin v = \frac{1}{3}$. Evaluate c.

SOLUTION We have

$$\frac{1}{3} = \sin v = \frac{a}{c}.$$

Thus

$$c = 3a = 3\sqrt{2}.$$

67. Suppose $\sin v = \frac{1}{3}$. Evaluate $\cos u$.

SOLUTION

$$\cos u = \frac{a}{c} = \sin v = \frac{1}{3}$$

69. Suppose $\sin v = \frac{1}{3}$. Evaluate $\sin u$.

SOLUTION

$$\sin u = \sqrt{1 - (\cos u)^2} = \sqrt{1 - \left(\tfrac{1}{3}\right)^2} = \sqrt{\frac{8}{9}} = \frac{2\sqrt{2}}{3}$$

71. Suppose $\sin v = \frac{1}{3}$. Evaluate $\tan u$.

SOLUTION

$$\tan u = \frac{\sin u}{\cos u} = \frac{\frac{2\sqrt{2}}{3}}{\frac{1}{3}} = 2\sqrt{2}$$

73. Suppose $\sin v = \frac{1}{3}$. Evaluate $\cos v$.

SOLUTION

$$\cos v = \sqrt{1 - (\sin v)^2} = \sqrt{1 - \left(\tfrac{1}{3}\right)^2} = \sqrt{\frac{8}{9}} = \frac{2\sqrt{2}}{3}$$

75. Suppose $\sin v = \frac{1}{3}$. Evaluate $\tan v$.

SOLUTION

$$\tan v = \frac{\sin v}{\cos v} = \frac{\frac{1}{3}}{\frac{2\sqrt{2}}{3}} = \frac{1}{2\sqrt{2}} = \frac{\sqrt{2}}{4}$$

77. Find the perimeter of a right triangle that has hypotenuse of length 6 and a 40° angle.

SOLUTION The side adjacent to the 40° angle has length $6\cos 40°$. The side opposite the 40° angle has length $6\sin 40°$. Thus the perimeter of the triangle is

$$6 + 6\cos 40° + 6\sin 40°,$$

which is approximately 14.453.

79. Suppose a 25-foot ladder is leaning against a wall, making a 63° angle with the ground (as measured from a perpendicular line from the base of the ladder to the wall). How high up the wall is the end of the ladder?

SOLUTION

In the sketch here, the vertical line represents the wall and the hypotenuse represents the ladder. As labeled here, the ladder touches the wall at height b; thus we need to evaluate b.

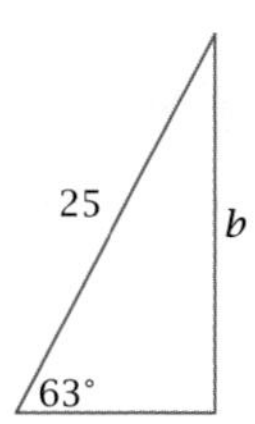

We have $\sin 63° = \frac{b}{25}$. Solving this equation for b, we get

$$b = 25\sin 63° \approx 22.28.$$

Thus the ladder touches the wall at a height of approximately 22.28 feet. Because $0.28 \times 12 = 3.36$, this is approximately 22 feet, 3 inches.

81. Suppose you need to find the height of a tall building. Standing 20 meters from the base of the building, you aim a laser pointer at the closest part of the top of the building. You measure that the laser pointer is 4° tilted from pointing straight up. The laser pointer is held 2 meters above the ground. How tall is the building?

SOLUTION

In the sketch here, the right-most vertical line represents the building and the hypotenuse represents the path of the laser beam. Because the laser pointer is 4° tilted from pointing straight up, the angle formed by the laser beam and a line parallel to the ground is 86°, as indicated in the figure (which is not drawn to scale).

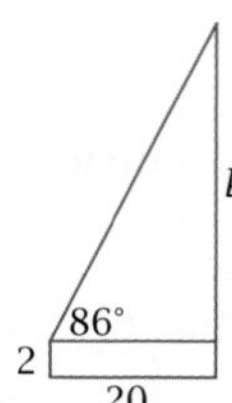

The side of the right triangle opposite the 86° angle has been labeled b. Thus the height of the building is $b + 2$.

We have $\tan 86° = \frac{b}{20}$. Solving this equation for b, we get

$$b = 20\tan 86° \approx 286.$$

Adding 2, we see that the height of the building is approximately 288 meters.

83. A surveyor wishes to measure the distance between points A and B, but buildings between A and B prevent a direct measurement. Thus the surveyor moves 50 meters perpendicular to the line AB to the point C and measures that angle BCA is 87°. What is the distance between the points A and B?

SOLUTION Let d denote the distance from A to B. From the figure (which is not drawn to scale) we have

$$\tan 87° = \frac{\text{opposite side}}{\text{adjacent side}} = \frac{d}{50}.$$

Solving this equation for d, we get

$$d = 50 \tan 87° \approx 954.$$

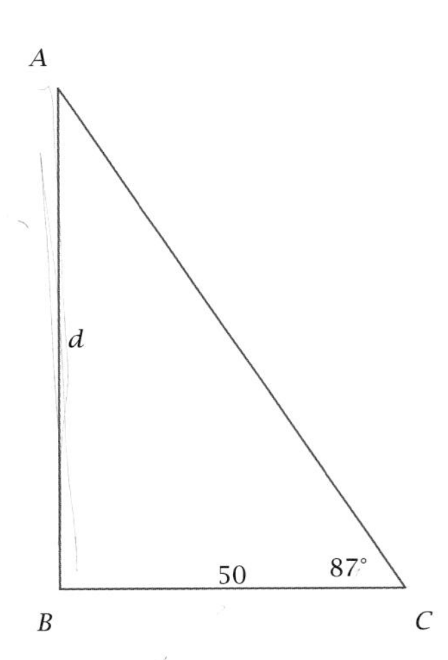

Thus the distance between A and B is approximately 954 meters.

For Exercises 85–90, assume that the surface of the earth is a sphere with radius 3963 miles. The* latitude *of a point P on the earth's surface is the angle between the line segment from the center of the earth to P and the line segment from the center of the earth to the point on the equator closest to P.

85. Dallas has latitude 32.8° north. Find the radius of the circle formed by the points with the same latitude as Dallas.

SOLUTION The figure below shows a cross section of the earth. The blue radius is the line segment from the center of the earth to Dallas, making a 32.8° angle with the line segment from the center of the earth to the point on the equator closest to Dallas.

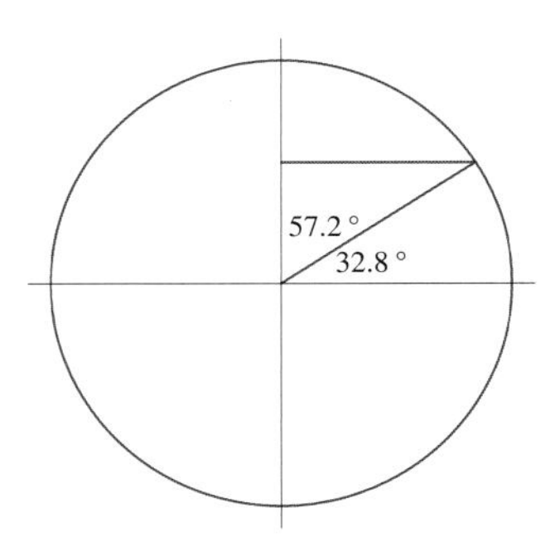

The red line segment above shows the radius of the circle formed by the points with the same latitude as Dallas. The angle opposite the red line segment in the right triangle is 57.2° (because 90 − 32.8 = 57.2). Thus we see from the figure that the length of the red line segment is 3963 sin 57.2° miles, which is approximately 3331.2 miles.

87. Suppose you travel east on the surface of the earth from Dallas (latitude 32.8° north, longitude 96.8° west), always staying at the same latitude as Dallas. You stop when reaching latitude 32.8° north, longitude 84.4° west (directly south of Atlanta). How far have your traveled?

SOLUTION Using the formula for the length of a circular arc from Section 9.1, we see that the distance traveled on the surface of the earth is $\frac{12.4\pi \cdot 3331.2}{180}$ miles (the number 12.4 is used because 12.4 = 96.8 − 84.4, and the radius of 3331.2 miles comes from the solution to Exercise 85). Because $\frac{12.4\pi \cdot 3331.2}{180} \approx 721$, you have traveled approximately 721 miles.

89. How fast is Dallas moving due to the daily rotation of the earth about its axis?

SOLUTION From the solution to Exercise 85, we see that Dallas travels $2\pi \cdot 3331.2$ miles, which is approximately 20,930 miles, each 24 hours due to the rotation of the earth about its axis. Thus the speed of Dallas due to the rotation of the earth is approximately $\frac{20930}{24}$ miles per hour, which is approximately 872 miles per hour.

9.6 Trigonometric Identities

LEARNING OBJECTIVES

By the end of this section you should be able to

- derive trigonometric identities and simplify trigonometric expressions;
- use the trigonometric identities for $-\theta$;
- use the trigonometric identities for $\frac{\pi}{2} - \theta$;
- use the trigonometric identities for $\theta + \pi$ and $\theta + 2\pi$.

Equations come in two flavors. One flavor is an equation such as

$$x^2 = 4x - 3,$$

which holds for only certain special values of the variable x. We can talk about solving such equations, which means finding the special values of the variable (or variables) that make the equations valid. For example, the equation above is valid only if $x = 1$ or $x = 3$.

A second flavor is an equation such as

$$(x + 3)^2 = x^2 + 6x + 9,$$

Throughout this book, identities generally appear in blue to distinguish them from equations of the first flavor.

which is valid for all numbers x. An equation such as this is called an **identity** because it is identically true without regard to the value of any variables. As another example, the logarithmic identity

$$\log(xy) = \log x + \log y$$

holds for all positive numbers x and y.

Do not memorize the many dozen useful trigonometric identities. Concentrate on understanding why these identities hold. Then you will be able to derive the ones you need in any particular situation.

In this section we focus on basic trigonometric identities, which are identities that involve trigonometric functions. Such identities are often useful for simplifying trigonometric expressions and for converting information about one trigonometric function into information about another trigonometric function. We will deal with additional trigonometric identities later, particularly in Sections 10.5 and 10.6.

The Relationship Among Cosine, Sine, and Tangent

We have already used the most important trigonometric identity, which is $(\cos\theta)^2 + (\sin\theta)^2 = 1$. Recall that this identity arises from the definition of $(\cos\theta, \sin\theta)$ as a point on the unit circle, whose equation is $x^2 + y^2 = 1$.

Usually the notation $\cos^2\theta$ is used instead of $(\cos\theta)^2$, and $\sin^2\theta$ is used instead of $(\sin\theta)^2$. We have been using the notation $(\cos\theta)^2$ and $(\sin\theta)^2$ to emphasize the meaning of these terms. Now it is time to switch to the more common notation. Keep in mind, however, that an expression such as $\cos^2\theta$ really means $(\cos\theta)^2$.

Notation for powers of cosine, sine, and tangent

If n is a positive integer, then

- $\cos^n \theta$ means $(\cos \theta)^n$;
- $\sin^n \theta$ means $(\sin \theta)^n$;
- $\tan^n \theta$ means $(\tan \theta)^n$.

With our new notation, the most important trigonometric identity can be rewritten as follows:

Relationship between cosine and sine

$$\cos^2 \theta + \sin^2 \theta = 1$$

The trigonometric identity above implies that

$$\cos \theta = \pm\sqrt{1 - \sin^2 \theta}$$

and

$$\sin \theta = \pm\sqrt{1 - \cos^2 \theta},$$

with the choices between the plus and minus signs depending on the quadrant in which the radius corresponding to θ lies.

Given either $\cos \theta$ *or* $\sin \theta$, *we can use these equations to evaluate the other provided that we also have enough information to choose between positive and negative values.*

The equations above can be used, for example, to write $\tan \theta$ solely in terms of $\cos \theta$, as follows:

EXAMPLE 1

Write $\tan \theta$ solely in terms of $\cos \theta$.

SOLUTION

$$\tan \theta = \frac{\sin \theta}{\cos \theta} = \pm\frac{\sqrt{1 - \cos^2 \theta}}{\cos \theta}$$

If both sides of the key trigonometric identity $\cos^2 \theta + \sin^2 \theta = 1$ are divided by $\cos^2 \theta$ and then we rewrite $\frac{\sin^2 \theta}{\cos^2 \theta}$ as $\tan^2 \theta$, we get another useful identity:

$$1 + \tan^2 \theta = \frac{1}{\cos^2 \theta}.$$

Using one of the three less common trigonometric functions, we could also write this identity in the form

$$1 + \tan^2 \theta = \sec^2 \theta,$$

where $\sec^2 \theta$ denotes, of course, $(\sec \theta)^2$.

Simplifying a trigonometric expression often involves doing a bit of algebraic manipulation and using an appropriate trigonometric identity, as in the following example.

EXAMPLE 2

Simplify the expression

$$(\tan^2\theta)\Big(\frac{1}{1-\cos\theta}+\frac{1}{1+\cos\theta}\Big).$$

SOLUTION

$$\begin{aligned}(\tan^2\theta)\Big(\frac{1}{1-\cos\theta}+\frac{1}{1+\cos\theta}\Big) &= (\tan^2\theta)\Big(\frac{(1+\cos\theta)+(1-\cos\theta)}{(1-\cos\theta)(1+\cos\theta)}\Big)\\ &= (\tan^2\theta)\Big(\frac{2}{1-\cos^2\theta}\Big)\\ &= (\tan^2\theta)\Big(\frac{2}{\sin^2\theta}\Big)\\ &= \Big(\frac{\sin^2\theta}{\cos^2\theta}\Big)\Big(\frac{2}{\sin^2\theta}\Big)\\ &= \frac{2}{\cos^2\theta}\end{aligned}$$

Trigonometric Identities for the Negative of an Angle

The first known table of values of trigonometric functions was compiled by the Greek astronomer Hipparchus over two thousand years ago.

By the definitions of cosine and sine, the endpoint of the radius of the unit circle corresponding to θ has coordinates $(\cos\theta, \sin\theta)$. Similarly, the endpoint of the radius of the unit circle corresponding to $-\theta$ has coordinates $(\cos(-\theta), \sin(-\theta))$, as shown in the figure below:

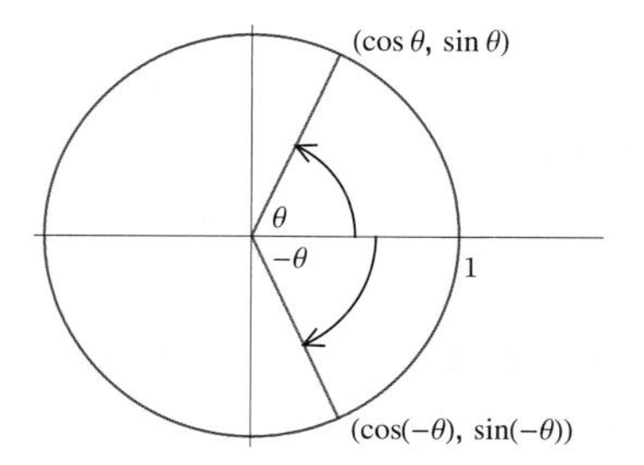

Flipping the radius corresponding to θ across the horizontal axis gives the radius corresponding to $-\theta$.

Each of the two radii in the figure above can be obtained by flipping the other radius across the horizontal axis. Thus the endpoints of the two radii in the figure above have the same first coordinate, and their second coordinates are the negative of each other. In other words, the figure above shows that

$$\cos(-\theta) = \cos\theta \quad \text{and} \quad \sin(-\theta) = -\sin\theta.$$

Using these equations and the definition of the tangent, we see that

$$\tan(-\theta) = \frac{\sin(-\theta)}{\cos(-\theta)} = \frac{-\sin\theta}{\cos\theta} = -\tan\theta.$$

An identity involving the tangent can often be derived from the corresponding identities for cosine and sine.

Collecting the three identities we have just derived gives the following:

Trigonometric identities with $-\theta$

$$\cos(-\theta) = \cos\theta$$

$$\sin(-\theta) = -\sin\theta$$

$$\tan(-\theta) = -\tan\theta$$

As we have just seen, the cosine of the negative of an angle is the same as the cosine of the angle. In other words, cosine is an even function (see Section 3.2 to review even functions). This explains why the graph of the cosine is symmetric about the vertical axis. Specifically, along with the typical point $(\theta, \cos\theta)$ on the graph of the cosine we also have the point $(-\theta, \cos(-\theta))$, which equals $(-\theta, \cos\theta)$.

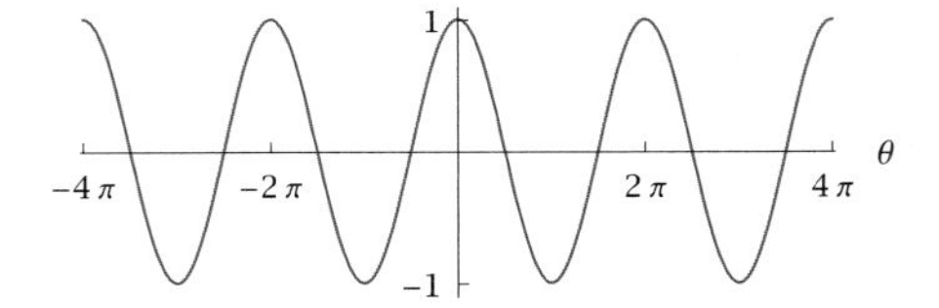

The graph of the even function cosine is symmetric about the vertical axis.

In contrast to the behavior of the cosine, the sine of the negative of an angle is the negative of the sine of the angle. In other words, sine is an odd function (see Section 3.2 to review odd functions). This explains why the graph of the sine is symmetric about the origin. Specifically, along with the typical point $(\theta, \sin\theta)$ on the graph of the sine we also have the point $(-\theta, \sin(-\theta))$, which equals $(-\theta, -\sin\theta)$.

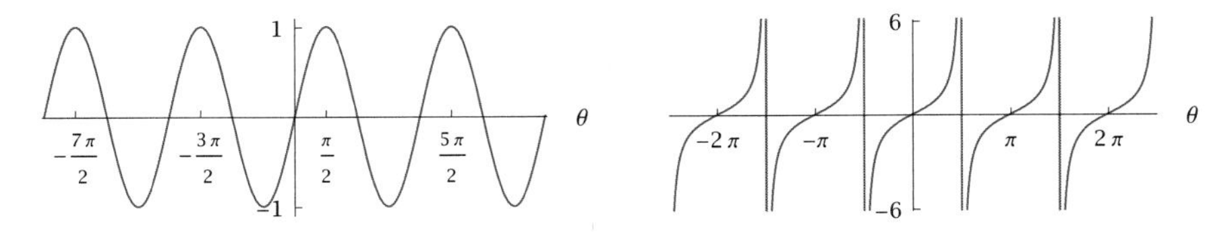

The graphs of the odd functions sine (left) and tangent (right) are symmetric about the origin.

Similarly, the graph of the tangent is also symmetric about the origin because the tangent of the negative of an angle is the negative of the tangent of the angle. In other words, tangent is an odd function.

Trigonometric Identities with $\frac{\pi}{2}$

A positive angle is called **acute** *if it is less than a right angle, which means less than $\frac{\pi}{2}$ radians or, equivalently, less than 90°.*

Suppose $0 < \theta < \frac{\pi}{2}$, and consider a right triangle with an angle of θ radians. Because the angles of a triangle add up to π radians, the triangle's other acute angle is $\frac{\pi}{2} - \theta$ radians, as shown in the figure below. If we were working in degrees rather than radians, then we would be stating that a right triangle with an angle of $\theta°$ also has an angle of $(90 - \theta)°$.

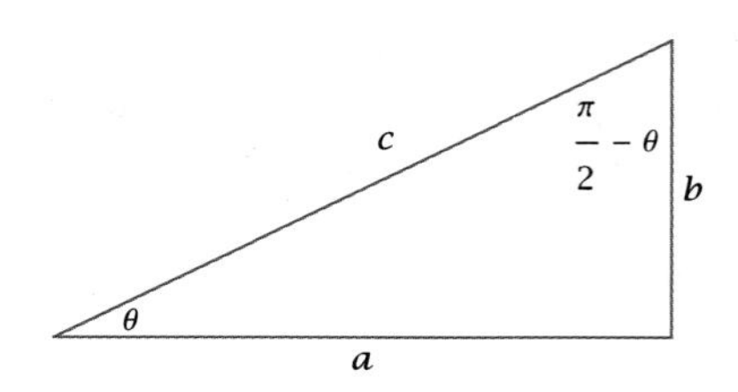

In a right triangle with an angle of θ radians, the other acute angle is $\frac{\pi}{2} - \theta$ radians.

In the triangle above, let c denote the length of the hypotenuse, let a denote the length of the side adjacent to the angle θ, and let b denote the length of the side opposite the angle θ. Focusing on the angle θ, our characterization from the previous section of cosine, sine, and tangent in terms of right triangles shows that

$$\cos\theta = \tfrac{a}{c} \quad \text{and} \quad \sin\theta = \tfrac{b}{c} \quad \text{and} \quad \tan\theta = \tfrac{b}{a}.$$

Now focusing instead on the angle $\frac{\pi}{2} - \theta$ in the triangle above, our right-triangle characterization of the trigonometric functions shows that

$$\cos(\tfrac{\pi}{2} - \theta) = \tfrac{b}{c} \quad \text{and} \quad \sin(\tfrac{\pi}{2} - \theta) = \tfrac{a}{c} \quad \text{and} \quad \tan(\tfrac{\pi}{2} - \theta) = \tfrac{a}{b}.$$

Comparing the last two sets of displayed equations, we get the following identities:

Trigonometric identities with $\frac{\pi}{2} - \theta$

$$\cos(\tfrac{\pi}{2} - \theta) = \sin\theta$$

$$\sin(\tfrac{\pi}{2} - \theta) = \cos\theta$$

$$\tan(\tfrac{\pi}{2} - \theta) = \frac{1}{\tan\theta}$$

Note that if $\tan\theta = 0$ (which happens when θ is an integer multiple of π), then $\tan(\frac{\pi}{2} - \theta)$ is undefined.

We have derived the identities above under the assumption that $0 < \theta < \frac{\pi}{2}$, but the first two identities hold for all values of θ. The third identity above holds for all values of θ except the integer multiples of $\frac{\pi}{2}$ [either $\tan(\frac{\pi}{2} - \theta)$ or $\tan\theta$ is not defined for such angles].

As an example of the formulas above, suppose $\theta = \frac{\pi}{6}$. Then

$$\tfrac{\pi}{2} - \theta = \tfrac{\pi}{2} - \tfrac{\pi}{6} = \tfrac{3\pi}{6} - \tfrac{\pi}{6} = \tfrac{2\pi}{6} = \tfrac{\pi}{3}.$$

For these angles $\frac{\pi}{6}$ and $\frac{\pi}{3}$ (which equal 30° and 60°), we already know the values of the trigonometric functions:

$$\cos\frac{\pi}{6} = \frac{\sqrt{3}}{2} \quad \text{and} \quad \sin\frac{\pi}{6} = \frac{1}{2} \quad \text{and} \quad \tan\frac{\pi}{6} = \frac{\sqrt{3}}{3}$$

and

$$\cos\frac{\pi}{3} = \frac{1}{2} \quad \text{and} \quad \sin\frac{\pi}{3} = \frac{\sqrt{3}}{2} \quad \text{and} \quad \tan\frac{\pi}{3} = \sqrt{3}.$$

Thus we see here the expected pattern when we consider an angle θ along with the angle $\frac{\pi}{2} - \theta$: the values of cosine and sine are interchanged, and the values of tangent are multiplicative inverses of each other (note that $\frac{1}{\sqrt{3}} = \frac{\sqrt{3}}{3}$).

Rewriting the identities in the box above in terms of degrees rather than radians, we obtain the following identities:

Trigonometric identities with $(90 - \theta)^\circ$

$$\cos(90 - \theta)^\circ = \sin\theta^\circ$$

$$\sin(90 - \theta)^\circ = \cos\theta^\circ$$

$$\tan(90 - \theta)^\circ = \frac{1}{\tan\theta^\circ}$$

These identities imply, for example, that $\cos 81^\circ = \sin 9^\circ$, $\sin 81^\circ = \cos 9^\circ$, *and* $\tan 81^\circ = \frac{1}{\tan 9^\circ}$.

Combining two or more trigonometric identities often leads to useful new identities, as shown in the following example.

EXAMPLE 3

Show that $\cos(\theta - \frac{\pi}{2}) = \sin\theta$.

SOLUTION Suppose θ is any real number. Then

$$\cos(\theta - \tfrac{\pi}{2}) = \cos(\tfrac{\pi}{2} - \theta)$$
$$= \sin\theta,$$

where the first equality above comes from the identity $\cos(-\theta) = \cos\theta$ (with θ replaced by $\frac{\pi}{2} - \theta$) and the second equality comes from one of the identities derived above.

The equation $\cos(\theta - \frac{\pi}{2}) = \sin\theta$ implies that the graph of sine is obtained by shifting the graph of cosine to the right by $\frac{\pi}{2}$ units (see Section 3.2 to review horizontal transformations of a function), as shown below:

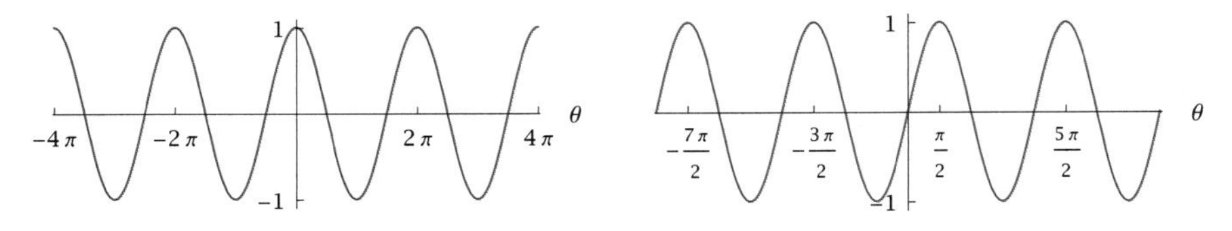

Shifting the graph of cosine (left) to the right by $\frac{\pi}{2}$ units produces the graph of sine (right).

Trigonometric Identities Involving a Multiple of π

Consider a typical angle θ radians and also the angle $\theta + \pi$ radians. Because π radians (which equals 180°) is a rotation halfway around the circle, the radius of the unit circle corresponding to $\theta + \pi$ radians forms a line with the radius corresponding to θ radians, as shown below.

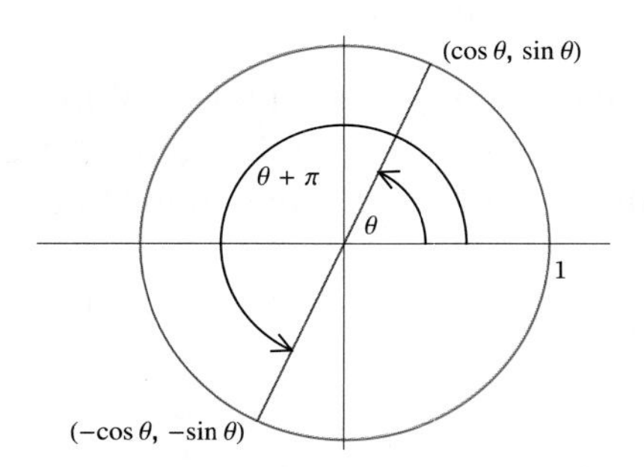

The radius corresponding to $\theta + \pi$ radians lies directly opposite the radius corresponding to θ radians. Thus the coordinates of the endpoints of these two radii are the negatives of each other.

By definition of the cosine and sine, the endpoint of the radius of the unit circle corresponding to θ has coordinates $(\cos\theta, \sin\theta)$, as shown above. The radius corresponding to $\theta+\pi$ lies directly opposite the radius corresponding to θ. Thus the endpoint of the radius corresponding to $\theta+\pi$ has coordinates $(-\cos\theta, -\sin\theta)$, as shown in the figure above.

By definition of the cosine and sine, the endpoint of the radius of the unit circle corresponding to $\theta+\pi$ has coordinates $(\cos(\theta+\pi), \sin(\theta+\pi))$. Thus $(\cos(\theta + \pi), \sin(\theta + \pi)) = (-\cos\theta, -\sin\theta)$. This implies that

$$\cos(\theta + \pi) = -\cos\theta \quad \text{and} \quad \sin(\theta + \pi) = -\sin\theta.$$

Recall that $\tan\theta$ equals the slope of the radius of the unit circle corresponding to θ. Similarly, $\tan(\theta + \pi)$ equals the slope of the radius corresponding to $\theta+\pi$. However, these two radii lie on the same line, as shown in the figure above. Thus these two radii have the same slope. Hence

$$\tan(\theta + \pi) = \tan\theta.$$

Another way to reach the same conclusion is to use the definition of the tangent as the ratio of the sine and cosine, along with the identities above:

$$\tan(\theta + \pi) = \frac{\sin(\theta + \pi)}{\cos(\theta + \pi)} = \frac{-\sin\theta}{-\cos\theta} = \frac{\sin\theta}{\cos\theta} = \tan\theta.$$

The first two identities hold for all values of θ. The third identity holds for all values of θ except odd multiples of $\frac{\pi}{2}$, which must be excluded because $\tan(\theta + \pi)$ and $\tan\theta$ are not defined for such angles.

Collecting the trigonometric identities involving $\theta + \pi$, we have:

Trigonometric identities with $\theta + \pi$

$$\cos(\theta + \pi) = -\cos\theta$$

$$\sin(\theta + \pi) = -\sin\theta$$

$$\tan(\theta + \pi) = \tan\theta$$

EXAMPLE 4

Using the result (which we will derive in Problem 101 in Section 10.5) that

$$\sin \tfrac{\pi}{10} = \frac{\sqrt{5}-1}{4},$$

find an exact expression for $\sin \frac{11\pi}{10}$.

SOLUTION

$$\begin{aligned}\sin \tfrac{11\pi}{10} &= \sin\left(\tfrac{\pi}{10} + \pi\right)\\ &= -\sin \tfrac{\pi}{10}\\ &= -\left(\frac{\sqrt{5}-1}{4}\right)\\ &= \frac{1-\sqrt{5}}{4}\end{aligned}$$

The trigonometric identity $\tan(\theta + \pi) = \tan\theta$ explains the periodic nature of the graph of the tangent, with the graph repeating the same shape after each interval of length π. This behavior is demonstrated in the graph below:

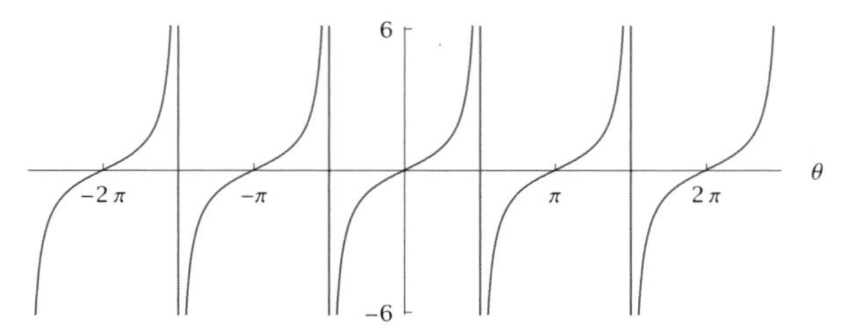

The graph of the tangent function. Because $\tan(\theta + \pi) = \tan\theta$*, this graph repeats the same shape after each interval of length* π*.*

This graph has been vertically truncated to show only values of the tangent that have absolute value less than 6.

Now we consider a typical angle θ radians and also the angle $\theta + 2\pi$. Because 2π radians (which equals 360°) is a complete rotation all the way around the circle, the radius of the unit circle corresponding to $\theta + 2\pi$ radians is the same as the radius corresponding to θ radians, as shown in the figure here.

By definition of the cosine and sine, the endpoint of the radius of the unit circle corresponding to θ has coordinates $(\cos\theta, \sin\theta)$, as shown here. Because the radius corresponding to $\theta + 2\pi$ is the same as the radius corresponding to θ, we see that $(\cos(\theta + 2\pi), \sin(\theta + 2\pi)) = (\cos\theta, \sin\theta)$. This implies that

$$\cos(\theta + 2\pi) = \cos\theta \quad \text{and} \quad \sin(\theta + 2\pi) = \sin\theta.$$

Recall that $\tan\theta$ equals the slope of the radius of the unit circle corresponding to θ. Similarly, $\tan(\theta + 2\pi)$ equals the slope of the radius corresponding to $\theta + 2\pi$. However, these two radii are the same. Thus

$$\tan(\theta + 2\pi) = \tan\theta.$$

Another way to reach the same conclusion is to use the definition of the tangent as the ratio of the sine and cosine, along with the identities above:

$$\tan(\theta + 2\pi) = \frac{\sin(\theta + 2\pi)}{\cos(\theta + 2\pi)} = \frac{\sin\theta}{\cos\theta} = \tan\theta.$$

Yet another way to reach the same conclusion is to use (twice!) the identity concerning the tangent of an angle plus π:

$$\tan(\theta + 2\pi) = \tan((\theta + \pi) + \pi) = \tan(\theta + \pi) = \tan\theta.$$

Collecting the trigonometric identities involving $\theta + 2\pi$, we have:

The first two identities hold for all values of θ. The third identity holds for all values of θ except odd multiples of $\frac{\pi}{2}$, which must be excluded because $\tan(\theta + 2\pi)$ and $\tan\theta$ are not defined for such angles.

Trigonometric identities with $\theta + 2\pi$

$$\cos(\theta + 2\pi) = \cos\theta$$

$$\sin(\theta + 2\pi) = \sin\theta$$

$$\tan(\theta + 2\pi) = \tan\theta$$

The trigonometric identities $\cos(\theta + 2\pi) = \cos\theta$ and $\sin(\theta + 2\pi) = \sin\theta$ explain the periodic nature of the graphs of cosine and sine, with the graphs repeating the same shape after each interval of length 2π. This behavior is demonstrated in the graphs below:

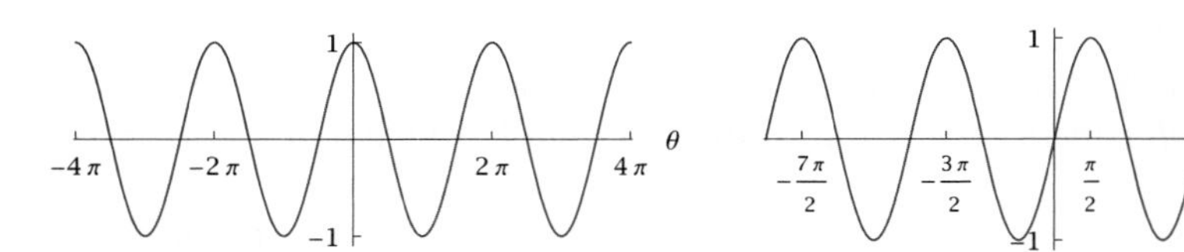

The graphs of the cosine (left) and sine (right).
Because $\cos(\theta + 2\pi) = \cos\theta$ and $\sin(\theta + 2\pi) = \sin\theta$,
these graphs repeat the same shape after each interval of length 2π.

In the box above, 2π could be replaced by any even multiple of π. For example, the radius corresponding to $\theta + 6\pi$ is obtained by starting with the radius corresponding to θ and then making three complete rotations around the circle, ending up with the same radius. Thus $\cos(\theta + 6\pi) = \cos\theta$, $\sin(\theta + 6\pi) = \sin\theta$, and $\tan(\theta + 6\pi) = \tan\theta$.

More generally, if n is an even integer, then the radius of the unit circle corresponding to $\theta + n\pi$ is the same as the radius corresponding to θ. Thus if n is an even integer, then the values of the trigonometric functions at $\theta + n\pi$ are the same as the values at θ.

Similarly, in our trigonometric formulas for $\theta + \pi$, we could replace π by any odd multiple of π. For example, the radius corresponding to $\theta + 5\pi$ is obtained by starting with the radius corresponding to θ and then making two-and-one-half rotations around the circle, ending up with the opposite radius. Thus $\cos(\theta + 5\pi) = -\cos\theta$, $\sin(\theta + 5\pi) = -\sin\theta$, and $\tan(\theta + 5\pi) = \tan\theta$.

More generally, if n is an odd integer, then the radius of the unit circle corresponding to $\theta + n\pi$ lies directly opposite the radius corresponding to θ. Thus if n is an odd integer, then the values of cosine and sine at $\theta + n\pi$ are the negatives of the values at θ, and the value of tangent at $\theta + n\pi$ is the same as the value at θ.

The trigonometric identities involving an integer multiple of π can be summarized as follows:

Trigonometric identities with $\theta + n\pi$

$$\cos(\theta + n\pi) = \begin{cases} \cos\theta & \text{if } n \text{ is an even integer} \\ -\cos\theta & \text{if } n \text{ is an odd integer} \end{cases}$$

$$\sin(\theta + n\pi) = \begin{cases} \sin\theta & \text{if } n \text{ is an even integer} \\ -\sin\theta & \text{if } n \text{ is an odd integer} \end{cases}$$

$$\tan(\theta + n\pi) = \tan\theta \quad \text{if } n \text{ is an integer}$$

The first two identities hold for all values of θ. The third identity holds for all values of θ except the odd multiples of $\frac{\pi}{2}$; these values must be excluded because $\tan(\theta + n\pi)$ and $\tan\theta$ are not defined for such angles.

EXERCISES

1. For $\theta = 7°$, evaluate each of the following:
   (a) $\cos^2\theta$ (b) $\cos(\theta^2)$

   [*Exercises 1 and 2 emphasize that* $\cos^2\theta$ *does not equal* $\cos(\theta^2)$.]

2. For $\theta = 5$ radians, evaluate each of the following:
   (a) $\cos^2\theta$ (b) $\cos(\theta^2)$

3. For $\theta = 4$ radians, evaluate each of the following:
   (a) $\sin^2\theta$ (b) $\sin(\theta^2)$

   [*Exercises 3 and 4 emphasize that* $\sin^2\theta$ *does not equal* $\sin(\theta^2)$.]

4. For $\theta = -8°$, evaluate each of the following:
   (a) $\sin^2\theta$ (b) $\sin(\theta^2)$

In Exercises 5–38, find exact expressions for the indicated quantities, given that

$$\cos\frac{\pi}{12} = \frac{\sqrt{2+\sqrt{3}}}{2} \quad \textit{and} \quad \sin\frac{\pi}{8} = \frac{\sqrt{2-\sqrt{2}}}{2}.$$

[These values for $\cos\frac{\pi}{12}$ and $\sin\frac{\pi}{8}$ will be derived in Examples 4 and 5 in Section 10.5.]

5. $\cos(-\frac{\pi}{12})$
6. $\sin(-\frac{\pi}{8})$
7. $\sin\frac{\pi}{12}$
8. $\cos\frac{\pi}{8}$
9. $\sin(-\frac{\pi}{12})$
10. $\cos(-\frac{\pi}{8})$
11. $\tan\frac{\pi}{12}$
12. $\tan\frac{\pi}{8}$
13. $\tan(-\frac{\pi}{12})$
14. $\tan(-\frac{\pi}{8})$
15. $\cos\frac{25\pi}{12}$
16. $\cos\frac{17\pi}{8}$
17. $\sin\frac{25\pi}{12}$
18. $\sin\frac{17\pi}{8}$
19. $\tan\frac{25\pi}{12}$
20. $\tan\frac{17\pi}{8}$
21. $\cos\frac{13\pi}{12}$
22. $\cos\frac{9\pi}{8}$
23. $\sin\frac{13\pi}{12}$
24. $\sin\frac{9\pi}{8}$
25. $\tan\frac{13\pi}{12}$
26. $\tan\frac{9\pi}{8}$
27. $\cos\frac{5\pi}{12}$
28. $\cos\frac{3\pi}{8}$
29. $\cos(-\frac{5\pi}{12})$
30. $\cos(-\frac{3\pi}{8})$
31. $\sin\frac{5\pi}{12}$
32. $\sin\frac{3\pi}{8}$
33. $\sin(-\frac{5\pi}{12})$
34. $\sin(-\frac{3\pi}{8})$
35. $\tan\frac{5\pi}{12}$
36. $\tan\frac{3\pi}{8}$
37. $\tan(-\frac{5\pi}{12})$
38. $\tan(-\frac{3\pi}{8})$

Suppose u ***and*** v ***are in the interval*** $(\frac{\pi}{2}, \pi)$***, with***

$$\tan u = -2 \quad \text{and} \quad \tan v = -3.$$

In Exercises 39–66, find exact expressions for the indicated quantities.

39. $\tan(-u)$
40. $\tan(-v)$
41. $\cos u$
42. $\cos v$
43. $\cos(-u)$
44. $\cos(-v)$
45. $\sin u$
46. $\sin v$
47. $\sin(-u)$
48. $\sin(-v)$
49. $\cos(u + 4\pi)$
50. $\cos(v - 6\pi)$
51. $\sin(u - 6\pi)$
52. $\sin(v + 10\pi)$
53. $\tan(u + 8\pi)$
54. $\tan(v - 4\pi)$
55. $\cos(u - 3\pi)$
56. $\cos(v + 5\pi)$
57. $\sin(u + 5\pi)$
58. $\sin(v - 7\pi)$
59. $\tan(u - 9\pi)$
60. $\tan(v + 3\pi)$
61. $\cos(\frac{\pi}{2} - u)$
62. $\cos(\frac{\pi}{2} - v)$
63. $\sin(\frac{\pi}{2} - u)$
64. $\sin(\frac{\pi}{2} - v)$
65. $\tan(\frac{\pi}{2} - u)$
66. $\tan(\frac{\pi}{2} - v)$

PROBLEMS

67. Show that
$$(\cos\theta + \sin\theta)^2 = 1 + 2\cos\theta\sin\theta$$
for every number θ.
[*Expressions such as* $\cos\theta \sin\theta$ *mean* $(\cos\theta)(\sin\theta)$, *not* $\cos(\theta\sin\theta)$.]

68. Show that
$$\frac{\sin x}{1 - \cos x} = \frac{1 + \cos x}{\sin x}$$
for every number x that is not an integer multiple of π.

69. Show that
$$\cos^3\theta + \cos^2\theta\sin\theta + \cos\theta\sin^2\theta + \sin^3\theta$$
$$= \cos\theta + \sin\theta$$
for every number θ.
[*Hint:* Try replacing the $\cos^2\theta$ term above with $1 - \sin^2\theta$ and replacing the $\sin^2\theta$ term above with $1 - \cos^2\theta$.]

70. Show that
$$\sin^2\theta = \frac{\tan^2\theta}{1 + \tan^2\theta}$$
for all θ except odd multiples of $\frac{\pi}{2}$.

71. Find a formula for $\cos\theta$ solely in terms of $\tan\theta$.

72. Find a formula for $\tan\theta$ solely in terms of $\sin\theta$.

73. Explain why $\sin 3° + \sin 357° = 0$.

74. Explain why $\cos 85° + \cos 95° = 0$.

75. Pretend that you are living in the time before calculators and computers existed, and that you have a table showing the cosines and sines of 1°, 2°, 3°, and so on, up to the cosine and sine of 45°. Explain how you would find the cosine and sine of 71°, which are beyond the range of your table.

76. Suppose n is an integer. Find formulas for $\sec(\theta + n\pi)$, $\csc(\theta + n\pi)$, and $\cot(\theta + n\pi)$ in terms of $\sec\theta$, $\csc\theta$, and $\cot\theta$.

77. Restate all the results in boxes in the subsection on *Trigonometric Identities Involving a Multiple of* π in terms of degrees instead of in terms of radians.

78. Show that
$$\cos(\pi - \theta) = -\cos\theta$$
for every angle θ.

79. Show that
$$\tan(\theta + \tfrac{\pi}{2}) = -\frac{1}{\tan\theta}$$
for every angle θ that is not an integer multiple of $\frac{\pi}{2}$. Interpret this result in terms of the characterization of the slopes of perpendicular lines.

80. Show that
$$\sin(\pi - \theta) = \sin\theta$$
for every angle θ.

81. Show that
$$\cos(x + \tfrac{\pi}{2}) = -\sin x,$$
for every number x.

82. Show that
$$\sin(t + \tfrac{\pi}{2}) = \cos t$$
for every number t.

83. Explain why
$$|\cos(x + n\pi)| = |\cos x|$$
for every number x and every integer n.

WORKED-OUT SOLUTIONS *to Odd-numbered Exercises*

1. For $\theta = 7°$, evaluate each of the following:

   (a) $\cos^2\theta$ (b) $\cos(\theta^2)$

   SOLUTION

(a) Using a calculator working in degrees, we have

$$\cos^2 7° = (\cos 7°)^2 \approx (0.992546)^2 \approx 0.985148.$$

(b) Note that $7^2 = 49$. Using a calculator working in degrees, we have

$$\cos 49° \approx 0.656059.$$

3. For $\theta = 4$ radians, evaluate each of the following:

   (a) $\sin^2\theta$ (b) $\sin(\theta^2)$

   SOLUTION

(a) Using a calculator working in radians, we have

$$\sin^2 4 = (\sin 4)^2 \approx (-0.756802)^2 \approx 0.57275.$$

(b) Note that $4^2 = 16$. Using a calculator working in radians, we have

$$\sin 16 \approx -0.287903.$$

In Exercises 5–38, find exact expressions for the indicated quantities, given that

$$\cos\tfrac{\pi}{12} = \frac{\sqrt{2+\sqrt{3}}}{2} \quad \textit{and} \quad \sin\tfrac{\pi}{8} = \frac{\sqrt{2-\sqrt{2}}}{2}.$$

5. $\cos(-\frac{\pi}{12})$

   SOLUTION

$$\cos(-\tfrac{\pi}{12}) = \cos\tfrac{\pi}{12} = \frac{\sqrt{2+\sqrt{3}}}{2}$$

7. $\sin\frac{\pi}{12}$

   SOLUTION We know that

$$\cos^2\tfrac{\pi}{12} + \sin^2\tfrac{\pi}{12} = 1.$$

Thus

$$\begin{aligned}\sin^2\tfrac{\pi}{12} &= 1 - \cos^2\tfrac{\pi}{12}\\ &= 1 - \Big(\frac{\sqrt{2+\sqrt{3}}}{2}\Big)^2\\ &= 1 - \frac{2+\sqrt{3}}{4}\\ &= \frac{2-\sqrt{3}}{4}.\end{aligned}$$

Because $\sin\frac{\pi}{12} > 0$, taking square roots of both sides of the equation above gives

$$\sin\tfrac{\pi}{12} = \frac{\sqrt{2-\sqrt{3}}}{2}.$$

9. $\sin(-\frac{\pi}{12})$

   SOLUTION

$$\sin(-\tfrac{\pi}{12}) = -\sin\tfrac{\pi}{12} = -\frac{\sqrt{2-\sqrt{3}}}{2}$$

11. $\tan\frac{\pi}{12}$

   SOLUTION

$$\begin{aligned}\tan\tfrac{\pi}{12} &= \frac{\sin\frac{\pi}{12}}{\cos\frac{\pi}{12}}\\ &= \frac{\sqrt{2-\sqrt{3}}}{\sqrt{2+\sqrt{3}}}\\ &= \frac{\sqrt{2-\sqrt{3}}}{\sqrt{2+\sqrt{3}}} \cdot \frac{\sqrt{2-\sqrt{3}}}{\sqrt{2-\sqrt{3}}}\\ &= \frac{2-\sqrt{3}}{\sqrt{4-3}}\\ &= 2-\sqrt{3}\end{aligned}$$

13. $\tan(-\frac{\pi}{12})$

   SOLUTION

$$\tan(-\tfrac{\pi}{12}) = -\tan\tfrac{\pi}{12} = -(2-\sqrt{3}) = \sqrt{3}-2$$

15. $\cos\frac{25\pi}{12}$

   SOLUTION Because $\frac{25\pi}{12} = \frac{\pi}{12} + 2\pi$, we have

$$\begin{aligned}\cos\tfrac{25\pi}{12} &= \cos(\tfrac{\pi}{12} + 2\pi)\\ &= \cos\tfrac{\pi}{12}\\ &= \frac{\sqrt{2+\sqrt{3}}}{2}.\end{aligned}$$

17. $\sin \frac{25\pi}{12}$

SOLUTION Because $\frac{25\pi}{12} = \frac{\pi}{12} + 2\pi$, we have

$$\begin{aligned}\sin \tfrac{25\pi}{12} &= \sin(\tfrac{\pi}{12} + 2\pi)\\ &= \sin \tfrac{\pi}{12}\\ &= \frac{\sqrt{2-\sqrt{3}}}{2}.\end{aligned}$$

19. $\tan \frac{25\pi}{12}$

SOLUTION Because $\frac{25\pi}{12} = \frac{\pi}{12} + 2\pi$, we have

$$\begin{aligned}\tan \tfrac{25\pi}{12} &= \tan(\tfrac{\pi}{12} + 2\pi)\\ &= \tan \tfrac{\pi}{12}\\ &= 2 - \sqrt{3}.\end{aligned}$$

21. $\cos \frac{13\pi}{12}$

SOLUTION Because $\frac{13\pi}{12} = \frac{\pi}{12} + \pi$, we have

$$\begin{aligned}\cos \tfrac{13\pi}{12} &= \cos(\tfrac{\pi}{12} + \pi)\\ &= -\cos \tfrac{\pi}{12}\\ &= -\frac{\sqrt{2+\sqrt{3}}}{2}.\end{aligned}$$

23. $\sin \frac{13\pi}{12}$

SOLUTION Because $\frac{13\pi}{12} = \frac{\pi}{12} + \pi$, we have

$$\begin{aligned}\sin \tfrac{13\pi}{12} &= \sin(\tfrac{\pi}{12} + \pi)\\ &= -\sin \tfrac{\pi}{12}\\ &= -\frac{\sqrt{2-\sqrt{3}}}{2}.\end{aligned}$$

25. $\tan \frac{13\pi}{12}$

SOLUTION Because $\frac{13\pi}{12} = \frac{\pi}{12} + \pi$, we have

$$\begin{aligned}\tan \tfrac{13\pi}{12} &= \tan(\tfrac{\pi}{12} + \pi)\\ &= \tan \tfrac{\pi}{12}\\ &= 2 - \sqrt{3}.\end{aligned}$$

27. $\cos \frac{5\pi}{12}$

SOLUTION

$$\cos \tfrac{5\pi}{12} = \sin(\tfrac{\pi}{2} - \tfrac{5\pi}{12}) = \sin \tfrac{\pi}{12} = \frac{\sqrt{2-\sqrt{3}}}{2}$$

29. $\cos(-\frac{5\pi}{12})$

SOLUTION

$$\cos(-\tfrac{5\pi}{12}) = \cos \tfrac{5\pi}{12} = \frac{\sqrt{2-\sqrt{3}}}{2}$$

31. $\sin \frac{5\pi}{12}$

SOLUTION

$$\sin \tfrac{5\pi}{12} = \cos(\tfrac{\pi}{2} - \tfrac{5\pi}{12}) = \cos \tfrac{\pi}{12} = \frac{\sqrt{2+\sqrt{3}}}{2}$$

33. $\sin(-\frac{5\pi}{12})$

SOLUTION

$$\sin(-\tfrac{5\pi}{12}) = -\sin \tfrac{5\pi}{12} = -\frac{\sqrt{2+\sqrt{3}}}{2}$$

35. $\tan \frac{5\pi}{12}$

SOLUTION

$$\begin{aligned}\tan \tfrac{5\pi}{12} &= \frac{1}{\tan(\frac{\pi}{2} - \frac{5\pi}{12})}\\ &= \frac{1}{\tan \frac{\pi}{12}}\\ &= \frac{1}{2-\sqrt{3}}\\ &= \frac{1}{2-\sqrt{3}} \cdot \frac{2+\sqrt{3}}{2+\sqrt{3}}\\ &= \frac{2+\sqrt{3}}{4-3}\\ &= 2+\sqrt{3}\end{aligned}$$

37. $\tan(-\frac{5\pi}{12})$

SOLUTION

$$\tan(-\tfrac{5\pi}{12}) = -\tan \tfrac{5\pi}{12} = -2 - \sqrt{3}$$

Suppose u and v are in the interval $(\frac{\pi}{2}, \pi)$, with

$$\tan u = -2 \quad \textit{and} \quad \tan v = -3.$$

In Exercises 39–66, find exact expressions for the indicated quantities.

39. $\tan(-u)$

SOLUTION $\tan(-u) = -\tan u = -(-2) = 2$

41. $\cos u$

SOLUTION We know that

$$-2 = \tan u$$
$$= \frac{\sin u}{\cos u}.$$

To find $\cos u$, make the substitution $\sin u = \sqrt{1 - \cos^2 u}$ in the equation above (this substitution is valid because $\frac{\pi}{2} < u < \pi$, which implies that $\sin u > 0$), getting

$$-2 = \frac{\sqrt{1 - \cos^2 u}}{\cos u}.$$

Now square both sides of the equation above, then multiply both sides by $\cos^2 u$ and rearrange to get the equation

$$5\cos^2 u = 1.$$

Thus $\cos u = -\frac{1}{\sqrt{5}}$ (the possibility that $\cos u$ equals $\frac{1}{\sqrt{5}}$ is eliminated because $\frac{\pi}{2} < u < \pi$, which implies that $\cos u < 0$). This can be written as $\cos u = -\frac{\sqrt{5}}{5}$.

43. $\cos(-u)$

SOLUTION $\cos(-u) = \cos u = -\dfrac{\sqrt{5}}{5}$

45. $\sin u$

SOLUTION

$$\sin u = \sqrt{1 - \cos^2 u}$$
$$= \sqrt{1 - \frac{1}{5}}$$
$$= \sqrt{\frac{4}{5}}$$
$$= \frac{2}{\sqrt{5}}$$
$$= \frac{2\sqrt{5}}{5}$$

47. $\sin(-u)$

SOLUTION $\sin(-u) = -\sin u = -\dfrac{2\sqrt{5}}{5}$

49. $\cos(u + 4\pi)$

SOLUTION $\cos(u + 4\pi) = \cos u = -\dfrac{\sqrt{5}}{5}$

51. $\sin(u - 6\pi)$

SOLUTION $\sin(u - 6\pi) = \sin u = \dfrac{2\sqrt{5}}{5}$

53. $\tan(u + 8\pi)$

SOLUTION $\tan(u + 8\pi) = \tan u = -2$

55. $\cos(u - 3\pi)$

SOLUTION $\cos(u - 3\pi) = -\cos u = \dfrac{\sqrt{5}}{5}$

57. $\sin(u + 5\pi)$

SOLUTION $\sin(u + 5\pi) = -\sin u = -\dfrac{2\sqrt{5}}{5}$

59. $\tan(u - 9\pi)$

SOLUTION $\tan(u - 9\pi) = \tan u = -2$

61. $\cos(\frac{\pi}{2} - u)$

SOLUTION $\cos(\frac{\pi}{2} - u) = \sin u = \dfrac{2\sqrt{5}}{5}$

63. $\sin(\frac{\pi}{2} - u)$

SOLUTION $\sin(\frac{\pi}{2} - u) = \cos u = -\dfrac{\sqrt{5}}{5}$

65. $\tan(\frac{\pi}{2} - u)$

SOLUTION $\tan(\frac{\pi}{2} - u) = \dfrac{1}{\tan u} = -\dfrac{1}{2}$

CHAPTER SUMMARY

To check that you have mastered the most important concepts and skills covered in this chapter, make sure that you can do each item in the following list:

- Explain what it means for an angle to be negative.
- Explain how an angle can be larger than $360°$.
- Convert angles from radians to degrees.
- Convert angles from degrees to radians.
- Compute the length of a circular arc.
- Compute the cosine, sine, and tangent of any multiple of $30°$ or $45°$ ($\frac{\pi}{6}$ radians or $\frac{\pi}{4}$ radians).
- Explain why $\cos^2\theta + \sin^2\theta = 1$ for every angle θ.
- Give the domain and range of the cosine, sine, and tangent functions.
- Compute $\cos\theta$, $\sin\theta$, and $\tan\theta$ if given just one of these quantities and the location of the corresponding radius.
- Compute the cosine, sine, and tangent of any angle of a right triangle if given the lengths of two sides of the triangle.
- Compute the lengths of all three sides of a right triangle if given any angle (in addition to the right angle) and the length of any side.
- Use the basic trigonometric identities involving $-\theta$, $\frac{\pi}{2} - \theta$, $\theta + \pi$, and $\theta + 2\pi$.

To review a chapter, go through the list above to find items that you do not know how to do, then reread the material in the chapter about those items. Then try to answer the chapter review questions below without looking back at the chapter.

CHAPTER REVIEW QUESTIONS

1. Find all points where the line through the origin with slope 5 intersects the unit circle.
2. Sketch a unit circle and the radius of that circle corresponding to $-70°$.
3. Sketch a unit circle and the radius of that circle corresponding to $440°$.
4. Explain how to convert an angle from degrees to radians.
5. Convert $27°$ to radians.
6. Explain how to convert an angle from radians to degrees.
7. Convert $\frac{7\pi}{9}$ radians to degrees.
8. Give the domain and range of each of the following functions: cos, sin, and tan.
9. Find three distinct angles, expressed in degrees, whose cosine equals $\frac{1}{2}$.
10. Find three distinct angles, expressed in radians, whose sine equals $-\frac{1}{2}$.

For Questions 11–13, assume that the earth has a circular orbit around the sun and that the distance from the earth to the sun is 92.956 million miles (the actual orbit is an ellipse rather than a circle, but it is very close to a circle). Also assume that the earth orbits the sun once every 365.24 days (again, this is a close approximation).

11. How far does the earth travel in a week?
12. How far does the earth travel in the month of July?
13. What is the speed of the earth due to its rotation around the sun?
14. Find three distinct angles, expressed in radians, whose tangent equals 1.
15. Explain why $\cos^2\theta + \sin^2\theta = 1$ for every angle θ.
16. Explain why $\cos(\theta + 2\pi) = \cos\theta$ for every angle θ.
17. Suppose $\frac{\pi}{2} < x < \pi$ and $\tan x = -4$. Evaluate $\cos x$ and $\sin x$.

Use the right triangle below for Questions 18–36. This triangle is not drawn to scale corresponding to the data in the questions.

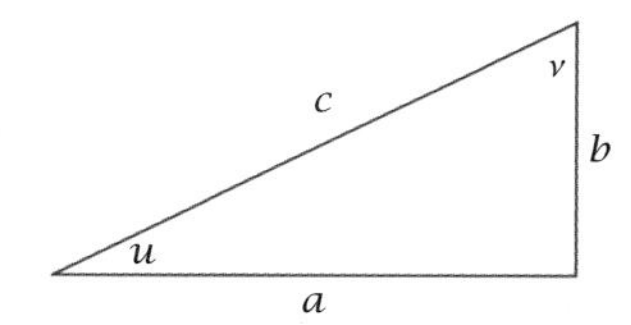

18. Suppose $a = 4$ and $b = 9$. Evaluate c.

19. Suppose $a = 4$ and $b = 9$. Evaluate $\cos u$.

20. Suppose $a = 4$ and $b = 9$. Evaluate $\sin u$.

21. Suppose $a = 4$ and $b = 9$. Evaluate $\tan u$.

22. Suppose $a = 4$ and $b = 9$. Evaluate $\cos v$.

23. Suppose $a = 4$ and $b = 9$. Evaluate $\sin v$.

24. Suppose $a = 4$ and $b = 9$. Evaluate $\tan v$.

25. Suppose $a = 3$ and $c = 8$. Evaluate b.

26. Suppose $a = 3$ and $c = 8$. Evaluate $\cos u$.

27. Suppose $a = 3$ and $c = 8$. Evaluate $\sin u$.

28. Suppose $a = 3$ and $c = 8$. Evaluate $\tan u$.

29. Suppose $a = 3$ and $c = 8$. Evaluate $\cos v$.

30. Suppose $a = 3$ and $c = 8$. Evaluate $\sin v$.

31. Suppose $a = 3$ and $c = 8$. Evaluate $\tan v$.

32. Suppose $b = 4$ and $u = 28°$. Evaluate a.

33. Suppose $b = 4$ and $u = 28°$. Evaluate c.

34. Suppose $u = 28°$. Evaluate $\cos v$.

35. Suppose $u = 28°$. Evaluate $\sin v$.

36. Suppose $u = 28°$. Evaluate $\tan v$.

37. Suppose θ is an angle such that $\cos\theta = \frac{3}{8}$. Evaluate $\cos(-\theta)$.

38. Suppose x is a number such that $\sin x = \frac{4}{7}$. Evaluate $\sin(-x)$.

39. Suppose y is a number such that $\tan y = -\frac{2}{9}$. Evaluate $\tan(-y)$.

40. Suppose u is a number such that $\cos u = -\frac{2}{5}$. Evaluate $\cos(u + \pi)$.

41. Suppose θ is an angle such that $\tan\theta = \frac{5}{6}$. Evaluate $\tan(\frac{\pi}{2} - \theta)$.

42. Find a formula for $\tan\theta$ solely in terms of $\cos\theta$.

43. Suppose $-\frac{\pi}{2} < x < 0$ and $\cos x = \frac{5}{13}$. Evaluate $\sin x$ and $\tan x$.

44. Show that

$$\frac{\sin x + \sin y}{\cos x + \cos y} = \frac{\cos x - \cos y}{\sin y - \sin x}$$

for all numbers x and y such that neither denominator is 0.

CHAPTER

10

The loonie, an 11-sided Canadian one-dollar coin. In this chapter you will learn how to use trigonometry to compute the area of a loonie (see Example 5 in Section 10.3).

Trigonometric Algebra and Geometry

This chapter begins by introducing the inverse trigonometric functions. These tremendously useful functions allow us to find angles from measurements of lengths. We will pay some attention to the inverse trigonometric identities, which are important in their own right and which also strengthen our understanding of trigonometry.

Then we will turn our attention to area, showing how trigonometry can be used to compute areas of various regions. We will also derive some important approximations of the trigonometric functions.

Our next subject will be the law of sines and the law of cosines. These results let us use trigonometry to compute all the angles and the lengths of the sides of a triangle given only some of this information.

The double-angle and half-angle formulas for the trigonometric functions will allow us to compute exact expressions for quantities such as $\sin 18^\circ$ and $\cos \frac{\pi}{32}$ (see Problems 101 and 104 in Section 10.5 for these beautiful expressions). Finally, the chapter concludes with the addition and subtraction formulas for the trigonometric functions, providing another group of useful identities.

10.1 *Inverse Trigonometric Functions*

LEARNING OBJECTIVES

By the end of this section you should be able to

- compute values of $\cos^{-1}$, $\sin^{-1}$, and $\tan^{-1}$;
- sketch the radius of the unit circle corresponding to the arccosine, arcsine, and arctangent of a number;
- use the inverse trigonometric functions to find angles in a right triangle, given the lengths of two sides;
- find the angles in an isosceles triangle, given the lengths of the sides;
- use $\tan^{-1}$ to find the angle a line with given slope makes with the horizontal axis.

Several of the most important functions in mathematics are defined as the inverse functions of familiar functions. For example, the cube root is defined as the inverse function of x^3, and the logarithm base 3 is defined as the inverse function of 3^x.

The inverse trigonometric functions provide a remarkably useful tool for solving many problems.

In this section, we will define the inverses of the cosine, sine, and tangent functions. These inverse functions are called the arccosine, the arcsine, and the arctangent. Neither cosine nor sine nor tangent is one-to-one when defined on its usual domain. Thus we will need to restrict the domains of these functions to obtain one-to-one functions that have inverses.

The Arccosine Function

As usual, we will assume throughout this section that all angles are measured in radians unless explicitly stated otherwise.

Consider the cosine function, whose domain is the entire real line. The cosine function is not one-to-one because, for example, $\cos 0 = \cos 2\pi$.

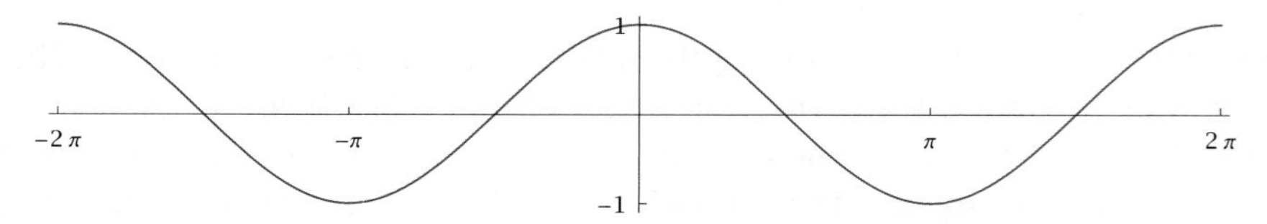

The graph of cosine on the interval $[-2\pi, 2\pi]$.

The graph above fails the horizontal line test—there are horizontal lines that intersect the graph in more than one point. Thus the cosine function is not one-to-one.

For example, suppose we are told that x is a number such that $\cos x = 0$, and we are asked to find the value of x. Of course $\cos \frac{\pi}{2} = 0$, but also $\cos \frac{3\pi}{2} = 0$; we also have $\cos(-\frac{\pi}{2}) = 0$ and $\cos(-\frac{3\pi}{2}) = 0$ and so on. Thus with the information given we have no way to determine a unique value of x such that $\cos x = 0$. Hence the cosine function does not have an inverse.

We faced a similar dilemma when we wanted to define the square root function as the inverse of the function x^2. The domain of the function x^2 is the entire real line. This function is not one-to-one; thus it does not have an inverse. For example, if we are told that $x^2 = 16$, then we cannot determine whether $x = 4$ or $x = -4$. We solved this problem by restricting the domain of x^2 to $[0, \infty)$; the resulting function is one-to-one, and its inverse is called the square root function. Roughly speaking, we say that the square root function is the inverse of x^2.

We will follow a similar process with the cosine. To decide how to restrict the domain of the cosine, we start by declaring 0 should be in the domain of the restricted function. Looking at the graph above of cosine, we see that starting at 0 and moving to the right, π is the farthest we can go while staying within an interval on which cosine is one-to-one. Once we decide that $[0, \pi]$ will be in the domain of our restricted cosine function, then we cannot move at all to the left from 0 and still have a one-to-one function. Thus $[0, \pi]$ is the natural domain to choose to get an inverse for the cosine.

If we restrict the domain of cosine to $[0, \pi]$, we obtain the one-to-one function whose graph is shown here. The inverse of this function is called the arccosine, which is abbreviated as $\cos^{-1}$. Here is the formal definition:

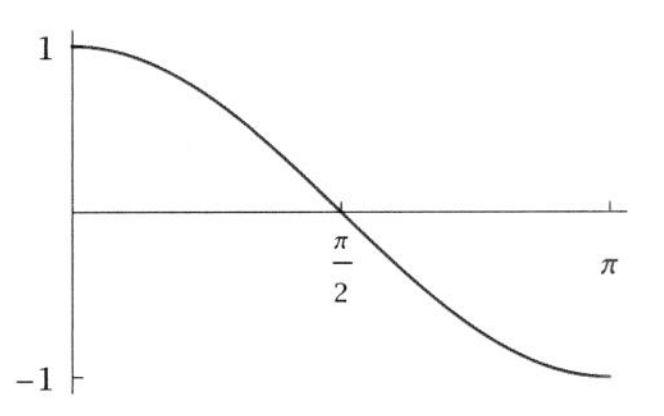

The graph of cosine on the interval $[0, \pi]$.

Arccosine

For t in $[-1, 1]$, the **arccosine** of t, denoted $\cos^{-1} t$, is the angle in $[0, \pi]$ whose cosine equals t. In other words, the equation

$$\cos^{-1} t = \theta$$

means θ is the angle in $[0, \pi]$ such that

$$\cos \theta = t.$$

In defining $\cos^{-1} t$, we must restrict t to be in the interval $[-1, 1]$ because otherwise there is no angle whose cosine equals t.

EXAMPLE 1

(a) Evaluate $\cos^{-1} 0$.

(b) Evaluate $\cos^{-1} 1$.

(c) Explain why the expression $\cos^{-1} 2$ makes no sense.

SOLUTION

(a) Because $\cos \frac{\pi}{2} = 0$ and $\frac{\pi}{2}$ is in the interval $[0, \pi]$, we have $\cos^{-1} 0 = \frac{\pi}{2}$.

(b) Because $\cos 0 = 1$ and 0 is in the interval $[0, \pi]$, we have $\cos^{-1} 1 = 0$.

(c) The expression $\cos^{-1} 2$ makes no sense because there is no angle whose cosine equals 2.

Do not confuse $\cos^{-1} t$ with $(\cos t)^{-1}$. Confusion can arise due to inconsistency in common notation. For example, $\cos^2 t$ is indeed equal to $(\cos t)^2$. However, we defined $\cos^n t$ to equal $(\cos t)^n$ only when n is a positive integer (see Section 9.6). This restriction concerning $\cos^n t$ was made precisely so that $\cos^{-1} t$ could be defined with $\cos^{-1}$ interpreted as an inverse function.

Be sure you understand that $\cos^{-1} t$ *is not equal to* $\frac{1}{\cos t}$.

The notation $\cos^{-1}$ to denote the arccosine function is consistent with our notation f^{-1} to denote the inverse of a function f. Even here a bit of explanation helps. The usual domain of the cosine function is the real line. However, when we write $\cos^{-1}$ we do mean not the inverse of the usual cosine function (which has no inverse because it is not one-to-one). Instead, $\cos^{-1}$ means the inverse of the cosine function whose domain is restricted to the interval $[0, \pi]$.

Some books use the notation arccos t *instead of* $\cos^{-1} t$.

The three different solutions to the three parts of the example below show why you need to pay careful attention to the meaning of notation.

EXAMPLE 2

For $x = 0.2$, evaluate each of the following:

(a) $\cos^{-1} x$ (b) $(\cos x)^{-1}$ (c) $\cos(x^{-1})$

SOLUTION The calculations below were done with a calculator. When checking them on your calculator, be sure that it is set to work in radians rather than degrees. Remember that the arccosine function $\cos^{-1}$ is the inverse of the cosine function where the input is assumed to represent radians (and the domain is restricted to the interval $[0, \pi]$).

(a) $\cos^{-1} 0.2 \approx 1.36944$ (note that $\cos 1.36944 \approx 0.2$ and 1.36944 is in $[0, \pi]$)

(b) $(\cos 0.2)^{-1} = \frac{1}{\cos 0.2} \approx 1.02034$

(c) $\cos(0.2^{-1}) = \cos \frac{1}{0.2} = \cos 5 \approx 0.283662$

The next example should help solidify your understanding of the arccosine function.

EXAMPLE 3

Sketch the radius of the unit circle corresponding to the angle $\cos^{-1} 0.3$.

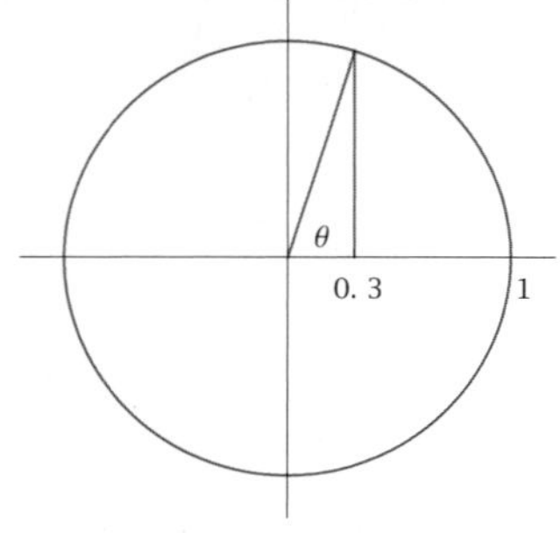

Here $\theta = \cos^{-1} 0.3$, *or equivalently* $\cos\theta = 0.3$.

SOLUTION We seek an angle in $[0, \pi]$ whose cosine equals 0.3. This means that the first coordinate of the endpoint of the corresponding radius will equal 0.3. Thus we start with 0.3 on the horizontal axis, as shown here, and extend a line upward until it intersects the unit circle. That point of intersection is the endpoint of the radius corresponding to the angle $\cos^{-1} 0.3$, as shown here.

A calculator shows that

$$\cos^{-1} 0.3 \approx 1.266.$$

Thus the angle θ shown in the figure here is approximately 1.266 radians, which is approximately 72.5°.

Recall that the inverse of a function interchanges the domain and range of the original function. Thus we have the following:

Domain and range of arccosine

- The domain of $\cos^{-1}$ is $[-1, 1]$.
- The range of $\cos^{-1}$ is $[0, \pi]$.

The graph of $\cos^{-1}$ can be obtained in the usual way when dealing with inverse functions. Specifically, flip the graph of the cosine (restricted to the interval $[0, \pi]$) across the line with slope 1 that contains the origin, getting the graph shown here.

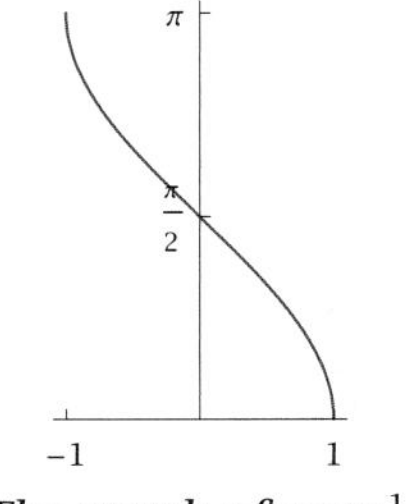

The graph of $\cos^{-1}$.

The inverse trigonometric functions are spectacularly useful in finding the angles of a right triangle when given the lengths of two of the sides. The example below gives our first illustration of this procedure.

EXAMPLE 4

Suppose a 13-foot ladder is leaned against a building, reaching to the bottom of a second-floor window 12 feet above the ground. What angle does the ladder make with the building?

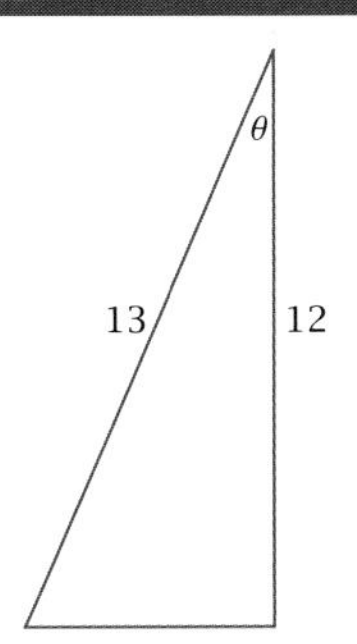

SOLUTION Let θ denote the angle the ladder makes with the building. Because the cosine of an angle in a right triangle equals the length of the adjacent side divided by the length of the hypotenuse, the figure here shows that $\cos\theta = \frac{12}{13}$. Thus

$$\theta = \cos^{-1}\tfrac{12}{13} \approx 0.3948.$$

Hence θ is approximately 0.3948 radians, which is approximately 22.6°.

The Arcsine Function

Now we consider the sine function, whose graph is shown below:

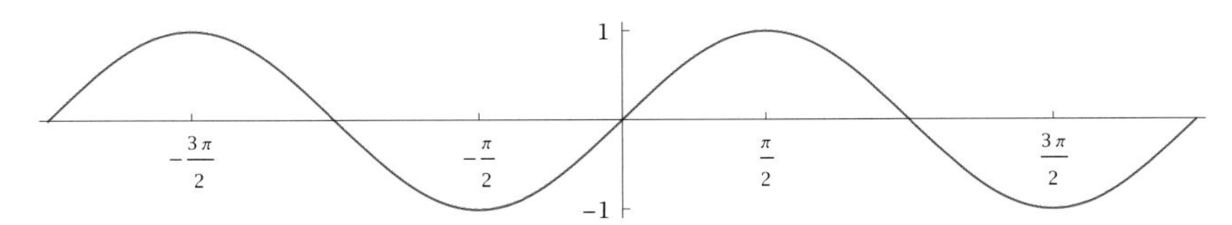

The graph of sine on the interval $[-2\pi, 2\pi]$.

Again, we need to restrict the domain to obtain a one-to-one function. We again start by declaring that 0 should be in the domain of the restricted function. Looking at the graph above of sine, we see that $[-\frac{\pi}{2}, \frac{\pi}{2}]$ is the largest interval containing 0 on which sine is one-to-one.

If we restrict the domain of sine to $[-\frac{\pi}{2}, \frac{\pi}{2}]$, we obtain the one-to-one function whose graph is shown here. The inverse of this function is called the arcsine, which is abbreviated as $\sin^{-1}$. Here is the formal definition:

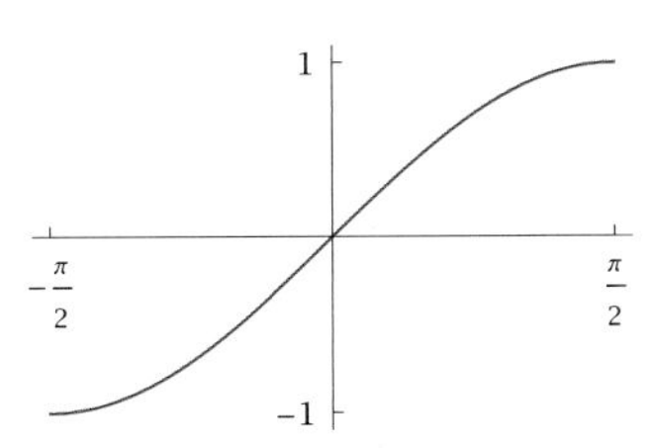

The graph of sine on the interval $[-\frac{\pi}{2}, \frac{\pi}{2}]$.

Arcsine

For t in $[-1, 1]$, the **arcsine** of t, denoted $\sin^{-1} t$, is the angle in $[-\frac{\pi}{2}, \frac{\pi}{2}]$ whose sine equals t. In other words, the equation

$$\sin^{-1} t = \theta$$

means θ is the angle in $[-\frac{\pi}{2}, \frac{\pi}{2}]$ such that

$$\sin\theta = t.$$

In defining $\sin^{-1} t$, we must restrict t to be in the interval $[-1, 1]$ because otherwise there is no angle whose sine equals t.

EXAMPLE 5

(a) Evaluate $\sin^{-1} 0$.

(b) Evaluate $\sin^{-1}(-1)$.

(c) Explain why the expression $\sin^{-1}(-3)$ makes no sense.

SOLUTION

(a) Because $\sin 0 = 0$ and 0 is in the interval $[-\frac{\pi}{2}, \frac{\pi}{2}]$, we have $\sin^{-1} 0 = 0$.

(b) Because $\sin(-\frac{\pi}{2}) = -1$ and $-\frac{\pi}{2}$ is in $[-\frac{\pi}{2}, \frac{\pi}{2}]$, we have $\sin^{-1}(-1) = -\frac{\pi}{2}$.

(c) The expression $\sin^{-1}(-3)$ makes no sense because there is no angle whose sine equals -3.

Be sure you understand that $\sin^{-1} t$ is not equal to $\frac{1}{\sin t}$.

Do not confuse $\sin^{-1} t$ with $(\sin t)^{-1}$. The same comments that were made earlier about the notation $\cos^{-1}$ apply to $\sin^{-1}$. Specifically, $\sin^2 t$ means $(\sin t)^2$, but $\sin^{-1} t$ involves an inverse function.

The next example should help solidify your understanding of the arcsine function.

EXAMPLE 6

Sketch the radius of the unit circle corresponding to the angle $\sin^{-1} 0.3$.

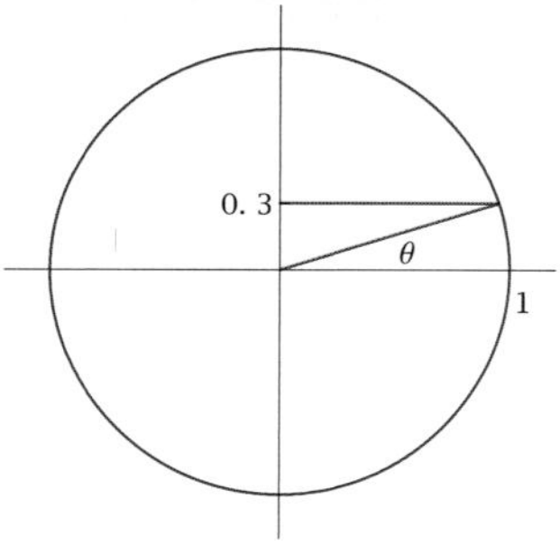

Here $\theta = \sin^{-1} 0.3$, or equivalently $\sin\theta = 0.3$.

SOLUTION We seek an angle in $[-\frac{\pi}{2}, \frac{\pi}{2}]$ whose sine equals 0.3. This means that the second coordinate of the endpoint of the corresponding radius will equal 0.3. Thus we start with 0.3 on the vertical axis, as shown here, and extend a line to the right until it intersects the unit circle. That point of intersection is the endpoint of the radius corresponding to the angle $\sin^{-1} 0.3$, as shown here.

A calculator shows that

$$\sin^{-1} 0.3 \approx 0.3047.$$

Thus the angle θ shown in the figure here is approximately 0.3047 radians, which is approximately 17.5°.

Because the inverse of a function interchanges the domain and range of the original function, we have the following:

Domain and range of arcsine

- The domain of $\sin^{-1}$ is $[-1, 1]$.
- The range of $\sin^{-1}$ is $[-\frac{\pi}{2}, \frac{\pi}{2}]$.

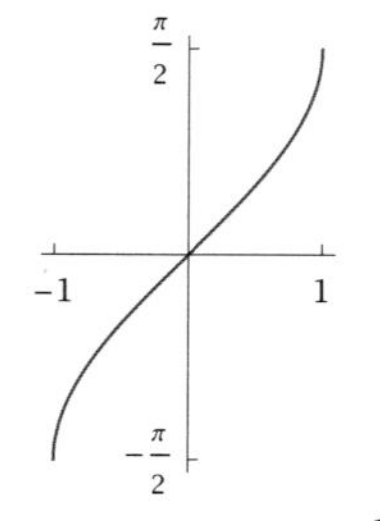

The graph of $\sin^{-1}$.

The graph of $\sin^{-1}$ can be obtained in the usual way when dealing with inverse functions. Specifically, flip the graph of the sine (restricted to the interval $[-\frac{\pi}{2}, \frac{\pi}{2}]$) across the line with slope 1 that contains the origin, getting the graph shown here.

Given the lengths of the hypotenuse and another side of a right triangle, you can use the arcsine function to determine the angle opposite the nonhypotenuse side. The example below illustrates the procedure.

EXAMPLE 7

Suppose your altitude goes up by 150 feet when driving one-half mile on a straight road. What is the angle of elevation of the road?

1/2 mile

150 feet

θ

This sketch is not drawn to scale.

SOLUTION The first step toward solving a problem like this is to sketch the situation. Thus we begin by constructing the sketch shown here, which is not drawn to scale. We want to find the angle of elevation of the road; this angle has been denoted by θ in the sketch.

As usual, we must use consistent units throughout a problem. The information we have been given uses both feet and miles. Thus we convert one-half mile to feet: because one mile equals 5280 feet, one-half mile equals 2640 feet.

Because the sine of an angle in a right triangle equals the length of the opposite side divided by the length of the hypotenuse, the sketch shows that $\sin\theta = \frac{150}{2640}$. Thus

$$\theta = \sin^{-1}\tfrac{150}{2640} \approx 0.057.$$

Hence θ is approximately 0.057 radians, which is approximately 3.3°.

The next example shows how to find the angles in an isosceles triangle, given the lengths of the sides.

EXAMPLE 8

Find the angle between the two sides of length 7 in an isosceles triangle that has one side of length 8 and two sides of length 7.

SOLUTION The trick here is create a right triangle by dropping a perpendicular from the vertex to the base, as shown in the figure below.

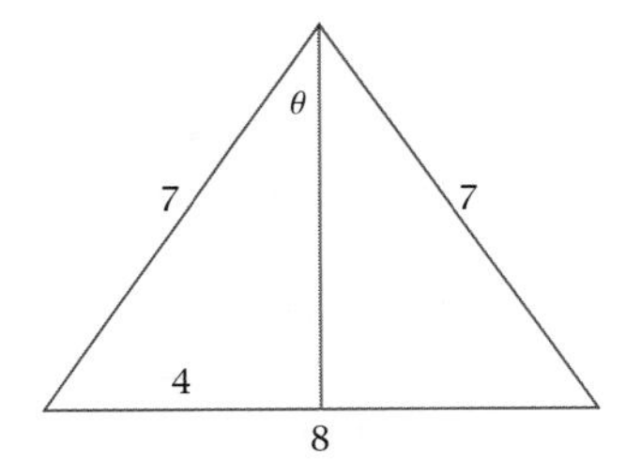

Let θ denote the angle between the perpendicular and a side of length 7. Because the base of the isosceles triangle has length 8, the side of the right triangle opposite the angle θ has length 4. Thus $\sin\theta = \frac{4}{7}$. Hence

$$\theta = \sin^{-1}\tfrac{4}{7} \approx 0.60825.$$

Thus the angle between the two sides of length 7 is approximately 1.2165 radians ($1.2165 = 2 \times 0.60825$), which is approximately 69.7°.

The Arctangent Function

Now we consider the tangent function, whose graph is shown below:

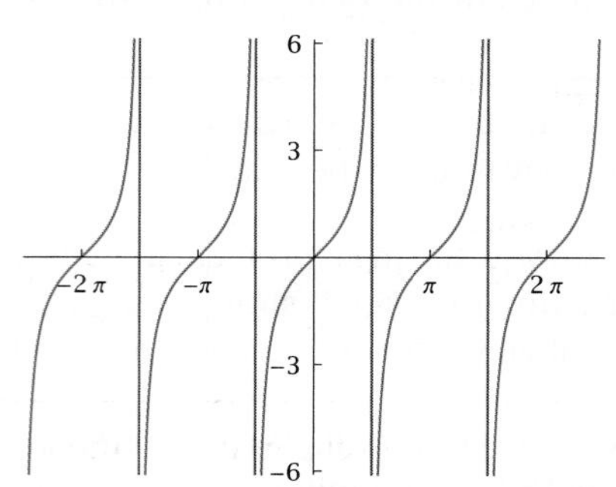

The graph of tangent on most of the interval $(-\frac{5}{2}\pi, \frac{5}{2}\pi)$*.*
Because $|\tan\theta|$ *gets very large for* θ *close to odd multiples of* $\frac{\pi}{2}$*,*
it is not possible to show the entire graph on this interval.

Again, we need to restrict the domain to obtain a one-to-one function. We again start by declaring that 0 should be in the domain of the restricted function. Looking at the graph above of tangent, we see that $(-\frac{\pi}{2}, \frac{\pi}{2})$ is the largest interval containing 0 on which tangent is one-to-one. This is an open interval that excludes the two endpoints $\frac{\pi}{2}$ and $-\frac{\pi}{2}$. Recall that the tangent function is not defined at $\frac{\pi}{2}$ or at $-\frac{\pi}{2}$; thus these numbers cannot be included in the domain.

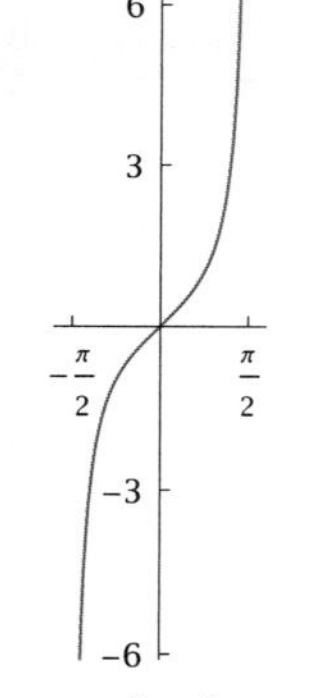

The graph of tangent on most of the interval $(-\frac{\pi}{2}, \frac{\pi}{2})$*.*

If we restrict the domain of tangent to $(-\frac{\pi}{2}, \frac{\pi}{2})$, we obtain the one-to-one function most of whose graph is shown here. The inverse of this function is called the arctangent, which is abbreviated as $\tan^{-1}$. Here is the formal definition:

Arctangent

The **arctangent** of t, denoted $\tan^{-1} t$, is the angle in $(-\frac{\pi}{2}, \frac{\pi}{2})$ whose tangent equals t. In other words, the equation

$$\tan^{-1} t = \theta$$

means θ is the angle in $(-\frac{\pi}{2}, \frac{\pi}{2})$ such that

$$\tan\theta = t.$$

EXAMPLE 9

(a) Evaluate $\tan^{-1} 0$.

(b) Evaluate $\tan^{-1} 1$.

(c) Evaluate $\tan^{-1} \sqrt{3}$.

SOLUTION

(a) Because $\tan 0 = 0$ and 0 is in the interval $(-\frac{\pi}{2}, \frac{\pi}{2})$, we have $\tan^{-1} 0 = 0$.

(b) Because $\tan \frac{\pi}{4} = 1$ and $\frac{\pi}{4}$ is in the interval $(-\frac{\pi}{2}, \frac{\pi}{2})$, we have $\tan^{-1} 1 = \frac{\pi}{4}$.

(c) Because $\tan \frac{\pi}{3} = \sqrt{3}$ and $\frac{\pi}{3}$ is in the interval $(-\frac{\pi}{2}, \frac{\pi}{2})$, we have $\tan^{-1} \sqrt{3} = \frac{\pi}{3}$.

Do not confuse $\tan^{-1} t$ with $(\tan t)^{-1}$. The same comments that were made earlier about the notation $\cos^{-1}$ and $\sin^{-1}$ apply to $\tan^{-1}$. Specifically, $\tan^2 t$ means $(\tan t)^2$, but $\tan^{-1} t$ involves an inverse function.

Be sure you understand that $\tan^{-1} t$ is not equal to $\frac{1}{\tan t}$.

EXAMPLE 10

Sketch the radius of the unit circle corresponding to the angle $\tan^{-1}(-3)$.

SOLUTION We seek an angle in $(-\frac{\pi}{2}, \frac{\pi}{2})$ whose tangent equals -3. This means that the slope of the corresponding radius will equal -3. The unit circle has two radii with slope -3; one of them is the radius shown here and the other is the radius in the opposite direction. But of these two radii, only the one shown here has a corresponding angle in the interval $(-\frac{\pi}{2}, \frac{\pi}{2})$. Notice that the indicated angle is negative because of the clockwise direction of the arrow.

A calculator shows that

$$\tan^{-1}(-3) \approx -1.249.$$

Thus the angle θ shown here is approximately -1.249 radians, which is approximately $-71.6°$.

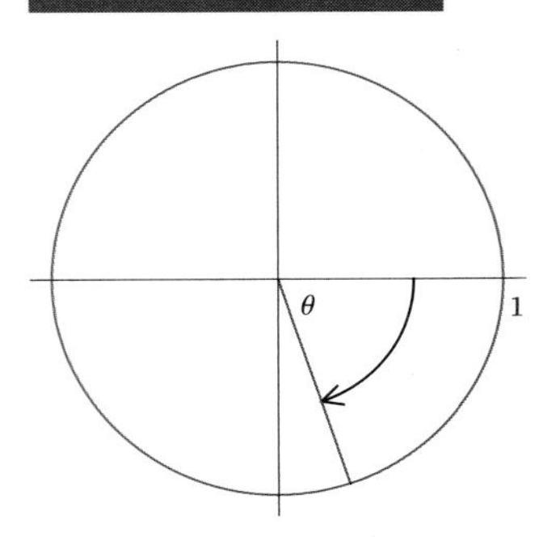

This radius has slope -3 and thus corresponds to $\tan^{-1}(-3)$.

Unlike $\cos^{-1} t$ and $\sin^{-1} t$, which make sense only when t is in $[-1, 1]$, $\tan^{-1} t$ makes sense for every real number t (because for every real number t there is an angle whose tangent equals t). Because the inverse of a function interchanges the domain and range of the original function, we have the following:

Domain and range of arctangent

- The domain of $\tan^{-1}$ is the set of real numbers.
- The range of $\tan^{-1}$ is $(-\frac{\pi}{2}, \frac{\pi}{2})$.

The graph of $\tan^{-1}$ can be obtained in the usual way when dealing with inverse functions. Specifically, flip the graph of the tangent (restricted to an interval slightly smaller than $(-\frac{\pi}{2}, \frac{\pi}{2})$) across the line with slope 1 that contains the origin, getting the graph shown below.

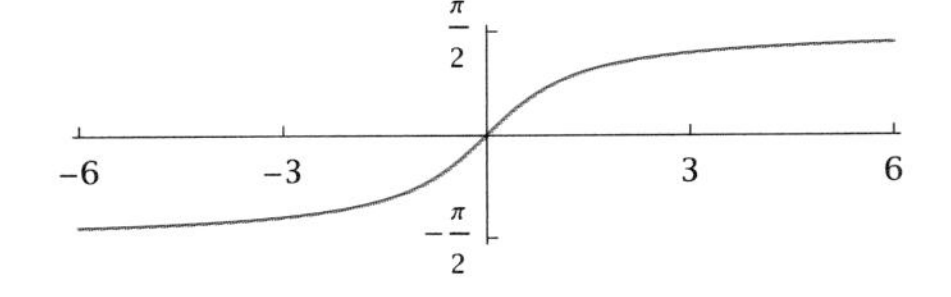

The graph of $\tan^{-1}$ on the interval $[-6, 6]$.

Given the lengths of the two nonhypotenuse sides of a right triangle, you can use the arctangent function to determine the angles of the triangle. The example below illustrates the procedure.

EXAMPLE 11

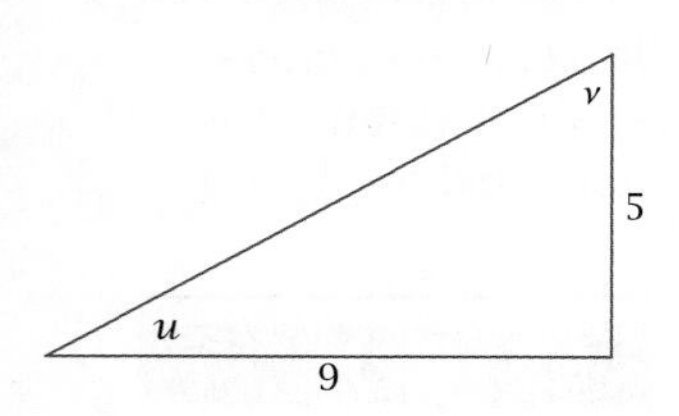

(a) In this right triangle, use the arctangent function to evaluate the angle u.

(b) In this right triangle, use the arctangent function to evaluate the angle v.

(c) As a check, compute the sum of the angles u and v obtained in parts (a) and (b). Does this sum have the expected value?

SOLUTION

(a) Because the tangent of an angle in a right triangle equals the length of the opposite side divided by the length of the adjacent side, we have $\tan u = \frac{5}{9}$. Thus

$$u = \tan^{-1} \tfrac{5}{9} \approx 0.5071.$$

Hence u is approximately 0.5071 radians, which is approximately 29.1°.

(b) Because the tangent of an angle in a right triangle equals the length of the opposite side divided by the length of the adjacent side, we have $\tan v = \frac{9}{5}$. Thus

$$v = \tan^{-1} \tfrac{9}{5} \approx 1.0637.$$

Hence v is approximately 1.0637 radians, which is approximately 60.9°.

(c) We have

$$u + v \approx 29.1^\circ + 60.9^\circ = 90^\circ.$$

Thus the sum of the two acute angles in this right triangle is 90°, as expected.

Given the slope of a line, the arctangent function allows us to find the angle the line makes with the positive horizontal axis, as shown in the next example.

EXAMPLE 12

What angle does the line $y = \frac{2}{3}x$ in the xy-plane make with the positive x-axis?

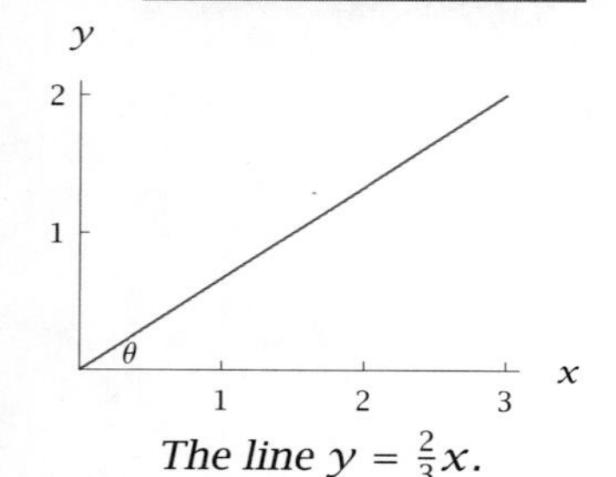

The line $y = \frac{2}{3}x$.

SOLUTION We seek the angle θ shown in the figure here. Because the line $y = \frac{2}{3}x$ has slope $\frac{2}{3}$, we have $\tan\theta = \frac{2}{3}$. Thus

$$\theta = \tan^{-1} \tfrac{2}{3} \approx 0.588.$$

Hence θ is approximately 0.588 radians, which is approximately 33.7°.

EXERCISES

You should be able to do Exercises 1–4 without using a calculator.

1. Evaluate $\cos^{-1} \frac{1}{2}$.
2. Evaluate $\sin^{-1} \frac{1}{2}$.
3. Evaluate $\tan^{-1}(-1)$.
4. Evaluate $\tan^{-1}(-\sqrt{3})$.

Exercises 5–16 emphasize the importance of understanding inverse notation as well as the importance of parentheses in determining the order of operations.

5. For $x = 0.3$, evaluate each of the following:
   (a) $\cos^{-1} x$ (c) $\cos(x^{-1})$
   (b) $(\cos x)^{-1}$ (d) $(\cos^{-1} x)^{-1}$
6. For $x = 0.4$, evaluate each of the following:
   (a) $\cos^{-1} x$ (c) $\cos(x^{-1})$
   (b) $(\cos x)^{-1}$ (d) $(\cos^{-1} x)^{-1}$
7. For $x = \frac{1}{7}$, evaluate each of the following:
   (a) $\sin^{-1} x$ (c) $\sin(x^{-1})$
   (b) $(\sin x)^{-1}$ (d) $(\sin^{-1} x)^{-1}$
8. For $x = \frac{1}{8}$, evaluate each of the following:
   (a) $\sin^{-1} x$ (c) $\sin(x^{-1})$
   (b) $(\sin x)^{-1}$ (d) $(\sin^{-1} x)^{-1}$
9. For $x = 2$, evaluate each of the following:
   (a) $\tan^{-1} x$ (c) $\tan(x^{-1})$
   (b) $(\tan x)^{-1}$ (d) $(\tan^{-1} x)^{-1}$
10. For $x = 3$, evaluate each of the following:
    (a) $\tan^{-1} x$ (c) $\tan(x^{-1})$
    (b) $(\tan x)^{-1}$ (d) $(\tan^{-1} x)^{-1}$
11. For $x = 4$, evaluate each of the following:
    (a) $(\cos(x^{-1}))^{-1}$ (c) $(\cos^{-1}(x^{-1}))^{-1}$
    (b) $\cos^{-1}(x^{-1})$
12. For $x = 5$, evaluate each of the following:
    (a) $(\cos(x^{-1}))^{-1}$ (c) $(\cos^{-1}(x^{-1}))^{-1}$
    (b) $\cos^{-1}(x^{-1})$
13. For $x = 6$, evaluate each of the following:
    (a) $(\sin(x^{-1}))^{-1}$ (c) $(\sin^{-1}(x^{-1}))^{-1}$
    (b) $\sin^{-1}(x^{-1})$
14. For $x = 9$, evaluate each of the following:
    (a) $(\sin(x^{-1}))^{-1}$ (c) $(\sin^{-1}(x^{-1}))^{-1}$
    (b) $\sin^{-1}(x^{-1})$
15. For $x = 0.1$, evaluate each of the following:
    (a) $(\tan(x^{-1}))^{-1}$ (c) $(\tan^{-1}(x^{-1}))^{-1}$
    (b) $\tan^{-1}(x^{-1})$
16. For $x = 0.2$, evaluate each of the following:
    (a) $(\tan(x^{-1}))^{-1}$ (c) $(\tan^{-1}(x^{-1}))^{-1}$
    (b) $\tan^{-1}(x^{-1})$

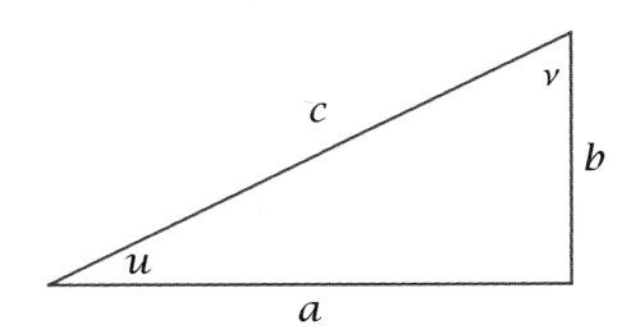

Use the right triangle above for Exercises 17–24. This triangle is not drawn to scale corresponding to the data in the exercises.

17. Suppose $a = 2$ and $c = 3$. Evaluate u in radians.
18. Suppose $a = 3$ and $c = 4$. Evaluate u in radians.
19. Suppose $a = 2$ and $c = 5$. Evaluate v in radians.
20. Suppose $a = 3$ and $c = 5$. Evaluate v in radians.
21. Suppose $a = 5$ and $b = 4$. Evaluate u in degrees.
22. Suppose $a = 5$ and $b = 6$. Evaluate u in degrees.
23. Suppose $a = 5$ and $b = 7$. Evaluate v in degrees.
24. Suppose $a = 7$ and $b = 6$. Evaluate v in degrees.
25. Find the angle between the two sides of length 9 in an isosceles triangle that has one side of length 14 and two sides of length 9.
26. Find the angle between the two sides of length 8 in an isosceles triangle that has one side of length 7 and two sides of length 8.
27. Find the angle between a side of length 6 and the side with length 10 in an isosceles triangle that has one side of length 10 and two sides of length 6.

28. Find the angle between a side of length 5 and the side with length 9 in an isosceles triangle that has one side of length 9 and two sides of length 5.
29. Find the smallest positive number θ such that $10^{\cos\theta} = 6$.
30. Find the smallest positive number θ such that $10^{\sin\theta} = 7$.
31. Find the smallest positive number θ such that $e^{\tan\theta} = 15$.
32. Find the smallest positive number θ such that $e^{\tan\theta} = 500$.
33. Find the smallest positive number y such that $\cos(\tan y) = 0.2$.
34. Find the smallest positive number y such that $\sin(\tan y) = 0.6$.
35. Find the smallest positive number x such that
$$\sin^2 x - 3\sin x + 1 = 0.$$
36. Find the smallest positive number x such that
$$\sin^2 x - 4\sin x + 2 = 0.$$
37. Find the smallest positive number x such that
$$\cos^2 x - 0.5\cos x + 0.06 = 0.$$
38. Find the smallest positive number x such that
$$\cos^2 x - 0.7\cos x + 0.12 = 0.$$
39. Find the smallest positive number θ such that $\sin\theta = -0.4$.
[*Hint:* Careful, the answer is not $\sin^{-1}(-0.4)$.]
40. Find the smallest positive number θ such that $\tan\theta = -5$.
[*Hint:* Careful, the answer is not $\tan^{-1}(-5)$.]
41. What angle does the line $y = \frac{2}{5}x$ in the xy-plane make with the positive x-axis?
42. What angle does the line $y = 4x$ in the xy-plane make with the positive x-axis?
43. What is the angle between the positive horizontal axis and the line containing the points $(3, 1)$ and $(5, 4)$?
44. What is the angle between the positive horizontal axis and the line containing the points $(2, 5)$ and $(6, 2)$?

For Exercises 45–48: Hilly areas often have road signs giving the percentage grade for the road. A 5% grade, for example, means that the altitude changes by 5 feet for each 100 feet of horizontal distance.

45. What percentage grade should be put on a road sign where the angle of elevation of the road is 3°?
46. What percentage grade should be put on a road sign where the angle of elevation of the road is 4°?
47. Suppose an uphill road sign indicates a road grade of 6%. What is the angle of elevation of the road?
48. Suppose an uphill road sign indicates a road grade of 8%. What is the angle of elevation of the road?

PROBLEMS

Some problems require considerably more thought than the exercises.
Unlike exercises, problems often have more than one correct answer.

49. Explain why
$$\cos^{-1}\tfrac{3}{5} = \sin^{-1}\tfrac{4}{5} = \tan^{-1}\tfrac{4}{3}.$$
[*Hint:* Take $a = 3$ and $b = 4$ in the triangle used for Exercises 17–24. Then find c and consider various ways to express u.]
50. Explain why
$$\cos^{-1}\tfrac{5}{13} = \sin^{-1}\tfrac{12}{13} = \tan^{-1}\tfrac{12}{5}.$$
51. Suppose a and b are numbers such that
$$\cos^{-1} a = \tfrac{\pi}{7} \quad \text{and} \quad \sin^{-1} b = \tfrac{\pi}{7}.$$
Explain why $a^2 + b^2 = 1$.
52. Without using a calculator, sketch the unit circle and the radius corresponding to $\cos^{-1} 0.1$.
53. Without using a calculator, sketch the unit circle and the radius corresponding to $\sin^{-1}(-0.1)$.

54. Without using a calculator, sketch the unit circle and the radius corresponding to $\tan^{-1} 4$.

55. Find all numbers t such that

$$\cos^{-1} t = \sin^{-1} t.$$

56. There exist angles θ such that $\cos\theta = -\sin\theta$ (for example, $-\frac{\pi}{4}$ and $\frac{3\pi}{4}$ are two such angles). However, explain why there do not exist any numbers t such that

$$\cos^{-1} t = -\sin^{-1} t.$$

57. Show that in an isosceles triangle with two sides of length b and a side of length c, the angle between the two sides of length b is

$$2\sin^{-1}\frac{c}{2b}.$$

58. Show that in an isosceles triangle with two sides of length b and a side of length c, the angle between a side of length b and the side of length c is

$$\cos^{-1}\frac{c}{2b}.$$

59. Suppose you are asked to find the angles in an isosceles triangle that has two sides of length 5 and one side of length 11. Using the two previous problems, this would require you to compute $\sin^{-1}\frac{11}{10}$ and $\cos^{-1}\frac{11}{10}$, neither of which makes sense. What is wrong here?

WORKED-OUT SOLUTIONS *to Odd-numbered Exercises*

Do not read these worked-out solutions before first attempting to do the exercises yourself. Otherwise you may merely mimic the techniques shown here without understanding the ideas.

Best way to learn: Carefully read the section of the textbook, then do all the odd-numbered exercises (even if they have not been assigned) and check your answers here. If you get stuck on an exercise, reread the section of the textbook—then try the exercise again. If you are still stuck, then look at the worked-out solution here.

You should be able to do Exercises 1–4 without using a calculator.

1. Evaluate $\cos^{-1}\frac{1}{2}$.

SOLUTION $\cos\frac{\pi}{3} = \frac{1}{2}$; thus $\cos^{-1}\frac{1}{2} = \frac{\pi}{3}$.

3. Evaluate $\tan^{-1}(-1)$.

SOLUTION $\tan(-\frac{\pi}{4}) = -1$; thus

$$\tan^{-1}(-1) = -\frac{\pi}{4}.$$

Exercises 5–16 emphasize the importance of understanding inverse notation as well as the importance of parentheses in determining the order of operations.

5. For $x = 0.3$, evaluate each of the following:

(a) $\cos^{-1} x$

(b) $(\cos x)^{-1}$

(c) $\cos(x^{-1})$

(d) $(\cos^{-1} x)^{-1}$

SOLUTION

(a) $\cos^{-1} 0.3 \approx 1.2661$

(b) $(\cos 0.3)^{-1} = \frac{1}{\cos 0.3} \approx 1.04675$

(c) $\cos(0.3^{-1}) = \cos\frac{1}{0.3} \approx -0.981674$

(d) $(\cos^{-1} 0.3)^{-1} = \frac{1}{\cos^{-1} 0.3} \approx 0.789825$

7. For $x = \frac{1}{7}$, evaluate each of the following:

(a) $\sin^{-1} x$

(b) $(\sin x)^{-1}$

(c) $\sin(x^{-1})$

(d) $(\sin^{-1} x)^{-1}$

SOLUTION

(a) $\sin^{-1}\frac{1}{7} \approx 0.143348$

(b) $(\sin\frac{1}{7})^{-1} = \frac{1}{\sin\frac{1}{7}} \approx 7.02387$

(c) $\sin\left((\frac{1}{7})^{-1}\right) = \sin 7 \approx 0.656987$

(d) $\left(\sin^{-1}\frac{1}{7}\right)^{-1} = \frac{1}{\sin^{-1}\frac{1}{7}} \approx 6.97605$

9. For $x = 2$, evaluate each of the following:

(a) $\tan^{-1} x$ (c) $\tan(x^{-1})$

(b) $(\tan x)^{-1}$ (d) $(\tan^{-1} x)^{-1}$

SOLUTION

(a) $\tan^{-1} 2 \approx 1.10715$

(b) $(\tan 2)^{-1} = \frac{1}{\tan 2} \approx -0.457658$

(c) $\tan(2^{-1}) = \tan \frac{1}{2} \approx 0.546302$

(d) $(\tan^{-1} 2)^{-1} = \frac{1}{\tan^{-1} 2} \approx 0.903221$

11. For $x = 4$, evaluate each of the following:

(a) $(\cos(x^{-1}))^{-1}$ (c) $(\cos^{-1}(x^{-1}))^{-1}$

(b) $\cos^{-1}(x^{-1})$

SOLUTION

(a) $(\cos(4^{-1}))^{-1} = (\cos \frac{1}{4})^{-1} = \frac{1}{\cos \frac{1}{4}} \approx 1.03209$

(b) $\cos^{-1}(4^{-1}) = \cos^{-1} \frac{1}{4} \approx 1.31812$

(c)
$$\begin{aligned}(\cos^{-1}(4^{-1}))^{-1} &= (\cos^{-1} \tfrac{1}{4})^{-1}\\ &= \frac{1}{\cos^{-1} \frac{1}{4}}\\ &\approx 0.758659\end{aligned}$$

13. For $x = 6$, evaluate each of the following:

(a) $(\sin(x^{-1}))^{-1}$ (c) $(\sin^{-1}(x^{-1}))^{-1}$

(b) $\sin^{-1}(x^{-1})$

SOLUTION

(a) $(\sin(6^{-1}))^{-1} = (\sin \frac{1}{6})^{-1} = \frac{1}{\sin \frac{1}{6}} \approx 6.02787$

(b) $\sin^{-1}(6^{-1}) = \sin^{-1} \frac{1}{6} \approx 0.167448$

(c)
$$\begin{aligned}(\sin^{-1}(6^{-1}))^{-1} &= (\sin^{-1} \tfrac{1}{6})^{-1}\\ &= \frac{1}{\sin^{-1} \frac{1}{6}}\\ &\approx 5.972\end{aligned}$$

15. For $x = 0.1$, evaluate each of the following:

(a) $(\tan(x^{-1}))^{-1}$ (c) $(\tan^{-1}(x^{-1}))^{-1}$

(b) $\tan^{-1}(x^{-1})$

SOLUTION

(a) $(\tan(0.1^{-1}))^{-1} = (\tan 10)^{-1} = \frac{1}{\tan 10} \approx 1.54235$

(b) $\tan^{-1}(0.1^{-1}) = \tan^{-1} 10 \approx 1.47113$

(c)
$$\begin{aligned}(\tan^{-1}(0.1^{-1}))^{-1} &= (\tan^{-1} 10)^{-1}\\ &= \frac{1}{\tan^{-1} 10}\\ &\approx 0.679751\end{aligned}$$

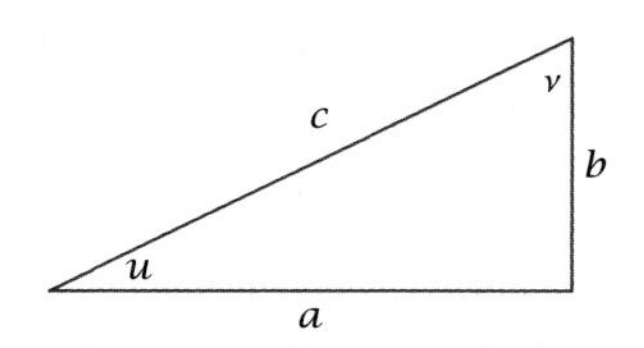

Use the right triangle above for Exercises 17–24. This triangle is not drawn to scale corresponding to the data in the exercises.

17. Suppose $a = 2$ and $c = 3$. Evaluate u in radians.

SOLUTION Because the cosine of an angle in a right triangle equals the length of the adjacent side divided by the length of the hypotenuse, we have $\cos u = \frac{2}{3}$. Using a calculator working in radians, we then have

$$u = \cos^{-1} \tfrac{2}{3} \approx 0.841 \text{ radians}.$$

19. Suppose $a = 2$ and $c = 5$. Evaluate v in radians.

SOLUTION Because the sine of an angle in a right triangle equals the length of the opposite side divided by the length of the hypotenuse, we have $\sin v = \frac{2}{5}$. Using a calculator working in radians, we then have

$$v = \sin^{-1} \tfrac{2}{5} \approx 0.412 \text{ radians}.$$

21. Suppose $a = 5$ and $b = 4$. Evaluate u in degrees.

SOLUTION Because the tangent of an angle in a right triangle equals the length of the opposite side divided by the length of the adjacent side, we have $\tan u = \frac{4}{5}$. Using a calculator working in degrees, we then have

$$u = \tan^{-1} \tfrac{4}{5} \approx 38.7^\circ.$$

23. Suppose $a = 5$ and $b = 7$. Evaluate v in degrees.

SOLUTION Because the tangent of an angle in a right triangle equals the length of the opposite side divided by the length of the adjacent side, we have $\tan \nu = \frac{5}{7}$. Using a calculator working in degrees, we then have

$$\nu = \tan^{-1} \tfrac{5}{7} \approx 35.5°.$$

25. Find the angle between the two sides of length 9 in an isosceles triangle that has one side of length 14 and two sides of length 9.

SOLUTION Create a right triangle by dropping a perpendicular from the vertex to the base, as shown in the figure below.

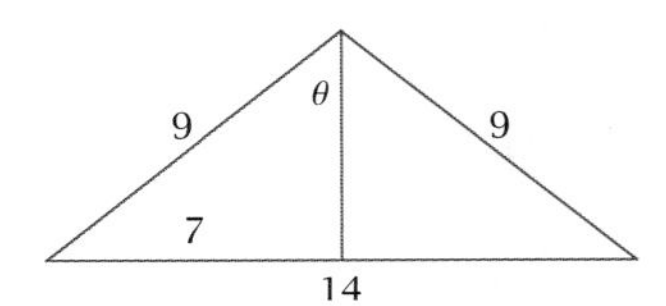

Let θ denote the angle between the perpendicular and a side of length 9. Because the base of the isosceles triangle has length 14, the side of the right triangle opposite the angle θ has length 7. Thus $\sin \theta = \frac{7}{9}$. Hence

$$\theta = \sin^{-1} \tfrac{7}{9} \approx 0.8911.$$

Thus the angle between the two sides of length 9 is approximately 1.7822 radians ($1.7822 = 2 \times 0.8911$), which is approximately 102.1°.

27. Find the angle between a side of length 6 and the side with length 10 in an isosceles triangle that has one side of length 10 and two sides of length 6.

SOLUTION Create a right triangle by dropping a perpendicular from the vertex to the base, as shown in the figure below.

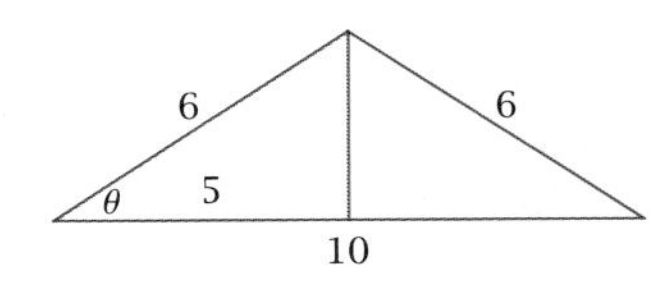

Let θ denote the angle between a side of length 10 and a side of length 6. Because the base of the isosceles triangle has length 10, the side of the right triangle adjacent to the angle θ has length 5. Thus $\cos \theta = \frac{5}{6}$. Hence

$$\theta = \cos^{-1} \tfrac{5}{6} \approx 0.58569.$$

Thus the angle between a side of length 6 and the side with length 10 is approximately 0.58569 radians, which is approximately 33.6°.

29. Find the smallest positive number θ such that $10^{\cos \theta} = 6$.

SOLUTION The equation above implies that $\cos \theta = \log 6$. Thus we take $\theta = \cos^{-1}(\log 6) \approx 0.67908$.

31. Find the smallest positive number θ such that $e^{\tan \theta} = 15$.

SOLUTION The equation above implies that $\tan \theta = \ln 15$. Thus we take $\theta = \tan^{-1}(\ln 15) \approx 1.21706$.

33. Find the smallest positive number y such that $\cos(\tan y) = 0.2$.

SOLUTION The equation above implies that we should choose $\tan y = \cos^{-1} 0.2 \approx 1.36944$. Thus we should choose $y \approx \tan^{-1} 1.36944 \approx 0.94007$.

35. Find the smallest positive number x such that

$$\sin^2 x - 3 \sin x + 1 = 0.$$

SOLUTION Write $y = \sin x$. Then the equation above can be rewritten as

$$y^2 - 3y + 1 = 0.$$

Using the quadratic formula, we find that the solutions to this equation are

$$y = \frac{3 + \sqrt{5}}{2} \approx 2.61803$$

and

$$y = \frac{3 - \sqrt{5}}{2} \approx 0.38197.$$

Thus $\sin x \approx 2.61803$ or $\sin x \approx 0.381966$. However, there is no real number x such that $\sin x \approx 2.61803$ (because $\sin x$ is at most 1 for every real number x), and thus we must have $\sin x \approx 0.381966$. Thus $x \approx \sin^{-1} 0.381966 \approx 0.39192$.

37. Find the smallest positive number x such that

$$\cos^2 x - 0.5\cos x + 0.06 = 0.$$

SOLUTION Write $y = \cos x$. Then the equation above can be rewritten as

$$y^2 - 0.5y + 0.06 = 0.$$

Using the quadratic formula or factorization, we find that the solutions to this equation are

$$y = 0.2 \quad \text{and} \quad y = 0.3.$$

Thus $\cos x = 0.2$ or $\cos x = 0.3$, which suggests that we choose $x = \cos^{-1} 0.2$ or $x = \cos^{-1} 0.3$. Because arccosine is a decreasing function, $\cos^{-1} 0.3$ is smaller than $\cos^{-1} 0.2$. Because we want to find the smallest positive value of x satisfying the original equation, we choose $x = \cos^{-1} 0.3 \approx 1.2661$.

39. Find the smallest positive number θ such that $\sin\theta = -0.4$.
[*Hint:* Careful, the answer is not $\sin^{-1}(-0.4)$.]

SOLUTION The answer is not $\sin^{-1}(-0.4)$ because $\sin^{-1}(-0.4)$ is a negative number and we need to find the smallest positive number θ such that $\sin\theta = -0.4$. In the figure below, the blue radius corresponds to the negative angle $\sin^{-1}(-0.4)$. The arrow and the red radius show the angle we seek, which is $\pi + \sin^{-1} 0.4$, which is approximately 3.55311 radians (which is approximately 203.6°).

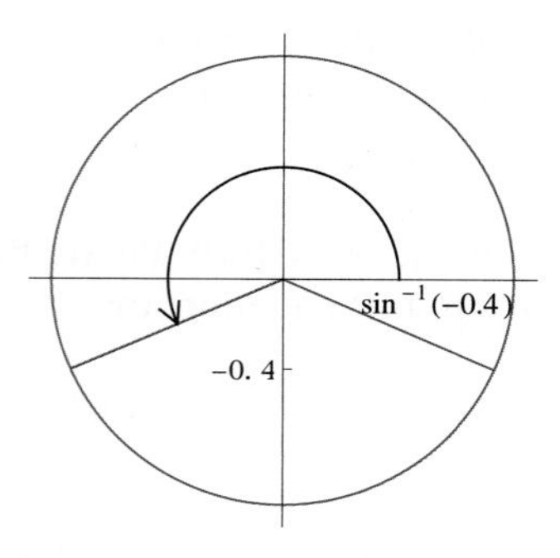

41. What angle does the line $y = \frac{2}{5}x$ in the xy-plane make with the positive x-axis?

SOLUTION We seek the angle θ shown in the figure here. Because the line $y = \frac{2}{5}x$ has slope $\frac{2}{5}$, we have $\tan\theta = \frac{2}{5}$. Thus

$$\theta = \tan^{-1}\tfrac{2}{5} \approx 0.3805.$$

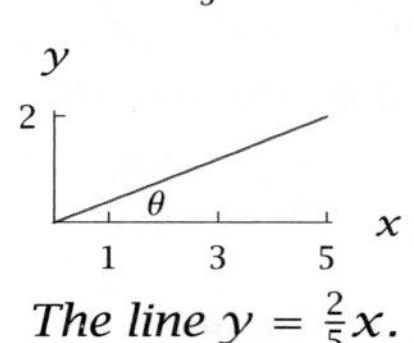

The line $y = \frac{2}{5}x$.

Hence θ is approximately 0.3805 radians, which is approximately 21.8°.

43. What is the angle between the positive horizontal axis and the line containing the points $(3, 1)$ and $(5, 4)$?

SOLUTION Let θ denote the angle between the positive horizontal axis and the line containing $(3, 1)$ and $(5, 4)$. The line containing $(3, 1)$ and $(5, 4)$ has slope $\frac{4-1}{5-3}$, which equals $\frac{3}{2}$. Thus $\tan\theta = \frac{3}{2}$. Thus

$$\theta = \tan^{-1}\tfrac{3}{2} \approx 0.982794.$$

Hence θ is approximately 0.982794 radians, which is approximately 56.3°.

For Exercises 45–48: Hilly areas often have road signs giving the percentage grade for the road. A 5% grade, for example, means that the altitude changes by 5 feet for each 100 feet of horizontal distance.

45. What percentage grade should be put on a road sign where the angle of elevation of the road is 3°?

SOLUTION The grade of a portion of a road is the change in altitude divided by the change in horizontal distance. Thus the grade is the slope of the road. A road with a 3° angle of elevation has slope $\tan 3°$, which is approximately 0.052. Thus a road sign should indicate a 5% grade.

47. Suppose an uphill road sign indicates a road grade of 6%. What is the angle of elevation of the road?

SOLUTION Let θ denote the angle of elevation of the road. A road grade of 6% means that $\tan\theta = 0.06$. Thus

$$\theta = \tan^{-1} 0.06 \approx 0.0599.$$

Hence θ is approximately 0.0599 radians, which is approximately 3.4°.

10.2 *Inverse Trigonometric Identities*

LEARNING OBJECTIVES

By the end of this section you should be able to

- compute the composition, in either order, of a trigonometric function and its inverse function;
- use the inverse trigonometric identities for $-t$;
- use the identity for $\tan^{-1} \frac{1}{t}$;
- compute the composition of a trigonometric function with the inverse of a different trigonometric function.

Inverse trigonometric identities are identities that involve inverse trigonometric functions.

Composition of Trigonometric Functions and Their Inverses

Recall that if f is a one-to-one function, then $f \circ f^{-1}$ is the identity function on the range of f, meaning that $f(f^{-1}(t)) = t$ for every t in the range of f. In the case of the trigonometric functions (or more precisely, the trigonometric functions restricted to the appropriate domain) and their inverses, this gives the following set of equations:

Trigonometric functions composed with their inverses

$$\cos(\cos^{-1} t) = t \quad \text{for every } t \text{ in } [-1, 1]$$

$$\sin(\sin^{-1} t) = t \quad \text{for every } t \text{ in } [-1, 1]$$

$$\tan(\tan^{-1} t) = t \quad \text{for every real number } t$$

For example, $\cos(\cos^{-1} 0.29) = 0.29$.

The left sides of the first two equations above make no sense unless t is in $[-1, 1]$ because $\cos^{-1}$ and the $\sin^{-1}$ are only defined on the interval $[-1, 1]$.

Recall also that if f is a one-to-one function, then $f^{-1} \circ f$ is the identity function on the domain of f, meaning that $f^{-1}(f(\theta)) = \theta$ for every θ in the domain of f. In the case of the trigonometric functions (or more precisely, the trigonometric functions restricted to the appropriate domain) and their inverses, this gives the following set of equations:

Inverse trigonometric functions composed with their inverses

$$\cos^{-1}(\cos\theta) = \theta \quad \text{for every } \theta \text{ in } [0, \pi]$$

$$\sin^{-1}(\sin\theta) = \theta \quad \text{for every } \theta \text{ in } [-\tfrac{\pi}{2}, \tfrac{\pi}{2}]$$

$$\tan^{-1}(\tan\theta) = \theta \quad \text{for every } \theta \text{ in } (-\tfrac{\pi}{2}, \tfrac{\pi}{2})$$

For example, $\cos^{-1}(\cos \frac{\pi}{17}) = \frac{\pi}{17}$.

The next example shows why the restrictions above on θ are necessary.

EXAMPLE 1 Evaluate $\cos^{-1}(\cos(2\pi))$.

SOLUTION The key point here is that the equation $\cos^{-1}(\cos\theta) = \theta$ is not valid here because 2π is not in the allowable range for θ. However, we can evaluate this expression directly. Because $\cos(2\pi) = 1$, we have

$$\cos^{-1}(\cos(2\pi)) = \cos^{-1} 1 = 0.$$

The example above shows that $\cos^{-1}(\cos\theta)$ does not equal θ if $\theta = 2\pi$. The next example shows how to deal with these compositions when θ is not in the required range.

EXAMPLE 2 Evaluate $\sin^{-1}(\sin 380°)$.

Here $\sin 380°$ is rewritten as $\sin\frac{\pi}{9}$ because we need to work in radians and we need an angle in the interval $[-\frac{\pi}{2}, \frac{\pi}{2}]$ so that we can apply the identity involving the composition of $\sin^{-1}$ and sin.

SOLUTION Note that $\sin 380° = \sin 20° = \sin\frac{\pi}{9}$. Thus

$$\sin^{-1}(\sin 380°) = \sin^{-1}(\sin\tfrac{\pi}{9}) = \tfrac{\pi}{9}.$$

REMARK Here is why the solution is $\frac{\pi}{9}$ rather than 20°: The input to a trigonometric function has units of either degrees or radians; if no units are specified then we assume that the units are radians. However, the output of a trigonometric function is a number with no units. Thus $\sin 380°$, which equals $\sin\frac{\pi}{9}$, is a number with no units. When this number is input into the $\sin^{-1}$ function, the $\sin^{-1}$ function does not know whether this number arose from a computation involving degrees or a computation involving radians. The $\sin^{-1}$ function produces the angle (in radians!) whose sine equals the given input.

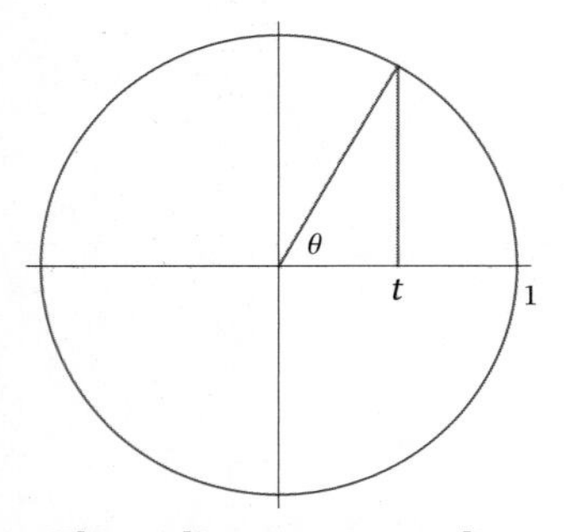

The radius corresponding to $\cos^{-1} t$.

The Arccosine, Arcsine, and Arctangent of $-t$: Graphical Approach

We begin by finding a formula for $\cos^{-1}(-t)$ in terms of $\cos^{-1} t$. To do this, suppose $0 < t < 1$. Let $\theta = \cos^{-1} t$, which implies that $\cos\theta = t$. Consider the radius of the unit circle corresponding to θ. The first coordinate of the endpoint of this radius will equal t, as shown in the first figure here.

To find $\cos^{-1}(-t)$, we need to find a radius whose first coordinate equals $-t$. To do this, flip the radius above across the vertical axis, giving the second figure here.

From the second figure, we see that the radius whose endpoint has first coordinate equal to $-t$ forms an angle of θ with the negative horizontal axis; thus this radius corresponds to $\pi - \theta$. In other words, we have $\cos^{-1}(-t) = \pi - \theta$, which we can rewrite as

$$\cos^{-1}(-t) = \pi - \cos^{-1} t.$$

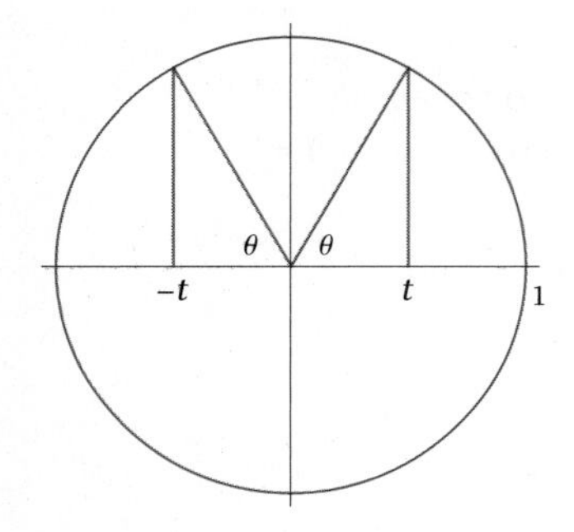

Note that $\pi - \theta$ is in $[0, \pi]$ whenever θ is in $[0, \pi]$. Thus $\pi - \cos^{-1} t$ is in the right interval to be the arccosine of some number.

EXAMPLE 3

Evaluate $\cos^{-1}(-\cos\frac{\pi}{7})$.

SOLUTION Using the formula above with $t = \cos\frac{\pi}{7}$, we have

$$\begin{aligned}\cos^{-1}(-\cos\tfrac{\pi}{7}) &= \pi - \cos^{-1}(\cos\tfrac{\pi}{7})\\ &= \pi - \tfrac{\pi}{7}\\ &= \tfrac{6\pi}{7}.\end{aligned}$$

We now turn to the problem of finding a formula for $\sin^{-1}(-t)$ in terms of $\sin^{-1} t$. To do this, suppose $0 < t < 1$. Let $\theta = \sin^{-1} t$, which implies that $\sin\theta = t$. Consider the radius of the unit circle corresponding to θ. The second coordinate of the endpoint of this radius will equal t, as shown here.

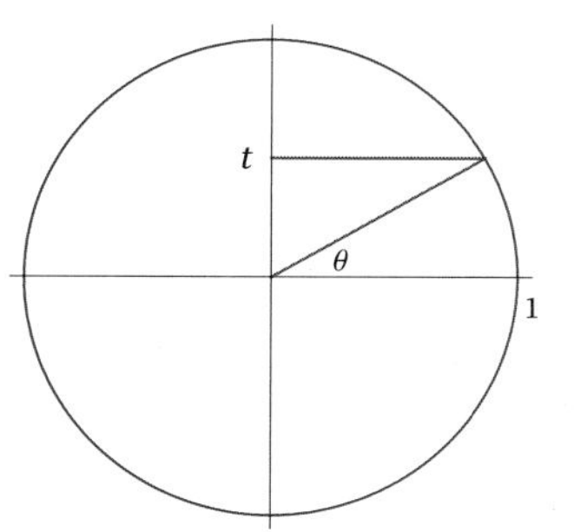

The radius corresponding to $\sin^{-1} t$.

To find $\sin^{-1}(-t)$, we need to find a radius whose second coordinate equals $-t$. To do this, flip the radius above across the horizontal axis, giving the figure here. From this figure, we see that the radius whose endpoint has second coordinate equal to $-t$ corresponds to $-\theta$.

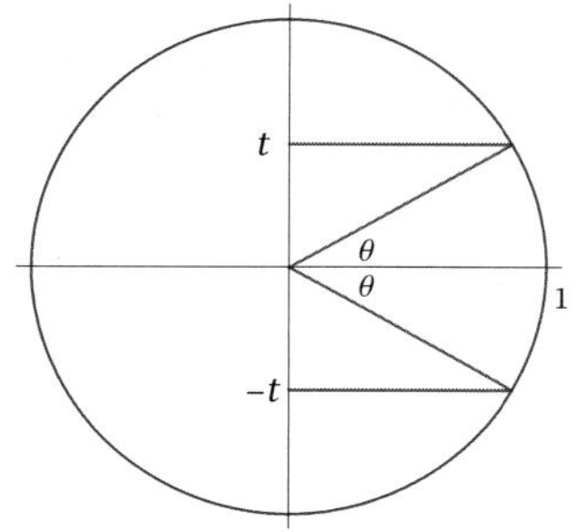

In other words, we have $\sin^{-1}(-t) = -\theta$, which we can rewrite as

$$\mathbf{sin^{-1}}(-t) = -\,\mathbf{sin^{-1}}\,t.$$

Note that $-\theta$ is in $[-\frac{\pi}{2}, \frac{\pi}{2}]$ whenever θ is in $[-\frac{\pi}{2}, \frac{\pi}{2}]$. Thus $-\sin^{-1} t$ is in the right interval to be the arcsine of some number.

EXAMPLE 4

Evaluate $\sin^{-1}(-\sin\frac{\pi}{7})$.

SOLUTION Using the formula above with $t = \sin\frac{\pi}{7}$, we have

$$\begin{aligned}\sin^{-1}(-\sin\tfrac{\pi}{7}) &= -\sin^{-1}(\sin\tfrac{\pi}{7})\\ &= -\tfrac{\pi}{7}.\end{aligned}$$

We now turn to the problem of finding a formula for $\tan^{-1}(-t)$ in terms of $\tan^{-1} t$. To do this, suppose $t > 0$. Let $\theta = \tan^{-1} t$, which implies that $\tan\theta = t$. Consider the radius of the unit circle corresponding to θ. This radius, which is shown here, has slope t.

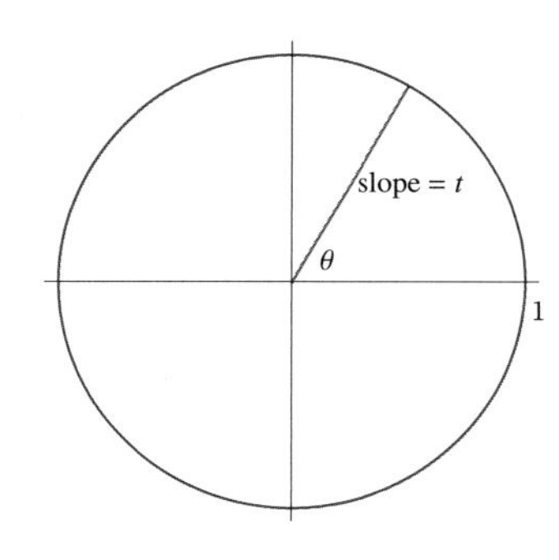

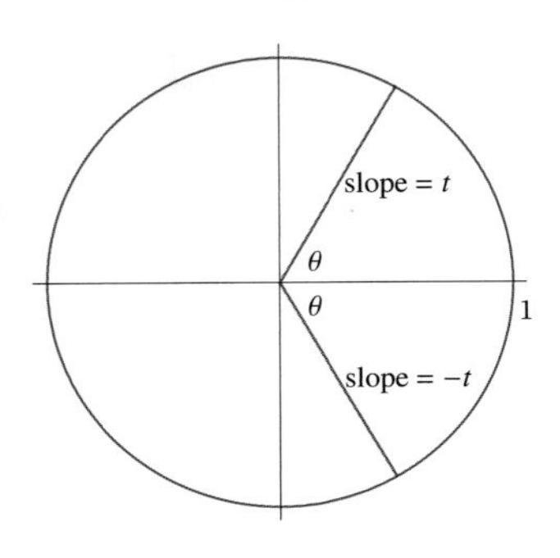

To find $\tan^{-1}(-t)$, we need to find a radius whose slope equals $-t$. To do this, flip the radius with slope t across the horizontal axis, which leaves the first coordinate of the endpoint unchanged and multiplies the second coordinate by -1, as shown here.

From the figure here, we see that the radius with slope $-t$ corresponds to $-\theta$. In other words, we have $\tan^{-1}(-t) = -\theta$, which we can rewrite as

$$\tan^{-1}(-t) = -\tan^{-1} t.$$

Note that $-\theta$ is in $(-\frac{\pi}{2}, \frac{\pi}{2})$ whenever θ is in $(-\frac{\pi}{2}, \frac{\pi}{2})$. Thus $-\tan^{-1} t$ is in the right interval to be the arctangent of some number.

In summary, we have found the following identities for computing the inverse trigonometric functions of $-t$:

Inverse trigonometric identities for $-t$

$$\cos^{-1}(-t) = \pi - \cos^{-1} t$$

$$\sin^{-1}(-t) = -\sin^{-1} t$$

$$\tan^{-1}(-t) = -\tan^{-1} t$$

We derived the first two identities above from figures using the assumption that $0 < t < 1$; for the last identity above our illustration assumed that $t > 0$. However, the algebraic approach, to which we now turn, shows that the first two identities above are actually valid whenever $-1 \le t \le 1$, and the last identity above is valid for all values of t.

The Arccosine, Arcsine, and Arctangent of $-t$: Algebraic Approach

Sometimes a second approach to a subject leads to better understanding.

In this subsection we will again derive the inverse trigonometric identities above, but this time using an algebraic approach that uses our previous trigonometric identities. We begin with an algebraic derivation of the identity for $\cos^{-1}(-t)$.

Suppose $-1 \le t \le 1$. Let $\theta = \cos^{-1} t$. Thus $\cos\theta = t$ and θ is in $[0, \pi]$ (which implies that $\pi - \theta$ is in $[0, \pi]$). Furthermore

$$\cos(\pi - \theta) = \cos(\theta - \pi) = -\cos\theta = -t,$$

where the first equality above comes from our identity for the cosine of the negative of an angle (Section 9.6) and the second equality comes from our identity for $\cos(\theta + n\pi)$ (also Section 9.6; here we are taking $n = -1$). Because $\pi - \theta$ is an angle in $[0, \pi]$ whose cosine equals $-t$ (by the equation above), we conclude that $\cos^{-1}(-t) = \pi - \theta$. This can be rewritten as

$$\cos^{-1}(-t) = \pi - \cos^{-1} t,$$

completing our second derivation of this identity.

Now we turn to an algebraic derivation of the identity for $\sin^{-1}(-t)$. Suppose $-1 \le t \le 1$. Let $\theta = \sin^{-1} t$. Thus θ is in $[-\frac{\pi}{2}, \frac{\pi}{2}]$ (which implies that $-\theta$ is in $[-\frac{\pi}{2}, \frac{\pi}{2}]$) and $\sin\theta = t$. Furthermore

Note that these inverse trigonometric identities follow from the corresponding trigonometric identities.

$$\sin(-\theta) = -\sin\theta = -t,$$

where the first equality above comes from our identity for $\sin(-\theta)$ (Section 9.6). Because $-\theta$ is an angle in $[-\frac{\pi}{2}, \frac{\pi}{2}]$ whose sine equals $-t$ (by the equation above), we conclude that $\sin^{-1}(-t) = -\theta$. This can be rewritten as

$$\sin^{-1}(-t) = -\sin^{-1} t,$$

completing our second derivation of this identity.

Finally, we turn to an algebraic derivation of the identity for $\tan^{-1}(-t)$. Suppose t is any real number. Let $\theta = \tan^{-1} t$. Thus θ is in $(-\frac{\pi}{2}, \frac{\pi}{2})$ [which implies that $-\theta$ is in $(-\frac{\pi}{2}, \frac{\pi}{2})$] and $\tan\theta = t$. Furthermore

$$\tan(-\theta) = -\tan\theta = -t,$$

where the first equality above comes from our identity for $\tan(-\theta)$ (Section 9.6). Because $-\theta$ is an angle in $(-\frac{\pi}{2}, \frac{\pi}{2})$ whose tangent equals $-t$ (by the equation above), we conclude that $\tan^{-1}(-t) = -\theta$. This can be rewritten as

$$\tan^{-1}(-t) = -\tan^{-1} t,$$

completing our second derivation of this identity.

Arccosine Plus Arcsine

Suppose $-1 \le t \le 1$ and $\theta = \cos^{-1} t$. Thus θ is in $[0, \pi]$ and $\cos\theta = t$. Now

$$\sin(\tfrac{\pi}{2} - \theta) = \cos\theta = t,$$

where the first equality comes from one of our identities in Section 9.6. The equation above shows that $\frac{\pi}{2} - \theta$ is an angle whose sine equals t. Furthermore, $\frac{\pi}{2} - \theta$ is in $[-\frac{\pi}{2}, \frac{\pi}{2}]$ (because θ is in $[0, \pi]$). Thus $\sin^{-1} t = \frac{\pi}{2} - \theta$, which can be rewritten as $\sin^{-1} t = \frac{\pi}{2} - \cos^{-1} t$. Adding $\cos^{-1} t$ to both sides of this equation produces a more symmetric version of this important identity.

Arccosine plus arcsine

$$\cos^{-1} t + \sin^{-1} t = \frac{\pi}{2}$$

for all t in $[-1, 1]$.

EXAMPLE 5 Verify the identity above when $t = \frac{1}{2}$.

SOLUTION We have $\cos^{-1}\frac{1}{2} = \frac{\pi}{3}$ and $\sin^{-1}\frac{1}{2} = \frac{\pi}{6}$. Adding these together, we have

$$\cos^{-1}\tfrac{1}{2} + \sin^{-1}\tfrac{1}{2} = \tfrac{\pi}{3} + \tfrac{\pi}{6} = \tfrac{\pi}{2},$$

as expected from the identity.

The Arctangent of $\frac{1}{t}$

Suppose $t > 0$ and $\theta = \tan^{-1} t$. Thus θ is in $(0, \frac{\pi}{2})$ and $\tan\theta = t$. Now

$$\tan(\tfrac{\pi}{2} - \theta) = \frac{1}{\tan\theta} = \frac{1}{t},$$

where the first equality comes from one of our identities in Section 9.6. The equation above shows that $\frac{\pi}{2} - \theta$ is an angle whose tangent equals $\frac{1}{t}$. Furthermore, $\frac{\pi}{2} - \theta$ is in $(0, \frac{\pi}{2})$ [because θ is in $(0, \frac{\pi}{2})$]. Thus $\tan^{-1}\frac{1}{t} = \frac{\pi}{2} - \theta$, which can be rewritten as

$$\tan^{-1}\tfrac{1}{t} = \tfrac{\pi}{2} - \tan^{-1} t.$$

Somewhat surprisingly, the formula derived in the paragraph above for $\tan^{-1}\frac{1}{t}$ does not hold when t is negative. To find the correct formula in this case, suppose $t < 0$. Then

$$\begin{aligned}\tan^{-1}\tfrac{1}{t} &= \tan^{-1}\big(-\tfrac{1}{(-t)}\big)\\ &= -\tan^{-1}\big(\tfrac{1}{(-t)}\big)\\ &= -\big(\tfrac{\pi}{2} - \tan^{-1}(-t)\big)\\ &= -\big(\tfrac{\pi}{2} + \tan^{-1} t\big)\\ &= -\tfrac{\pi}{2} - \tan^{-1} t,\end{aligned}$$

where the second and fourth equalities above come from the identity we found earlier in this section for the arctangent of the negative of a number and the third identity above comes from applying the result of the previous paragraph to the positive number $-t$.

Putting together the results from the last two paragraphs, we have the following identity for $\tan^{-1}\frac{1}{t}$:

This formula for $\tan^{-1}\frac{1}{t}$ is unusual because it depends upon whether $t > 0$ or $t < 0$. Problem 38 gives a formula for $\tan^{-1}\frac{1}{t}$ that does not depend upon whether $t > 0$ or $t < 0$.

Arctangent of $\frac{1}{t}$

$$\tan^{-1}\tfrac{1}{t} = \begin{cases} \tfrac{\pi}{2} - \tan^{-1} t & \text{if } t > 0\\ -\tfrac{\pi}{2} - \tan^{-1} t & \text{if } t < 0\end{cases}$$

EXAMPLE 6

Verify the identity above when $t = 4$.

SOLUTION Using a calculator, we have $\tan^{-1} 4 \approx 1.3258$. Thus

$$\tfrac{\pi}{2} - \tan^{-1} 4 \approx 1.5708 - 1.3258 = 0.245.$$

A calculator shows that $\tan^{-1} \frac{1}{4} \approx 0.245$, and thus the calculation above agrees with the identity.

We can also verify the identity without using a calculator by using a figure.

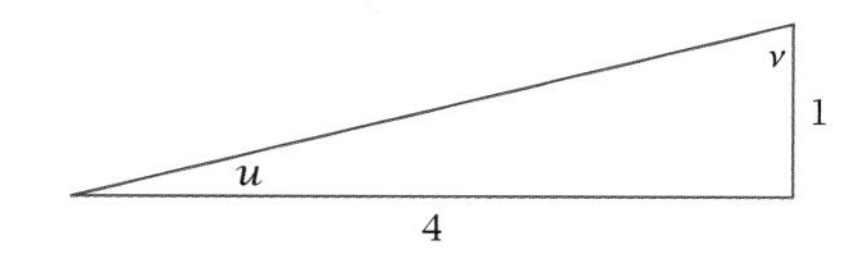

Here $\tan v = 4$ *and* $\tan u = \frac{1}{4}$.

Looking at the figure above, we see that $\tan v = 4$ and $\tan u = \frac{1}{4}$. Thus $v = \tan^{-1} 4$ and $u = \tan^{-1} \frac{1}{4}$. Because this is a right triangle, we have $u = \frac{\pi}{2} - v$, which translates to the equation

$$\tan^{-1} \tfrac{1}{4} = \tfrac{\pi}{2} - \tan^{-1} 4.$$

More Compositions with Inverse Trigonometric Functions

In the previous subsection we discussed the composition of a trigonometric function with its inverse function. In this subsection we will discuss the composition of a trigonometric function with the inverse of a different trigonometric function.

For example, consider the problem of evaluating $\cos(\sin^{-1} \frac{2}{3})$. One way to approach this problem would be to evaluate $\sin^{-1} \frac{2}{3}$, then evaluate the cosine of that angle. However, no one knows how to find an exact expression for $\sin^{-1} \frac{2}{3}$.

A calculator could give an approximate answer. A calculator working in radians shows that

$$\sin^{-1} \tfrac{2}{3} \approx 0.729728.$$

Using a calculator again to take cosine of the number above, we see that

$$\cos(\sin^{-1} \tfrac{2}{3}) \approx 0.745356.$$

The exact answer that we will obtain in the next two examples is more satisfying than this approximation.

When working with trigonometric functions, an accurate numerical approximation such as computed above is sometimes the best that can be done. However, for compositions of the type discussed above, exact answers can be found. The example below shows how to do this.

EXAMPLE 7 Evaluate $\cos(\sin^{-1}\frac{2}{3})$.

SOLUTION Let $\theta = \sin^{-1}\frac{2}{3}$. Thus θ is in $[-\frac{\pi}{2}, \frac{\pi}{2}]$ and $\sin\theta = \frac{2}{3}$. Now

In general we know that $\cos\theta = \pm\sqrt{1-\sin^2\theta}$. Here we can choose the plus sign because this θ is in $[-\frac{\pi}{2}, \frac{\pi}{2}]$, which implies that $\cos\theta \geq 0$.

$$\begin{aligned}\cos(\sin^{-1}\tfrac{2}{3}) &= \cos\theta \\ &= \sqrt{1-\sin^2\theta} \\ &= \sqrt{1-(\tfrac{2}{3})^2} \\ &= \sqrt{\tfrac{5}{9}} \\ &= \tfrac{\sqrt{5}}{3}.\end{aligned}$$

A calculator shows that $\frac{\sqrt{5}}{3} \approx 0.745356$. Thus the exact value we just obtained for $\cos(\sin^{-1}\frac{2}{3})$ is consistent with the approximate value obtained earlier.

The method used in the example above might be called the algebraic approach. The example below solves the same problem using a right-triangle approach. Some people prefer the algebraic approach; others prefer the right-triangle approach. Use whichever method seems clearer to you.

EXAMPLE 8 Evaluate $\cos(\sin^{-1}\frac{2}{3})$.

SOLUTION Let $\theta = \sin^{-1}\frac{2}{3}$; thus $\sin\theta = \frac{2}{3}$. Recall that

$$\sin\theta = \frac{\text{opposite side}}{\text{hypotenuse}}$$

in a right triangle with an angle of θ, where *opposite side* means the length of the side opposite the angle θ. The easiest choices for side lengths to have $\sin\theta = \frac{2}{3}$ are shown in the triangle below:

We could also have chosen sides of length 4 and 6, or $\frac{2}{3}$ and 1, or any pair of numbers whose ratio equals $\frac{2}{3}$. But choosing sides of length 2 and 3 is the simplest choice.

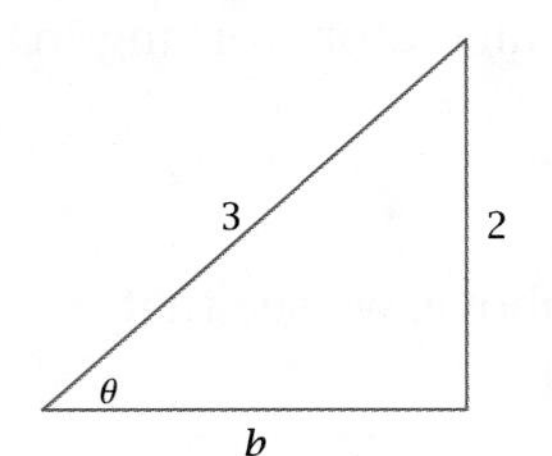

A right triangle with $\sin\theta = \frac{2}{3}$.

We need to evaluate $\cos\theta$. In terms of the figure above, we have

$$\cos\theta = \frac{\text{adjacent side}}{\text{hypotenuse}} = \frac{b}{3}.$$

Applying the Pythagorean Theorem to the triangle above, we have $b^2 + 4 = 9$, which implies that $b = \sqrt{5}$. Thus $\cos\theta = \frac{\sqrt{5}}{3}$. In other words, $\cos(\sin^{-1}\frac{2}{3}) = \frac{\sqrt{5}}{3}$.

The procedures used in the examples above can be used to find identities for the composition of a trigonometric function and the inverse of another trigonometric function. We first illustrate this procedure using the algebraic approach.

EXAMPLE 9

Find a formula for $\tan(\cos^{-1} t)$.

SOLUTION Suppose $-1 \le t \le 1$ with $t \neq 0$ (we are excluding $t = 0$ because in that case we would have $\cos^{-1} t = \frac{\pi}{2}$, but $\tan \frac{\pi}{2}$ is undefined). Let $\theta = \cos^{-1} t$. Thus θ is in $[0, \pi]$ and $\cos\theta = t$. Now

$$\begin{aligned}\tan(\cos^{-1} t) &= \tan\theta \\ &= \frac{\sin\theta}{\cos\theta} \\ &= \frac{\sqrt{1-\cos^2\theta}}{\cos\theta} \\ &= \frac{\sqrt{1-t^2}}{t}.\end{aligned}$$

In general we know that $\sin\theta = \pm\sqrt{1-\cos^2\theta}$*. Here we can choose the plus sign because this* θ *is in* $[0, \pi]$*, which implies that* $\sin\theta \ge 0$*.*

Thus the formula we seek is

$$\tan(\cos^{-1} t) = \frac{\sqrt{1-t^2}}{t}.$$

Next, we derive the same identity using the right-triangle approach. Again, you should use whichever method you find clearer.

EXAMPLE 10

Find a formula for $\tan(\cos^{-1} t)$.

SOLUTION Let $\theta = \cos^{-1} t$; thus $\cos\theta = t$. Recall that

$$\cos\theta = \frac{\text{adjacent side}}{\text{hypotenuse}}$$

in a right triangle with an angle of θ, where *adjacent side* means the length of the (nonhypotenuse) side adjacent to the angle θ. The easiest choices for side lengths to have $\cos\theta = t$ are shown in the triangle here.

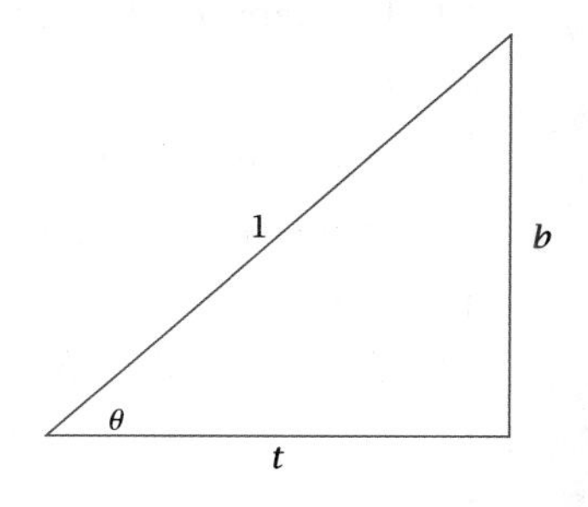

A right triangle with $\cos\theta = t$*.*

We need to evaluate $\tan\theta$. In terms of the figure here, we have

$$\tan\theta = \frac{\text{opposite side}}{\text{adjacent side}} = \frac{b}{t}.$$

Applying the Pythagorean Theorem to the triangle above, we have $t^2 + b^2 = 1$, which implies that $b = \sqrt{1-t^2}$. Thus $\tan\theta = \frac{\sqrt{1-t^2}}{t}$. In other words, we have the identity

$$\tan(\cos^{-1} t) = \frac{\sqrt{1-t^2}}{t}.$$

In the examples above, we derived the identity

$$\tan(\cos^{-1} t) = \frac{\sqrt{1-t^2}}{t},$$

which holds whenever $-1 \le t \le 1$ with $t \ne 0$. There are five more such identities, involving the composition of a trigonometric function and the inverse of another trigonometric function. The problems in this section ask you to derive those five additional identities, which can be done using the same methods as for the identity above. Memorizing these identities is not a good use of your mental energy, but be sure that you understand how to derive them.

See Problems 39–43.

EXERCISES

1. Evaluate $\cos(\cos^{-1} \frac{1}{4})$.
2. Evaluate $\tan(\tan^{-1} 5)$.
3. Evaluate $\tan(\tan^{-1}(e + \pi))$.
4. Evaluate $\sin(\sin^{-1}(\frac{1}{e} - \frac{1}{\pi}))$.
5. Evaluate $\sin^{-1}(\sin \frac{2\pi}{7})$.
6. Evaluate $\cos^{-1}(\cos \frac{1}{2})$.
7. Evaluate $\cos^{-1}(\cos 3\pi)$.
8. Evaluate $\sin^{-1}(\sin \frac{9\pi}{4})$.
9. Evaluate $\tan^{-1}(\tan \frac{11\pi}{5})$.
10. Evaluate $\tan^{-1}(\tan \frac{17\pi}{7})$.
11. Evaluate $\sin^{-1}(\sin \frac{6\pi}{7})$.
12. Evaluate $\cos^{-1}(\cos \frac{10\pi}{9})$.
13. Evaluate $\cos^{-1}(\cos 40°)$.
14. Evaluate $\sin^{-1}(\sin 70°)$.
15. Evaluate $\tan^{-1}(\tan 340°)$.
16. Evaluate $\tan^{-1}(\tan 310°)$.
17. Evaluate $\sin(-\sin^{-1} \frac{3}{13})$.
18. Evaluate $\tan(-\tan^{-1} \frac{7}{11})$.
19. Suppose t is such that $\cos^{-1} t = 2$. Evaluate the following:
    (a) $\cos^{-1}(-t)$
    (b) $\sin^{-1} t$
    (c) $\sin^{-1}(-t)$
20. Suppose t is such that $\cos^{-1} t = \frac{3}{2}$. Evaluate the following:
    (a) $\cos^{-1}(-t)$
    (b) $\sin^{-1} t$
    (c) $\sin^{-1}(-t)$
21. Suppose t is such that $\sin^{-1} t = \frac{3\pi}{8}$. Evaluate the following:
    (a) $\sin^{-1}(-t)$
    (b) $\cos^{-1} t$
    (c) $\cos^{-1}(-t)$
22. Suppose t is such that $\sin^{-1} t = -\frac{2\pi}{7}$. Evaluate the following:
    (a) $\sin^{-1}(-t)$
    (b) $\cos^{-1} t$
    (c) $\cos^{-1}(-t)$
23. Suppose t is such that $\tan^{-1} t = \frac{3\pi}{7}$. Evaluate the following:
    (a) $\tan^{-1} \frac{1}{t}$
    (b) $\tan^{-1}(-t)$
    (c) $\tan^{-1}(-\frac{1}{t})$
24. Suppose t is such that $\tan^{-1} t = -\frac{4\pi}{11}$. Evaluate the following:
    (a) $\tan^{-1} \frac{1}{t}$
    (b) $\tan^{-1}(-t)$
    (c) $\tan^{-1}(-\frac{1}{t})$
25. Evaluate $\sin(\cos^{-1} \frac{1}{3})$.
26. Evaluate $\cos(\sin^{-1} \frac{2}{5})$.
27. Evaluate $\tan(\cos^{-1} \frac{1}{3})$.
28. Evaluate $\tan(\sin^{-1} \frac{2}{5})$.
29. Evaluate $\cos(\tan^{-1}(-4))$.
30. Evaluate $\sin(\tan^{-1}(-9))$.
31. Evaluate $\sin^{-1}(\cos \frac{2\pi}{5})$.
32. Evaluate $\cos^{-1}(\sin \frac{4\pi}{9})$.
33. Evaluate $\cos^{-1}(\sin \frac{6\pi}{7})$.
34. Evaluate $\sin^{-1}(\cos \frac{10\pi}{9})$.

PROBLEMS

35. Is arccosine an even function, an odd function, or neither?

36. Is arcsine an even function, an odd function, or neither?

37. Is arctangent an even function, an odd function, or neither?

38. Show that
$$\tan^{-1}\frac{1}{t} = \frac{t}{|t|}\frac{\pi}{2} - \tan^{-1} t$$
for all $t \neq 0$.

39. Show that
$$\cos(\sin^{-1} t) = \sqrt{1-t^2}$$
whenever $-1 \le t \le 1$.

40. Find an identity expressing $\sin(\cos^{-1} t)$ as a nice function of t.

41. Find an identity expressing $\tan(\sin^{-1} t)$ as a nice function of t.

42. Show that
$$\cos(\tan^{-1} t) = \frac{1}{\sqrt{1+t^2}}$$
for every number t.

43. Find an identity expressing $\sin(\tan^{-1} t)$ as a nice function of t.

44. Explain why
$$\cos^{-1} t = \sin^{-1}\sqrt{1-t^2}$$
whenever $0 \le t \le 1$.

45. Explain why
$$\cos^{-1} t = \tan^{-1}\frac{\sqrt{1-t^2}}{t}$$
whenever $0 < t \le 1$.

46. Explain why
$$\sin^{-1} t = \tan^{-1}\frac{t}{\sqrt{1-t^2}}$$
whenever $-1 < t < 1$.

47. Explain what is wrong with the following "proof" that $\theta = -\theta$:

Let θ be any angle. Then
$$\cos\theta = \cos(-\theta).$$
Apply $\cos^{-1}$ to both sides of the equation above, getting
$$\cos^{-1}(\cos\theta) = \cos^{-1}(\cos(-\theta)).$$
Because $\cos^{-1}$ is the inverse of cos, the equation above implies that
$$\theta = -\theta.$$

WORKED-OUT SOLUTIONS *to Odd-numbered Exercises*

1. Evaluate $\cos(\cos^{-1}\frac{1}{4})$.

SOLUTION Let $\theta = \cos^{-1}\frac{1}{4}$. Thus θ is the angle in $[0, \pi]$ such that $\cos\theta = \frac{1}{4}$. Thus $\cos(\cos^{-1}\frac{1}{4}) = \cos\theta = \frac{1}{4}$.

3. Evaluate $\tan(\tan^{-1}(e+\pi))$.

SOLUTION Let $\theta = \tan^{-1}(e+\pi)$. Thus θ is the angle in $(-\frac{\pi}{2}, \frac{\pi}{2})$ such that $\tan\theta = e+\pi$. Thus $\tan(\tan^{-1}(e+\pi)) = \tan\theta = e+\pi$.

5. Evaluate $\sin^{-1}(\sin\frac{2\pi}{7})$.

SOLUTION Let $\theta = \sin^{-1}(\sin\frac{2\pi}{7})$. Thus θ is the angle in the interval $[-\frac{\pi}{2}, \frac{\pi}{2}]$ such that
$$\sin\theta = \sin\frac{2\pi}{7}.$$
Because $-\frac{1}{2} \le \frac{2}{7} \le \frac{1}{2}$, we see that $\frac{2\pi}{7}$ is in $[-\frac{\pi}{2}, \frac{\pi}{2}]$. Thus the equation above implies that $\theta = \frac{2\pi}{7}$.

7. Evaluate $\cos^{-1}(\cos 3\pi)$.

SOLUTION Because $\cos 3\pi = -1$, we see that
$$\cos^{-1}(\cos 3\pi) = \cos^{-1}(-1).$$
Because $\cos\pi = -1$, we have $\cos^{-1}(-1) = \pi$ ($\cos 3\pi$ also equals -1, but $\cos^{-1}(-1)$ must be in the interval $[0, \pi]$). Thus $\cos^{-1}(\cos 3\pi) = \pi$.

9. Evaluate $\tan^{-1}(\tan\frac{11\pi}{5})$.

SOLUTION Because $\tan^{-1}$ is the inverse of tan, it may be tempting to think that $\tan^{-1}(\tan\frac{11\pi}{5})$ equals $\frac{11\pi}{5}$. However, the values of $\tan^{-1}$ must be between $-\frac{\pi}{2}$ and $\frac{\pi}{2}$. Because $\frac{11\pi}{5} > \frac{\pi}{2}$, we conclude that $\tan^{-1}(\tan\frac{11\pi}{5})$ cannot equal $\frac{11\pi}{5}$.

Note that

$$\tan\tfrac{11\pi}{5} = \tan(2\pi + \tfrac{\pi}{5}) = \tan\tfrac{\pi}{5}.$$

Because $\frac{\pi}{5}$ is in $(-\frac{\pi}{2}, \frac{\pi}{2})$, we have $\tan^{-1}(\tan\frac{\pi}{5}) = \frac{\pi}{5}$. Thus

$$\tan^{-1}(\tan\tfrac{11\pi}{5}) = \tan^{-1}(\tan\tfrac{\pi}{5}) = \tfrac{\pi}{5}.$$

11. Evaluate $\sin^{-1}(\sin\frac{6\pi}{7})$.

SOLUTION Because $\sin^{-1}$ is the inverse of sin, it may be tempting to think that $\sin^{-1}(\sin\frac{6\pi}{7})$ equals $\frac{6\pi}{7}$. However, the values of $\sin^{-1}$ lie in the interval $[-\frac{\pi}{2}, \frac{\pi}{2}]$. Because $\frac{6\pi}{7} > \frac{\pi}{2}$, we conclude that $\sin^{-1}(\sin\frac{6\pi}{7})$ cannot equal $\frac{6\pi}{7}$.

Note that

$$\sin\tfrac{6\pi}{7} = \sin(\pi - \tfrac{6\pi}{7}) = \sin\tfrac{\pi}{7}.$$

Because $\frac{\pi}{7}$ is in $[-\frac{\pi}{2}, \frac{\pi}{2}]$, we have $\sin^{-1}(\sin\frac{\pi}{7}) = \frac{\pi}{7}$. Thus

$$\sin^{-1}(\sin\tfrac{6\pi}{7}) = \sin^{-1}(\sin\tfrac{\pi}{7}) = \tfrac{\pi}{7}.$$

13. Evaluate $\cos^{-1}(\cos 40°)$.

SOLUTION

$$\cos^{-1}(\cos 40°) = \cos^{-1}(\cos\tfrac{2\pi}{9}) = \tfrac{2\pi}{9}$$

15. Evaluate $\tan^{-1}(\tan 340°)$.

SOLUTION Note that

$$\tan 340° = \tan(-20°) = \tan(-\tfrac{\pi}{9}).$$

Thus

$$\tan^{-1}(\tan 340°) = \tan^{-1}(\tan(-\tfrac{\pi}{9})) = -\tfrac{\pi}{9}.$$

REMARK The expression $\tan 340°$ is rewritten above as $\tan(-\frac{\pi}{9})$ because (1) we need to work in radians and (2) we need an angle in the interval $(-\frac{\pi}{2}, \frac{\pi}{2})$ so that we can apply the identity involving the composition of $\tan^{-1}$ and tan.

17. Evaluate $\sin(-\sin^{-1}\frac{3}{13})$.

SOLUTION

$$\sin(-\sin^{-1}\tfrac{3}{13}) = -\sin(\sin^{-1}\tfrac{3}{13})$$
$$= -\tfrac{3}{13}$$

19. Suppose t is such that $\cos^{-1} t = 2$. Evaluate the following:

(a) $\cos^{-1}(-t)$

(b) $\sin^{-1} t$

(c) $\sin^{-1}(-t)$

SOLUTION

(a) $\cos^{-1}(-t) = \pi - \cos^{-1} t = \pi - 2$

(b) $\sin^{-1} t = \frac{\pi}{2} - \cos^{-1} t = \frac{\pi}{2} - 2$

(c) $\sin^{-1}(-t) = -\sin^{-1} t = 2 - \frac{\pi}{2}$

21. Suppose t is such that $\sin^{-1} t = \frac{3\pi}{8}$. Evaluate the following:

(a) $\sin^{-1}(-t)$

(b) $\cos^{-1} t$

(c) $\cos^{-1}(-t)$

SOLUTION

(a) $\sin^{-1}(-t) = -\sin^{-1} t = -\frac{3\pi}{8}$

(b) $\cos^{-1} t = \frac{\pi}{2} - \sin^{-1} t = \frac{\pi}{2} - \frac{3\pi}{8} = \frac{\pi}{8}$

(c) $\cos^{-1}(-t) = \pi - \cos^{-1} t = \pi - \frac{\pi}{8} = \frac{7\pi}{8}$

23. Suppose t is such that $\tan^{-1} t = \frac{3\pi}{7}$. Evaluate the following:

(a) $\tan^{-1}\frac{1}{t}$

(b) $\tan^{-1}(-t)$

(c) $\tan^{-1}(-\frac{1}{t})$

SOLUTION

(a) Because $t = \tan\frac{3\pi}{7}$, we see that $t > 0$. Thus

$$\tan^{-1}\tfrac{1}{t} = \tfrac{\pi}{2} - \tan^{-1} t = \tfrac{\pi}{2} - \tfrac{3\pi}{7} = \tfrac{\pi}{14}.$$

(b) $\tan^{-1}(-t) = -\tan^{-1} t = -\frac{3\pi}{7}$

(c) $\tan^{-1}(-\frac{1}{t}) = -\tan^{-1}(\frac{1}{t}) = -\frac{\pi}{14}$

25. Evaluate $\sin(\cos^{-1}\frac{1}{3})$.

SOLUTION We give two ways to work this exercise: the algebraic approach and the right-triangle approach.

Algebraic approach: Let $\theta = \cos^{-1}\frac{1}{3}$. Thus θ is the angle in $[0, \pi]$ such that $\cos\theta = \frac{1}{3}$. Note that $\sin\theta \geq 0$ because θ is in $[0, \pi]$. Thus

$$\begin{aligned}\sin(\cos^{-1}\tfrac{1}{3}) &= \sin\theta\\ &= \sqrt{1-\cos^2\theta}\\ &= \sqrt{1-\tfrac{1}{9}}\\ &= \sqrt{\tfrac{8}{9}}\\ &= \tfrac{2\sqrt{2}}{3}.\end{aligned}$$

Right-triangle approach: Let $\theta = \cos^{-1}\frac{1}{3}$; thus $\cos\theta = \frac{1}{3}$. Because

$$\cos\theta = \frac{\text{adjacent side}}{\text{hypotenuse}}$$

in a right triangle with an angle of θ, the following figure (which is not drawn to scale) illustrates the situation:

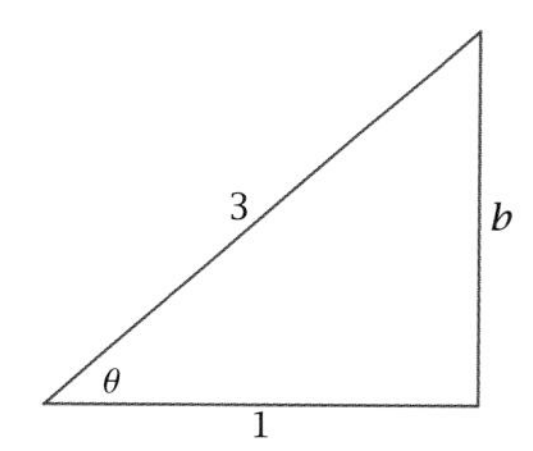

We need to evaluate $\sin\theta$. In terms of the figure above, we have

$$\sin\theta = \frac{\text{opposite side}}{\text{hypotenuse}} = \frac{b}{3}.$$

Applying the Pythagorean Theorem to the triangle above, we have $b^2 + 1 = 9$, which implies that $b = \sqrt{8} = 2\sqrt{2}$. Thus $\sin\theta = \frac{2\sqrt{2}}{3}$. In other words, $\sin(\cos^{-1}\frac{1}{3}) = \frac{2\sqrt{2}}{3}$.

27. Evaluate $\tan(\cos^{-1}\frac{1}{3})$.

SOLUTION We give two ways to work this exercise: the algebraic approach and the right-triangle approach.

Algebraic approach: From Exercise 25, we already know that

$$\sin(\cos^{-1}\tfrac{1}{3}) = \tfrac{2\sqrt{2}}{3}.$$

Thus

$$\begin{aligned}\tan(\cos^{-1}\tfrac{1}{3}) &= \frac{\sin(\cos^{-1}\frac{1}{3})}{\cos(\cos^{-1}\frac{1}{3})}\\ &= \frac{\frac{2\sqrt{2}}{3}}{\frac{1}{3}}\\ &= 2\sqrt{2}.\end{aligned}$$

Right-triangle approach: Let $\theta = \cos^{-1}\frac{1}{3}$; thus $\cos\theta = \frac{1}{3}$. Because

$$\cos\theta = \frac{\text{adjacent side}}{\text{hypotenuse}}$$

in a right triangle with an angle of θ, the following figure (which is not drawn to scale) illustrates the situation:

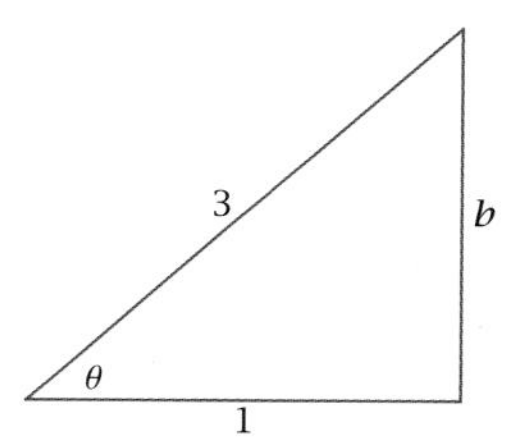

We need to evaluate $\tan\theta$. In terms of the figure above, we have

$$\tan\theta = \frac{\text{opposite side}}{\text{adjacent side}} = b.$$

Applying the Pythagorean Theorem to the triangle above, we have $b^2 + 1 = 9$, which implies that $b = \sqrt{8} = 2\sqrt{2}$. Thus $\tan\theta = 2\sqrt{2}$. In other words, $\tan(\cos^{-1}\frac{1}{3}) = 2\sqrt{2}$.

29. Evaluate $\cos(\tan^{-1}(-4))$.

SOLUTION We give two ways to work this exercise: the algebraic approach and the right-triangle approach.

Algebraic approach: Let $\theta = \tan^{-1}(-4)$. Thus θ is the angle in $(-\frac{\pi}{2}, \frac{\pi}{2})$ such that $\tan\theta = -4$. Note that $\cos\theta > 0$ because θ is in $(-\frac{\pi}{2}, \frac{\pi}{2})$.

Recall that dividing both sides of the identity $\cos^2\theta + \sin^2\theta = 1$ by $\cos^2\theta$ produces the equation $1 + \tan^2\theta = \frac{1}{\cos^2\theta}$. Solving this equation for $\cos\theta$ gives the following:

$$\begin{aligned}\cos(\tan^{-1}(-4)) &= \cos\theta\\ &= \frac{1}{\sqrt{1+\tan^2\theta}}\\ &= \frac{1}{\sqrt{1+(-4)^2}}\\ &= \tfrac{1}{\sqrt{17}} = \tfrac{\sqrt{17}}{17}.\end{aligned}$$

Right-triangle approach: Sides with negative length make no sense in a right triangle. Thus first we use some identities to get rid of the minus sign, as follows:

$$\begin{aligned}\cos(\tan^{-1}(-4)) &= \cos(-\tan^{-1}4)\\ &= \cos(\tan^{-1}4).\end{aligned}$$

Thus we need to evaluate $\cos(\tan^{-1}4)$.

Now let $\theta = \tan^{-1}4$; thus $\tan\theta = 4$. Because

$$\tan\theta = \frac{\text{opposite side}}{\text{adjacent side}}$$

in a right triangle with an angle of θ, the following figure (which is not drawn to scale) illustrates the situation:

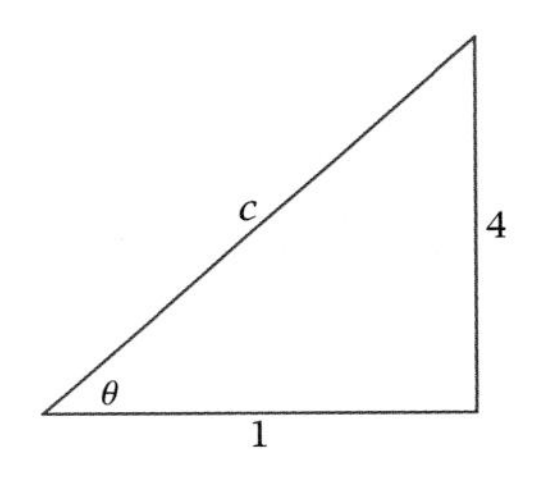

We need to evaluate $\cos\theta$. In terms of the figure above, we have

$$\cos\theta = \frac{\text{adjacent side}}{\text{hypotenuse}} = \frac{1}{c}.$$

Applying the Pythagorean Theorem to the triangle above, we have $c^2 = 1 + 16$, which implies that $c = \sqrt{17}$. Thus $\cos\theta = \frac{1}{\sqrt{17}} = \frac{\sqrt{17}}{17}$. In other words, $\cos(\tan^{-1}4) = \frac{\sqrt{17}}{17}$. Thus $\cos(\tan^{-1}(-4)) = \frac{\sqrt{17}}{17}$.

31. Evaluate $\sin^{-1}(\cos\frac{2\pi}{5})$.

SOLUTION

$$\begin{aligned}\sin^{-1}(\cos\tfrac{2\pi}{5}) &= \tfrac{\pi}{2} - \cos^{-1}(\cos\tfrac{2\pi}{5})\\ &= \tfrac{\pi}{2} - \tfrac{2\pi}{5} = \tfrac{\pi}{10}\end{aligned}$$

33. Evaluate $\cos^{-1}(\sin\frac{6\pi}{7})$.

SOLUTION

$$\begin{aligned}\cos^{-1}(\sin\tfrac{6\pi}{7}) &= \tfrac{\pi}{2} - \sin^{-1}(\sin\tfrac{6\pi}{7})\\ &= \tfrac{\pi}{2} - \tfrac{\pi}{7}\\ &= \tfrac{5\pi}{14},\end{aligned}$$

where the value of $\sin^{-1}(\sin\frac{6\pi}{7})$ comes from the solution to Exercise 11.

10.3 *Using Trigonometry to Compute Area*

LEARNING OBJECTIVES

By the end of this section you should be able to

- compute the area of a triangle given the lengths of two sides and the angle between them;
- deal with the ambiguous angle problem that sometimes arises when trying to find the angle between two sides of a triangle or a parallelogram;
- compute the area of a parallelogram given the lengths of two adjacent sides and the angle between them;
- compute the area of a regular polygon;
- use the trigonometric approximations.

The Area of a Triangle via Trigonometry

Suppose we know the lengths of two sides of a triangle and the angle between those two sides. How can we find the area of the triangle? The example below shows how a knowledge of trigonometry helps solve this problem.

EXAMPLE 1

Find the area of a triangle that has sides of length 4 and 7 and a 49° angle between those two sides.

SOLUTION We consider the side of length 7 to be the base of the triangle. Let h denote the corresponding height of the triangle, as shown here.

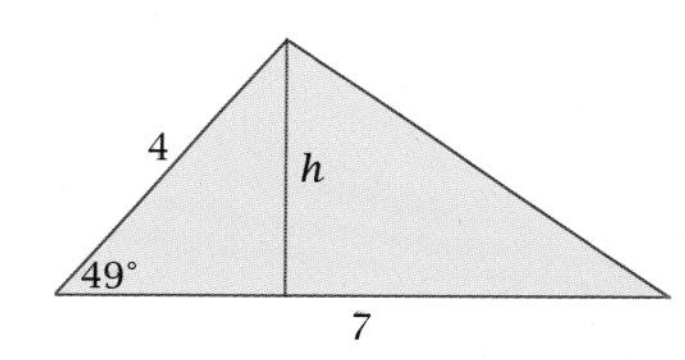

Looking at the figure above, we see that $\sin 49^\circ = \frac{h}{4}$. Solving for h, we have $h = 4\sin 49^\circ$. Thus the triangle has area

$$\tfrac{1}{2} \cdot 7h = \tfrac{1}{2} \cdot 7 \cdot 4\sin 49^\circ = 14\sin 49^\circ \approx 10.566.$$

To find a formula for the area of a triangle given the lengths of two sides and the angle between those two sides, we repeat the process used in the example above. Thus consider a triangle with sides of length a and b and angle θ between those two sides. We consider b to be the base of the triangle. Let h denote the corresponding height of this triangle.

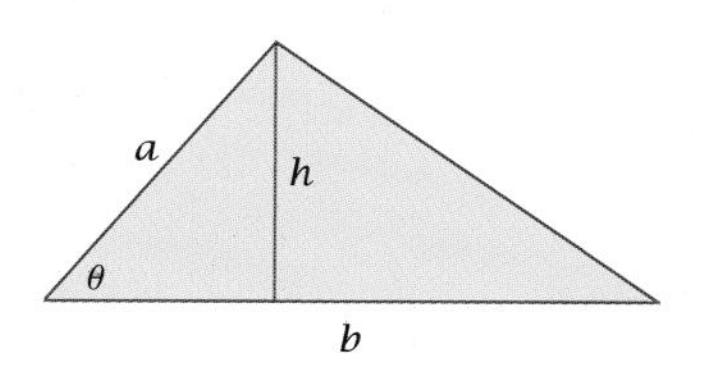

A triangle with base b and height h.

We want to write the height h in terms of the known measurements of the triangle, which are a, b, and θ. Looking at the figure here, we see that $\sin\theta = \frac{h}{a}$. Solving for h, we have

$$h = a\sin\theta.$$

The area of the triangle above is $\frac{1}{2}bh$. Substituting $a\sin\theta$ for h shows that the area of the triangle equals $\frac{1}{2}ab\sin\theta$. Thus we have arrived at our desired formula giving the area of a triangle in terms of the lengths of two sides and the angle between those sides:

> ***Area of a triangle***
>
> A triangle with sides of length a and b and with angle θ between those two sides has area $\frac{1}{2}ab\sin\theta$.

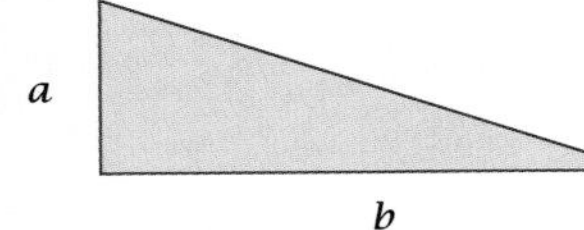

This right triangle has area $\frac{1}{2}ab$.

Whenever we encounter a new formula we should check that it agrees with previously known formulas in cases where both formulas apply. The formula above allows us to compute the area of a triangle whenever we know the lengths of two sides and the angle between those sides. We already knew how to do this when the angle in question is a right angle. Specifically, we already knew that the right triangle shown here has area $\frac{1}{2}ab$ (which is half the area of a rectangle with sides of length a and b).

To apply our new formula to the right triangle above, we take $\theta = \frac{\pi}{2}$. Because $\sin\frac{\pi}{2} = 1$, the expression $\frac{1}{2}ab\sin\theta$ becomes $\frac{1}{2}ab$. In other words, our new formula for the area of a triangle gives the same result as our previous formula for the area of a right triangle. Thus the two formulas are consistent (if they had been inconsistent, then we would know that one of them was incorrect).

Ambiguous Angles

Suppose a triangle has sides of lengths a and b, an angle θ between those sides, and area R. Given any three of a, b, θ, and R, we can use the equation

$$R = \tfrac{1}{2}ab\sin\theta$$

to solve for the other quantity. This process is mostly straightforward—the exercises at the end of this section provide some practice in this procedure.

However, a subtlety arises when we know the lengths a, b and the area R and we need to find the angle θ. Solving the equation above for $\sin\theta$, we get

$$\sin\theta = \frac{2R}{ab}.$$

Thus θ is an angle whose sine equals $\frac{2R}{ab}$, and it would seem that we finish by taking $\theta = \sin^{-1}\frac{2R}{ab}$. Sometimes this is correct, but not always. Let's look at an example to see what can happen.

EXAMPLE 2 Suppose a triangle with area 6 has sides of lengths 3 and 8. Find the angle between those two sides.

SOLUTION Solving for $\sin\theta$ as above, we have

$$\sin\theta = \frac{2R}{ab} = \frac{2\cdot 6}{3\cdot 8} = \frac{1}{2}.$$

Now $\sin^{-1}\frac{1}{2}$ equals $\frac{\pi}{6}$ radians, which equals 30°. Thus it appears that our triangle should look like this:

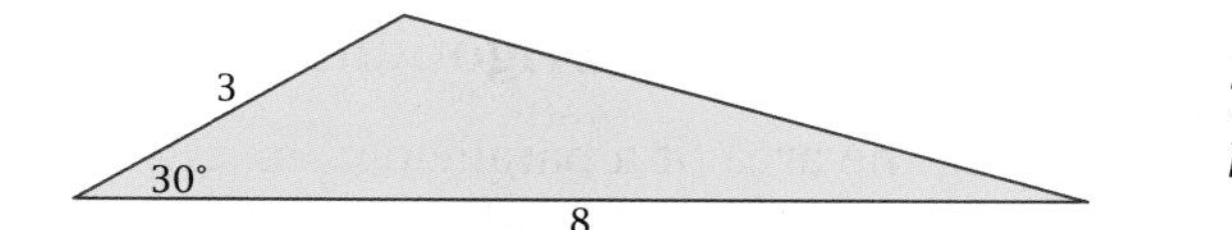

This triangle has area 6.

However, the sine of 150° also equals $\frac{1}{2}$. Thus the following triangle with sides of lengths 3 and 8 also has area 6:

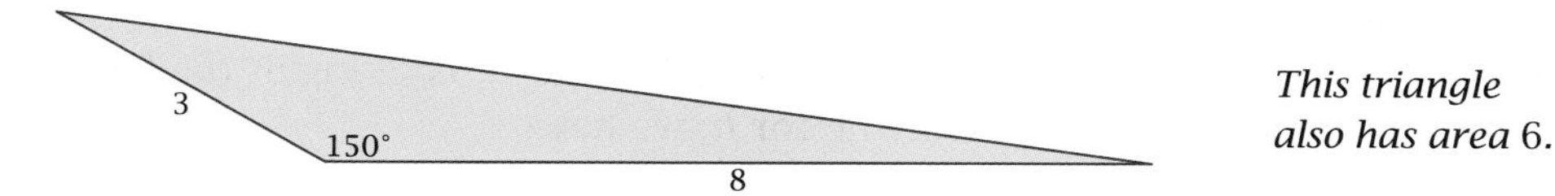

This triangle also has area 6.

If the only information available is that the triangle has area 6 and sides of length 3 and 8, then there is no way to decide which of the two possibilities above truly represents the triangle.

Do not mistakenly think that because $\sin^{-1}\frac{1}{2}$ is defined to equal $\frac{\pi}{6}$ radians (which equals 30°), the preferred solution in the example above is to choose $\theta = 30°$. We defined the arcsine of a number to be in the interval $[-\frac{\pi}{2}, \frac{\pi}{2}]$ because some choice needed to be made in order to obtain a well-defined inverse for the sine. However, remember that given a number t in $[-1, 1]$, there are angles other than $\sin^{-1} t$ whose sine equals t (although there is only one such angle in the interval $[-\frac{\pi}{2}, \frac{\pi}{2}]$).

In the example above, we had 30° and 150° as two angles whose sine equals $\frac{1}{2}$. More generally, given any number t in $[-1, 1]$ and an angle θ such that $\sin\theta = t$, we also have $\sin(\pi - \theta) = t$. This follows from the identity $\sin(\pi - \theta) = \sin\theta$, which can be derived as follows:

$$\begin{aligned}\sin(\pi - \theta) &= -(\sin(\theta - \pi)) \\ &= -(-\sin\theta) \\ &= \sin\theta,\end{aligned}$$

where the first identity above follows from our identity for the sine of the negative of an angle (Section 9.6) and the second identity follows from our formula for the sine of $\theta + n\pi$, with $n = -1$ (also Section 9.6).

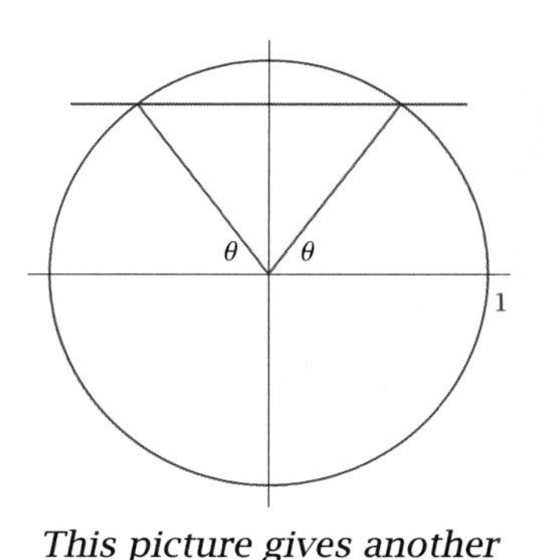

This picture gives another proof that $\sin(\pi - \theta) = \sin\theta$. *Here one radius corresponds to the angle* θ*; the other radius corresponds to the angle* $\pi - \theta$. *The two radii have endpoints with the same second coordinate; thus the corresponding angles have the same sine.*

When working in degrees instead of radians, the result in the paragraph above should be restated to say that the angles $\theta°$ and $(180 - \theta)°$ have the same sine.

Returning to the example above, note that in addition to 30° and 150°, there are other angles whose sine equals $\frac{1}{2}$. For example, −330° and 390° are two such angles. But a triangle cannot have a negative angle, and a triangle cannot have an angle larger than 180°. Thus neither −330° nor 390° is a viable possibility for the angle θ in the triangle in question.

The Area of a Parallelogram via Trigonometry

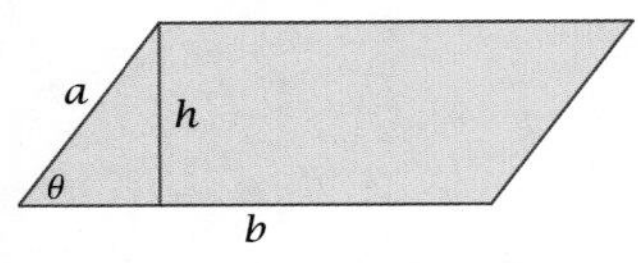

A parallelogram with base b and height h.

The procedure for finding the area of a parallelogram, given an angle of the parallelogram and the lengths of the two adjacent sides, is the same as the procedure followed for a triangle. Consider a parallelogram with sides of length a and b and angle θ between those two sides, as shown here. We consider b to be the base of the parallelogram, and we let h denote the height of the parallelogram.

We want to write the height h in terms a, b, and θ. Looking at the figure above, we see that $\sin\theta = \frac{h}{a}$. Solving for h, we have

$$h = a\sin\theta.$$

The area of the parallelogram above is bh. Substituting $a\sin\theta$ for h shows that the area equals $ab\sin\theta$. Thus we have the following formula:

Area of a parallelogram

A parallelogram with adjacent sides of length a and b and with angle θ between those two sides has area $ab\sin\theta$.

Suppose a parallelogram has adjacent sides of lengths a and b, an angle θ between those sides, and area R. Given any three of a, b, θ, and R, we can use the equation

$$R = ab\sin\theta$$

to solve for the other quantity. As with the case of a triangle, if we know the lengths a and b and the area R, then there can be two possible choices for θ. For a parallelogram, both choices can be correct, as illustrated in the example below.

EXAMPLE 3

A parallelogram has area 40 and pairs of sides with lengths 5 and 10, as shown here. Find the angle between the sides of lengths 5 and 10.

SOLUTION Solving the area formula above for $\sin\theta$, we have

$$\sin\theta = \frac{R}{ab} = \frac{40}{5\cdot 10} = \frac{4}{5}.$$

A calculator shows that $\sin^{-1}\frac{4}{5} \approx 0.927$ radians, which is approximately $53.1°$. An angle of $\pi - \sin^{-1}\frac{4}{5}$, which is approximately $126.9°$, also has a sine equal to $\frac{4}{5}$.

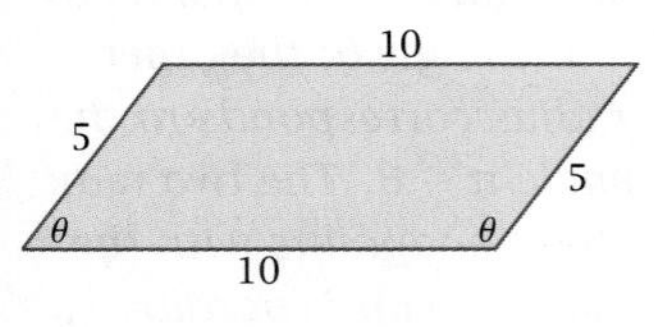

An angle θ measured in degrees is **obtuse** *if 90° < θ < 180°.*

To determine whether $\theta \approx 53.1°$ or whether $\theta \approx 126.9°$, we need to look at the figure here. As you can see, two angles have been labeled θ—both angles are between sides of length 5 and 10, reflecting the ambiguity in the statement of the problem. Although the ambiguity makes this a poorly stated problem, our formula has found both possible answers!

Specifically, if what was meant was the acute angle θ above (the leftmost angle labeled θ), then $\theta \approx 53.1°$; if what was meant was the obtuse angle θ above (the rightmost angle labeled θ), then $\theta \approx 126.9°$.

The Area of a Polygon

One way to find the area of a polygon is to decompose the polygon into triangles and then compute the sum of the areas of the triangles. This procedure works particularly well for a **regular polygon**, which is a polygon all of whose sides have the same length and all of whose angles are equal. For example, a regular polygon with four sides is a square. As another example, the figure here shows a regular octagon.

The following example illustrates the procedure for finding the area of a regular polygon.

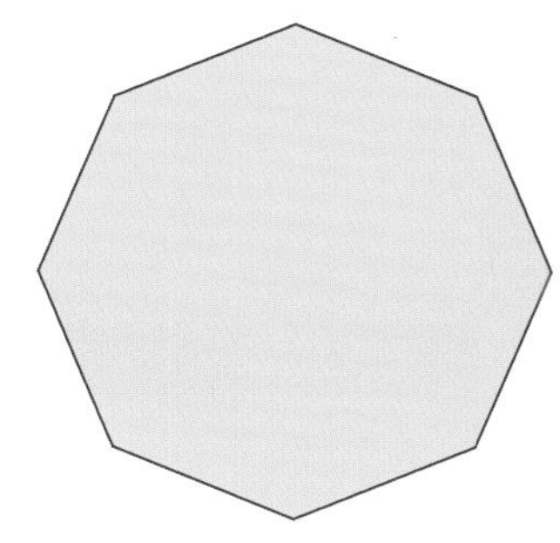

A regular octagon, which has 8 sides.

EXAMPLE 4

Find the area of a regular octagon whose vertices are eight equally spaced points on the unit circle.

SOLUTION The figure here shows how the octagon can be decomposed into triangles by drawing line segments from the center of the circle to the vertices.

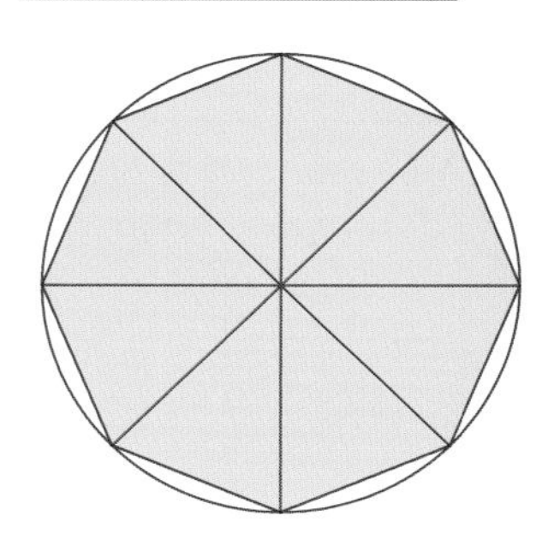

A regular octagon inscribed in the unit circle and decomposed into 8 triangles.

Each triangle shown here has two sides that are radii of the unit circle; thus those two sides of the triangle each have length 1. The angle between those two radii is $\frac{2\pi}{8}$ radians (because one rotation around the entire circle is an angle of 2π radians and each of the eight triangles has an angle that takes up one-eighth of the total). Now $\frac{2\pi}{8}$ radians equals $\frac{\pi}{4}$ radians (or 45°). Thus each of the eight triangles has area

$$\tfrac{1}{2} \cdot 1 \cdot 1 \cdot \sin \tfrac{\pi}{4},$$

which equals $\frac{\sqrt{2}}{4}$. Thus the sum of the areas of the eight triangles equals $8 \cdot \frac{\sqrt{2}}{4}$, which equals $2\sqrt{2}$. In other words, the octagon has area $2\sqrt{2}$.

The next example shows how to compute the area of a regular polygon that is not inscribed in the unit circle.

EXAMPLE 5

Assume that the face of a loonie, the 11-sided Canadian one-dollar coin shown below and on the opening page of this chapter, is a regular 11-sided polygon (this is not quite true, because the edges of the loonie are slightly curved). The distance from the center of a loonie to one of the vertices is 1.325 centimeters. Find the area of the face of a loonie.

SOLUTION Decompose a regular 11-sided polygon representing the loonie into triangles by drawing line segments from the center to the vertices, as shown here. Each triangle has two sides of length 1.325 centimeters. The angle between those two sides is $\frac{2\pi}{11}$ radians (because one full rotation is an angle of 2π radians and each of the 11 triangles has an angle that takes up one-eleventh of the total). Thus each of the 11 triangles has area

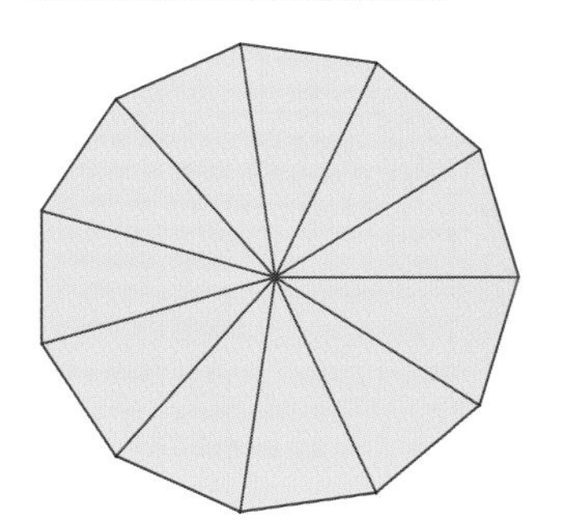

A regular 11-sided polygon, decomposed into 11 triangles.

$$\tfrac{1}{2} \cdot 1.325 \cdot 1.325 \cdot \sin \tfrac{2\pi}{11}$$

square centimeters. The area of the 11-sided polygon is the sum of the areas of the 11 triangles, which equals $11 \cdot \frac{1}{2} \cdot 1.325 \cdot 1.325 \cdot \sin \frac{2\pi}{11}$ square centimeters, which is approximately 5.22 square centimeters.

The technique for finding the area of a regular polygon changes slightly when we know the length of each side instead of the distance from the center to a vertex. The next example illustrates this procedure.

EXAMPLE 6

Each side of the Pentagon that houses the U.S. Department of Defense has length 921 feet. Find the area of the Pentagon.

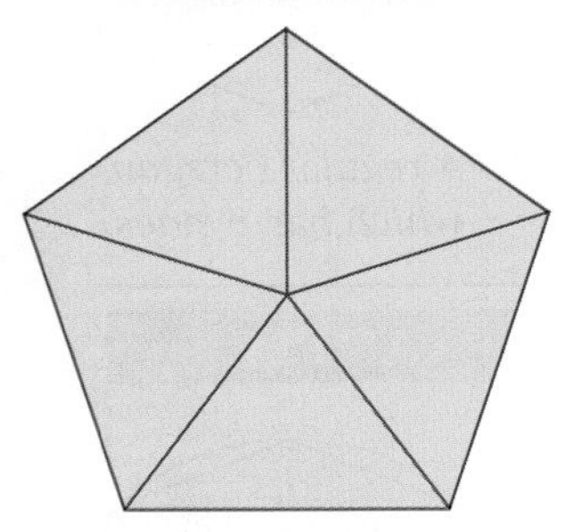

A regular pentagon, decomposed into 5 triangles.

SOLUTION Decompose a regular pentagon representing the Pentagon into triangles by drawing line segments from the center to the vertices, as shown here. In each triangle, there is a $\frac{2\pi}{5}$ radian angle between the two sides of equal length (because we have divided the circle into five equal angles), with the side opposite that angle having length 921 feet.

We do not know the distance from the center of the pentagon to a vertex. Thus to find the area of each triangle, consider the bottom triangle, as shown here (with a different scale than in the pentagon).

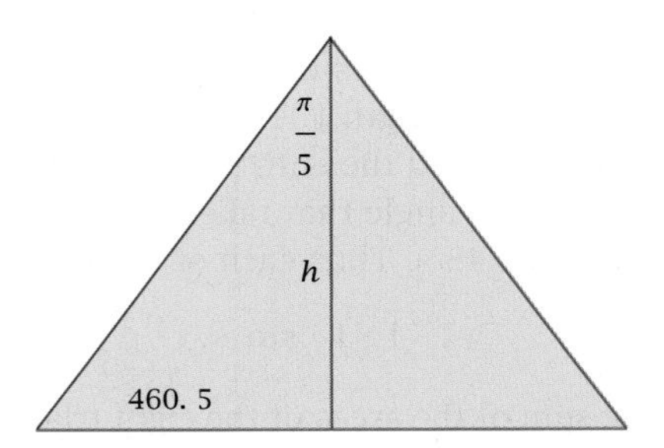

A perpendicular has been dropped from the vertex to the base. The length of this perpendicular has been labeled h, the height of the triangle. Because the angle at the vertex is $\frac{2\pi}{5}$ radians, the top angle in the right triangle is $\frac{\pi}{5}$ radians, as shown in the figure above. Because each side of the Pentagon has length 921 feet, the side of the right triangle opposite this angle has length $\frac{921}{2}$, which equals 460.5 as shown above.

The figure above shows that $\tan\frac{\pi}{5} = \frac{460.5}{h}$. Thus

$$h = \frac{460.5}{\tan\frac{\pi}{5}}.$$

Hence the area (one-half base times height) of the large triangle is

$$\frac{1}{2} \cdot 921 \cdot \frac{460.5}{\tan\frac{\pi}{5}}$$

square feet, which equals

$$\frac{212060.25}{\tan\frac{\pi}{5}}$$

square feet. Thus the total area of the Pentagon is

$$5 \cdot \frac{212060.25}{\tan\frac{\pi}{5}}$$

square feet, which is approximately 1.46 million square feet.

Trigonometric Approximations

Consider the following table showing $\sin\theta$ and $\tan\theta$ (rounded off to an appropriate number of digits) for some small values of θ:

θ	$\sin\theta$	$\tan\theta$
0.5	0.48	0.55
0.05	0.04998	0.05004
0.005	0.00499998	0.00500004
0.0005	0.00049999998	0.00050000004
0.00005	0.00004999999998	0.00005000000004

Here $\sin\theta$ and $\tan\theta$ are computed assuming that θ is an angle measured in radians.

Note that $\sin\theta \approx \tan\theta$ for the small values of θ in this table, with the approximate equality of $\sin\theta$ and $\tan\theta$ becoming closer to an equality as θ gets smaller. This approximation happens because $\tan\theta = \frac{\sin\theta}{\cos\theta}$, and with $\cos\theta$ getting very close to 1 as θ gets very close to 0, we have $\tan\theta \approx \sin\theta$ when θ is small.

The table above also shows the more surprising approximations $\sin\theta \approx \theta$ and $\tan\theta \approx \theta$, again with these approximations becoming closer to equalities as θ gets smaller.

To see why these last approximations hold, consider the figure shown here. You may want to review the formula for the length of a circular arc given in Section 9.2. Think of θ as being even smaller than what is shown in the figure. Then the red vertical line segment and the red circular arc have approximately the same length. Thus $\sin\theta \approx \theta$ if θ is small.

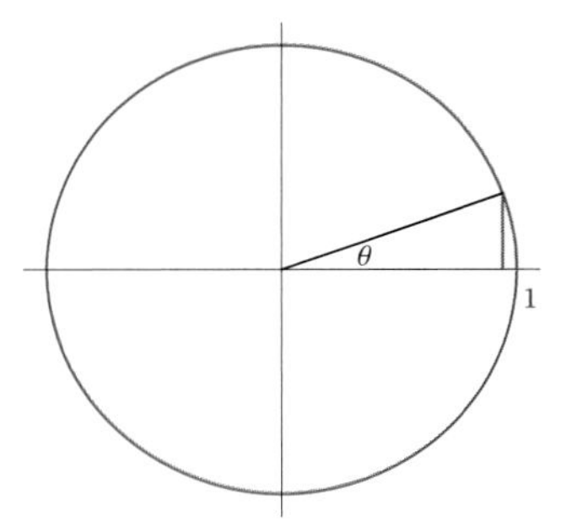

The red vertical line segment has length $\sin\theta$; the red circular arc has length θ. Thus $\sin\theta \approx \theta$ if θ is small.

The approximations $\tan\theta \approx \sin\theta$ and $\sin\theta \approx \theta$, both holding if θ is small, now show that $\tan\theta \approx \theta$ if θ is small. In summary, the paragraph above explains the behavior we noticed in the table above. We have the following result:

Approximation of $\sin\theta$ *and* $\tan\theta$

If $|\theta|$ is small then

$$\sin\theta \approx \theta \quad \text{and} \quad \tan\theta \approx \theta.$$

The pretty approximation above holds only when θ is measured in radians. This result shows again why radians are the natural unit for angles.

In the figure we used above, the red vertical line segment is shorter than the red circular arc. Thus we conclude that $\sin\theta < \theta$.

To get an inequality involving $\tan\theta$, consider the figure shown here. The yellow region has area $\frac{\theta}{2}$ (because the area inside the entire circle is π and the fraction of the area inside the entire circle occupied by the yellow region is $\frac{\theta}{2\pi}$; see the material titled "Area of a Slice" in Section 9.2). The red vertical line segment has length $\tan\theta$. Thus the right triangle, which has base 1 and height $\tan\theta$, has area $\frac{\tan\theta}{2}$. Because the yellow region lies inside the triangle, we have $\frac{\theta}{2} < \frac{\tan\theta}{2}$. Thus $\theta < \tan\theta$.

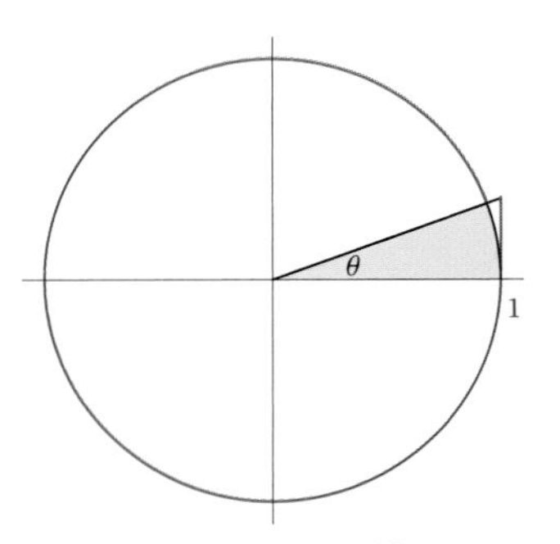

The red vertical line segment has length $\tan\theta$. The yellow region has area $\frac{\theta}{2}$; the triangle has area $\frac{\tan\theta}{2}$. Thus $\theta < \tan\theta$.

Putting together the inequalities derived in the two previous paragraphs gives the following beautiful result:

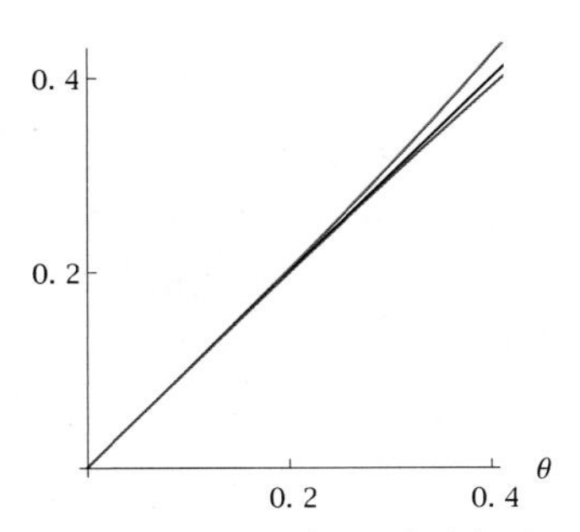

The graphs of $\sin\theta$ *(blue),* θ *(black), and* $\tan\theta$ *(red) on the interval* $[0, 0.4]$*. These functions are so close together on the first half of this interval that their graphs cannot be distinguished there.*

Inequality between $\sin\theta$*,* θ*, and* $\tan\theta$

If $0 < \theta < \frac{\pi}{2}$ then

$$\sin\theta < \theta < \tan\theta.$$

The inequalities above agree with the table at the beginning of this subsection, where we saw that $\sin\theta$ is slightly less than θ and $\tan\theta$ is slightly larger than θ for small values of θ.

Sometimes the result in the box above is more useful if it is rewritten in a slightly different form. To do this, start with the inequality $\theta < \frac{\sin\theta}{\cos\theta}$, which follows from the box above because the last expression equals $\tan\theta$. Multiply both sides of this inequality by $\cos\theta$, then divide both sides by θ to produce the inequality $\cos\theta < \frac{\sin\theta}{\theta}$.

Next, divide both sides of the inequality $\sin\theta < \theta$ from the box above by θ to produce the inequality $\frac{\sin\theta}{\theta} < 1$.

Putting together the inequalities from the two previous paragraphs shows that $\cos\theta < \frac{\sin\theta}{\theta} < 1$. This result has been derived under the assumption that $0 < \theta < \frac{\pi}{2}$, but replacing θ by $-\theta$ does not change either $\cos\theta$ or $\frac{\sin\theta}{\theta}$. Thus we have the following result:

Inequality for $\frac{\sin\theta}{\theta}$

This inequality holds only if we are working with radians.

If $0 < |\theta| < \frac{\pi}{2}$, then

$$\cos\theta < \frac{\sin\theta}{\theta} < 1.$$

Suppose $|\theta|$ is small. Then $\cos\theta$ is very close to 1. Thus the inequality in the box above sandwiches $\frac{\sin\theta}{\theta}$ between 1 and a number very close to 1. This implies the following result:

Approximation of $\frac{\sin\theta}{\theta}$

For example, $\frac{\sin 0.001}{0.001}$ *rounded off to seven digits after the decimal point equals* 0.9999998.

If $|\theta|$ is small but nonzero, then

$$\frac{\sin\theta}{\theta} \approx 1.$$

Now we will see how to approximate $\cos\theta$ for small values of θ. The key to this result is the next manipulation, which allows us to use the approximation for $\sin\theta$ that we have already derived.

Suppose $|\theta|$ is small. Then

$$\begin{aligned} 1 - \cos\theta &= (1 - \cos\theta) \cdot \frac{1 + \cos\theta}{1 + \cos\theta} \\ &= \frac{1 - \cos^2\theta}{1 + \cos\theta} \\ &= \frac{\sin^2\theta}{1 + \cos\theta} \\ &\approx \frac{\theta^2}{1 + \cos\theta} \\ &\approx \frac{\theta^2}{2}. \end{aligned}$$

The graphs of $1 - \cos\theta$ *(blue) and* $\frac{\theta^2}{2}$ *(red) on the interval* $[-0.8, 0.8]$*. These functions are so close together on most of this interval that their graphs cannot be distinguished there.*

We have just shown that if $|\theta|$ is small, then $1 - \cos\theta \approx \frac{\theta^2}{2}$. Solving for $\cos\theta$ now gives the following result:

For example, $\cos 0.01$ *rounded off to ten digits after the decimal point equals* 0.9999500004*. Because* $1 - \frac{0.01^2}{2}$ *equals exactly* 0.99995*, the approximation is quite good.*

Approximation of $\cos\theta$

If $|\theta|$ is small then

$$\cos\theta \approx 1 - \frac{\theta^2}{2}.$$

We conclude this section with an amusing trick.

EXAMPLE 7

Consider the following trick:

With an audience containing at least one person with a scientific calculator, ask everyone to set their calculators to work in degrees. Ask for a number with two digits before the decimal point and two digits after the decimal point (you may want to let four different people give you four different digits so that the audience sees that there is no collusion). Let's say the number given to you by the audience is 69.23. Now ask the people with calculators to compute

$$\tan(\cos(\sin 69.23^\circ))$$

but caution them not to say the result. Meanwhile, you (pretend to) calculate this quantity in your head. After the people with calculators have had enough time to get the answer, announce that the result is 0.01745 plus some additional digits. Ask the people with calculators to confirm that this is correct, and accept applause for being able to do this calculation in your head.

(a) How did you do this trick?

(b) Why does this trick work?

SOLUTION

(a) The answer is always 0.01745 plus some additional digits, regardless of the original input.

(b) Call the original input θ. The first step in this calculation evaluates $\sin\theta$, producing a number in the interval $[-1, 1]$.

This explanation requires a good understanding of radians.

The next step in this calculation evaluates $\cos(\sin\theta)$. But remember that you asked for the calculators to be set to work in degrees. So we are evaluating the cosine of an angle between $-1°$ and $1°$. In terms of radians, this is an angle between $-\frac{\pi}{180}$ radians and $\frac{\pi}{180}$ radians, which are small numbers. Thus according to our estimate for $\cos\theta$ when θ is small, $\cos(\sin\theta)$ should differ from 1 by at most $\frac{1}{2}\cdot(\frac{\pi}{180})^2$, which is about 0.00015. Because 0.00015 is a small number, we have $\cos(\sin\theta)\approx 1$.

Some people in your audience are likely to experiment and discover that the result is always 0.01745 plus some additional digits. You might ask them to try to figure out why this happens. A good hint to give them is that $\frac{\pi}{180}\approx 0.01745$.

The final step in this calculation evaluates $\tan(\cos(\sin\theta))$. Again, remember that the calculators are working in degrees. Thus the paragraph above shows that the number we are evaluating is approximately $\tan 1°$. Convert to radians, showing that the number we are evaluating is approximately $\tan\frac{\pi}{180}$. Now $\frac{\pi}{180}$ is approximately 0.01745, which is a fairly small number. Our estimate for the tangent of a small number now shows that

$$\begin{aligned}\tan(\cos(\sin\theta)) &\approx \tan 1°\\ &= \tan\tfrac{\pi}{180}\\ &\approx \tan 0.01745\\ &\approx 0.01745.\end{aligned}$$

A small amount of experimentation shows that the estimates above are good enough to produce a result of 0.01745 plus some additional digits regardless of the original angle.

EXERCISES

1. Find the area of a triangle that has sides of length 3 and 4, with a 37° angle between those sides.
2. Find the area of a triangle that has sides of length 4 and 5, with a 41° angle between those sides.
3. Find the area of a triangle that has sides of length 2 and 7, with a 3 radian angle between those sides.
4. Find the area of a triangle that has sides of length 5 and 6, with a 2 radian angle between those sides.

For Exercises 5–12 use the following figure (which is not drawn to scale):

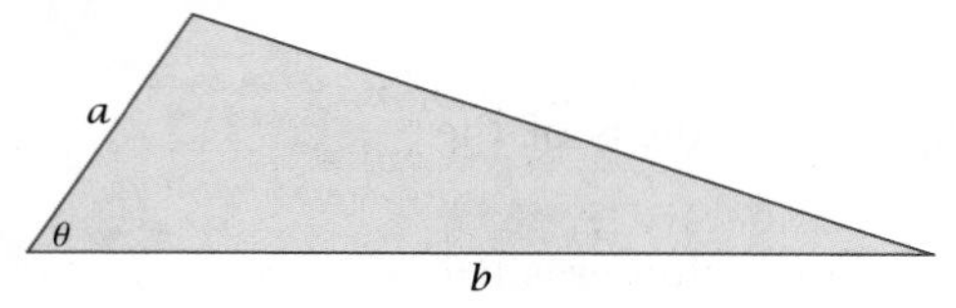

5. Find the value of b if $a = 3$, $\theta = 30°$, and the area of the triangle equals 5.
6. Find the value of a if $b = 5$, $\theta = 45°$, and the area of the triangle equals 8.
7. Find the value of a if $b = 7$, $\theta = \frac{\pi}{4}$, and the area of the triangle equals 10.
8. Find the value of b if $a = 9$, $\theta = \frac{\pi}{3}$, and the area of the triangle equals 4.
9. Find the value of θ (in radians) if $a = 7$, $b = 6$, the area of the triangle equals 15, and $\theta < \frac{\pi}{2}$.
10. Find the value of θ (in radians) if $a = 5$, $b = 4$, the area of the triangle equals 3, and $\theta < \frac{\pi}{2}$.
11. Find the value of θ (in degrees) if $a = 6$, $b = 3$, the area of the triangle equals 5, and $\theta > 90°$.

12. Find the value of θ (in degrees) if $a = 8$, $b = 5$, and the area of the triangle equals 12, and $\theta > 90°$.

13. Find the area of a parallelogram that has pairs of sides of lengths 6 and 9, with an 81° angle between two of those sides.

14. Find the area of a parallelogram that has pairs of sides of lengths 5 and 11, with a 28° angle between two of those sides.

15. Find the area of a parallelogram that has pairs of sides of lengths 4 and 10, with a $\frac{\pi}{6}$ radian angle between two of those sides.

16. Find the area of a parallelogram that has pairs of sides of lengths 3 and 12, with a $\frac{\pi}{3}$ radian angle between two of those sides.

For Exercises 17–24, use the following figure (which is not drawn to scale except that u is indeed meant to be an acute angle and v is indeed meant to be an obtuse angle):

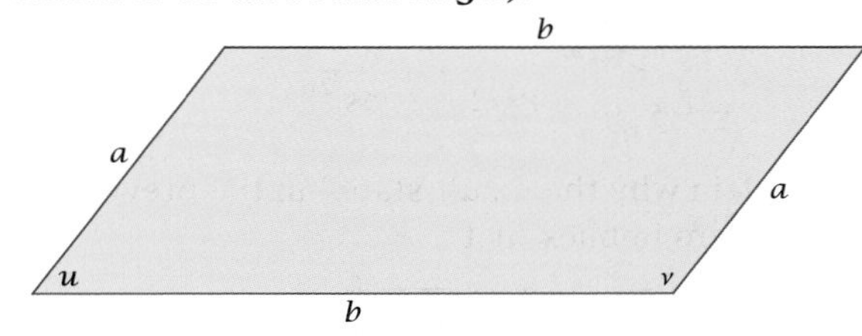

17. Find the value of b if $a = 4$, $v = 135°$, and the area of the parallelogram equals 7.

18. Find the value of a if $b = 6$, $v = 120°$, and the area of the parallelogram equals 11.

19. Find the value of a if $b = 10$, $u = \frac{\pi}{3}$, and the area of the parallelogram equals 7.

20. Find the value of b if $a = 5$, $u = \frac{\pi}{4}$, and the area of the parallelogram equals 9.

21. Find the value of u (in radians) if $a = 3$, $b = 4$, and the parallelogram has area 10.

22. Find the value of u (in radians) if $a = 4$, $b = 6$, and the parallelogram has area 19.

23. Find the value of v (in degrees) if $a = 6$, $b = 7$, and the parallelogram has area 31.

24. Find the value of v (in degrees) if $a = 8$, $b = 5$, and the parallelogram has area 12.

25. What is the largest possible area for a triangle that has one side of length 4 and one side of length 7?

26. What is the largest possible area for a parallelogram that has pairs of sides with lengths 5 and 9?

27. Sketch the regular hexagon whose vertices are six equally spaced points on the unit circle, with one of the vertices at the point $(1, 0)$.

28. Sketch the regular dodecagon whose vertices are twelve equally spaced points on the unit circle, with one of the vertices at the point $(1, 0)$. [*A* **dodecagon** *is a twelve-sided polygon.*]

29. Find the coordinates of all six vertices of the regular hexagon whose vertices are six equally spaced points on the unit circle, with $(1, 0)$ as one of the vertices. List the vertices in counterclockwise order starting at $(1, 0)$.

30. Find the coordinates of all twelve vertices of the dodecagon whose vertices are twelve equally spaced points on the unit circle, with $(1, 0)$ as one of the vertices. List the vertices in counterclockwise order starting at $(1, 0)$.

31. Find the area of a regular hexagon whose vertices are six equally spaced points on the unit circle.

32. Find the area of a regular dodecagon whose vertices are twelve equally spaced points on the unit circle.

33. Find the perimeter of a regular hexagon whose vertices are six equally spaced points on the unit circle.

34. Find the perimeter of a regular dodecagon whose vertices are twelve equally spaced points on the unit circle.

35. Find the area of a regular hexagon with sides of length s.

36. Find the area of a regular dodecagon with sides of length s.

37. Find the area of a regular 13-sided polygon whose vertices are 13 equally spaced points on a circle of radius 4.

38. Find the area of a regular 15-sided polygon whose vertices are 15 equally spaced points on a circle of radius 7.

39. The one-dollar coin in the Pacific island country Tuvalu is a regular 9-sided polygon. The distance from the center of the face of this coin to a vertex is 1.65 centimeters. Find the area of a face of the Tuvalu one-dollar coin.

40. The 50-pence coin in Great Britain is a regular 7-sided polygon (the edges are actually slightly curved, but ignore that small curvature for this exercise). The distance from the center of the face of this coin to a vertex is 1.4 centimeters. Find the area of a face of a British 50-pence coin.

PROBLEMS

41. What is the area of a triangle whose sides all have length r?

42. Explain why there does not exist a triangle with area 15 having one side of length 4 and one side of length 7.

43. Show that if a triangle has area R, sides of length A, B, and C, and angles a, b, and c, then

$$R^3 = \tfrac{1}{8}A^2B^2C^2(\sin a)(\sin b)(\sin c).$$

[*Hint:* Write three formulas for the area R, and then multiply these formulas together.]

44. Suppose a trapezoid has bases b_1 and b_2, another side with length c, and an angle θ between the side with length c and the base side with length b_1. Show that the area of the trapezoid is

$$\tfrac{1}{2}(b_1 + b_2)c \sin\theta.$$

45. Find numbers b and c such that an isosceles triangle with sides of length b, b, and c has perimeter and area that are both integers.

46. Explain why the solution to Exercise 32 is somewhat close to π.

47. Use a calculator to evaluate numerically the exact solution you obtained to Exercise 34. Then explain why this number is somewhat close to 2π.

48. Explain why a regular polygon with n sides whose vertices are n equally spaced points on the unit circle has area $\frac{n}{2}\sin\frac{2\pi}{n}$.

49. Explain why the result stated in the previous problem implies

$$\sin\frac{2\pi}{n} \approx \frac{2\pi}{n}$$

for large positive integers n.

50. Show that each edge of a regular polygon with n sides whose vertices are n equally spaced points on the unit circle has length

$$\sqrt{2 - 2\cos\tfrac{2\pi}{n}}.$$

51. Explain why a regular polygon with n sides, each with length s, has area

$$\frac{n\sin\frac{2\pi}{n}}{4(1-\cos\frac{2\pi}{n})}s^2.$$

52. Verify that for $n = 4$, the formula given by the previous problem reduces to the usual formula for the area of a square.

53. Explain why a regular polygon with n sides whose vertices are n equally spaced points on the unit circle has perimeter

$$n\sqrt{2 - 2\cos\tfrac{2\pi}{n}}.$$

54. Explain why the result stated in the previous problem implies that

$$n\sqrt{2 - 2\cos\tfrac{2\pi}{n}} \approx 2\pi$$

for large positive integers n.

55. Choose three large values of n, and use a calculator to verify that $n\sqrt{2 - 2\cos\frac{2\pi}{n}} \approx 2\pi$ for each of those three large values of n.

56. We derived the inequality $\sin\theta < \theta$ using a figure that assumed that $0 < \theta < \frac{\pi}{2}$. Does the inequality $\sin\theta < \theta$ hold for all positive values of θ?

57. We derived the inequality $\theta < \tan\theta$ using a figure that assumed that $0 < \theta < \frac{\pi}{2}$. Does the inequality $\theta < \tan\theta$ hold for all positive values of θ?

58. Show that if $\theta \approx \frac{\pi}{2}$, then $\cos\theta \approx \frac{\pi}{2} - \theta$.

59. Show that if $\theta \approx \frac{\pi}{2}$, then $\sin\theta \approx 1 - \frac{1}{2}(\frac{\pi}{2} - \theta)^2$.

60. Show that if $\theta \approx \frac{\pi}{2}$, then $\tan\theta \approx \dfrac{1}{\frac{\pi}{2} - \theta}$.

61. Suppose $|x|$ is small but nonzero. Explain why the slope of the line containing the point $(x, \sin x)$ and the origin is approximately 1.

WORKED-OUT SOLUTIONS *to Odd-numbered Exercises*

1. Find the area of a triangle that has sides of length 3 and 4, with a $37°$ angle between those sides.

   SOLUTION The area of this triangle equals $\frac{3 \cdot 4 \cdot \sin 37°}{2}$, which equals $6 \sin 37°$. A calculator shows that this is approximately 3.61 (make sure that your calculator is computing in degrees, or first convert to radians, when doing this calculation).

3. Find the area of a triangle that has sides of length 2 and 7, with a 3 radian angle between those sides.

   SOLUTION The area of this triangle equals $\frac{2 \cdot 7 \cdot \sin 3}{2}$, which equals $7 \sin 3$. A calculator shows that this is approximately 0.988 (make sure that your calculator is computing in radians, or first convert to degrees, when doing this calculation).

For Exercises 5–12 use the following figure (which is not drawn to scale):

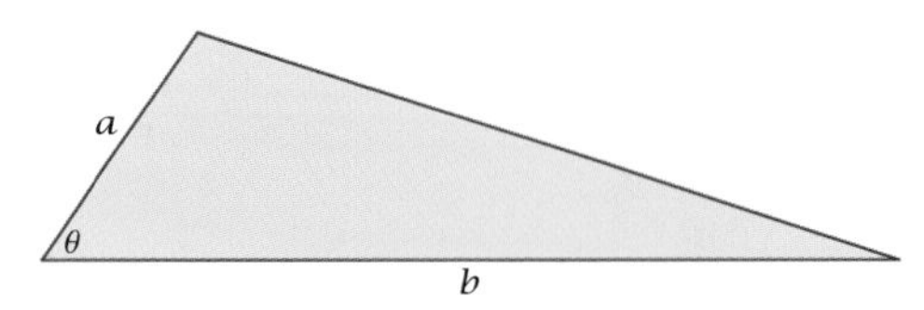

5. Find the value of b if $a = 3$, $\theta = 30°$, and the area of the triangle equals 5.

   SOLUTION Because the area of the triangle equals 5, we have

$$5 = \frac{ab \sin \theta}{2} = \frac{3b \sin 30°}{2} = \frac{3b}{4}.$$

   Solving the equation above for b, we get $b = \frac{20}{3}$.

7. Find the value of a if $b = 7$, $\theta = \frac{\pi}{4}$, and the area of the triangle equals 10.

   SOLUTION Because the area of the triangle equals 10, we have

$$10 = \frac{ab \sin \theta}{2} = \frac{7a \sin \frac{\pi}{4}}{2} = \frac{7a}{2\sqrt{2}}.$$

   Solving the equation above for a, we get $a = \frac{20\sqrt{2}}{7}$.

9. Find the value of θ (in radians) if $a = 7$, $b = 6$, the area of the triangle equals 15, and $\theta < \frac{\pi}{2}$.

   SOLUTION Because the area of the triangle equals 15, we have

$$15 = \frac{ab \sin \theta}{2} = \frac{7 \cdot 6 \cdot \sin \theta}{2} = 21 \sin \theta.$$

   Solving the equation above for $\sin \theta$, we get $\sin \theta = \frac{5}{7}$. Thus $\theta = \sin^{-1} \frac{5}{7} \approx 0.7956$.

11. Find the value of θ (in degrees) if $a = 6$, $b = 3$, the area of the triangle equals 5, and $\theta > 90°$.

   SOLUTION Because the area of the triangle equals 5, we have

$$5 = \frac{ab \sin \theta}{2} = \frac{6 \cdot 3 \cdot \sin \theta}{2} = 9 \sin \theta.$$

   Solving the equation above for $\sin \theta$, we get $\sin \theta = \frac{5}{9}$. Thus θ equals $\pi - \sin^{-1} \frac{5}{9}$ radians. Converting this to degrees, we have

$$\theta = 180° - (\sin^{-1} \tfrac{5}{9}) \tfrac{180}{\pi}° \approx 146.25°.$$

13. Find the area of a parallelogram that has pairs of sides of lengths 6 and 9, with an $81°$ angle between two of those sides.

   SOLUTION The area of this parallelogram equals $6 \cdot 9 \cdot \sin 81°$, which equals $54 \sin 81°$. A calculator shows that this is approximately 53.34.

15. Find the area of a parallelogram that has pairs of sides of lengths 4 and 10, with a $\frac{\pi}{6}$ radian angle between two of those sides.

   SOLUTION The area of this parallelogram equals $4 \cdot 10 \cdot \sin \frac{\pi}{6}$, which equals 20.

For Exercises 17–24, use the following figure (which is not drawn to scale except that u is indeed meant to be an acute angle and v is indeed meant to be an obtuse angle):

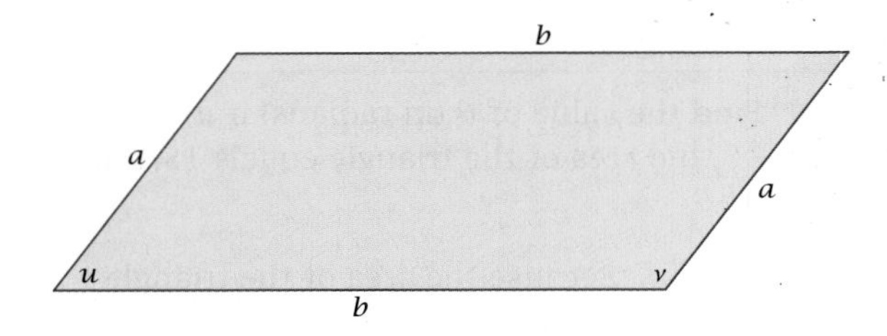

17. Find the value of b if $a = 4$, $v = 135°$, and the area of the parallelogram equals 7.

SOLUTION Because the area of the parallelogram equals 7, we have

$$7 = ab \sin v = 4b \sin 135° = 2\sqrt{2}b.$$

Solving the equation above for b, we get $b = \frac{7}{2\sqrt{2}} = \frac{7\sqrt{2}}{4}$.

19. Find the value of a if $b = 10$, $u = \frac{\pi}{3}$, and the area of the parallelogram equals 7.

SOLUTION Because the area of the parallelogram equals 7, we have

$$7 = ab \sin u = 10a \sin \tfrac{\pi}{3} = 5a\sqrt{3}.$$

Solving the equation above for a, we get $a = \frac{7}{5\sqrt{3}} = \frac{7\sqrt{3}}{15}$.

21. Find the value of u (in radians) if $a = 3$, $b = 4$, and the parallelogram has area 10.

SOLUTION Because the area of the parallelogram equals 10, we have

$$10 = ab \sin u = 3 \cdot 4 \cdot \sin u = 12 \sin u.$$

Solving the equation above for $\sin u$, we get $\sin u = \frac{5}{6}$. Thus

$$u = \sin^{-1} \tfrac{5}{6} \approx 0.9851.$$

23. Find the value of v (in degrees) if $a = 6$, $b = 7$, and the parallelogram has area 31.

SOLUTION Because the area of the parallelogram equals 31, we have

$$31 = ab \sin v = 6 \cdot 7 \cdot \sin v = 42 \sin v.$$

Solving the equation above for $\sin v$, we get $\sin v = \frac{31}{42}$. Because v is an obtuse angle, we thus have $v = \pi - \sin^{-1} \frac{31}{42}$ radians. Converting this to degrees, we have $v = 180° - (\sin^{-1} \frac{31}{42})\frac{180°}{\pi} \approx 132.43°$.

25. What is the largest possible area for a triangle that has one side of length 4 and one side of length 7?

SOLUTION In a triangle that has one side of length 4 and one side of length 7, let θ denote the angle between those two sides. Thus the area of the triangle will equal

$$14 \sin \theta.$$

We need to choose θ to make this area as large as possible. The largest possible value of $\sin \theta$ is 1, which occurs when $\theta = \frac{\pi}{2}$ (or $\theta = 90°$ if we are working in degrees). Thus we choose $\theta = \frac{\pi}{2}$, which gives us a right triangle with sides of length 4 and 7 around the right angle.

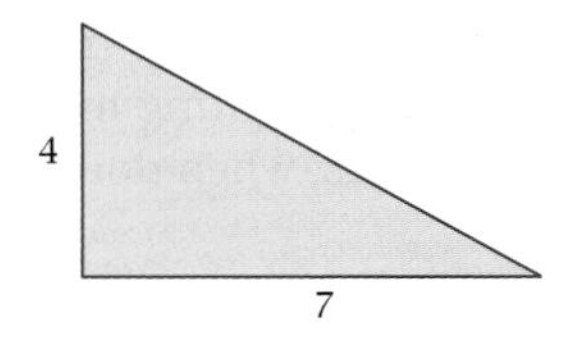

This right triangle has area 14, which is the largest area of any triangle with sides of length 4 and 7.

27. Sketch the regular hexagon whose vertices are six equally spaced points on the unit circle, with one of the vertices at the point $(1, 0)$.

SOLUTION

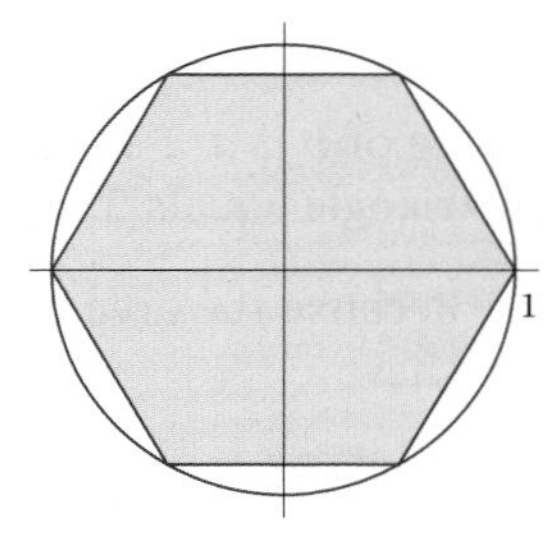

29. Find the coordinates of all six vertices of the regular hexagon whose vertices are six equally spaced points on the unit circle, with $(1, 0)$ as one of the vertices. List the vertices in counterclockwise order starting at $(1, 0)$.

SOLUTION The coordinates of the six vertices, listed in counterclockwise order starting at $(1, 0)$, are $(\cos \frac{2\pi m}{6}, \sin \frac{2\pi m}{6})$, with m going from 0 to 5. Evaluating the trigonometric functions, we get the following list of coordinates of vertices: $(1,0)$, $(\frac{1}{2}, \frac{\sqrt{3}}{2})$, $(-\frac{1}{2}, \frac{\sqrt{3}}{2})$, $(-1,0)$, $(-\frac{1}{2}, -\frac{\sqrt{3}}{2})$, $(\frac{1}{2}, -\frac{\sqrt{3}}{2})$.

31. Find the area of a regular hexagon whose vertices are six equally spaced points on the unit circle.

SOLUTION Decompose the hexagon into triangles by drawing line segments from the center of the circle (the origin) to the vertices. Each triangle has two sides that are radii of the unit circle; thus those two sides of the triangle each have length 1. The angle between those two radii is $\frac{2\pi}{6}$ radians (because one rotation around the entire circle is an angle of 2π radians, and each of the six triangles has an angle that takes up one-sixth of the total). Now $\frac{2\pi}{6}$ radians equals $\frac{\pi}{3}$ radians (or 60°). Thus each of the six triangles has area

$$\tfrac{1}{2} \cdot 1 \cdot 1 \cdot \sin \tfrac{\pi}{3},$$

which equals $\frac{\sqrt{3}}{4}$. Thus the sum of the areas of the six triangles equals $6 \cdot \frac{\sqrt{3}}{4}$, which equals $\frac{3\sqrt{3}}{2}$. In other words, the hexagon has area $\frac{3\sqrt{3}}{2}$.

33. Find the perimeter of a regular hexagon whose vertices are six equally spaced points on the unit circle.

SOLUTION If we assume that one of the vertices of the hexagon is the point $(1,0)$, then the next vertex in the counterclockwise direction is the point $(\frac{1}{2}, \frac{\sqrt{3}}{2})$. Thus the length of each side of the hexagon equals the distance between $(1,0)$ and $(\frac{1}{2}, \frac{\sqrt{3}}{2})$, which equals

$$\sqrt{(1 - \tfrac{1}{2})^2 + (\tfrac{\sqrt{3}}{2})^2},$$

which equals 1. Thus the perimeter of the hexagon equals $6 \cdot 1$, which equals 6.

35. Find the area of a regular hexagon with sides of length s.

SOLUTION This computation could be done by using the technique shown in Example 6. For variety, another method will be shown here.

By the Area Stretch Theorem (see Section 2.4), there is a constant c such that a regular hexagon with sides of length s has area cs^2. From Exercises 31 and 33, we know that the area equals $\frac{3\sqrt{3}}{2}$ if $s = 1$. Thus

$$\tfrac{3\sqrt{3}}{2} = c \cdot 1^2 = c.$$

Thus a regular hexagon with sides of length s has area $\frac{3\sqrt{3}}{2}s^2$.

37. Find the area of a regular 13-sided polygon whose vertices are 13 equally spaced points on a circle of radius 4.

SOLUTION Decompose the 13-sided polygon into triangles by drawing line segments from the center of the circle to the vertices. Each triangle has two sides that are radii of the circle of radius 4; thus those two sides of the triangle each have length 4. The angle between those two radii is $\frac{2\pi}{13}$ radians (because one rotation around the entire circle is an angle of 2π radians, and each of the 13 triangles has an angle that takes up one-thirteenth of the total). Thus each of the 13 triangles has area

$$\tfrac{1}{2} \cdot 4 \cdot 4 \cdot \sin \tfrac{2\pi}{13},$$

which equals $8 \sin \frac{2\pi}{13}$. The area of the 13-sided polygon is the sum of the areas of the 13 triangles, which equals $13 \cdot 8 \sin \frac{2\pi}{13}$, which is approximately 48.3.

39. The one-dollar coin in the Pacific island country Tuvalu is a regular 9-sided polygon. The distance from the center of the face of this coin to a vertex is 1.65 centimeters. Find the area of a face of the Tuvalu one-dollar coin.

SOLUTION Decompose a regular 9-sided polygon representing the Tuvalu one-dollar coin into triangles by drawing line segments from the center to the vertices. Each triangle has two sides of length 1.65 centimeters. The angle between those two sides is 40° (because one full rotation is an angle of 360° and each of the 9 triangles has an angle that takes up one-ninth of the total). Thus each of the 9 triangles has area

$$\tfrac{1}{2} \cdot 1.65 \cdot 1.65 \cdot \sin 40°$$

square centimeters. The area of the 9-sided polygon is the sum of the areas of the 9 triangles, which equals $9 \cdot \frac{1}{2} \cdot 1.65 \cdot 1.65 \cdot \sin 40°$ square centimeters, which is approximately 7.875 square centimeters.

10.4 *The Law of Sines and the Law of Cosines*

LEARNING OBJECTIVES

By the end of this section you should be able to

- find angles or side lengths in a triangle using the law of sines;
- find angles or side lengths in a triangle using the law of cosines;
- determine when to use which of these two laws.

In this section we will learn how to find all the angles and the lengths of all the sides of a triangle given only some of this data.

The Law of Sines

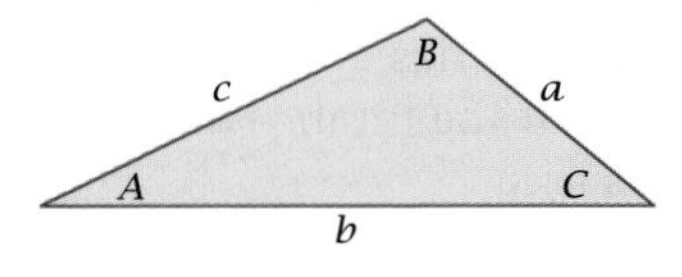

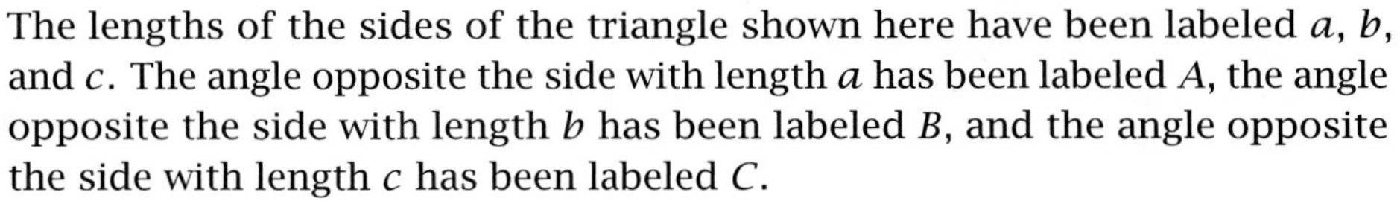

The lengths of the sides of the triangle shown here have been labeled a, b, and c. The angle opposite the side with length a has been labeled A, the angle opposite the side with length b has been labeled B, and the angle opposite the side with length c has been labeled C.

We know from the previous section that the area of the triangle equals one-half the product of the lengths of any two sides times the sine of the angle between those two sides. Different choices of the two sides of the triangle will lead to different formulas for the area of the triangle. As we are about to see, setting those different formulas for the area equal to each other leads to an interesting result.

Using the sides with lengths b and c, we see that the area of the triangle equals

$$\tfrac{1}{2}bc \sin A.$$

Using the sides with lengths a and c, we see that the area of the triangle equals

$$\tfrac{1}{2}ac \sin B.$$

Using the sides with lengths a and b, we see that the area of the triangle equals

$$\tfrac{1}{2}ab \sin C.$$

Setting the three formulas obtained above for the area of the triangle equal to each other, we get

$$\tfrac{1}{2}bc \sin A = \tfrac{1}{2}ac \sin B = \tfrac{1}{2}ab \sin C.$$

Multiplying all three expressions above by 2 and then dividing all three expressions by abc gives a result called the **law of sines**:

Law of sines

$$\frac{\sin A}{a} = \frac{\sin B}{b} = \frac{\sin C}{c}$$

in a triangle with sides whose lengths are a, b, and c, with corresponding angles A, B, and C opposite those sides.

Using the Law of Sines

The following example shows how the law of sines can be used to find the lengths of all three sides of a triangle given only two angles of the triangle and the length of one side.

EXAMPLE 1

Find the lengths of all three sides of the triangle shown here in the margin.

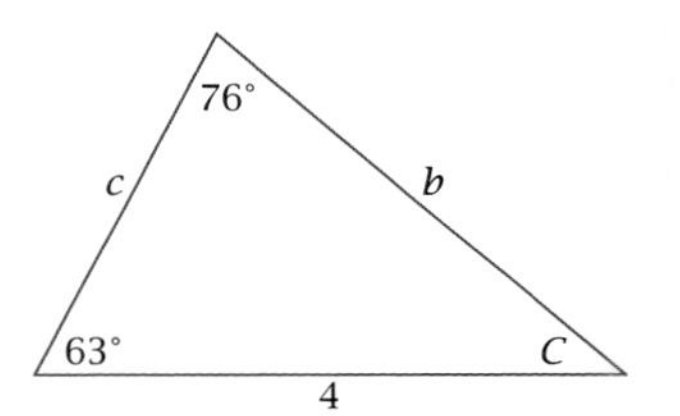

SOLUTION Applying the law of sines to this triangle, we have

$$\frac{\sin 76°}{4} = \frac{\sin 63°}{b}.$$

Solving for b, we get

$$b = 4\frac{\sin 63°}{\sin 76°} \approx 3.67,$$

where the approximate value was obtained with the use of a calculator.

To find the length c, we want to apply the law of sines. Thus we first find the angle C. We have

$$C = 180° - 63° - 76° = 41°.$$

Now applying the law of sines again to the triangle above, we have

$$\frac{\sin 76°}{4} = \frac{\sin 41°}{c}.$$

Solving for c, we get

$$c = 4\frac{\sin 41°}{\sin 76°} \approx 2.70.$$

When using the law of sines, sometimes the same ambiguity arises as we saw in the previous section, as illustrated in the following example.

EXAMPLE 2

Find all the angles in a triangle that has one side of length 8, one side of length 5, and a 30° angle opposite the side of length 5.

SOLUTION Labeling the triangle as on the first page of this section, we take $b = 8$, $c = 5$, and $C = 30°$. Applying the law of sines, we have

$$\frac{\sin B}{8} = \frac{\sin 30°}{5}.$$

Using the information that $\sin 30° = \frac{1}{2}$, we can solve the equation above for $\sin B$, getting

$$\sin B = \tfrac{4}{5}.$$

Now $\sin^{-1}\frac{4}{5}$, when converted from radians to degrees, is approximately 53°, which suggests that $B \approx 53°$. However, 180° minus this angle also has a sine equal to $\frac{4}{5}$, which suggests that $B \approx 127°$. There is no way to distinguish between these two choices, which are shown below, unless we have some additional information (for example, we might know that B is an obtuse angle, and in that case we would choose $B \approx 127°$).

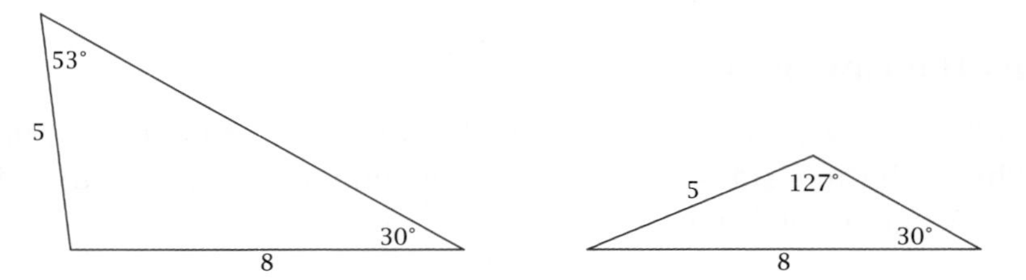

Both these triangles have one side of length 8, one side of length 5, and a 30° angle opposite the side of length 5.

Once we decide between the two possible choices of approximately 53° or 127° for the angle opposite the side of length 8, the other angle in the triangle is forced upon us by the requirement that the sum of the angles in a triangle equals 180°. Thus if we make the choice on the left above, then the unlabeled angle is approximately 97°, but if we make the choice on the right, then the unlabeled angle is approximately 23°.

The law of sines does not always lead to an ambiguity when given the lengths of two sides of a triangle and the angle opposite one of the sides, as shown in the following example.

EXAMPLE 3

Find all the angles in a triangle that has one side of length 5, one side of length 7, and a 100° angle opposite the side of length 7.

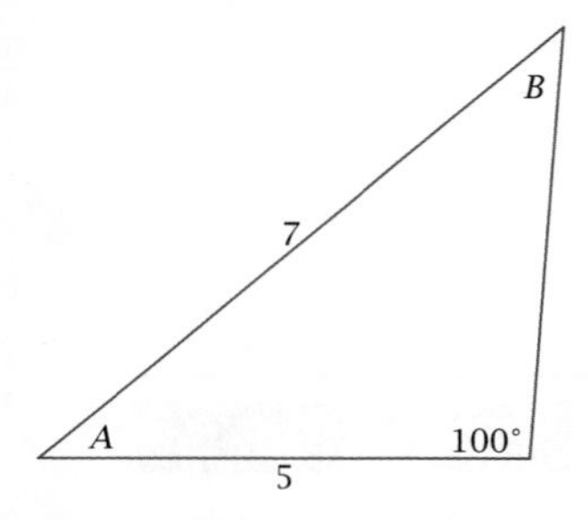

A triangle that has one side of length 5, one side of length 7, and a 100° angle opposite the side of length 7 must look like this, where $A \approx 35.3°$ and $B \approx 44.7°$.

SOLUTION Labeling the angles of the triangle as shown here and applying the law of sines, we have

$$\frac{\sin B}{5} = \frac{\sin 100°}{7}.$$

Thus

$$\sin B = \frac{5 \sin 100°}{7} \approx 0.703.$$

Now $\sin^{-1} 0.703$, when converted from radians to degrees, is approximately 44.7°, which suggests that $B \approx 44.7°$. Note that 180° minus this angle also has a sine equal to 0.703, which suggests that $B \approx 135.3°$ might be another possible choice for B. However, that choice would give us a triangle with angles of 100° and 135.3°, which adds up to more than 180°. Thus this second choice is not possible. Hence there is no ambiguity here—we must have $B \approx 44.7°$.

Because $180° - 100° - 44.7° = 35.3°$, the angle A in the triangle is approximately 35.3°. Now that we know all three angles of the triangle, we could use the law of sines to find the length of the side opposite A.

The Law of Cosines

The law of sines is a wonderful tool for finding the lengths of all three sides of a triangle when we know two of the angles of the triangle (which means that we know all three angles) and the length of at least one side of the triangle. Also, if we know the lengths of two sides of a triangle and one of the angles other than the angle between those two sides, then the law of sines allows us to find the other angles and the length of the other side, although it may produce two possible choices rather than a unique solution.

However, the law of sines is of no use if we know the lengths of all three sides of a triangle and want to find the angles of the triangle. Similarly, the law of sines cannot help us if the only information we know about a triangle is the length of two sides and the angle between those sides. Fortunately the law of cosines, our next topic, provides the necessary tools for these tasks.

As we will see, the law of cosines is a generalization to all triangles of the Pythagorean Theorem, which applies only to right triangles.

Consider a triangle with sides of lengths a, b, and c and an angle of C opposite the side of length c, as shown here.

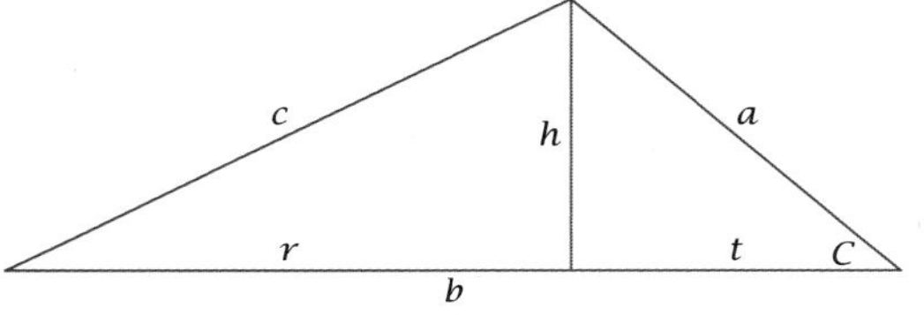

Drop a perpendicular line segment from the vertex opposite the side of length b to the side of length b, as shown above. The length of this line segment is the height of the triangle; label it h. The endpoint of this line segment of length h divides the side of the triangle of length b into two smaller line segments, which we have labeled r and t above.

The line segment of length h shown above divides the original larger triangle into two smaller right triangles. Looking at the right triangle on the right, we see that $\sin C = \frac{h}{a}$. Thus

$$h = a \sin C.$$

Furthermore, looking at the same right triangle, we see that $\cos C = \frac{t}{a}$. Thus

$$t = a \cos C.$$

The figure above also shows that $r = b - t$. Using the equation above for t, we thus have

$$r = b - a \cos C.$$

For convenience, we now redraw the figure above, replacing h, t, and r with the values we have just found for them.

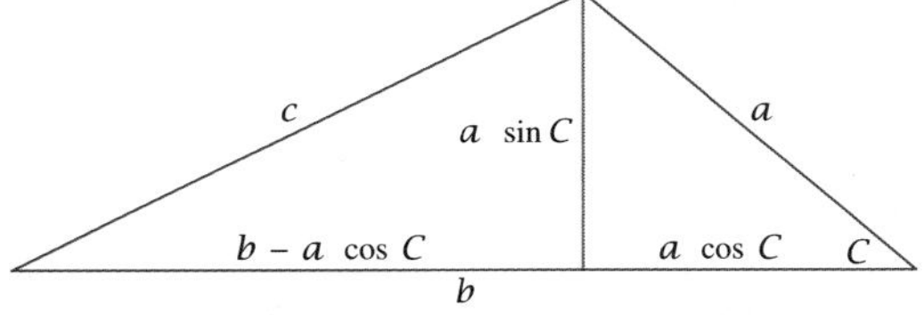

In the figure above, consider the right triangle on the left. This right triangle has a hypotenuse of length c and sides of length $a \sin C$ and $b - a \cos C$. By the Pythagorean Theorem, we have

$$\begin{aligned} c^2 &= (a \sin C)^2 + (b - a \cos C)^2 \\ &= a^2 \sin^2 C + b^2 - 2ab \cos C + a^2 \cos^2 C \\ &= a^2(\sin^2 C + \cos^2 C) + b^2 - 2ab \cos C \\ &= a^2 + b^2 - 2ab \cos C. \end{aligned}$$

Thus we have shown that

$$c^2 = a^2 + b^2 - 2ab \cos C.$$

This result is called the **law of cosines.**

> ***Law of cosines***
>
> $$c^2 = a^2 + b^2 - 2ab \cos C$$
>
> in a triangle with sides whose lengths are a, b, and c, with an angle of C opposite the side with length c.

This reformulation allows use of the law of cosines regardless of the labels used for sides and angles.

The law of cosines can be restated without symbols as follows: In any triangle, the length squared of one side equals the sum of the squares of the lengths of the other two sides minus twice the product of those two lengths times the cosine of the angle opposite the first side.

Suppose we have a right triangle, with hypotenuse of length c and sides of lengths a and b. In this case we have $C = \frac{\pi}{2}$ (or $C = 90°$ if we want to work in degrees). Thus $\cos C = 0$. Hence the law of cosines in this case becomes

$$c^2 = a^2 + b^2,$$

which is the familiar Pythagorean Theorem.

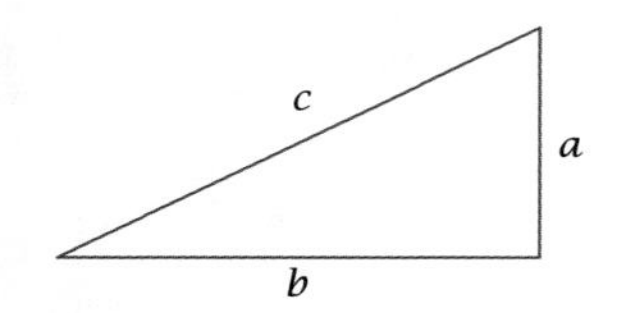

For a right triangle, the law of cosines reduces to the Pythagorean Theorem.

Using the Law of Cosines

The following example shows how the law of cosines can be used to find all three angles of a triangle given only the lengths of the three sides. The idea is to use the law of cosines to solve for the cosine of each angle of the triangle. Unlike the situation that sometimes arises with the law of sines, there will be no ambiguity because no two angles between 0 radians and π radians (or between 0° and 180° if we work in degrees) have the same cosine.

EXAMPLE 4

Find all three angles of the triangle shown here in the margin..

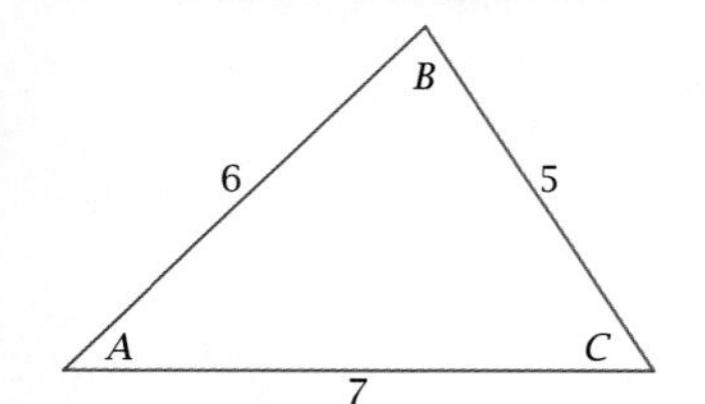

SOLUTION Here we know that the triangle has sides of lengths 5, 6, and 7, but we do not know any of the angles. The angles have been labeled in this figure. Applying the law of cosines, we have

$$6^2 = 5^2 + 7^2 - 2 \cdot 5 \cdot 7 \cos C.$$

Solving the equation above for $\cos C$, we get

$$\cos C = \tfrac{19}{35}.$$

Thus $C = \cos^{-1} \frac{19}{35}$, which is approximately 0.997 radians (or, equivalently, approximately 57.1°).

Now we apply the law of cosines again, this time focusing on the angle B, getting

$$7^2 = 5^2 + 6^2 - 2 \cdot 5 \cdot 6 \cos B.$$

Solving the equation above for $\cos B$, we get

$$\cos B = \tfrac{1}{5}.$$

Thus $B = \cos^{-1} \frac{1}{5}$, which is approximately 1.37 radians (or, equivalently, approximately 78.5°).

To find the third angle A, we could simply subtract from π (or from 180° if we are using degrees) the sum of the other two angles. But as a check that we have not made any errors, we will instead use the law of cosines again, this time focusing on the angle A. We have

$$5^2 = 7^2 + 6^2 - 2 \cdot 7 \cdot 6 \cos A.$$

Solving the equation above for $\cos A$, we get

$$\cos A = \tfrac{5}{7}.$$

Thus $A = \cos^{-1} \frac{5}{7}$, which is approximately 0.775 radians (or, equivalently, approximately 44.4°).

As a check, we can add up our approximate solutions in angles. Because

$$78.5 + 57.1 + 44.4 = 180,$$

all is well.

The next example shows how the law of cosines can be used to find the lengths of all the sides of a triangle given the lengths of two sides and the angle between them.

EXAMPLE 5

Find the lengths of all three sides of the triangle shown here in the margin.

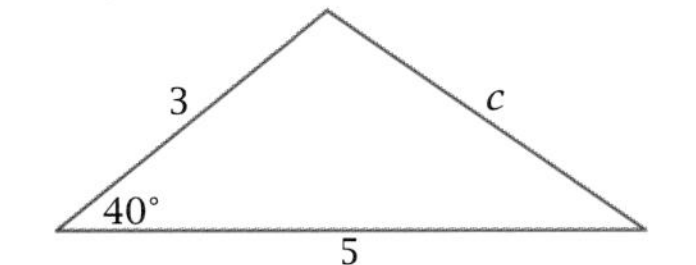

SOLUTION Here we know that the triangle has sides of lengths 3 and 5 and that the angle between them equals 40°. The side opposite that angle has been labeled c. By the law of cosines, we have

$$c^2 = 3^2 + 5^2 - 2 \cdot 3 \cdot 5 \cos 40°.$$

Thus

$$c = \sqrt{34 - 30 \cos 40°} \approx 3.32.$$

Now that we know the lengths of all three sides of the triangle, we could use the law of cosines twice more to find the other two angles, using the same procedure as in the last example.

When to Use Which Law

A triangle has three angles and three side lengths. If you know some of these six pieces of data, you can often use either the law of sines or the law of cosines to determine the remainder of the data about the triangle. To determine which law to use, think about how to come up with an equation that has only one unknown:

- If you know only the lengths of the three sides of a triangle, then the law of sines is not useful because it involves two angles, both of which are unknown. Thus if you know only the lengths of the three sides of a triangle, use the law of cosines.

- If you know only the lengths of two sides of a triangle and the angle between them, then any use of the law of sines leads to an equation with either an unknown side and an unknown angle or an equation with two unknown angles. Either way, with two unknowns you will not be able to solve the equation; thus the law of sines is not useful in this situation. Hence use the law of cosines if you know the lengths of two sides of a triangle and the angle between them.

The term law *is unusual in mathematics. The law of sines and law of cosines could have been called the* sine theorem *and* cosine theorem.

Sometimes you have enough data so that either the law of sines or the law of cosines could be used, as discussed below:

- Suppose you start by knowing the lengths of all three sides of a triangle. The only possibility in this situation is first to use the law of cosines to find one of the angles. Then, knowing the lengths of all three sides of the triangle and one angle, you could use either the law of cosines or the law of sines to find another angle. However, the law of sines may lead to two choices for the angle rather than a unique choice; thus it is better to use the law of cosines in this situation.

- Another case where you could use either law is when you know the length of two sides of a triangle and an angle other than the angle between those two sides. With the notation from the beginning of this section, suppose we know a, c, and C. We could use either the law of sines or the law of cosines to get an equation with only one unknown:

$$\frac{\sin A}{a} = \frac{\sin C}{c} \quad \text{or} \quad c^2 = a^2 + b^2 - 2ab\cos C.$$

  The first equation above, where A is the unknown, may lead to two possible choices for A. Similarly, the second equation above, where b is the unknown and we need to use the quadratic formula to solve for b, may lead to two possible choices for b. Thus both laws may give us two choices. The law of sines is probably a bit simpler to apply.

The box below summarizes when to use which law. As usual, you will be better off understanding how these guidelines arise (you can then always reconstruct them) rather than memorizing them.

When to use which law

Use the law of cosines if you know

- the lengths of all three sides of a triangle;
- the lengths of two sides of a triangle and the angle between them.

Use the law of sines if you know

- two angles of a triangle and the length of one side;
- the length of two sides of a triangle and an angle other than the angle between those two sides.

If you know two angles of a triangle, then finding the third angle is easy because the sum of the angles of a triangle equals π *radians or* 180°.

EXAMPLE 6

A surveyor needs to measure the distance to a landmark on the inaccessible side of a large canyon. There are two observation posts on the accessible side of the canyon, separated by 563 meters. From observation post 1, the surveyor uses a surveying instrument to find that the angle landmark–observation post 1–observation post 2 is 84°. From observation post 2, the surveyor finds that the angle landmark–observation post 2–observation post 1 is 87°.

(a) What is the distance from observation post 1 to the landmark?

(b) What is the distance from observation post 2 to the landmark?

SOLUTION

(a) The figure here shows a sketch of the situation. We need to find a, which is the distance from observation post 1 to the landmark. Thus we need an equation in which a is the only unknown. The law of cosines involves all three lengths in a triangle; here we know only one of those lengths, and hence the law of cosines will have at least two unknowns.

Thus we need to use the law of sines. We could set $\frac{\sin 87°}{a}$ equal to $\frac{\sin 84°}{b}$, but this does not help us because we do not know b. Hence we set $\frac{\sin 87°}{a}$ equal to $\frac{\sin C}{563}$. Fortunately we can easily find C, because the equation $84 + 87 + C = 180$ implies that $C = 9°$. Thus we have

$$\frac{\sin 87°}{a} = \frac{\sin 9°}{563},$$

which implies that

$$a = 563\frac{\sin 87°}{\sin 9°} \approx 3594.$$

Thus the distance from observation post 1 to the landmark is approximately 3594 meters.

(b) Using the same reasoning as in the solution to part (a), we have

$$b = 563\frac{\sin 84°}{\sin 9°} \approx 3579.$$

Thus the distance from observation post 2 to the landmark is approximately 3579 meters.

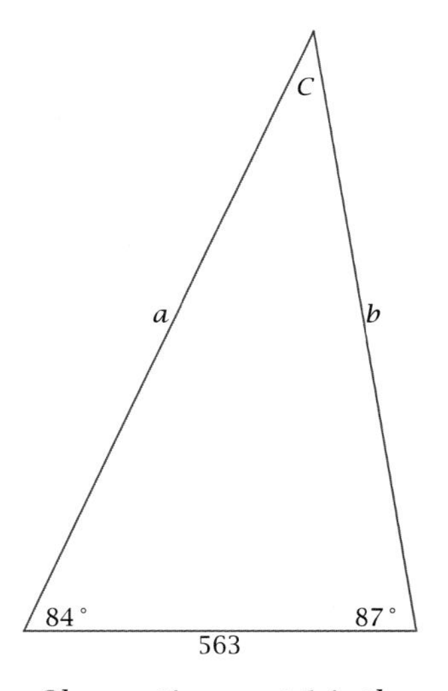

Observation post 1 is the vertex at the 84° angle. Observation post 2 is the vertex at the 87° angle. The landmark is the other vertex. This sketch is not drawn to scale.

EXERCISES

In Exercises 1–16 use the following figure (which is not drawn to scale). When an exercise requests that you evaluate an angle, give answers in both radians and degrees.

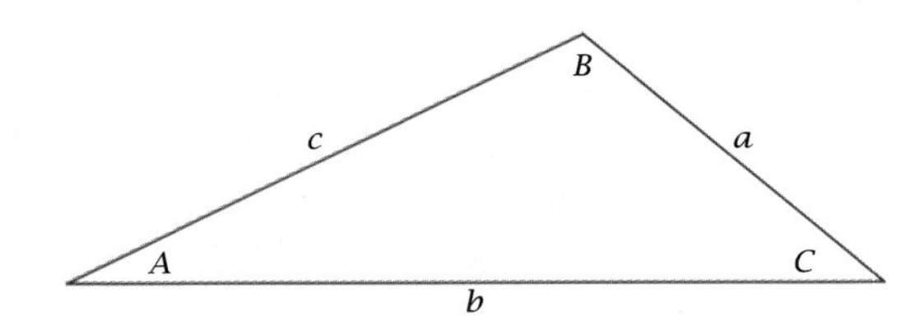

1. Suppose $a = 6$, $B = 25°$, and $C = 40°$. Evaluate:
   (a) A (b) b (c) c

2. Suppose $a = 7$, $B = 50°$, and $C = 35°$. Evaluate:
   (a) A (b) b (c) c

3. Suppose $a = 6$, $A = \frac{\pi}{7}$ radians, and $B = \frac{4\pi}{7}$ radians. Evaluate:
   (a) C (b) b (c) c

4. Suppose $a = 4$, $B = \frac{2\pi}{11}$ radians, and $C = \frac{3\pi}{11}$ radians. Evaluate:
   (a) A (b) b (c) c

5. Suppose $a = 3$, $b = 5$, and $c = 6$. Evaluate:
   (a) A (b) B (c) C

6. Suppose $a = 4$, $b = 6$, and $c = 7$. Evaluate:
   (a) A (b) B (c) C

7. Suppose $a = 5$, $b = 6$, and $c = 9$. Evaluate:
   (a) A (b) B (c) C

8. Suppose $a = 6$, $b = 7$, and $c = 8$. Evaluate:
   (a) A (b) B (c) C

9. Suppose $a = 2$, $b = 3$, and $C = 37°$. Evaluate:
   (a) c (b) A (c) B

10. Suppose $a = 5$, $b = 7$, and $C = 23°$. Evaluate:
    (a) c (b) A (c) B

11. Suppose $a = 3$, $b = 4$, and $C = 1$ radian. Evaluate:
    (a) c (b) A (c) B

12. Suppose $a = 4$, $b = 5$, and $C = 2$ radians. Evaluate:
    (a) c (b) A (c) B

13. Suppose $a = 4$, $b = 3$, and $B = 30°$. Evaluate:
    (a) A (assume that $A < 90°$)
    (b) C
    (c) c

14. Suppose $a = 14$, $b = 13$, and $B = 60°$. Evaluate:
    (a) A (assume that $A < 90°$)
    (b) C
    (c) c

Exercises 15 and 16 should be compared with Exercises 13 and 14.

15. Suppose $a = 4$, $b = 3$, and $B = 30°$. Evaluate:
    (a) A (assume that $A > 90°$)
    (b) C
    (c) c

16. Suppose $a = 14$, $b = 13$, and $B = 60°$. Evaluate:
    (a) A (assume that $A > 90°$)
    (b) C
    (c) c

17. Find the lengths of both diagonals in a parallelogram that has two sides of length 4, two sides of length 7, and two angles of 38°.

18. Find the lengths of both diagonals in a parallelogram that has two sides of length 5, two sides of length 11, and two angles of 41°.

19. Find the angle between the shorter diagonal of the parallelogram in Exercise 17 and a side with length 4.

20. Find the angle between the shorter diagonal of the parallelogram in Exercise 18 and a side with length 5.

PROBLEMS

21. The famous Flatiron Building in New York City often appears in popular culture (for example, in the *Spider-Man* movies) because of its unusual triangular shape. The base of the Flatiron Building is a triangle whose sides have lengths 190 feet, 173 feet, and 87 feet. Find the angles of the Flatiron Building.

22. Suppose the minute hand of a clock is 5 inches long, and the hour hand is 3 inches long. Suppose the angle formed by the minute hand and hour hand is 68°.

    (a) Find the distance between the endpoint of the minute hand and the endpoint of the hour hand by using the law of cosines.

    (b) Find the distance between the endpoint of the minute hand and the endpoint of the hour hand by assuming that the center of the clock is located at the origin, choosing a convenient location for the minute hand and finding the coordinates of its endpoint, then finding the coordinates of the hour hand in a position that makes a 68° angle with the minute hand, and finally using the usual distance formula to find the distance between the endpoint of the minute hand and the endpoint of the hour hand.

    (c) Make sure that your answers for parts (a) and (b) are the same. Which method did you find easier?

23. Write the law of sines in the special case of a right triangle.

24. Show how the previous problem gives the familiar characterization of the sine of an angle in a right triangle as the length of the opposite side divided by the length of the hypotenuse.

25. Show how Problem 23 gives the familiar characterization of the tangent of an angle in a right triangle as the length of the opposite side divided by the length of the adjacent side.

26. In a triangle whose sides and angles are labeled as in the instructions before Exercise 1, let h denote the height as measured from the vertex of angle B to the side with length b.

    (a) Express h in terms of A and c.

    (b) Express h in terms of C and a.

    (c) Setting the two values of h obtained in parts (a) and (b) equal to each other gives an equation. What result from this section leads to the same equation?

27. A surveyor on the south bank of a river needs to measure the distance from a boulder on the south bank of the river to a tree on the north bank of the river. The surveyor measures the distance from the boulder to a small hill on the south bank of the river and finds that distance to be 413 feet. From the boulder, the surveyor uses a surveying instrument to find that the angle tree–boulder–hill is 71°. From the hill, the surveyor finds that the angle tree–hill–boulder is 93°.

    (a) What is the distance from the boulder to the tree?

    (b) What is the distance from the hill to the tree?

28. Suppose a triangle has sides of length a, b, and c satisfying the equation

$$a^2 + b^2 = c^2.$$

Show that this triangle is a right triangle.

29. Show that in a triangle whose sides have lengths a, b, and c, the angle between the sides of length a and b is an acute angle if and only if

$$a^2 + b^2 > c^2.$$

30. Show that

$$p = \frac{r}{\sqrt{2}\sqrt{1 - \cos\theta}}$$

in an isosceles triangle that has two sides of length p, an angle of θ between these two sides, and a third side of length r.

31. Use the law of cosines to show that if a, b, and c are the lengths of the three sides of a triangle, then

$$c^2 > a^2 + b^2 - 2ab.$$

32. Use the previous problem to show that in every triangle, the sum of the lengths of any two sides is greater than the length of the third side.

33. Suppose one side of a triangle has length 5 and another side of the triangle has length 8. Let c denote the length of the third side of the triangle. Show that $3 < c < 13$.

34. Suppose you need to walk from a point P to a point Q. You can either walk in a line from P to Q, or you can walk in a line from P to another point R and then walk in a line from R to Q. Use the previous problem to determine which of these two paths is shorter.

35. Suppose you are asked to find the angle C formed by the sides of length 2 and 3 in a triangle whose sides have length 2, 3, and 7.

    (a) Show that in this situation the law of cosines leads to the equation $\cos C = -3$.

    (b) There is no angle whose cosine equals -3. Thus part (a) seems to give a counterexample to the law of cosines. Explain what is happening here.

36. The law of cosines is stated in this section using the angle C. Using the labels of the triangle just before Exercise 1, write two versions of the law of cosines, one involving the angle A and one involving the angle B.

37. Use one of the examples from this section to show that

$$\cos^{-1}\tfrac{1}{5} + \cos^{-1}\tfrac{5}{7} + \cos^{-1}\tfrac{19}{35} = \pi.$$

38. Discover another equation similar to the one given in the previous problem by choosing a triangle whose side lengths are all integers and using the law of cosines.

39. Show that

$$a(\sin B - \sin C) + b(\sin C - \sin A) + c(\sin A - \sin B) = 0$$

in a triangle with sides whose lengths are a, b, and c, with corresponding angles A, B, and C opposite those sides.

40. Show that

$$a^2 + b^2 + c^2 = 2(bc\cos A + ac\cos B + ab\cos C)$$

in a triangle with sides whose lengths are a, b, and c, with corresponding angles A, B, and C opposite those sides.

41. Show that

$$c = b\cos A + a\cos B$$

in a triangle with sides whose lengths are a, b, and c, with corresponding angles A, B, and C opposite those sides.
[*Hint:* Add together the equations $a^2 = b^2 + c^2 - 2bc\cos A$ and $b^2 = a^2 + c^2 - 2ac\cos B$.]

WORKED-OUT SOLUTIONS *to Odd-numbered Exercises*

In Exercises 1–16 use the following figure (which is not drawn to scale). When an exercise requests that you evaluate an angle, give answers in both radians and degrees.

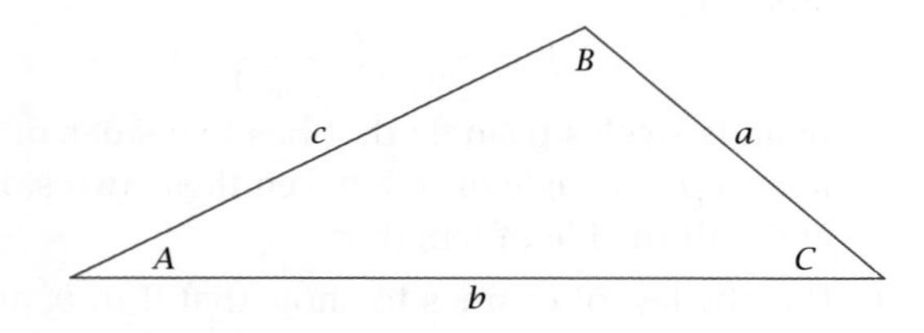

1. Suppose $a = 6$, $B = 25°$, and $C = 40°$. Evaluate:

   (a) A (b) b (c) c

SOLUTION

(a) The angles in a triangle add up to 180°. Thus $A + B + C = 180°$. Solving for A, we have

$$A = 180° - B - C = 180° - 25° - 40° = 115°.$$

Multiplying by $\frac{\pi}{180°}$ to convert to radians gives

$$A = 115° = \frac{23\pi}{36} \text{ radians} \approx 2.007 \text{ radians}.$$

(b) Use the law of sines in the form

$$\frac{\sin A}{a} = \frac{\sin B}{b},$$

which in this case becomes the equation

$$\frac{\sin 115°}{6} = \frac{\sin 25°}{b}.$$

Solve the equation above for b, getting

$$b = \frac{6\sin 25°}{\sin 115°} \approx 2.80.$$

(c) Use the law of sines in the form

$$\frac{\sin A}{a} = \frac{\sin C}{c},$$

which in this case becomes the equation

$$\frac{\sin 115°}{6} = \frac{\sin 40°}{c}.$$

Solve the equation above for c, getting

$$c = \frac{6\sin 40°}{\sin 115°} \approx 4.26.$$

3. Suppose $a = 6$, $A = \frac{\pi}{7}$ radians, and $B = \frac{4\pi}{7}$ radians. Evaluate:

(a) C (b) b (c) c

SOLUTION

(a) The angles in a triangle add up to π radians. Thus $A + B + C = \pi$. Solving for C, we have

$$C = \pi - A - B = \pi - \tfrac{\pi}{7} - \tfrac{4\pi}{7} = \tfrac{2\pi}{7}.$$

Multiplying by $\frac{180°}{\pi}$ to convert to radians gives

$$C = \tfrac{2\pi}{7} \text{ radians} = \tfrac{360}{7}°.$$

Using a calculator to obtain decimal approximations, we have

$$C \approx 0.8976 \text{ radians} \approx 51.429°.$$

(b) Use the law of sines in the form

$$\frac{\sin A}{a} = \frac{\sin B}{b},$$

which in this case becomes the equation

$$\frac{\sin \frac{\pi}{7}}{6} = \frac{\sin \frac{4\pi}{7}}{b}.$$

Solve the equation above for b, getting

$$b = \frac{6\sin \frac{4\pi}{7}}{\sin \frac{\pi}{7}} \approx 13.48.$$

(c) Use the law of sines in the form

$$\frac{\sin A}{a} = \frac{\sin C}{c},$$

which in this case becomes the equation

$$\frac{\sin \frac{\pi}{7}}{6} = \frac{\sin \frac{2\pi}{7}}{c}.$$

Solve the equation above for c, getting

$$c = \frac{6\sin \frac{2\pi}{7}}{\sin \frac{\pi}{7}} \approx 10.81.$$

5. Suppose $a = 3$, $b = 5$, and $c = 6$. Evaluate:

(a) A (b) B (c) C

SOLUTION The law of cosines allows us to solve for the angles of the triangle when we know the lengths of all the sides. Note the check that is performed below after part (c).

(a) To find A, use the law of cosines in the form

$$a^2 = b^2 + c^2 - 2bc\cos A,$$

which in this case becomes the equation

$$3^2 = 5^2 + 6^2 - 2 \cdot 5 \cdot 6 \cdot \cos A,$$

which can be rewritten as

$$9 = 61 - 60\cos A.$$

Solve the equation above for $\cos A$, getting

$$\cos A = \tfrac{13}{15}.$$

Thus $A = \cos^{-1}\frac{13}{15}$. Use a calculator to evaluate $\cos^{-1}\frac{13}{15}$ in radians, and then multiply that result by $\frac{180°}{\pi}$ to convert to degrees, getting

$$A = \cos^{-1}\tfrac{13}{15} \approx 0.522 \text{ radians} \approx 29.9°.$$

(b) To find B, use the law of cosines in the form

$$b^2 = a^2 + c^2 - 2ac\cos B,$$

which in this case becomes the equation

$$25 = 45 - 36\cos B.$$

Solve the equation above for $\cos B$, getting

$$\cos B = \tfrac{5}{9}.$$

Thus $B = \cos^{-1}\frac{5}{9}$. Use a calculator to evaluate $\cos^{-1}\frac{5}{9}$ in radians, and then multiply that result by $\frac{180°}{\pi}$ to convert to degrees, getting

$$B = \cos^{-1}\tfrac{5}{9} \approx 0.982 \text{ radians} \approx 56.3°.$$

(c) To find C, use the law of cosines in the form

$$c^2 = a^2 + b^2 - 2ab\cos C,$$

which in this case becomes the equation

$$36 = 34 - 30\cos C.$$

Solve the equation above for $\cos C$, getting

$$\cos C = -\tfrac{1}{15}.$$

Thus $C = \cos^{-1}(-\frac{1}{15})$. Use a calculator to evaluate $\cos^{-1}(-\frac{1}{15})$ in radians, and then multiply that result by $\frac{180°}{\pi}$ to convert to degrees, getting

$$C = \cos^{-1}(-\tfrac{1}{15}) \approx 1.638 \text{ radians} \approx 93.8°.$$

CHECK The angles in a triangle add up to 180°. Thus we can check for mistakes by seeing if our values of A, B, and C add up to 180°:

$$A + B + C \approx 29.9° + 56.3° + 93.8° = 180.0°.$$

Because the sum above equals 180.0°, this check uncovers no problems. If the sum had differed from 180.0° by more than 0.1° (a small difference might arise due to using approximate values rather than exact values), then we would know that an error had been made.

7. Suppose $a = 5$, $b = 6$, and $c = 9$. Evaluate:

(a) A (b) B (c) C

SOLUTION The law of cosines allows us to solve for the angles of the triangle when we know the lengths of all the sides. Note the check that is performed below after part (c).

(a) To find A, use the law of cosines in the form

$$a^2 = b^2 + c^2 - 2bc\cos A,$$

which in this case becomes the equation

$$5^2 = 6^2 + 9^2 - 2 \cdot 6 \cdot 9 \cdot \cos A,$$

which can be rewritten as

$$25 = 117 - 108\cos A.$$

Solve the equation above for $\cos A$, getting

$$\cos A = \tfrac{23}{27}.$$

Thus $A = \cos^{-1}\frac{23}{27}$. Use a calculator to evaluate $\cos^{-1}\frac{23}{27}$ in radians, and then multiply that result by $\frac{180°}{\pi}$ to convert to degrees, getting

$$A = \cos^{-1}\tfrac{23}{27} \approx 0.551 \text{ radians} \approx 31.6°.$$

(b) To find B, use the law of cosines in the form

$$b^2 = a^2 + c^2 - 2ac\cos B,$$

which in this case becomes the equation

$$36 = 106 - 90\cos B.$$

Solve the equation above for $\cos B$, getting

$$\cos B = \tfrac{7}{9}.$$

Thus $B = \cos^{-1}\frac{7}{9}$. Use a calculator to evaluate $\cos^{-1}\frac{7}{9}$ in radians, and then multiply that result by $\frac{180°}{\pi}$ to convert to degrees, getting

$$B = \cos^{-1}\tfrac{7}{9} \approx 0.680 \text{ radians} \approx 38.9°.$$

(c) To find C, use the law of cosines in the form

$$c^2 = a^2 + b^2 - 2ab\cos C,$$

which in this case becomes the equation

$$81 = 61 - 60\cos C.$$

Solve the equation above for $\cos C$, getting

$$\cos C = -\tfrac{1}{3}.$$

Thus $C = \cos^{-1}(-\frac{1}{3})$. Use a calculator to evaluate $\cos^{-1}(-\frac{1}{3})$ in radians, and then multiply that result by $\frac{180°}{\pi}$ to convert to degrees, getting

$$C = \cos^{-1}(-\tfrac{1}{3}) \approx 1.911 \text{ radians} \approx 109.5°.$$

CHECK The angles in a triangle add up to 180°. Thus we can check for mistakes by seeing if our values of A, B, and C add up to 180°:

$$A + B + C \approx 31.6° + 38.9° + 109.5° = 180.0°.$$

Because the sum above equals 180.0°, this check uncovers no problems. If the sum had differed from 180.0° by more than 0.1° (a small difference might arise due to using approximate values rather than exact values), then we would know that an error had been made.

9. Suppose $a = 2$, $b = 3$, and $C = 37°$. Evaluate:

   (a) c (b) A (c) B

SOLUTION Note the check that is performed below after part (c).

(a) To find c, use the law of cosines in the form

$$c^2 = a^2 + b^2 - 2ab\cos C,$$

which in this case becomes the equation

$$c^2 = 2^2 + 3^2 - 2 \cdot 2 \cdot 3 \cdot \cos 37°,$$

which can be rewritten as

$$c^2 = 13 - 12\cos 37°.$$

Thus

$$c = \sqrt{13 - 12\cos 37°} \approx 1.848.$$

(b) To find A, use the law of cosines in the form

$$a^2 = b^2 + c^2 - 2bc\cos A,$$

which in this case becomes the approximate equation

$$4 \approx 12.415 - 11.088\cos A,$$

where we have an approximation rather than an exact equality because we have used an approximate value for c. Solve the equation above for $\cos A$, getting

$$\cos A \approx 0.7589.$$

Thus $A \approx \cos^{-1} 0.7589$. Use a calculator to evaluate $\cos^{-1} 0.7589$ in radians, and then multiply that result by $\frac{180°}{\pi}$ to convert to degrees, getting

$$A \approx \cos^{-1} 0.7589 \approx 0.7092 \text{ radians} \approx 40.6°.$$

(c) The angles in a triangle add up to 180°. Thus $A + B + C = 180°$. Solving for B, we have

$$B = 180° - A - C \approx 180° - 40.6° - 37° = 102.4°.$$

Multiplying by $\frac{\pi}{180°}$ to convert to radians gives

$$B \approx 102.4° \approx 1.787 \text{ radians}.$$

CHECK We will check our results by computing B by a different method. Specifically, we will use the law of cosines rather than the simpler method used above in part (c).

We use the law of cosines in the form

$$b^2 = a^2 + c^2 - 2ac\cos B,$$

which in this case becomes the approximate equation

$$9 \approx 7.4151 - 7.392\cos B,$$

where we have an approximation rather than an exact equality because we have used the approximate value 1.848 for c. Solve the equation above for $\cos B$, getting

$$\cos B \approx -0.2144.$$

Thus $B \approx \cos^{-1}(-0.2144)$. Use a calculator to evaluate $\cos^{-1}(-0.2144)$ in radians, and then multiply that result by $\frac{180°}{\pi}$ to convert to degrees, getting

$$B \approx \cos^{-1}(-0.2144) \approx 1.787 \text{ radians} \approx 102.4°.$$

In part (c) above, we also obtained a value of 102.4° for B. Thus this check uncovers no problems. If the two methods for computing B had produced results differing by more than 0.1° (a small difference might arise due to using approximate values rather than exact values), then we would know that an error had been made.

11. Suppose $a = 3$, $b = 4$, and $C = 1$ radian. Evaluate:

   (a) c (b) A (c) B

SOLUTION Note the check that is performed below after part (c).

(a) To find c, use the law of cosines in the form

$$c^2 = a^2 + b^2 - 2ab\cos C,$$

which in this case becomes the equation

$$c^2 = 3^2 + 4^2 - 2 \cdot 3 \cdot 4 \cdot \cos 1,$$

which can be rewritten as

$$c^2 = 25 - 24\cos 1.$$

Thus

$$c = \sqrt{25 - 24\cos 1} \approx 3.469.$$

(b) To find A, use the law of cosines in the form

$$a^2 = b^2 + c^2 - 2bc\cos A,$$

which in this case becomes the approximate equation

$$9 \approx 28.034 - 27.752\cos A,$$

where we have an approximation rather than an exact equality because we have used an approximate value for c. Solve the equation above for $\cos A$, getting

$$\cos A \approx 0.6859.$$

Thus $A \approx \cos^{-1} 0.6859$. Use a calculator to evaluate $\cos^{-1} 0.6859$ in radians, and then multiply that result by $\frac{180°}{\pi}$ to convert to degrees, getting

$$A \approx \cos^{-1} 0.6859 \approx 0.8150 \text{ radians} \approx 46.7°.$$

(c) The angles in a triangle add up to π radians. Thus $A + B + C = \pi$. Solving for B, we have

$$B = \pi - A - C \approx \pi - 0.8150 - 1 \approx 1.3266.$$

Multiplying by $\frac{180°}{\pi}$ to convert to radians gives

$$B \approx 1.3266 \text{ radians} \approx 76.0°.$$

CHECK We will check our results by computing B by a different method. Specifically, we will use the law of cosines rather than the simpler method used above in part (c).

We use the law of cosines in the form

$$b^2 = a^2 + c^2 - 2ac\cos B,$$

which in this case becomes the approximate equation

$$16 \approx 21.034 - 20.814\cos B,$$

where we have an approximation rather than an exact equality because we have used the approximate value 3.469 for c. Solve the equation above for $\cos B$, getting

$$\cos B \approx 0.2419.$$

Thus $B \approx \cos^{-1} 0.2419$. Use a calculator to evaluate $\cos^{-1} 0.2419$ in radians, getting

$$B \approx \cos^{-1} 0.2419 \approx 1.3265 \text{ radians}.$$

In part (c) above, we obtained a value of 1.3266 radians for B. Thus the two methods for computing B differed by only 0.0001 radians. This tiny difference is almost certainly due to using approximate values rather than exact values. Thus this check uncovers no problems.

13. Suppose $a = 4$, $b = 3$, and $B = 30°$. Evaluate:

(a) A (assume that $A < 90°$)

(b) C

(c) c

SOLUTION

(a) Use the law of sines in the form

$$\frac{\sin A}{a} = \frac{\sin B}{b},$$

which in this case becomes the equation

$$\frac{\sin A}{4} = \frac{\frac{1}{2}}{3}.$$

Solve the equation above for $\sin A$, getting

$$\sin A = \tfrac{2}{3}.$$

The assumption that $A < 90°$ now implies that

$$A = \sin^{-1} \tfrac{2}{3} \approx 0.7297 \text{ radians} \approx 41.8°.$$

(b) The angles in a triangle add up to 180°. Thus $A + B + C = 180°$. Solving for C, we have

$$C = 180° - A - B \approx 180° - 41.8° - 30° = 108.2°.$$

Multiplying by $\frac{\pi}{180°}$ to convert to radians gives

$$C \approx 108.2° \approx 1.888 \text{ radians}.$$

(c) Use the law of sines in the form

$$\frac{\sin A}{a} = \frac{\sin C}{c},$$

which in this case becomes the equation

$$\frac{\frac{2}{3}}{4} \approx \frac{\sin 108.2°}{c},$$

where we have an approximation rather than an exact equality because we have used the approximate value 108.2° for C (our solution in part (a) showed that $\sin A$ has the exact value $\frac{2}{3}$; thus the left side above is not an approximation). Solving the equation above for c, we get

$$c \approx 5.70.$$

Exercises 15 and 16 should be compared with Exercises 13 and 14.

15. Suppose $a = 4$, $b = 3$, and $B = 30°$. Evaluate:

 (a) A (assume that $A > 90°$)

 (b) C

 (c) c

SOLUTION

(a) Use the law of sines in the form

$$\frac{\sin A}{a} = \frac{\sin B}{b},$$

which in this case becomes the equation

$$\frac{\sin A}{4} = \frac{\frac{1}{2}}{3}.$$

Solve the equation above for $\sin A$, getting

$$\sin A = \tfrac{2}{3}.$$

The assumption that $A > 90°$ now implies that

$$A = \pi - \sin^{-1} \tfrac{2}{3} \approx 2.4119 \text{ radians} \approx 138.2°.$$

(b) The angles in a triangle add up to 180°. Thus $A + B + C = 180°$. Solving for C, we have

$$C = 180° - A - B \approx 180° - 138.2° - 30° = 11.8°.$$

Multiplying by $\frac{\pi}{180°}$ to convert to radians gives

$$C \approx 11.8° \approx 0.206 \text{ radians}.$$

(c) Use the law of sines in the form

$$\frac{\sin A}{a} = \frac{\sin C}{c},$$

which in this case becomes the equation

$$\frac{\frac{2}{3}}{4} \approx \frac{\sin 11.8°}{c},$$

where we have an approximation rather than an exact equality because we have used the approximate value 11.8° for C (our solution in part (a) showed that $\sin A$ has the exact value $\frac{2}{3}$; thus the left side above is not an approximation). Solving the equation above for c, we get

$$c \approx 1.23.$$

17. Find the lengths of both diagonals in a parallelogram that has two sides of length 4, two sides of length 7, and two angles of 38°.

SOLUTION The diagonal opposite the 38° angles is the third side of a triangle that has a side of length 4, a side of length 7, and a 38° angle between those two sides. Let c denote the length of this diagonal. Then by the law of cosines we have

$$\begin{aligned} c^2 &= 4^2 + 7^2 - 2 \cdot 4 \cdot 7 \cos 38° \\ &= 65 - 56 \cos 38° \\ &\approx 20.87. \end{aligned}$$

Thus $c \approx \sqrt{20.87} \approx 4.57$.

The parallelogram has two 38° angles; thus the other two angles of the parallelogram must be 142° (because all four angles of the parallelogram add up to 360°). The diagonal opposite the 142° angles is the third side of a triangle that has a side of length 4, a side of length 7, and a 142° angle between those two sides. Let d denote the length of this diagonal. Then by the law of cosines we have

$$\begin{aligned} d^2 &= 4^2 + 7^2 - 2 \cdot 4 \cdot 7 \cos 142° \\ &= 65 - 56 \cos 142° \\ &\approx 109.13. \end{aligned}$$

Thus $d \approx \sqrt{109.13} \approx 10.45$.

19. Find the angle between the shorter diagonal of the parallelogram in Exercise 17 and a side with length 4.

SOLUTION We have enough information to use either the Law of Sines or the Law of Cosines. However, to avoid the ambiguity that can arise from using the Law of Sines to solve for an angle, we will use the Law of Cosines.

Let A denote the angle between the shorter diagonal (which has length approximately 4.57) and a side of the parallelogram of length 4. By the Law of Cosines, we have

$$\begin{aligned} \cos A &\approx \frac{4^2 + 4.57^2 - 7^2}{2 \cdot 4 \cdot 4.57} \\ &\approx -0.331. \end{aligned}$$

Thus $A \approx \cos^{-1}(-0.331) \approx 1.91$. Hence the angle in question is approximately 1.91 radians, which is approximately 109°.

10.5 *Double-Angle and Half-Angle Formulas*

LEARNING OBJECTIVES

By the end of this section you should be able to

- compute $\cos(2\theta)$, $\sin(2\theta)$, and $\tan(2\theta)$ from the values of $\cos\theta$, $\sin\theta$, and $\tan\theta$;
- compute $\cos\frac{\theta}{2}$, $\sin\frac{\theta}{2}$, and $\tan\frac{\theta}{2}$ from the values of $\cos\theta$, $\sin\theta$, and $\tan\theta$.

How are the values of $\cos(2\theta)$ and $\sin(2\theta)$ and $\tan(2\theta)$ related to the values of $\cos\theta$ and $\sin\theta$ and $\tan\theta$? What about the values of $\cos\frac{\theta}{2}$ and $\sin\frac{\theta}{2}$ and $\tan\frac{\theta}{2}$? In this section we will see how to answer these questions. We will begin with the double-angle formulas involving 2θ and then use those formulas to find the half-angle formulas involving $\frac{\theta}{2}$.

The Cosine of 2θ

Suppose $0 < \theta < \frac{\pi}{2}$, and consider a right triangle with a hypotenuse of length 1 and an angle of θ radians. The other angle of this right triangle will be $\frac{\pi}{2} - \theta$ radians. The side opposite the angle θ has length $\sin\theta$, as shown below (the unlabeled side of the triangle has length $\cos\theta$, but that side is not of interest right now):

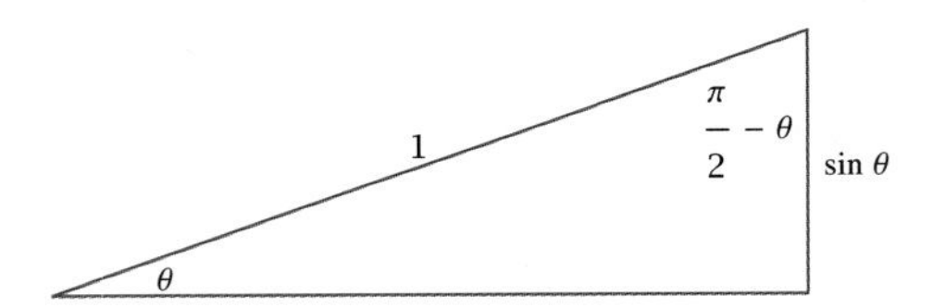

Flip the triangle above across the horizontal side, producing another right triangle with a hypotenuse of length 1 and angles of θ and $\frac{\pi}{2} - \theta$ radians, as shown below:

The triangle formed by the outer edges is an isosceles triangle with two sides of length 1 and an angle of 2θ between those two sides.

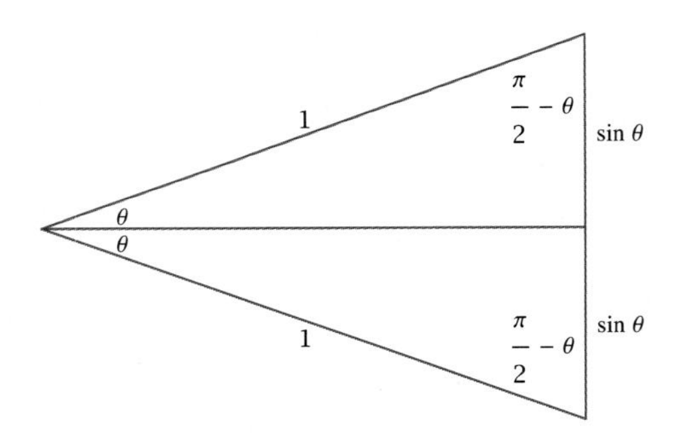

Now consider the isosceles triangle above formed by the union of the two right triangles. Two sides of this isosceles triangle have length 1. As can be seen above, the angle between these two sides is 2θ. As can also be seen above, the side opposite this angle has length $2\sin\theta$. Thus applying the law of cosines to this isosceles triangle gives

$$(2\sin\theta)^2 = 1^2 + 1^2 - 2\cdot 1\cdot 1\cdot\cos(2\theta),$$

which can be rewritten as

$$4\sin^2\theta = 2 - 2\cos(2\theta).$$

Solving this equation for $\cos(2\theta)$ gives the equation

$$\mathbf{cos(2\theta) = 1 - 2\sin^2\theta.}$$

We just found a formula for $\cos(2\theta)$ in terms of $\sin\theta$. Sometimes we need a formula expressing $\cos(2\theta)$ in terms of $\cos\theta$. To obtain such a formula, replace $\sin^2\theta$ by $1-\cos^2\theta$ in the equation above, getting

$$\mathbf{cos(2\theta) = 2\cos^2\theta - 1.}$$

Yet another formula for $\cos(2\theta)$ arises if we replace 1 in the formula above by $\cos^2\theta + \sin^2\theta$, getting

$$\mathbf{cos(2\theta) = \cos^2\theta - \sin^2\theta.}$$

Thus we have found three formulas for $\cos(2\theta)$, which are collected below:

Double-angle formulas for cosine

$$\cos(2\theta) = 1 - 2\sin^2\theta = 2\cos^2\theta - 1 = \cos^2\theta - \sin^2\theta$$

Never, ever, make the mistake of thinking that $\cos(2\theta)$ *equals* $2\cos\theta$*.*

In practice, use whichever of the three formulas is most convenient, as shown in the next example.

EXAMPLE 1

Suppose θ is an angle such that $\cos\theta = \frac{3}{4}$. Evaluate $\cos(2\theta)$.

SOLUTION Because we know the value of $\cos\theta$, we use the second of the formulas given above for $\cos(2\theta)$:

$$\cos(2\theta) = 2\cos^2\theta - 1 = 2\left(\tfrac{3}{4}\right)^2 - 1 = 2\cdot\tfrac{9}{16} - 1 = \tfrac{9}{8} - 1 = \tfrac{1}{8}.$$

The Sine of 2θ

To find a formula for $\sin(2\theta)$, we will apply the law of sines to the isosceles triangle in the last figure above. As we have already noted, this triangle has an angle of 2θ, with a side of length $2\sin\theta$ opposite this angle. The uppermost angle in the isosceles triangle is $\frac{\pi}{2} - \theta$ radians, with a side of length 1 opposite this angle. The law of sines now tells us that

$$\frac{\sin(2\theta)}{2\sin\theta} = \frac{\sin(\frac{\pi}{2}-\theta)}{1}.$$

Recall that $\sin(\frac{\pi}{2}-\theta) = \cos\theta$ (see Section 9.6). Thus the equation above can be rewritten as

$$\frac{\sin(2\theta)}{2\sin\theta} = \cos\theta.$$

Solving this equation for $\sin(2\theta)$ gives the following formula:

Expressions such as $\cos\theta\sin\theta$ *should be interpreted to mean* $(\cos\theta)(\sin\theta)$, *not* $\cos(\theta\sin\theta)$.

Double-angle formula for sine

$$\sin(2\theta) = 2\cos\theta\sin\theta$$

EXAMPLE 2 Using the information that $\cos 30° = \frac{\sqrt{3}}{2}$ and $\sin 30° = \frac{1}{2}$, use the double-angle formula for sine to evaluate $\sin 60°$.

SOLUTION Using the double-angle formula for $\sin(2\theta)$ with $\theta = 30°$, we have

$$\sin 60° = 2\cos 30°\sin 30° = 2\cdot\frac{\sqrt{3}}{2}\cdot\frac{1}{2} = \frac{\sqrt{3}}{2}.$$

REMARK This is a truly terrible method for evaluating $\sin 60°$. Once we know that $\cos 30° = \frac{\sqrt{3}}{2}$, the information that $\sin 60° = \frac{\sqrt{3}}{2}$ follows immediately from the identity $\sin(90° - \theta) = \cos\theta$. The double-angle formula is used here to evaluate $\sin 60°$ only to help you get comfortable with the meaning of the double-angle formula.

The Tangent of 2θ

Now that we have found formulas for $\cos(2\theta)$ and $\sin(2\theta)$, we can find a formula for $\tan(2\theta)$ in the usual fashion of writing the tangent as a ratio of a sine and cosine. In doing so, we will find it more convenient to use the last of the three formulas we found for $\cos(2\theta)$. Specifically, we have

$$\begin{aligned}\tan(2\theta) &= \frac{\sin(2\theta)}{\cos(2\theta)}\\ &= \frac{2\cos\theta\sin\theta}{\cos^2\theta - \sin^2\theta}.\end{aligned}$$

In the last expression, divide numerator and denominator by $\cos^2\theta$, getting

$$\tan(2\theta) = \frac{2\frac{\sin\theta}{\cos\theta}}{1 - \frac{\sin^2\theta}{\cos^2\theta}}.$$

Now replace $\frac{\sin\theta}{\cos\theta}$ above by $\tan\theta$, getting the following nice formula:

Double-angle formula for tangent

$$\tan(2\theta) = \frac{2\tan\theta}{1 - \tan^2\theta}$$

EXAMPLE 3

Suppose θ is an angle such that $\tan\theta = 5$. Evaluate $\tan(2\theta)$.

SOLUTION Because $\tan\theta = 5$, the formula above tells us that

$$\tan(2\theta) = \frac{2 \cdot 5}{1 - 5^2} = -\frac{10}{24} = -\frac{5}{12}.$$

We derived the double-angle formulas for cosine, sine, and tangent starting with the figure on the first page of this section. That figure assumes that θ is between 0 and $\frac{\pi}{2}$. Actually these double-angle formulas are valid for all values of θ, except that in the formula for $\tan(2\theta)$ we must exclude values of θ for which $\tan\theta$ or $\tan(2\theta)$ is undefined.

The Cosine and Sine of $\frac{\theta}{2}$

Now we are ready to find the half-angle formulas for evaluating $\cos\frac{\theta}{2}$ and $\sin\frac{\theta}{2}$. We start with the double-angle formula

$$\cos(2\theta) = 2\cos^2\theta - 1.$$

This formula allows us to find the value of $\cos(2\theta)$ if we know the value of $\cos\theta$. If instead we start out knowing the value of $\cos(2\theta)$, then the equation above could be solved for $\cos\theta$. The example below illustrates this procedure.

EXAMPLE 4

Find an exact expression for $\cos 15°$.

SOLUTION We know that $\cos 30° = \frac{\sqrt{3}}{2}$. We want to find the cosine of half of 30°. Thus we set $\theta = 15°$ in the identity above, getting

$$\cos 30° = 2\cos^2 15° - 1.$$

In this equation, replace $\cos 30°$ with its value, getting

$$\frac{\sqrt{3}}{2} = 2\cos^2 15° - 1.$$

Now solve the equation above for $\cos 15°$, getting

$$\cos 15° = \sqrt{\frac{1 + \frac{\sqrt{3}}{2}}{2}} = \sqrt{\frac{(1 + \frac{\sqrt{3}}{2}) \cdot 2}{2 \cdot 2}} = \frac{\sqrt{2 + \sqrt{3}}}{2}.$$

This value for $\cos 15°$ *(or* $\cos\frac{\pi}{12}$ *if we work in radians) was used in exercises in Sections 9.4 and 9.6.*

REMARK In the first equality in the last line above, we did not need to worry about choosing a plus or minus sign associated with the square root because we know that $\cos 15°$ is positive.

To find a general formula for $\cos\frac{\theta}{2}$ in terms of $\cos\theta$, we will carry out the procedure followed in the example above. The key idea is that we can substitute any value for θ in the identity

$$\cos(2\theta) = 2\cos^2\theta - 1,$$

provided that we make the same substitution on both sides of the equation. We want to find a formula for $\cos\frac{\theta}{2}$. Thus we replace θ by $\frac{\theta}{2}$ on both sides of the equation above, getting

$$\cos\theta = 2\cos^2\frac{\theta}{2} - 1.$$

Now solve this equation for $\cos\frac{\theta}{2}$, getting the following half-angle formula:

Never, ever, make the mistake of thinking that $\cos\frac{\theta}{2}$ equals $\frac{\cos\theta}{2}$.

Half-angle formula for cosine

$$\cos\frac{\theta}{2} = \pm\sqrt{\frac{1+\cos\theta}{2}}$$

The choice of the plus or minus sign in the formula above will need to depend on knowledge of the sign of $\cos\frac{\theta}{2}$. For example, if $0 < \theta < \pi$, then $0 < \frac{\theta}{2} < \frac{\pi}{2}$, which implies that $\cos\frac{\theta}{2}$ is positive (thus we would choose the plus sign in the formula above). As another example, if $\pi < \theta < 3\pi$, then $\frac{\pi}{2} < \frac{\theta}{2} < \frac{3\pi}{2}$, which implies that $\cos\frac{\theta}{2}$ is negative (thus we would choose the minus sign in the formula above).

To find a formula for $\sin\frac{\theta}{2}$, we start with the double-angle formula

$$\cos(2\theta) = 1 - 2\sin^2\theta.$$

In the identity above, replace θ by $\frac{\theta}{2}$ on both sides of the equation, getting

$$\cos\theta = 1 - 2\sin^2\frac{\theta}{2}.$$

Now solve this equation for $\sin\frac{\theta}{2}$, getting the following half-angle formula:

Half-angle formula for sine

$$\sin\frac{\theta}{2} = \pm\sqrt{\frac{1-\cos\theta}{2}}$$

The choice of the plus or minus sign in the formula above will need to depend on knowledge of the sign of $\sin\frac{\theta}{2}$. The examples below illustrate this procedure.

EXAMPLE 5

Find an exact expression for $\sin\frac{\pi}{8}$.

SOLUTION We already know how to evaluate $\sin\frac{\pi}{4}$. Thus we take $\theta = \frac{\pi}{4}$ in the half-angle formula for sine, getting

$$\sin\frac{\pi}{8} = \sqrt{\frac{1-\cos\frac{\pi}{4}}{2}} = \sqrt{\frac{1-\frac{\sqrt{2}}{2}}{2}} = \sqrt{\frac{2-\sqrt{2}}{4}} = \frac{\sqrt{2-\sqrt{2}}}{2}.$$

In the first equality above, we chose the plus sign in the half-angle formula because we know that $\sin\frac{\pi}{8}$ is positive.

This value for $\sin\frac{\pi}{8}$ (or $\sin 22.5°$ if we work in degrees) was used in exercises in Sections 9.4 and 9.6.

The next example shows that sometimes the minus sign must be chosen when using a half-angle formula.

EXAMPLE 6

Suppose $-\frac{\pi}{2} < \theta < 0$ and $\cos\theta = \frac{2}{3}$. Evaluate $\sin\frac{\theta}{2}$.

SOLUTION Because $-\frac{\pi}{4} < \frac{\theta}{2} < 0$, we see that $\sin\frac{\theta}{2} < 0$. Thus we need to choose the negative sign in the identity above. We have

$$\sin\frac{\theta}{2} = -\sqrt{\frac{1-\cos\theta}{2}} = -\sqrt{\frac{1-\frac{2}{3}}{2}} = -\sqrt{\frac{1}{6}} = -\frac{\sqrt{6}}{6}.$$

The Tangent of $\frac{\theta}{2}$

We start with the equation

$$\tan\theta = \frac{\sin\theta}{\cos\theta}.$$

Because we seek a formula involving $\sin(2\theta)$, we multiply numerator and denominator above by $2\cos\theta$, getting

$$\tan\theta = \frac{\sin\theta}{\cos\theta} = \frac{2\cos\theta\sin\theta}{2\cos^2\theta}.$$

The numerator of the last term above equals $\sin(2\theta)$. Furthermore, the identity $\cos(2\theta) = 2\cos^2\theta - 1$ shows that the denominator of the last term above equals $1 + \cos(2\theta)$. Making these substitutions in the equation above gives

$$\tan\theta = \frac{\sin(2\theta)}{1+\cos(2\theta)}.$$

In the equation above, replace θ by $\frac{\theta}{2}$ on both sides of the equation, getting the half-angle formula

$$\tan\frac{\theta}{2} = \frac{\sin\theta}{1+\cos\theta}.$$

We could find a formula for $\tan\frac{\theta}{2}$ by writing $\tan\frac{\theta}{2}$ as $\sin\frac{\theta}{2}$ divided by $\cos\frac{\theta}{2}$ and using the half-angle formulas for cosine and sine. However, the process used here leads to a simpler formula.

This formula is valid for all values of θ, except that we must exclude odd multiples of π (because we need to exclude cases where $\cos\theta = -1$ to avoid division by 0).

To find another formula for $\tan\frac{\theta}{2}$, note that

$$\begin{aligned}\frac{\sin\theta}{1+\cos\theta} &= \frac{\sin\theta}{1+\cos\theta}\cdot\frac{1-\cos\theta}{1-\cos\theta}\\ &= \frac{(\sin\theta)(1-\cos\theta)}{1-\cos^2\theta}\\ &= \frac{(\sin\theta)(1-\cos\theta)}{\sin^2\theta}\\ &= \frac{1-\cos\theta}{\sin\theta}.\end{aligned}$$

Thus our identity above for $\tan\frac{\theta}{2}$ can be rewritten to give the half-angle formula

This formula is valid whenever $\sin\theta \neq 0$.

$$\tan\frac{\theta}{2} = \frac{1-\cos\theta}{\sin\theta}.$$

For convenience, we now collect the half-angle formulas for tangent.

Half-angle formulas for tangent

$$\tan\frac{\theta}{2} = \frac{1-\cos\theta}{\sin\theta} = \frac{\sin\theta}{1+\cos\theta}$$

EXERCISES

1. For $\theta = 23°$, evaluate each of the following:

   (a) $\cos(2\theta)$ (b) $2\cos\theta$

   [*This exercise and the next one emphasize that* $\cos(2\theta)$ *does not equal* $2\cos\theta$.]

2. For $\theta = 7$ radians, evaluate each of the following:

   (a) $\cos(2\theta)$ (b) $2\cos\theta$

3. For $\theta = -5$ radians, evaluate each of the following:

   (a) $\sin(2\theta)$ (b) $2\sin\theta$

   [*This exercise and the next one emphasize that* $\sin(2\theta)$ *does not equal* $2\sin\theta$.]

4. For $\theta = 100°$, evaluate each of the following:

   (a) $\sin(2\theta)$ (b) $2\sin\theta$

5. For $\theta = 6$ radians, evaluate each of the following:

   (a) $\cos\frac{\theta}{2}$ (b) $\frac{\cos\theta}{2}$

   [*This exercise and the next one emphasize that* $\cos\frac{\theta}{2}$ *does not equal* $\frac{\cos\theta}{2}$.]

6. For $\theta = -80°$, evaluate each of the following:

   (a) $\cos\frac{\theta}{2}$ (b) $\frac{\cos\theta}{2}$

7. For $\theta = 65°$, evaluate each of the following:

   (a) $\sin\frac{\theta}{2}$ (b) $\frac{\sin\theta}{2}$

   [*This exercise and the next one emphasize that* $\sin\frac{\theta}{2}$ *does not equal* $\frac{\sin\theta}{2}$.]

8. For $\theta = 9$ radians, evaluate each of the following:

   (a) $\sin\frac{\theta}{2}$ (b) $\frac{\sin\theta}{2}$

9. Given that $\sin 18^\circ = \frac{\sqrt{5}-1}{4}$, find an exact expression for $\cos 36^\circ$.
[*The value used here for* $\sin 18^\circ$ *is derived in Problem 101 in this section.*]

10. Given that $\sin \frac{3\pi}{10} = \frac{\sqrt{5}+1}{4}$, find an exact expression for $\cos \frac{3\pi}{5}$.
[*Problem 71 asks you to explain how the value for* $\sin \frac{3\pi}{10}$ *used here follows from the solution to Exercise 9.*]

For Exercises 11–26, evaluate the given quantities assuming that u and v are both in the interval $(0, \frac{\pi}{2})$ and

$$\cos u = \tfrac{1}{3} \quad \text{and} \quad \sin v = \tfrac{1}{4}.$$

11. $\sin u$
12. $\cos v$
13. $\tan u$
14. $\tan v$
15. $\cos(2u)$
16. $\cos(2v)$
17. $\sin(2u)$
18. $\sin(2v)$
19. $\tan(2u)$
20. $\tan(2v)$
21. $\cos \frac{u}{2}$
22. $\cos \frac{v}{2}$
23. $\sin \frac{u}{2}$
24. $\sin \frac{v}{2}$
25. $\tan \frac{u}{2}$
26. $\tan \frac{v}{2}$

For Exercises 27–42, evaluate the given quantities assuming that u and v are both in the interval $(\frac{\pi}{2}, \pi)$ and

$$\sin u = \tfrac{1}{5} \quad \text{and} \quad \sin v = \tfrac{1}{6}.$$

27. $\cos u$
28. $\cos v$
29. $\tan u$
30. $\tan v$
31. $\cos(2u)$
32. $\cos(2v)$
33. $\sin(2u)$
34. $\sin(2v)$
35. $\tan(2u)$
36. $\tan(2v)$
37. $\cos \frac{u}{2}$
38. $\cos \frac{v}{2}$
39. $\sin \frac{u}{2}$
40. $\sin \frac{v}{2}$
41. $\tan \frac{u}{2}$
42. $\tan \frac{v}{2}$

For Exercises 43–58, evaluate the given quantities assuming that u and v are both in the interval $(-\frac{\pi}{2}, 0)$ and

$$\tan u = -\tfrac{1}{7} \quad \text{and} \quad \tan v = -\tfrac{1}{8}.$$

43. $\cos u$
44. $\cos v$
45. $\sin u$
46. $\sin v$
47. $\cos(2u)$
48. $\cos(2v)$
49. $\sin(2u)$
50. $\sin(2v)$
51. $\tan(2u)$
52. $\tan(2v)$
53. $\cos \frac{u}{2}$
54. $\cos \frac{v}{2}$
55. $\sin \frac{u}{2}$
56. $\sin \frac{v}{2}$
57. $\tan \frac{u}{2}$
58. $\tan \frac{v}{2}$

59. Suppose $0 < \theta < \frac{\pi}{2}$ and $\sin \theta = 0.4$.
    (a) Without using a double-angle formula, evaluate $\sin(2\theta)$.
    (b) Without using an inverse trigonometric function, evaluate $\sin(2\theta)$ again.

[*Your solutions to (a) and (b), which are obtained through different methods, should be the same, although they might differ by a tiny amount due to using approximations rather than exact amounts.*]

60. Suppose $0 < \theta < \frac{\pi}{2}$ and $\sin \theta = 0.2$.
    (a) Without using a double-angle formula, evaluate $\sin(2\theta)$.
    (b) Without using an inverse trigonometric function, evaluate $\sin(2\theta)$ again.

61. Suppose $-\frac{\pi}{2} < \theta < 0$ and $\cos \theta = 0.3$.
    (a) Without using a double-angle formula, evaluate $\cos(2\theta)$.
    (b) Without using an inverse trigonometric function, evaluate $\cos(2\theta)$ again.

62. Suppose $-\frac{\pi}{2} < \theta < 0$ and $\cos \theta = 0.8$.
    (a) Without using a double-angle formula, evaluate $\cos(2\theta)$.
    (b) Without using an inverse trigonometric function, evaluate $\cos(2\theta)$ again.

63. Find an exact expression for $\sin 15^\circ$.

64. Find an exact expression for $\cos 22.5^\circ$.

65. Find an exact expression for $\sin \frac{\pi}{24}$.

66. Find an exact expression for $\cos \frac{\pi}{16}$.

67. Find a formula for $\sin(4\theta)$ in terms of $\cos \theta$ and $\sin \theta$.

68. Find a formula for $\cos(4\theta)$ in terms of $\cos \theta$.

69. Find constants a, b, and c such that

$$\cos^4\theta = a + b\cos(2\theta) + c\cos(4\theta)$$

for all θ.

70. Find constants a, b, and c such that

$$\sin^4\theta = a + b\cos(2\theta) + c\cos(4\theta)$$

for all θ.

PROBLEMS

71. Explain how the equation $\sin\frac{3\pi}{10} = \frac{\sqrt{5}+1}{4}$ follows from the solution to Exercise 9.

72. Show that
$$(\cos x + \sin x)^2 = 1 + \sin(2x)$$
for every number x.

73. Show that
$$\cos(2\theta) \le \cos^2\theta$$
for every angle θ.

74. Show that
$$|\sin(2\theta)| \le 2|\sin\theta|$$
for every angle θ.

75. Do not ever make the mistake of thinking that
$$\frac{\sin(2\theta)}{2} = \sin\theta$$
is a valid identity. Although the equation above is false in general, it is true for some special values of θ. Find all values of θ that satisfy the equation above.

76. Explain why there does not exist an angle θ such that $\cos\theta\sin\theta = \frac{2}{3}$.

77. Show that
$$|\cos\theta\sin\theta| \le \tfrac{1}{2}$$
for every angle θ.

78. Do not ever make the mistake of thinking that
$$\frac{\cos(2\theta)}{2} = \cos\theta$$
is a valid identity.

(a) Show that the equation above is false whenever $0 < \theta < \frac{\pi}{2}$.

(b) Show that there exists an angle θ in the interval $(\frac{\pi}{2}, \pi)$ satisfying the equation above.

79. Without doing any algebraic manipulations, explain why
$$(2\cos^2\theta - 1)^2 + (2\cos\theta\sin\theta)^2 = 1$$
for every angle θ.

80. Find angles u and v such that $\cos(2u) = \cos(2v)$ but $\cos u \ne \cos v$.

81. Show that if $\cos(2u) = \cos(2v)$, then $|\cos u| = |\cos v|$.

82. Find angles u and v such that $\sin(2u) = \sin(2v)$ but $|\sin u| \ne |\sin v|$.

83. Show that
$$\sin^2(2\theta) = 4(\sin^2\theta - \sin^4\theta)$$
for all θ.

84. Find a formula that expresses $\sin^2(2\theta)$ only in terms of $\cos\theta$.

85. Show that
$$(\cos\theta + \sin\theta)^2(\cos\theta - \sin\theta)^2 + \sin^2(2\theta) = 1$$
for all angles θ.

86. Suppose θ is not an integer multiple of π. Explain why the point $(1, 2\cos\theta)$ is on the line containing the point $(\sin\theta, \sin(2\theta))$ and the origin.

87. Show that
$$\tan^2(2x) = \frac{4(\cos^2 x - \cos^4 x)}{(2\cos^2 x - 1)^2}$$
for all numbers x except odd multiples of $\frac{\pi}{4}$.

88. Find a formula that expresses $\tan^2(2\theta)$ only in terms of $\sin\theta$.

89. Find all numbers t such that
$$\frac{\cos^{-1} t}{2} = \sin^{-1} t.$$

90. Find all numbers t such that
$$\cos^{-1} t = \frac{\sin^{-1} t}{2}.$$

91. Show that
$$\tan\frac{\theta}{2} = \pm\sqrt{\frac{1-\cos\theta}{1+\cos\theta}}$$
for all θ except odd multiples of π.

92. Find a formula that expresses $\tan\frac{\theta}{2}$ only in terms of $\tan\theta$.

93. Suppose θ is an angle such that $\cos\theta$ is rational. Explain why $\cos(2\theta)$ is rational.

94. Give an example of an angle θ such that $\sin\theta$ is rational but $\sin(2\theta)$ is irrational.

95. Give an example of an angle θ such that both $\sin\theta$ and $\sin(2\theta)$ are rational.

Problems 96–101 will lead you to the discovery of an exact expression for the value of* $\sin 18°$*. For convenience, throughout these problems let

$$t = \sin 18°.$$

96. Using a double-angle formula, show that $\cos 36° = 1 - 2t^2$.

97. Using a double-angle formula and the previous problem, show that

$$\cos 72° = 8t^4 - 8t^2 + 1.$$

98. Explain why $\sin 18° = \cos 72°$. Then using the previous problem, explain why

$$8t^4 - 8t^2 - t + 1 = 0.$$

99. Verify that

$$8t^4 - 8t^2 - t + 1 = (t-1)(2t+1)(4t^2+2t-1).$$

100. Explain why the two previous problems imply that

$$t = 1, \quad t = -\frac{1}{2}, \quad t = \frac{-\sqrt{5}-1}{4}, \text{ or } t = \frac{\sqrt{5}-1}{4}.$$

101. Explain why the first three values in the previous problem are not possible values for $\sin 18°$. Conclude that

$$\sin 18° = \frac{\sqrt{5}-1}{4}.$$

[*This value for* $\sin 18°$ *(or* $\sin \frac{\pi}{10}$ *if we work in radians) was used in Exercise 9.*]

102. Use the result from the previous problem to show that

$$\cos 18° = \sqrt{\frac{\sqrt{5}+5}{8}}.$$

103. Show that

$$\tan \tfrac{x}{4} = \frac{\sqrt{2 - 2\cos x} - \sin x}{1 - \cos x}$$

for all x in the interval $(0, 2\pi)$.
[*Hint:* Start with a half-angle formula for tangent to express $\tan \frac{x}{4}$ in terms of $\sin \frac{x}{2}$ and $\cos \frac{x}{2}$. Then use half-angle formulas for cosine and sine, along with algebraic manipulations.]

104. Show that

$$\cos \tfrac{\pi}{32} = \frac{\sqrt{2 + \sqrt{2 + \sqrt{2 + \sqrt{2}}}}}{2}.$$

[*Hint:* First do Exercise 66.]

105. Show that

$$\sin \tfrac{\pi}{32} = \frac{\sqrt{2 - \sqrt{2 + \sqrt{2 + \sqrt{2}}}}}{2}.$$

WORKED-OUT SOLUTIONS *to Odd-numbered Exercises*

1. For $\theta = 23°$, evaluate each of the following:
   (a) $\cos(2\theta)$ (b) $2\cos\theta$

SOLUTION

(a) Note that $2 \times 23 = 46$. Using a calculator working in degrees, we have

$$\cos 46° \approx 0.694658.$$

(b) Using a calculator working in degrees, we have

$$2\cos 23° \approx 2 \times 0.920505 = 1.841010.$$

3. For $\theta = -5$ radians, evaluate each of the following:
   (a) $\sin(2\theta)$ (b) $2\sin\theta$

SOLUTION

(a) Note that $2 \times (-5) = -10$. Using a calculator working in radians, we have

$$\sin(-10) \approx 0.544021.$$

(b) Using a calculator working in radians, we have

$$2\sin(-5) \approx 2 \times 0.9589 = 1.9178.$$

5. For $\theta = 6$ radians, evaluate each of the following:
   (a) $\cos \frac{\theta}{2}$ (b) $\frac{\cos\theta}{2}$

SOLUTION

(a) Using a calculator working in radians, we have

$$\cos \tfrac{6}{2} = \cos 3 \approx -0.989992.$$

(b) Using a calculator working in radians, we have

$$\frac{\cos 6}{2} \approx \frac{0.96017}{2} = 0.480085.$$

7. For $\theta = 65°$, evaluate each of the following:
(a) $\sin \frac{\theta}{2}$ (b) $\frac{\sin \theta}{2}$

SOLUTION

(a) Using a calculator working in degrees, we have

$$\sin \tfrac{65°}{2} = \sin 32.5° \approx 0.537300.$$

(b) Using a calculator working in degrees, we have

$$\frac{\sin 65°}{2} \approx \frac{0.906308}{2} = 0.453154.$$

9. Given that $\sin 18° = \frac{\sqrt{5}-1}{4}$, find an exact expression for $\cos 36°$.

SOLUTION To evaluate $\cos 36°$, use one of the double-angle formulas for $\cos(2\theta)$ with $\theta = 18°$:

$$\begin{aligned}\cos 36° &= 1 - 2\sin^2 18° \\ &= 1 - 2\left(\tfrac{\sqrt{5}-1}{4}\right)^2 = 1 - 2\left(\tfrac{3-\sqrt{5}}{8}\right) = \tfrac{\sqrt{5}+1}{4}.\end{aligned}$$

For Exercises 11–26, evaluate the given quantities assuming that u and v are both in the interval $(0, \frac{\pi}{2})$ and

$$\cos u = \tfrac{1}{3} \quad \textit{and} \quad \sin v = \tfrac{1}{4}.$$

11. $\sin u$

SOLUTION Because $0 < u < \frac{\pi}{2}$, we know that $\sin u > 0$. Thus

$$\sin u = \sqrt{1 - \cos^2 u} = \sqrt{1 - \tfrac{1}{9}} = \sqrt{\tfrac{8}{9}} = \tfrac{2\sqrt{2}}{3}.$$

13. $\tan u$

SOLUTION To evaluate $\tan u$, use its definition as a ratio:

$$\tan u = \frac{\sin u}{\cos u} = \frac{\frac{2\sqrt{2}}{3}}{\frac{1}{3}} = 2\sqrt{2}.$$

15. $\cos(2u)$

SOLUTION To evaluate $\cos(2u)$, use one of the double-angle formulas for cosine:

$$\cos(2u) = 2\cos^2 u - 1 = \tfrac{2}{9} - 1 = -\tfrac{7}{9}.$$

17. $\sin(2u)$

SOLUTION To evaluate $\sin(2u)$, use the double-angle formula for sine:

$$\sin(2u) = 2\cos u \sin u = 2 \cdot \tfrac{1}{3} \cdot \tfrac{2\sqrt{2}}{3} = \tfrac{4\sqrt{2}}{9}.$$

19. $\tan(2u)$

SOLUTION To evaluate $\tan(2u)$, use its definition as a ratio:

$$\tan(2u) = \frac{\sin(2u)}{\cos(2u)} = \frac{\frac{4\sqrt{2}}{9}}{-\frac{7}{9}} = -\tfrac{4\sqrt{2}}{7}.$$

Alternatively, we could have used the double-angle formula for tangent, which will produce the same answer.

21. $\cos \frac{u}{2}$

SOLUTION Because $0 < \frac{u}{2} < \frac{\pi}{4}$, we know that $\cos \frac{u}{2} > 0$. Thus

$$\begin{aligned}\cos \tfrac{u}{2} &= \sqrt{\frac{1 + \cos u}{2}} \\ &= \sqrt{\frac{1 + \frac{1}{3}}{2}} = \sqrt{\frac{\frac{4}{3}}{2}} = \sqrt{\tfrac{2}{3}} = \tfrac{\sqrt{6}}{3}.\end{aligned}$$

23. $\sin \frac{u}{2}$

SOLUTION Because $0 < \frac{u}{2} < \frac{\pi}{4}$, we know that $\sin \frac{u}{2} > 0$. Thus

$$\begin{aligned}\sin \tfrac{u}{2} &= \sqrt{\frac{1 - \cos u}{2}} \\ &= \sqrt{\frac{1 - \frac{1}{3}}{2}} = \sqrt{\frac{\frac{2}{3}}{2}} = \sqrt{\tfrac{1}{3}} = \tfrac{1}{\sqrt{3}} = \tfrac{\sqrt{3}}{3}.\end{aligned}$$

25. $\tan \frac{u}{2}$

SOLUTION To evaluate $\tan \frac{u}{2}$, use its definition as a ratio:

$$\tan \tfrac{u}{2} = \frac{\sin \frac{u}{2}}{\cos \frac{u}{2}} = \frac{\frac{\sqrt{3}}{3}}{\frac{\sqrt{6}}{3}} = \frac{\sqrt{3}}{\sqrt{6}} = \frac{1}{\sqrt{2}} = \frac{\sqrt{2}}{2}.$$

Alternatively, we could have used the half-angle formula for tangent, which will produce the same answer.

For Exercises 27–42, evaluate the given quantities assuming that u and v are both in the interval $(\frac{\pi}{2}, \pi)$ and

$$\sin u = \tfrac{1}{5} \quad \textit{and} \quad \sin v = \tfrac{1}{6}.$$

27. $\cos u$

SOLUTION Because $\frac{\pi}{2} < u < \pi$, we know that $\cos u < 0$. Thus

$$\cos u = -\sqrt{1 - \sin^2 u} = -\sqrt{1 - \tfrac{1}{25}} = -\sqrt{\tfrac{24}{25}}$$
$$= -\tfrac{2\sqrt{6}}{5}.$$

29. $\tan u$

SOLUTION To evaluate $\tan u$, use its definition as a ratio:

$$\tan u = \frac{\sin u}{\cos u} = \frac{\frac{1}{5}}{-\frac{2\sqrt{6}}{5}} = -\frac{1}{2\sqrt{6}} = -\frac{\sqrt{6}}{12}.$$

31. $\cos(2u)$

SOLUTION To evaluate $\cos(2u)$, use one of the double-angle formulas for cosine:

$$\cos(2u) = 1 - 2\sin^2 u = 1 - \tfrac{2}{25} = \tfrac{23}{25}.$$

33. $\sin(2u)$

SOLUTION To evaluate $\sin(2u)$, use the double-angle formula for sine:

$$\sin(2u) = 2\cos u \sin u = 2 \cdot \left(-\tfrac{2\sqrt{6}}{5}\right) \cdot \tfrac{1}{5} = -\tfrac{4\sqrt{6}}{25}.$$

35. $\tan(2u)$

SOLUTION To evaluate $\tan(2u)$, use its definition as a ratio:

$$\tan(2u) = \frac{\sin(2u)}{\cos(2u)} = \frac{-\frac{4\sqrt{6}}{25}}{\frac{23}{25}} = -\frac{4\sqrt{6}}{23}.$$

Alternatively, we could have used the double-angle formula for tangent, which will produce the same answer.

37. $\cos \frac{u}{2}$

SOLUTION Because $\frac{\pi}{4} < \frac{u}{2} < \frac{\pi}{2}$, we know that $\cos \frac{u}{2} > 0$. Thus

$$\cos \tfrac{u}{2} = \sqrt{\frac{1 + \cos u}{2}}$$
$$= \sqrt{\frac{1 - \frac{2\sqrt{6}}{5}}{2}} = \sqrt{\frac{\frac{5-2\sqrt{6}}{5}}{2}} = \sqrt{\frac{5 - 2\sqrt{6}}{10}}.$$

39. $\sin \frac{u}{2}$

SOLUTION Because $\frac{\pi}{4} < \frac{u}{2} < \frac{\pi}{2}$, we know that $\sin \frac{u}{2} > 0$. Thus

$$\sin \tfrac{u}{2} = \sqrt{\frac{1 - \cos u}{2}}$$
$$= \sqrt{\frac{1 + \frac{2\sqrt{6}}{5}}{2}} = \sqrt{\frac{\frac{5+2\sqrt{6}}{5}}{2}} = \sqrt{\frac{5 + 2\sqrt{6}}{10}}.$$

41. $\tan \frac{u}{2}$

SOLUTION To evaluate $\tan \frac{u}{2}$, use one of the half-angle formulas for tangent:

$$\tan \tfrac{u}{2} = \frac{1 - \cos u}{\sin u} = \frac{1 + \frac{2\sqrt{6}}{5}}{\frac{1}{5}} = 5 + 2\sqrt{6}.$$

We could also have evaluated $\tan \frac{u}{2}$ by using its definition as the ratio of $\sin \frac{u}{2}$ and $\cos \frac{u}{2}$, but in this case that procedure would lead to a more complicated algebraic expression.

For Exercises 43–58, evaluate the given quantities assuming that u and v are both in the interval $(-\frac{\pi}{2}, 0)$ and

$$\tan u = -\tfrac{1}{7} \quad \textit{and} \quad \tan v = -\tfrac{1}{8}.$$

43. $\cos u$

SOLUTION Because $-\frac{\pi}{2} < u < 0$, we know that $\cos u > 0$ and $\sin u < 0$. Thus

$$-\tfrac{1}{7} = \tan u = \frac{\sin u}{\cos u} = \frac{-\sqrt{1 - \cos^2 u}}{\cos u}.$$

Squaring the first and last entries above gives

$$\tfrac{1}{49} = \frac{1 - \cos^2 u}{\cos^2 u}.$$

Multiplying both sides by $\cos^2 u$ and then by 49 gives

$$\cos^2 u = 49 - 49\cos^2 u.$$

Thus $50\cos^2 u = 49$, which implies that

$$\cos u = \sqrt{\tfrac{49}{50}} = \tfrac{7}{5\sqrt{2}} = \tfrac{7\sqrt{2}}{10}.$$

45. $\sin u$

SOLUTION Solve the equation $\tan u = \frac{\sin u}{\cos u}$ for $\sin u$:

$$\sin u = \cos u \tan u = \tfrac{7\sqrt{2}}{10} \cdot \left(-\tfrac{1}{7}\right) = -\tfrac{\sqrt{2}}{10}.$$

47. $\cos(2u)$

SOLUTION To evaluate $\cos(2u)$, use one of the double-angle formulas for cosine:

$$\cos(2u) = 2\cos^2 u - 1 = 2 \cdot \tfrac{49}{50} - 1 = \tfrac{24}{25}.$$

49. $\sin(2u)$

SOLUTION To evaluate $\sin(2u)$, use the double-angle formula for sine:

$$\sin(2u) = 2\cos u \sin u = 2 \cdot \tfrac{7\sqrt{2}}{10} \cdot \left(-\tfrac{\sqrt{2}}{10}\right) = -\tfrac{7}{25}.$$

51. $\tan(2u)$

SOLUTION To evaluate $\tan(2u)$, use the double-angle formula for tangent:

$$\tan(2u) = \frac{2\tan u}{1 - \tan^2 u} = \frac{-\frac{2}{7}}{\frac{48}{49}} = -\tfrac{7}{24}.$$

Alternatively, we could have evaluated $\tan(2u)$ by using its definition as a ratio of $\sin(2u)$ and $\cos(2u)$, producing the same answer.

53. $\cos\frac{u}{2}$

SOLUTION Because $-\frac{\pi}{4} < \frac{u}{2} < 0$, we know that $\cos\frac{u}{2} > 0$. Thus

$$\cos\tfrac{u}{2} = \sqrt{\frac{1+\cos u}{2}}$$

$$= \sqrt{\frac{1+\frac{7\sqrt{2}}{10}}{2}} = \sqrt{\frac{\frac{10+7\sqrt{2}}{10}}{2}} = \sqrt{\frac{10+7\sqrt{2}}{20}}.$$

55. $\sin\frac{u}{2}$

SOLUTION Because $-\frac{\pi}{4} < \frac{u}{2} < 0$, we know that $\sin\frac{u}{2} < 0$. Thus

$$\sin\tfrac{u}{2} = -\sqrt{\frac{1-\cos u}{2}}$$

$$= -\sqrt{\frac{1-\frac{7\sqrt{2}}{10}}{2}} = -\sqrt{\frac{\frac{10-7\sqrt{2}}{10}}{2}} = -\sqrt{\frac{10-7\sqrt{2}}{20}}.$$

57. $\tan\frac{u}{2}$

SOLUTION To evaluate $\tan\frac{u}{2}$, use one of the half-angle formulas for tangent:

$$\tan\tfrac{u}{2} = \frac{1-\cos u}{\sin u} = \frac{1-\frac{7\sqrt{2}}{10}}{-\frac{\sqrt{2}}{10}} = 7 - \tfrac{10}{\sqrt{2}} = 7 - 5\sqrt{2}.$$

We could also have evaluated $\tan\frac{u}{2}$ by using its definition as the ratio of $\sin\frac{u}{2}$ and $\cos\frac{u}{2}$, but in this case that procedure would lead to a more complicated algebraic expression.

59. Suppose $0 < \theta < \frac{\pi}{2}$ and $\sin\theta = 0.4$.

(a) Without using a double-angle formula, evaluate $\sin(2\theta)$.

(b) Without using an inverse trigonometric function, evaluate $\sin(2\theta)$ again.

SOLUTION

(a) Because $0 < \theta < \frac{\pi}{2}$ and $\sin\theta = 0.4$, we see that

$$\theta = \sin^{-1} 0.4 \approx 0.411517 \text{ radians}.$$

Thus

$$2\theta \approx 0.823034 \text{ radians}.$$

Hence

$$\sin(2\theta) \approx \sin(0.823034) \approx 0.733212.$$

(b) To use the double-angle formula to evaluate $\sin(2\theta)$, we must first evaluate $\cos\theta$. Because $0 < \theta < \frac{\pi}{2}$, we know that $\cos\theta > 0$. Thus

$$\cos\theta = \sqrt{1-\sin^2\theta} = \sqrt{1-0.16} = \sqrt{0.84}$$
$$\approx 0.916515.$$

Now

$$\sin(2\theta) = 2\cos\theta\sin\theta \approx 2(0.916515)(0.4)$$
$$= 0.733212.$$

61. Suppose $-\frac{\pi}{2} < \theta < 0$ and $\cos\theta = 0.3$.

(a) Without using a double-angle formula, evaluate $\cos(2\theta)$.

(b) Without using an inverse trigonometric function, evaluate $\cos(2\theta)$ again.

SOLUTION

(a) Because $-\frac{\pi}{2} < \theta < 0$ and $\cos\theta = 0.3$, we see that

$$\theta = -\cos^{-1} 0.3 \approx -1.2661 \text{ radians.}$$

Thus

$$2\theta \approx -2.5322 \text{ radians.}$$

Hence

$$\cos(2\theta) \approx \cos(-2.5322) \approx -0.82.$$

(b) Using a double-angle formula, we have

$$\begin{aligned}\cos(2\theta) &= 2\cos^2\theta - 1 = 2(0.3)^2 - 1\\ &= -0.82.\end{aligned}$$

63. Find an exact expression for $\sin 15°$.

SOLUTION Use the half-angle formula for $\sin\frac{\theta}{2}$ with $\theta = 30°$ (choose the plus sign associated with the square root because $\sin 15°$ is positive), getting

$$\begin{aligned}\sin 15° &= \sqrt{\frac{1-\cos 30°}{2}}\\ &= \sqrt{\frac{1-\frac{\sqrt{3}}{2}}{2}} = \sqrt{\frac{(1-\frac{\sqrt{3}}{2})\cdot 2}{2\cdot 2}} = \frac{\sqrt{2-\sqrt{3}}}{2}.\end{aligned}$$

65. Find an exact expression for $\sin\frac{\pi}{24}$.

SOLUTION Using the half-angle formula for $\sin\frac{\theta}{2}$ with $\theta = \frac{\pi}{12}$ (and choosing the plus sign associated with the square root because $\sin\frac{\pi}{24}$ is positive), we have

$$\sin\frac{\pi}{24} = \sqrt{\frac{1-\cos\frac{\pi}{12}}{2}}.$$

Note that $\frac{\pi}{12}$ radians equals $15°$. Substituting for $\cos\frac{\pi}{12}$ the value for $\cos 15°$ from Example 4 gives

$$\sin\frac{\pi}{24} = \sqrt{\frac{1-\frac{\sqrt{2+\sqrt{3}}}{2}}{2}} = \frac{\sqrt{2-\sqrt{2+\sqrt{3}}}}{2}.$$

67. Find a formula for $\sin(4\theta)$ in terms of $\cos\theta$ and $\sin\theta$.

SOLUTION Use the double-angle formula for sine, with θ replaced by 2θ, getting

$$\sin(4\theta) = 2\cos(2\theta)\sin(2\theta).$$

Now use the double-angle formulas for the expressions on the right side, getting

$$\begin{aligned}\sin(4\theta) &= 2(2\cos^2\theta - 1)(2\cos\theta\sin\theta)\\ &= 4(2\cos^2\theta - 1)\cos\theta\sin\theta.\end{aligned}$$

There are also other correct ways to express a solution. For example, replacing $\cos^2\theta$ with $1-\sin^2\theta$ in the expression above leads to the formula

$$\sin(4\theta) = 4(1 - 2\sin^2\theta)\cos\theta\sin\theta.$$

69. Find constants a, b, and c such that

$$\cos^4\theta = a + b\cos(2\theta) + c\cos(4\theta)$$

for all θ.

SOLUTION One of the double-angle formulas for $\cos(2\theta)$ can be written in the form

$$\cos^2\theta = \frac{1+\cos(2\theta)}{2}.$$

Squaring both sides, we get

$$\cos^4\theta = \frac{1 + 2\cos(2\theta) + \cos^2(2\theta)}{4}.$$

We now see that we need an expression for $\cos^2(2\theta)$, which we can obtain by replacing θ by 2θ in the formula above for $\cos^2\theta$:

$$\cos^2(2\theta) = \frac{1+\cos(4\theta)}{2}.$$

Substituting this expression into the expression above for $\cos^4\theta$ gives

$$\begin{aligned}\cos^4\theta &= \frac{1 + 2\cos(2\theta) + \frac{1+\cos(4\theta)}{2}}{4}\\ &= \tfrac{3}{8} + \tfrac{1}{2}\cos(2\theta) + \tfrac{1}{8}\cos(4\theta).\end{aligned}$$

Thus $a = \frac{3}{8}$, $b = \frac{1}{2}$, and $c = \frac{1}{8}$.

10.6 *Addition and Subtraction Formulas*

LEARNING OBJECTIVES

By the end of this section you should be able to

- find the cosine of the sum and difference of two angles;
- find the sine of the sum and difference of two angles;
- find the tangent of the sum and difference of two angles.

The Cosine of a Sum and Difference

Consider the figure below, which shows the unit circle along with the radius corresponding to u and the radius corresponding to $-v$.

This figure has been carefully chosen to lead us to an easy derivation of the formula for $\cos(u+v)$.

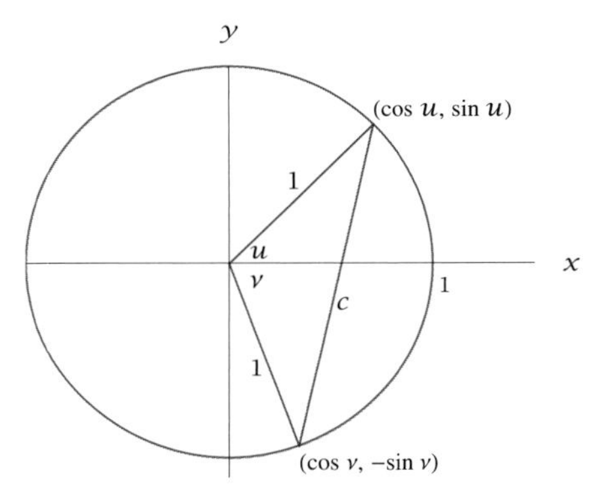

We defined the cosine and sine so that the endpoint of the radius corresponding to u has coordinates $(\cos u, \sin u)$. The endpoint of the radius corresponding to $-v$ has coordinates $(\cos(-v), \sin(-v))$, which we know equals $(\cos v, -\sin v)$, as shown above.

The large triangle in the figure above has two sides that are radii of the unit circle and thus have length 1. The angle between these two sides is $u + v$. The length of the third side of this triangle has been labeled c. The idea now is that we can compute c^2 in two different ways: first by using the formula for the distance between two points, and second by using the law of cosines. We will then set these two computed values of c^2 equal to each other, obtaining a formula for $\cos(u + v)$.

To carry out the plan discussed in the paragraph above, note that one endpoint of the line segment above with length c has coordinates $(\cos u, \sin u)$ and the other endpoint has coordinates $(\cos v, -\sin v)$. Recall that the distance between two points is the square root of the sum of the squares of the differences of the coordinates. Thus

$$c = \sqrt{(\cos u - \cos v)^2 + (\sin u + \sin v)^2}.$$

Squaring both sides of this equation, we have

$$\begin{aligned} c^2 &= (\cos u - \cos v)^2 + (\sin u + \sin v)^2 \\ &= \cos^2 u - 2\cos u \cos v + \cos^2 v \\ &\quad + \sin^2 u + 2\sin u \sin v + \sin^2 v \\ &= (\cos^2 u + \sin^2 u) + (\cos^2 v + \sin^2 v) \\ &\quad - 2\cos u \cos v + 2\sin u \sin v \\ &= 2 - 2\cos u \cos v + 2\sin u \sin v. \end{aligned}$$

To compute c^2 by another method, apply the law of cosines to the large triangle in the figure above, getting $c^2 = 1^2 + 1^2 - 2 \cdot 1 \cdot 1 \cos(u + v)$, which can be rewritten as

$$c^2 = 2 - 2\cos(u + v).$$

We have now found two expressions that equal c^2. Setting those expressions equal to each other, we have

$$2 - 2\cos(u + v) = 2 - 2\cos u \cos v + 2\sin u \sin v.$$

Subtracting 2 from both sides of the equation above and then dividing both sides by -2 gives the following result:

Addition formula for cosine

$$\cos(u + v) = \cos u \cos v - \sin u \sin v$$

Never, ever, make the mistake of thinking that $\cos(u + v)$ *equals* $\cos u + \cos v$.

We derived this formula using the figure above, which assumes that u and v are between 0 and $\frac{\pi}{2}$. However, the formula above is valid for all values of u and v.

EXAMPLE 1

Find an exact expression for $\cos 75°$.

SOLUTION Note that $75° = 45° + 30°$, and we already know how to evaluate the cosine and sine of 45° and 30°. Using the addition formula for cosine, we have

$$\begin{aligned} \cos 75° &= \cos(45° + 30°) \\ &= \cos 45° \cos 30° - \sin 45° \sin 30° \\ &= \tfrac{\sqrt{2}}{2} \cdot \tfrac{\sqrt{3}}{2} - \tfrac{\sqrt{2}}{2} \cdot \tfrac{1}{2} \\ &= \tfrac{\sqrt{6}-\sqrt{2}}{4}. \end{aligned}$$

Notice that if $v = u$, the addition formula for cosine becomes

$$\cos(2u) = \cos^2 u - \sin^2 u,$$

which agrees with one of our previous double-angle formulas.

We can now find a formula for the cosine of the difference of two angles. In the formula for $\cos(u+v)$, replace v by $-v$ on both sides of the equation and use the identities $\cos(-v) = \cos v$ and $\sin(-v) = -\sin(v)$ to get the following result:

Subtraction formula for cosine

$$\cos(u - v) = \cos u \cos v + \sin u \sin v$$

EXAMPLE 2

Find an exact expression for $\cos 15°$.

SOLUTION Note that $15° = 45° - 30°$, and we already know how to evaluate the cosine and sine of 45° and 30°. Using the subtraction formula for cosine, we have

$$\begin{aligned}\cos 15° &= \cos(45° - 30°)\\ &= \cos 45° \cos 30° + \sin 45° \sin 30°\\ &= \tfrac{\sqrt{2}}{2} \cdot \tfrac{\sqrt{3}}{2} + \tfrac{\sqrt{2}}{2} \cdot \tfrac{1}{2}\\ &= \tfrac{\sqrt{6}+\sqrt{2}}{4}.\end{aligned}$$

Note that the expression produced by the subtraction formula for cosine is simpler than the expression produced by the half-angle formula for cosine.

REMARK Using a half-angle formula, in Example 4 in Section 10.5 we showed that

$$\cos 15° = \frac{\sqrt{2+\sqrt{3}}}{2}.$$

Thus we have two seemingly different exact expressions for $\cos 15°$, one produced by the subtraction formula for cosine and the other produced by the half-angle formula for cosine. Problem 39 in this section asks you to verify that these two expressions for $\cos 15°$ are equal.

The Sine of a Sum and Difference

To find the formula for the sine of the sum of two angles, we will make use of the identities

$$\sin\theta = \cos(\tfrac{\pi}{2} - \theta) \quad \text{and} \quad \sin(\tfrac{\pi}{2} - \theta) = \cos\theta,$$

which you can review in Section 9.6. We begin by converting the sine into a cosine and then we use the identity just derived above:

$$\begin{aligned}\sin(u+v) &= \cos(\tfrac{\pi}{2} - u - v)\\ &= \cos((\tfrac{\pi}{2} - u) - v)\\ &= \cos(\tfrac{\pi}{2} - u)\cos v + \sin(\tfrac{\pi}{2} - u)\sin v.\end{aligned}$$

The equation above and the identities above now imply the following result:

Addition formula for sine

$$\sin(u+v) = \sin u \cos v + \cos u \sin v$$

Never, ever, make the mistake of thinking that $\sin(u+v)$ *equals* $\sin u + \sin v$.

Notice that if $v = u$, the addition formula for sine becomes

$$\sin(2u) = 2\cos u \sin u,$$

which agrees with our previous double-angle formula for sine.

We can now find a formula for the sine of the difference of two angles. In the formula for $\sin(u+v)$, replace v by $-v$ on both sides of the equation and use the identities $\cos(-v) = \cos v$ and $\sin(-v) = -\sin(v)$ to get the following result:

Subtraction formula for sine

$$\sin(u-v) = \sin u \cos v - \cos u \sin v$$

EXAMPLE 3

Verify that the subtraction formula for sine gives the expected identity for $\sin(\frac{\pi}{2}-\theta)$.

SOLUTION Using the subtraction formula for sine, we have

$$\begin{aligned}\sin(\tfrac{\pi}{2}-\theta) &= \sin\tfrac{\pi}{2}\cos\theta - \cos\tfrac{\pi}{2}\sin\theta\\ &= 1\cdot\cos\theta - 0\cdot\sin\theta\\ &= \cos\theta.\end{aligned}$$

The Tangent of a Sum and Difference

Now that we have found formulas for the cosine and sine of the sum of two angles, we can find a formula for the tangent of the sum of two angles in the usual fashion by writing the tangent as a ratio of a sine and cosine. Specifically, we have

$$\begin{aligned}\tan(u+v) &= \frac{\sin(u+v)}{\cos(u+v)}\\ &= \frac{\sin u \cos v + \cos u \sin v}{\cos u \cos v - \sin u \sin v}\\ &= \frac{\frac{\sin u}{\cos u} + \frac{\sin v}{\cos v}}{1 - \frac{\sin u \sin v}{\cos u \cos v}}.\end{aligned}$$

The last equality is obtained by dividing the numerator and denominator of the previous expression by $\cos u \cos v$.

Using the definition of the tangent, rewrite the equation above as follows:

Addition formula for tangent

$$\tan(u+v) = \frac{\tan u + \tan v}{1 - \tan u \tan v}$$

In this section we derive six addition and subtraction formulas. Memorizing all six would not be a good use of your time or mental energy. Instead, concentrate on learning the formulas for $\cos(u+v)$ *and* $\sin(u+v)$ *and on understanding how the other formulas follow from those two.*

The identity above is valid for all u, v such that $\tan u$, $\tan v$, and $\tan(u+v)$ are defined (in other words, avoid odd multiples of $\frac{\pi}{2}$).

Notice that if $v = u$, the addition formula for tangent becomes

$$\tan(2u) = \frac{2\tan u}{1 - \tan^2 u},$$

which agrees with our previous double-angle formula for tangent.

We can now find a formula for the tangent of the difference of two angles. In the formula for $\tan(u+v)$, replace v by $-v$ on both sides of the equation and use the identity $\tan(-v) = -\tan v$ to get the following result:

Subtraction formula for tangent

$$\tan(u-v) = \frac{\tan u - \tan v}{1 + \tan u \tan v}$$

EXAMPLE 4

Use the subtraction formula for tangent to find a formula for $\tan(\pi - \theta)$.

SOLUTION Using the subtraction formula for tangent, we have

$$\begin{aligned}\tan(\pi - \theta) &= \frac{\tan \pi - \tan\theta}{1 + \tan\pi \tan\theta} \\ &= \frac{0 - \tan\theta}{1 + 0 \cdot \tan\theta} \\ &= -\tan\theta.\end{aligned}$$

We could have skipped the derivation of the double-angle formulas in the previous section and instead we could have obtained the double-angle formulas as consequences of the addition formulas (in fact, your instructor may have done this). However, sometimes additional understanding comes from seeing multiple derivations of a formula.

EXERCISES

1. For $x = 19°$ and $y = 13°$, evaluate each of the following:

   (a) $\cos(x+y)$ (b) $\cos x + \cos y$

   [*This exercise and the next one emphasize that* $\cos(x+y)$ *does not equal* $\cos x + \cos y$.]

2. For $x = 1.2$ radians and $y = 3.4$ radians, evaluate each of the following:

   (a) $\cos(x+y)$ (b) $\cos x + \cos y$

3. For $x = 5.7$ radians and $y = 2.5$ radians, evaluate each of the following:

   (a) $\sin(x - y)$ (b) $\sin x - \sin y$

   [*This exercise and the next one emphasize that* $\sin(x - y)$ *does not equal* $\sin x - \sin y$.]

4. For $x = 79°$ and $y = 33°$, evaluate each of the following:

   (a) $\sin(x - y)$ (b) $\sin x - \sin y$

For Exercises 5–12, find exact expressions for the indicated quantities. The following information will be useful:

$$\cos 22.5° = \frac{\sqrt{2+\sqrt{2}}}{2} \quad \textit{and} \quad \sin 22.5° = \frac{\sqrt{2-\sqrt{2}}}{2};$$

$$\cos 18° = \sqrt{\frac{\sqrt{5}+5}{8}} \quad \textit{and} \quad \sin 18° = \frac{\sqrt{5}-1}{4}.$$

[The value for $\sin 22.5°$ used here was derived in Example 5 in Section 10.5; the other values were derived in Exercise 64 and Problems 101 and 102 in Section 10.5.]

5. $\cos 82.5°$
6. $\cos 48°$ [*Hint:* $48 = 30 + 18$]
7. $\sin 82.5°$
8. $\sin 48°$
9. $\cos 37.5°$
10. $\cos 12°$ [*Hint:* $12 = 30 - 18$]
11. $\sin 37.5°$
12. $\sin 12°$

For Exercises 13–24, evaluate the indicated expressions assuming that

$$\cos x = \tfrac{1}{3} \quad \textit{and} \quad \sin y = \tfrac{1}{4},$$

$$\sin u = \tfrac{2}{3} \quad \textit{and} \quad \cos v = \tfrac{1}{5}.$$

Assume also that x and u are in the interval $(0, \frac{\pi}{2})$, that y is in the interval $(\frac{\pi}{2}, \pi)$, and that v is in the interval $(-\frac{\pi}{2}, 0)$.

13. $\cos(x + y)$
14. $\cos(u + v)$
15. $\cos(x - y)$
16. $\cos(u - v)$
17. $\sin(x + y)$
18. $\sin(u + v)$
19. $\sin(x - y)$
20. $\sin(u - v)$
21. $\tan(x + y)$
22. $\tan(u + v)$
23. $\tan(x - y)$
24. $\tan(u - v)$
25. Evaluate $\cos(\frac{\pi}{6} + \cos^{-1}\frac{3}{4})$.
26. Evaluate $\sin(\frac{\pi}{3} + \sin^{-1}\frac{2}{5})$.
27. Evaluate $\sin(\cos^{-1}\frac{1}{4} + \tan^{-1} 2)$.
28. Evaluate $\cos(\cos^{-1}\frac{2}{3} + \tan^{-1} 3)$.
29. Find a formula for $\cos(\theta + \frac{\pi}{2})$.
30. Find a formula for $\sin(\theta + \frac{\pi}{2})$.
31. Find a formula for $\cos(\theta + \frac{\pi}{4})$.
32. Find a formula for $\sin(\theta - \frac{\pi}{4})$.
33. Find a formula for $\tan(\theta + \frac{\pi}{4})$.
34. Find a formula for $\tan(\theta - \frac{\pi}{4})$.
35. Find a formula for $\tan(\theta + \frac{\pi}{2})$.
36. Find a formula for $\tan(\theta - \frac{\pi}{2})$.

PROBLEMS

37. Show (without using a calculator) that

$$\sin 10° \cos 20° + \cos 10° \sin 20° = \tfrac{1}{2}.$$

38. Show (without using a calculator) that

$$\sin\tfrac{\pi}{7}\cos\tfrac{4\pi}{21} + \cos\tfrac{\pi}{7}\sin\tfrac{4\pi}{21} = \tfrac{\sqrt{3}}{2}.$$

39. Show that

$$\frac{\sqrt{6}+\sqrt{2}}{4} = \frac{\sqrt{2+\sqrt{3}}}{2}.$$

Do this without using a calculator and without using the knowledge that both expressions above are equal to $\cos 15°$ (see Example 2).

40. Show that

$$\cos(3\theta) = 4\cos^3\theta - 3\cos\theta$$

for all θ.
[*Hint:* $\cos(3\theta) = \cos(2\theta + \theta)$.]

41. Show that $\cos 20°$ is a zero of the polynomial $8x^3 - 6x - 1$.
[*Hint:* Set $\theta = 20°$ in the identity from the previous problem.]

42. Show that

$$\sin(3\theta) = 3\sin\theta - 4\sin^3\theta$$

for all θ.

43. Show that $\sin \frac{\pi}{18}$ is a zero of the polynomial $8x^3 - 6x + 1$.
[*Hint:* Use the identity from the previous problem.]

44. Show that
$$\cos(5\theta) = 16\cos^5\theta - 20\cos^3\theta + 5\cos\theta$$
for all θ.

45. Find a nice formula for $\sin(5\theta)$ in terms of $\sin\theta$.

46. Show that
$$\cos u \cos v = \frac{\cos(u+v) + \cos(u-v)}{2}$$
for all u, v.
[*Hint:* Add together the formulas for $\cos(u+v)$ and $\cos(u-v)$.]

47. Show that
$$\sin u \sin v = \frac{\cos(u-v) - \cos(u+v)}{2}$$
for all u, v.

48. Show that
$$\cos u \sin v = \frac{\sin(u+v) - \sin(u-v)}{2}$$
for all u, v.

49. Show that
$$\cos x + \cos y = 2\cos\tfrac{x+y}{2}\cos\tfrac{x-y}{2}$$
for all x, y.
[*Hint:* Take $u = \frac{x+y}{2}$ and $v = \frac{x-y}{2}$ in the formula given by Problem 46.]

50. Show that
$$\cos x - \cos y = 2\sin\tfrac{x+y}{2}\sin\tfrac{y-x}{2}$$
for all x, y.

51. Show that
$$\sin x - \sin y = 2\cos\tfrac{x+y}{2}\sin\tfrac{x-y}{2}$$
for all x, y.

52. Find a formula for $\sin x + \sin y$ analogous to the formula in the previous problem.

53. Show that
$$\tan\tfrac{x+y}{2} = \frac{\cos x - \cos y}{\sin y - \sin x}$$
for all numbers x and y such that both sides make sense.
[*Hint:* Divide the result in Exercise 50 by the result in Exercise 51.]

54. Show that if $|t|$ is small but nonzero, then
$$\frac{\sin(x+t) - \sin x}{t} \approx \cos x.$$

55. Show that if $|t|$ is small but nonzero, then
$$\frac{\cos(x+t) - \cos x}{t} \approx -\sin x.$$

56. Show that if $|t|$ is small but nonzero and x is not an odd multiple of $\frac{\pi}{2}$, then
$$\frac{\tan(x+t) - \tan x}{t} \approx 1 + \tan^2 x.$$

57. Suppose $u = \tan^{-1} 2$ and $v = \tan^{-1} 3$. Show that $\tan(u+v) = -1$.

58. Suppose $u = \tan^{-1} 2$ and $v = \tan^{-1} 3$. Using the previous problem, explain why $u + v = \frac{3\pi}{4}$.

59. Using the previous problem, derive the beautiful equation
$$\tan^{-1} 1 + \tan^{-1} 2 + \tan^{-1} 3 = \pi.$$
[*Problem 39 in Section 11.4 gives another derivation of the equation above.*]

WORKED-OUT SOLUTIONS *to Odd-numbered Exercises*

1. For $x = 19°$ and $y = 13°$, evaluate each of the following:

(a) $\cos(x+y)$ (b) $\cos x + \cos y$

SOLUTION

(a) Using a calculator working in degrees, we have
$$\cos(19° + 13°) = \cos 32° \approx 0.84805.$$

(b) Using a calculator working in degrees, we have
$$\cos 19° + \cos 13° \approx 0.94552 + 0.97437$$
$$= 1.91989.$$

3. For $x = 5.7$ radians and $y = 2.5$ radians, evaluate each of the following:

   (a) $\sin(x - y)$ (b) $\sin x - \sin y$

   SOLUTION

(a) Using a calculator working in radians, we have

$$\sin(5.7 - 2.5) = \sin 3.2 \approx -0.05837.$$

(b) Using a calculator working in radians, we have

$$\begin{aligned}\sin 5.7 - \sin 2.5 &\approx -0.55069 - 0.59847\\ &= -1.14916.\end{aligned}$$

For Exercises 5–12, find exact expressions for the indicated quantities. The following information will be useful:

$$\cos 22.5^\circ = \frac{\sqrt{2+\sqrt{2}}}{2} \quad \textit{and} \quad \sin 22.5^\circ = \frac{\sqrt{2-\sqrt{2}}}{2};$$

$$\cos 18^\circ = \sqrt{\frac{\sqrt{5}+5}{8}} \quad \textit{and} \quad \sin 18^\circ = \frac{\sqrt{5}-1}{4}.$$

5. $\cos 82.5^\circ$

   SOLUTION

$$\begin{aligned}\cos 82.5^\circ &= \cos(60^\circ + 22.5^\circ)\\ &= \cos 60^\circ \cos 22.5^\circ - \sin 60^\circ \sin 22.5^\circ\\ &= \frac{1}{2}\cdot\frac{\sqrt{2+\sqrt{2}}}{2} - \frac{\sqrt{3}}{2}\cdot\frac{\sqrt{2-\sqrt{2}}}{2}\\ &= \frac{\sqrt{2+\sqrt{2}} - \sqrt{3}\sqrt{2-\sqrt{2}}}{4}\end{aligned}$$

7. $\sin 82.5^\circ$

   SOLUTION

$$\begin{aligned}\sin 82.5^\circ &= \sin(60^\circ + 22.5^\circ)\\ &= \sin 60^\circ \cos 22.5^\circ + \cos 60^\circ \sin 22.5^\circ\\ &= \frac{\sqrt{3}}{2}\cdot\frac{\sqrt{2+\sqrt{2}}}{2} + \frac{1}{2}\cdot\frac{\sqrt{2-\sqrt{2}}}{2}\\ &= \frac{\sqrt{3}\sqrt{2+\sqrt{2}} + \sqrt{2-\sqrt{2}}}{4}\end{aligned}$$

9. $\cos 37.5^\circ$

   SOLUTION

$$\begin{aligned}\cos 37.5^\circ &= \cos(60^\circ - 22.5^\circ)\\ &= \cos 60^\circ \cos 22.5^\circ + \sin 60^\circ \sin 22.5^\circ\\ &= \frac{1}{2}\cdot\frac{\sqrt{2+\sqrt{2}}}{2} + \frac{\sqrt{3}}{2}\cdot\frac{\sqrt{2-\sqrt{2}}}{2}\\ &= \frac{\sqrt{2+\sqrt{2}} + \sqrt{3}\sqrt{2-\sqrt{2}}}{4}\end{aligned}$$

11. $\sin 37.5^\circ$

   SOLUTION

$$\begin{aligned}\sin 37.5^\circ &= \sin(60^\circ - 22.5^\circ)\\ &= \sin 60^\circ \cos 22.5^\circ - \cos 60^\circ \sin 22.5^\circ\\ &= \frac{\sqrt{3}}{2}\cdot\frac{\sqrt{2+\sqrt{2}}}{2} - \frac{1}{2}\cdot\frac{\sqrt{2-\sqrt{2}}}{2}\\ &= \frac{\sqrt{3}\sqrt{2+\sqrt{2}} - \sqrt{2-\sqrt{2}}}{4}\end{aligned}$$

For Exercises 13–24, evaluate the indicated expressions assuming that

$$\cos x = \tfrac{1}{3} \quad \textit{and} \quad \sin y = \tfrac{1}{4},$$

$$\sin u = \tfrac{2}{3} \quad \textit{and} \quad \cos v = \tfrac{1}{5}.$$

Assume also that x and u are in the interval $(0, \frac{\pi}{2})$, that y is in the interval $(\frac{\pi}{2}, \pi)$, and that v is in the interval $(-\frac{\pi}{2}, 0)$.

13. $\cos(x + y)$

    SOLUTION To use the addition formula for $\cos(x + y)$, we will need to know the cosine and sine of both x and y. Thus first we find those values, beginning with $\sin x$. Because $0 < x < \frac{\pi}{2}$, we know that $\sin x > 0$. Thus

$$\begin{aligned}\sin x = \sqrt{1 - \cos^2 x} = \sqrt{1 - \tfrac{1}{9}} = \sqrt{\tfrac{8}{9}} &= \tfrac{\sqrt{4}\sqrt{2}}{\sqrt{9}}\\ &= \tfrac{2\sqrt{2}}{3}.\end{aligned}$$

Because $\frac{\pi}{2} < y < \pi$, we know that $\cos y < 0$. Thus

$$\begin{aligned}\cos y = -\sqrt{1 - \sin^2 y} = -\sqrt{1 - \tfrac{1}{16}} &= -\sqrt{\tfrac{15}{16}}\\ &= -\tfrac{\sqrt{15}}{4}.\end{aligned}$$

Thus

$$\begin{aligned}\cos(x+y) &= \cos x \cos y - \sin x \sin y \\ &= \tfrac{1}{3}\cdot(-\tfrac{\sqrt{15}}{4}) - \tfrac{2\sqrt{2}}{3}\cdot\tfrac{1}{4} \\ &= \tfrac{-\sqrt{15}-2\sqrt{2}}{12}.\end{aligned}$$

15. $\cos(x-y)$

SOLUTION

$$\begin{aligned}\cos(x-y) &= \cos x \cos y + \sin x \sin y \\ &= \tfrac{1}{3}\cdot(-\tfrac{\sqrt{15}}{4}) + \tfrac{2\sqrt{2}}{3}\cdot\tfrac{1}{4} \\ &= \tfrac{2\sqrt{2}-\sqrt{15}}{12}\end{aligned}$$

17. $\sin(x+y)$

SOLUTION

$$\begin{aligned}\sin(x+y) &= \sin x \cos y + \cos x \sin y \\ &= \tfrac{2\sqrt{2}}{3}\cdot(-\tfrac{\sqrt{15}}{4}) + \tfrac{1}{3}\cdot\tfrac{1}{4} \\ &= \tfrac{1-2\sqrt{30}}{12}\end{aligned}$$

19. $\sin(x-y)$

SOLUTION

$$\begin{aligned}\sin(x-y) &= \sin x \cos y - \cos x \sin y \\ &= \tfrac{2\sqrt{2}}{3}\cdot(-\tfrac{\sqrt{15}}{4}) - \tfrac{1}{3}\cdot\tfrac{1}{4} \\ &= \tfrac{-1-2\sqrt{30}}{12}\end{aligned}$$

21. $\tan(x+y)$

SOLUTION To use the addition formula for $\tan(x+y)$, we will need to know the tangent of both x and y. Thus first we find those values, beginning with $\tan x$:

$$\tan x = \frac{\sin x}{\cos x} = \frac{\frac{2\sqrt{2}}{3}}{\frac{1}{3}} = 2\sqrt{2}.$$

Also,

$$\tan y = \frac{\sin y}{\cos y} = \frac{\frac{1}{4}}{-\frac{\sqrt{15}}{4}} = -\frac{1}{\sqrt{15}} = -\frac{\sqrt{15}}{15}.$$

Thus

$$\begin{aligned}\tan(x+y) &= \frac{\tan x + \tan y}{1 - \tan x \tan y} \\ &= \frac{2\sqrt{2} - \frac{\sqrt{15}}{15}}{1 + 2\sqrt{2}\cdot\frac{\sqrt{15}}{15}} \\ &= \frac{30\sqrt{2} - \sqrt{15}}{15 + 2\sqrt{30}},\end{aligned}$$

where the last expression is obtained by multiplying the numerator and denominator of the previous expression by 15.

23. $\tan(x-y)$

SOLUTION

$$\begin{aligned}\tan(x-y) &= \frac{\tan x - \tan y}{1 + \tan x \tan y} \\ &= \frac{2\sqrt{2} + \frac{\sqrt{15}}{15}}{1 - 2\sqrt{2}\cdot\frac{\sqrt{15}}{15}} \\ &= \frac{30\sqrt{2} + \sqrt{15}}{15 - 2\sqrt{30}},\end{aligned}$$

where the last expression is obtained by multiplying the numerator and denominator of the middle expression by 15.

25. Evaluate $\cos(\frac{\pi}{6} + \cos^{-1}\frac{3}{4})$.

SOLUTION To use the addition formula for cosine, we will need to evaluate the cosine and sine of $\cos^{-1}\frac{3}{4}$. Thus we begin by computing those values.

The definition of $\cos^{-1}$ implies that

$$\cos(\cos^{-1}\tfrac{3}{4}) = \tfrac{3}{4}.$$

Evaluating $\sin(\cos^{-1}\frac{3}{4})$ takes a bit more work. Let $v = \cos^{-1}\frac{3}{4}$. Thus v is the angle in $[0,\pi]$ such that $\cos v = \frac{3}{4}$. Note that $\sin v \ge 0$ because v is in $[0,\pi]$. Thus

$$\begin{aligned}\sin(\cos^{-1}\tfrac{3}{4}) = \sin v &= \sqrt{1-\cos^2 v} \\ &= \sqrt{1-\tfrac{9}{16}} = \sqrt{\tfrac{7}{16}} = \tfrac{\sqrt{7}}{4}.\end{aligned}$$

Using the addition formula for cosine, we now have

$$\cos(\tfrac{\pi}{6} + \cos^{-1}\tfrac{3}{4})$$
$$= \cos\tfrac{\pi}{6}\cos(\cos^{-1}\tfrac{3}{4}) - \sin\tfrac{\pi}{6}\sin(\cos^{-1}\tfrac{3}{4})$$
$$= \tfrac{\sqrt{3}}{2}\cdot\tfrac{3}{4} - \tfrac{1}{2}\cdot\tfrac{\sqrt{7}}{4}$$
$$= \tfrac{3\sqrt{3}-\sqrt{7}}{8}.$$

27. Evaluate $\sin(\cos^{-1}\frac{1}{4} + \tan^{-1}2)$.

SOLUTION To use the addition formula for sine, we will need to evaluate the cosine and sine of $\cos^{-1}\frac{1}{4}$ and $\tan^{-1}2$. Thus we begin by computing those values.

The definition of $\cos^{-1}$ implies that

$$\cos(\cos^{-1}\tfrac{1}{4}) = \tfrac{1}{4}.$$

Evaluating $\sin(\cos^{-1}\frac{1}{4})$ takes a bit more work. Let $u = \cos^{-1}\frac{1}{4}$. Thus u is the angle in $[0,\pi]$ such that $\cos u = \frac{1}{4}$. Note that $\sin u \geq 0$ because u is in $[0,\pi]$. Thus

$$\sin(\cos^{-1}\tfrac{1}{4}) = \sin u = \sqrt{1-\cos^2 u}$$
$$= \sqrt{1-\tfrac{1}{16}} = \sqrt{\tfrac{15}{16}} = \tfrac{\sqrt{15}}{4}.$$

Now let $v = \tan^{-1}2$. Thus v is the angle in $(0,\frac{\pi}{2})$ such that $\tan v = 2$ (the range of $\tan^{-1}$ is the interval $(-\frac{\pi}{2},\frac{\pi}{2})$, but for this particular v we know that $\tan v$ is positive, which excludes the interval $(-\frac{\pi}{2},0]$ from consideration). We have

$$2 = \tan v = \frac{\sin v}{\cos v} = \frac{\sqrt{1-\cos^2 v}}{\cos v}.$$

Squaring the first and last terms above, we get

$$4 = \frac{1-\cos^2 v}{\cos^2 v}.$$

Solving the equation above for $\cos v$ now gives

$$\cos(\tan^{-1}2) = \cos v = \tfrac{\sqrt{5}}{5}.$$

The identity $\sin v = \sqrt{1-\cos^2 v}$ now implies that

$$\sin(\tan^{-1}2) = \sin v = \tfrac{2\sqrt{5}}{5}.$$

Using the addition formula for sine, we now have

$$\sin(\cos^{-1}\tfrac{1}{4} + \tan^{-1}2)$$
$$= \sin(\cos^{-1}\tfrac{1}{4})\cos(\tan^{-1}2)$$
$$+ \cos(\cos^{-1}\tfrac{1}{4})\sin(\tan^{-1}2)$$
$$= \tfrac{\sqrt{15}}{4}\cdot\tfrac{\sqrt{5}}{5} + \tfrac{1}{4}\cdot\tfrac{2\sqrt{5}}{5}$$
$$= \tfrac{5\sqrt{3}+2\sqrt{5}}{20}.$$

29. Find a formula for $\cos(\theta + \frac{\pi}{2})$.

SOLUTION

$$\cos(\theta + \tfrac{\pi}{2}) = \cos\theta\cos\tfrac{\pi}{2} - \sin\theta\sin\tfrac{\pi}{2}$$
$$= -\sin\theta.$$

31. Find a formula for $\cos(\theta + \frac{\pi}{4})$.

SOLUTION

$$\cos(\theta + \tfrac{\pi}{4}) = \cos\theta\cos\tfrac{\pi}{4} - \sin\theta\sin\tfrac{\pi}{4}$$
$$= \tfrac{\sqrt{2}}{2}(\cos\theta - \sin\theta)$$

33. Find a formula for $\tan(\theta + \frac{\pi}{4})$.

SOLUTION

$$\tan(\theta + \tfrac{\pi}{4}) = \frac{\tan\theta + \tan\frac{\pi}{4}}{1 - \tan\theta\tan\frac{\pi}{4}}$$
$$= \frac{\tan\theta + 1}{1 - \tan\theta}$$

35. Find a formula for $\tan(\theta + \frac{\pi}{2})$.

SOLUTION Because $\tan\frac{\pi}{2}$ is undefined, we cannot use the formula for the tangent of the sum of two angles. But the following calculation works:

$$\tan(\theta + \tfrac{\pi}{2}) = \frac{\sin(\theta + \frac{\pi}{2})}{\cos(\theta + \frac{\pi}{2})}$$
$$= \frac{\sin\theta\cos\frac{\pi}{2} + \cos\theta\sin\frac{\pi}{2}}{\cos\theta\cos\frac{\pi}{2} - \sin\theta\sin\frac{\pi}{2}}$$
$$= \frac{\cos\theta}{-\sin\theta}$$
$$= -\frac{1}{\tan\theta}.$$

CHAPTER SUMMARY

To check that you have mastered the most important concepts and skills covered in this chapter, make sure that you can do each item in the following list:

- Give the domain and range of $\cos^{-1}$, $\sin^{-1}$, and $\tan^{-1}$.
- Find the angle a line with given slope makes with the horizontal axis.
- Compute the composition of a trigonometric function and an inverse trigonometric function.
- Compute the area of a triangle or a parallelogram given the lengths of two adjacent sides and the angle between them.
- Compute the area of a regular polygon.
- Explain why knowing the sine of an angle of a triangle is sometimes not enough information to determine the angle.
- Find all the angles and the lengths of all the sides of a triangle given only some of this data.
- Use the double-angle and half-angle formulas for cosine, sine, and tangent.
- Use the addition and subtraction formulas for cosine, sine, and tangent.

To review a chapter, go through the list above to find items that you do not know how to do, then reread the material in the chapter about those items. Then try to answer the chapter review questions below without looking back at the chapter.

CHAPTER REVIEW QUESTIONS

1. Find the angles (in radians) in a right triangle with a hypotenuse of length 7 and another side with length 4.

2. Find the angles (in degrees) in a right triangle whose nonhypotenuse sides have lengths 6 and 7.

3. Find the lengths of both circular arcs of the unit circle connecting the points $(\frac{3}{5}, \frac{4}{5})$ and $(\frac{5}{13}, \frac{12}{13})$.

4. Give the domain and range of each of the following functions: $\cos^{-1}$, $\sin^{-1}$, and $\tan^{-1}$.

5. Evaluate $\cos^{-1} \frac{\sqrt{3}}{2}$.

6. Evaluate $\sin^{-1} \frac{\sqrt{3}}{2}$.

7. Evaluate $\cos(\cos^{-1} \frac{2}{5})$.

8. Without using a calculator, sketch the radius of the unit circle corresponding to $\cos^{-1}(-0.8)$.

9. Explain why your calculator is likely to be unhappy if you ask it to evaluate $\cos^{-1} 3$.

10. Find the smallest positive number x such that
$$3\sin^2 x - 4\sin x + 1 = 0.$$

11. Evaluate $\sin^{-1}(\sin \frac{19\pi}{8})$.

12. Evaluate $\cos(\tan^{-1} 5)$.

13. Find the area of a triangle that has sides of length 7 and 10, with a 29° angle between those sides.

14. Find the area of a regular 9-sided polygon whose vertices are nine equally spaced points on a circle of radius 2.

15. Find the angles in a rhombus (a parallelogram whose four sides have equal length) that has area 19 and sides of length 5.

16. Find the perimeter of a regular 13-sided polygon whose vertices are 13 equally spaced points on a circle of radius 5.

17. Suppose $\sin u = \frac{3}{7}$. Evaluate $\cos(2u)$.

18. Suppose θ is an angle such that $\sin\theta$ is a rational number. Explain why $\cos(2\theta)$ is a rational number.

19. Suppose θ is an angle such that $\tan\theta$ is a rational number other than 1 or -1. Explain why $\tan(2\theta)$ is a rational number.

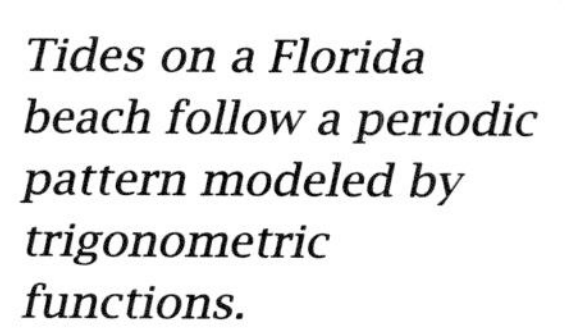

Tides on a Florida beach follow a periodic pattern modeled by trigonometric functions.

Applications of Trigonometry

This chapter begins with a discussion of parametric curves. We will see that this technique for describing paths in the coordinate plane gives new insight into several previously discussed topics, including the cosine and sine functions, the graphs of inverse functions, and the graphs arising from function transformations.

Then we turn to transformations of trigonometric functions. Such transformations are used to model periodic events. Redoing function transformations in the context of trigonometric functions will also help us review the key concepts of function transformations from Chapter 3.

Our next subject will be polar coordinates, which provide an alternative method of locating points in the coordinate plane. As we will see, converting between polar coordinates and rectangular coordinates requires a good understanding of the trigonometric functions.

Polar coordinates will be useful in the last two sections of the chapter, as we deal with vectors and the complex plane.

11.1 *Parametric Curves*

LEARNING OBJECTIVES

By the end of this section you should be able to

- use parametric curves to describe the motion of a point moving in the coordinate plane;
- use parametric curves to graph the inverse of a function;
- find formulas that shift, stretch, or flip a parametric curve.

Curves in the Coordinate Plane

Parametric curves can be used to describe the path of a point moving in the coordinate plane. A more formal definition of a parametric curve will be given soon, but first we look at an example.

EXAMPLE 1

Suppose a point moving in the coordinate plane has coordinates

$$(\cos t, \sin t)$$

at time t seconds for t in the interval $[0, 2\pi]$.

(a) What are the coordinates of the point at time $t = 0$ seconds?

(b) What are the coordinates of the point at time $t = 3$ seconds?

(c) What are the coordinates of the point at time $t = 2\pi$ seconds?

(d) Describe the path followed by the point over the time interval $[0, 2\pi]$ seconds.

(e) How far does the point travel in the time interval $[0, 3]$ seconds?

SOLUTION

(a) At time 0 seconds, the point has coordinates $(\cos 0, \sin 0)$, which equals $(1, 0)$. Thus the location of the point at time 0 is shown by the blue dot in the figure.

(b) At time 3 seconds, the point has coordinates $(\cos 3, \sin 3)$. The location of the point at time 3 seconds is shown by the red dot in the figure. The red dot is near the point $(-1, 0)$ because 3 is a bit less than π (and because π radians equals $180°$).

(c) At time 2π seconds, the point has coordinates $(\cos 2\pi, \sin 2\pi)$, which equals $(1, 0)$.

(d) The point starts at $(1, 0)$ at time 0. It travels counterclockwise on the unit circle, as shown in the figure, coming back to its starting point at time 2π.

(e) We need to find the length of the arc on the unit circle from $(1, 0)$ to $(\cos 3, \sin 3)$. Because we are working in radians, the length of an arc on the unit circle equals the number of radians in the corresponding angle. Thus this arc has length 3.

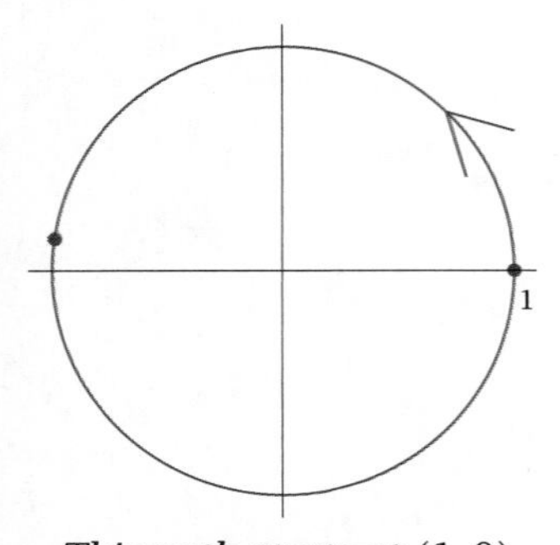

This path starts at $(1, 0)$ (blue dot) at time 0 and ends at the same location at time 2π seconds. The arrow shows the direction of motion. The red dot shows the location of the point at time 3 seconds.

We can think of the path of the point in the example above as described by two functions f and g defined by

$$f(t) = \cos t \quad \text{and} \quad g(t) = \sin t,$$

where the domain of both f and g is the interval $[0, 2\pi]$. The location of the point at time t is $(f(t), g(t))$, where t is in the domain of f and g. These considerations lead to the following definition.

We could have defined $\cos t$ and $\sin t$ as the first and second coordinates at time t seconds of a point traveling counterclockwise on the unit circle, starting at $(1, 0)$ at time 0 and moving at a speed of one unit of distance per second.

Parametric curves

A **parametric curve** is a path in the coordinate plane described by an ordered pair of functions that have an interval as their common domain.

EXAMPLE 2

Consider the parametric curve described by

$$(t^3 - 6t^2 + 10t, 2t^2 - 8t + 9)$$

for t in the interval $[0, 4]$.

(a) Using a computer or a graphing calculator, sketch this parametric curve.

(b) Which point of this curve corresponds to $t = 0$?

(c) Which point of this curve corresponds to $t = 1$?

(d) Which point of this curve corresponds to $t = 4$?

(e) Is this parametric curve the graph of some function?

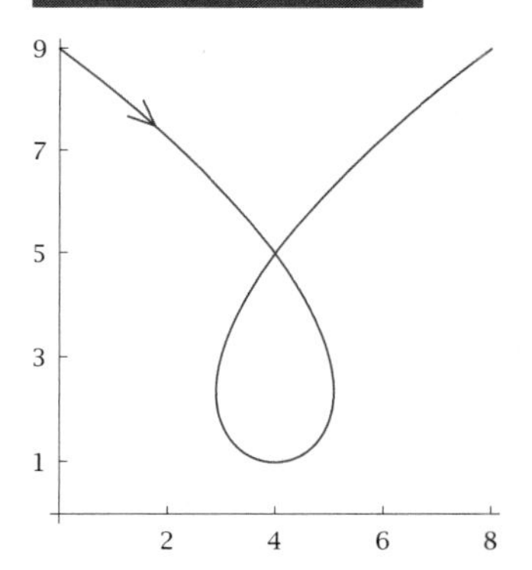

The parametric curve described by $(t^3 - 6t^2 + 10t, 2t^2 - 8t + 9)$ for t in the interval $[0, 4]$.

SOLUTION

(a) Enter

```
parametric plot (t^3 - 6t^2 + 10t, 2t^2 - 8t + 9) for t=0 to 4
```

in a Wolfram|Alpha entry box, or use the appropriate input in other software or on a graphing calculator that can produce parametric curves. Using any of these methods, you should see a curve that looks like the figure shown here.

(b) If $t = 0$ then $(t^3 - 6t^2 + 10t, 2t^2 - 8t + 9)$ equals $(0, 9)$. Make sure you can locate the point $(0, 9)$, which is where the path starts, in the figure.

(c) If $t = 1$ then $(t^3 - 6t^2 + 10t, 2t^2 - 8t + 9)$ equals $(5, 3)$. Make sure you can locate the point $(5, 3)$ in the figure.

(d) If $t = 4$ then $(t^3 - 6t^2 + 10t, 2t^2 - 8t + 9)$ equals $(8, 9)$. Make sure you can locate the point $(8, 9)$, which is where the path ends, in the figure.

(e) Looking at the curve in the margin above, we see that there are vertical lines that intersect the curve in more than one point. Because this curve fails the vertical line test, it is not the graph of any function (see Section 3.1).

The arrow in the curve above (and in other figures in this section) indicates the direction of motion when considering a parametric curve as the path of a point moving in the plane.

The dynamic representation with a moving point tracing out the path may give you a better understanding of a parametric curve.

The figure above shows only a static representation of the parametric curve. To see a dynamic representation with a moving point tracing out the path, point your web browser to `algebraTrig.axler.net/parametric.html`. This web site contains dynamic representations with moving points tracing out the path for each parametric curve shown in this book.

The graph of every function whose domain is an interval can be thought of as a parametric curve. Specifically, if f is a function whose domain is some interval, then the parametric curve described by $(t, f(t))$ at time t is the graph of f.

EXAMPLE 3

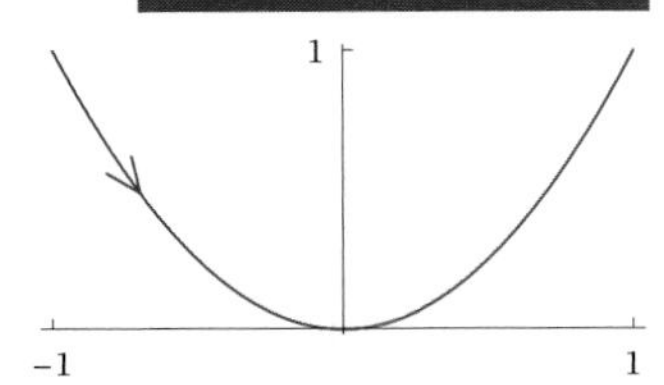

The parametric curve described by (t, t^2) for t in $[-1, 1]$. The same path is also described by the parametric curve (t^3, t^6) for t in $[-1, 1]$. As this example shows, a path can be described by more than one pair of functions.

Suppose f is the function defined by $f(x) = x^2$, with the domain of f the interval $[-1, 1]$.

(a) Describe the graph of f as a parametric curve.

(b) Consider the parametric curve described by (t^3, t^6) for t in $[-1, 1]$. What is the relationship of this parametric curve to the parametric curve in part (a)?

SOLUTION

(a) The graph of f is the parametric curve described by (t, t^2) for t in $[-1, 1]$, as shown in the figure. The initial point of this path, corresponding to $t = -1$, is $(-1, 1)$. The endpoint of this path, corresponding to $t = 1$, is $(1, 1)$. This parametric curve is a familiar parabola.

(b) Let $u = t^3$. As t varies from -1 to 1, then u also varies from -1 to 1. Furthermore, $t^6 = (t^3)^2$, and thus $(t^3, t^6) = (u, u^2)$. Thus the parametric curve described by (t^3, t^6) for t in $[-1, 1]$ is the same as the parametric curve described by (u, u^2) for u in $[-1, 1]$, which is the same as the parametric curve described by (t, t^2) for t in $[-1, 1]$ (because the name of the variable does not matter).

Parametric curves provide a useful method for describing the motion of an object thrown in the air until gravity brings it back to the ground.

These formulas ignore air friction and thus should be used with caution if air friction becomes an important factor.

These formulas are valid only until the object hits the ground or another object.

Motion under the influence of gravity

Suppose an object is thrown from height H feet at time 0 with initial horizontal velocity U feet per second and initial vertical velocity V feet per second. Then at time t seconds

- the object is Ut feet (in the horizontal direction) from the thrower;
- the height of the object is $-16.1t^2 + Vt + H$ feet.

A few comments about the last formula above:

- If using meters instead of feet, replace 16.1 by 4.9.
- The number 16.1 is valid on Earth but must be replaced on other planets by a suitable constant depending upon the mass and radius of the planet.

EXAMPLE 4

Suppose a football is thrown from height 6 feet with initial horizontal velocity 50 feet per second and initial vertical velocity 40 feet per second.

(a) How long after the football is thrown does it hit the ground?

(b) Describe the path of the football as a parametric curve, and sketch the path.

(c) How far away from the thrower (in the horizontal direction) is the football when it hits the ground?

(d) How high is the football at its maximum height?

SOLUTION

(a) The football hits the ground when its height is 0, which happens when

$$-16.1t^2 + 40t + 6 = 0.$$

Using the quadratic formula to solve for t gives $t \approx -0.142$ or $t \approx 2.63$. A negative value for t makes no sense for this problem. Thus we conclude that the football hits the ground about 2.63 seconds after being thrown.

(b) In an appropriate coordinate plane, at time t the football has coordinates

$$(50t, -16.1t^2 + 40t + 6).$$

Thus we have the following computer-generated sketch of the football's path:

The path of the football is part of a parabola.

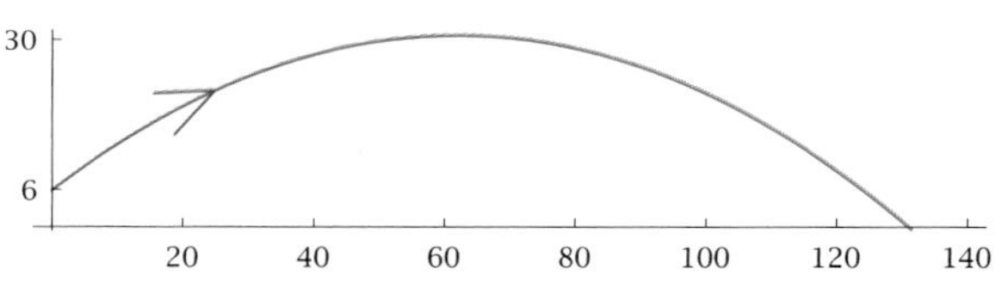

The parametric curve described by $(50t, -16.1t^2 + 40t + 6)$ *for* t *in* $[0, 2.63]$.

(c) The football hits the ground approximately 2.63 seconds after being thrown. Its horizontal distance from the thrower at that time is 50×2.63 feet, which is 131.5 feet.

(d) To find the maximum height of the football, we complete the square:

$$\begin{aligned} -16.1t^2 + 40t + 6 &= -16.1\left[t^2 - \tfrac{40}{16.1}t\right] + 6 \\ &= -16.1\left[\left(t - \tfrac{20}{16.1}\right)^2 - \left(\tfrac{20}{16.1}\right)^2\right] + 6 \\ &= -16.1\left(t - \tfrac{20}{16.1}\right)^2 + \tfrac{20^2}{16.1} + 6 \end{aligned}$$

The expression above shows that the maximum height of the football is attained when $t = \frac{20}{16.1}$ and that the football's height at that time will be $\frac{20^2}{16.1} + 6$ feet, which is approximately 30.8 feet. Note that a maximum height of 30.8 feet is consistent with the figure above.

Parametric curves can easily be graphed by computers and most graphing calculators. Indeed, parametric curves are often the easiest way to get such machines to graph some familiar curves, as shown by the next two examples.

EXAMPLE 5 Explain why the parametric curve described by

$$(5\cos t, 3\sin t)$$

for t in $[0, 2\pi]$ is an ellipse.

SOLUTION Note that

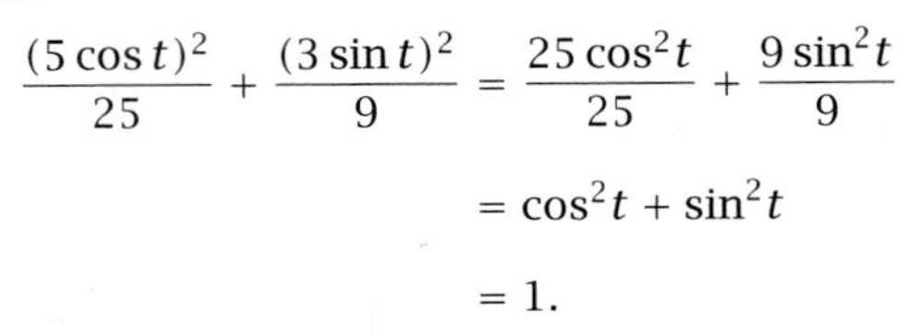

$$\frac{(5\cos t)^2}{25} + \frac{(3\sin t)^2}{9} = \frac{25\cos^2 t}{25} + \frac{9\sin^2 t}{9}$$
$$= \cos^2 t + \sin^2 t$$
$$= 1.$$

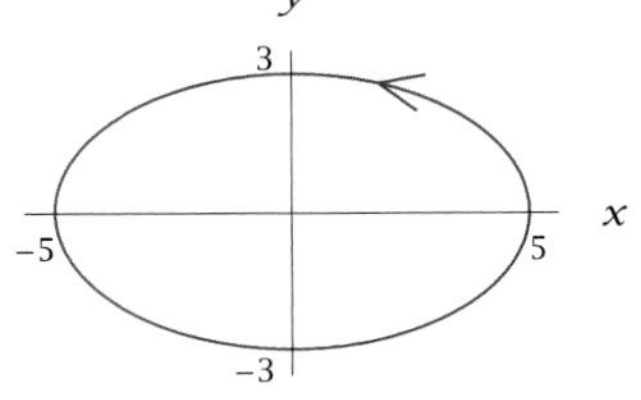

The parametric curve described by $(5\cos t, 3\sin t)$ *for t in* $[0, 2\pi]$ *is the ellipse* $\frac{x^2}{25} + \frac{y^2}{9} = 1$.

Thus setting $x = 5\cos t$ and $y = 3\sin t$, we have $\frac{x^2}{25} + \frac{y^2}{9} = 1$, which is the equation of the ellipse shown here.

The graph of this ellipse can be obtained by entering

parametric plot (5 cos t, 3 sin t) for t=0 to 2 pi

in a Wolfram|Alpha entry box, or use the appropriate input in other software or on a graphing calculator.

EXAMPLE 6 Explain why the parametric curve described by

$$\left(\tfrac{3}{\cos t}, 2\tan t\right)$$

for t in $[-\frac{\pi}{3}, \frac{\pi}{3}]$ is part of a hyperbola.

SOLUTION Note that

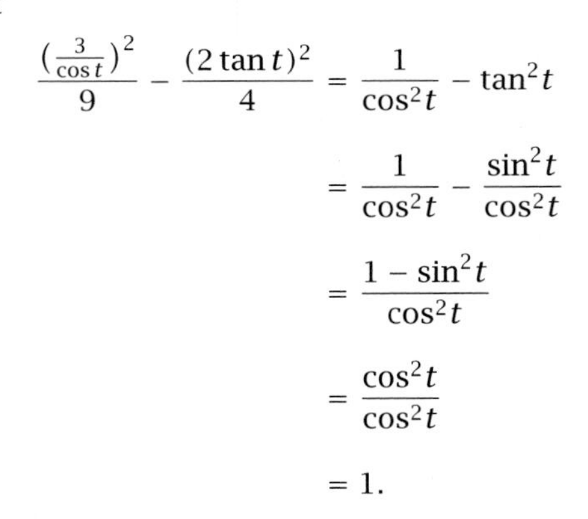

$$\frac{\left(\frac{3}{\cos t}\right)^2}{9} - \frac{(2\tan t)^2}{4} = \frac{1}{\cos^2 t} - \tan^2 t$$
$$= \frac{1}{\cos^2 t} - \frac{\sin^2 t}{\cos^2 t}$$
$$= \frac{1 - \sin^2 t}{\cos^2 t}$$
$$= \frac{\cos^2 t}{\cos^2 t}$$
$$= 1.$$

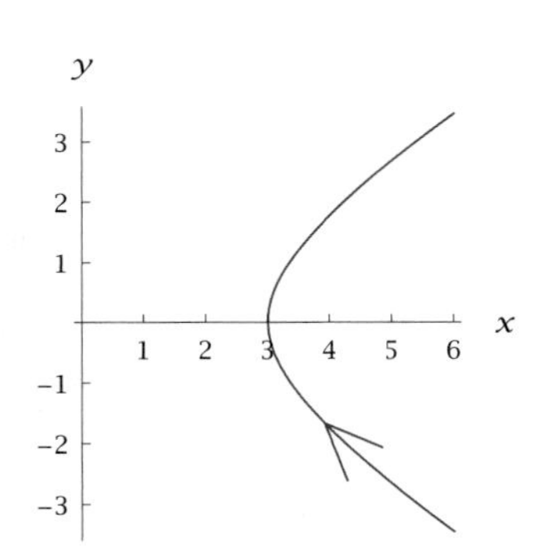

The parametric curve described by $(\frac{3}{\cos t}, 2\tan t)$ *for t in* $[-\frac{\pi}{3}, \frac{\pi}{3}]$ *is part of the hyperbola* $\frac{x^2}{9} - \frac{y^2}{4} = 1$.

Thus setting $x = \frac{3}{\cos t}$ and $y = 2\tan t$, we have $\frac{x^2}{9} - \frac{y^2}{4} = 1$, which is the equation of a hyperbola.

The part of this hyperbola shown here can be obtained by entering

parametric plot (3/cos t, 2 tan t) for t=-pi/3 to pi/3

in a Wolfram|Alpha entry box, or use the appropriate input in other software or on a graphing calculator.

Graphing Inverse Functions as Parametric Curves

Even when it is not possible to find a formula for the inverse of a function, the graph of an inverse function can be constructed as a parametric curve.

Graphing a function and its inverse

Suppose f is a function whose domain is an interval.

- The graph of f is described by the parametric curve $(t, f(t))$ as t varies over the domain of f.
- If f is one-to-one, then the graph of the inverse function f^{-1} is described by the parametric curve $(f(t), t)$ as t varies over the domain of f.

EXAMPLE 7

Suppose f is the function with domain $[0, 1]$ defined by $f(t) = \frac{1}{2}t^5 + \frac{3}{2}t^3$. Describe the graphs of f and f^{-1} as parametric curves, and use technology to sketch these graphs.

SOLUTION A computer or graphing calculator can draw the graph of f as the parametric curve described by $(t, \frac{1}{2}t^5 + \frac{3}{2}t^3)$ for t in $[0, 1]$ (see the blue curve).

Even a computer cannot find a formula for f^{-1} because it it not possible to solve the equation

$$\tfrac{1}{2}t^5 + \tfrac{3}{2}t^3 = y$$

for t in terms of y.

However, the graph of f^{-1} is obtained by flipping the graph of f across the line through the origin with slope 1. This operation is the same as interchanging the two coordinates of each point on the graph of f. In other words, the graph of f^{-1} is the parametric curve described by $(\frac{1}{2}t^5 + \frac{3}{2}t^3, t)$ for t in $[0, 1]$. A computer or graphing calculator can now easily draw that parametric curve (see the red curve). Thus even though we do not have a formula for f^{-1}, we do have its graph!

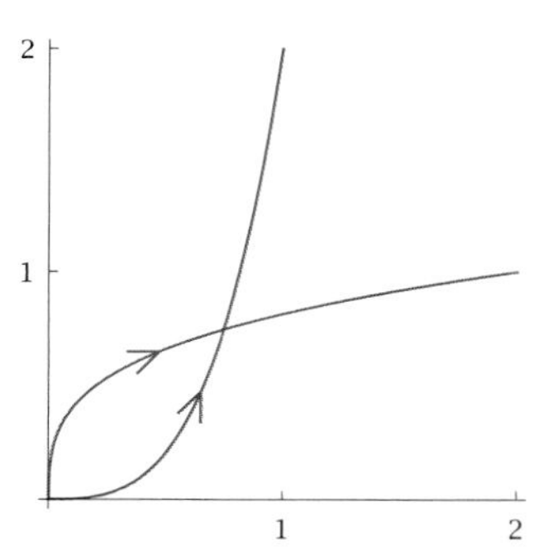

The parametric curve described by $(t, \frac{1}{2}t^5 + \frac{3}{2}t^3)$ (blue) and the parametric curve described by $(\frac{1}{2}t^5 + \frac{3}{2}t^3, t)$ (red), both for t in $[0, 1]$.

Enter

```
parametric plot { ( t, (1/2)t^5 + (3/2)t^3 ), ( (1/2)t^5 + (3/2)t^3, t ) } for t=0 to 1
```

in a Wolfram|Alpha entry box to obtain the graphs of f and f^{-1} in the same figure as shown here, or use the appropriate input in other software or on a graphing calculator.

Note that by thinking about parametric curves, we have turned an impossibly hard task [find a formula for f^{-1} and then graph that function] into an easy task [graph f^{-1} as the parametric curve described by $(f(t), t)$].

The example above is the same as Example 2 in Section 3.5, but now more explanation has been provided of how a computer was instructed to generate the graph of f^{-1}.

Shifting, Stretching, or Flipping a Parametric Curve

We now turn to the topic of transformations of parametric curves. This subject is analogous to the function transformations that we studied in Section 3.2. However, the ideas and the results are easier with parametric curves. Going through this material on transformations of parametric curves may help solidify your understanding of Section 3.2. We will also use this topic as an excuse to look at some new parametric curves.

We begin with a horizontal transformation of a parametric curve.

EXAMPLE 8

(a) Sketch the parametric curves described by (t^2, t^3) and $(t^2 + 1, t^3)$ for t in the interval $[-1, 1]$.

(b) Explain how the parametric curve described by $(t^2 + 1, t^3)$ is obtained from the parametric curve described by (t^2, t^3).

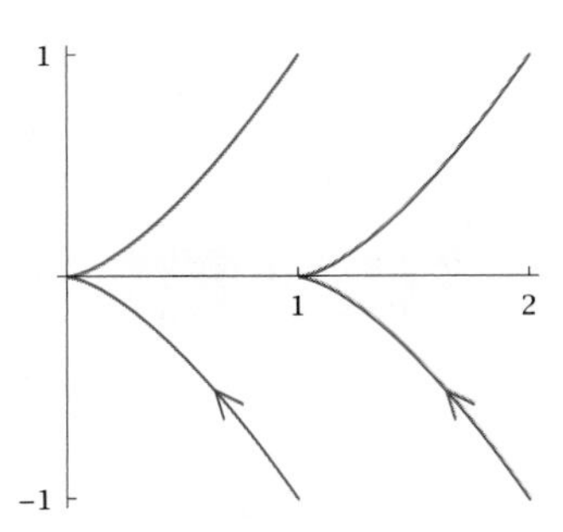

The parametric curve described by (t^2, t^3) (blue) and the parametric curve described by $(t^2 + 1, t^3)$ (red), both for t in $[-1, 1]$.

SOLUTION

(a) Sketching these parametric curves is best done by a computer or graphing calculator. If you use Wolfram|Alpha, then the following entry

```
parametric plot { (t^2, t^3), (t^2 + 1, t^3) } for t=-1 to 1
```

produces both parametric curves in the same figure, as shown here.

(b) The figure here shows that the parametric curve described by $(t^2 + 1, t^3)$ is obtained by shifting the parametric curve described by (t^2, t^3) right 1 unit.

The conclusion above about shifting right 1 unit is no surprise because 1 is added to the first coordinate of (t^2, t^3) to obtain $(t^2 + 1, t^3)$. The same reasoning works with any parametric curve and any positive number b in place of 1. Similarly, subtracting a positive number b in the first coordinate shifts left. Thus we have the following general result.

Shifting a parametric curve right or left

- Adding a positive constant b to the first coordinate of every point on a parametric curve shifts the curve right b units.
- Subtracting a positive constant b from the first coordinate of every point on a parametric curve shifts the curve left b units.

The next example involves a vertical transformation of a parametric curve.

EXAMPLE 9

(a) Sketch the parametric curves described by (t^3, t^4) and $(t^3, t^4 + 6)$ for t in the interval $[-2, 2]$.

(b) Explain how the parametric curve described by $(t^3, t^4 + 6)$ is obtained from the parametric curve described by (t^3, t^4).

SOLUTION

(a) A computer produces the parametric curves shown here.

(b) The figure here shows that the parametric curve described by $(t^3, t^4 + 6)$ is obtained by shifting the parametric curve described by (t^3, t^4) up 6 units.

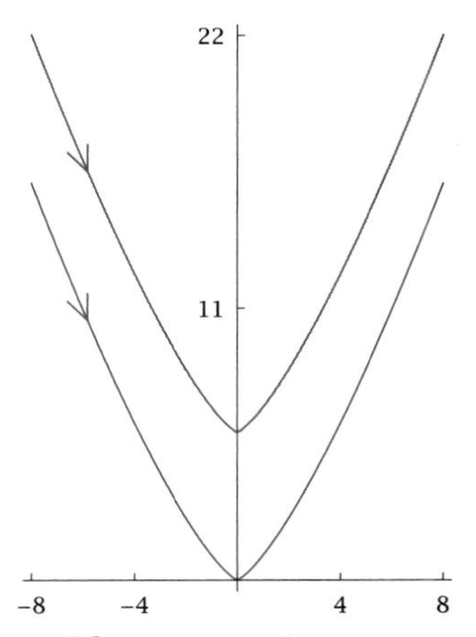

The parametric curve described by (t^3, t^4) (blue) and the parametric curve described by $(t^3, t^4 + 6)$ (red), both for t in $[-2, 2]$.

The conclusion above about shifting up 6 units is no surprise because 6 is added to the second coordinate of (t^3, t^4) to obtain $(t^3, t^4 + 6)$. The same reasoning works with any parametric curve and any positive number b in place of 1. Similarly, subtracting a positive number b in the second coordinate shifts down. Thus we have the following general result.

Shifting a parametric curve up or down

- Adding a positive constant b to the second coordinate of every point on a parametric curve shifts the curve up b units.
- Subtracting a positive constant b from the second coordinate of every point on a parametric curve shifts the curve down b units.

Now we look at a transformation that stretches a parametric curve.

EXAMPLE 10

(a) Sketch the parametric curves described by $(t^2 - t^3, t - t^2)$ and $(2t^2 - 2t^3, t - t^2)$ for t in the interval $[0, 1]$.

(b) Explain how the parametric curve described by $(2t^2 - 2t^3, t - t^2)$ is obtained from the parametric curve described by $(t^2 - t^3, t - t^2)$.

SOLUTION

(a) A computer produces the parametric curves shown here.

(b) The figure here shows that the parametric curve described by

$$(2t^2 - 2t^3, t - t^2)$$

is obtained by horizontally stretching the parametric curve described by

$$(t^2 - t^3, t - t^2)$$

by a factor of 2.

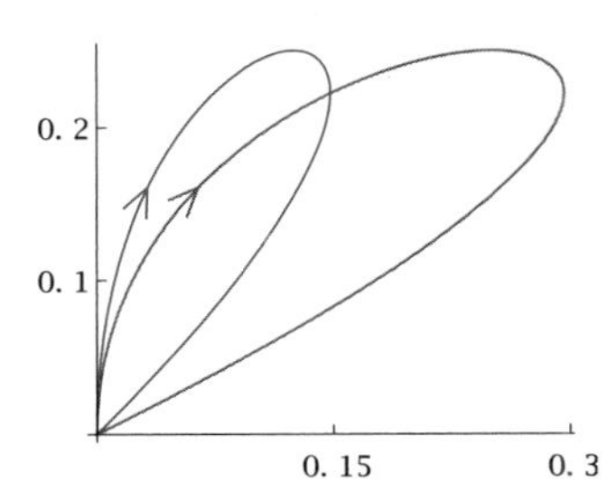

The parametric curve described by $(t^2 - t^3, t - t^2)$ (blue) and the parametric curve described by $(2t^2 - 2t^3, t - t^2)$ (red), both for t in $[0, 1]$.

The stretching above by a factor of 2 occurs because the first coordinate of $(t^2 - t^3, t - t^2)$ is multiplied by 2 to obtain $(2t^2 - 2t^3, t - t^2)$. The same reasoning works with any parametric curve and any positive number b in place of 2. Similarly, multiplying the second coordinate stretches vertically.

Stretching a parametric curve horizontally or vertically

- Multiplying the first coordinate of every point on a parametric curve by a positive constant b horizontally stretches the curve by a factor of b.
- Multiplying the second coordinate of every point on a parametric curve by a positive constant b vertically stretches the curve by a factor of b.

Our next transformation flips a parametric curve.

EXAMPLE 11

(a) Sketch the parametric curves described by

$$(t^4 + \sqrt{1+t} - t^2 - \tfrac{3}{4}, t - t^5) \quad \text{and} \quad (-t^4 - \sqrt{1+t} + t^2 + \tfrac{3}{4}, t - t^5)$$

for t in the interval $[0, 1]$.

(b) Explain how the parametric curve described by $(-t^4 - \sqrt{1+t} + t^2 + \frac{3}{4}, t - t^5)$ is obtained from the parametric curve described by $(t^4 + \sqrt{1+t} - t^2 - \frac{3}{4}, t - t^5)$.

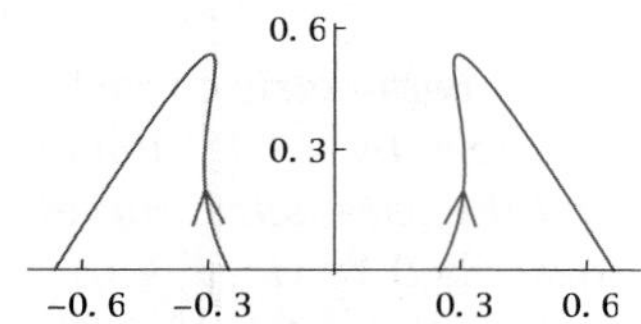

The parametric curve described by $(t^4 + \sqrt{1+t} - t^2 - \frac{3}{4}, t - t^5)$ *(blue) and the parametric curve described by* $(-t^4 - \sqrt{1+t} + t^2 + \frac{3}{4}, t - t^5)$ *(red), both for* t *in* $[0, 1]$.

SOLUTION

(a) A computer produces the parametric curves shown here.

(b) The figure here shows that the parametric curve described by

$$(-t^4 - \sqrt{1+t} + t^2 + \tfrac{3}{4}, t - t^5)$$

is obtained by flipping the parametric curve described by

$$(t^4 + \sqrt{1+t} - t^2 - \tfrac{3}{4}, t - t^5)$$

across the vertical axis.

The conclusion above about flipping across the vertical axis is no surprise because the first coordinate of $(t^4 + \sqrt{1+t} - t^2 - \frac{3}{4}, t - t^5)$ is multiplied by -1 to obtain $(-t^4 - \sqrt{1+t} + t^2 + \frac{3}{4}, t - t^5)$. The same reasoning works with any parametric curve. Similarly, multiplying the second coordinate by -1 flips across the horizontal axis. Thus we have the following general result.

Flipping a parametric curve vertically or horizontally

- Multiplying the first coordinate of every point on a parametric curve by -1 flips the curve across the vertical axis.
- Multiplying the second coordinate of every point on a parametric curve by -1 flips the curve across the horizontal axis.

EXERCISES

For Exercises 1–4, consider the parametric curve described by the given ordered pair of functions defined on the given interval.

(a) ***What is the initial point of the parametric curve?***

(b) ***What is the endpoint of the parametric curve?***

(c) ***Sketch the parametric curve.***

(d) ***Is the parametric curve the graph of some function?***

1. $(2t^2 - 8t + 5, t^3 - 6t^2 + 10t)$ for t in $[0, 3]$

2. $(2t^2 - 8t + 5, t^3 - 6t^2 + 10t)$ for t in $[1, 4]$

3. $(t^3 + 1, \frac{3t^2 - t + 1}{2})$ for t in $[-1, 1]$

4. $(t^3 - 1, 4t^2 - 2t + 1)$ for t in $[-1, 1]$

For Exercises 5–6, answer the following questions using the given information.

(a) ***Describe the path of the basketball as a parametric curve.***

(b) ***How long after the basketball is thrown does it hit the ground?***

(c) ***How far away from the thrower (in the horizontal direction) is the basketball when it hits the ground?***

(d) ***How high is the basketball at its maximum height?***

5. A basketball is thrown from height 5 feet with initial horizontal velocity 20 feet per second and initial vertical velocity 15 feet per second.

6. A basketball is thrown from height 7 feet with initial horizontal velocity 25 feet per second and initial vertical velocity 17 feet per second.

For Exercises 7–8, answer the following questions using the given information.

(a) ***Describe the path of the basketball as a parametric curve.***

(b) ***How long after the basketball is thrown does it hit the wall?***

(c) ***How high is the basketball when it hits the wall?***

7. A basketball is thrown at a wall 40 feet away from height 6 feet with initial horizontal velocity 35 feet per second and initial vertical velocity 20 feet per second.

8. A basketball is thrown at a wall 50 feet away from height 5 feet with initial horizontal velocity 55 feet per second and initial vertical velocity 30 feet per second.

9. Find a function f such that the parametric curve described by $(t - 1, t^2)$ for t in $[0, 2]$ is the graph of f.

10. Find a function f such that the parametric curve described by $(t + 2, 4t^3)$ for t in $[2, 5]$ is the graph of f.

11. (a) Sketch the parametric curve described by (t^2, t) for t in the interval $[-1, 1]$.

    (b) Find a function f such that the parametric curve in part (a) could be obtained by flipping the graph of f across the line with slope 1 that goes through the origin.

12. (a) Sketch the parametric curve with coordinates (t^6, t^3) for t in the interval $[-1, 1]$.

    (b) Find a function f such that the parametric curve in part (a) could be obtained by flipping the graph of f across the line with slope 1 that goes through the origin.

13. Use a parametric curve to sketch the inverse of the function f defined by $f(x) = x^3 + x$ on the interval $[0, 1]$.

14. Use a parametric curve to sketch the inverse of the function f defined by $f(x) = x^5 + x^3 + 1$ on the interval $[0, 1]$.

15. Find the equation of the ellipse in the xy-plane given by the parametric curve described by $(2 \cos t, 7 \sin t)$ for t in $[0, 2\pi]$.

16. Find the equation of the ellipse in the xy-plane given by the parametric curve described by $(5 \sin t, 6 \cos t)$ for t in $[-\pi, \pi]$.

17. Write the ellipse $\frac{x^2}{16} + \frac{y^2}{81} = 1$ as a parametric curve.

18. Write the ellipse $\frac{x^2}{100} + \frac{y^2}{64} = 1$ as a parametric curve.

19. Write the ellipse $7x^2 + 5y^2 = 3$ as a parametric curve.

20. Write the ellipse $2x^2 + 3y^2 = 1$ as a parametric curve.

For Exercises 21-36, explain how the parametric curve described by the given ordered pair of functions could be obtained from the parametric curve shown in Example 2; assume that t is in [0,4] in all cases. Then sketch the given parametric curve.

The coordinates in Exercises 21-36 can be easily obtained from the coordinates

$$(t^3 - 6t^2 + 10t, 2t^2 - 8t + 9)$$

in Example 2. The intention here is that you shift, stretch, or flip the parametric curve in Example 2 to obtain your results. Do not use a computer or graphing calculator to sketch the graphs in these exercises, because doing so would deprive you of a good understanding of transformations of parametric curves.

21. $(t^3 - 6t^2 + 10t + 1, 2t^2 - 8t + 9)$

22. $(t^3 - 6t^2 + 10t + 3, 2t^2 - 8t + 9)$

23. $(t^3 - 6t^2 + 10t - 1, 2t^2 - 8t + 9)$

24. $(t^3 - 6t^2 + 10t - 4, 2t^2 - 8t + 9)$

25. $(t^3 - 6t^2 + 10t, 2t^2 - 8t + 10)$

26. $(t^3 - 6t^2 + 10t, 2t^2 - 8t + 12)$

27. $(t^3 - 6t^2 + 10t, 2t^2 - 8t + 6)$

28. $(t^3 - 6t^2 + 10t, 2t^2 - 8t + 4)$

29. $(\frac{t^3 - 6t^2 + 10t}{2}, 2t^2 - 8t + 9)$

30. $(2(t^3 - 6t^2 + 10t), 2t^2 - 8t + 9)$

31. $(t^3 - 6t^2 + 10t, \frac{3}{2}(2t^2 - 8t + 9))$

32. $(t^3 - 6t^2 + 10t, \frac{2t^2 - 8t + 9}{2})$

33. $(-(t^3 - 6t^2 + 10t + 1), 2t^2 - 8t + 9)$

34. $(-(t^3 - 6t^2 + 10t - 4), 2t^2 - 8t + 9)$

35. $(t^3 - 6t^2 + 10t, -(2t^2 - 8t + 6))$

36. $(t^3 - 6t^2 + 10t, -(2t^2 - 8t + 4))$

PROBLEMS

Some problems require considerably more thought than the exercises. Unlike exercises, problems often have more than one correct answer.

37. Explain why the parametric curve described by $(t^3, 5t^3)$ for t in the interval $[-2, 2]$ is a line segment.

38. Explain why the parametric curve described by $(t^3, 2t^3)$ for t a real number is a line.

39. Explain why the parametric curve described by $(t^2, 3t^2)$ for t a real number is a ray.

40. Explain why the parametric curve described by $(\cos^2 t, \sin^2 t)$ for t in the interval $[0, \pi]$ is a line segment. Describe this line segment.

41. Suppose a star is located at the origin of a coordinate plane and a planet orbits the star, with the planet located at

$$(\cos(2\pi t), \sin(2\pi t))$$

at time t. Here t is measured in years and distance is measured in astronomical units (one astronomical unit is approximately the average distance from the Earth to the sun, which is approximately 93 million miles). Explain why the planet orbits the star once per year in a circular orbit.

42. Using the same units as in Problem 41, now suppose that due to the gravitational effect of a moon the planet's location at time t is actually

$$(\cos(2\pi(1 + 10^{-10})t), \sin(2\pi t)).$$

Show that after one million years, the planet is about 18 miles [a tiny distance in terms of astronomical units] from where it would have been if following the orbit given by Problem 41.

43. Assuming that the planet follows the orbit given in Problem 42, show that after five billion years the planet crashes into the star. [*This problem shows that a tiny change in the formula describing the orbit can result in a serious change after a long time period. Currently we do not know the formula for the orbit of the Earth sufficiently accurately to know whether or not it will eventually crash into the sun.*]

44. The parametric curve described by $((\sin t)(1 - \cos t), (\cos t)(1 - \cos t))$ for t in $[0, 2\pi]$ is called a **cardiod**. Sketch a graph of this parametric curve and discuss the reason for its name.

WORKED-OUT SOLUTIONS *to Odd-numbered Exercises*

Do not read these worked-out solutions before first attempting to do the exercises yourself. Otherwise you may merely mimic the techniques shown here without understanding the ideas.

Best way to learn: Carefully read the section of the textbook, then do all the odd-numbered exercises (even if they have not been assigned) and check your answers here. If you get stuck on an exercise, reread the section of the textbook—then try the exercise again. If you are still stuck, then look at the worked-out solution here.

For Exercises 1–4, consider the parametric curve described by the given ordered pair of functions defined on the given interval.

(a) ***What is the initial point of the parametric curve?***

(b) ***What is the endpoint of the parametric curve?***

(c) ***Sketch the parametric curve.***

(d) ***Is the parametric curve the graph of some function?***

1. $(2t^2 - 8t + 5, t^3 - 6t^2 + 10t)$ for t in $[0, 3]$

SOLUTION

(a) Taking $t = 0$, we see that the initial point of this parametric curve is $(5, 0)$, as shown in the figure below.

(b) Taking $t = 3$, we see that the endpoint of this parametric curve is $(-1, 3)$, as shown in the figure below.

(c) If you are using Wolfram|Alpha, see Example 2 for the kind of input that produces this figure.

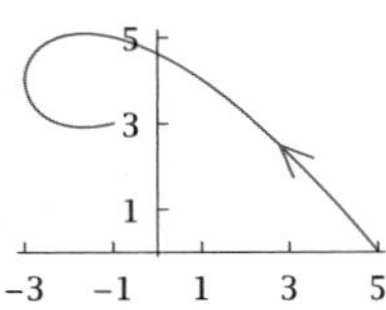

(d) This curve fails the vertical line test, and thus it is not the graph of any function.

3. $(t^3 + 1, \frac{3t^2 - t + 1}{2})$ for t in $[-1, 1]$

SOLUTION

(a) Taking $t = -1$, we see that the initial point of this parametric curve is $(0, \frac{5}{2})$, as shown in the figure below.

(b) Taking $t = 1$, we see that the endpoint of this parametric curve is $(2, \frac{3}{2})$, as shown in the figure below.

(c)

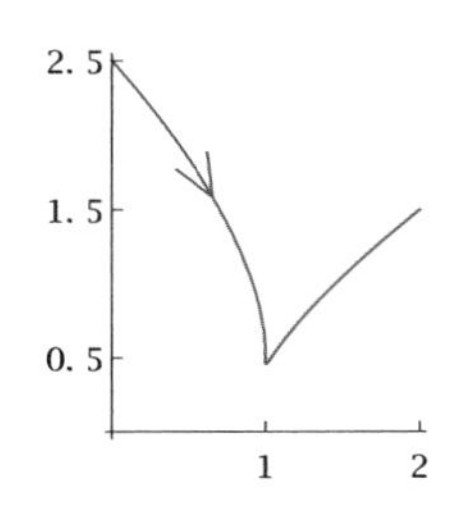

(d) This curve passes the vertical line test, and thus it is the graph of some function.

For Exercises 5–6, answer the following questions using the given information.

(a) ***Describe the path of the basketball as a parametric curve.***

(b) ***How long after the basketball is thrown does it hit the ground?***

(c) ***How far away from the thrower (in the horizontal direction) is the basketball when it hits the ground?***

(d) ***How high is the basketball at its maximum height?***

5. A basketball is thrown from height 5 feet with initial horizontal velocity 20 feet per second and initial vertical velocity 15 feet per second.

SOLUTION

(a) The parametric curve described by $(20t, -16.1t^2 + 15t + 5)$ gives the path of the basketball. Here we are assuming that the basketball is thrown at time $t = 0$. The first coordinate $20t$ is the horizontal distance of the basketball from the thrower at time t seconds. The second coordinate $-16.1t^2 + 15t + 5$ is the height of the basketball at time t seconds.

Part (b) shows that this description of the basketball's path is valid only for t in the interval $[0, 1.19]$.

(b) The basketball hits the ground when $-16.1t^2 + 15t + 5 = 0$. The quadratic formula shows that this happens when $t \approx 1.19$ seconds. Thus the formula for the parametric curve given in part (a) is valid only for t in the interval $[0, 1.19]$.

(c) When the basketball hits the ground at time 1.19, its horizontal distance from the thrower is 20×1.19 feet, which is 23.8 feet.

(d) To find the maximum height of the basketball, we complete the square:

$$\begin{aligned}-16.1t^2 &+ 15t + 5 \\ &= -16.1[t^2 - \tfrac{15}{16.1}t] + 5 \\ &= -16.1[(t - \tfrac{15}{32.1})^2 - (\tfrac{15}{32.2})^2] + 5 \\ &= -16.1(t - \tfrac{15}{32.2})^2 + \tfrac{15^2}{64.4} + 5\end{aligned}$$

The expression above shows that the maximum height of the basketball is attained when $t = \frac{15}{32.2}$ and that the basketball's height at that time will be $\frac{15^2}{64.4} + 5$ feet, which is approximately 8.5 feet.

For Exercises 7–8, answer the following questions using the given information.

(a) ***Describe the path of the basketball as a parametric curve.***

(b) ***How long after the basketball is thrown does it hit the wall?***

(c) ***How high is the basketball when it hits the wall?***

7. A basketball is thrown at a wall 40 feet away from height 6 feet with initial horizontal velocity 35 feet per second and initial vertical velocity 20 feet per second.

SOLUTION

(a) The parametric curve $(35t, -16.1t^2 + 20t + 6)$ describes the path of the basketball. Here we are assuming that the basketball is thrown at time $t = 0$. The first coordinate $35t$ is the horizontal distance of the basketball from the thrower at time t seconds. The second coordinate $-16.1t^2 + 20t + 6$ is the height of the basketball at time t seconds. Part (b) shows that this description of the basketball's path is valid only for t in the interval $[0, 1.14]$.

(b) The basketball hits the wall when $35t = 40$, which means $t = \frac{40}{35} \approx 1.14$ seconds. Thus the formula for the parametric curve given in part (a) is valid only for t in the interval $[0, 1.14]$.

(c) When the basketball hits the wall at time 1.14, its height is $-16.1 \times 1.14^2 + 20 \times 1.14 + 6$ feet, which is approximately 7.9 feet.

9. Find a function f such that the parametric curve described by $(t - 1, t^2)$ for t in $[0, 2]$ is the graph of f.

SOLUTION Let $s = t - 1$. As t varies over the interval $[0, 2]$, clearly s varies over the interval $[-1, 1]$.

Because $t = s + 1$, we have

$$(t - 1, t^2) = (s, (s + 1)^2).$$

Thus if we define a function f with domain $[-1, 1]$ by $f(s) = (s + 1)^2$, then the graph of f is the parametric curve described by $(t - 1, t^2)$ for t in $[0, 2]$.

11. (a) Sketch the parametric curve described by (t^2, t) for t in the interval $[-1, 1]$.

(b) Find a function f such that the parametric curve in part (a) could be obtained by flipping the graph of f across the line with slope 1 that goes through the origin.

SOLUTION

(a)

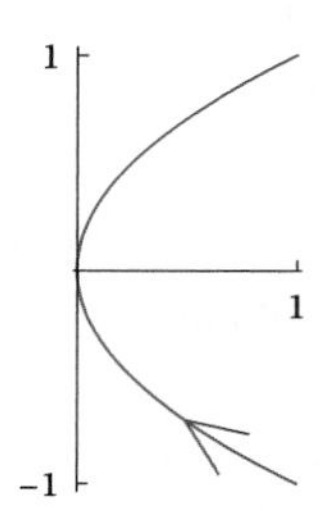

(b) Flipping a parametric curve across the line with slope 1 that goes through the origin interchanges the coordinates. Thus we are looking for a function f whose graph is the parametric curve described by (t, t^2) for t in $[-1, 1]$. Thus we define f by $f(t) = t^2$, with the domain of f equal to $[-1, 1]$.

13. Use a parametric curve to sketch the inverse of the function f defined by $f(x) = x^3 + x$ on the interval $[0, 1]$.

SOLUTION

The graph of f is the parametric curve described by $(t, t^3 + t)$ for t in $[0, 1]$. Thus the graph of f^{-1} is the parametric curve described by $(t^3 + t, t)$ for t in $[0, 1]$, as shown here.

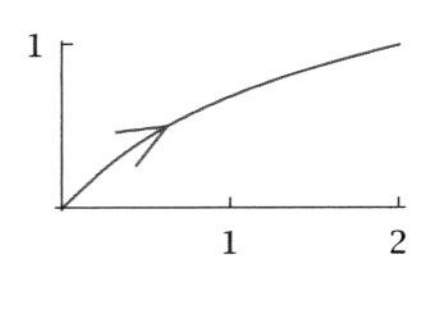

15. Find the equation of the ellipse in the xy-plane given by the parametric curve described by $(2\cos t, 7\sin t)$ for t in $[0, 2\pi]$.

SOLUTION Note that

$$\begin{aligned}\frac{(2\cos t)^2}{4} + \frac{(7\sin t)^2}{49} &= \frac{4\cos^2 t}{4} + \frac{49\sin^2 t}{49}\\ &= \cos^2 t + \sin^2 t\\ &= 1.\end{aligned}$$

Thus setting $x = 2\cos t$ and $y = 7\sin t$, we have the equation of the ellipse $\frac{x^2}{4} + \frac{y^2}{49} = 1$.

17. Write the ellipse $\frac{x^2}{16} + \frac{y^2}{81} = 1$ as a parametric curve.

SOLUTION The parametric curve described by $(4\cos t, 9\sin t)$ for t in $[0, 2\pi]$ gives the desired ellipse. There are also many other correct solutions, such as the parametric curve described by $(4\sin t, 9\cos t)$ for t in $(-\pi, \pi]$.

19. Write the ellipse $7x^2 + 5y^2 = 3$ as a parametric curve.

SOLUTION The equation $7x^2 + 5y^2 = 3$ can be rewritten as

$$\left(\sqrt{\tfrac{7}{3}}x\right)^2 + \left(\sqrt{\tfrac{5}{3}}y\right)^2 = 1.$$

The equation above shows that a good way to write this ellipse as a parametric curve is to take

$$\cos t = \sqrt{\tfrac{7}{3}}x \quad \text{and} \quad \sin t = \sqrt{\tfrac{5}{3}}y.$$

Solving these equations for x and y leads to the parametric curve described by

$$\left(\sqrt{\tfrac{3}{7}}\cos t, \sqrt{\tfrac{3}{5}}\sin t\right)$$

for t in $[0, 2\pi)$. There are also many other correct solutions.

For Exercises 21–36, explain how the parametric curve described by the given ordered pair of functions could be obtained from the parametric curve shown in Example 2; assume that t is in [0, 4] in all cases. Then sketch the given parametric curve.

The coordinates in Exercises 21–36 can be easily obtained from the coordinates

$$(\mathbf{t^3 - 6t^2 + 10t, 2t^2 - 8t + 9})$$

in Example 2. The intention here is that you shift, stretch, or flip the parametric curve in Example 2 to obtain your results. Do not use a computer or graphing calculator to sketch the graphs in these exercises, because doing so would deprive you of a good understanding of transformations of parametric curves.

21. $(t^3 - 6t^2 + 10t + 1, 2t^2 - 8t + 9)$

SOLUTION These coordinates are obtained by adding 1 to the first coordinate of every point on the parametric curve from Example 2.

Thus the parametric curve with these coordinates is obtained by shifting the parametric curve from Example 2 right 1 unit, giving the curve shown here.

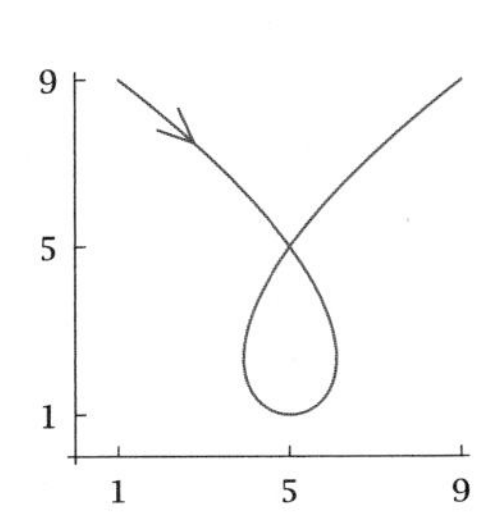

23. $(t^3 - 6t^2 + 10t - 1, 2t^2 - 8t + 9)$

SOLUTION These coordinates are obtained by subtracting 1 from the first coordinate of every point on the parametric curve from Example 2.

Thus the parametric curve with these coordinates is obtained by shifting the parametric curve from Example 2 left 1 unit, giving the curve shown here.

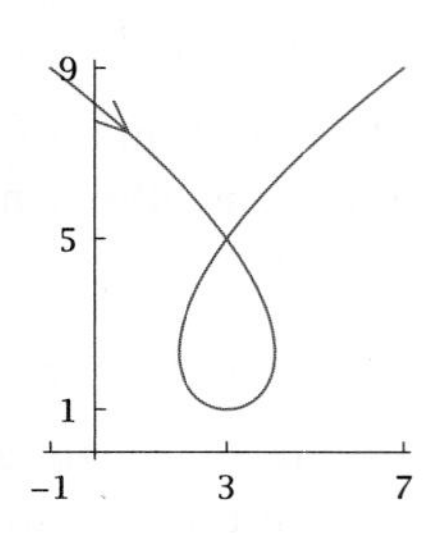

25. $(t^3 - 6t^2 + 10t, 2t^2 - 8t + 10)$

SOLUTION These coordinates are obtained by adding 1 to the second coordinate of every point on the parametric curve from Example 2.

Thus the parametric curve with these coordinates is obtained by shifting the parametric curve from Example 2 up 1 unit, giving the curve shown here.

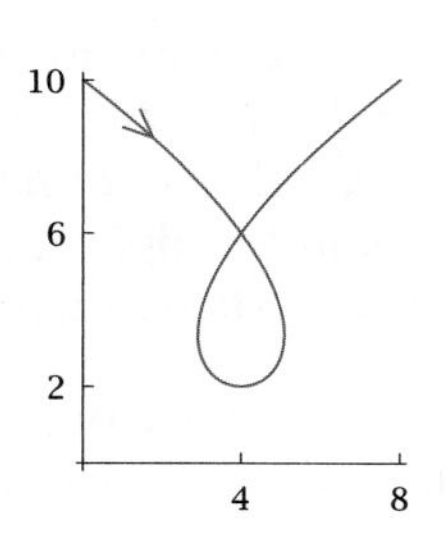

27. $(t^3 - 6t^2 + 10t, 2t^2 - 8t + 6)$

SOLUTION These coordinates are obtained by subtracting 3 from the second coordinate of every point on the parametric curve from Example 2.

Thus the parametric curve with these coordinates is obtained by shifting the parametric curve from Example 2 down 3 units, giving the curve shown here.

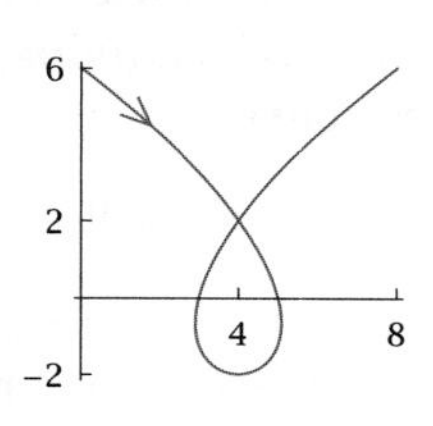

29. $\left(\dfrac{t^3 - 6t^2 + 10t}{2}, 2t^2 - 8t + 9\right)$

SOLUTION These coordinates are obtained by multiplying the first coordinate of every point on the parametric curve from Example 2 by $\frac{1}{2}$.

Thus the parametric curve with these coordinates is obtained by horizontally stretching the parametric curve from Example 2 by a factor of $\frac{1}{2}$, giving the curve shown here.

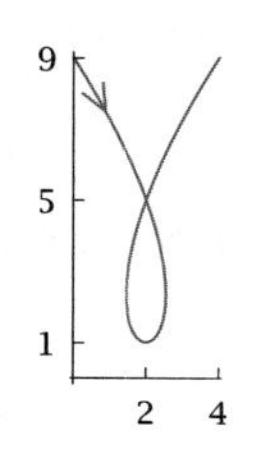

31. $(t^3 - 6t^2 + 10t, \frac{3}{2}(2t^2 - 8t + 9))$

SOLUTION These coordinates are obtained by multiplying the second coordinate of every point on the parametric curve from Example 2 by $\frac{3}{2}$.

Thus the parametric curve with these coordinates is obtained by vertically stretching the parametric curve from Example 2 by a factor of $\frac{3}{2}$, giving the curve shown here.

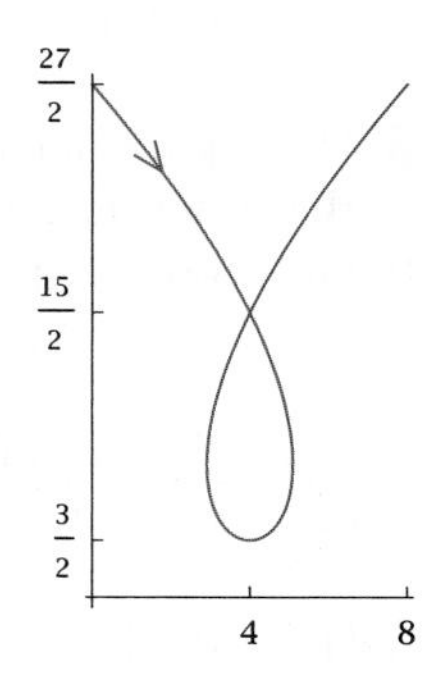

33. $(-(t^3 - 6t^2 + 10t + 1), 2t^2 - 8t + 9)$

SOLUTION These coordinates are obtained by multiplying the first coordinate of every point on the parametric curve from Exercise 21 by -1.

Thus the parametric curve with these coordinates is obtained by flipping the parametric curve from Exercise 21 across the vertical axis, giving the curve shown here.

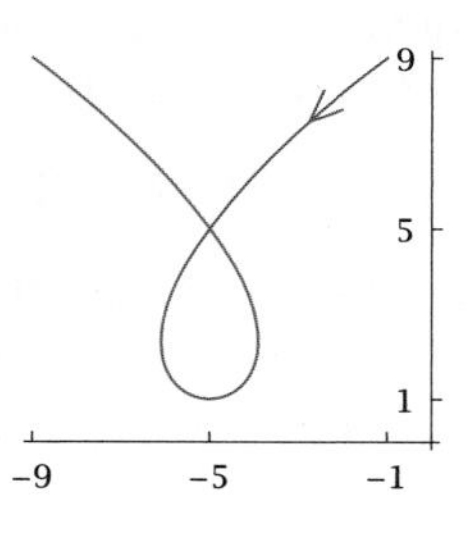

35. $(t^3 - 6t^2 + 10t, -(2t^2 - 8t + 6))$

SOLUTION These coordinates are obtained by multiplying the second coordinate of every point on the parametric curve from Exercise 27 by -1.

Thus the parametric curve with these coordinates is obtained by flipping the parametric curve from Exercise 27 across the horizontal axis, giving the curve shown here.

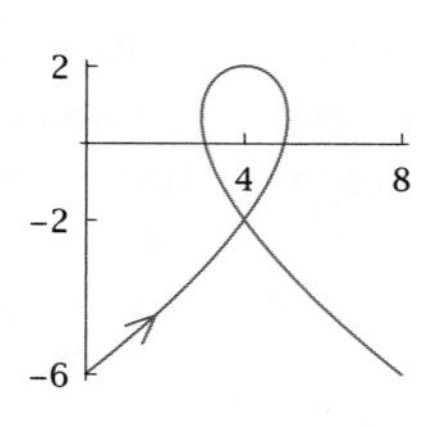

11.2 *Transformations of Trigonometric Functions*

LEARNING OBJECTIVES

By the end of this section you should be able to

- compute the amplitude of a function;
- compute the period of a function;
- graph a phase shift;
- find transformations of trigonometric functions that adjust the amplitude, period, and/or phase shift to desired values.

Some events have patterns that repeat roughly periodically, such as tides (approximately daily), total daily nationwide ridership on mass transit (approximately weekly, with decreases on weekends as compared to weekdays), phases of the moon (approximately monthly), and the noon temperature in Chicago (approximately yearly, as the seasons change).

The cosine and sine functions are periodic functions and thus are particularly well suited for modeling such events. However, values of the cosine and sine, which are between -1 and 1, and the period of the cosine and sine, which is 2π, rarely fit the events being modeled. Thus transformations of these functions are needed.

In Section 3.2 we discussed various transformations of a function that could stretch the graph of the function vertically or horizontally, shift the graph to the left or right, or flip the graph across the vertical or horizontal axis. In this section we will revisit function transformations, this time using trigonometric functions. Thus this section will help you review and solidify the concepts of function transformations introduced in Section 3.2 while also deepening your understanding of the behavior of the key trigonometric functions.

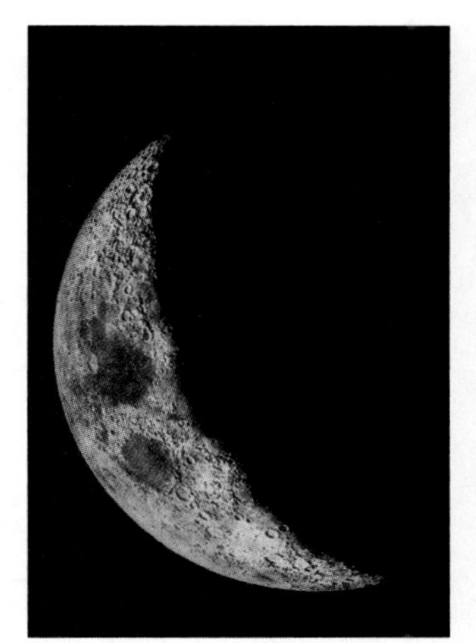

The phases of the moon, which repeat approximately monthly, provide an excellent example of periodic behavior.

Amplitude

Recall that if f is a function, c is a positive number, and a function g is defined by $g(x) = cf(x)$, then the graph of g is obtained by vertically stretching the graph of f by a factor of c (see Section 3.2).

EXAMPLE 1

Here we are informally using $\cos x$ *as an abbreviation for the function whose value at a number* x *equals* $\cos x$.

(a) Sketch the graphs of the functions $\cos x$ and $3\cos x$ on the interval $[-4\pi, 4\pi]$.

(b) What is the range of the function $3\cos x$?

SOLUTION

(a) The graph of $3\cos x$ is obtained by vertically stretching the graph of $\cos x$ by a factor of 3:

For convenience, throughout this section different scales are used on the horizontal and vertical axes.

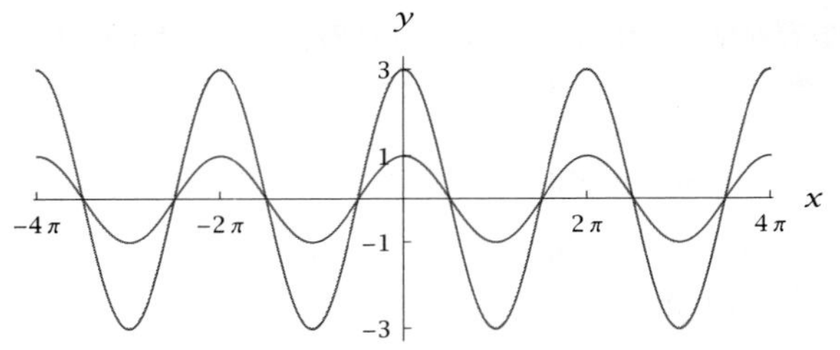

The graphs of $\cos x$ *(blue) and* $3\cos x$ *(red) on the interval* $[-4\pi, 4\pi]$.

(b) The range of $3\cos x$ is obtained by multiplying each number in the range of $\cos x$ by 3. Thus the range of $3\cos x$ is the interval $[-3, 3]$.

We say that $3\cos x$ has amplitude 3. Here is the formal definition:

Not every function has an amplitude. For example, the tangent function defined on the interval $[0, \frac{\pi}{2})$ *does not have a maximum value and thus does not have an amplitude.*

Amplitude

The **amplitude** of a function is one-half the difference between the maximum and minimum values of the function.

For example, the function $3\cos x$ has a maximum value of 3 and a minimum value of −3. Thus the difference between the maximum and minimum values of $3\cos x$ is 6. Half of 6 is 3, and hence the function $3\cos x$ has amplitude 3.

The next example illustrates the effect of multiplying a trigonometric function by a negative number.

EXAMPLE 2

(a) Sketch the graphs of the functions $\sin x$ and $-3\sin x$ on the interval $[-4\pi, 4\pi]$.

(b) What is the range of the function $-3\sin x$?

(c) What is the amplitude of the function $-3\sin x$?

SOLUTION

(a) The graph of $-3\sin x$ is obtained by vertically stretching the graph of $\sin x$ by a factor of 3 and then flipping across the horizontal axis:

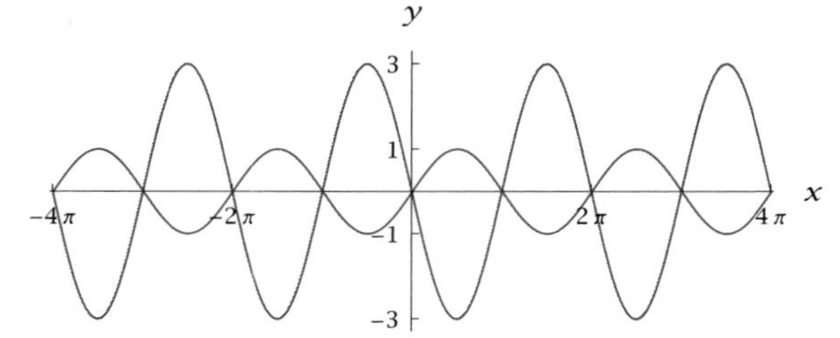

The graphs of $\sin x$ *(blue) and* $-3\sin x$ *(red) on the interval* $[-4\pi, 4\pi]$.

(b) The range of $-3\sin x$ is obtained by multiplying each number in the range of $\sin x$ by −3. Thus the range of $-3\sin x$ is the interval $[-3, 3]$.

(c) The function $-3\sin x$ has a maximum value of 3 and a minimum value of −3. Thus the difference between the maximum and minimum values of $-3\sin x$ is 6. Half of 6 is 3, and hence the function $-3\sin x$ has amplitude 3.

Recall that if f is a function, a is a positive number, and a function g is defined by $g(x) = f(x) + a$, then the graph of g is obtained by shifting the graph of f up a units (see Section 3.2). The next example illustrates a function whose graph is obtained from the graph of the cosine function by stretching vertically and shifting up.

EXAMPLE 3

(a) Sketch the graphs of $\cos x$ and $2 + 0.3\cos x$ on the interval $[-4\pi, 4\pi]$.

(b) What is the range of the function $2 + 0.3\cos x$?

(c) What is the amplitude of the function $2 + 0.3\cos x$?

SOLUTION

(a) The graph of $2 + 0.3\cos x$ is obtained by vertically stretching the graph of $\cos x$ by a factor of 0.3 and then shifting up by 2 units:

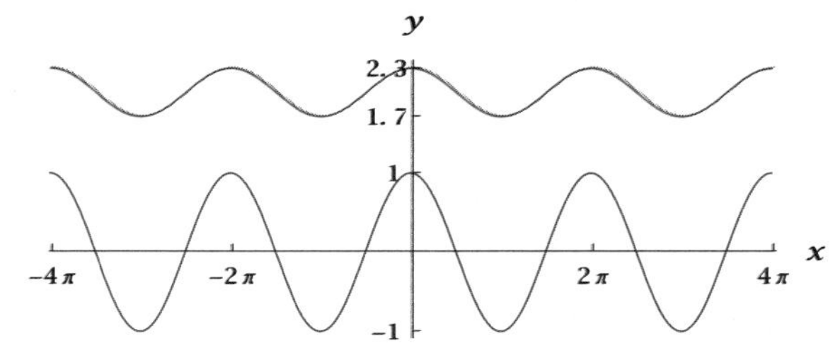

The graphs of $\cos x$ *(blue) and* $2 + 0.3\cos x$ *(red) on the interval* $[-4\pi, 4\pi]$.

(b) The range of $2 + 0.3\cos x$ is obtained by multiplying each number in the range of $\cos x$ by 0.3, which produces the interval $[-0.3, 0.3]$, and then adding 2 to each number. Thus the range of $2 + 0.3\cos x$ is the interval $[1.7, 2.3]$, as shown by the graph above.

(c) The function $2 + 0.3\cos x$ has a maximum value of 2.3 and a minimum value of 1.7. Thus the difference between the maximum and minimum values of $2 + 0.3\cos x$ is 0.6. Half of 0.6 is 0.3, and hence the function $2 + 0.3\cos x$ has amplitude 0.3.

Even though $2 + 0.3\cos x$ *is larger than* $\cos x$ *for every real number* x*, the function* $2 + 0.3\cos x$ *has a smaller amplitude than the function* $\cos x$.

Period

The graphs of the cosine and sine functions are periodic, meaning that they repeat their behavior at regular intervals. More specifically,

$$\cos(x + 2\pi) = \cos x \quad \text{and} \quad \sin(x + 2\pi) = \sin x$$

for every number x. In the equations above, we could have replaced 2π with 4π or 6π or 8π, and so on, but no positive number smaller than 2π would make these equations valid for all values of x. Thus we say that the cosine and sine functions have **period** 2π. Here is the formal definition:

Although the cosine and sine functions have period 2π, the tangent function has period π (see Section 9.6).

Period

Suppose f is a function and $p > 0$. We say that f has **period** p if p is the smallest positive number such that

$$f(x + p) = f(x)$$

for every real number x in the domain of f.

Some functions do not repeat their behavior at regular intervals and thus do not have a period. For example, the function f defined by $f(x) = x^2$ does not have a period. A function is called **periodic** if it has a period.

Recall that if f is a function, c is a positive number, and a function h is defined by $h(x) = f(cx)$, then the graph of h is obtained by horizontally stretching the graph of f by a factor of $\frac{1}{c}$ (see Section 3.2). This implies that if f has period p, then h has period $\frac{p}{c}$, as illustrated by the next example.

EXAMPLE 4

(a) Sketch the graphs of $3 + \cos x$ and $\cos(2x)$ on the interval $[-4\pi, 4\pi]$.

(b) What is the range of the function $\cos(2x)$?

(c) What is the amplitude of the function $\cos(2x)$?

(d) What is the period of the function $\cos(2x)$?

SOLUTION

(a) The graph of $3 + \cos x$ is obtained by shifting the graph of $\cos x$ up by 3 units. The graph of $\cos(2x)$ is obtained by horizontally stretching the graph of $\cos x$ by a factor of $\frac{1}{2}$:

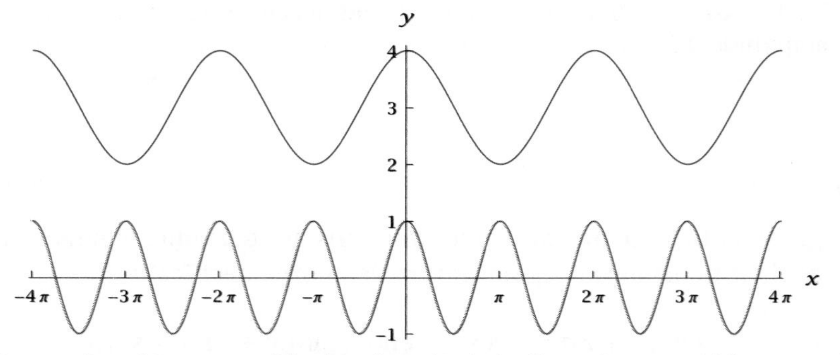

The graphs of $3 + \cos x$ *(blue) and* $\cos(2x)$ *(red) on the interval* $[-4\pi, 4\pi]$.

(b) As x varies over the real numbers, $\cos x$ and $\cos(2x)$ take on the same values. Thus the range of $\cos(2x)$ is the interval $[-1, 1]$.

(c) The function $\cos(2x)$ has a maximum value of 1 and a minimum value of -1. Thus the difference between the maximum and minimum values of $\cos(2x)$ is 2. Half of 2 is 1, and hence the function $\cos(2x)$ has amplitude 1.

(d) To find the period of $\cos(2x)$, we need to find the smallest positive number p such that

$$\cos(2(x+p)) = \cos(2x)$$

for every number x. The equation above can be rewritten as

$$\cos(2x+2p) = \cos(2x).$$

Because the cosine function has period 2π, to find the smallest positive number p that satisfies the equation above for all numbers x we need to solve the simple equation $2p = 2\pi$. Thus $p = \pi$, which means that $\cos(2x)$ has period π.

Another way to compute that $\cos(2x)$ has period π is to recall that the graph of $\cos(2x)$ is obtained by horizontally stretching the graph of $\cos x$ by a factor of $\frac{1}{2}$, as discussed in the solution to part (a) above. Because the graph of $\cos x$ repeats its behavior in intervals of size 2π (and not in any intervals of smaller size), this means that the graph of $\cos(2x)$ repeats its behavior in intervals of size $\frac{1}{2}(2\pi)$ (and not in any intervals of smaller size); see the figure above. Thus $\cos(2x)$ has period π.

If we think of the horizontal axis in the graph in part (a) above as representing time, then we can say that the graph of $\cos(2x)$ oscillates twice as fast as the graph of $\cos x$.

The next example illustrates a transformation of the sine function that changes both the amplitude and the period.

EXAMPLE 5

(a) Sketch the graph of the function $7\sin(2\pi x)$ on the interval $[-3, 3]$.

(b) What is the range of the function $7\sin(2\pi x)$?

(c) What is the amplitude of the function $7\sin(2\pi x)$?

(d) What is the period of the function $7\sin(2\pi x)$?

SOLUTION

(a) The graph of $7\sin(2\pi x)$ is obtained from the graph of $\sin x$ by stretching horizontally by a factor of $\frac{1}{2\pi}$ and stretching vertically by a factor of 7:

$\frac{2\pi}{2} = \pi$

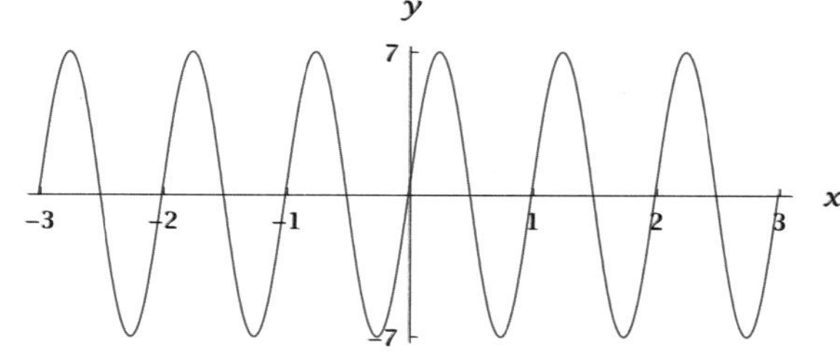

The graph of $7\sin(2\pi x)$ on the interval $[-3, 3]$.

(b) As x varies over the real numbers, $\sin(2\pi x)$ takes on the same values as $\sin x$. Hence the range of the function $\sin(2\pi x)$ is the interval $[-1, 1]$. The range of $7\sin(2\pi x)$ is obtained by multiplying each number in the range of $\sin(2\pi x)$ by 7. Thus the range of $7\sin(2\pi x)$ is the interval $[-7, 7]$.

(c) The function $7\sin(2\pi x)$ has a maximum value of 7 and a minimum value of -7. Thus the difference between the maximum and minimum values of $7\sin(2\pi x)$ is 14. Half of 14 is 7, and hence the function $7\sin(2\pi x)$ has amplitude 7.

(d) To find the period of $7\sin(2\pi x)$, we need to find the smallest positive number p such that

$$7\sin(2\pi(x+p)) = 7\sin(2\pi x)$$

for every number x. After dividing both sides by 7, we can rewrite the equation above as

$$\sin(2\pi x + 2\pi p) = \sin(2\pi x).$$

As this example shows, multiplying a function by a constant (in this case 7) changes the amplitude but has no effect on the period.

Because the sine function has period 2π, to find the smallest positive number p that satisfies the equation above for all numbers x we need to solve the simple equation $2\pi p = 2\pi$. Thus $p = 1$, which means that $7\sin(2\pi x)$ has period 1.

Another way to compute that $7\sin(2\pi x)$ has period 1 is to recall that the graph of $7\sin(2\pi x)$ is obtained from the graph of $\sin x$ by stretching horizontally by a factor of $\frac{1}{2\pi}$ and stretching vertically by a factor of 7, as discussed in the solution to part (a) above. Because the graph of $\sin x$ repeats its behavior in intervals of size 2π (and not in any intervals of smaller size), this means that the graph of $7\sin(2\pi x)$ repeats its behavior in intervals of size $\frac{1}{2\pi}(2\pi)$ (and not in any intervals of smaller size); see the figure above. Thus $7\sin(2\pi x)$ has period 1.

Phase Shift

Recall that if f is a function, b is a positive number, and a function g is defined by $g(x) = f(x - b)$, then the graph of g is obtained by shifting the graph of f right b units (see Section 3.2).

EXAMPLE 6

(a) Sketch the graphs of $\cos x$ and $\cos(x - \frac{\pi}{3})$ on the interval $[-4\pi, 4\pi]$.

(b) What is the range of the function $\cos(x - \frac{\pi}{3})$?

(c) What is the amplitude of the function $\cos(x - \frac{\pi}{3})$?

(d) What is the period of the function $\cos(x - \frac{\pi}{3})$?

(e) By what fraction of the period of $\cos x$ has the graph been shifted right to obtain the graph of $\cos(x - \frac{\pi}{3})$?

SOLUTION

(a) The graph of $\cos(x - \frac{\pi}{3})$ is obtained by shifting the graph of $\cos x$ right by $\frac{\pi}{3}$ units:

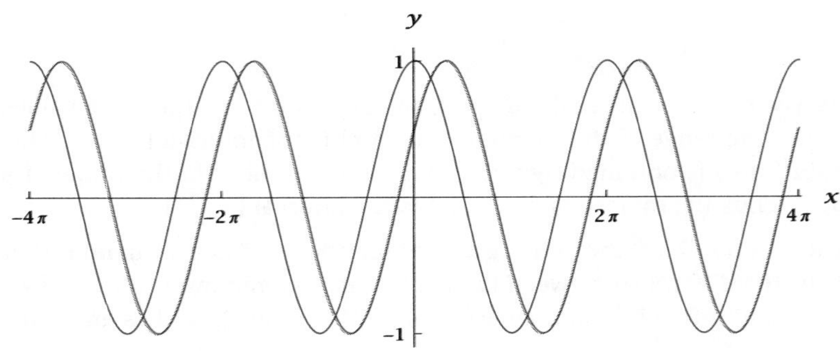

The graphs of $\cos x$ *(blue) and* $\cos(x - \frac{\pi}{3})$ *(red) on the interval* $[-4\pi, 4\pi]$.

(b) As x varies over the real numbers, $\cos x$ and $\cos(x - \frac{\pi}{3})$ take on the same values. Thus the range of $\cos(x - \frac{\pi}{3})$ is the interval $[-1, 1]$.

(c) The function $\cos(x - \frac{\pi}{3})$ has a maximum value of 1 and a minimum value of -1. Thus the difference between the maximum and minimum values of $\cos(x - \frac{\pi}{3})$ is 2. Half of 2 is 1, and hence the function $\cos(x - \frac{\pi}{3})$ has amplitude 1.

As this example shows, shifting the graph of a function to the right or the left changes neither the range nor the amplitude nor the period.

(d) Because the graph of $\cos(x - \frac{\pi}{3})$ is obtained by shifting the graph of $\cos x$ right by $\frac{\pi}{3}$ units, the graph of $\cos x$ repeats its behavior in intervals of the same size as the graph of $\cos x$. Because $\cos x$ has period 2π, this implies that $\cos(x - \frac{\pi}{3})$ also has period 2π.

(e) The graph of $\cos x$ is shifted right by $\frac{\pi}{3}$ units to obtain the graph of $\cos(x - \frac{\pi}{3})$. The period of $\cos x$ is 2π. Thus the fraction of the period of $\cos x$ by which the graph has been shifted is $\frac{\pi/3}{2\pi}$, which equals $\frac{1}{6}$.

In the solution to part (e) above, we saw that the graph of $\cos x$ is shifted right by one-sixth of a period to obtain the graph of $\cos(x - \frac{\pi}{3})$. Shifting the graph of a periodic function to the right or the left is often called a **phase shift** because the original function and the new function have the same period and the same behavior, although they are out of phase.

Here is how the $\cos x$ behaves with phase shifts of one-fourth its period, one-half its period, and all of its period:

- If the graph of $\cos x$ is shifted right by $\frac{\pi}{2}$ units, which is one-fourth of its period, then we obtain the graph of $\sin x$; this happens because $\cos(x - \frac{\pi}{2}) = \sin x$ (see Example 3 in Section 9.6).

- If the graph of $\cos x$ is shifted right by π units, which is one-half of its period, then we obtain the graph of $-\cos x$; this happens because $\cos(x - \pi) = -\cos x$ (using the formula in Section 9.6 for $\cos(\theta + n\pi)$, with $n = -1$).

- If the graph of $\cos x$ is shifted right by 2π units, which is its period, then we obtain the graph of $\cos x$; this happens because $\cos(x - 2\pi) = \cos x$.

The next example shows how to deal with a change in amplitude and a change in period and a phase shift.

EXAMPLE 7

(a) Sketch the graphs of the functions $5\sin\frac{x}{2}$ and $5\sin(\frac{x}{2} - \frac{\pi}{3})$ on the interval $[-4\pi, 4\pi]$.

(b) What is the range of the function $5\sin(\frac{x}{2} - \frac{\pi}{3})$?

(c) What is the amplitude of the function $5\sin(\frac{x}{2} - \frac{\pi}{3})$?

(d) What is the period of the function $5\sin(\frac{x}{2} - \frac{\pi}{3})$?

(e) By what fraction of the period of $5\sin\frac{x}{2}$ has the graph been shifted right to obtain the graph of $5\sin(\frac{x}{2} - \frac{\pi}{3})$?

SOLUTION

(a) The graph of $5\sin\frac{x}{2}$ is obtained from the graph of $\sin x$ by stretching vertically by a factor of 5 and stretching horizontally by a factor of 2, as shown below.

You may be surprised that the graph is shifted right by $\frac{2\pi}{3}$ units, not $\frac{\pi}{3}$ units. Take extra care with problems that involve both a change of period and a phase shift.

To see how to construct the graph of $5\sin(\frac{x}{2}-\frac{\pi}{3})$, define a function f by

$$f(x) = 5\sin\tfrac{x}{2}.$$

Now

$$5\sin(\tfrac{x}{2}-\tfrac{\pi}{3}) = 5\sin\frac{x-\frac{2\pi}{3}}{2} = f(x-\tfrac{2\pi}{3}).$$

Thus the graph of $5\sin(\frac{x}{2}-\frac{\pi}{3})$, shown below, is obtained by shifting the graph of $5\sin\frac{x}{2}$ right by $\frac{2\pi}{3}$ units:

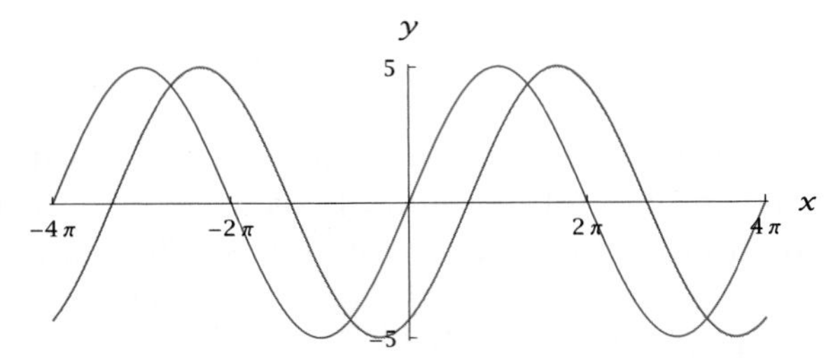

The graphs of $5\sin\frac{x}{2}$ *(blue) and* $5\sin(\frac{x}{2}-\frac{\pi}{3})$ *(red) on the interval* $[-4\pi, 4\pi]$.

(b) The range of $5\sin(\frac{x}{2}-\frac{\pi}{3})$ is obtained by multiplying each number in the range of $\sin(\frac{x}{2}-\frac{\pi}{3})$ by 5. Thus the range of $5\sin(\frac{x}{2}-\frac{\pi}{3})$ is the interval $[-5, 5]$.

(c) The function $5\sin(\frac{x}{2}-\frac{\pi}{3})$ has a maximum value of 5 and a minimum value of -5. Thus the difference between the maximum and minimum values of $5\sin(\frac{x}{2}-\frac{\pi}{3})$ is 10. Half of 10 is 5, and hence the function $5\sin(\frac{x}{2}-\frac{\pi}{3})$ has amplitude 5.

(d) Because the graph of $5\sin(\frac{x}{2}-\frac{\pi}{3})$ is obtained by shifting the graph of $5\sin\frac{x}{2}$ right by $\frac{2\pi}{3}$ units, the graph of $5\sin(\frac{x}{2}-\frac{\pi}{3})$ repeats its behavior in intervals of the same size as the graph of $5\sin\frac{x}{2}$. Thus the period of $5\sin(\frac{x}{2}-\frac{\pi}{3})$ equals the period of $5\sin\frac{x}{2}$, which equals the period of $\sin\frac{x}{2}$ (because changing the amplitude does not change the period). The function $\sin\frac{x}{2}$ has period 4π, because its graph is obtained by horizontally stretching the graph of $\sin x$ (which has period 2π) by a factor of 2. Thus $5\sin(\frac{x}{2}-\frac{\pi}{3})$ has period 4π.

(e) The graph of the function $5\sin\frac{x}{2}$ is shifted right by $\frac{2\pi}{3}$ units to obtain the graph of $5\sin(\frac{x}{2}-\frac{\pi}{3})$. The period of $5\sin\frac{x}{2}$ is 4π. Thus the fraction of the period of $5\sin\frac{x}{2}$ by which the graph has been shifted is $\frac{2\pi/3}{4\pi}$, which equals $\frac{1}{6}$.

Fitting Transformations of Trigonometric Functions to Data

We now know how to modify a trigonometric function by modifying its amplitude, period, and/or phase shift. These tools give us enough flexibility to model periodic events using transformations of trigonometric functions. The next example illustrates the ideas and the techniques.

Although we use a transformation of the cosine in the next example, we could just have easily used the sine with a different phase shift. The identities $\sin x = \cos(x-\frac{\pi}{2})$ and $\cos x = \sin(\frac{\pi}{2}-x)$ allow us to switch easily between cosine and sine.

EXAMPLE 8

The graph below shows the average monthly temperature in Chicago for the five years ending in January 2011 (the twelve data points for each year have been joined by line segments). Not surprisingly, the graph appears somewhat periodic. Find a function of the form

$$a\cos(bx+c)+d$$

that models the temperature in Chicago, where x is measured in years (thus $x = 2010.5$ would correspond to the time halfway through the year 2010).

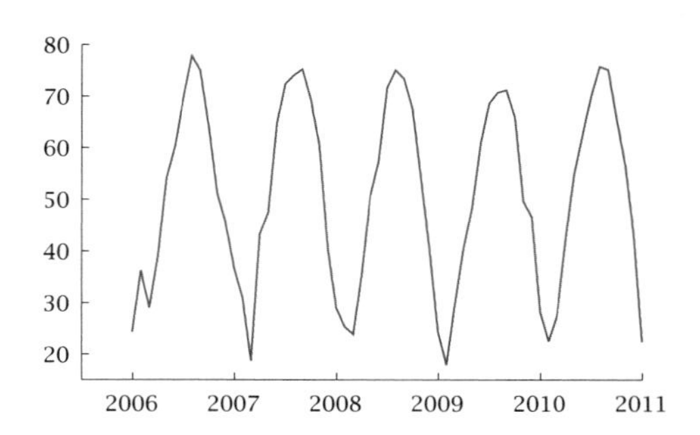

Average monthly temperature in Chicago.

SOLUTION As we have seen from examples, the period of the function above is $\frac{2\pi}{b}$ (we plan to choose $b > 0$). Because this is temperature data, it is reasonable to assume that the period is one year, as suggested by the graph above. Thus we want $\frac{2\pi}{b} = 1$, which means that $b = 2\pi$.

As we have seen, the amplitude of the function above is a (we plan to choose $a > 0$). The yearly maximum temperature on this graph seems to average about 75; the yearly minimum temperature seems to average about 21. Thus our function should have amplitude $\frac{75-21}{2}$, which equals 27. Thus we take $a = 27$.

Here we have no data, only a graph from which to make approximations.

As we have seen, d should be chosen to be halfway between the minimum and maximum values of our function. Thus we should take $d = \frac{21+75}{2} = 48$.

In the graph above, we see that the annual minimum value of this function seems to occur about one-twelfth of the way into the year (which is the end of January, typically the coldest time of the year). Thus we want to choose c so that $\cos(2\pi x + c) = -1$ when $x = \frac{1}{12}$. Thus we choose c so that $\frac{2\pi}{12} + c = \pi$. In other words, we take $c = \frac{5\pi}{6}$.

Putting all this together, our function that models the temperature in Chicago is

$$27\cos(2\pi x + \tfrac{5\pi}{6}) + 48,$$

whose graph is shown below.

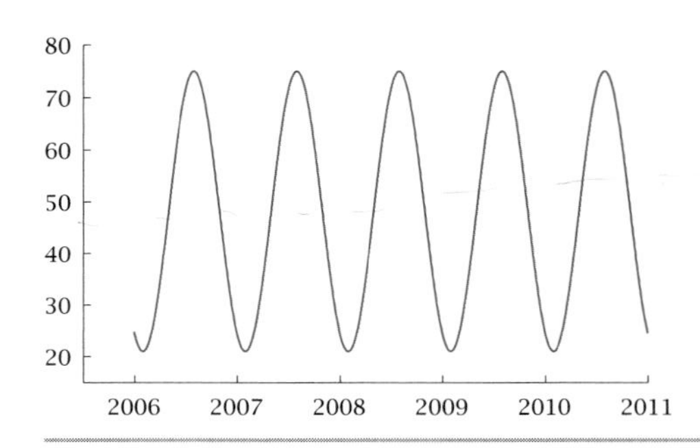

The graph of $27\cos(2\pi x + \frac{5\pi}{6}) + 48$ *on the interval* $[2006, 2011]$.

Real data is messy, so we cannot expect a mathematical model to match real data exactly. Although the graph here does not perfectly reproduce the actual data, it gives a pretty good approximation to the graph above.

EXERCISES

1. Sketch the graphs of the functions $4 \sin x$ and $\sin(4x)$ on the interval $[-\pi, \pi]$ (use the same coordinate axes for both graphs).
2. Sketch the graphs of the functions $-5 \sin x$ and $\sin(-5x)$ on the interval $[-\pi, \pi]$ (use the same coordinate axes for both graphs).
3. What is the range of the function $4 \sin x$?
4. What is the range of the function $-5 \sin x$?
5. What is the range of the function $\sin(4x)$?
6. What is the range of the function $\sin(-5x)$?
7. What is the amplitude of the function $4 \sin x$?
8. What is the amplitude of the function $-5 \sin x$?
9. What is the amplitude of the function $\sin(4x)$?
10. What is the amplitude of the function $\sin(-5x)$?
11. What is the period of the function $4 \sin x$?
12. What is the period of the function $-5 \sin x$?
13. What is the period of the function $\sin(4x)$?
14. What is the period of the function $\sin(-5x)$?

Use the following graph for Exercises 15–22:

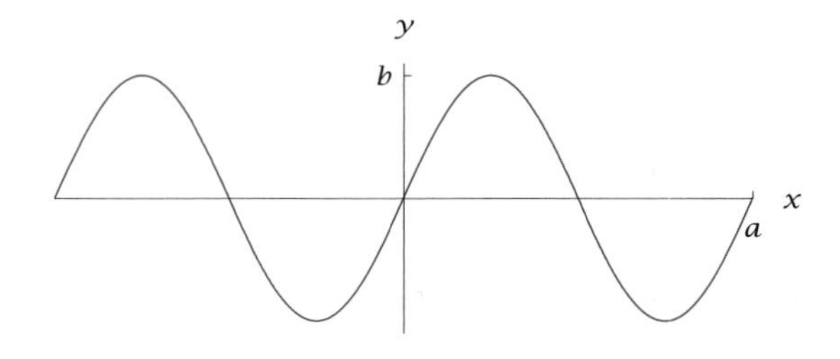

15. Suppose the figure above is part of the graph of the function $3 \sin x$. What is the value of b?
16. Suppose the figure above is part of the graph of the function $4 \sin(5x)$. What is the value of b?
17. Suppose the figure above is part of the graph of the function $\sin(7x)$. What is the value of a?
18. Suppose the figure above is part of the graph of the function $9 \sin(6x)$. What is the value of a?
19. Find the smallest positive number c such that the figure above is part of the graph of the function $\sin(x + c)$.
20. Find the smallest positive number c such that the figure above is part of the graph of the function $\sin(x - c)$.
21. Find the smallest positive number c such that the figure above is part of the graph of the function $\cos(x - c)$.
22. Find the smallest positive number c such that the figure above is part of the graph of the function $\cos(x + c)$.
[*Hint:* The correct answer is not $\frac{\pi}{2}$.]
23. Sketch the graphs of the functions $2 + \cos x$ and $\cos(2 + x)$ on the interval $[-3\pi, 3\pi]$ (use the same coordinate axes for both graphs).
24. Sketch the graphs of the functions $4 - \cos x$ and $\cos(4 - x)$ on the interval $[-3\pi, 3\pi]$ (use the same coordinate axes for both graphs).
25. What is the range of the function $\cos(2 + x)$?
26. What is the range of the function $\cos(4 - x)$?
27. What is the range of the function $2 + \cos x$?
28. What is the range of the function $4 - \cos x$?
29. What is the amplitude of the function $\cos(2 + x)$?
30. What is the amplitude of the function $\cos(4 - x)$?
31. What is the amplitude of the function $2 + \cos x$?
32. What is the amplitude of the function $4 - \cos x$?
33. What is the period of the function $\cos(2 + x)$?
34. What is the period of the function $\cos(4 - x)$?
35. What is the period of the function $2 + \cos x$?
36. What is the period of the function $4 - \cos x$?
37. Sketch the graph of the function $5 \cos(\pi x)$ on the interval $[-4, 4]$.
38. Sketch the graph of the function $4 \cos(3\pi x)$ on the interval $[-2, 2]$.
39. What is the range of the function $5 \cos(\pi x)$?
40. What is the range of the function $4 \cos(3\pi x)$?
41. What is the amplitude of the function $5 \cos(\pi x)$?
42. What is the amplitude of the function $4 \cos(3\pi x)$?
43. What is the period of the function $5 \cos(\pi x)$?
44. What is the period of the function $4 \cos(3\pi x)$?

45. Sketch the graph of the function $7\cos(\frac{\pi}{2}x + \frac{6\pi}{5})$ on the interval $[-8, 8]$.

46. Sketch the graph of the function $6\cos(\frac{\pi}{3}x + \frac{8\pi}{5})$ on the interval $[-9, 9]$.

47. What is the range of the function $7\cos(\frac{\pi}{2}x + \frac{6\pi}{5})$?

48. What is the range of the function $6\cos(\frac{\pi}{3}x + \frac{8\pi}{5})$?

49. What is the amplitude of the function $7\cos(\frac{\pi}{2}x + \frac{6\pi}{5})$?

50. What is the amplitude of the function $6\cos(\frac{\pi}{3}x + \frac{8\pi}{5})$?

51. What is the period of the function $7\cos(\frac{\pi}{2}x + \frac{6\pi}{5})$?

52. What is the period of the function $6\cos(\frac{\pi}{3}x + \frac{8\pi}{5})$?

53. By what fraction of the period of $7\cos(\frac{\pi}{2}x)$ has the graph been shifted left to obtain the graph of $7\cos(\frac{\pi}{2}x + \frac{6\pi}{5})$?

54. By what fraction of the period of $6\cos(\frac{\pi}{3}x)$ has the graph been shifted left to obtain the graph of $6\cos(\frac{\pi}{3}x + \frac{8\pi}{5})$?

55. Sketch the graph of the function $7\cos(\frac{\pi}{2}x + \frac{6\pi}{5}) + 3$ on the interval $[-8, 8]$.

56. Sketch the graph of the function $6\cos(\frac{\pi}{3}x + \frac{8\pi}{5}) + 7$ on the interval $[-9, 9]$.

For Exercises 57–66, assume that f is the function defined by

$$f(x) = a\cos(bx + c) + d,$$

where a, b, c, and d are constants.

57. Find two distinct values for a so that f has amplitude 3.

58. Find two distinct values for a so that f has amplitude $\frac{17}{5}$.

59. Find two distinct values for b so that f has period 4.

60. Find two distinct values for b so that f has period $\frac{7}{3}$.

61. Find values for a and d, with $a > 0$, so that f has range $[3, 11]$.

62. Find values for a and d, with $a > 0$, so that f has range $[-8, 6]$.

63. Find values for a, d, and c, with $a > 0$ and $0 \le c \le \pi$, so that f has range $[3, 11]$ and $f(0) = 10$.

64. Find values for a, d, and c, with $a > 0$ and $0 \le c \le \pi$, so that f has range $[-8, 6]$ and $f(0) = -2$.

65. Find values for a, d, c, and b, with $a > 0$ and $b > 0$ and $0 \le c \le \pi$, so that f has range $[3, 11]$, $f(0) = 10$, and f has period 7.

66. Find values for a, d, c, and b, with $a > 0$ and $b > 0$ and $0 \le c \le \pi$, so that f has range $[-8, 6]$, $f(0) = -2$, and f has period 8.

67. What is the range of the function $\sin^2 x$?

68. What is the range of the function $\cos^2(3x)$?

69. What is the amplitude of the function $\sin^2 x$?

70. What is the amplitude of the function $\cos^2(3x)$?

71. What is the period of the function $\sin^2 x$?

72. What is the period of the function $\cos^2(3x)$?

73. Sketch the graph of the function $\sin^2 x$ on the interval $[-3\pi, 3\pi]$.

74. Sketch the graph of the function $\cos^2(3x)$ on the interval $[-2\pi, 2\pi]$.

For Exercises 75–76, use the following information: In the northern hemisphere, the day with the longest daylight is June 21. Also, "day x of the year" means that $x = 1$ on January 1, $x = 2$ on January 2, $x = 32$ on February 1, etc.

75. Anchorage, Alaska, receives 19.37 hours of daylight on June 21. Six months later, on the day with the shortest daylight, Anchorage receives only 5.45 hours of daylight. Find a function of the form

$$a\cos(bx + c) + d$$

that models the number of hours of daylight in Anchorage on day x of the year.

76. Phoenix, Arizona, receives 14.37 hours of daylight on June 21. Six months later, on the day with the shortest daylight, Phoenix receives only 9.93 hours of daylight. Find a function of the form

$$a\cos(bx + c) + d$$

that models the number of hours of daylight in Phoenix on day x of the year.

PROBLEMS

Use the following graph for Problems 77–79. Note that no scale is shown on the coordinate axes here. Do not assume that the scale is the same on the two coordinate axes.

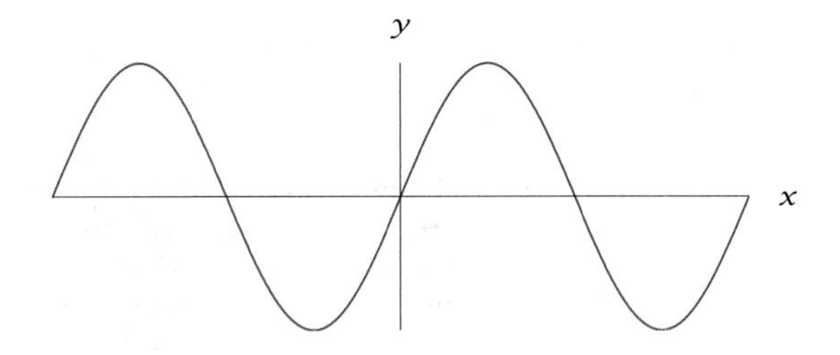

77. Explain why, with no scale on either axis, it is not possible to determine whether the figure above is the graph of $\sin x$, $3\sin x$, $\sin(5x)$, or $3\sin(5x)$.

78. Suppose you are told that the function graphed above is either $\sin x$ or $3\sin x$. To narrow the choice down to just one of these two functions, for which axis would you want to know the scale?

79. Suppose you are told that the function graphed above is either $\sin x$ or $\sin(5x)$. To narrow the choice down to just one of these two functions, for which axis would you want to know the scale?

80. Suppose f is the function whose value at x is the cosine of x degrees. Explain how the graph of f is obtained from the graph of $\cos x$.

81. Explain why a function of the form
$$-5\cos(bx + c),$$
where b and c are constants, can be rewritten in the form
$$5\cos(bx + \tilde{c}),$$
where $\tilde{c}$ is a constant. What is the relationship between $\tilde{c}$ and c?

82. Explain why a function of the form
$$a\cos(-7x + c),$$
where a and c are constants, can be rewritten in the form
$$a\cos(7x + \tilde{c}),$$
where $\tilde{c}$ is a constant. What is the relationship between $\tilde{c}$ and c?

83. Explain why a function of the form
$$a\cos(bx - 4),$$
where a and b are constants, can be rewritten in the form
$$a\cos(bx + \tilde{c}),$$
where $\tilde{c}$ is a positive constant.

84. Explain why a function of the form
$$a\cos(bx + c),$$
where a, b, and c are constants, can be rewritten in the form
$$\tilde{a}\cos(\tilde{b}x + \tilde{c}),$$
where $\tilde{a}$, $\tilde{b}$, and $\tilde{c}$ are nonnegative constants. What is the relationship between $\tilde{c}$ and c?

85. Explain why a function of the form
$$a\sin(bx + c),$$
where a, b, and c are constants, can be rewritten in the form
$$a\cos(bx + \tilde{c}),$$
where $\tilde{c}$ is a constant. What is the relationship between $\tilde{c}$ and c?

86. Explain why a function of the form
$$a\sin(bx + c),$$
where a, b, and c are constants, can be rewritten in the form
$$\tilde{a}\cos(\tilde{b}x + \tilde{c}),$$
where $\tilde{a}$, $\tilde{b}$, and $\tilde{c}$ are nonnegative constants.

87. Suppose f is a function with period p. Explain why
$$f(x + 2p) = f(x)$$
for every number x in the domain of f.

88. Suppose f is a function with period p. Explain why
$$f(x - p) = f(x)$$
for every number x such that $x - p$ is in the domain of f.

89. Suppose f is the function defined by $f(x) = \sin^4 x$. Is f a periodic function? Explain.

90. Suppose g is the function defined by $g(x) = \sin(x^4)$. Is g a periodic function? Explain.

91. Explain how the sine behaves with phase shifts of one-fourth its period, one-half its period, and all of its period, similarly to what was done for the cosine in the bulleted list that appears between Examples 6 and 7.

92. Find a graph of real data that suggests periodic behavior and then find a function that models this behavior (as in Example 8).

93. Find some real data involving periodic behavior and then find a function that models this behavior (as in Exercise 75).

WORKED-OUT SOLUTIONS *to Odd-numbered Exercises*

1. Sketch the graphs of the functions $4\sin x$ and $\sin(4x)$ on the interval $[-\pi, \pi]$ (use the same coordinate axes for both graphs).

SOLUTION The graph of $4\sin x$ is obtained by vertically stretching the graph of $\sin x$ by a factor of 4. The graph of $\sin(4x)$ is obtained by horizontally stretching the graph of $\sin x$ by a factor of $\frac{1}{4}$:

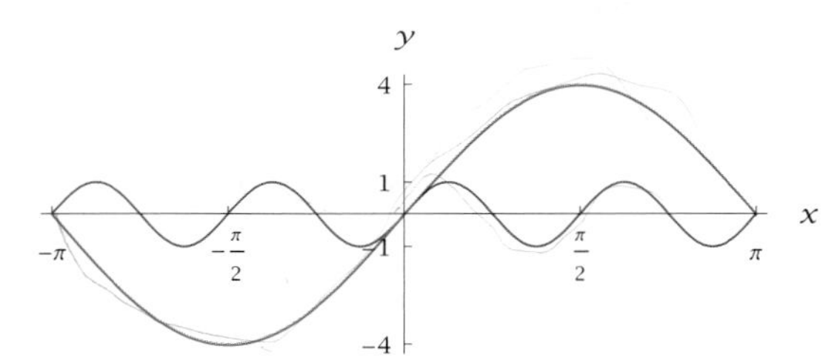

The graphs of $4\sin x$ *(blue) and* $\sin(4x)$ *(red) on the interval* $[-\pi, \pi]$.

3. What is the range of the function $4\sin x$?

SOLUTION The range of $4\sin x$ is obtained by multiplying each number in the range of $\sin x$ by 4. Thus the range of $4\sin x$ is the interval $[-4, 4]$.

5. What is the range of the function $\sin(4x)$?

SOLUTION As x ranges over the real numbers, $\sin x$ and $\sin(4x)$ take on the same values. Thus the range of $\sin(4x)$ is the interval $[-1, 1]$.

7. What is the amplitude of the function $4\sin x$?

SOLUTION The function $4\sin x$ has a maximum value of 4 and a minimum value of -4. Thus the difference between the maximum and minimum values of $4\sin x$ is 8. Half of 8 is 4, and hence the function $4\sin x$ has amplitude 4.

9. What is the amplitude of the function $\sin(4x)$?

SOLUTION The function $\sin(4x)$ has a maximum value of 1 and a minimum value of -1. Thus the difference between the maximum and minimum values of $\sin(4x)$ is 2. Half of 2 is 1, and hence the function $\sin(4x)$ has amplitude 1.

11. What is the period of the function $4\sin x$?

SOLUTION The period of $4\sin x$ is the same as the period of $\sin x$. Thus $4\sin x$ has period 2π.

13. What is the period of the function $\sin(4x)$?

SOLUTION The period of $\sin(4x)$ is the period of $\sin x$ divided by 4. Thus $\sin(4x)$ has period $\frac{2\pi}{4}$, which equals $\frac{\pi}{2}$. The figure above shows that $\sin(4x)$ indeed has period $\frac{\pi}{2}$.

Use the following graph for Exercises 15–22:

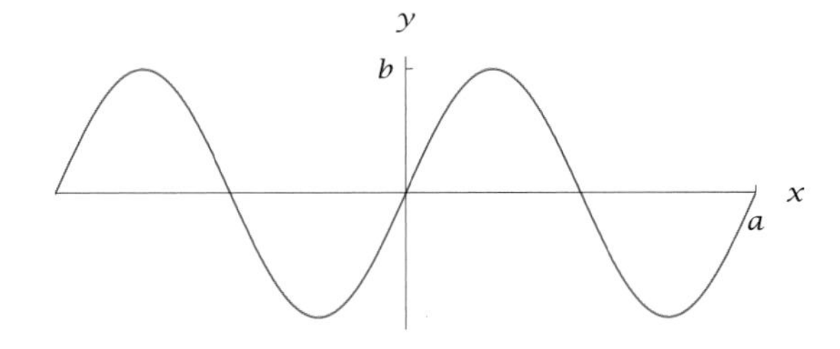

15. Suppose the figure above is part of the graph of the function $3\sin x$. What is the value of b?

SOLUTION The function shown in the graph has a maximum value of b. The function $3\sin x$ has a maximum value of 3. Thus $b = 3$.

17. Suppose the figure above is part of the graph of the function $\sin(7x)$. What is the value of a?

SOLUTION The function $\sin x$ has period 2π; thus the function $\sin(7x)$ has period $\frac{2\pi}{7}$. The function shown in the graph above has period a. Thus $a = \frac{2\pi}{7}$.

19. Find the smallest positive number c such that the figure above is part of the graph of the function $\sin(x + c)$.

SOLUTION The graph of $\sin(x + c)$ is obtained by shifting the graph of $\sin x$ left by c units. The graph above looks like the graph of $\sin x$ (for example, the graph goes through the origin and depicts a function that is increasing on an interval centered at 0).

The graph above is indeed the graph of $\sin x$ if we take $a = 2\pi$ and $b = 1$. Because $\sin x$ has period 2π, taking $c = 2\pi$ gives the smallest positive number such that the figure above is part of the graph of the function $\sin(x + c)$.

21. Find the smallest positive number c such that the figure above is part of the graph of the function $\cos(x - c)$.

SOLUTION The graph of $\cos(x - c)$ is obtained by shifting the graph of $\cos x$ right by c units. Shifting the graph of $\cos x$ right by $\frac{\pi}{2}$ units gives the graph of $\sin x$; in other words, $\cos(x - \frac{\pi}{2}) = \sin x$, as can be verified from the subtraction formula for cosine.

The graph above is indeed the graph of $\sin x$ if we take $a = 2\pi$ and $b = 1$. No positive number smaller than $\frac{\pi}{2}$ produces a graph of $\cos(x - c)$ that goes through the origin. Thus we must have $c = \frac{\pi}{2}$.

23. Sketch the graphs of the functions $2 + \cos x$ and $\cos(2 + x)$ on the interval $[-3\pi, 3\pi]$ (use the same coordinate axes for both graphs).

SOLUTION The graph of $2 + \cos x$ is obtained by shifting the graph of $\cos x$ up by 2 units. The graph of $\cos(2 + x)$ is obtained by shifting the graph of $\cos x$ left by 2 units:

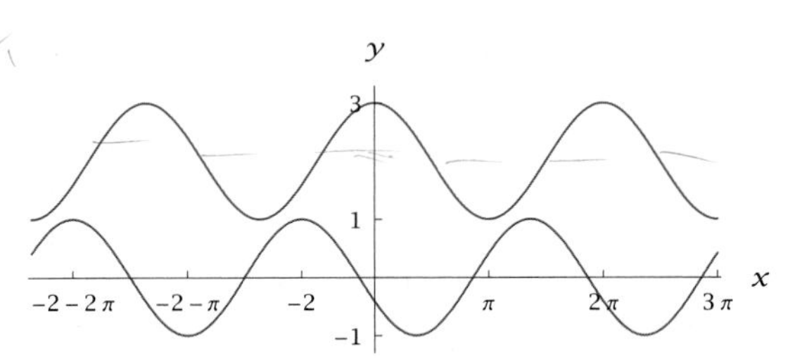

The graphs of $2 + \cos x$ *(blue) and* $\cos(2 + x)$ *(red) on the interval* $[-3\pi, 3\pi]$.

25. What is the range of the function $\cos(2 + x)$?

SOLUTION As x ranges over the real numbers, $\cos(2 + x)$ and $\cos x$ take on the same values. Thus the range of $\cos(2 + x)$ is the interval $[-1, 1]$.

27. What is the range of the function $2 + \cos x$?

SOLUTION The range of $2 + \cos x$ is obtained by adding 2 to each number in the range of $\cos x$. Thus the range of $2 + \cos x$ is the interval $[1, 3]$.

29. What is the amplitude of the function $\cos(2 + x)$?

SOLUTION The function $\cos(2 + x)$ has a maximum value of 1 and a minimum value of -1. Thus the difference between the maximum and minimum values of $\cos(2+x)$ is 2. Half of 2 is 1, and hence the function $\cos(2+x)$ has amplitude 1.

31. What is the amplitude of the function $2 + \cos x$?

SOLUTION The function $2 + \cos x$ has a maximum value of 3 and a minimum value of 1. Thus the difference between the maximum and minimum values of $2 + \cos x$ is 2. Half of 2 is 1, and hence the function $2 + \cos x$ has amplitude 1.

33. What is the period of the function $\cos(2 + x)$?

SOLUTION The period of $\cos(2 + x)$ is the same as the period of $\cos x$. Thus $\cos(2 + x)$ has period 2π.

35. What is the period of the function $2 + \cos x$?

SOLUTION The period of $2 + \cos x$ is the same as the period of $\cos x$. Thus $2 + \cos x$ has period 2π.

37. Sketch the graph of the function $5\cos(\pi x)$ on the interval $[-4, 4]$.

SOLUTION The graph of $5\cos(\pi x)$ is obtained by vertically stretching the graph of $\cos x$ by a factor of 5 and horizontally stretching by a factor of $\frac{1}{\pi}$:

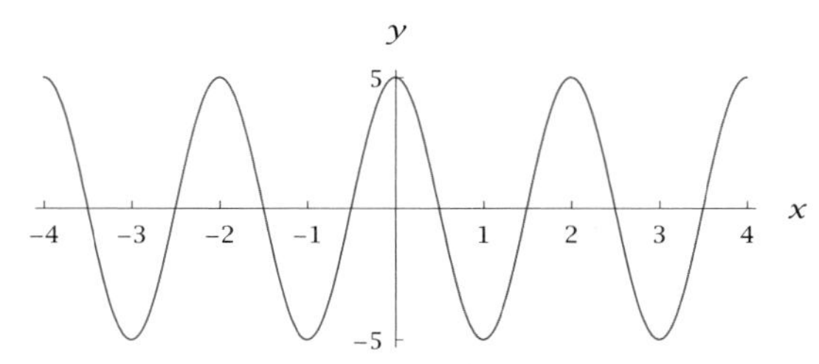

The graph of $5\cos(\pi x)$ *on the interval* $[-4, 4]$.

39. What is the range of the function $5\cos(\pi x)$?

SOLUTION The range of $5\cos(\pi x)$ is obtained by multiplying each number in the range of $\cos(\pi x)$ by 5. Thus the range of $5\cos(\pi x)$ is the interval $[-5, 5]$.

41. What is the amplitude of the function $5\cos(\pi x)$?

SOLUTION The function $5\cos(\pi x)$ has a maximum value of 5 and a minimum value of -5. Thus the difference between the maximum and minimum values of $5\cos(\pi x)$ is 10. Half of 10 is 5, and hence the function $5\cos(\pi x)$ has amplitude 5.

43. What is the period of the function $5\cos(\pi x)$?

SOLUTION The period of $5\cos(\pi x)$ is the period of $\cos x$ divided by π. Thus $5\cos(\pi x)$ has period $\frac{2\pi}{\pi}$, which equals 2. The figure above shows that $5\cos(\pi x)$ indeed has period 2.

45. Sketch the graph of the function $7\cos(\frac{\pi}{2}x + \frac{6\pi}{5})$ on the interval $[-8, 8]$.

SOLUTION The graph of $7\cos(\frac{\pi}{2}x)$ is obtained by vertically stretching the graph of $\cos x$ by a factor of 7 and horizontally stretching by a factor of $\frac{2}{\pi}$:

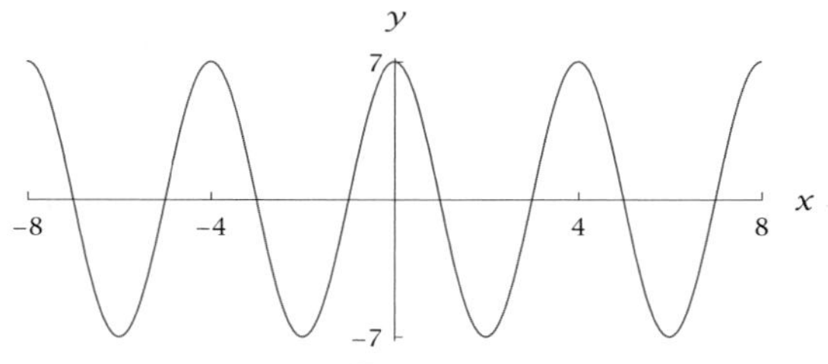

The graph of $7\cos(\frac{\pi}{2}x)$ *on the interval* $[-8, 8]$.

To see how to construct the graph of $7\cos(\frac{\pi}{2}x + \frac{6\pi}{5})$, define a function f by

$$f(x) = 7\cos(\tfrac{\pi}{2}x).$$

Now

$$7\cos(\tfrac{\pi}{2}x + \tfrac{6\pi}{5}) = 7\cos(\tfrac{\pi}{2}(x + \tfrac{12}{5})) = f(x + \tfrac{12}{5}).$$

Thus the graph of $7\cos(\frac{\pi}{2}x + \frac{6\pi}{5})$ is obtained by shifting the graph of $7\cos(\frac{\pi}{2}x)$ left by $\frac{12}{5}$ units:

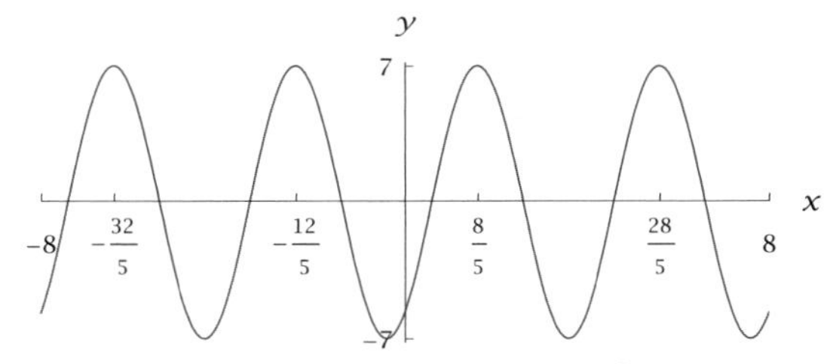

The graph of $7\cos(\frac{\pi}{2}x + \frac{6\pi}{5})$ *on the interval* $[-8, 8]$.

Note that the peaks of the graph of the function $7\cos(\frac{\pi}{2}x)$ that occur at $x = -4$, at $x = 0$, at $x = 4$, and at $x = 8$ have been shifted by $\frac{12}{5}$ units to the left, now occurring in the graph at $x = -4 - \frac{12}{5}$ (which equals $-\frac{32}{5}$), at $x = 0 - \frac{12}{5}$ (which equals $-\frac{12}{5}$), at $x = 4 - \frac{12}{5}$ (which equals $\frac{8}{5}$), and at $x = 8 - \frac{12}{5}$ (which equals $\frac{28}{5}$).

47. What is the range of the function $7\cos(\frac{\pi}{2}x + \frac{6\pi}{5})$?

SOLUTION The range of the function $7\cos(\frac{\pi}{2}x + \frac{6\pi}{5})$ is obtained by multiplying each number in the range of $\cos(\frac{\pi}{2}x + \frac{6\pi}{5})$ by 7. Thus the range of the function $7\cos(\frac{\pi}{2}x + \frac{6\pi}{5})$ is the interval $[-7, 7]$.

49. What is the amplitude of the function $7\cos(\frac{\pi}{2}x + \frac{6\pi}{5})$?

SOLUTION The function $7\cos(\frac{\pi}{2}x + \frac{6\pi}{5})$ has a maximum value of 7 and a minimum value of -7. Thus the difference between the maximum and minimum values of $7\cos(\frac{\pi}{2}x + \frac{6\pi}{5})$ is 14. Half of 14 is 7, and hence the function $7\cos(\frac{\pi}{2}x + \frac{6\pi}{5})$ has amplitude 7.

51. What is the period of the function $7\cos(\frac{\pi}{2}x + \frac{6\pi}{5})$?

SOLUTION The period of $7\cos(\frac{\pi}{2}x + \frac{6\pi}{5})$ is the period of $\cos x$ divided by $\frac{\pi}{2}$. Thus $7\cos(\frac{\pi}{2}x + \frac{6\pi}{5})$ has period $(2\pi)/(\frac{\pi}{2})$, which equals 4. The figure above shows that $7\cos(\frac{\pi}{2}x + \frac{6\pi}{5})$ indeed has period 4.

53. By what fraction of the period of $7\cos(\frac{\pi}{2}x)$ has the graph been shifted left to obtain the graph of $7\cos(\frac{\pi}{2}x + \frac{6\pi}{5})$?

SOLUTION The graph of $7\cos(\frac{\pi}{2}x)$ is shifted left by $\frac{12}{5}$ units to obtain the graph of $7\cos(\frac{\pi}{2}x + \frac{6\pi}{5})$. The period of $7\cos(\frac{\pi}{2}x)$ is 4. Thus the fraction of the period of $7\cos(\frac{\pi}{2}x)$ by which the graph has been shifted is $\frac{12}{5}/4$, which equals $\frac{3}{5}$.

55. Sketch the graph of the function $7\cos(\frac{\pi}{2}x + \frac{6\pi}{5}) + 3$ on the interval $[-8, 8]$.

SOLUTION The graph of $7\cos(\frac{\pi}{2}x + \frac{6\pi}{5}) + 3$ is obtained by shifting the graph of $7\cos(\frac{\pi}{2}x + \frac{6\pi}{5})$ up by 3 units. Fortunately, we already graphed the function $7\cos(\frac{\pi}{2}x + \frac{6\pi}{5})$ in Exercise 45. Shifting the graph obtained there up by 3 units, we obtain the following graph:

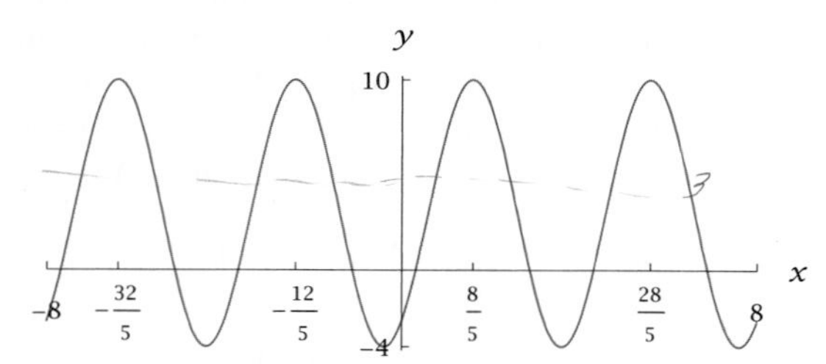

The graph of $7\cos(\frac{\pi}{2}x + \frac{6\pi}{5}) + 3$ *on the interval* $[-8, 8]$.

For Exercises 57–66, assume that f is the function defined by

$$f(x) = a\cos(bx + c) + d,$$

where a, b, c, and d are constants.

57. Find two distinct values for a so that f has amplitude 3.

SOLUTION The amplitude of a function is half the difference between its maximum and minimum values. The function $\cos(bx + c)$ has a maximum value of 1 and a minimum value of -1 (regardless of the values of b and c).

Thus the function $a\cos(bx + c)$ has a maximum value of $|a|$ and a minimum value of $-|a|$. Hence the function $a\cos(bx + c) + d$ has a maximum value of $|a| + d$ and a minimum value of $-|a| + d$. The difference between this maximum value and this minimum value is $2|a|$. Thus the amplitude of $a\cos(bx + c) + d$ is $|a|$ (notice that the value of d does not affect the amplitude).

Hence the function f has amplitude 3 if $|a| = 3$. Thus we can take $a = 3$ or $a = -3$.

59. Find two distinct values for b so that f has period 4.

SOLUTION The function $\cos x$ has period 2π. If $b > 0$, then the graph of $\cos(bx)$ is obtained by horizontally stretching the graph of $\cos x$ by a factor of $\frac{1}{b}$. Thus $\cos(bx)$ has period $\frac{2\pi}{b}$.

The graph of $\cos(bx + c)$ differs from the graph of $\cos(bx)$ only by a phase shift, which does not change the period. Thus $\cos(bx + c)$ also has period $\frac{2\pi}{b}$.

The graph of $a\cos(bx + c)$ is obtained from the graph of $\cos(bx + c)$ by stretching vertically, which changes the amplitude but not the period. Thus $a\cos(bx + c)$ also has period $\frac{2\pi}{b}$.

The graph of $a\cos(bx + c) + d$ is obtained by shifting the graph of $a\cos(bx + c)$ up or down (depending on whether d is positive or negative). Adding d changes neither the period nor the amplitude. Thus $a\cos(bx + c) + d$ also has period $\frac{2\pi}{b}$.

We want $a\cos(bx + c) + d$ to have period 4. Thus we solve the equation $\frac{2\pi}{b} = 4$, getting $b = \frac{\pi}{2}$. In other words, $a\cos(\frac{\pi}{2}x + c) + d$ has period 4, regardless of the values of a, c, and d.

Note that

$$a\cos(-\tfrac{\pi}{2}x + c) + d = a\cos(\tfrac{\pi}{2}x - c) + d,$$

and thus $a\cos(-\frac{\pi}{2}x + c) + d$ also has period 4. Hence to make f have period 4, we can take $b = \frac{\pi}{2}$ or $b = -\frac{\pi}{2}$.

61. Find values for a and d, with $a > 0$, so that f has range $[3, 11]$.

SOLUTION Because f has range $[3, 11]$, the maximum value of f is 11 and the minimum value of f is 3. Thus the difference between the maximum and minimum values of f is 8. Thus the amplitude of f is half of 8, which equals 4. Reasoning as in the solution to Exercise 57, we see that this implies $a = 4$ or $a = -4$. This exercise requires that $a > 0$, and thus we must take $a = 4$.

The function $4\cos(bx + c)$ has range $[-4, 4]$ (regardless of the values of b and c). Note that $[-4, 4]$ is an interval of length 8, just as $[3, 11]$ is an interval of length 8. We want to find a number d such that each number in the interval $[3, 11]$ is obtained by adding d to a number in the interval $[-4, 4]$. To find d, we can subtract either the left endpoints or the right endpoints of these two intervals. In other words, we can find d by evaluating $3 - (-4)$ or $11 - 4$. Either way, we obtain $d = 7$.

Thus the function $4\cos(bx + c) + 7$ has range $[3, 11]$ (regardless of the values of b and c).

63. Find values for a, d, and c, with $a > 0$ and $0 \le c \le \pi$, so that f has range $[3, 11]$ and $f(0) = 10$.

SOLUTION From the solution to Exercise 61, we see that we need to choose $a = 4$ and $d = 7$. Thus we have

$$f(x) = 4\cos(bx + c) + 7,$$

and we need to choose c so that $0 \le c \le \pi$ and $f(0) = 10$. Hence we need to choose c so that $0 \le c \le \pi$ and

$$4\cos c + 7 = 10.$$

Thus $\cos c = \frac{3}{4}$. Because $0 \le c \le \pi$, this means that $c = \cos^{-1}\frac{3}{4}$.

Thus the function $4\cos(bx + \cos^{-1}\frac{3}{4}) + 7$ has range $[3, 11]$ and $f(0) = 10$ (regardless of the value of b).

65. Find values for a, d, c, and b, with $a > 0$ and $b > 0$ and $0 \le c \le \pi$, so that f has range $[3, 11]$, $f(0) = 10$, and f has period 7.

SOLUTION From the solution to Exercise 63, we see that we need to choose $a = 4$, $c = \cos^{-1}\frac{3}{4}$, and $d = 7$. Thus we have

$$f(x) = 4\cos(bx + \cos^{-1}\tfrac{3}{4}) + 7,$$

and we need to choose $b > 0$ so that f has period 7. Because the cosine function has period 2π, this means that we need to choose $b = \frac{2\pi}{7}$. Thus the function $4\cos(\frac{2\pi}{7}x + \cos^{-1}\frac{3}{4}) + 7$ has range $[3, 11]$, equals 10 when $x = 0$, and has period 7.

67. What is the range of the function $\sin^2 x$?

SOLUTION The sine function takes on all the values in the interval $[-1, 1]$; squaring the numbers in this interval gives the numbers in the interval $[0, 1]$. Thus the range of $\sin^2 x$ is the interval $[0, 1]$.

69. What is the amplitude of the function $\sin^2 x$?

SOLUTION The function $\sin^2 x$ has a maximum value of 1 and a minimum value of 0. The difference between this maximum value and this minimum value is 1. Thus the amplitude of $\sin^2 x$ is $\frac{1}{2}$.

71. What is the period of the function $\sin^2 x$?

SOLUTION We know that $\sin(x + \pi) = -\sin x$ for every number x (see Section 9.6). Squaring both sides of this equation, we get

$$\sin^2(x + \pi) = \sin^2 x.$$

No positive number p smaller than π can produce the identity

$$\sin^2(x + p) = \sin^2 x,$$

as can be seen by taking $x = 0$, in which case the equation above becomes $\sin^2 p = 0$. The smallest positive number p satisfying this last equation is π. Putting all this together, we conclude that the function $\sin^2 x$ has period π.

73. Sketch the graph of the function $\sin^2 x$ on the interval $[-3\pi, 3\pi]$.

SOLUTION The function $\sin^2 x$ takes on values between 0 and 1, has period π, equals 0 when x is an integer multiple of π, and equals 1 when x is halfway between two zeros of this function. Thus a sketch of the graph of $\sin^2 x$ should resemble the figure below:

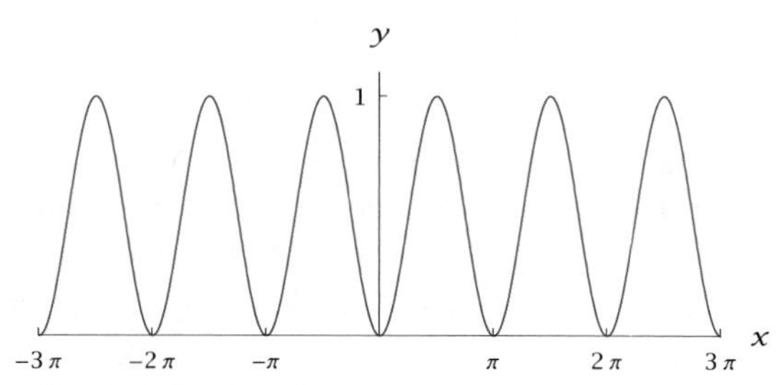

The graph of $\sin^2 x$ *on the interval* $[-3\pi, 3\pi]$.

For Exercises 75–76, use the following information: In the northern hemisphere, the day with the longest daylight is June 21. Also, "day x ***of the year" means that*** $x = 1$ ***on January 1,*** $x = 2$ ***on January 2,*** $x = 32$ ***on February 1, etc.***

75. Anchorage, Alaska, receives 19.37 hours of daylight on June 21. Six months later, on the day with the shortest daylight, Anchorage receives only 5.45 hours of daylight. Find a function of the form

$$a\cos(bx + c) + d$$

that models the number of hours of daylight in Anchorage on day x of the year.

SOLUTION As we have seen from examples, the period of the function above is $\frac{2\pi}{b}$ (here we are assuming that $b > 0$, which we can do because if $b < 0$ then we could replace b and c by $-b$ and $-c$ and end up with the same function). Because a year has 365 days (we will ignore leap years), we want the period of our function to be 365. In other words, we want $\frac{2\pi}{b} = 365$, which means that $b = \frac{2\pi}{365}$.

As we have seen from examples, the amplitude of the function above is a (here we are assuming that $a > 0$, which we can do because if $a < 0$ we could replace a by $-a$ and c by $c + \pi$ and end up with the same function). The maximum daylight in Anchorage is 19.37 hours, and the minimum daylight is 5.45 hours. Thus our function should have amplitude $\frac{19.37-5.45}{2}$, which equals 6.96. In other words, we take $a = 6.96$.

As we have seen from examples, d should be chosen to be halfway between the minimum and maximum values of our function. Thus we should take $d = \frac{19.37+5.45}{2} = 12.41$.

June 21 is day 172 of the year. Thus we want to choose c so that $\cos(bx + c)$ is largest when $x = 172$. We already know that $b = \frac{2\pi}{365}$, so we want to chose c so that $\cos(\frac{2\pi}{365}172 + c)$ is as large as possible. The largest value of cosine is 1, and $\cos 0 = 1$. Thus we choose c so that $\frac{2\pi}{365}172 + c = 0$. Thus we take $c = -\frac{2\pi}{365}172$.

Putting all this together, our function that models the number of daylight hours in Anchorage on day x of the year is

$$6.96\cos(2\pi\tfrac{x-172}{365}) + 12.41.$$

How well does this function model the actual behavior? For January 21 ($x = 21$), the function above predicts 6.45 hours of daylight in Anchorage. The actual number of hours of daylight in Anchorage on January 21 is 6.87. Thus the model is off by about 6%, a decent but not spectacular performance.

For March 21 ($x = 80$), the model above predicts 12.32 hours of daylight in Anchorage. The actual amount is 12.33 hours. Thus in this case the model is extremely accurate.

11.3 *Polar Coordinates*

LEARNING OBJECTIVES

By the end of this section you should be able to

- locate a point from its polar coordinates;
- convert from polar to rectangular coordinates;
- convert from rectangular to polar coordinates;
- graph an equation given in polar coordinates.

The usual rectangular coordinates (x, y) of a point in the coordinate plane tell us the horizontal and vertical displacement of the point from the origin. In this section we discuss another useful coordinate system, called polar coordinates, that focuses more directly on the line segment from the origin to a point. The first polar coordinate tells us the length of this line segment; the second polar coordinate tells us the angle this line segment makes with the positive horizontal axis.

Defining Polar Coordinates

The two polar coordinates of a point are traditionally called r and θ. These coordinates have a simple geometric description in terms of the line segment from the origin to the point.

Polar coordinates

The **polar coordinates** (r, θ) of a point in the coordinate plane are characterized as follows:

- The first polar coordinate r is the distance from the origin to the point.
- The second polar coordinate θ is the angle between the positive horizontal axis and the line segment from the origin to the point.

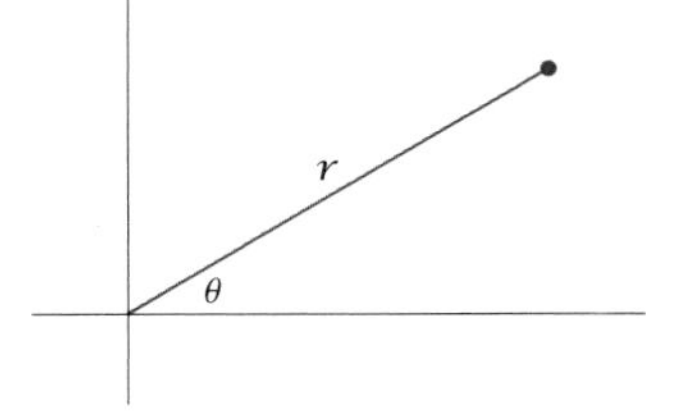

As usual, positive angles are measured in the counterclockwise direction starting at the positive horizontal axis and negative angles are measured in the clockwise direction.

EXAMPLE 1

Sketch the line segment from the origin to the point with polar coordinates $(3, \frac{\pi}{4})$.

SOLUTION The line segment is shown in the following figure. The length of the line segment is 3, and the line segment makes an angle of $\frac{\pi}{4}$ radians (which equals $45°$) with the positive horizontal axis.

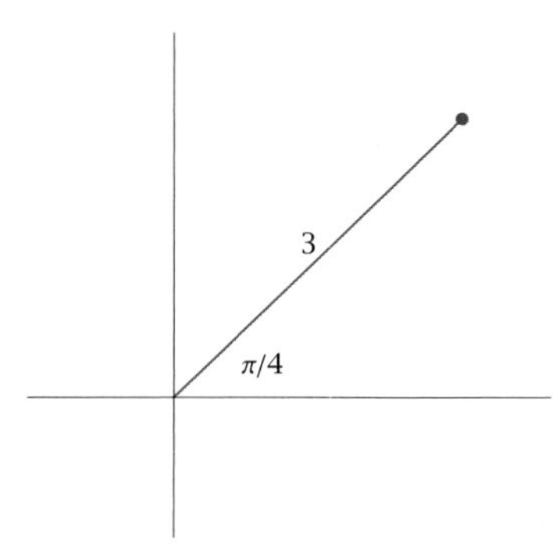

The endpoint of this line segment has first polar coordinate $r = 3$ and second polar coordinate $\theta = \frac{\pi}{4}$.

As another example, a point whose second polar coordinate equals $\frac{\pi}{2}$ is on the positive vertical axis (because the positive vertical axis makes an angle of $\frac{\pi}{2}$ with the positive horizontal axis). A point whose second polar coordinate equals $-\frac{\pi}{2}$ is on the negative vertical axis (because the negative vertical axis makes an angle of $-\frac{\pi}{2}$ with the positive horizontal axis).

Converting from Polar to Rectangular Coordinates

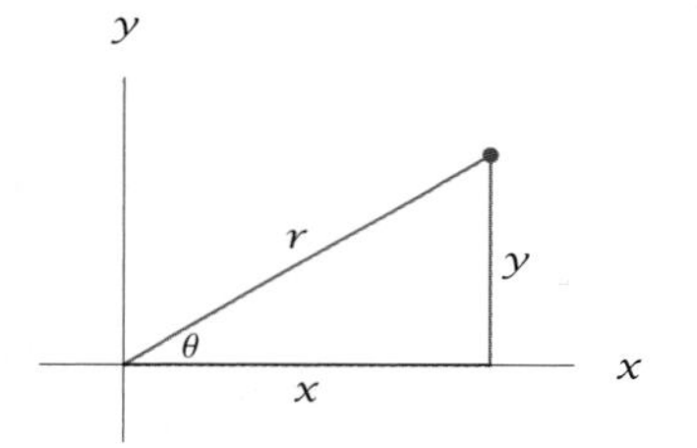

The term coordinates *with no adjective in front refers to rectangular coordinates. Always use the phrase* polar coordinates *when an ordered pair should be interpreted as polar coordinates.*

To obtain a formula for converting from polar to rectangular coordinates in the xy-plane, draw the line segment from the origin to the point in question, and then form the right triangle shown in the figure here.

Looking at this right triangle, we see that $\cos\theta = \frac{x}{r}$ and $\sin\theta = \frac{y}{r}$. Solving for x and y gives the following formulas:

Converting from polar to rectangular coordinates

A point with polar coordinates (r, θ) has rectangular coordinates

$$(r\cos\theta, r\sin\theta).$$

In the xy-plane, this conversion is expressed by the following equations:

$$x = r\cos\theta \quad \text{and} \quad y = r\sin\theta.$$

EXAMPLE 2 Find the rectangular coordinates of the point with polar coordinates $(5, \frac{\pi}{3})$.

SOLUTION This point has rectangular coordinates $(5\cos\frac{\pi}{3}, 5\sin\frac{\pi}{3})$, which equals $(\frac{5}{2}, \frac{5\sqrt{3}}{2})$.

The point with polar coordinates $(6, 0)$ has rectangular coordinates $(6, 0)$, as does the point with polar coordinates $(6, 2\pi)$. More generally, adding any integer multiple of 2π to an angle does not change the cosine or sine of the angle. Thus the polar coordinates of a point are not unique.

Converting from Rectangular to Polar Coordinates

We know how to convert from polar coordinates to rectangular coordinates. Now we take up the question of converting in the other direction. In other words, given rectangular coordinates (x, y), how do we find the polar coordinates (r, θ)?

Recall that the first polar coordinate r is the distance from the origin to the point (x, y). Thus

$$r = \sqrt{x^2 + y^2}.$$

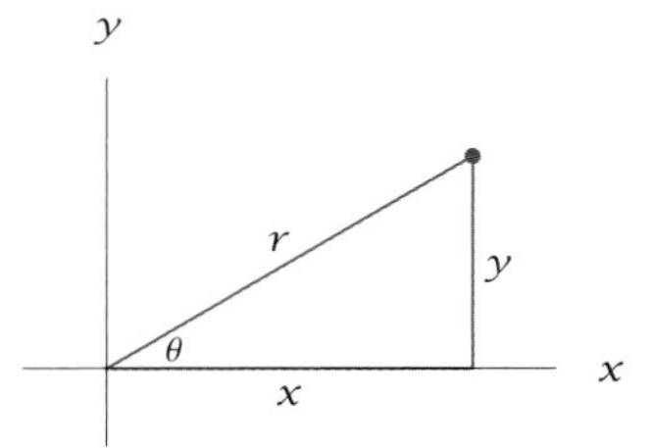

This figure shows that $\tan\theta = \frac{y}{x}$.

To see how to choose the second polar coordinate θ given the rectangular coordinates (x, y), let's look once again at the standard figure showing the relationship between polar coordinates and rectangular coordinates.

Looking at the right triangle shown here, we see that $\tan\theta = \frac{y}{x}$. Thus it is tempting to choose $\theta = \tan^{-1}\frac{y}{x}$. However, there are two problems with the formula $\theta = \tan^{-1}\frac{y}{x}$. We now turn to a discussion of these problems.

The first problem involves the lack of uniqueness for the polar coordinate θ, as shown by the following example.

EXAMPLE 3

Find polar coordinates for the point with rectangular coordinates $(1, 1)$.

SOLUTION There is no choice about the first polar coordinate r for this point—we must take

$$r = \sqrt{1^2 + 1^2} = \sqrt{2}.$$

If we use the formula $\theta = \tan^{-1}\frac{y}{x}$ to obtain the second polar coordinate θ for the point with rectangular coordinates $(1, 1)$, we get

$$\theta = \tan^{-1}\tfrac{1}{1} = \tan^{-1} 1 = \tfrac{\pi}{4}.$$

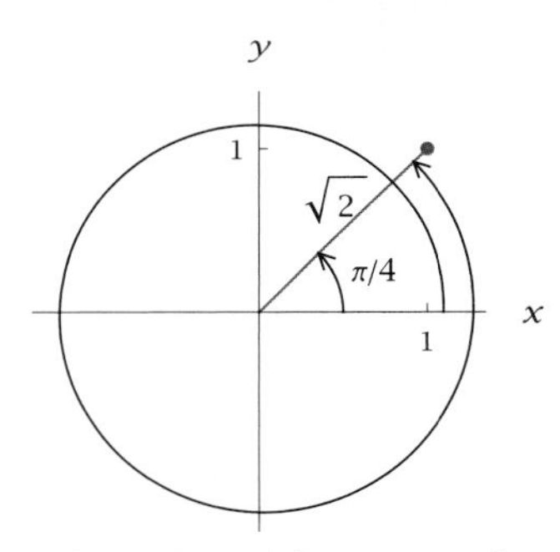

The point with rectangular coordinates $(1, 1)$ *has polar coordinates* $r = \sqrt{2}$ *and* $\theta = \frac{\pi}{4}$, *but* $\theta = \frac{\pi}{4} + 2\pi$ *is also a valid choice.*

The point with polar coordinates $(\sqrt{2}, \frac{\pi}{4})$ indeed has rectangular coordinates $(1, 1)$, so it seems that all is well.

However, the point with polar coordinates $(\sqrt{2}, \frac{\pi}{4} + 2\pi)$ also has rectangular coordinates $(1, 1)$, as shown here, as does the point with polar coordinates $(\sqrt{2}, \frac{\pi}{4} + 4\pi)$. Or we could have chosen the second polar coordinate to be $\frac{\pi}{4} + 2\pi n$ for any integer n. Thus using the arctangent formula for the second polar coordinate θ produced a correct answer in this case, but if we had been seeking one of the other correct choices for θ, the arctangent formula would not have provided it.

The second problem with the formula $\theta = \tan^{-1}\frac{y}{x}$ is more serious. To see how this problem arises, we will look at a few more examples.

EXAMPLE 4

Find polar coordinates for the point with rectangular coordinates $(1, -1)$.

SOLUTION Using the formula for the first polar coordinate r, we get

$$r = \sqrt{1^2 + (-1)^2} = \sqrt{2}.$$

If we use the formula $\theta = \tan^{-1}\frac{y}{x}$ to obtain the second polar coordinate θ for the point with rectangular coordinates $(1, -1)$, we get

$$\theta = \tan^{-1}\left(\tfrac{-1}{1}\right) = \tan^{-1}(-1) = -\tfrac{\pi}{4}.$$

The point with polar coordinates $(\sqrt{2}, -\frac{\pi}{4})$ indeed has rectangular coordinates $(1, -1)$. Thus in this case the formula $\theta = \tan^{-1}\frac{y}{x}$ has worked (although it ignored other possible correct choices for θ).

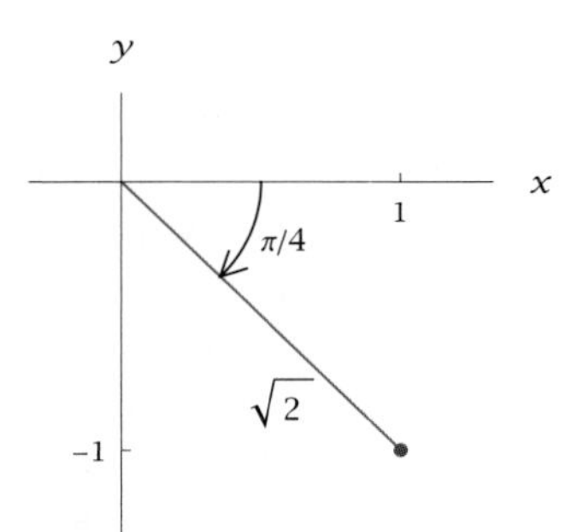

The point with rectangular coordinates $(1, -1)$ has polar coordinates $(\sqrt{2}, -\frac{\pi}{4})$ (along with other possible correct choices for the second polar coordinate).

The formula $\theta = \tan^{-1}\frac{y}{x}$ can be wrong, as shown in the next example.

EXAMPLE 5

Find polar coordinates for the point with rectangular coordinates $(-1, 1)$.

SOLUTION Using the formula for the first polar coordinate r, we get

$$r = \sqrt{(-1)^2 + 1^2} = \sqrt{2}.$$

If we use the formula $\theta = \tan^{-1}\frac{y}{x}$ to obtain the second polar coordinate θ, we get

$$\theta = \tan^{-1}\left(\tfrac{1}{-1}\right) = \tan^{-1}(-1) = -\tfrac{\pi}{4}.$$

In this example the formula $\theta = \tan^{-1}\frac{y}{x}$ gives an incorrect result.

However, the point with polar coordinates $(\sqrt{2}, -\frac{\pi}{4})$ has rectangular coordinates $(1, -1)$, not $(-1, 1)$, which is what we seek now. The figure below shows that the correct choice of the second polar coordinate θ for the point $(-1, 1)$ is $\theta = \frac{3\pi}{4}$ (or $\theta = \frac{3\pi}{4} + 2\pi n$ for any integer n).

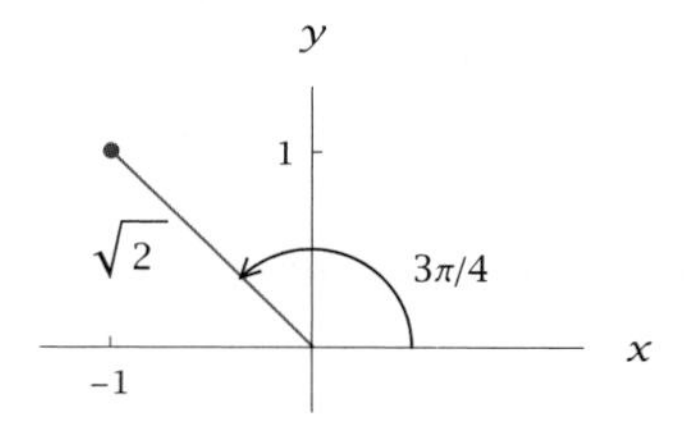

The point with rectangular coordinates $(-1, 1)$ has polar coordinates $(\sqrt{2}, \frac{3\pi}{4})$ (along with other possible correct choices for θ).

The formula $\theta = \tan^{-1}\frac{y}{x}$ produced an incorrect answer when applied to the point $(-1, 1)$ in the example above. To understand why this happened, recall that $\tan^{-1}\frac{y}{x}$ is the angle in the interval $(-\frac{\pi}{2}, \frac{\pi}{2})$ whose tangent equals $\frac{y}{x}$. Note that $r\cos\theta > 0$ if $r > 0$ and θ is in the interval $(-\frac{\pi}{2}, \frac{\pi}{2})$. Thus the formula $\theta = \tan^{-1}\frac{y}{x}$ cannot produce a correct polar coordinate θ if $x < 0$.

In the case of the point with rectangular coordinates $(-1, 1)$, the formula $\theta = \tan^{-1} \frac{y}{x} = \tan^{-1}(-1)$ produces the angle $-\frac{\pi}{4}$, which indeed has tangent equal to -1. However, the angle $-\frac{\pi}{4}$ is an incorrect choice for the second polar coordinate of $(-1, 1)$, as shown in the figure in Example 5.

The problem here is that although $\tan^{-1} \frac{y}{x}$ is an angle whose tangent equals $\frac{y}{x}$, there are also other angles whose tangent equals $\frac{y}{x}$.

The angle $\frac{3\pi}{4}$ (which equals $-\frac{\pi}{4} + \pi$) also has tangent equal to -1. The figure in Example 5 shows that $\frac{3\pi}{4}$ (or $\frac{3\pi}{4} + 2\pi n$ for any integer n) is the angle that we need for the second polar coordinate of $(-1, 1)$.

Another example may help show what is happening here.

EXAMPLE 6

Find polar coordinates for the point with rectangular coordinates $(-1, -1)$.

SOLUTION Using the formula for the first polar coordinate r, we get

$$r = \sqrt{(-1)^2 + (-1)^2} = \sqrt{2}.$$

If we use the formula $\theta = \tan^{-1} \frac{y}{x}$ to obtain the second polar coordinate θ, we get

$$\theta = \tan^{-1}\left(\tfrac{-1}{-1}\right) = \tan^{-1} 1 = \tfrac{\pi}{4}.$$

However, the point with polar coordinates $(\sqrt{2}, \frac{\pi}{4})$ has rectangular coordinates $(1, 1)$, not $(-1, -1)$, which is what we seek now. The figure below shows that the correct choice of the second polar coordinate θ for the point $(-1, -1)$ is $\theta = -\frac{3\pi}{4}$ (or $\theta = -\frac{3\pi}{4} + 2\pi n$ for any integer n). Note that $-\frac{3\pi}{4} = \frac{\pi}{4} - \pi$; thus the incorrect formula $\theta = \tan^{-1} \frac{y}{x}$ was off by an odd multiple of π.

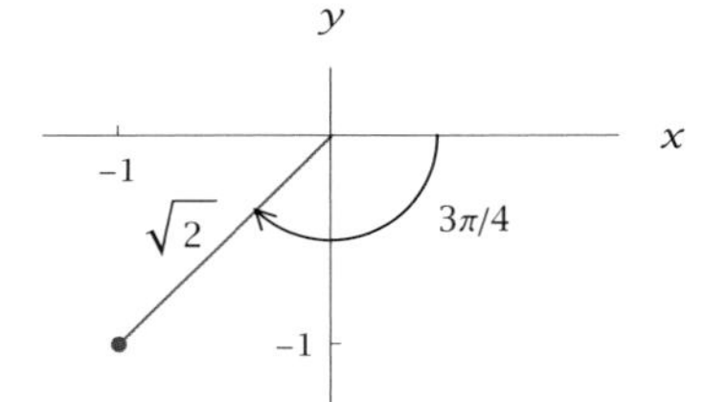

The point with rectangular coordinates $(-1, -1)$ has polar coordinates $(\sqrt{2}, -\frac{3\pi}{4})$ (along with other possible correct choices for θ).

The example above shows that the incorrect formula $\theta = \tan^{-1} \frac{y}{x}$ does not distinguish between $\tan^{-1} \frac{1}{1}$ and $\tan^{-1}\left(\frac{-1}{-1}\right)$, giving a result of $\frac{\pi}{4}$ in both cases. Thus we will state the formula for converting from rectangular to polar coordinates in terms of a requirement on $\tan \theta$ rather than a formula involving $\tan^{-1}$.

The examples presented in this section should help reinforce the idea that the equation $\tan \theta = t$ is not equivalent to the equation $\theta = \tan^{-1} t$.

Although there are many angles θ satisfying $\tan \theta = \frac{y}{x}$, we will need to have $x = r \cos \theta$ and $y = r \sin \theta$. Because r is positive, this means that $\cos \theta$ will need to have the same sign as x and $\sin \theta$ will need to have the same sign as y. If we pick θ accordingly from among the angles whose tangent equals $\frac{y}{x}$, then we will have a correct choice of polar coordinates.

Converting from rectangular to polar coordinates

A point with rectangular coordinates (x, y), with $x \neq 0$, has polar coordinates (r, θ) that satisfy the equations

$$r = \sqrt{x^2 + y^2} \quad \text{and} \quad \tan\theta = \tfrac{y}{x},$$

where θ must be chosen so that $\cos\theta$ has the same sign as x and $\sin\theta$ has the same sign as y.

For example, the point with rectangular coordinates $(0, 5)$ *has polar coordinates* $(5, \frac{\pi}{2})$, *and the point with rectangular coordinates* $(0, -5)$ *has polar coordinates* $(5, -\frac{\pi}{2})$.

In the box above, we excluded the case where $x = 0$ (in other words, points on the vertical axis) to avoid division by 0 in the formula $\tan\theta = \frac{y}{x}$. To convert $(0, y)$ to polar coordinates, you can choose $\theta = \frac{\pi}{2}$ if $y > 0$ and $\theta = -\frac{\pi}{2}$ if $y < 0$.

The box below provides a convenient summary of how to choose the second polar coordinate θ to be in the interval $(-\pi, \pi]$:

Choosing the polar coordinate θ in $(-\pi, \pi]$

The second polar coordinate θ corresponding to a point with rectangular coordinates (x, y) can be chosen as follows:

- If $x > 0$, then $\theta = \tan^{-1}\frac{y}{x}$.
- If $x < 0$ and $y \geq 0$, then $\theta = \tan^{-1}\frac{y}{x} + \pi$.
- If $x < 0$ and $y < 0$, then $\theta = \tan^{-1}\frac{y}{x} - \pi$.
- If $x = 0$ and $y > 0$, then $\theta = \frac{\pi}{2}$.
- If $x = 0$ and $y < 0$, then $\theta = -\frac{\pi}{2}$.

Do not memorize this procedure. Instead, focus on understanding the meaning of polar coordinates. With that understanding, this procedure will be clear.

In the box above, none of the cases covers the origin, whose rectangular coordinates are $(0, 0)$. The origin has polar coordinates $(0, \theta)$, where θ can be chosen to be any number.

EXAMPLE 7

Find polar coordinates for the point with rectangular coordinates $(-4, 3)$. For the second polar coordinate θ, use radians and choose θ to be in the interval $(-\pi, \pi]$.

SOLUTION Using the formula for the first polar coordinate r, we have

$$r = \sqrt{(-4)^2 + 3^2} = \sqrt{16 + 9} = \sqrt{25} = 5.$$

Because the first coordinate of $(-4, 3)$ is negative and the second coordinate is positive, for the second polar coordinate θ we have

$$\theta = \tan^{-1}\left(\tfrac{3}{-4}\right) + \pi \approx 2.498.$$

Thus $(-4, 3)$ has polar coordinates approximately $(5, 2.498)$.

Graphs of Polar Equations

Some curves or regions in the coordinate plane can be described more simply by using polar coordinates instead of rectangular coordinates, as shown by the following example.

EXAMPLE 8

(a) Find an equation in rectangular coordinates for the circle of radius 3 centered at the origin.

(b) Find an equation in polar coordinates for the circle of radius 3 centered at the origin.

SOLUTION

(a) In rectangular coordinates in the xy-plane, this circle can be described by the equation

$$x^2 + y^2 = 9.$$

(b) The first polar coordinate r measures the distance to the origin. Because the circle of radius 3 centered at the origin equals the set of points whose distance to the origin equals 3, this circle above can be described in polar coordinates (r, θ) by the simpler equation

$$r = 3.$$

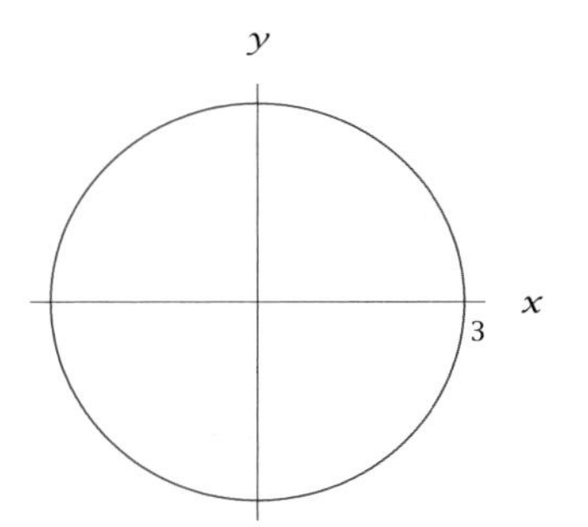

The circle described by the polar equation $r = 3$.

More generally, we have the following result.

Polar equation of a circle

If c is a positive number, then the polar equation $r = c$ describes the circle of radius c centered at the origin.

EXAMPLE 9

(a) Find inequalities in rectangular coordinates describing the region between the circles of radius 2 and 5, both centered at the origin.

(b) Find inequalities in polar coordinates describing the region between the circles of radius 2 and 5, both centered at the origin.

SOLUTION

(a) In rectangular coordinates in the xy-plane, this region can be described by the inequalities

$$4 < x^2 + y^2 < 25.$$

(b) In polar coordinates (r, θ), this region can be described by the inequalities

$$2 < r < 5.$$

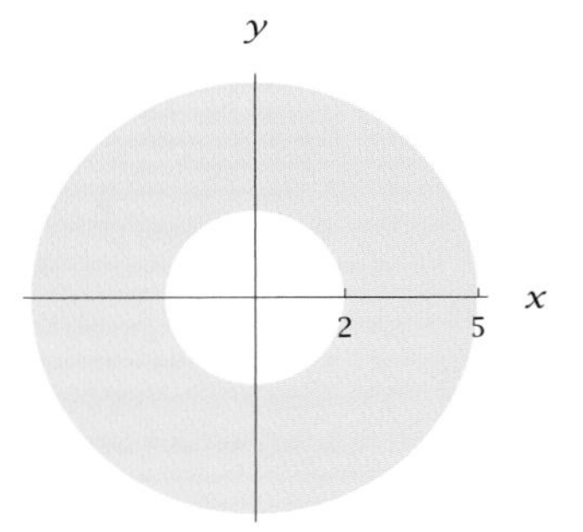

The region described by the polar inequalities $2 < r < 5$.

As we have seen, the set of points where the first polar coordinate r equals some constant is a circle. Now we look at the set of points where the second polar coordinate θ equals some constant.

EXAMPLE 10

Describe the set of points whose second polar coordinate θ equals $\frac{\pi}{4}$.

SOLUTION A point in the coordinate plane has second polar coordinate θ equal to $\frac{\pi}{4}$ if and only if the line segment from the origin to the point makes an angle of $\frac{\pi}{4}$ radians (or 45°) with the positive horizontal axis. Thus the equation

$$\theta = \frac{\pi}{4}$$

describes the ray shown here.

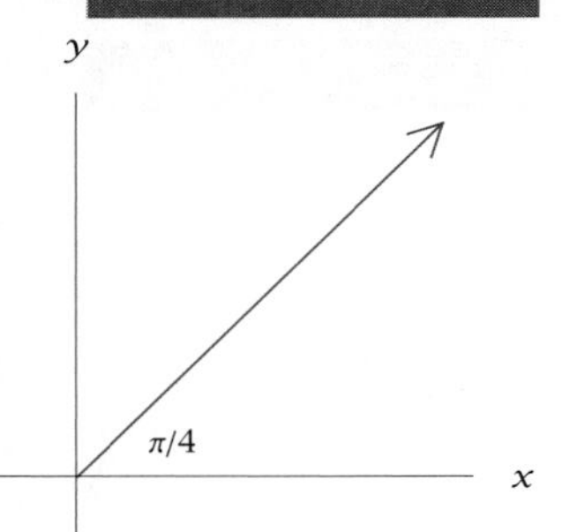

The ray described by the polar equation $\theta = \frac{\pi}{4}$. The ray continues without end; only part of it is shown.

More generally, we have the following result; here c is an arbitrary constant.

Polar equation of a ray

The polar equation $\theta = c$ describes the ray starting at the origin and making angle c with the positive horizontal axis.

So far our examples of equations in polar coordinates have involved only one of the two polar coordinates. Now we look at an example of a polar equation using both polar coordinates.

EXAMPLE 11

Consider the polar equation

$$r = \sin\theta.$$

(a) Plot enough points satisfied by this polar equation to see the shape of its graph.

(b) Convert this polar equation to an equation in rectangular coordinates and describe its graph.

SOLUTION

(a) The table below shows a few values of the second polar coordinate θ and the corresponding values of the first polar coordinate r that satisfy the equation $r = \sin\theta$:

θ	$r = \sin\theta$
0	0
$\frac{\pi}{4}$	$\frac{\sqrt{2}}{2}$
$\frac{\pi}{2}$	1
$\frac{3\pi}{4}$	$\frac{\sqrt{2}}{2}$
π	0

Some values of θ and $r = \sin\theta$.

The five points listed here (actually four points, because two of them are the origin) are plotted in the figure.

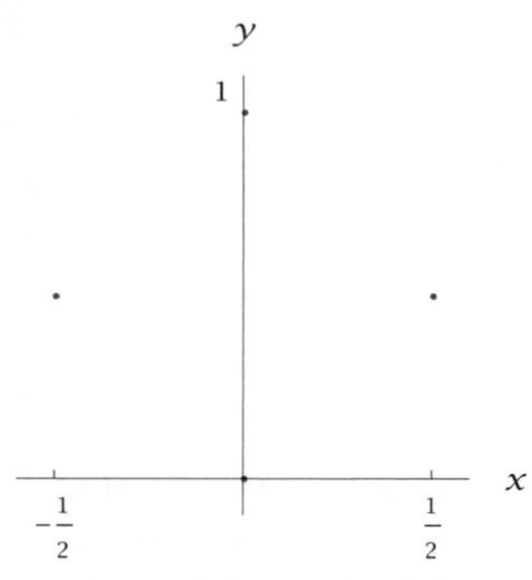

A partial graph of the polar equation $r = \sin\theta$.

The first point in the table above is the origin (because this point has $r = 0$). The second point in the table above is on the ray $\theta = \frac{\pi}{4}$, a distance of $\frac{\sqrt{2}}{2}$ from the origin. The third point in the table above is on the positive vertical axis (corresponding to $\theta = \frac{\pi}{2}$), a distance of 1 from the origin, and so on.

The figure above using just five points (two of which are the same) does not provide enough information to guess the shape of the graph. Thus instead of considering values of θ in the interval $[0, \pi]$ separated by $\frac{\pi}{4}$ as in the table above, we consider values of θ in the interval $[0, \pi]$ separated by $\frac{\pi}{50}$. Plotting the resulting points leads to the figure shown here.

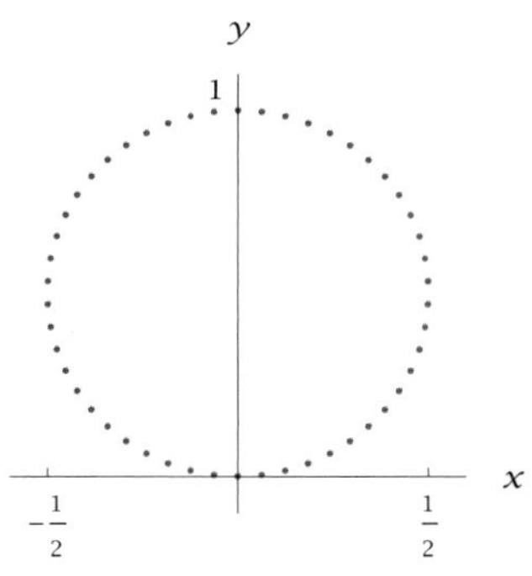

A partial graph of the polar equation $r = \sin\theta$.

The graph shown here appears to be part of a circle. In part (b), we will verify that the graph is indeed a circle.

(b) To convert our polar equation to an equation in rectangular coordinates, multiply both sides of the equation $r = \sin\theta$ by r, obtaining

$$r^2 = r\sin\theta.$$

Converting this equation to rectangular coordinates in the xy-plane gives

$$x^2 + y^2 = y.$$

Subtract y from both sides, getting

$$x^2 + y^2 - y = 0.$$

Completing the square, we can rewrite this equation as

$$x^2 + (y - \tfrac{1}{2})^2 = \tfrac{1}{4}.$$

Thus we see that the polar equation $r = \sin\theta$ describes a circle centered at $(0, \frac{1}{2})$ with radius $\frac{1}{2}$, which is consistent with the figure above.

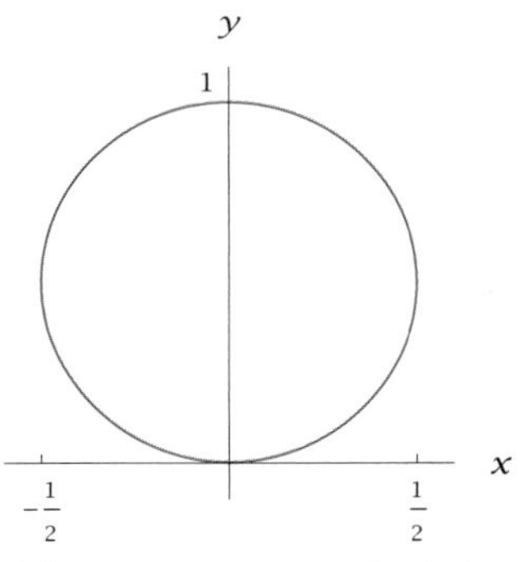

The complete graph of the polar equation $r = \sin\theta$.

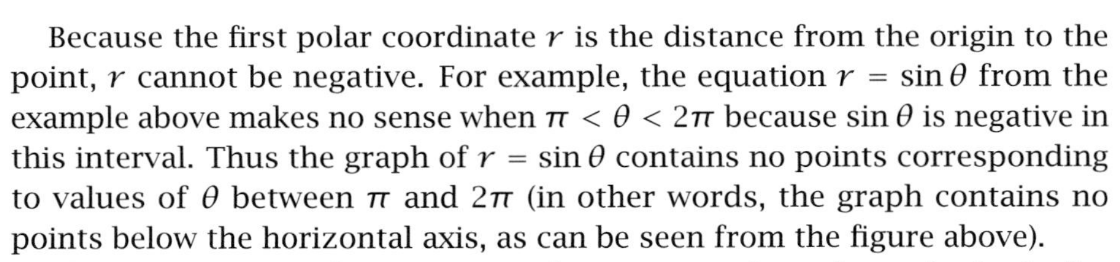

Because the first polar coordinate r is the distance from the origin to the point, r cannot be negative. For example, the equation $r = \sin\theta$ from the example above makes no sense when $\pi < \theta < 2\pi$ because $\sin\theta$ is negative in this interval. Thus the graph of $r = \sin\theta$ contains no points corresponding to values of θ between π and 2π (in other words, the graph contains no points below the horizontal axis, as can be seen from the figure above).

The restriction on θ to correspond to nonnegative values of r is similar to what happens when we graph the equation $y = \sqrt{x - 3}$. In graphing this equation, we do not consider values of x less than 3 because the equation $y = \sqrt{x - 3}$ makes no sense when $x < 3$. Similarly, the equation $r = \sin\theta$ makes no sense when $\pi < \theta < 2\pi$.

Some books allow r to be negative, which is contrary to the notion of r as the distance from the origin.

Graph of polar equations can often be described as parametric curves, as shown by the next example.

EXAMPLE 12

Describe the graph of the polar equation

$$r = 1 - \cos\theta$$

for θ in $[0, 2\pi]$ as a parametric curve.

SOLUTION A point with polar coordinates (r, θ) has rectangular coordinates $(r\cos\theta, r\sin\theta)$. Thus to graph the polar equation $r = 1 - \cos\theta$, we replace r by $1 - \cos\theta$ in $(r\cos\theta, r\sin\theta)$. This substitution shows that the graph of the polar equation $r = 1 - \cos\theta$ for θ in $[0, 2\pi]$ is the set of points of the form

$$((1 - \cos\theta)\cos\theta, (1 - \cos\theta)\sin\theta)$$

for θ in $[0, 2\pi]$. Thus we have described the graph of the polar equation $r = 1 - \cos\theta$ as a parametric curve (using θ rather than the usual variable t that is commonly used for parametric curves).

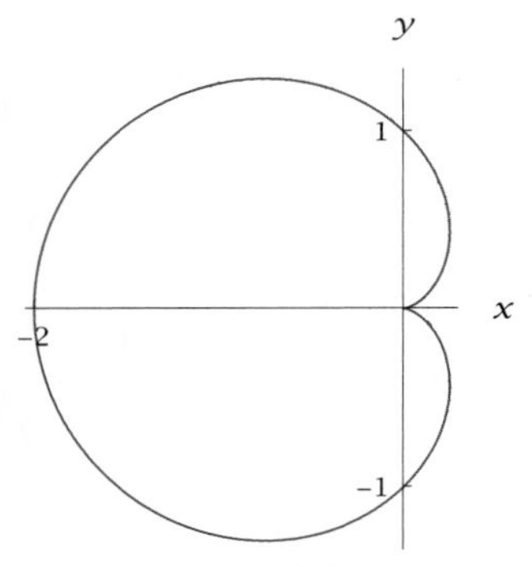

The graph of the polar equation $r = 1 - \cos\theta$.

Now that the graph of the polar equation $r = 1 - \cos\theta$ has been described as a parametric curve, a graphing calculator or software that can plot parametric curves can produce the graph. For example, the expression

```
parametric plot ( (1 - cos t) cos t, (1 - cos t) sin t ) for t=0 to 2pi
```

could be entered in a Wolfram|Alpha entry box (note that for convenience θ has been replaced by t), producing the graph shown here.

As shown in the next example, some graphing calculators can directly, without converting them to

EXAMPLE 13

Graph the polar equation $r = \theta$ for θ in $[0, 6\pi]$.

SOLUTION The graph of this polar equation can be obtained by entering

```
polar plot r = theta for theta=0 to 6pi
```

in a Wolfram|Alpha entry box, or use the appropriate input in other software or on a graphing calculator that can handle polar equations. Doing so should produce the spiral graph shown here.

The graph of the polar equation $r = \theta$ for θ in $[0, 6\pi]$.

Polar equations can lead to beautiful curves, as shown in the two previous examples. You should use Wolfram|Alpha or other software or a graphing calculator to experiment. Surprisingly complex curves can arise from simple polar equations, as shown below.

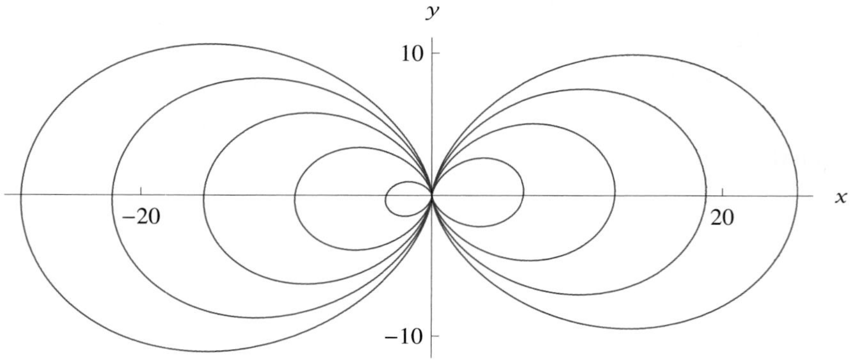

The graph of the polar equation $r = \theta\cos^2\theta$ for θ in $[\frac{\pi}{2}, \frac{19\pi}{2}]$.

EXERCISES

In Exercises 1–12, convert the point with the given polar coordinates to rectangular coordinates (x, y).

1. polar coordinates $(\sqrt{19}, 5\pi)$
2. polar coordinates $(3, 2^{1000}\pi)$
3. polar coordinates $(4, \frac{\pi}{2})$
4. polar coordinates $(5, -\frac{\pi}{2})$
5. polar coordinates $(6, -\frac{\pi}{4})$
6. polar coordinates $(7, \frac{\pi}{4})$
7. polar coordinates $(8, \frac{\pi}{3})$
8. polar coordinates $(9, -\frac{\pi}{3})$
9. polar coordinates $(10, \frac{\pi}{6})$
10. polar coordinates $(11, -\frac{\pi}{6})$
11. polar coordinates $(12, \frac{11\pi}{4})$
12. polar coordinates $(13, \frac{8\pi}{3})$

In Exercises 13–28, convert the point with the given rectangular coordinates to polar coordinates (r, θ)***. Use radians, and always choose the angle*** θ ***to be in the interval*** $(-\pi, \pi]$.

13. $(2, 0)$
14. $(-\sqrt{3}, 0)$
15. $(0, -\pi)$
16. $(0, 2\pi)$
17. $(3, 3)$
18. $(4, -4)$
19. $(-5, 5)$
20. $(-6, -6)$
21. $(3, 2)$
22. $(4, 7)$
23. $(3, -7)$
24. $(6, -5)$
25. $(-4, 1)$
26. $(-2, 5)$
27. $(-5, -2)$
28. $(-3, -6)$
29. Find the center and radius of the circle whose equation in polar coordinates is $r = 3\cos\theta$.
30. Find the center and radius of the circle whose equation in polar coordinates is $r = 10\sin\theta$.

PROBLEMS

31. Sketch the graph of the polar equation $r = 4$.
32. Sketch the graph of the polar equation $\theta = -\frac{\pi}{4}$.
33. Sketch the graph of the polar equation $r = \cos\theta + \sin\theta$.
34. Sketch the graph of the polar equation $r = 1 + \sin(15\theta)$.
35. Use the law of cosines to find a formula for the distance (in the usual rectangular coordinate plane) between the point with polar coordinates (r_1, θ_1) and the point with polar coordinates (r_2, θ_2).
36. Describe the set of points whose polar coordinates are equal to their rectangular coordinates.
37. What is the relationship between the point with polar coordinates $(5, 0.2)$ and the point with polar coordinates $(5, -0.2)$?
38. What is the relationship between the point with polar coordinates $(5, 0.2)$ and the point with polar coordinates $(5, 0.2 + \pi)$?
39. Explain why the second polar coordinate θ corresponding to a point with rectangular coordinates (x, y) can be chosen as follows:
    - If $x > 0$, then $\theta = \tan^{-1}\frac{y}{x}$.
    - If $x < 0$, then $\theta = \tan^{-1}\frac{y}{x} + \pi$.
    - If $x = 0$ and $y \geq 0$, then $\theta = \frac{\pi}{2}$.
    - If $x = 0$ and $y < 0$, then $\theta = -\frac{\pi}{2}$.

    Furthermore, explain why the formula above always leads to a choice of θ in the interval $[-\frac{\pi}{2}, \frac{3\pi}{2})$.
40. Give a formula for the second polar coordinate θ corresponding to a point with rectangular coordinates (x, y), similar in nature to the formula in the previous problem, that always leads to a choice of θ in the interval $[0, 2\pi)$.

WORKED-OUT SOLUTIONS *to Odd-numbered Exercises*

In Exercises 1–12, convert the point with the given polar coordinates to rectangular coordinates (x, y).

1. polar coordinates $(\sqrt{19}, 5\pi)$

SOLUTION We have

$$x = \sqrt{19}\cos(5\pi) \quad \text{and} \quad y = \sqrt{19}\sin(5\pi).$$

Subtracting even multiples of π does not change the value of cosine and sine. Because $5\pi - 4\pi = \pi$, we have $\cos(5\pi) = \cos\pi = -1$ and $\sin(5\pi) = \sin\pi = 0$. Thus the point in question has rectangular coordinates $(-\sqrt{19}, 0)$.

3. polar coordinates $(4, \frac{\pi}{2})$

SOLUTION We have

$$x = 4\cos\tfrac{\pi}{2} \quad \text{and} \quad y = 4\sin\tfrac{\pi}{2}.$$

Because $\cos\frac{\pi}{2} = 0$ and $\sin\frac{\pi}{2} = 1$, the point in question has rectangular coordinates $(0, 4)$.

5. polar coordinates $(6, -\frac{\pi}{4})$

SOLUTION We have

$$x = 6\cos(-\tfrac{\pi}{4}) \quad \text{and} \quad y = 6\sin(-\tfrac{\pi}{4}).$$

Because $\cos(-\frac{\pi}{4}) = \frac{\sqrt{2}}{2}$ and $\sin(-\frac{\pi}{4}) = -\frac{\sqrt{2}}{2}$, the point in question has rectangular coordinates $(3\sqrt{2}, -3\sqrt{2})$.

7. polar coordinates $(8, \frac{\pi}{3})$

SOLUTION We have

$$x = 8\cos\tfrac{\pi}{3} \quad \text{and} \quad y = 8\sin\tfrac{\pi}{3}.$$

Because $\cos\frac{\pi}{3} = \frac{1}{2}$ and $\sin\frac{\pi}{3} = \frac{\sqrt{3}}{2}$, the point in question has rectangular coordinates $(4, 4\sqrt{3})$.

9. polar coordinates $(10, \frac{\pi}{6})$

SOLUTION We have

$$x = 10\cos\tfrac{\pi}{6} \quad \text{and} \quad y = 10\sin\tfrac{\pi}{6}.$$

Because $\cos\frac{\pi}{6} = \frac{\sqrt{3}}{2}$ and $\sin\frac{\pi}{6} = \frac{1}{2}$, the point in question has rectangular coordinates $(5\sqrt{3}, 5)$.

11. polar coordinates $(12, \frac{11\pi}{4})$

SOLUTION We have

$$x = 12\cos\tfrac{11\pi}{4} \quad \text{and} \quad y = 12\sin\tfrac{11\pi}{4}.$$

Because $\cos\frac{11\pi}{4} = -\frac{\sqrt{2}}{2}$ and $\sin\frac{11\pi}{4} = \frac{\sqrt{2}}{2}$, the point in question has rectangular coordinates $(-6\sqrt{2}, 6\sqrt{2})$.

In Exercises 13–28, convert the point with the given rectangular coordinates to polar coordinates (r, θ). Use radians, and always choose the angle θ to be in the interval $(-\pi, \pi]$.

13. $(2, 0)$

SOLUTION The point $(2, 0)$ is on the positive x-axis, 2 units from the origin. Thus this point has polar coordinates $(2, 0)$.

15. $(0, -\pi)$

SOLUTION The point $(0, -\pi)$ is on the negative y-axis, π units from the origin. Thus this point has polar coordinates $(\pi, -\frac{\pi}{2})$.

17. $(3, 3)$

SOLUTION We have

$$r = \sqrt{3^2 + 3^2} = \sqrt{3^2 \cdot 2} = \sqrt{3^2}\sqrt{2} = 3\sqrt{2}.$$

The point $(3, 3)$ is on the portion of the line $y = x$ that makes a $45°$ angle with the positive x-axis. Thus $\theta = \frac{\pi}{4}$.

Hence this point has polar coordinates $(3\sqrt{2}, \frac{\pi}{4})$.

19. $(-5, 5)$

SOLUTION We have

$$r = \sqrt{5^2 + (-5)^2} = \sqrt{5^2 \cdot 2} = \sqrt{5^2}\sqrt{2} = 5\sqrt{2}.$$

The point $(-5, 5)$ is on the portion of the line $y = -x$ that makes a $135°$ angle with the positive x-axis. Thus $\theta = \frac{3\pi}{4}$.

Hence this point has polar coordinates $(5\sqrt{2}, \frac{3\pi}{4})$.

21. (3, 2)

SOLUTION We have

$$r = \sqrt{3^2 + 2^2} = \sqrt{13} \approx 3.61.$$

Because both coordinates of $(3, 2)$ are positive, we have

$$\theta = \tan^{-1} \tfrac{2}{3} \approx 0.588 \text{ radians.}$$

Hence this point has polar coordinates approximately $(3.61, 0.588)$.

23. (3, −7)

SOLUTION We have

$$r = \sqrt{3^2 + (-7)^2} = \sqrt{58} \approx 7.62.$$

Because the first coordinate of $(3, -7)$ is positive, we have

$$\theta = \tan^{-1}(\tfrac{-7}{3}) \approx -1.166 \text{ radians.}$$

Hence this point has polar coordinates approximately $(7.62, -1.166)$.

25. (−4, 1)

SOLUTION We have

$$r = \sqrt{(-4)^2 + 1^2} = \sqrt{17} \approx 4.12.$$

Because the first coordinate of $(-4, 1)$ is negative and the second coordinate is positive, we have

$$\begin{aligned}\theta = \tan^{-1}(\tfrac{1}{-4}) + \pi &= \tan^{-1}(-\tfrac{1}{4}) + \pi \\ &\approx 2.897 \text{ radians.}\end{aligned}$$

Hence this point has polar coordinates approximately $(4.12, 2.897)$.

27. (−5, −2)

SOLUTION We have

$$r = \sqrt{(-5)^2 + (-2)^2} = \sqrt{29} \approx 5.39.$$

Because both coordinates of $(-5, -2)$ are negative, we have

$$\begin{aligned}\theta = \tan^{-1}(\tfrac{-2}{-5}) - \pi &= \tan^{-1} \tfrac{2}{5} - \pi \\ &\approx -2.761 \text{ radians.}\end{aligned}$$

Hence this point has polar coordinates approximately $(5.39, -2.761)$.

29. Find the center and radius of the circle whose equation in polar coordinates is $r = 3\cos\theta$.

SOLUTION Multiply both sides of the equation $r = 3\cos\theta$ by r, obtaining

$$r^2 = 3r\cos\theta.$$

Now convert the equation above to rectangular coordinates in the xy-plane, getting

$$x^2 + y^2 = 3x.$$

Subtract $3x$ from both sides, getting

$$x^2 - 3x + y^2 = 0.$$

Completing the square, we can rewrite this equation as

$$(x - \tfrac{3}{2})^2 + y^2 = \tfrac{9}{4}.$$

Thus we see that the polar equation $r = 3\cos\theta$ describes a circle centered at $(\tfrac{3}{2}, 0)$ with radius $\tfrac{3}{2}$.

11.4 *Vectors*

LEARNING OBJECTIVES

By the end of this section you should be able to

- determine whether two vectors are equal;
- find the magnitude and direction of a vector from its coordinates;
- add and subtract two vectors algebraically and geometrically;
- compute the product of a number and a vector algebraically and geometrically;
- compute the dot product of two vectors;
- compute the angle between two vectors.

An Algebraic and Geometric Introduction to Vectors

To see how vectors naturally arise, consider weather data at a specific location and at a specific time. One key item of weather data is the temperature, which is a number that could be positive or negative (for example, 14 degrees Fahrenheit or -10 degrees Celsius, depending on the units used). Another key item of weather data is the wind velocity, which consists of a magnitude that must be a nonnegative number (for example, 10 miles per hour) and a direction (for example, northwest).

Measurements that have both a magnitude and a direction are common enough to deserve their own terminology:

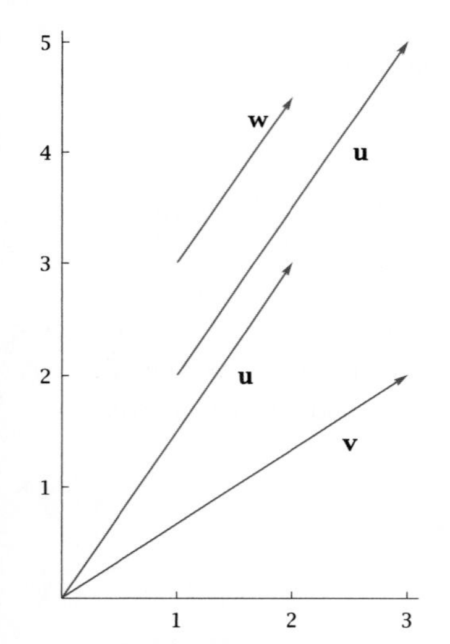

Vector

A **vector** is characterized by its magnitude and its direction. Usually a vector is drawn as an arrow:

- the length of the arrow is the **magnitude** of the vector;
- the direction of the arrowhead indicates the **direction** of the vector;
- two vectors are equal if and only if they have the same magnitude and the same direction.

EXAMPLE 1

Symbols denoting vectors appear in boldface in this book to emphasize that they denote vectors, not numbers.

The figure above shows vectors **u**, **v**, and **w**.

(a) Explain why the two vectors labeled **u** are equal to each other.

(b) Explain why $\mathbf{v} \neq \mathbf{u}$.

(c) Explain why $\mathbf{w} \neq \mathbf{u}$.

SOLUTION

(a) The two vectors with the label **u** have the same length and their arrows are parallel and point in the same direction. Because these two vectors have the same magnitude and the same direction, they are equal vectors and thus it is appropriate to give them the same label **u**.

(b) The vector **v** shown above has the same magnitude as **u** but points in a different direction (the arrows are not parallel); thus $\mathbf{v} \neq \mathbf{u}$.

(c) The vector **w** shown above has the same direction as **u** (parallel arrows pointing in the same direction) but has a different magnitude; thus $\mathbf{w} \neq \mathbf{u}$.

A vector is determined by its initial point and its endpoint. For example, the vector **v** shown above has initial point the origin $(0,0)$ and has endpoint $(3,2)$. One version of the vector **u** shown above has initial point the origin $(0,0)$ and has endpoint $(2,3)$; the other version of the vector **u** shown above has initial point $(1,2)$ and has endpoint $(3,5)$.

Sometimes a vector is specified by giving only the endpoint, with the assumption that the initial point is the origin. For example, the vector **v** shown above can be identified as $(3,2)$, with the understanding that the origin is the initial point. In other words, sometimes we think of $(3,2)$ as a point in the coordinate plane, and sometimes we think of $(3,2)$ as the vector from the origin to that point.

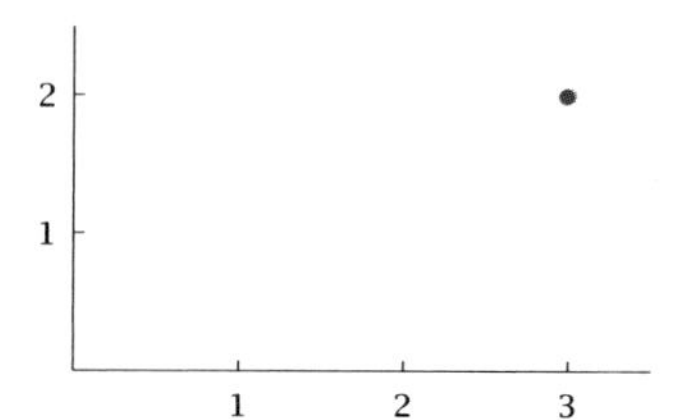

The notation $(3,2)$ can be used to denote the point shown above.

Notation for vectors with initial point at the origin

If a and b are real numbers, then (a,b) can denote either a point or a vector, depending on the context. In other words, (a,b) can be used as notation for either of the two following objects:

- the point in the coordinate plane whose first coordinate is a and whose second coordinate is b;
- the vector whose initial point is the origin and whose endpoint has first coordinate a and second coordinate b.

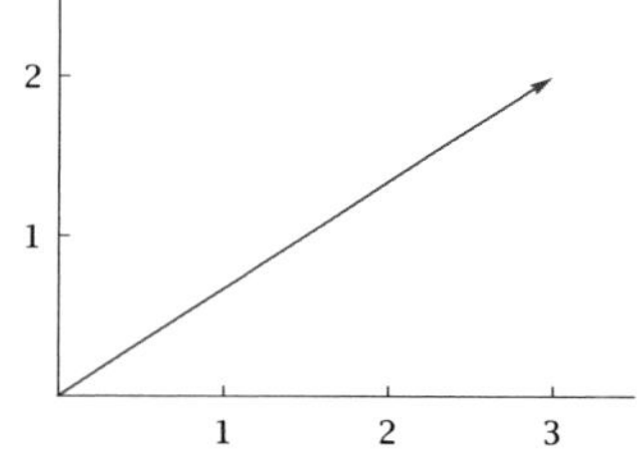

The notation $(3,2)$ can also be used to denote the vector shown above.

Polar coordinates allow us to be more precise about what we mean by the magnitude and direction of a vector:

Magnitude and direction of a vector

Suppose a vector **u** is positioned with its initial point at the origin. If the endpoint of **u** has polar coordinates (r,θ), then

- the **magnitude** of **u**, denoted $|\mathbf{u}|$, is defined to equal r;
- the **direction** of **u** is determined by θ, which is the angle that **u** makes with the positive horizontal axis.

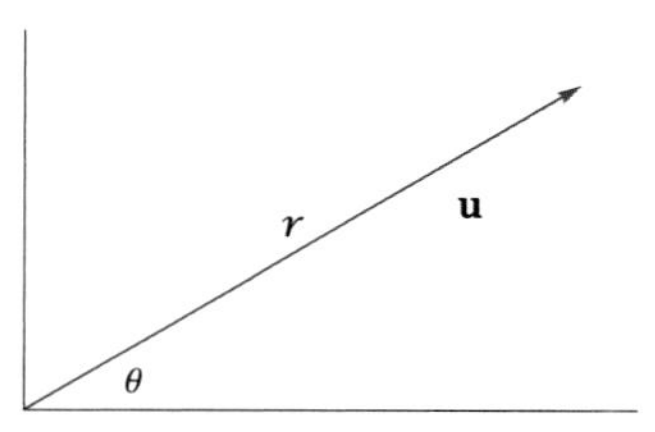

*The endpoint of this vector **u** has polar coordinates (r,θ).*

The result below simply repeats the conversion from rectangular to polar coordinates that we saw in the previous section:

Computing the magnitude and direction of a vector

If $\mathbf{u} = (a, b)$, then

- $|\mathbf{u}| = \sqrt{a^2 + b^2}$;
- an angle θ that determines the direction of $\mathbf{u}$ satisfies the equation $\tan\theta = \frac{b}{a}$, where θ must be chosen so that $\cos\theta$ has the same sign as a and $\sin\theta$ has the same sign as b.

In the last bullet item above, as usual we must exclude the case where $a = 0$ to avoid division by 0.

EXAMPLE 2

Suppose the vector u shown above has endpoint $(5, 3)$. Find the magnitude of u and an angle that determines the direction of u.

SOLUTION We have

$$|\mathbf{u}| = \sqrt{5^2 + 3^2} = \sqrt{34} \approx 5.83$$

and

$$\theta = \tan^{-1}\tfrac{3}{5} \approx 0.54.$$

Because the second polar coordinate θ is not unique, we could also add any integer multiple of 2π to the value of θ chosen above.

Vector Addition

Two vectors can be added, producing another vector. The following definition presents vector addition from the viewpoint of a vector as an arrow and also from the viewpoint of identifying a vector with its endpoint (assuming that the initial point is the origin):

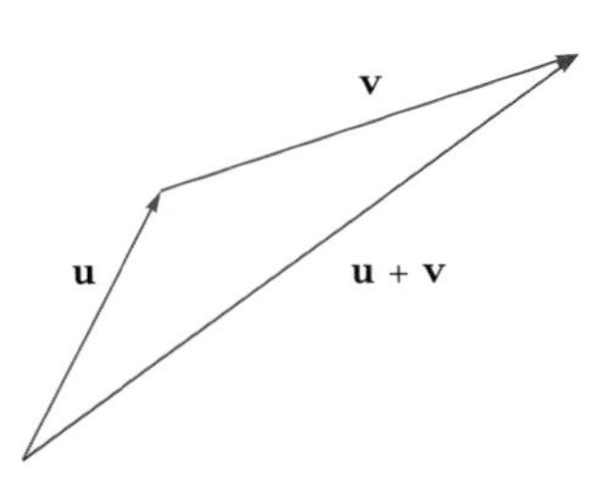

Vector addition

- If the endpoint of a vector $\mathbf{u}$ coincides with the initial point of a vector $\mathbf{v}$, then the vector $\mathbf{u} + \mathbf{v}$ has the same initial point as $\mathbf{u}$ and the same endpoint as $\mathbf{v}$.
- If $\mathbf{u} = (a, b)$ and $\mathbf{v} = (c, d)$, then $\mathbf{u} + \mathbf{v} = (a + c, b + d)$.

EXAMPLE 3

Suppose $\mathbf{u} = (1, 2)$ and $\mathbf{v} = (3, 1)$.

(a) Draw a figure illustrating the sum of $\mathbf{u}$ and $\mathbf{v}$ as arrows.

(b) Compute the sum $\mathbf{u} + \mathbf{v}$ using coordinates.

SOLUTION

(a) The figure below on the left shows the two vectors **u** and **v**, both with their initial point at the origin. In the figure below in the middle, the vector **v** has been moved parallel to its original position so that its initial point now coincides with the endpoint of **u**. The figure below on the right shows that the vector **u** + **v** is the vector with the same initial point as **u** and the same endpoint as the second version of **v**.

To add two vectors as arrows, move one vector parallel to itself so that its initial point coincides with the endpoint of the other vector.

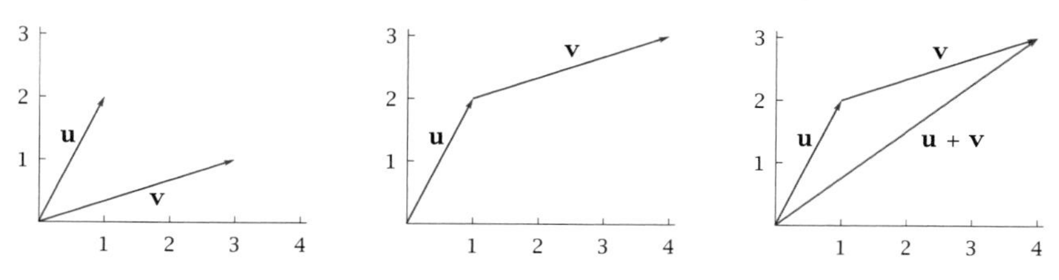

Here **v** *is moved parallel to itself so that the initial point of* **v** *coincides with the endpoint of* **u**.

(b) The coordinates of **u** + **v** are obtained by adding the corresponding coordinates of **u** and **v**. Thus **u** + **v** = $(1, 2) + (3, 1) = (4, 3)$. Note that $(4, 3)$ is the endpoint of the red vector above on the right.

Vector addition satisfies the usual commutative and associative properties that are expected for an operation of addition. In other words,

$$\mathbf{u} + \mathbf{v} = \mathbf{v} + \mathbf{u} \quad \text{and} \quad (\mathbf{u} + \mathbf{v}) + \mathbf{w} = \mathbf{u} + (\mathbf{v} + \mathbf{w})$$

for all vectors **u**, **v**, and **w**. The figure in the margin shows why vector addition is commutative.

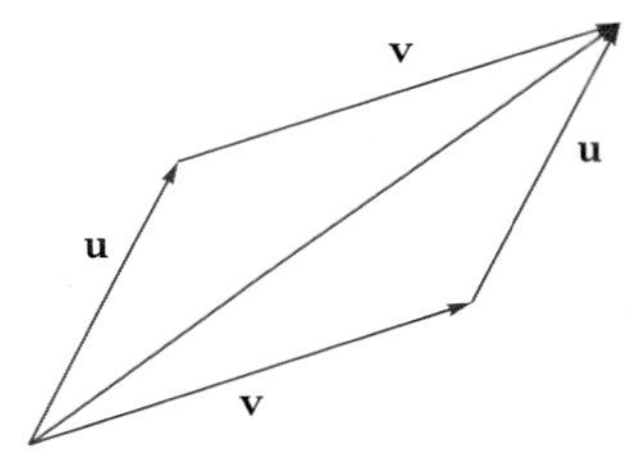

The vector shown here as the red diagonal of the parallelogram equals **u** + **v** *and also equals* **v** + **u**.

The zero vector, denoted with boldface **0**, is the vector whose magnitude is 0. The direction of the zero vector can be chosen to be anything convenient and is irrelevant because this vector has magnitude 0. In terms of coordinates, the zero vector equals $(0, 0)$. For every vector **u**, we have

$$\mathbf{u} + \mathbf{0} = \mathbf{0} + \mathbf{u} = \mathbf{u}.$$

An important application of vector addition is the computation of the effect of wind upon an airplane. Consider, for example, an airplane whose engines allow the airplane to travel at 500 miles per hour when there is no wind. With a 50 miles per hour headwind (winds are much stronger at airplane heights than at ground level), the airplane can travel only 450 miles per hour relative to the ground. With a 50 miles per hour tailwind, the airplane can travel 550 miles per hour relative to the ground. This behavior explains, for example, why it usually takes an hour longer to fly from Miami to Los Angeles than to fly from Los Angeles to Miami (the wind at airplane heights is usually from west to east).

If the wind direction is neither exactly in the same direction as the airplane's flight nor exactly in the opposite direction, then vector addition is used to determine the effect of the wind upon the airplane. Specifically, the wind vector is added to the airspeed vector (the vector giving the airplane's speed and direction relative to the wind) to produce the groundspeed vector (the vector giving the airplane's speed and direction relative to the ground). The next example illustrates this idea.

EXAMPLE 4

Suppose the wind at airplane heights is 50 miles per hour (relative to the ground) moving 20° south of east. Relative to the wind, an airplane is flying at 300 miles per hour in a direction 35° north of the wind. Find the speed and direction of the airplane relative to the ground.

SOLUTION Let's call the wind vector **w**. Thus **w** has magnitude 50 and direction −20°. This implies that the wind vector **w** has coordinates $(50\cos(-20°), 50\sin(-20°))$.

Let **a** denote the vector giving the motion of the airplane relative to the wind (this vector is called the **airspeed vector**). Relative to the wind, the airplane is heading 35° north, which means that **a** has magnitude 300 and direction 15°. Thus the airspeed vector **a** has coordinates $(300\cos 15°, 300\sin 15°)$.

This figure is drawn to scale for this data. One version of **w** *has its initial point at the same location as the initial point of* **a** *(think of this point as the origin). The second version of* **w** *has its initial point at the endpoint of* **a** *so that we can visualize the sum* $\mathbf{g} = \mathbf{a} + \mathbf{w}$.

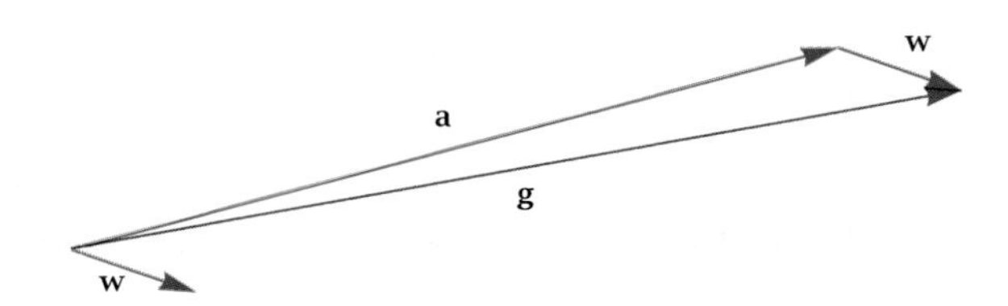

Let **g** denote the vector giving the motion of the airplane relative to the ground (this vector is called the **groundspeed vector**). The properties of motion imply that the groundspeed vector equals the airspeed vector plus the wind vector. In other words, $\mathbf{g} = \mathbf{a} + \mathbf{w}$.

In terms of coordinates, we have

$$\begin{aligned}\mathbf{g} &= \mathbf{a} + \mathbf{w}\\ &= (300\cos 15°, 300\sin 15°) + (50\cos 20°, -50\sin 20°)\\ &= (300\cos 15° + 50\cos 20°, 300\sin 15° - 50\sin 20°)\\ &\approx (336.762, 60.5447).\end{aligned}$$

Now the speed of the airplane relative to the ground is the magnitude of **g**:

$$\begin{aligned}|\mathbf{g}| &\approx \sqrt{336.762^2 + 60.5447^2}\\ &\approx 342.2.\end{aligned}$$

The direction of the airplane relative to the ground is the direction of g, which is approximately $\tan^{-1}\frac{60.5447}{336.762}$, which is approximately 0.178 radians, which is approximately 10.2°.

Conclusion: Relative to the ground, the airplane is traveling at approximately 342.2 miles per hour in a direction 10.2° north of east.

Vector Subtraction

Vectors have additive inverses, just as numbers do. The following definition presents the additive inverse from the viewpoint of a vector as an arrow and from the viewpoint of identifying a vector with its coordinates:

Additive inverse

- If $\mathbf{u}$ is a vector, then $-\mathbf{u}$ has the same magnitude as $\mathbf{u}$ and has the opposite direction.
- If $\mathbf{u}$ has polar coordinates (r, θ), then $-\mathbf{u}$ has polar coordinates $(r, \theta + \pi)$.
- If $\mathbf{u} = (a, b)$, then $-\mathbf{u} = (-a, -b)$.

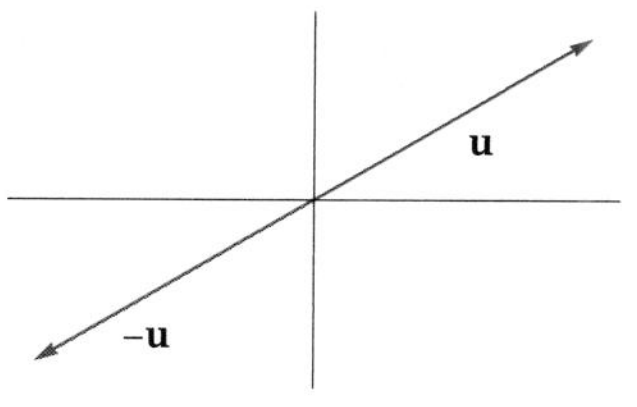

A vector $\mathbf{u}$ *and its additive inverse* $-\mathbf{u}$.

Make sure you understand why the definition above implies that

$$\mathbf{u} + (-\mathbf{u}) = \mathbf{0}$$

for every vector $\mathbf{u}$.

Two vectors can be subtracted, producing another vector. The following definition presents vector subtraction from the viewpoint of a vector as an arrow and from the viewpoint of identifying a vector with its endpoint (assuming that the initial point is the origin):

Vector subtraction

- If $\mathbf{u}$ and $\mathbf{v}$ are vectors, then the difference $\mathbf{u} - \mathbf{v}$ is defined by

$$\mathbf{u} - \mathbf{v} = \mathbf{u} + (-\mathbf{v}).$$

- If vectors $\mathbf{u}$ and $\mathbf{v}$ are positioned to have the same initial point, then $\mathbf{u} - \mathbf{v}$ is the vector whose initial point is the endpoint of $\mathbf{v}$ and whose endpoint is the endpoint of $\mathbf{u}$.
- If $\mathbf{u} = (a, b)$ and $\mathbf{v} = (c, d)$, then $\mathbf{u} - \mathbf{v} = (a - c, b - d)$.

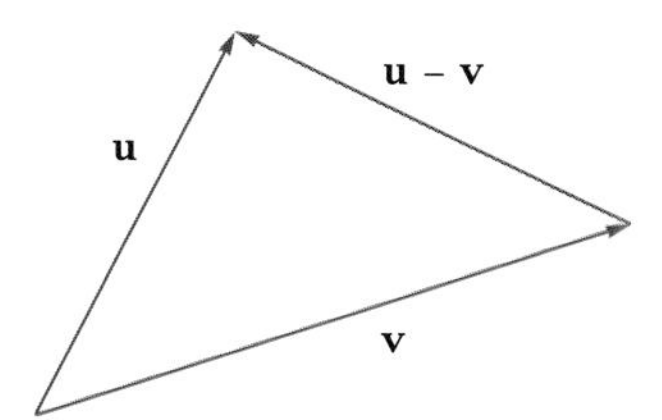

To figure out in which direction the arrow for $\mathbf{u} - \mathbf{v}$ *points, choose the direction that makes* $\mathbf{v} + (\mathbf{u} - \mathbf{v})$ *equal to* $\mathbf{u}$.

EXAMPLE 5

Suppose $\mathbf{u} = (1, 2)$ and $\mathbf{v} = (3, 1)$.

(a) Draw a figure using arrows illustrating the difference $\mathbf{u} - \mathbf{v}$.

(b) Compute the difference $\mathbf{u} - \mathbf{v}$ using coordinates.

SOLUTION

(a) The figure below on the left shows the two vectors $\mathbf{u}$ and $\mathbf{v}$, both with their initial point at the origin. The figure below in the center shows that the vector $\mathbf{u} - \mathbf{v}$ is the vector whose initial point is the endpoint of $\mathbf{v}$ and whose endpoint is the endpoint of $\mathbf{u}$.

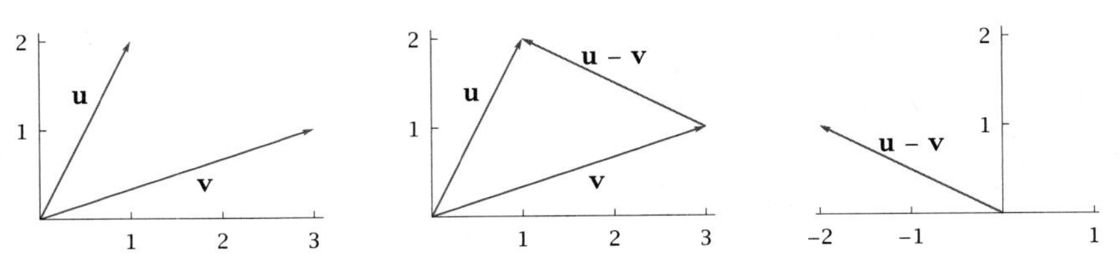

To subtract two vectors, position them to have the same initial point.

(b) The coordinates of $\mathbf{u}-\mathbf{v}$ are obtained by subtracting the corresponding coordinates of $\mathbf{u}$ and $\mathbf{v}$. Thus $\mathbf{u}-\mathbf{v}=(1,2)-(3,1)=(-2,1)$. The figure above on the right shows $\mathbf{u}-\mathbf{v}$ with its initial point at the origin and its endpoint at $(-2,1)$.

The next example shows a practical application of vector subtraction.

EXAMPLE 6

Suppose the wind at airplane heights is 60 miles per hour (relative to the ground) moving 25° south of east. An airplane wants to fly directly west at 400 miles per hour relative to the ground. Find the speed and direction that the airplane must fly relative to the wind.

SOLUTION The wind vector $\mathbf{w}$ has coordinates $(60\cos(-25°), 60\sin(-25°))$.

We want the groundspeed vector $\mathbf{g}$ to have coordinates $(-400, 0)$.

Let $\mathbf{a}$ denote the airspeed vector, which shows the motion of the airplane relative to the wind. As usual, $\mathbf{g}=\mathbf{a}+\mathbf{w}$. Solving for the airspeed vector $\mathbf{a}$, we have

$$\begin{aligned}\mathbf{a} &= \mathbf{g}-\mathbf{w}\\ &= (-400,0)-(60\cos(-25°),60\sin(-25°))\\ &= (-400-60\cos 25°, 60\sin 25°)\\ &\approx (-454.38, 25.357).\end{aligned}$$

This figure is drawn to scale for this data.

g

w

a

The speed of the airplane relative to the wind is the magnitude of $\mathbf{a}$:

$$\begin{aligned}|\mathbf{a}| &\approx \sqrt{(-454.38)^2+25.357^2}\\ &\approx 455.1.\end{aligned}$$

Because $\tan^{-1}\frac{25.357}{-454.38}$ is in the interval $(-\frac{\pi}{2},0)$, we add π to it to obtain an angle in the appropriate interval with the same tangent.

The direction of the airplane relative to the wind is the direction of $\mathbf{a}$. As can be seen in the figure, the angle that $\mathbf{a}$ makes with the positive horizontal axis is in the interval $(\frac{\pi}{2},\pi)$. Thus this angle is approximately $\pi+\tan^{-1}\frac{25.357}{-454.38}$, which is approximately 3.0858 radians, which is approximately 176.8°, which is 3.2° north of west.

Conclusion: Relative to the wind, the airplane must fly at approximately 455.1 miles per hour in a direction that makes an angle of 201.8° with the wind direction (because $25°+176.8°=201.8°$), as measured counterclockwise from the wind vector.

Scalar Multiplication

The word **scalar** is simply a fancy word for *number*. The term **scalar multiplication** refers to the operation defined below of multiplying a vector by a scalar, producing a vector.

Often the word scalar *is used to emphasize that a quantity is a number rather than a vector.*

Scalar multiplication

Suppose t is a real number and $\mathbf{u}$ is a vector.

- The vector $t\mathbf{u}$ has magnitude $|t|$ times the magnitude of $\mathbf{u}$; thus $|t\mathbf{u}| = |t||\mathbf{u}|$.
  - If $t > 0$, then $t\mathbf{u}$ has the same direction as $\mathbf{u}$.
  - If $t < 0$, then $t\mathbf{u}$ has the opposite direction of $\mathbf{u}$.
- Suppose $\mathbf{u}$ has polar coordinates (r, θ).
  - If $t > 0$, then tu has polar coordinates (tr, θ).
  - If $t < 0$, then tu has polar coordinates $(-tr, \theta + \pi)$.
- If $\mathbf{u} = (a, b)$, then $t\mathbf{u} = (ta, tb)$.

EXAMPLE 7

Suppose $\mathbf{u} = (2, 1)$.

(a) Draw a figure showing $\mathbf{u}$, $2\mathbf{u}$, and $-2\mathbf{u}$.

(b) Compute $2\mathbf{u}$ and $-2\mathbf{u}$ using coordinates.

SOLUTION

(a) The figure below on the left shows $\mathbf{u}$. The figure below in the middle shows that $2\mathbf{u}$ is the vector having twice the magnitude of $\mathbf{u}$ and having the same direction as $\mathbf{u}$. The figure below on the right shows that $-2\mathbf{u}$ is the vector having twice the magnitude of $\mathbf{u}$ and having the opposite direction of $\mathbf{u}$.

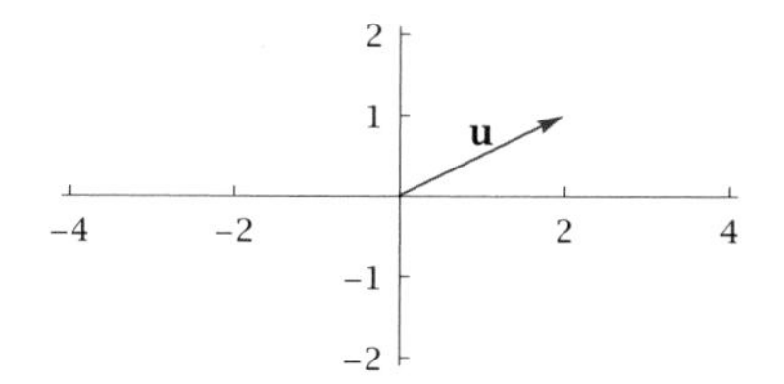

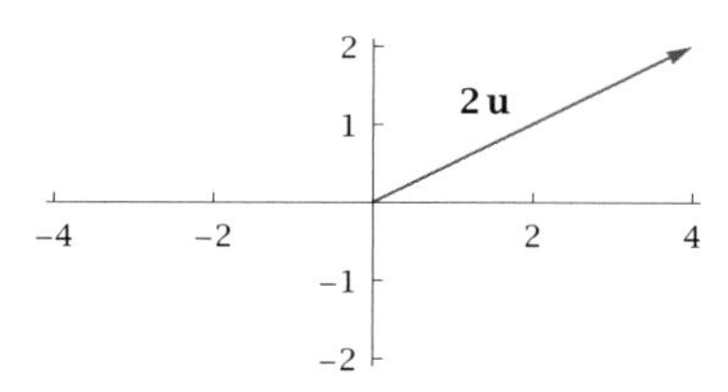

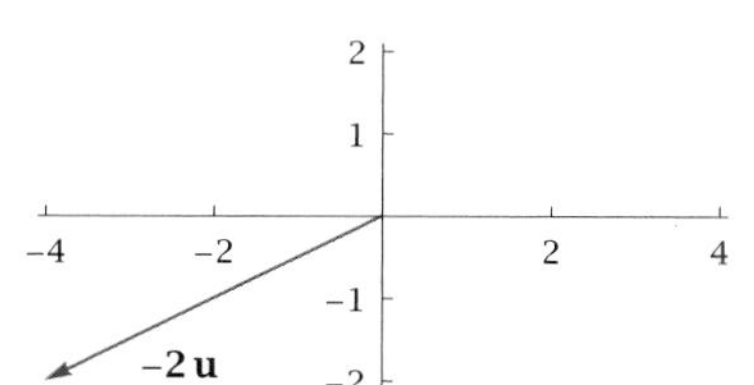

(b) The coordinates of $2\mathbf{u}$ are obtained by multiplying the corresponding coordinates of $\mathbf{u}$ by 2, and the coordinates of $-2\mathbf{u}$ are obtained by multiplying the corresponding coordinates of $\mathbf{u}$ by -2. Thus we have $2\mathbf{u} = (4, 2)$ and $-2\mathbf{u} = (-4, -2)$.

Dividing a vector by a nonzero scalar c is the same as multiplying by $\frac{1}{c}$. For example, $\frac{\mathbf{u}}{2}$ should be interpreted to mean $\frac{1}{2}\mathbf{u}$. Note that division by a vector is not defined.

The Dot Product

We have defined the sum and difference of two vectors, and the scalar product of a number and a vector. Each of those operations produces another vector. Now we turn to another operation, called the dot product, that produces a number from two vectors. We begin with a definition in terms of coordinates; soon we will also see a formula from the viewpoint of vectors as arrows.

Always remember that the dot product of two vectors is a number, not a vector.

Dot product

Suppose $\mathbf{u} = (a, b)$ and $\mathbf{v} = (c, d)$. Then the **dot product** of **u** and **v**, denoted $\mathbf{u} \cdot \mathbf{v}$, is defined by

$$\mathbf{u} \cdot \mathbf{v} = ac + bd.$$

Thus to compute the dot product of two vectors, multiply together the first coordinates, multiply together the second coordinates, and then add these two products.

EXAMPLE 8 Suppose $\mathbf{u} = (2, 3)$ and $\mathbf{v} = (5, 4)$. Compute $\mathbf{u} \cdot \mathbf{v}$.

SOLUTION Using the formula above, we have

$$\mathbf{u} \cdot \mathbf{v} = 2 \cdot 5 + 3 \cdot 4 = 10 + 12 = 22.$$

The dot product has the following pleasant algebraic properties:

Algebraic properties of the dot product

Suppose **u**, **v**, and **w** are vectors and t is a real number. Then

- $\mathbf{u} \cdot \mathbf{v} = \mathbf{v} \cdot \mathbf{u}$ (commutativity);
- $\mathbf{u} \cdot (\mathbf{v} + \mathbf{w}) = \mathbf{u} \cdot \mathbf{v} + \mathbf{u} \cdot \mathbf{w}$ (distributive property);
- $(t\mathbf{u}) \cdot \mathbf{v} = \mathbf{u} \cdot (t\mathbf{v}) = t(\mathbf{u} \cdot \mathbf{v})$;
- $\mathbf{u} \cdot \mathbf{u} = |\mathbf{u}|^2$.

To verify the last property above, suppose $\mathbf{u} = (a, b)$. Then

$$\mathbf{u} \cdot \mathbf{u} = a^2 + b^2 = \left(\sqrt{a^2 + b^2}\right)^2 = |\mathbf{u}|^2,$$

as desired. The verifications of the first three properties are left as similarly easy problems for the reader.

The next result gives a remarkably useful formula for computing $\mathbf{u} \cdot \mathbf{v}$ in terms of the magnitude of **u**, the magnitude of **v**, and the angle between these two vectors.

Computing the dot product geometrically

If **u** and **v** are vectors with the same initial point, then

$$\mathbf{u} \cdot \mathbf{v} = |\mathbf{u}|\,|\mathbf{v}| \cos \theta,$$

where θ is the angle between **u** and **v**.

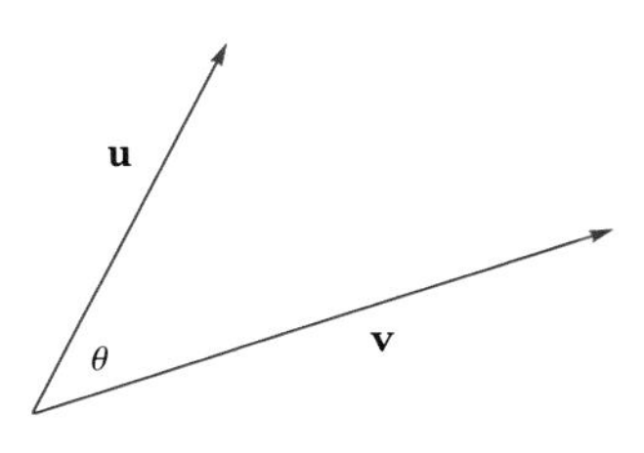

To verify the formula above, first draw the vector $\mathbf{u} - \mathbf{v}$, whose initial point is the endpoint of **v** and whose endpoint is the endpoint of **u**, as shown here. Next, use the algebraic properties of the dot product to compute a formula for $|\mathbf{u} - \mathbf{v}|^2$ as follows:

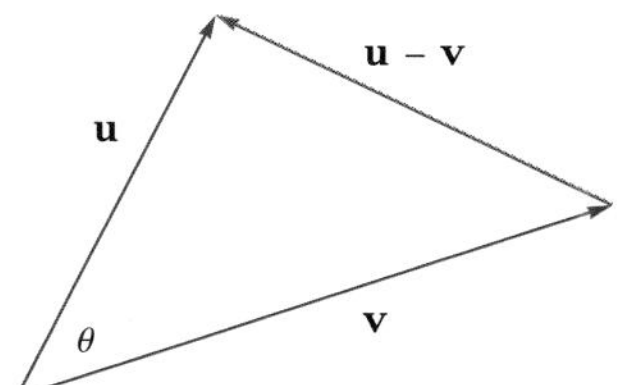

$$\begin{aligned} |\mathbf{u} - \mathbf{v}|^2 &= (\mathbf{u} - \mathbf{v}) \cdot (\mathbf{u} - \mathbf{v}) \\ &= \mathbf{u} \cdot (\mathbf{u} - \mathbf{v}) - \mathbf{v} \cdot (\mathbf{u} - \mathbf{v}) \\ &= \mathbf{u} \cdot \mathbf{u} - \mathbf{u} \cdot \mathbf{v} - \mathbf{v} \cdot \mathbf{u} + \mathbf{v} \cdot \mathbf{v} \\ &= |\mathbf{u}|^2 - 2\mathbf{u} \cdot \mathbf{v} + |\mathbf{v}|^2. \end{aligned}$$

Now apply the law of cosines to the triangle above, getting

$$|\mathbf{u} - \mathbf{v}|^2 = |\mathbf{u}|^2 + |\mathbf{v}|^2 - 2|\mathbf{u}|\,|\mathbf{v}| \cos \theta.$$

Finally, set the two expressions that we have obtained for $|\mathbf{u} - \mathbf{v}|^2$ equal to each other, getting

$$|\mathbf{u}|^2 - 2\mathbf{u} \cdot \mathbf{v} + |\mathbf{v}|^2 = |\mathbf{u}|^2 + |\mathbf{v}|^2 - 2|\mathbf{u}|\,|\mathbf{v}| \cos \theta.$$

Subtract $|\mathbf{u}|^2 + |\mathbf{v}|^2$ from both sides of the equation above, and then divide both sides by -2, getting $\mathbf{u} \cdot \mathbf{v} = |\mathbf{u}|\,|\mathbf{v}| \cos \theta$, completing our derivation of this remarkable formula.

EXAMPLE 9

Find the angle between the vectors $(1, 2)$ and $(3, 1)$.

SOLUTION Let $\mathbf{u} = (1, 2)$, let $\mathbf{v} = (3, 1)$, and let θ denote the angle between these two vectors, as shown here.

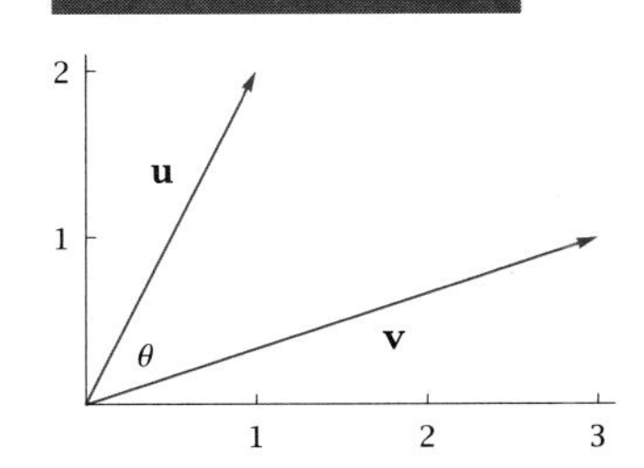

We could solve this problem without using the dot product by noting that the angle between the positive horizontal axis and **u** equals $\tan^{-1} 2$ and the angle between the positive horizontal axis and **v** equals $\tan^{-1} \frac{1}{3}$; thus $\theta = \tan^{-1} 2 - \tan^{-1} \frac{1}{3}$. Neither $\tan^{-1} 2$ nor $\tan^{-1} \frac{1}{3}$ can be evaluated exactly, and thus it appears that this expression for θ cannot be simplified.

However, we have another way to compute θ. Specifically, from the formula above we have

$$\cos \theta = \frac{\mathbf{u} \cdot \mathbf{v}}{|\mathbf{u}|\,|\mathbf{v}|} = \frac{5}{\sqrt{5}\sqrt{10}} = \frac{5}{\sqrt{5}\sqrt{5}\sqrt{2}} = \frac{1}{\sqrt{2}} = \frac{\sqrt{2}}{2}.$$

The equation above now implies that $\theta = \frac{\pi}{4}$.

Problem 39 gives a nice application of this example.

The formula for computing the dot product geometrically gives us an easy method to determine if two vectors are perpendicular.

Suppose θ is the angle between two nonzero vectors **u** and **v** with the same initial point. The vectors **u** and **v** are perpendicular if and only if θ equals $\frac{\pi}{2}$ radians, which happens if and only if $\cos\theta = 0$, which happens (according to our formula for computing the dot product geometrically) if and only if $\mathbf{u} \cdot \mathbf{v} = 0$. Thus we have the following result.

Perpendicular vectors

Two nonzero vectors with the same initial point are perpendicular if and only if their dot product equals 0.

EXAMPLE 10 Find a number t such that the vectors $(3, -5)$ and $(4, 2^t)$ are perpendicular.

SOLUTION The dot product of $(3, -5)$ and $(4, 2^t)$ is

$$12 - 5 \cdot 2^t.$$

For the vectors to be perpendicular, we want this dot product to equal 0. Thus we must solve the equation $12 - 5 \cdot 2^t = 0$, which is equivalent to the equation $2^t = \frac{12}{5}$. Taking logs of both sides gives $t \log 2 = \log \frac{12}{5}$. Thus

$$t = \frac{\log \frac{12}{5}}{\log 2} \approx 1.26303.$$

EXERCISES

1. Find the magnitude of the vector $(-3, 2)$.
2. Find the magnitude of the vector $(-5, -2)$.
3. Find an angle that determines the direction of the vector $(-3, 2)$.
4. Find an angle that determines the direction of the vector $(-5, -2)$.
5. Find two distinct numbers t such that $|t(1, 4)| = 5$.
6. Find two distinct numbers b such that $|b(3, -7)| = 4$.
7. Suppose $\mathbf{u} = (2, 1)$ and $\mathbf{v} = (3, 1)$.
   (a) Draw a figure illustrating the sum of **u** and **v** as arrows.
   (b) Compute the sum $\mathbf{u} + \mathbf{v}$ using coordinates.
8. Suppose $\mathbf{u} = (-3, 2)$ and $\mathbf{v} = (-2, -1)$.
   (a) Draw a figure illustrating the sum of **u** and **v** as arrows.
   (b) Compute the sum $\mathbf{u} + \mathbf{v}$ using coordinates.
9. Suppose $\mathbf{u} = (2, 1)$ and $\mathbf{v} = (3, 1)$.
   (a) Draw a figure using arrows illustrating the difference $\mathbf{u} - \mathbf{v}$.
   (b) Compute the difference $\mathbf{u} - \mathbf{v}$ using coordinates.
10. Suppose $\mathbf{u} = (-3, 2)$ and $\mathbf{v} = (-2, -1)$.
    (a) Draw a figure using arrows illustrating the difference $\mathbf{u} - \mathbf{v}$.
    (b) Compute the difference $\mathbf{u} - \mathbf{v}$ using coordinates.

11. Suppose the wind at airplane heights is 40 miles per hour (relative to the ground) moving $15°$ north of east. Relative to the wind, an airplane is flying at 450 miles per hour in a direction $20°$ south of the wind. Find the speed and direction of the airplane relative to the ground.

12. Suppose the wind at airplane heights is 70 miles per hour (relative to the ground) moving $17°$ south of east. Relative to the wind, an airplane is flying at 500 miles per hour in a direction $200°$ measured counterclockwise from the wind direction. Find the speed and direction of the airplane relative to the ground.

13. Suppose $\mathbf{u} = (3, 2)$ and $\mathbf{v} = (4, 5)$. Compute $\mathbf{u} \cdot \mathbf{v}$.

14. Suppose $\mathbf{u} = (-4, 5)$ and $\mathbf{v} = (2, -6)$. Compute $\mathbf{u} \cdot \mathbf{v}$.

15. Use the dot product to find the angle between the vectors $(2, 3)$ and $(3, 4)$.

16. Use the dot product to find the angle between the vectors $(3, -5)$ and $(-4, 3)$.

17. Find a number t such that the vectors $(6, -7)$ and $(2, \tan t)$ are perpendicular.

18. Find a number t such that the vectors $(2 \cos t, 4)$ and $(10, 3)$ are perpendicular.

PROBLEMS

19. Find coordinates for five different vectors $\mathbf{u}$, each of which has magnitude 5.

20. Find coordinates for three different vectors $\mathbf{u}$, each of which has a direction determined by an angle of $\frac{\pi}{6}$.

21. Suppose $\mathbf{u}$ and $\mathbf{v}$ are vectors with the same initial point. Explain why $|\mathbf{u} - \mathbf{v}|$ equals the distance between the endpoint of $\mathbf{u}$ and the endpoint of $\mathbf{v}$.

22. Suppose the wind at airplane heights is 55 miles per hour (relative to the ground) moving in the direction of $22°$ south of east. An airplane wants to fly directly north at 400 miles per hour relative to the ground. Find the speed and direction that the airplane must fly relative to the wind.

23. Using coordinates, show that if t is a scalar and $\mathbf{u}$ and $\mathbf{v}$ are vectors, then
$$t(\mathbf{u} + \mathbf{v}) = t\mathbf{u} + t\mathbf{v}.$$

24. Using coordinates, show that if s and t are scalars and $\mathbf{u}$ is a vector, then
$$(s + t)\mathbf{u} = s\mathbf{u} + t\mathbf{u}.$$

25. Using coordinates, show that if s and t are scalars and $\mathbf{u}$ is a vector, then
$$(st)\mathbf{u} = s(t\mathbf{u}).$$

26. Show that if $\mathbf{u}$ and $\mathbf{v}$ are vectors, then
$$2(|\mathbf{u}|^2 + |\mathbf{v}|^2) = |\mathbf{u} + \mathbf{v}|^2 + |\mathbf{u} - \mathbf{v}|^2.$$
[*This equality is often called the Parallelogram Equality, for reasons that are explained by the next problem.*]

27. Draw an appropriate figure and explain why the result in the problem above implies the following result: In any parallelogram, the sum of the squares of the lengths of the four sides equals the sum of the squares of the lengths of the two diagonals.

28. Suppose $\mathbf{v}$ is a vector other than $\mathbf{0}$. Explain why the vector $\frac{\mathbf{v}}{|\mathbf{v}|}$ has magnitude 1.

29. Show that if $\mathbf{u}$ and $\mathbf{v}$ are vectors, then
$$\mathbf{u} \cdot \mathbf{v} = \mathbf{v} \cdot \mathbf{u}.$$

30. Show that if $\mathbf{u}$, $\mathbf{v}$ and $\mathbf{w}$ are vectors, then
$$\mathbf{u} \cdot (\mathbf{v} + \mathbf{w}) = \mathbf{u} \cdot \mathbf{v} + \mathbf{u} \cdot \mathbf{w}.$$

31. Show that if $\mathbf{u}$ and $\mathbf{v}$ are vectors and t is a real number, then
$$(t\mathbf{u}) \cdot \mathbf{v} = \mathbf{u} \cdot (t\mathbf{v}) = t(\mathbf{u} \cdot \mathbf{v}).$$

32. Suppose $\mathbf{u}$ and $\mathbf{v}$ are vectors, neither of which is $\mathbf{0}$. Show that $\mathbf{u} \cdot \mathbf{v} = |\mathbf{u}|\,|\mathbf{v}|$ if and only if $\mathbf{u}$ and $\mathbf{v}$ have the same direction.

33. Suppose **u** and **v** are vectors. Show that

$$|\mathbf{u} \cdot \mathbf{v}| \le |\mathbf{u}|\,|\mathbf{v}|.$$

[*This result is called the Cauchy-Schwarz Inequality. Although this problem asks for a proof only in the setting of vectors in the plane, a similar inequality is true in many other settings and has important uses throughout mathematics.*]

34. Show that if **u** and **v** are vectors, then

$$|\mathbf{u}+\mathbf{v}|^2 = |\mathbf{u}|^2 + 2\mathbf{u}\cdot\mathbf{v} + |\mathbf{v}|^2.$$

35. Show that if **u** and **v** are vectors, then

$$|\mathbf{u}+\mathbf{v}| \le |\mathbf{u}| + |\mathbf{v}|.$$

[*Hint:* Square both sides and use the two previous problems.]

36. Interpret the inequality in the previous problem (which is often called the Triangle Inequality) as saying something interesting about triangles.

37. In Exercise 9, draw the vector from the endpoint of **v** to the endpoint of **u**. Show that the resulting triangle (whose edges are **u**, **v**, and **u** − **v**) is a right triangle.

38. Explain why there does not exist a number t such that the vectors $(2, t)$ and $(3, t)$ are perpendicular.

39. In Example 9 we found that the angle θ equals $\tan^{-1} 2 - \tan^{-1} \frac{1}{3}$ and also that θ equals $\frac{\pi}{4}$. Thus

$$\tan^{-1} 2 - \tan^{-1} \tfrac{1}{3} = \tfrac{\pi}{4}.$$

(a) Use one of the inverse trigonometric identities from Section 10.2 to show that the equation above can be rewritten as

$$\tan^{-1} 2 + \tan^{-1} 3 = \tfrac{3\pi}{4}.$$

(b) Explain how adding $\frac{\pi}{4}$ to both sides of the equation above leads to the beautiful equation

$$\tan^{-1} 1 + \tan^{-1} 2 + \tan^{-1} 3 = \pi.$$

[*Problem 59 in Section 10.6 gives another derivation of the equation above.*]

WORKED-OUT SOLUTIONS *to Odd-numbered Exercises*

1. Find the magnitude of the vector $(-3, 2)$.

SOLUTION $|(-3,2)| = \sqrt{(-3)^2 + 2^2} = \sqrt{13}$

3. Find an angle that determines the direction of the vector $(-3, 2)$.

SOLUTION We want an angle θ such that $\tan\theta = -\frac{2}{3}$ and also such that $\cos\theta$ is negative and $\sin\theta$ is positive. The angle $\tan^{-1}(-\frac{2}{3})$ has the correct tangent, but its cosine and sine have the wrong sign. Thus the angle we seek is $\tan^{-1}(-\frac{2}{3}) + \pi$.

5. Find two distinct numbers t such that $|t(1,4)| = 5$.

SOLUTION Because $t(1,4) = (t, 4t)$, we have

$$|t(1,4)| = |(t,4t)| = \sqrt{t^2 + 16t^2} = \sqrt{17t^2}.$$

We want this to equal 5, which means $17t^2 = 25$. Thus $t^2 = \frac{25}{17}$, which implies that $t = \pm\frac{5}{\sqrt{17}}$, which can be rewritten as $t = \pm\frac{5\sqrt{17}}{17}$.

7. Suppose $\mathbf{u} = (2, 1)$ and $\mathbf{v} = (3, 1)$.

(a) Draw a figure illustrating the sum of **u** and **v** as arrows.

(b) Compute the sum **u** + **v** using coordinates.

SOLUTION

(a) The figure below on the left shows **u** and **v** with their initial point at the origin. In the figure below on the right, **v** has been moved parallel to its original position so that its initial point now coincides with the endpoint of **u**. The last figure below shows that the vector **u** + **v** is the vector with the same initial point as **u** and the same endpoint as the second version of **v**.

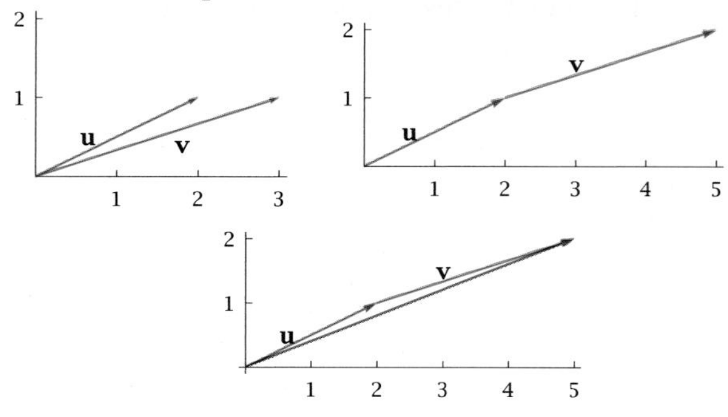

(b) The coordinates of $\mathbf{u} + \mathbf{v}$ are obtained by adding the corresponding coordinates of $\mathbf{u}$ and $\mathbf{v}$. Thus

$$\mathbf{u} + \mathbf{v} = (2, 1) + (3, 1) = (5, 2).$$

9. Suppose $\mathbf{u} = (2, 1)$ and $\mathbf{v} = (3, 1)$.

   (a) Draw a figure using arrows illustrating the difference $\mathbf{u} - \mathbf{v}$.

   (b) Compute the difference $\mathbf{u} - \mathbf{v}$ using coordinates.

SOLUTION

(a) The figure below shows the two vectors $\mathbf{u}$ and $\mathbf{v}$, both with their initial point at the origin. The vector $\mathbf{u} - \mathbf{v}$ is the vector whose initial point is the endpoint of $\mathbf{v}$ and whose endpoint is the endpoint of $\mathbf{u}$.

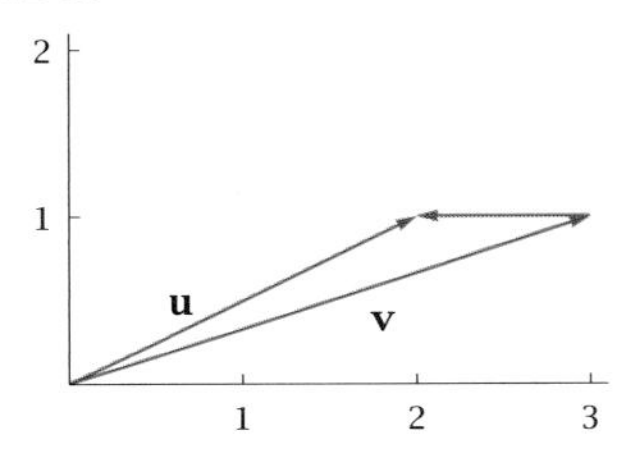

(b) The coordinates of $\mathbf{u} - \mathbf{v}$ are obtained by subtracting the corresponding coordinates of $\mathbf{u}$ and $\mathbf{v}$. Thus

$$\mathbf{u} - \mathbf{v} = (2, 1) - (3, 1) = (-1, 0).$$

11. Suppose the wind at airplane heights is 40 miles per hour (relative to the ground) moving 15° north of east. Relative to the wind, an airplane is flying at 450 miles per hour in a direction 20° south of the wind. Find the speed and direction of the airplane relative to the ground.

SOLUTION Let $\mathbf{w}$ denote the wind vector. Because $\mathbf{w}$ has magnitude 40 and direction 15°, the wind vector $\mathbf{w}$ has coordinates $(40 \cos 15°, 40 \sin 15°)$.

Let $\mathbf{a}$ denote the airspeed vector giving the motion of the airplane relative to the wind. The airspeed vector $\mathbf{a}$ has coordinates $(450 \cos(-5°), 450 \sin(-5°))$.

Let $\mathbf{g}$ denote the groundspeed vector giving the motion of the airplane relative to the ground. The properties of motion imply that the groundspeed vector equals the airspeed vector plus the wind vector. In other words, $\mathbf{g} = \mathbf{a} + \mathbf{w}$.

In terms of coordinates, we have

$$\begin{aligned} \mathbf{g} &= \mathbf{a} + \mathbf{w} \\ &= (450 \cos(-5°), 450 \sin(-5°)) \\ &\qquad + (40 \cos 15°, 40 \sin 15°) \\ &\approx (486.925, -28.8673). \end{aligned}$$

Now the speed of the airplane relative to the ground is the magnitude of $\mathbf{g}$:

$$\begin{aligned} |\mathbf{g}| &\approx \sqrt{486.925^2 + (-28.8673)^2} \\ &\approx 487.8. \end{aligned}$$

The direction of the airplane relative to the ground is the direction of g, which is approximately $\tan^{-1} \frac{-28.8673}{486.925}$, which is approximately -0.0592 radians, which is approximately $-3.4°$.

Thus relative to the ground, the airplane is traveling at approximately 487.8 miles per hour in a direction 3.4° south of east.

13. Suppose $\mathbf{u} = (3, 2)$ and $\mathbf{v} = (4, 5)$. Compute $\mathbf{u} \cdot \mathbf{v}$.

SOLUTION $\mathbf{u} \cdot \mathbf{v} = 3 \cdot 4 + 2 \cdot 5 = 12 + 10 = 22$

15. Use the dot product to find the angle between the vectors $(2, 3)$ and $(3, 4)$.

SOLUTION Note that $|(2, 3)| = \sqrt{13}$, $|(3, 4)| = 5$, and $(2, 3) \cdot (3, 4) = 18$. Thus the angle between $(2, 3)$ and $(3, 4)$ is

$$\cos^{-1} \frac{(2, 3) \cdot (3, 4)}{|(2, 3)|\,|(3, 4)|} = \cos^{-1} \frac{18}{5\sqrt{13}} \approx 0.0555.$$

17. Find a number t such that the vectors $(6, -7)$ and $(2, \tan t)$ are perpendicular.

SOLUTION The dot product of these two vectors is $12 - 7 \tan t$. To make these two vectors perpendicular, we want $12 - 7 \tan t = 0$, which implies that we want $\tan t = \frac{12}{7}$. Thus we take $t = \arctan \frac{12}{7}$.

There are also other correct answers in addition to $\arctan \frac{12}{7}$ because there are other angles whose tangent equals $\frac{12}{7}$. Specifically, we could choose $t = (\arctan \frac{12}{7}) + n\pi$ for any integer n.

11.5 *The Complex Plane*

LEARNING OBJECTIVES

By the end of this section you should be able to

- compute the absolute value of a complex number;
- write a complex number in polar form;
- compute large powers of complex numbers using De Moivre's Theorem;
- compute roots of complex numbers using De Moivre's Theorem.

Before starting this section, you should review the introduction to complex numbers given in Section 4.4.

Complex Numbers as Points in the Plane

Recall that a complex number has the form $a + bi$, where a and b are real numbers and $i^2 = -1$. We turn now to an interpretation of the coordinate plane that will help us better understand the complex number system.

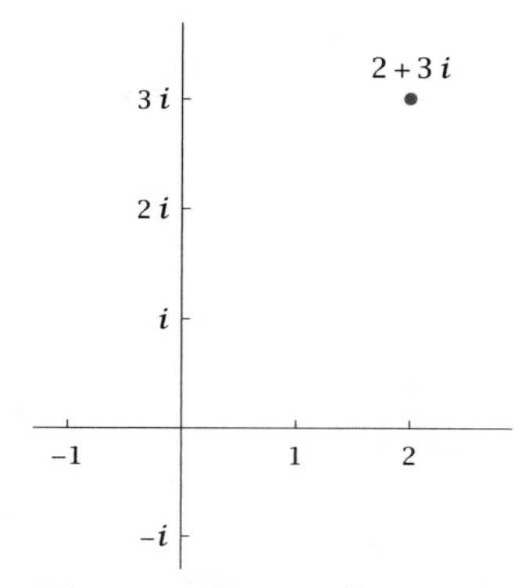

The complex number $2 + 3i$ as a point in the complex plane.

The complex plane

- Label the horizontal axis of a coordinate plane in the usual fashion but label the vertical axis with multiples of i, as shown in the figure.
- Identify the complex number $a+bi$, where a and b are real numbers, with the point (a, b).
- The coordinate plane with this interpretation of points as complex numbers is called the **complex plane**.

We can think of the system of complex numbers as being represented by the complex plane, just as we can think of the system of real numbers as being represented by the real line.

Each real number is also a complex number. For example, the real number 3 is also the complex number $3 + 0i$, which we usually write as just 3.

When we think of the real numbers also being the complex numbers, then the real numbers correspond to the horizontal axis in the complex plane. Thus the horizontal axis of the complex plane is sometimes called the **real axis** and the vertical axis is sometimes called the **imaginary axis**.

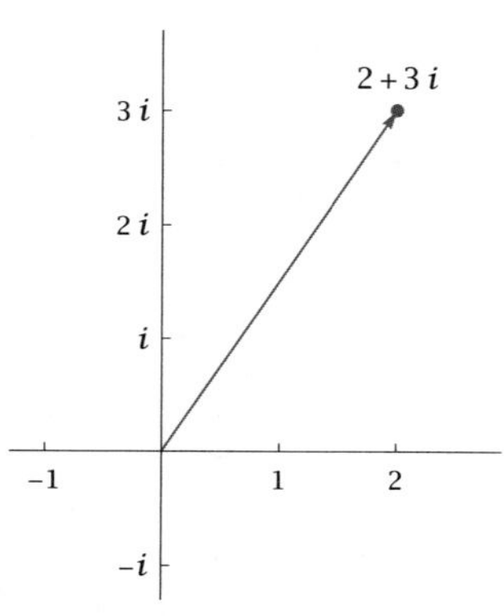

The complex number $2 + 3i$ as a vector in the complex plane.

Sometimes we identify a complex number $a + bi$ with the vector whose initial point is the origin and whose endpoint is located at the point corresponding to $a + bi$ in the complex plane, as shown here. Because complex numbers are added and subtracted by adding and subtracting their real and imaginary parts separately, complex addition and subtraction have the same geometric interpretation as vector addition and subtraction.

Recall that the absolute value of a real number is the distance from 0 to the number (when thinking of numbers as points on the real line). Similarly, the absolute value of a complex number is defined to be the distance from the origin to the complex number (when thinking of complex numbers as points on the complex plane). Here is the formal definition:

Absolute value of a complex number

If $z = a + bi$, where a and b are real numbers, then the **absolute value** of z, denoted $|z|$, is defined by

$$|z| = \sqrt{a^2 + b^2}.$$

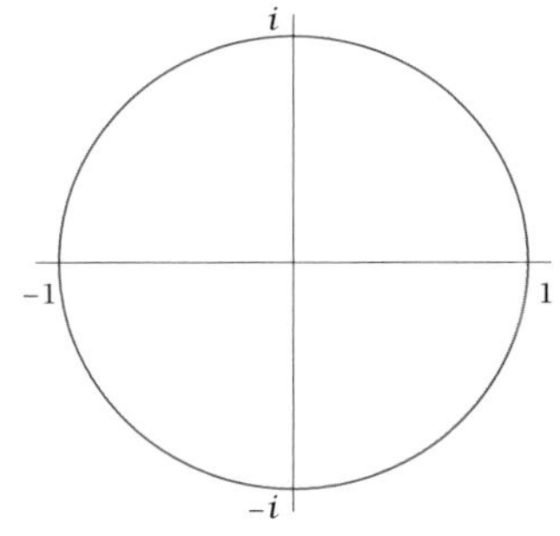

The unit circle in the complex plane is described by the equation $|z| = 1$.

When thinking of complex numbers as vectors in the complex plane, the absolute value of a complex number is simply the magnitude of the corresponding vector.

EXAMPLE 1

Evaluate $|2 + 3i|$.

SOLUTION $|2 + 3i| = \sqrt{2^2 + 3^2} = \sqrt{13} \approx 3.60555$

EXAMPLE 2

Show that

$$|\cos\theta + i\sin\theta| = 1$$

for every real number θ.

Some authors use the term modulus *instead of* absolute value.

SOLUTION $|\cos\theta + i\sin\theta| = \sqrt{\cos^2\theta + \sin^2\theta} = \sqrt{1} = 1$

Recall that the complex conjugate of a complex number $a + bi$, where a and b are real numbers, is denoted by $\overline{a + bi}$ and is defined by

$$\overline{a + bi} = a - bi.$$

In terms of the complex plane, the operation of complex conjugation is the same as flipping across the real axis. The figure here shows a complex number and its complex conjugate.

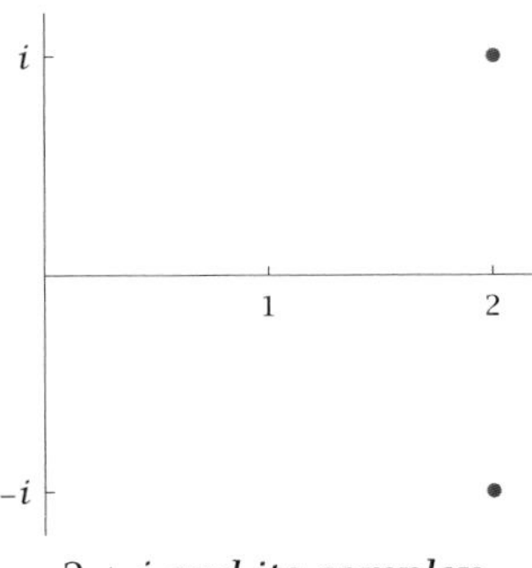

$2 + i$ and its complex conjugate $2 - i$.

A nice formula connects the complex conjugate and the absolute value of a complex number. To derive this formula, suppose $z = a + bi$, where a and b are real numbers. Then

$$\begin{aligned} z\overline{z} &= (a + bi)(a - bi) \\ &= a^2 - b^2 i^2 \\ &= a^2 + b^2 \\ &= |z|^2. \end{aligned}$$

We record this result as follows:

Complex conjugates and absolute values

If z is a complex number, then

$$z\bar{z} = |z|^2.$$

Geometric Interpretation of Complex Multiplication and Division

As we will soon see, using polar coordinates with complex numbers can bring extra insight into the operations of multiplication, division, and raising a complex number to a power.

The idea here is to think of a complex number as a point in the complex plane and then use the polar coordinates that were developed in Section 11.3.

Suppose $z = x + yi$, where x and y are real numbers. We identify z with the point (x, y) in the complex plane. If (x, y) has polar coordinates (r, θ), then $x = r\cos\theta$ and $y = r\sin\theta$. Thus

$$\begin{aligned} z &= x + yi \\ &= r\cos\theta + ir\sin\theta \\ &= r(\cos\theta + i\sin\theta). \end{aligned}$$

The equation above leads to the following definition.

Polar form of a complex number

The **polar form** of a complex number z is an expression of the form

$$z = r(\cos\theta + i\sin\theta),$$

where $r = |z|$ and θ, which is called the **argument** of z, is the angle that z (thought of as a vector) makes with the positive horizontal axis.

When writing a complex number z in polar form, there is only one correct choice for the number $r \geq 0$ in the expression above: we must choose $r = |z|$. However, given any correct choice for the argument θ, another correct choice can be found by adding an integer multiple of 2π.

EXAMPLE 3

Write the following complex numbers in polar form:

(a) 2

(b) $3i$

(c) $1 + i$

(d) $\sqrt{3} - i$

(e) $-\sqrt{3} + i$

SOLUTION

(a) We have $|2| = 2$. Also, because 2 is on the positive horizontal axis, the argument of 2 is 0. Thus the polar form of 2 is

$$2 = 2(\cos 0 + i \sin 0).$$

We could also write

$$2 = 2(\cos(2\pi) + i \sin(2\pi)) \quad \text{or} \quad 2 = 2(\cos(4\pi) + i \sin(4\pi))$$

or use any integer multiple of 2π as the argument.

(b) We have $|3i| = 3$. Also, because $3i$ is on the positive vertical axis, the argument of $3i$ is $\frac{\pi}{2}$. Thus the polar form of $3i$ is

$$3i = 3(\cos \tfrac{\pi}{2} + i \sin \tfrac{\pi}{2}).$$

As usual, we could add any integer multiple of 2π to the argument $\frac{\pi}{2}$ to obtain other arguments.

(c) We have $|1 + i| = \sqrt{1^2 + 1^2} = \sqrt{2}$. Also, the figure here shows that the argument θ of $1 + i$ is $\frac{\pi}{4}$. Thus the polar form of $1 + i$ is

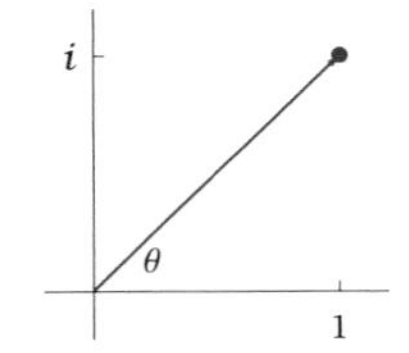

$$1 + i = \sqrt{2}(\cos \tfrac{\pi}{4} + i \sin \tfrac{\pi}{4}).$$

(d) We have $|\sqrt{3} - i| = \sqrt{\sqrt{3}^2 + (-1)^2} = \sqrt{3 + 1} = 2$. Note that $\tan^{-1}(\frac{-1}{\sqrt{3}}) = -\frac{\pi}{6}$. Thus the figure here shows that the argument θ of $\sqrt{3} - i$ is $-\frac{\pi}{6}$. Thus the polar form of $\sqrt{3} - i$ is

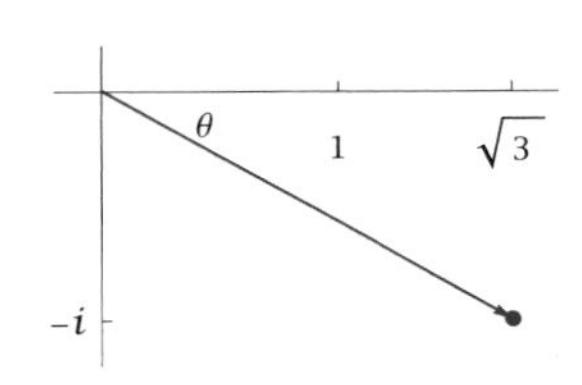

$$\sqrt{3} - i = 2(\cos(-\tfrac{\pi}{6}) + i \sin(-\tfrac{\pi}{6})).$$

(e) We have $|-\sqrt{3} + i| = \sqrt{(-\sqrt{3})^2 + 1^2} = \sqrt{3 + 1} = 2$. Note that $\tan^{-1}(\frac{1}{-\sqrt{3}}) = -\frac{\pi}{6}$. Although the angle $-\frac{\pi}{6}$ has the correct tangent, its cosine and sine have the wrong sign. As can be seen in the figure, the angle we seek is $-\frac{\pi}{6} + \pi$, which equals $\frac{5\pi}{6}$. Thus the polar form of $-\sqrt{3} + i$ is

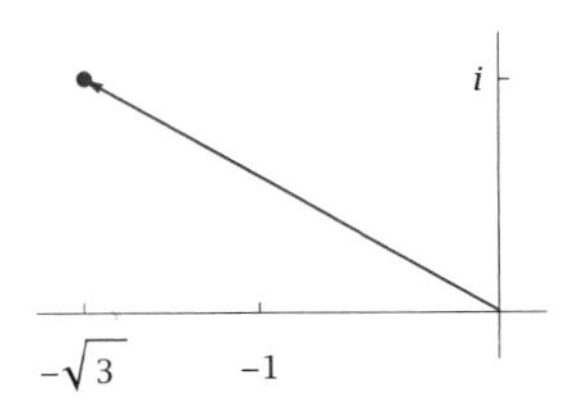

This example shows that extra care must be used in finding the argument of a complex number whose real part is negative.

$$-\sqrt{3} + i = 2(\cos \tfrac{5\pi}{6} + i \sin \tfrac{5\pi}{6}).$$

The multiplicative inverse of a complex number has a nice interpretation in polar form. Suppose $z = r(\cos\theta + i \sin\theta)$ is a nonzero complex number (from now on, whenever we write an expression like this, we will assume that $r = |z|$ and that θ is a real number). We know that $|z|^2 = z\bar{z}$. Dividing both sides of this equation by $z|z|^2$ shows that

$$\frac{1}{z} = \frac{\bar{z}}{|z|^2} = \frac{r(\cos\theta - i \sin\theta)}{r^2} = \frac{\cos\theta - i \sin\theta}{r}.$$

We record this result as follows:

This result states that the polar form of $\frac{1}{z}$ is obtained from the polar form $r(\cos\theta + i\sin\theta)$ of z by replacing r by $\frac{1}{r}$ and replacing θ by $-\theta$.

Multiplicative inverse of a complex number in polar form

If $z = r(\cos\theta + i\sin\theta)$ is a nonzero complex number, then

$$\frac{1}{z} = \frac{1}{r}(\cos\theta - i\sin\theta) = \frac{1}{r}(\cos(-\theta) + i\sin(-\theta)).$$

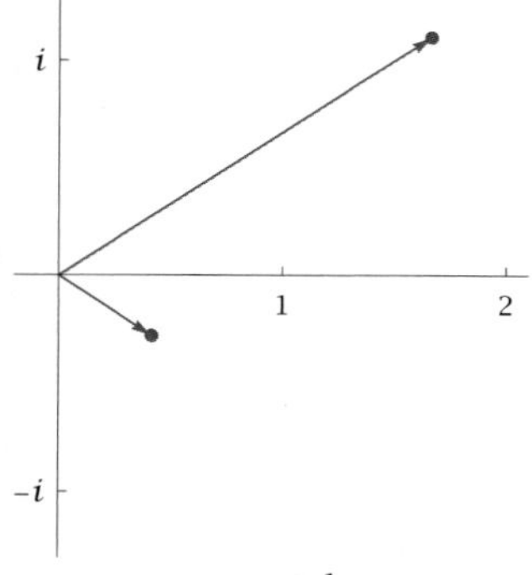

z and $\frac{1}{z}$.

The figure here illustrates the formula above. The longer vector represents a complex number z with $|z| = 2$. The shorter vector represents the complex number $\frac{1}{z}$; it has absolute value $\frac{1}{2}$, and its argument is the negative of the argument of z. Thus the angle from the shorter vector to the positive horizontal axis equals the angle from the positive horizontal axis to the longer vector.

Complex multiplication also has a pretty expression in terms of polar form. Suppose

$$z_1 = r_1(\cos\theta_1 + i\sin\theta_1) \quad \text{and} \quad z_2 = r_2(\cos\theta_2 + i\sin\theta_2).$$

Then

$$\begin{aligned} z_1 z_2 &= r_1 r_2(\cos\theta_1 + i\sin\theta_1)(\cos\theta_2 + i\sin\theta_2) \\ &= r_1 r_2\big((\cos\theta_1\cos\theta_2 - \sin\theta_1\sin\theta_2) + i(\sin\theta_1\cos\theta_2 + \cos\theta_1\sin\theta_2)\big) \\ &= r_1 r_2(\cos(\theta_1 + \theta_2) + i\sin(\theta_1 + \theta_2)), \end{aligned}$$

where the addition formulas for cosine and sine from Section 10.6 gave the last simplification.

Thus we have the following result, expressed first as a formula and then in words:

Complex multiplication in polar form

- If $z_1 = r_1(\cos\theta_1 + i\sin\theta_1)$ and $z_2 = r_2(\cos\theta_2 + i\sin\theta_2)$, then

$$z_1 z_2 = r_1 r_2(\cos(\theta_1 + \theta_2) + i\sin(\theta_1 + \theta_2)).$$

- The absolute value of the product of two complex numbers is the product of their absolute values.
- The argument of the product of two complex numbers is the sum of their arguments.

EXAMPLE 4

Suppose $z_1 = 3(\cos\frac{\pi}{7} + i\sin\frac{\pi}{7})$ and $z_2 = 4(\cos\frac{3\pi}{7} + i\sin\frac{3\pi}{7})$. Find the polar form of $z_1 z_2$.

SOLUTION Using the formula above, we have

$$\begin{aligned} z_1 z_2 &= 3 \cdot 4\left(\cos\left(\tfrac{\pi}{7} + \tfrac{3\pi}{7}\right) + i \sin\left(\tfrac{\pi}{7} + \tfrac{3\pi}{7}\right)\right) \\ &= 12\left(\cos \tfrac{4\pi}{7} + i \sin \tfrac{4\pi}{7}\right). \end{aligned}$$

The quotient $\frac{z_1}{z_2}$ can be computed by thinking of division by z_2 as multiplication by $\frac{1}{z_2}$. We already know that the polar form of $\frac{1}{z_2}$ is obtained by replacing r_2 by $\frac{1}{r_2}$ and replacing θ_2 by $-\theta_2$. Thus combining our result on the multiplicative inverse and our result on complex multiplication gives the following result:

Complex division in polar form

- If $z_1 = r_1(\cos\theta_1 + i \sin\theta_1)$ and $z_2 = r_2(\cos\theta_2 + i \sin\theta_2)$, then

$$\frac{z_1}{z_2} = \frac{r_1}{r_2}(\cos(\theta_1 - \theta_2) + i \sin(\theta_1 - \theta_2)).$$

- The absolute value of the quotient of two complex numbers is the quotient of their absolute values.
- The argument of the quotient of two complex numbers is the difference of their arguments.

Here we assume that $z_2 \neq 0$.

De Moivre's Theorem

Suppose $z = r(\cos\theta + i \sin\theta)$. Take $z_1 = z$ and $z_2 = z$ in the formula just derived for complex multiplication in polar form, getting

$$z^2 = r^2(\cos(2\theta) + i \sin(2\theta)).$$

Now apply the formula again, this time with $z_1 = z^2$ and $z_2 = z$, getting

$$z^3 = r^3(\cos(3\theta) + i \sin(3\theta)).$$

If we apply the formula once more, this time with $z_1 = z^3$ and $z_2 = z$, we get

$$z^4 = r^4(\cos(4\theta) + i \sin(4\theta)).$$

This pattern continues, leading to the beautiful result called De Moivre's Theorem:

De Moivre's Theorem

If $z = r(\cos\theta + i \sin\theta)$ and n is a positive integer, then

$$z^n = r^n(\cos(n\theta) + i \sin(n\theta)).$$

Abraham de Moivre first published this result in 1722.

De Moivre's Theorem is a wonderful tool for evaluating large powers of complex numbers.

EXAMPLE 5

Evaluate $(\sqrt{3} - i)^{100}$.

One way to solve this problem would be to multiply $\sqrt{3} - i$ times itself 100 times. But that process would be tedious, it would take a long time, and errors can easily creep into such long calculations.

SOLUTION As the first step in using De Moivre's Theorem, we must write $(\sqrt{3} - i)$ in polar form. However, we already did that in Example 3, getting

$$\sqrt{3} - i = 2(\cos(-\tfrac{\pi}{6}) + i\sin(-\tfrac{\pi}{6})).$$

De Moivre's Theorem tells us that

$$(\sqrt{3} - i)^{100} = 2^{100}(\cos(-\tfrac{100}{6}\pi) + i\sin(-\tfrac{100}{6}\pi)).$$

Now $-\frac{100}{6} = -\frac{50}{3} = -16 - \frac{2}{3}$. Because even multiples of π can be discarded when computing values of cosine and sine, we thus have

$$\begin{aligned}(\sqrt{3} - i)^{100} &= 2^{100}(\cos(-\tfrac{2}{3}\pi) + i\sin(-\tfrac{2}{3}\pi))\\ &= 2^{100}(-\tfrac{1}{2} - \tfrac{\sqrt{3}}{2}i)\\ &= 2^{99}\cdot 2(-\tfrac{1}{2} - \tfrac{\sqrt{3}}{2}i)\\ &= -2^{99}(1 + \sqrt{3}i).\end{aligned}$$

Finding Complex Roots

De Moivre's Theorem also allows us to find roots of complex numbers.

EXAMPLE 6

Find three distinct complex numbers z such that $z^3 = 1$.

SOLUTION Taking $z = 1$ is one choice of a complex number such that $z^3 = 1$, but the other two choices are not obvious. To find them, suppose $z = r(\cos\theta + i\sin\theta)$. Then

$$z^3 = r^3(\cos(3\theta) + i\sin(3\theta)).$$

We want z^3 to equal 1. Thus we take $r = 1$. Now we must find values of θ such that $\cos(3\theta) = 1$ and $\sin(3\theta) = 0$. One choice is to take $\theta = 0$, which gives us

$$z = 1,$$

which we already knew was one choice for z.

Another choice of θ that satisfies $\cos(3\theta) = 1$ and $\sin(3\theta) = 0$ can be obtained by choosing $3\theta = 2\pi$, which means $\theta = \frac{2\pi}{3}$. This choice of θ gives

$$z = -\tfrac{1}{2} + \tfrac{\sqrt{3}}{2}i.$$

Yet another choice of θ that satisfies $\cos(3\theta) = 1$ and $\sin(3\theta) = 0$ can be obtained by choosing $3\theta = 4\pi$, which means $\theta = \frac{4\pi}{3}$. This choice of θ gives

$$z = -\tfrac{1}{2} - \tfrac{\sqrt{3}}{2}i.$$

You should verify that $(-\frac{1}{2} \pm \frac{\sqrt{3}}{2}i)^3 = 1$.

Thus three distinct values of z such that $z^3 = 1$ are 1, $-\frac{1}{2} + \frac{\sqrt{3}}{2}i$, and $-\frac{1}{2} - \frac{\sqrt{3}}{2}i$.

EXERCISES

1. Evaluate $|4 - 3i|$.

2. Evaluate $|7 + 12i|$.

3. Find two real numbers b such that $|3 + bi| = 7$.

4. Find two real numbers a such that $|a - 5i| = 9$.

5. Write $2 - 2i$ in polar form.

6. Write $-3 + 3\sqrt{3}i$ in polar form.

7. Write
$$\frac{1}{6(\cos\frac{\pi}{11} + i\sin\frac{\pi}{11})}$$
in polar form.

8. Write
$$\frac{1}{7(\cos\frac{\pi}{9} + i\sin\frac{\pi}{9})}$$
in polar form.

9. Write
$$(\cos\tfrac{\pi}{7} + i\sin\tfrac{\pi}{7})(\cos\tfrac{\pi}{9} + i\sin\tfrac{\pi}{9})$$
in polar form.

10. Write
$$(\cos\tfrac{\pi}{5} + i\sin\tfrac{\pi}{5})(\cos\tfrac{\pi}{11} + i\sin\tfrac{\pi}{11})$$
in polar form.

11. Evaluate $(2 - 2i)^{333}$.

12. Evaluate $(-3 + 3\sqrt{3}i)^{555}$.

13. Find four distinct complex numbers z such that $z^4 = -2$.

14. Find three distinct complex numbers z such that $z^3 = 4i$.

PROBLEMS

15. Explain why there does not exist a real number b such that $|5 + bi| = 3$.

16. Show that if z is a complex number, then the real part of z is in the interval $[-|z|, |z|]$.

17. Show that if z is a complex number, then the imaginary part of z is in the interval $[-|z|, |z|]$.

18. Show that
$$\frac{1}{\cos\theta + i\sin\theta} = \cos\theta - i\sin\theta$$
for every real number θ.

19. Suppose z is a nonzero complex number. Show that $\overline{z} = \frac{1}{z}$ if and only if $|z| = 1$.

20. Suppose z is a complex number whose real part has absolute value equal to $|z|$. Show that z is a real number.

21. Suppose z is a complex number whose imaginary part has absolute value equal to $|z|$. Show that the real part of z equals 0.

22. Suppose w and z are complex numbers. Show that
$$|wz| = |w|\,|z|.$$

23. Suppose w and z are complex numbers. Show that
$$|w + z| \le |w| + |z|.$$

24. Describe the subset of the complex plane consisting of the complex numbers z such that z^3 is a real number.

25. Describe the subset of the complex plane consisting of the complex numbers z such that z^3 is a positive number.

26. Describe the subset of the complex plane consisting of the complex numbers z such that the real part of z^3 is a positive number.

27. Explain why $(\cos 1^\circ + i\sin 1^\circ)^{360} = 1$.

28. Explain why 360 is the smallest positive integer n such that $(\cos 1^\circ + i\sin 1^\circ)^n = 1$.

29. In Example 6, we found cube roots of 1 by finding numbers θ such that
$$\cos(3\theta) = 1 \quad\text{and}\quad \sin(3\theta) = 0.$$
The three choices $\theta = 0$, $\theta = \frac{2\pi}{3}$, and $\theta = \frac{4\pi}{3}$ gave us three distinct cube roots of 1. Other choices of θ, such as $\theta = 2\pi$, $\theta = \frac{8\pi}{3}$, and $\theta = \frac{10\pi}{3}$, also satisfy the equations above. Explain why these choices of θ do not give us additional cube roots of 1.

30. Explain why the six distinct complex numbers that are sixth roots of 1 are the vertices of a regular hexagon inscribed in the unit circle.

WORKED-OUT SOLUTIONS *to Odd-numbered Exercises*

1. Evaluate $|4 - 3i|$.

SOLUTION

$$|4 - 3i| = \sqrt{4^2 + (-3)^2} = \sqrt{16 + 9} = \sqrt{25} = 5$$

3. Find two real numbers b such that $|3 + bi| = 7$.

SOLUTION We have

$$7 = |3 + bi| = \sqrt{9 + b^2}.$$

Squaring both sides, we have $49 = 9 + b^2$. Thus $b^2 = 40$, which implies that

$$b = \pm\sqrt{40} = \pm\sqrt{4 \cdot 10} = \pm\sqrt{4} \cdot \sqrt{10} = \pm 2\sqrt{10}.$$

5. Write $2 - 2i$ in polar form.

SOLUTION First we compute $|2 - 2i|$:

$$|2 - 2i| = \sqrt{2^2 + (-2)^2} = \sqrt{8} = \sqrt{4 \cdot 2} = 2\sqrt{2}.$$

The vector whose initial point is the origin and whose endpoint is $2 - 2i$ in the complex plane makes an angle of $-\frac{\pi}{4}$ with the positive horizontal axis. Thus

$$2 - 2i = 2\sqrt{2}(\cos(-\tfrac{\pi}{4}) + i\sin(-\tfrac{\pi}{4}))$$

gives the polar form of $2 - 2i$.

7. Write

$$\frac{1}{6(\cos\frac{\pi}{11} + i\sin\frac{\pi}{11})}$$

in polar form.

SOLUTION From the formula for the polar form of the multiplicative inverse of a complex number, we have

$$\frac{1}{6(\cos\frac{\pi}{11} + i\sin\frac{\pi}{11})} = \tfrac{1}{6}(\cos(-\tfrac{\pi}{11}) + i\sin(-\tfrac{\pi}{11})).$$

9. Write

$$(\cos\tfrac{\pi}{7} + i\sin\tfrac{\pi}{7})(\cos\tfrac{\pi}{9} + i\sin\tfrac{\pi}{9})$$

in polar form.

SOLUTION Because $\frac{\pi}{7} + \frac{\pi}{9} = \frac{16\pi}{63}$, the product above equals

$$\cos\tfrac{16\pi}{63} + i\sin\tfrac{16\pi}{63}.$$

11. Evaluate $(2 - 2i)^{333}$.

SOLUTION From Exercise 5 we know that

$$2 - 2i = 2\sqrt{2}(\cos(-\tfrac{\pi}{4}) + i\sin(-\tfrac{\pi}{4})).$$

Thus

$$(2 - 2i)^{333} = (2\sqrt{2})^{333}(\cos(-\tfrac{333\pi}{4}) + i\sin(-\tfrac{333\pi}{4})).$$

Now

$$(2\sqrt{2})^{333} = 2^{333}2^{333/2} = 2^{333}2^{166}\sqrt{2} = 2^{499}\sqrt{2}.$$

Also,

$$-\tfrac{333}{4} = -83 - \tfrac{1}{4} = -82 - \tfrac{5}{4}.$$

Thus

$$\begin{aligned}(2\sqrt{2})^{333} &= 2^{499}\sqrt{2}(\cos(-\tfrac{5}{4}\pi) + i\sin(-\tfrac{5}{4}\pi)) \\ &= 2^{499}\sqrt{2}(-\tfrac{\sqrt{2}}{2} + i\tfrac{\sqrt{2}}{2}) \\ &= 2^{499}(-1 + i).\end{aligned}$$

13. Find four distinct complex numbers z such that $z^4 = -2$.

SOLUTION Suppose $z = r(\cos\theta + i\sin\theta)$. Then

$$z^4 = r^4(\cos(4\theta) + i\sin(4\theta)).$$

We want z^4 to equal -2. Thus we need $r^4 = 2$, which implies that $r = 2^{1/4}$.

Now we must find values of θ such that $\cos(4\theta) = -1$ and $\sin(4\theta) = 0$. One choice is to take $4\theta = \pi$, which implies that $\theta = \frac{\pi}{4}$, which gives us

$$z = 2^{1/4}(\tfrac{\sqrt{2}}{2} + \tfrac{\sqrt{2}}{2}i).$$

Another choice of θ that satisfies $\cos(4\theta) = -1$ and $\sin(4\theta) = 0$ can be obtained by choosing $4\theta = 3\pi$, which means $\theta = \frac{3\pi}{4}$, which gives us

$$z = 2^{1/4}(-\tfrac{\sqrt{2}}{2} + \tfrac{\sqrt{2}}{2}i).$$

Yet another choice of θ that satisfies $\cos(4\theta) = -1$ and $\sin(4\theta) = 0$ can be obtained by choosing $4\theta = 5\pi$, which means $\theta = \frac{5\pi}{4}$, which gives us

$$z = 2^{1/4}(-\tfrac{\sqrt{2}}{2} - \tfrac{\sqrt{2}}{2}i).$$

Yet another choice of θ that satisfies $\cos(4\theta) = -1$ and $\sin(4\theta) = 0$ can be obtained by choosing $4\theta = 7\pi$, which means $\theta = \frac{7\pi}{4}$, which gives us

$$z = 2^{1/4}(\tfrac{\sqrt{2}}{2} - \tfrac{\sqrt{2}}{2}i).$$

Thus four distinct values of z such that $z^4 = -2$ are $2^{1/4}(\frac{\sqrt{2}}{2} + \frac{\sqrt{2}}{2}i)$, $2^{1/4}(-\frac{\sqrt{2}}{2} + \frac{\sqrt{2}}{2}i)$, $2^{1/4}(-\frac{\sqrt{2}}{2} - \frac{\sqrt{2}}{2}i)$, and $2^{1/4}(\frac{\sqrt{2}}{2} - \frac{\sqrt{2}}{2}i)$.

CHAPTER SUMMARY

To check that you have mastered the most important concepts and skills covered in this chapter, make sure that you can do each item in the following list:

- Use parametric curves to graph the inverse of a function.
- Find formulas that shift, stretch, or flip a parametric curve.
- Graph transformations of trigonometric functions that change the amplitude, period, and/or phase shift.
- Convert from polar to rectangular coordinates.
- Convert from rectangular to polar coordinates.
- Graph a curve described by polar coordinates.
- Compute the sum, difference, and dot product of two vectors.
- Determine when two vectors are perpendicular.
- Find the polar form of a complex number.
- Use De Moivre's Theorem to compute powers and roots of complex numbers.

To review a chapter, go through the list above to find items that you do not know how to do, then reread the material in the chapter about those items. Then try to answer the chapter review questions below without looking back at the chapter.

CHAPTER REVIEW QUESTIONS

1. If a cannonball is shot from the ground with initial horizontal velocity 700 feet per second and an initial vertical velocity 600 feet per second, how far away does it land?

2. Explain how parametric curves can be used to graph the inverse of a function even in cases where no formula can be found for the inverse.

3. How should the coordinates of a parametric curve be changed to shift it right 4 units?

4. How should the coordinates of a parametric curve be changed to shift it down 3 units?

5. How should the coordinates of a parametric curve be changed to stretch it horizontally by a factor of 6?

6. How should the coordinates of a parametric curve be changed to stretch it vertically by a factor of 5?

7. How should the coordinates of a parametric curve be changed to flip it across the horizontal axis?

8. Give an example of a function that has amplitude 5 and period 3.

9. Give an example of a function that has period 3π and range $[2, 12]$.

10. Sketch the graph of the function
$$4\sin(2x+1)+5$$
on the interval $[-3\pi, 3\pi]$.

For Questions 11–15, assume that g is the function defined by

$$g(x) = a\sin(bx+c)+d,$$

where a, b, c, and d are constants with $a \neq 0$ and $b \neq 0$.

11. Find two distinct values for a so that g has amplitude 4.

12. Find two distinct values for b so that g has period $\frac{\pi}{2}$.

13. Find values for a and d, with $a > 0$, so that g has range $[-3, 4]$.

14. Find values for a, d, and c, with $a > 0$ and $0 \le c \le \pi$, so that g has range $[-3, 4]$ and $g(0) = 2$.

15. Find values for a, d, c, and b, with $a > 0$ and $b > 0$ and $0 \leq c \leq \pi$, so that g has range $[-3, 4]$, $g(0) = 2$, and g has period 5.

16. Sketch the line segment from the origin to the point with polar coordinates $(3, \frac{\pi}{3})$.

17. Find the rectangular coordinates of the point whose polar coordinates are $(4, 2.1)$.

18. Find the polar coordinates of the point whose rectangular coordinates are $(5, 9)$.

19. Sketch the graph of the polar equation $r = 7$.

20. Sketch the graph of the polar equation $\theta = \frac{5\pi}{6}$.

21. Sketch the graph of the polar equation $r = 5\cos\theta$.

22. Evaluate $\left|\cos\frac{\pi}{11} - i\sin\frac{\pi}{11}\right|$.

23. Write
$$(\cos\tfrac{\pi}{4} + i\tfrac{\pi}{4})(\cos\tfrac{\pi}{7} + i\sin\tfrac{\pi}{7})$$
in polar form.

24. Write
$$\frac{2(\cos\frac{\pi}{3} + i\frac{\pi}{3})}{5(\cos\frac{\pi}{7} + i\sin\frac{\pi}{7})}$$
in polar form.

25. Evaluate $(1 + i)^{500}$.

26. Find three distinct complex numbers z such that $z^3 = -5i$.

Photo Credits

- page v: Jonathan Shapiro
- page 1: Goodshot/SuperStock
- page 41: Public domain image from Wikipedia
- page 61: Brand X/SuperStock
- page 81: Public domain image from Wikipedia
- page 85: NASA Jet Propulsion Laboratory/UCLA
- page 103: Christine Balderas/iStockphoto
- page 104: Manuela Miller/iStockphoto
- page 106: Sheldon Axler
- page 117: Public domain image from Wikipedia
- page 213: Mostafa Azizi/Public domain image from Wikimedia
- page 279: SuperStock, Inc./SuperStock
- page 295: Age Fotostock/SuperStock
- page 296: Public domain image from Wikipedia
- page 299: Culver Pictures, Inc./SuperStock
- page 300: Sheldon Axler
- page 309: Sheldon Axler
- page 328: iStockphoto
- page 333: Center for Disease Control and Prevention, U.S. Department of Health and Human Services/Evangeline Sowers, Janice Haney Carr
- page 334: Alex Slobodkin/iStockphoto
- page 335: © Scott Adams/United Feature Syndicate, Inc.
- page 338: Adam Kazmierski/iStockphoto
- page 349: iStockphoto
- page 376: Corbis/SuperStock
- page 387: Ingram Publishing/SuperStock
- page 412: Public domain image from Wikipedia
- page 447: Christie's Images/SuperStock
- page 455: Public domain image from Wikipedia
- page 471: Hamelin de Guettelet/Wikimedia
- page 476: Public domain image from Wikipedia
- page 492: Jaime Abecasis/SuperStock
- page 497: Age Fotostock/SuperStock
- page 501: Paul Kline/iStockphoto
- page 521: Tetra Images/SuperStock
- page 523: FoodCollection/SuperStock
- page 527: FoodCollection/SuperStock
- page 583: NewsCom
- page 617: NewsCom
- page 671: PhotoAlto/SuperStock
- page 687: Photodisc/SuperStock

Index